FEATURES AND BENEFITS
PRENTICE HALL TRIGONOMETRY

Prentice Hall Trigonometry **is a comprehensive and cohesive course written to ensure that students will master trigonometric skills and concepts for use in future mathematics courses and in related disciplines.**

CONTENTS

1 *Comprehensive coverage* begins with the development of the Fundamental Functions of Trigonometry; Degree and Radian Measure; Special Angles; and Evaluation of Trigonometric Functions, followed by chapters on Graphing Trigonometric Functions; Right Triangle Trigonometry and Basic Identities; Oblique Triangles; Trigonometric Identities; Inverse Trigonometric Functions; Complex Numbers; Exponential and Logarithmic Functions; and Sequences and Series. (See Contents, pp. iii–vi.)

2 *The Preview* at the outset of the lesson prepares the students for the lesson, using a variety of techniques including reviewing the prerequisite skill needed for the lesson or informally investigating the lesson concept before it is formally developed. (See pp. 156, 255, 398.)

3 *Applications* are included in the Practice Exercises in the lessons, and as special features in the chapters to underscore the relevancy of Trigonometry in the real world and in related disciplines. (See pp. 198–199, 278–279.)

4 *Calculator and Computer* activities guide students in investigating trigonometric concepts and in using technology as a problem-solving tool. (See pp. 145–146, 327.)

5 *Carefully graded Practice Exercises* provide a wealth of practice for students of all ability levels. Although labeled A, B, C for you in the Teacher's Edition, they are unlabeled in the Student Text, thereby encouraging students to try all problems. (See pp. 121–123, 383–385.)

6 *Review and maintenance* features include 2-page Chapter Summary and Reviews, Cumulative Reviews, Maintaining Skills, and College Entrance Exam Reviews. (See pp. 280–284, 326–332.)

7 *A complete testing program* consists of Test Yourself midway and at the end of each chapter, and a Chapter Test at the end of each chapter. (See pp. 56–58, 104–110.)

8 *Special features* include Critical Thinking, Careers, and Did You Know? (See pp. 185, 317.)

- **The Teacher's Edition** includes a Chapter Overview immediately preceding each chapter, and a complete Lesson Plan for every lesson consisting of background information, teaching suggestions, follow-up activities, and assignment guides to help you plan effectively and efficiently.
- **The Teacher's Resource Book** contains Practice and Enrichment organized by chapter, plus Tests, Preparing for College Entrance Exams, Critical Thinking Activities, Algebra and Geometry Review, Teaching Aids, Transparencies, Technology Activities (with diskettes), and Answers.
- **The Solutions Manual** contains completely worked-out solutions to Student Text exercises.

THE APPROACH THAT DELIVERS THE POWER...

BRAND NEW!

TEACHER'S RESOURCE BOOK

PRENTICE HALL

Trigonometry

TEACHER'S EDITION

PRENTICE HALL

Trigonometry

SOLUTION MANUAL

PRENTICE HALL

Trigonometry

PRENTICE HALL

Trigonometry

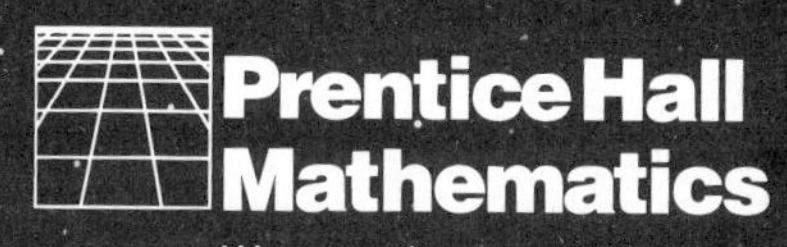

STUDENT TEXT

Distinctive...

Over 3000 Trigonometry teachers have told us that they want a program for high school students that builds understanding better than current programs do. Whereas most current programs have been developed for college students, our program has been developed specifically for high school students. Our Preview in every lesson is *the approach that delivers the power* of understanding by investigating new concepts in a real-world setting, or by reviewing important prerequisite skills before the concepts are formally developed.

The Preview at the outset of the lesson prepares the students for the lesson using a variety of techniques including reviewing the prerequisite skill needed for the lesson or informally investigating the lesson concept before it is formally developed.

3.1 Solving Right Triangles

Objectives: To solve right triangles, given the measures of one angle and one side or the measures of two sides

In this chapter, trigonometric functions will sometimes be referred to as trigonometric *ratios*, since they can be thought of as the ratios of lengths of sides of right triangles.

Preview

Recall that a calculator can be used to find both a value of a trigonometric function and an angle measure for a given trigonometric-function value.

EXAMPLE 1 **Find $\cot \frac{\pi}{5}$ to four decimal places.**

$\cot \frac{\pi}{5} = 1.3764$ *Use radian mode. Enter $\frac{\pi}{5}$ and use the tan key and the reciprocal key.*

EXAMPLE 2 **Find θ to the nearest tenth of one degree if $\sin \theta = 0.7083$ and $0° < \theta < 90°$.**

$\theta = 45.1°$ *Use degree mode. Enter 0.7083 and use the inverse key and the sine key.*

Find each value to four decimal places.

1. $\sec 203.8°$
2. $\tan \frac{\pi}{7}$
3. $\sin \frac{7\pi}{4}$

Find θ to the nearest tenth of one degree, where $0° < \theta < 90°$.

4. $\cos \theta = 0.6128$
5. $\cot \theta = 1.2341$
6. $\csc \theta = 3.1215$

The table on page 113 compares trigonometric *functions* and trigonometric *ratios*. Right triangle ABC can be used to express the values of the trigonometric ratios in terms of the sides of the triangle, where

$$a = \text{length of leg opposite } \theta$$
$$b = \text{length of leg adjacent to } \theta$$
$$c = \sqrt{a^2 + b^2} = \text{hypotenuse}$$

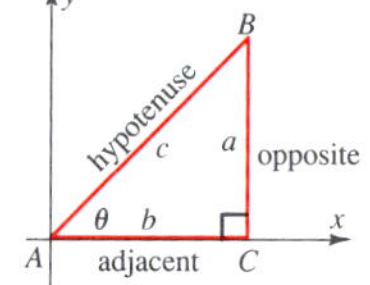

112 Chapter 3 Right Triangle Trigonometry and Basic Identities

Trigonometric Functions	Right Triangle Ratios ($0° < \theta < 90°$)
$\sin \theta = \frac{y}{r}$	$\sin \theta = \frac{a}{c} = \frac{\text{length of leg opposite } \theta}{\text{length of hypotenuse}}$
$\cos \theta = \frac{x}{r}$	$\cos \theta = \frac{b}{c} = \frac{\text{length of leg adjacent to } \theta}{\text{length of hypotenuse}}$
$\tan \theta = \frac{y}{x}$	$\tan \theta = \frac{a}{b} = \frac{\text{length of leg opposite } \theta}{\text{length of leg adjacent to } \theta}$
$\csc \theta = \frac{r}{y}$	$\csc \theta = \frac{c}{a} = \frac{\text{length of hypotenuse}}{\text{length of leg opposite } \theta}$
$\sec \theta = \frac{r}{x}$	$\sec \theta = \frac{c}{b} = \frac{\text{length of hypotenuse}}{\text{length of leg adjacent to } \theta}$
$\cot \theta = \frac{x}{y}$	$\cot \theta = \frac{b}{a} = \frac{\text{length of leg adjacent to } \theta}{\text{length of leg opposite } \theta}$

Capital letters are usually used to represent the angles of triangles, or their measures. Lowercase letters refer to the sides opposite their respective angles, or to their measures. The right triangle ratios can be used to *solve* a right triangle, that is, to find the unknown measures of the sides and angles.

EXAMPLE 1 **Solve right triangle *ABC* if $b = 32$, $\angle A = 25°$, and $\angle C = 90°$. Find *a* and *c* to the nearest unit.**

To find a, use $\tan 25°$.

$\tan 25° = \frac{a}{32}$ — *$\tan A = \frac{a}{b}$*

$a = 32 \tan 25°$ — *Calculation-ready form*

$a = 15$ — *To the nearest unit*

B
c
a
25°
A
32
C

To find c, use $\cos 25°$.

$\cos 25° = \frac{32}{c}$ — *$\cos A = \frac{b}{c}$*

$c = \frac{32}{\cos 25°} = 35$ — *To the nearest unit*

Since angles A and B are complementary, $\angle B = 90° - 25° = 65°$.

A *significant digit* is any nonzero digit or any zero that serves a purpose other than to locate the decimal point. Consider the following examples:

0.00*304*	Three significant digits
29.40	Four significant digits
30.305	Five significant digits

Step-by-step solutions to examples with the reason for each step build students' understanding of the new skill or concept.

STUDENT TEXT

Complete...

The goal of every Trigonometry teacher is to build a solid foundation of trigonometric skills and concepts that will be used in Calculus and other advanced courses. *Our approach delivers the power* to lay that foundation by providing a comprehensive and cohesive course that systematically develops the skills and concepts needed to select a successful approach to solve any problem.

Numerous models and examples provide important learning aids as students do homework assignments.

Technology is incorporated throughout. Within the lessons, technology is used in the development of trigonometric skills and concepts and as problem solving and discovery tools. Within the Practice Exercises, technology activities allow students to experiment and apply what they have learned. More detailed development of the use of technology as it applies to trigonometric skills and concepts is given in technology lessons within the chapters.

In general, the greater the number of significant digits in a measurement, the greater the accuracy. Use the table below as a guide to decide how accurate your answers should be.

Angle measures accurate to nearest . . .	1°	10′ or 0.1°	1′ or 0.01°
Significant digits in lengths of sides	2	3	4

To illustrate, if angle measures are given to the nearest degree, compute the lengths of the sides to two significant digits. If lengths are given to three significant digits, compute angle measures to the nearest 0.1°.

EXAMPLE 2 **Solve right triangle ABC if $\angle C = 90°$, $c = 7.25$, and $b = 4.37$.**

To find the value of $\angle A$, use the cosine function.

$$\cos A = \frac{4.37}{7.25}$$

$$A = 52.9°$$ *Since b and c are given to 3 significant digits, round to the nearest 0.1°.*

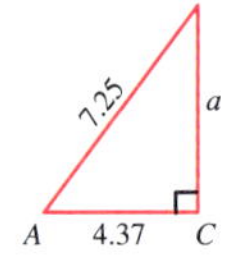

Since angles A and B are complementary, $\angle B = 90° - 52.9° = 37.1°$.

To find the value of a, use sin 52.9°.

$$\sin 52.9° = \frac{a}{7.25}$$

$$a = 7.25 \sin 52.9°$$

$$a = 5.78$$ *Round to 3 significant digits.*

In Example 3, the angle measures are expressed in degrees and minutes. If your calculator has a key that converts from degrees and minutes to decimal degrees, you can omit the next-to-last steps shown in the calculation of c and in the calculation of b.

EXAMPLE 3 **Solve triangle ABC if $a = 22.14$, $\angle B = 32°12'$, and $\angle C = 90°$.**

To find c, use sec 32°12′.

$$\sec 32°12' = \frac{c}{22.14}$$

$$c = 22.14 \sec 32°12'$$

$$c = 22.14 \sec \left(32 + \frac{12}{60}\right)^{\circ}$$

$$c = 26.16$$ *Round to 4 significant digits.*

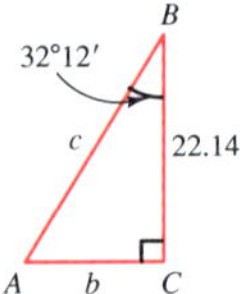

114 Chapter 3 Right Triangle Trigonometry and Basic Identities

To find b, use $\tan 32°12'$.

$$\tan 32°12' = \frac{b}{22.14}$$

$$b = 22.14 \tan 32°12'$$

$$b = 22.14 \tan\left(32 + \frac{12}{60}\right)^\circ$$

$$b = 13.94 \qquad \textit{Round to 4 significant digits.}$$

Since angles A and B are complementary, $\angle A = 90° - 32°12' = 57°48'$.

In order to solve a right triangle, it is helpful to draw a reasonably accurate figure and label the parts of known measure. You can usually choose which side or angle measure to determine first, and more than one trigonometric function may be appropriately applied. The Pythagorean theorem can also be used when the measures of the two sides are known.

To illustrate, it is possible to solve first for c, b, or $\angle A$ in Example 3. In order to find b first, $\tan B$ or $\cot B$ can be used. After b is found, $\sin B$, $\cos B$, $\sec B$, $\csc B$, or the Pythagorean theorem can be used to determine c.

CLASS EXERCISES

For triangle *ABC* at the right, express the given trigonometric function in terms of *a*, *b*, and *c*.

1. $\sin B$ **2.** $\sin A$ **3.** $\cot A$

4. $\sec A$ **5.** $\cos B$ **6.** $\tan B$

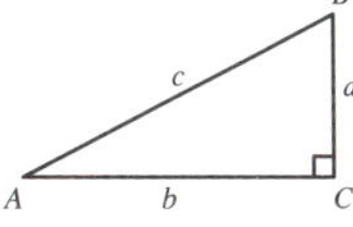

State two equations, using reciprocal functions, that could be used to find *x* or *θ*.

7.

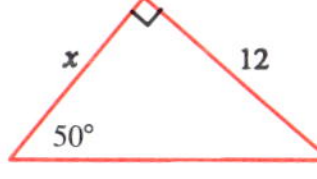

8.

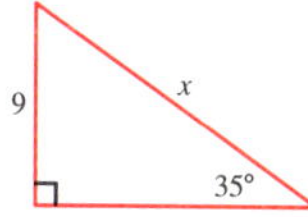

9.

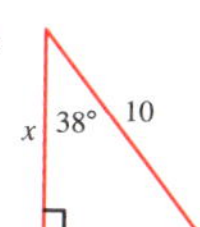

10.

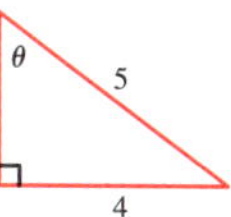

11.

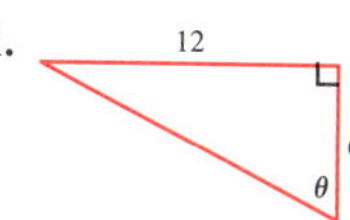

12.

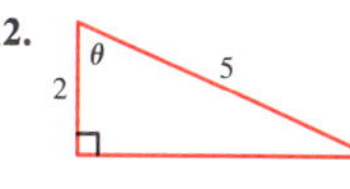

Class Exercises assess students' understanding of lesson concepts and readiness for the Practice Exercises that follow.

The Practice Exercises contain a wealth of practice for all ability levels. Although the exercise sets are labeled A, B, C in the Teacher's Edition to indicate their relative difficulty, they are unlabeled in the Student Text, thereby encouraging students to try all problems.

STUDENT TEXT

Relevant...

One of the most frequently asked questions by students is, "Why do I have to learn this?" Our chapter themes, applications at the end of the lessons, and featured applications throughout the text is *the approach that delivers the power* of trigonometric skills and concepts in the real world and in related disciplines.

Applications are included in the Practice Exercises and as special features to underscore the relevancy of trigonometric skills and concepts in the real world and in related desciplines.

PRACTICE EXERCISES

Solve each right triangle *ABC*.

1. 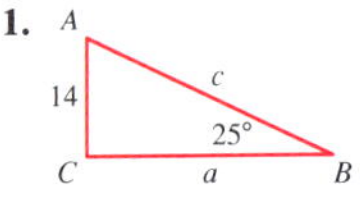

2. 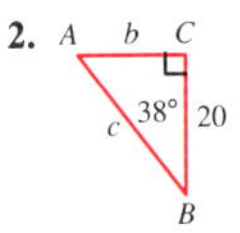

3. 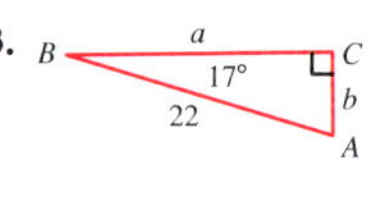

4. 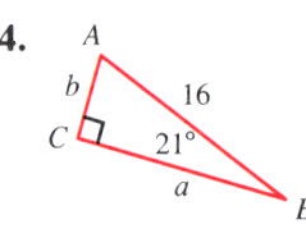

5. 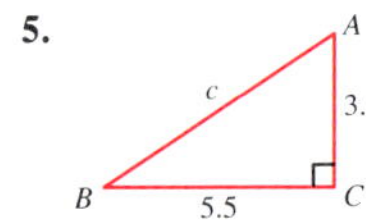

6. 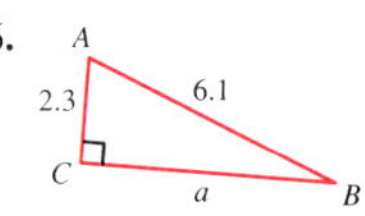

Solve each right triangle *ABC* ($\angle C = 90°$) given the measure of one angle and the length of one side.

7. $\angle A = 58°$, $c = 27$
8. $\angle B = 29°$, $c = 14$
9. $\angle B = 15.1°$, $c = 10.7$
10. $\angle A = 17.2°$, $c = 29.4$
11. $\angle A = 63°$, $a = 11$
12. $\angle B = 24°$, $b = 36$
13. $\angle B = 42.5°$, $a = 188$
14. $\angle A = 70.5°$, $b = 276$

Solve each right triangle *ABC* ($\angle C = 90°$) given the measure of one angle and the length of one side.

15. $\angle A = 36°41'$, $a = 19.32$
16. $\angle B = 42°35'$, $a = 71.22$
17. $\angle B = 72°28'$, $a = 84.84$
18. $\angle A = 80°12'$, $a = 36.22$

Solve each right triangle *ABC* ($\angle C = 90°$) given the lengths of two sides.

19. $b = 17.62$, $c = 23.91$
20. $b = 13.42$, $c = 26.31$
21. $a = 18.65$, $b = 14.22$
22. $a = 7.613$, $c = 14.05$
23. $a = 1632$, $c = 2015$
24. $a = 1503$, $b = 1635$

Use the given information for triangle *ABC* ($\angle C = 90°$) to express the other five trigonometric functions of $\angle A$ in terms of t.

25. $\sin A = \frac{3}{t}$
26. $\tan A = \frac{1}{t}$
27. $\sec A = t$
28. For any right triangle *DEF* ($\angle F = 90°$), find the numerical value of $(\cos D)^2 + (\cos E)^2 + (\cos F)^2$.

116 Chapter 3 Right Triangle Trigonometry and Basic Identities

Applications

29. Construction A ladder rests against a building at a point that is 63 ft from the ground. If the ladder makes a 47° angle with the ground, what is the length of the ladder?

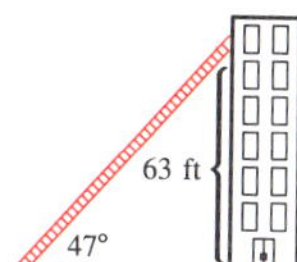

30. Geometry The base of an isosceles triangle is 14 cm in length and the angle opposite the base measures 86°. Find the length of each of the congruent sides.

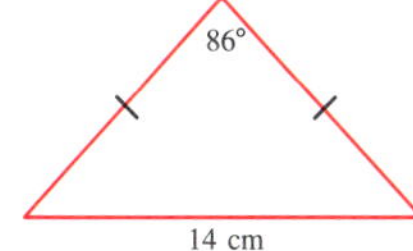

31. Architecture The Leaning Tower of Pisa is 179 ft high, making a 5.1° angle with the vertical. Find x.

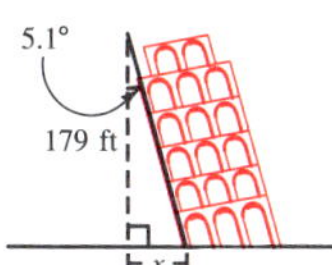

32. Sports A ski slope has a 42° incline and is 170 yd long. Find the vertical drop d.

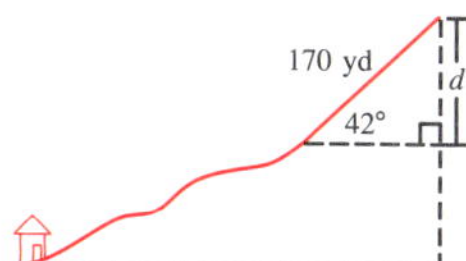

HISTORICAL NOTE

The Pythagorean Theorem There is evidence that the Pythagorean theorem was known and used long before the time of the Greek mathematician Pythagoras, who lived in the sixth century B.C. In fact, the Babylonians knew of the theorem more than 1000 years earlier. The ancient Egyptians may have been familiar with the concept also. It has been speculated that Egyptian surveyors used Pythagorean triples to construct right angles. To do this, a rope divided by knots into segments in the ratio 3:4:5 was stretched out into a triangle. The desired angle was opposite the longest segment.

Since the time of Pythagoras, this theorem has been proved in many ways. Locate two proofs of the theorem and write them in your own words.

3.1 Solving Right Triangles **117**

Special end-of-lesson features include Trigonometry in Careers, Historical Note, Extra, and Test Yourself.

TEACHER'S EDITION

More support than ever before...

The Teacher's Edition is so easy to use that planning and assessment are never a problem.

A Chapter Overview preceding each chapter capsulizes the key features of the chapter. It includes an Assignment Guide that correlates supplementary materials on a lesson-by-lesson basis to enable you to meet the needs of students of all ability levels.

OVERVIEW • Chapter 3

SUMMARY

Right-triangle trigonometry is introduced and then used to find unknown side lengths or angle measures. The concepts of angles of elevation or depression are developed and used to solve a variety of real-world problems.

Then, the fundamental reciprocal, ratio, Pythagorean, and odd-even identities are defined and proved. These identities are then used to simplify complex trigonometric expressions. This chapter concludes by using the technology of a graphing calculator or a computer to prove identities as well as the traditional method of simplifying independently one or both sides of the identity.

After this chapter is completed, students should be able to solve right triangles, and they should be able to solve a variety of problems involving angles of depression or angles of elevation. Students should be able to define and prove the fundamental identities and use these identities to write equivalent trigonometric expressions and to prove other identities. They should also be able to use a graphing calculator or computer to determine whether or not an equation is an identity.

CHAPTER OBJECTIVES

- To solve right triangles given the measures of one angle and one side or given the measures of two sides
- To define and use angles of elevation and depression
- To solve real-world problems using trigonometry
- To introduce and prove reciprocal, ratio, Pythagorean, and odd-even identities
- To use the fundamental identities to write equivalent trigonometric expressions
- To use the fundamental identities to prove other identities
- To check identities by graphing both sides of the equations

CHAPTER HIGHLIGHTS

The *theme* of Chapter 3 is lasers. A discussion of the use of lasers in compact disc players is only one of the chapter's special features.

APPLICATIONS

Right triangle trigonometry lends itself to an abundance of applications from many diverse areas. Problems from such fields as construction, architecture, sports, and engineering apply trigonometric functions and the Pythagorean theorem. Basic trigonometric identities are used in problems from genetics and computer science. Lesson 3.3 provides situations from navigation and surveying which require the use of trigonometry.

TECHNOLOGY

Calculator

Calculators are extremely useful tools for solving problems involving trigonometric functions. Students should be encouraged to use them once problems are in calculation-ready form.

Computer

Programs for evaluating trigonometric functions of specific angles and for evaluating a trigonometric expression for given angles are presented.

RESOURCES

Teacher's Resource Book

- Teaching Aid 3
- Transparencies 5 and 6

110A

T GUIDE Meeting Student Needs

	Basic	Average	Enriched	TEACHER'S RESOURCE BOOK P	E
nt	D: 116/1–23 odd, 29	D: 116/7–25 odd, 29, 31	D: 116/11–33 odd	1	2
Angles of nd	D: 121/1–15 odd R: 116/5, 8, 16	D: 121/5–19 odd R: 116/10, 16, 20	D: 121/5–19 odd R: 116/14, 16, 22	3	4
s	D: 126/1–11 odd R: 121/2, 6, 8 129/TY	D: 127/7–17 odd R: 121/6, 10, 14 129/TY	D: 127/7–17 odd R: 121/8, 12, 16 129/TY	5	6
al	D: 133/1–23 odd, 37 R: 126/2, 6, 10	D: 134/13–33 odd, 37 R: 127/8, 10, 14	D: 134/19–39 odd R: 127/8, 10, 14	7	8
ric s	D: 137/1–21 odd, 33 R: 133/6, 12, 20	D: 138/9–29 odd, 33 R: 134/18, 26, 30	D: 138/11–33 odd R: 134/22, 26, 30	9	10
ntities	D: 143/1–21 odd R: 137/4, 8, 18	D: 143/11–31 odd R: 138/14, 18, 26	D: 143/15–35 odd R: 138/18, 22, 28	11	12
Representa- tities	D: 146/1–13 odd R: 143/2, 10, 20 147/TY	D: 146/11–23 odd, 29 R: 143/12, 20, 28 147/TY	D: 146/15–29 odd R: 143/16, 20, 30 147/TY	13	14

D = Daily R = Review TY = Test Yourself P = Practice E = Enrichment

	DENT TEXT				TEACHER'S RESOURCE BOOK	
and Testing	Yourself	129, 147	College Ent. Exam Rev.	153	Tests	
	pter Sum. and Rev.	150	Maintaining Skills	154	• Quizzes	25–28
	Chapter Test	152			• Chapter Test (Form A)	29–30
					• Chapter Test (Form B)	31–32
Special Features	Historical Note	117	Extra	139	Applications—Chapter 3	15
	Challenge	123	Historical Note	144	Critical Thinking	2
	Logical Reasoning	135	Application	148	Alg. and Geom. Review	9–12
					Technology	3

110B

...Makes it one of a kind!

The convenient format accommodates easy-to-read student pages with answers in place. Plus, a wealth of teaching aids appear right next to each student page...where you need them, when you need them.

A Complete Lesson Plan for every lesson helps you meet the challenges of teaching. It includes Background Information, Teaching Suggestions, and...

...Follow-Up Activities with references to pages from the Teacher's Resource Book that relate to the lesson.

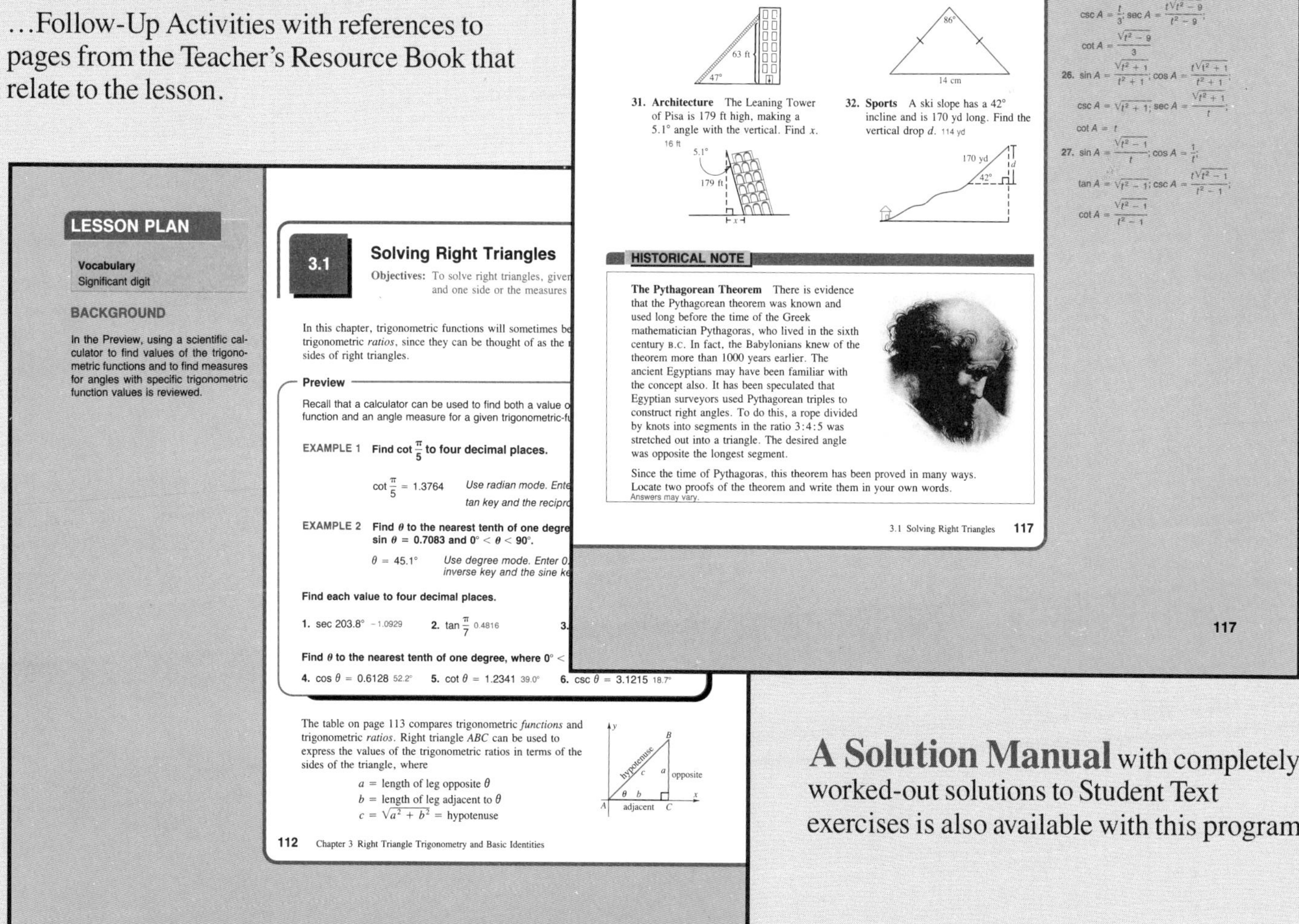

LESSON PLAN

Vocabulary
Significant digit

BACKGROUND

In the Preview, using a scientific calculator to find values of the trigonometric functions and to find measures for angles with specific trigonometric function values is reviewed.

3.1 Solving Right Triangles

Objectives: To solve right triangles, given ... and one side or the measures ...

In this chapter, trigonometric functions will sometimes be ... trigonometric *ratios*, since they can be thought of as the ... sides of right triangles.

Preview

Recall that a calculator can be used to find both a value o... function and an angle measure for a given trigonometric-f...

EXAMPLE 1 **Find $\cot \frac{\pi}{5}$ to four decimal places.**

$\cot \frac{\pi}{5} = 1.3764$ *Use radian mode. Ente... tan key and the recipro...*

EXAMPLE 2 **Find θ to the nearest tenth of one degre... $\sin \theta = 0.7083$ and $0° < \theta < 90°$.**

$\theta = 45.1°$ *Use degree mode. Enter 0... inverse key and the sine ke...*

Find each value to four decimal places.

1. sec 203.8° −1.0929 **2.** $\tan \frac{\pi}{7}$ 0.4816 **3.** ...

Find θ to the nearest tenth of one degree, where 0° < ...

4. $\cos \theta = 0.6128$ 52.2° **5.** $\cot \theta = 1.2341$ 39.0° **6.** $\csc \theta = 3.1215$ 18.7°

The table on page 113 compares trigonometric *functions* and trigonometric *ratios*. Right triangle *ABC* can be used to express the values of the trigonometric ratios in terms of the sides of the triangle, where

a = length of leg opposite θ
b = length of leg adjacent to θ
$c = \sqrt{a^2 + b^2}$ = hypotenuse

112 Chapter 3 Right Triangle Trigonometry and Basic Identities

112

Applications

29. Construction A ladder rests against a building at a point that is 63 ft from the ground. If the ladder makes a 47° angle with the ground, what is the length of the ladder? 86 ft

30. Geometry The base of an isosceles triangle is 14 cm in length and the angle opposite the base measures 86°. Find the length of each of the congruent sides. 10 cm

31. Architecture The Leaning Tower of Pisa is 179 ft high, making a 5.1° angle with the vertical. Find x. 16 ft

32. Sports A ski slope has a 42° incline and is 170 yd long. Find the vertical drop d. 114 yd

HISTORICAL NOTE

The Pythagorean Theorem There is evidence that the Pythagorean theorem was known and used long before the time of the Greek mathematician Pythagoras, who lived in the sixth century B.C. In fact, the Babylonians knew of the theorem more than 1000 years earlier. The ancient Egyptians may have been familiar with the concept also. It has been speculated that Egyptian surveyors used Pythagorean triples to construct right angles. To do this, a rope divided by knots into segments in the ratio 3:4:5 was stretched out into a triangle. The desired angle was opposite the longest segment.

Since the time of Pythagoras, this theorem has been proved in many ways. Locate two proofs of the theorem and write them in your own words. Answers may vary.

3.1 Solving Right Triangles 117

Teacher's Resource Book
Practice—Chapter 3, p. 1
Enrichment—Chapter 3, p. 2

Additional Answers

25. $\cos A = \frac{\sqrt{t^2 - 9}}{t}$; $\tan A = \frac{3\sqrt{t^2 - 9}}{t^2 - 9}$; $\csc A = \frac{t}{3}$; $\sec A = \frac{t\sqrt{t^2 - 9}}{t^2 - 9}$; $\cot A = \frac{\sqrt{t^2 - 9}}{3}$

26. $\sin A = \frac{\sqrt{t^2 + 1}}{t^2 + 1}$; $\cos A = \frac{t\sqrt{t^2 + 1}}{t^2 + 1}$; $\csc A = \sqrt{t^2 + 1}$; $\sec A = \frac{\sqrt{t^2 + 1}}{t}$; $\cot A = t$

27. $\sin A = \frac{\sqrt{t^2 - 1}}{t}$; $\cos A = \frac{1}{t}$; $\tan A = \sqrt{t^2 - 1}$; $\csc A = \frac{t\sqrt{t^2 - 1}}{t^2 - 1}$; $\cot A = \frac{\sqrt{t^2 - 1}}{t^2 - 1}$

117

A Solution Manual with completely worked-out solutions to Student Text exercises is also available with this program.

TEACHER'S RESOURCE BOOK

With a total teaching package...

Tabbed 3-ring binder contains a wealth of supplementary material organized by chapter, plus special aids and teaching resources including transparencies and computer diskettes.

Follow-Up activities for every lesson aid you in reaching all students.

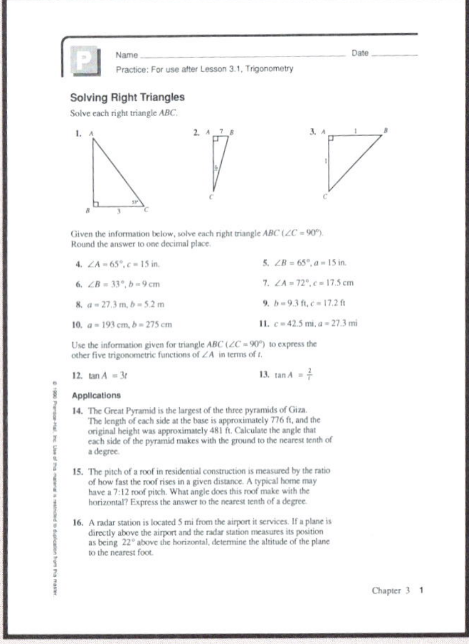

P Name ________________ Date ______

Practice: For use after Lesson 3.1, Trigonometry

Solving Right Triangles

Solve each right triangle ABC.

1.

2.

3.

Given the information below, solve each right triangle ABC ($\angle C = 90°$). Round the answer to one decimal place.

4. $\angle A = 65°$, $c = 15$ in.
5. $\angle B = 65°$, $a = 15$ in.
6. $\angle B = 33°$, $b = 9$ cm
7. $\angle A = 72°$, $c = 17.5$ cm
8. $a = 27.3$ m, $b = 5.2$ m
9. $b = 9.3$ ft, $c = 17.2$ ft
10. $a = 193$ cm, $b = 275$ cm
11. $c = 42.5$ mi, $a = 27.3$ mi

Use the information given for triangle ABC ($\angle C = 90°$) to express the other five trigonometric functions of $\angle A$ in terms of t.

12. $\tan A = 3t$
13. $\tan A = \frac{2}{t}$

Applications

14. The Great Pyramid is the largest of the three pyramids of Giza. The length of each side at the base is approximately 776 ft, and the original height was approximately 481 ft. Calculate the angle that each side of the pyramid makes with the ground to the nearest tenth of a degree.
15. The pitch of a roof in residential construction is measured by the ratio of how fast the roof rises in a given distance. A typical home may have a 7:12 roof pitch. What angle does this roof make with the horizontal? Express the answer to the nearest tenth of a degree.
16. A radar station is located 5 mi from the airport it services. If a plane is directly above the airport and the radar station measures its position as being 22° above the horizontal, determine the altitude of the plane to the nearest foot.

Chapter 3 1

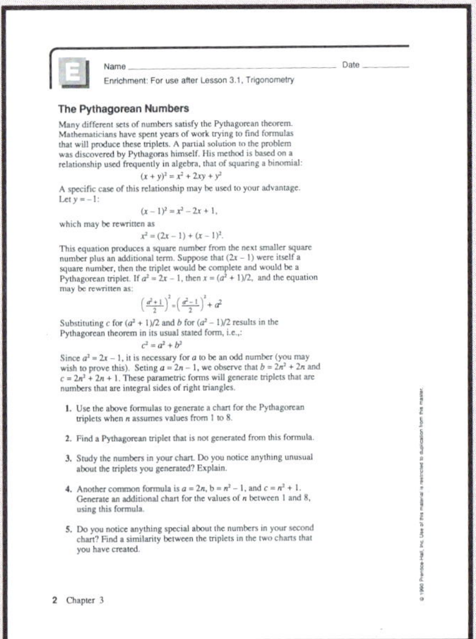

E Name ________________ Date ______

Enrichment: For use after Lesson 3.1, Trigonometry

The Pythagorean Numbers

Many different sets of numbers satisfy the Pythagorean theorem. Mathematicians have spent years of work trying to find formulas that will produce these triplets. A partial solution to the problem was discovered by Pythagoras himself. His method is based on a relationship used frequently in algebra, that of squaring a binomial:

$$(x + y)^2 = x^2 + 2xy + y^2$$

A specific case of this relationship may be used to your advantage. Let $y = -1$:

$$(x - 1)^2 = x^2 - 2x + 1,$$

which may be rewritten as

$$x^2 = (2x - 1) + (x - 1)^2.$$

This equation produces a square number from the next smaller square number plus an additional term. Suppose that $(2x - 1)$ were itself a square number, then the triplet would be complete and would be a Pythagorean triplet. If $a^2 = 2x - 1$, then $x = (a^2 + 1)/2$, and the equation may be rewritten as:

$$\left(\frac{a^2+1}{2}\right)^2 = \left(\frac{a^2-1}{2}\right)^2 + a^2$$

Substituting c for $(a^2 + 1)/2$ and b for $(a^2 - 1)/2$ results in the Pythagorean theorem in its usual stated form, i.e.,:

$$c^2 = a^2 + b^2$$

Since $a^2 = 2x - 1$, it is necessary for a to be an odd number (you may wish to prove this). Seting $a = 2n - 1$, we observe that $b = 2n^2 + 2n$ and $c = 2n^2 + 2n + 1$. These parametric forms will generate triplets that are numbers that are integral sides of right triangles.

1. Use the above formulas to generate a chart for the Pythagorean triplets when n assumes values from 1 to 8.
2. Find a Pythagorean triplet that is not generated from this formula.
3. Study the numbers in your chart. Do you notice anything unusual about the triplets you generated? Explain.
4. Another common formula is $a = 2n$, $b = n^2 - 1$, and $c = n^2 + 1$. Generate an additional chart for the values of n between 1 and 8, using this formula.
5. Do you notice anything special about the numbers in your second chart? Find a similarity between the triplets in the two charts that you have created.

2 Chapter 3

...that's the ultimate in completeness!

Special features support, strengthen, and enrich development in key areas.

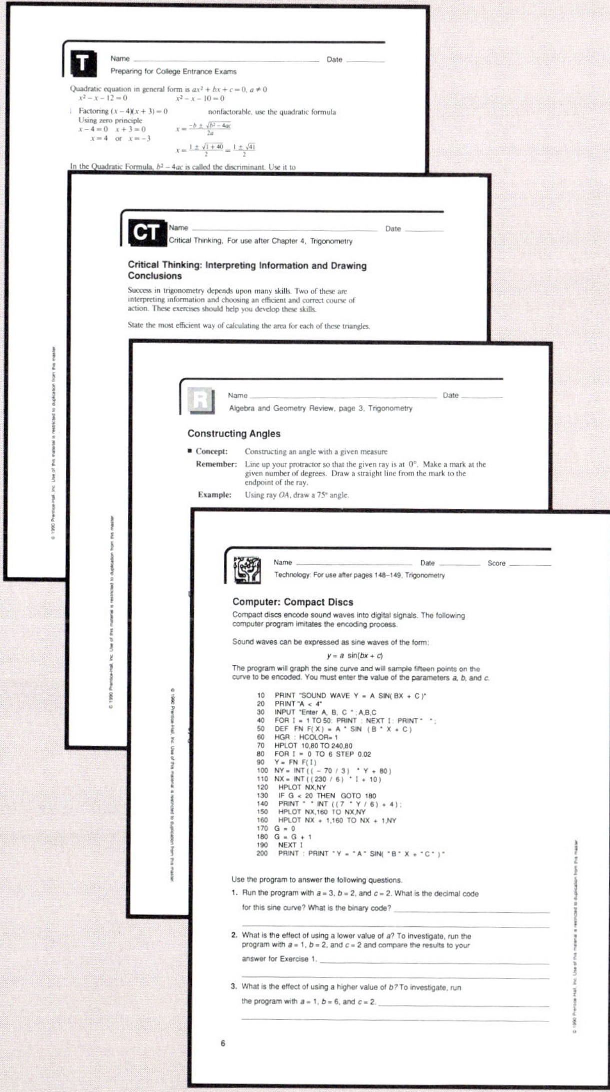

Name ______ Date ______
Preparing for College Entrance Exams

Quadratic equation in general form is $ax^2 + bx + c = 0,\ a \neq 0$

$x^2 - x - 12 = 0$ $\qquad$ $x^2 - x - 10 = 0$

Factoring $(x-4)(x+3) = 0$ $\qquad$ nonfactorable, use the quadratic formula

Using zero principle

$x - 4 = 0 \quad x + 3 = 0$ $\qquad$ $x = \frac{-b \pm \sqrt{b^2 - 4ac}}{2a}$

$x = 4 \text{ or } x = -3$ $\qquad$ $x = \frac{1 \pm \sqrt{1+40}}{2} = \frac{1 \pm \sqrt{41}}{2}$

In the Quadratic Formula, $b^2 - 4ac$ is called the discriminant. Use it to

Name ______ Date ______
Critical Thinking, For use after Chapter 4, Trigonometry

Critical Thinking: Interpreting Information and Drawing Conclusions

Success in trigonometry depends upon many skills. Two of these are interpreting information and choosing an efficient and correct course of action. These exercises should help you develop these skills.

State the most efficient way of calculating the area for each of these triangles.

Name ______ Date ______
Algebra and Geometry Review, page 3, Trigonometry

Constructing Angles

- **Concept:** Constructing an angle with a given measure
- **Remember:** Line up your protractor so that the given ray is at 0°. Make a mark at the given number of degrees. Draw a straight line from the mark to the endpoint of the ray.
- **Example:** Using ray OA, draw a 75° angle.

Name ______ Date ______ Score ______
Technology: For use after pages 148–149, Trigonometry

Computer: Compact Discs

Compact discs encode sound waves into digital signals. The following computer program imitates the encoding process.

Sound waves can be expressed as sine waves of the form:

$$y = a \sin(bx + c)$$

The program will graph the sine curve and will sample fifteen points on the curve to be encoded. You must enter the value of the parameters a, b, and c.

```
10  PRINT "SOUND WAVE Y = A SIN( BX + C )"
20  PRINT "A < 4"
30  INPUT "Enter A, B, C ";A,B,C
40  FOR I = 1 TO 50: PRINT : NEXT I: PRINT " ";
50  DEF FN F(X) = A * SIN (B * X + C)
60  HGR : HCOLOR= 1
70  HPLOT 10,80 TO 240,80
80  FOR I = 0 TO 6 STEP 0.02
90  Y = FN F(I)
100 NY = INT((- 70 / 3) * Y + 80)
110 NX = INT((230 / 6) * I + 10)
120  HPLOT NX,NY
130  IF G < 20 THEN GOTO 180
140  PRINT " " INT ((7 * Y / 6) + 4);
150  HPLOT NX,160 TO NX,NY
160  HPLOT NX + 1,160 TO NX + 1,NY
170 G = 0
180 G = G + 1
190  NEXT I
200  PRINT : PRINT "Y = "A" SIN( "B" X + "C" )"
```

Use the program to answer the following questions.

1. Run the program with $a = 3$, $b = 2$, and $c = 2$. What is the decimal code for this sine curve? What is the binary code? ______
2. What is the effect of using a lower value of a? To investigate, run the program with $a = 1$, $b = 2$, and $c = 2$ and compare the results to your answer for Exercise 1. ______
3. What is the effect of using a higher value of b? To investigate, run the program with $a = 1$, $b = 6$, and $c = 2$. ______

6

A complete testing program assesses student progress.

Name ______ Date ______
Chapter 3 Test, Form A, page 1, Trigonometry

Chapter Test

Solve each triangle ABC ($\angle C = 90°$). Round your answers to one decimal place. **ANSWERS**

1. $\angle A = 73°,\ b = 19$ cm $\qquad$ 2. $c = 21.5$ in., $a = 12.7$ in. $\qquad$ 1. ______ 2. ______
3. A monkey on a tropical island is climbing a tree to get a coconut. Because of the high winds on the coast of the island the tree has an angle of elevation of 78° 27'. If the monkey climbs 52 feet up the trunk of the tree to get the coconut, how high is he from the ground? 3. ______
4. Looking below, the monkey spots his buddy at an angle of depression of 48° 49'. How far would the monkey in the tree have to throw the coconut in order for his buddy to catch it? (For this problem, you may ignore the effects of gravity.) 4. ______
5. A plane has lost all of its power for instruments except the radio emergency power. The pilot radios airport A and finds out from radar that he is due east at a distance of 14 miles. If he wishes to land at airport B which is 137 miles due north of airport A, calculate the bearing and distance from the plane to airport B. (Note: a compass requires no power.) 5. ______

Verify each identity below for the given angle measure.

6. $\frac{1}{\sin 300°} + \frac{\tan^2 300°}{\sec^2 300°} = \frac{\sin^2 300° + 1}{\sin 300°}$ 6. ______
7. $(1 - \tan^2 240°)(\sec^2 240°) = \frac{2\cos^2 240° - 1}{\cos^4 240°}$ 7. ______

Convert the first trigonometric expression to the second expression.

8. $\frac{1}{1 - \cos x} + \frac{1}{1 + \cos x};\ 2\csc^2 x$ 8. ______
9. $\frac{\csc \gamma}{\cos \gamma} + \frac{\cos \gamma}{\sin \gamma};\ \tan \gamma$ 9. ______
10. $\frac{1 - 3\cos\beta - 4\cos^2\beta}{\sin^2\beta};\ \frac{1 - 4\cos\beta}{1 - \cos\beta}$ 10. ______

Name ______ Date ______
Chapter 3 Test, Form B, page 1, Trigonometry

Chapter Test

Solve each triangle ABC ($\angle C = 90°$). Round your answers to one decimal place. **ANSWERS**

1. $\angle B = 54°,\ b = 37$ mm $\qquad$ 2. $b = 7$ yds, $a = 2.3$ yds $\qquad$ 1. ______ 2. ______
3. A hot air balloon is being held to the ground by a 100 meter rope. The wind is blowing slightly and so the rope makes an angle of elevation of 78° 13' with the ground. If you are in the basket of the balloon, how high above ground are you? 3. ______
4. You would like to release the rope, but you cannot untie the knot at the basket. However, your friend is located on the ground at an angle of depression of 38°37'. How far must you yell to get his attention in order for him to untie the rope at the ground level? 4. ______
5. Your ship has lost all electrical power except the radio which operates on its own battery back-up power. You contact another ship and they determine by radar that you are 17 miles due west of them while they are 55 miles due south of port. Calculate the bearing and distance that you must sail in order to reach port. (Note: a compass requires no electrical power) 5. ______

Verify each identity below for the given angle measure.

6. $\frac{\sin 30°}{\csc 30°} + \frac{\cos 30°}{\sec 30°} = 1$ 6. ______
7. $\csc^2 135° = \frac{1 + \tan^2 135}{\tan^2 135°}$ 7. ______

Convert the first trigonometric expression to the second expression.

8. $\frac{\cot \gamma}{\csc \gamma + 1};\ \frac{\csc \gamma - 1}{\cot \gamma}$ 8. ______
9. $\sec x + \csc x - \cos x - \sin x;\ \sin x \tan x + \cos x \cot x$ 9. ______
10. $\frac{\sin\alpha\cos\beta + \cos\alpha\sin\beta}{\cos\alpha\cos\beta - \sin\alpha\sin\beta};\ \frac{\tan\alpha + \tan\beta}{1 - \tan\alpha\tan\beta}$ 10. ______

Tests 31

The approach that delivers the power...

	STUDENT TEXT	TEACHER'S EDITION	TEACHER'S RESOURCE BOOK
Distinctive...	• Preview at the outset of every skill lesson	• Background information provides the purpose for the skill being reviewed or the concept being investigated	• Algebra and Geometry Review Activities
	• On-going maintenance and review includes: • Chapter Summary and Review • Test-Yourself • Maintaining Skills • Preparing for College Entrance Exams • Cumulative Reviews	• Quiz for every lesson	• Complete Testing Program
Complete...	• Numerous models and examples	• Materials/Manipulatives • Additional chalkboard examples • Common Error(s) to identify often repeated errors	• Transparencies • Teaching Aids
	• Highlighted key concepts	• Lesson Vocabulary	• Critical Thinking Activities
	• Use of calculator and computer as problem-solving tools	• Teaching Suggestions for incorporating technology into the lesson	• Technology Activities • Computer diskettes
	• A wealth of carefully graded Practice Exercises	• Assignment Guide for different ability levels • Enrichment Problems	• Practice • Enrichment
Relevant...	• Theme-centered chapters • Applications at the end of the lessons • Featured Applications throughout the text	• Background information on the chapter's theme • Suggested assignments • Teaching Suggestions	• Follow-Up Applications

For more information, please write: **or call TOLL FREE: 1-800-848-9500**

PRENTICE HALL
School Division of Simon & Schuster
4343 Equity Drive, P.O. Box 2649
Columbus, OH 43216-2649

 ISBN 136-96973-9

TEACHER'S EDITION

PRENTICE HALL

Trigonometry

Jerome D. Hayden

Bettye C. Hall

1991 Printing

Printed in the United States of America.

ISBN 0-13-930843-1

10 9 8 7 6 5 4 3

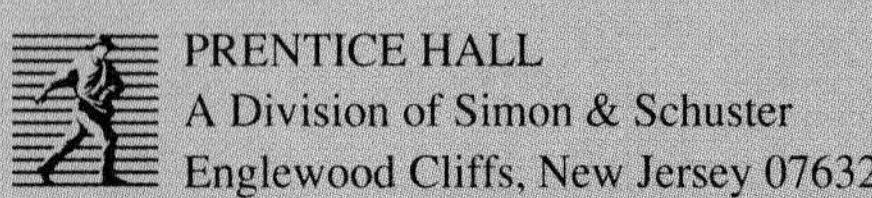
PRENTICE HALL
A Division of Simon & Schuster
Englewood Cliffs, New Jersey 07632

CONTENTS

About the Program	T16
Bibliography	T18
Pacing Chart	T19
Chapter 1 Overview	A
Chapter 2 Overview	58A
Chapter 3 Overview	110A
Chapter 4 Overview	154A
Chapter 5 Overview	206A
Chapter 6 Overview	248A
Chapter 7 Overview	284A
Chapter 8 Overview	332A
Chapter 9 Overview	370A
Additional Answers	473

ABOUT THE PROGRAM

As a teacher of higher level mathematics, you tackle a difficult job every day. Not only do you have to frequently adapt college texts for use in your high school courses, you also have to do without the support material generally available in other mathematics courses.

Prentice Hall Trigonometry has been specifically designed for high school students and offers complete teaching support. Along with a fully annotated Teacher's Edition and a Solution Manual, our program also includes a Teacher's Resource Book that provides a wealth of materials to supplement your classroom needs.

Because our program is specifically designed for high school students, it builds understanding better than current programs do. The Preview in every lesson prepares students for the lesson. Completely worked–out examples and an abundant supply of exercises and applications in every lesson ensure that students will gain a solid foundation of trigonometric skills and concepts.

The flexibility of this program allows you to choose from many approaches to make Trigonometry work. Real world situations, teaching suggestions involving a variety of techniques, and the inclusion of technology are but a few of the approaches from which you may choose.

Structure of the Student Text

Prentice Hall Trigonometry is a comprehensive and cohesive course that lays a solid foundation of trigonometric skills and concepts and prepares students to use them confidently in mathematics and in related disciplines. Our systematic development of the skills necessary to select a successful approach to solve any problem will enable students to become proficient in these skills and to go on to study other advanced topics.

In order to succeed in mathematics, you must first comprehend the language of mathematics. The Student Text accomplishes this with a readable text that includes the following:

- *Lesson Objective(s)* tell the students at the outset what is expected of them.
- *The Preview* at the outset of a lesson prepares the students for the lesson, using a variety of techniques such as reviewing the prerequisite skill needed for the lesson or informally investigating the lesson concept before it is formally developed.
- *Highlighted key concepts* are boxed in red to underscore their importance.
- *Completely worked–out* examples provide important learning aids as students do their homework assignments. Solutions to examples are given step by step along with the reason for each step.
- *Class Exercises* assess students' understanding of lesson concepts and readiness for the Practice Exercises that follow.
- *Practice Exercises* encompass a wealth of carefully graded practice including real world applications to underscore the relevancy of the topic.
- *Special Features* at the end of lessons include Critical Thinking, Careers, Did You Know?, and Test Yourself.
- *Applications* are special chapter features that focus on trigonometric concepts applied to real world situations.

The complete testing and review program consists of:

- *Test Yourself.* A test midway and at the end of each chapter
- *Chapter Test.* One-page test at the end of each chapter
- *Chapter Summary and Review.* Two-page summary and review of chapter's content including important vocabulary

- *Cumulative Review.* One- to three-page review after all even-numbered chapters
- *Maintaining Skills Review.* One-page review of Algebra and Geometry skills after all chapters that do not contain a Cumulative Review
- *College Entrance Exam Review.* A review patterned to reflect the actual Scholastic Aptitude Test (SAT) and Advanced Placement Test (Levels I and II) in mathematics

Structure of the Teacher's Edition

The Teacher's Edition is designed to provide both the experienced and novice Trigonometry teacher with more features than ever before to plan effectively and assess student understanding. Among the many features is a *Pacing Chart* that suggests how many days might be allocated to each lesson for three possible courses of study in a one-semester course.
Each chapter is preceded by a *Chapter Overview* which includes the following key features:

- *Summary.* A brief description of the chapter's content
- *Chapter Objectives.* A complete listing of the lesson objectives
- *Chapter Highlights.* Key features contained within the chapter including the chapter theme, problem solving and applications in the chapter, calculator and computer features, and available resources
- *Assignment Guide.* A chart that provides suggested assignments and related resources (found in the Teacher's Resource Book) on a lesson by lesson basis, plus review, testing and other special features in the chapter and in the Teacher's Resource Book

Throughout the book, nearly full–size student text pages have answers in place, plus teaching aids for every lesson right alongside. These aids include the following:

- *Vocabulary.* A listing of important terms introduced in the lesson
- *Materials/Manipulatives.* Concrete materials that may be used to present lesson concepts
- *Background.* Suggestions for using the Preview
- *Teaching Suggestions.* Helpful recommendations on how to present the lesson including Critical Thinking Skills, extra Chalkboard Examples, and Common Errors
- *Follow–Up.* Activities that provide practice and enrichment for each lesson, a quiz on the lesson content, plus references to pages in the Teacher's Resource Book that relate to the lesson
- *Labeled Exercises.* Labels indicating degree of difficulty of exercise sets noted only in the Teacher's Edition

Technology: Using Calculators and Computers

The integration of calculators and computers in the high school mathematics curriculum is a critical issue that concerns teachers. You are the only one who can determine the extent to which technology can be integrated into your classroom. Our approach to using the calculator and computer in developing trigonometric concepts provides you with all the options necessary for you to be most effective.

In the Student Text technology is incorporated throughout. Within the lessons, technology is used in the development of trigonometric skills and concepts as a problem solving and discovery tool. Within the Practice Exercises, technology activities allow students to experiment and apply what they have learned. More detailed development of the use of technology as it applies to trigonometric skills and concepts is given in technology lessons within the chapters.

In the Teacher's Edition you are alerted to the use of technology within a chapter in the Chapter Overview immediately preceding the chapter. Here you will also find references to related Technology Activities that may be found in the Teacher's Resource Book. Also included in the Teacher's Edition are Teaching Suggestions on how to integrate technology within your daily Lesson Plan.

In the Teacher's Resource Book additional activities involving technology are provided in a special section, together with a computer diskette that contains all the programs written exclusively for *Prentice Hall Trigonometry*. This diskette is duplicable, so you may make as many copies as you wish. Students with no previous programming experience can use the activities along with the diskette that has been provided for them. Students with some experience can extend these programs using Apple BASIC and Appleworks.

BIBLIOGRAPHY

BOOKLETS AND PERIODICALS

Commission on Standards for School Mathematics, National Council of Teachers of Mathematics. *Curriculum and Evaluation Standards for School Mathematics.* Reston, VA: NCTM, 1988.

National Council of Teachers of Mathematics. *An Agenda for Action : Recommendations for School Mathematics of the 1980's.* Reston, VA: NCTM, 1980.

National Council of Teachers of Mathematics. *Journal for Research in Mathematics Education*, January, March, May, July, and November. Reston, VA:NCTM.

National Council of Teachers of Mathematics. *The Mathematics Teacher.* Monthly September through May. Reston, VA: NCTM.

BOOKS

Barnsley, Michael. *Fractals Everywhere.* San Diego, CA: Academic Press, Inc., 1988.

Bell, Eric Temple. *Mathematics: Queen & Servant of Science.* Mathematical Association of America. Washington, DC: 1979.

Boles, Martha, and Newman, Rochelle. *The Golden Relationship : Art, Math, Nature.* Bradford, MA: Pythagorean Press, 1983.

College Entrance Examination Board. *Academic Preparation in Mathematics: Teaching for Transition from High School to College.* New York: CEEB, 1988.

Eves, Howard. *An Introduction to the History of Mathematics.* New York: Holt Rinehart and Winston, 1976.

Gardner, Martin. *The Second Scientific American Book of Mathematical Puzzles and Diversions.* Chicago: University of Chicago Press, 1987.

Graham, Neill. *Computers and Computing.* St. Paul, MN: West Publishing Company, 1983.

Grinstein, Louise S., and Paul J. Campbell. *Women of Mathematics:* A Bibliographic Sourcebook. Westport, CT: Greenwood Press, 1987.

The Ideas of Algebra, K–12. *The 1988 Yearbook of the Council of Teachers of Mathematics.* Reston, VA: NCTM, 1988.

Jacobs, Harold R. *Mathematics: A Human Endeavor,* 2nd ed. San Francisco: W. H. Freeman & Company, 1982.

James, Glenn, and Robert C. eds. *Mathematics Dictionary.* New York: Van Nostrand Reinhold, 1976.

Krulik, Stephen, ed. *Problem Solving in School Mathematics.* 1980 Yearbook of the National Council of Teachers of Mathematics. Reston, VA: NCTM, 1980.

Olds, C. D. *Continued Fractions.* Vol. 9 of New Mathematical Library, Mathematical Association of America, Washington, DC.

Papert, Seymour. *Mindstorms: Children, Computers, and Powerful Ideas.* New York: Basic Books, 1980.

Perl, T. *Math Equals—Biographies of Women Mathematicians and Related Activities.* Menlo Park, CA: Addison Wesley Publishing Company, 1978.

Peterson, Ivars. *The Mathematical Tourist.* New York: W. H. Freeman, 1988.

Polya, George. *How to Solve It: A New Aspect of Mathematical Method.* 2nd ed. Princeton, NJ: Princeton University Press, 1973.

Sobel, Max, and Evan M. Maletsky. *Teaching Mathematics; A Sourcebook of Aids, Activities, and Strategies.* 2nd ed. Englewood Cliffs, NJ: Prentice Hall, 1988.

Tobias, Sheila. *Succeed with Math: Every Student's Guide to Conquering Math Anxiety.* New York: College Board Publications, 1987.

West, Beverly H., et al. *The Prentice– Hall Encyclopedia of Mathematics.* Englewood Cliffs, NJ: Prentice Hall, 1982.

PACING TRIGONOMETRY

The following three charts suggest how a total of 85 class days may be allocated by chapter for three different levels of ability.

BASIC

Chapter	1	2	3	4	5	6	7	8	9
Days	16	13	11	12	7	7	12	0	7
Section	all	1–4, 6, 7	all	1–5, 7	1–5	1–5	1–7	none	1–6

AVERAGE

Chapter	1	2	3	4	5	6	7	8	9
Days	12	14	10	12	8	8	13	0	8
Section	all	all	all	all	all	all	all	none	all

ENRICHED

Chapter	1	2	3	4	5	6	7	8	9
Days	11	12	9	11	7	7	11	9	8
Section	all	all	all	all	all	all	all	all	all

A more detailed pacing chart is provided on the following pages. It is based on the number of days allocated per chapter suggested in the above three charts. Each of the 85 class days is represented (including reviews and tests) with text pages and practice exercises to be assigned.

D = Daily R = Review TY = Test Yourself Cum.Rev. = Cumulative Review
S&R = Summary and Review

TRIGONOMETRY PACING CHART

Day	Basic	Average	Enriched
1	**1.1** 5/1–19 odd	**1.1** 6/21–51 odd, 57	**1.1** 6/29–59 odd
2	**1.1** 6/21–37 odd, 57	**1.2** 10/9–35 odd, 39, 41 R 6/30, 42, 54	**1.2** 10/11, 17, 23–43 odd R 6/44, 48, 52
3	**1.2** 10/1–23 odd, 39 R 5/4, 12, 24	**1.3** 15/9–41 odd, 51 R 10/4, 16, 30	**1.3** 15/11, 19, 29–51 odd R 10/12, 18, 38
4	**1.3** 15/1–27 odd, 51 R 10/4, 14, 22	**1.4** 21/19, 31, 43–57 odd R 15/10, 32	**1.4** 21/23, 41, 45, 53, 59–87 odd R 15/12, 20, 36
5	**1.4** 21/1–23 odd R 15/2, 24	**1.4** 21/59–75 odd, 85 R 15/18	**1.5** 27/5, 11, 17, 23–33 odd R 21/30, 50, 72 28/Test Yourself
6	**1.4** 25–41 odd, 85 R 15/16	**1.5** 27/3, 9, 15, 21–31 R 21/26, 48, 70 28/Test Yourself	**1.6** 32/11, 15, 17–31 odd R 27/12, 18, 24
7	**1.5** 26/1–11 odd R 21/4, 22	**1.6** 32/5–27 odd R 27/10, 16, 26	**1.7** 37/11, 21–43 odd R 32/12, 22, 26
8	**1.5** 27/13–23 odd R 21/10 28/Test Yourself	**1.7** 37/9, 17–37 odd, 43 R 32/6, 10, 24	**1.8** 44/23–45 odd R 37/12, 26
9	**1.6** 32/1–21 odd R 26/4, 8, 14	**1.8** 44/19–37 odd R 37/10, 24	**1.8** 45/47–71 odd R 38/32
10	**1.7** 37/1–15 odd R 32/2, 10	**1.8** 44/39–61 odd, 69 R 37/30	**1.9** 50/21–59 odd R 44/42, 46, 54 51/Test Yourself
11	**1.7** 37/17–27 odd, 43 R 32/30	**1.9** 50/13–55 odd R 44/32, 42, 40 51/Test Yourself	54/Chap. 1 S&R, Test

Day	Basic	Average	Enriched
12	**1.8** 44/1–25 odd R 37/4, 14	54/Chap. 1 S&R, Test	**2.1** 64/3, 11, 19–35 odd, 39, 41
13	**1.8** 44/27–41 odd, 69 R 37/22	**2.1** 64/3, 9, 13–33 odd, 39	**2.2** 71/11, 13, 17–23 odd R 64/4, 22
14	**1.9** 50/1–23 odd R 44/20, 32	**2.2** 71/11, 17, 21–27 odd R 64/4, 16	**2.2** 71/25–41 odd R 64/28
15	**1.9** 50/25–37 odd R 44/36 51/Test Yourself	**2.2** 71/29–37 odd, 41 R 64/28	**2.3** 75/1–39 odd R 71/14, 24
16	54/Chap. 1 S&R, Test	**2.3** 75/1–15 odd R 71/12	**2.4** 82/5–15 odd R 75/10, 12
17	**2.1** 64/1–11 odd	**2.3** 75/17–29 odd, 37 R 71/22	**2.4** 82/17–31 odd R 75/16 83/Test Yourself
18	**2.1** 64/13–27 odd, 39	**2.4** 82/5–17 odd R 75/6, 8	**2.5** 86/9–21 odd R 82/6, 8, 22
19	**2.2** 71/1–13 odd R 64/4, 16	**2.4** 82/19–25 odd, 29 R 75/10 83/Test Yourself	**2.6** 92/9–17 odd R 86/10
20	**2.2** 71/15–23 odd, 41 R 64/26	**2.5** 86/5–15 odd, 21 R 82/4, 6, 22	**2.6** 92/19–27 odd R 86/16, 20
21	**2.3** 75/1–13 odd R 71/4, 10	**2.6** 92/7–17 odd R 86/6, 10	**2.7** 98/9–27 odd R 92/10, 14, 16
22	**2.3** 75/15–21 odd, 37 R 71/22	**2.6** 92/19–25 odd R 86/14	**2.8** 103/5–15 odd R 98/10, 12, 18 104/Test Yourself
23	**2.4** 82/1–13 odd R 75/4, 6	**2.7** 98/5–13 odd R 92/8, 14	106/Chap. 2 S&R, Test, Cum, Rev.

Day	Basic	Average	Enriched
24	**2.4** 82/15–21 odd, 29 R 75/8 83/Test Yourself	**2.7** 98/15–25 odd R 92/16	**3.1** 116/11–33 odd
25	**2.6** 92/1–9 odd R 82/4, 10	**2.8** 103/5–15 odd R 98/10, 12, 18 104/Test Yourself	**3.2** 121/5–19 odd R 116/14, 16, 22
26	**2.6** 92/11–19 odd R 82/14	106/Chap. 2 S&R, Test, Cum. Rev.	**3.3** 127/7–17 odd R 121/8, 12, 16 129/Test Yourself
27	**2.7** 98/1–9 odd R 92/2, 12	**3.1** 116/7–25 odd, 29, 31	**3.4** 134/19–39 odd R 127/8, 10, 14
28	**2.7** 98/11–19 odd R 92/20	**3.2** 121/5–19 odd R 116/10, 16, 20	**3.5** 138/11–33 odd R 134.22, 26, 30
29	106/Chap. 2 S&R, Test, Cum. Rev.	**3.3** 127/7–17 odd R 121/6, 10, 14 129/Test Yourself	**3.6** 143/15–25 odd R 138/18, 28
30	**3.1** 116/1–13 odd	**3.4** 134/13–23 odd R 127/8, 10	**3.6** 143/27–35 odd R 138/22
31	**3.1** 116/15–23 odd, 29	**3.4** 134/25–33 odd, 37 R 127/14	**3.7** 146/15–29 odd R 143/16, 20, 30 147/Test Yourself
32	**3.2** 121/1–15 odd R 116/5, 8, 16	**3.5** 138/9–29 odd, 33 R 134/18, 26, 30	150/Chap. 3 S&R, Test
33	**3.3** 126/1–11 odd R 121/2, 6, 8 129/Test Yourself	**3.6** 143/11–21 odd R 138/14, 18	**4.1** 161/11–23 odd
34	**3.4** 133/1–13 odd R 126/2, 6	**3.6** 143/23–31 odd R 138/26	**4.1** 161/25–39 odd

Day	Basic	Average	Enriched
35	**3.4** 134/15–23 odd, 37 R 126/10	**3.7** 146/11–23 odd, 29 R 143/12, 20, 28 147/Test Yourself	**4.2** 166/13–25 odd R 161/14, 16, 24
36	**3.5** 137/1–21 odd, 33 R 133/6, 12, 20	150/Chap. 3 S&R, Test	**4.3** 173/17–27 odd R 166/12, 18
37	**3.6** 143/1–11 odd R 137/4, 8	**4.1** 160/7–17 odd	**4.3** 173/29–39 odd R 166/26
38	**3.6** 143/13–21 odd R 137/18	**4.1** 161/19–33 odd, 37	**4.4** 170/5–23 odd R 172/16, 20, 24 180/Test Yourself
39	**3.7** 146/1–13 odd R 143/2, 10, 20 147/Test Yourself	**4.2** 166/9–31 odd, 35 R 161/10, 16, 24	**4.5** 184/13–35 odd R 179/6, 12, 16
40	150/Chap. 3 S&R, Test	**4.3** 172/7–21 odd R 166/10, 18	**4.6** 189/7–25 odd R 184/16, 18, 24
41	**4.1** 160/1–13 odd	**4.3** 173/23–31 odd, 35, 37 R 166/26	**4.7** 195/3, 9, 15–31 odd R 189/10, 16
42	**4.1** 161/15–25 odd, 37	**4.4** 179/3–11 odd R 172/10, 14	**4.7** 196/33–45 odd R 189/18 197/Test Yourself
43	**4.2** 166/1–15 odd R 160/4, 16	**4.4** 179/13–23 odd R 172/20 180/Test Yourself	200/Chap. 4 S&R, Test, Cum. Rev.
44	**4.2** 166/17–23 odd, 35 R 160/20	**4.5** 184/11–29 odd, 33 R 179/6, 12, 16	**5.1** 212/7, 11–31 odd
45	**4.3** 172/1–15 odd R 166/4, 12	**4.6** 189/7–25 odd R 184/12, 18, 22	**5.2** 218/5, 11, 15, 17–31 odd R 212/8, 14, 20

Day	Basic	Average	Enriched
46	**4.3** 173/17–23 odd, 35 R 166/18	**4.7** 195/3, 9, 13–25 odd R 189/8, 14	**5.3** 223/7, 11, 15, 17–33 odd R 218/8, 12, 20 225/Test Yourself
47	**4.4** 179/1–11 odd R 172/4, 12	**4.7** 195/27–39 odd, 45 R 189/18 197/Test Yourself	**5.4** 229/11–33 odd R 223/8, 12, 20
48	**4.4** 179/13–17 odd, 23 R 172/18	200/Chap. 4 S&R, Test, Cum. Rev.	**5.5** 235/9–33 odd R 229/12, 18, 24
49	**4.5** 184/1–19 odd, 33 R 179/2, 8, 14	**5.1** 212/5, 9–19 odd	**5.6** 240/3, 7, 11, 15–29 odd R 235/12, 22, 28 241/Test Yourself
50	**4.7** 195/1–17 odd R 184/2, 10 179/10	**5.1** 213/21–27 odd, 31	244/Chap. 5 S&R, Test
51	**4.7** 195/19–27 odd R 184/18 197/Test Yourself	**5.2** 218/5, 9–27 odd R 212/4, 14, 20	**6.1** 253/9–37 odd
52	200/Chap. 4 S&R, Test, Cum. Rev.	**5.3** 223/5, 9–29 odd, 33 R 218/6, 10, 18 225/Test Yourself	**6.2** 257/7, 15–49 odd R 253/14, 18, 20
53	**5.1** 212/1–11 odd	**5.4** 229/9–29 odd, 33	**6.3** 263/3, 7, 13–29 odd 264.Test Yourself
54	**5.1** 212/13–29 odd, 31	**5.5** 235/7–29 odd, 33 R 229/10, 18, 20	**6.4** 267/5. 11, 17–41 odd R 263/4, 16, 24
55	**5.2** 218/1–19 odd R 212/4, 10, 18	**5.6** 240/3, 7, 11 R 235/8, 18, 26 241/Test Yourself	**6.5** 271/17–37 odd R 267/6, 24, 34
56	**5.3** 223/1–19 odd R 218/2, 14, 18 225/Test Yourself	244/Chap. 5 S&R, Test	**6.6** 276/7, 13, 29 odd R 271/22, 26, 30 277/Test Yourself

Day	Basic	Average	Enriched
57	**5.4** 228/1–21 odd, 33 R 223/4, 10, 18	**6.1** 253/7–25 odd	280/Chap. 6 S&R, Test, Cum. Rev.
58	**5.5** 235/1–23 odd R 229/6, 14, 20	**6.1** 254/27–33 odd, 37	**7.1** 290/7–17 odd
59	244/Chap. 5 S&R, Test	**6.2** 257/7, 13–43 odd, 49 R 253/10, 16, 22	**7.1** 291/29–45 odd
60	**6.1** 253/1–13 odd	**6.3** 263/3, 7–25 odd, 29 R 257/8, 14, 34 264/Test Yourself	**7.2** 296/1, 5, 11–21 odd R 290/10, 16
61	**6.1** 253/15–27 odd, 37	**6.4** 267/3, 9, 13–35 odd R 263/2, 14, 22	**7.2** 296/23–35 odd R 290/28, 32
62	**6.2** 257/1–31 odd R 253/2, 6, 20	**6.5** 271/11–31 odd, 37 R 267/4, 16, 22	**7.3** 301/13, 19–59 odd R 296/2, 6, 16
63	**6.3** 263/1–21 odd R 257/4, 10, 26 264/Test Yourself	**6.6** 276/5, 11–25 odd, 29 R 271/12, 22, 28 277/Test Yourself	**7.4** 305/15–59 odd R 301/20, 34, 44 307/Test Yourself
64	**6.4** 267/1–25 odd R 263/2, 6, 22	280/Chap. 6 S&R, Test, Cum. Rev.	**7.5** 311/7, 11, 19–43 odd R 305/18, 28, 36
65	**6.5** 271/1–23 odd R 267/2, 8, 14	**7.1** 290/7–25 odd	**7.6** 316/9, 15–33 odd R 311/12, 20, 30
66	280/Chap. 6 S&R, Test, Cum. Rev.	**7.1** 290/27–41 odd, 45	**7.7** 321/11–29 odd R 316/10, 14, 20
67	**7.1** 290/1–15 odd	**7.2** 296/1, 3, 9–17 odd R 290/8, 16	**7.8** 325/9, 13, 17–39 odd R 321/12, 22, 24 326/Test Yourself
68	**7.1** 290/17–35 odd	**7.2** 296/19–29 odd, 33 R 290/26, 30	328/Chap. 7 S&R, Test

Day	Basic	Average	Enriched
69	**7.2** 296/1–15 odd R 290/2, 14	**7.3** 301/11–33 odd R 296/2, 4	**8.1** 336/3, 9, 17–41 odd
70	**7.2** 296/17–23 odd R 290/24, 30	**7.3** 301/35–49 odd, 57 R 296/16	**8.2** 341/3, 7, 17–33 odd R 336/4, 10, 30
71	**7.3** 301/1–19 odd R 296/2, 4	**7.4** 305/13–35 odd R 301/14, 24	**8.3** 345/7, 15, 21–59 odd R 341/4, 16, 24 346/Test Yourself
72	**7.3** 301/21–37 odd, 57 R 296/14	**7.4** 306/37–51 odd, 57 R 301/42 307/Test Yourself	**8.4** 351/17–47 odd R 345/8, 16
73	**7.4** 305/1–27 odd R 301/4, 22	**7.5** 311/5, 9, 15–35 odd, 41 R 305/14, 24, 32	**8.4** 352/49–67 odd R 345/28
74	**7.4** 306/29–39 odd, 57 R 301/36	**7.6** 316/5, 9, 15–29 odd R 311/8, 16, 24	**8.5** 357/5, 11, 17, 23–39 R 351/16, 30
75	**7.5** 311/1–25 odd, 41 R 305/2, 20, 44	**7.7** 321/7–25 odd, 29 R 316/6, 16, 20	**8.5** 358/41–65 odd R 351/46
76	**7.6** 315/1–19 odd R 311/2, 14, 22	**7.8** 325/9–35 odd R 321/6, 20, 24 326/Test Yourself	**8.6** 363/5–19 odd R 357/24, 44, 50 364/Test Yourself
77	**7.7** 320/1–19 odd, 29 R 315/2, 12, 18	328/Chap. 7 S&R, Test	366/Chap. 8 S&R, Test, Cum. Rev.
78	828/Chap. 7 S&R, Test	**9.1** 378/3, 9, 17–43 odd, 49	**9.1** 378/5, 11, 17, 23–49 odd
79	**9.1** 378/1–29 odd, 49	**9.2** 383/3, 9, 13–37 odd, 41 R 378/4, 10, 28	**9.2** 383/5, 11, 15–43 odd R 378/6, 12, 30
80	**9.2** 383/1–27 odd, 41 R 378/2, 8, 20	**9.3** 389/5, 15–43 odd, 49 R 383/10, 16, 26	**9.3** 389/7, 19–51 odd R 383/12, 18, 28

Day	Basic	Average	Enriched
81	**9.3** 389/1–31 odd, 49 R 383/2, 8, 20	**9.4** 396/396/9–39 odd, 45 R 389/16, 26, 32 397/Test Yourself	**9.4** 396/15–47 odd R 390/18, 28, 34 397/Test Yourself
82	**9.4** 395/1–29 odd, 45 R 389/4, 14, 26 397/Test Yourself	**9.5** 402/9–33 odd, 37, 39 R 396/14, 24, 28	**9.5** 402/9–41 odd R 396;14, 26, 30
83	**9.5** 402/1–25 odd, 37 R 395/4, 14, 24	**9.6** 407/3, 7, 11–37 odd R 402/10, 16, 20	**9.6** 407/3, 7, 11–47 odd R 402/12, 18, 22
84	**9.6** 407/1–27 odd R 402/6, 14, 20	**9.7** 412/3, 9–21 odd, 25 R 407/18, 26, 30 413/Test Yourself	**9.7** 412/3, 9–25 odd R 407/20, 28, 32 413/Test Yourself
85	416/Chap. 9 S&R, Test	416/Chap. 9 S&R, Test	416/Chap. 9 S&R, Test

PRENTICE HALL

Trigonometry

Jerome D. Hayden

Bettye C. Hall

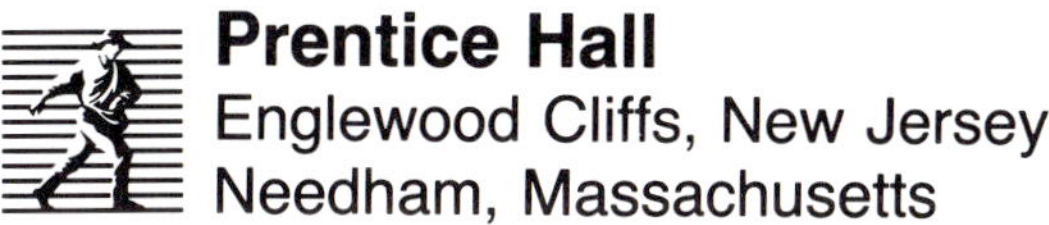
Prentice Hall
Englewood Cliffs, New Jersey
Needham, Massachusetts

Prentice Hall Trigonometry

Student Text
Teacher's Edition
Solution Manual
Teacher's Resource Book

AUTHORS

Jerome D. Hayden
Mathematics Department Chairman
McLean County District 5
Normal, Illinois

Bettye C. Hall
Director of Mathematics
Houston Independent School District
Houston, Texas

REVIEWERS

Calvin T. Long
Professor of Mathematics
Washington State University
Pullman, Washington

John F. Lamb
Professor of Mathematics
East Texas State University
Commerce, Texas

Francis Yu-Chaw Meng
School of Computer Science
Rochester Institute of Technology
Rochester, New York

CONSULTANTS

Herbert Hollister
Professor of Mathematics
Bowling Green State University
Bowling Green, Ohio

Glenn Miller
Mathematics Instructor
Interboro Institute
New York, New York

Photo credits begin on page 472.
Front cover: Stock Market.
Back cover: Dan McCoy/Rainbow; Hank Morgan/Rainbow.

1991 Printing

Printed in the United States of America.

ISBN 0-13-930835-0

10 9 8 7 6 5 4 3

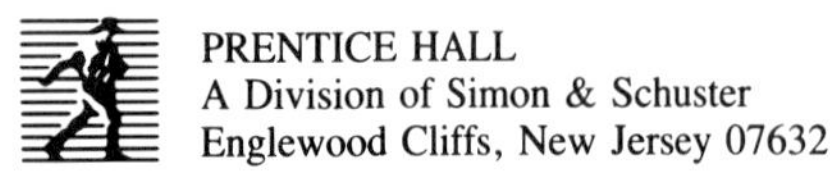

PRENTICE HALL
A Division of Simon & Schuster
Englewood Cliffs, New Jersey 07632

CONTENTS

1 TRIGONOMETRIC FUNCTIONS

1.1 Functions and the Coordinate Plane 2
Applications: Meteorology, Geometry

1.2 The Distance Formula 8
Applications: Navigation, Construction, Engineering

1.3 Angles in the Coordinate Plane 12
Applications: Entertainment

1.4 Angle Measure in Degrees and Radians 17
Applications: Horology, Meteorology

1.5 Applications: Angular and Linear Velocity 23

1.6 Cosine and Sine Functions 29
Applications: Computer

1.7 The Trigonometric Functions 34
Applications: Surveying, Construction

1.8 Trigonometric Functions of Special Angles 39
Applications: Construction, Routing

1.9 Evaluating Trigonometric Functions 46
Applications: Computer

Featured Application: Space Exploration 52
Chapter Summary and Review 54, Chapter Test 56
College Entrance Exam Review 57, Maintaining Skills 58

2 GRAPHING TRIGONOMETRIC FUNCTIONS

2.1 Periodic Functions and Symmetry 60
Applications: Landscaping, Algebra, Architecture

2.2 Graphs of the Sine and Cosine Functions 67
Applications: Algebra

2.3 Amplitude and Period 73
Applications: Physics

2.4 Phase Shift and Vertical Shift 78
Applications: Music

2.5 Graphing by Addition of Ordinates 84
Applications: Music, Oceanography

2.6 Graphs of the Tangent and Cotangent Functions 88
Applications: Graphing Calculator

2.7 Graphs of the Secant and Cosecant Functions 94
Applications: Engineering

2.8 Application: Harmonic Motion 100

Featured Application: Magnetic Resonance Imaging 105
Chapter Summary and Review 106, Chapter Test 108
College Entrance Exam Review 109, Cumulative Review 110

3 RIGHT TRIANGLE TRIGONOMETRY AND BASIC IDENTITIES

3.1 Solving Right Triangles 112
Applications: Construction, Geometry, Architecture, Sports
3.2 Application: Angles of Elevation and Depression 118
3.3 Applications 124
3.4 Fundamental Identities 130
Applications: Engineering, Aviation, Computer
3.5 Equivalent Trigonometric Expressions 136
Applications: Genetics, Computer
3.6 Proving Identities 140
Applications: Computer
3.7 Graphical Representation of Identities 145
Applications: Physics, Metallurgy

Featured Application: Compact Discs 148
Chapter Summary and Review 150, Chapter Test 152
College Entrance Exam Review 153, Maintaining Skills 154

4 OBLIQUE TRIANGLES

4.1 The Law of Sines 156
Applications: Surveying
4.2 Law of Sines: The Ambiguous Case 163
Applications: Engineering, Surveying
4.3 The Law of Cosines 168
Applications: Surveying, Navigation
4.4 The Law of Tangents 175
Applications: Surveying, Geometry
4.5 The Area of a Triangle 181
Applications: Surveying
4.6 Heron's Formula 186
Applications: Construction, Landscaping

4.7 Vectors in the Plane 190
Applications: Physics

Featured Application: Industrial Robots 198
Chapter Summary and Review 200, Chapter Test 202
College Entrance Exam Review 203, Cumulative Review 204

5 TRIGONOMETRIC IDENTITIES

5.1 Cosine: Sum and Difference Identities 208
Applications: Surveying, Aviation

5.2 Sine: Sum and Difference Identities 214
Applications: Computer

5.3 Tangent: Sum and Difference Identities 220
Applications: Coordinate Geometry

5.4 Double-Angle Identities 226
Applications: Indirect Measurement

5.5 Half-Angle Identities 231
Applications: Geometry

5.6 Product/Sum Identities 237
Applications: Computer

Featured Application: Aviation 242
Chapter Summary and Review 244, Chapter Test 246
College Entrance Exam Review 247, Maintaining Skills 248

6 INVERSE TRIGONOMETRIC FUNCTIONS

6.1 Inverse Relations and Functions 250
Applications: Science, Geometry

6.2 The Inverse Sine and Cosine Functions 255
Applications: Mechanics, Engineering

6.3 Other Inverse Trigonometric Functions 260
Applications: Physics, Construction

6.4 Solving Trigonometric Equations: Using Special Angles 265
Applications: Physics

6.5 Trigonometric Equations: Approximate Solutions 269
Applications: Physics

6.6 Rotation of Axes 273
Applications: Engineering, Geometry

Featured Application: The Rainbow 278
Chapter Summary and Review 280, Chapter Test 282
College Entrance Exam Review 283, Cumulative Review 284

7 COMPLEX NUMBERS

7.1 Polar Coordinates 286
Applications: Physics

7.2 Graphs of Polar Equations 292
Applications: Physics, Coordinate Geometry

7.3 Sums and Differences of Complex Numbers 298
Applications: Algebra, Electricity

7.4 Products and Quotients of Complex Numbers 303
Applications: Physics, Computer

7.5 Complex Numbers in Polar Form 308
Applications: Geometry

7.6 Multiplying and Dividing Complex Numbers in Polar Form 313
Applications: Computer

7.7 De Moivre's Theorem 318
Applications: Algebra

7.8 Roots of Complex Numbers 322
Applications: Computer

Technology: Fractals 327
Chapter Summary and Review 328, Chapter Test 330
College Entrance Exam Review 331, Maintaining Skills 332

8 EXPONENTIAL AND LOGARITHMIC FUNCTIONS

8.1 Real Exponents 334
Applications: Biology, Archaeology

8.2 Exponential Functions 338
Applications: Algebra, Optics

8.3 Logarithmic Functions 342
Applications: Physics, Computer

8.4 Properties of Logarithms 347
Applications: Physics

8.5 Evaluating Logarithms and Solving Exponential Equations 353
Applications: Astronomy, Business, Algebra

8.6 Applications: Exponential and Logarithmic Equations 359

Technology: The Richter Scale 365
Chapter Summary and Review 366, Chapter Test 368
College Entrance Exam Review 369, Cumulative Review 370

9 SEQUENCES AND SERIES

9.1 Arithmetic Sequences 372
Applications: Consumerism, Business

9.2 Geometric Sequences 378
Applications: Business, Investment, Geometry

9.3 Arithmetic Series 384
Applications: Building Design, Sports, Business, Number Theory

9.4 Geometric Series 390
Applications: Construction, Entertainment, Finance

9.5 Infinite Geometric Series 396
Applications: Physics, Geometry

9.6 Power Series and Trigonometric Functions 402
Applications: Statistics

9.7 Hyperbolic Functions 407
Applications: Civil Engineering

Technology: Spirals 412
Chapter Summary and Review 414, Chapter Test 416
College Entrance Exam Review 417, Cumulative Review 418

Table of Squares, Cubes, Square, and Cube Roots 421

Using a Table of Trigonometric Values 422

Linear Interpolation 424

Using a Table of Common Logarithms 426

Computation Using Common Logarithms 428

Table of Common Logarithms 430

Table of Natural Logarithms 432

Table of Values of the Trigonometric Functions 434

Answers to Selected Exercises 439

Glossary 462

Index 466

To the Student

The word "trigonometry" comes from two Greek words interpreted to mean "the measurement of triangles." Early trigonometry was developed in ancient times as a device to aid in solving problems with right triangles—problems of the kind that still arise in surveying and navigation. Today trigonometry applies to every kind of triangle, including spherical ones, and also to quantities that rise and fall such as vibrations, alternating current, and business cycles.

This course is designed to extend your fundamental mathematical skills so that you may enjoy topics in other advanced mathematical and scientific disciplines. In order for you to become more proficient, this text provides you with completely worked out examples and an abundant supply of exercises in each lesson. Key concepts are highlighted and many applications are included.

This book should help prepare you to understand and therefore enjoy your future work in mathematics. The responsibility for learning and understanding advanced academic disciplines lies with you. Always study with pencil and paper at hand; concentrate, plan well, work hard, and take pleasure in your accomplishments. Remember, it is important to gain knowledge because, as you can see from the applications at the end of each lesson, knowledge gives you the power to solve problems related to your future schoolwork and career choices, and also enables you to better understand the world around you.

OVERVIEW • Chapter 1

SUMMARY

Relations, functions, and the distance formula are discussed and related to graphs in the coordinate plane. A smooth transition is thus provided for defining the six trigonometric functions using the concept of rotation around the unit circle. Trigonometric functions of special angles as well as evaluating trigonometric functions using a calculator are presented. A number of application problems are introduced and solved throughout the chapter.

After completing this chapter, students should be able to evaluate the trigonometric functions in degrees and radians using appropriate technology. They should also be able to apply this knowledge to solve problems involving arc length, angular velocity, and linear velocity.

CHAPTER OBJECTIVES

- To define relations and functions
- To graph functions in the coordinate plane
- To define and use the distance formula
- To measure angles in rotations and in degrees
- To find the measures of coterminal angles
- To measure angles and arcs in degrees and in radians
- To change from degree measure to radian measure and from radian measure to degree measure
- To solve problems involving arc length, angular velocity, and linear velocity
- To define the cosine and sine functions
- To evaluate the sine and cosine functions of an angle given a point on its terminal side
- To define the tangent, cotangent, secant, and cosecant functions
- To evaluate the trigonometric functions of an angle
- To find the values of the six trigonometric functions of special angles
- To find decimal approximations for the values of the six trigonometric functions for all angles
- To find the measure of an angle given the value of one of its trigonometric functions

CHAPTER HIGHLIGHTS

The *theme* of Chapter 1 is satellite communications. The chapter's special features show why different methods of angle measurement provide greater accuracy in solving particular problems.

APPLICATIONS

Applications form an integral part of each lesson. Students are given applications from many diverse fields to give them a better understanding of the many different problems that contain radian measurement and trigonometric functions in today's world. In Lesson 1.5, students apply skills learned in previous lessons to solve problems involving arc length and linear and angular velocity.

TECHNOLOGY

Calculator

The calculator is used in Lesson 1.4 for changing radian measure to degree measure and back again. In Lesson 1.9, the calculator is used to find the values of trigonometric functions in radians and in degrees.

Computer

Computer usage involves a program for finding the sine and cosine of specific angles using the Pythagorean theorem. Students are asked why the program gives the correct sign for each function and why the program gives angle measures within a particular interval.

RESOURCES

Teacher's Resource Book

- Teaching Aid 1
- Transparencies 1 and 2

ASSIGNMENT GUIDE Meeting Student Needs

STUDENT TEXT					TEACHER'S RESOURCE BOOK	
Chapter Content		**Basic**	**Average**	**Enriched**	**P**	**E**
1.1	Functions and the Coordinate Plane	D: 5/1–37 odd, 57	D: 6/21–51 odd, 57	D: 6/29–59 odd	1	2
1.2	The Distance Formula	D: 10/1–23 odd, 39 R: 5/4, 12, 24	D: 10/9–35 odd, 39, 41 R: 6/30, 42, 54	D: 10/11, 17, 23–43 odd R: 6/44, 48, 52	3	4
1.3	Angles in the Coordinate Plane	D: 15/1–27 odd, 51 R: 10/4, 14, 22	D: 15/9–41 odd, 51 R: 10/4, 16, 30	D: 15/11, 19, 29–51 odd R: 10/12, 18, 38	5	6
1.4	Angle Measures in Degrees and Radians	D: 21/1–41 odd, 85 R: 15/2, 16, 24	D: 21/19, 31, 43–75 odd, 85 R: 15/10, 18, 32	D: 21/23, 41, 45, 53, 59–87 odd R: 15/12, 20, 36	7	8
1.5	Applications: Angular and Linear Velocity	D: 26/1–23 odd R: 21/4, 10, 22 28/TY	D: 27/3, 9, 15, 21–31 odd R: 21/26, 48, 70 28/TY	D: 27/5, 11, 17, 23–33 odd R: 21/30, 50, 72 28/TY	9	10
1.6	Cosine and Sine Functions	D: 32/1–21 odd R: 26/4, 8, 14	D: 32/5–27 odd R: 27/10, 16, 26	D: 32/11, 15, 16–31 odd R: 27/12, 18, 24	11	12
1.7	The Trigonometric Functions	D: 37/1–27 odd, 43 R: 32/2, 10, 20	D: 37/9, 17–37 odd, 43 R: 32/6, 10, 24	D: 37/11, 21–43 odd R: 32/12, 22, 26	13	14
1.8	Trigonometric Functions of Special Angles	D: 44/1–41 odd, 69 R: 37/4, 14, 22	D: 44/19–61 odd, 69 R: 37/10, 24, 30	D: 44/23–71 odd R: 37/12, 26, 32	15	16
1.9	Evaluating Trigonometric Functions	D: 50/1–37 odd R: 44/20, 32, 36 51/TY	D: 50/13–55 odd R: 44/32, 42, 50 51/TY	D: 50/21–59 odd R: 44/42, 46, 54 51/TY	17	18

D = Daily R = Review TY = Test Yourself P = Practice E = Enrichment

	STUDENT TEXT				TEACHER'S RESOURCE BOOK	
Review and Testing	Test Yourself	28, 51	College Ent. Exam Rev.	57	Tests	
	Chapter Sum. and Rev.	54	Maintaining Skills	58	• Quizzes	5–8
	Chapter Test	56			• Chapter Test (Form A)	9–10
					• Chapter Test (Form B)	11–12
Special Features	Historical Note	7	Trigonometry in the Machine Shop	33	Applications—Chapter 1	19–20
	Extra	11, 16	Challenge	38, 45	Critical Thinking	1
	Trigonometry in Surveying	22	Application	52	Alg. and Geom. Review	1–4
					Technology	1

1 Trigonometric Functions

Satellites in orbit are programmed to transmit data to earth from various points in their orbit. Small errors in measuring an angle may produce large errors over long distances. Trigonometric concepts play an important part in the calculations that make sure transmission signals do not miss their targets on earth.

BACKGROUND

The primary vehicle for research and exploration in the United States space program is the space shuttle. Satellites launched into orbit from the shuttle provide radio transmissions that include television pictures and instrument-recorded scientific data.

Ask students to find out about different units of angle measure and why some units provide more accurate solutions than others in navigation, physics, or other related disciplines.

LESSON PLAN

Vocabulary
Coordinate plane
Domain
Function
Origin
Quadrant
Range
Relation
x – axis
y – axis

BACKGROUND

In the Preview, a formula from physics, $d = 16t^2$, is used to introduce the concept of a relation. Before beginning the exercises, review the definition of exponent and the procedure for evaluating powers. As the exercises are discussed, the concept of a one-to-one correspondence could also be developed.

Critical Thinking

Predicting Consequences Ask students to predict whether a baseball or a feather will hit the ground first if both are dropped from the same height at the same time, assuming that no outside forces, other than gravity, are acting on the objects. Then ask them to give reasons for their answers. The time it takes an object to fall when it is dropped depends upon the distance it has to travel and not its weight. Therefore, the baseball and the feather will hit the ground at the same time.

1.1 Functions and the Coordinate Plane

Objectives: To define relations and functions
To graph functions in the coordinate plane

Mathematical relationships often exist between sets of numbers. Formulas and equations are frequently used to define such relationships.

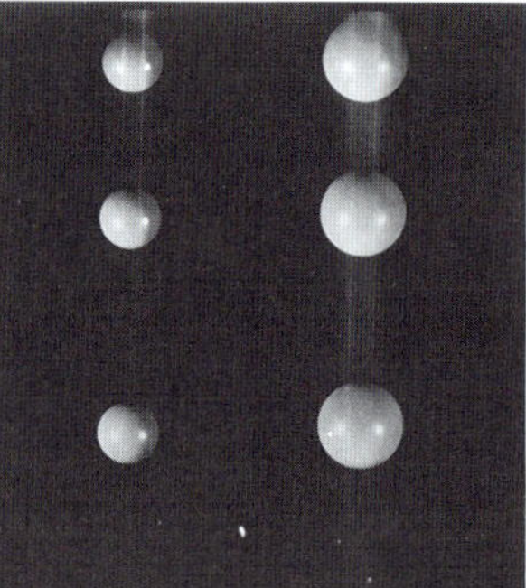

Preview

When an object is dropped, the distance it travels depends upon the amount of time elapsed since its release. For a given time t, the approximate distance d is given by the equation

$$d = 16t^2$$

where d is in feet and t is in seconds.

Find the height of a building if an object dropped from the roof takes the given amount of time to hit the ground.

1. 3 s 144 ft **2.** 5 s 400 ft **3.** 0.1 min 576 ft

Find the number of seconds it takes an object to hit the ground if it is dropped from the roof of a building with the given height.

4. 64 ft 2 s **5.** 256 ft 4 s **6.** 100 ft 2.5 s

The solutions to the equation $d = 16t^2$ can be written as ordered pairs (t, d), such as

$$\{(1, 16), (2, 64), (3, 144), \ldots\}$$

A set of ordered pairs of numbers is called a **relation.** The set of replacements for the first variable is called the **domain** of the relation. The domain may be limited to a specific set of numbers. In the equation $d = 16t^2$, it would be reasonable to limit the time t to numbers that are greater than or equal to zero. The domain of this relation is the set of nonnegative real numbers.

The set of replacements for the second variable is called the **range** of the relation. In the equation $d = 16t^2$, the distance d will always be greater than or equal to zero, because t^2 is always greater than or equal to zero. The range of this relation is the set of nonnegative real numbers.

EXAMPLE 1 **Find the domain and range of the relation: {(0, 0), (1, 2), (2, 4), (3, 6), (4, 8)}**

The domain is the set of replacements for the first variable:

{0, 1, 2, 3, 4}

The range is the set of replacements for the second variable:

{0, 2, 4, 6, 8}

Each ordered pair in a relation is associated with exactly one point on the **coordinate plane.** The set of all points (x, y) associated with a relation can be graphed on a rectangular coordinate system, as shown at the right.

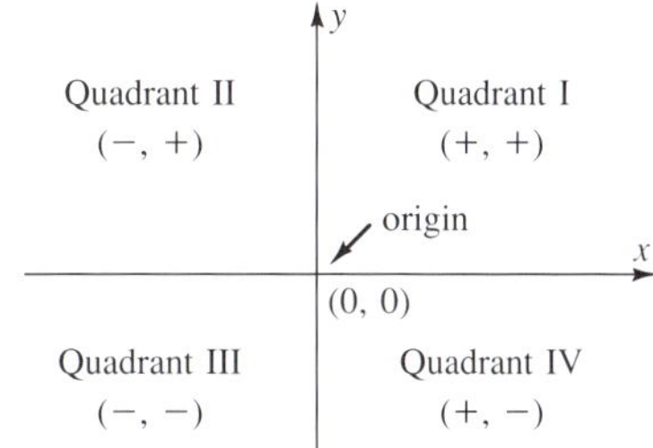

The horizontal number line is called the ***x*-axis** (sometimes called the axis of *abscissas*), where x represents the first number in the ordered pair. The vertical number line is called the ***y*-axis** (sometimes called the axis of *ordinates*), where y represents the second number in the ordered pair. The point of intersection of the axes is called the **origin.** The axes divide the plane into four regions, called **quadrants.** Every point on the plane can be represented by an ordered pair (x, y). When you plot ordered pairs, or points, it is important to label the axes. Although it is often convenient to choose the same scale for both axes, it is not necessary to do so.

EXAMPLE 2 **Find the domain and range of each relation:**

a.

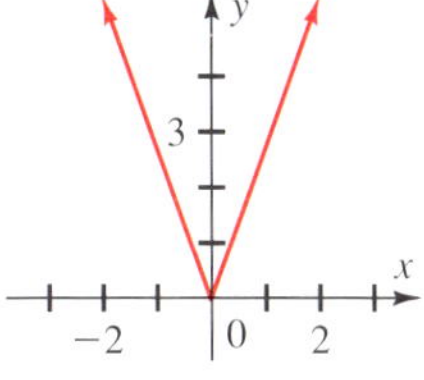

b.

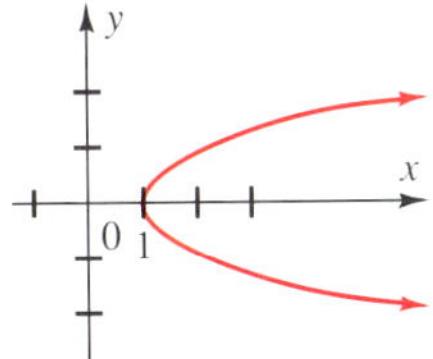

c.

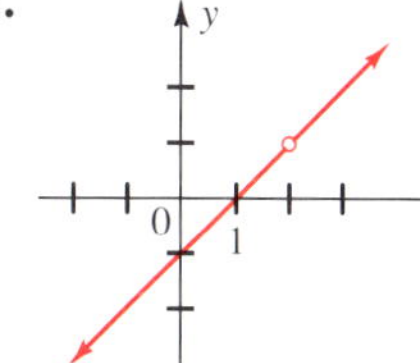

a. Domain {all real numbers}
Range $\{y: y \geq 0\}$

b. Domain $\{x: x \geq 1\}$
Range $\{y: y$ is a real number$\}$

c. Domain $\{x: x \neq 2\}$
Range $\{y: y \neq 1\}$

In the relation given in Example 1, notice that each member x of the domain is paired with exactly one member y of the range. Because of this unique pairing, the ordered pairs represent a special type of relation called a *function*. A **function** is a relation in which each member of the domain is paired with exactly one member of the range. In this book, special types of functions called *trigonometric functions* will be studied.

TEACHING SUGGESTIONS

- Emphasize that functional notation $f(x)$ does not mean multiplication. Note that the symbol $f(x)$ indicates the second component of an ordered pair and that the symbol f represents a set of ordered pairs.
- Stress that for a function each member of the domain is paired with exactly one member of the range, but that this is not necessarily true for a relation.

CHALKBOARD EXAMPLES

- **For Example 1**
  Find the domain and range of each relation.
  1. {(1, 2), (3, 4), (3, 5)}
  D = {1, 3}; R = {2, 4, 5}
  2. {(−3, 7), (−2, 3), (−1, −1), (0, 2), (2, 8)}
  D = {−3, −2, −1, 0, 2}; R = {7, 3, −1, 2, 8}
- **For Example 2**
  Find domain and range of each relation.

3.

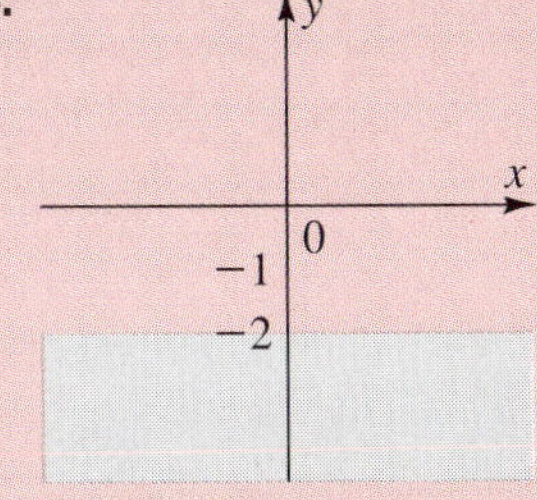

D = {all real numbers};
R = $\{y{:}y \leq -2\}$

4.

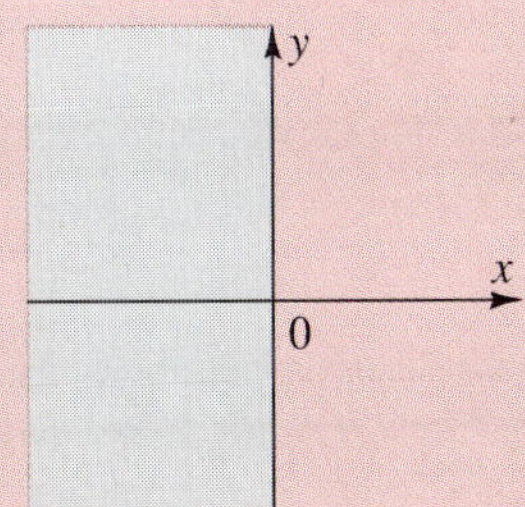

D = $\{x{:}x \leq 0\}$;
R = {all real numbers}

- **For Example 3**
  State whether or not each relation is a function.

5. $\{(2, 3), (3, -4), (4, 1), (1, 3)\}$ function

6. $\{(4, 4), (-2, 3), (4, 2), (3, 4)\}$ not a function

- **For Example 4**
  State whether or not each graph represents a function.

7.

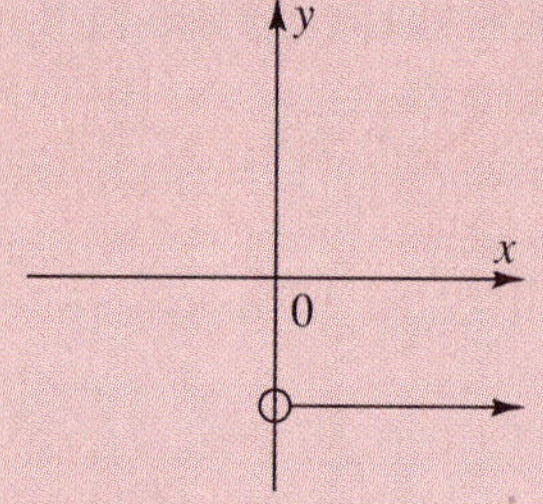

function

8.

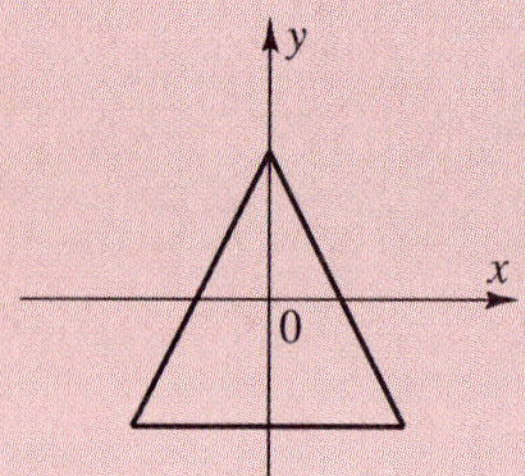

not a function

- **For Example 5**
  For the function $f(x) = \frac{7}{x - 2}$, find each value.

9. $f(16)$ $\frac{1}{2}$ **10.** $f\left(\frac{2}{3}\right)$ $-\frac{21}{4}$

EXAMPLE 3 **State whether or not each relation is a function.**

a. $\{(1, 2), (2, 3), (3, 4)\}$ **b.** $\{(1, 4), (2, 5), (1, 6)\}$ **c.** $\{(1, 6), (2, 7), (3, 6), (4, 7)\}$

a. a function **b.** not a function **c.** a function

A quick way to decide if a graph represents a function is to use the *vertical line test*: If any vertical line can be drawn that intersects the graph of a relation in more than one point, the graph does *not* represent a function.

EXAMPLE 4 **State whether or not each graph represents a function.**

a.

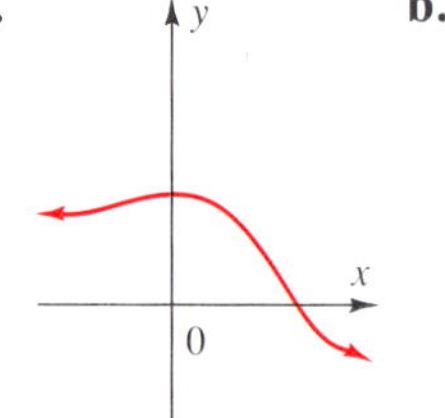

b.

c.

d.

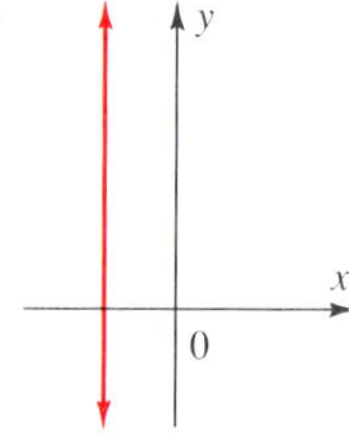

a. function **b.** not a function **c.** function **d.** not a function

Special notation is used to denote a function. Symbols such as $f(x)$, $g(t)$, and $F(w)$ are used. For example, the function $d = 16t^2$ can be written

$$f(t) = 16t^2 \qquad \textit{f(t) is read f of t.}$$

Specific values of this function are written as follows:

$$f(1) = 16(1)^2 = 16(1) = 16$$
$$f(2) = 16(2)^2 = 16(4) = 64$$

EXAMPLE 5 **For the function $f(x) = 6x + 7$, find:**

a. $f(-3)$ **b.** $f(0)$ **c.** $f(3.1)$

a.
$$\begin{aligned} f(x) &= 6x + 7 \\ f(-3) &= 6(-3) + 7 \\ &= -18 + 7 \\ &= -11 \end{aligned}$$

b.
$$\begin{aligned} f(x) &= 6x + 7 \\ f(0) &= 6(0) + 7 \\ &= 0 + 7 \\ &= 7 \end{aligned}$$

c.
$$\begin{aligned} f(x) &= 6x + 7 \\ f(3.1) &= 6(3.1) + 7 \\ &= 18.6 + 7 \\ &= 25.6 \end{aligned}$$

If the domain of a function is not stated, assume that it is the greatest set of real numbers for which the value of the function is also a real number. You must exclude from the domain any numbers that make a denominator equal to zero, since the value of the function would be undefined. Numbers that lead to the square root of a negative quantity must also be excluded, since the value of the function would not be a real number.

EXAMPLE 6 **Determine which real numbers must be excluded from the domain of each function:**

a. $f(x) = \frac{2}{x}$ **b.** $f(x) = \frac{2x}{3x - 6}$ **c.** $g(x) = \frac{5}{x^2 - 9}$ **d.** $g(x) = \sqrt{x}$

a. Exclude 0, since $\frac{2}{0}$ is undefined. **b.** Exclude 2, since $\frac{4}{0}$ is undefined.

c. Exclude 3 and -3, since $\frac{5}{0}$ is undefined. **d.** Exclude all real numbers less than 0.

CLASS EXERCISES

Find the domain and range of each relation, and state whether or not the relation is a function.

1. $\{(2, 4), (3, 6), (4, 8), (5, 10)\}$ d: {2, 3, 4, 5}; r: {4, 6, 8, 10}; func.

2. $\{(-2, -1), (-1, 0), (0, 1), (1, 2)\}$ d: {−2, −1, 0, 1}; r: {−1, 0, 1, 2}; func.

3. $\{(6, 1), (6, 2), (6, 3), (6, 4)\}$ d: {6}; r: {1, 2, 3, 4}; not a func.

4. $\{(2, 5), (3, 5), (4, 5), (5, 5), (6, 5)\}$ d: {2, 3, 4, 5, 6}; r: {5}; func.

5. $\{(2, 4), (4, 2), (3, 6), (6, 3)\}$ d: {2, 4, 3, 6}; r: {4, 2, 6, 3}; func.

6. $\{(0, 2), (1, 3), (1, 4), (2, 5)\}$ d: {0, 1, 2}; r: {2, 3, 4, 5}; not a func.

7. $\{(-2, -2), (1, -1), (-1, 1), (-2, 2)\}$ d: {−2, −1, 1}; r: {−2, −1, 1, 2}; not a func.

8. $\{(-2, -2), (-1, -1), (1, -1), (2, 2)\}$ d: {−2, −1, 1, 2}; r: {−2, −1, 2}; func.

For the function $f(x) = 2x + 4$, find each value.

9. $f(1)$ 6 **10.** $f(10)$ 24 **11.** $f(0)$ 4 **12.** $f(-2)$ 0 **13.** $f(a)$ $2a + 4$

Determine which real numbers must be excluded from the domain of each function.

14. $\frac{3}{x + 5}$ $x \neq -5$ **15.** $\frac{-2x}{5x - 15}$ $x \neq 3$ **16.** $\frac{4}{x^2 - 16}$ $x \neq \pm 4$ **17.** $\frac{1}{\sqrt{8x}}$ $x \leq 0$

PRACTICE EXERCISES

Find the domain and range of each relation.

A

1. $\{(1, 0), (2, 1), (5, 3), (7, 5), (16, 9)\}$ d: {1, 2, 5, 7, 16}; r: {0, 1, 3, 5, 9}

2. $\{(2, 1), (3, 3), (5, 5), (10, 6)\}$ d: {2, 3, 5, 10}; r: {1, 3, 5, 6}

3. $\{(-2, -10), (-1, -1), (0, 8), (1, 11)\}$ d: {−2, −1, 0, 1}; r: {−10, −1, 8, 11}

4. $\{(-3, -2), (-2, -1), (0, 2), (8, 11)\}$ d: {−3, −2, 0, 8}; r: {−2, −1, 2, 11}

5. $\{(-3, 1), (2, 2), (4, 3), (-3, 3)\}$ d: {−3, 2, 4}; r: {1, 2, 3}

6. $\{(-100, 100), (-100, 5), (1, -100)\}$ d: {−100, 1}; r: {100, 5, −100}

7. $\{(4, 25), (6, 25), (8, 25), (10, 25)\}$ d: {4, 6, 8, 10}; r: {25}

8. $\{(7, -3), (6, 2), (5, 6), (7, -3)\}$ d: {7, 6, 5}; r: {−3, 2, 6}

State whether or not each relation is a function.

9. $\{(2, 3), (2, 4), (3, 5)\}$ not a func.

10. $\{(3, 1), (4, 1), (5, 2)\}$ func.

11. $\{(6, 4), (7, 4), (8, 5)\}$ func.

12. $\{(6, 2), (6, 3), (6, 4)\}$ not a func.

13. $\{(2, 6), (3, 6), (4, 6)\}$ func.

14. $\{(6, 1), (5, 1), (4, 1)\}$ func.

- **For Example 6**

  Determine which real numbers must be excluded from the domain of each function.

  11. $g(x) = \frac{4}{x^3 - 8}$ $x \neq 2$

  12. $g(x) = \frac{7}{5x - 3}$ $x \neq \frac{3}{5}$

Common Errors

- Students often interchange the domain and range elements. Emphasize that the set of first elements is the domain and the set of second elements is the range.
- Some students misinterpret $f(x)$. Stress that $f(x)$ means the value of the function of f at x, where x is a real number, not f times x.
- See *Teacher's Resource Book* for additional remediation.

LESSON FOLLOW-UP

Discussion

Describe the domain of the function $f(x) = \sqrt{x^2 - 9x + 14}$. The domain is the set of all real numbers except $2 < x < 7$, and $-7 < x < -2$, since for each of these numbers $x^2 - 9x + 14 < 0$.

Assignment Guide

See p. B for assignments.

Historical Note

The present Cartesian coordinate system is compared to that originally used by Descartes.

Lesson Quiz

Find the domain and range of each relation and state whether or not the relation is a function.

1. $\{(-1, 2), (3, 4), (-1, 6)\}$
   D = $\{-1, 3\}$; R = $\{2, 4, 6\}$; not a function
2. $\{(0, 3), (1, -4), (2, -4), (3, 3)\}$
   D = $\{0, 1, 2, 3\}$; R = $\{-4, 3\}$; function

For the function $g(x) = 3x - 2$, find each value.

3. $g(0)$ -2 4. $g(-1)$ -5

Determine which real numbers must be excluded from the domain of each function.

5. $\frac{13}{x - 4}$ $x \neq 4$

6. $-\frac{x}{x^2 - 25}$ $x \neq \pm 5$

State whether or not each graph represents a function.

6. 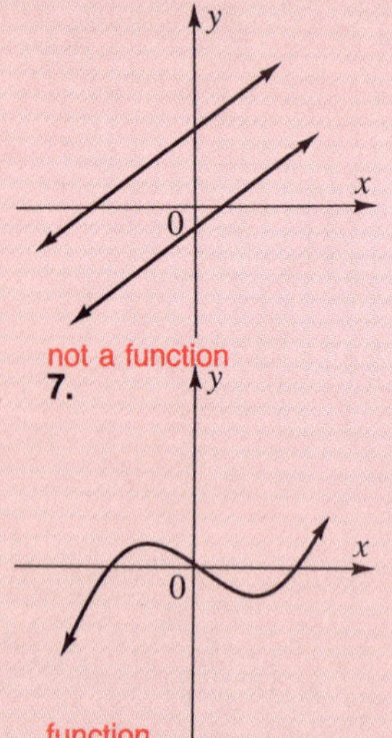

not a function

7. function

Enrichment

If $f(x) = 2x$ and $g(x) = x^2 + 4$, find $f(g(1))$ and $g(f(1))$. $g(1) = 5$, then $f(g(1))$ or $f(5) = 10$; $f(1) = 2$, then $g(f(1))$ or $g(2) = 8$.

State whether or not each graph represents a function.

15. not a func.

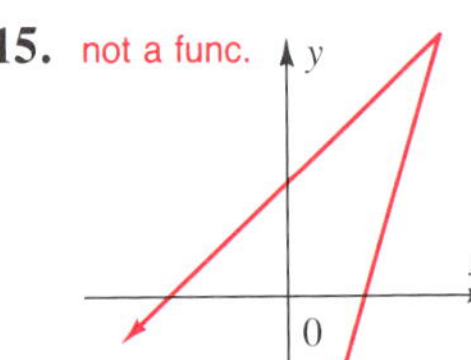

16. func.

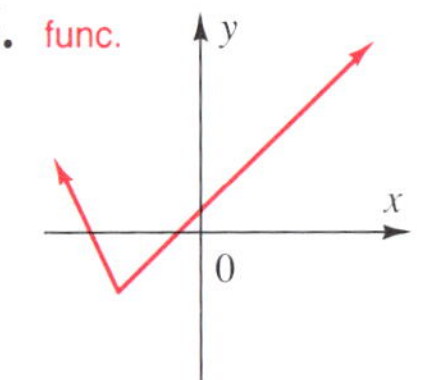

17. func.

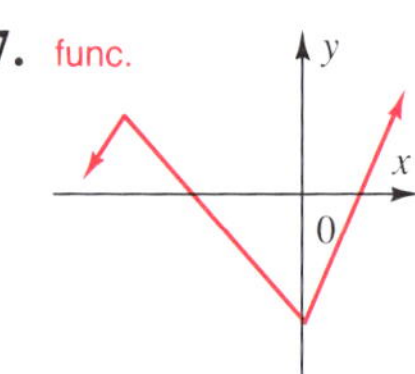

18. not a func.

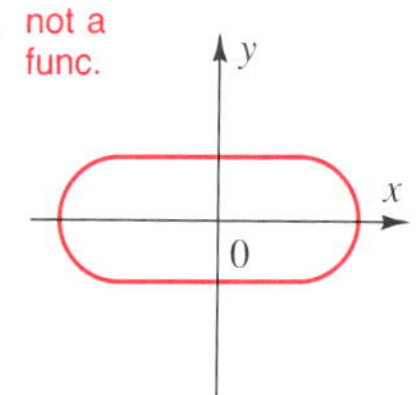

19. not a func.

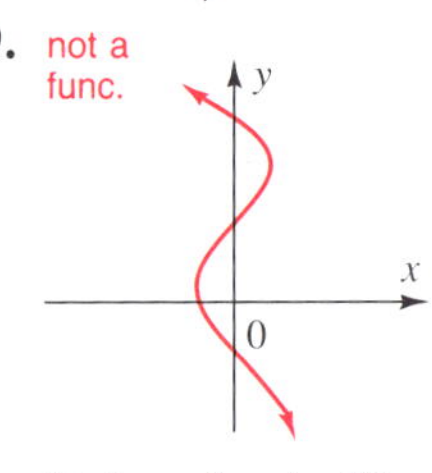

20. func.

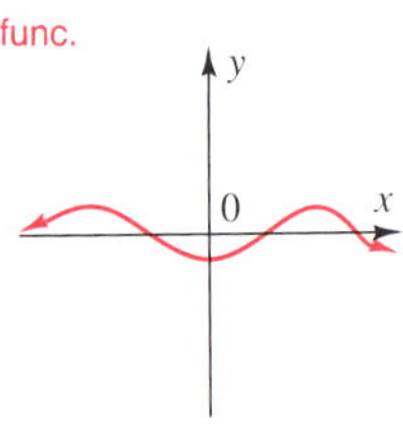

Find each value for the function $f(x) = 3x + 10$.

21. $f(5)$ 25 **22.** $f(6)$ 28 **23.** $f(-4)$ -2 **24.** $f(-5)$ -5

Find each value for the function $g(x) = 2x - 5$.

25. $g\left(\frac{1}{2}\right)$ -4 **26.** $g\left(\frac{3}{4}\right)$ $-\frac{7}{2}$ **27.** $g(1.7)$ -1.6 **28.** $g(2.5)$ 0

Find the domain and range of each relation, and state whether or not the relation is a function. See side column page 7.

B **29.** $\{(5, -1), (5, 3), (-1, -10), (8, 8)\}$ **30.** $\{(3, 3), (5, 3), (4, 3), (-3, 3)\}$

31. $f(x) = x^2$ **32.** $f(x) = -x^2$ **33.** $f(x) = -x$ **34.** $f(x) = x^3$

35. 

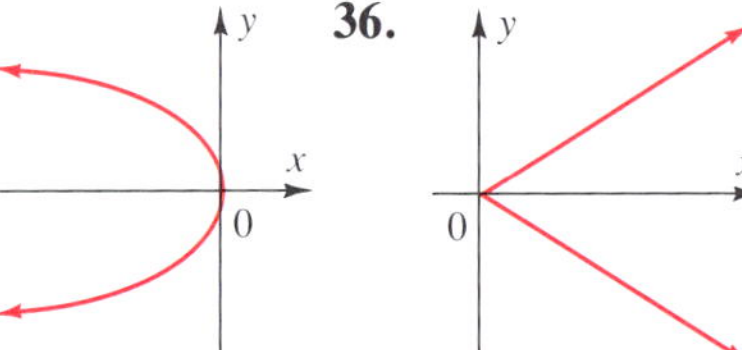

36.

37. 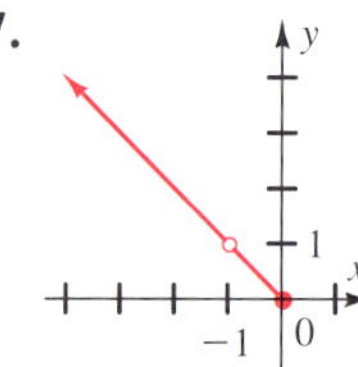

38. 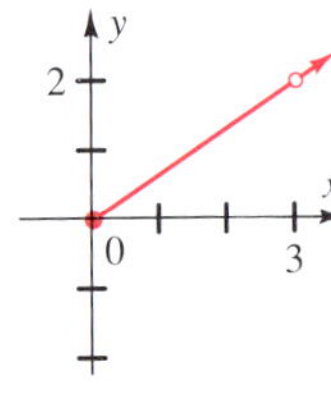

Find each value for the function $f(x) = x^2 - 1$

39. $f(2)$ 3 **40.** $f(3)$ 8 **41.** $f(-5)$ 24 **42.** $f(-6)$ 35 **43.** $f(\sqrt{5})$ 4 **44.** $f(\sqrt{7})$ 6

Determine which real numbers must be excluded from the domain of each function.

45. $f(x) = \frac{x}{x}$ $x \neq 0$ **46.** $g(x) = \frac{3x}{-3x + 9}$ $x \neq 3$ **47.** $f(x) = \frac{-x}{x^2 - 49}$ $x \neq \pm 7$ **48.** $g(x) = \frac{1}{\sqrt{2x}}$ 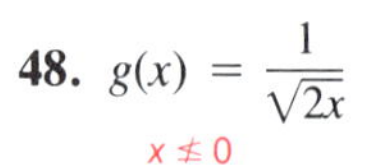 $x \leq 0$

Find each value for the function $f(x) = -\frac{1}{x^2}$.

C **49.** $f(-3)$ $-\frac{1}{9}$ **50.** $f(-4)$ $-\frac{1}{16}$ **51.** $f\left(\frac{1}{9}\right)$ -81 **52.** $f\left(\frac{1}{10}\right)$ -100

Find each value for the functions $f(x) = \sqrt{x} - x$ and $g(x) = \sqrt{x} + 1$.

53. $f(4) + g(16)$ 3 **54.** $f(3) + g(9)$ $1 + \sqrt{3}$ **55.** $f(25) - g(25)$ -26 **56.** $f(36) - g(36)$ -37

Applications

57. Meteorology Temperatures given in degrees Celsius can be converted to degrees Fahrenheit using the function $f(t) = \frac{9}{5}t + 32°$, where t is the temperature in degrees Celsius. If the temperature is 35°C, determine the temperature in degrees Fahrenheit. 95°F

58. Meteorology Temperatures given in degrees Fahrenheit can be converted to degrees Celsius by the function $g(t) = \frac{5}{9}(t - 32)°$, where t is the temperature in degrees Fahrenheit. If the temperature is 86°F, determine the temperature in degrees Celsius. 30°C

59. Geometry The surface area of a cube is given by the function $A = 6e^2$, where e is the length of an edge. Find the surface area of a cube whose edges each measure 3 in. 54 in.^{2}

60. Geometry The volume of a cube is given by the function $V = e^3$, where e is the length of an edge. Find the length of an edge of a cube with volume 512 cm^3. 8 cm

HISTORICAL NOTE

Descartes and the Coordinate Plane The French mathematician and philosopher René Descartes (1596–1650) is usually considered the founder of modern coordinate geometry. It is interesting to note, however, that there are differences between the original coordinate system Descartes devised and today's *Cartesian* coordinate system, which was named in his honor. Descartes used a single horizontal axis and only positive values. At each point on the axis, he imagined the construction of a line segment whose length represented the value of the related second coordinate. These segments were not always perpendicular to the axis, but they were parallel to each other. Investigate Descartes.
Answers may vary.

Teacher's Resource Book
Practice—Chapter 1, p. 1
Enrichment—Chapter 1, p. 2

Additional Answers
Practice Exercises

29. D: {−1, 5, 8}; R: {−10, −1, 3, 8}; not a func.
30. D: {−3, 3, 4, 5}; R: {3}; func.
31. D: {x: x is a real number} R: {$f(x)$: $f(x) \geq 0$}; func.
32. D: {x: x is a real number}; R: {$f(x)$: $f(x) \leq 0$}; func.
33. D:{x: x is a real number}; R: {$f(x)$: $f(x)$ is a real number}; func.
34. D: {x: x is a real number}; R: {$f(x)$: $f(x)$ is a real number}; func.
35. D: {x: $x \leq 0$}; R: {y: y is a real number}; not a func.
36. D: {x: $x \geq 0$}; R: {y: y is a real number}; not a func.
37. D: {x: $x \leq 0$, $x \neq -1$}; R: {y: $y \geq 0$, $y \neq -1$}; func.
38. D: {x: $x \geq 0$, $x \neq 3$}; R: {y: $y \geq 0$, $y \neq 2$}; func.

LESSON PLAN

Vocabulary
Distance formula
Pythagorean theorem

BACKGROUND

In the Preview, simplifying radicals is reviewed. Students will use this skill when finding the distance between two points. A list of "perfect squares" may also be developed at this time.

Critical Thinking

Causal Explanation Ask students to explain whether or not $\sqrt{-72}$ can be simplified in the system of real numbers. Reasoning that the square of a real number is always nonnegative, students should conclude that $\sqrt{-72}$ cannot be simplified in the system of real numbers.

1.2 The Distance Formula

Objective: To define and use the distance formula

It is important to be able to find the distance between two points in the coordinate plane. When distances are represented by irrational numbers, it is customary to simplify radicals.

Preview

To simplify a square root, factor the radicand into as many perfect squares as possible. Then take the square root of each square factor and leave the remaining factors under the square root sign.

EXAMPLE **Simplify:** $\sqrt{150}$

$\sqrt{150} = \sqrt{25 \cdot 6} = 5\sqrt{6}$

Simplify each expression.

1. $\sqrt{90}$ $3\sqrt{10}$ **2.** $\sqrt{200}$ $10\sqrt{2}$ **3.** $\sqrt{98}$ $7\sqrt{2}$ **4.** $\sqrt{125}$ $5\sqrt{5}$ **5.** $\sqrt{108}$ $6\sqrt{3}$ **6.** $\sqrt{112}$ $4\sqrt{7}$

The distance formula is based on the **Pythagorean theorem,** which states that a and b are the lengths of the legs of a right triangle and c is the length of its hypotenuse if and only if

$$c^2 = a^2 + b^2$$

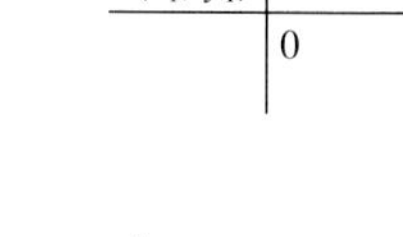

Consider two points $E(x_2, y_2)$ and $F(x_1, y_1)$. The distance between E and F is found by applying the Pythagorean theorem to the triangle at the right.

$$d^2 = e^2 + f^2$$
$$= (x_2 - x_1)^2 + (y_2 - y_1)^2 \qquad e = x_2 - x_1;\ f = y_2 - y_1$$

The distance formula is found by taking the square root of each side.

Distance Formula The distance between two points $E(x_2, y_2)$ and $F(x_1, y_1)$ is

$$d = \sqrt{(x_2 - x_1)^2 + (y_2 - y_1)^2}$$

EXAMPLE 1 **Find the distance between each pair of points:**

a. $A(-5, 2)$ and $B(4, 2)$ **b.** $P(-2, -6)$ and $Q(-5, 3)$

a. $d = \sqrt{[4 - (-5)]^2 + (2 - 2)^2}$
$= \sqrt{9^2 + 0^2}$
$= \sqrt{81} = 9$

b. $d = \sqrt{[-5 - (-2)]^2 + [3 - (-6)]^2}$
$= \sqrt{(-3)^2 + 9^2}$
$= \sqrt{90} = 3\sqrt{10}$

The distance formula can be used to find the equation of a circle of radius r with center O at the origin. If $P(x, y)$ represents any point on the circle, then the distance between P and O is r.

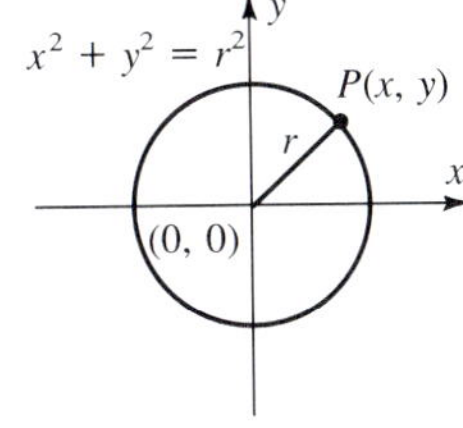

$$\sqrt{(x - 0)^2 + (y - 0)^2} = r$$
$$\sqrt{x^2 + y^2} = r$$
$$x^2 + y^2 = r^2 \quad \textit{Square both sides.}$$

If a circle of radius r has its center at the origin, then its equation is $x^2 + y^2 = r^2$. Conversely, if the equation $x^2 + y^2 = r^2$ is given, then its graph is a circle of radius r with center at the origin.

EXAMPLE 2 **Find the radius r of the circle with the equation $x^2 + y^2 = 25$.**

$r = \sqrt{25} = 5$

CLASS EXERCISES

For Discussion

1. The formula for the distance between two points $E(x_2, y_2)$ and $F(x_1, y_1)$ is $d = \sqrt{(x_2 - x_1)^2 + (y_2 - y_1)^2}$. Would you get a different value for d if you interchanged x_1 and x_2? if you interchanged y_1 and y_2? no; no

State whether or not each triangle is a right triangle.

2. 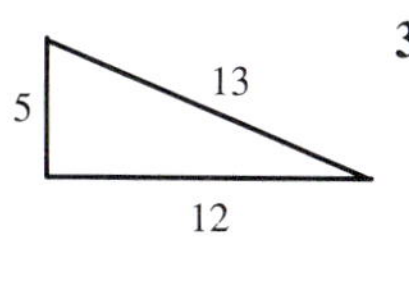

yes

3. 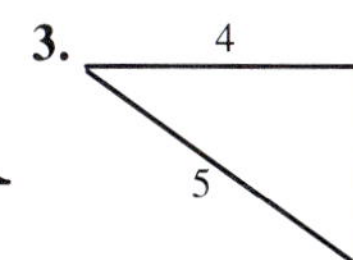

yes

4. 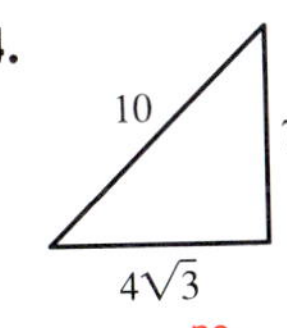

no

5. 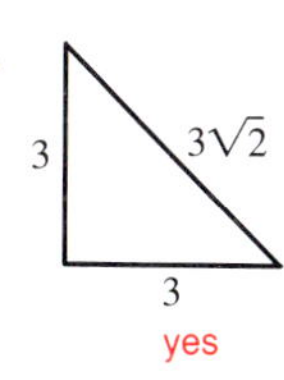

yes

Match the equation of each circle with its description.

6. center at the origin, radius 4 d
7. center at the origin, radius 2 a
8. center at the origin, radius 7 c
9. center at the origin, radius $3\sqrt{6}$ b

a. $x^2 + y^2 = 4$
b. $x^2 + y^2 = 54$
c. $x^2 + y^2 - 49 = 0$
d. $2x^2 + 2y^2 = 32$

TEACHING SUGGESTIONS

- Emphasize that $\sqrt{100 - 36} = \sqrt{64} = 8$, not $\sqrt{100 - 36} = \sqrt{100} - \sqrt{36} = 10 - 6 = 4$.
- When deriving the equation of a circle, emphasize that the coordinates of the origin are (0, 0).

CHALKBOARD EXAMPLES

- **For Example 1**
Find the distance between each pair of points.

1. $(-1, 6), (5, -4)$ $2\sqrt{34}$
2. $\left(\frac{1}{2}, \frac{1}{3}\right), (2, -1)$ $\frac{\sqrt{145}}{6}$

- **For Example 2**
Find the radius r of the circle with the given equation.

3. $x^2 + y^2 = 16$ 4
4. $x^2 = 9 - y^2$ 3

Common Error

- When writing the equation of a circle that contains a given point, some students may forget to square the r, yielding $x^2 + y^2 = r$. Remind them that $3^2 + 4^2 = 5^2$, not 5.
- See *Teacher's Resource Book* for additional remediation.

LESSON FOLLOW-UP

Assignment Guide

See p. B for assignments.

Extra

The distance formula in 3-space is discussed.

Lesson Quiz

Find the distance between each pair of points.

1. (5, −2), (11, 6) 10
2. (2, 4), (−3, 6) $\sqrt{29}$

Find the radius of each circle.

3. $x^2 + y^2 = 225.$ 15
4. $y^2 = 144 - x^2$ 12
5. A 9-m ramp must be 1 m tall. How far is the base of the ramp from its vertical support? Give your answer to the nearest hundredth of a meter. 8.94 m

Enrichment

An equilateral triangle has vertices at (−4, 6) (−4, 2), and (x, y). Find two possible pairs of coordinates for (x, y).

$\sqrt{(x+4)^2 + (y-6)^2}$
$= \sqrt{(x+4)^2 + (y-2)^2} = 4$
$y = 4, x = -4 \pm 2\sqrt{3}$; thus, the coordinates are $(-4 + 2\sqrt{3}, 4)$ or $(-4 - 2\sqrt{3}, 4)$.

PRACTICE EXERCISES

A **Find the distance between each pair of points.**

1. (1, 1), (4, 5) 5
2. (2, 2), (8, 10) 10
3. (−2, 5), (5, −2) $7\sqrt{2}$
4. (−7, 2), (3, −3) $5\sqrt{5}$
5. (0, 0), (6, −6) $6\sqrt{2}$
6. (0, 0), (5, −5) $5\sqrt{2}$
7. (6, −5), (0, −4) $\sqrt{37}$
8. (3, 0), (−2, 2) $\sqrt{29}$
9. $\left(\frac{1}{2}, -1\right), \left(\frac{1}{2}, 4\right)$ 5
10. $\left(3, \frac{2}{3}\right), \left(12, \frac{2}{3}\right)$ 9
11. (0.5, 1.2), (1.5, −1.8) $\sqrt{10}$
12. (−3, 1.4), (−2, 1.4) 1

Find the radius of each circle.

13. $x^2 + y^2 = 49$ 7
14. $x^2 + y^2 = 81$ 9
15. $x^2 + y^2 - 121 = 0$ 11
16. $x^2 + y^2 - 225 = 0$ 15
17. $3x^2 + 3y^2 = 180$ $2\sqrt{15}$
18. $\frac{1}{3}x^2 + \frac{1}{3}y^2 = 21$ $3\sqrt{7}$

B **Find the distance between each pair of points.**

19. (3.2, 5.6), (4.7, 6.1) $\sqrt{2.5}$
20. (1.4, −4.3), (−0.6, −2.7) $\sqrt{6.56}$
21. $(\sqrt{2}, -3), (2\sqrt{2}, 1)$ $3\sqrt{2}$
22. $(5, \sqrt{2}), (9, -\sqrt{2})$ $2\sqrt{6}$
23. $(\sqrt{5}, 4\sqrt{7}), (3\sqrt{5}, 3\sqrt{7})$ $3\sqrt{3}$
24. $(-\sqrt{2}, \sqrt{3}), (2\sqrt{2}, 2\sqrt{3})$ $\sqrt{21}$
25. $(2a, b), (4a, -b)$ $2\sqrt{a^2 + b^2}$
26. $(-4r, 3s), (-r, -3s)$ $3\sqrt{r^2 + 4s^2}$

Write the equation of the circle that contains the given point P and has its center at the origin.

27. $P(-3, 4)$ $x^2 + y^2 = 25$
28. $P(6, -8)$ $x^2 + y^2 = 100$
29. $P(-3, -6)$ $x^2 + y^2 = 45$
30. $P(-9, -3)$ $x^2 + y^2 = 90$
31. $P(\sqrt{2}, 5)$ $x^2 + y^2 = 27$
32. $P(-3\sqrt{2}, -4)$ $x^2 + y^2 = 34$

C

33. Find x if the distance between (1, 3) and (x, 9) is 10. $x = 9$ or $x = -7$
34. Find y if the distance between (2, y) and (7, 8) is 13. $y = 20$ or $y = -4$
35. Find the area of the rectangle with vertices A (−5, 1), B (−3, −1), C (3, 5), and D (1, 7). 24 units²
36. Find the area of the rectangle with vertices A (−3, 0), B (−2, −1), C (1, 2), and D (0, 3). 6 units²
37. A quadrilateral is a parallelogram if both pairs of opposite sides are equal in length. If $A(-1, 0)$, $B(3, 8)$, $C(3, 2)$, and D are consecutive vertices of a parallelogram, what are the coordinates of point D? (−1, −6)
38. A quadrilateral is a rhombus if all four sides are equal in length. If $P(1, 2)$, $Q(4, -2)$, $R(1, -6)$, and S are consecutive vertices of a rhombus, what are the coordinates of point S? (−2, −2)

Teacher's Resource Book
Practice—Chapter 1, p. 3
Enrichment—Chapter 1, p. 4

Applications

39. Navigation Airplane A is located 3 mi west and 1 mi south of airport O. Airplane B is located 5 mi due north of the airport at the same altitude as airplane A. Draw a diagram using a coordinate system with origin O. How far apart are the two airplanes? Express the answer in radical form. $3\sqrt{5}$ mi

40. Navigation A rowboat is located 5 mi due east of buoy O. A sailboat is located 3 mi west and 6 mi north of the same buoy. Draw a diagram using a coordinate system with origin O. How far apart are the two boats? 10 mi

41. Construction A stringer is a board that supports a set of stairs. If a staircase is 8 ft high and extends 10 ft horizontally, find the length of the stringer to the nearest foot. 13 ft

42. Engineering A bridge crosses a river that is 2140 ft wide. The vertical arch support at the center of the bridge is 560 ft tall. Find the length of the longest spanning wire to the nearest foot. (Assume that the wire is taut.) 1208 ft

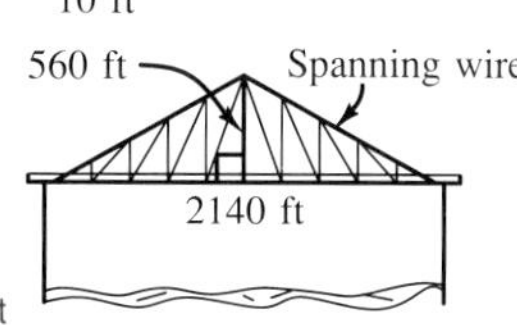

43. Engineering A guy wire is used to support or steady an object. A trapeze artist is standing on her landing platform, which is attached to a vertical post. A guy wire supporting the post is 87 ft long, and the base of the post is 45 ft from the wire's stake. Find the height of the platform to the nearest foot. 74 ft

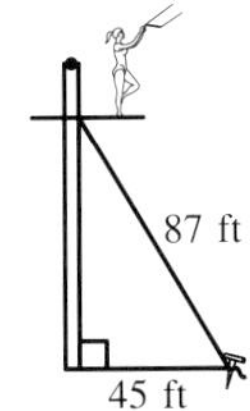

EXTRA

The distance between two points in space
In the coordinate system shown at the right, points in three-dimensional space are represented by ordered triples of the form (x, y, z). The following formula gives the distance d between points $P(x_2, y_2, z_2)$ and $Q(x_1, y_1, z_1)$:

$$d = \sqrt{(x_2 - x_1)^2 + (y_2 - y_1)^2 + (z_2 - z_1)^2}$$

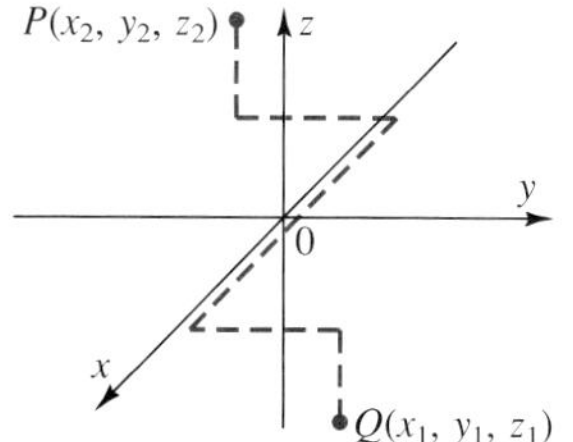

Use the formula above to find the distance between each pair of points.

1. $A(0, 1, 2)$, $B(5, -3, 1)$ $\sqrt{42}$

2. $C(1, 1, 1)$, $D(1, -1, 2)$ $\sqrt{5}$

3. Could a distance formula be written for "four-dimensional space"? If so, how? What applications might there be for this?
yes; $d = \sqrt{(w_2 - w_1)^2 + (x_2 - x_1)^2 + (y_2 - y_1)^2 + (z_2 - z_1)^2}$; answers may vary.

LESSON PLAN

Vocabulary

Angle
Coterminal angles
Degree
Initial side
Standard position
Terminal side

BACKGROUND

In the Preview, the Coriolis force is discussed. This force is produced by a combination of the Earth's rotation and major wind patterns. This causes a general clockwise water swirl in northern oceans and a counterclockwise water swirl in oceans in the Southern Hemisphere.

Critical Thinking

Discovering Patterns Ask students to explain in which direction the water swirls when it drains out of a sink at school. The water would swirl clockwise in the northern hemisphere, and counterclockwise in the southern hemisphere.

Additional Answers

1. The water would drain straight down with little or no swirling.
2. If you stirred the water with enough force, you could cause it to temporarily swirl in the opposite direction.

1.3 Angles in the Coordinate Plane

Objectives: To measure angles in rotations and in degrees
To find the measures of coterminal angles

In trigonometry, there is a distinction between the measures of angles formed by a ray rotating in a clockwise direction and the measures of angles formed by a ray rotating in a counterclockwise direction.

Preview

The Coriolis force is the inertial force caused by the earth's rotation. It affects many phenomena, including water draining from a bathtub. Above the equator, water draining from a bathtub swirls clockwise. Below the equator, it swirls counterclockwise.
See side column.

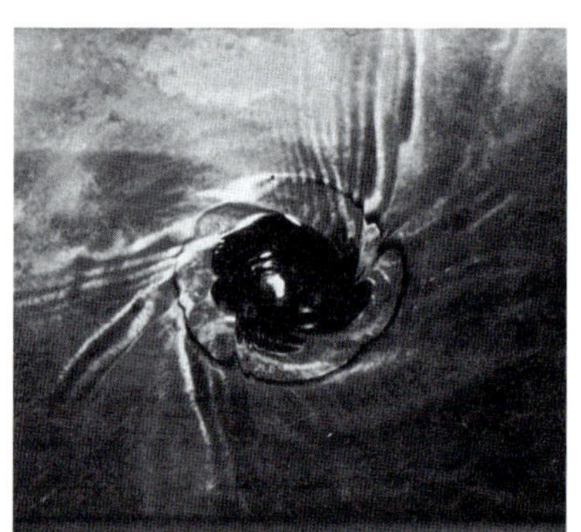

1. How do you think water draining from a bathtub would behave *at* the equator?
2. If you stirred draining water with your finger, do you think you could cause it to swirl in the opposite direction?

An **angle** can be defined as the union of two rays with a common endpoint, the *vertex* of the angle. In trigonometry, an angle is formed by rotating a ray about its vertex from one position, called the **initial side** of the angle, to another, called the **terminal side.**

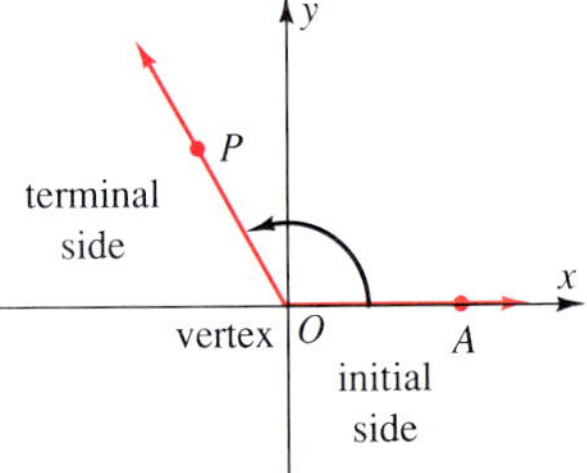

When the vertex of an angle is the origin of the coordinate plane and its initial side coincides with the positive x-axis, the angle is said to be in **standard position.** Angle AOP ($\angle AOP$) is in standard position. Its measure is the amount of rotation of $\overrightarrow{OA}$ (ray OA), the initial side, to the position of $\overrightarrow{OP}$, the terminal side of the angle.

If an angle is formed by a counterclockwise rotation, its measure is *positive*. If an angle is formed by a clockwise rotation, its measure is *negative*.

One unit of angle measure is the **degree.** One full counterclockwise rotation measures 360° (360 degrees), and one full clockwise rotation measures −360°. Note that the terminal and initial sides of the angle coincide under a full rotation.

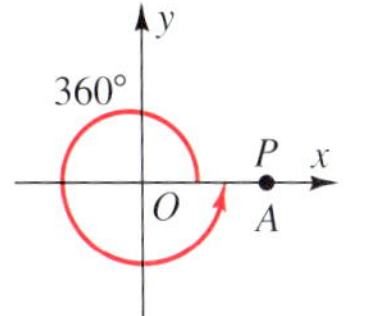

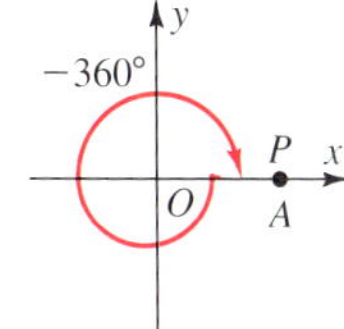

EXAMPLE 1 **Find the degree measure of the angle for each rotation. Sketch the angle in standard position.**

a. $\frac{1}{2}$ rotation, counterclockwise

b. $\frac{2}{3}$ rotation, clockwise

a. $\frac{1}{2}(360°) = 180°$

b. $\frac{2}{3}(-360°) = -240°$

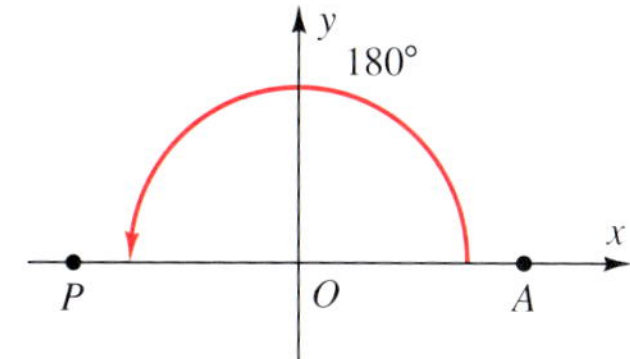

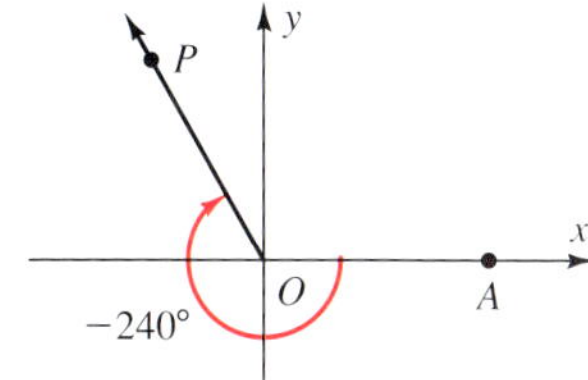

If an angle is in standard position, it is said to lie in the quadrant in which its terminal side falls. For instance, $\angle AOP$ in Example 1b lies in quadrant II.

Coterminal angles are angles in standard position whose terminal sides coincide.

EXAMPLE 2 **Find the measure of the angle coterminal with $\angle BOC$ under the given rotation from the position of $\overrightarrow{OC}$.**

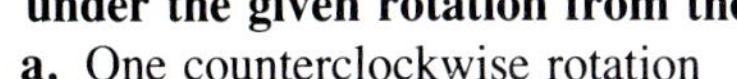

a. One counterclockwise rotation

b. One clockwise rotation

c. A counterclockwise rotation other than the one used in part a.

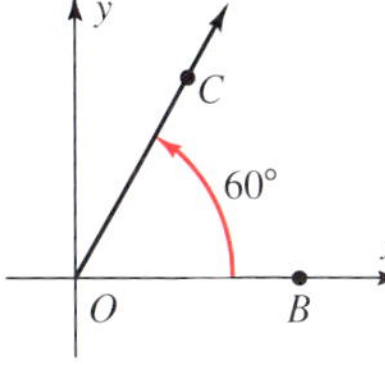

a.

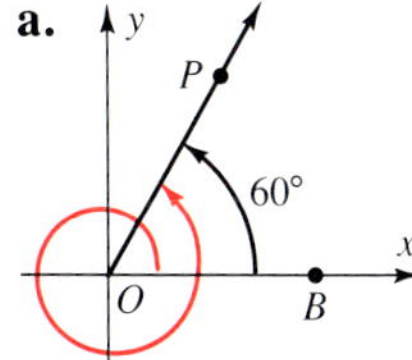

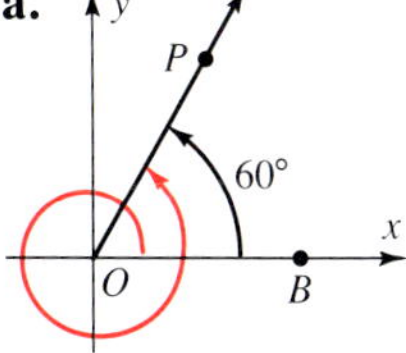

$60° + 360° = 420°$
1 counterclockwise rotation

b.

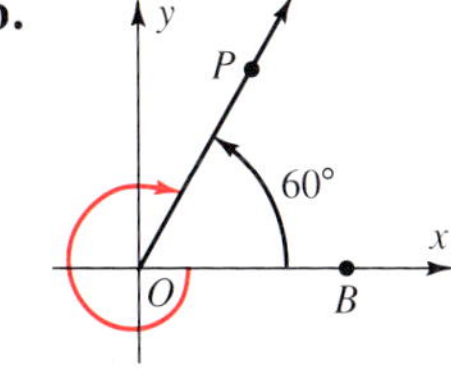

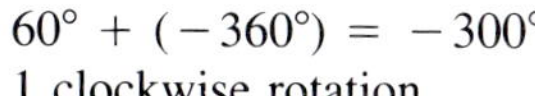

$60° + (-360°) = -300°$
1 clockwise rotation

c.

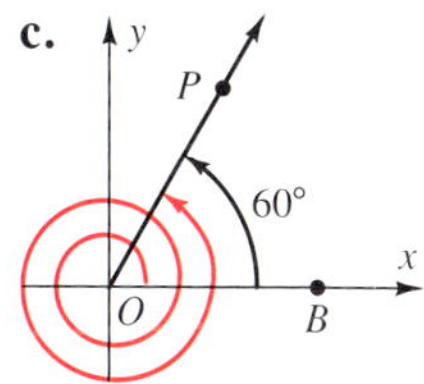

$60° + 2(360°) = 780°$
2 counterclockwise rotations

TEACHING SUGGESTIONS

- Emphasize that for an angle to be in standard position, its vertex must be at the origin and its initial side must correspond to the positive x-axis.
- Point out that in geometry, the measure of an angle is always positive, but in trigonometry the measure of an angle may be positive or negative.
- Coterminal angles can be defined as angles that share a terminal side. That is, $\frac{1}{2}$ rotation clockwise and $\frac{3}{2}$ rotations counterclockwise form coterminal angles.

CHALKBOARD EXAMPLES

- **For Example 1**

  Find the degree measure of the angle for each rotation.

  1. $\frac{1}{4}$ rotation, counterclockwise 90°

  2. $\frac{2}{5}$ rotation, clockwise −144°

- **For Example 2**

  Find the measure of the angle coterminal with $\angle BOC$ under the given rotation from the position of $\overrightarrow{OC}$.

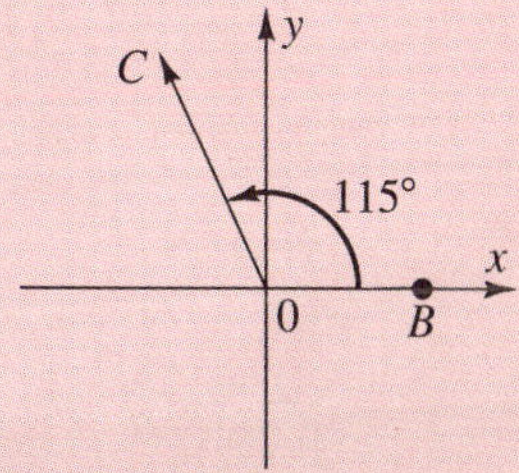

3. Two clockwise rotations −605°

4. Three counterclockwise rotations 1195°

- **For Example 3**

  Given $g(n) = (n \cdot 360 + 150)°$, find:

  5. $g(-1)$ $-210°$

  6. $g\left(\frac{1}{2}\right)$ $330°$

Common Error

- Some students will incorrectly say that angles of 510° and 150° are equal since they are coterminal angles. Have those students practice sketching pairs of coterminal angles.
- See *Teacher's Resource Book* for additional remediation.

LESSON FOLLOW-UP

Assignment Guide

- See p. B for assignments.
- Although Exercises 29–32 and 37–40 are B exercises, you may wish to assign them to all students.

If two angles are coterminal, their measures differ by an integral multiple of 360°.

$420° = 1 \cdot 360° + 60°$ $\quad -300° = (-1) \cdot 360° + 60°$ $\quad 780° = 2 \cdot 360° + 60°$

The measure of any angle coterminal with an angle of degree measure θ (the Greek letter *theta*) is found by evaluating $n \cdot 360 + \theta$, where n is an integer. That is, the measures of all angles coterminal with a given angle of measure θ may be expressed as a function of the number n of complete rotations of the terminal side

$$f(n) = (n \cdot 360 + \theta)°$$

EXAMPLE 3 **Given $f(n) = (n \cdot 360 + 48)°$, find:**

a. $f(-1)$ **b.** $f(0)$ **c.** $f(1)$ **d.** $f(2)$

a. $f(-1) = (-1 \cdot 360 + 48)° = -312°$

b. $f(0) = (0 \cdot 360 + 48)° = 48°$

c. $f(1) = (1 \cdot 360 + 48)° = 408°$

d. $f(2) = (2 \cdot 360 + 48)° = 768°$

CLASS EXERCISES

Find the degree measure of the angle for each rotation.

1. $\frac{1}{3}$ rotation, clockwise $-120°$

2. $\frac{1}{5}$ rotation, clockwise $-72°$

3. $\frac{13}{12}$ rotation, counterclockwise $390°$

4. $\frac{11}{12}$ rotation, counterclockwise $330°$

In which quadrant does the terminal side of each angle lie when it is in standard position?

5. 95° II **6.** $-30°$ IV **7.** $-40°$ IV **8.** 75° I

9. 210° III **10.** $-340°$ I **11.** $-350°$ I **12.** 295° IV

For each angle shown in standard position, name the initial and terminal sides and estimate the degree measure.

13.
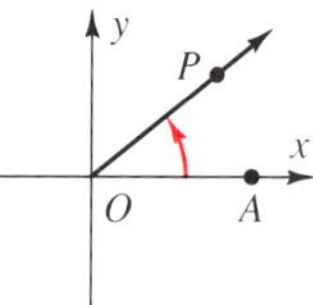

$\overrightarrow{OA}$; $\overrightarrow{OP}$; 40°

14.
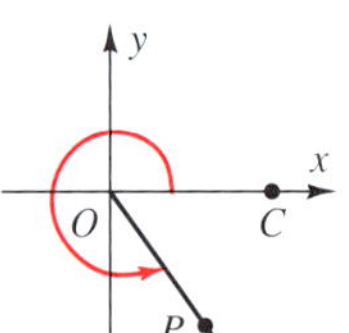

$\overrightarrow{OC}$; $\overrightarrow{OP}$; 310°

15.
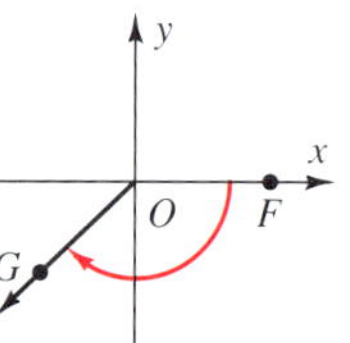

$\overrightarrow{OF}$; $\overrightarrow{OG}$; $-135°$

16.
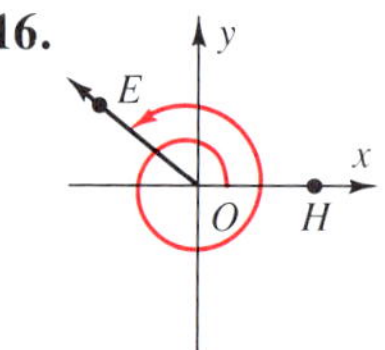

$\overrightarrow{OH}$; $\overrightarrow{OE}$; 500°

Name two angles that are coterminal with the given angle. Answers may vary.

17. 30° 390°; $-330°$

18. 150° 510°; $-210°$

19. $-30°$ 330°; $-390°$

20. $-120°$ 240°; $-480°$

PRACTICE EXERCISES

Find the degree measure of the angle for each rotation. Then sketch each angle in standard position. Indicate its position by a curved arrow.

A **1.** $\frac{1}{4}$ rotation, clockwise $-90°$ **2.** $\frac{3}{4}$ rotation, clockwise $-270°$

3. $\frac{1}{2}$ rotation, counterclockwise $180°$ **4.** $\frac{1}{8}$ rotation, counterclockwise $45°$

5. $\frac{3}{5}$ rotation, clockwise $-216°$ **6.** $\frac{1}{3}$ rotation, counterclockwise $120°$

7. $\frac{1}{6}$ rotation, clockwise $-60°$ **8.** $\frac{5}{6}$ rotation, clockwise $-300°$

9. $\frac{4}{3}$ rotation, counterclockwise $480°$ **10.** $\frac{3}{10}$ rotation, clockwise $-108°$

11. $\frac{5}{12}$ rotation, clockwise $-150°$ **12.** $\frac{17}{12}$ rotation, counterclockwise $510°$

Name three angles (two positive and one negative) that are coterminal with the given angle. Answers may vary.

13. 40° 400°; 760°; −320° **14.** 70° 430°; 790°; −290° **15.** 183° 543°; 903°; −177° **16.** 97° 457°; 817°; −263°

17. −34° 326°; 686°; −394° **18.** −62° 298°; 658°; −422° **19.** 151° 511°; 871°; −209° **20.** 209° 569°; 929°; −151°

Given $f(n) = (n \cdot 360 + 115)°$, find each value.

21. $f(2)$ 835° **22.** $f(1)$ 475° **23.** $f(0)$ 115° **24.** $f(4)$ 1555°

B **25.** $f(-8.4)$ −2909° **26.** $f(8.4)$ 3139° **27.** $f(1.8)$ 763° **28.** $f(-1.8)$ −533°

Start at the terminal side of the given angle θ in standard position. Find the measure of the resulting angle (in standard position) after the given number of rotations.

29. $\theta = 210°$, 1 clockwise −150° **30.** $\theta = 240°$, 1 clockwise −120°

31. $\theta = 18°$, 1 counterclockwise 378° **32.** $\theta = -16°$, 1 counterclockwise 344°

33. $\theta = -18°$, 2 counterclockwise 702° **34.** $\theta = 16°$, 2 counterclockwise 736°

35. $\theta = 30°$, 1.5 counterclockwise 570° **36.** $\theta = 60°$, 2.5 counterclockwise 960°

37. $\theta = 310°$, 2 clockwise −410° **38.** $\theta = -150°$, 1.5 clockwise −690°

For each of the following angles, find a coterminal angle with measure θ such that $0° \le \theta < 360°$.

39. −100° 260° **40.** −215° 145° **41.** 534° 174° **42.** −386° 334°

C **43.** 900° 180° **44.** 1250° 170° **45.** −3610° 350° **46.** −7201° 359°

Extra

The advantages of naming angles with Greek letters are developed.

Lesson Quiz

In which quadrant does the angle lie? Assume that the angle is in standard position.

1. 220° III
2. −75° IV

For each rotation, find the degree measure of the angle.

3. $\frac{1}{2}$ rotation, clockwise −180°
4. $\frac{7}{12}$ rotation, counterclockwise 210°

Given $f(n) = (n \cdot 360 + 125)°$, find each value.

5. $f(-2)$ −595°
6. $f(3)$ 1205°

For each of the following angles, find a coterminal angle with measure θ such that $0° \le \theta \le 360°$.

7. 529° 169°
8. 1048° 328°

Starting at the terminal side of the given angle θ in standard position, find the measure of the resulting angle (in standard position) after the given number of rotations.

9. $\theta = 39°$, 2 counterclockwise 759°
10. $\theta = -105°$, 1.5 clockwise −645°

Enrichment

Angles *AOB* and *AOC* are coterminal. The measure of the $\angle AOB$ is $\frac{1}{9}$ the measure of $\angle AOC$. If the sum of the angles is 450°, find the measure of each angle. $\angle AOB + \angle AOC = 450°$ or $\frac{1}{9}x + x = 450°$; $\angle AOB = 45°$; $\angle AOC = 405°$

Teacher's Resource Book
Practice—Chapter 1, p. 5
Enrichment—Chapter 1, p. 6

The angles measuring $(n \cdot 360 + 20)°$ are coterminal for all integers n. There are two possible positions of the terminal sides for the angles $(n \cdot 180 + 20)°$ in standard position. How many different positions of the terminal sides are possible for angles in standard position given by each of the following expressions?

47. $(n \cdot 120 + 20)°$ 3 **48.** $(n \cdot 72 + 20)°$ 5 **49.** $(n \cdot 100 + 20)°$ 18 **50.** $(n \cdot 150 + 20)°$ 12

Applications

51. Entertainment Many records make $33\frac{1}{3}$ revolutions per minute. Through how many degrees does a point on the record turn during one minute? 12,000°

52. Entertainment Singles are usually available on records that make 45 revolutions per minute. Through how many degrees does a point on the record turn during one minute? during 30 seconds? 16,200°; 8100°

EXTRA

In geometry, angles are named using capital letters of the English alphabet. For example, this angle could be called $\angle Q$ or $\angle PQR$. As mentioned previously in this lesson, Greek letters are often used in trigonometry to name angles and their measures.

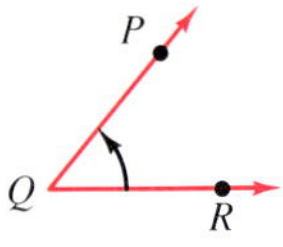

Greek Alphabet

A	α	Alpha	B	β	Beta	Γ	γ	Gamma	Δ	δ	Delta
E	ϵ	Epsilon	Z	ζ	Zeta	H	η	Eta	Θ	θ	Theta
I	ι	Iota	K	κ	Kappa	Λ	λ	Lambda	M	μ	Mu
N	ν	Nu	Ξ	ξ	Xi	O	o	Omicron	Π	π	Pi
P	ρ	Rho	Σ	σ	Sigma	T	τ	Tao	Υ	υ	Upsilon
ϑ	φ	Phi	X	χ	Chi	Ψ	ψ	Psi	Ω	ω	Omega

It is particularly helpful to use Greek letters to name angles between 180° and 360°. To illustrate, "$\angle ABC$" could refer to *either* of the two angles shown at the right. However, the letters α and β name the angles uniquely.

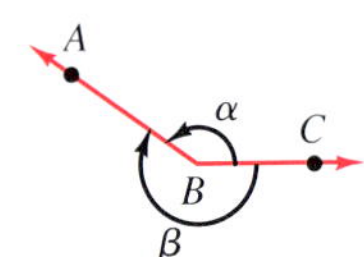

Determine the degree measure of each angle labeled with a Greek letter.

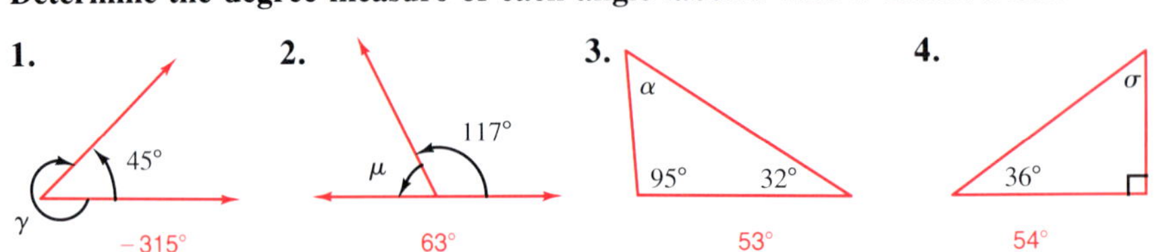

1.4 Angle Measure in Degrees and Radians

Objectives: To measure angles and arcs in degrees and radians
To change from degree measure to radian measure and from radian measure to degree measure

If a ray makes a complete 360° rotation about its endpoint, the path of a point on that ray is a circle.

Preview

The circumference C of a circle of radius r is given by the formula

$$C = 2\pi r$$

Find the circumference (in terms of π) of each circle of given radius r.

1. $r = 3$ 6π **2.** $r = 6.8$ 13.6π **3.** $r = 8\frac{1}{2}$ 17π **4.** $r = 4.25$ 8.5π

Find the radius of each circle of given circumference C.

5. $C = 8\pi$ 4 **6.** $C = 2\pi$ 1 **7.** $C = 12.4\pi$ 6.2 **8.** $C = 3.5\pi$ 1.75

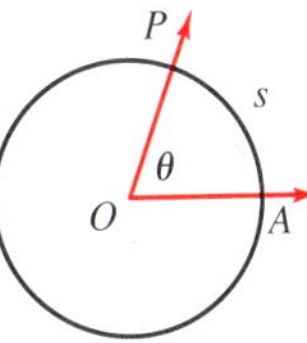

Letters of the Greek alphabet, such as θ (theta), α (alpha), and β (beta), are used to name angles as well as to represent their measures. Angle AOP, or angle θ, is called a *central angle*. A **central angle** of a circle is an angle whose vertex is the center of the circle. The sides of angle θ intercept arc AP with length s.

Angles and arcs can be measured in different ways. One common unit of measure is the degree. An arc of 1 degree is $\frac{1}{360}$ of the circumference of a circle. The degree measure of a central angle is equal to that of its intercepted arc. That is, a central angle that intercepts an arc of 1° also has measure 1°. A degree is subdivided into **minutes,** where $1° = 60'$ (60 min). A minute is further subdivided into **seconds,** where $1' = 60''$ (60 s).

EXAMPLE 1 **Express each angle using degrees, minutes, and seconds:**

a. 10.5° **b.** $13° \, 3\frac{1}{3}'$

a. $10.5° = 10° + (0.5°)\left(\frac{60'}{1°}\right)$ *Multiply by 1 in the form $\frac{60'}{1°}$.*

$= 10° + 30'$ *Notice how the units of degrees in the numerator and denominator divide out.*

$= 10°30'$

LESSON PLAN

Vocabulary
Central angle
Minute
Radian
Second

Materials/Manipulatives
Calculators
Degree protractor
Radian protractor

BACKGROUND

In the Preview, the formula for the circumference of a circle is presented.

TEACHING SUGGESTIONS

- Review the fact that the measure of a central angle is the same as the measure of its intercepted arc.
- When changing from degree measure to radian measure and the reverse, emphasize that the sign remains the same.
- It would be helpful if students would memorize the radian equivalents for 0°, 30°, 45°, 60°, 90°, and 180°.
- You may wish to have students use a calculator to check solutions to exercises, and a protractor to measure angles.

CHALKBOARD EXAMPLES

- **For Example 1**

  Express each angle using degrees, minutes, and seconds.

  1. 20.6° 20°36″

  2. $5°2\frac{1}{2}'$ 5°02′30″

- **For Example 2**

  Express each angle in decimal degrees, to the nearest hundredth of a degree.

  3. 10°30′15″ 10.50°

  4. 210°40′45″ 210.68°

b. $13°\,3\frac{1'}{3} = 13° + 3' + \left(\frac{1'}{3}\right)\left(\frac{60''}{1'}\right)$ *Multiply by 1 in the form $\frac{60''}{1'}$.*

$= 13° + 3' + 20''$ *Notice how the units of minutes in the numerator and denominator divide out.*

$= 13°3'20''$

For most applications, decimal parts of degrees are preferred to minutes and seconds. That is, 32.5° is used instead of 32°30′.

EXAMPLE 2 **Express 127°18′36″ in decimal degrees.**

$$127°18'36'' = 127° + 18'\left(\frac{1°}{60'}\right) + 36''\left(\frac{1'}{60''}\right)\left(\frac{1°}{60'}\right)$$
$$= 127° + 0.3° + 0.01°$$ *Units divide out.*
$$= 127.31°$$

Thus, 127°18′36″ equals 127.31°.

Your calculator may have special keys for changing degrees, minutes, and seconds to decimal degrees and the reverse. If so, you can solve the first two examples by entering the given measures and pressing the appropriate keys.

Angle measures can also be given in *radians*. When a central angle of a circle intercepts an arc that has the same length as the radius of the circle, the measure of this angle is defined to be one **radian,** abbreviated 1 rad.

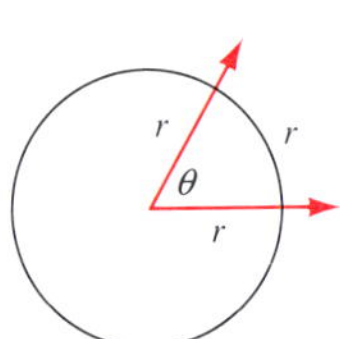

A central angle of 1 radian intercepts an arc of length r regardless of the length of the radius r of the circle. However, if the radius of one circle is longer than that of another, the intercepted arc of the larger circle is longer than that of the smaller circle.

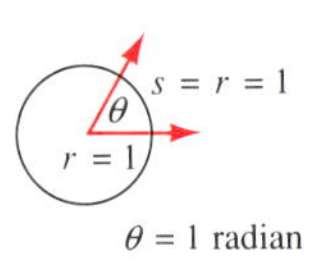

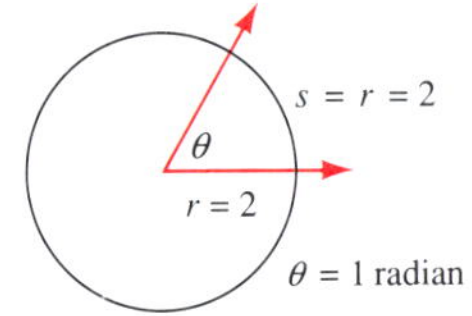

Let r be the radius of a circle and s the length of an arc intercepted by central angle θ. The radian measure of θ can be found by dividing s by r. That is,

$$\theta = \frac{s}{r}$$

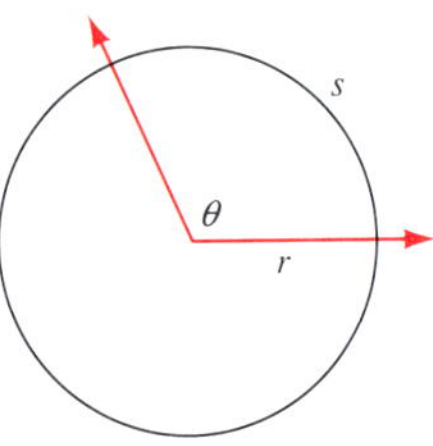

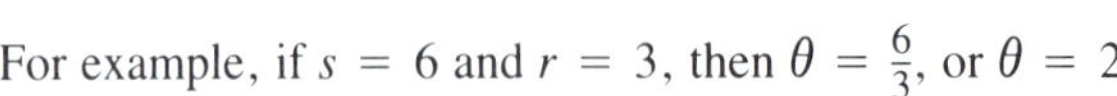

For example, if $s = 6$ and $r = 3$, then $\theta = \frac{6}{3}$, or $\theta = 2$.

Notice that no unit is given. When no unit is specified, assume that radian measure is being used. Thus, $\theta = 2$ means that the measure of angle θ is 2 radians. Similarly, $\theta = 3\pi$ means that the measure of angle θ is 3π radians.

If θ is a complete counterclockwise revolution, $s = 2\pi r$ and thus

$$\theta = \frac{2\pi r}{r} = 2\pi$$

Since 1 revolution is equal to 360° and is also equal to 2π rad,

$$2\pi \text{ rad} = 360° \qquad \text{or} \qquad \pi \text{ rad} = 180°$$

Thus,

$$1° = \frac{\pi}{180} \qquad 1° \approx 0.0174533 \text{ rad}$$

$$1 \text{ rad} = \frac{180°}{\pi} \qquad 1 \text{ rad} \approx 57.2958°$$

To express a given number of degrees in radians, multiply the number of degrees by $\frac{\pi}{180}$.

EXAMPLE 3 **Express each angle measure in radians:** **a.** 45° **b.** −120°

a. $45° = 45 \cdot \frac{\pi}{180} \text{ rad} = \frac{45\pi}{180} \text{ rad} = \frac{\pi}{4} \text{ rad}$

b. $-120° = -120 \cdot \frac{\pi}{180} \text{ rad} = \frac{-120\pi}{180} \text{ rad} = -\frac{2\pi}{3} \text{ rad}$

To express a given number of radians in degrees, multiply the number of radians by $\frac{180}{\pi}$.

EXAMPLE 4 **Express each angle measure in degrees:** **a.** $\frac{\pi}{12}$ **b.** -5π

a. $\frac{\pi}{12} \text{ rad} = \frac{\pi}{12} \cdot \frac{180°}{\pi} = 15°$

b. $-5\pi \text{ rad} = -5\pi \cdot \frac{180°}{\pi} = -900°$

It is a good idea to know the following equivalent degree and radian measures. The fact that $360° = 2\pi$ radians should help you to recall the other entries in the table. For example, $120° = \frac{1}{3}(360°)$, so $120° = \frac{1}{3}(2\pi) = \frac{2\pi}{3}$ rad.

- **For Example 3**

Express each angle measure in radians.

5. 60° $\frac{\pi}{3}$

6. −150° $-\frac{5\pi}{6}$

7. 450° $\frac{5\pi}{2}$

8. −110° $-\frac{11\pi}{18}$

- **For Example 4**

Express each angle measure in degrees.

9. $\frac{4\pi}{3}$ 240°

10. $-\frac{7\pi}{4}$ −315°

11. $-\frac{7\pi}{18}$ −70°

12. 5π 900°

- **For Example 5**

  In which quadrant does the terminal side of each angle lie when it is in standard position?

13. $\frac{3}{4}\pi$ II

14. $-\frac{11\pi}{6}$ I

Common Error

- Some students will use the wrong conversion factor when changing degrees to radians and vice versa. Stress that when changing from degrees to radians, students should use the conversion factor with π in the numerator, and when changing from radians to degrees, students should use the conversion factor with π in the denominator.
- See *Teacher's Resource Book* for additional remediation.

LESSON FOLLOW-UP

Assignment Guide

See p. B for assignments.

Critical Thinking

Comparing-Contrasting Ask students to discuss the similarities and/or differences between the radian measures of the central angles of a circle with radius of 5 which intercepts an arc of length 5 and a circle with radius of 10 which intercepts an arc of length 10. Students should reason that both central angles would measure 1 radian, because when a central angle intercepts an arc that has the same length as the radius of the circle, the measure of this angle is defined to be 1 radian.

Degrees	Radians
0°	0
30°	$\frac{\pi}{6}$
45°	$\frac{\pi}{4}$
60°	$\frac{\pi}{3}$
90°	$\frac{\pi}{2}$
120°	$\frac{2\pi}{3}$
135°	$\frac{3\pi}{4}$
150°	$\frac{5\pi}{6}$
180°	π
270°	$\frac{3\pi}{2}$
360°	2π

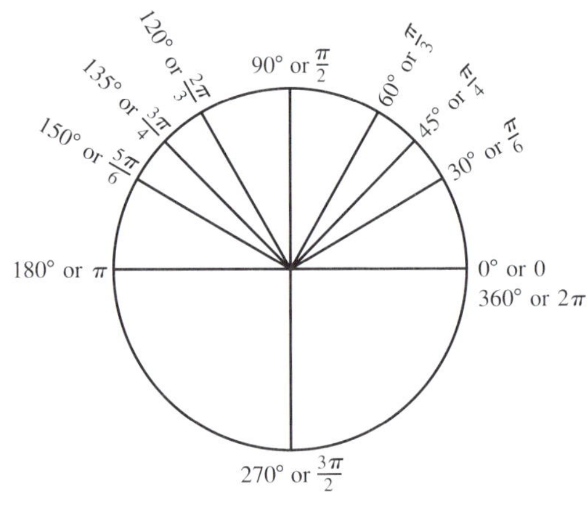

Many scientific calculators have special keys that can be used to change from degree measure to radian measure and the reverse.

EXAMPLE 5 **In which quadrant does the terminal side of each angle lie when it is in standard position?**

a. $\frac{2\pi}{3}$ **b.** $\frac{5\pi}{4}$ **c.** $\frac{\pi}{6}$ **d.** $-\frac{\pi}{3}$

a. Quadrant II **b.** Quadrant III **c.** Quadrant I **d.** Quadrant IV

Remember that for any angle θ expressed in degrees, the measure of any coterminal angle is $n \cdot 360° + \theta$, where n is an integer. Thus, for any angle θ expressed in radians, the measure of any coterminal angle is $n \cdot 2\pi + \theta$, where n is an integer.

CLASS EXERCISES

Express each angle measure in radians.

1. 90° $\frac{\pi}{2}$ **2.** 180° π **3.** 360° 2π **4.** −720° -4π **5.** −180° $-\pi$ **6.** 270° $\frac{3\pi}{2}$

Express each angle measure in degrees.

7. 0 0° **8.** 3π 540° **9.** $-\frac{\pi}{2}$ −90° **10.** $-\frac{3\pi}{2}$ −270° **11.** $\frac{\pi}{4}$ 45° **12.** $\frac{3\pi}{4}$ 135°

In which quadrant does the terminal side of each angle lie when it is in standard position?

13. $\frac{3\pi}{4}$ II **14.** $\frac{\pi}{3}$ I **15.** $\frac{5\pi}{4}$ III **16.** $\frac{11\pi}{12}$ II

PRACTICE EXERCISES

In which quadrant does the terminal side of each angle lie when it is in standard position?

A **1.** $\frac{4\pi}{5}$ II **2.** $\frac{2\pi}{3}$ II **3.** $-\frac{2\pi}{3}$ III **4.** $-\frac{3\pi}{4}$ III **5.** $-\frac{\pi}{12}$ IV **6.** $-\frac{7\pi}{3}$ IV

Express each angle measure in degrees and minutes.

7. 27.5° 27°30′ **8.** 108.75° 108°45′ **9.** 20.2° 20°12′ **10.** 150.1° 150°6′ **11.** 134.3° 134°18′ **12.** 126.6° 126°36′

Express each angle measure in decimal degrees.

13. 135°30′ 135.5° **14.** 75°30′ 75.5° **15.** 220°18′ 220.3° **16.** 300°24′ 300.4° **17.** 25°12′ 25.2° **18.** 100°36′ 100.6°

Express each angle measure in radians. Give answers in terms of π.

19. 60° $\frac{\pi}{3}$ **20.** 30° $\frac{\pi}{6}$ **21.** 120° $\frac{2\pi}{3}$ **22.** 135° $\frac{3\pi}{4}$ **23.** 150° $\frac{5\pi}{6}$ **24.** 210° $\frac{7\pi}{6}$
25. 54° $\frac{3\pi}{10}$ **26.** 28° $\frac{7\pi}{45}$ **27.** 335° $\frac{67\pi}{36}$ **28.** −310° $-\frac{31\pi}{18}$ **29.** −115° $-\frac{23\pi}{36}$ **30.** 155° $\frac{31\pi}{36}$

Express each angle measure in degrees.

31. $\frac{4\pi}{3}$ 240° **32.** $\frac{2\pi}{3}$ 120° **33.** $\frac{5\pi}{12}$ 75° **34.** $\frac{7\pi}{12}$ 105° **35.** $-\frac{13\pi}{18}$ −130° **36.** $-\frac{4\pi}{9}$ −80°
37. $\frac{7\pi}{6}$ 210° **38.** $\frac{11\pi}{6}$ 330° **39.** $-\frac{3\pi}{4}$ −135° **40.** $-\frac{\pi}{4}$ −45° **41.** $\frac{5\pi}{3}$ 300° **42.** $\frac{5\pi}{4}$ 225°

Express each angle measure in degrees, minutes, and seconds, to the nearest second.

B **43.** 21.56° 21°33′36″ **44.** 35.54° 35°32′24″ **45.** 160.11° 160°6′36″ **46.** 280.99° 280°59′24″
47. −313.53° −313°31′48″ **48.** −404.13° −404°7′48″ **49.** −381.87° −381°52′12″ **50.** −555.66° −555°39′36″

Express each angle measure in decimal degrees, to the nearest hundredth of a degree.

51. 18°15′15″ 18.25° **52.** 25°45′45″ 25.76° **53.** 315°48′03″ 315.80° **54.** 333°33′33″ 333.56°
55. −102°02′02″ −102.03° **56.** −103°44′04″ −103.73° **57.** −515°15′15″ −515.25° **58.** −602°12′22″ −602.21°

Express each angle measure in radians. Give answers in terms of π.

59. 450° $\frac{5\pi}{2}$ **60.** 630° $\frac{7\pi}{2}$ **61.** −405° $-\frac{9\pi}{4}$ **62.** −675° $-\frac{15\pi}{4}$ **63.** −1035° $-\frac{23\pi}{4}$ **64.** −1125° $-\frac{25\pi}{4}$

Express each angle measure in degrees.

65. $\frac{21\pi}{4}$ 945° **66.** $\frac{19\pi}{4}$ 855° **67.** $-\frac{18\pi}{3}$ −1080° **68.** $-\frac{21\pi}{3}$ −1260°
69. $-\frac{20\pi}{3}$ −1200° **70.** $-\frac{10\pi}{3}$ −600° **71.** $\frac{16\pi}{5}$ 576° **72.** $\frac{2\pi}{15}$ 24°

Trigonometry in Surveying

Directions as they apply to surveying are used to give yet another example of angle-measure applications.

Lesson Quiz

Express each angle measure in radians.

1. 480° $\frac{8\pi}{3}$

2. 270° $\frac{3\pi}{2}$

3. −225° $-\frac{5\pi}{4}$

Express each angle measure in degrees.

4. $\frac{5\pi}{6}$ 150°

5. 3π 540°

6. $-\frac{5\pi}{12}$ −75°

7. Express 27.11° in degrees, minutes, and seconds, to the nearest second. 27°06′36″

8. Express 87°45′45″ in decimal degrees, to the nearest hundredth of a degree. 87.76°

Enrichment

Find the radian measure of the angle formed by the hands of a clock at 4:30. $\frac{\pi}{4}$

Teacher's Resource Book
Practice—Chapter 1, p. 7
Enrichment—Chapter 1, p. 8

Start at the terminal side of the given angle θ in standard position. Find the radian measure of the resulting angle, in standard position, after the given number of rotations. Give answers in terms of π.

C **73.** $\theta = \frac{\pi}{3}$, 1 clockwise $-\frac{5\pi}{3}$

74. $\theta = \frac{\pi}{3}$, 1 counterclockwise $\frac{7\pi}{3}$

75. $\theta = -\frac{\pi}{3}$, 1 counterclockwise $\frac{5\pi}{3}$

76. $\theta = -\frac{\pi}{3}$, 1 clockwise $-\frac{7\pi}{3}$

77. $\theta = -\frac{\pi}{6}$, $1\frac{1}{2}$ clockwise $-\frac{19\pi}{6}$

78. $\theta = -\frac{\pi}{6}$, $1\frac{1}{2}$ counterclockwise $\frac{17\pi}{6}$

79. $\theta = \frac{3\pi}{2}$, 2 counterclockwise $\frac{11\pi}{2}$

80. $\theta = \frac{3\pi}{2}$, 2 clockwise $-\frac{5\pi}{2}$

81. $\theta = \frac{3\pi}{4}$, $2\frac{1}{2}$ clockwise $-\frac{17\pi}{4}$

82. $\theta = \frac{3\pi}{4}$, $2\frac{1}{2}$ counterclockwise $\frac{23\pi}{4}$

83. $\theta = -\frac{5\pi}{6}$, $1\frac{1}{3}$ counterclockwise $\frac{11\pi}{6}$

84. $\theta = -\frac{5\pi}{6}$, $1\frac{1}{3}$ clockwise $-\frac{7\pi}{2}$

Applications

85. Horology Through how many *degrees* does the minute hand of a clock rotate in 5 minutes? in $1\frac{1}{2}$ hours? $-30°$; $-540°$

86. Horology Through how many *radians* does the minute hand of a clock rotate in 45 minutes? in $2\frac{1}{4}$ hours? $-\frac{3\pi}{2}$, $-\frac{9\pi}{2}$

87. Meteorology Through how many *degrees* and how many *radians* does a weather vane rotate if a north wind changes to an east wind? $-90°$; $-\frac{\pi}{2}$

TRIGONOMETRY IN SURVEYING

Surveyors use trigonometry to find and represent boundaries of regions. An instrument called a transit is used to measure angles. The direction of a ray is sometimes given as the measure of the acute angle formed by the intersection of the ray with the true north-south line.

To write a direction in this way, first write N or S, then the measure of the acute angle, and finally E or W. The direction of $\overrightarrow{OA}$ is written N30°15′W, and is read 30 degrees 15 minutes west of north.

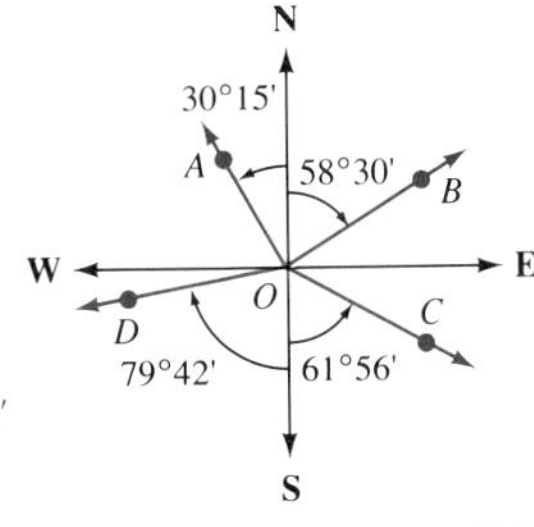

1. Write the directions for $\overrightarrow{OB}$, $\overrightarrow{OC}$, and $\overrightarrow{OD}$. N58°30′E; S61°56′E; S79°42′W
2. Find the measures of $\angle AOB$ and $\angle COD$. 88°45′; 141°38′

1.5 Applications: Angular and Linear Velocity

Objective: To solve problems involving arc length, angular velocity, and linear velocity

In order to solve problems involving rotary motion, it is often necessary to be able to find arc length.

Preview

The first Ferris wheel built was 250 ft in diameter. Modern Ferris wheels are somewhat smaller. For example, a large Ferris wheel located in Japan is 208 ft in diameter.

1. Suppose the 250-ft Ferris wheel could make one rotation in 45 seconds and the 208-ft wheel can make one rotation in 25 seconds. On which Ferris wheel would a rider move faster? Japanese Ferris wheel
2. Why do you think Ferris wheels are smaller today? safety
3. When a Ferris wheel is in motion, do all riders move at the same rate? yes

You have learned that the radian measure of a central angle θ can be found by dividing the length s of the intercepted arc by the radius r of the circle.

$$\theta = \frac{s}{r}$$

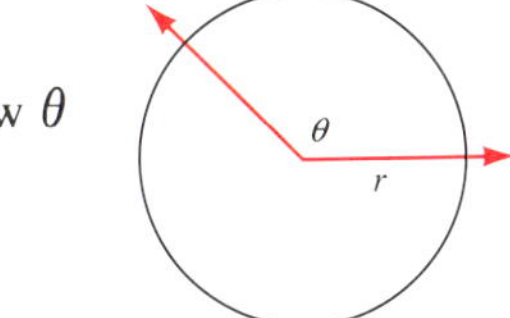

You can also use the formula to find s if you know θ (in radians) and r.

$$s = r \cdot \theta$$

As the figure at the right shows, s is a fraction of the circumference of the circle.

EXAMPLE 1 **Find the length of the arc intercepted by a central angle of 2.5 radians in a circle of radius 10 cm.**

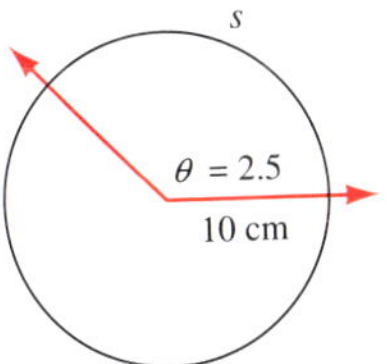

$$\begin{aligned} s &= r \cdot \theta \\ &= (10)(2.5) \\ &= 25 \end{aligned}$$

The arc has a length of 25 cm.

LESSON PLAN

Vocabulary
Angular displacement
Angular velocity
Linear velocity

Materials/Manipulatives
Calculators

BACKGROUND

In the Preview, the concept of angular velocity is informally introduced. It may be advisable to review the formula for finding the circumference of a circle before discussing the three questions.

TEACHING SUGGESTIONS

- It may be necessary to give an example of axes when discussing pulleys turning about their axes.
- In Example 4, demonstrate the canceling of units.
- Emphasize how problems such as the one in Example 5 may be checked.
- A calculator may be used to do any necessary calculations.

Critical Thinking

Comparing-Contrasting Ask students to discuss the similarities and/or differences between linear and angular velocity. Give examples of each in everyday situations. Answers may vary.

CHALKBOARD EXAMPLES

- **For Example 1**
  1. Find the length of the arc intercepted by a central angle of 2.5 radians in a circle of radius 5 cm. 12.5 cm
  2. Find the length of the arc intercepted by a central angle of 3 radians in a circle of diameter 13 mm. 19.5 mm
- **For Example 2**
  3. If $\overrightarrow{OA}$ on the record in Example 2 makes a $\frac{2}{3}$ revolution about its axis, find the angular displacement θ of point A. Express your answer in radians. $\frac{4\pi}{3}$
  4. If $\overrightarrow{OB}$ has an angular displacement of $\theta = \frac{5\pi}{3}$ rad, find the number of revolutions $\overrightarrow{OB}$ makes about its axis. $\frac{5}{6}$

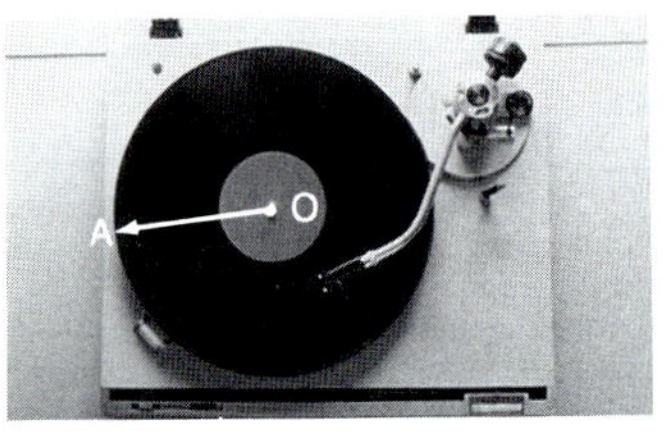

Objects such as Ferris wheels, phonograph records, and pulleys turn about their axes. This is called *rotary motion*, since a point on such an object follows a circular path. Imagine point A on the edge of a record spinning about its axis O. The **angular displacement** of point A is the angle θ through which $\overrightarrow{OA}$ rotates.

EXAMPLE 2 **If $\overrightarrow{OA}$ on the record pictured above makes a $\frac{4}{5}$ revolution about its axis, find the angular displacement θ of point A in radians.**

$\overrightarrow{OA}$ travels through 2π radians during each complete revolution. Thus, $\theta = \frac{4}{5} \cdot 2\pi = \frac{8\pi}{5}$ rad.

The **angular velocity** of a point on a revolving ray is the angular displacement per unit of time. The Greek letter ω (omega) is usually used to represent angular velocity.

$$\omega = \frac{\theta}{t}$$

Angular velocity can be expressed in various units, such as revolutions per minute (rpm) or radians per second (rad/s). Sometimes you will have to change from one unit of measure to another.

EXAMPLE 3 **Find the angular velocity ω in radians per second of a wheel turning at 25 rpm (revolutions per minute).**

$$\omega = \frac{25 \text{ revolutions}}{1 \text{ minute}} \cdot \frac{2\pi \text{ radians}}{1 \text{ revolution}} \qquad \textit{1 revolution} = 2\pi \textit{ radians}$$

$$= 50\pi \text{ rad/min}$$

$$= \frac{50\pi \text{ radians}}{1 \text{ minute}} \cdot \frac{1 \text{ minute}}{60 \text{ seconds}} \qquad \textit{1 minute} = \textit{60 seconds}$$

$$= \frac{5\pi}{6} \text{ rad/s}$$

The **linear velocity** V of a point on a revolving ray is defined to be the linear distance traveled by the point per unit of time: $V = \frac{s}{t}$. Since $s = r \cdot \theta$ and $\omega = \frac{\theta}{t}$, the linear velocity can be expressed as the product of the angular velocity ω and the radius r.

$$V = \frac{s}{t}$$

$$V = \frac{r \cdot \theta}{t} = \frac{\theta}{t} \cdot r$$

$$V = \omega \cdot r$$

EXAMPLE 4 **The wheel of a truck is turning at 6 rps (revolutions per second). The wheel is 4 ft in diameter.**

a. Find the angular velocity ω of the wheel in radians per second.

b. Find the linear velocity V, in feet per second, of a point on the rim of the wheel.

a. $\omega = \dfrac{6 \text{ revolutions}}{1 \text{ second}} \cdot \dfrac{2\pi \text{ radians}}{1 \text{ revolution}} = 12\pi \text{ rad/s}$

b. Since the diameter is 4 ft, a point on the rim is 2 ft from the center. That is, $r = 2$ ft.

$$V = \omega \cdot r = \frac{12\pi \text{ radians}}{1 \text{ second}} \cdot 2 \text{ ft} = 24\pi \text{ ft/s}$$

Note that the word *radian* is not included in the answer in Example 4b. Technically, a radian is the ratio of two lengths and has no dimension. However, to avoid confusion in calculations and to indicate that angle measure is involved, the word radian is sometimes retained.

Problems dealing with pulleys often involve linear velocity and angular velocity.

EXAMPLE 5 **Two pulleys, one 6 in. and the other 2 ft in diameter, are connected by a belt. The larger pulley revolves at the rate of 60 rpm.**

a. Find the linear velocity of the belt in feet per minute.

b. Calculate the angular velocity of the smaller pulley in radians per minute.

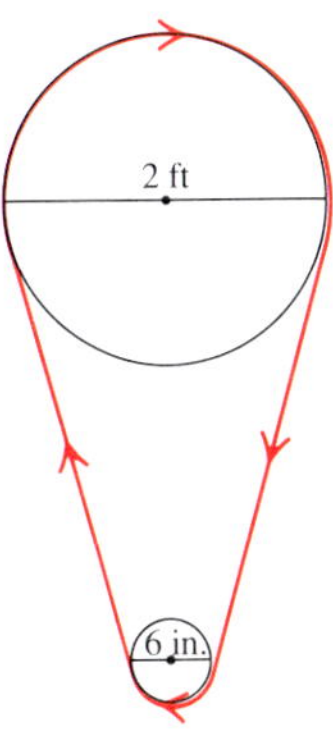

a. The linear velocity V of the belt equals the linear velocity of a point on the rim of either pulley. You know the angular velocity of the larger pulley in revolutions per minute. Change this to radians per minute.

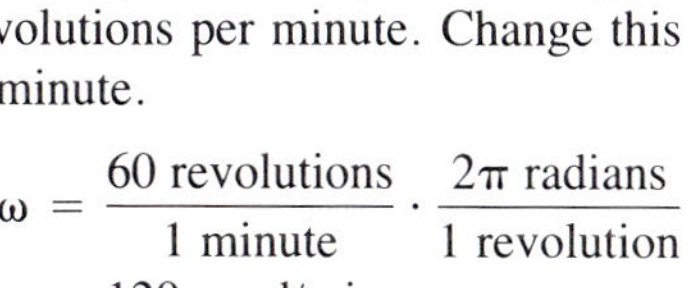

$$\omega = \frac{60 \text{ revolutions}}{1 \text{ minute}} \cdot \frac{2\pi \text{ radians}}{1 \text{ revolution}}$$
$$= 120\pi \text{ rad/min}$$

A point on the rim of the larger pulley is 1 ft from its center. Therefore,

$$V = \omega \cdot r$$
$$= \frac{120\pi \text{ radians}}{1 \text{ minute}} \cdot 1 \text{ ft} = 120\pi \text{ ft/min}$$

- **For Example 3**

Find the angular velocity ω in radians per second for a wheel turning at the given number of revolutions.

5. 40 rpm $\frac{4\pi}{3}$ rad/s

6. 65 rpm $\frac{13\pi}{6}$ rad/s

- **For Example 4**

7. The wheel of a car is turning at 15 rps. The wheel has a radius of 1.5 m.

a. Find the angular velocity of the wheel in rad/s. 30π rad/s

b. Find the linear velocity in feet per second of a point 1 ft from the center (axis) of the wheel. 30π ft/s

8. The minute hand of a clock is 7 cm long. How far does the top of the minute hand move in 2 h 30 min? Give your answer to the nearest centimeter. 110 cm

- **For Example 5**

The radius of the Earth is approximately 3840 mi long, and the Earth revolves on its axis once every 24 h.

9. Determine the angular velocity of the Earth in radians per day and radians per week.
2π rad/da 14π rad/wk

10. Determine the linear velocity of a person standing on the equator as opposed to a person standing on the North Pole.
At equator: 7680π mi/da
At North Pole: 0

Common Error

- Some students do not correctly distinguish between linear and angular velocity. Stress the difference between the two.
- See *Teacher's Resource Book* for additional remediation.

LESSON FOLLOW-UP

Discussion

Explain how you would calculate the linear velocity in miles per hour of a car wheel that turns completely around 14 times every second and has a diameter 2 ft long. First find the angular velocity of the wheel: $14 \cdot 60 \cdot 60 \cdot 2\pi = 100{,}800\pi$ rad/h. Next find the radius of the wheel in miles: 1 ft $= \frac{1}{5280}$ mi. Then find the linear velocity in miles per hour: $100{,}800\pi \cdot \frac{1}{5280} = 60$ mi/h, to the nearest mi/h.

Assignment Guide

See p. B for assignments.

b. The linear velocity of the smaller pulley is 120π ft/min and the radius of that pulley is 3 in., or $\frac{1}{4}$ ft. Substitute the known information into the formula $V = \omega \cdot r$.

$$V = \omega \cdot r$$
$$120\pi = \omega \cdot \frac{1}{4}$$
$$\omega = 480\pi \text{ rad/min} \qquad \textit{Angular velocity of smaller pulley}$$

To check the answer, note that the radius of the smaller pulley is $\frac{1}{4}$ that of the larger pulley. In order to keep up, the angular velocity of the smaller pulley must be 4 times that of the larger one. Consequently, $\omega = 4(120\pi) = 480\pi$.

CLASS EXERCISES

Express answers to Exercises 1–8 in terms of π.

1. Find the length of the arc intercepted by a central angle of $\frac{5\pi}{4}$ if the radius of the circle is 20 mm. 25π mm
2. Find the length of the arc intercepted by a central angle of $\frac{7\pi}{4}$ if the radius of the circle is 12 ft. 21π ft

Find the angular velocity, in radians per second, of a wheel turning at each of the following numbers of revolutions per minute.

3. 200 $\frac{20\pi}{3}$ rad/s
4. 100 $\frac{10\pi}{3}$ rad/s
5. 144 $\frac{24\pi}{5}$ rad/s
6. 180 6π rad/s

For Exercises 7–8, the wheel of a bicycle is turning at 5 rpm. The wheel is 24 in. in diameter.

7. Find the angular velocity of the wheel in radians per second. $\frac{\pi}{6}$ rad/s
8. Find the linear velocity in inches per second of a point on the rim. 2π in./s
9. Name at least three things in real life that have angular velocity. Answers may vary.

PRACTICE EXERCISES

Express answers to Exercises 1–18 in terms of π.

For Exercises 1–6, find the length of the arc intercepted by the given central angle in a circle with the given radius.

A

1. $\theta = \frac{\pi}{2}$, $r = 10$ in. 5π in.
2. $\theta = \frac{\pi}{4}$, $r = 8$ m 2π m
3. $\theta = \frac{5\pi}{6}$, $r = 15$ in. $\frac{25\pi}{2}$ in.
4. $\theta = \frac{3\pi}{4}$, $r = 12$ m 9π m
5. $\theta = \frac{9\pi}{4}$, $r = 20$ in. 45π in.
6. $\theta = \frac{7\pi}{6}$, $r = 25$ cm $\frac{175\pi}{6}$ in.

Point A is on the rim of a rotating wheel. Find the angular displacement of point A for the given number of revolutions. Express answers in radians.

7. $\frac{1}{2}$ π rad. **8.** $\frac{1}{4}$ $\frac{\pi}{2}$ rad. **9.** $\frac{1}{3}$ $\frac{2\pi}{3}$ rad. **10.** $\frac{2}{3}$ $\frac{4\pi}{3}$ rad. **11.** $\frac{3}{5}$ $\frac{6\pi}{5}$ rad. **12.** $\frac{5}{6}$ $\frac{5\pi}{3}$ rad.

Find the angular velocity for each angular displacement in Exercises 7–12 for the given values of t. Express answers in radians per minute.

13. Exercise 7, $t = 2$ min $\frac{\pi}{2}$ rad/min

14. Exercise 8, $t = 10$ min $\frac{\pi}{20}$ rad/min

15. Exercise 9, $t = 5$ min $\frac{2\pi}{15}$ rad/min

16. Exercise 10, $t = 3$ min $\frac{4\pi}{9}$ rad/min

17. Exercise 11, $t = 12$ min $\frac{\pi}{10}$ rad/min

18. Exercise 12, $t = 15$ min $\frac{\pi}{9}$ rad/min

Calculate the linear velocity V of an object rotating at angular velocity ω at a distance r from the center. Express answers in centimeters per second.

B **19.** $r = 10$ cm, $\omega = 20\pi$ rad/s 200π cm/s

20. $r = 3$ cm, $\omega = 20\pi$ rad/s 60π cm/s

21. $r = 15$ cm, $\omega = 5\pi$ rad/s 75π cm/s

22. $r = 25$ cm, $\omega = 8\pi$ rad/s 200π cm/s

23. Find the angular velocity in radians per second of the second hand of a clock. Express the answer in terms of π. $\frac{\pi}{30}$ rad/s

24. Find the angular velocity in radians per second of the minute hand of a clock. Express the answer in terms of π. $\frac{\pi}{1800}$ rad/s

25. Calculate the angular velocity in radians per minute of a Ferris wheel 250 ft in diameter that takes 45 s to rotate once. Express the answer in terms of π. $\frac{8\pi}{3}$ rad/min

26. If you sat on the rim of the Ferris wheel in Exercise 25, what would your linear velocity be, to the nearest foot per minute? 1047 ft/min

27. Calculate the angular velocity in radians per minute of a Ferris wheel 208 ft in diameter that takes 25 s to rotate once. Express the answer in terms of π. $\frac{24\pi}{5}$ rad/min

28. If you sat on the rim of the Ferris wheel in Exercise 27, what would your linear velocity be, to the nearest foot per minute? 1568 ft/min

29. The Earth spins on its axis every 24 hours. If the Earth's circumference is 24,800 miles, find the linear velocity of a person standing on the equator in miles per hour. $1033\frac{1}{3}$ mph

C **30.** A car is traveling at a speed of 45 miles per hour. Find the angular velocity of a tire in revolutions per minute (rpm), if the diameter of each rim is 13 in. with 4 in. of tire between the road surface and each rim. 720 rpm

31. Assume the car tire in Exercise 30 above is not properly inflated so that the measurement between the road surface and the rim is only 2 in. Find the angular velocity of the underinflated tire in rpm. 890 rpm

Test Yourself

See *Teacher's Resource Book,* Tests, pp. 5–6.

Lesson Quiz

1. Find the length of an arc intercepted by a central angle of $\frac{2\pi}{3}$ if the radius of the circle is 12 in. 8π in.
2. Point B is on the edge of the rotating wheel. Find the angular displacement of point B for $\frac{3}{4}$ revolution. Give your answer in radians. $\frac{3\pi}{2}$ rad
3. Find the linear velocity of an object rotating at angular velocity 6π rad/s at a distance 15 cm from the center. Give your answer to the nearest tenth of a centimeter per second. 282.7 cm/s

Enrichment

A satellite traveling in a circular orbit 2000 km above the Earth completes one orbit every 3 h. If the Earth's radius is 6400 km, find the speed of the satellite in kilometers per hour.
$\frac{2\pi}{3} \cdot 8400 = 5600\pi$ km/h

Teacher's Resource Book
Practice—Chapter 1, p. 9
Enrichment—Chapter 1, p. 10

Two pulleys are connected by a belt. The smaller pulley has a diameter of 8 ft and the larger pulley has a diameter of 12 ft. The smaller pulley is revolving at the rate of 60 rpm.

32. Find the angular velocity of each pulley in radians per second. Express the answers in terms of π. 2π rad/s; $\frac{4\pi}{3}$ rad/s

33. Find the linear velocity of a point on the rim of each pulley in feet per second. Express the answers in terms of π. 8π ft/s; 8π ft/s

TEST YOURSELF

Give the domain and range of each relation and state whether the relation *is* or *is not* a function.

1. $\{(1, 3), (2, 3), (4, 5), (6, 7)\}$ d:{1, 2, 4, 6}; r:{3, 5, 7}; function

2. $\{(2, 6), (3, 7), (3, 8), (4, 9)\}$ d:{2, 3, 4}; r:{6, 7, 8, 9}; not a function — 1.1

For the function $f(x) = \frac{1}{x+2}$, find each value if it is defined.

3. $f(5)$ $\frac{1}{7}$

4. $f\left(\frac{1}{2}\right)$ $\frac{2}{5}$

5. $f(-2)$ undefined

6. Find the lengths of the sides of triangle ABC with vertices $A(-4, 1)$, $B(0, 1)$, and $C(3, -2)$. $AB = 4$; $BC = 3\sqrt{2}$; $AC = \sqrt{58}$ — 1.2

In which quadrant does the terminal side of each angle lie? Assume that the angles are in standard position.

7. 132° II

8. −78° IV

9. 400° I — 1.3

10. For a $\frac{13}{12}$ clockwise rotation, find the measure of the angle in degrees. −390°

11. Express 75° in radians. Give the answer in terms of π. $\frac{5\pi}{12}$ — 1.4

12. Express $\frac{5\pi}{36}$ in degrees. 25°

For Exercises 13–15, express answers in terms of π. — 1.5

13. Find the length of the arc intercepted by a central angle of $\frac{2\pi}{3}$ if the radius of the circle is 18 m. 12π m

14. Find the angular velocity ω in radians per second for a pulley turning at 1800 rpm. 60π rad/s

15. Calculate the linear velocity V of a point located 12 in. from the center of a disk rotating at 7π rad/s. 84π in./s

1.6 Cosine and Sine Functions

Objectives: To define the cosine and sine functions
To evaluate the sine and cosine functions of an angle given a point on its terminal side

Sometimes the Pythagorean theorem is used to find the length of a side of a right triangle in order to determine the sine and cosine of an angle.

Preview

If triangle ABC is a right triangle with hypotenuse c, the lengths of sides a, b, and c are related by the Pythagorean theorem: $c^2 = a^2 + b^2$.

EXAMPLE **Find b if $a = 4.5$ and $c = 7.5$.**

$$(7.5)^2 = (4.5)^2 + b^2 \qquad c^2 = a^2 + b^2$$
$$56.25 = 20.25 + b^2$$
$$36 = b^2$$
$$\sqrt{36} = b \qquad \textit{Length is positive.}$$
$$6 = b$$

Find the length of the third side of each right triangle ABC with hypotenuse c.

1. $a = 12, b = 9, c = \underline{?}$ 15
2. $b = 5, c = 5\sqrt{2}, a = \underline{?}$ 5
3. $a = 1, c = 2, b = \underline{?}$ $\sqrt{3}$
4. $a = 6, c = 15.6, b = \underline{?}$ 14.4
5. $a = 11, b = 11, c = \underline{?}$ $11\sqrt{2}$
6. $b = 1.2, c = 1.3, a = \underline{?}$ 0.5

The radius of a **unit circle** is 1. Consider the origin-centered unit circle shown at the right together with an angle θ in standard position drawn on the same coordinate plane. The initial side of the angle intersects the circle at (1, 0) and the terminal side intersects the circle at the point $P(x, y)$.

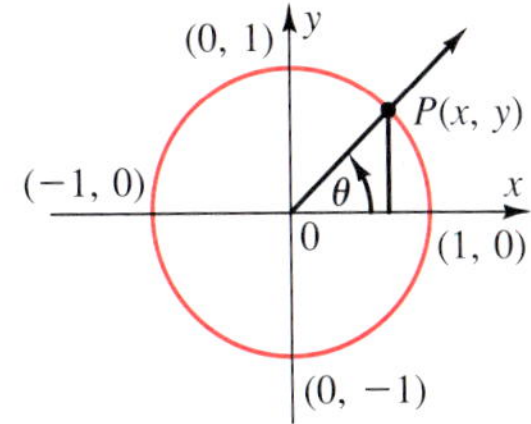

If the terminal side of θ is rotated counterclockwise about the origin, the values of x and y associated with point P will change. For each value of θ, however, there is exactly one corresponding value for x. Because of this correspondence, the set of ordered pairs (θ, x) is a function. Also, for each value of θ, there is exactly one value for y. Therefore, the set of ordered pairs (θ, y) is also a function. These two functions are called the **cosine** and the **sine** functions, respectively.

LESSON PLAN

Vocabulary

Cosine
Sine
Radius vector
Reference triangle
Unit circle

Materials/Manipulatives

Calculators
Computer

BACKGROUND

In the Preview, the Pythagorean theorem is reviewed. The Pythagorean theorem is often used to find sine and cosine values.

TEACHING SUGGESTIONS

- When developing reference triangles, you may want to do examples that involve 30°- 60°- 90° triangles and 45°- 45°- 90° triangles.
- In Example 1, emphasize that *r* is actually the hypotenuse of the right triangle.
- In Example 2, you may want to determine the sign of the cosine first. That is, since the terminal side lies in the third quadrant, *x* must be negative, so the cosine will be negative.
- Point out that there is a close relationship between the circular functions and the corresponding trigonometric functions. In fact, the circular functions and angle functions have identical properties.
- Calculators and computers may be used to find the sine and cosine of given angles.

Critical Thinking

Observations Ask students to explain whether or not sin θ and cos θ can be found if θ is not in standard position and how, if possible, they would find these values. Students should reason that as long as the measure of θ is known, approximations for sin θ and cos θ can be found by using a scientific calculator.

If θ is an angle in standard position whose terminal side intersects the unit circle at (x, y), then

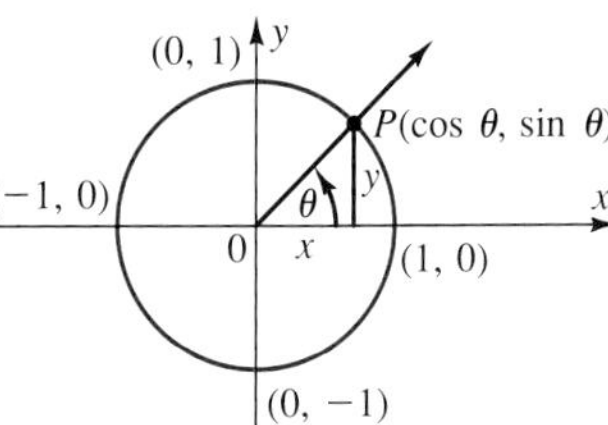

$x =$ cosine θ *abbreviated* $x = \cos\theta$

and $y =$ sine θ *abbreviated* $y = \sin\theta$

The ordered pair (x, y) can be written as $(\cos\theta, \sin\theta)$, which are the coordinates of point P in the figure.

Since the unit circle has radius 1, the values of x and y can range from -1 to 1. That is,

$$-1 \leq \cos\theta \leq 1 \quad \text{and} \quad -1 \leq \sin\theta \leq 1$$

Given an arbitrary angle θ, $\cos\theta$ and $\sin\theta$ can be found if a point (x, y) on the terminal side of θ is given, even if (x, y) is not on the unit circle. A right triangle is formed when a perpendicular is drawn from *any* point on the terminal side of θ to the x-axis. Such a triangle is called a **reference triangle.** In the figure, (x, y) and (x', y') are the coordinates of corresponding vertices of similar right triangles. The lengths of the sides are x, y, and r, and x', y', and 1, respectively. The lengths of corresponding sides of similar triangles are proportional, so $\frac{x'}{1} = \frac{x}{r}$ and $\frac{y'}{1} = \frac{y}{r}$. Thus, since $x' = \cos\theta$ and $y' = \sin\theta$,

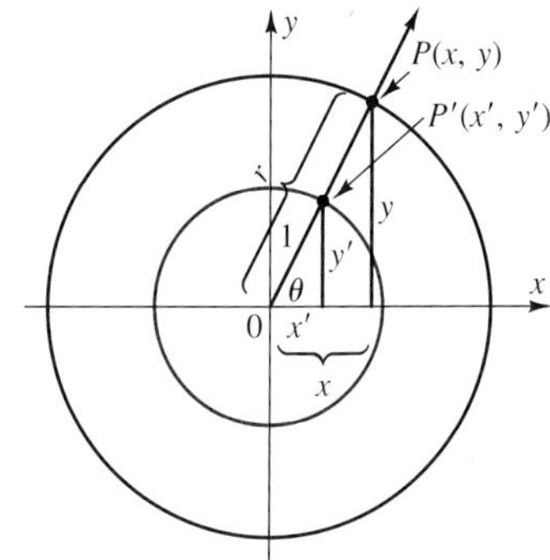

$$\cos\theta = \frac{x}{r} \quad \text{and} \quad \sin\theta = \frac{y}{r}$$

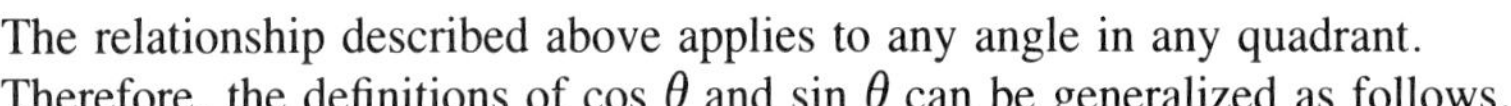

The relationship described above applies to any angle in any quadrant. Therefore, the definitions of $\cos\theta$ and $\sin\theta$ can be generalized as follows.

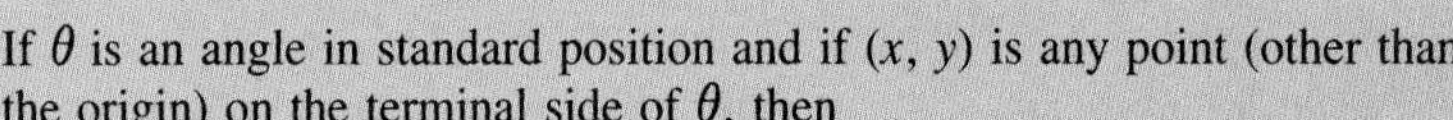

> If θ is an angle in standard position and if (x, y) is any point (other than the origin) on the terminal side of θ, then
>
> $$\cos\theta = \frac{x}{r} \quad \text{and} \quad \sin\theta = \frac{y}{r} \quad \text{where } r = \sqrt{x^2 + y^2}$$

Note that θ can be measured in degrees or in radians. Since r, sometimes called the **radius vector,** is the distance from point P to the origin, it is always positive. However, x and y can be positive or negative, depending upon the location of the terminal side of the angle. If x is negative and y is positive, θ is in the second quadrant; if x is negative and y is negative, θ is in the third quadrant; and if x is positive and y is negative, θ is in the fourth quadrant.

When you are asked to find *exact* values of the trigonometric functions of an angle (as in Example 1 below), leave irrational answers in radical form.

EXAMPLE 1 **Find the exact values of the sine and cosine of an angle θ in standard position if point Q is on its terminal side. Sketch the reference triangle.**
a. $Q(3, -4)$ **b.** $Q(-3, -6)$

a. $r = \sqrt{3^2 + (-4)^2} = \sqrt{25} = 5$ *$r = \sqrt{x^2 + y^2}$*

Therefore, $\cos \theta = \frac{3}{5}$ *$\cos \theta = \frac{x}{r}$*

$\sin \theta = \frac{-4}{5} = -\frac{4}{5}$ *$\sin \theta = \frac{y}{r}$*

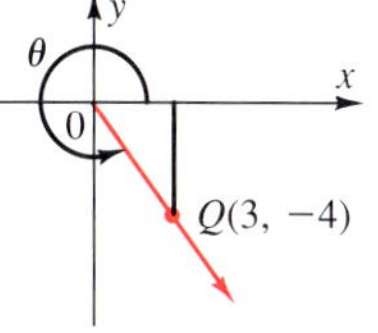

b. $r = \sqrt{(-3)^2 + (-6)^2} = \sqrt{45} = 3\sqrt{5}$ *$r = \sqrt{x^2 + y^2}$*

Therefore, $\cos \theta = \frac{-3}{3\sqrt{5}} = -\frac{1}{\sqrt{5}} = -\frac{\sqrt{5}}{5}$

$\sin \theta = \frac{-6}{3\sqrt{5}} = -\frac{2}{\sqrt{5}} = -\frac{2\sqrt{5}}{5}$

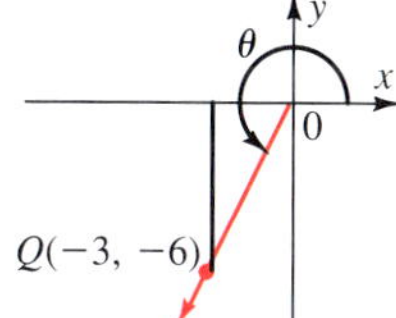

If you are given either the sine or the cosine, then you can find the other function, provided you know the quadrant in which the angle terminates.

EXAMPLE 2 **Angle θ is in standard position with its terminal side in the third quadrant. Find the exact value of $\cos \theta$ if $\sin \theta = -\frac{3}{5}$.**

Since $\sin \theta = -\frac{3}{5}$, $r = 5$ and $y = -3$ for some point (x, y) on the terminal side of θ.   *r is always positive and y is negative in the third quadrant.*

$$\begin{aligned} x^2 + (-3)^2 &= 5^2 \\ x^2 + 9 &= 25 \\ x^2 &= 16 \\ x &= 4 \quad \text{or} \quad x = -4 \end{aligned}$$ *$x^2 + y^2 = r^2$*

Since θ is in the third quadrant, $x < 0$. Thus, $x = -4$ and $\cos \theta = \frac{-4}{5} = -\frac{4}{5}$.

The set of real numbers is the domain for both the cosine and the sine functions. These functions can be associated with the measures of angles and they can also be associated with the lengths of arcs on the unit circle. For this reason, these functions are sometimes called **circular functions.**

CHALKBOARD EXAMPLES

- **For Example 1**
  Find the exact values of the sine and cosine of an angle θ in standard position, if point A is on its terminal side.

  1. $A(4, 3)$ $\sin \theta = \frac{3}{5}$; $\cos \theta = \frac{4}{5}$

  2. $A(-3, 4)$ $\sin \theta = \frac{4}{5}$; $\cos \theta = -\frac{3}{5}$

- **For Example 2**
  Angle θ is in standard position in the first quadrant. Find the exact value of $\sin \theta$ if $\cos \theta$ is given.

  3. $\cos \theta = \frac{7}{25}$ $\frac{24}{25}$

  4. $\cos \theta = \frac{12}{13}$ $\frac{5}{13}$

Common Error

- Some students use the incorrect sign for $\sin \theta$ and/or $\cos \theta$. Emphasize that the quadrant in which the terminal side of angle θ lies must be considered.
- See *Teacher's Resource Book* for additional remediation.

LESSON FOLLOW-UP

Discussion

If an angle is not in standard position, how could you find the values of sine and cosine using the methods discussed in this lesson? You could not find the values of the sine and cosine of the angle using the methods discussed in the lesson, unless you were given a point on the terminal side of the angle or the measure of the angle and the sides of its reference triangle.

Assignment Guide

See p. B for assignments.

Computer

The complete form of the program that uses the Pythagorean theorem to find values of the sine and cosine is on the disk provided with the *Teacher's Resource Book*.

Lesson Quiz

1. In Quadrant II, is $\sin\theta$ positive or negative? positive
2. In Quadrant IV, is $\cos\theta$ positive or negative? positive

Find the exact values of the sine and cosine of an angle θ in standard position if the point with the given coordinates is on its terminal side.

3. $(-9, 12)$ $\sin\theta = \frac{4}{5}$; $\cos\theta = -\frac{3}{5}$
4. $(4, -4)$ $\sin\theta = -\frac{\sqrt{2}}{2}$; $\cos\theta = \frac{\sqrt{2}}{2}$
5. Find the exact value of $\cos\theta$ if $\sin\theta = -\frac{12}{13}$ and θ is an angle in standard position in Quadrant III. $-\frac{5}{13}$
6. Find the exact value of $\sin\theta$ if $\cos\theta = \frac{4}{5}$ and θ is an angle in standard position in Quadrant IV. $-\frac{3}{5}$

CLASS EXERCISES

Let θ be an angle in standard position. For each given value of $\sin\theta$ or $\cos\theta$, state the quadrants in which θ can lie.

1. $\sin\theta = \frac{1}{2}$ I; II **2.** $\sin\theta = -\frac{\sqrt{3}}{2}$ III; IV **3.** $\cos\theta = \frac{3}{5}$ I; IV **4.** $\cos\theta = -\frac{4}{5}$ II; III

Complete the table below by writing the sign (+ or −) of $\cos\theta$ and of $\sin\theta$ for each quadrant in the coordinate plane.

Quadrant	I	II	III	IV
$\cos\theta$	**5.** +	**6.** −	**7.** −	**8.** +
$\sin\theta$	**9.** +	**10.** +	**11.** −	**12.** −

13. What is the minimum value of $\sin\theta$? of $\cos\theta$? −1; −1

14. What is the maximum value of $\sin\theta$? of $\cos\theta$? 1; 1

PRACTICE EXERCISES

Find the exact values of the sine and cosine of an angle θ in standard position if the point with the given coordinates is on the terminal side of the angle. Sketch the reference triangle. 6. $-\frac{\sqrt{2}}{2}$; $-\frac{\sqrt{2}}{2}$ 3. $-\frac{12}{13}$; $-\frac{5}{13}$ 4. $-\frac{2\sqrt{5}}{5}$; $\frac{\sqrt{5}}{5}$

A **1.** $(3, 4)$ $\frac{4}{5}$; $\frac{3}{5}$ **2.** $(-4, 3)$ $\frac{3}{5}$; $-\frac{4}{5}$ **3.** $(-5, -12)$ **4.** $(3, -6)$

5. $(-3, 0)$ 0; −1 **6.** $(-2, -2)$ **7.** $(-6, 3)$ $\frac{\sqrt{5}}{5}$; $-\frac{2\sqrt{5}}{5}$ **8.** $(8, 6)$ $\frac{3}{5}$; $\frac{4}{5}$

Find the exact value of $\sin\theta$ if $\cos\theta$ is given and θ is an angle in standard position in Quadrant I.

9. $\cos\theta = \frac{4}{5}$ $\frac{3}{5}$ **10.** $\cos\theta = \frac{3}{5}$ $\frac{4}{5}$ **11.** $\cos\theta = \frac{12}{13}$ $\frac{5}{13}$ **12.** $\cos\theta = \frac{5}{13}$ $\frac{12}{13}$

Find the exact value of $\cos\theta$ if $\sin\theta$ is given and θ is an angle in standard position in Quadrant II.

13. $\sin\theta = \frac{\sqrt{3}}{2}$ $-\frac{1}{2}$ **14.** $\sin\theta = \frac{1}{2}$ $-\frac{\sqrt{3}}{2}$ **15.** $\sin\theta = \frac{3}{5}$ $-\frac{4}{5}$ **16.** $\sin\theta = \frac{5}{13}$ $-\frac{12}{13}$

Find the exact values of the sine and cosine of an angle θ in standard position if the point with the given coordinates is on the terminal side of the angle. Sketch the reference triangle.

B **17.** $(5, 7)$ $\frac{7\sqrt{74}}{74}$; $\frac{5\sqrt{74}}{74}$ **18.** $(-2, 4)$ $\frac{2\sqrt{5}}{5}$; $-\frac{\sqrt{5}}{5}$ **19.** $(3, -3)$ $-\frac{\sqrt{2}}{2}$; $\frac{\sqrt{2}}{2}$ **20.** $(4, 6)$ $\frac{3\sqrt{13}}{13}$; $\frac{2\sqrt{13}}{13}$ **21.** $(-6, 10)$ $\frac{5\sqrt{34}}{34}$; $-\frac{3\sqrt{34}}{34}$ **22.** $(12, -18)$ $\frac{-3\sqrt{13}}{13}$; $\frac{2\sqrt{13}}{13}$

Angle θ is in standard position and lies in the given quadrant. When $\sin\theta$ is given, find the exact value of $\cos\theta$. When $\cos\theta$ is given, find the exact value of $\sin\theta$.

23. $\sin\theta = \frac{1}{2}$; I $\frac{\sqrt{3}}{2}$ **24.** $\cos\theta = \frac{1}{2}$; I $\frac{\sqrt{3}}{2}$ **25.** $\sin\theta = -\frac{\sqrt{3}}{2}$; III $-\frac{1}{2}$

26. $\sin\theta = -\frac{12}{13}$; IV $\frac{5}{13}$ **27.** $\cos\theta = \frac{5}{11}$; IV $-\frac{4\sqrt{6}}{11}$ **28.** $\cos\theta = -\frac{5}{6}$; II $\frac{\sqrt{11}}{6}$

Let θ be an angle in standard position. In which quadrant or quadrants can θ lie under the given conditions?

C **29.** $\sin\theta > 0$ I; II **30.** $\sin\theta < 0$ and $\cos\theta > 0$ IV

31. $\sin\theta = \cos\theta$ I; III **32.** $\sin\theta = -\cos\theta$ II; IV

Applications

Computer In the program shown, the Pythagorean theorem is used to calculate specific values of the cosine and sine functions. Line 30 allows the user to enter the coordinates of a point on the terminal side of an angle. Lines 40–60 calculate the sine and cosine, and lines 70 and 80 display the results.

```
10 PRINT "ENTER THE COORDINATES OF A POINT"
20 PRINT "ON THE TERMINAL SIDE OF THE ANGLE"
30 PRINT: INPUT "ENTER X THEN Y"; X,Y
40 R = SQR(X^2 + Y^2)
50 S = Y/R
60 C = X/R
70 HOME: PRINT "THE SINE OF THE ANGLE IS"S
80 PRINT: "THE COSINE OF THE ANGLE IS"C
90 END
```

33. Why does this program give the correct sign for each cosine and sine value? See side column.

34. In the program, within what interval are degree measures of the angles assumed to be? $0° \le \theta \le 360°$

TRIGONOMETRY IN THE MACHINE SHOP

The *sine bar* is a precision tool used to measure angles. It is composed of two parallel cylinders of equal radii with axes 25 cm apart attached to a flat surface parallel to a plane through the axes of the cylinders. The cylinders are placed on blocks set on a horizontal table, and the bar is tilted. The sine of angle β is given by the formula

$$\sin\beta = \frac{b - a}{25}$$

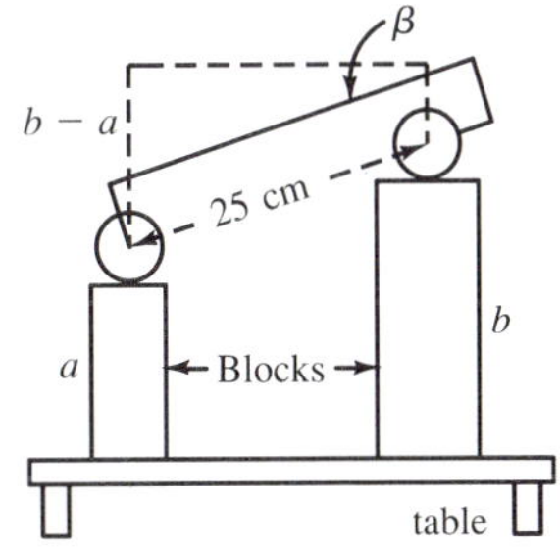

1. If $a = 1.0$ cm and $b = 3.5$ cm, find the exact value of $\sin\beta$. 0.1

2. If $a = 2.0$ cm and $b = 5.2$ cm, find $\sin\beta$ to the nearest hundredth. 0.13

Teacher's Resource Book

Practice—Chapter 1, p. 11

Enrichment—Chapter 1, p. 12

Additional Answers

Practice Exercises

33. The correct sign is entered in line 30 and then used to compute the sine and cosine in lines 50–60.

LESSON PLAN

Vocabulary
Cosecant
Cotangent
Reciprocal trigonometric functions
Secant
Tangent
Trigonometric functions

BACKGROUND

In the Preview, the sine and cosine functions are reviewed and the tangent, cotangent, secant, and cosecant functions are introduced but are not labeled as such.

1.7 The Trigonometric Functions

Objectives: To define the tangent, cotangent, secant, and cosecant functions
To evaluate the trigonometric functions of an angle

In lesson 1.6, the sine and cosine functions of an angle θ in standard position were defined to be $\sin\theta = \frac{y}{r}$ and $\cos\theta = \frac{x}{r}$, where (x, y) is a point other than the origin on the terminal side of the angle and $r = \sqrt{x^2 + y^2}$. In this lesson, the *tangent*, *cotangent*, *secant*, and *cosecant* functions will be defined.

Preview

Angle θ is in standard position. The points $(4, -2)$ and $(6, -3)$ are on the terminal side of θ.

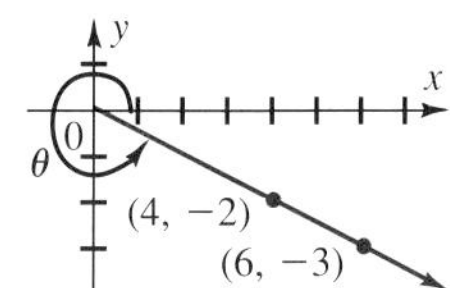

Let $r = \sqrt{x^2 + y^2}$ and calculate the following six ratios for each of the two points.

1. $\frac{x}{r}$ $\frac{2\sqrt{5}}{5}$ **2.** $\frac{y}{r}$ $-\frac{\sqrt{5}}{5}$ **3.** $\frac{y}{x}$ $-\frac{1}{2}$ **4.** $\frac{r}{x}$ $\frac{\sqrt{5}}{2}$ **5.** $\frac{r}{y}$ $-\sqrt{5}$ **6.** $\frac{x}{y}$ -2

7. What are the values of $\sin\theta$ and $\cos\theta$? Do they differ when they are calculated using the point $(6, -3)$ instead of $(4, -2)$? Explain. $\sin\theta = -\frac{\sqrt{5}}{5}$; $\cos\theta = \frac{2\sqrt{5}}{5}$; no, same ratios for $(6, -3)$ and $(4, -2)$

Let $P(x, y)$ be any point other than the origin on the terminal side of an angle θ in standard position. The six *trigonometric functions* are defined as follows, where $r = \sqrt{x^2 + y^2}$.

Trigonometric Functions

sine	$\sin\theta = \frac{y}{r}$	**cosecant**	$\csc\theta = \frac{r}{y}, y \neq 0$
cosine	$\cos\theta = \frac{x}{r}$	**secant**	$\sec\theta = \frac{r}{x}, x \neq 0$
tangent	$\tan\theta = \frac{y}{x}, x \neq 0$	**cotangent**	$\cot\theta = \frac{x}{y}, y \neq 0$

The values of these functions depend only on the size of angle θ. They do not depend on the choice of point P on the terminal side of the angle.

Notice that three pairs of reciprocal relationships exist between the six trigonometric functions. Note that no denominator may equal zero.

Reciprocal Trigonometric Functions

$$\csc\theta = \frac{1}{\sin\theta} \qquad \sec\theta = \frac{1}{\cos\theta} \qquad \cot\theta = \frac{1}{\tan\theta}$$

Remember that if two numbers are reciprocals of each other, their product is 1. Therefore, the reciprocal relationships may also be written as follows, where neither factor on the left is equal to 0:

$$\sin\theta \cdot \csc\theta = 1 \qquad \cos\theta \cdot \sec\theta = 1 \qquad \tan\theta \cdot \cot\theta = 1$$

EXAMPLE 1 **The terminal side of angle θ in standard position passes through point $P(-3, -4)$. Draw θ and find the values of the six trigonometric functions of θ.**

If P has the coordinates $(-3, -4)$, then $x = -3$, and $y = -4$, and

$$\begin{aligned} r &= \sqrt{(-3)^2 + (-4)^2} \\ &= \sqrt{9 + 16} \\ &= \sqrt{25} = 5 \end{aligned}$$

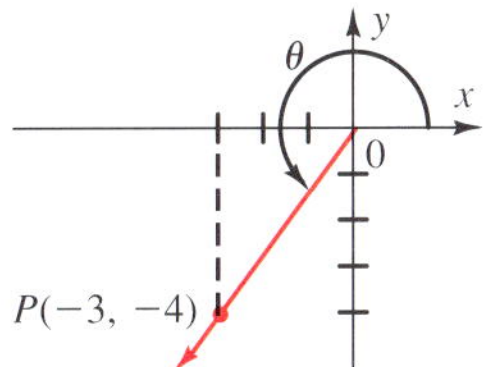

Therefore,

$$\sin\theta = \frac{-4}{5} = -\frac{4}{5} \qquad \sin\theta = \frac{y}{r} \qquad \csc\theta = \frac{5}{-4} = -\frac{5}{4} \qquad \csc\theta = \frac{r}{y}$$

$$\cos\theta = \frac{-3}{5} = -\frac{3}{5} \qquad \cos\theta = \frac{x}{r} \qquad \sec\theta = \frac{5}{-3} = -\frac{5}{3} \qquad \sec\theta = \frac{r}{x}$$

$$\tan\theta = \frac{-4}{-3} = \frac{4}{3} \qquad \tan\theta = \frac{y}{x} \qquad \cot\theta = \frac{-3}{-4} = \frac{3}{4} \qquad \cot\theta = \frac{x}{y}$$

Recall that r is always positive. Thus, the signs of the values of the trigonometric functions of θ are determined by the signs of x and y, which depend upon the quadrant in which the terminal side of θ lies. Using the reciprocal properties, you can see that the signs of the values of $\csc\theta$ are the same as those for $\sin\theta$, the signs for $\sec\theta$ are the same as those for $\cos\theta$, and the signs for $\cot\theta$ are the same as those for $\tan\theta$. The chart at the right summarizes this information.

Signs of Trigonometric Functions

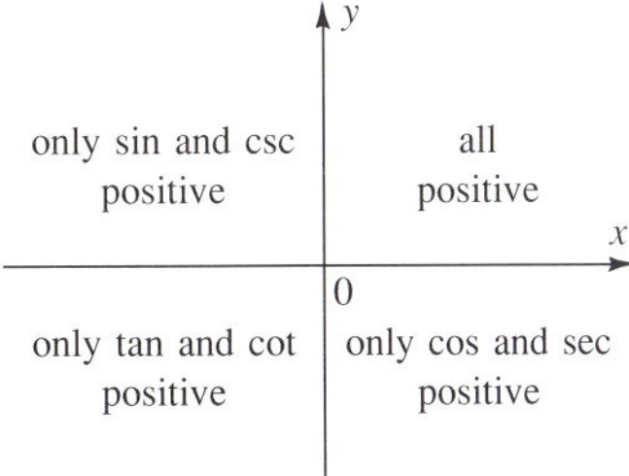

TEACHING SUGGESTIONS

- Emphasize that the quadrant of the terminal side determines the sign of the trigonometric function.
- Since r is always positive, the signs of the trigonometric functions are determined by the signs of x and y.

Critical Thinking

Discovering Patterns Ask students to tell what other information they must be given in order to find the coordinates of a point on the terminal side of an angle θ in standard position if the value of one of the six trigonometric functions is given. Ask them to explain their answers. Reasoning from the definitions of the trigonometric functions of angle θ, students should conclude that in order to find the coordinates of a point on the terminal side of θ, given the value of one of the six trigonometric functions of θ, they must also be given the quadrant in which θ lies, because there is more than one angle that will yield the given ratio. For example, if they are given $\sin\theta = 0.5$, θ could be 30° or 150°, so the point (x,y) could lie in quadrant I or quadrant II.

CHALKBOARD EXAMPLES

- **For Example 1**
  1. The terminal side of angle θ in standard position passes through $(5, -12)$. Find the values of the six trigonometric functions of θ.
  $\sin\theta = -\frac{12}{13}$; $\cos\theta = \frac{5}{13}$; $\tan\theta = -\frac{12}{5}$; $\csc\theta = -\frac{13}{12}$; $\sec\theta = \frac{13}{5}$; $\cot\theta = -\frac{5}{12}$
  2. The terminal side of angle θ in standard position passes through $(2, 1)$. Find the values of the six trigonometric functions of θ.
  $\sin\theta = \frac{\sqrt{5}}{5}$; $\cos\theta = \frac{2\sqrt{5}}{5}$; $\tan\theta = \frac{1}{2}$; $\csc\theta = \sqrt{5}$; $\sec\theta = \frac{\sqrt{5}}{2}$; $\cot\theta = 2$

- **For Example 2**
  In what quadrant or quadrants does θ lie under the given conditions?
  3. Cos θ and tan θ have opposite signs. III and IV
  4. Sin θ and sec θ have the same signs. I and III
- **For Example 3**
  Let θ be an angle in standard position. Evaluate each of the following if θ lies in Quadrant III and cos $\theta = -\frac{1}{2}$.
  5. sin θ $-\frac{\sqrt{3}}{2}$
  6. tan θ $\sqrt{3}$

Common Error

- Some students confuse the signs of the trigonometric functions. You may wish to have those students use *A Simple Trig Chart* to point out the signs of the trigonometric functions. That is, *A*ll positive (first quadrant), *S*ine positive (2nd quadrant), *T*angent positive (3rd quadrant), *C*osine positive (4th quadrant).
- See *Teacher's Resource Book* for additional remediation.

LESSON FOLLOW-UP

Assignment Guide

See p. B for assignments.

EXAMPLE 2 **In which quadrants do sin θ and tan θ have opposite signs?**

The tangent is negative and the sine is positive in quadrant II; the tangent is positive and the sine is negative in quadrant III. Therefore, sin θ and tan θ have opposite signs in quadrants II and III.

If you know the value of one of the six trigonometric functions of θ and the quadrant in which it lies, you can find the values of the other five.

EXAMPLE 3 **Let θ be an angle in standard position. Evaluate cos θ, tan θ, cot θ, sec θ, and csc θ if θ lies in quadrant IV and sin $\theta = -\frac{5}{13}$.**

Since $\sin\theta = -\frac{5}{13}$ and θ is in quadrant IV, $y = -5$ and $r = 13$. Find x.

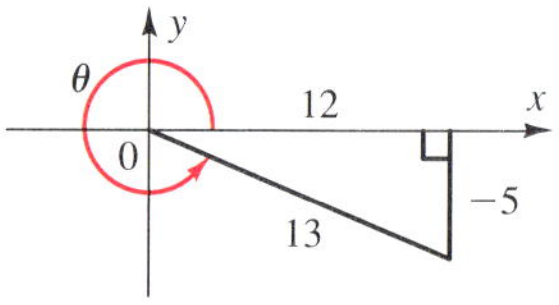

$$x^2 + (-5)^2 = 13^2 \qquad x^2 + y^2 = r^2$$
$$x^2 + 25 = 169$$
$$x^2 = 144$$
$$x = 12 \quad \text{or} \quad x = -12$$

Since θ lies in quadrant IV, $x > 0$. Therefore, $x = 12$ and

$$\cos\theta = \frac{12}{13} \qquad \tan\theta = \frac{-5}{12} = -\frac{5}{12} \qquad \csc\theta = \frac{13}{-5} = -\frac{13}{5}$$
$$\sec\theta = \frac{13}{12} \qquad \cot\theta = \frac{12}{-5} = -\frac{12}{5}$$

CLASS EXERCISES

State the values of all six trigonometric functions of θ. 3. $-\frac{\sqrt{2}}{2}$; $-\frac{\sqrt{2}}{2}$; 1; $-\sqrt{2}$; $-\sqrt{2}$; 1
Answers are in the following order: sin; cos; tan; csc; sec; cot.

1.
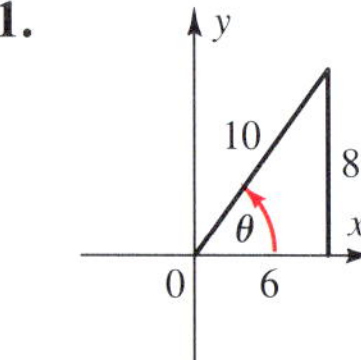

$\frac{4}{5}$; $\frac{3}{5}$; $\frac{4}{3}$; $\frac{5}{4}$; $\frac{5}{3}$; $\frac{3}{4}$

2.
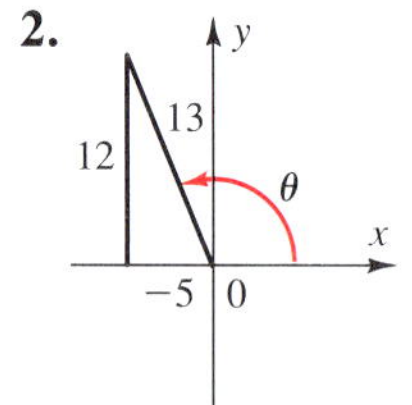

$\frac{12}{13}$; $-\frac{5}{13}$; $-\frac{12}{5}$; $\frac{13}{12}$; $-\frac{13}{5}$; $-\frac{5}{12}$

3.
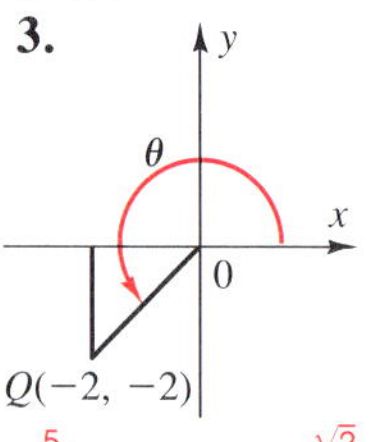

4.
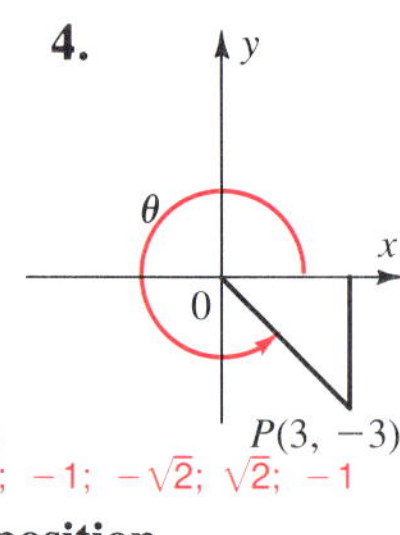

$-\frac{\sqrt{2}}{2}$; $\frac{\sqrt{2}}{2}$; -1; $-\sqrt{2}$; $\sqrt{2}$; -1

Determine in which quadrant θ lies if θ is an angle in standard position.

5. $\sin\theta > 0, \cos\theta < 0$ II **6.** $\sin\theta < 0, \cos\theta > 0$ IV **7.** $\csc\theta < 0, \sec\theta < 0$ III

8. $\csc\theta > 0, \sec\theta > 0$ I **9.** $\sec\theta < 0, \cot\theta > 0$ III **10.** $\sec\theta < 0, \cot\theta < 0$ II

Find the value of each reciprocal function.

11. If $\sin\theta = \frac{1}{2}$, $\csc\theta = \underline{?}$ 2
12. If $\cos\theta = -\frac{3}{4}$, $\sec\theta = \underline{?}$ $-\frac{4}{3}$
13. If $\tan\theta = -4$, $\cot\theta = \underline{?}$ $-\frac{1}{4}$
14. If $\sec\theta = 1.5$, $\cos\theta = \underline{?}$ $0.\overline{6}$
15. If $\sec\theta = -2.5$, $\cos\theta = \underline{?}$ -0.4
16. If $\cot\theta = 3$, $\tan\theta = \underline{?}$ $\frac{1}{3}$

PRACTICE EXERCISES

The terminal side of angle θ in standard position passes through point P. Draw θ and find the exact values of the six trigonometric functions of θ.

Answers are in the following order: sin; cos; tan; csc; sec; cot

See side column.

A

1. $P(3, 4)$ $\frac{4}{5}; \frac{3}{5}; \frac{4}{3}; \frac{5}{4}; \frac{5}{3}; \frac{3}{4}$
2. $P(6, 8)$ $\frac{4}{5}; \frac{3}{5}; \frac{4}{3}; \frac{5}{4}; \frac{5}{3}; \frac{3}{4}$
3. $P(12, -5)$
4. $P(-5, 12)$
5. $P(-3, -3)$
6. $P(2, -2)$
7. $P(2, -\sqrt{3})$
8. $P(-\sqrt{3}, -2)$
9. $P(7, -24)$
10. $P(24, -7)$
11. $P(-2, 2\sqrt{3})$
12. $P(3, -3\sqrt{3})$

Let θ be an angle in standard position. In which quadrant or quadrants can θ lie under the given conditions?

13. Sin θ and tan θ have the same sign. I, IV
14. Cos θ and tan θ have the same sign. I, II
15. Sin θ is negative and cos θ is positive. IV
16. Sin θ, cos θ, and tan θ all have the same sign. I
17. Sin θ and cos θ have opposite signs. II, IV
18. Tan θ and cot θ have the same sign. I, II, III, IV
19. Cot θ and sec θ have opposite signs. III, IV
20. Sec θ and csc θ have the same sign. I, III

Find the exact values of the other five trigonometric functions for an angle θ in standard position lying in the given quadrant.

B

21. $\sin\theta = -\frac{4}{5}$; III $-\frac{3}{5}; \frac{4}{3}; -\frac{5}{4}; -\frac{5}{3}; \frac{3}{4}$
22. $\cos\theta = -\frac{3}{5}$; II $\frac{4}{5}; -\frac{4}{3}; \frac{5}{4}; -\frac{5}{3}; -\frac{3}{4}$
23. $\tan\theta = \frac{4}{3}$; I $\frac{4}{5}; \frac{3}{5}; \frac{5}{4}; \frac{5}{3}; \frac{3}{4}$
24. $\tan\theta = -\frac{3}{4}$; IV $-\frac{3}{5}; \frac{4}{5}; -\frac{5}{3}; \frac{5}{4}; -\frac{4}{3}$
25. $\sec\theta = -2$; III $-\frac{\sqrt{3}}{2}; -\frac{1}{2}; \sqrt{3}; -\frac{2\sqrt{3}}{3}; \frac{\sqrt{3}}{3}$
26. $\csc\theta = \frac{4}{3}$; II $\frac{3}{4}; -\frac{\sqrt{7}}{4}; -\frac{3\sqrt{7}}{7}; -\frac{4\sqrt{7}}{7}; -\frac{\sqrt{7}}{3}$
27. $\cot\theta = \frac{24}{7}$; III $-\frac{7}{25}; -\frac{24}{25}; \frac{7}{24}; -\frac{25}{7}; -\frac{25}{24}$
28. $\cot\theta = -\frac{12}{5}$; IV $-\frac{5}{13}; \frac{12}{13}; -\frac{5}{12}; -\frac{13}{5}; \frac{13}{12}$

Challenge

Sin θ and cos θ are shown as lengths of certain line segments of a unit circle. Then students are challenged to show that the other named line segments represent the other four trigonometric functions of θ.

Lesson Quiz

Determine in which quadrant the terminal side of θ will lie if θ is an angle in standard position.

1. $\sin\theta < 0$, $\cos\theta < 0$ III
2. $\cot\theta < 0$, $\sec\theta > 0$ IV
3. If $\sin\theta = -\frac{7}{8}$, what is the value of $\csc\theta$? $-\frac{8}{7}$
4. The terminal side of angle θ in standard position passes through point $P(8, -6)$. Find the exact values of the six trigonometric functions of θ. $\sin\theta = -\frac{3}{5}$; $\cos\theta = \frac{4}{5}$; $\tan\theta = -\frac{3}{4}$; $\csc\theta = -\frac{5}{3}$; $\sec\theta = \frac{5}{4}$; $\cot\theta = -\frac{4}{3}$
5. In which quadrant or quadrants do $\csc\theta$ and $\tan\theta$ have the same sign, if θ is in standard position? I and IV

Additional Answers
Practice Exercises

3. $-\frac{5}{13}; \frac{12}{13}; -\frac{5}{12}; -\frac{13}{5}, \frac{13}{12}, -\frac{12}{5}$
4. $\frac{12}{13}; -\frac{5}{13}; -\frac{12}{5}; \frac{13}{12}, -\frac{13}{5}, -\frac{5}{12}$
5. $-\frac{\sqrt{2}}{2}; -\frac{\sqrt{2}}{2}; 1; -\sqrt{2}; -\sqrt{2}; 1$
6. $-\frac{\sqrt{2}}{2}; \frac{\sqrt{2}}{2}; -1; -\sqrt{2}; \sqrt{2}; -1$
7. $-\frac{\sqrt{21}}{7}; \frac{2\sqrt{7}}{7}; -\frac{\sqrt{3}}{2}; -\frac{\sqrt{21}}{3}; \frac{\sqrt{7}}{2}; -\frac{2\sqrt{3}}{3}$
8. $-\frac{2\sqrt{7}}{7}; -\frac{\sqrt{21}}{7}; \frac{2\sqrt{3}}{3}; -\frac{\sqrt{7}}{2}; -\frac{\sqrt{21}}{3}; \frac{\sqrt{3}}{2}$
9. $-\frac{24}{25}; \frac{7}{25}; -\frac{24}{7}; -\frac{25}{24}; \frac{25}{7}; -\frac{7}{24}$
10. $-\frac{7}{25}; \frac{24}{25}; -\frac{7}{24}; -\frac{25}{7}; \frac{25}{24}; -\frac{24}{7}$
11. $\frac{\sqrt{3}}{2}; -\frac{1}{2}; -\sqrt{3}, \frac{2\sqrt{3}}{3}; -2, -\frac{\sqrt{3}}{3}$
12. $-\frac{\sqrt{3}}{2}; \frac{1}{2}; -\sqrt{3}; -\frac{2\sqrt{3}}{3}; 2, -\frac{\sqrt{3}}{3}$

Teacher's Resource Book

Practice—Chapter 1, p. 13
Enrichment—Chapter 1, p. 14

Additional Answers

Challenge

Since $\triangle OQP \sim \triangle OSR$,

$\tan\theta = \frac{QP}{OP} = \frac{RS}{OR} = \frac{RS}{1} = RS$

$\sec\theta = \frac{OQ}{OP} = \frac{OS}{OR} = \frac{OS}{1} = OS$

Since $\triangle OPQ \sim \triangle TUO$ ($\angle T = \theta$),

$\cot\theta = \frac{OP}{PQ} = \frac{TU}{UO} = \frac{TU}{1} = TU$

$\csc\theta = \frac{OQ}{QP} = \frac{TO}{UO} = \frac{TO}{1} = TO$

A point $P(x, y)$ is on the unit circle with equation $x^2 + y^2 = 1$. Show that the following points are on the unit circle. Then find the exact values of $\sin\theta$, $\cos\theta$, and $\tan\theta$ if P is on the terminal side of θ.

29. $P\left(-\frac{3}{4}, \frac{\sqrt{7}}{4}\right)$ $\frac{\sqrt{7}}{4}; -\frac{3}{4}; -\frac{\sqrt{7}}{3}$ **30.** $P\left(\frac{\sqrt{2}}{2}, -\frac{\sqrt{2}}{2}\right)$ $-\frac{\sqrt{2}}{2}; \frac{\sqrt{2}}{2}; -1$ **31.** $P\left(\frac{1}{2}, -\frac{\sqrt{3}}{2}\right)$ $-\frac{\sqrt{3}}{2}; \frac{1}{2}; -\sqrt{3}$

32. $P\left(\frac{\sqrt{3}}{2}, -\frac{1}{2}\right)$ $-\frac{1}{2}; \frac{\sqrt{3}}{2}; -\frac{\sqrt{3}}{3}$ **33.** $P\left(-\frac{5}{13}, -\frac{12}{13}\right)$ $-\frac{12}{13}; -\frac{5}{13}; \frac{12}{5}$ **34.** $P\left(-\frac{3}{5}, -\frac{4}{5}\right)$ $-\frac{4}{5}; -\frac{3}{5}; \frac{4}{3}$

Determine the exact coordinates of the point at the given distance from the origin in the given quadrant, if θ is in standard position.

C **35.** 10; II; $\sin\theta = \frac{3}{5}$ $(-8, 6)$

36. 4; IV; $\csc\theta = -2$ $(2\sqrt{3}, -2)$

37. $\sqrt{2}$; IV; $\cos\theta = \frac{\sqrt{2}}{2}$ $(1, -1)$

38. 6; III; $\cot\theta = 1$ $(-3\sqrt{2}, -3\sqrt{2})$

39. 8; IV; $\tan\theta = -1$ $(4\sqrt{2}, -4\sqrt{2})$

40. $\sqrt{5}$; II; $\sin\theta = \frac{\sqrt{5}}{5}$ $(-2, 1)$

Applications

41. Surveying A surveyor needs to use the cotangent of an angle but does not have a calculator with a cot key. What function of the same angle could the surveyor use to find the cotangent? tangent

42. Construction A 15-ft ladder forms an angle θ with the side of a building such that $\sec\theta = \frac{5}{3}$. The distance d from the base of the ladder to the base of the building can be found using the equation $\cos\theta = \frac{d}{15}$. Find the value of d. 9 ft

43. Construction The wooden truss pictured at the right forms an isosceles triangle constructed in such a way that the length of its base is 5 times its height. Find the exact value of $\tan\theta$. $\tan\theta = \frac{2}{5}$

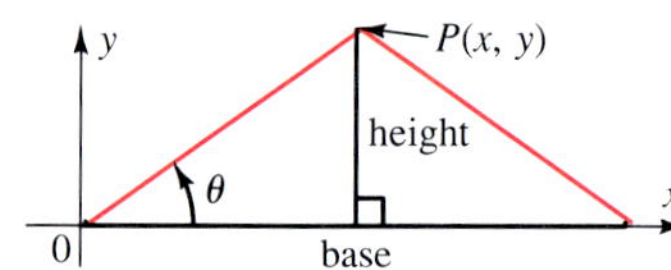

CHALLENGE

Geometric Interpretations of Trigonometric Functions In the unit circle figure at the right, the values of the trigonometric functions of angle θ are given by the lengths of certain line segments. For example, $\sin\theta = QP$ and $\cos\theta = OP$. Show that $\tan\theta = RS$, $\sec\theta = OS$, $\cot\theta = UT$, and $\csc\theta = OT$. (Use similar triangles and write proportions.) See side column.

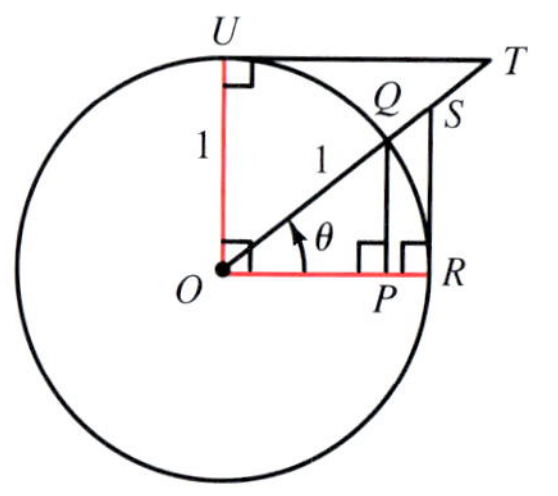

1.8 Trigonometric Functions of Special Angles

Objective: To find the values of the six trigonometric functions of special angles

Triangles with special angle-side relationships, introduced in geometry, are useful in evaluating the trigonometric functions of 30°, 45°, and 60°.

Preview

In an isosceles right triangle (45°-45°-90°), the length of the hypotenuse is $\sqrt{2}$ times the length of a leg. In a 30°-60°-90° triangle, the length of the hypotenuse is two times the length of the leg opposite the 30° angle. The length of the leg opposite the 60° angle is $\sqrt{3}$ times the length of the leg opposite the 30° angle.

EXAMPLE 1 **In the triangle at the right, $a = 5$. Find b and c.**

$b = a = 5$

$c = a\sqrt{2} = 5\sqrt{2}$

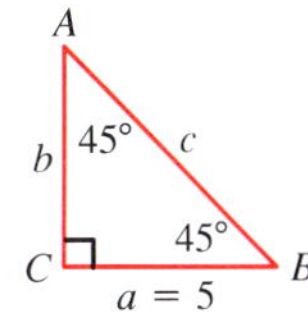

EXAMPLE 2 **In the triangle at the right, $a = 14$. Find b and c.**

$b = a\sqrt{3} = 14\sqrt{3}$

$c = 2a = 2(14) = 28$

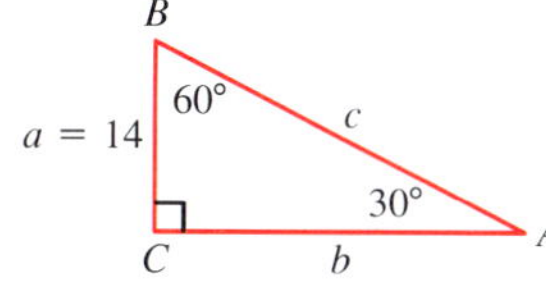

Find the indicated lengths of the sides of a 45°-45°-90° triangle *ABC* with right angle *C*.

1. If $c = 2$, find a and b. $a = b = \sqrt{2}$
2. If $b = 3$, find a and c. $a = 3;\ c = 3\sqrt{2}$
3. If $a = 0.5$, find b and c. $b = 0.5;\ c = 0.5\sqrt{2}$

Find the indicated lengths of the sides of a 30°-60°-90° triangle *ABC* with right angle *C* and angle *A* equal to 30°.

4. If $a = 2$, find b and c. $b = 2\sqrt{3};\ c = 4$
5. If $c = 6$, find a and b. $a = 3;\ b = 3\sqrt{3}$
6. If $b = \frac{3\sqrt{2}}{2}$, find a and c. $a = \frac{\sqrt{6}}{2};\ c = \sqrt{6}$

An angle in standard position whose terminal side lies on the x- or y-axis is called a **quadrantal angle.** In the figure at the right, angles *POR*, *POD*, *POQ*, and *POP* are quadrantal angles.

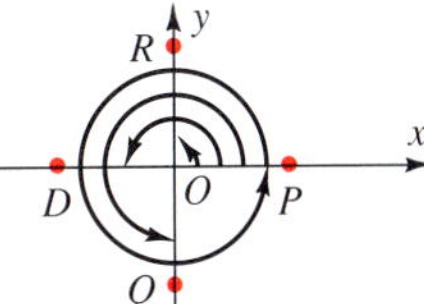

LESSON PLAN

Vocabulary

Quadrantal angle

Reference angle

BACKGROUND

In the Preview, 30°- 60°- 90° and 45°- 45°- 90° triangle relationships are discussed. Rationalizing radical expressions may also be reviewed at this time. These skills will be used in finding the values of the trigonometric functions for angles of 30°, 60°, 45°, and 90°.

TEACHING SUGGESTIONS

- Interchange radian measure with degree measure so that quadrantal angles and the other special angles (30°, 45°, 60°) are recognized both ways.
- Be sure that students understand that the values of r used in Examples 1–3 on text pp. 40–41 were convenient and that the values of the trigonometric functions would have been the same for different values of r.

Critical Thinking

Causal Explanation Ask students to explain why division by zero is undefined. Students should reason that if $\frac{n}{0}$ is defined, then there is some number x such that $0 \cdot x = n$, which is not possible by the zero product property.

CHALKBOARD EXAMPLES

- **For Example 1**
  Given an angle θ in standard position, evaluate $\sin \theta$, $\cos \theta$, and $\tan \theta$ for each value of θ. Use $r = 3$.

  1. $\theta = 90°$ $\sin 90° = \frac{3}{3} = 1$; $\cos 90° = \frac{0}{3} = 0$; $\tan 90° = \frac{3}{0}$: undefined

  2. $\theta = 180°$ $\sin 180° = \frac{0}{3} = 0$; $\cos 180° = -\frac{3}{3} = -1$; $\tan 180° = -\frac{0}{3} = 0$

For an angle θ in standard position, the terminal side of θ intersects a circle of radius r at a unique point $P(x, y)$. The definitions of the trigonometric functions can be used to evaluate the trigonometric functions of $\theta = 0°$, $90°$, $180°$, $270°$, and $360°$ by setting r equal to 1.

EXAMPLE 1 **Given angle θ in standard position, evaluate $\sin \theta$, $\cos \theta$, and $\tan \theta$ for each value of θ. Use $r = 1$.**

a. $\theta = 0°$ **b.** $\theta = 90°$ **c.** $\theta = 180°$ **d.** $\theta = 270°$ **e.** $\theta = 360°$

a. Since $\theta = 0°$ and $r = 1$, $x = 1$ and $y = 0$.

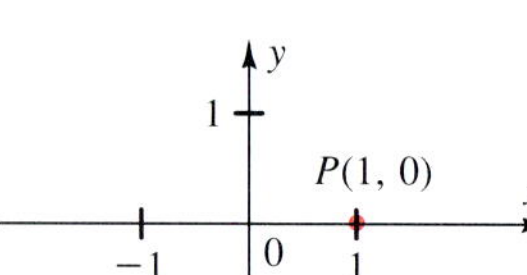

Therefore, $\sin 0° = \frac{0}{1} = 0$

$\cos 0° = \frac{1}{1} = 1$

$\tan 0° = \frac{0}{1} = 0$

b. Since $\theta = 90°$ and $r = 1$, $x = 0$ and $y = 1$.

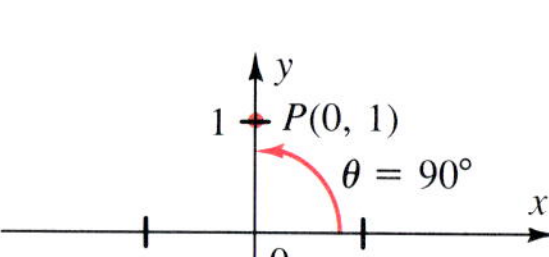

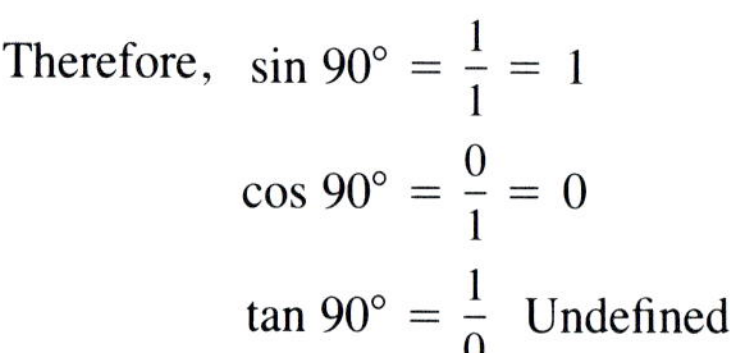

Therefore, $\sin 90° = \frac{1}{1} = 1$

$\cos 90° = \frac{0}{1} = 0$

$\tan 90° = \frac{1}{0}$ Undefined

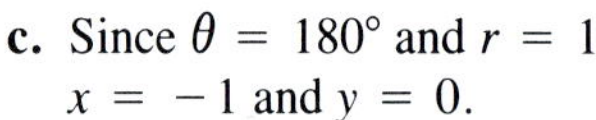

c. Since $\theta = 180°$ and $r = 1$, $x = -1$ and $y = 0$.

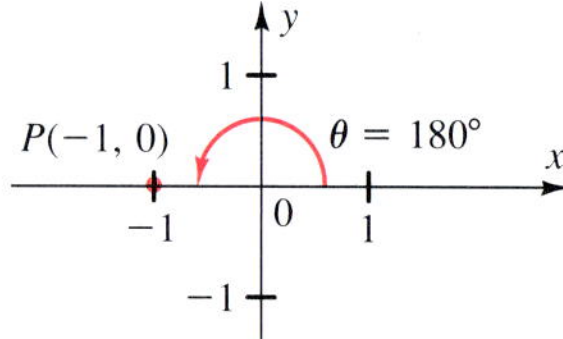

Therefore, $\sin 180° = \frac{0}{1} = 0$

$\cos 180° = \frac{-1}{1} = -1$

$\tan 180° = \frac{0}{-1} = 0$

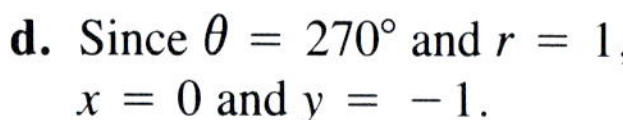

d. Since $\theta = 270°$ and $r = 1$, $x = 0$ and $y = -1$.

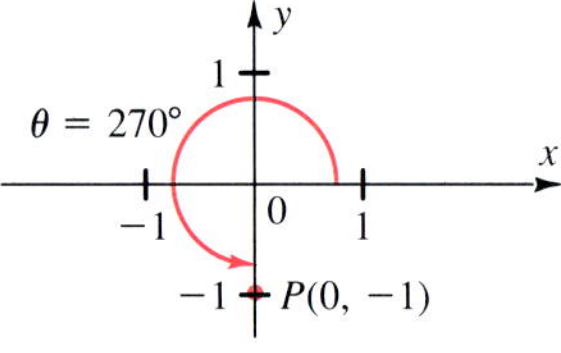

Therefore, $\sin 270° = \frac{-1}{1} = -1$

$\cos 270° = \frac{0}{1} = 0$

$\tan 270° = \frac{-1}{0}$ Undefined

e. Since 0° and 360° angles are coterminal, their trigonometric functions have the same values. Therefore, $\sin 360° = 0$, $\cos 360° = 1$, and $\tan 360° = 0$.

Because division by 0 is not defined, some values of the trigonometric functions are not defined. Example 1 illustrates this.

Choosing a convenient value for r makes it easier to find the trigonometric values of special angles such as 30°, 45°, and 60°. Since a 45°-45°-90° triangle has sides in the ratio of $1:1:\sqrt{2}$, where $\sqrt{2}$ corresponds to the hypotenuse, a convenient value for r is $\sqrt{2}$.

EXAMPLE 2 **If $\theta = 45°$ and $r = \sqrt{2}$, find $\sin\theta$, $\cos\theta$, and $\tan\theta$.**

Since legs OB and PB have the same length, $OB = PB = 1$. Thus, $x = 1$ and $y = 1$. Therefore,

$$\sin 45° = \frac{1}{\sqrt{2}} = \frac{\sqrt{2}}{2}$$

$$\cos 45° = \frac{1}{\sqrt{2}} = \frac{\sqrt{2}}{2}$$

$$\tan 45° = \frac{1}{1} = 1$$

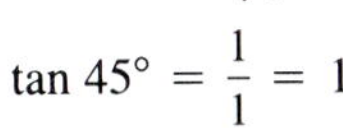

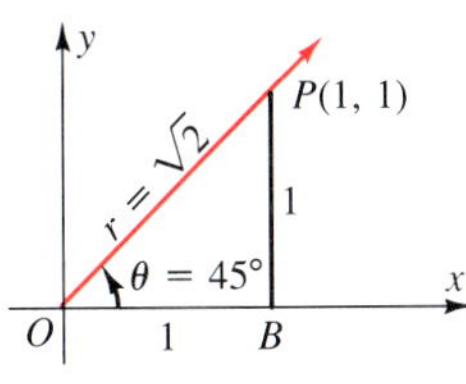

In a 30°-60°-90° triangle, the sides are in the ratio $1:\sqrt{3}:2$, where 2 corresponds to the hypotenuse and $\sqrt{3}$ corresponds to the side opposite the 60° angle. Thus, a convenient value for r is 2.

EXAMPLE 3 **Evaluate $\sin\theta$, $\cos\theta$, $\tan\theta$, $\csc\theta$, $\sec\theta$, and $\cot\theta$ for each value of r and θ:** **a.** $\theta = 60°$; $r = 2$ **b.** $\theta = \frac{\pi}{6}$; $r = 2$

a. $x = OB = 1$ and $y = PB = \sqrt{3}$, so

$$\sin 60° = \frac{\sqrt{3}}{2} \qquad \csc 60° = \frac{2}{\sqrt{3}} = \frac{2\sqrt{3}}{3}$$

$$\cos 60° = \frac{1}{2} \qquad \sec 60° = \frac{2}{1} = 2$$

$$\tan 60° = \frac{\sqrt{3}}{1} = \sqrt{3} \qquad \cot 60° = \frac{1}{\sqrt{3}} = \frac{\sqrt{3}}{3}$$

b. $x = OB = \sqrt{3}$ and $y = PB = 1$, so

$$\sin\frac{\pi}{6} = \frac{1}{2} \qquad \csc\frac{\pi}{6} = \frac{2}{1} = 2$$

$$\cos\frac{\pi}{6} = \frac{\sqrt{3}}{2} \qquad \sec\frac{\pi}{6} = \frac{2}{\sqrt{3}} = \frac{2\sqrt{3}}{3}$$

$$\tan\frac{\pi}{6} = \frac{1}{\sqrt{3}} = \frac{\sqrt{3}}{3} \qquad \cot\frac{\pi}{6} = \frac{\sqrt{3}}{1} = \sqrt{3}$$

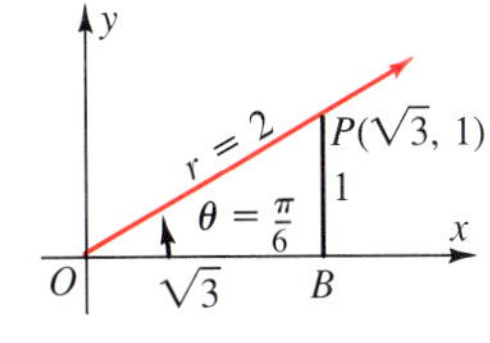

- **For Example 2**

Evaluate $\sin\theta$, $\cos\theta$, and $\tan\theta$ for each value of r and θ.

3. $\theta = 45°$; $r = 1$ $\sin 45° = \frac{\sqrt{2}}{2}$; $\cos 45° = \frac{\sqrt{2}}{2}$; $\tan 45° = 1$

4. $\theta = \frac{\pi}{4}$; $r = 3\sqrt{2}$ $\sin\frac{\pi}{4} = \frac{\sqrt{2}}{2}$; $\cos\frac{\pi}{4} = \frac{\sqrt{2}}{2}$; $\tan\frac{\pi}{4} = 1$

- **For Example 3**

Find $\sin\theta$, $\cos\theta$, $\tan\theta$, $\csc\theta$, $\sec\theta$, and $\cot\theta$ for each value θ and r.

5. $\theta = 30°$; $r = 1$ $\sin 30° = \frac{1}{2}$; $\cos 30° = \frac{\sqrt{3}}{2}$; $\tan 30° = \frac{\sqrt{3}}{3}$; $\csc 30° = 2$; $\sec 30° = \frac{2\sqrt{3}}{3}$; $\cot 30° = \sqrt{3}$

6. $\theta = \frac{\pi}{6}$; $r = 6$ $\sin\frac{\pi}{6} = \frac{1}{2}$; $\cos\frac{\pi}{6} = \frac{\sqrt{3}}{2}$; $\tan\frac{\pi}{6} = \frac{\sqrt{3}}{3}$; $\csc\frac{\pi}{6} = 2$; $\sec\frac{\pi}{6} = \frac{2\sqrt{3}}{3}$; $\cot\frac{\pi}{6} = \sqrt{3}$

- **For Example 4**

  θ is an angle in standard position. Find the measure of its reference angle θ'.

  7. 150° $\quad \theta' = 180^\circ - 150^\circ = 30^\circ$

  8. $\frac{11\pi}{6}$ $\quad \theta' = 2\pi - \frac{11\pi}{6} = \frac{\pi}{6}$

  9. $\frac{4\pi}{3}$ $\quad \theta' = \frac{4\pi}{3} - \pi = \frac{\pi}{3}$

  10. -240° $\quad -240^\circ + 360^\circ = 120^\circ$; $\theta' = 180^\circ - 120^\circ = 60^\circ$

You can also use the methods introduced in this lesson to evaluate trigonometric functions of angles whose measures are integral multiples of 30°, 45°, and 60°. In order to do so, the concept of a *reference angle* is needed. The **reference angle** θ' of a given angle θ in standard position is the smallest positive acute angle determined by the x-axis and the terminal side of θ. Recall that an acute angle is an angle whose measure is between 0° and 90°.

The following diagrams illustrate the relationship between θ and θ' in each of the four quadrants:

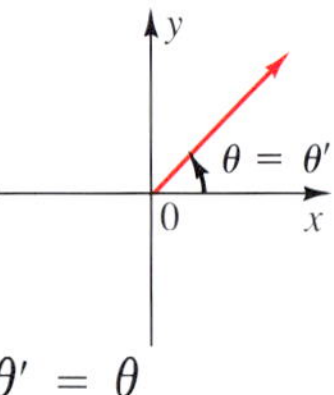

$\theta' = \theta$

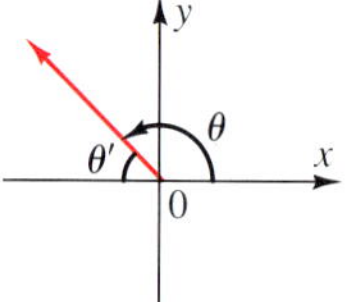

$\theta' = 180^\circ - \theta$

$\theta' = \pi - \theta$

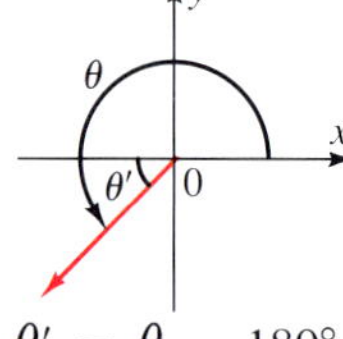

$\theta' = \theta - 180^\circ$

$\theta' = \theta - \pi$

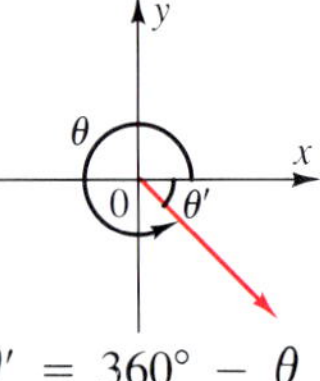

$\theta' = 360^\circ - \theta$

$\theta' = 2\pi - \theta$

To find the reference angle for an angle with negative measure or for an angle greater than 360°, first find a coterminal angle whose measure is between 0° and 360°.

EXAMPLE 4 **θ is an angle in standard position. Find the measure of its reference angle θ'.**

a. 250° **b.** $\frac{5\pi}{6}$ **c.** $\frac{23\pi}{12}$ **d.** −120° **e.** 460°

a. A 250° angle is in quadrant III.
$\theta' = 250^\circ - 180^\circ = 70^\circ$

b. An angle measuring $\frac{5\pi}{6}$ is in quadrant II.

$$\theta' = \pi - \frac{5\pi}{6} = \frac{\pi}{6}$$

c. An angle measuring $\frac{23\pi}{12}$ is in quadrant IV.

$$\theta' = 2\pi - \frac{23\pi}{12} = \frac{\pi}{12}$$

d. Since $-120^\circ + 360^\circ = 240^\circ$, an angle of −120° is coterminal with a 240° angle, which is in quadrant III.
$\theta' = 240^\circ - 180^\circ = 60^\circ$

e. Since $460^\circ - 360^\circ = 100^\circ$, a 460° angle is coterminal with a 100° angle, which is in quadrant II.
$\theta' = 180^\circ - 100^\circ = 80^\circ$

The possible values of the trigonometric functions of angles greater than 90° $\left(\frac{\pi}{2}\text{ radians}\right)$ are the same as those of angles between 0° and 90°, except for some of the signs.

EXAMPLE 5 **Evaluate sin θ, cos θ, and tan θ for each value of θ. Sketch θ and θ'.**
a. 150° **b.** 210° **c.** 330°

a.

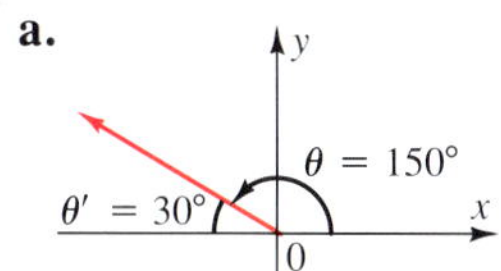

b.

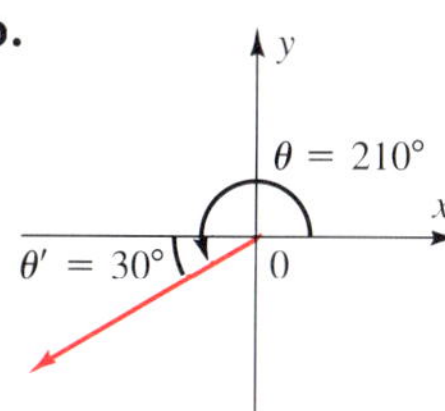

c.

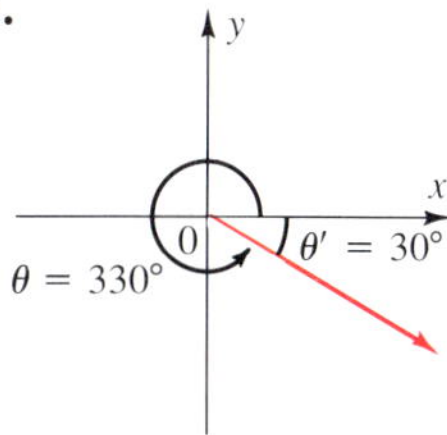

$\sin 150° = \frac{1}{2}$ $\qquad$ $\sin 210° = -\frac{1}{2}$ $\qquad$ $\sin 330° = -\frac{1}{2}$

$\cos 150° = -\frac{\sqrt{3}}{2}$ $\qquad$ $\cos 210° = -\frac{\sqrt{3}}{2}$ $\qquad$ $\cos 330° = \frac{\sqrt{3}}{2}$

$\tan 150° = -\frac{\sqrt{3}}{3}$ $\qquad$ $\tan 210° = \frac{\sqrt{3}}{3}$ $\qquad$ $\tan 330° = -\frac{\sqrt{3}}{3}$

Thus, the value of each trigonometric function of an angle is equal to the value of the same function of its reference angle, or to the opposite of that value. For example, either $\sin\theta = \sin\theta'$ or $\sin\theta = -\sin\theta'$.

There are an infinite number of values of θ for which $\sin\theta = \frac{1}{2}$. For example, θ could be 30°, 150°, 390°, or −330°. However, if you know that θ is between 0° and 360°, then the value of θ must be either 30° or 150° for $\sin\theta = \frac{1}{2}$.

EXAMPLE 6 **If $0° \le \theta \le 360°$, find all values of θ for which $\tan\theta = -1$.**

Tan θ is negative in quadrants II and IV. Since $\tan\theta = \frac{y}{x}$, let $y = 1$ and $x = -1$, or let $y = -1$ and $x = 1$. Draw the reference triangles in quadrants II and IV. The values of θ for which $\tan\theta = -1$ are 135° and 315°.

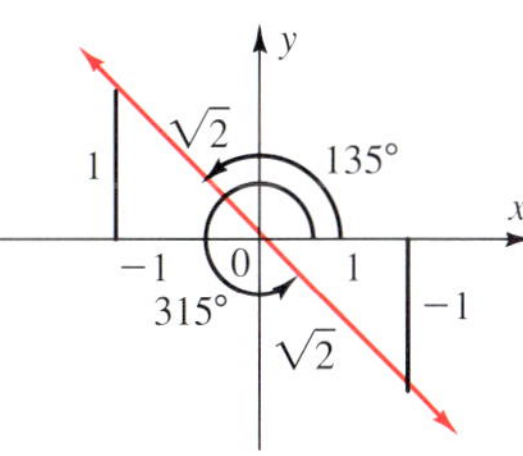

- **For Example 5**
  Evaluate sin θ, cos θ, and tan θ for each value of θ. Give the measure of θ'.

11. 315° $\sin 315° = -\frac{\sqrt{2}}{2}$; $\cos 315° = \frac{\sqrt{2}}{2}$; $\tan 315° = -1$; 45°

12. $\frac{4\pi}{3}$ $\sin\frac{4\pi}{3} = -\frac{\sqrt{3}}{2}$; $\cos\frac{4\pi}{3} = -\frac{1}{2}$; $\tan\frac{4\pi}{3} = \sqrt{3}$; $\frac{\pi}{3}$

- **For Example 6**
  If $0 \le \theta < 2\pi$, find all values of θ for which each of the following is true.

13. $\tan\theta = 1$ $\quad\frac{\pi}{4}, \frac{5\pi}{4}$

14. $\tan\theta = -\sqrt{3}$ $\quad\frac{2\pi}{3}, \frac{5\pi}{3}$

Common Error

- The relationships of the lengths of the sides of a 30°- 60°- 90° triangle are often confused with those of a 45°- 45°- 90° triangle. Emphasize that the ratio of the lengths of the legs in a 30°- 60°- 90° triangle is $1:\sqrt{3}$, and the ratio of the lengths of the legs in a 45°- 45°- 90° triangle is 1:1.
- See *Teacher's Resource Book* for additional remediation.

LESSON FOLLOW-UP

Assignment Guide

See p. B for assignments.

Lesson Quiz

Find the exact value of each trigonometric function.

1. cos 360° 1
2. tan 180° 0
3. sin 270° −1
4. sec 90° undefined
5. csc 0° undefined

Give the measure of the reference angle θ' for each angle θ in standard position.

6. 162° 18°
7. −59° 59°
8. 246° 66°

Find the exact values of the six trigonometric functions for each angle.

9. 315° $\sin 315° = -\frac{\sqrt{2}}{2}$; $\cos 315° = \frac{\sqrt{2}}{2}$; $\tan 315° = -1$; $\csc 315° = -\sqrt{2}$; $\sec 315° = \sqrt{2}$; $\cot 315° = -1$

10. $\frac{5\pi}{6}$ $\sin\frac{5\pi}{6} = \frac{1}{2}$; $\cos\frac{5\pi}{6} = -\frac{\sqrt{3}}{2}$; $\tan\frac{5\pi}{6} = -\frac{\sqrt{3}}{3}$; $\csc\frac{5\pi}{6} = 2$; $\sec\frac{5\pi}{6} = -\frac{2\sqrt{3}}{3}$; $\cot\frac{5\pi}{6} = -\sqrt{3}$

CLASS EXERCISES

Complete the following chart. Write *undefined* if a value is undefined.

	$\sin\theta$	$\cos\theta$	$\tan\theta$	$\csc\theta$	$\sec\theta$	$\cot\theta$
$\theta = 0° = 0$	**1.** 0	**2.** 1	**3.** 0	**4.** undef.	**5.** 1	**6.** undef.
$\theta = 90° = \frac{\pi}{2}$	**7.** 1	**8.** 0	**9.** undef.	**10.** 1	**11.** undef.	**12.** 0
$\theta = 180° = \pi$	**13.** 0	**14.** −1	**15.** 0	**16.** undef.	**17.** −1	**18.** undef.
$\theta = 270° = \frac{3\pi}{2}$	**19.** −1	**20.** 0	**21.** undef.	**22.** −1	**23.** undef.	**24.** 0

If $0° \le \theta \le 360°$, find the value(s) of θ that make each statement true. Express your answers in degrees.

25. $\sin\theta = 0$ 0°; 180°; 360°
26. $\cos\theta = 0$ 90°; 270°
27. $\sec\theta = 2$ 60°; 300°
28. $\csc\theta = -2$ 210°; 330°

PRACTICE EXERCISES

A **Complete the following chart.**

	$\sin\theta$	$\cos\theta$	$\tan\theta$	$\csc\theta$	$\sec\theta$	$\cot\theta$
$\theta = 30° = \frac{\pi}{6}$	**1.** $\frac{1}{2}$	**2.** $\frac{\sqrt{3}}{2}$	**3.** $\frac{\sqrt{3}}{3}$	**4.** 2	**5.** $\frac{2\sqrt{3}}{3}$	**6.** $\sqrt{3}$
$\theta = 45° = \frac{\pi}{4}$	**7.** $\frac{\sqrt{2}}{2}$	**8.** $\frac{\sqrt{2}}{2}$	**9.** 1	**10.** $\sqrt{2}$	**11.** $\sqrt{2}$	**12.** 1
$\theta = 60° = \frac{\pi}{3}$	**13.** $\frac{\sqrt{3}}{2}$	**14.** $\frac{1}{2}$	**15.** $\sqrt{3}$	**16.** $\frac{2\sqrt{3}}{3}$	**17.** 2	**18.** $\frac{\sqrt{3}}{3}$

Give the measure of the reference angle θ' for each angle θ in standard position.

19. 171° 9°
20. 133° 47°
21. 305° 55°
22. 325° 35°
23. 412° 52°
24. 505° 35°
25. −110° 70°
26. −200° 20°

Find the exact values of the six trigonometric functions for each angle.
See side column.

27. 150°
28. 210°
29. 135°
30. 120°
31. 225°
32. $\frac{5\pi}{6}$
33. $\frac{7\pi}{6}$
34. $\frac{7\pi}{4}$

B
35. 390°
36. 420°
37. 495°
38. 585°
39. −30°
40. −135°
41. −240°
42. 1350°
43. $-\frac{\pi}{3}$
44. $\frac{9\pi}{4}$
45. $\frac{10\pi}{3}$
46. $-\frac{15\pi}{4}$

If $0 \le \theta \le 2\pi$, find the values of θ that make each statement true. Express your answers in radians.

47. $\cos \theta = \frac{1}{2}$ $\frac{\pi}{3}; \frac{5\pi}{3}$ **48.** $\sin \theta = \frac{1}{2}$ $\frac{\pi}{6}; \frac{5\pi}{6}$ **49.** $\tan \theta = -1$ $\frac{3\pi}{4}; \frac{7\pi}{4}$ **50.** $\cot \theta = 1$ $\frac{\pi}{4}; \frac{5\pi}{4}$

51. $\sec \theta = \sqrt{2}$ $\frac{\pi}{4}; \frac{7\pi}{4}$ **52.** $\csc \theta = -\sqrt{2}$ $\frac{5\pi}{4}; \frac{7\pi}{4}$ **53.** $\cot \theta = -\sqrt{3}$ $\frac{5\pi}{6}; \frac{11\pi}{6}$ **54.** $\tan \theta = \sqrt{3}$ $\frac{\pi}{3}; \frac{4\pi}{3}$

Give the measure of the reference angle θ' for each angle θ in standard position.

55. $1002°$ 78° **56.** $1500°$ 60° **57.** $-907°$ 7° **58.** $-950°$ 50°

If $0° \le \theta \le 360°$, find the value of θ that makes each statement true. Express your answer in degrees.

C **59.** $\tan \theta = 1$ and $\sin \theta \ge 0$ 45° **60.** $\cos \theta = \frac{\sqrt{3}}{2}$ and $\tan \theta \ge 0$ 30°

61. $\sin \theta = \frac{1}{2}$ and $\cos \theta \le 0$ 150° **62.** $\cos \theta = -\frac{1}{2}$ and $\sin \theta \ge 0$ 120°

63. $\sin \theta = -\frac{\sqrt{3}}{2}$ and $\cos \theta \le 0$ 240° **64.** $\cos \theta = -\frac{\sqrt{3}}{2}$ and $\sin \theta \le 0$ 210°

65. $\sin \theta = -1$ and $\cos \theta = 0$ 270° **66.** $\sin \theta = 0$ and $\cos \theta = -1$ 180°

67. $\sec \theta = \sqrt{2}$ and $\csc \theta = -\sqrt{2}$ 315° **68.** $\cot \theta = -\sqrt{3}$ and $\sec \theta = \frac{2\sqrt{3}}{3}$ 330°

Applications

69. Construction A 30-ft ladder resting against a building makes a 60° angle with the ground. The height h from the ground at which the ladder touches the building can be found using the equation $\sin 60° = \frac{h}{30}$. Find h to the nearest foot. 26 ft

70. Routing A square city block is 135 m on each side. Determine the distance from the NW corner to the SE corner via two of the bordering streets. Then calculate the distance between the corners if a diagonal shortcut is taken. Round your answers to the nearest meter. 270 m; 191 m

CHALLENGE

A rectangular piece of paper 12 cm by 24 cm is folded along line segment AB, as shown.

Find $\csc \theta$. $\frac{5}{3}$

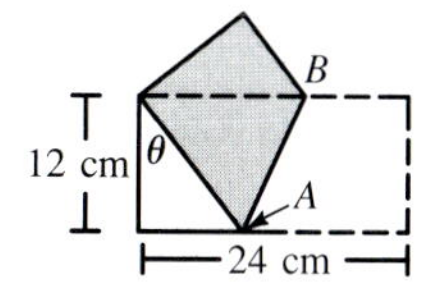

Teacher's Resource Book

Practice—Chapter 1, p. 15

Enrichment—Chapter 1, p. 16

Additional Answers

27. $\frac{1}{2}, -\frac{\sqrt{3}}{2}, -\frac{\sqrt{3}}{3}, 2, -\frac{2\sqrt{3}}{3}, -\sqrt{3}$

28. $-\frac{1}{2}, -\frac{\sqrt{3}}{2}, \frac{\sqrt{3}}{3}, -2, -\frac{2\sqrt{3}}{3}, \sqrt{3}$

29. $\frac{\sqrt{2}}{2}, -\frac{\sqrt{2}}{2}, -1, \sqrt{2}, -\sqrt{2}, -1$

30. $\frac{\sqrt{3}}{2}, -\frac{1}{2}, -\sqrt{3}, \frac{2\sqrt{3}}{3}, -2, -\frac{\sqrt{3}}{3}$

31. $-\frac{\sqrt{2}}{2}, -\frac{\sqrt{2}}{2}, 1, -\sqrt{2}, -\sqrt{2}, 1$

32. $\frac{1}{2}, -\frac{\sqrt{3}}{2}, -\frac{\sqrt{3}}{3}, 2, -\frac{2\sqrt{3}}{3}, -\sqrt{3}$

33. $-\frac{1}{2}, -\frac{\sqrt{3}}{2}, \frac{\sqrt{3}}{3}, -2, -\frac{2\sqrt{3}}{3}, \sqrt{3}$

34. $-\frac{\sqrt{2}}{2}, \frac{\sqrt{2}}{2}, -1, -\sqrt{2}, \sqrt{2}, -1$

35. $\frac{1}{2}, \frac{\sqrt{3}}{2}, \frac{\sqrt{3}}{3}, 2, \frac{2\sqrt{3}}{3}, \sqrt{3}$

36. $\frac{\sqrt{3}}{2}, \frac{1}{2}, \sqrt{3}, \frac{2\sqrt{3}}{3}, 2, \frac{\sqrt{3}}{3}$

37. $\frac{\sqrt{2}}{2}, -\frac{\sqrt{2}}{2}, -1, \sqrt{2}, -\sqrt{2}, -1$

38. $-\frac{\sqrt{2}}{2}, -\frac{\sqrt{2}}{2}, 1, -\sqrt{2}, -\sqrt{2}, 1$

39. $-\frac{1}{2}, \frac{\sqrt{3}}{2}, -\frac{\sqrt{3}}{3}, -2, \frac{2\sqrt{3}}{3}, -\sqrt{3}$

40. $-\frac{\sqrt{2}}{2}, -\frac{\sqrt{2}}{2}, 1, -\sqrt{2}, -\sqrt{2}, 1$

41. $\frac{\sqrt{3}}{2}, -\frac{1}{2}, -\sqrt{3}, \frac{2\sqrt{3}}{3}, -2, -\frac{\sqrt{3}}{3}$

42. -1, 0, undef., -1, undef, 0

43. $-\frac{\sqrt{3}}{2}, \frac{1}{2}, -\sqrt{3}, -\frac{2\sqrt{3}}{3}, 2, -\frac{\sqrt{3}}{3}$

44. $\frac{\sqrt{2}}{2}, \frac{\sqrt{2}}{2}, 1, \sqrt{2}, \sqrt{2}, 1$

45. $-\frac{\sqrt{3}}{2}, -\frac{1}{2}, \sqrt{3}, -\frac{2\sqrt{3}}{3}, -2, \frac{\sqrt{3}}{3}$

46. $\frac{\sqrt{2}}{2}, \frac{\sqrt{2}}{2}, 1, \sqrt{2}, \sqrt{2}, 1$

LESSON PLAN

Materials/Manipulatives
Calculators
Computer

BACKGROUND

In the Preview, the concept of nautical mile is discussed. The Example uses a latitude of 60°. You may wish to have a student look up the latitude of your school and then calculate the approximate length of a nautical mile for that latitude.

1.9 Evaluating Trigonometric Functions

Objectives: To find decimal approximations for the values of the six trigonometric functions for all angles
To find the measure of an angle given the value of one of its trigonometric functions

Using the values of the trigonometric functions of angles, navigators and surveyors are able to calculate distances that they cannot measure directly.

Preview

A nautical mile is a unit of distance used on ships and aircraft. One *nautical mile* is equal to $\frac{1}{60}$ degree of arc length on a meridian.

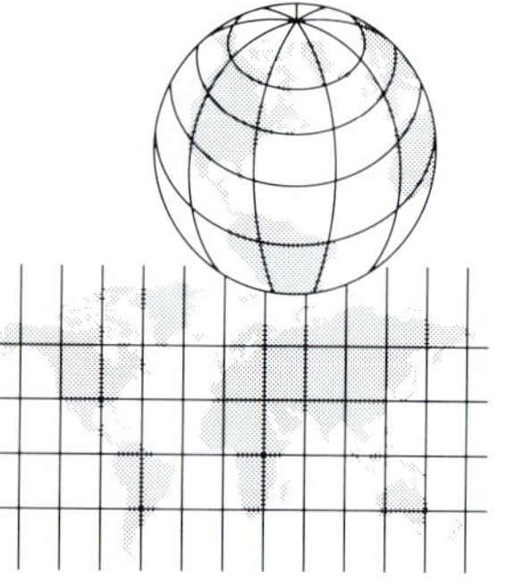

The earth is not a perfect sphere, so the actual length of a nautical mile varies with the latitude. The length can be approximated using the formula

$$1 \text{ nautical mile} = (6077 - 31 \cos 2\theta) \text{ ft}$$

where θ is the latitude in degrees.

The latitude of Anchorage, Alaska is about 61°. If this is rounded to 60°, you can calculate the length of one nautical mile at that location, since you know the value of cos 2(60°), or cos 120°.

1. Round 61° to 60° and calculate the approximate length of a nautical mile at Anchorage. Round the answer to the nearest foot. 6093 ft
2. Which of the following is the best approximation of the length of a nautical mile at a location with latitude 29°? b.

   a. 6055 ft **b.** 6062 ft **c.** 6068 ft **d.** 6075 ft

A scientific calculator is an invaluable tool when you wish to find the values of trigonometric functions of an angle such as 61° that is not an integral multiple of a special angle. You can also use a table of trigonometric values. See pages 434–438 for the table and pages 422–423 for a discussion of its use.

A scientific calculator has at least two modes. Make sure that it is in *degree mode* when you are asked to find the values of the trigonometric functions of angles expressed in degrees, and in *radian mode* when you must find the values of the trigonometric functions of angles expressed in radians.

EXAMPLE 1 **Find each value. Round to four decimal places.**

a. cos 56.2° **b.** tan 15°23′ **c.** $\sin \frac{\pi}{9}$

a. Place your calculator in degree mode. Since the angle measure is given in decimal degrees, simply enter 56.2. Use the cos key.

$$\cos 56.2° = 0.556295615$$
$$= 0.5563 \quad \text{to four decimal places}$$

b. Change 15°23′ to decimal degrees, unless your calculator accepts angle measures given in degrees and minutes. Use the tan key.

$$\tan 15°23' = \tan\left(15 + \frac{23}{60}\right)^{\circ}$$
$$= \tan 15.38333333°$$
$$= 0.275132957$$
$$= 0.2751 \quad \text{to four decimal places}$$

c. Place your calculator in radian mode.

$$\sin \frac{\pi}{9} = \sin 0.34906585 \quad \textit{Use the } \pi \textit{ and } \div \textit{ keys.}$$
$$= 0.342020143 \quad \textit{Use the sin key.}$$
$$= 0.3420 \quad \text{to four decimal places}$$

In Example 1, cos 56.2° is not exactly equal to 0.556295615. The exact values of tan 15°23′ and sin $\frac{\pi}{9}$ are not given either. Most of the values of the trigonometric functions are irrational numbers and cannot be written exactly as decimals. Nevertheless, it is customary to use the equals sign in examples of this type. For exercises in this book, values of the trigonometric functions should be rounded to four decimal places, unless otherwise indicated.

When you use a calculator to find the values of the cosecant, secant, and cotangent functions, remember that csc θ, sec θ, and cot θ are the reciprocal functions of sin θ, cos θ, and tan θ, respectively. Therefore, use the reciprocal key, which is usually labeled $\frac{1}{x}$.

EXAMPLE 2 **Find sec 24.37°.**

Make sure that your calculator is in degree mode.
Enter 24.37. Use the cos key and the reciprocal key.

$$\sec 24.37° = 1.097815544$$
$$= 1.0978 \quad \text{to four decimal places}$$

TEACHING SUGGESTIONS

- Point out that calculators are only effective, if the correct keys are used.
- Stress that since $\sec \theta = \frac{1}{\cos \theta}$, $\csc \theta = \frac{1}{\sin \theta}$, and $\cot \theta = \frac{1}{\tan \theta}$, the cos key followed by the $\frac{1}{x}$ key is used to obtain secant, and so on.
- You may wish to point out that the inverse key is the second function key on some calculators.
- When finding an angle measure from the values of trigonometric functions of the angle, be sure that students realize that there may be more than one angle measure that satisfies the condition.

Critical Thinking

Observations Without using calculators, ask the students how they would evaluate cot y, sec x, and csc θ, given the values: $\sin \theta = \frac{1}{2}$, $\cos x = \frac{\sqrt{3}}{2}$, $\tan y = -\frac{5}{6}$. Students should observe that sin θ and csc θ, cos x and sec x, and tan y and cot y are reciprocals of each other.

CHALKBOARD EXAMPLES

- **For Example 1**

  Find each value. Round to four decimal places.

  1. sin 31.8° 0.5270
  2. tan 10° 45′ 0.1899
  3. $\cos \frac{\pi}{9}$ 0.9397
  4. $\sin \frac{\pi}{5}$ 0.5878

- **For Example 2**

  Find each value. Round to four decimal places.

  5. csc 36.20° 1.6932
  6. cot 83°25′ 0.1154
  7. $\sec \frac{\pi}{12}$ 1.0353
  8. $\csc \frac{\pi}{9}$ 2.9238

- **For Example 3**

  Find each value.

  9. cos 310.57° 0.6504

  10. sec (−227°33′) −1.4816

  11. csc $\frac{(-5\pi)}{6}$ −2.0000

  12. tan $\frac{8\pi}{9}$ −0.3640

- **For Example 4**

  Find the measure of each angle θ, $0° \le \theta < 360°$, to the nearest tenth of one degree.

  13. $\cos\theta = 0.8746$ 29.0°; 331.0°

  14. $\sec\theta = 1.197$ 33.3°; 326.7°

  15. $\sin\theta = -0.5324$ 212.2°; 327.8°

  16. $\cot\theta = -1.2103$ 140.4°; 320.4°

Common Errors

- Some students may make incorrect entries. Have those students check the signs of their answers to help verify correct entries.
- When finding angle measures, some students may forget to find all possible angle values within the required interval. Point out that if the upper limit of the interval is 360°, two angles are possible since each function has the same sign in two quadrants.
- See *Teacher's Resource Book* for additional remediation.

The angles in the first two examples are all acute. You can also use a calculator to find the values of the trigonometric functions of angles whose measures are greater than 90° or less than 0°. If you make the entries correctly, the calculator will give the correct signs for the values of the trigonometric functions of these angles. However, you should check the results against your knowledge of the signs of the various functions in each quadrant.

EXAMPLE 3 **Find each value: a.** sin 324.72° **b.** $\tan\left(-\frac{7\pi}{10}\right)$

a. Make sure that your calculator is in degree mode.

$\sin 324.72° = -0.5776$ *Enter 324.72 and use the sin key.*

Check that the sign is correct: $270° < 324.72° < 360°$, so the terminal side of 324.72° is in quadrant IV where sine is negative.

b. Make sure that your calculator is in radian mode.

$\tan\left(-\frac{7\pi}{10}\right) = 1.3764$ *Enter* $-\frac{7\pi}{10}$. *Use the keys for* π, ÷, +/−, *tan.*

Check that the sign is correct: $-\frac{7\pi}{10}$ is coterminal with $\frac{13\pi}{10}$, whose terminal side is in quadrant III where tangent is positive.

If you know the value of a trigonometric function of an angle, a calculator will give you the measure of one angle with that value. You must then use your knowledge of the signs of the various trigonometric functions in each quadrant to decide whether there is a second angle that has the same value.

Finding the measure of an angle with a known sine and finding the sine of an angle of known measure are *inverse operations*. Therefore, you will probably have to use the inverse key (usually labeled inv) as shown in Example 4.

EXAMPLE 4 **Find the measure of each angle θ, where $0° \le \theta \le 360°$, to the nearest tenth of one degree.**

a. $\sin\theta = 0.5358$ **b.** $\cos\theta = -0.5663$ **c.** $\cot\theta = -1.3182$

a. $\sin\theta = 0.5358$ *Enter 0.5358. Make sure calculator is in degree mode.*

$\theta = 32.39818172°$ *Use the inv and the sin keys.*

$\theta = 32.4°$ to the nearest tenth

Since $\sin\theta$ is positive, there is also an angle in quadrant II whose sine is 0.5358 and whose reference angle is 32.4°.

$$180° - 32.4° = 147.6°$$

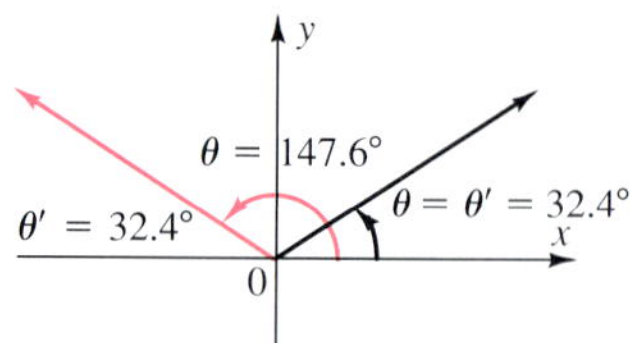

Check to show that sin 147.6° = 0.5358.
The solutions are $\theta = 32.4°$ and $\theta = 147.6°$.

b. $\cos\theta = -0.5663$ *Enter −0.5663. Use the +/− key.*

$\theta = 124.4926144°$ *Use the inv and cos keys.*

$\theta = 124.5°$ to the nearest tenth

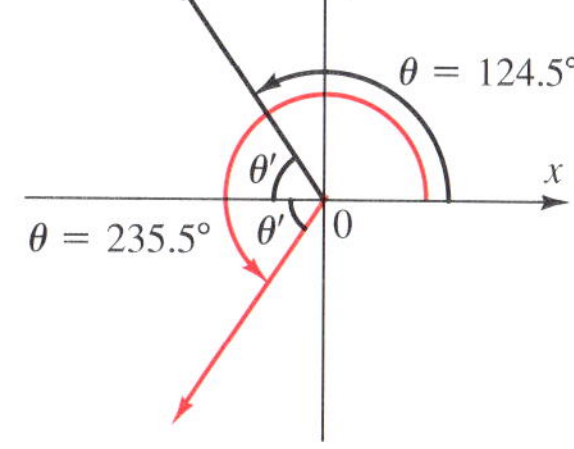

Since $\cos\theta$ is negative, there is also an angle in quadrant III whose cosine is −0.5663 and whose reference angle is the same as that of 124.5°.

$180° - 124.5° = 55.5°$ *Reference angle for 124.5°*

$180° + 55.5° = 235.5°$ *A 235.5° angle also has reference angle 55.5°.*

Check to show that cos 235.5° = −0.5663. (A calculator gives cos 235.5° as −0.5664, to four decimal places. The discrepancy is due to the fact that 124.4926144° was rounded to 124.5°.)

The solutions are $\theta = 124.5°$ and $\theta = 235.5°$.

c. $\cot\theta = -1.3182$ *Enter −1.3182. Use the +/− key.*

$\theta = -37.18432593°$ *Use the reciprocal, inv and tan keys.*

$\theta = -37.2°$ to the nearest tenth

Since $0° \le \theta \le 360°$, angle θ cannot be negative. Find a positive angle coterminal with −37.2°.

$$360° + (-37.2°) = 322.8°$$

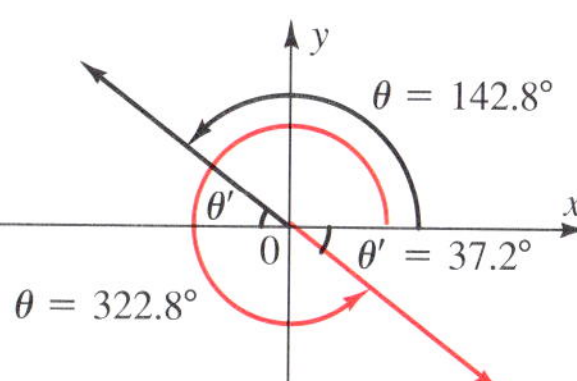

Since $\cot\theta$ is negative, there is also an angle in quadrant II whose cotangent is −1.3182 and whose reference angle is the same as that of 322.8°.

$360° - 322.8° = 37.2°$ *Reference angle for 322.8°*

$180° - 37.2° = 142.8°$ *A 142.8° angle also has reference angle 37.2°.*

Check to show that cot 142.8° is approximately equal to −1.3182. (The calculator gives −1.3175. The discrepancy is due to the rounding procedures used.) The solutions are $\theta = 322.8°$ and $\theta = 142.8°$.

CLASS EXERCISES

Find each value. Round to four decimal places.

1. $\sin 24.8°$ 0.4195
2. $\cos 36.9°$ 0.7997
3. $\tan 12°18'$ 0.2180
4. $\cos(-101.7°)$ −0.2028
5. $\sin(-82.5°)$ −0.9914
6. $\sec\frac{13\pi}{12}$ −1.0353
7. $\csc\frac{17\pi}{12}$ −1.0353
8. $\cot\left(-\frac{2\pi}{5}\right)$ −0.3249

Find θ, where $0° \le \theta \le 360°$, to the nearest tenth of one degree.

9. $\sin\theta = 0.7218$ 46.2°, 133.8°
10. $\cos\theta = -0.5864$ 125.9°, 234.1°
11. $\csc\theta = -1.1111$ 244.2°, 295.8°
12. $\sec\theta = 1.1111$ 25.8°, 334.2°
13. $\cot\theta = -1.2222$ 140.7°, 320.7°
14. $\cot\theta = -2.3456$ 156.9°, 336.9°

LESSON FOLLOW-UP

Discussion

Discuss the advantages of using a scientific calculator as opposed to a table of values to find the values of the trigonometric functions of an angle or the reverse. Answers may vary. One major advantage is that a calculator saves time by eliminating the need for interpolation.

Assignment Guide

See p. B for assignments.

Computer

The complete form of the program for determining the values of trigonometric functions is on the disk provided with the *Teacher's Resource Book*.

Test Yourself

See *Teacher's Resource Book*, Tests, pp. 7–8.

Lesson Quiz

Find each value. Round to four decimal places.

1. $\sin 47.2°$ 0.7337
2. $\tan 14°48'$ 0.2642
3. $\cos \frac{2\pi}{3}$ −0.5000
4. $\cot 318.6°$ −1.1343
5. $\csc(-206.7°)$ 2.2256
6. $\sec \frac{(-5\pi)}{3}$ 2.0000

Find θ, where $0° \le \theta \le 360°$, to the nearest tenth of a degree.

7. $\cos\theta = -0.2672$ 105.5°, 254.5°
8. $\tan\theta = 12.5135$ 85.4°, 265.4°
9. $\sin\theta = -0.9681$ 284.5°, 255.5°
10. $\sec\theta = -2.5990$ 112.6°, 247.4°

Enrichment

Find θ if $\cos\theta = 0.7\sin\theta$.

$\cos\theta = 0.7\sin\theta$

$\frac{\cos\theta}{\sin\theta} = 0.7$

$\frac{\frac{x}{r}}{\frac{y}{r}} = 0.7$

$\frac{x}{y} = 0.7$

$\cot\theta = 0.7$

$\theta = 55°, 235°, 415°, 595°, \ldots$

PRACTICE EXERCISES

Find each value. Round to four decimal places.

A

1. $\sin 18.6°$ 0.3190
2. $\cos 21.5°$ 0.9304
3. $\tan 83.7°$ 9.0579
4. $\sin 85.8°$ 0.9973
5. $\cos 12.7°$ 0.9755
6. $\tan 48.4°$ 1.1263
7. $\sin 18°32'$ 0.3179
8. $\cos 24°25'$ 0.9106
9. $\tan 87°18'$ 21.2049
10. $\sin 73°37'$ 0.9594
11. $\cos\frac{\pi}{5}$ 0.8090
12. $\tan\frac{\pi}{6}$ 0.5774
13. $\sin\frac{2\pi}{3}$ 0.8660
14. $\cos\frac{3\pi}{5}$ −0.3090
15. $\tan\frac{3\pi}{4}$ −1
16. $\tan\frac{4\pi}{5}$ −0.7265
17. $\csc 20.7°$ 2.8291
18. $\sec 23.9°$ 1.0938
19. $\cot 41.4°$ 1.1343
20. $\cot 60.6°$ 0.5635
21. $\sec 145.5°$ −1.2134
22. $\cot 160.6°$ −2.8397
23. $\csc 98.9°$ 1.0122
24. $\sec 175.4°$ −1.0032

Find θ, where $0° \le \theta \le 360°$, to the nearest tenth of one degree.

25. $\sin\theta = 0.7777$ 51.1°, 128.9°
26. $\cos\theta = 0.8888$ 27.3°, 332.7°
27. $\tan\theta = 0.6765$ 34.1°, 214.1°
28. $\sin\theta = 0.3535$ 20.7°, 159.3°
29. $\cos\theta = 0.6161$ 52.0°, 308.0°
30. $\tan\theta = 0.9808$ 44.4°, 224.4°

Find each value. Round to four decimal places.

B

31. $\sin 310°16'$ −0.7630
32. $\cos 311°55'$ 0.6680
33. $\tan 610.7°$ 2.8556
34. $\sin 511.1°$ 0.4833
35. $\cos(-111.8°)$ −0.3714
36. $\tan(-200.8°)$ −0.3799
37. $\csc 321.2°$ −1.5959
38. $\sec 315.6°$ 1.3996
39. $\sin\frac{3\pi}{7}$ 0.9749
40. $\tan\frac{9\pi}{8}$ 0.4142
41. $\cot\frac{\pi}{11}$ 3.4057
42. $\csc\frac{25\pi}{8}$ −2.6131
43. $\cot(-77°16')$ −0.2260
44. $\csc\left(-\frac{3\pi}{7}\right)$ −1.0257
45. $\sec\left(-\frac{7\pi}{4}\right)$ 1.4142
46. $\cot\left(-\frac{6\pi}{5}\right)$ −1.3764

Find θ, where $0° \le \theta \le 360°$, to the nearest tenth of one degree.

47. $\csc\theta = 2.5012$ 23.6°, 156.4°
48. $\sec\theta = 3.1909$ 71.7°, 288.3°
49. $\cot\theta = 4.4444$ 12.7°, 192.7°
50. $\csc\theta = -1.3737$ 313.3°, 226.7°
51. $\cot\theta = -6.6175$ 351.4°, 171.4°
52. $\sec\theta = -1.7172$ 125.6°, 234.4°

C

53. $\tan\theta = 1.1213$ and $\sin\theta$ is positive. 48.3°
54. $\sin\theta = 0.5151$ and $\cos\theta$ is negative. 149.0°
55. $\cot\theta = 0.6878$ and θ is in quadrant III. 235.5°
56. $\csc\theta = 2.5151$ and θ is in quadrant I. 23.4°
57. $\sin\theta = -0.5267$ and $270° \le \theta \le 360°$ 328.2°
58. $\cos\theta = -0.0655$ and $90° \le \theta \le 180°$ 93.8°
59. $\csc\theta = 1.0222$ and $90° \le \theta \le 180°$ 102.0°
60. $\sec\theta = 1.0222$ and $270° \le \theta \le 360°$ 348.0°

Applications

Computer The following program calculates and displays the values of the six trigonometric functions of an angle whose measure is given in degrees. Line 40 calculates the sine and cosine, lines 50 and 60 calculate the other functions, and lines 70–100 display the results.

```
 10 PRINT "PLEASE ENTER THE MEASURE OF AN ANGLE"
 20 INPUT M
 30 M = M*0.01745329
 40 S = SIN(M): C = COS(M)
 50 T = S/C: CS = 1/S
 60 SE = 1/C: CT = 1/T
 70 HOME: PRINT "THE TRIG FUNCTION VALUES OF "M" DEGREES ARE"
 80 PRINT: PRINT "SINE-"S, "COSINE-"C
 90 PRINT "TANGENT-"T, "COSECANT-"CS
100 PRINT "SECANT-"SE, "COTANGENT-"CT
110 END
```

61. Line 30 converts the degree measure to radians. How is the constant 0.01745329 derived? $\pi \div 180$

62. CT represents the cotangent. State another way that CT could be found.
C/S or CS/SE

TEST YOURSELF

1. Angle θ is in standard position and lies in quadrant II. If $\sin\theta = \frac{\sqrt{3}}{2}$, find the exact value of $\cos\theta$. $\cos\theta = -\frac{1}{2}$ **1.6**

2. Find the exact values of $\sin\theta$ and $\cos\theta$ if θ is an angle in standard position whose terminal side passes through $(-2, -2)$. Sketch the reference triangle.
$\sin\theta = -\frac{\sqrt{2}}{2}$; $\cos\theta = -\frac{\sqrt{2}}{2}$

Find the exact values of the other five trigonometric functions for an angle in standard position lying in the given quadrant.
Answers are in the order sin, cos, tan, csc, sec, cot.

3. $\sin\theta = -\frac{12}{13}$, III
$-\frac{5}{13}$; $\frac{12}{5}$; $-\frac{13}{12}$; $-\frac{13}{5}$; $\frac{5}{12}$

4. $\cos\theta = \frac{4}{5}$, I
$\frac{3}{5}$; $\frac{3}{4}$; $\frac{5}{3}$; $\frac{5}{4}$; $\frac{4}{3}$

5. $\sec\theta = -\frac{12}{7}$, II **1.7**
$\frac{\sqrt{95}}{12}$; $-\frac{7}{12}$; $-\frac{\sqrt{95}}{7}$; $\frac{12\sqrt{95}}{95}$; $-\frac{7\sqrt{95}}{95}$

Give the measure of the reference angle θ' for each angle θ in standard position.

6. 179° 1°

7. 312° 48°

8. $-\frac{4\pi}{3}$ $\frac{\pi}{3}$

9. $\frac{\pi}{4}$ $\frac{\pi}{4}$ **1.8**

10. Find the exact values of the six trigonometric functions of a 240° angle.
$\sin\theta = -\frac{\sqrt{3}}{2}$; $\cos\theta = -\frac{1}{2}$; $\tan\theta = \sqrt{3}$; $\csc\theta = -\frac{2\sqrt{3}}{3}$; $\sec\theta = -2$; $\cot\theta = \frac{\sqrt{3}}{3}$

Find each value. Round to four decimal places.

11. sin 212.2°
−0.5329

12. tan (−63.5°)
−2.0057

13. $\csc\frac{\pi}{7}$ 2.3048

14. $\cot\frac{2\pi}{5}$ **1.9**
0.3249

Teacher's Resource Book
Practice—Chapter 1, p. 17
Enrichment—Chapter 1, p. 18

Application

The purpose of this application is to give students an idea of how scientists might use trigonometry in their work. This application illustrates how space scientists use trigonometric functions to determine when a booster rocket should launch (release) its payload into its own orbit.

See *Teacher's Resource Book*, Chapter 1, p. 19.

See *Teacher's Resource Book*, Technology, p.1.

APPLICATION: Space Exploration

Did you know that space scientists prefer to launch a payload, such as a communications satellite, into Earth-orbit in more than one stage? The reason for this is that while a large, heavy fuel module is required to lift a heavy rocket through the lower atmosphere, relatively little fuel is required to propel a light payload into its own trajectory through the thin atmosphere of high altitudes. As the diagram below shows, a booster rocket releases its payload at a previously selected altitude and then follows a parabolic trajectory back to Earth. Then the payload projectile starts a new trajectory that will eventually carry it into an orbit around the Earth. The trajectories of both the booster rocket and the payload projectile are curves. However, the path of the payload may be approximated by a straight line for a brief period of time at the start of its independent flight.

In the diagram, a booster rocket has reached altitude y_1 with the direction of its path forming an angle θ with an imaginary horizontal line. The value of θ has been steadily diminishing since the launch began. The space scientists on the ground now wish to fire the payload into its own trajectory, provided that θ has diminished to a certain critical value that will put the payload into Earth-orbit. This value of θ can be calculated by solving the equations

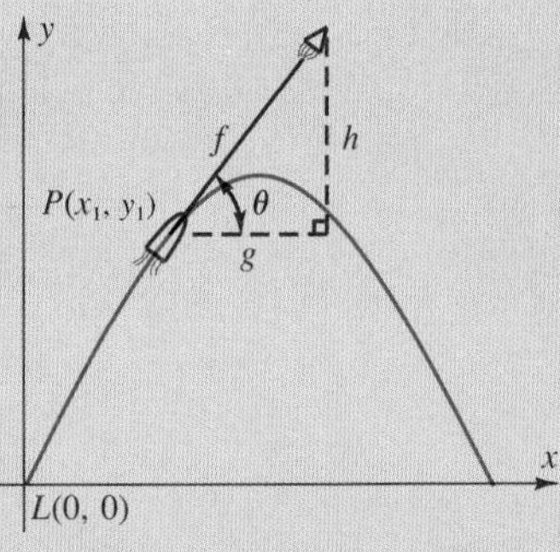

$y = ax^2 + bx$ *Path of the booster rocket*

$\tan \theta = 2ax + b$ *Direction of the booster rocket*

where x and y are the horizontal and vertical coordinates, respectively, of the position of the rocket. However, before the equations can be used, the values of constants a and b must be determined.

EXAMPLE 1 The equation of the trajectory of a booster rocket is $y = -x^2 + 3x$. Space scientists plan to fire a payload from the booster at a horizontal distance of 1 km from the launch site. Find the height and direction at which the payload will be fired.

$$y = -x^2 + 3x$$
$$= -1^2 + 3(1) = 2$$ The payload will be fired when the height is 2 km.

Next, evaluate a and b in the equation for the direction.

$$\tan\theta = 2(-1)x + 3$$ *$a = -1, b = 3$*
$$= -2x + 3$$
$$= -2(1) + 3 = 1$$ *Substitute 1 for x_1.*

Since $\tan 45° = 1$, the initial direction of the projectile in independent flight is at an angle of 45° with the horizontal.

EXAMPLE 2 For the flight of Example 1, use $\sin\theta = \frac{h}{f}$ to find the vertical distance h traveled by the projectile after it has traveled on an approximately linear path of 10 km.

$$\sin 45° = \frac{h}{10}$$ *Substitute 10 for f.*
$$\frac{\sqrt{2}}{2} = \frac{h}{10}$$
$$h = 10 \cdot \frac{\sqrt{2}}{2} \approx 7.1$$ The vertical distance is about 7.1 km.

EXERCISES

1. Find the altitude of the payload in Example 2 after 10 km of its free flight. 9.1 km
2. Use $\cos\theta = \frac{g}{f}$ to find the horizontal distance g traveled by the payload in Example 2. 7.1 km
3. In Examples 1 and 2, suppose that the payload is fired when $\theta = 58°$ rather than 45°. What will be the altitude of the rocket? 1.6 km
4. A booster rocket with a weather satellite payload is launched in a trajectory with equation $y = -2x^2 + 1.9x$. The satellite is fired into its own trajectory when the horizontal distance reaches 0.3 km. Find, to the nearest degree, the free-flight direction of the satellite. 35°
5. Find the maximum altitude attained by the booster rocket of Exercise 4. *Hint*: What must be the value of θ at the vertex of the parabola? 0.45 km

CHAPTER 1 SUMMARY AND REVIEW

Vocabulary

angle (12)
angular displacement (24)
angular velocity (24)
central angle (17)
coordinate plane (3)
cosecant (34)
cosine (29)
cotangent (34)
coterminal angles (13)
distance formula (8)
degree (13)
domain (2)
function (3)
initial side (12)
linear velocity (24)
minute (17)
origin (3)
Pythagorean theorem (8)
quadrant (3)
quadrantal angle (39)
radian (18)
range (2)
reciprocal functions (35)
reference angle (42)
reference triangle (30)
relation (2)
secant (34)
second (17)
sine (29)
standard position (12)
tangent (34)
terminal side (12)
trigonometric functions (34)
unit circle (29)
x-axis, y-axis (3)

Functions and the Coordinate Plane A function is a relation that pairs each member of the domain with exactly one member of the range. 1.1

1. Find the domain and range of the relation $\{(1, 3), (5, 7), (5, 9)\}$. Is the relation a function? domain: {1, 5}; range: {3, 7, 9}; no
2. Find $f(-3)$ if $f(x) = 2x - 7$. -13

The Distance Formula The distance d between two points with coordinates (x_2, y_2) and (x_1, y_1) is given by $d = \sqrt{(x_2 - x_1)^2 + (y_2 - y_1)^2}$. 1.2

3. Find the distance between the points $(3, 0)$ and $(-5, -6)$. 10

Angles in the Coordinate Plane An angle is in standard position if its vertex is the origin and its initial side lies on the positive x-axis. 1.3

4. Find the degree measure of the angle formed by a $\frac{1}{3}$ counterclockwise rotation. Sketch the angle in standard position. 120°
5. Find the measures of two angles that are coterminal with a 110° angle. 470°, $-250°$ Answers may vary.

Angle Measures in Degrees and Radians To change from degrees to radians, multiply the number of degrees by $\frac{\pi}{180}$. To change from radians to degrees, multiply the number of radians by $\frac{180}{\pi}$. 1.4

6. Express 240° in radians. Give the answer in terms of π. $\frac{4\pi}{3}$
7. Express $\frac{11\pi}{4}$ in degrees. 495°
8. Express 100°33′ in decimal degrees. 100.55°

Applications: Angular and Linear Velocity The formulas for arc length s, angular velocity ω, and linear velocity V are: 1.5

$$s = r\theta \qquad \omega = \frac{\theta}{t} \qquad V = \frac{s}{t} = \omega r$$

9. Find the length of the arc intercepted by a central angle of $\frac{\pi}{3}$, if the radius r of the circle is 12 in. Express the answer in terms of π. 4π in.

10. Find the angular velocity in radians per second and the linear velocity in inches per second of a point on the rim of a wheel with a diameter of 20 in., rotating at 150 revolutions per minute. Express the answers in terms of π. 5π rad/s; 50π in./s

Cosine and Sine Functions The cosine and sine functions of an angle θ are determined by the coordinates $(x, y) = (\cos\theta, \sin\theta)$ of the intersection of the terminal side of the angle with the unit circle. If the circle has radius r, then $\cos\theta = \frac{x}{r}$ and $\sin\theta = \frac{y}{r}$. 1.6

11. If $\sin\theta = -\frac{15}{17}$ and θ is in quadrant III, find the exact value of $\cos\theta$. $-\frac{8}{17}$

The Trigonometric Functions The remaining functions are $\tan\theta = \frac{y}{x}\ (x \neq 0)$, $\csc\theta = \frac{r}{y}\ (y \neq 0)$, $\sec\theta = \frac{r}{x}\ (x \neq 0)$, and $\cot\theta = \frac{x}{y}\ (y \neq 0)$. 1.7

12. Find the exact values of the six trigonometric functions of angle θ in standard position, if the point $(4, -3)$ is on its terminal side.
$\sin\theta = -\frac{3}{5}$; $\cos\theta = \frac{4}{5}$; $\tan\theta = -\frac{3}{4}$; $\csc\theta = -\frac{5}{3}$; $\sec\theta = \frac{5}{4}$; $\cot\theta = -\frac{4}{3}$

Trigonometric Functions of Special Angles Angles with measures 30°, 45°, and 60°, as well as quadrantal angles, are special angles in trigonometry. Reference angles are used to find the values of trigonometric functions for angles that are not in the first quadrant. Answers are in the order sin, cos, tan, csc, sec, cot. 1.8

13. Find the exact values of the six trigonometric functions of 180°.
0; -1; 0; undef.; -1; undef.

14. Find the exact values of the six trigonometric functions of $-30°$.
$-\frac{1}{2}$; $\frac{\sqrt{3}}{2}$; $-\frac{\sqrt{3}}{3}$; -2; $\frac{2\sqrt{3}}{3}$; $-\sqrt{3}$

Evaluating Trigonometric Functions A scientific calculator can be used to find values of the trigonometric functions of an angle. It can also be used to find the measure of an angle if the value of one of its trigonometric functions is given. 1.9

15. Find $\sin\frac{\pi}{4}$ and $\sec -215.7°$. Round to four decimal places. 0.7071; -1.2314

16. If $0° \leq \theta \leq 360°$ and $\csc\theta = -2.2023$, find θ to the nearest tenth of one degree. 333.0°, 207.0°

See *Teacher's Resource Book, Tests*, pp. 9–12.

CHAPTER TEST

1. Write the domain and the range for the relation {(1, 3), (3, 8), (0, 8)}. Is the relation a function? domain: {1, 3, 0}; range: {3, 8}; yes
2. If $f(x) = 2x - 5$, find $f(-3)$. -11
3. Find the distance between the points $(-3, 4)$ and $(9, -1)$. 13

Find the degree measure of the angle for each rotation.

4. $\frac{17}{6}$ clockwise rotation $-1020°$
5. $\frac{2}{3}$ counterclockwise rotation 240°
6. Express 225° in radians. Give the answer in terms of π. $\frac{5\pi}{4}$
7. Express $-\frac{5\pi}{3}$ in degrees. $-300°$

Express the answers to Exercises 8–10 in terms of π.

8. Find the length of the arc intercepted by a central angle of $\frac{9\pi}{4}$ if the radius of the circle is 28 in. 63π in.
9. Find the angular velocity in radians per minute of an object rotating at 315 revolutions per minute. 630π rad/min
10. An object rotates at 4 revolutions per second at a distance of 15 ft from a point. Find the linear velocity of the object in feet per second. 120π ft/s
11. If $\sin\theta = \frac{\sqrt{3}}{2}$ and θ is in quadrant II, find the exact value of $\cos\theta$. $-\frac{1}{2}$
12. Point P $(3, -3)$ is on the terminal side of θ, an angle in standard position. Find the exact values of the six trigonometric functions of θ.
13. Find the exact value of $\cot(-585°)$. -1
14. Find the value of cos 37.9°. Round to four decimal places. 0.7891
15. If $0° \le \theta \le 360°$ and $\tan\theta = 1.3535$, find θ to the nearest tenth of one degree. 53.5°, 233.5°

Challenge

12. $\sin\theta = -\frac{\sqrt{2}}{2}$; $\cos\theta = \frac{\sqrt{2}}{2}$; $\tan\theta = -1$; $\csc\theta = -\sqrt{2}$; $\sec\theta = \sqrt{2}$; $\cot\theta = -1$

Find the exact coordinates of a point that separates a semicircular arc of the unit circle (from 0° to 180°) into two parts, the ratio of whose lengths is 1:3. Is there more than one such point? $\left(\frac{\sqrt{2}}{2}, \frac{\sqrt{2}}{2}\right)$; yes, $\left(-\frac{\sqrt{2}}{2}, \frac{\sqrt{2}}{2}\right)$

COLLEGE ENTRANCE EXAM REVIEW

Select the best choice for each question.

1. If $\triangle ABC$ is a right triangle, find the length of side x.
B
A. -34 **B.** $\sqrt{119}$
C. $\sqrt{34}$ **D.** $-\sqrt{34}$
E. 119

2. Find $\cot \theta$ if $\sin \theta = -\frac{1}{3}$ and $\tan \theta$ is positive.
A
A. $2\sqrt{2}$ **B.** -3 **C.** $\frac{2\sqrt{2}}{3}$
D. $-2\sqrt{2}$ **E.** $2\sqrt{3}$

3. Solve for x.
D
$$\frac{3}{x-1} + 1 = \frac{x+4}{x^2-1} - \frac{x}{1-x}$$
A. 4 **B.** -2 **C.** $\frac{3}{13}$
D. 2 **E.** No solution

4. If line m is parallel to line n, then $c - a$ equals
E

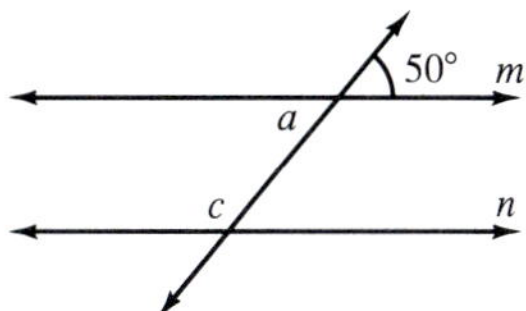

A. 180° **B.** 130° **C.** 0°
D. 50° **E.** 80°

5. Which of the following relations is not a function?
D
A. $\{(3, 4), (-1, 0), (2, 2)\}$
B. $\{(1, 2), (3, 2), (5, 2), (6, 2)\}$
C. $\{(-1, -6), (-6, -1)\}$
D. $\{(3, 1), (0, 1), (3, 2)\}$
E. $\{(2, 4), (4, 6), (6, 8)\}$

6. $\left(\frac{x-2}{x+4} \div \frac{x+1}{x+4}\right) = \frac{x-2}{x+1}$ where
D
A. $x \neq 2$ **B.** $x \neq 0$ **C.** $x \neq -1$
D. $x \neq -1, x \neq -4$ **E.** $x \neq -4$

7. The graphs of $4x + y = 4$ and $4x - 3y = 2$ are
A
A. equivalent **B.** parallel
C. perpendicular **D.** intersecting
E. not straight

8. If $\cos \theta = \frac{\sqrt{3}}{2}$ and θ is in quadrant IV, then $\sin \theta$ equals
C
A. $\frac{1}{2}$ **B.** 2 **C.** $-\frac{1}{2}$
D. $\frac{\sqrt{3}}{3}$ **E.** $-\sqrt{3}$

9. For what value(s) of θ does $\sin 2\theta = -1$, where $0 \leq \theta \leq 2\pi$?
C
I. $\frac{\pi}{2}$ II. $\frac{\pi}{4}$ III. $\frac{3\pi}{4}$
A. I only **B.** II only **C.** III only
D. I and III only **E.** II and III only

10. If f and g are inverse functions and $g(x) = 3x - 1$, then $f(x)$ equals
B
A. $3x - 1$ **B.** $\frac{1}{3}x + \frac{1}{3}$ **C.** $x + \frac{1}{3}$
D. $x - 3$ **E.** $\frac{1}{3}x + 1$

11. Which of the following is the smallest?
D
A. $(4x^2)^0$ **B.** $\left(\frac{2}{3}\right)^{-2}$ **C.** $\frac{\sqrt{2}}{2}$
D. $\left(\frac{1}{2}\right)^3$ **E.** $\sin 30°$

12. Which of the following is (are) true?
C
I. $\frac{\pi}{2}$ is a quadrantal angle.
II. If $\theta = \frac{\pi}{3}$, then $0° \leq \theta \leq 45°$.
III. $\frac{3\pi}{4} = 135°$
A. I only **B.** I and II only
C. I and III only **D.** II and III only
E. I, II, and III

Maintaining Skills

The following skills and concepts are reviewed:

Solving equations with rational expressions
Factoring polynomials
Solving polynomial equations
Graphing functions

Additional Answers

19.

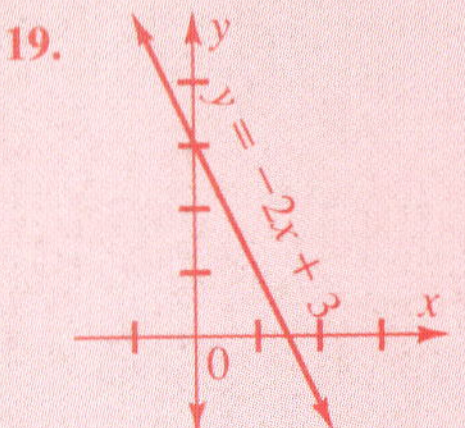

20.

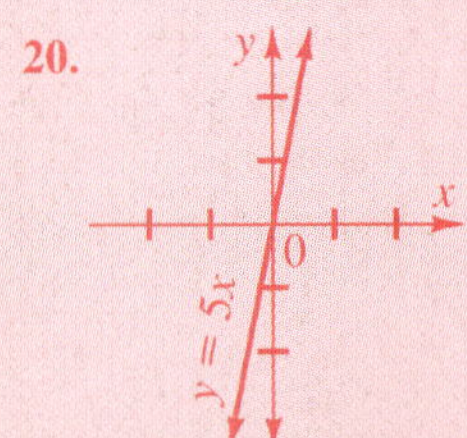

21.

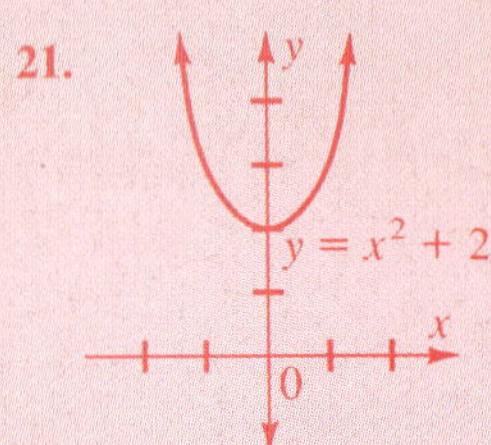

22.

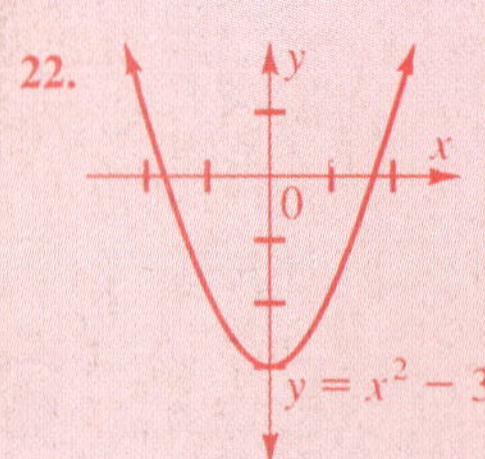

MAINTAINING SKILLS

Solve for *x*.

Example $\dfrac{x-5}{10} = \dfrac{2x+6}{15}$

$(x-5)(15) = 10(2x+6)$ *If $\frac{a}{b} = \frac{c}{d}$, then $ad = bc$.*

$15x - 75 = 20x + 60$ *Distributive property*

$-135 = 5x$ *Subtract 60 and 15x from both sides.*

$-27 = x$

1. $-\dfrac{3}{5} = \dfrac{6}{x}$ -10 **2.** $\dfrac{x+2}{5} = \dfrac{x-3}{7}$ $-\frac{29}{2}$ **3.** $-\dfrac{x}{7} = \dfrac{x+3}{5}$ $-\frac{7}{4}$

4. $\dfrac{x+3}{x-4} = \dfrac{x+5}{x-7}$ $-\frac{1}{5}$ **5.** $\dfrac{x+2}{x+7} = \dfrac{x+3}{x-2}$ $-\frac{5}{2}$ **6.** $\dfrac{2}{x+3} = \dfrac{2x-4}{x^2+3x+2}$ -4

Factor each polynomial.

Example $4x^2 + 12x + 9 = (2x+3)^2$ $a^2 + 2ab + b^2 = (a+b)^2$

7. $4x^2 - 9$ $(2x+3)(2x-3)$ **8.** $x^2 - 9y^2$ $(x+3y)(x-3y)$ **9.** $8 - x^3$ $(2-x)(4+2x+x^2)$

10. $9x^2 - 24x + 16$ $(3x-4)^2$ **11.** $x^3 - 27y^3$ $(x-3y)(x^2+3xy+9y^2)$ **12.** $x^8 - 16$ $(x^4+4)(x^2+2)(x^2-2)$

Solve each equation.

Example $0 = x^3 - 4x$

$0 = x(x^2 - 4)$

$0 = x(x+2)(x-2)$ $a^2 - b^2 = (a+b)(a-b)$

$x = 0, -2, 2$ The solutions are 0, -2, and 2.

13. $0 = 2x^3 - 4x$ $0, \sqrt{2}, -\sqrt{2}$ **14.** $0 = -x^3 + x$ $0, 1, -1$ **15.** $0 = x^3 - x^2 - 2x$ $0, 2, -1$

16. $x^3 = 4x$ $0, 2, -2$ **17.** $-x^3 + 4x^2 = -5x$ $0, 5, -1$ **18.** $x^3 = x^2 + 6x$ $0, -2, 3$

Graph each equation. See side column.

Example $y = x^2 - 1$

Make a table of values.

x	-2	-1	0	1	2
y	3	0	-1	0	3

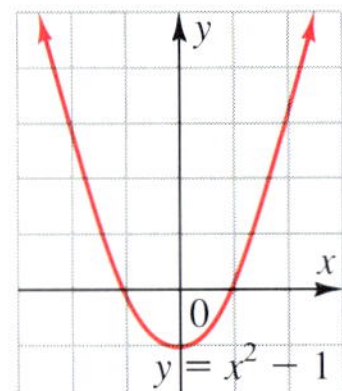

19. $y = -2x + 3$ **20.** $y = 5x$ **21.** $y = x^2 + 2$

22. $y = x^2 - 3$ **23.** $y = -x^2 + 1$ **24.** $y = -x^2 - 2$

23. 24.

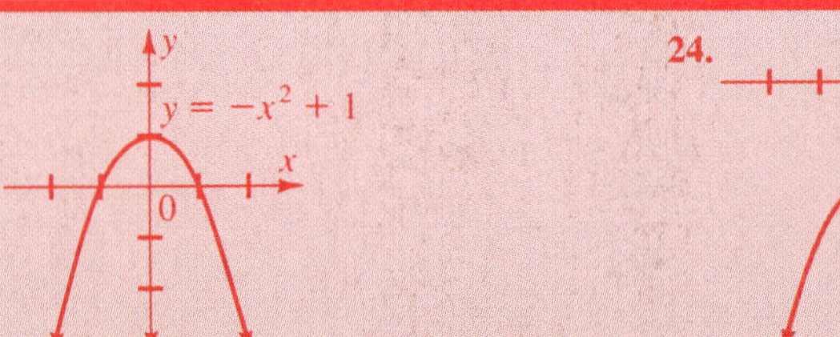

OVERVIEW • Chapter 2

SUMMARY

The chapter begins with a development of the graphs of the sine and cosine functions, including those involving a change in amplitude and/or period, a phase shift, and/or a vertical shift. Then the graphs of the other four trigonometric functions are introduced and developed. An application that addresses the phenomenon of simple harmonic motion rounds out the chapter.

After completing this chapter, students should be able to graph any one of the six trigonometric functions with any amplitude, period, phase shift, or vertical shift and find the sum or difference of two functions. They should be able to solve a variety of problems using the properties of the trigonometric functions.

CHAPTER OBJECTIVES

- To define periodic functions and odd and even functions
- To determine the symmetry of a graph
- To develop the properties of the sine and cosine functions
- To graph the sine and cosine functions
- To find the amplitude and period of a trigonometric function from its equation
- To graph the sine and cosine functions with varied amplitudes and periods
- To find the phase shift and vertical shift of graphs of sine and cosine functions from their equations
- To graph sine and cosine functions with varied phase shifts and vertical shifts
- To graph functions that are the sums or differences of two sine and/or cosine functions
- To determine the properties of the tangent and cotangent functions
- To graph the tangent and cotangent functions
- To determine the properties of the secant and cosecant functions
- To graph the secant and cosecant functions
- To solve problems involving simple harmonic motion

CHAPTER HIGHLIGHTS

The *theme* of Chapter 2 is medical technology. Included in the chapter's special features is a discussion on magnetic resonance imaging, a contemporary technique for viewing tissues in the human body.

APPLICATIONS

Applications in Chapter 2 illustrate the symmetric and periodic natures of many real-world phenomena. Problems from fields such as music, oceanography, and architecture provide vivid illustrations of such concepts as period, amplitude, and addition of ordinates. Physics problems involving simple harmonic motion are presented in Lesson 2.8.

TECHNOLOGY

Calculator

A calculator is recommended for evaluating trigonometric functions, especially when graphing by addition of ordinates. Graphing calculators should be used to show the effects of parameter changes on trigonometric graphs.

Computer

A computer may be used to demonstrate the graphing of trigonometric functions. A computer program may also be helpful for finding equations in problems involving simple harmonic motion.

RESOURCES

Teacher's Resource Book

- Teaching Aid 2
- Transparencies 3 and 4

ASSIGNMENT GUIDE Meeting Student Needs

STUDENT TEXT					TEACHER'S RESOURCE BOOK	
Chapter Content		**Basic**	**Average**	**Enriched**	**P**	**E**
2.1	Periodic Functions and Symmetry	D: 64/1–27 odd, 39	D: 64/3, 9, 13–33 odd, 39	D: 64/3, 11, 19–35 odd, 39, 41	1	2
2.2	Graphs of the Sine and Cosine Functions	D: 71/1–23 odd, 41 R: 64/4, 16, 26	D: 71/11, 17, 21–37 odd, 41 R: 64/4, 16, 28	D: 71/14, 20, 21–41 odd R: 64/4, 22, 28	3	4
2.3	Amplitude and Period	D: 75/1–21 odd, 37 R: 71/4, 10, 22	D: 75/1–29 odd, 37 R: 71/12, 22	D: 75/1–39 odd R: 71/14, 24	5	6
2.4	Phase Shift and Vertical Shift	D: 82/1–21 odd, 29 R: 75/4, 6, 8 83/TY	D: 82/5–25 odd, 29 R: 75/6, 8, 10 83/TY	D: 82/5–31 odd R: 75/10, 12, 16 83/TY	7	8
2.5	Graphing by Addition of Ordinates	Omit	D: 86/5–15 odd, 21 R: 82/4, 6, 22	D: 86/9–21 odd R: 82/6, 8, 22	9	10
2.6	Graphs of the Tangent and Cotangent Functions	D: 92/1–19 odd R: 82/4, 10, 14	D: 92/7–25 odd R: 86/6, 10, 14	D: 92/9–27 odd R: 86/10, 16, 20	11	12
2.7	Graphs of the Secant and Cosecant Functions	D: 98/1–19 odd R: 92/2, 12, 20	D: 98/5–25 odd R: 92/8, 14, 16	D: 98/9–27 odd R: 92/10, 14, 16	13	14
2.8	Application: Harmonic Motion	Omit	D: 103/5–15 odd R: 98/10, 12, 18 104/TY	D: 103/5–15 odd R: 98/10, 12, 18 104/TY	15	16

D = Daily R = Review TY = Test Yourself P = Practice E = Enrichment

	STUDENT TEXT				TEACHER'S RESOURCE BOOK	
Review and Testing	Test Yourself	83, 104	College Ent. Exam Rev.	109	Tests	
	Chapter Sum. and Rev.	106	Cumulative Review	110	• Quizzes	13–16
	Chapter Test	108			• Chapter Test (Form A)	17–18
					• Chapter Test (Form B)	19–20
Special Features	Historical Note	66	Extra	93	Applications—Chapter 2	17
	Career	72	Biography	99	Alg. and Geom. Review	5–8
	Trigonometry and the Calendar	87	Application	105	Technology	2

2 Graphing Trigonometric Functions

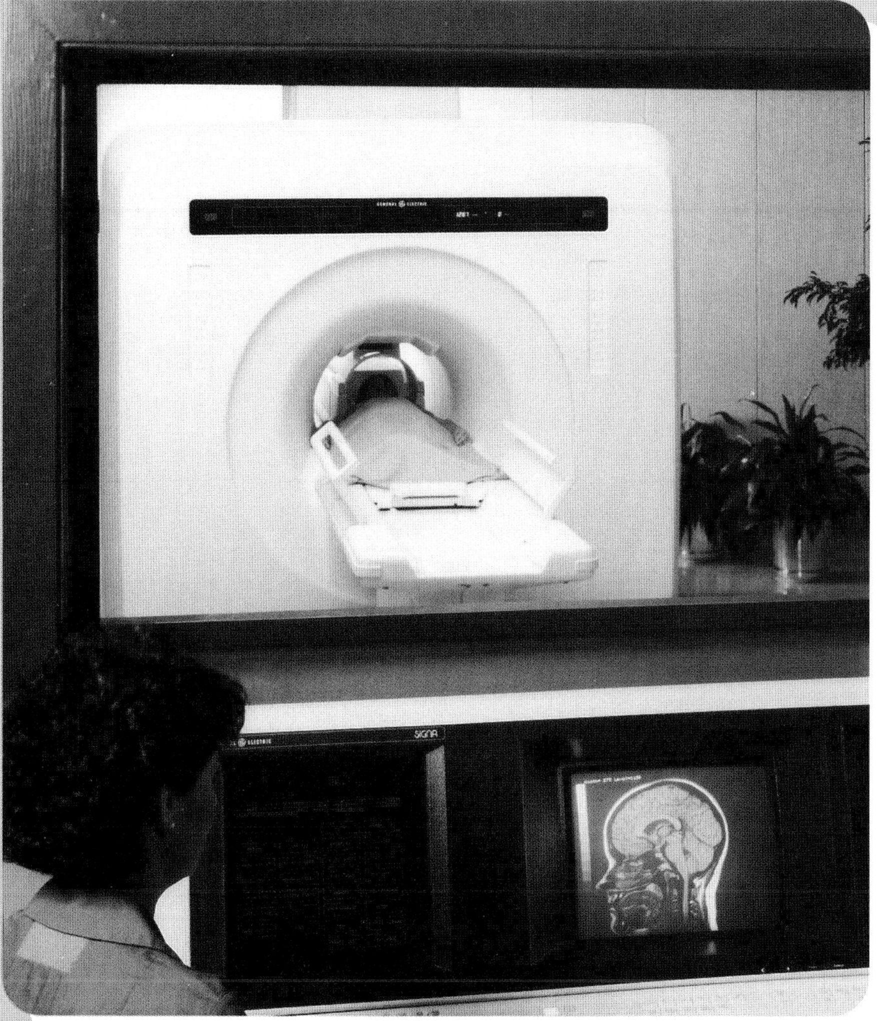

Magnetic Resonance Imaging (MRI) has been heralded as the most significant advance in medical technology since the advent of the x-ray. It enables physicians to take images of body organs from any angle on any plane without moving the patient.

BACKGROUND

The use of trigonometry and algebra in the life sciences is quite extensive. Health care is just one area that has greatly improved in recent years, benefitting from scientific and mathematical advances.

LESSON PLAN

Vocabulary

Cycle
Even function
Odd function
Period
Periodic
Symmetric with respect to the origin
Symmetric with respect to the *x*-axis
Symmetric with respect to the *y*-axis

Materials/Manipulatives

Unruled paper
Clear acetate or plastic

BACKGROUND

In the Preview, determining whether or not a geometric figure is symmetric with respect to a given line or point is discussed.

2.1 Periodic Functions and Symmetry

Objectives: To define periodic functions and odd and even functions
To determine the symmetry of a graph

Many functions are *periodic*; that is, their graphs repeat in cycles. Also, many functions have graphs that are symmetric with respect to one of the axes or to the origin.

Preview

If a figure can be "folded" along a line so that the two parts coincide, the figure is said to be *symmetric* with respect to that line. For example, square *WXYZ* is symmetric with respect to either diagonal *WY* or diagonal *XZ*.

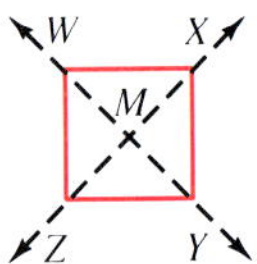

If a figure appears to be in exactly the same position after a 180° rotation about a point, it is said to be *symmetric* with respect to that point. For example, square *WXYZ* is symmetric with respect to point *M*.

***E*, *F*, *G*, and *H* are the midpoints of the sides of rectangle *ABCD*, and *AB* > *BC*. Tell whether or not each figure is symmetric with respect to the given line or point.**

1. Line segment *AB*, to line *FH* yes
2. Line segment *AD*, to point *E* yes
3. Rectangle *ABCD*, to line *FH* yes
4. Rectangle *ABCD*, to point *I* yes
5. Rectangle *ABGE*, to line *AG* no
6. Triangle *AGD*, to line *EG* yes
7. Triangle *AGD*, to point *I* no

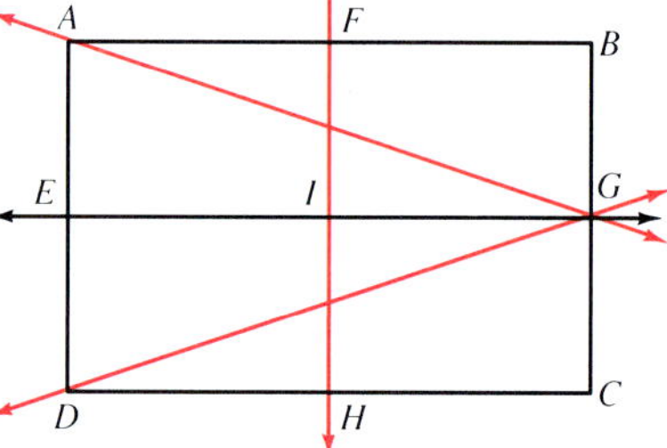

A periodic function is a function that repeats itself. That is, a function $f(x)$ is **periodic** if there exists a smallest positive number p such that

$$f(x + p) = f(x)$$

whenever both $f(x + p)$ and $f(x)$ are defined. The number p is said to be the **period** of the function, and one **cycle** of the graph is completed in each period.

The graphs of two periodic functions are shown below. The arrowheads indicate that the graphs continue the same way, without bound, to the left and to the right. The period of $f(x)$ is 2 and the period of $g(x)$ is 4.

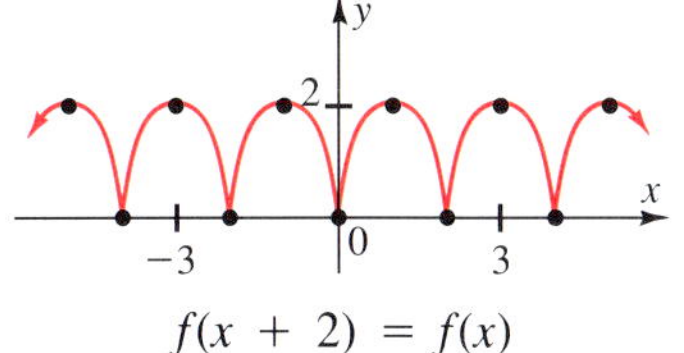

$f(x + 2) = f(x)$

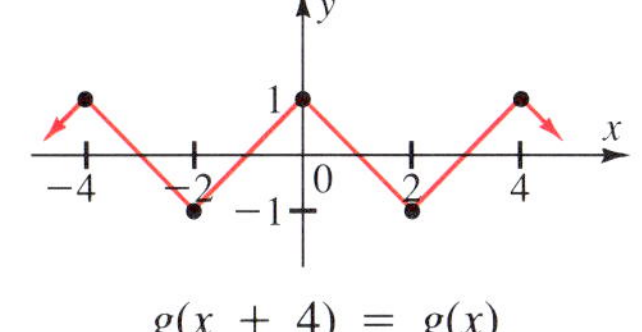

$g(x + 4) = g(x)$

Recall that a graph represents a function if no vertical line can be drawn that intersects the graph in more than one point. Given the graph of a relation, you can determine if the graph is a function, if it is periodic, and what the period is.

EXAMPLE 1 **Examine the graph at the right.**

a. Does it represent a function?
b. Is it periodic?
c. If it is periodic, what is the period?

y, x, −5, 0, 5, 10, 15, −5

a. The vertical line test shows that the graph represents a function.

b. The graph repeats in cycles, so the function is periodic.

c. The graph repeats every 5 units; that is, $f(x + 5) = f(x)$. The period is 5.

If you know the pattern of a periodic function, as well as its period, you can extend the graph.

EXAMPLE 2 **The figure at the left below shows one cycle of the graph of a function with period 3. Extend the graph over the interval $-3 < x \leq 9$.**

To extend the graph, repeat the pattern every 3 units.

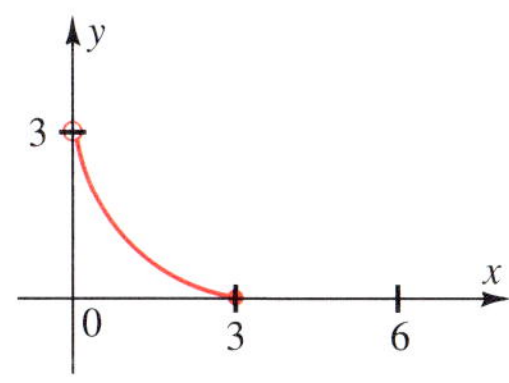

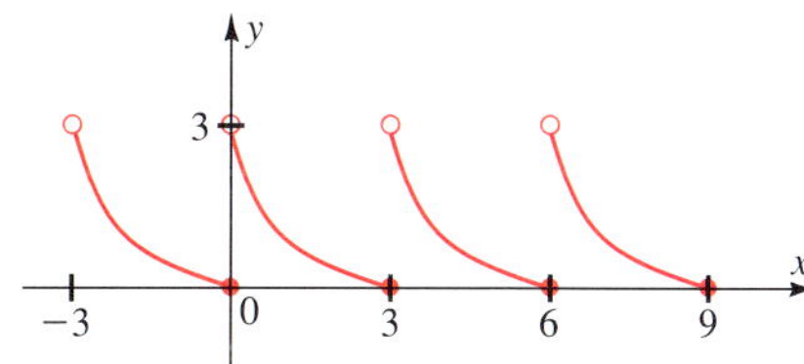

If $f(-x) = f(x)$ for every value in its domain, then the function is said to be an **even function.** If $f(-x) = -f(x)$ for every value in its domain, then it is said to be an **odd function.**

TEACHING SUGGESTIONS

- Odd, even, and periodic functions may be new concepts for most of the students. More time should be spent on these than on the other ideas in this section.
- To emphasize the notion of a periodic function, copy or have students copy one section of a periodic function onto a sheet of acetate or other clear plastic and slide it along the x-axis.
- When you discuss the concepts of even and odd functions, you may wish to point out that a polynomial with only even powers of the variables is an even function, while a polynomial with only odd powers of the variable is an odd function. Also point out that some even functions and some odd functions are not polynomials.
- Have students copy symmetric graphs onto unruled paper, and use paper folding to stress the concept of symmetry.

CHALKBOARD EXAMPLES

- **For Example 1**

Examine each graph. (a) Does it represent a function? (b) Is the graph periodic? (c) If it is periodic, what is the period?

1.

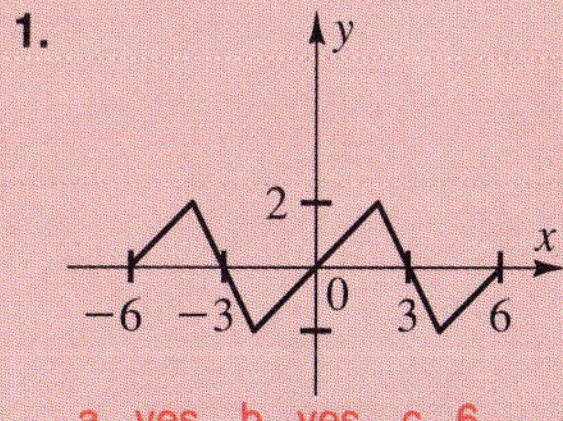

a. yes b. yes c. 6

2.

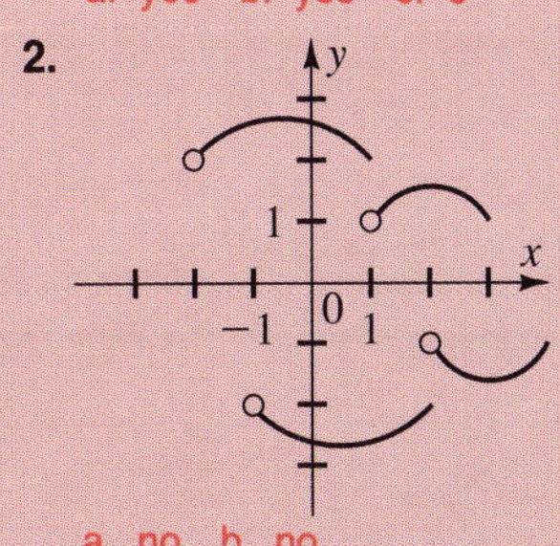

a. no b. no

- **For Example 2**

  Each figure below shows one cycle of the graph of a function. Extend the graph over the internal $-6 \leq x < 6$.

3.

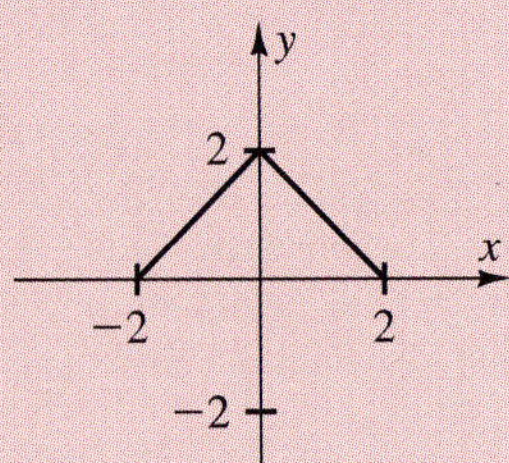

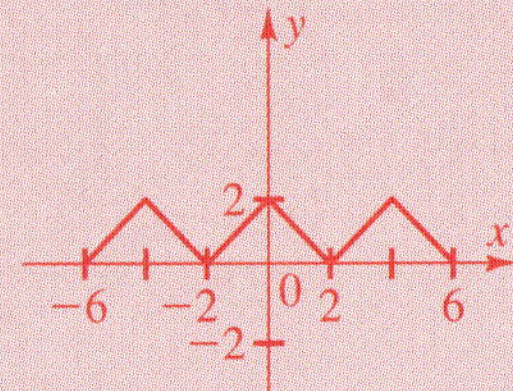

4.

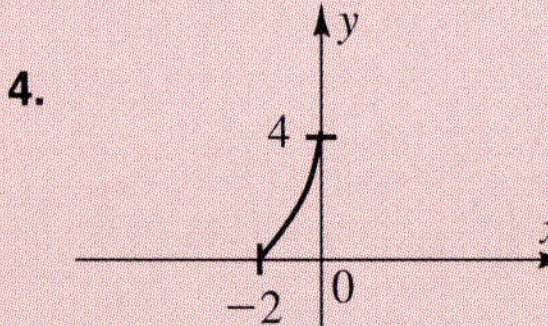

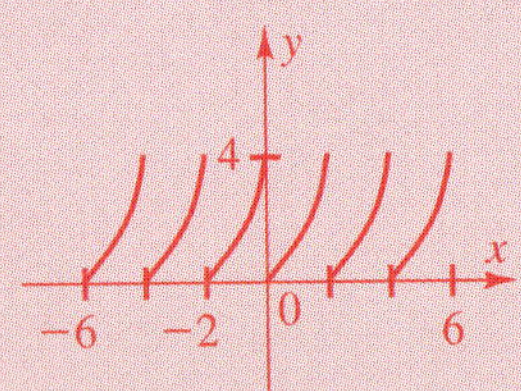

- **For Example 3**

  Determine whether each function is odd, even, or neither.

  5. f(x) = 4x + 6 neither
  6. f(x) = x odd
  7. f(x) = x + x neither
  8. f(x) = x + 1 even

EXAMPLE 3 **Determine whether each function is even, odd, or neither.**

a. $f(x) = x^2 + 2$ **b.** $f(x) = x^3$ **c.** $f(x) = x^2 + x$

a. For any value of x,

$$f(x) = x^2 + 2 \quad \text{and} \quad f(-x) = (-x)^2 + 2 = x^2 + 2$$

Since $f(-x) = f(x)$, the function is even.

b. For any value of x,

$$f(x) = x^3 \quad \text{and} \quad f(-x) = (-x)^3 = -x^3$$

Since $f(-x) = -f(x)$, the function is odd.

c. For any value of x,

$$f(x) = x^2 + x \quad \text{and} \quad f(-x) = (-x)^2 + (-x) = x^2 - x$$

Since $f(-x) \neq f(x)$ and $f(-x) \neq -f(x)$, the function is neither even nor odd.

A graph may be symmetric with respect to the x-axis, the y-axis, or the origin. Knowing that the graph of a relation is symmetric with respect to one of the axes or to the origin makes it easier to sketch.

For every point (a, b) on the graph of a relation, if the point $(-a, b)$ is also on the graph, then the graph is **symmetric with respect to the *y*-axis.** An *even* function is symmetric with respect to the y-axis.

For every point (a, b) on the graph of a relation, if the point $(a, -b)$ is also on the graph, then the graph is **symmetric with respect to the *x*-axis.**

For every point (a, b) on the graph of a relation, if the point $(-a, -b)$ is also on the graph, then the graph is **symmetric with respect to the origin.** An *odd* function is symmetric with respect to the origin.

The figures below provide examples of each type of symmetry.

y-axis symmetry

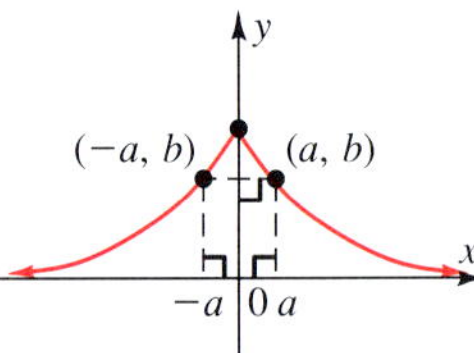

x-axis symmetry

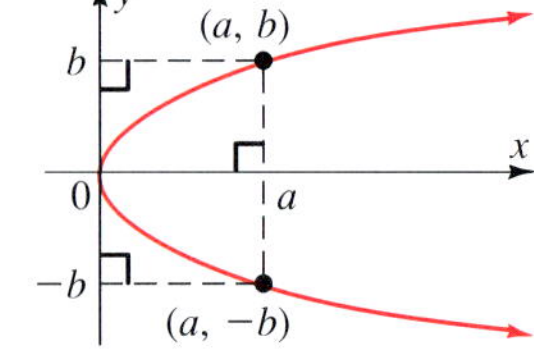

origin symmetry

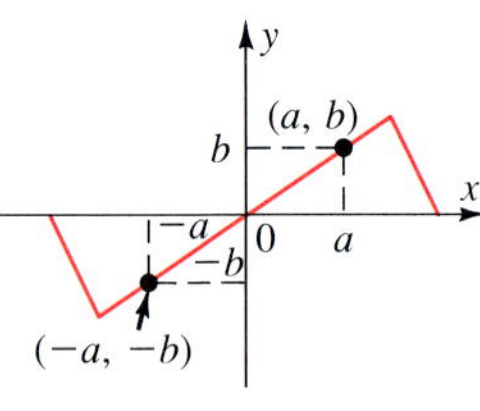

EXAMPLE 4 **Determine whether each graph is symmetric with respect to the x-axis, the y-axis, the origin, or to none of these.**

a.

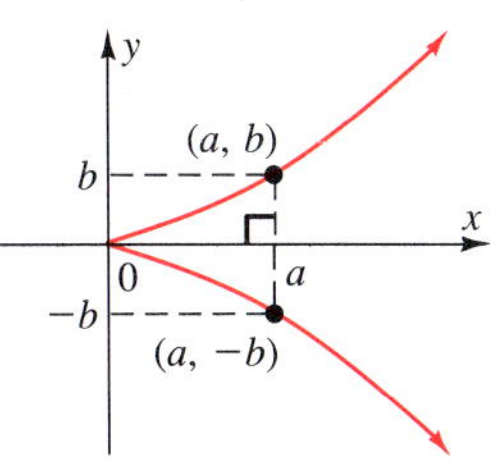

b.

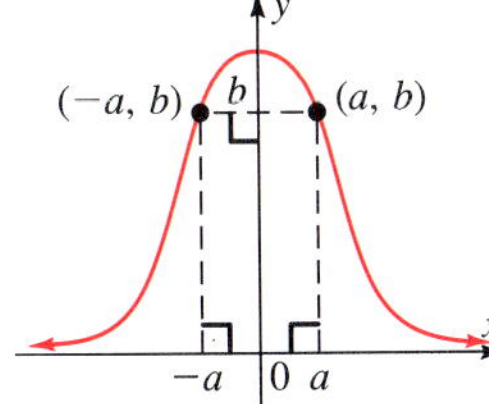

c.

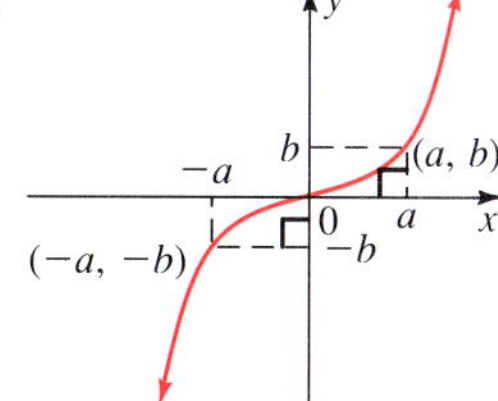

a. For any point (a, b) on the graph, $(a, -b)$ is also on the graph. The graph is symmetric with respect to the x-axis.

b. For any point (a, b) on the graph, $(-a, b)$ is also on the graph. The graph is symmetric with respect to the y-axis.

c. For any point (a, b) on the graph, $(-a, -b)$ is also on the graph. The graph is symmetric with respect to the origin.

CLASS EXERCISES

Which of these graphs represent periodic functions?

1.

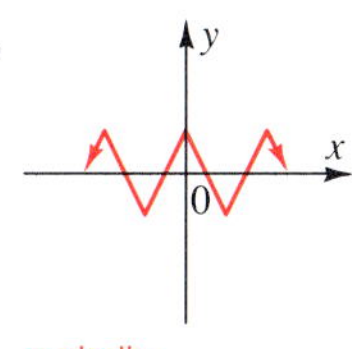

periodic

2.

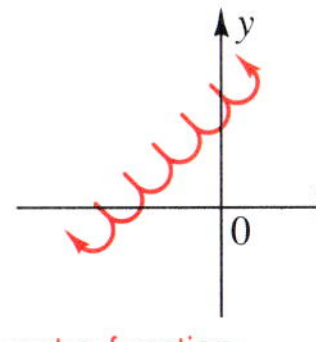

not a function

3.

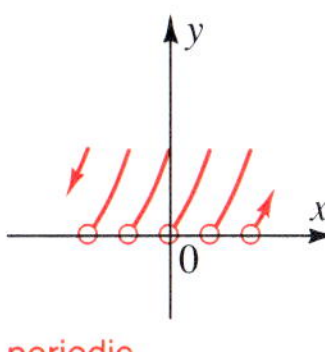

periodic

4.

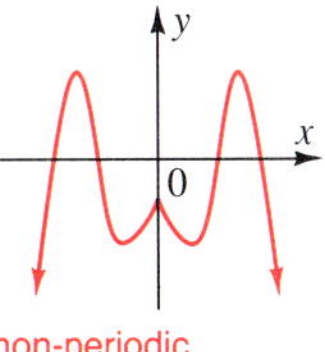

non-periodic

Determine whether each graph is symmetric with respect to the x-axis, the y-axis, the origin, or to none of these.

5.

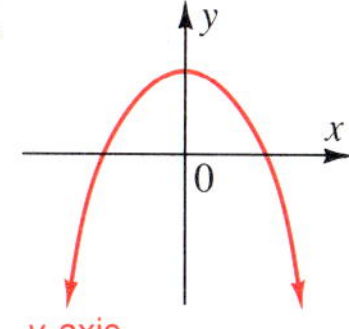

y-axis

6.

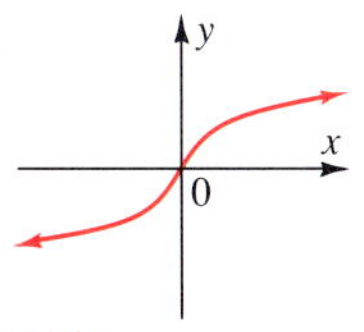

origin

7.

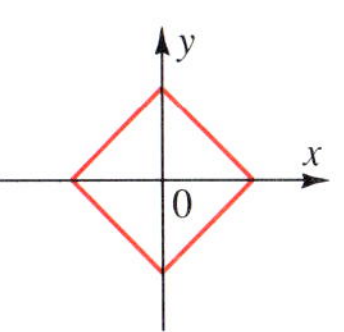

y-axis; x-axis; origin

8.

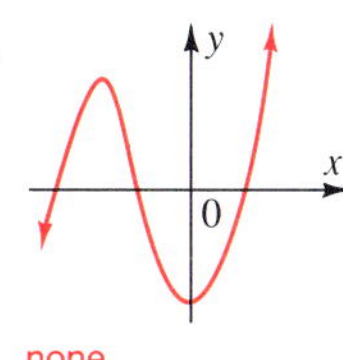

none

Determine whether each function is even, odd, or neither.

9.

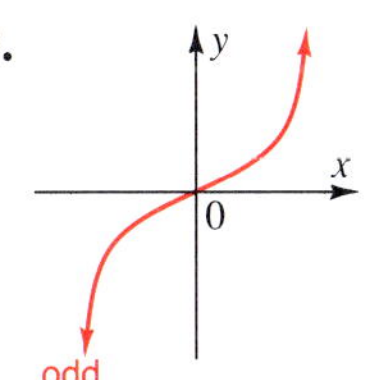

odd

10.

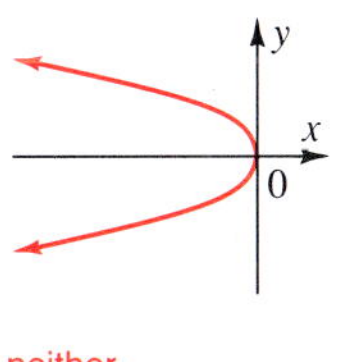

neither

11.

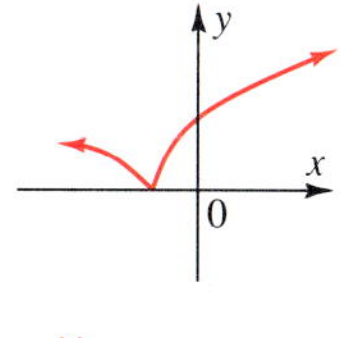

neither

12.

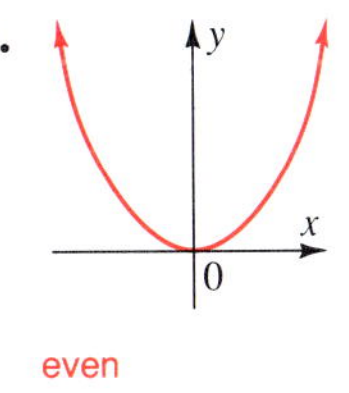

even

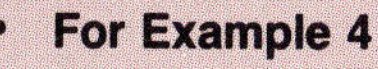

- **For Example 4**

Determine whether each graph is symmetric with respect to the x-axis, the y-axis, the origin, or to none of these.

9.

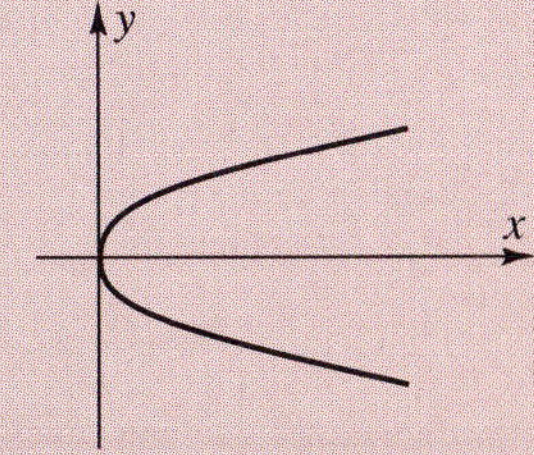

x-axis

10.

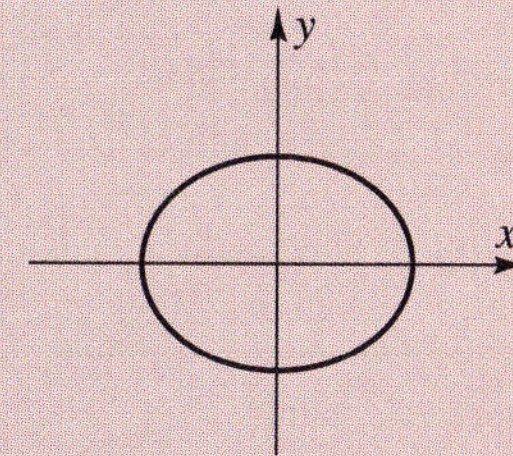

y-axis, x-axis, origin

11.

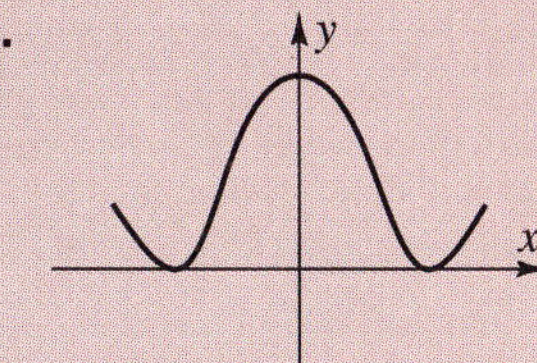

y-axis

12.

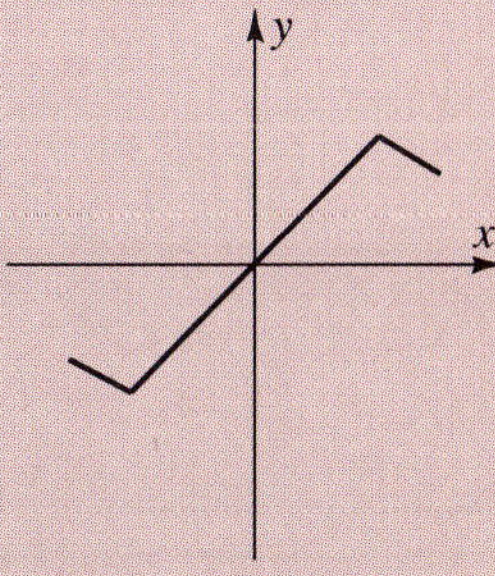

origin

Common Error

- When determining if a function is odd or even, some students fail to realize that they must substitute $(-x)$ for x and simplify each term. Have those students review what is meant by the additive inverse of a function.
- See *Teacher's Resource Book* for additional remediation.

LESSON FOLLOW-UP

Discussion

Which type of function, odd or even, has a graph that is symmetric to the y-axis? Give a reason for your answer. An even function has a graph that is symmetric to the y-axis, because by the definition of an even function, $f(x) = f(-x)$. Both (a, b) and $(-a, b)$ are on the graph of an even function.

Critical Thinking

Application Ask students to sketch the graph of a function that is even, periodic, and symmetric with respect to the y-axis. Then ask them to justify the fact that their sketches meet the criteria specified. Answers may vary.

Assignment Guide

See p. 58B for assignments.

Historical Note

Some of the major contributions to the development of trigonometry are presented in chart form.

Additional Answers

39. $x^2 + y^2 = 36$

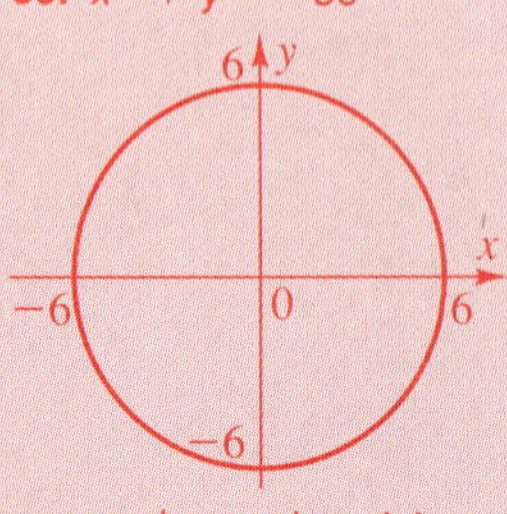

x-axis, y-axis, origin

PRACTICE EXERCISES

For each of the graphs in Exercises 1–6, answer the following questions:

a. Does the graph represent a function?

b. Is the graph periodic?

c. If the graph is periodic, what is the period?

A

1.

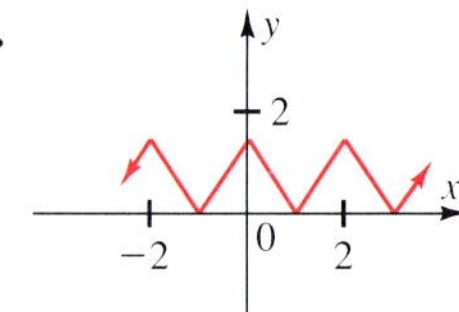

yes; yes; 2

2.

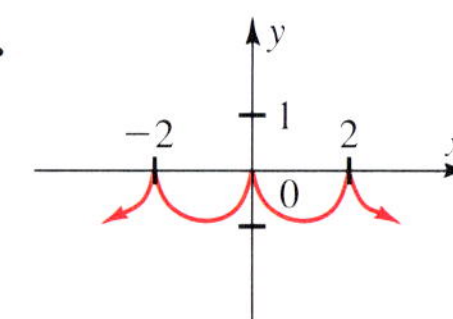

yes; yes; 2

3.

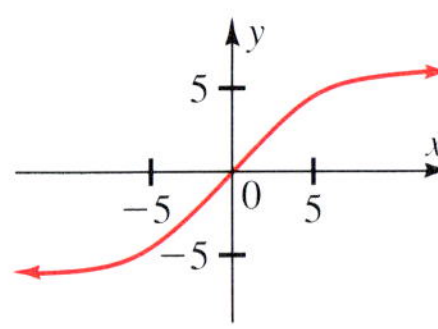

yes; no

4.

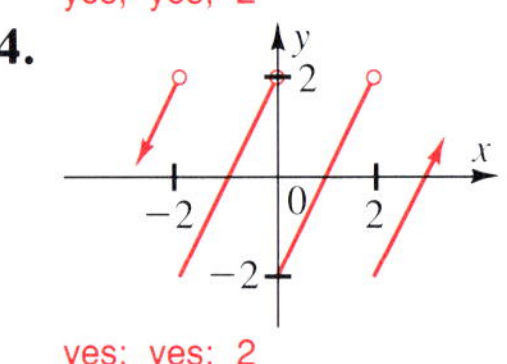

yes; yes; 2

5.

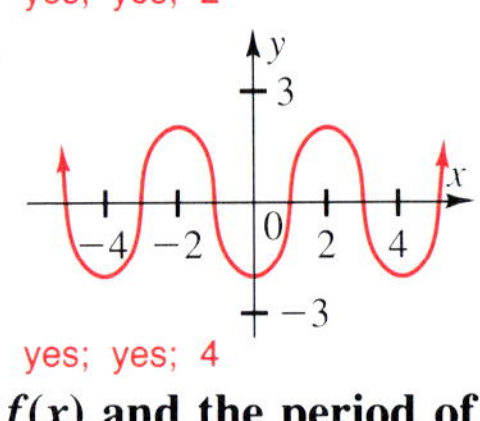

yes; yes; 4

6.

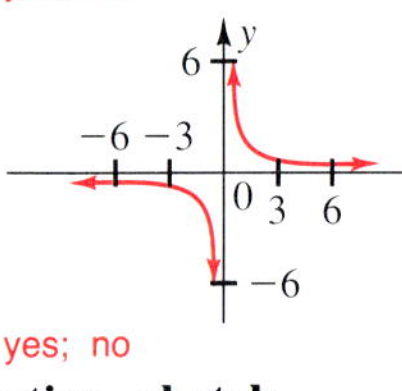

yes; no

Given one cycle of the graph of $f(x)$ and the period of the function, sketch the graph of the function over the specified domain. See page 473.

7. period 2; $0 \leq x \leq 8$

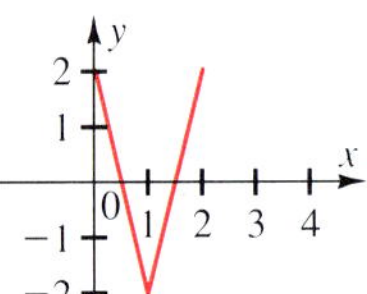

8. period 3; $0 \leq x \leq 6$

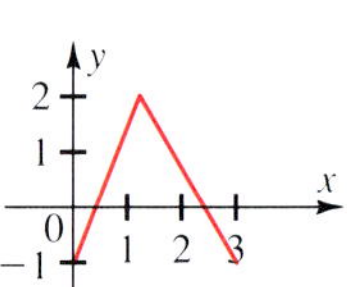

9. period 1; $-2 \leq x \leq 3$

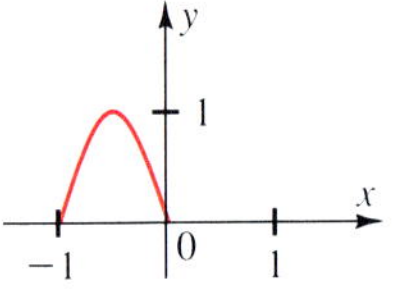

10. period 2; $-1 \leq x \leq 5$

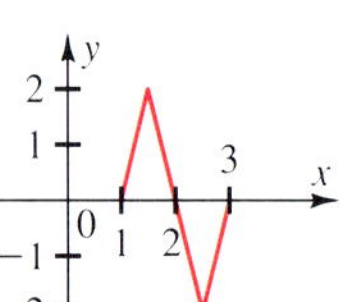

11. period 2; $-4 \leq x \leq 6$

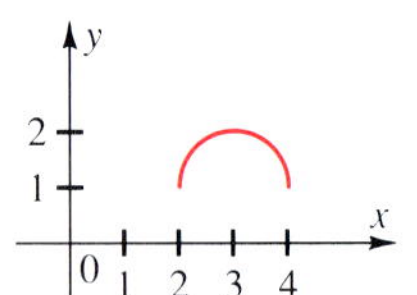

12. period 4; $4 \leq x \leq 8$

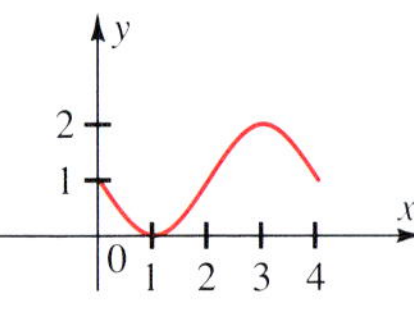

Determine whether each function is even, odd, or neither.

13. $f(x) = 2x^2$ even

14. $f(x) = 5x^3$ odd

15. $f(x) = x$ odd

16. $f(x) = 3x^2 - 7$ even

17. $f(x) = 3x^2 + 5$ even

18. $f(x) = 2x + 1$ neither

B

19. $f(x) = x^3 + x$ odd

20. $f(x) = x^5$ odd

21. $f(x) = -x^2$ even

22. $f(x) = -2x^2 + 1$ even

23. $f(x) = -x^3$ odd

24. $f(x) = -3x^3 - x$ odd

Determine whether each graph is symmetric to the x-axis, the y-axis, the origin, or to none of these.

25.

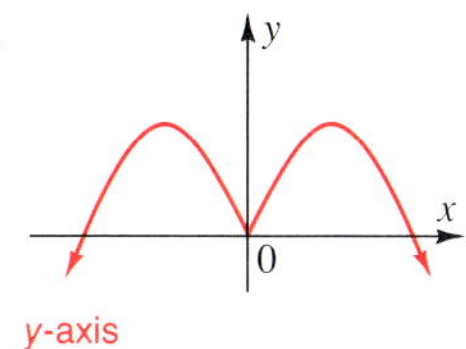

y-axis

26.

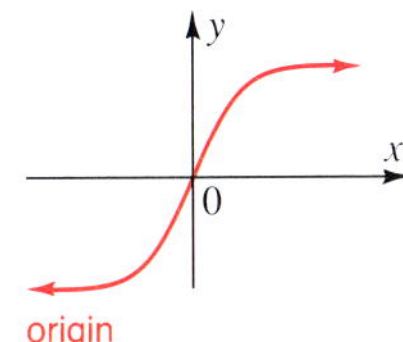

origin

27.

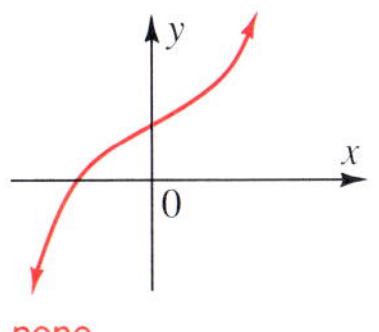

none

28.

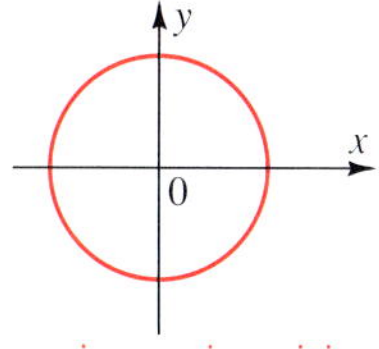

x-axis; y-axis; origin

29.

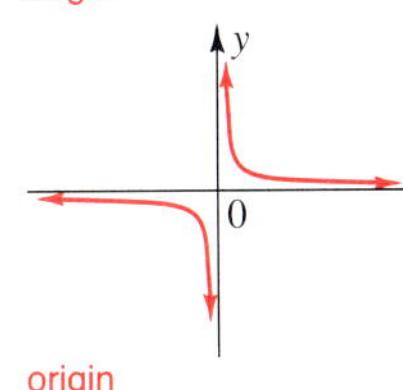

origin

30.

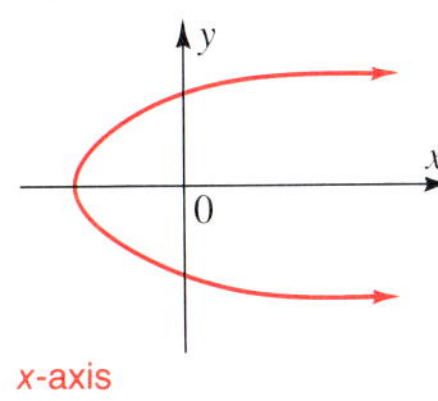

x-axis

Make three copies of each figure below. Extend the first graph so that it is symmetric to the x-axis. Extend the second graph so that it is symmetric to the y-axis. Extend the third graph so that it is symmetric to the origin.
See page 473.

C **31.**

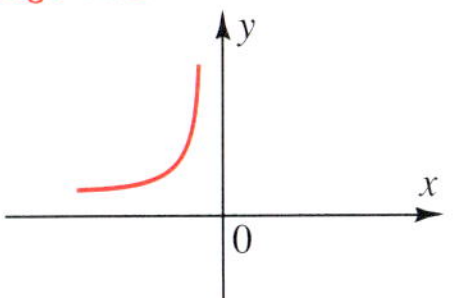

32.

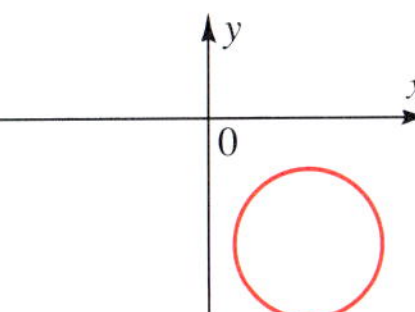

33.

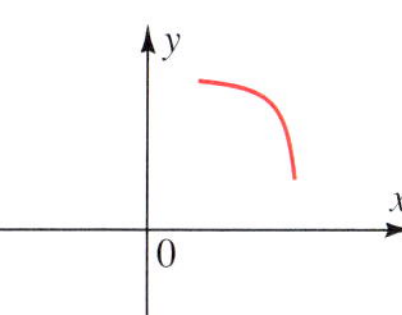

34.

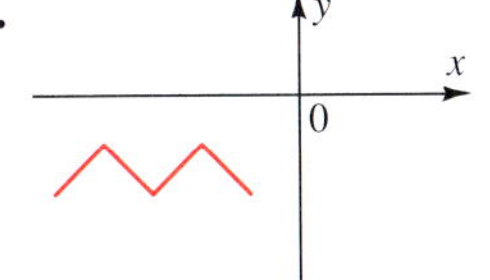

35.

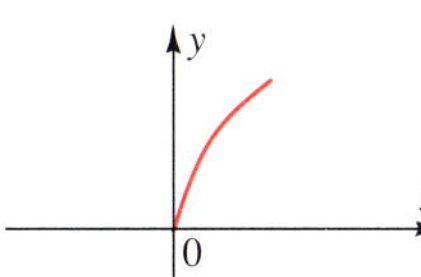

36.

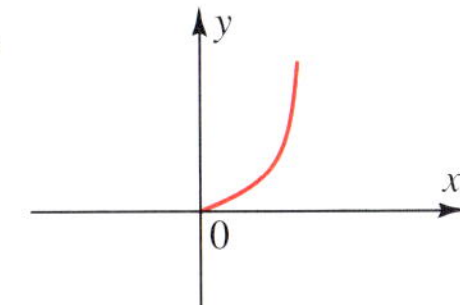

37. Which of the graphs constructed in Exercises 31–36 represent functions?
31–2nd, 3rd; 33–2nd, 3rd; 34–2nd, 3rd; 35–2nd, 3rd; 36–2nd, 3rd

Applications

38. Landscaping If a rectangular garden is plotted on a coordinate system so that its boundary is symmetric with respect to both the x-axis and the y-axis, and the area of the portion in the third quadrant is 187 square units, what is the area of the garden? 748 square units

39. Algebra Graph $x^2 + y^2 = 36$. Describe the symmetry of the graph.
See side column, page 64.

Lesson Quiz

For each graph, answer the following questions.

a. Does the graph represent a function? If so, is it even or odd?
b. Is the graph periodic? If so, what is the period?
c. Is the graph symmetric? If so, with respect to what?

1.

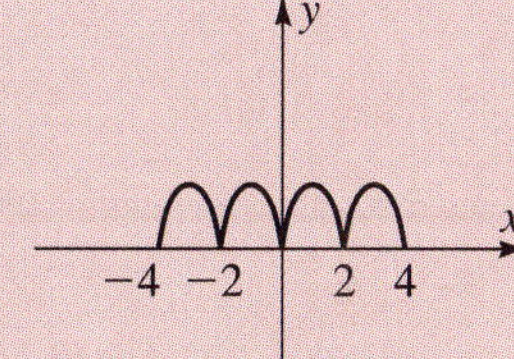

a. yes; even b. yes; 2 c. yes; y-axis

2.

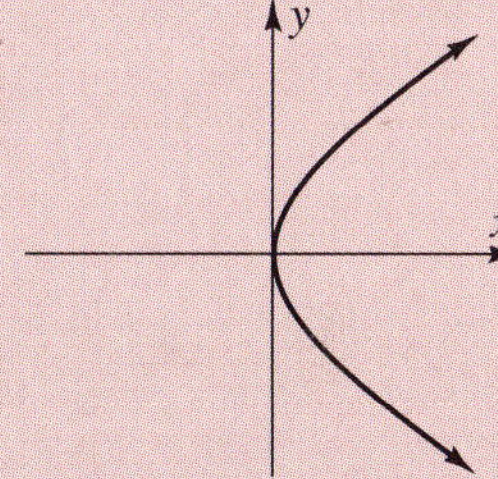

a. no b. no c. yes; x-axis

3.

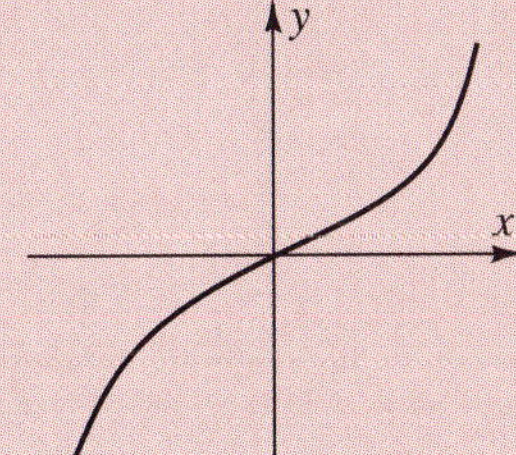

a. yes; odd b. no c. yes; origin

4.

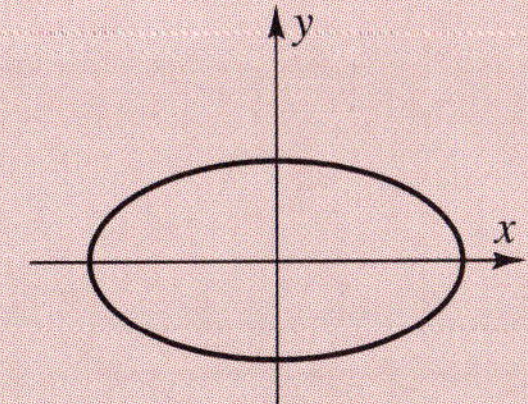

a. no b. no c. yes; x-axis, y-axis, origin

Teacher's Resource Book

Practice—Chapter 2, p. 1
Enrichment—Chapter 2, p. 2

Additional Answers

40.

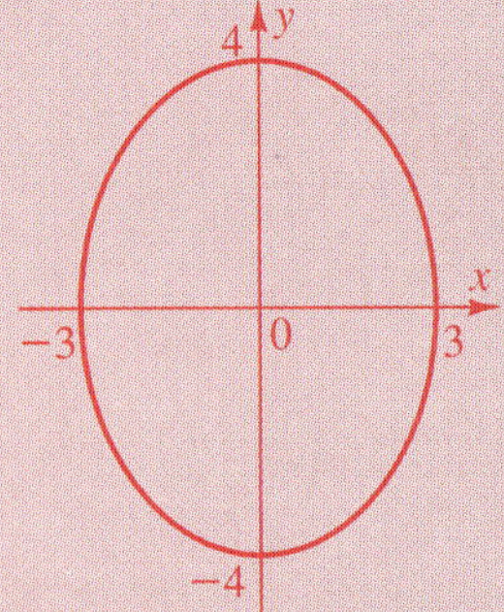

symmetric about the x-axis and the y-axis

41.

40. Algebra Graph $\frac{x^2}{9} + \frac{y^2}{16} = 1$. Describe the symmetry of the graph. See side column.

41. Architecture The front of a building is to be symmetric with respect to a vertical line through the center of its base. The left half of a diagram of the front of the building is shown at the right. Copy the figure and complete the drawing of the front of the building. See side column.

HISTORICAL NOTE

Trigonometric concepts were developed in ancient times, often for use in the study of astronomy. However, it was not until the 17th century, when a satisfactory system of algebraic symbols had been devised, that trigonometry was thought of as a unique branch of mathematics. Listed below are some of the major contributions to the development of trigonometry.

Century	Mathematicians	Contributions
2nd, BC	Hipparchus	Related certain functions of angle measures to astronomy
	Claudius Ptolemy	Summarized in the *Almagest* the theorems known to Hipparchus Computed entries for a table of chords of circles that was roughly equivalent to a sine table
15th	Regiomontanus (Johann Müller)	Consolidated known material dealing with trigonometry Did much to establish trigonometry as a science independent of astronomy
16th	François Viète	Computed entries for more extensive trigonometric tables Obtained many of the trigonometric identities algebraically Used all six trigonometric functions to develop methods of solving plane and spherical triangles
18th	Joseph Fourier	Developed Fourier series, used to approximate values of any function

Find information on Ptolemy's *table of chords*. Explain why such a table is similar to a table of sines. Answers may vary.

2.2 Graphs of the Sine and Cosine Functions

Objectives: To develop the properties of the sine and cosine functions
To graph the sine and cosine functions

The sine and cosine functions are defined for all real numbers, and these functions have many real-world applications.

Preview

An oscilloscope is an electronic instrument used to display changing electrical signals. It displays electrical or sound waves as patterns on a fluorescent screen. Many of the graphs shown on an oscilloscope are sine or cosine waves.

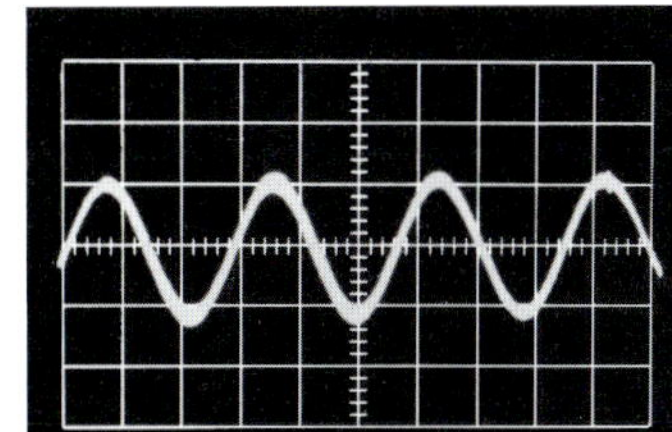

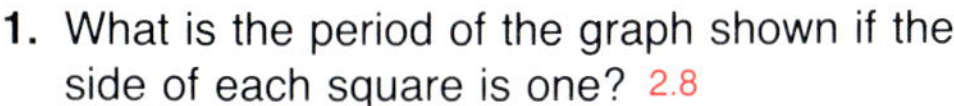

1. What is the period of the graph shown if the side of each square is one? 2.8

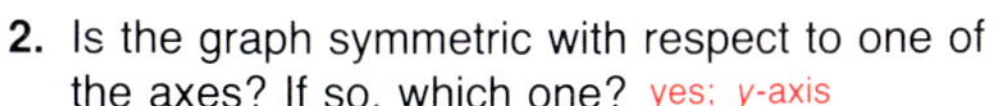

2. Is the graph symmetric with respect to one of the axes? If so, which one? yes; *y*-axis

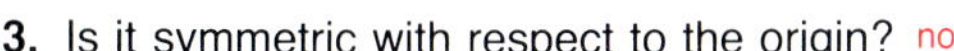

3. Is it symmetric with respect to the origin? no
4. What is the maximum value of the function? the minimum value? 1.2; −1.2
5. How do you think an oscilloscope can be used by engineers or physicians? Answers may vary.

Two acute angles are **complementary** if the sum of their measures is 90°. For example, 40° and 50° are complementary angles. The sine, tangent, and secant functions are associated with the **cofunctions** cosine, cotangent, and cosecant, respectively. The value of any defined trigonometric function of an acute angle is equal to the value of the cofunction of its complement.

Cofunctions

sine and cosine	tangent and cotangent	secant and cosecant
$\sin x = \cos(90° - x)$ $\cos x = \sin(90° - x)$	$\tan x = \cot(90° - x)$ $\cot x = \tan(90° - x)$	$\sec x = \csc(90° - x)$ $\csc x = \sec(90° - x)$

Thus, $\sin 30° = \cos(90° - 30°) = \cos 60° = \frac{1}{2}$

$\cos 30° = \sin(90° - 30°) = \sin 60° = \frac{\sqrt{3}}{2}$

2 60° 1 30° 0 $\sqrt{3}$

LESSON PLAN

Vocabulary
Cofunctions
Complementary angles

Materials/Manipulatives
Scientific calculators
Graphing calculators

BACKGROUND

In the Preview, the oscilloscope is introduced and students examine the graph shown on the photograph of the oscilloscope.

TEACHING SUGGESTIONS

- It may be necessary to review the radian measure of the special angles in quadrants other than the first quadrant.
- When sketching the graphs of sin *x* and cos *x*, remind students to locate the zeros and the *y*-intercepts.
- Encourage students to use their scientific calculators to find the values of the sine and cosine for specific angles. Remind them that they must check to be sure the calculator is in the proper mode, degree or radian, depending on the conditions stated in the problem. Emphasize that radian mode is usually used when graphing the functions.

Critical Thinking

Analysis Ask students to determine what effect taking the absolute value of the sine and cosine functions would have on the graphs of $y = \sin x$ and $y = \cos x$. After making a table of value for $y = |\sin x|$ and $y = |\cos x|$ and plotting the points, students should conclude that the curves for $y = |\sin x|$ and $y = |\cos x|$ are always on or above the *x*-axis.

CHALKBOARD EXAMPLES

- **For Example 1**

  Express each function in terms of its cofunction.

  1. cos 58° sin 32°
  2. sec 43° csc 47°
  3. tan 35° cot 55°
  4. csc 78° sec 12°

- **For Example 2**

  5. Use the fact that cosine is an even function to find cos (−263°). $\cos(-263°) = \cos 263° = -0.1219$

$$\tan 30° = \cot(90° - 30°) = \cot 60° = \frac{\sqrt{3}}{3}$$

$$\cot 30° = \tan(90° - 30°) = \tan 60° = \sqrt{3}$$

$$\sec 30° = \csc(90° - 30°) = \csc 60° = \frac{2\sqrt{3}}{3}$$

$$\csc 30° = \sec(90° - 30°) = \sec 60° = 2$$

EXAMPLE 1 **Express each function in terms of its cofunction:**

a. sin 47° **b.** cot 78°

a. $\sin 47° = \cos(90° - 47°) = \cos 43°$

b. $\cot 78° = \tan(90° - 78°) = \tan 12°$

The sine and cosine functions exhibit a number of the properties discussed in the previous lesson. They are periodic. One is odd and the other is even. One is symmetric with respect to the origin, and one is symmetric with respect to the *y*-axis. Knowledge of these properties makes it easier to sketch the graphs of $y = \sin x$ and $y = \cos x$.

Recall that a function is odd if $f(-x) = -f(x)$, and even if $f(-x) = f(x)$. To determine whether the sine function is odd, even, or neither, check the values of sin *x* for $x = \pm\frac{\pi}{6}$ and a few other values of *x*.

$$\sin\frac{\pi}{6} = \frac{1}{2} \qquad \sin\left(-\frac{\pi}{6}\right) = -\frac{1}{2} \qquad \sin\frac{2\pi}{3} = \frac{\sqrt{3}}{2} \qquad \sin\left(-\frac{2\pi}{3}\right) = -\frac{\sqrt{3}}{2}$$

$$\sin\frac{4\pi}{3} = -\frac{\sqrt{3}}{2} \qquad \sin\left(-\frac{4\pi}{3}\right) = \frac{\sqrt{3}}{2} \qquad \sin\frac{11\pi}{6} = -\frac{1}{2} \qquad \sin\left(-\frac{11\pi}{6}\right) = \frac{1}{2}$$

It appears that $\sin(-\theta) = -\sin\theta$. The figure below demonstrates that $\sin\theta = \frac{y}{r}$ and $\sin(-\theta) = -\frac{y}{r} = -\sin\theta$. Thus, $\sin(-\theta) = -\sin\theta$. The sine is an odd function, and its graph is symmetric with respect to the origin.

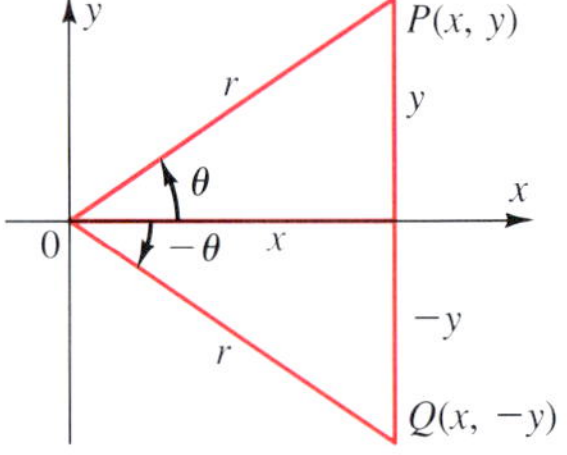

The figure also shows that $\cos\theta = \frac{x}{r}$ and $\cos(-\theta) = \frac{x}{r}$. Thus, $\cos(-\theta) = \cos\theta$. The cosine is an even function, and its graph is symmetric with respect to the *y*-axis.

EXAMPLE 2 **Use the fact that sine is an odd function to find sin (−46°) if sin 46° = 0.7193.**

$$\sin(-46°) = -\sin 46° = -0.7193$$

In Chapter 1 it was shown that after one complete clockwise or counterclockwise rotation the values of the sine and cosine repeat, since angles with measures less than 0 or greater than 2π are coterminal with angles with measures between 0 and 2π. Thus, the sine function is periodic, with period 2π. That is, $\sin(\theta + 2\pi) = \sin\theta$. The cosine function is also periodic, with period 2π. That is, $\cos(\theta + 2\pi) = \cos\theta$.

EXAMPLE 3 **Use the fact that the sine is a periodic function to find $\sin \frac{13\pi}{6}$.**

The sine function is periodic with period 2π, so $\sin(\theta + 2\pi) = \sin\theta$.

$$\sin\frac{13\pi}{6} = \sin\left(\frac{\pi}{6} + 2\pi\right) = \sin\frac{\pi}{6} = 0.5$$

Before $y = \sin x$ and $y = \cos x$ are graphed for the first time, it is helpful to construct a table of values for $0 \le x \le 2\pi$. The properties of symmetry, along with the fact that $\sin x$ and $\cos x$ have periods of 2π, can be used to extend the graphs in either direction.

When $y = \sin x$ is graphed on the traditional xy-coordinate plane, each point on the graph has coordinates of the form $(x, \sin x)$, where x is given in radians. In other words, x is used in place of θ.

EXAMPLE 4 **Sketch the graph of $y = \sin x$ over the interval $-2\pi \le x \le 2\pi$.**

To graph $y = \sin x$ over the interval $0 \le x \le 2\pi$, use radian measure and make a table of values. Round the values of $\sin x$ to the nearest hundredth.

x	0	$\frac{\pi}{6}$	$\frac{\pi}{4}$	$\frac{\pi}{3}$	$\frac{\pi}{2}$	$\frac{2\pi}{3}$	$\frac{3\pi}{4}$	$\frac{5\pi}{6}$
$\sin x$	0	0.5	0.71	0.87	1	0.87	0.71	0.5

x	π	$\frac{7\pi}{6}$	$\frac{5\pi}{4}$	$\frac{4\pi}{3}$	$\frac{3\pi}{2}$	$\frac{5\pi}{3}$	$\frac{7\pi}{4}$	$\frac{11\pi}{6}$	2π
$\sin x$	0	−0.5	−0.71	−0.87	−1	−0.87	−0.71	−0.5	0

Plot the points and sketch the curve over the interval $0 \le x \le 2\pi$. To sketch the curve over the interval $-2\pi \le x \le 0$, use the facts that $y = \sin x$ has a period of 2π and that the graph is symmetric with respect to the origin.

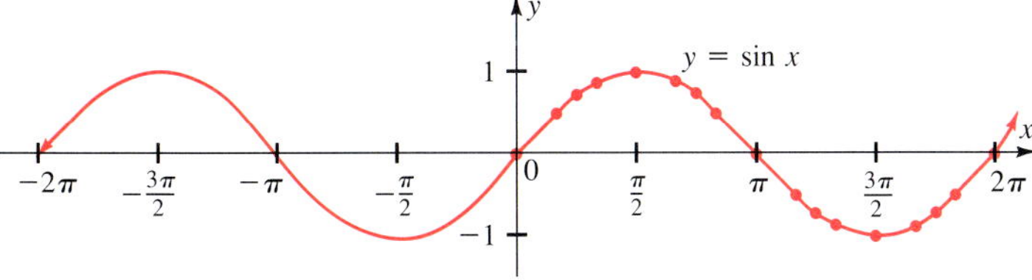

- **For Example 3**

6. Use the fact that cosine is a periodic function to find $\cos\frac{11\pi}{4}$.

$\cos\frac{11\pi}{4} = \cos\left(\frac{3\pi}{4} + 2\pi\right) = -0.7071$

- **For Example 4**

7. Sketch the graph of $y = \sin x$ over the interval $-\pi \le x \le \pi$.

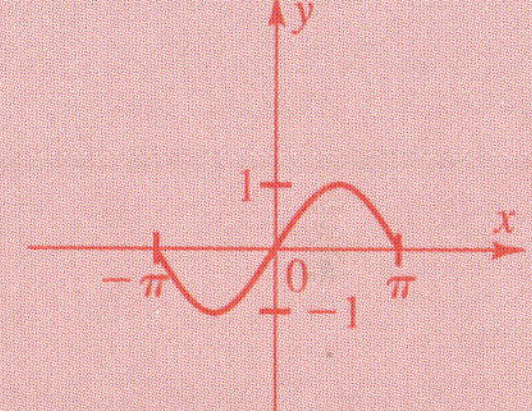

- **For Example 5**

8. Sketch the graph of $y = \cos x$ over the interval $-\pi \le x \le \pi$.

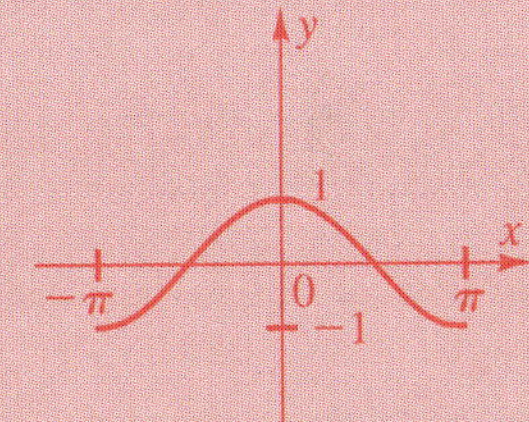

Common Error

- When using a calculator to find the values of the trigonometric functions for given angles, some students fail to read the negative sign when it occurs in the display. Have those students get in the habit of thinking of the sign of the functions in the various quadrants to ensure that they have the correct sign.
- See *Teacher's Resource Book* for additional remediation.

LESSON FOLLOW-UP

Discussion

Compare and contrast the graphs of the sine and cosine functions. Comparisons: Both graphs have the same maximum and minimum points. Both graphs are smooth continuous curves. Both graphs have a period of 2π. Contrasts: The maximum value of the sine occurs when $x = \frac{\pi}{2}$, while the maximum value of the cosine occurs when $x = 0$. The minimum value of the sine occurs when $x = \frac{3\pi}{2}$, while the minimum value of the cosine occurs when $x = \pi$.

Critical Thinking

Discovering patterns Ask students to sketch the graph of $y = \sin x$ on a sheet of clear acetate and then to explain how this graph could be used to produce the graph of $y = \cos x$. After sketching the graph on the acetate and experimenting, students should discover that the graph of $y = \cos x$ can be produced by sliding the graph of $y = \sin x$ to the left $\frac{\pi}{2}$ units.

Assignment Guide

See p. 58B for assignments.

The graph of $y = \cos x$ can be constructed in a similar manner.

EXAMPLE 5 **Sketch the graph of $y = \cos x$ over the interval $-2\pi \leq x \leq 2\pi$.**

Begin by constructing a table of values for the interval $0 \leq x \leq 2\pi$.

x	0	$\frac{\pi}{6}$	$\frac{\pi}{4}$	$\frac{\pi}{3}$	$\frac{\pi}{2}$	$\frac{2\pi}{3}$	$\frac{3\pi}{4}$	$\frac{5\pi}{6}$
$\cos x$	1	0.87	0.71	0.5	0	−0.5	−0.71	−0.87

x	π	$\frac{7\pi}{6}$	$\frac{5\pi}{4}$	$\frac{4\pi}{3}$	$\frac{3\pi}{2}$	$\frac{5\pi}{3}$	$\frac{7\pi}{4}$	$\frac{11\pi}{6}$	2π
$\cos x$	−1	−0.87	−0.71	−0.5	0	0.5	0.71	0.87	1

Plot the points and sketch the curve over the interval $0 \leq x \leq 2\pi$. To extend the curve over the interval $-2\pi \leq x \leq 0$, use the facts that $y = \cos x$ has a period of 2π and that the graph is symmetric with respect to the y-axis.

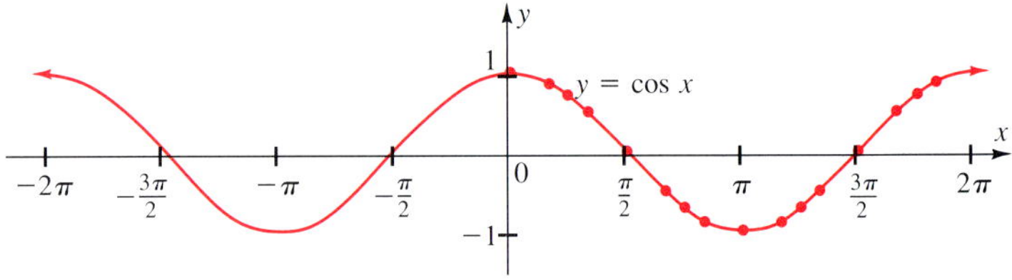

A scientific calculator can be used to find specified values of the functions being graphed. Graphing calculators or computers can be used to sketch trigonometric functions. Their displays are helpful when checking the accuracy of graphs that you have sketched.

CLASS EXERCISES

Use properties of the sine and cosine functions to match each value in the column on the left with one on the right.

1. cos 65° d	**a.** cos 25°
2. cos (−25°) a	**b.** sin (−55°)
3. cos 75° e	**c.** sin (−75°)
4. sin 35° g	**d.** sin 25°
5. cos (−35°) f	**e.** sin 15°
6. −cos (−15°) c	**f.** sin 55°
7. sin (−25°) h	**g.** cos 55°
8. −sin 55° b	**h.** −cos 65°

PRACTICE EXERCISES

Express each function in terms of its cofunction.

A 1. sin 28° cos 62° 2. tan 32° cot 58° 3. csc 85° sec 5° 4. cos 14° sin 76°
5. sec 27° csc 63° 6. sec 10° csc 80° 7. cot 53° tan 37° 8. csc 42° sec 48°

Use the fact that the sine is an odd function and the cosine is an even function to determine the following values.

9. If sin 32° = 0.5299, find sin (−32°). −0.5299
10. If sin 129° = 0.7771, find sin (−129°). −0.7771
11. If cos 48° = 0.6691, find cos (−48°). 0.6691
12. If cos 117° = −0.4540, find cos (−117°). −0.4540
13. If sin 238° = −0.8480, find sin (−238°). 0.8480
14. If sin 306° = −0.8090, find sin (−306°). 0.8090

Use the fact that the sine and cosine are periodic functions to determine the following values.

15. If sin 54° = 0.8090, find sin 414°. 0.8090
16. If sin 134° = 0.7193, find sin 494°. 0.7193
17. If cos 62° = 0.4695, find cos 422°. 0.4695
18. If cos 153° = −0.8910, find cos 513°. −0.8910
19. If sin 222° = −0.6691, find sin 582°. −0.6691
20. If sin 312° = −0.7431, find sin 672°. −0.7431

Graph each function over the given interval. See page 474.

B 21. $y = \sin x,\ 2\pi \le x \le 4\pi$ 22. $y = \cos x,\ 2\pi \le x \le 4\pi$
23. $y = \cos x,\ -6\pi \le x \le 0$ 24. $y = \sin x,\ -6\pi \le x \le 0$

Use the fact that the sine and cosine are periodic functions, and your knowledge of cofunctions, to determine the following values.

25. If cos 68° = 0.3746, find sin 22°. 0.3746
26. If cos 54° = 0.5878, find sin 36°. 0.5878
27. If sin 27° = 0.4540, find cos 63°. 0.4540
28. If cos 74° = 0.2756, find sin 16°. 0.2756
29. If sin 141° = 0.6293, find sin 861°. 0.6293

Career: Air Traffic Control Specialist

Students may be interested in researching the trigonometric concepts used by air traffic controllers.

Lesson Quiz

Use the properties of the sine and the cosine functions to match each value in Column A with its equivalent value in Column B.

Column A		*Column B*
1. $\sin\frac{\pi}{3}$	C	A. $\cos\frac{\pi}{4}$
2. sin 35°	D	B. cos 16°
3. $\sin\frac{4\pi}{3}$	F	C. $\cos\frac{\pi}{6}$
4. $\sin\frac{5\pi}{6}$	E	D. cos 55°
5. $\sin\frac{\pi}{4}$	A	E. $\sin\frac{\pi}{6}$
6. sin 74°	B	F. $\cos\frac{5\pi}{6}$

Use the fact that the sine and the cosine are periodic functions and your knowledge of cofunctions to determine the following values.

7. If cos 56° = 0.5592, find sin 394° and cos (−124°). 0.5592; −0.5592
8. If sin (−48°) = −0.7431, find cos 138° and sin 132°. −0.7431; 0.7431

Enrichment

If each angle value were increased by π units, what would the graph of the new function $y = \sin(x + \pi)$ look like over the interval $0 \le x \le 2\pi$?

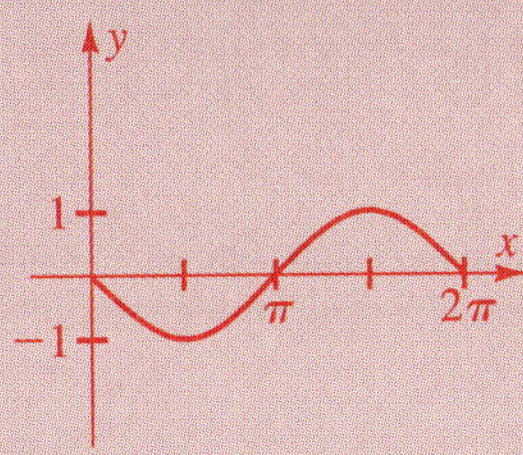

Teacher's Resource Book
Practice—Chapter 2, p. 3
Enrichment—Chapter 2, p. 4

30. If $\cos 123° = -0.5446$, find $\cos 843°$. -0.5446

31. If $\sin 227° = -0.7314$, find $\sin 947°$. -0.7314

32. If $\sin 34° = 0.5592$, find $\sin(-326°)$. 0.5592

33. If $\sin 139° = 0.6561$, find $\sin(-221°)$. 0.6561

34. If $\cos 162° = -0.9511$, find $\cos(-198°)$. -0.9511

Graph each function over the interval $-2\pi \le x \le 2\pi$. See page 474.

C **35.** $y = \sin x + 2$ **36.** $y = \cos x - 3$ **37.** $y = 2 \cos x$

38. $y = 3 \sin x$ **39.** $y = \sin 2x$ **40.** $y = \cos 2x$

Applications

See page 474.

41. Algebra Sketch the graphs of $y = x$ and $y = \sin x$ on the same coordinate plane. Find approximate solutions to $x = \sin x$, where x is in radians.
For $0 \le x \le 0.6$, $x \approx \sin x$ (to the nearest tenth).

42. Algebra Sketch the graphs of $y = x$ and $y = \cos x$ on the same coordinate plane. Find approximate solutions to $x = \cos x$, where x is in radians. $x \approx 0.7$

CAREER

Air traffic control specialists are guardians of the airways. They keep track of planes in the airspace surrounding an airport, monitor them to ensure a safe and orderly flow of traffic, and provide pilots with take off and landing directions as well as weather information.

Individuals must have the ability to analyze what is shown on the radar screen, make predictions, and be prepared to make split-second decisions. Thus, they must exercise common sense, patience, and emotional control, and be able to enunciate clearly. In addition, they must pass a physical exam and possess 20/20 vision or vision that can be corrected to 20/20.

Educational requirements include a degree in air traffic control, at least 3 years of college, 2 years of college combined with work experience, or military experience. They must also train for $2\frac{1}{2}$ months in Oklahoma City.

2.3

Amplitude and Period

Objectives: To find the amplitude and period of a trigonometric function from its equation
To graph sine and cosine functions with various amplitudes and periods

Amplitude and frequency are important concepts that are essential to the understanding of many fields, including music.

Preview

The loudness, or *amplitude*, of a musical sound is a comparative measure of the strength of the sound that is heard. Loudness is related to the distance from the source and the energy of the vibration.

Pitch is determined by the number of vibrations per second, or the *frequency* of the vibrations that yield a sound wave. Frequency is the reciprocal of the period. For example, if the period of a sound wave is 0.01 second, then the frequency is $\frac{1}{0.01}$, or 100 vibrations per second.

1. The period of a sound wave is 0.002 seconds. What is the frequency? 500 vibrations per second
2. The lowest frequency of the human voice is about 60 vibrations per second. What is the period of a sound wave with this frequency? 0.0167 second
3. The highest frequency of the human voice is about 1300 vibrations per second. What is the period of a sound wave with this frequency? 0.00077 second
4. The periods of sound waves produced by a piano range from about 0.0002416 to 0.03676 seconds. What is the range of the frequencies? 27 to 4139 vibrations per second

Several observations can be made about the basic sine and cosine functions, $y = \sin x$ and $y = \cos x$.

- Their graphs are *continuous*; that is, they have no breaks.
- They each have a period of 2π; that is, $\sin(x + 2\pi) = \sin x$ and $\cos(x + 2\pi) = \cos x$ for all real values of x.
- Their domains are the set of all real numbers.
- For all values of x, $-1 \leq \sin x \leq 1$ and $-1 \leq \cos x \leq 1$. Thus, the range of both functions is the set of all real numbers from -1 through 1.

The maximum value of each function is 1, and the minimum value is -1. The functions are said to be *bounded* by these values. The **amplitude** of a periodic function is defined to be half the difference of the maximum value M and the

LESSON PLAN

Vocabulary
Amplitude

Materials/Manipulatives
Scientific calculators
Graphs of the sine and the cosine functions on clear acetate
Computer
Graphing calculators

BACKGROUND

In the Preview, the amplitude and period of sound waves are discussed. Various physical phenomena, such as sound waves, are applications of the graphs of the sine and the cosine functions.

TEACHING SUGGESTIONS

- Use the graph of $y = \cos x$ to obtain the graph of $y = 3 \cos x$. Relate the effect of the constant in $y = \cos x$ and $y = 3 \cos x$ to the effect of the constant in $y = x$ and $y = 3x$.
- In Example 2 on text page 75, help students to see that the graph of $y = \sin 2x$ is a contraction of the graph of $y = \sin x$.
- Encourage students to use their scientific calculators to find the values of the functions and to do any calculations. Students may use a graphing calculator or a computer to see the effects of the constants on the graphs.

Critical Thinking

Analysis Ask students to explain if the four graphs $y = \sin x$, $y = \sin 3x$, $y = 3 \sin x$, and $y = \sin x + 3$ have any points in common. Also ask them to explain if any three of the graphs have points in common. Comparing $y = \sin x$ and $y = \sin x + 3$ to $y = x$ and $y = x + 3$, students should conclude that the four graphs have no points in common, since the graph of $y = \sin x + 3$ is three units above the graph of $y = \sin x$. The three graphs $y = \sin x$, $y = \sin 3x$, and $y = 3 \sin x$ do have points in common, since $y = 0$ if $x = 0$ or $x = 2\pi$.

CHALKBOARD EXAMPLES

- **For Example 1**
  Sketch the graphs of $y = \sin x$, $y = \frac{1}{2} \sin x$, and $y = 2 \sin x$ on the same coordinate plane. Then determine their amplitudes.

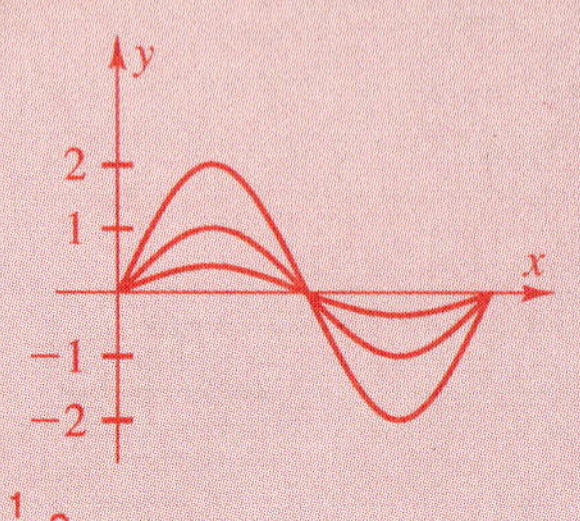

$1; \frac{1}{2}; 2$

minimum value m. Thus, the amplitude is equal to $\frac{1}{2}(M - m)$. The amplitude is related to the height of the graph.

EXAMPLE 1 **Sketch the graphs of $y = \cos x$, $y = -2 \cos x$, and $y = \frac{1}{2} \cos x$ on the same coordinate plane. Then determine their amplitudes.**

x	0	$\frac{\pi}{6}$	$\frac{\pi}{3}$	$\frac{\pi}{2}$	$\frac{2\pi}{3}$	$\frac{5\pi}{6}$	π	$\frac{3\pi}{2}$	2π
$\cos x$	1	0.87	0.5	0	−0.5	−0.87	−1	0	1
$-2 \cos x$	−2	−1.73	−1	0	1	1.73	2	0	−2
$\frac{1}{2} \cos x$	0.5	0.43	0.25	0	−0.25	−0.43	−0.5	0	0.5

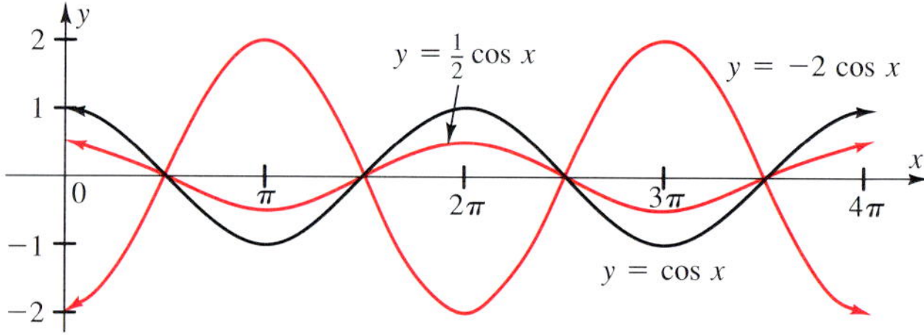

For $y = \cos x$, $M = 1$ and $m = -1$, so the amplitude is $\frac{1}{2}[1 - (-1)] = 1$.

For $y = -2 \cos x$, $M = 2$ and $m = -2$, so the amplitude is $\frac{1}{2}[2 - (-2)] = 2$.

For $y = \frac{1}{2} \cos x$, $M = \frac{1}{2}$ and $m = -\frac{1}{2}$, so the amplitude is $\frac{1}{2}\left[\frac{1}{2} - \left(-\frac{1}{2}\right)\right] = \frac{1}{2}$.

The amplitude of each of the functions in Example 1 is equal to the absolute value of a, the coefficient of $\cos x$. This is, in fact, the case for all functions of the form $y = a \sin bx$ or $y = a \cos bx$. Since the maximum value of the sine or cosine is 1 and the minimum value is -1, the maximum and mimimum values of $y = a \sin bx$ and $y = a \cos bx$ are $|a|$ and $-|a|$, respectively. Thus, the amplitude of $y = a \sin bx$ and $y = a \cos bx$ is $\frac{1}{2}[|a| - (-|a|)] = \frac{1}{2}(2|a|) = |a|$.

Note that the graph of $y = -a \cos x$ is the reflection of the graph of $y = a \cos x$ about the x-axis. This is illustrated at the right for $y = -\cos x$ and $y = \cos x$. Similarly, the graph of $y = -a \sin x$ is the reflection of the graph of $y = a \sin x$ about the x-axis.

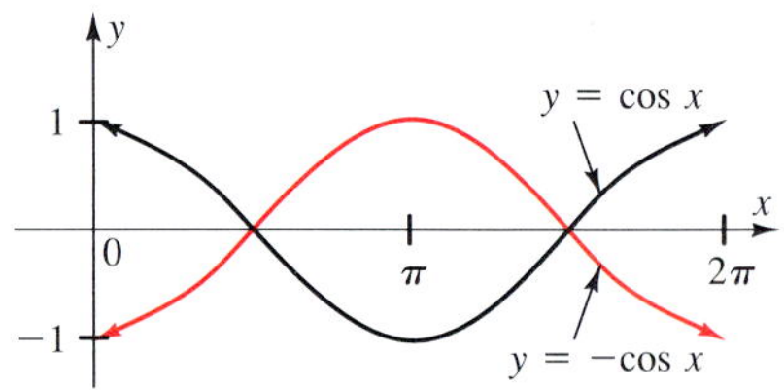

Sometimes it is convenient to use different scales on the x- and y-axes of a trigonometric graph. This is illustrated in Example 1 and in later examples.

EXAMPLE 2 **Sketch the graphs of $y = \sin x$ and $y = \sin 2x$ on the same coordinate plane.**

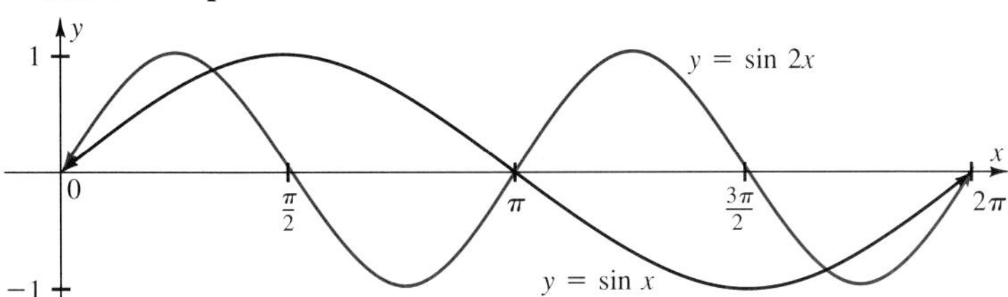

Example 2 shows that $y = \sin x$ has a period of 2π, but $y = \sin 2x$ has a period of π. That is, the period of $y = \sin 2x$ is half that of $y = \sin x$. In general, the period of a function of the form $y = a \sin bx$ or $y = a \cos bx$, where $b \neq 0$, is $\frac{2\pi}{|b|}$.

EXAMPLE 3 **Find an equation of the graph shown below.**

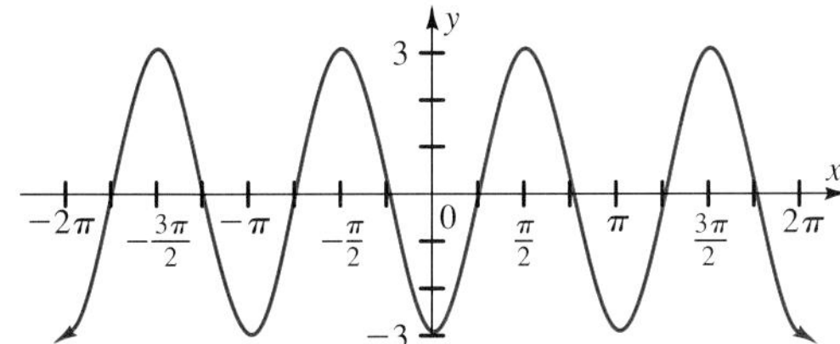

First determine whether the graph represents a sine or cosine function. Note that the curve has y-axis symmetry and when $x = 0$, $y \neq 0$. These two facts are true for cosine graphs but not for sine graphs. Therefore, use $y = a \cos bx$.

Find the period: $\frac{2\pi}{|b|} = \pi$, so $|b| = 2$ and $b = \pm 2$. Choose $b = 2$, so $y = a \cos 2x$.

Find the amplitude: $|a| = 3$, so $a = \pm 3$. From the graph, if $x = 0$, $y = -3$. Therefore, $-3 = a \cos 2(0) = a \cos 0 = a \cdot 1 = a$.

Thus, an equation of the graph is $y = -3 \cos 2x$.

> For $y = a \sin bx$ and $y = a \cos bx$
>
> The amplitude is $|a|$. The period is $\frac{2\pi}{|b|}$.

- **For Example 2**

  Sketch the graphs of $y = \cos x$ and $y = \cos 2x$ on the same coordinate plane. Then determine the periods.

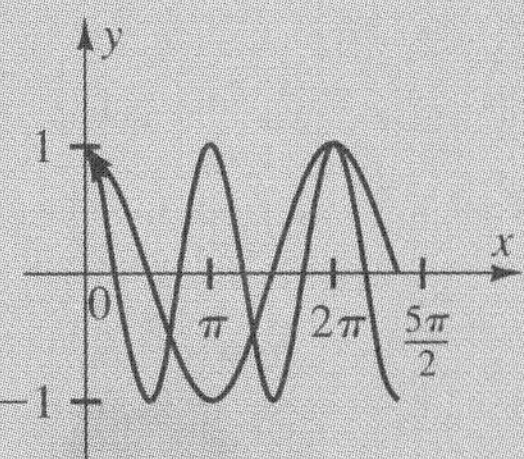

  2π; π

- **For Example 3**

  Find an equation of the graph shown below.

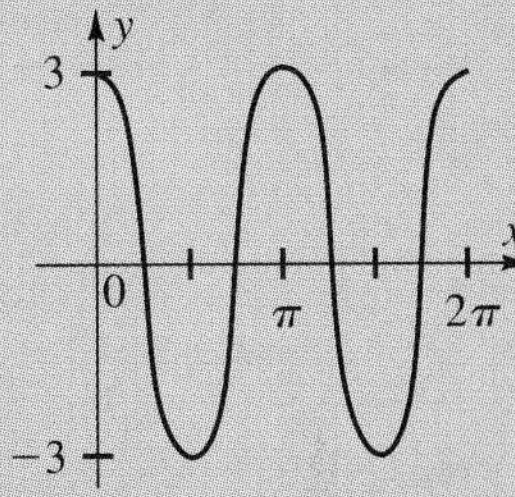

  $y = 3 \cos 2x$

LESSON FOLLOW-UP

Assignment Guide

See p. 58B for assignments.

Challenge

Given a triangular array of the letters of the word trigonometry (one letter per row), students are challenged to find the number of ways that trigonometry can be spelled.

Lesson Quiz

Determine the amplitude and period of each function.

1. $y = \cos 4x$ $1; \frac{\pi}{2}$
2. $y = \sin 5x$ $1; \frac{2\pi}{5}$
3. $y = 5 \sin 2x$ $5; \pi$
4. $y = -3 \cos 2x$ $3; \pi$
5. $y = \frac{3}{2} \cos x$ $\frac{3}{2}; 2\pi$
6. $y = \sin \frac{3}{2}x$ $1; \frac{4\pi}{3}$
7. Sketch the graph of $y = 4 \sin 2x$ over the interval $0 \le x \le 2\pi$.

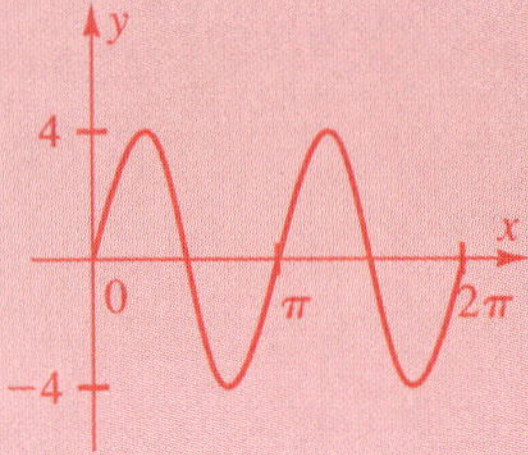

Enrichment

If you were to combine the graphs of $y = 3 \sin 2x$ and $y = -3 \sin 2x$, what would the graph of the resulting function be? $y = 0$, or the x-axis

CLASS EXERCISES

Determine the amplitude and the period of each function.

1. $y = -\frac{1}{2} \cos x$ $\frac{1}{2}; 2\pi$
2. $y = -\sin x$ $1; 2\pi$
3. $y = 2 \cos 6x$ $2; \frac{1}{3}\pi$
4. $y = \frac{3}{2} \cos \frac{1}{2}x$ $\frac{3}{2}; 4\pi$
5. $y = -6 \sin(-4x)$ $6; \frac{1}{2}\pi$
6. $y = -\frac{1}{5} \cos x$ $\frac{1}{5}; 2\pi$
7. $y = \sin 3x$ $1; \frac{2}{3}\pi$
8. $y = \cos 2x$ $1; \pi$
9. $y = \cos\left(-\frac{1}{2}x\right)$ $1; 4\pi$
10. $y = 3 \sin \frac{3}{2}x$ $3; \frac{4\pi}{3}$
11. $y = -\cos x$ $1; 2\pi$
12. $y = -4 \sin \frac{1}{4}x$ $4; 8\pi$

PRACTICE EXERCISES

Determine the amplitude and period of each function.

A

1. $y = \sin 4x$ $1; \frac{\pi}{2}$
2. $y = \cos 5x$ $1; \frac{2\pi}{5}$
3. $y = \sin x$ $1; 2\pi$
4. $y = \cos 6x$ $1; \frac{\pi}{3}$
5. $y = 3 \sin x$ $3; 2\pi$
6. $y = 2 \cos x$ $2; 2\pi$
7. $y = 4 \cos x$ $4; 2\pi$
8. $y = -2 \sin x$ $2; 2\pi$
9. $y = \sin(-2x)$ $1; \pi$
10. $y = 2 \sin(-4x)$ $2; \frac{\pi}{2}$
11. $y = -4 \cos 5x$ $4; \frac{2}{5}\pi$
12. $y = 3 \cos(-2x)$ $3; \pi$
13. $y = 4 \cos \frac{1}{2}x$ $4; 4\pi$
14. $y = 3 \sin \frac{2}{3}x$ $3; 3\pi$
15. $y = 2 \cos \frac{3}{2}x$ $2; \frac{4\pi}{3}$
16. $y = 5 \sin \frac{5}{3}x$ $5; \frac{6\pi}{5}$
17. $y = -2 \cos \frac{5}{4}x$ $2; \frac{8}{5}\pi$
18. $y = -3 \sin \frac{3}{5}x$ $3; \frac{10}{3}\pi$

Give the amplitude and period of each function graphed below. Then write an equation of each graph.

19.

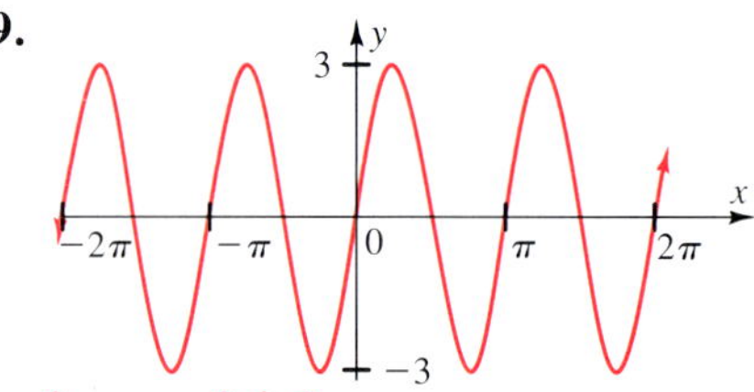

$3; \pi; y = 3 \sin 2x$

20.

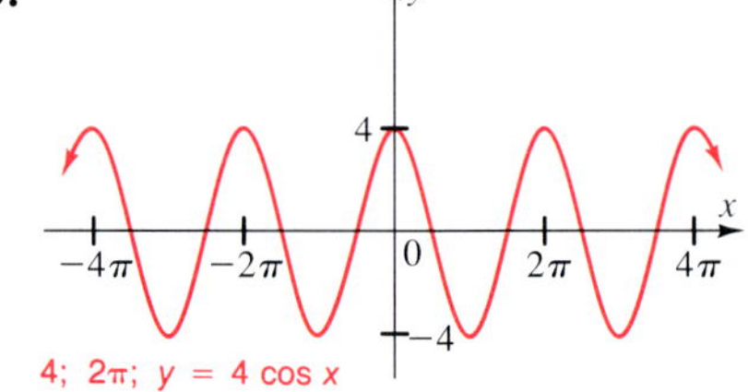

$4; 2\pi; y = 4 \cos x$

21.

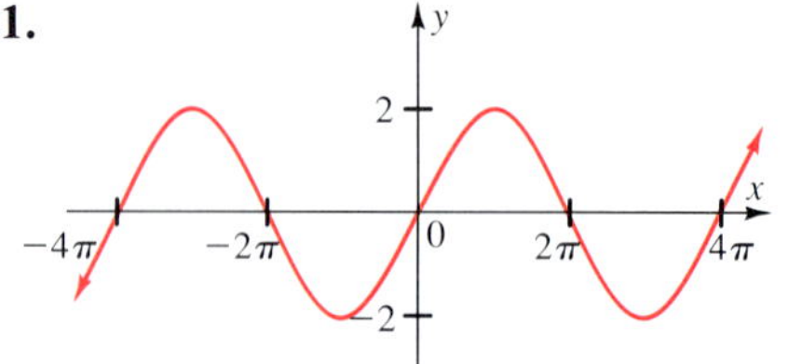

$2; 4\pi; y = 2 \sin \frac{1}{2}x$

22.

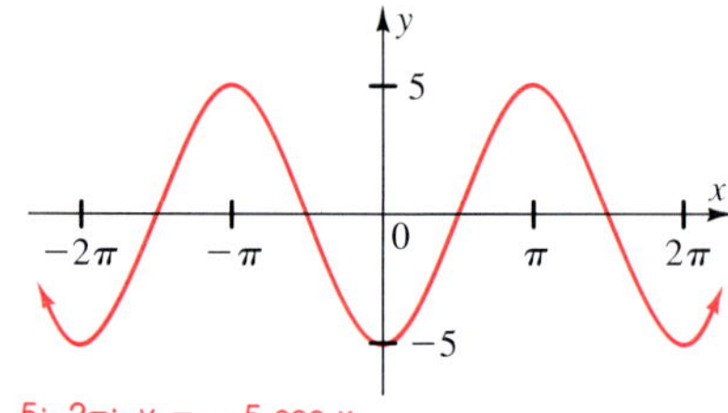

$5; 2\pi; y = -5 \cos x$

Give the amplitude and period of each function. Then sketch the graph of the function over the interval $-2\pi \le x \le 2\pi$. See Solutions Manual for graphs.

B **23.** $y = 3 \sin x$ 3; 2π **24.** $y = 2 \cos x$ 2; 2π **25.** $y = \sin 2x$ 1; π

26. $y = \cos 2x$ 1; π **27.** $y = 3 \sin 2x$ 3; π **28.** $y = 4 \cos 2x$ 4; π

29. $y = 3 \cos \frac{1}{2}x$ 3; 4π **30.** $y = 2 \sin \frac{1}{3}x$ 2; 6π **31.** $y = 2 \cos \frac{3}{2}x$ 2; $\frac{4}{3}\pi$

32. $y = 3 \sin \frac{2}{3}x$ 3; 3π **33.** $y = -3 \cos x$ 3; 2π **34.** $y = -2 \sin x$ 2; 2π

C **35.** $y = \cos(-x)$ 1; 2π **36.** $y = \sin(-x)$ 1; 2π

37. $y = -2 \sin(-2x)$ 2; π **38.** $y = -3 \cos(-3x)$ 3; $\frac{2\pi}{3}$

39. Write four equations of the form $y = a \cos bx$ with amplitude 5 and period 3.

40. Write four equations of the form $y = a \sin bx$ with amplitude $\frac{2}{3}$ and period $\frac{3}{2}\pi$.

$y = \frac{2}{3} \sin \frac{4}{3}x,\ y = -\frac{2}{3} \sin \frac{4}{3}x,\ y = \frac{2}{3} \sin\left(-\frac{4}{3}x\right),\ y = -\frac{2}{3} \sin\left(-\frac{4}{3}x\right)$

Applications 39. $y = 5 \cos \frac{2}{3}\pi x,\ y = -5 \cos \frac{2}{3}\pi x,\ y = 5 \cos\left(-\frac{2}{3}\pi\right)x,\ y = -5 \cos\left(-\frac{2}{3}\pi\right)x$

Physics The *frequency* of a periodic function is defined to be the reciprocal of the period.

41. Find the frequency of the function $y = \cos 2x$. $\frac{1}{\pi}$

42. Find the frequency of the function $y = -4 \sin \frac{1}{4}x$. $\frac{1}{8\pi}$

43. The voltage V in a particular electrical circuit is given by the formula $V = 6 \cos 90\pi t$, where t is measured in seconds. Find the amplitude, the period, and the frequency of this function. 6; $\frac{1}{45}$ seconds; 45 cycles per second

CHALLENGE

In how many ways can *TRIGONOMETRY* be spelled using adjacent letters in this triangle? One possible solution is shown in red. 2^{11}
Hint: Look for a pattern.

```
                      T
                     R R
                    I I I
                   G G G G
                  O O O O O
                 N N N N N N
                O O O O O O O
               M M M M M M M M
              E E E E E E E E E
             T T T T T T T T T T
            R R R R R R R R R R R
           Y Y Y Y Y Y Y Y Y Y Y Y
```


Teacher's Resource Book
Practice—Chapter 2, p. 5
Enrichment—Chapter 2, p. 6

LESSON PLAN

Vocabulary

Phase shift
Vertical shift

Materials/Manipulatives

Scientific calculators
Graphs of $y = \sin x$, $y = \cos x$, $y = a \sin bx$, and $y = a \cos bx$ for various values of a and b on acetate sheets
Overhead projector
Graphing calculators
Computer

BACKGROUND

In the Preview, horizontal and vertical shifts in the graphs of the quadratic equations are reviewed.

2.4 Phase Shift and Vertical Shift

Objectives: To find the phase shift and vertical shift of sine and cosine functions from their equations
To graph sine and cosine functions with various phase shifts and vertical shifts

Before shifts in the graphs of the sine and cosine functions are introduced, it will be helpful to review shifts in the graphs of quadratic functions.

Preview

The graph of an equation of the form $y = (x + a)^2 + b$ is the same shape as that of $y = x^2$. However, it is shifted a units to the left if $a > 0$, $|a|$ units to the right if $a < 0$, b units up if $b > 0$, and $|b|$ units down if $b < 0$.

To illustrate, the graph of $y = (x - 1)^2 + 3$ is 1 unit to the right of the graph of $y = x^2$ and 3 units above it.

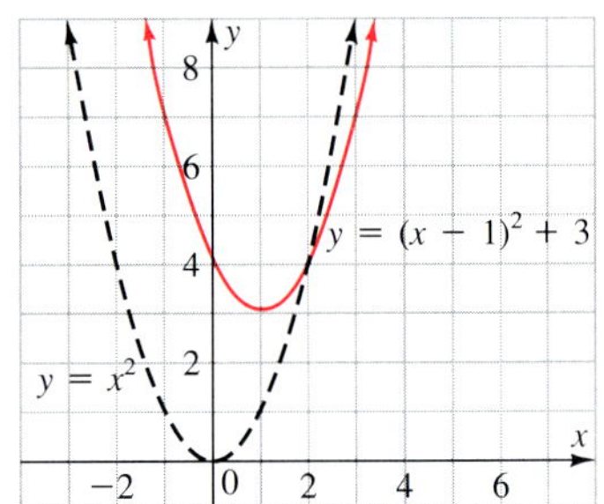

State how the graph of each equation is shifted from the graph of $y = x^2$. Check by graphing.

1. $y = (x + 3)^2$ 3 left **2.** $y = x^2 - 4$ 4 down **3.** $y = (x + 2)^2 - 1$ 2 left; 1 down **4.** $y = (x - 4)^2 + 5$ 4 right; 5 up

Consider the graphs of the sine and cosine shown on the same coordinate plane. If the graph of the cosine were shifted to the right $\frac{\pi}{2}$ units, it would coincide with the graph of the sine. This relationship implies that $\sin x = \cos\left(x - \frac{\pi}{2}\right)$. The distance that one graph must be shifted to the right or to the left to coincide with another graph is called the **phase shift.**

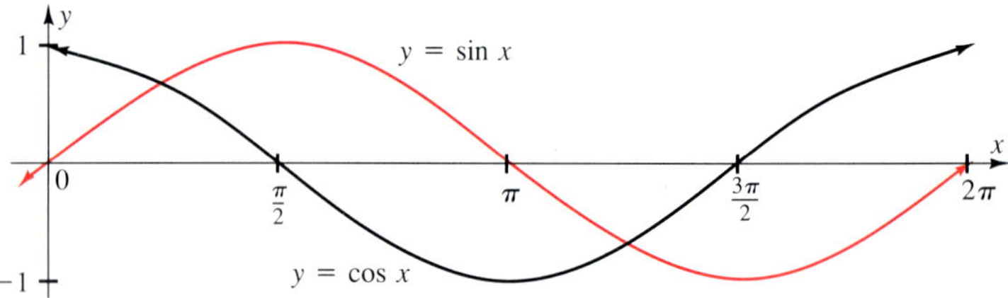

EXAMPLE 1 **Sketch the graph of $y = \sin\left(x - \frac{\pi}{4}\right)$.**

Construct a table of values, rounding to the nearest hundredth. Then graph the points and join them with a smooth curve.

x	0	$\frac{\pi}{4}$	$\frac{\pi}{2}$	$\frac{3\pi}{4}$	π	$\frac{5\pi}{4}$	$\frac{3\pi}{2}$	$\frac{7\pi}{4}$	2π	$\frac{9\pi}{4}$
$x - \frac{\pi}{4}$	$-\frac{\pi}{4}$	0	$\frac{\pi}{4}$	$\frac{\pi}{2}$	$\frac{3\pi}{4}$	π	$\frac{5\pi}{4}$	$\frac{3\pi}{2}$	$\frac{7\pi}{4}$	2π
$\sin\left(x - \frac{\pi}{4}\right)$	-0.71	0	0.71	1	0.71	0	-0.71	-1	-0.71	0

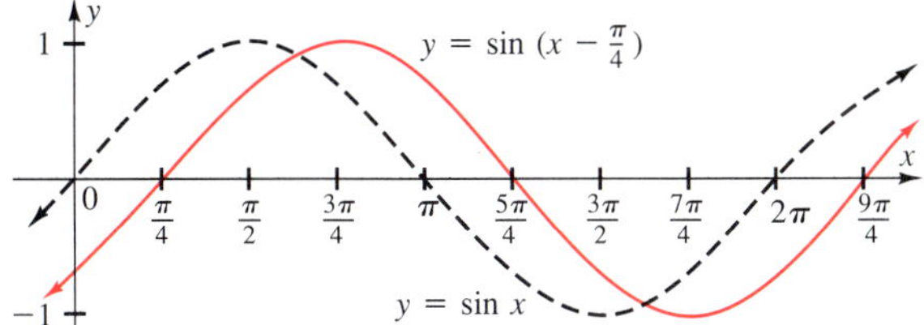

The graph of $y = \sin\left(x - \frac{\pi}{4}\right)$ is the same as that of $y = \sin x$, except that it is shifted $\frac{\pi}{4}$ units to the right. The phase shift of $y = \sin\left(x - \frac{\pi}{4}\right)$ is $\frac{\pi}{4}$, the period is 2π, and the amplitude is 1.

In general, the graph of a function of the form $y = a \sin b(x + c)$ or $y = a \cos b(x + c)$ is the same as that of $y = a \sin bx$ or $y = a \cos bx$, but it is shifted c units to the left if $c > 0$, and $|c|$ units to the right if $c < 0$. The period of a function of the form $y = a \sin b(x + c)$ or $y = a \cos b(x + c)$ is $\frac{2\pi}{|b|}$, and the amplitude is $|a|$.

EXAMPLE 2 **Determine the amplitude, period, and phase shift of $y = 3 \cos\left(x + \frac{\pi}{6}\right)$. Then sketch its graph.**

Amplitude $= |3| = 3$ *Amplitude* $= |a|$

Period $= \frac{2\pi}{|1|} = 2\pi$ *Period* $= \frac{2\pi}{|b|}$

Phase shift $= \left|\frac{\pi}{6}\right|$

$= \frac{\pi}{6}$ units to the left, since $c > 0$

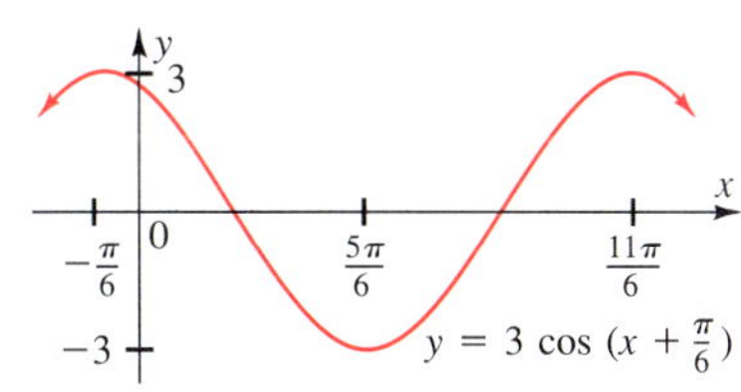

TEACHING SUGGESTIONS

- Use the graphs of $y = \sin x$ and $y = \cos x$ to obtain the graphs of $y = \sin x + 3$ and $y = \cos x - 2$. Relate the effect of the constants in these equations to the effect of the constants in $y = x$, $y = x + 3$, and $y = x - 1$.
- Use the graphs of $y = \sin x$ and $y = \cos x$ to obtain the graphs of $y = \sin(x + \pi)$ and $y = \cos(x - \pi)$. Relate the effect of the constants in these equations to the effect of the constants $y = x^2$, $y = (x - 3)^2$, and $y = (x + 3)^2$.
- Encourage students to use graphing calculators or computers to show the effects of the constants.

CHALKBOARD EXAMPLES

- **For Example 1**

  1. Sketch the graph of $y = \cos\left(x - \frac{\pi}{2}\right)$.

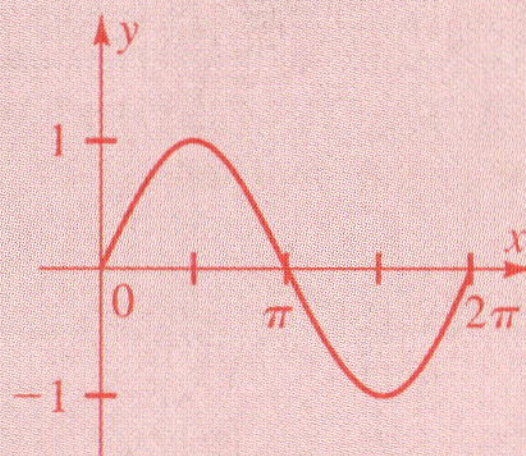

- **For Example 2**

  2. Give the amplitude, period, and phase shift of $y = 2 \sin\left(x - \frac{2\pi}{3}\right)$. Then sketch the graph. ampl.: 2; period: 2π; shift: $\frac{2\pi}{3}$ to the right

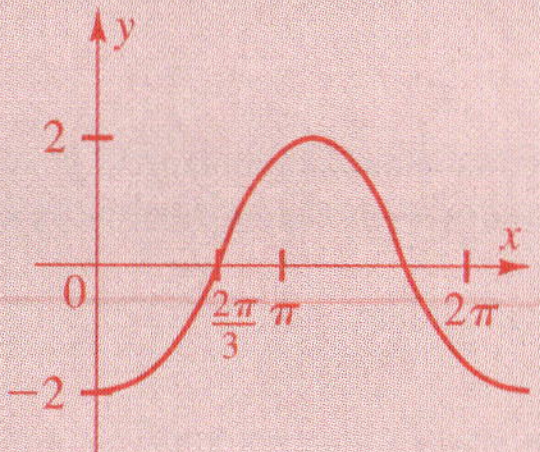

- **For Example 3**

3. Give the amplitude, period, and phase shift of $y = \cos(4x + \pi)$. Then sketch the graph.

ampl.: 1; period: $\frac{\pi}{2}$; shift: $\frac{\pi}{4}$ to the left

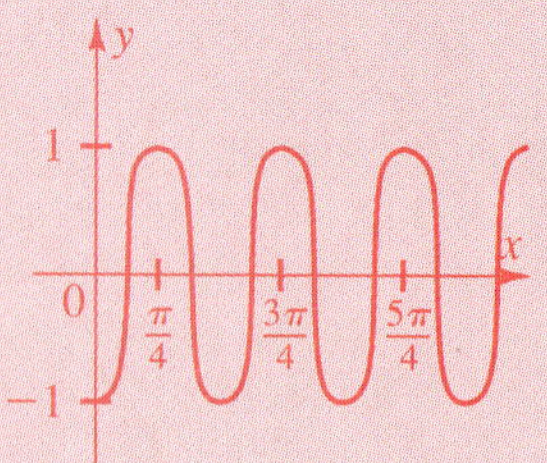

- **For Example 4**

Sketch each graph.

4. $y = \sin x - 4$

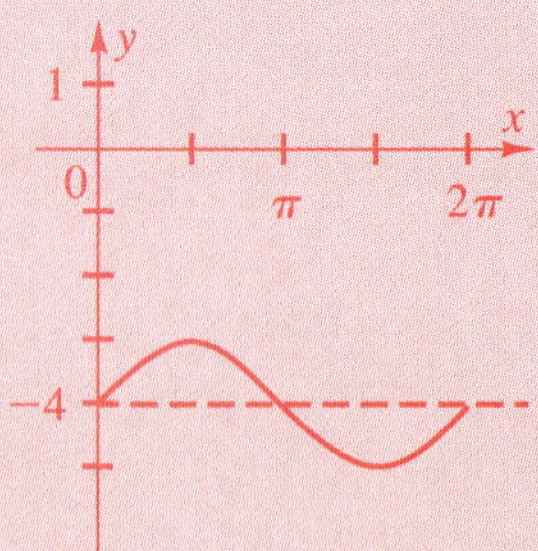

5. $y = \cos x + 2$

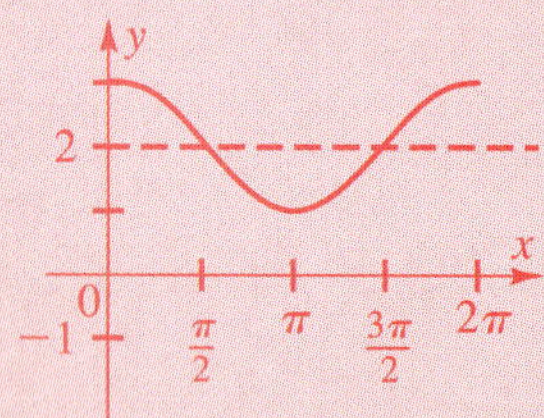

In Example 3, the equation must be factored before the phase shift can be determined.

EXAMPLE 3 **Determine the amplitude, period, and phase shift of $y = \sin(2x + \pi)$. Then sketch its graph.**

$y = \sin(2x + \pi)$

$y = \sin 2\left(x + \frac{\pi}{2}\right)$ *Factor.*

Amplitude $= |1| = 1$ *Amplitude* $= |a|$

Period $= \frac{2\pi}{|2|} = \pi$ *Period* $= \frac{2\pi}{|b|}$

Phase shift $= \left|\frac{\pi}{2}\right| = \frac{\pi}{2}$ units to the left, since $c > 0$

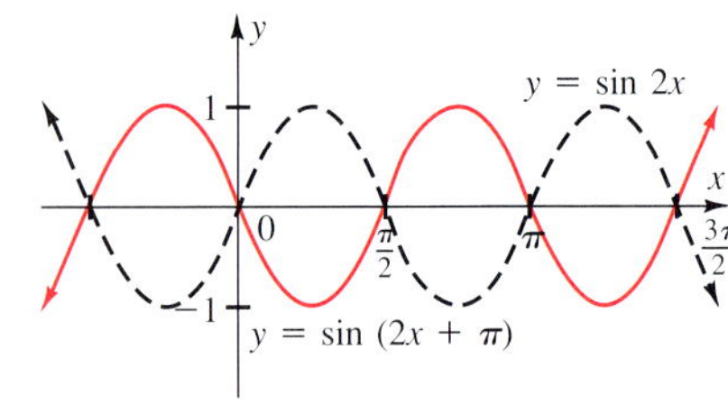

Phase shifts are horizontal shifts. However, if a constant term d is added to the right side of an equation of the form $y = a \sin b(x + c)$ or $y = a \cos b(x + c)$, a **vertical shift** results. You can verify graphs using a graphing calculator or computer.

EXAMPLE 4 **Sketch each graph: a.** $y = \sin x + 3$ **b.** $y = \cos x - 2$

a. The graph of $y = \sin x$ is shifted up 3 units.

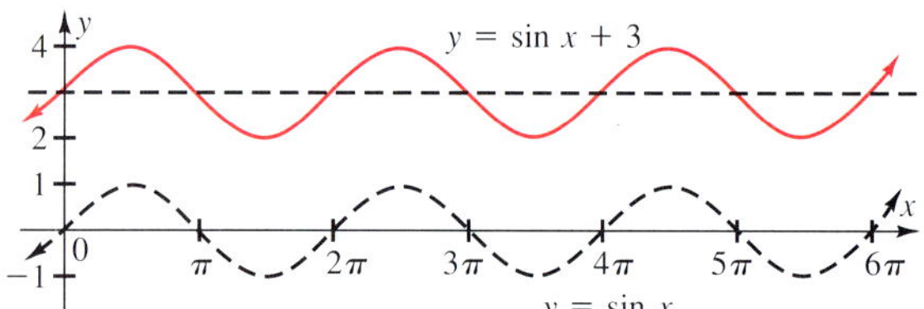

b. The graph of $y = \cos x$ is shifted down 2 units.

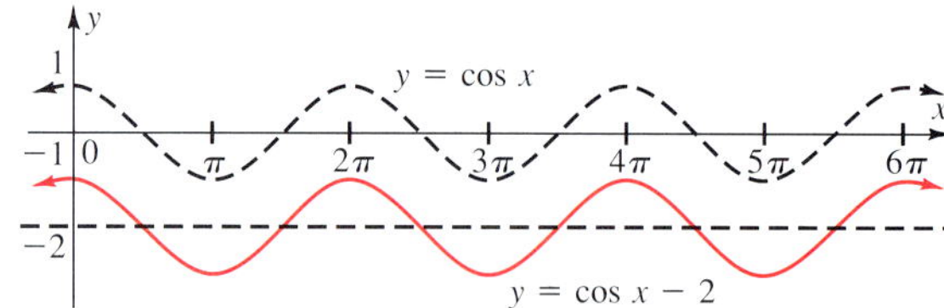

In general, the graphs of $y = a \sin b(x + c) + d$ and $y = a \cos b(x + c) + d$ are shifted d units upward from those of $y = a \sin bx$ and $y = a \cos bx$ if $d > 0$, and $|d|$ units downward if $d < 0$. Combinations of phase shifts and vertical shifts also occur, as illustrated in Example 5.

EXAMPLE 5 **Determine the amplitude, period, phase shift, and vertical shift for the graph of $y = \sin\left(x - \frac{\pi}{4}\right) + 2$. Then sketch the graph.**

Amplitude $= |1| = 1$ *Amplitude* $= |a|$

Period $= \frac{2\pi}{|1|} = 2\pi$ *Period* $= \frac{2\pi}{|b|}$

Phase shift $= \left|-\frac{\pi}{4}\right| = \frac{\pi}{4}$ units to the right

Vertical shift $= |2| = 2$ units up

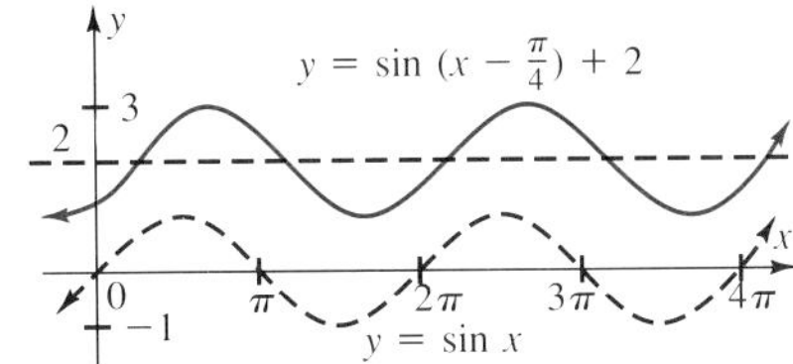

For $y = a \sin b(x + c) + d$ and $y = a \cos b(x + c) + d$

The amplitude is $|a|$.   The period is $\frac{2\pi}{|b|}$.

The phase shift from $y = a \sin bx$ or $y = \cos bx$ is c units to the left if $c > 0$, and $|c|$ units to the right if $c < 0$.

The vertical shift from $y = a \sin bx$ or $y = a \cos bx$ is d units upward if $d > 0$, and $|d|$ units downward if $d < 0$.

EXAMPLE 6 **Sketch the graph of $y = 3 \cos\left(\frac{1}{2}x - \frac{\pi}{2}\right) - 2$.**

$$y = 3 \cos\left(\frac{1}{2}x - \frac{\pi}{2}\right) - 2$$

$$y = 3 \cos \tfrac{1}{2}(x - \pi) - 2$$

Amplitude $= |3| = 3$

Period $= \frac{2\pi}{\left|\frac{1}{2}\right|} = 4\pi$

Phase shift $= |-\pi| = \pi$ units to the right

Vertical shift $= |-2| = 2$ units down

CLASS EXERCISES

Determine the amplitude, period, phase shift, and vertical shift for each.

1. $y = 3 \sin 2x$ 3; π; none; none
2. $y = \cos(x - \pi)$ 1; 2π; right π; none
3. $y = 2 \sin 3x - 5$ 3. 2; $\frac{2\pi}{3}$; none; down 5
4. $y = \cos 2x + 2$ 1; π; none; up 2
5. $y = 4 \sin(2x + 2\pi)$ 4; π; left π; none
6. $y = 2 \cos \frac{\pi}{2}x - 6$ 2; 4; none; down 6
7. $y = 4 \sin\left(2x - \frac{\pi}{2}\right) - 3$ 4; π; $\frac{\pi}{4}$ right; down 3
8. $y = -2 \sin\left(3x - \frac{3\pi}{4}\right) + 2$ 2; $\frac{2\pi}{3}$; right $\frac{\pi}{4}$; up 2

- **For Example 5**

6. Give the amplitude, period, phase shift, and vertical shift for $y = 3 \cos(2x - \pi) + 4$. Then sketch the graph. 3; π; $\frac{\pi}{2}$ to the right; up 4

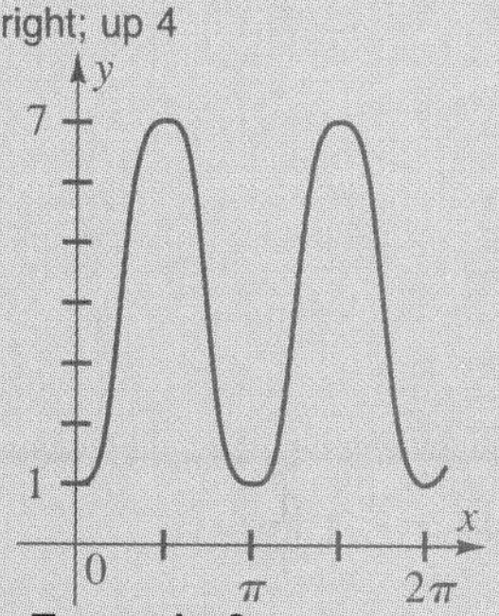

- **For Example 6**

7. Sketch the graph of $y = \frac{3}{2} \sin(2x - 2\pi) + 2$

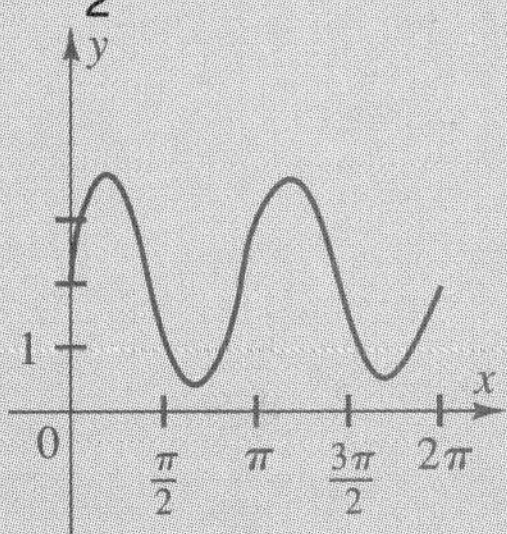

Common Error

- Some students sketch the graph of the sine for the cosine when they do a number of exercises that involve phase shifts. Encourage those students to double check by looking at the value at the origin of the new system.
- See *Teacher's Resource Book* for additional remediation.

LESSON FOLLOW-UP

Discussion

How would you determine where to locate a new origin if you wanted to graph the function $y = \sin b(x + c) + d$? Give the coordinates of the origin of this graph. Determine the phase and vertical shifts from the equation; the origin is at $(-c, d)$.

Assignment Guide
See p. 58B for assignments.

Critical Thinking
Analysis Ask students to contrast the graphs of $y = \sin x$ and $y = 3 \sin (2x + 2\pi) - 4$. Students note that the second graph has an amplitude 3 times as great and a period $\frac{1}{2}$ that of $y = \sin x$. Thus, they conclude that the second graph is shifted to the left π units and down 4 units.

Test Yourself
See *Teacher's Resource Book*, *Tests*, pp. 13–14.

Lesson Quiz
Determine the amplitude, period, phase shift, and vertical shift for each function.

1. $y = 3 \cos 2x$ 3; π; 0; 0
2. $y = \sin 3x + 2$ 1; $\frac{2\pi}{3}$; 0; 2 up
3. $y = 4 \cos (2x + 2\pi)$ 4; π; π left; 0
4. $y = 2 \sin (3x - \pi) + 1$ 2; $\frac{2\pi}{3}$; $\frac{\pi}{3}$ right; 1 up
5. Sketch the graph of $y = \sin 2x + 1$ over the interval $0 \le x \le 2\pi$.

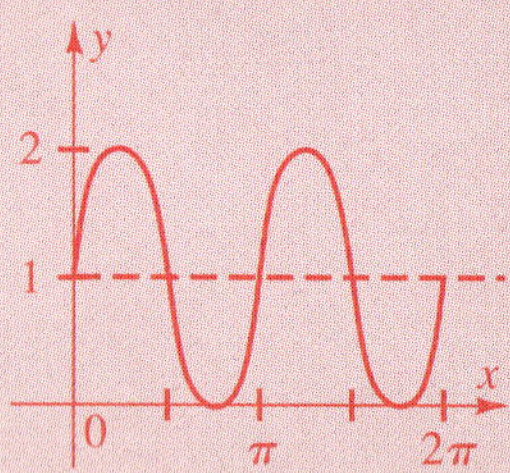

Enrichment
If the origin of a new coordinate system is located at $(\pi, 3)$, and the graph is a regular cosine graph on this system, what did the equation look like before the translation? $y = \cos (x - \pi) + 3$

PRACTICE EXERCISES

Determine the amplitude, period, phase shift, and vertical shift for each.

A

1. $y = 2 \sin 3x$ 2; $\frac{2\pi}{3}$; none; none
2. $y = \sin (x - \pi)$ 1; 2π; right π; none
3. $y = 3 \cos 4x$ 3; $\frac{\pi}{2}$; none; none
4. $y = 3 \sin 6x - 3$ 3; $\frac{\pi}{3}$; none; down 3
5. $y = \cos 2x - 5$ 1; π; none; down 5
6. $y = 2 \sin (3x + 3\pi)$ 2; $\frac{2\pi}{3}$; left π; none
7. $y = \frac{1}{4} \sin 2x$ $\frac{1}{4}$; π; none; none
8. $y = 3 \cos \frac{1}{2}x + 4$ 3; 4π; none; up 4
9. $y = 4 \sin \left(3x - \frac{\pi}{3}\right) + 2$ 4; $\frac{2\pi}{3}$; right $\frac{\pi}{9}$; up 2
10. $y = -2 \cos \left(3x - \frac{4\pi}{3}\right) - 3$ 2; $\frac{2\pi}{3}$; right $\frac{4\pi}{9}$; down 3

Sketch the graph of each function over the interval $0 \le x \le 4\pi$.

11. $y = \sin 2x + 3$ See page 475.
12. $y = \cos (x - \pi)$
13. $y = \cos 2x - 1$
14. $y = 3 \sin x - 1$

B

15. $y = \sin (x - \pi)$
16. $y = \sin (x - \pi) - 1$
17. $y = \cos 3x - 5$
18. $y = 2 \sin (2x + 2\pi)$
19. $y = \cos 2(x - \pi)$
20. $y = \sin \frac{1}{2}(x - \pi)$
21. $y = \frac{1}{2} \sin 2x$
22. $y = 2 \cos \frac{1}{2}x + 4$
23. $y = 3 \sin \left(2x - \frac{\pi}{2}\right) + 1$
24. $y = -2 \cos \left(3x - \frac{4\pi}{3}\right) - 2$

C

25. Find an equation for a sine function which has an amplitude of 4, a period of 180°, and a y-intercept of -3. $y = 4 \sin 2x - 3$ Other answers are possible.
26. Find an equation for a sine function which has an amplitude of 3, a period of 90°, and a y-intercept of 2. $y = 3 \sin 4x + 2$ Other answers are possible.
27. Find an equation for a cosine function which has an amplitude of $\frac{3}{5}$, a period of 270°, and y-intercept of 5. $y = \frac{3}{5} \cos \frac{4}{3}x + \frac{22}{5}$ Other answers are possible.
28. Find an equation for a cosine function which has an amplitude of $\frac{3}{2}$, a period of 90°, and a y-intercept of -4. $y = \frac{3}{2} \cos 4x - \frac{11}{2}$ Other answers are possible.

Applications

Music Sound waves can be represented by sine or cosine functions.

29. A tuning fork for G (below middle C) has a frequency of 196 vibrations per second $\left(\text{frequency} = \frac{1}{\text{period}}\right)$. If the amplitude is 0.04 mm, find an equation for the resulting sound wave. $y = \frac{1}{25} \sin 2 \cdot 196\ \pi x$ or $y = \frac{1}{25} \cos 2 \cdot 196\pi x$

30. The vibrations may not be exactly in phase when two instruments are played together. Suppose the sound wave from one instrument can be represented by $y_1 = 3 \sin (3\pi x)$, and the sound wave from another instrument can be represented by $y_2 = 3 \sin \left(3\pi x + \frac{3\pi}{2}\right)$. Sketch the graphs of both. See side column.

31. A musical sound wave is represented by $y = 0.004 \sin (300\pi x + 150\pi^2)$. Determine its amplitude, period, and phase shift. Sketch the graph of the sound wave for $0 \leq x \leq 0.01$, letting x increase in steps of 0.0005 radians. $0.004; \frac{1}{150}; \frac{\pi}{2}$

TEST YOURSELF

Determine whether each function is even, odd, or neither.

1. $f(x) = 2x^2$ even **2.** $f(x) = 3x^3 + 1$ neither **3.** $f(x) = x^7$ odd **2.1**

Determine whether each graph is symmetric with respect to the x-axis, the y-axis, the origin, or to none of these.

4.

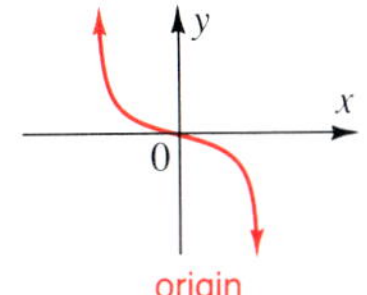

origin

5.

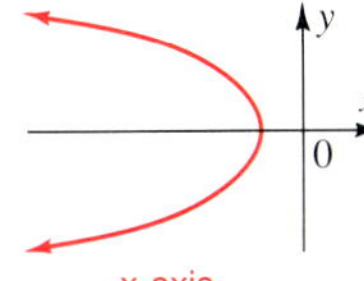

x-axis

6.

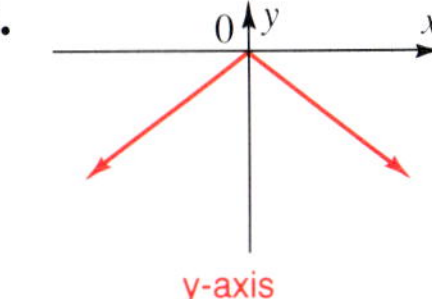

y-axis

Use the fact that the sine is an odd function and the cosine is an even function to determine the following values.

7. If sine 121° = 0.8572, find sin (−121°). −0.8572 **2.2**

8. If cos 121° = −0.5150, find cos (−121°). −0.5150

Determine the amplitude and period of each function.

9. $y = -5 \cos x$ $5; 2\pi$ **10.** $y = \sin \frac{3}{4}x$ $1; \frac{8\pi}{3}$ **2.3**

11. $y = 4 \cos \frac{1}{3}x$ $4; 6\pi$ **12.** $y = \frac{1}{2} \sin (-2x)$ $\frac{1}{2}; \pi$

Determine the amplitude, period, phase shift, and vertical shift of each.

13. $y = 2 \cos \left(x - \frac{\pi}{2}\right) + 5$ $2; 2\pi; \frac{\pi}{2}$ right; 5 up

14. $y = \frac{1}{2} \cos \left(x + \frac{\pi}{2}\right) - 7$ $\frac{1}{2}; 2\pi; \frac{\pi}{2}$ left; 7 down

15. $y = \sin (2x + 3\pi) - 1$ $1; \pi; \frac{3\pi}{2}$ left; 1 down

16. $y = \sin (3x + \pi) + 2$ $1; \frac{2\pi}{3}; \frac{\pi}{3}$ left; 2 up

Sketch the graph of each function over the interval $0 \leq x \leq 4\pi$.
See page 476.

17. $y = 2 \sin \left(x - \frac{\pi}{2}\right) + 1$ **18.** $y = 3 \cos \left(x + \frac{\pi}{4}\right) - 2$

Teacher's Resource Book
Practice—Chapter 2, p. 7
Enrichment—Chapter 2, p. 8

Additional Answers

30.

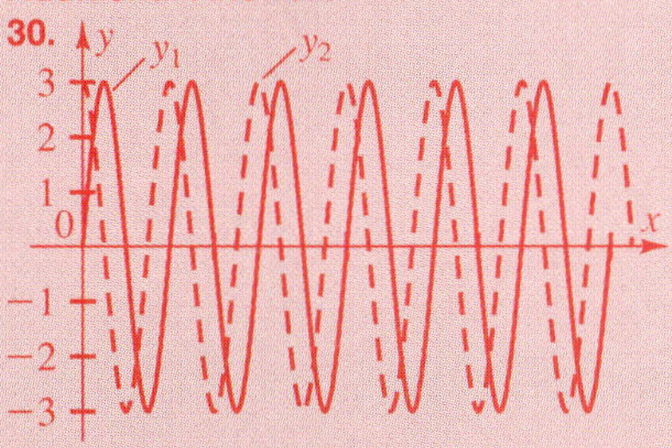

31.

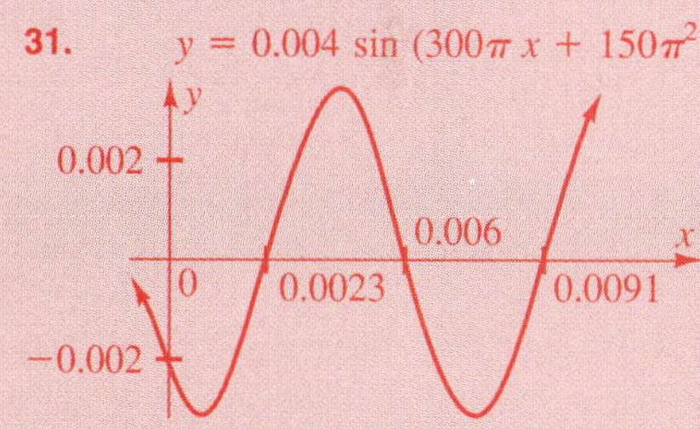

LESSON PLAN

Vocabulary

Addition of ordinates

Materials/Manipulatives

Graphs of varied forms of the sine and cosine functions on acetate sheets
Graphing calculators
Computer

BACKGROUND

In the Preview, the graphs of $y = \sin(\pi x)$, $y = \sin(2\pi x)$ and their sum, $y = \sin(\pi x) + \sin(2\pi x)$ are discussed and compared.

Critical Thinking

Analyzing Relationships Ask the students if addition of ordinates is a commutative operation. Have students explain their answers. Yes. The ordinates may be added in either order and the result is the same.

2.5 Graphing by Addition of Ordinates

Objectives: To graph functions that are the sums or differences of two sine functions, two cosine functions, or a sine function and a cosine function

The function $y = 2x + 4$ is actually the sum of two functions, $y = 2x$ and $y = 4$. For every value in the domain, there is a corresponding value in the range so that the point $(x, 2x + 4)$ is on the graph of $y = 2x + 4$. The sums of trigonometric functions can also be graphed.

Preview

If two strings of a piano have the same frequency, they are in tune with each other. If they are off by even a small number of vibrations per second, the sound has *beats*. The figure at the right shows the graphs of $y_1 = \sin(\pi x)$, $y_2 = \sin(2\pi x)$, and their sum,

$$y = \sin(\pi x) + \sin(2\pi x)$$

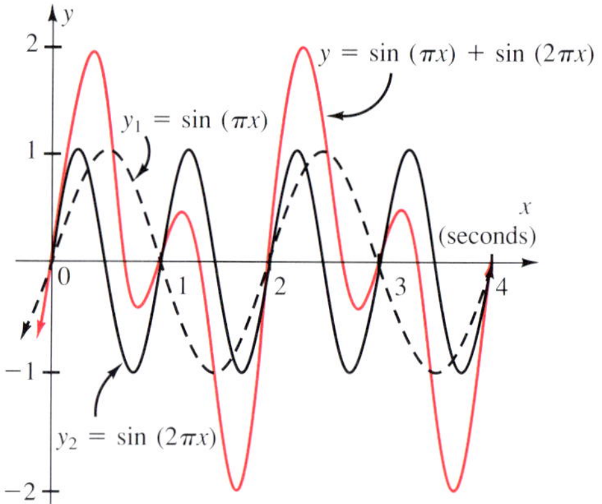

1. How many cycles does y_1 complete in four seconds? 2
2. How many cycles does y_2 complete in four seconds? 4
3. The graphs of y_1 and y_2 coincide at $x = 0$ and $x = 2$. Then both graphs rise for a fraction of a second. What number does the amplitude of y approach during this time? 2

On page 80, it was shown that the graph of $y = \sin x + 3$ is the same as the graph of $y = \sin x$ shifted up 3 units. In other words, each value of the function $y = \sin x + 3$ is the sum of the values $y = \sin x$ and $y = 3$. Similarly, the graph of $y = \cos x - 2$ is the same as the graph of $y = \cos x$ shifted down 2 units. That is, each value of the function $y = \cos x - 2$ is the sum of the values $y = \cos x$ and $y = -2$.

A function that is the sum of the two trigonometric functions can be sketched by first graphing the two functions and then adding the two y-values for each value of x. A graphing calculator is useful for checking the results.

EXAMPLE 1 **Graph $y = \sin x + \cos x$ over the interval $0 \le x \le 4\pi$.**

First, sketch the graphs of $y = \sin x$ and $y = \cos x$.

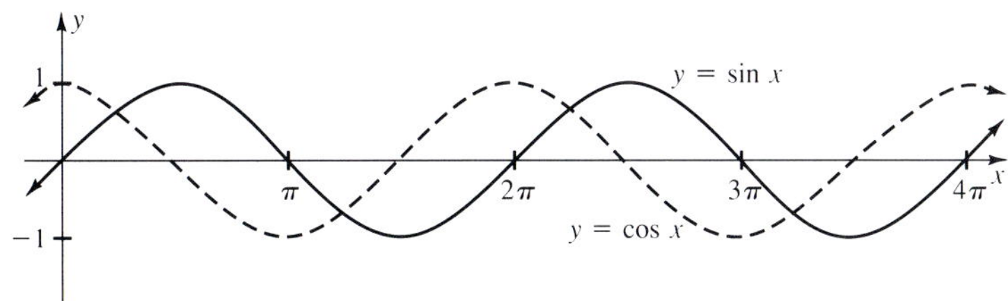

For each value of x, add the two values of y and sketch their sum.

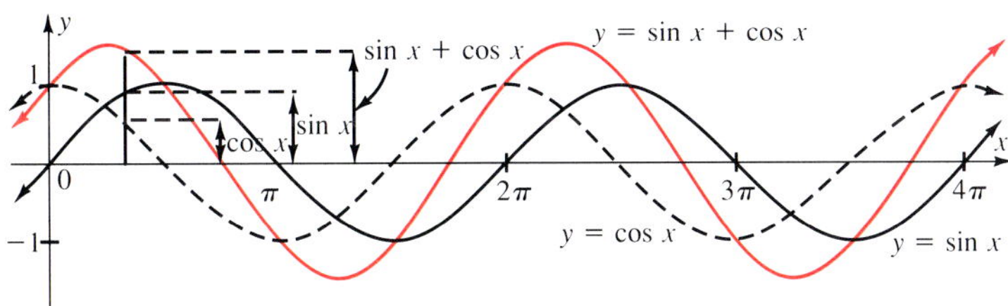

The y-coordinates of a point are sometimes called *ordinates*, and the method illustrated in Example 1 is called graphing by **addition of ordinates.** Check the graph using a graphing calculator or computer.

EXAMPLE 2 **Graph $y = 2 \sin x + \sin \frac{1}{2}x$ over the interval $0 \le x \le 4\pi$.**

First, sketch the graphs of $y = 2 \sin x$ and $y = \sin \frac{1}{2}x$.
Then, for each value of x, add the two values of y to sketch $y = 2 \sin x + \sin \frac{1}{2}x$.

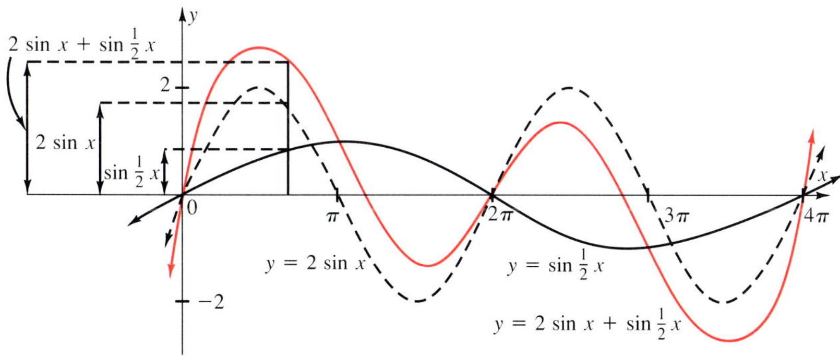

TEACHING SUGGESTIONS

- Be sure that students understand that in $(x, f(x))$, $f(x)$ is the distance the point is from the x-axis.
- Have students use different colors to graph each function and the sum of the functions.
- Point out a quick way to get points on the sum is to locate all points where the value of one of the functions is zero.
- Encourage students to use their graphing calculators or computers to find the sum of two functions.

CHALKBOARD EXAMPLES

- **For Example 1**

  1. Graph $y = \sin x - \cos x$.

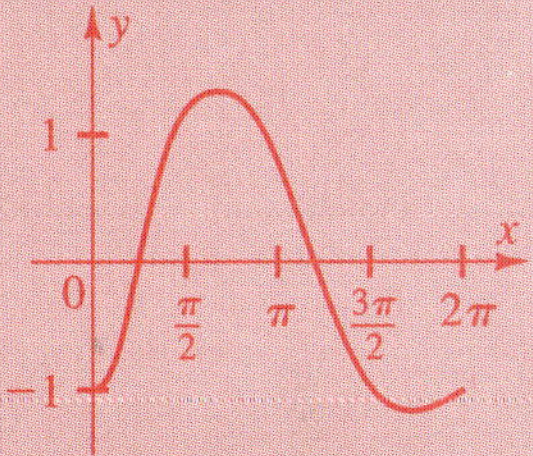

- **For Example 2**

  2. Graph $y = \sin x + 2 \sin x$.

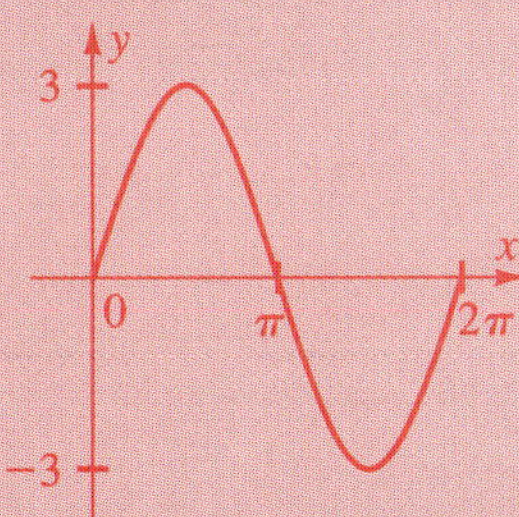

Common Error

- Some students do not watch for places where the functions have opposite signs and thus graph the sum as if both have the same sign. Have those students locate the zeros of both functions and find the sum for these points to get an idea of the shape of the graph.
- See *Teacher's Resource Book* for additional remediation.

LESSON FOLLOW-UP

Assignment Guide

See p. 58B for assignments.

Lesson Quiz

1. Make a table of values for $y = \cos x$, $y = 2 \cos x$, and $y = \cos x + 2 \cos x$.

x	$\cos x$	$2 \cos x$	$\cos x + 2 \cos x$
0	1	2	3
$\frac{\pi}{6}$	0.9	1.8	2.7
$\frac{\pi}{4}$	0.7	1.4	2.1
$\frac{\pi}{3}$	0.5	1.0	1.5
$\frac{\pi}{2}$	0	0	0
$\frac{2\pi}{3}$	−0.5	−1.0	−1.5
$\frac{3\pi}{4}$	−0.7	−1.4	−2.1
$\frac{5\pi}{6}$	−0.9	−1.8	−2.7
π	−1	−2	−3

2. Graph $y = \cos x - \cos 2x$.

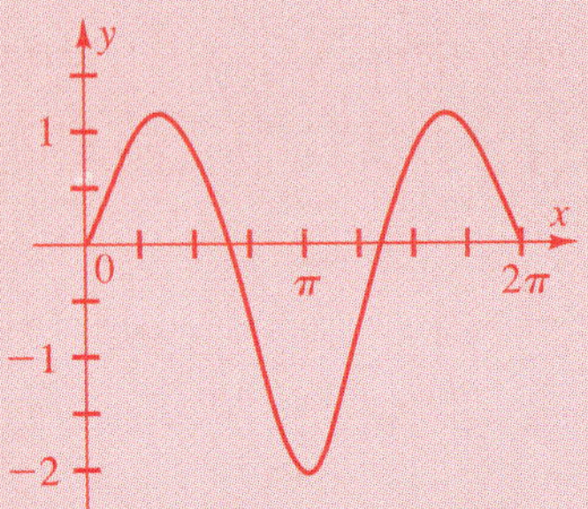

3. Graph $y = \sin x + \cos \frac{x}{2}$.

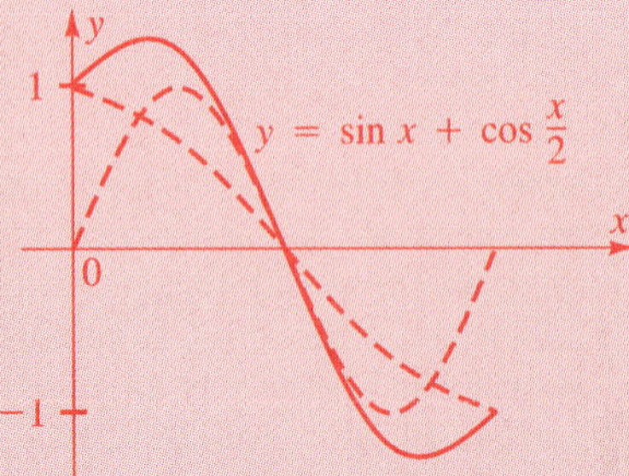

CLASS EXERCISES

1. Complete the table below for the function $y = \sin x + 2 \sin x$.

x	0	$\frac{\pi}{6}$	$\frac{\pi}{4}$	$\frac{\pi}{3}$	$\frac{\pi}{2}$	$\frac{2\pi}{3}$	$\frac{3\pi}{4}$	$\frac{5\pi}{6}$	π
$\sin x$	0	0.5	0.71	0.87	1	0.87	0.71	0.5	0
$2 \sin x$	0	1	1.41	1.73	2	1.73	1.41	1	0
$\sin x + 2 \sin x$	0	1.5	2.12	2.60	3	2.60	2.12	1.5	0

2. Complete the table below for the function $y = \cos x + \cos 2x$.

x	0	$\frac{\pi}{6}$	$\frac{\pi}{4}$	$\frac{\pi}{3}$	$\frac{\pi}{2}$	$\frac{2\pi}{3}$	$\frac{3\pi}{4}$	$\frac{5\pi}{6}$	π
$\cos x$	1	0.87	0.71	0.5	0	−0.5	−0.71	−0.87	−1
$2x$	0	$\frac{\pi}{3}$	$\frac{\pi}{2}$	$\frac{2\pi}{3}$	π	$\frac{4\pi}{3}$	$\frac{3\pi}{2}$	$\frac{5\pi}{3}$	2π
$\cos 2x$	1	0.5	0	−0.5	−1	−0.5	0	0.5	1
$\cos x + \cos 2x$	2	1.37	0.71	0	−1	−1	−0.71	−0.37	0

PRACTICE EXERCISES See pages 476–477.

Graph each by addition of ordinates over at least a one-period interval.

A
1. $y = 2 \sin x + \sin x$
2. $y = 3 \sin x + \sin x$
3. $y = \sin x + 2 \cos x$
4. $y = 2 \sin x + \cos x$
5. $y = 3 \sin x + \sin 2x$
6. $y = 4 \sin x + \sin 2x$
7. $y = 2 \sin x - \cos x$
8. $y = 2 \cos x - \sin x$
9. $y = \sin 2x + \cos 2x$
10. $y = \sin 3x + \cos 3x$

B
11. $y = \sin 2x - \cos 2x$
12. $y = \cos 2x - \sin 2x$
13. $y = \sin 3x + \cos 2x$
14. $y = \sin x + \cos 2x$
15. $y = -2 \sin x - \cos x$
16. $y = -\sin x - 2 \cos x$

C
17. $y = \cos\left(x - \frac{\pi}{4}\right) + \sin\left(x - \frac{\pi}{4}\right)$
18. $y = \sin\left(x + \frac{\pi}{3}\right) + \cos\left(x + \frac{\pi}{2}\right)$
19. $y = \sin x + \cos x + \sin 2x$
20. $y = 2 \sin x + \sin \frac{1}{2}x + \sin 2x$

Applications

21. **Music** Two strings of a piano are out of tune. Their sound waves can be represented by $y_1 = \sin 2\pi x$ and $y_2 = \sin 3\pi x$. Graph $y_1 = \sin 2\pi x$ and $y_2 = \sin 3\pi x$. Then graph $y = \sin 2\pi x + \sin 3\pi x$ on the same plane.
See page 477.

22. Oceanography The graph of a tide function can be represented by the equation $y = 2 \sin\left(x - \frac{\pi}{4}\right) + 5 \sin x$. Graph the function.

TRIGONOMETRY AND THE CALENDAR

The table below lists times of sunrise and sunset at Lancaster, Pennsylvania, which is located at approximately 40°N latitude, on various dates during a year. The number of hours of daylight is shown in the last column.

Date	Sunrise (AM)	Sunset (PM)	Hours of Daylight	Date	Sunrise (AM)	Sunset (PM)	Hours of Daylight
Jan 1	7:22	4:44	9.37	Jul 13	4:42	7:29	14.78
15	7:20	4:58	9.63	30	4:57	7:16	14.32
30	7:11	5:16	10.08	Aug 15	5:12	6:56	13.73
Feb 15	6:54	5:35	10.68	29	5:25	6:36	13.18
28	6:36	5:50	11.23	Sep 15	5:41	6:09	12.47
Mar 15	6:11	6:07	11.93	30	5:55	5:44	11.28
30	5:48	6:22	12.57	Oct 15	6:10	5:20	11.17
Apr 14	5:24	6:38	13.23	30	6:27	5:06	10.65
29	5:03	6:53	13.83	Nov 15	6:45	4:43	9.97
May 14	4:46	7:08	14.37	30	7:02	4:36	9.57
29	4:34	7:20	14.77	Dec 15	7:15	4:36	9.35
Jun 15	4:30	7:31	15.02	30	7:21	4:44	9.38
30	4:34	7:33	14.98				

A graph of the hours of daylight is shown in red below. The graph is periodic and approximates a sine curve.

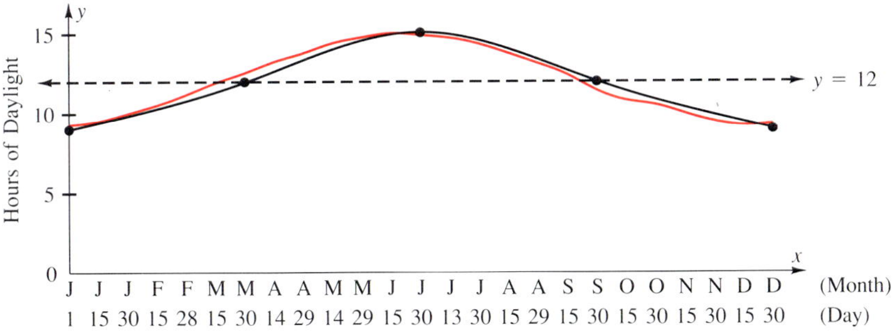

1. Find the equation of the sine function shown by the black curve. $y = 3 \sin\left(x - \frac{\pi}{2}\right) + 12$
2. Find the times of sunrise and sunset for the area in which you live. (You must know the approximate latitude of your area to find the times.) Make a table and a graph similar to those shown above. Answers may vary.

Enrichment

If you were to add the two given functions $y = x^2$ and $y = 2x - 8$, what would the graph of the sum look like? Where would the graph cross the x-axis? A parabola, opening up, with vertex at $(-1, -9)$ and axis of symmetry with equation $x = -1$; $(-4, 0)$ and $(2, 0)$

Teacher's Resource Book

Practice—Chapter 2, p. 9
Enrichment—Chapter 2, p. 10

LESSON PLAN

Vocabulary
Asymptotes
Discontinuous

Materials/Manipulatives
Scientific calculators
Graphs of tangent and cotangent functions on acetate sheets
Graphing calculators
Computer

BACKGROUND

In the Preview, finding the asymptotes of a hyperbola is reviewed.

Critical Thinking

Analysis Ask students to examine the figure below showing angle θ in standard position, a unit circle P with its center at the origin, and $\overline{QT}$ and $\overline{RS}$ tangent to the circle at Q and R, respectively. Then ask them to draw conclusions from the figure by answering the following questions:

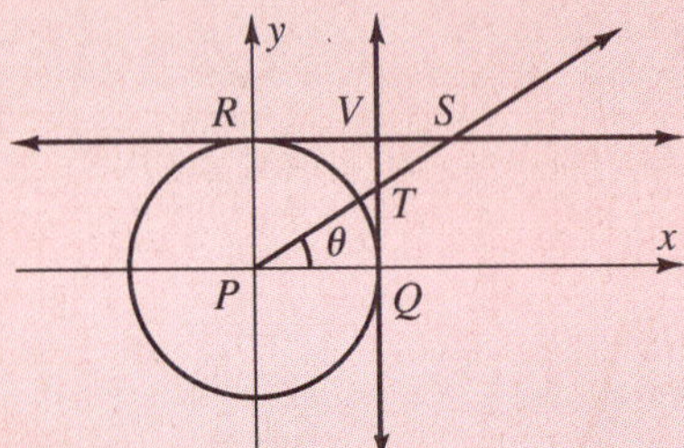

Which trigonometric functions of θ are represented by the lengths of $\overline{QT}$ and $\overline{RS}$? As θ increases from 0 to $\frac{\pi}{2}$, what happens to the length of $\overline{QT}$? As θ decreases from $\frac{\pi}{2}$ to 0, what happens to the length of $\overline{RS}$? What principle does this figure illustrate? $\overline{QT}$ represents the tangent of θ and $\overline{RS}$ represents the cotangent of θ. The length of $\overline{QT}$ increases; the length of $\overline{RS}$ decreases; tangent is undefined at $\frac{\pi}{2}$ and cotangent is undefined at 0.

2.6 Graphs of the Tangent and Cotangent Functions

Objectives: To determine the properties of the tangent and cotangent functions
To graph the tangent and cotangent functions

The tangent and cotangent functions are not defined for all real numbers, and their graphs are not like the graphs of the sine and cosine functions.

Preview

The graph of a hyperbola is discontinuous, and it has two **asymptotes,** which are lines that the graph approaches but never crosses as it gets infinitely far from the origin. The asymptotes of a hyperbola with an equation of the form $\frac{x^2}{a^2} - \frac{y^2}{b^2} = 1$ have equations of the form

$$y = \frac{b}{a}x \quad \text{and} \quad y = -\frac{b}{a}x$$

Similarly, the asymptotes of a hyperbola with an equation of the form $\frac{y^2}{a^2} - \frac{x^2}{b^2} = 1$ have equations of the form

$$y = \frac{a}{b}x \quad \text{and} \quad y = -\frac{a}{b}x$$

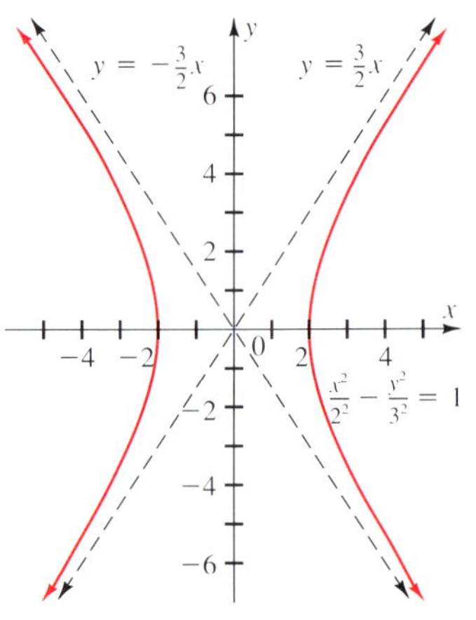

Find the equations of the asymptotes of each hyperbola. Then sketch the graph of the hyperbola. See page 477.

1. $\frac{x^2}{5^2} - \frac{y^2}{2^2} = 1$ $y = \pm\frac{2}{5}x$ **2.** $\frac{y^2}{3^2} - \frac{x^2}{4^2} = 1$ $y = \pm\frac{3}{4}x$ **3.** $x^2 - \frac{y^2}{36} = 1$ $y = \pm 6x$

In order to graph the tangent and cotangent functions, it will be useful to construct a table of values. Recall that the tangent is not defined for $\pm\frac{\pi}{2}, \pm\frac{3\pi}{2}, \ldots$. Near these values of x, the value of tan x increases or decreases without bound. Consider the values of tan x as x approaches $\frac{\pi}{2}$ from 0. Round answers to the nearest hundredth.

x	0	$\frac{\pi}{6}$	$\frac{\pi}{4}$	$\frac{\pi}{3}$	$\frac{\pi}{2.1}$	$\frac{\pi}{2.01}$	$\frac{\pi}{2.001}$	$\frac{\pi}{2}$
tan x	0	0.58	1	1.73	13.34	127.96	1273.88	undefined

The value of tan x increases without bound as x approaches $\frac{\pi}{2}$ from the left. The tangent of $\frac{\pi}{2}$ is not defined. Similarly, as x decreases from π to $\frac{\pi}{2}$, tan x decreases without bound. This can be seen by reading the table below from right to left.

x	$\frac{\pi}{2}$	$\frac{\pi}{1.999}$	$\frac{\pi}{1.99}$	$\frac{\pi}{1.9}$	$\frac{2\pi}{3}$	$\frac{3\pi}{4}$	$\frac{5\pi}{6}$	π
tan x	undefined	−1272.6	−126.68	−12.07	−1.73	−1.00	−0.58	0

After x increases from 0 to π, the values of tan x begin to repeat. That is, $\tan (x + \pi) = \tan x$. Thus, the period of the tangent function is π.

EXAMPLE 1 **Graph $y = \tan x$ over the interval $-\pi \leq x \leq 2\pi$.**

Use the tables of values shown above to find the coordinates of the points over the interval $0 \leq x \leq \pi$, and sketch the curve over that interval. Then use the fact that the period of the function is π to extend the graph in both directions.

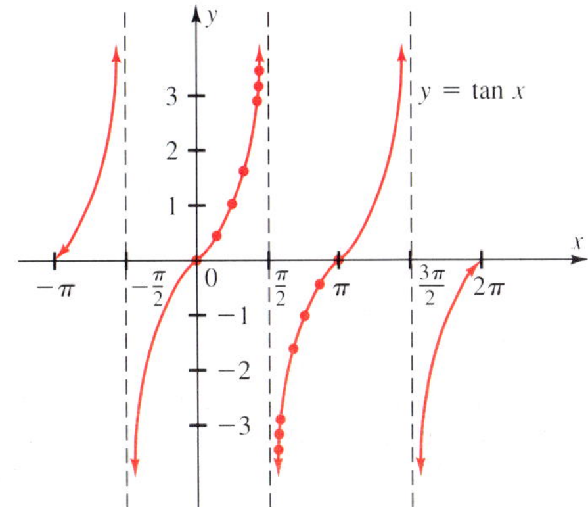

Tan x is undefined for some values of x. Therefore, the tangent function is said to be **discontinuous,** since its graph has breaks. Like the hyperbola, the graph of the tangent function approaches asymptotes at certain values in the domain of the function. The asymptotes are lines with equations of the form $x = \left(\frac{2n + 1}{2}\right)\pi$, where n is an integer. The range of $y = \tan x$ is the set of all real numbers. The amplitude is not defined because tan x increases or decreases without bound as x approaches the asymptotes.

Since the tangent and cotangent functions are reciprocal functions, the value of cot x is undefined when $\tan x = 0$, and it is zero when tan x is undefined.

TEACHING SUGGESTIONS

- One way of having students get the feel for the discontinuity of the tangent and cotangent functions is to use the scientific calculator and take values of the angles closer and closer to the points of discontinuity so that students actually see the values increasing and decreasing.
- Encourage students to use their scientific calculators to find the values of the functions.
- Have students use their graphing calculators or a computer to see the effects of the various constants on the tangent and cotangent functions.

CHALKBOARD EXAMPLES

- **For Example 1**

1. Graph $y = \tan x$ over the interval $0 \leq x \leq \pi$.

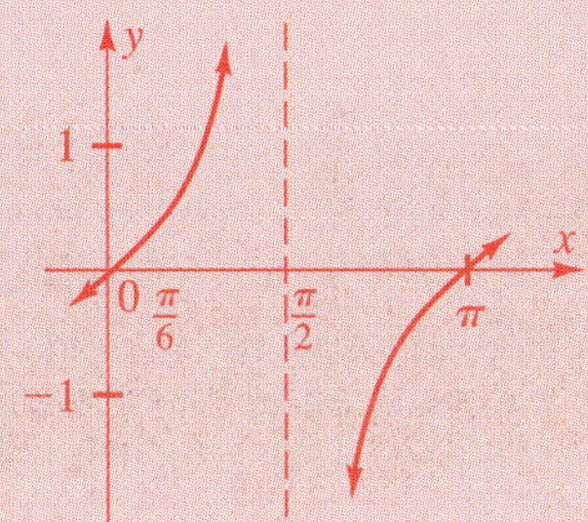

- **For Example 2**

2. Graph $y = \cot x$ over the interval $-\pi \le x \le 0$.

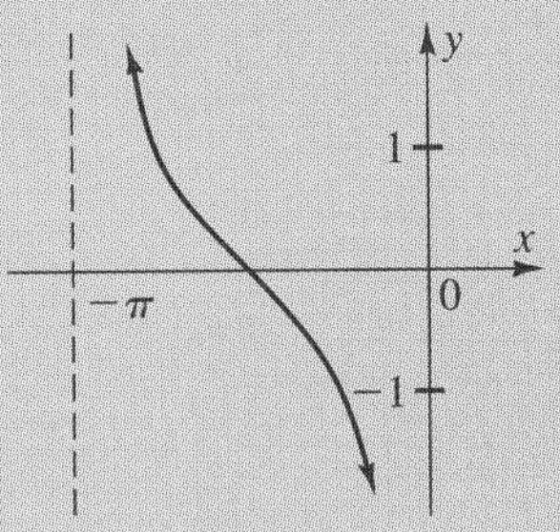

EXAMPLE 2 **Graph $y = \cot x$ over the interval $0 < x < 2\pi$.**

First, make a table of values for the interval $0 \le x \le \pi$.

x	0	$\frac{\pi}{6}$	$\frac{\pi}{4}$	$\frac{\pi}{3}$	$\frac{\pi}{2}$	$\frac{2\pi}{3}$	$\frac{3\pi}{4}$	$\frac{5\pi}{6}$	π
$\cot x$	undefined	1.73	1	0.58	0	-0.58	-1	-1.73	undefined

The asymptotes for $y = \cot x$ have equations of the form $x = n\pi$, where n is an integer. It is helpful to sketch the asymptotes of a tangent or cotangent function before drawing the graph. Then use the table of values to plot points and sketch the curve over the interval $0 < x < \pi$. Use the fact that the period is π to extend the graph. The range of $y = \cot x$ is the set of all real numbers, and the amplitude is not defined.

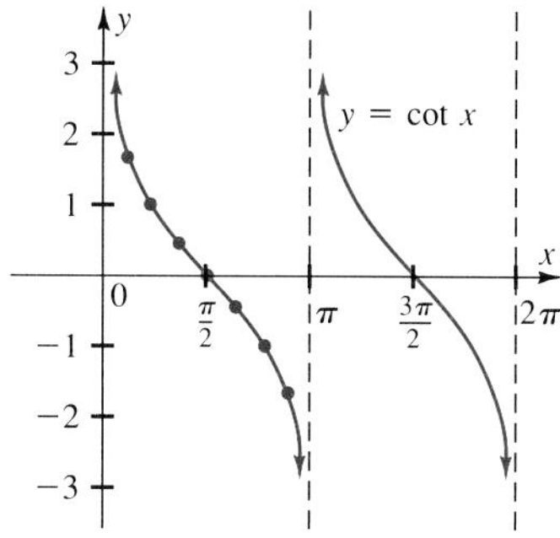

Even though the tangent and cotangent are cofunctions, the graph of the cotangent function is not a phase shift of the tangent function. The functions are reciprocals, and this is reflected in their graphs. When the value of the tangent is very small, the value of the cotangent is very large. The reverse is also true.

Constants affect the graphs of the tangent and cotangent functions in many of the same ways that they affect the sine and cosine functions.

> For $y = a \tan b(x + c) + d$ and $y = a \cot b(x + c) + d$
>
> The constant a vertically stretches or shrinks the graph compared to the graph of $y = \tan x$ or $y = \cot x$.
>
> The period is $\frac{\pi}{|b|}$.
>
> The phase shift from $y = a \tan bx$ or $y = a \cot bx$ is c units to the left if $c > 0$, and $|c|$ units to the right if $c < 0$.
>
> The vertical shift from $y = a \tan bx$ or $y = a \cot bx$ is d units upward if $d > 0$, and $|d|$ units downward if $d < 0$.

Verify your graphs using a graphing calculator or computer.

EXAMPLE 3 **Determine the period, the phase shift, and the vertical shift of $y = 3\tan 2\left(x + \frac{\pi}{4}\right) - 5$. Describe any stretching or shrinking.**

Period: $\frac{\pi}{|2|} = \frac{\pi}{2}$ $\quad$ *Period* $= \frac{\pi}{|b|}$

Phase shift: $\left|\frac{\pi}{4}\right| = \frac{\pi}{4}$ units to the left, since $c > 0$

Vertical shift: $|-5| = 5$ units down, since $d < 0$

Since $a = 3$, the graph is stretched vertically by a factor of 3.

You can use your knowledge of the period, the phase shift, and the vertical shift to graph the tangent and cotangent functions.

EXAMPLE 4 **Sketch each graph over the given interval:**

a. $y = \tan 3x$; $-\pi \le x \le \pi$ $\qquad$ **b.** $y = \cot\left(x - \frac{\pi}{4}\right)$; $-\frac{3\pi}{4} \le x \le \frac{17\pi}{4}$

a. Period: $\frac{\pi}{|3|} = \frac{\pi}{3}$ $\quad$ $b = 3$

Phase shift: none $\quad$ $c = 0$

Vertical shift: none $\quad$ $d = 0$

The asymptotes occur when $\tan 3x$ is undefined, so the asymptotes have the equations $x = -\frac{5\pi}{6}$, $x = -\frac{\pi}{2}$, $x = -\frac{\pi}{6}$, $x = \frac{\pi}{6}$, $x = \frac{\pi}{2}$, and $x = \frac{5\pi}{6}$.

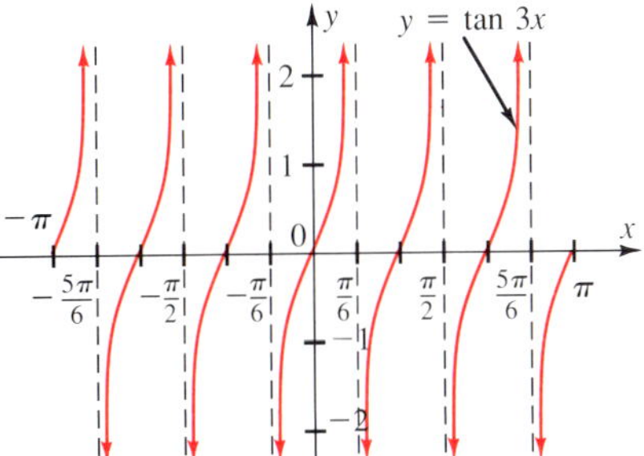

b. Period: $\frac{\pi}{|1|} = \pi$ $\quad$ $b = 1$

Phase shift: $\left|-\frac{\pi}{4}\right| = \frac{\pi}{4}$ units to the right, since $c < 0$

Vertical shift: none $\quad$ $d = 0$

The asymptotes occur when $\cot\left(x - \frac{\pi}{4}\right)$ is undefined, so the asymptotes have the equations $x = -\frac{3\pi}{4}$, $x = \frac{\pi}{4}$, $x = \frac{5\pi}{4}$, $x = \frac{9\pi}{4}$, $x = \frac{13\pi}{4}$, and $x = \frac{17\pi}{4}$.

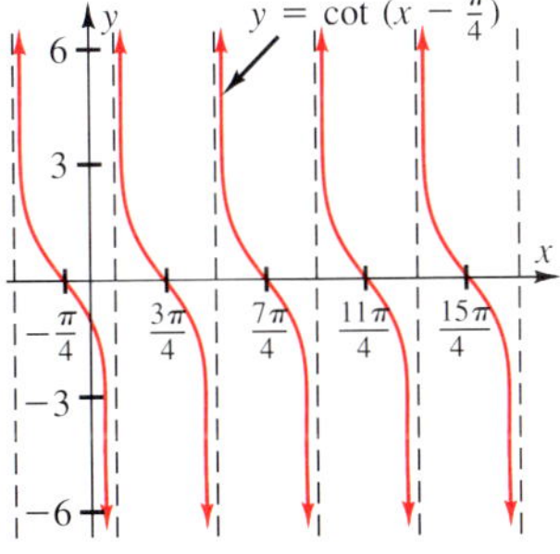

- **For Example 3**

3. Find the period, the phase shift, and the vertical shift of the graph of $y = 2\cot 3\left(x - \frac{\pi}{2}\right) + 4$. Describe the stretching or shrinking of the graph. $\frac{\pi}{3}$; $\frac{\pi}{2}$ to right; 4 units up; stretched vertically by a factor of 2

- **For Example 4**

4. Graph $y = \tan 2\left(x - \frac{\pi}{4}\right)$ over the interval $-\pi \le x \le \pi$.

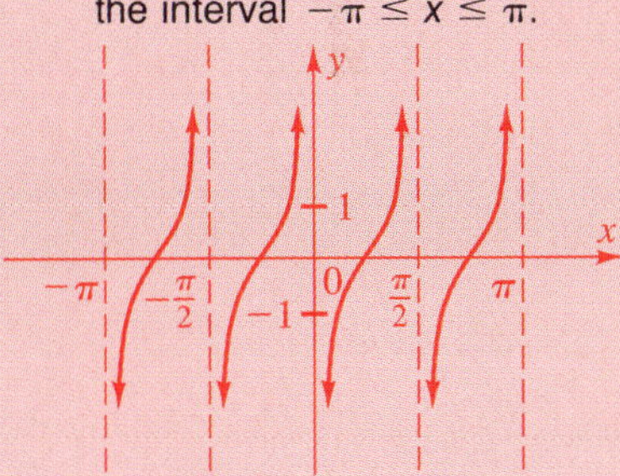

Common Errors

- Some students sketch the cotangent for the tangent. Remind those students that as the measure of the angle increases the value of the tangent increases while the value of the cotangent decreases.
- When graphing various forms of the tangent and cotangent functions, some students use incorrect asymptotes. Post a copy of the graphs of $y = \tan x$ and $y = \cot x$ so that the students can refer to them.
- See *Teacher's Resource Book* for additional remediation.

LESSON FOLLOW-UP

Discussion

Why is it easy to see the difference between the graphs of $y = \cot x$ and $y = \cot x + c$? The constant c shifts the entire graph up or down from the x-axis.

Assignment Guide

See p. 58B for assignments.

Lesson Quiz

Find the period, the phase shift, and the vertical shift, if any, for each function.

1. $y = \tan \frac{1}{3}x$ 3π; 0; 0

2. $y = \cot \frac{3}{4}x + 2$ $\frac{4\pi}{3}$; 0; 2 up

3. $y = \tan \frac{1}{2}(x - \pi) - 1$ 2π; π right; 1 down

4. $y = \cot \frac{2}{3}\left(x + \frac{\pi}{4}\right)$ $\frac{3\pi}{2}$; $\frac{\pi}{4}$ left; 0

5. Graph $y = \cot \frac{1}{2}x$ over the interval $-\pi \leq x \leq \pi$.

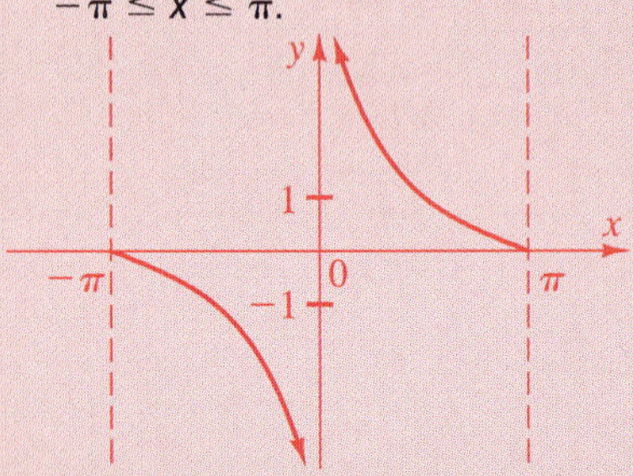

Enrichment

The tangent and cotangent are reciprocal functions. What does this fact tell you about the location of the asymptotes of one graph in relation to the other?
Tangent is undefined where cotangent is zero; cotangent is undefined where tangent is zero.

CLASS EXERCISES

Determine the period of each function.

1. $y = \tan \frac{1}{4}x$ 4π **2.** $y = \tan \frac{1}{2}x$ 2π **3.** $y = \cot \frac{2}{3}x$ $\frac{3\pi}{2}$ **4.** $y = \cot \frac{3}{4}x$ $\frac{4}{3}\pi$

Determine the phase shift of each function.

5. $y = \tan\left(x - \frac{\pi}{4}\right)$ $\frac{\pi}{4}$ right

6. $y = \tan\left(x + \frac{\pi}{3}\right)$ $\frac{\pi}{3}$ left

7. $y = \cot\left(x - \frac{2\pi}{3}\right)$ $\frac{2\pi}{3}$ right

8. $y = \cot(2x + \pi)$ $\frac{\pi}{2}$ left

PRACTICE EXERCISES

Determine the period, the phase shift, and the vertical shift, if any, of each function. Describe any vertical stretching or shrinking.

A

1. $y = \tan 3(x - \pi)$ $\frac{\pi}{3}$; π right

2. $y = \cot 2(x + \pi)$ $\frac{\pi}{2}$; π left

3. $y = \tan\left(x - \frac{\pi}{4}\right)$ π; $\frac{\pi}{4}$ right

4. $y = \tan\left(x + \frac{\pi}{2}\right)$ π; $\frac{\pi}{2}$ left

5. $y = \cot\left(x + \frac{2\pi}{3}\right)$ π; $\frac{2\pi}{3}$ left

6. $y = \cot\left(x - \frac{3\pi}{4}\right)$ π; $\frac{3\pi}{4}$ right

7. $y = \tan 2(x + \pi) + 3$ $\frac{\pi}{2}$; π left; 3 up

8. $y = \cot 3(x - \pi) - 5$ $\frac{\pi}{3}$; π right; 5 down

9. $y = \tan\left(x - \frac{4\pi}{3}\right) - 1$ π; $\frac{4\pi}{3}$ right; 1 down

10. $y = \cot\left(x - \frac{5\pi}{6}\right) + 4$ π; $\frac{5\pi}{6}$ right; 4 up

11. $y = 5 \cot\left(x + \frac{2\pi}{3}\right) - 2$ π; $\frac{2\pi}{3}$ left; 2 down; 5

12. $y = 4 \tan\left(x - \frac{3\pi}{4}\right) + 3$ π; $\frac{3\pi}{4}$ right; 3 up; 4

13. $y = 3 \tan\left(x - \frac{\pi}{6}\right) + 2$ π; $\frac{\pi}{6}$ right; 2 up; 3

14. $y = 3 \cot\left(x + \frac{\pi}{6}\right) + 1$ π; $\frac{\pi}{6}$ left; 1 up; 3

Graph each function over a two-period interval. See page 478.

B

15. $y = \cot 2x$

16. $y = \tan 2x$

17. $y = 3 \cot 2x$

18. $y = 4 \cot 2x$

19. $y = \tan(x - \pi)$

20. $y = \cot(x + \pi)$

21. $y = \cot 3\left(x + \frac{2\pi}{3}\right)$

22. $y = \cot 2\left(x - \frac{3\pi}{4}\right)$

23. $y = \tan\left(x - \frac{2\pi}{3}\right) + 2$

24. $y = \cot\left(x + \frac{\pi}{6}\right) - 2$

C

25. $y = \sin x + \tan x$

26. $y = \cos x + \tan x$

27. $y = \sin x + \cot x$

28. $y = \cos x + \cot x$

Applications

Graphing Calculator To use a graphing calculator to graph trigonometric functions, set the mode in radians and choose values for the x- and y-axes. The following values were used for the graph of $y = \tan 2x$ shown at the left below: $-1 \le x \le 3.2$, scale: 0.5 and $-15 \le y \le 15$, scale: 5.

When entering a function to be graphed, use parentheses and operations as necessary. For the graph of $y = 2 \sin 2x$ shown at the right, the following was entered: $y = 2 \times \sin (2 \times x)$.

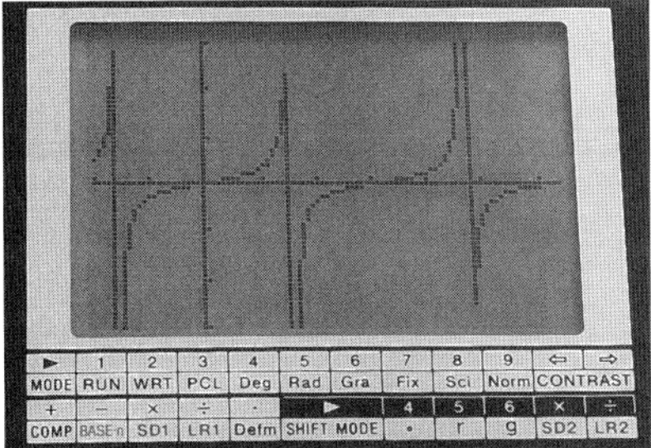

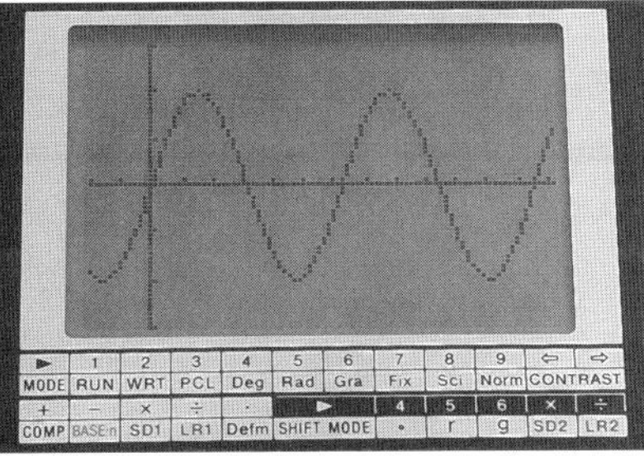

29. Use a graphing calculator to graph $y = 5 \cot 3x$. Answers may vary.

EXTRA

The values of x, $\sin x$, and $\tan x$ are very nearly the same for small positive values of x.

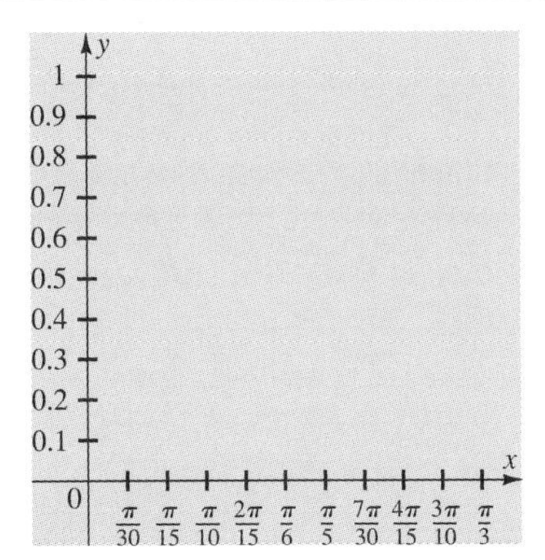

1. Graph $y = x$ and $y = \sin x$ carefully on the same coordinate plane over the interval $0 \le x \le \frac{\pi}{3}$. Label the vertical axis 0, 0.1, 0.2, . . . , 0.8, 0.9, and 1, as shown. Over approximately what interval do the nonnegative values of x and $\sin x$ differ by less than 0.05? $0 \le x \le 0.65$; See side column.
2. Without graphing, give the approximate interval over which the negative values of x and $\sin x$ differ by less than 0.05. $-0.65 \le x \le 0$
3. Graph $y = x$ and $y = \tan x$ carefully on the same coordinate plane over the interval $0 \le x \le \frac{\pi}{4}$. Over approximately what interval do the nonnegative values of x and $\tan x$ differ by less than 0.05? $0 \le x \le 0.50$; See side column.
4. Without graphing, give the approximate interval over which the negative values of x and $\tan x$ differ by less than 0.05. $-0.50 \le x \le 0$

Teacher's Resource Book
Practice—Chapter 2, p. 11
Enrichment—Chapter 2, p. 12

Applications
Note that when plotting graphs of functions with asymptotes you may obtain asymptote-like lines on a graphing calculator as shown in the photo on the left.

Additional Answers

Extra

1.

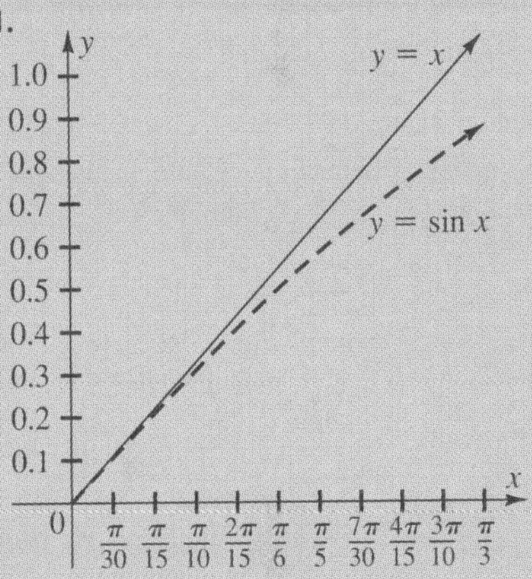

3.

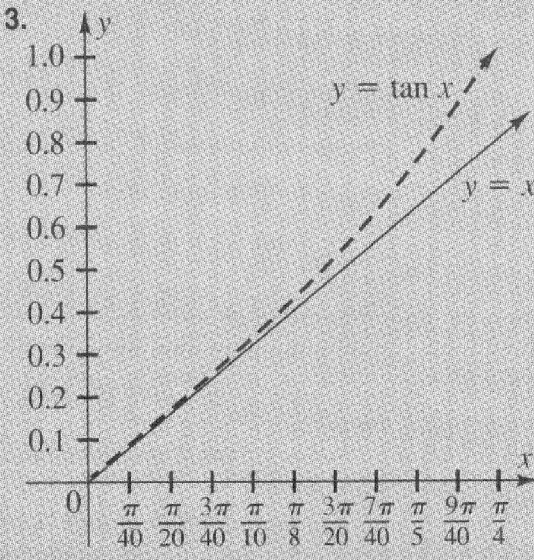

LESSON PLAN

Materials/Manipulatives

Scientific calculators
Graphs of the sine, cosine, cosecant, and secant functions on acetate sheets
Graphing calculators
Computer

BACKGROUND

In the Preview, evaluating secants and cosecants of the special angles measured in radians is reviewed.

Critical Thinking

Application Ask students to combine what they know about discontinuous functions, the fact that the secant and cosecant functions are reciprocals of the sine and cosine functions, and the shape of the graphs of the sine and cosine, to sketch the graphs of the secant and cosecant. Then ask them to compare their sketches to the actual graphs in Examples 1 and 2 on text pages 95 and 96. Answers may vary.

2.7 Graphs of the Secant and Cosecant Functions

Objectives: To determine the properties of the secant and cosecant functions
To graph the secant and cosecant functions

The sine and cosecant are reciprocal functions, as are the cosine and secant. Use these relationships to evaluate specific values of the cosecant and secant functions.

Preview

EXAMPLE **Evaluate:** $\sec \frac{\pi}{3}$

$$\sec \frac{\pi}{3} = \frac{1}{\cos \frac{\pi}{3}} = \frac{1}{\frac{1}{2}} = 2$$

Find each value. Write *undefined*, where appropriate.

1. $\csc \frac{\pi}{3}$ $\frac{2\sqrt{3}}{3}$ **2.** $\sec \frac{\pi}{2}$ undefined **3.** $\csc \pi$ undefined **4.** $\sec \frac{2\pi}{3}$ -2

5. $\sec \frac{\pi}{4}$ $\sqrt{2}$ **6.** $\csc \frac{7\pi}{4}$ $-\sqrt{2}$ **7.** $\sec \pi$ -1 **8.** $\csc 2\pi$ undefined

Since the secant and cosecant functions are not defined when the values of the cosine and sine functions, respectively, are equal to zero, their graphs are discontinuous and have asymptotes. Consider the values of $\sec x$ as x approaches $\frac{\pi}{2}$.

x	0	$\frac{\pi}{6}$	$\frac{\pi}{4}$	$\frac{\pi}{3}$	$\frac{\pi}{2.1}$	$\frac{\pi}{2.01}$	$\frac{\pi}{2}$
$\sec x$	1	1.15	1.41	2	13.38	127.96	undefined

x	$\frac{\pi}{2}$	$\frac{\pi}{1.99}$	$\frac{\pi}{1.9}$	$\frac{2\pi}{3}$	$\frac{3\pi}{4}$	$\frac{5\pi}{6}$	π
$\sec x$	undefined	-126.69	-12.11	-2	-1.41	-1.15	-1

The value of $\sec x$ increases without bound as x approaches $\frac{\pi}{2}$ from the left. The secant of $\frac{\pi}{2}$ is not defined. As x decreases from π to $\frac{\pi}{2}$, $\sec x$ decreases without bound. Thus, $x = \frac{\pi}{2}$ is an asymptote.

x	π	$\frac{7\pi}{6}$	$\frac{5\pi}{4}$	$\frac{4\pi}{3}$	$\frac{3\pi}{2.1}$	$\frac{3\pi}{2.01}$	$\frac{3\pi}{2.001}$	$\frac{3\pi}{2}$
$\sec x$	-1	-1.15	-1.41	-2	-4.49	-42.66	-424.63	undefined

x	$\frac{3\pi}{2}$	$\frac{3\pi}{1.999}$	$\frac{3\pi}{1.99}$	$\frac{3\pi}{1.9}$	$\frac{5\pi}{3}$	$\frac{7\pi}{4}$	$\frac{11\pi}{6}$	2π
$\sec x$	undefined	424.20	42.23	4.07	2	1.41	1.15	1

As x increases from π to $\frac{3\pi}{2}$, $\sec x$ decreases without bound. As x decreases from 2π to $\frac{3\pi}{2}$, $\sec x$ increases without bound. Thus, $x = \frac{3\pi}{2}$ is an asymptote. Other asymptotes occur at $x = \left(\frac{2n+1}{2}\right)\pi$, where n is an integer.

To graph $y = \sec x$, first sketch the asymptotes. It is also helpful to sketch the graph of $y = \cos x$. Since the period of $y = \cos x$ is 2π, the period of its reciprocal function, $y = \sec x$, is also 2π. The graphs of $y = \cos x$ and $y = \sec x$ intersect at the points where both $\cos x$ and $\sec x$ equal 1, and at the points where both $\cos x$ and $\sec x$ equal -1.

EXAMPLE 1 **Graph $y = \sec x$ over the interval $-\frac{\pi}{2} < x < \frac{7\pi}{2}$.**

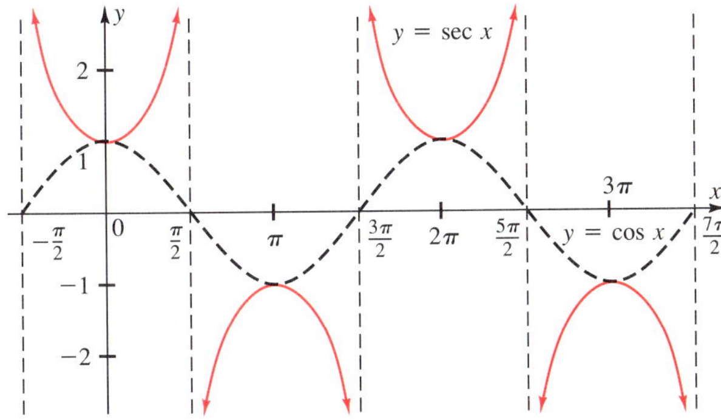

The amplitude is not defined because $y = \sec x$ increases or decreases without bound as x approaches the asymptotes. The range of $y = \sec x$ is the set of all real numbers y such that $|y| \geq 1$.

The graph of $y = \csc x$ can be constructed in a similar manner. Asymptotes occur when $\sin x = 0$, since $\csc x = \frac{1}{\sin x}$. Since the period of $y = \sin x$ is 2π, the period of $y = \csc x$ is also 2π. The asymptotes for $y = \csc x$ occur when $x = n\pi$, where n is any integer.

TEACHING SUGGESTIONS

- Remind students of the form that the sine and cosine functions took when there was a change in amplitude or period, a phase shift or a vertical shift. Show them how these changes are illustrated with the secant and cosecant functions.
- Encourage students to use their scientific calculators to find the values of the functions.
- Have students use their graphing calculators or a computer to see the effects of the various constants on the secant and cosecant functions.

CHALKBOARD EXAMPLES

- **For Example 1**

  1. Graph $y = \sec x$ over the interval $-\pi \leq x \leq \pi$.

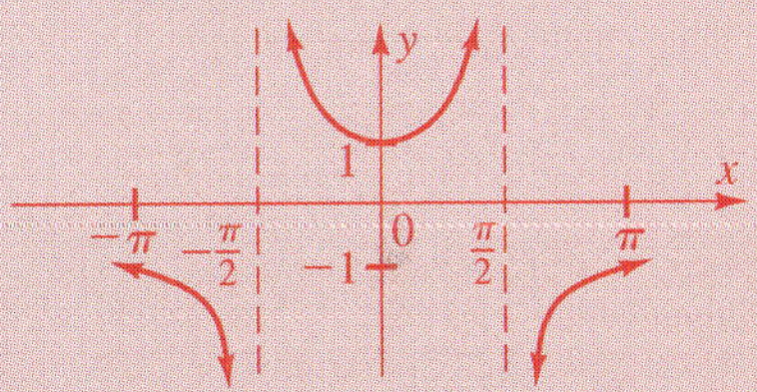

- **For Example 2**

  2. Graph $y = \csc x$ over the interval $-\pi \leq x \leq \pi$.

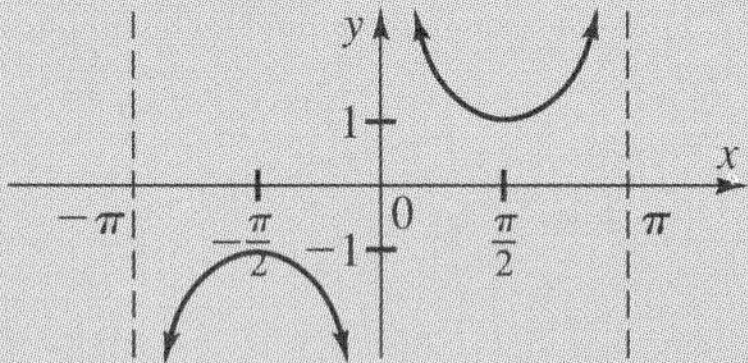

EXAMPLE 2 **Graph $y = \csc x$ over the interval $0 < x < 4\pi$.**

Sketch the asymptotes and the graph of $y = \sin x$. Then graph the cosecant function.

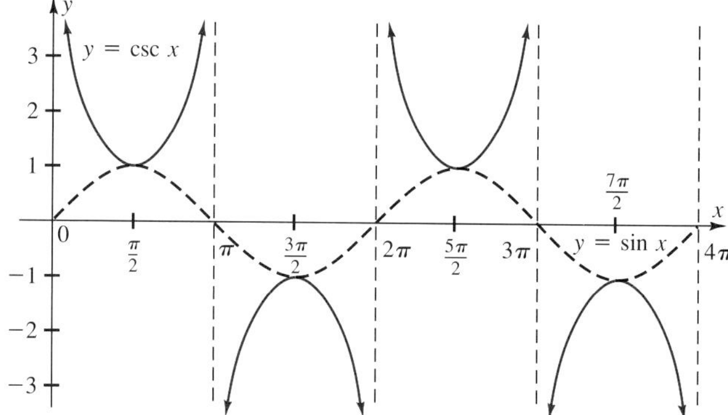

The amplitude is not defined because $y = \csc x$ increases or decreases without bound as x approaches the asymptotes. The range of $y = \csc x$ is the set of all real numbers y such that the $|y| \geq 1$. The graph of the cosecant function is that of the secant function shifted $\frac{\pi}{2}$ units to the right.

Constants affect the graphs of the secant and cosecant functions in many of the same ways that they affect the graphs of the sine and cosine functions.

For $y = a \sec b(x + c) + d$ and $y = a \csc b(x + c) + d$

The constant a vertically stretches or shrinks the graph compared to the graph of $y = \sec x$ or $y = \csc x$.

The period is $\frac{2\pi}{|b|}$.

The phase shift from $y = a \sec bx$ or $y = a \csc bx$ is c units to the left if $c > 0$, and $|c|$ units to the right if $c < 0$.

The vertical shift from $y = a \sec bx$ or $y = a \csc bx$ is d units upward if $d > 0$, and $|d|$ units downward if $d < 0$.

You can also use a graphing calculator or computer to verify the graphs of the secant and cosecant functions.

EXAMPLE 3 **Determine the period, the phase shift, and the vertical shift of $y = 4 \sec 3\left(x + \frac{\pi}{3}\right) - 6$. Describe any stretching or shrinking.**

The period is $\frac{2\pi}{|3|} = \frac{2\pi}{3}$. The phase shift is $\frac{\pi}{|3|} = \frac{\pi}{3}$ units to the left. The vertical shift is $|-6| = 6$ units down, and the graph will be stretched vertically by a factor of 4.

The properties are also useful when you graph a secant or cosecant function.

EXAMPLE 4 **Sketch each graph over the given interval:**

a. $y = \sec \frac{4}{3}x;\ -\frac{9\pi}{8} \le x \le \frac{15\pi}{8}$ **b.** $y = \csc\left(x - \frac{\pi}{4}\right);\ -\frac{3\pi}{4} \le x \le \frac{13\pi}{4}$

a. $y = \sec \frac{4}{3}x$

Period: $\frac{2\pi}{\left|\frac{4}{3}\right|} = \frac{3}{2}\pi.$

Phase shift: none

Vertical shift: none

No vertical stretching or shrinking

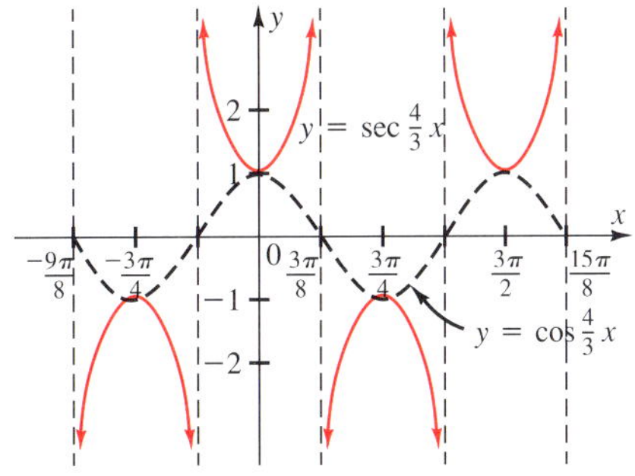

Asymptotes: $\frac{4}{3}x = \left(\frac{2n+1}{2}\right)\pi$

$x = \frac{3}{4}\left(\frac{2n+1}{2}\right)\pi, = \left(\frac{6n+3}{8}\right)\pi$

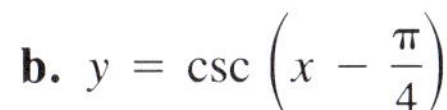

b. $y = \csc\left(x - \frac{\pi}{4}\right)$

Period: $\frac{2\pi}{|1|} = 2\pi$

Phase shift: $\left|-\frac{\pi}{4}\right| =$

$\frac{\pi}{4}$ units to the right

Vertical shift: none

No vertical stretching or shrinking

Asymptotes: $x - \frac{\pi}{4} = n\pi$

$x = \frac{\pi}{4} + n\pi$

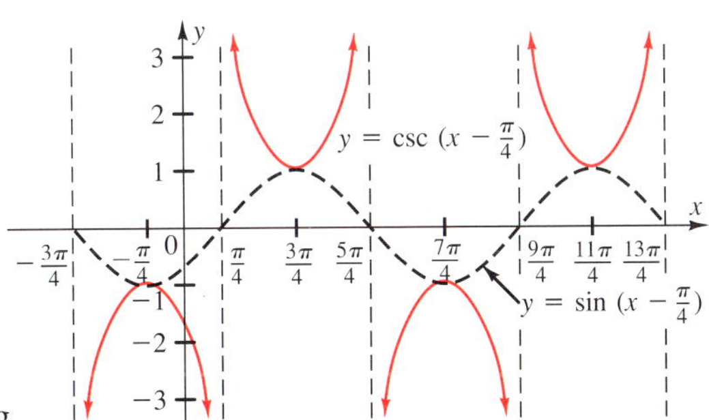

- **For Example 3**
  3. Find the period, the phase shift, and the vertical shift of the graph of $y = 3 \sec 2(x + \pi) - 4$. Describe any stretching or shrinking of the graph. π; π left; 4 down; stretched vertically by a factor of 3

- **For Example 4**
  4. Sketch the graph of $y = \csc \frac{2}{3}(x - \pi) - 1$ over the interval $-\pi \le x \le \pi$.

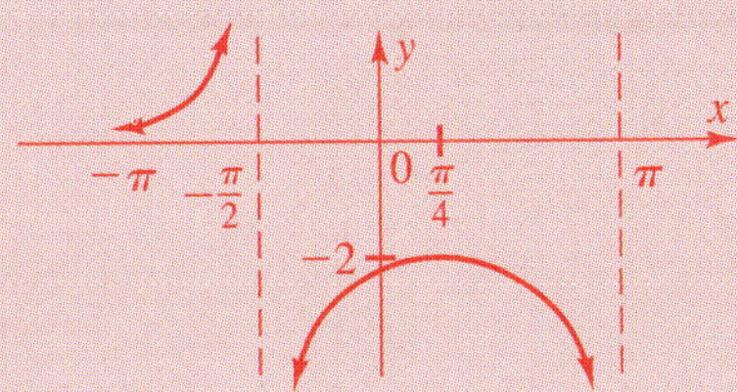

Common Error

- When graphing various forms of the secant and cosecant functions, some students use incorrect asymptotes. Remind those students that the secant and cosecant functions are undefined where the values of the cosine and sine, respectively, are zero. Also, post a copy of the graphs of $y = \sec x$ and $y = \csc x$ so that the students can refer to them.
- See *Teacher's Resource Book* for additional remediation.

LESSON FOLLOW-UP

Discussion

Why is it easy to see the difference between the graphs of $y = \csc x$ and $y = \csc x + c$? The constant c shifts the entire graph up or down from the x-axis.

Critical Thinking

Analysis Ask students to compare the pair of equations $y = \tan x$ and $y = a \tan x$ to the pair of equations $y = \sec x$ and $y = a \sec x$ and to explain in which pair the graphs are more easily distinguishable. Students note that $y = a \tan x$ is continuous with no maximum or minimum values and that in $y = a \sec x$, a affects the value of the maximum and minimum values; thus they conclude that the graphs of $y = \sec x$ and $y = a \sec x$ are more easily distinguishable.

Assignment Guide

See p. 58B for assignments.

Lesson Quiz

Find the period, the phase shift, and the vertical shift, if any, of each function.

1. $y = \sec 2x + 3$ π; 0; 3 up
2. $y = 3 \csc 4x - 2$ $\frac{\pi}{2}$; 0; 2 down
3. $y = \frac{2}{3} \sec \frac{1}{2}(x - \pi)$ 4π; π right; 0
4. $y = \frac{1}{2} \sec \left(x - \frac{\pi}{3}\right) + 1$ 2π; $\frac{\pi}{3}$ right; 1 up
5. $y = \frac{4}{3} \csc \frac{1}{2}(x + \pi)$ 4π; π left; 0

Enrichment

Write the equations for the two horizontal tangents for the graphs of the secant and cosecant.
$y = 1$ and $y = -1$

CLASS EXERCISES

For discussion

1. Why are there no values of $\sec x$ and $\csc x$ between -1 and 1?
Neither $\sin x$ nor $\cos x$ is greater than 1 or less than -1.

Determine the period of each function.

2. $y = \sec \frac{1}{4} x$ 8π
3. $y = \sec \frac{1}{2} x$ 4π
4. $y = \csc \frac{2}{3} x$ 3π
5. $y = \csc \frac{3}{4} x$ $\frac{8\pi}{3}$

Find the phase shift of each function.

6. $y = \sec \left(x - \frac{\pi}{2}\right)$ $\frac{\pi}{2}$ right
7. $y = \sec \left(x + \frac{\pi}{4}\right)$ $\frac{\pi}{4}$ left
8. $y = \csc (x + \pi)$ π left
9. $y = \csc \left(x - \frac{2\pi}{3}\right)$ $\frac{2\pi}{3}$ right

PRACTICE EXERCISES

Determine the period, the phase shift, and the vertical shift, if any, of each function. Describe any vertical stretching or shrinking.

A

1. $y = \sec 3(x - \pi)$ $\frac{2\pi}{3}$; π right; none
2. $y = \csc 2(x + \pi)$ π; π left; none
3. $y = \sec \left(x - \frac{\pi}{4}\right)$ 2π; $\frac{\pi}{4}$ right; none
4. $y = \sec \left(x + \frac{\pi}{2}\right)$ 2π; $\frac{\pi}{2}$ left; none
5. $y = \csc \left(x + \frac{2\pi}{3}\right)$ 2π; $\frac{2\pi}{3}$ left; none
6. $y = \csc \left(x - \frac{3\pi}{4}\right)$ 2π; $\frac{3\pi}{4}$ right; none
7. $y = \sec 2(x + \pi) + 3$ π; π left ; 3 up
8. $y = \csc 3(x - \pi) - 5$ $\frac{2\pi}{3}$; π right; 5 down
9. $y = \sec \left(x - \frac{4\pi}{3}\right) - 1$ 2π; $\frac{4\pi}{3}$ right; 1 down
10. $y = \csc \left(x + \frac{5\pi}{6}\right) + 4$ 2π; $\frac{5\pi}{6}$ left; 4 up
11. $y = \csc \left(3x + \frac{2\pi}{3}\right) - 2$ $\frac{2\pi}{3}$; $\frac{2\pi}{9}$ left; 2 down
12. $y = \sec \left(4x - \frac{3\pi}{4}\right) + 3$ $\frac{\pi}{2}$; $\frac{3\pi}{16}$ right; 3 up
13. $y = \sec \left(2x + \frac{\pi}{2}\right) - 1$ π; $\frac{\pi}{4}$ left; 1 down
14. $y = \csc \left(2x - \frac{3\pi}{2}\right) + 1$ π; $\frac{3\pi}{4}$ right; 1 up

Graph each function over a two-period interval. See page 479.

B

15. $y = \csc 2x$
16. $y = \sec 2x$
17. $y = 3 \csc 2x$
18. $y = 4 \csc 2x$
19. $y = \sec (x - \pi)$
20. $y = \csc (x + \pi)$
21. $y = \csc 2\left(x + \frac{2\pi}{3}\right)$
22. $y = \csc 3\left(x - \frac{3\pi}{4}\right)$

23. $y = \sec(2x - \pi)$ See page 479.

24. $y = \sec(2x + \pi)$

C 25. $y = \sin x + \sec x$

26. $y = \cos x + \sec x$

27. $y = \sin x + \csc x$

28. $y = \cos x + \csc x$

Applications

29. **Engineering** The tension at a given point T in a cable supporting a distributed load can be represented by the equation $T = T_h \sec \theta$, where T_h is the tension at which the cable is horizontal, and θ is the angle between the cable and the horizontal at any point. Sketch the graph when $T_h = 400$ lb, letting θ vary from 0° to 90° in intervals of 10°. See side column.

BIOGRAPHY: Grace Chisholm Young

Grace Chisholm received the first doctorate that was ever granted to a woman in Germany. Her dissertation was entitled "The algebraic groups of spherical trigonometry." Spherical trigonometry is the study of the relations between the sides and angles of triangles drawn on spheres.

She was born in England in 1868 when Great Britain was the center of technological advance. It was also a time of educational reform. It was only at the end of the 19th century that girls were given the opportunity to attend day and boarding schools which would prepare them to qualify for university entrance exams. Grace received an informal education at home until she was ten.

At seventeen, she passed the Cambridge Senior Exam. Because she was a woman, she was not allowed to apply for university entrance. Her family wanted her to be a social worker to pass her time until she married. However, she was determined to continue her studies. Her mother forbade her to study medicine, her first choice, so she chose mathematics. She won a scholarship to Girton College, a new women's college at Cambridge. Since women were not admitted to graduate school in England, she decided to go to Germany to continue her studies after receiving her degree from Cambridge. At the University of Gottingen in Germany, she became a student of Felix Klein, a leading mathematician.

A year after her graduation, Grace Chisholm married Will Young, a former tutor of hers at Girton College. They collaborated on over 220 articles and books, and Grace Chisholm spent her life raising six children and writing mathematics. In 1906 she and her husband published a book entitled *The Theory of Sets of Points*. This was the first textbook on set theory. She also spoke six languages and wrote two children's books on biology. Investigate Grace Chisholm Young.
Answers may vary.

Teacher's Resource Book

Practice—Chapter 2, p. 13

Enrichment—Chapter 2, p. 14

Additional Answers

29.

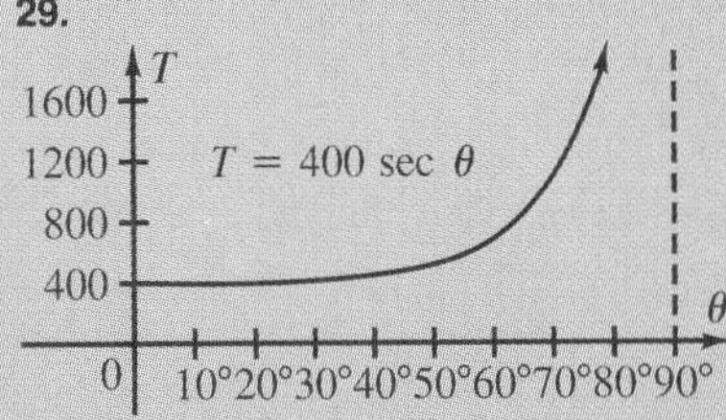

LESSON PLAN

Vocabulary
Displacement
Frequency
Simple harmonic motion
Sinusoidal

Materials/Manipulatives
Scientific calculators
Spring with a weight attached
Oscilloscope (optional)

BACKGROUND

In the Preview, the graph of the motion of a tuning fork is described as sinusoidal. Any motion that can be modeled by a sine-like curve is said to be sinusoidal.

2.8 Application: Harmonic Motion

Objective: To solve problems involving simple harmonic motion

Many motions, including the oscillations of a pendulum, the orbit of a satellite or a space shuttle, and the vibrations of a tuning fork, are periodic in nature. Such motions can often be modeled by sine or cosine curves and are examples of **simple harmonic motion.** Because the graphs of these motions are similar to the graph of the sine, they are said to be **sinusoidal.**

Preview

An oscilloscope can convert sound waves into electrical impulses and display them on a fluorescent screen. The sound of a tuning fork produces a sine-like wave on the screen when used with a microphone.

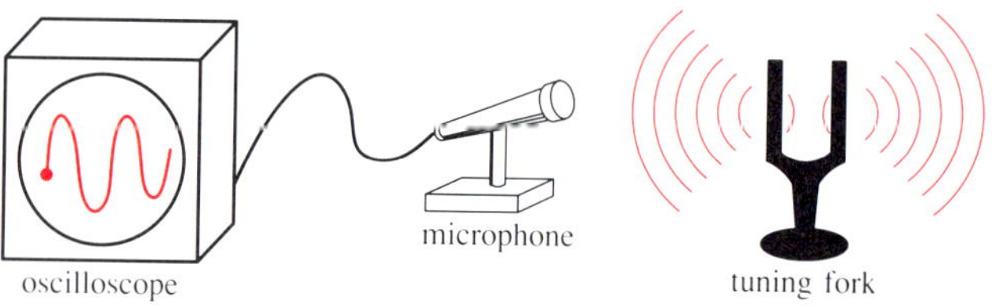

Suppose that the equation for the instantaneous movement of a particular tuning fork is $y = 2 \sin (400\pi t)$, where t is the elapsed time in seconds.

1. What is the amplitude of $y = 2 \sin (400\pi t)$? 2
2. What is the period? $\frac{1}{200}$
3. What is the phase shift? none
4. What is the vertical shift? none

The equations representing sinusoidal graphs can be expressed in the form $y = a \sin b(t + c) + d$ or $y = a \cos b(t + c) + d$, where a, b, c, and d are constants and t represents time. The period of each function is $\frac{2\pi}{|b|}$, the amplitude is $|a|$, the horizontal shift is $|c|$, and the vertical shift is $|d|$. The **frequency** is the reciprocal of the period and is therefore equal to $\frac{|b|}{2\pi}$.

Simple harmonic motion is the motion that occurs when an object is shifted from its rest position and the force attempting to return it to rest is directly proportional to the shift. One example of simple harmonic motion is the vertical oscillation of a weight suspended from a spring. If the weight is pulled down a certain distance, it will oscillate up and down. The maximum distance the weight is pulled from its rest position is called the **displacement,** a. The time the weight takes to reach a given position is represented by t.

EXAMPLE 1 **A weight is at rest hanging from a spring. It is then pulled down 6 cm and released. The weight bounces back and forth through its rest position, making one full cycle from −6 to 6 and then back to −6 every 3 s. This motion can be described by a sine function. Find an equation for the function. Then sketch the graph.**

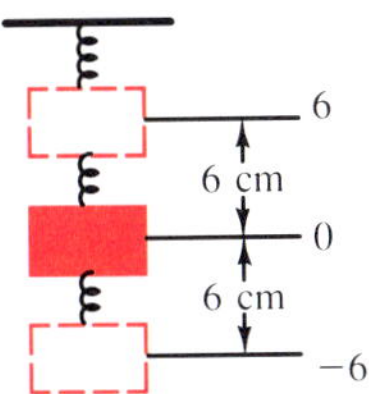

The weight moves up and down between a minimum of −6 and a maximum of 6. So, its amplitude is 6. Its period is 3, since a complete cycle takes 3 s. There is no vertical shift, since the rest position is at 0; that is, $d = 0$.

The motion can be modeled by either the sine or the cosine curve, but the sine has been selected. Thus, $y = a \sin b(t + c)$. The period is $\frac{2\pi}{|b|}$, so $3 = \frac{2\pi}{|b|}$. Therefore, letting b be positive, $b = \frac{2\pi}{3}$.

Substituting $a = 6$, $b = \frac{2\pi}{3}$, and $d = 0$ into $y = a \sin b(t + c) + d$ gives

$$y = 6 \sin \frac{2\pi}{3}(t + c)$$

Remember that $y = -6$ when $t = 0$, since the spring was pulled down 6 cm to start a cycle. Therefore,

$$-6 = 6 \sin \frac{2\pi}{3}(0 + c)$$

$$-6 = 6 \sin \frac{2\pi}{3} c$$

$$-1 = \sin \frac{2\pi}{3} c$$

$$\frac{2\pi}{3} c = -\frac{\pi}{2}$$ *If* $\sin \frac{2\pi}{3} c = -1$, *one possible value for* $\frac{2\pi}{3} c$ *is* $-\frac{\pi}{2}$.

$$c = -\frac{3}{4}$$

Note that other possible values for c would also result in a correct equation. Substituting $-\frac{3}{4}$ for c yields $y = 6 \sin \frac{2\pi}{3}\left(t - \frac{3}{4}\right)$.

TEACHING SUGGESTIONS

- If an oscilloscope can be obtained from the physics department, show the students the results of the vibrations of the tuning fork as described in the preview.
- Use a spring with a weight attached to actually show the students the type of motion described in Example 1. A diagram such as the one below will show how this translated into a graph.

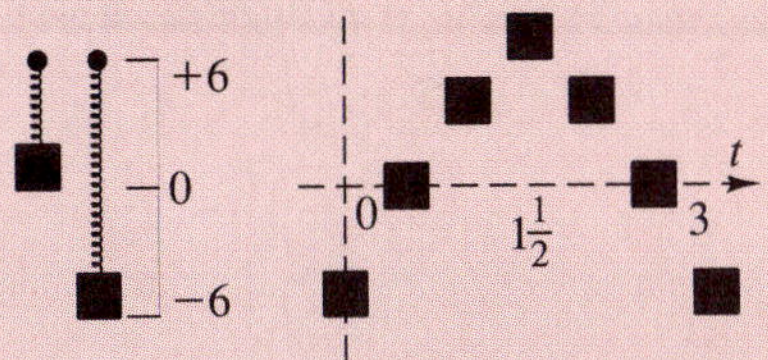

- NASA can be most helpful when talking about space flight. Contact them for information that can be used to illustrate the flight of the space shuttle.
- Encourage students to use their scientific calculators to do any necessary calculations.

Critical Thinking

Application Ask students to illustrate how a child swinging on a swing would represent harmonic motion. Observing a child swinging on a swing, students should note that such motion is sinusoidal and thus represents simple harmonic motion.

CHALKBOARD EXAMPLES

- **For Example 1**

1. A weight is at rest hanging from a spring. The weight is pulled down 5 cm and released and makes a full cycle in 1 s. Write an equation for the function. Then sketch the graph.

$y = 5 \sin 2\pi\left(t - \frac{1}{4}\right)$

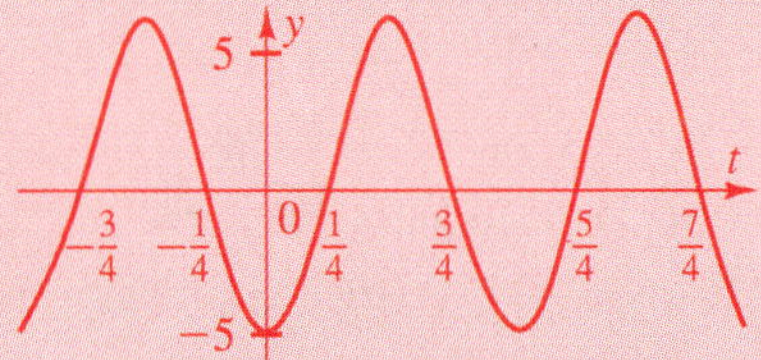

- **For Example 2**

2. The pendulum of a clock swings so that the distance that the tip of the pendulum travels from its vertical position, as a function of time, is sinusoidal. If the displacement of the pendulum on a clock is 12 cm and the pendulum makes one cycle in 3 s, write an equation for its motion. $y = 12 \sin \frac{2\pi}{3} t$

Common Error

- Some students start the graph of a simple harmonic motion at the origin, failing to take into account the location of the object at $t = 0$. Emphasize that students should ask themselves where the object was at the beginning of the time frame.
- See *Teacher's Resource Book* for additional remediation.

Before sketching the graph, note that the phase shift is $\frac{3}{4}$ units to the right.

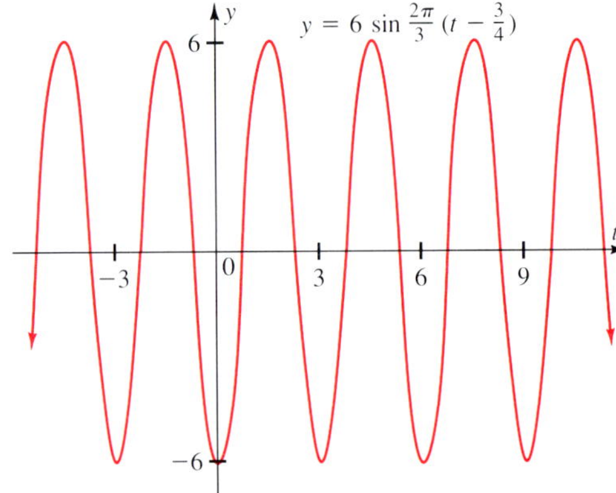

The motion of the tides in the Bay of Fundy is well known around the world. It has been shown that such motion is sinusoidal.

EXAMPLE 2 **At high tide in the Bay of Fundy, the water level is approximately 8 m above the level at low tide, and the time between high tides is 12.4 h. Write an equation for the motion of these tides.**

Since the time between high tides is 12.4 h, the period is 12.4. Therefore, $12.4 = \frac{2\pi}{|b|}$, and $b = \frac{2\pi}{12.4} = \frac{\pi}{6.2}$.

The water level at high tide is approximately 8 m above the level at low tide, so $a = \frac{8}{2} = 4$. Also, there is no vertical shift, so $d = 0$.

$$y = a \sin b(t + c) + d$$

$$y = 4 \sin \frac{\pi}{6.2}(t + c)$$ *Substitute $a = 4$, $b = \frac{\pi}{6.2}$, and $d = 0$.*

Assume that a period starts at high tide. Then $y = 4$ when $t = 0$.

$$4 = 4 \sin \frac{\pi}{6.2}(0 + c)$$

$$\sin \frac{\pi}{6.2} c = 1$$

$$\frac{\pi}{6.2} c = \frac{\pi}{2}$$

$$c = 3.1$$

Note that other possible values of c would also result in a correct equation.

$$y = 4 \sin \frac{\pi}{6.2}(t + 3.1)$$ *Substitute 3.1 for c.*

CLASS EXERCISES

Use the information given in Examples 1 and 2 to answer each of the following questions.

1. What is the distance of the weight from its rest position after 1 s? 3 cm
2. When is the first time that the weight will be 6 cm above its rest position? 1.5 s
3. How many meters has the water level dropped 2 hours after a high tide in the Bay of Fundy? 2.12 m
4. How many hours after a high tide does the water level in the Bay of Fundy first reach a level halfway between that of high tide and low tide? 3.1 h

PRACTICE EXERCISES

A weight is hanging at rest from a spring. At $t = 0$, the weight is pulled down q units and released. It makes r full cycles every second. Write an equation using each of the following values of q and r, and then graph the equation. See Solutions Manual for graphs.

A

1. $q = 3$ cm, $r = 1$ $y = 3 \sin 2\pi\left(t - \frac{1}{4}\right)$
2. $q = 6$ cm, $r = 1$ $y = 6 \sin 2\pi\left(t - \frac{1}{4}\right)$
3. $q = 5$ cm, $r = \frac{1}{2}$ $y = 5 \sin \pi\left(t - \frac{1}{2}\right)$
4. $q = 8$ cm, $r = \frac{1}{2}$ $y = 8 \sin \pi\left(t - \frac{1}{2}\right)$
5. $q = 5$ in., $r = 2$ $y = 5 \sin 4\pi\left(t - \frac{1}{8}\right)$
6. $q = 8$ in., $r = 2$ $y = 8 \sin 4\pi\left(t - \frac{1}{8}\right)$
7. $q = 20$ in., $r = \frac{1}{4}$ $y = 20 \sin \frac{\pi}{2}(t - 1)$
8. $q = 16$ in., $r = \frac{1}{4}$ $y = 16 \sin \frac{\pi}{2}(t - 1)$
9. $q = 20$ in., $r = 4$ $y = 20 \sin 8\pi\left(t - \frac{1}{16}\right)$
10. $q = 16$ in., $r = 4$ $y = 16 \sin 8\pi\left(t - \frac{1}{16}\right)$

For Exercises 11–15, assume that the equations are of the form $y = a \sin b(t + c) + d$.

B

11. If the amplitude of the sound wave produced by middle C is 1 and the frequency is 264 cycles per second, write an equation for the graph seen on an oscilloscope when middle C is sounded. $y = \sin 528\pi t$
12. If the amplitude of the wave produced by A below middle C is 1 and the frequency is 220 cycles per second, what is an equation for the sound wave? $y = \sin 440\pi t$
13. Although the tides in Fishfin are not as famous as those in the Bay of Fundy, the high tide of 3.2 m is quite impressive. If the tides occur every 12.4 h in Fishfin, what is an equation for the changes in the tides? $y = 3.2 \sin \frac{\pi}{6.2}(t + 3.1)$
14. The pendulum of a clock swings so that the distance that its tip travels from its vertical position as a function of time is sinusoidal. If the displacement of the pendulum on a grandfather clock is 5 in. and the pendulum makes one complete cycle in 2 s, write an equation for its motion. $y = 5 \sin \pi\left(t + \frac{1}{2}\right)$

LESSON FOLLOW-UP

Discussion

The tip of a pendulum of a clock moves in simple harmonic motion. Do you think the length of the pendulum has any effect on the motion? Give a reason for your answer. Yes; it affects the displacement.

Assignment Guide

See p. 58B for assignments.

Test Yourself

See *Teacher's Resource Book, Tests*, pp. 15–16.

Lesson Quiz

1. What do we mean when we say that an object moves in simple harmonic motion? a periodic displacement that is a function of time
2. Describe the graph of a weight hanging on the end of a spring when the weight is pulled down and released. sinusoidal
3. If the amplitude of a sound wave produced by A above middle C is 1 and the frequency is 440 cycles per second, what is the equation for its graph as it would look on an oscilloscope? $y = \sin 880\pi t$

Enrichment

Write an equation for the simple harmonic motion shown on the graph below. $y = -4 \cos \frac{3}{2}\pi t$

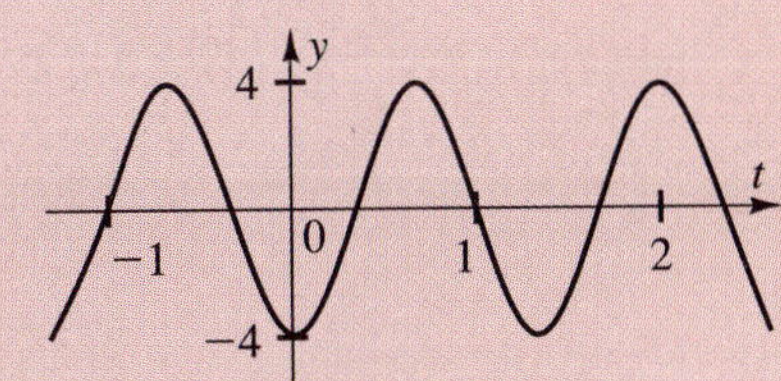

Teacher's Resource Book
Practice—Chapter 2, p. 15
Enrichment—Chapter 2, p. 16

C **15.** A point A is moving at a constant speed in a clockwise direction around the circumference of a circle. The circle is located on the coordinate plane as shown. The projection of A on the x-axis is the point B. B moves back and forth between the points $(6, 0)$ and $(-6, 0)$. Assume that the position x of B is a sinusoidal function of the time t. If A makes one complete rotation in 4 s and starts at $(0,6)$ when $t = 0$, what is an equation for the position of B? $x = 6 \sin \frac{\pi}{2} t$
Other equations are also possible.

(0, 6)
A
(−6, 0)
0
B
(6, 0)
(0, −6)

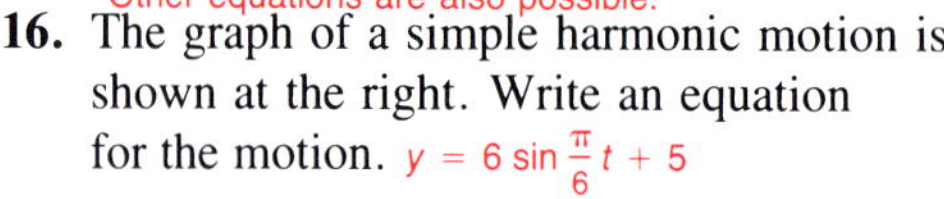

16. The graph of a simple harmonic motion is shown at the right. Write an equation for the motion. $y = 6 \sin \frac{\pi}{6} t + 5$

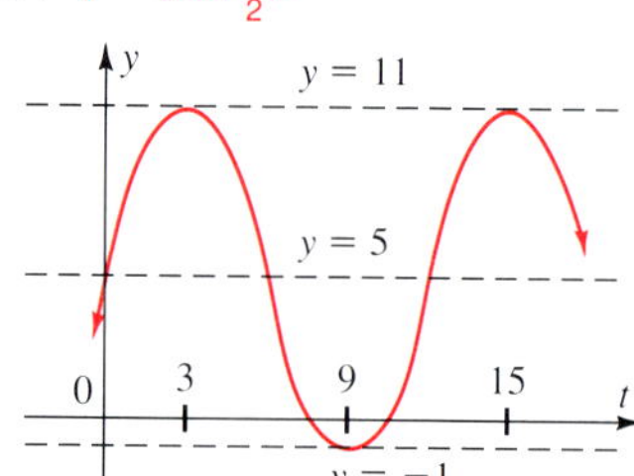

TEST YOURSELF

Graph each by addition of ordinates over at least a one-period interval.
See page 480.

1. $y = \cos x + \cos 3x$
2. $y = \sin x + \sin \frac{1}{2} x$ **2.5**

Determine the period, phase shift, and vertical shift, if any, of each.

3. $y = \tan 4(x + \pi)$ $\frac{\pi}{4}$; π left; none
4. $y = \tan 2(x - \pi)$ $\frac{\pi}{2}$; π right; none **2.6**
5. $y = \cot\left(x + \frac{\pi}{4}\right) - 1$ π; $\frac{\pi}{4}$ left; 1 down
6. $y = \cot\left(x - \frac{\pi}{4}\right) + 2$ π; $\frac{\pi}{4}$ right; 2 up
7. $y = \csc\left(x + \frac{\pi}{2}\right) + 1$ 2π; $\frac{\pi}{2}$ left; 1 up
8. $y = \csc\left(x - \frac{\pi}{2}\right) + 3$ 2π; $\frac{\pi}{2}$ right; 3 up **2.7**
9. $y = \sec 2\left(x + \frac{\pi}{4}\right) - 3$ π; $\frac{\pi}{4}$ left; 3 down
10. $y = \sec 3\left(x - \frac{\pi}{4}\right) - 2$ $\frac{2\pi}{3}$; $\frac{\pi}{4}$ right; 2 down

Graph each function over a two-period interval. See page 480.

11. $y = \tan \frac{1}{2} x$
12. $y = \cot\left(x + \frac{\pi}{4}\right)$ **2.6**
13. $y = \sec \frac{1}{2} x$
14. $y = \csc\left(x - \frac{\pi}{2}\right)$ **2.7**
15. A weight is hanging at rest from a spring. It is pulled down 8 in. and released when $t = 0$. The weight oscillates up and down from its rest position and makes one full cycle every 4 s. Write an equation for the function and then sketch its graph. $y = 8 \sin \frac{\pi}{2}(t - 1)$ **2.8**

APPLICATION: Magnetic Resonance Imaging

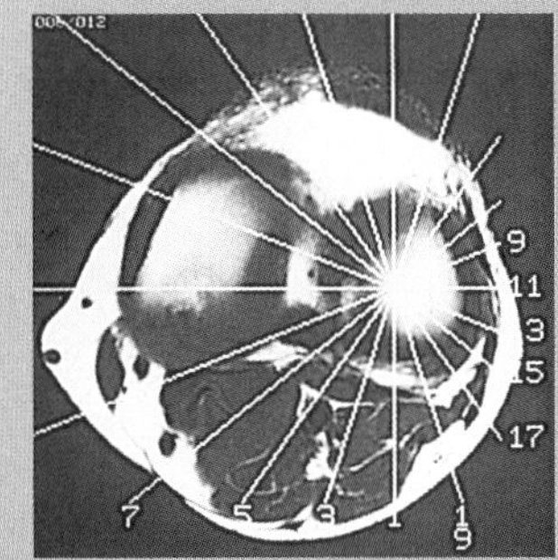

Did you know that Magnetic Resonance Imaging (MRI) is a technique used to view tissues in the human body? This is done by using radio signals to effect temporary electromagnetic changes in hydrogen (or other) atoms. These weak changes are detected by magnetic fields. Using computer technology, these magnetic impressions are converted into visual images.

An MRI system has three magnetic *gradient coils*. Their axes (x, y, and z) are mutually perpendicular. If the plane that is to contain the image (the *slice plane*) is perpendicular to one of the coil axes, then the magnetic field, measured in gauss units, will be needed from only the gradient coil for that axis. Otherwise, contributions will be needed from more than one coil.

EXAMPLE A slice plane forms a 30° angle with the axis of the y-gradient coil. The field strength of the slice plane is composed of the contributions of the y-gradient coil and the z-gradient coil. Find the magnetic field components from each coil.

The two components are added as *vectors*, not as real numbers (see Lesson 4.7). The fields are related by the equations

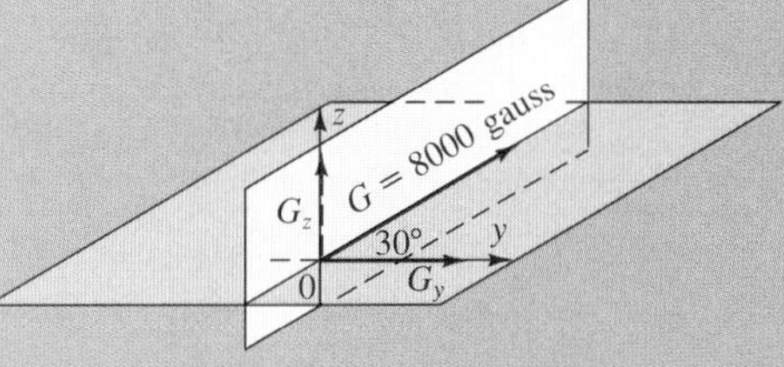

$G_y = G \cos \theta$ $\qquad$ $G_z = G \sin \theta$

$G_y = 8000 \cos 30° = 6928$ gauss $\qquad$ $G_z = 8000 \sin 30° = 4000$ gauss

EXERCISES

1. For a vertebral disk image, a slice plane must form a 20° angle with the y-coil and requires a magnetic field of 10,000 gauss. Find the contributions of the y- and z-coils. 9397 gauss; 3420 gauss
2. For a slice plane image of a shoulder bone, an angle of 15° with the z-axis is required. The contribution of the z-coil is to be 12,000 gauss. Find the magnetic field in the slice plane and the contribution from the y-coil. 12,423 gauss; 3215 gauss

Application

A sophisticated technique for viewing tissues in the body is introduced. Magnetic Resonance Imaging makes use of vector algebra and trigonometric functions to project clear images of human systems.

See *Teacher's Resource Book*, Chapter 2, Application, p. 17.

See *Teacher's Resource Book* for Technology, p. 2.

Additional Answers

4.

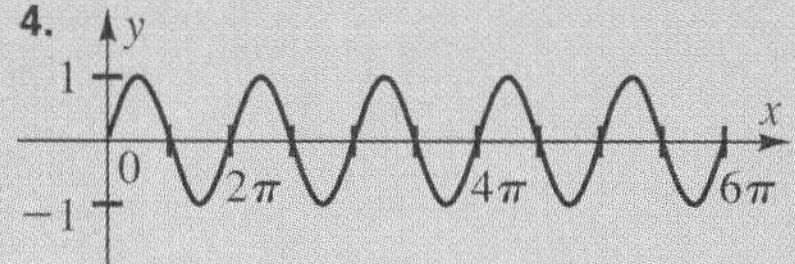

6.

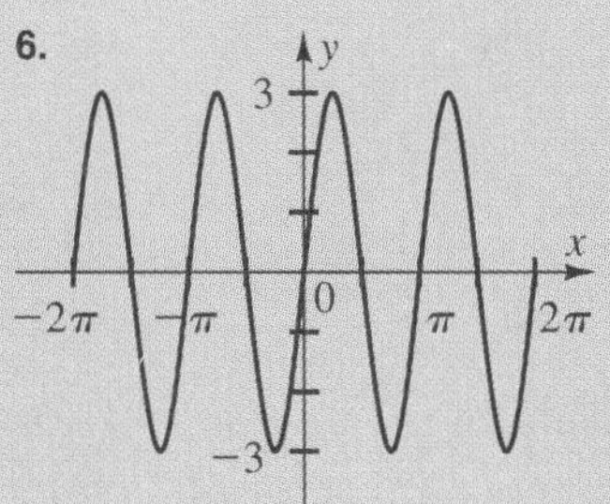

CHAPTER 2 SUMMARY AND REVIEW

Vocabulary

addition of ordinates (85)
amplitude (73)
asymptotes (88)
cofunctions (67)
complementary angles (67)
cycle (60)
discontinuous (89)
displacement (101)
even function (61)
frequency (100)
odd function (61)
period (60)
periodic (60)
phase shift (78)
simple harmonic motion (100)
sinusoidal (100)
symmetric with respect to the origin (62)
symmetric with respect to the x-axis (62)
symmetric with respect to the y-axis (62)
vertical shift (80)

Periodic Functions and Symmetry A function $f(x)$ is periodic (with period p) if there exists a smallest positive number p such that $f(x + p) = f(x)$ when both $f(x + p)$ and $f(x)$ are defined. If $f(-x) = f(x)$ for every value x in the domain of a function, $f(x)$ is an even function. If $f(-x) = -f(x)$ for every value of x, $f(x)$ is an odd function. Even functions are symmetric with respect to the y-axis; odd functions are symmetric with respect to the origin. 2.1

1. Determine if the graph at the right represents a periodic function. If so, state the period. Also state whether the graph is symmetric with respect to the y-axis, the x-axis, the origin, or to none of these.
periodic with period 6; y-axis symmetry

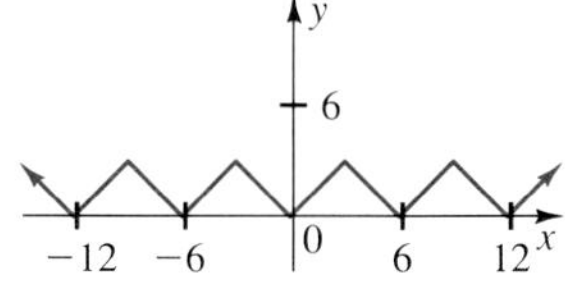

Graphs of the Sine and Cosine Functions The sine and cosine functions are periodic with period 2π. Sine is an odd function. Cosine is an even function. 2.2

2. Express $\cos 26°$ in terms of its cofunction. $\sin 64°$
3. Use the fact that the cosine function is even to find $\cos(-505°)$, if $\cos 505° = -0.8192$. $\cos(-505°) = -0.8192$
4. Graph $y = \sin x$ over the interval $2\pi \leq x \leq 6\pi$. See side column.

Amplitude and Period For $y = a \sin bx$ and $y = a \cos bx$, the amplitude is $|a|$ and the period is $\frac{2\pi}{|b|}$, $b \neq 0$. 2.3

5. Determine the amplitude and period of $y = -4 \cos 3x$. $4; \frac{2\pi}{3}$
6. Graph $y = 3 \sin 2x$ over the interval $-2\pi \leq x \leq 2\pi$. See side column.

Phase Shift and Vertical Shift For $y = a \sin b(x + c) + d$ and $y = a \cos b(x + c) + d$, the amplitude is $|a|$, the period is $\frac{2\pi}{|b|}$, the phase shift is $|c|$ (to the left if $c > 0$, to the right if $c < 0$), and the vertical shift is $|d|$ (up if $d > 0$, down if $d < 0$). 2.4

Determine the amplitude, period, phase shift, and vertical shift for each. Then sketch the graph over a one-period interval. See page 480.

7. $y = 3 \cos 2(x + \pi) - 4$
   3; π; π left; 4 down

8. $y = 4 \sin 3(x - \pi) + 2$
   4; $\frac{2\pi}{3}$; π right; 2 up

Addition of Ordinates To graph a function that is the sum of two functions, graph the two functions. Then add the two y-values. 2.5

Graph each by addition of ordinates over a one-period interval. See page 480.

9. $y = 3 \cos x + \cos x$

10. $y = \sin 2x + \cos x$

Graphs of the Tangent and Cotangent Functions The tangent and cotangent functions are discontinuous and have vertical asymptotes. For $y = a \tan b(x + c) + d$ and $y = a \cot b(x + c) + d$, the period is $\frac{\pi}{|b|}$, and the phase shift and vertical shift are determined by the constants c and d. 2.6

Determine the period, the phase shift, and the vertical shift, if any, for each function. Then graph the function over a two-period interval. See page 480.

11. $y = \tan 3x$ $\frac{\pi}{3}$; none; none

12. $y = 2 \cot \left(x + \frac{\pi}{2}\right)$ π; $\frac{\pi}{2}$ left; none

Graphs of the Secant and Cosecant Functions The secant and cosecant are discontinuous functions and have vertical asymptotes. For $y = a \csc b(x + c) + d$ and $y = a \sec b(x + c) + d$, the period is $\frac{2\pi}{|b|}$, and the phase shift and vertical shift are determined by c and d. 2.7

Determine the period, the phase shift, and the vertical shift, if any, for each function. Then graph the function over a two-period interval. See page 480.

13. $y = 2 \sec 3x$ $\frac{2\pi}{3}$; none; none

14. $y = \csc (x - \pi) + 1$ 2π; π right; 1 up

Harmonic Motion Simple harmonic motion can be modeled by equations of the form $y = a \sin b(t + c) + d$ and $y = a \cos b(t + c) + d$. 2.8

15. The amplitude of the sound wave for C below middle C is 1, and the frequency is 132 cycles per second. Find an equation for the graph.
    $y = \sin 264 \pi t$

See *Teacher's Resource Book, Tests*, pp. 17–20.

CHAPTER TEST

1. State whether this graph represents a periodic function. If it does, give the period. yes; 2

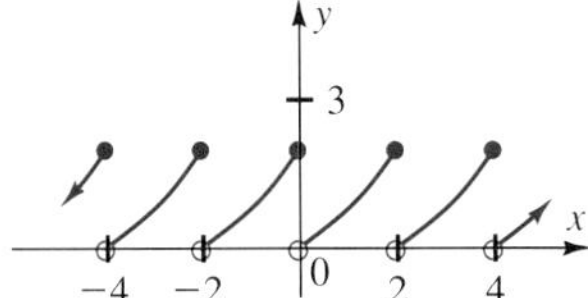

2. State whether this graph is symmetric to the x-axis, the y-axis, or the origin. origin symmetry

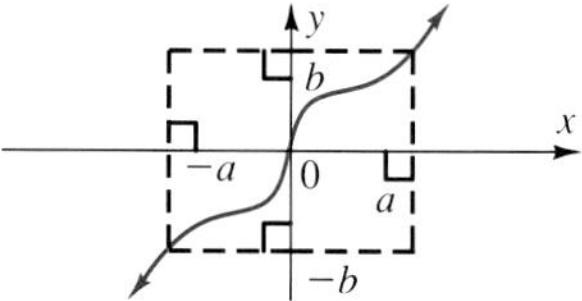

Express each function in terms of its cofunction.

3. $\cos 88°$ $\sin 2°$

4. $\csc 54°$ $\sec 36°$

5. $\tan 8°$ $\cot 82°$

6. Use the properties of the cosine function to find $\cos 123°$, given that $\cos(-123°) = -0.5446$. $\cos(123°) = -0.5446$

Determine the amplitude, the period, the phase shift, and the vertical shift of each function. Then sketch the graph over the interval $0 \le x \le 2\pi$.

7. $y = 4 \sin x$
   4; 2π; none; none

8. $y = \cos 3x$
   1; $\frac{2\pi}{3}$; none; none

9. $y = \sin\left(x - \frac{\pi}{2}\right) + 1$
   1; 2π; $\frac{\pi}{2}$ right; 1 up

10. Graph the function $y = 2 \sin x + 2 \cos x$ by addition of ordinates over the interval $0 \le x \le 2\pi$. See page 480.

Determine the period, the phase shift, and the vertical shift of each.

11. $y = \tan 4x$ $\frac{\pi}{4}$; none; none

12. $y = \cot\left(x + \frac{\pi}{4}\right)$ π; $\frac{\pi}{4}$ left; none

13. $y = 2 \sec \frac{1}{2}x$ 4π; none; none

14. $y = \csc\left(2x - \frac{2\pi}{3}\right) - 2$ π; $\frac{\pi}{3}$ right; 2 down

15. A weight hanging from a spring is pulled down 10 cm and released at time $t = 0$. The weight oscillates up and down from its rest position, making one full cycle every 2 seconds. Write an equation to describe its motion. $y = 10 \sin \pi\left(t - \frac{1}{2}\right)$

Challenge

The graph of the function $y = \sin x \cos x$ is a sine curve. What is its period? π

COLLEGE ENTRANCE EXAM REVIEW

Select the best choice for each question.

1. For $0 \le x \le 2\pi$, the following represents the graph of
D

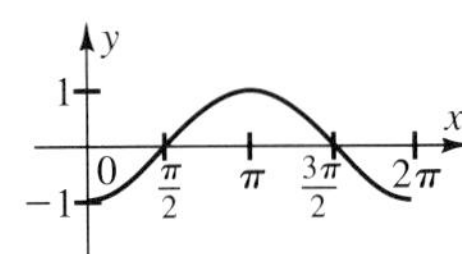

A. $y = \cos x$ **B.** $y = \sec x$
C. $y = \sin x$ **D.** $y = -\cos x$
E. $y = -\sin x$

2. The amplitude of $y = \sin 2\theta$ is
B
A. 2π **B.** 1 **C.** 2 **D.** $\frac{1}{2}$ **E.** π

3. Evaluate $4x^3 - 3x^2 + 2x - 1$ when $x = 1$.
D
A. -2 **B.** -10 **C.** -4 **D.** 2
E. -12

4. In the figure, $\sin\theta \cdot \tan\theta$ equals
C
A. $\frac{x}{r}$ **B.** $\frac{y}{r}$ **C.** $\frac{x^2}{ry}$
D. $\frac{y^2}{rx}$ **E.** $\frac{xy}{r^2}$

5. $y = -2 \sin 4x$ has
E
A. amplitude -2 and period 4
B. amplitude -2 and period π
C. amplitude $\frac{1}{2}$ and period $\frac{\pi}{2}$
D. amplitude -8 and period 4
E. amplitude 2 and period $\frac{\pi}{2}$

6. The length of a rectangle is $2x + 3$ and the width is $x + 5$. The area of the rectangle expressed in terms of x is
E
A. $3x + 8$ **B.** $2x^2 + 2x + 15$
C. $3x^2 + 8x + 8$ **D.** $2x^2 + 2x + 8$
E. $2x^2 + 13x + 15$

7. $\frac{3}{4 - \sqrt{5}}$ equals
D
A. $\frac{3(4 + \sqrt{5})}{-1}$ **B.** $\frac{11}{4 + \sqrt{5}}$
C. $\frac{12 + \sqrt{5}}{11}$ **D.** $\frac{12 + 3\sqrt{5}}{11}$
E. $\frac{12 - 3\sqrt{5}}{11}$

8. In $\triangle ABC$, x equals
E
A. 10 **B.** 5 **C.** 90 **D.** 45 **E.** 20

9. The graph of $y = \sin x$
E
I. rises from $x = 0$ to $x = \frac{\pi}{2}$
II. falls from $x = \frac{\pi}{2}$ to $x = \pi$
III. rises from $x = \pi$ to $x = 2\pi$
A. I only **B.** II only **C.** III only
D. II and III only **E.** I and II only

10. If $2p + q = 4$, then $6p + 3q$ equals
B
A. $-\frac{2}{3}$ **B.** 12 **C.** 24 **D.** 8 **E.** 2

11. Twelve times the first of two consecutive even integers equals eight times the second. The sum of the integers is
C
A. 22 **B.** 26 **C.** 10 **D.** 14 **E.** 3

12. $\sin 90°$ equals
A
A. 1 **B.** 0 **C.** -1
D. $\frac{1}{2}$ **E.** undefined

13. If $MN = NP$, then
D
A. $MO - NP = MN$
B. $MO = MO + NP$
C. $MN + NO = MN + OP$
D. $MN - NO = NP - NO$
E. $MN - OP = MO$

See *Teacher's Resource Book, Tests,* pp. 21–24.

Additional answers

7. $y = 6 \sin \pi \left(t - \frac{1}{2}\right)$

14.

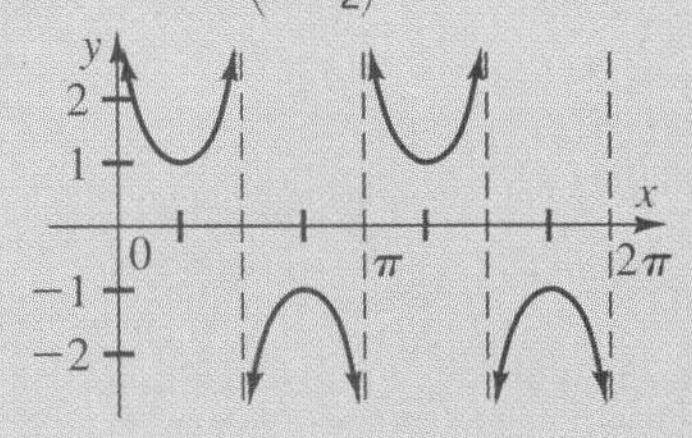

CUMULATIVE REVIEW

1. Sketch the graph of $y = \sin\left(x - \frac{\pi}{3}\right)$ over the interval $0 \le x \le 4\pi$. See below. 2.4

2. Angle θ is in standard position with its terminal side in the second quadrant. Find the exact value of $\sin\theta$ if $\cos\theta = -\frac{1}{2}$. $\frac{\sqrt{3}}{2}$ 1.6

3. Determine whether the function $f(x) = x^2 + 3$ is odd, even, or neither. even 2.1

State whether or not each relation is a function. 1.1

4. $\{(2, 3), (3, 6), (4, 6)\}$ yes

5. $\{(1, 4), (4, 1), (1, 1)\}$ no

6. Determine the amplitude, period, phase shift, and vertical shift for the function $y = 4\cos(2x + \pi)$. 4; π; $\frac{\pi}{2}$ left; none 2.4

7. A weight is hanging at rest from a spring. It is pulled down 6 cm and released. The weight oscillates up and down from its rest position and makes one full cycle every 2 s. Write an equation for the function. See side column. 2.8

8. Find the exact values of the six trigonometric functions of $\theta = -150°$. 1.7
$-\frac{1}{2}$; $-\frac{\sqrt{3}}{2}$; $\frac{\sqrt{3}}{3}$; -2; $-\frac{2\sqrt{3}}{3}$; $\sqrt{3}$

9. Express the function cos 49° in terms of its cofunction. sin 41° 2.2

10. Name two angles that are coterminal with an angle of 135°. −225°, 495° 1.3

11. Graph $y = \sin x - \cos x$ over the interval $0 \le x \le 4\pi$. See below. 2.5

12. If $0° < \theta < 360°$, find all values of θ for which $\cos\theta = -\frac{\sqrt{3}}{2}$. 150°, 210° 1.8

13. The wheel of a car is turning at 8 rps. It is 3 ft in diameter. Find the linear velocity, in feet per second, of a point on the rim of the wheel. 24π ft/s 1.5

14. Graph $y = \sec\left(2x - \frac{\pi}{2}\right)$ over a two-period interval. See side column. 2.7

15. Express 315° in radians. $\frac{7\pi}{4}$ 1.4

16. In which quadrants do $\sin\theta$ and $\cos\theta$ have opposite signs? II, IV 1.7

17. Express $\frac{11\pi}{6}$ in degrees. 330° 1.4

18. Find the radius r of a circle with the equation $2x^2 + 2y^2 = 36$. $3\sqrt{2}$ 1.2

Find θ, where $0° \le \theta \le 360°$, to the nearest tenth of a degree. 1.9

19. $\cos\theta = 0.6587$ 48.8°, 311.2°

20. $\cot\theta = 1.4019$ 35.5°, 215.5°

1.

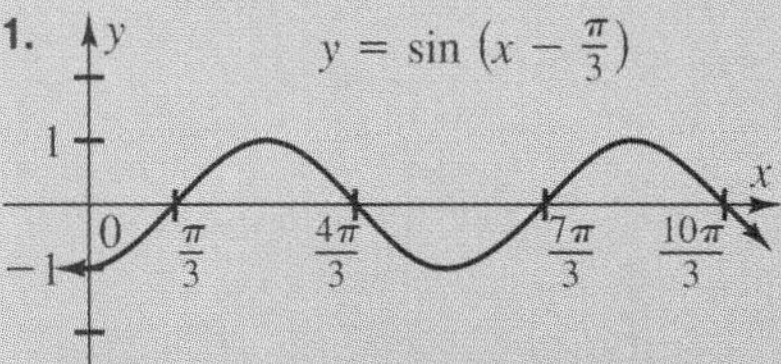

11.

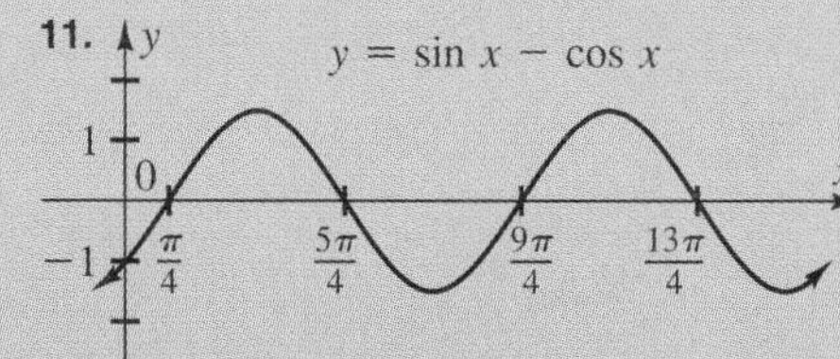

OVERVIEW • Chapter 3

SUMMARY

Right-triangle trigonometry is introduced and then used to find unknown side lengths or angle measures. The concepts of angles of elevation or depression are developed and used to solve a variety of real-world problems.

Then, the fundamental reciprocal, ratio, Pythagorean, and odd-even identities are defined and proved. These identities are then used to simplify complex trigonometric expressions. This chapter concludes by using the technology of a graphing calculator or a computer to prove identities as well as the traditional method of simplifying independently one or both sides of the identity.

After this chapter is completed, students should be able to solve right triangles, and they should be able to solve a variety of problems involving angles of depression or angles of elevation. Students should be able to define and prove the fundamental identities and use these identities to write equivalent trigonometric expressions and to prove other identities. They should also be able to use a graphing calculator or computer to determine whether or not an equation is an identity.

CHAPTER OBJECTIVES

- To solve right triangles given the measures of one angle and one side or given the measures of two sides
- To define and use angles of elevation and depression
- To solve real-world problems using trigonometry
- To introduce and prove reciprocal, ratio, Pythagorean, and odd-even identities
- To use the fundamental identities to write equivalent trigonometric expressions
- To use the fundamental identities to prove other identities
- To check identities by graphing both sides of the equations

CHAPTER HIGHLIGHTS

The *theme* of Chapter 3 is lasers. A discussion of the use of lasers in compact disc players is only one of the chapter's special features.

APPLICATIONS

Right triangle trigonometry lends itself to an abundance of applications from many diverse areas. Problems from such fields as construction, architecture, sports, and engineering apply trigonometric functions and the Pythagorean theorem. Basic trigonometric identities are used in problems from genetics and computer science. Lesson 3.3 provides situations from navigation and surveying which require the use of trigonometry.

TECHNOLOGY

Calculator

Calculators are extremely useful tools for solving problems involving trigonometric functions. Students should be encouraged to use them once problems are in calculation-ready form.

Computer

Programs for evaluating trigonometric functions of specific angles and for evaluating a trigonometric expression for given angles are presented.

RESOURCES

Teacher's Resource Book

- Teaching Aid 3
- Transparencies 5 and 6

ASSIGNMENT GUIDE Meeting Student Needs

STUDENT TEXT					TEACHER'S RESOURCE BOOK	
Chapter Content		**Basic**	**Average**	**Enriched**	**P**	**E**
3.1	Solving Right Triangles	D: 116/1–23 odd, 29	D: 116/7–25 odd, 29, 31	D: 116/11–33 odd	1	2
3.2	Application: Angles of Elevation and Depression	D: 121/1–15 odd R: 116/5, 8, 16	D: 121/5–19 odd R: 116/10, 16, 20	D: 121/5–19 odd R: 116/14, 16, 22	3	4
3.3	Applications	D: 126/1–11 odd R: 121/2, 6, 8 129/TY	D: 127/7–17 odd R: 121/6, 10, 14 129/TY	D: 127/7–17 odd R: 121/8, 12, 16 129/TY	5	6
3.4	Fundamental Identities	D: 133/1–23 odd, 37 R: 126/2, 6, 10	D: 134/13–33 odd, 37 R: 127/8, 10, 14	D: 134/19–39 odd R: 127/8, 10, 14	7	8
3.5	Equivalent Trigonometric Expressions	D: 137/1–21 odd, 33 R: 133/6, 12, 20	D: 138/9–29 odd, 33 R: 134/18, 26, 30	D: 138/11–33 odd R: 134/22, 26, 30	9	10
3.6	Proving Identities	D: 143/1–21 odd R: 137/4, 8, 18	D: 143/11–31 odd R: 138/14, 18, 26	D: 143/15–35 odd R: 138/18, 22, 28	11	12
3.7	Graphical Representation of Identities	D: 146/1–13 odd R: 143/2, 10, 20 147/TY	D: 146/11–23 odd, 29 R: 143/12, 20, 28 147/TY	D: 146/15–29 odd R: 143/16, 20, 30 147/TY	13	14

D = Daily R = Review TY = Test Yourself P = Practice E = Enrichment

	STUDENT TEXT				TEACHER'S RESOURCE BOOK	
Review and Testing	Test Yourself	129, 147	College Ent. Exam Rev.	153	Tests	
	Chapter Sum. and Rev.	150	Maintaining Skills	154	• Quizzes	25–28
	Chapter Test	152			• Chapter Test (Form A)	29–30
					• Chapter Test (Form B)	31–32
Special Features	Historical Note	117	Extra	139	Applications—Chapter 3	15
	Challenge	123	Historical Note	144	Critical Thinking	2
	Logical Reasoning	135	Application	148	Alg. and Geom. Review	9–12
					Technology	3

3 Right Triangle Trigonometry and Basic Identities

A laser (an acronym for light amplification by stimulated emission of radiation) is a device that produces a nearly perfect plane wave of light. Lasers are used as surgical tools, to treat cancer, to align surveying equipment, and in compact disc players.

BACKGROUND

Lasers form an integral part of much of today's technology. Have students research some of the areas in which lasers are used and the scientific advances to which they have contributed.

LESSON PLAN

Vocabulary
Significant digit

BACKGROUND

In the Preview, using a scientific calculator to find values of the trigonometric functions and to find measures for angles with specific trigonometric function values is reviewed.

112

3.1 Solving Right Triangles

Objectives: To solve right triangles, given the measures of one angle and one side or the measures of two sides

In this chapter, trigonometric functions will sometimes be referred to as trigonometric *ratios*, since they can be thought of as the ratios of lengths of sides of right triangles.

Preview

Recall that a calculator can be used to find both a value of a trigonometric function and an angle measure for a given trigonometric-function value.

EXAMPLE 1 **Find $\cot \frac{\pi}{5}$ to four decimal places.**

$\cot \frac{\pi}{5} = 1.3764$ *Use radian mode. Enter $\frac{\pi}{5}$ and use the tan key and the reciprocal key.*

EXAMPLE 2 **Find θ to the nearest tenth of one degree if $\sin \theta = 0.7083$ and $0° < \theta < 90°$.**

$\theta = 45.1°$ *Use degree mode. Enter 0.7083 and use the inverse key and the sine key.*

Find each value to four decimal places.

1. $\sec 203.8°$ −1.0929 **2.** $\tan \frac{\pi}{7}$ 0.4816 **3.** $\sin \frac{7\pi}{4}$ −0.7071

Find θ to the nearest tenth of one degree, where $0° < \theta < 90°$.

4. $\cos \theta = 0.6128$ 52.2° **5.** $\cot \theta = 1.2341$ 39.0° **6.** $\csc \theta = 3.1215$ 18.7°

The table on page 113 compares trigonometric *functions* and trigonometric *ratios*. Right triangle ABC can be used to express the values of the trigonometric ratios in terms of the sides of the triangle, where

a = length of leg opposite θ
b = length of leg adjacent to θ
$c = \sqrt{a^2 + b^2}$ = hypotenuse

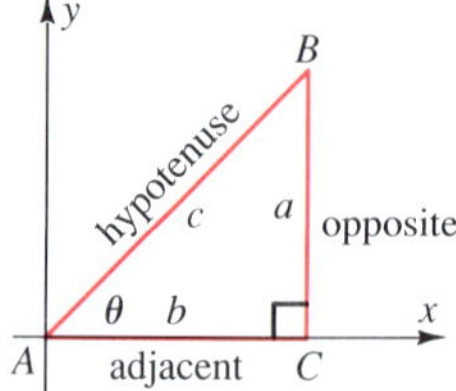

Trigonometric Functions	Right Triangle Ratios ($0° < \theta < 90°$)
$\sin\theta = \frac{y}{r}$	$\sin\theta = \frac{a}{c} = \frac{\text{length of leg opposite } \theta}{\text{length of hypotenuse}}$
$\cos\theta = \frac{x}{r}$	$\cos\theta = \frac{b}{c} = \frac{\text{length of leg adjacent to } \theta}{\text{length of hypotenuse}}$
$\tan\theta = \frac{y}{x}$	$\tan\theta = \frac{a}{b} = \frac{\text{length of leg opposite } \theta}{\text{length of leg adjacent to } \theta}$
$\csc\theta = \frac{r}{y}$	$\csc\theta = \frac{c}{a} = \frac{\text{length of hypotenuse}}{\text{length of leg opposite } \theta}$
$\sec\theta = \frac{r}{x}$	$\sec\theta = \frac{c}{b} = \frac{\text{length of hypotenuse}}{\text{length of leg adjacent to } \theta}$
$\cot\theta = \frac{x}{y}$	$\cot\theta = \frac{b}{a} = \frac{\text{length of leg adjacent to } \theta}{\text{length of leg opposite } \theta}$

Capital letters are usually used to represent the angles of triangles, or their measures. Lowercase letters refer to the sides opposite their respective angles, or to their measures. The right triangle ratios can be used to *solve* a right triangle, that is, to find the unknown measures of the sides and angles.

EXAMPLE 1 **Solve right triangle ABC if $b = 32$, $\angle A = 25°$, and $\angle C = 90°$. Find a and c to the nearest unit.**

To find a, use tan 25°.

$$\tan 25° = \frac{a}{32} \qquad \textit{tan } A = \frac{a}{b}$$

$a = 32 \tan 25°$ *Calculation-ready form*

$a = 15$ *To the nearest unit*

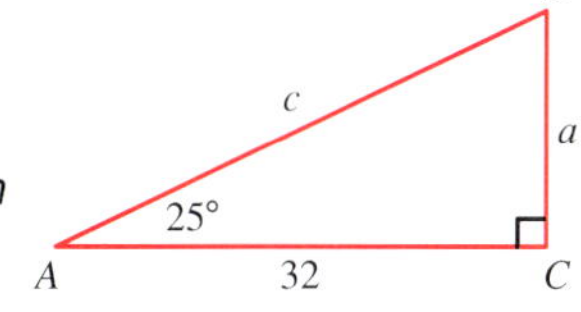

To find c, use cos 25°.

$$\cos 25° = \frac{32}{c} \qquad \textit{cos } A = \frac{b}{c}$$

$c = \frac{32}{\cos 25°} = 35$ *To the nearest unit*

Since angles A and B are complementary, $\angle B = 90° - 25° = 65°$.

A **significant digit** is any nonzero digit or any zero that serves a purpose other than to locate the decimal point. Consider the following examples:

0.00*304*	Three significant digits
29.40	Four significant digits
30.305	Five significant digits

TEACHING SUGGESTIONS

- Review the Pythagorean theorem and its use in solving right triangles.
- Be sure that students know the trigonometric ratios in terms of x, y, r, a, b, c; and opposite, adjacent, hypotenuse.
- Emphasize that solving a right triangle means finding the measures of *all* unknown sides and angles.
- Review significant digits and how they are determined.
- When discussing Examples 1–3 in the text, stress alternate methods of finding the unknown measures (that is, the use of other trigonometric ratios). However, point out that the ratio that involves the simplest calculation should be used.

Critical Thinking

Synthesis Ask students to outline a set of steps that could be followed when solving right triangles. Then ask them to point out where students are most likely to make mistakes.
Answers may vary.

CHALKBOARD EXAMPLES

- **For Example 1**
  1. Solve right triangle ABC if $a = 25$, $\angle B = 42°$, and $\angle C = 90°$. Find b and c to the nearest unit.
  $\angle A = 48°$, $b = 23$, $c = 34$
  2. Solve right triangle ABC if $\angle A = 33°$, $\angle C = 90°$, and $c = 42$. Find a and b to the nearest unit.
  $\angle B = 57°$, $a = 23$, $b = 35$

- **For Example 2**
  3. Solve right triangle ABC if $\angle C = 90°$, $b = 8.15$ m, and $a = 3.78$ m. $\angle B = 65.1°$, $\angle A = 24.9°$, $c = 8.98$ m
  4. Solve right triangle ABC if $\angle C = 90°$, $c = 9.45$ ft, and $a = 4.23$ ft. $\angle A = 26.6°$, $\angle B = 63.4°$, $b = 8.45$ ft

- **For Example 3**
  5. Solve right triangle ABC if $a = 14.70$ ft, $\angle B = 61°15'$, and $\angle C = 90°$. $\angle A = 28°45'$, $b = 26.79$ ft, $c = 30.56$ ft
  6. Solve right triangle ABC if $b = 12.24$ cm, $\angle A = 36°14'$, and $\angle C = 90°$. $\angle B = 53°46'$, $c = 15.17$ cm, $a = 8.97$ cm

Common Errors

- Students often confuse a, b, and c with A, B, and C. Be sure to stress that the lowercase letters represent side lengths, while the capital letters represent angle measures.
- When using a calculator, some students will forget to check the mode of the calculator. Have those students do several exercises asking them to use radians for some and degrees for others.
- Some students will not express answers to the correct number of significant digits. Have those students refer to the table on text p. 114.
- See *Teacher's Resource Book* for additional remediation.

In general, the greater the number of significant digits in a measurement, the greater the accuracy. Use the table below as a guide to decide how accurate your answers should be.

Angle measures accurate to nearest . . .	1°	10′ or 0.1°	1′ or 0.01°
Significant digits in lengths of sides	2	3	4

To illustrate, if angle measures are given to the nearest degree, compute the lengths of the sides to two significant digits. If lengths are given to three significant digits, compute angle measures to the nearest 0.1°.

EXAMPLE 2 **Solve right triangle ABC if $\angle C = 90°$, $c = 7.25$, and $b = 4.37$.**

To find the value of $\angle A$, use the cosine function.

$$\cos A = \frac{4.37}{7.25}$$

$$A = 52.9°$$ *Since b and c are given to 3 significant digits, round to the nearest 0.1°.*

Since angles A and B are complementary, $\angle B = 90° - 52.9° = 37.1°$.

To find the value of a, use $\sin 52.9°$.

$$\sin 52.9° = \frac{a}{7.25}$$

$$a = 7.25 \sin 52.9°$$

$$a = 5.78$$ *Round to 3 significant digits.*

In Example 3, the angle measures are expressed in degrees and minutes. If your calculator has a key that converts from degrees and minutes to decimal degrees, you can omit the next-to-last steps shown in the calculation of c and in the calculation of b.

EXAMPLE 3 **Solve triangle ABC if $a = 22.14$, $\angle B = 32°12'$, and $\angle C = 90°$.**

To find c, use $\sec 32°12'$.

$$\sec 32°12' = \frac{c}{22.14}$$

$$c = 22.14 \sec 32°12'$$

$$c = 22.14 \sec\left(32 + \frac{12}{60}\right)^{\circ}$$

$$c = 26.16$$ *Round to 4 significant digits.*

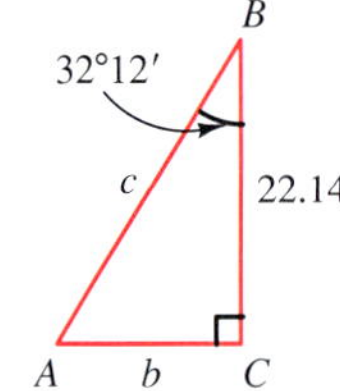

To find b, use $\tan 32°12'$.

$$\tan 32°12' = \frac{b}{22.14}$$

$$b = 22.14 \tan 32°12'$$

$$b = 22.14 \tan \left(32 + \frac{12}{60}\right)^{\circ}$$

$$b = 13.94 \qquad \textit{Round to 4 significant digits.}$$

Since angles A and B are complementary, $\angle A = 90° - 32°12' = 57°48'$.

In order to solve a right triangle, it is helpful to draw a reasonably accurate figure and label the parts of known measure. You can usually choose which side or angle measure to determine first, and more than one trigonometric function may be appropriately applied. The Pythagorean theorem can also be used when the measures of the two sides are known.

To illustrate, it is possible to solve first for c, b, or $\angle A$ in Example 3. In order to find b first, $\tan B$ or $\cot B$ can be used. After b is found, $\sin B$, $\cos B$, $\sec B$, $\csc B$, or the Pythagorean theorem can be used to determine c.

CLASS EXERCISES

For triangle *ABC* at the right, express the given trigonometric function in terms of *a*, *b*, and *c*.

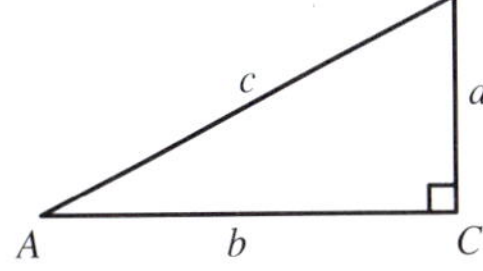

1. $\sin B$ $\frac{b}{c}$
2. $\sin A$ $\frac{a}{c}$
3. $\cot A$ $\frac{b}{a}$
4. $\sec A$ $\frac{c}{b}$
5. $\cos B$ $\frac{a}{c}$
6. $\tan B$ $\frac{b}{a}$

State two equations, using reciprocal functions, that could be used to find *x* or *θ*.

7. 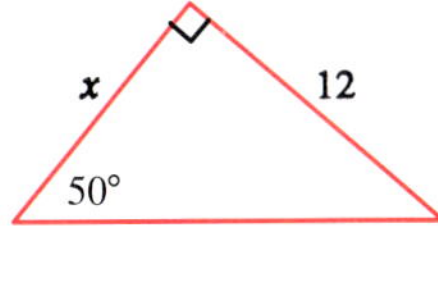

$\tan 50° = \frac{12}{x}$; $\cot 50° = \frac{x}{12}$

8. 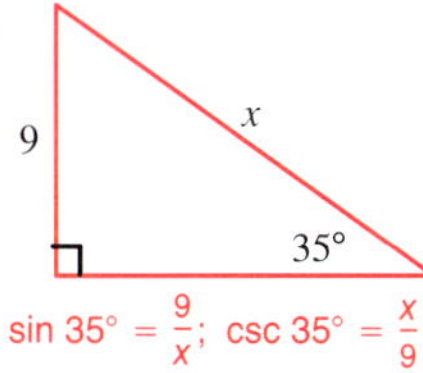

$\sin 35° = \frac{9}{x}$; $\csc 35° = \frac{x}{9}$

9. 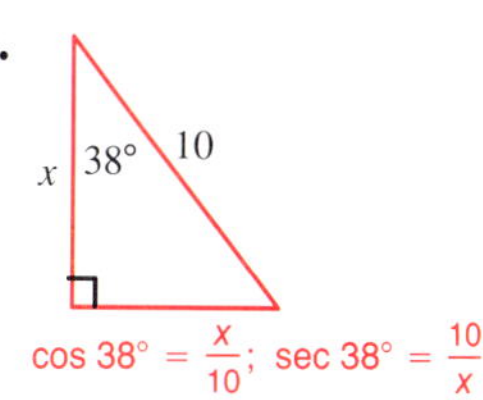

$\cos 38° = \frac{x}{10}$; $\sec 38° = \frac{10}{x}$

10. 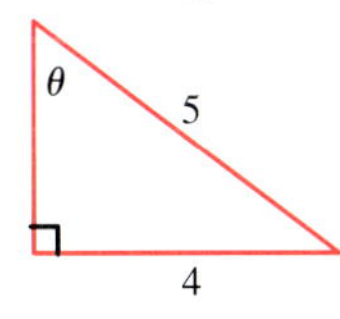

$\sin \theta = \frac{4}{5}$; $\csc \theta = \frac{5}{4}$

11. 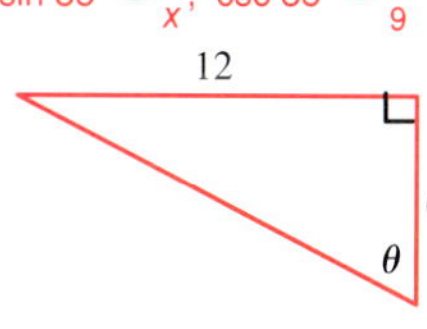

$\tan \theta = \frac{12}{6}$; $\cot \theta = \frac{6}{12}$

12. θ, 2, 5

$\sec \theta = \frac{5}{2}$; $\cos \theta = \frac{2}{5}$

LESSON FOLLOW-UP

Extension

Using the principles of solving right triangles, find as many of the missing measurements as possible.

Assignment Guide

See p. 110B for assignments.

Historical Note

The Pythagorean theorem has been useful to numerous cultures for a variety of reasons, beginning with the Babylonians (1540 BC) and continuing through the present day.

Lesson Quiz

Solve for x in each right triangle.

1.

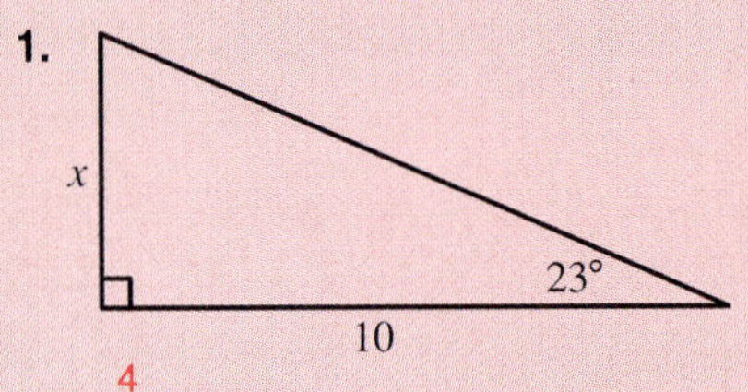

4

2.

12.1, 18.2, x

48.3°

3.

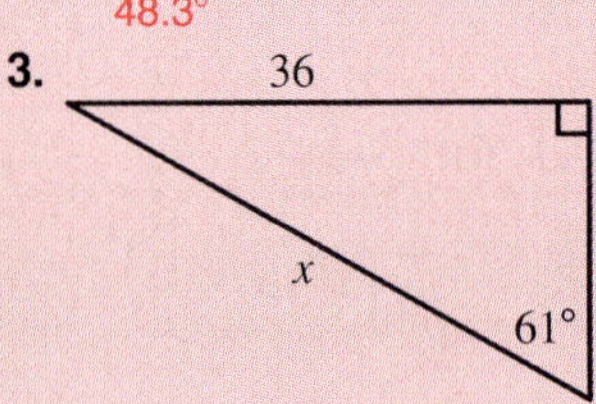

41

4. Solve right triangle ABC if $\angle C = 90°$, $\angle B = 48°$, and $c = 15$. $\angle A = 42°$, $a = 10$, $b = 11$

5. Solve right triangle ABC if $\angle C = 90°$, $a = 12.1$, and $b = 14.3$ in. $\angle A = 40.2°$, $\angle B = 49.8°$, $c = 18.7$

6. An 8.6-ft ladder leans against a building and it makes a 78° angle with the ground. How high up the building does it reach? 8.4 ft

PRACTICE EXERCISES

Solve each right triangle *ABC*.

A **1.** (A, C, B; 14, c, a, 25°)

$a = 30$; $c = 33$; $\angle A = 65°$

2.

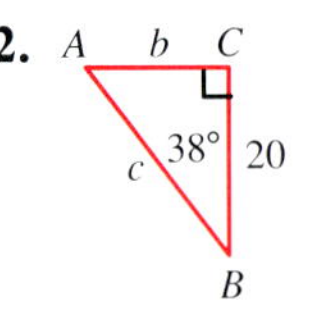

$b = 16$; $c = 25$; $\angle A = 52°$

3. (B, C, A; a, 17°, 22, b)

$a = 21$; $b = 6.4$; $\angle A = 73°$

4. (A, C, B; b, 16, 21°, a)

$a = 15$; $b = 5.7$; $\angle A = 69°$

5.

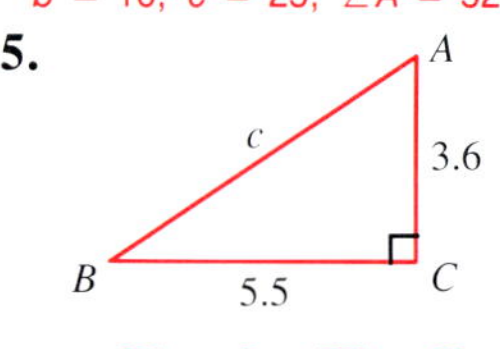

$c = 6.6$; $\angle A = 57°$; $\angle B = 33°$

6. (A, C, B; 2.3, 6.1, a)

$a = 5.6$; $\angle A = 68°$; $\angle B = 2$

Solve each right triangle *ABC* ($\angle C = 90°$) given the measure of one angle and the length of one side.

7. $\angle A = 58°$, $c = 27$
$\angle B = 32°$; $a = 23$; $b = 14$

8. $\angle B = 29°$, $c = 14$
$\angle A = 61°$; $a = 12$; $b = 6.8$

9. $\angle B = 15.1°$, $c = 10.7$
$\angle A = 74.9°$; $a = 10.3$; $b = 2.79$

10. $\angle A = 17.2°$, $c = 29.4$
$\angle B = 72.8°$; $a = 8.69$; $b = 28.1$

11. $\angle A = 63°$, $a = 11$
$\angle B = 27°$; $b = 5.6$; $c = 12$

12. $\angle B = 24°$, $b = 36$
$\angle A = 66°$; $a = 81$; $c = 89$

13. $\angle B = 42.5°$, $a = 188$
$\angle A = 47.5°$; $b = 172$; $c = 255$

14. $\angle A = 70.5°$, $b = 276$
$\angle B = 19.5°$; $a = 779$; $c = 827$

Solve each right triangle *ABC* ($\angle C = 90°$) given the measure of one angle and the length of one side.

B **15.** $\angle A = 36°41'$, $a = 19.32$
$\angle B = 53°19'$; $b = 25.94$; $c = 32.34$

16. $\angle B = 42°35'$, $a = 71.22$
$\angle A = 47°25'$; $b = 65.45$; $c = 96.73$

17. $\angle B = 72°28'$, $a = 84.84$
$\angle A = 17°32'$; $b = 268.5$; $c = 281.6$

18. $\angle A = 80°12'$, $a = 36.22$
$\angle B = 9°48'$; $b = 6.26$; $c = 36.76$

Solve each right triangle *ABC* ($\angle C = 90°$) given the lengths of two sides.

19. $b = 17.62$, $c = 23.91$
$\angle A = 42.53°$; $\angle B = 47.47°$; $a = 16.16$

20. $b = 13.42$, $c = 26.31$
$\angle A = 59.33°$; $\angle B = 30.67°$; $a = 22.63$

21. $a = 18.65$, $b = 14.22$
$\angle A = 52.68°$; $\angle B = 37.32°$; $c = 23.45$

22. $a = 7.613$, $c = 14.05$
$\angle A = 32.81°$; $\angle B = 57.19°$; $b = 11.81$

23. $a = 1632$, $c = 2015$
$\angle A = 54.09°$; $\angle B = 35.91°$; $b = 1182$

24. $a = 1503$, $b = 1635$
$\angle A = 42.59°$; $\angle B = 47.41°$; $c = 2221$

Use the given information for triangle *ABC* ($\angle C = 90°$) to express the other five trigonometric functions of $\angle A$ in terms of t. See side column.

C **25.** $\sin A = \frac{3}{t}$ **26.** $\tan A = \frac{1}{t}$ **27.** $\sec A = t$

28. For any right triangle DEF ($\angle F = 90°$), find the numerical value of $(\cos D)^2 + (\cos E)^2 + (\cos F)^2$. 1

Applications

29. Construction A ladder rests against a building at a point that is 63 ft from the ground. If the ladder makes a 47° angle with the ground, what is the length of the ladder? 86 ft

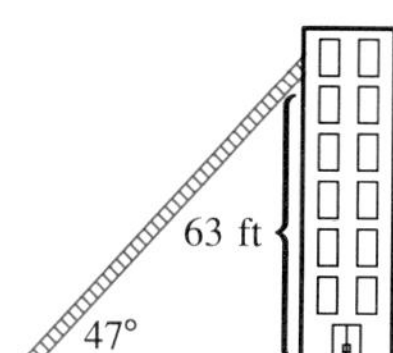

30. Geometry The base of an isosceles triangle is 14 cm in length and the angle opposite the base measures 86°. Find the length of each of the congruent sides. 10 cm

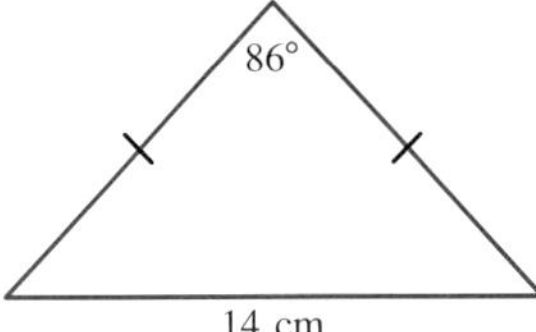

31. Architecture The Leaning Tower of Pisa is 179 ft high, making a 5.1° angle with the vertical. Find x.
16 ft

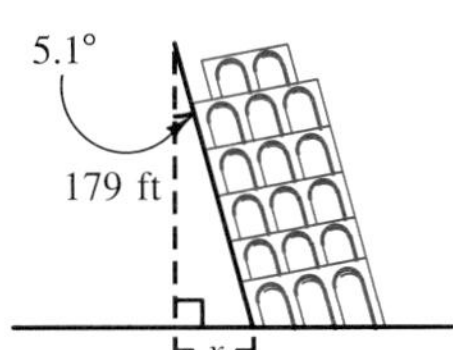

32. Sports A ski slope has a 42° incline and is 170 yd long. Find the vertical drop d. 114 yd

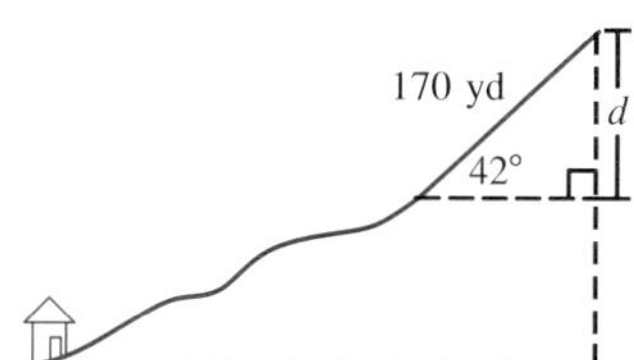

HISTORICAL NOTE

The Pythagorean theorem There is evidence that the Pythagorean theorem was known and used long before the time of the Greek mathematician Pythagoras, who lived in the sixth century B.C. In fact, the Babylonians knew of the theorem more than 1000 years earlier. The ancient Egyptians may have been familiar with the concept also. It has been speculated that Egyptian surveyors used Pythagorean triples to construct right angles. To do this, a rope divided by knots into segments in the ratio 3:4:5 was stretched out into a triangle. The desired angle was opposite the longest segment.

Since the time of Pythagoras, this theorem has been proved in many ways. Locate two proofs of the theorem and write them in your own words.
Answers may vary.

Teacher's Resource Book
Practice—Chapter 3, p. 1
Enrichment—Chapter 3, p. 2

Additional Answers

25. $\cos A = \frac{\sqrt{t^2 - 9}}{t}$; $\tan A = \frac{3\sqrt{t^2 - 9}}{t^2 - 9}$; $\csc A = \frac{t}{3}$; $\sec A = \frac{t\sqrt{t^2 - 9}}{t^2 - 9}$; $\cot A = \frac{\sqrt{t^2 - 9}}{3}$

26. $\sin A = \frac{\sqrt{t^2 + 1}}{t^2 + 1}$; $\cos A = \frac{t\sqrt{t^2 + 1}}{t^2 + 1}$; $\csc A = \sqrt{t^2 + 1}$; $\sec A = \frac{\sqrt{t^2 + 1}}{t}$; $\cot A = t$

27. $\sin A = \frac{\sqrt{t^2 - 1}}{t}$; $\cos A = \frac{1}{t}$; $\tan A = \sqrt{t^2 - 1}$; $\csc A = \frac{t\sqrt{t^2 - 1}}{t^2 - 1}$; $\cot A = \frac{\sqrt{t^2 - 1}}{t^2 - 1}$

LESSON PLAN

Vocabulary

Angle of depression
Angle of elevation

BACKGROUND

In the Preview, the use of trigonometry in landscaping and solar heating is discussed.

3.2 Application: Angles of Elevation and Depression

Objective: To define and use angles of elevation and depression

Solar heating panels are placed on roofs facing due south, if possible, in order to absorb maximum energy. Trigonometry can be used to determine where trees should be planted so that they do not block sunlight striking the panels.

Preview

The angle of the sun at noon varies, with the smallest angle occurring on December 21, the winter solstice. Richard lives in Winnipeg, where the angle of the sun at noon on December 21 is 14°. He has solar panels on his roof with the bottom edge of the panels 25 ft above the ground. Richard wants to plant a tree 110 ft from the panels. What is the maximum height the tree can attain if it is not to block sunlight to the panels?

$$\tan 14^\circ = \frac{x}{110}$$

$$x = 110 \tan 14^\circ$$

$$x = 27 \text{ ft} \qquad \textit{Two significant digits}$$

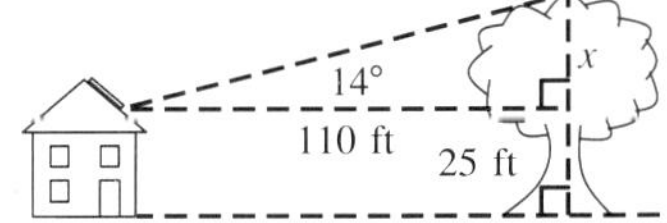

$$\text{maximum height} = 25 \text{ ft} + 27 \text{ ft} = 52 \text{ ft}$$

1. In Philadelphia, the angle of the sun on December 21 is 24°. Solar panels are placed on a roof 42 ft above the ground. How far from the house should a tree be planted if the tree is to attain a maximum height of 75 ft? 74 ft
2. On December 21, the angle of the sun at noon in Toronto is 18°. How far from the ground should solar panels be placed so that sunlight will not be blocked by a fully grown, 33-ft tree 68 ft from the house? 11 ft

If an observer sights an object, the angle formed between a horizontal line and his or her *line of sight* is called the **angle of elevation** if the line of sight is above the horizontal and the **angle of depression** if the line of sight is below the horizontal. In the figure at the right, lines n and p are parallel. Therefore, the angle of elevation from the person on the ground to the airplane is equal in measure to the angle of depression from the plane to the person on the ground.

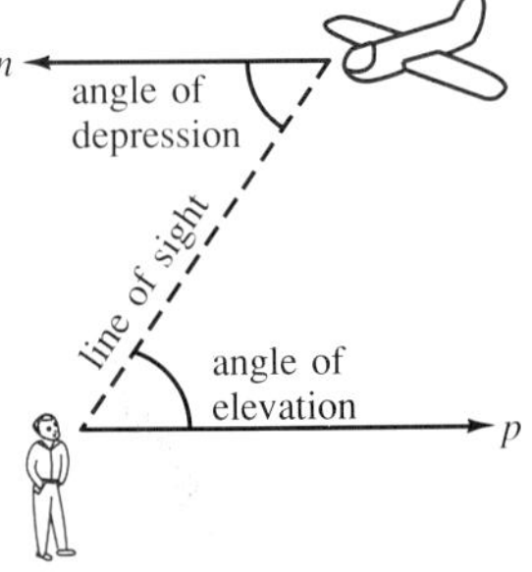

EXAMPLE 1 **A 98-ft extension ladder rests on top of a hook and ladder truck with its base 11 ft from the ground. When the angle of elevation of the ladder is 73°, how high up the building will it reach?**

$$\sin 73° = \frac{h}{98}$$
$$h = 98 \sin 73°$$
$$h = 94 \text{ ft} \qquad \textit{Two significant digits}$$

The vertical reach is about 11 ft + 94 ft, or 105 ft.

The altimeter of an airplane records the height at which the plane is flying. The altimeter reading, along with the angle of depression from the plane to a point on the ground, can be used to find the distance to that point from a point directly below the plane.

EXAMPLE 2 **The altimeter of a jet airplane approaching O'Hare Airport in Chicago records 5900 ft as it passes over the Sears Tower. At the same time, the angle of depression from the plane to the near end of the runway is 5°. How far is it from the base of the Sears Tower to the end of the runway?**

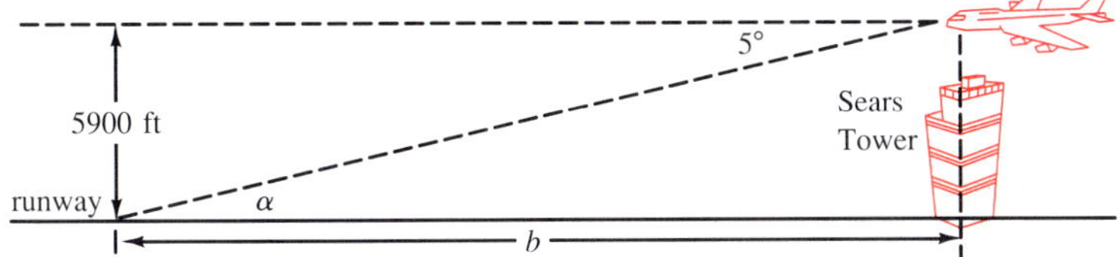

Since the angle of depression from the plane to the runway is equal to the angle of elevation from the runway to the plane, $\alpha = 5°$. The required distance is b.

$$\cot 5° = \frac{b}{5900}$$
$$b = 5900 \cot 5°$$
$$b = 67{,}000 \text{ ft} \qquad \textit{Two significant digits}$$

The base of the Sears Tower is about 67,000 ft from the runway. Since 5280 ft = 1 mi, the distance is about 13 mi.

In the next example, two known distances are used to find an angle of depression.

TEACHING SUGGESTIONS

- Emphasize that angles of depression and elevation are formed with the horizontal.
- Show that any angle of elevation problem can be solved as an angle of depression problem.
- Stress that if a drawing is not given, one should always be made to depict given information.

Critical Thinking

Classifying Ask students to list areas of interest in their lives where they could use trigonometry and angles of elevation and depression to solve problems. Answers may vary.

CHALKBOARD EXAMPLES

- **For Example 1**
  1. From a point on the ground 4.0 m from the base of a tree, the angle of elevation of the top of the tree is 62°. Determine the height of the tree. 7.5 m
  2. The Gateway Arch in St. Louis is 630 ft high. If a person is walking away from the Arch until the angle of elevation to the top is 55°, how far from the Arch is the person? 440 ft
- **For Example 2**
  3. Determine an airplane's altimeter reading if the angle of depression to the runway is 40° when the plane is 12 mi from the runway. 10 mi
  4. The angle of depression of an airplane to the airport is 30.1° when the plane is 10.3 km from the airport. Determine the altimeter reading inside the plane. 5.97 km

- **For Example 3**
  5. A stranded boater tries to gain attention of a rescue helicopter that is 30 ft away. If the helicopter is hovering at 70 ft, what is the angle of depression from the pilot to the boater? $67°$
  6. A tower with a height of 45 m is 25 m from point A. Find the angle of elevation from point A to the top of the tower. $61°$

Common Error

- Students often confuse angles of depression and elevation. Be sure to stress their differences.
- See *Teacher's Resource Book* for additional remediation.

LESSON FOLLOW-UP

Discussion

Find a general expression for solving problems such as the one in Example 1.

Assignment Guide

See p. 110B for assignments.

EXAMPLE 3 **A passenger in a helicopter shines a light on a car stranded 45 ft from a point just below the helicopter. If the helicopter is hovering at 85 ft, what is the angle of depression from the light source to the car?**

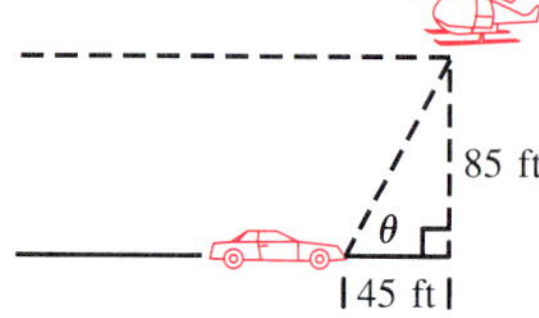

The angle of depression from the helicopter to the car is equal to θ, the angle of elevation from the car to the helicopter.

$$\tan \theta = \frac{85}{45}$$

$$\theta = 62° \quad \textit{To the nearest degree}$$

The angle of depression is approximately 62°.

CLASS EXERCISES

For each figure, give two trigonometric equations that could be used to find x.

1. 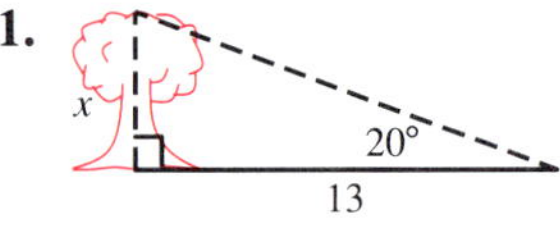

$\tan 20° = \frac{x}{13}$; $\cot 20° = \frac{13}{x}$

2. 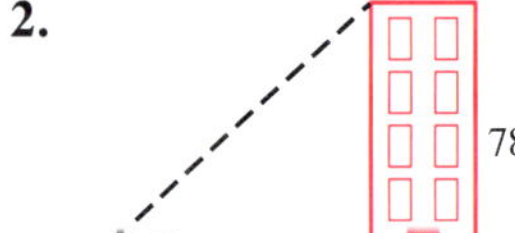

$\tan 42° = \frac{78}{x}$; $\cot 42° = \frac{x}{78}$

3. 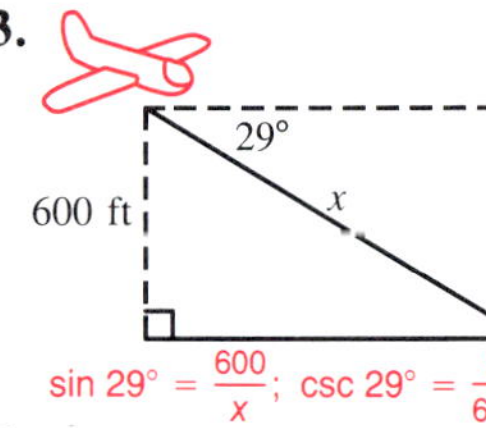

$\sin 29° = \frac{600}{x}$; $\csc 29° = \frac{x}{600}$

4. 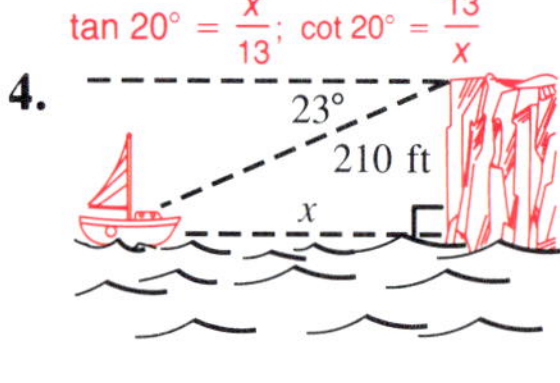

$\tan 23° = \frac{210}{x}$; $\cot 23° = \frac{x}{210}$

5. 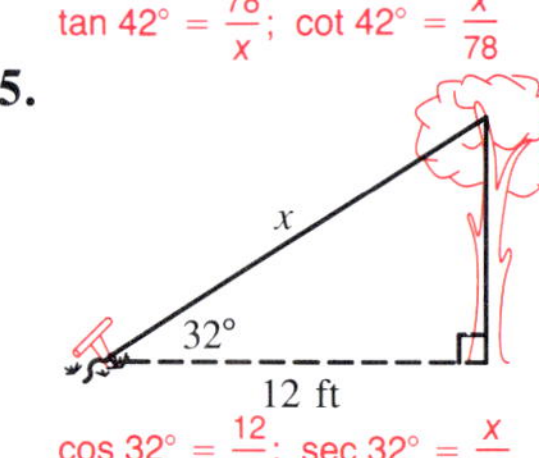

$\cos 32° = \frac{12}{x}$; $\sec 32° = \frac{x}{12}$

6. 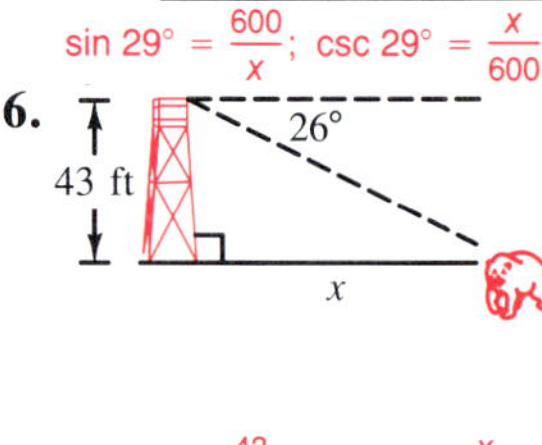

$\tan 26° = \frac{43}{x}$; $\cot 26° = \frac{x}{43}$

7. If a guy wire for a tree is 14 ft long, making a 41° angle with the ground, how far is the base of the tree from the stake anchoring the wire? 11 ft

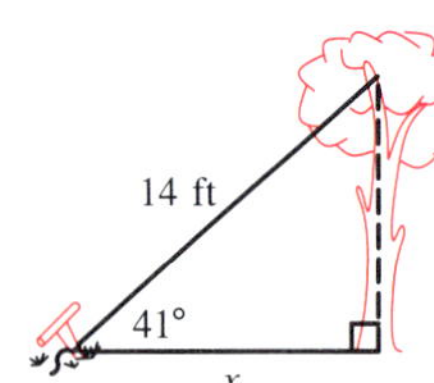

8. From a height of 38 m above sea level, two ships are sighted due west. The angles of depression are 53° and 28°. How far apart are the ships? 43 m

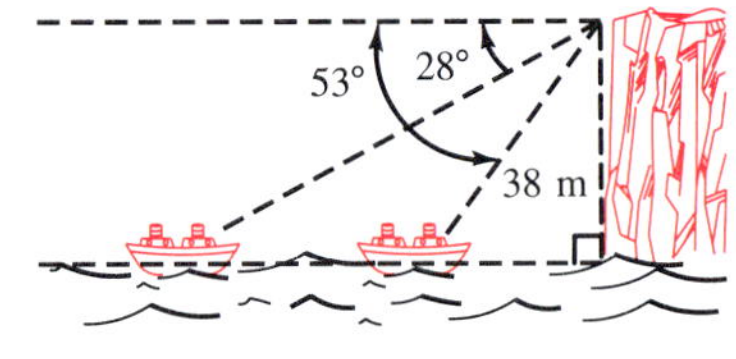

PRACTICE EXERCISES

A

1. The angle of depression is measured from the top of a 43-ft tower to a reference point on the ground. Its value is found to be 63°. How far is the base of the tower from the point on the ground? 22 ft

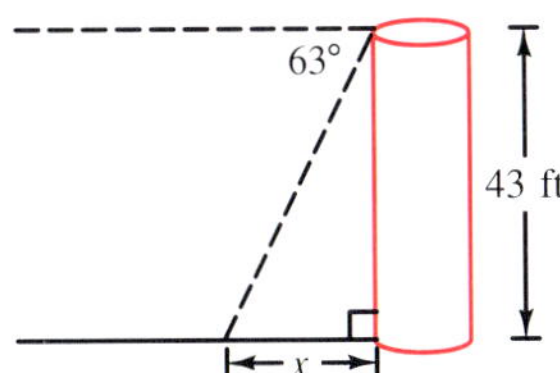

2. The angle of depression from a searchlight to its target is 58°. How long is the beam of light, if the searchlight is 26 ft above ground? 31 ft

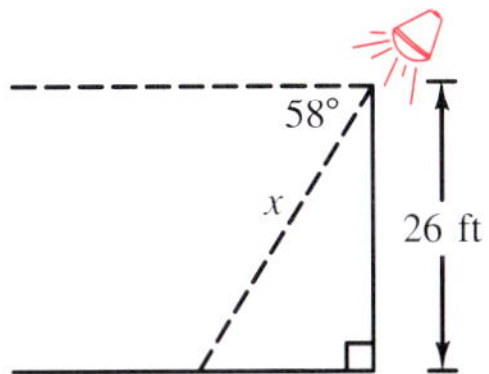

3. A NASA recovery helicopter hovers 75 ft above a space capsule. If the angle of depression to the recovery ship is 40°, how far is the ship from the space capsule? 89 ft

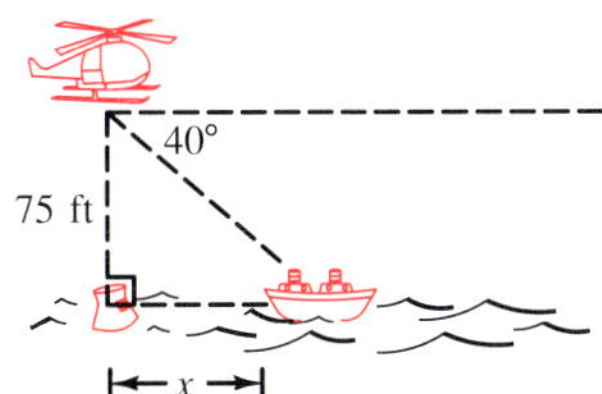

4. The entrance of the old town library is 2.3 ft above ground level. A ramp from the ground level to the library entrance is scheduled to be built. The angle of elevation from the base of the ramp to its top is to be 15°. Find the length of the ramp. 8.9 ft

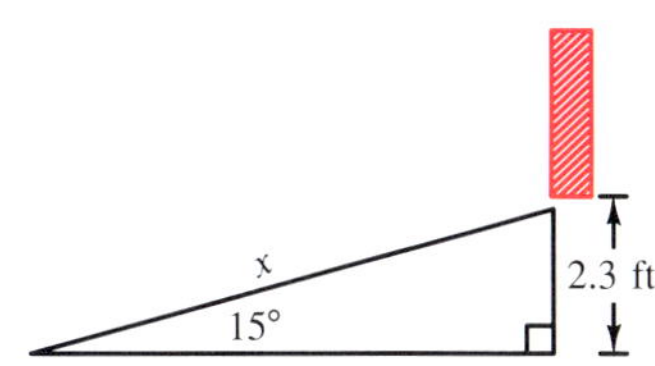

5. A closed-circuit television camera is mounted on a wall 7.4 ft above a security desk in an office building. It is used to view an entrance door 9.3 ft from the desk. Find the angle of depression from the camera lens to the entrance door. 39°

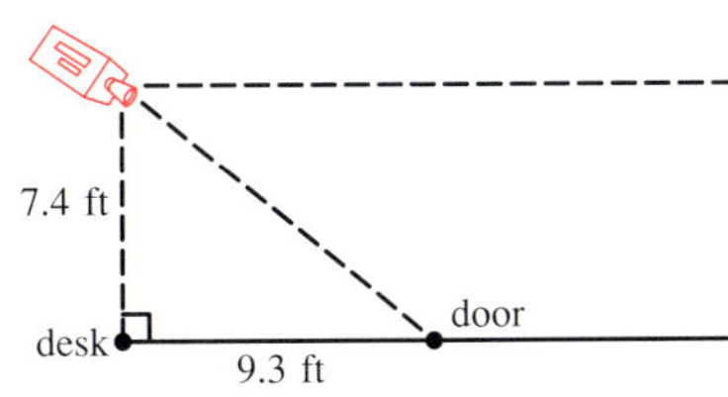

6. To illuminate the entrance of an apartment building, a night light is mounted on a 6.6-m pole. If the base of the pole is 24 m from the entrance, find the angle of depression from the light. 15°

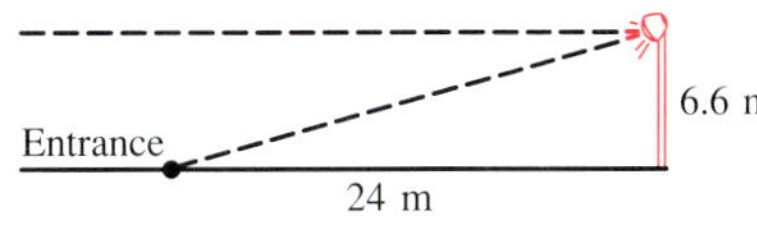

Challenge

Students are challenged to find formulas for the areas of a parallelogram and a rectangle using trigonometry.

Lesson Quiz

1. Name the angle of elevation in the figure below. 5
2. Name the angle of depression in the figure below. 2

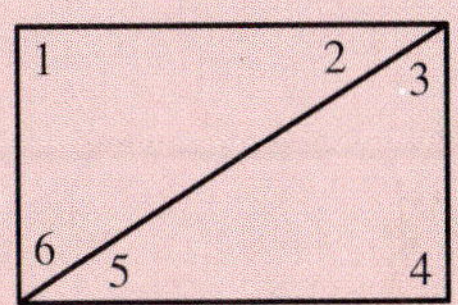

3. From a sailboat, the angle of elevation to the top of a lighthouse 311 ft away is 17°. Find the height of the lighthouse. 95.1 ft
4. An airplane is flying at an altitude of 1.4 mi and is 10.7 mi from the runway. Find the angle of depression that the airplane must make to land safely. 7.5°
5. An antenna stands on top of a 140-ft building. From a point 112 ft from the building, the angle of elevation to the top of the antenna is 62.1°. Find the height of the antenna. 71.5 ft

Enrichment

Find θ if $x = 50$ cm and $y = 40$ cm.

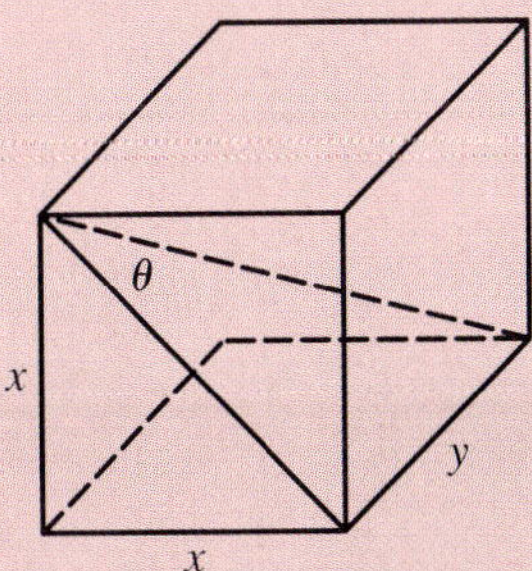

$\tan \theta = \frac{40}{50\sqrt{2}}$; $\theta = 29°$

7. The angle of elevation from the bottom of the world's longest slide, located in Peru, Vermont, is approximately 10.3°. The slide has a vertical drop of 821 ft. Find the length of the slide. 4590 ft

8. The largest doors in the world are located in the Vehicle Assembly Building near Cape Canaveral, Florida. If the angle of elevation from a point on the ground 199 ft from the base of the doors is 66.6°, how high are the doors? 460 ft

9. The world's longest escalator is at the Leningrad Underground in Lenin Square. The escalator has an angle of elevation of 10.36° and a vertical rise of 195.8 ft. Find the length of the escalator. 1089 ft

10. The world's tallest fountain is in Fountain Hills, Arizona. If the angle of elevation to the top of the fountain from a point 755 ft from its base is 36.5°, find the height of the fountain. 559 ft

11. The extension ladder on top of a 6.0-ft high hook and ladder truck is 150 ft long. If the angle of elevation of the ladder is 70°, to what height on a building will the ladder reach? 147 ft

12. A child holds the end of a kite string 36 in. above the ground. The string is taut and it makes a 68° angle with the horizontal. How high off the ground is the kite if 540 in. of string are out? 537 in.

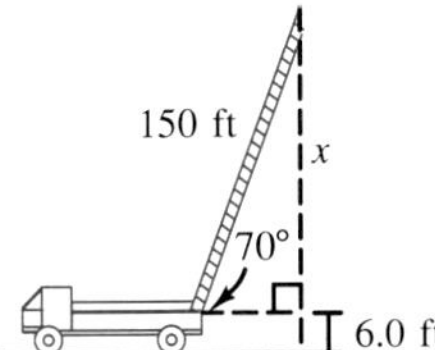

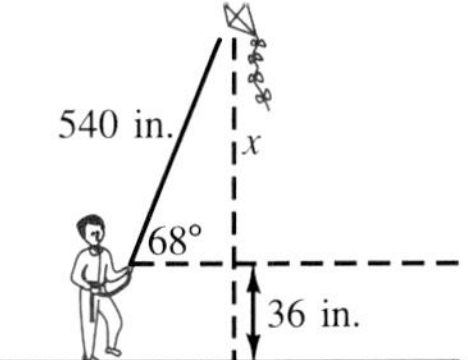

B 13. The Statue of Liberty stands on a 150-ft pedestal. From a point 280 ft from the base of the pedestal, the angle of elevation to the top of Liberty's torch is 47°. Find the height of the statue. 150 ft

14. A television antenna stands on the edge of the top of a 52-story building. From a point 320 ft from the base of the building, the angle of elevation to the top of the antenna is 64°. If each story is 12 ft high, find the height of the antenna. 32 ft

15. A ranger's tower is located 44 m from a tall tree. From the top of the tower, the angle of elevation to the top of the tree is 28°, and the angle of depression to the base of the tree is 36°. How tall is the tree? 55 m

16. A building is 16.3 m from a television tower. From the top of the building, the angle of depression to the base of the tower is 43.5°, and the angle of elevation to the top of the tower is 23.8°. Find the height of the tower. 22.7 ft

17. A jet took off at a rate of 260 ft/s and climbed in a straight path for 3.2 min. What was the angle of elevation of its path if its final altitude was 12,000 ft? 13.91°

18. A jet took off at a rate of 280 ft/s and climbed in a straight path for 5.1 min. What was the angle of elevation of its path if its final altitude was 15,000 ft? 10.08°

C 19. An engineer determines that the angle of elevation from her position to the top of a tower is 52°. She measures the angle of elevation again from a point 47 m farther from the tower and finds it to be 31°. Both positions are due east of the town. Find the height of the tower. 53 m

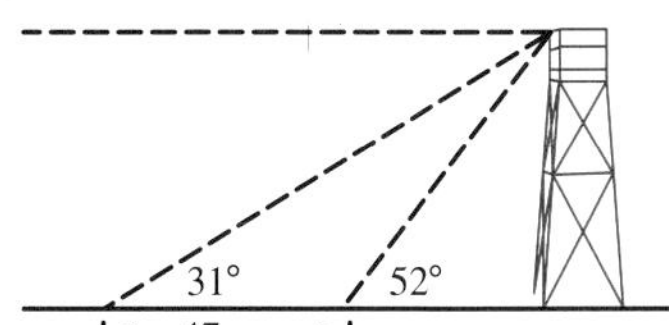

20. From one point on the ground, the angle of elevation of the peak of a mountain is 10.38°, and from a point 15,860 ft closer to the mountain, the angle of elevation is 14.67°. Both points are due south of the mountain. Find the height of the mountain. 9674 ft

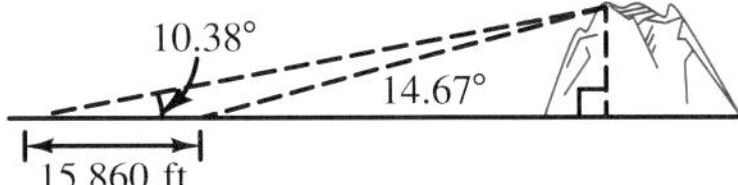

CHALLENGE

1. The area of a parallelogram can be given in terms of the lengths x and y of two adjacent sides and a trigonometric function of the angle formed by those two sides. Which of the following formulas gives the area A of this parallelogram? c

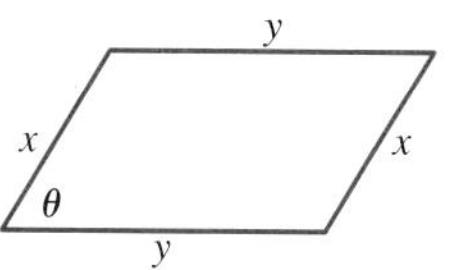

a. $A = xy \cos \theta$ **b.** $A = xy \tan \theta$ **c.** $A = xy \sin \theta$ **d.** $A = xy \cot \theta$

2. Would the answer to Exercise 1 be the same if θ were one of the other three angles of the parallelogram? yes

3. The area of a rectangle can be given in terms of the length p of a diagonal and two trigonometric functions of an angle formed by the diagonal and a side. Which of the following formulas gives the area A of this rectangle? d

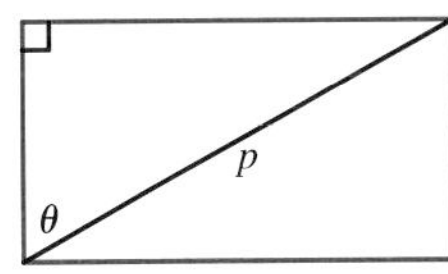

a. $A = p^2 \sec \theta$ **b.** $A = p^2 \cot \theta$ **c.** $A = p^2 \sin \theta \tan \theta$ **d.** $A = p^2 \sin \theta \cos \theta$

4. Would the answer to Exercise 3 be the same if θ were one of the other three angles formed by the diagonal and a side of the rectangle? yes

Teacher's Resource Book

Practice—Chapter 3, p. 3

Enrichment—Chapter 3, p. 4

LESSON PLAN

Vocabulary

Bearing
Course

BACKGROUND

In the Preview, the terms navigation, course, and bearing as they relate to directions are discussed.

3.3 Applications

Objective: To solve real-world problems using trigonometry

Right triangle trigonometry can be used to solve problems in navigation, surveying, and construction.

Preview

In air and sea navigation, the angle measured clockwise from north to the line of travel is the **course** of the plane or ship. The clockwise angle from north to the line of sight to a point of reference is called the **bearing** of the point.

Directions are also written by referring to north or south, using an acute angle. In the figure at the right below, the direction of $\overrightarrow{OA}$ is N 60°E and that of $\overrightarrow{OB}$ is S15°E.

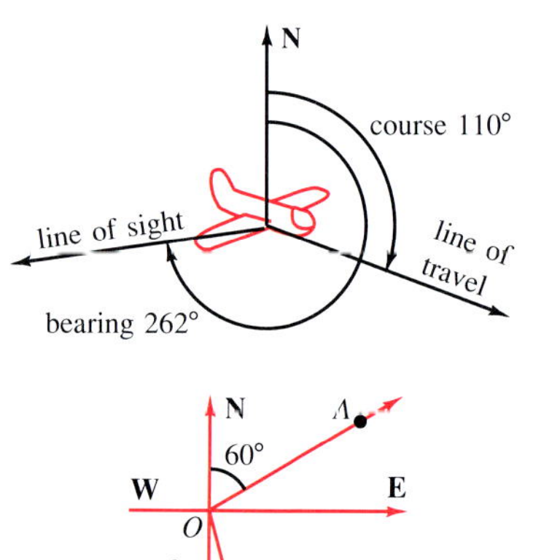

Write the bearing for each direction.

1. N15°E 15° **2.** N 20°W 340° **3.** S18°E 162°

4. S 45°W 225° **5.** N 72°W 288° **6.** S 59°E 121°

In order to solve navigation problems, it is often necessary to find or to interpret bearings.

EXAMPLE 1 **The pilot of a plane traveling on a course of 30° sights the San Francisco International Airport. The pilot's line of sight forms a right angle with the plane's line of travel, as shown at the right. Find the bearing of the airport.**

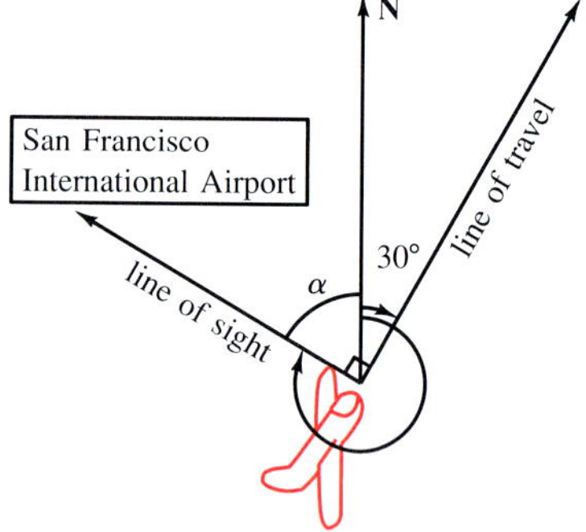

$$\alpha = 90° - 30° = 60°$$
$$\text{bearing} = 360° - \alpha$$
$$= 360° - 60° = 300°$$

EXAMPLE 2 **After traveling 65 mi on the same course, the new bearing of the airport from the plane in Example 1 is 238°. Find the distance from the plane to San Francisco International Airport.**

β = bearing − 180° − 30°
= 238° − 210° = 28°

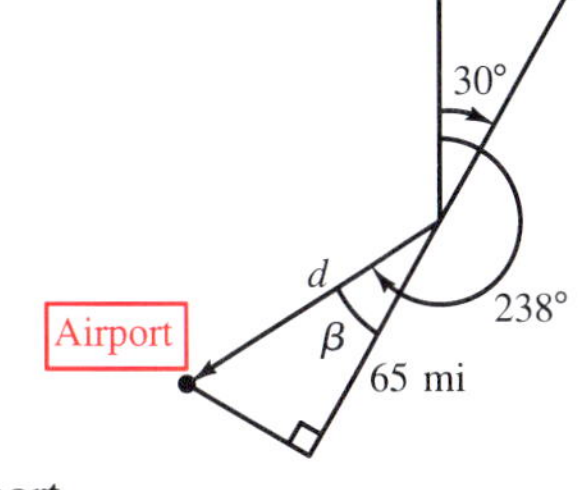

Find the distance using the secant ratio.

$$\sec 28° = \frac{d}{65}$$

$d = 65 \sec 28°$ *Calculation-ready form*

$d = 74$ *Two significant digits*

The plane is approximately 74 mi from the airport.

Small errors in angle measurement can cause large errors in computed distances. To minimize such errors, surveyors measure angles using an extremely accurate instrument called a *transit*.

EXAMPLE 3 **A bridge is being built across a river, from *B* to *C*. A surveyor using a transit determines that $\angle A = 43°20'$. It is also known that the distance from *A* to *C* is 465 m. Find *BC*, the distance across the river.**

$$\tan 43°20' = \frac{BC}{465}$$

$BC = 465 \tan 43°20'$

$BC = 439$ *Three significant digits*

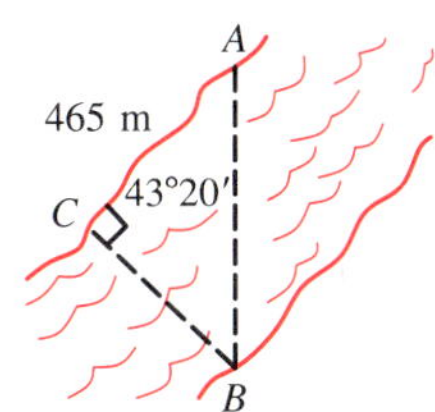

The distance across the river is 439 m.

Trigonometric ratios can also be used to solve construction problems.

EXAMPLE 4 **This hip roof is in the form of four congruent isosceles triangles with a common vertex that forms the peak of the roof. If each triangle has a base of 26 ft and a height of 14 ft, find the measure of each base angle.**

Since side *AC* is bisected by the altitude, *AD* = 13 ft.

$$\tan \alpha = \frac{14}{13}$$

$\alpha = 47°$ *To the nearest degree*

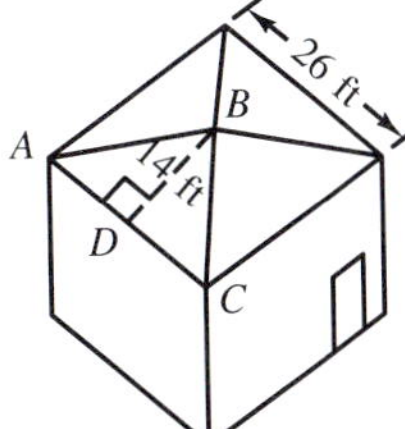

Each base angle measures approximately 47°.

TEACHING SUGGESTIONS

Use the Preview exercises to further develop the idea of bearing. Use students' drawings to determine a bearing given a particular direction or line of travel.

CHALKBOARD EXAMPLES

- **For Example 1**

  1. A pilot is traveling on a course of 36° and sights an airport. The pilot's line of sight forms a right angle with the plane's line of travel. Find the bearing of the airport. 306°

- **For Example 2**

  2. After traveling 45 min at 25 km/h on a course of 30°, a tugboat finds that its new bearing to a lighthouse is 214°. How far is the boat from the lighthouse? 19 km

- **For Example 3**

  3. A surveyor needs to determine the distance across a river. If $\angle X = 50°30'$ and $y = 415$ in., find the distance across the river. 503 in.

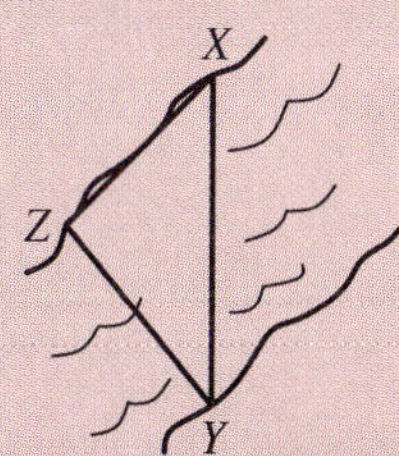

- **For Example 4**

  4. A hip roof on a garage consists of four congruent isosceles triangles with a common vertex that forms the peak of the roof. If each triangle has base 32 ft and each base angle measures 50°, find the height of each triangle. 19 ft

Common Error

- Students often confuse course and bearing. Be sure to emphasize that course is the angle formed by the line of travel with due north.
- See *Teacher's Resource Book* for additional remediation.

LESSON FOLLOW-UP

Critical Thinking

Analysis Ask students to explain how trigonometry can be used to determine the "ceiling" (an aviation term that indicates the height of the cloud cover) at night. Students should reason that they could shine a searchlight vertically upward, from a point at a measured distance to the right or left of the clouds, and then use a trigonometric ratio to find the ceiling by using the measure of the angle of the searchlight as the angle of elevation and the distance from the clouds as one of the other measurements.

Assignment Guide

See p. 110B for assignments.

CLASS EXERCISES

Give the course of each ship or airplane and its bearing to the lighthouse or tower.

1. N; 43°

43°; 270°

2. N; 132°; 110°; Tower

242°; 132°

3. N; 33°

0°; 57°

4. To find the distance AC across the western corner of Ocean Pond, a surveyor determined the measurements shown. Find AC. 39 m

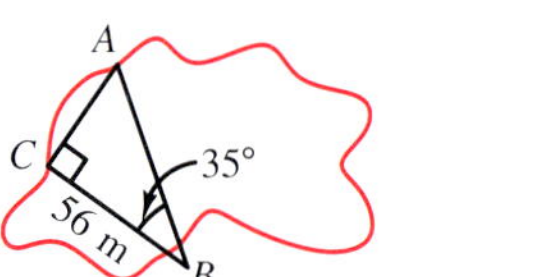

5. A hip roof is in the form of four congruent isosceles triangles. The base of each triangle is 24 ft and the height is 11 ft. Find the measure of each base angle. 43°

6. Each of the four congruent isosceles triangles that form a hip roof has a height of 12 ft and base angles of 36°. What is the length of the base of each triangle? 33 ft

PRACTICE EXERCISES

A **1.** A tugboat is 36 km due north of lighthouse A. Lighthouse B has a bearing of 90° from lighthouse A. The lighthouses are 53 km apart. Find the bearing of lighthouse B from the tugboat and the distance from lighthouse B to the tugboat. 124°; 64 km

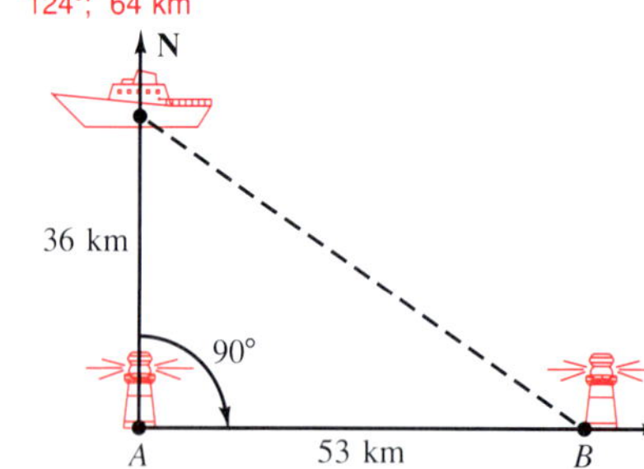

2. A boat is 23 mi due west of lighthouse A. Lighthouse B is 14 mi due north of lighthouse A. Find the bearing of lighthouse B from the boat and the distance from lighthouse B to the boat. 59°; 27 mi

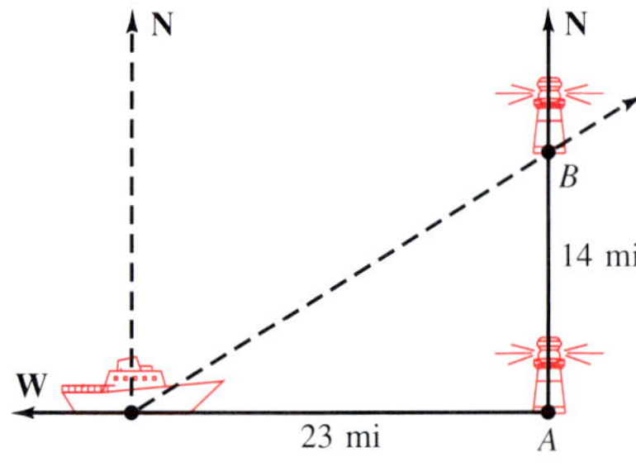

3. Two ships leave the same port at 7 AM. The first ship sails towards Europe on a 54° course at a constant rate of 36 mi/h. The second ship, with a tropical destination, sails on a 144° course at a constant speed of 42 mi/h. Find the distance between the ships at 11 AM. 221 mi

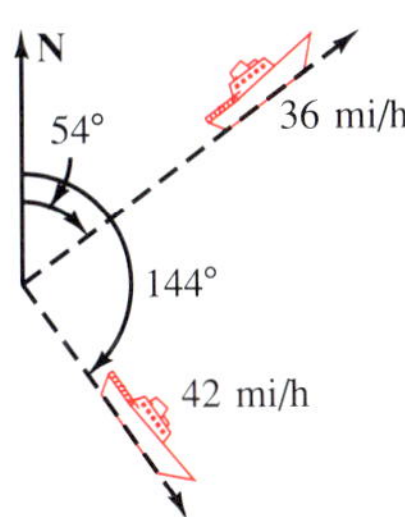

4. A jet flew 140 mi on a course of 196° and then 120 mi on a course of 106°. Then the jet returned to its starting point via the shortest route possible. Find the total distance that the jet traveled. 444 mi

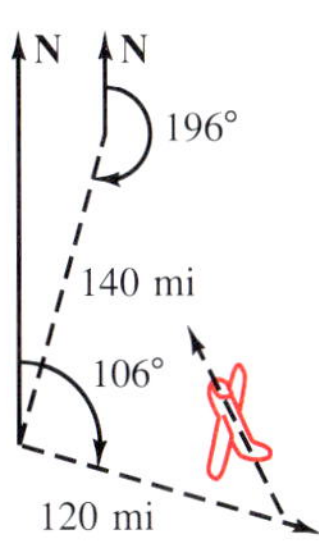

5. To find the length of a diagonal of a rectangular city block, a surveyor determined the measurements shown below. How long is the diagonal? 916 ft

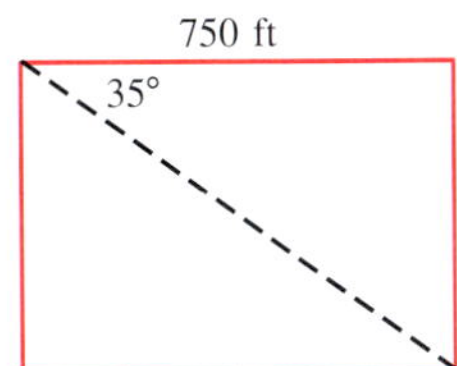

6. To find the distance d across a pond, a surveyor determined the measurements shown below. How far is it across the pond? 1.3 mi

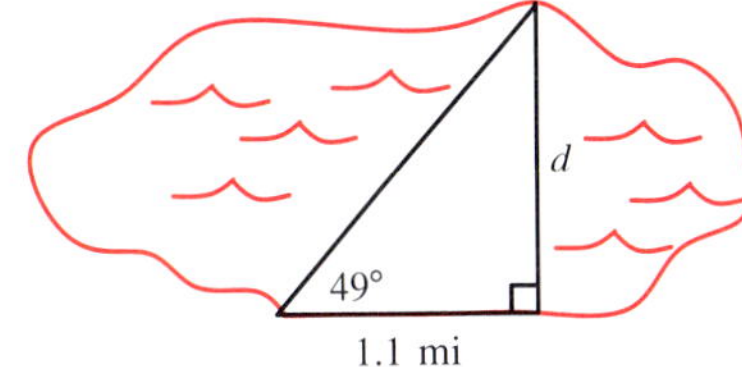

7. To find the distance AC across a canyon, a hiker walks 29 yd along one side from A to B. If segments AB and AC are perpendicular and $\angle B = 73°$, find AC. 95 yd

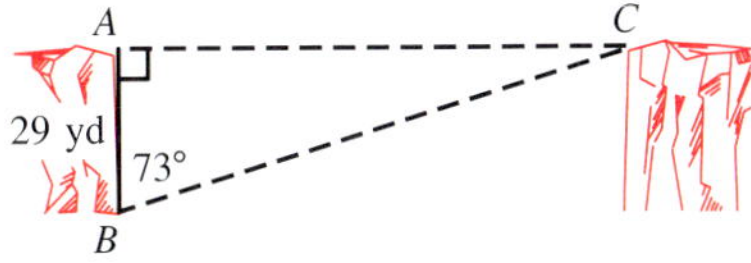

8. To find the width of a river between points P and R, a surveyor locates a point Q such that segments PQ and PR are perpendicular and $PQ = 420$ yd. Using a transit, the surveyor finds that $\angle Q = 66°$. Find PR, the width of the river. 943 yd

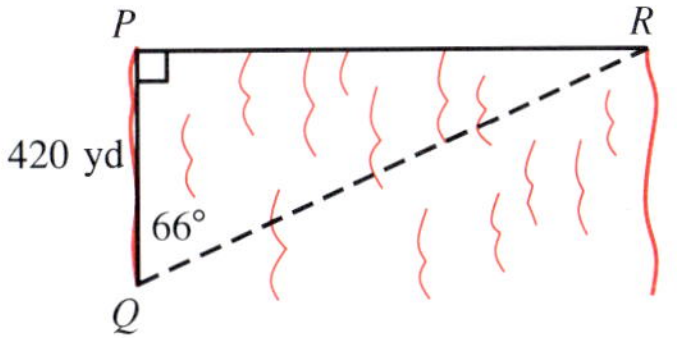

Test Yourself

See *Teacher's Resource Book*, *Tests*, pp. 25–26.

Lesson Quiz

1. To find the diagonal across a rectangular plot, a surveyor determined the measurements shown. Find the length of the diagonal. 83 ft

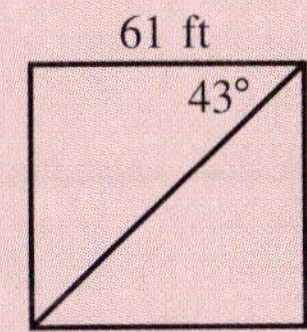

2. A boy walking on a course of 87° sights a tree. The line of sight forms a 103°-angle with the line of travel. Find the bearing of the tree. 190°
3. In a cross-country race, a horse and rider are on a course of 38° and are 63 km from the starting point. A second rider is on a course of 128° and 46 km from the starting point. Find the distance between the riders. 78 km
4. The rise of symmetrical roof is 8 ft and the run 15 ft. Find the pitch. $\frac{4}{15}$

Enrichment

Standing at the midpoint of the Golden Gate Bridge you observe a speed boat coming directly toward you. Suppose your eyes are 270 ft above the water. The angle of depression, your horizontal line of sight, changes from 20° to 45° while you are watching the boat. Approximate the distance the boat traveled during the observation period. Then estimate the speed, to the nearest mile per hour, of the boat if it took 10 s to travel that distance.

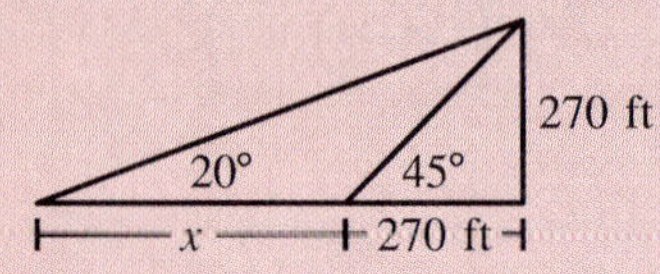

$\tan 20° = \frac{270}{270 + x}$, 427 ft;
$472 \div 10 \cdot 3600 \div 5280$, 32 mi/h

9. A hip roof is in the form of four congruent isosceles triangles with a common vertex that forms the peak of the roof. If each triangle has an 18-ft base and a height of 16 ft, find the measure of each base angle. 61°

10. A hip roof is in the form of four congruent isosceles triangles with base angles of 30° and heights of 15 ft. Find the length of the base of each triangle. 52 ft

B 11. The bearing of a buoy from a ship 8.7 mi away is 64°. The ship is headed due north, and the navigator plans to change course when the buoy has a bearing of 154°. How much farther will the ship travel before a change of course is needed? 19.8 mi

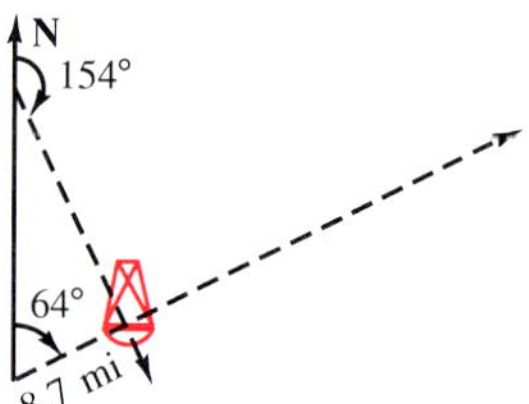

12. A pilot of a San Antonio-to-Houston express plane traveling on a course of 79° sights the Austin airport. His line of sight forms a right angle with the plane's line of travel, as shown below. Find the bearing of the Austin airport. 349°

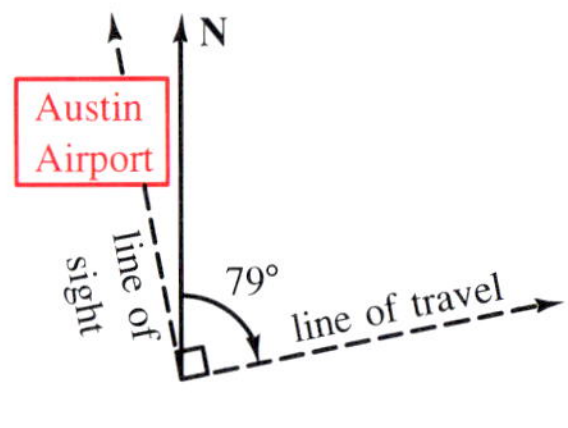

13. After the plane in Exercise 12 travels 45 min at 180 mi/h along the same course, the airport has a new bearing of 280°. How far is the plane from the airport then? 145 mi

14. The navigator of a ship on a 44° course sights a buoy with bearing 134°. After the ship sails 15 km along the same course, the navigator sights the same buoy with bearing 168°. Find the distance between the ship and the buoy at the time of each sighting. first sighting: 22.2 km; second sighting: 26.8 km

15. The two sections of the Sault Sainte Marie railroad bridge in Michigan are each 210 ft in length. Suppose that the maximum angle of elevation of each section is 75°. When the bridge is closed, the water level is normally 13 ft below the bridge. When the bridge is fully opened, what is the distance from the water to point *A* on the upper left corner of the right section? 216 ft

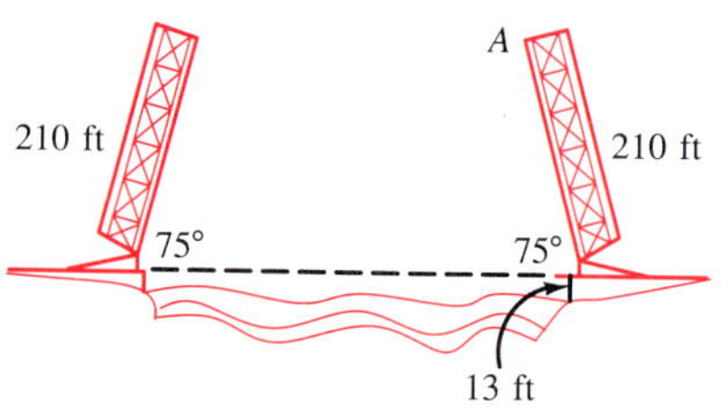

16. When the Sault Sainte Marie bridge is fully opened, find the distance between the separated ends of the sections (see Exercise 15). 311 ft

C **17.** A space-shuttle pilot flying toward the Suez Canal finds that the angle of depression to one end of the canal is 38.25° and the angle of depression to the other end is 52.75°. If the canal is 100.6 mi long, find the altitude of the space shuttle. 198.0 mi

18. A helicopter is hovering near the Fort Pitt Tunnel in Pittsburgh, Pennsylvania. The tunnel is 3560 ft long and the angles of depression to the two ends of the tunnel are 22.2° and 68.5°. Find the altitude of the helicopter. 1730 ft

Teacher's Resource Book
Practice—Chapter 3, p. 5
Enrichment—Chapter 3, p. 6

TEST YOURSELF

Solve each right triangle *ABC* ($\angle C = 90°$) using the given information.

3.1

1. $\angle A = 43°$, $c = 19$
$\angle B = 47°$; $a = 13$; $b = 14$

2. $\angle B = 21°$, $a = 7.6$
$b = 2.9$; $c = 8.1$; $\angle A = 69°$

3. $a = 15$, $b = 19$
$\angle A = 38°$; $\angle B = 52°$; $c = 24$

4. $a = 23.2$, $c = 31.6$
$b = 21.5$; $\angle A = 47.2°$; $\angle B = 42.8°$

3.2

5. A 22-ft flagpole casts a 62-ft shadow. Calculate the angle of elevation to the sun. 20°

6. The angle of depression from a helicopter to its landing port is 64°. If the altitude of the helicopter is 1600 m, find the direct distance from the helicopter to the landing port. 1780 m

3.3

7. An ocean liner is 19 mi due west of lighthouse *A*. Lighthouse *B* is 32 mi due south of lighthouse *A*. Find the bearing of lighthouse *B* from the ocean liner and the distance from lighthouse *B* to the liner. 149°; 37 mi

8. Two planes take off from an airport at 9:15 AM. One flies on a course of 86° at a constant rate of 190 mi/h. The second one flies on a course of 176° at 160 mi/h. How far apart are they at 10:45 AM? 373 mi

9. To find the distance across a pond, a surveyor measured 72 m along a line perpendicular to the line of sight across the pond, and also measured an angle of 52°, as shown below. What is the distance *d* across the pond? 92 m

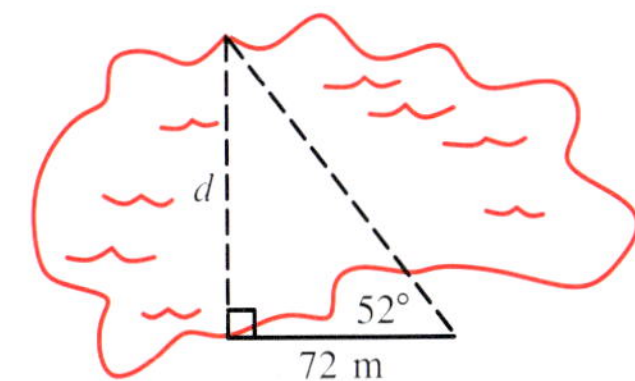

10. The sloping edges *AB* and *AC* of a symmetrical roof each measure 27 ft and form 35° angles with the horizontal. Find the distance *BC* between the edges of the roof. 44 ft

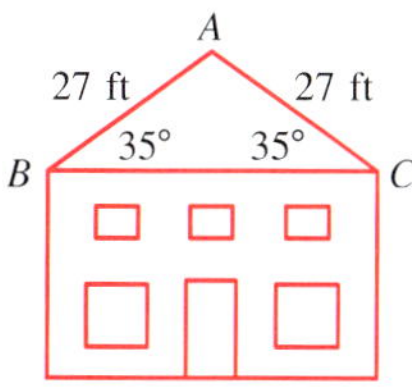

LESSON PLAN

Vocabulary

Odd-even identities
Pythagorean identities
Ratio identities
Reciprocal identities
Trigonometric identity

BACKGROUND

In the Preview, algebraic sentences and their relative truth values are reviewed.

3.4 Fundamental Identities

Objective: To introduce and prove the reciprocal, ratio, Pythagorean, and odd-even identities

Many fundamental relationships between trigonometric functions can be derived directly from definitions. These relationships are useful in simplifying more complicated trigonometric expressions.

Preview

In both algebra and trigonometry, sentences in one variable may be true for all values of the variable, false for all values, or true for just some values.

True for *all* values	False for *all* values	True for *some* values
$ab = ba$	$3(x + 3) = 3x + 5$	$7x + 2 = 5$
$2x^2 + 4x^2 = 6x^2$	$x + 3 = 2 + x$	$3(x + 2) = 16$
$-1 \le \sin\theta \le 1$	$\sin\theta = 3\sqrt{3}$	$\cos\theta = 1$

Classify each sentence as *always true*, *always false*, or *sometimes true*.

1. $-x = -(x)$ always true
2. $x - 5 = x + (-5)$ always true
3. $(x + 2) - 4 = x + 2$ always false
4. $\frac{x^2}{2x} = \frac{1}{2}$ sometimes true
5. $\frac{1}{x^2} = \frac{1}{x}$ sometimes true
6. $\frac{1}{2}(2x) = 3x\left(\frac{1}{3}\right)$ always true

A **trigonometric identity** is a trigonometric equation that is true for all values in the domain of its variable. Examples are the **reciprocal identities** (Chapter 1).

$$\csc\theta = \frac{1}{\sin\theta} \qquad \sec\theta = \frac{1}{\cos\theta} \qquad \cos\theta = \frac{1}{\tan\theta}$$

An identity may be proved using the basic definitions from Chapter 1.

EXAMPLE 1 **Use the definitions $\sin\theta = \frac{y}{r}$ and $\csc\theta = \frac{r}{y}$ to prove $\csc\theta = \frac{1}{\sin\theta}$.**

$$\frac{1}{\sin\theta} = \frac{1}{\frac{y}{r}} = \frac{1 \cdot r}{\frac{y}{r} \cdot r} = \frac{r}{y} = \csc\theta$$

Therefore, $\csc\theta = \frac{1}{\sin\theta}$.

In Example 1, $\csc\theta$ must have no value of θ for which $\sin\theta = 0$. Similar restrictions exist for $\sec\theta$ ($\cos\theta \neq 0$) and $\cot\theta$ ($\tan\theta \neq 0$). Assume that a denominator in a trigonometric expression is not zero, even if not explicitly stated.

A proof of an identity may be written using a vertical line to separate the expressions to be proved equivalent. Use known algebraic and trigonometric identities to replace one of the expressions with an equivalent expression. Repeat the process until an expression is obtained that is the same as that on the other side of the line. The procedure is illustrated in Example 2.

The **ratio identities** can also be proved using definitions.

$$\tan\theta = \frac{\sin\theta}{\cos\theta},\quad \cos\theta \neq 0 \qquad \cot\theta = \frac{\cos\theta}{\sin\theta},\quad \sin\theta \neq 0$$

EXAMPLE 2 **Prove: $\tan\theta = \dfrac{\sin\theta}{\cos\theta}$, $\cos\theta \neq 0$**

$\tan\theta$	$\dfrac{\sin\theta}{\cos\theta}$	
	$\dfrac{\frac{y}{r}}{\frac{x}{r}}$	*Definitions of $\sin\theta$ and $\cos\theta$ ($x \neq 0$, since $\cos\theta \neq 0$.)*
	$\dfrac{y}{x}$	*Multiply numerator and denominator by r.*
	$= \tan\theta$	Therefore, $\tan\theta = \dfrac{\sin\theta}{\cos\theta}$, $\cos\theta \neq 0$.

The expression $(\sin x)^2$ is usually written as $\sin^2 x$, and similarly for the other trigonometric functions. Such expressions are found in the three **Pythagorean identities,** so called because they can be derived from the Pythagorean theorem.

$$\sin^2\theta + \cos^2\theta = 1 \qquad 1 + \cot^2\theta = \csc^2\theta \qquad 1 + \tan^2\theta = \sec^2\theta$$

EXAMPLE 3 **Prove: $\sin^2\theta + \cos^2\theta = 1$**

$\sin^2\theta + \cos^2\theta$	1	
$\left(\dfrac{y}{r}\right)^2 + \left(\dfrac{x}{r}\right)^2$		*Definitions of $\sin\theta$ and $\cos\theta$*
$\dfrac{y^2}{r^2} + \dfrac{x^2}{r^2}$		
$\dfrac{x^2 + y^2}{r^2}$		
$\dfrac{r^2}{r^2}$		$x^2 + y^2 = r^2$
1	$=$	Thus, $\sin^2\theta + \cos^2\theta = 1$.

TEACHING SUGGESTIONS

Show, or have students express, the alternate forms of the reciprocal and Pythagorean identities. Answers may vary. For example: $\sin\theta = \dfrac{1}{\csc\theta}$ or $\sin\theta\csc\theta = 1$; $\sin^2\theta + \cos^2\theta = 1$ or $\sin^2\theta = 1 - \cos^2\theta$ or $\cos^2\theta = 1 - \sin^2\theta$

Critical Thinking

Comprehension Ask students why it is not permissible to work across the equal sign when proving identities. To do so would mean assuming the identity is true.

CHALKBOARD EXAMPLES

- **For Example 1**

1. Prove: $\sec\theta\cos\theta = 1$

$\sec\theta\cos\theta$	1
$\left(\dfrac{r}{x}\right)\left(\dfrac{x}{r}\right)$	
$\dfrac{rx}{xr}$	
1	$=$

- **For Example 2**

2. Prove: $\cos\theta = \cot\theta\sin\theta$

$\cos\theta$	$\cot\theta\sin\theta$
	$\left(\dfrac{x}{y}\right)\left(\dfrac{y}{r}\right)$
	$\dfrac{x}{r}$
$=$	$\cos\theta$

- **For Example 3**

3. Prove: $1 = \sec^2\theta - \tan^2\theta$

1	$\sec^2\theta - \tan^2\theta$
	$\dfrac{r^2}{x^2} - \dfrac{y^2}{x^2}$
	$\dfrac{r^2 - y^2}{x^2}$
	$\dfrac{x^2}{x^2}$
$=$	1

- **For Example 4**

4. Prove: $\cos(-\theta) = \cos\theta$

$\cos(-\theta)$	$\cos\theta$
$\frac{x}{r}$	
$\cos\theta$	$=$

5. $\tan(-\theta) = -\tan\theta$

$\tan(-\theta)$	$-\tan\theta$
$\frac{x}{-y}$	
$-\frac{x}{y}$	
$-\tan\theta$	$=$

- **For Example 5**

Verify the following identities for $\theta = 45°$.

6. $\sin^2\theta + \cos^2\theta = 1$

$\sin^2\theta + \cos^2\theta$	1
$\sin^2 45 + \cos^2 45$	
$\left(\frac{\sqrt{2}}{2}\right)^2 + \left(\frac{\sqrt{2}}{2}\right)^2$	
$\frac{2}{4} + \frac{2}{4}$	
1	$=$

7. $\cot\theta = \frac{\cos\theta}{\sin\theta}$

$\cot\theta$	$\frac{\cos\theta}{\sin\theta}$
$\cot 45°$	$\frac{\cos 45°}{\sin 45°}$
1	$\frac{\frac{\sqrt{2}}{2}}{\frac{\sqrt{2}}{2}}$
1	$=$ 1

Common Errors

- In Example 1 and 2, students are often confused by the simplifying of complex fractions. Be sure to show that $\frac{1}{\frac{y}{r}}$ is $1 \cdot \frac{1}{\frac{y}{r}} = 1\left(\frac{r}{y}\right) = \frac{r}{y}$.
- Students often confuse proving with verifying. Emphasize that verifying involves the use of specific angle measures.
- See *Teacher's Resource Book* for additional remediation.

Recall that a function $f(x)$ is *even* if $f(-x) = f(x)$ for all real values of x in its domain and *odd* if $f(-x) = -f(x)$ for all real values of x in its domain. The last two identities of this lesson will be the **odd-even identities.**

$$\sin(-\theta) = -\sin\theta \qquad \cos(-\theta) = \cos\theta \qquad \tan(-\theta) = -\tan\theta$$

EXAMPLE 4 **Prove: $\sin(-\theta) = -\sin\theta$**

Let θ be an angle that intersects a circle of radius r at a point (x, y). Then $-\theta$ will intersect the circle at the point $(x, -y)$.

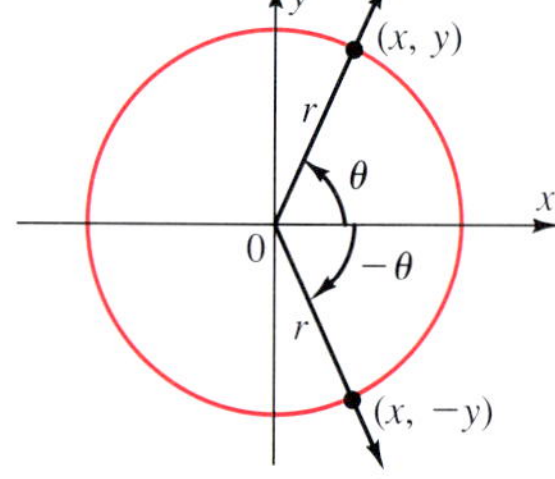

$\sin(-\theta)$	$-\sin\theta$
$\frac{-y}{r}$	
$-\frac{y}{r}$	
$-\sin\theta$	$=$

Therefore, $\sin(-\theta) = -\sin\theta$.

Any identity can be *verified* for a specific value of the variable. Keep in mind that verification for a specific value does not constitute a proof of the identity.

EXAMPLE 5 **Verify the following identities for $\theta = 30°$.**

a. $\tan\theta = \frac{\sin\theta}{\cos\theta}$ **b.** $\sin^2\theta + \cos^2\theta = 1$

a.

$\tan 30°$	$\frac{\sin 30°}{\cos 30°}$
	$\frac{\frac{1}{2}}{\frac{\sqrt{3}}{2}}$
	$\frac{1}{\sqrt{3}}$
	$\frac{\sqrt{3}}{3}$
	$=$ $\tan 30°$

Therefore, $\tan 30° = \frac{\sin 30°}{\cos 30°}$.

b.

$\sin^2(30°) + \cos^2(30°)$	1
$\left(\frac{1}{2}\right)^2 + \left(\frac{\sqrt{3}}{2}\right)^2$	
$\frac{1}{4} + \frac{3}{4}$	
1	$=$

Thus, $\sin^2(30°) + \cos^2(30°) = 1$.

Identities may assume alternate forms. For example, $\sin\theta = \cos\theta\tan\theta$ is an alternate form of $\tan\theta = \frac{\sin\theta}{\cos\theta}$.

Here is a summary of the **fundamental identities.**

Reciprocal Identities

$\csc\theta = \frac{1}{\sin\theta}, \quad \sin\theta \neq 0$

$\sec\theta = \frac{1}{\cos\theta}, \quad \cos\theta \neq 0$

$\cot\theta = \frac{1}{\tan\theta}, \quad \tan\theta \neq 0$

Ratio Identities

$\tan\theta = \frac{\sin\theta}{\cos\theta}, \quad \cos\theta \neq 0$

$\cot\theta = \frac{\cos\theta}{\sin\theta}, \quad \sin\theta \neq 0$

Pythagorean Identities

$\sin^2\theta + \cos^2\theta = 1$

$1 + \cot^2\theta = \csc^2\theta$

$1 + \tan^2\theta = \sec^2\theta$

Odd-Even Identities

$\sin(-\theta) = -\sin\theta$

$\cos(-\theta) = \cos\theta$

$\tan(-\theta) = -\tan\theta$

CLASS EXERCISES

Complete by writing 1 or −1 to make each sentence an identity.

1. $\sin\theta\csc\theta = \underline{?}$ 1
2. $\sec^2\theta - \tan^2\theta = \underline{?}$ 1
3. $\frac{\sin(-\theta)}{\sin\theta} = \underline{?}$ $(\sin\theta \neq 0)$ −1
4. $\cot\theta\tan\theta = \underline{?}$ 1
5. $\tan^2\theta - \sec^2\theta = \underline{?}$ −1
6. $\cos^2\theta + \sin^2\theta = \underline{?}$ 1
7. $\cot^2\theta - \csc^2\theta = \underline{?}$ −1
8. $\sec\theta\cos\theta = \underline{?}$ 1
9. $\frac{\cos\theta}{\cos(-\theta)} = \underline{?}$ $(\cos(-\theta) \neq 0)$ 1
10. $\frac{\sin\theta}{\cos\theta\tan\theta} = \underline{?}$ $(\cos\theta\tan\theta \neq 0)$ 1

PRACTICE EXERCISES

Write two equivalent ways to express each identity. Answers 1–8 may vary. Examples are listed.

A

1. $\csc\theta = \frac{1}{\sin\theta}, \quad \sin\theta \neq 0$
   $\sin\theta = \frac{1}{\csc\theta}$; $\csc\theta\sin\theta = 1$
2. $\sec\theta = \frac{1}{\cos\theta}, \quad \cos\theta \neq 0$
   $\cos\theta = \frac{1}{\sec\theta}$; $\sec\theta\cos\theta = 1$
3. $\tan\theta = \frac{\sin\theta}{\cos\theta}, \quad \cos\theta \neq 0$
   $\cos\theta = \frac{\sin\theta}{\tan\theta}$; $\sin\theta = \cos\theta\tan\theta$
4. $\cot\theta = \frac{\cos\theta}{\sin\theta}, \quad \sin\theta \neq 0$
   $\cos\theta = \sin\theta\cot\theta$; $\sin\theta = \frac{\cos\theta}{\cot\theta}$
5. $\sin^2\theta + \cos^2\theta = 1$ $\sin^2\theta = 1 - \cos^2\theta$; $\cos^2\theta = 1 - \sin^2\theta$
6. $1 + \cot^2\theta = \csc^2\theta$
   $\csc^2\theta - \cot^2\theta = 1$; $\cot^2\theta = \csc^2\theta - 1$
7. $1 + \tan^2\theta = \sec^2\theta$
   $\sec^2\theta - \tan^2\theta = 1$; $\tan^2\theta = \sec^2\theta - 1$
8. Use the two ratio identities to show that $(\tan\theta)(\cot\theta) = 1$.
   $\tan\theta\cot\theta = \left(\frac{\sin\theta}{\cos\theta}\right)\left(\frac{\cos\theta}{\sin\theta}\right) = 1$

LESSON FOLLOW-UP

Discussion

Prove $\cos\theta\sin\theta = \frac{\cos\theta}{\csc\theta}$ and $\sec\theta\sin\theta\cot\theta = 1$, then discuss the difference in the proofs.

$\cos\theta\sin\theta$	$\frac{\cos\theta}{\csc\theta}$
$\frac{x}{r}\cdot\frac{y}{r}$	$\frac{\frac{x}{r}}{\frac{r}{y}}$
$\frac{xy}{r^2}$ =	$\frac{xy}{r^2}$

$\sec\theta\sin\theta\cot\theta$	1
$\frac{r}{x}\cdot\frac{y}{r}\cdot\frac{x}{y}$	
1 =	1

In the first proof, the expressions on both sides of the equal sign were replaced with known identities, while in the second proof only the expressions on the left side were replaced with known identities.

Assignment Guide

See p. 110B for assignments.

Additional Answers

11. $\left(\frac{\sqrt{2}}{2}\right)^2 + \left(\frac{\sqrt{2}}{2}\right)^2 = 1$

12. $1 = \frac{\frac{\sqrt{2}}{2}}{\frac{\sqrt{2}}{2}}$

13. $-\sqrt{3} = \frac{\frac{-\sqrt{3}}{2}}{\frac{1}{2}}$

14. $2 = \frac{1}{\frac{1}{2}}$

Logical Reasoning

Students are asked to find the error in the given algebraic proof that 1 = 2.

Lesson Quiz

1. Use the ratio identities to show that $(\sin\theta)(\csc\theta) = 1$.
   $\sin\theta\left(\frac{1}{\sin\theta}\right) = \frac{\sin\theta}{\sin\theta} = 1$

Verify each identity for the given angle measure.

2. $1 - \sin 45° \cos 45° = \cos^2 45°$

$1 - \sin 45° \cos 45°$	$\cos^2 45°$
$1 - \frac{\sqrt{2}}{2}\cdot\frac{\sqrt{2}}{2}\cdot 1$	$\left(\frac{\sqrt{2}}{2}\right)^2$
$1 - \frac{2}{4}$	$\frac{2}{4}$
$\frac{1}{2}$	$= \frac{1}{2}$

3. $1 = \frac{\sec^2 30°}{1 + \tan^2 30°}$

1	$\frac{\sec^2 30°}{1 + \tan^2 30°}$
	$\frac{\left(\frac{2}{\sqrt{3}}\right)^2}{1 + \left(\frac{1}{\sqrt{3}}\right)^2}$
	$\frac{\frac{4}{3}}{1 + \frac{1}{3}}$
	$\frac{\frac{4}{3}}{\frac{4}{3}}$
=	1

Prove each identity.

4. $\csc\theta = \sec\theta \cot\theta$

$\csc\theta$	$\sec\theta \cot\theta$
	$\frac{1}{\cos\theta}\cdot\frac{\cos\theta}{\sin\theta}$
	$\frac{1}{\sin\theta}$
=	$\csc\theta$

5. $\cos A \sec A = 1$

$\cos A \sec A$	1
$\cos A \cdot \frac{1}{\cos A}$	
1	=

Verify each identity for the given angle measure. 10. $1 + \left(\frac{\sqrt{3}}{3}\right)^2 = \left(\frac{2\sqrt{3}}{3}\right)^2$

9. $1 + \cot^2 30° = \csc^2 30°$ $\quad 1 + (\sqrt{3})^2 = (2)^2$

10. $1 + \tan^2 30° = \sec^2 30°$

11. $\sin^2 45° + \cos^2 45° = 1$ See side column on page 133.

12. $\cot 45° = \frac{\cos 45°}{\sin 45°}$

13. $\tan(-60°) = \frac{\sin(-60°)}{\cos(-60°)}$

14. $\sec(-60°) = \frac{1}{\cos(-60°)}$

15. $\sin(-120°) = -\sin 120°$ $\quad \frac{-\sqrt{3}}{2} = -\left(\frac{\sqrt{3}}{2}\right)$

16. $\cos(-120°) = \cos 120°$ $\quad \frac{-1}{2} = -\frac{1}{2}$

17. $1 - \sin^2\left(\frac{3\pi}{4}\right) = \cos^2\left(\frac{3\pi}{4}\right)$ $\quad 1 - \left(\frac{\sqrt{2}}{2}\right)^2 = \left(\frac{-\sqrt{2}}{2}\right)^2$

18. $\tan^2\left(\frac{3\pi}{4}\right) = \sec^2\left(\frac{3\pi}{4}\right) - 1$ $\quad \left(\frac{1}{-1}\right)^2 = \left(\frac{\sqrt{2}}{-1}\right)^2 - 1$

Use the definitions of the trigonometric functions to prove each identity.
See side column.

B

19. $\sec\theta = \frac{1}{\cos\theta}, \quad \cos\theta \neq 0$

20. $\cot\theta = \frac{1}{\tan\theta}, \quad \tan\theta \neq 0$

21. $\cot\theta = \frac{\cos\theta}{\sin\theta}, \quad \sin\theta \neq 0$

22. $1 + \cot^2\theta = \csc^2\theta$

23. $1 + \tan^2\theta = \sec^2\theta$

24. $\cos(-\theta) = \cos\theta$

25. $\sin\theta \csc\theta = 1$

26. $\sin\theta = \cos\theta \tan\theta$

27. $\cos^2\theta = 1 - \sin^2\theta$

28. $\sin^2\theta = 1 - \cos^2\theta$

Verify each identity for the given angle measure.

29. $\sin 45° \cot 45° = \cos 45°$ $\quad \left(\frac{\sqrt{2}}{2}\right)(1) = \frac{\sqrt{2}}{2}$

30. $\tan 60° = \sin 60° \sec 60°$ $\quad \sqrt{3} = \frac{\sqrt{3}}{2}\cdot 2$

31. $\sin 135° \cos 135° \tan 135° = 1 - \cos^2 135°$ $\quad \left(\frac{\sqrt{2}}{2}\right)\left(-\frac{\sqrt{2}}{2}\right)(-1) = 1 - \left(-\frac{\sqrt{2}}{2}\right)^2$

32. $\cos 30° \cot 30° = \csc 30° - \sin 30°$ $\quad \left(\frac{\sqrt{3}}{2}\right)(\sqrt{3}) = 2 - \frac{1}{2}$

In triangle *ABC*, sides *a*, *b*, and *c* measure 9, 40, and 41, respectively.

C

33. Show that triangle ABC is a right triangle. $9^2 + 40^2 = 41^2$; $1681 = 1681$

34. Verify the identity $\cos A = \frac{\sin A}{\tan A}$ for triangle ABC. $\quad \frac{40}{41} = \frac{\frac{9}{41}}{\frac{9}{40}}$

35. Verify the identity $\csc B \cos B \tan B = 1$ for triangle ABC. $\quad \left(\frac{41}{40}\right)\left(\frac{9}{41}\right)\left(\frac{40}{9}\right) = 1$

36. Verify the identity $\sec B \sin B \cot B = 1$ for triangle ABC. $\quad \left(\frac{41}{9}\right)\left(\frac{40}{41}\right)\left(\frac{9}{40}\right) = 1$

Applications

37. **Engineering** A 20.2-m walkway between two buildings is higher at one end than at the other. The sine of the angle of elevation is 0.1386. Use a Pythagorean identity to find the cosine of that angle. Then find the distance between the buildings. 0.9903; 20.0 m

38. Aviation A plane is flying 8100 m above the ground. The secant of the angle of depression to the end of the runway is 5.7588. Find the tangent of the angle of elevation from the end of the runway to the plane. Then find the ground distance from the plane to the runway. 5.6713; 1428.2 m

Computer The program at the right prints the values of the sine, cosine, and tangent functions for 0, $\frac{\pi}{6}$, $\frac{\pi}{4}$, $\frac{\pi}{3}$, and $\frac{\pi}{2}$. Lines 70 and 90 round the values to four decimal places. An identity is used to calculate the tangent.

```
1    HOME:RESTORE
10   PRINT"ANGLE",TAB(16); "SINE";
     TAB(20); "COSINE";
     TAB(36)"TANGENT"
20   FOR R = 1 TO 5
30   READ ANG
40   DATA 0, 30, 45, 60, 90
50   RAD = ANG*0.0174533
60   Y = SIN(RAD) :X = COS(RAD)
70   Y = (INT(Y*10000))/10000:
     X = (INT(X*10000))/10000
80   IF X = 0 THEN T$ = "UNDEF": GOTO 100
90   T = Y/X: T = (INT(T*10000))/10000:
     T$ = STR$(T)
100  PRINT ANG;TAB(10)Y;TAB(20)X;
     TAB(30)T$
110  NEXT R
120  END
```

39. Run the program and verify the results.

40. Line 90 calculates the tangent function. Which fundamental identity is used? $\tan\theta = \frac{\sin\theta}{\cos\theta}$

41. Line 80 checks to see if the cosine of the angle is 0. Why is this line necessary? It prevents division by 0.

LOGICAL REASONING

When you prove a trigonometric identity, or any other statement in mathematics, every step must be mathematically sound. Here is an algebraic "proof" that 1 = 2. Can you find the error in the reasoning? Step 5 is in error. Since $s = t$, $t - s = 0$. Therefore you cannot divide by $t - s$.

Given: s and t are positive real numbers, and $s = t$.

Proof:

1.	$s = t$	Given.
2.	$st = t^2$	Multiply both sides by t.
3.	$st - s^2 = t^2 - s^2$	Subtract s^2 from both sides.
4.	$s(t - s) = (t + s)(t - s)$	Factor both sides.
5.	$s = t + s$	Divide both sides by $t - s$.
6.	$t = t + t$	Substitute t for s.
7.	$t = 2t$	$t + t = t(1 + 1) = 2t$
8.	$1 = 2$	Divide both sides by t. ($t \neq 0$, since it was given that t is a positive real number.)

Enrichment

Use reciprocal identities to find the numerical value of x to the nearest hundredth of a degree.

$$\frac{1}{\tan x} + \frac{1}{\cot x} = \frac{144}{25}$$ 79.84°, 10.16°

Computer

The complete form of the program for finding the tangent of specific angles is on the disk provided with the *Teacher's Resource Book*.

Teacher's Resource Book

Practice—Chapter 3, p. 7
Enrichment—Chapter 3, p. 8

Additional Answers

19. $\frac{1}{\cos\theta} = \frac{1}{\frac{x}{r}} = \frac{r}{x} = \sec\theta$

20. $\frac{1}{\tan\theta} = \frac{1}{\frac{y}{x}} = \frac{x}{y} = \cot\theta$

21. $\frac{\cos\theta}{\sin\theta} = \frac{\frac{x}{r}}{\frac{y}{r}} = \frac{x}{y} = \cot\theta$

22. $1 + \cot^2\theta = 1 + \left(\frac{x}{y}\right)^2 = 1 + \frac{x^2}{y^2} = \frac{x^2 + y^2}{y^2} = \frac{r^2}{y^2} = \left(\frac{r}{y}\right)^2 = \csc^2\theta$

23. $1 + \tan^2\theta = 1 + \left(\frac{y}{x}\right)^2 = 1 + \frac{y^2}{x^2} = \frac{x^2 + y^2}{x^2} = \frac{r^2}{x^2} = \left(\frac{r}{x}\right)^2 = \sec^2\theta$

24. $\cos(-\theta) = \frac{x}{r} = \cos\theta$

25. $\sin\theta\csc\theta = \left(\frac{y}{r}\right)\left(\frac{r}{y}\right) = \frac{yr}{ry} = 1$

26. $\cos\theta\tan\theta = \left(\frac{x}{r}\right)\left(\frac{y}{x}\right) = \frac{y}{r} = \sin\theta$

27. $1 - \sin^2\theta = 1 - \left(\frac{y}{r}\right)^2 = 1 - \frac{y^2}{r^2} = \frac{r^2 - y^2}{r^2} = \frac{x^2}{r^2} = \left(\frac{x}{r}\right)^2 = \cos^2\theta$

28. $1 - \cos^2\theta = 1 - \left(\frac{x}{r}\right)^2 = 1 - \frac{x^2}{r^2} = \frac{r^2 - x^2}{r^2} = \frac{y^2}{r^2} = \left(\frac{y}{r}\right)^2 = \sin^2\theta$

LESSON PLAN

BACKGROUND

In the Preview, special methods of factoring are reviewed. Greatest common factor, perfect square trinomials, the difference of squares and cubes, and the sum of cubes are recalled.

Critical Thinking

Analysis Ask students to explain how to find the value of $\frac{\csc\theta}{\cos\theta\tan\theta}$ if $\sin\theta = \frac{1}{4}$. Students should reason that since no specific method is implied, the expression can be evaluated in two ways: (1) Draw θ in standard position and find x. Then

$$\frac{\csc\theta}{\cos\theta\tan\theta} = \frac{\frac{r}{y}}{\frac{x}{r}\cdot\frac{y}{x}} = \frac{\frac{4}{1}}{\frac{\sqrt{15}}{4}\cdot\frac{1}{\sqrt{15}}} = \frac{\frac{4}{1}}{\frac{1}{4}} = 16.$$

(2) Write $\frac{\csc\theta}{\cos\theta\tan\theta}$ in terms of sine only. Then substitute:

$$\frac{\csc\theta}{\cos\theta\tan\theta} = \frac{\csc\theta}{\cos\theta\left(\frac{\sin\theta}{\cos\theta}\right)} = \frac{\csc\theta}{\sin\theta} = \frac{1}{\sin\theta\sin\theta} = \frac{1}{\sin^2\theta} = \frac{1}{\left(\frac{1}{4}\right)^2} = \frac{1}{\frac{1}{16}} = 16$$

3.5 Equivalent Trigonometric Expressions

Objective: To use the fundamental identities to write equivalent trigonometric expressions

The skills used to manipulate and simplify algebraic expressions can also be used with trigonometric expressions.

Preview

Recall the following patterns that are used to factor polynomials:

$$am + bm + cm = m(a + b + c) \qquad a^2 - b^2 = (a + b)(a - b)$$
$$a^2 + 2ab + b^2 = (a + b)(a + b) \qquad a^3 + b^3 = (a + b)(a^2 - ab + b^2)$$
$$a^2 - 2ab + b^2 = (a - b)(a - b) \qquad a^3 - b^3 = (a - b)(a^2 + ab + b^2)$$

EXAMPLE Simplify: $\frac{x^2 - 25}{x^2 + 10x + 25}$

$$\frac{x^2 - 25}{x^2 + 10x + 25} = \frac{(x + 5)(x - 5)}{(x + 5)(x + 5)} = \frac{x - 5}{x + 5}$$

Simplify.

1. $\frac{9x^2 - 4}{9x^2 + 12x + 4}$ $\frac{3x-2}{3x+2}$
2. $\frac{x^2 - 36}{x^2 - 12x + 36}$ $\frac{x+6}{x-6}$
3. $\frac{27x^3 - 8}{9x^2 - 4}$ $\frac{9x^2+6x+4}{3x+2}$
4. $\frac{x^2 - 1}{x^3 + 1}$ $\frac{x-1}{x^2-x+1}$

The patterns for factoring polynomials can be used along with the fundamental identities to convert one trigonometric expression to another.

EXAMPLE 1 **Convert the first trigonometric expression to the second.**

a. $(1 - \cos\theta)(1 + \cos\theta)$; $\sin^2\theta$

b. $\frac{9\sin^2\theta - 24\sin\theta + 16}{9\sin^2\theta - 16}$; $\frac{3\sin\theta - 4}{3\sin\theta + 4}$

a. $(1 - \cos\theta)(1 + \cos\theta) = 1 - \cos^2\theta$
$= \sin^2\theta$ *$\sin^2\theta + \cos^2\theta = 1$, so $1 - \cos^2\theta = \sin^2\theta$.*

b. $$\frac{9\sin^2\theta - 24\sin\theta + 16}{9\sin^2\theta - 16} = \frac{(3\sin\theta - 4)(3\sin\theta - 4)}{(3\sin\theta + 4)(3\sin\theta - 4)} = \frac{3\sin\theta - 4}{3\sin\theta + 4}$$

Algebraic manipulations can be used together with the fundamental trigonometric identities to write a trigonometric expression in terms of a single function.

EXAMPLE 2 **Simplify each expression by writing it in terms of cosine only.**

a. $\sin\theta\cot\theta$ **b.** $(\csc^2\theta - 1)(\sin^2\theta)$ **c.** $\dfrac{1+\sin\theta}{\cos\theta} + \dfrac{\cos\theta}{1+\sin\theta}$

a. $\sin\theta\cot\theta = (\sin\theta)\left(\dfrac{\cos\theta}{\sin\theta}\right) = \cos\theta$

b. $(\csc^2\theta - 1)(\sin^2\theta) = \cot^2\theta\sin^2\theta = \left(\dfrac{\cos^2\theta}{\sin^2\theta}\right)(\sin^2\theta) = \cos^2\theta$

c.
$$\frac{1+\sin\theta}{\cos\theta} + \frac{\cos\theta}{1+\sin\theta} = \frac{(1+\sin\theta)(1+\sin\theta)}{(\cos\theta)(1+\sin\theta)} + \frac{\cos\theta\cos\theta}{(\cos\theta)(1+\sin\theta)}$$
$$= \frac{1+2\sin\theta+\sin^2\theta}{(\cos\theta)(1+\sin\theta)} + \frac{\cos^2\theta}{(\cos\theta)(1+\sin\theta)}$$
$$= \frac{1+2\sin\theta+(\sin^2\theta+\cos^2\theta)}{(\cos\theta)(1+\sin\theta)}$$
$$= \frac{1+2\sin\theta+1}{(\cos\theta)(1+\sin\theta)} \qquad \sin^2\theta+\cos^2\theta = 1$$
$$= \frac{2+2\sin\theta}{(\cos\theta)(1+\sin\theta)} = \frac{2(1+\sin\theta)}{(\cos\theta)(1+\sin\theta)} = \frac{2}{\cos\theta}$$

CLASS EXERCISES

Write each expression in terms of sine or cosine only.

1. $\tan\theta\cos\theta$ $\sin\theta$
2. $\cot\theta\cos\theta$ $\dfrac{1-\sin^2\theta}{\sin\theta}$
3. $\dfrac{\cot\theta}{\cos\theta}$ $\dfrac{1}{\sin\theta}$
4. $\tan\theta\sin\theta$ $\dfrac{1-\cos^2\theta}{\cos\theta}$
5. $\dfrac{\sec\theta}{\cos\theta}$ $\dfrac{1}{\cos^2\theta}$
6. $\dfrac{\csc\theta}{\sin\theta}$ $\dfrac{1}{\sin^2\theta}$
7. $\dfrac{\cot\theta}{\sin\theta}$ $\dfrac{\cos\theta}{1-\cos^2\theta}$
8. $\dfrac{\tan\theta}{\cos\theta}$ $\dfrac{\sin\theta}{1-\sin^2\theta}$

PRACTICE EXERCISES

Convert the first expression to the second expression. In Exercises 1–16, the intermediate step is shown.

A

1. $(1-\sin\theta)(1+\sin\theta);\ \cos^2\theta$ $1-\sin^2\theta$
2. $(\csc\alpha+1)(\csc\alpha-1);\ \cot^2\alpha$ $\csc^2\alpha - 1$
3. $(\sec\beta+1)(\sec\beta-1);\ \tan^2\beta$ $\sec^2\beta - 1$
4. $(1+\tan\alpha)(1-\cot\alpha);\ \tan\alpha-\cot\alpha$ $1-\cot\alpha+\tan\alpha-1$
5. $\dfrac{\sin^2\alpha-25}{\sin^2\alpha+10\sin\alpha+25};\ \dfrac{\sin\alpha-5}{\sin\alpha+5}$ See below.
6. $\dfrac{\sin^2\alpha-36}{\sin^2\alpha+12\sin\alpha+36};\ \dfrac{\sin\alpha-6}{\sin\alpha+6}$
7. $\dfrac{\cos^2\alpha-14\cos\alpha+49}{\cos^2\alpha-49};\ \dfrac{\cos\alpha-7}{\cos\alpha+7}$
8. $\dfrac{\cos^2\alpha+16\cos\alpha+64}{\cos^2\alpha-64};\ \dfrac{\cos\alpha+8}{\cos\alpha-8}$

TEACHING SUGGESTIONS

- Before discussing the examples, do several problems like those in the Preview, but replace the variables with trigonometric functions.
- Emphasize that simplifying an expression does not always reduce the number of terms in the expression.

CHALKBOARD EXAMPLES

- **For Example 1**

1. Convert the first trigonometric expression to the second: $\dfrac{\sin^3\theta-64}{\sin^2\theta-16};\ \dfrac{\sin^2\theta+4\sin\theta+16}{\sin\theta+4}$

$\dfrac{\sin^3\theta-64}{\sin^2\theta-16} = \dfrac{(\sin\theta-4)(\sin^2\theta+4\sin\theta+16)}{(\sin\theta-4)(\sin\theta+4)} = \dfrac{\sin^2\theta+4\sin\theta+16}{\sin\theta+4}$

- **For Example 2**

Write each expression in terms of cosine only.

2. $1+\tan^2\theta$ $\quad 1+\tan^2\theta = 1+\dfrac{\sin^2\theta}{\cos^2\theta} = \dfrac{\cos^2\theta}{\cos^2\theta}+\dfrac{\sin^2\theta}{\cos^2\theta} = \dfrac{\cos^2\theta+\sin^2\theta}{\cos^2\theta} = \dfrac{1}{\cos^2\theta}$

3. $\dfrac{1+\tan\theta}{\cos\theta+\sin\theta}$ $\quad \dfrac{1+\tan\theta}{\cos\theta+\sin\theta} = \dfrac{1+\dfrac{\sin\theta}{\cos\theta}}{\cos\theta+\sin\theta} = \dfrac{\dfrac{\cos\theta+\sin\theta}{\cos\theta}}{\cos\theta+\sin\theta} = \dfrac{1}{\cos\theta}$

Common Error

- Some students are afraid to make substitutions that expand a trigonometric expression. Emphasize that it is often necessary to expand an expression in order to simplify it.
- See *Teacher's Resource Book* for additional remediation.

Additional Answers:

Practice Exercises

5. $\dfrac{(\sin\alpha+5)(\sin\alpha-5)}{(\sin\alpha+5)^2}$
6. $\dfrac{(\sin\alpha+6)(\sin\alpha-6)}{(\sin\alpha+6)^2}$
7. $\dfrac{(\cos\alpha-7)^2}{(\cos\alpha+7)(\cos\alpha-7)}$
8. $\dfrac{(\cos\alpha+8)^2}{(\cos\alpha+8)(\cos\alpha-8)}$

LESSON FOLLOW-UP

Discussion

How are the expressions $\cot^2 \theta \sec^2 \theta$ and $1 + \cot^2 \theta$ related? The expressions are equivalent because $\cot^2 \theta \sec^2 \theta = \frac{1}{\sin^2 \theta} = \csc^2 \theta$ and $1 + \cot^2 \theta = \csc^2 \theta$.

Assignment Guide

See p. 110B for assignments.

Extra

Calculus for finding sin x and cos x, where x is in radians, is introduced.

Lesson Quiz

1. Write $\frac{\sin x}{\csc x}$ in terms of sine or cosine only. $\sin^2 x$ or $1 - \cos^2 x$

Convert the first expression to the second expression.

2. $\frac{\sin^2 \theta - 36}{\sin^2 \theta + 12 \sin \theta + 36}; \frac{\sin \theta - 6}{\sin \theta + 6}$

$\frac{\sin^2 \theta - 36}{\sin^2 \theta + 12 \sin \theta + 36} = \frac{(\sin \theta + 6)(\sin \theta - 6)}{(\sin \theta + 6)(\sin \theta + 6)} = \frac{\sin \theta - 6}{\sin \theta + 6}$

3. $\frac{\sin^3 \alpha - 8}{\sin^2 \alpha - 4}; \frac{\sin^2 \alpha + 2 \sin \alpha + 4}{\sin \alpha + 2}$

$\frac{\sin^3 \alpha - 8}{\sin^2 \alpha - 4} = \frac{(\sin \alpha - 2)(\sin^2 \alpha + 2 \sin \alpha + 4)}{(\sin \alpha - 2)(\sin \alpha + 2)} = \frac{\sin^2 \alpha + 2 \sin \alpha + 4}{\sin \alpha + 2}$

4. Write $\frac{\csc x}{\sec x}$ in terms of tan x.

$\frac{\csc x}{\sec x} = \frac{\frac{1}{\sin x}}{\frac{1}{\cos x}} = \frac{\cos x}{\sin x} = \cot x = \frac{1}{\tan x}$

9. $\frac{\tan^3 \theta - 27}{\tan^2 \theta - 9}; \frac{\tan^2 \theta + 3 \tan \theta + 9}{\tan \theta + 3}$ $\frac{(\tan \theta - 3)(\tan^2 \theta + 3 \tan \theta + 9)}{(\tan \theta - 3)(\tan \theta + 3)}$

10. $\frac{\tan^3 \theta - 64}{\tan^2 \theta - 16}; \frac{\tan^2 \theta + 4 \tan \theta + 16}{\tan \theta + 4}$ $\frac{(\tan \theta - 4)(\tan^2 \theta + 4 \tan \theta + 16)}{(\tan \theta - 4)(\tan \theta + 4)}$

11. $\frac{\cot^3 \theta + 8}{\cot^2 \theta - 4}; \frac{\cot^2 \theta - 2 \cot \theta + 4}{\cot \theta - 2}$ $\frac{(\cot \theta + 2)(\cot^2 \theta - 2 \cot \theta + 4)}{(\cot \theta + 2)(\cot \theta - 2)}$

12. $\frac{\cot^3 \theta + 125}{\cot^2 \theta - 25}; \frac{\cot^2 \theta - 5 \cot \theta + 25}{\cot \theta - 5}$ $\frac{(\cot \theta + 5)(\cot^2 \theta - 5 \cot \theta + 25)}{(\cot \theta + 5)(\cot \theta - 5)}$

13. $\frac{1}{\sin^2 \theta} - \frac{1}{\tan^2 \theta}; 1$ $\csc^2 \theta - \cot^2 \theta$

14. $\frac{1}{\cos^2 \theta} - \frac{1}{\cot^2 \theta}; 1$ $\sec^2 \theta - \tan^2 \theta$

15. $\frac{1 - \cos^2 \theta}{\cos^2 \theta}; \tan^2 \theta$ $\frac{\sin^2 \theta}{\cos^2 \theta}$

16. $\frac{1 - \sin^2 \theta}{\sin^2 \theta}; \cot^2 \theta$ $\frac{\cos^2 \theta}{\sin^2 \theta}$

Write each expression in terms of sin θ.

B 17. $\tan \theta \sec \theta$ $\frac{\sin \theta}{1 - \sin^2 \theta}$

18. $(\tan^2 \theta)\left(\frac{1}{\sec^2 \theta}\right) + \frac{1}{\sin \theta}$ $\frac{\sin^3 \theta + 1}{\sin \theta}$

19. $\frac{1 + \cos \theta}{\sin \theta} + \frac{\sin \theta}{1 + \cos \theta}$ $\frac{2}{\sin \theta}$

20. $\frac{1}{1 - \cos \theta} + \frac{1}{1 + \cos \theta}$ $\frac{2}{\sin^2 \theta}$

Write each expression in terms of cos θ.

21. $\frac{\sin^2 \theta}{\sec^2 \theta - 1}$ $\cos^2 \theta$

22. $\frac{\sin^2 \theta}{\cos \theta} + \cos \theta$ $\frac{1}{\cos \theta}$

23. $\frac{1}{\sec \theta - \tan \theta} + \frac{1}{\sec \theta + \tan \theta}$ $\frac{2}{\cos \theta}$

24. $(1 - \tan^2 \theta)(\sec^2 \theta)$ $\frac{2 \cos^2 \theta - 1}{\cos^4 \theta}$

Write each expression in terms of tan θ.

25. $\frac{\sec \theta}{\csc \theta}$ $\tan \theta$

26. $\frac{1 + \tan^2 \theta}{\csc^2 \theta}$ $\tan^2 \theta$

27. $\frac{\sec \theta}{\sin \theta} - \frac{\sec \theta}{\csc \theta}$ $\frac{1}{\tan \theta}$

28. $\frac{\sec \theta - \cos \theta}{\sin \theta}$ $\tan \theta$

C 29. Convert $\frac{\sin^3 \theta - \cos^3 \theta}{\sin \theta - \cos \theta}$ to $\sin \theta \cos \theta + 1$. See below.

30. Convert $\frac{\sin^3 \theta + \cos^3 \theta}{\sin \theta + \cos \theta}$ to $1 - \sin \theta \cos \theta$.

31. If $s = a \sec \theta - b \tan \theta$ and $t = b \sec \theta - a \tan \theta$, find $s^2 - t^2$. $a^2 - b^2$

32. If $\sin \alpha = \frac{(a + b)^2}{a - b}$ and $\cos \alpha = \frac{a^2 + b^2}{a^2 - b^2}$, find tan α. $\frac{(a + b)^3}{a^2 + b^2}$

Additional Answers

29. $\frac{\sin^3 \theta - \cos^3 \theta}{\sin \theta - \cos \theta} = \frac{(\sin \theta - \cos \theta)(\sin^2 \theta + \sin \theta \cos \theta + \cos^2 \theta)}{\sin \theta - \cos \theta} = \sin \theta \cos \theta + \sin^2 \theta + \cos^2 \theta = \sin \theta \cos \theta + 1$

30. $\frac{\sin^3 \theta + \cos^3 \theta}{\sin \theta + \cos \theta} = \frac{(\sin \theta + \cos \theta)(\sin^2 \theta - \sin \theta \cos \theta + \cos^2 \theta)}{\sin \theta + \cos \theta} = \sin^2 \theta + \cos^2 \theta - \sin \theta \cos \theta = 1 - \sin \theta \cos \theta$

Applications

33. Genetics In an analysis of heredity, a genetic engineer was led to the expression $\frac{1}{1 + \sin \theta} + \frac{1}{1 - \sin \theta}$. Show that this expression is equivalent to $\frac{2}{\cos^2 \theta}$.
Write as one fraction: $\frac{1 - \sin \theta + 1 + \sin \theta}{1 - \sin^2 \theta}$, or $\frac{2}{\cos^2 \theta}$.

Computer The program at the right evaluates the expression $\frac{\tan^2 \theta - \sin^2 \theta}{\tan^2 \theta \sin^2 \theta}$ for any angle measure (in degrees) entered by the user. Line 20 changes the angle measure to radians. Lines 40 and 50 evaluate the expression.

```
1  HOME
10 INPUT"ENTER AN ANGLE
   MEASURE IN DEGREES";A
20 R = A*0.0174533
30 S = SIN(R):T = TAN(R)
40 S = S^2:T = T^2
50 E = (T - S)/(T * S)
60 PRINT E
70 END
```

34. Run the program several times, using a different angle measure each time.

35. The same result should be displayed each time the program is run. Explain.
The expression $\frac{\tan^2 \theta - \sin^2 \theta}{\tan^2 \theta \sin^2 \theta}$ is equivalent to 1.

EXTRA

Recall that $x!$ (x factorial) is equal to $x(x - 1)(x - 2)(x - 3) \ldots (3)(2)(1)$.

$5! = 5(4)(3)(2)(1) = 120 \qquad 8! = 8(7)(6)(5)(4)(3)(2)(1) = 40{,}320$

In calculus, $\sin x$ and $\cos x$ are each written as the sum of an infinite number of terms using the following formulas, where x is in radians:

$$\sin x = x - \frac{x^3}{3!} + \frac{x^5}{5!} - \frac{x^7}{7!} + \frac{x^9}{9!} - \cdots$$

$$\cos x = 1 - \frac{x^2}{2!} + \frac{x^4}{4!} - \frac{x^6}{6!} + \frac{x^8}{8!} - \cdots$$

EXAMPLE **Find sin 2, using the first five terms of the above formula.**

$$\sin 2 = 2 - \frac{2^3}{3!} + \frac{2^5}{5!} - \frac{2^7}{7!} + \frac{2^9}{9!}$$

$$= 2 - \frac{8}{6} + \frac{32}{120} - \frac{128}{5040} + \frac{512}{362880} = 0.9093$$

Check that this result is correct to the nearest ten-thousandth.

1. Find cos 2 using the first five terms of the formula for cos x. How many more terms are needed to obtain a result correct to the nearest ten-thousandth?
-0.4159; 2 more terms
2. Find sin 1 and cos 1 using the first four terms of the formulas. 0.8415; 0.5403
3. Use the results from Exercise 2 and a fundamental identity to approximate the value of tan 1. $\tan 1 = \frac{\sin 1}{\cos 1} = \frac{0.8415}{0.5403} = 1.557$

Enrichment

Express $\frac{\cos \theta}{\sec \theta + 1} + \frac{\cos \theta}{\sec \theta - 1}$ in terms of $\cot \theta$.

$\frac{\cos \theta}{\sec \theta + 1} + \frac{\cos \theta}{\sec \theta - 1} =$

$\frac{\cos \theta (\sec \theta - 1) + \cos \theta (\sec \theta + 1)}{(\sec \theta + 1)(\sec \theta - 1)} =$

$\frac{\cos \theta \sec \theta + \cos \theta \sec \theta}{\sec^2 \theta - 1} =$

$\frac{2 \cos \theta \sec \theta}{\tan^2 \theta} = \frac{2}{\tan^2 \theta} = 2 \cot^2 \theta$

Teacher's Resource Book

Practice—Chapter 3, p. 9
Enrichment—Chapter 3, p. 10

LESSON PLAN

BACKGROUND

In the Preview, simplifying complex fractions is discussed. This skill is sometimes used when proving identities.

3.6 Proving Identities

Objective: To use the fundamental identities to prove other identities

Recall that a *complex rational expression* has a rational expression in its numerator or denominator, or in both. Complex expressions occur in trigonometry as well as in algebra.

Preview

One method used to simplify complex rational expressions is to multiply the numerator and the denominator by the lowest common denominator (LCD) of all the rational expressions in the numerator and the denominator.

EXAMPLE **Simplify:** **a.** $\dfrac{\frac{1}{x}+\frac{1}{y}}{\frac{1}{x}-\frac{1}{y}}$ **b.** $\dfrac{\sin x - 1}{1-\frac{1}{\sin x}}$

a. $\dfrac{\frac{1}{x}+\frac{1}{y}}{\frac{1}{x}-\frac{1}{y}} = \dfrac{\left(\frac{1}{x}+\frac{1}{y}\right)(xy)}{\left(\frac{1}{x}-\frac{1}{y}\right)(xy)} = \dfrac{y+x}{y-x}$ $\quad LCD = xy$

b. $\dfrac{\sin x - 1}{1-\frac{1}{\sin x}} = \dfrac{(\sin x - 1)(\sin x)}{\left(1-\frac{1}{\sin x}\right)(\sin x)} = \dfrac{(\sin x - 1)(\sin x)}{\sin x - 1} = \sin x$ $\quad LCD = \sin x$

Simplify.

1. $\dfrac{1-\frac{1}{x}}{\frac{1}{y}}$ $\quad \dfrac{xy-y}{x}$

2. $\dfrac{1+\frac{x^2}{y^2}}{\frac{1}{x^2}}$ $\quad \dfrac{x^2y^2+x^4}{y^2}$

3. $\dfrac{3+\frac{1}{\sin^2 A}}{\frac{1}{\sin^2 A}}$ $\quad 3\sin^2 A + 1$

4. $\dfrac{1+\frac{1}{\cos A}}{2-\frac{1}{\sin A}}$ $\quad \dfrac{\sin A(\cos A + 1)}{\cos A(2\sin A - 1)}$

To prove that a trigonometric equation is an identity means to show that the equation is true for any permissible replacement of the variable. Recall that one way to write a proof of an identity is to continue to write equivalent expressions for one side of the equation until you arrive at an expression identical to the one on the other side of the equation.

EXAMPLE 1 **Prove: $\cot\theta = \cos\theta\csc\theta$**

$$\begin{array}{r|l} \cot\theta & \cos\theta\csc\theta \\ & (\cos\theta)\left(\frac{1}{\sin\theta}\right) \\ & \frac{\cos\theta}{\sin\theta} \\ = & \cot\theta \end{array}$$

Therefore, $\cot\theta = \cos\theta\csc\theta$.

Another way to write a proof of an identity is to replace expressions on *both* sides of the given equation with equivalent expressions.

EXAMPLE 2 **Prove: $\cos\theta\sec\theta = \tan\theta\cot\theta$**

$$\begin{array}{r|l} \cos\theta\sec\theta & \tan\theta\cot\theta \\ (\cos\theta)\left(\frac{1}{\cos\theta}\right) & (\tan\theta)\left(\frac{1}{\tan\theta}\right) \\ 1 & = 1 \end{array}$$

Therefore, $\cos\theta\sec\theta = \tan\theta\cot\theta$.

There are several strategies to use when you prove identities.

1. Know the fundamental identities and look for ways to apply them.
2. Write all expressions in terms of sines and cosines.
3. If you choose to work with only one side of an identity, continuously refer back to the other side to see what you are trying to obtain.
4. When one side contains only one trigonometric function, attempt to rewrite all the functions on the other side in terms of that function.
5. Use the Pythagorean identities to substitute for expressions equal to 1.
6. Perform algebraic operations.
   a. Factor.
   b. Simplify complex rational expressions.
   c. Find the LCD and combine fractions.
   d. Combine like terms.
   e. Multiply both numerator and denominator by the same expression to obtain an equivalent fraction.
   f. Replace a binomial with a monomial.

Proving an identity is not the same as solving an equation. The vertical line separating the left side from the right side is used to show that each side is done independently. Thus, such equation-solving techniques as operating on both sides are not to be used in the proof of an identity.

TEACHING SUGGESTIONS

- Stress that variables must be included to have identities.
- Stress that verifying identities using specific substitutions is different from proving identities.
- Stress that identities are true for all values for which the expression is defined.

Critical Thinking

Analysis Ask students to compare and contrast solving equations with proving identities. Answers may vary.

CHALKBOARD EXAMPLES

- **For Example 1**

1. Prove: $\cos\beta = \cot\beta\sin\beta$

$$\begin{array}{r|l} \cos\beta & \cot\beta\sin\beta \\ & \frac{\cos\beta}{\sin\beta}\sin\beta \\ = & \cos\beta \end{array}$$

2. Prove: $\frac{\sin\theta\cot\theta}{\cos\theta} = 1$

$$\begin{array}{l|r} \frac{\sin\theta\cot\theta}{\cos\theta} & 1 \\ \frac{\sin\theta\frac{\cos\theta}{\sin\theta}}{\cos\theta} & \\ \frac{\cos\theta}{\cos\theta} & \\ 1 & = \end{array}$$

- **For Example 2**

3. Prove: $\sin\theta\csc\theta = \cos\theta\sec\theta$

$$\begin{array}{l|l} \sin\theta\csc\theta & \cos\theta\sec\theta \\ \sin\theta\frac{1}{\sin\theta} & \cos\theta\frac{1}{\cos\theta} \\ 1 & = 1 \end{array}$$

4. Prove: $\frac{1}{\tan\beta} = \cos\beta\csc\beta$

$$\begin{array}{l|l} \frac{1}{\tan\beta} & \cos\beta\csc\beta \\ \frac{1}{\frac{\sin\beta}{\cos\beta}} & \cos\beta\frac{1}{\sin\beta} \\ \frac{\cos\beta}{\sin\beta} & \cot\beta \\ \cot\beta & = \cot\beta \end{array}$$

- **For Example 3**

5. Prove: $\cos\theta = \frac{1-\sin^2\theta}{\cos\theta}$

$$\begin{array}{l|l} \cos\theta & \frac{1-\sin^2\theta}{\cos\theta} \\ & \frac{\cos^2\theta}{\cos\theta} \\ = & \cos\theta \end{array}$$

6. Prove: $\frac{1}{\cos^2\alpha} - 1 = \tan^2\alpha$

$$\begin{array}{l|l} \frac{1}{\cos^2\alpha} - 1 & \tan^2\alpha \\ \frac{1}{\cos^2\alpha} - \frac{\cos^2\alpha}{\cos^2\alpha} & \\ \frac{1-\cos^2\alpha}{\cos^2\alpha} & \\ \frac{\sin^2\alpha}{\cos^2\alpha} & \\ \tan^2\alpha & = \end{array}$$

- **For Example 4**

7. Prove: $\frac{\sin^2\theta}{1+2\cos\theta+\cos^2\theta} = \frac{\csc\theta-\cot\theta}{\csc\theta+\cot\theta}$

$$\begin{array}{l|l} \frac{\sin^2\theta}{1+2\cos\theta+\cos^2\theta} & \frac{\csc\theta-\cot\theta}{\csc\theta+\cot\theta} \\ \frac{1-\cos^2\theta}{(1+\cos\theta)(1+\cos\theta)} & \frac{\frac{1}{\sin\theta}-\frac{\cos\theta}{\sin\theta}}{\frac{1}{\sin\theta}+\frac{\cos\theta}{\sin\theta}} \\ \frac{(1+\cos\theta)(1-\cos\theta)}{(1+\cos\theta)(1+\cos\theta)} & \frac{\frac{1-\cos\theta}{\sin\theta}}{\frac{1+\cos\theta}{\sin\theta}} \\ \frac{1-\cos\theta}{1+\cos\theta} & = \frac{1-\cos\theta}{1+\cos\theta} \end{array}$$

Common Errors

- Students often make incorrect substitutions. Review the fundamental identities.
- Students try to prove identities as if they were equations. Emphasize that proving identities requires simplifying each side separately.
- See *Teacher's Resource Book* for additional remediation.

LESSON FOLLOW-UP

Assignment Guide

See p. 110B for assignments.

EXAMPLE 3 **Prove:** $2\sec^2\theta = \frac{1}{1-\sin\theta} + \frac{1}{1+\sin\theta}$

$$\begin{array}{l|ll} 2\sec^2\theta & \frac{1}{1-\sin\theta} + \frac{1}{1+\sin\theta} & \\ & \frac{(1+\sin\theta)+(1-\sin\theta)}{(1-\sin\theta)(1+\sin\theta)} & LCD = (1-\sin\theta)(1+\sin\theta) \\ & \frac{2}{1-\sin^2\theta} & \\ & \frac{2}{\cos^2\theta} & 1-\sin^2\theta = \cos^2\theta \\ & = 2\sec^2\theta & \end{array}$$

Therefore, $2\sec^2\theta = \frac{1}{1-\sin\theta} + \frac{1}{1+\sin\theta}$.

In Example 4, a Pythagorean identity is used.

EXAMPLE 4 **Prove:** $\frac{\cos^2\theta}{1+2\sin\theta+\sin^2\theta} = \frac{\sec\theta-\tan\theta}{\sec\theta+\tan\theta}$

$$\begin{array}{ll|l} & \frac{\cos^2\theta}{1+2\sin\theta+\sin^2\theta} & \frac{\sec\theta-\tan\theta}{\sec\theta+\tan\theta} \\ \cos^2\theta = 1-\sin^2\theta & \frac{1-\sin^2\theta}{(1+\sin\theta)(1+\sin\theta)} & \frac{\frac{1}{\cos\theta}-\frac{\sin\theta}{\cos\theta}}{\frac{1}{\cos\theta}+\frac{\sin\theta}{\cos\theta}} \\ \text{Factor } 1-\sin^2\theta. & \frac{(1+\sin\theta)(1-\sin\theta)}{(1+\sin\theta)(1+\sin\theta)} & \frac{\left(\frac{1}{\cos\theta}-\frac{\sin\theta}{\cos\theta}\right)(\cos\theta)}{\left(\frac{1}{\cos\theta}+\frac{\sin\theta}{\cos\theta}\right)(\cos\theta)} \\ & \frac{1-\sin\theta}{1+\sin\theta} & = \frac{1-\sin\theta}{1+\sin\theta} \end{array}$$

Therefore, $\frac{\cos^2\theta}{1+2\sin\theta+\sin^2\theta} = \frac{\sec\theta-\tan\theta}{\sec\theta+\tan\theta}$.

CLASS EXERCISES

Factor each expression.

1. $\cot^2\theta + 2\cot\theta\csc\theta + \csc^2\theta$ $(\cot\theta + \csc\theta)^2$
2. $9 - \sec^2\theta$ $(3+\sec\theta)(3-\sec\theta)$
3. $1 - \sin^2\alpha$ $(1+\sin\alpha)(1-\sin\alpha)$
4. $\sin^2\alpha + 2\sin\alpha\cos\alpha + \cos^2\alpha$ $(\sin\alpha + \cos\alpha)^2$
5. $\cos^4\theta - \sin^4\theta$ $(\cos^2\theta + \sin^2\theta)(\cos\theta + \sin\theta)(\cos\theta - \sin\theta)$
6. $9\cot^4\theta + 6\cot^2\theta + 1$ $(3\cot^2\theta + 1)^2$
7. $\sec^4\beta + 6\sec^2\beta + 5$ $(\sec^2\beta + 5)(\sec^2\beta + 1)$
8. $\csc^3\beta + \sec^3\beta$ $(\csc\beta + \sec\beta)(\csc^2\beta - \csc\beta\sec\beta + \sec^2\beta)$

PRACTICE EXERCISES

Prove each identity. See pages 481–482.

A

1. $\frac{1-\cos^2\theta}{\cos^2\theta} = \left(\frac{\sin\theta}{\cos\theta}\right)^2$
2. $\frac{\sin^2\theta+\cos^2\theta}{\cos^2\theta} = \left(\frac{1}{\cos\theta}\right)^2$
3. $\frac{\cos\alpha}{1-\sin^2\alpha} = \frac{1}{\cos\alpha}$
4. $\frac{1-\cos^2\lambda}{\sin\lambda} = \sin\lambda$
5. $\frac{1}{\tan\alpha} = \frac{\cot\alpha}{\sin^2\alpha+\cos^2\alpha}$
6. $\frac{1+\cot^2\alpha}{\csc^2\alpha} = 1$
7. $\frac{1+\cos\mu}{\cos\mu} = \sec\mu + 1$
8. $\frac{\csc^2\mu - 1}{\csc^2\mu} = \cos^2\mu$
9. $\frac{\sin^2\alpha}{1-\sin^2\alpha} = \sec^2\alpha - 1$
10. $\cot\alpha - 1 = \frac{\csc\alpha - \sec\alpha}{\sec\alpha}$
11. $\sin^2\theta\cos\theta\sec\theta = 1-\cos^2\theta$
12. $\frac{1}{\sin\phi} = \sin\phi + \cos\phi\cot\phi$
13. $\sin^2\beta = 1 - \frac{1}{\sec^2\beta}$
14. $\frac{\cos\beta}{\sec\beta} - \frac{\cot\beta}{\tan\beta} = -\cos^2\beta\cot^2\beta$
15. $\frac{\cos\lambda}{\sec\lambda+\tan\lambda} = 1-\sin\lambda$
16. $\frac{1+\cot^2\alpha}{\csc\alpha} = \frac{1}{\sin\alpha}$
17. $\csc^2\beta = \cos^2\beta + \cot^2\beta + \sin^2\beta$
18. $\frac{1}{\tan\beta+\cot\beta} = \sin\beta\cos\beta$

B

19. $2\cos^2\alpha - 1 = \cos^4\alpha - \sin^4\alpha$
20. $(\sin\lambda - \tan\lambda)^2 = (\tan^2\lambda)(\cos\lambda - 1)^2$
21. $\frac{\cos\lambda}{1+\sin\lambda} + \frac{\cos\lambda}{1-\sin\lambda} = \frac{2}{\cos\lambda}$
22. $\frac{\sec\lambda}{\cos\lambda} - 1 = \frac{\sin^2\lambda}{\cos^2\lambda}$
23. $\frac{\sin^4\theta - 1}{\cos^2\theta} = \cos^2\theta - 2$
24. $\frac{\sin\theta - 1}{1-\sin^2\theta} = \frac{\csc\theta}{-\csc\theta - 1}$
25. $\tan^2\alpha + \cos^2\alpha - 1 = \tan^2\alpha - \sin^2\alpha$
26. $2\csc\lambda = \frac{1}{\csc\lambda - \cot\lambda} + \frac{1}{\csc\lambda + \cot\lambda}$
27. $\frac{1-\sin\alpha}{1+\sin\alpha} = (\tan\alpha - \sec\alpha)^2$
28. $(\sin\theta + \cos\theta)^2\tan\theta = \tan\theta + 2\sin^2\theta$
29. $(1+\sin\theta+\cos\theta)(1-\sin\theta-\cos\theta) = -2\sin\theta\cos\theta$
30. $(\sec\theta)(\sin\theta\sec\theta + 1) = \frac{\sec^2\theta - \tan^2\theta + \tan\theta}{\cos\theta}$

C

31. $\frac{\tan^2\alpha}{1-\cos^2\alpha} + \frac{\sin\alpha}{\sec^2\alpha - 1} = (\cos\alpha)(\sec^3\alpha + \cot\alpha)$

Historical Note

Students get a historical look at mechanical and electronic computing devices, beginning with the abacus and continuing up to our present day hand-held calculators.

Lesson Quiz

Prove each identity.

1. $1 - \sin\theta\cos\theta\cot\theta = \sin^2\theta$

$1 - \sin\theta\cos\theta\cot\theta$	$\sin^2\theta$
$1 - \sin\theta\cos\theta\frac{\cos\theta}{\sin\theta}$	
$1-\cos^2\theta$	
$\sin^2\theta$	=

2. $\frac{\cos x - 1}{\sin x} = \cot x - \csc x$

$\frac{\cos x - 1}{\sin x}$	$\cot x - \csc x$
$\frac{\cos x - 1}{\sin x}$	$\frac{\cos x}{\sin x} - \frac{1}{\sin x}$
$\frac{\cos x - 1}{\sin x}$ =	$\frac{\cos x - 1}{\sin x}$

3. $\frac{1}{1-\cos x} + \frac{1}{1+\cos x} = 2\csc^2 x$

$\frac{1}{1-\cos x} + \frac{1}{1+\cos x}$	$2\csc^2 x$
$\frac{(1+\cos x)+(1-\cos x)}{1-\cos^2 x}$	
$\frac{2}{\sin^2 x}$	
$2\csc^2 x$	=

4. $\frac{\sin\theta - \cos\theta - \sin^3\theta}{\sin\theta} = \cos^2\theta - \cot\theta$

$\frac{\sin\theta - \cos\theta - \sin^3\theta}{\sin\theta}$	$\cos^2\theta - \cot\theta$
$\frac{\sin\theta}{\sin\theta} - \frac{\cos\theta}{\sin\theta} - \frac{\sin^3\theta}{\sin\theta}$	
$1 - \cot\theta - \sin^2\theta$	
$\cos^2\theta - \cot\theta$	=

Enrichment

Prove: $\frac{1 + \sin x}{1 - \sin x} = (\sec x + \tan x)^2$

$\frac{1 + \sin x}{1 - \sin x}$	$(\sec x + \tan x)^2$
$\frac{\frac{1}{\cos x} + \frac{\sin x}{\cos x}}{\frac{1}{\cos x} - \frac{\sin x}{\cos x}}$	
$\frac{\sec x + \tan x}{\sec x - \tan x}$	
$\frac{(\sec x + \tan x)^2}{\sec^2 x - \tan^2 x}$	
$(\sec x + \tan x)^2$	$=$

Computer

The complete form of the program in the applications is on the disk provided with the *Teacher's Resource Book.*

Teacher's Resource Book

Practice—Chapter 3, p. 11
Enrichment—Chapter 3, p. 12

32. $\frac{\cot \alpha}{1 - \sin^2 \alpha} + \frac{\cos \alpha}{\csc^2 \alpha - 1} = (\sec \alpha)(\sin^2 \alpha + \csc \alpha)$

33. $\sqrt{(5 \cos \theta + 12 \sin \theta)^2 + (12 \cos \theta - 5 \sin \theta)^2} = 13$

34. $\frac{-2 \sin \alpha \cos \alpha}{1 - \sin \alpha - \cos \alpha} = 1 + \sin \alpha + \cos \alpha$

35. $\frac{2 \sin \theta \cos \theta}{\sin \theta + \cos \theta - 1} = \sin \theta + \cos \theta + 1$

Hint: Multiply the numerator and denominator on the left side by $(\sin \theta + \cos \theta + 1)$.

36. $\frac{\sin \alpha - \cos \alpha - 1}{\sin \alpha + \cos \alpha - 1} = -\frac{\cos \alpha + 1}{\sin \alpha}$

Hint: Multiply the numerator and denominator on the left side by $(\sin \theta + \cos \theta + 1)$.

Applications

Computer This program evaluates $\cos x + \sin x \tan x$ for $0° \leq x \leq 360°$. The value of the expression is displayed.

```
10 HOME
20 INPUT"ENTER AN ANGLE MEASURE
   FROM 0 TO 360 DEGREES";A
30 IF A < 0 OR A > 360 THEN 20
40 R = A*0.0174533
50 E = COS(R) + SIN(R)*TAN(R)
60 PRINT E
70 END
```

37. Run the program for several values of x.

38. For what angle measures from 0° through 360° will an error message be displayed? Why? 90°; 270°; tan *A* is undefined at these values.

39. Write $\cos x + \sin x \tan x$ in terms of a single function of x. $\sec x$

HISTORICAL NOTE

Milestones in the History of Computing Devices Human beings have been interested in building work-saving calculating devices since ancient times. Around 300 BC the first known calculating device, the abacus, was created by the Babylonians and/or the Egyptians. In the seventeenth century, mechanical devices were developed in France by Blaise Pascal and in Germany by Gottfried Leibniz.

During the next two centuries, mechanical, rotary, and electrical calculating devices continued to be invented and refined. Each refinement generally upgraded the capability and features of the devices. Through the 1960s, the devices were rather large. This changed with the invention of the electronic microchip, and the hand-held calculator became a reality in the early 1970s.

Find information on the mechanical calculating device invented by Pascal or Leibniz. Write a report on your findings. Answers may vary.

3.7 Graphical Representation of Identities

Objectives: To check identities by graphing both sides of the equation

You should be familiar with the graphs of the trigonometric functions when you use a computer or a graphing calculator to check identities. If you know the general characteristics of a graph, you are more likely to catch errors made while entering the equations in the computer or calculator.

Preview

Graph each function.
See page 482.

1. $y = \sin x$
2. $y = \tan x$
3. $y = \sec x$
4. $y = 2 \sin x$
5. $y = \cos 2x$
6. $y = \sin x + 2$

EXAMPLE **Graph: $y = \cos x$**

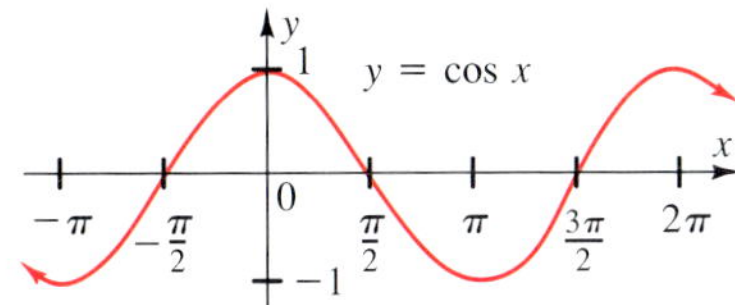

To check that a trigonometric equation is an identity, first graph the left side of the equation using a graphing calculator or a computer. Then, without clearing the screen, graph (overwrite) the right side. If the graphs coincide, you can be fairly sure that the original equation is an identity. Note that graphs are inexact, so this method of verification does not constitute a proof.

EXAMPLE 1 **Verify graphically that $\tan x = \dfrac{\sin x}{\cos x}$ is an identity.**

Graph the left side. $y = \tan x$

Without clearing the screen, graph the right side. $y = \dfrac{\sin x}{\cos x}$

The graphs coincide. This suggests that $\tan x = \dfrac{\sin x}{\cos x}$ is an identity.

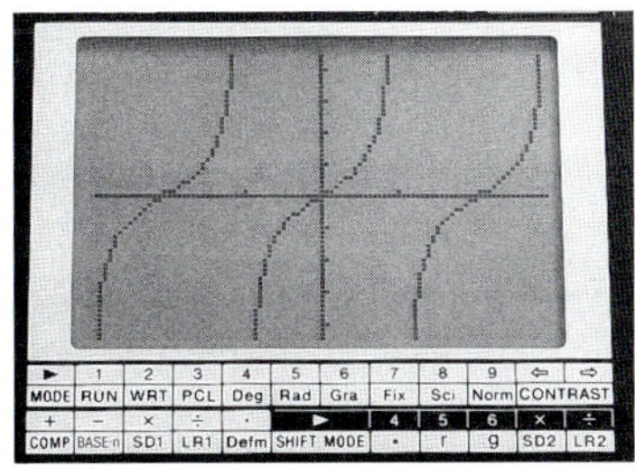

LESSON PLAN

Materials/Manipulatives
Graphing calculators
Computer

BACKGROUND

In the Preview, graphing trigonometric functions is reviewed. This skill will be used to determine if an equation is an identity.

TEACHING SUGGESTIONS

Unless students are familiar with graphing calculators or computers, it will be necessary to discuss how graphing is accomplished. Consult the owner's manual for specific problem areas.

Critical Thinking

Comprehension Ask students to explain how graphing calculators or computers could be used in other areas of trigonometry. Answers may vary.

CHALKBOARD EXAMPLES

- **For Example 1**
  1. Verify graphically that $\csc x = \dfrac{1}{\sin x}$ is an identity. When the left and right sides are graphed separately without clearing the screen, the graphs coincide. This implies that $\csc x = \dfrac{1}{\sin x}$ is an identity.
- **For Example 2**
  2. Check graphically to see whether $\csc \theta \tan^2 \theta = \tan \theta$ is an identity. not an identity

Common Error

- Some students may feel that they cannot use a calculator or a computer to graph secant, cosecant, and tangent. Remind those students to use the reciprocal identity.
- See *Teacher's Resource Book* for additional remediation.

LESSON FOLLOW-UP

Discussion

How could a graphing calculator or computer be used to find $\sin \theta$ given that $\cos \theta = \frac{3}{4}$. Graph $\sin \theta$ and $\cos \theta$ on the same display. Find where $\cos \theta = \frac{3}{4}$, then read the corresponding value on the graph of $\sin \theta$.

Assignment Guide

See p. 110B for assignments.

Test Yourself

See *Teacher's Resource Book*, *Tests*, pp. 27–28.

Lesson Quiz

Check graphically to see whether or not each equation is an identity.

1. $\frac{\sin^2 \theta}{\cos \theta} + \cos \theta = \sec \theta$ identity
2. $\frac{\csc x \cos x}{\sec x \sin x} = \frac{(1 + \csc x)}{(1 - \csc x)}$ not an identity
3. $\csc^4 \theta \cot^2 \theta = (\tan \theta + \cot \theta)^2$ not an identity
4. $\sec x = \frac{\cos x}{1 - \sin x} - \tan x$ identity
5. $\frac{1}{1 - \sin \theta} + \frac{1}{1 + \sin \theta} = \frac{2}{\cos^2 \theta}$ identity

Enrichment

Find the relationship between the graph of $\cot^2 \theta \sec^2 \theta$ and the graph of $2 + \cot^2 \theta$. The graphs do not coincide.

If the graphs are not identical and the equations have been graphed correctly, then the given equation is *not* an identity.

EXAMPLE 2 **Check graphically to see whether $\sec x \sin^2 x = \cos^2 x$ is an identity.**

Since the graphs of the two sides are not the same, this is *not* an identity.

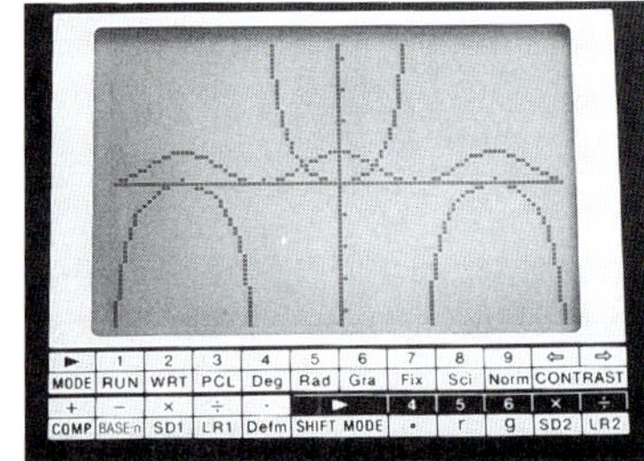

CLASS EXERCISES

Simplify each expression. Then check your result by graphing the given expression and your answer.

1. $\frac{1}{\sec \theta}$ $\cos \theta$
2. $\frac{1}{\cot x}$ $\tan x$
3. $\frac{1}{\tan B}$ $\cot B$
4. $\frac{1}{\csc x}$ $\sin x$
5. $\cot x \tan x$ 1
6. $\sin A \csc A$ 1

PRACTICE EXERCISES

Check graphically to see whether or not each equation is an identity.

A

1. $\frac{\sin A}{\csc A} = \sin^2 A$ identity
2. $\frac{\cos A}{\sec A} = \cos^2 A$ identity
3. $\frac{\tan x}{\sin x} = \frac{1}{\cos x}$ identity
4. $\frac{\cos x}{\sin x} = \cot x$ identity
5. $\cos B \csc B = \cot B$ identity
6. $\cos \beta \sec \beta = \tan \beta \cot \beta$ identity
7. $\sec A \cot A = \csc A$ identity
8. $\sin \alpha \cot \alpha = \cos \alpha$ identity
9. $\frac{\sin x \csc x}{\tan x} = \cot x$ identity
10. $\frac{\cot x \tan x}{\sin x} = \csc x$ identity
11. $\cos x \sin x \cot x = 2 \cos x$ not an identity
12. $\sin x + \cos x \tan x = 2 \sin x$ identity
13. $\frac{1 - \cos^2 \beta + \sin \beta}{1 + \sin \beta} = \sin \beta$ identity
14. $\frac{\sec \theta + 1}{\sin \theta} = \frac{1 + \cos \theta}{\sin \theta \cos \theta}$ identity

B

15. $(1 - \sin \theta)^2 = (\sec \theta - \tan \theta)^2$ not an identity
16. $(\sec^2 \theta)(\tan^2 \theta - 1) = (\tan^2 \theta)(\tan^2 \theta + 1)$ not an identity

17. $\tan^2 \beta \cos^2 \beta = 1 - \cot^2 \beta \sin^2 \beta$ identity

18. $\sin^2 \beta + \dfrac{\cos^2 \beta}{\sin^2 \beta} + \cos^2 \beta = \dfrac{1}{\cos^2 \beta}$ not an identity

19. $\dfrac{\sin^2 A + \cos^2 A}{\sin A \cos A} = \sec A \csc A$ identity

20. $\dfrac{\sec \lambda \sin \lambda + \tan \lambda}{\cos \lambda \cot \lambda} = \dfrac{2 \sin^2 \lambda}{\cos^3 \lambda}$ identity

21. $\dfrac{\sin \alpha}{\cos \alpha - 1} + \dfrac{\sin \alpha}{\cos \alpha + 1} = \dfrac{2 \cos \alpha}{-\sin \alpha}$ identity

22. $\dfrac{\cot \theta}{\cos \theta} - \dfrac{\csc^2 \theta}{\sec \theta} = \cot \theta$ not an identity

23. $\dfrac{\sin \mu}{\sec \mu - 1} = \dfrac{\sin \mu + 1}{\tan \mu}$ not an identity

24. $\dfrac{2 + 2\cos^2 \mu}{1 - \cos^2 \mu} = 4 \cot \mu \csc \mu$ not an identity

25. $\dfrac{2 \sin^2 x - 1}{\sin x \cos x} = \tan x - \cot x$ identity

26. $\dfrac{\tan \lambda + \sin \lambda}{\sin \lambda + \sec \lambda} = 1 + \sin \lambda$ not an identity

C 27. $\dfrac{2\cos^2 \beta + \cot \beta}{\tan \beta} = \dfrac{2 \sin \beta \cos^3 \beta + \cos^2 \beta}{\sin \beta}$ not an identity

28. $\dfrac{\sin \alpha - \sin^3 \alpha + \cos^2 \alpha}{\cos^2 \alpha} = \sin \alpha + 1$ identity

Applications

29. **Physics** A scientist working with the expression $4 \cos \theta + 3 \sin \theta$ guessed that it was about equivalent to $5 \cos (\theta - 36°52')$. Verify this graphically. See side column.

30. **Metallurgy** A metallurgist has to find the values of θ between 0° and 360° for which $\sec \theta + \csc \theta = 0$. Find the solutions graphically. 135°; 315°

TEST YOURSELF

Verify each trigonometric identity. 3.4

1. $\cot^2 \left(\dfrac{2\pi}{3}\right) = \csc^2 \left(\dfrac{2\pi}{3}\right) - 1$

$\left(-\dfrac{\sqrt{3}}{3}\right)^2 = \left(\dfrac{2\sqrt{3}}{3}\right)^2 - 1$

2. $\cos 210° \cot 210° = \csc 210° - \sin 210°$

$\left(-\dfrac{\sqrt{3}}{2}\right)\sqrt{3} = -2 - \left(-\dfrac{1}{2}\right)$

Convert the first expression to the second expression. 3.5

3. $\dfrac{\cot \theta}{\sin \theta}(\sec \theta - \cos \theta);\ 1$

4. $(\cos \alpha + 1)(\cos \alpha - 1) - \cos^2 \alpha;\ -1$

$\cos^2 \alpha - 1 - \cos^2 \alpha = -1$

3. $\dfrac{\frac{\cos \theta}{\sin \theta}}{\sin \theta}\left(\dfrac{1}{\cos \theta} - \cos \theta\right) = \dfrac{\cos \theta}{\sin^2 \theta}\left(\dfrac{1 - \cos^2 \theta}{\cos \theta}\right) = 1$

Prove each identity. 3.6

See side column.

5. $\dfrac{\cot \theta}{\sec \theta + 1} = \dfrac{\sec \theta - 1}{\tan^3 \theta}$

6. $\dfrac{\tan \beta}{\sec \beta} + \dfrac{\cot \beta}{\csc \beta} = \sin \beta + \cos \beta$

Check graphically to see whether or not each equation is an identity. 3.7

7. $(\cot x + \csc x)^2 = \dfrac{\sec x + 1}{\sec x - 1}$ identity

8. $2 \sin x = \dfrac{\sin x}{1 + \cos x} + \dfrac{1 + \cos x}{\sin x}$ not an identity

Teacher's Resource Book

Practice—Chapter 3, p. 13

Enrichment—Chapter 3, p. 14

Additional Answers

29.

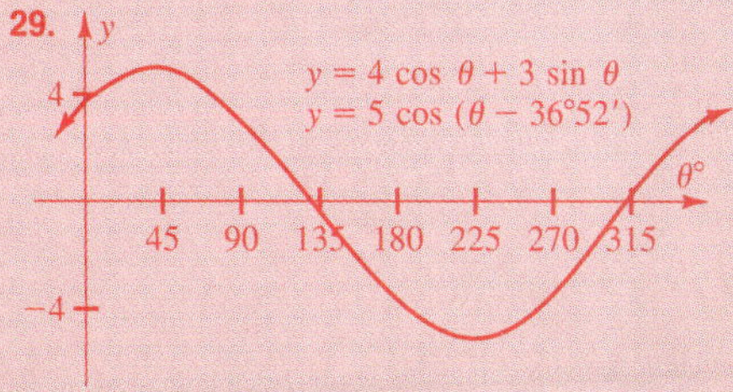

Test Yourself

5. $\dfrac{\cot \theta}{\sec \theta + 1}$ | $\dfrac{\sec \theta - 1}{\tan^3 \theta}$

$\dfrac{\sec \theta - 1}{\tan \theta(\sec^2 \theta - 1)}$

$\dfrac{\sec \theta - 1}{\tan \theta(\sec \theta - 1)(\sec \theta + 1)}$

$= \dfrac{\cot \theta}{\sec \theta + 1}$

6. $\dfrac{\tan \beta}{\sec \beta} + \dfrac{\cot \beta}{\csc \beta}$ | $\sin \beta + \cos \beta$

$\dfrac{\frac{\sin \beta}{\cos \beta}}{\frac{1}{\cos \beta}} + \dfrac{\frac{\cos \beta}{\sin \beta}}{\frac{1}{\sin \beta}}$

$\sin \beta + \cos \beta =$

Application

The benefits of digital electronics in recording are discussed in this feature on compact discs. Students are introduced to binary coding of audio waves in sound reproduction.

See *Teacher's Resource Book* for *Application*, Chapter 3, p. 15.

See Teacher's Resource Book, *Technology*, p. 3.

APPLICATION: Compact Discs

Did you know that a revolution in recorded music began with the introduction of the compact disc (CD) in 1983? Compact discs have captured approximately one-quarter of the audio market. It is likely that CDs will one day make conventional long-playing vinyl records obsolete. What is a CD, how is sound stored on one, and how does a CD player retrieve that sound?

A CD is a plastic coated metallic disc that is less than five inches in diameter. Despite its small size, it is capable of providing approximately one hour of uninterrupted playing time. Its tracks are so fine that over 300 of them could fit next to one another in a single groove of a standard vinyl record! There are no grooves on the surface of a CD, and the CD player has no stylus. CDs are scanned by a beam of laser light in the CD player that is focused deep into the disc. Because there is no physical contact between the CD and the sensing device of the player, there is virtually no wear on the CD as there is on a conventional record with a stylus. Thus, if handled with reasonable care, a CD will not wear out nor deteriorate with age.

As do other music reproduction systems, a CD system stores and retrieves sound waves. The sound waves are translated by a microphone into an alternating electrical current. The waveform of this current represents the original sound wave input and is called an *analog*. The electrical analog is then used to produce a permanent physical equivalent of the sound wave, which can then be stored in the grooves of a conventional vinyl record.

Unlike analog systems such as vinyl records, CDs are *digitally* encoded using data from audio waveforms. They do not reproduce and store the actual shape of the waveform. Instead, a digital system records frequent measurements of the height of the graph. Indeed, the measurements are made 44,100 times each second. The CD player uses a binary code to handle the numbers rapidly, using just the symbols 0 and 1. Electrically, 1 corresponds to "on" and 0 to "off." A compact disc is encoded using 16 binary digits. Thus, the audio waveforms are sampled $2^{16} = 65{,}536$ times, which leads to very precise plotting.

The following diagram demonstrates how information is recorded on a CD and how it is played back. The original sound wave is sampled, and the wave height is recorded as a binary number (for purposes of this example, only 3 binary digits are used). When a digitally recorded compact disc is played, the sampled points are recreated, giving the signal a stair-step appearance. If the points are connected with a smooth line, the curve is shaped like the graphs of the sine or cosine functions.

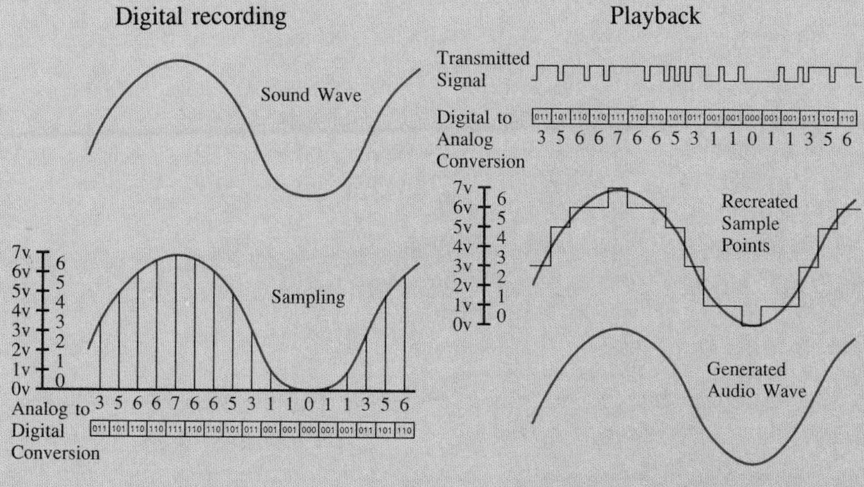

EXERCISES See side column.

1. Draw the graph of a sound wave with sample points 7, 6, 6, 5, 3, 2, 1, 0, 1, 2, 3, 5, 6, 6, 7.
2. Convert the sample points from Exercise 1 to binary notation.
3. Referring to the sound wave shown below, find the voltage of selected sample points and then convert them to binary notation.
   Answers may vary.

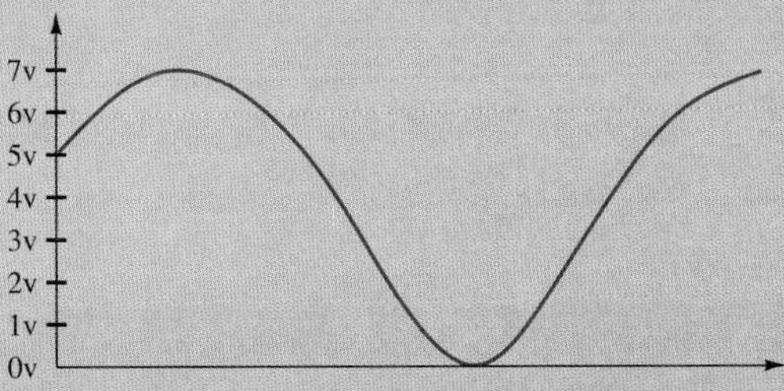

Additional Answers

1.

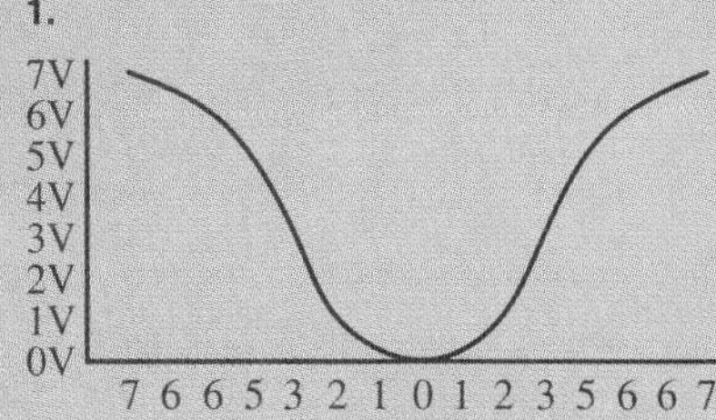

2. 111, 110, 110, 101, 011, 010, 001, 000, 001, 010, 011, 101, 110, 110, 111

CHAPTER 3 SUMMARY AND REVIEW

Vocabulary

angle of depression (118)
angle of elevation (118)
bearing (124)
course (124)
fundamental identities (133)
odd-even identities (132)
Pythagorean identities (131)
ratio identities (131)
reciprocal identities (133)
significant digit (113)
trigonometric identity (130)

Solving Right Triangles Right triangle trigonometry can be used to find lengths of missing sides and measures of missing angles. 3.1

Solve each triangle *ABC* ($\angle C = 90°$).

1. $\angle B = 21°$, $a = 170$
   $\angle A = 69°$; $b = 65$; $c = 180$
2. $a = 11.3$, $b = 20.5$
   $\angle A = 29°$; $\angle B = 61°$; $c = 23.4$

Angles of Elevation and Depression Angles of elevation and depression are used to find distances that cannot be measured directly. 3.2

3. A girl who is 54 in. tall casts a 140-in. shadow. Find the angle of elevation from the tip of the shadow to the sun. 21°
4. The angle of depression from an airplane to an airport is 25°. If the plane is flying at an altitude of 1200 m and is directly over a football stadium, how far is it from the stadium to the airport? 2573 m

Applications In navigation, the *course* is the angle measured clockwise from north to the line of travel, and the *bearing* is the angle measured clockwise from north to the line of sight to a point of reference. 3.3

5. An airplane is 24.3 km due east of radar station *A*, and radar station *B* is 37.4 km due south of radar station *A*. Find the bearing of radar station *B* from the plane and the distance between radar station *B* and the plane. 213°; 44.6 km
6. To find the distance *AB* across a lake, a surveyor made the following measurements:

   $AC = 420$ yd $\qquad \angle C = 57°$

   Find *AB*. 647 yd
7. A surveyor determined that the diagonal of a city block formed an angle of 28° with one of the streets forming the block. The side of the block on that street was 600 ft long. How long was the diagonal? 680 ft

Fundamental Identities The ratio, reciprocal, Pythagorean, and odd-even identities are considered the fundamental trigonometric identities. 3.4

Reciprocal Identities

$\csc\theta = \frac{1}{\sin\theta}$, $\sin\theta \neq 0$

$\sec\theta = \frac{1}{\cos\theta}$, $\cos\theta \neq 0$

$\cot\theta = \frac{1}{\tan\theta}$, $\tan\theta \neq 0$

Pythagorean Identities

$\sin^2\theta + \cos^2\theta = 1$

$1 + \cot^2\theta = \csc^2\theta$

$1 + \tan^2\theta = \sec^2\theta$

Ratio Identities

$\tan\theta = \frac{\sin\theta}{\cos\theta}$, $\cos\theta \neq 0$

$\cot\theta = \frac{\cos\theta}{\sin\theta}$, $\sin\theta \neq 0$

Odd-Even Identities

$\sin(-\theta) = -\sin\theta$

$\cos(-\theta) = \cos\theta$

$\tan(-\theta) = -\tan\theta$

Verify each identity for the given angle measure.

8. $1 + \cot^2 150° = \csc^2 150°$ $1 + (-\sqrt{3})^2 = (2)^2$

9. $\tan\frac{7\pi}{6} = \frac{\sin\frac{7\pi}{6}}{\cos\frac{7\pi}{6}}$ $\frac{\sqrt{3}}{3} = \frac{-\frac{1}{2}}{-\frac{\sqrt{3}}{2}}$

Equivalent Trigonometric Expressions The fundamental identities, along with algebraic manipulations, can be used to write a trigonometric expression equivalent to a given expression. 3.5

Convert the first trigonometric expression to the second expression.

10. $\frac{\csc^2\theta - 10\csc\theta + 25}{\csc^2\theta - 25}$; $\frac{\csc\theta - 5}{\csc\theta + 5}$ $\frac{(\csc\theta - 5)(\csc\theta - 5)}{(\csc\theta - 5)(\csc\theta + 5)}$

11. $\frac{\cos^2\theta}{1 - \cos^2\theta}$; $\cot^2\theta$ $\frac{\cos^2\theta}{\sin^2\theta}$

Proving Identities Trigonometric identities can be proved using algebraic manipulations and the fundamental identities. 3.6

Prove each identity. See side column.

12. $\sin\theta\sec\theta + \cot\theta = \frac{1}{\sin\theta\cos\theta}$

13. $\frac{\sec^2\beta - 1}{\sec\beta - 1} = 1 + \frac{1}{\cos\beta}$

Graphical Representation of Identities If the graphs of the two sides of an equation are identical, it is probable that the equation is an identity. If the graphs are not the same, the equation is not an identity. 3.7

Check graphically to see whether or not each equation is an identity.

14. $\frac{1}{\cos x} - \frac{\cos x}{1 + \sin x} = \tan x$ identity

15. $\frac{\csc^2 B - \cot B}{\sin^2 B - \tan B} = \tan^2 B$ not an identity

Additional Answers

12. $\sin\theta\sec\theta + \cot\theta \;\Big|\; \frac{1}{\sin\theta\cos\theta}$

$\sin\theta\cdot\frac{1}{\cos\theta} + \frac{\cos\theta}{\sin\theta}$

$\frac{\sin^2\theta + \cos^2\theta}{\cos\theta\sin\theta}$

$\frac{1}{\sin\theta\cos\theta} =$

13. $\frac{\sec^2\beta - 1}{\sec\beta - 1} \;\Big|\; 1 + \frac{1}{\cos\beta}$

$\frac{(\sec\beta + 1)(\sec\beta - 1)}{\sec\beta - 1} \;\Big|\; 1 + \sec\beta$

$\sec\beta + 1 = \sec\beta + 1$

See *Teacher's Resource Book, Tests*, pp. 29–32.

Additional Answers

12. $\sec\theta \;\Big|\; \cos\theta + \tan\theta\sin\theta$

$\cos\theta + \dfrac{\sin\theta}{\cos\theta}\cdot\sin\theta$

$\cos\theta + \dfrac{\sin^2\theta}{\cos\theta}$

$\dfrac{\cos^2\theta + \sin^2\theta}{\cos\theta}$

$= \sec\theta$

13. $\dfrac{\cos A}{\sec A} + \dfrac{\sin A}{\csc A} \;\Big|\; \sec^2 A - \tan^2 A$

$\cos^2 A + \sin^2 A \;\Big|\; 1$

$1 = 1$

14. $\sin\beta\cos\beta + \sin\beta \;\Big|\; \dfrac{\sin^3\beta}{1-\cos\beta}$

$\sin\beta(\cos\beta + 1) \;\Big|\; \sin\beta\cdot\dfrac{\sin^2\beta}{1-\cos\beta}$

$\sin\beta\dfrac{(1-\cos^2\beta)}{1-\cos\beta}$

$\sin\beta(1+\cos\beta)$

$= \sin\beta(\cos\beta + 1)$

Challenge

Area of $\triangle ABC = \frac{1}{2}bh = xh$; $x = a\cos A$; $h = a\sin A$. Area of $\triangle ABC = a\cos A\cdot a\sin A = a^2\cos A\sin A$.

CHAPTER TEST

Solve each triangle ABC ($\angle C = 90°$).

1. $\angle A = 56°$, $b = 14$
   $\angle B = 34°$; $a = 21$; $c = 25$
2. $a = 19.2$, $b = 42.1$
   $\angle A = 24.5°$; $\angle B = 65.5°$; $c = 46.3$
3. The angle of elevation of a 16-ft ladder leaning against a wall is 36°. How far is the base of the ladder from the wall? 13 ft
4. From an airplane flying over a bridge at an altitude of 640 m, the angle of depression to the airport is 26°. How far is it from the bridge to the airport? 1312 m
5. A boat is 21 m due west of buoy A. Buoy B is 38 m due south of buoy A. Find the bearing of buoy B from the boat and the distance between buoy B and the boat. 151°; 43 m

Verify each identity for the given angle measure.

6. $\dfrac{\cot 60°}{\sin^2 60° + \cos^2 60°} = \cot 60°$ $\dfrac{\frac{\sqrt{3}}{3}}{\frac{3}{4}+\frac{1}{4}} = \dfrac{\sqrt{3}}{3}$
7. $\dfrac{\sec^2 135°}{1 + \tan^2 135°} = 1$ $\dfrac{(\sqrt{2})^2}{1+1^2} = 1$

Convert the first trigonometric expression to the second expression.
In items 8–11, the middle step or steps are given.

8. $\dfrac{\cos^2\theta - 81}{\cos^2\theta + 18\cos\theta + 81}$; $\dfrac{\cos\theta - 9}{\cos\theta + 9}$ $\dfrac{(\cos\theta - 9)(\cos\theta + 9)}{(\cos\theta + 9)(\cos\theta + 9)}$
9. $\dfrac{\tan^2\theta - 100}{\tan^2\theta - 20\tan\theta + 100}$; $\dfrac{\tan\theta + 10}{\tan\theta - 10}$ $\dfrac{(\tan\theta - 10)(\tan\theta + 10)}{(\tan\theta - 10)(\tan\theta - 10)}$
10. $\dfrac{\sin^2\alpha}{\cos^2\alpha} + \sin\alpha\csc\alpha$; $\sec^2\alpha$
    $\tan^2\alpha + \sin\alpha\cdot\dfrac{1}{\sin\alpha} = \tan^2\alpha + 1$
11. $\dfrac{\sec^2\alpha - 1}{\csc^2\alpha - 1}$; $\tan^4\alpha$
    $\dfrac{\tan^2\alpha}{\cot^2\alpha} = \tan^2\alpha\cdot\tan^2\alpha$

Prove each identity. See side column.

12. $\sec\theta = \cos\theta + \tan\theta\sin\theta$
13. $\dfrac{\cos A}{\sec A} + \dfrac{\sin A}{\csc A} = \sec^2 A - \tan^2 A$
14. $\sin\beta\cos\beta + \sin\beta = \dfrac{\sin^3\beta}{1 - \cos\beta}$

Check graphically to see whether or not each equation is an identity.

15. $\tan 2x = \dfrac{2\sin x\cos x}{\sin^2 x - \cos^2 x}$
    not an identity
16. $\cos x + \sin x = \sqrt{2}\sin\left(x + \dfrac{\pi}{4}\right)$
    identity

Challenge

Show that the area of any isosceles triangle with legs of lengths a and base angles of measure θ is given by the expression $a^2\cos\theta\sin\theta$. See side column.

COLLEGE ENTRANCE EXAM REVIEW

In each item you are to compare a quantity in Column 1 with a quantity in Column 2. Write the letter of the correct answer from these choices:

A. The quantity in Column 1 is greater than the quantity in Column 2.
B. The quantity in Column 2 is greater than the quantity in Column 1.
C. The quantity in Column 1 is equal to the quantity in Column 2.
D. The relationship cannot be determined from the information given.

Notes: Information centered over both columns refers to one or both of the quantities being compared. A symbol that appears in both columns has the same meaning in each column. All variables represent real numbers. Most figures are not drawn to scale.

	Column 1	Column 2
1. C	$\sin A \cot A$	$\cos A$

$$3x - 2y = -2$$
$$2x - y = 5$$

	Column 1	Column 2
2. B	x	y
3. B	$9^0 + 9^{-\frac{1}{2}} + 16^{\frac{1}{4}}$	$-4 \sin 270°$

Use this diagram for 4–6.

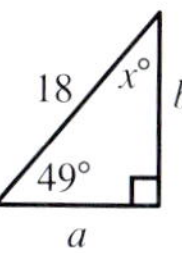

	Column 1	Column 2
4. A	x	$36°$
5. B	a	b
6. A	b	$8\sqrt{2}$
7. C	$\frac{1}{\cos 55°}$	$\sec 55°$
8. C	$\sin^2 \theta + \cos^2 \theta$	$\sec^2 \theta - \tan^2 \theta$
9. D	$\sin(-\theta)$	$\cos \theta$

$$ac > bc$$

	Column 1	Column 2
10. D	a	b

$$\cos^2 x = \tfrac{1}{2}, \text{ where } 0 \le x \le \tfrac{\pi}{2}$$

	Column 1	Column 2
11. B	x	$\frac{\pi}{3}$

$$x + y = 4$$

	Column 1	Column 2
12. D	The average of x, y, and z	The average of x, y, and p
13. C	$\frac{15d + 12}{3}$	$5d + 4$

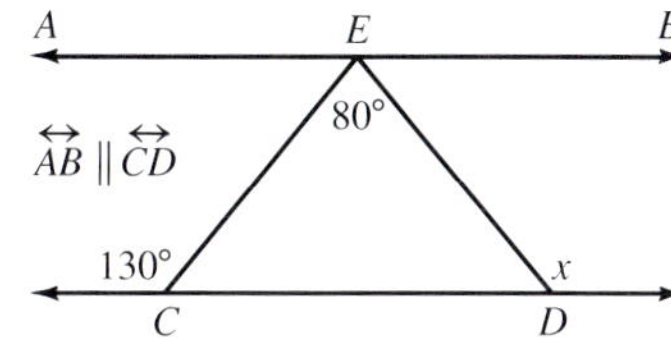

	Column 1	Column 2
14. B	$200° - x$	$80°$
15. A	$\tan 45°$	$\tan(-45°)$

A kite string measures 400 ft and makes an angle of 38° with the ground. Let h represent the height of the kite.

	Column 1	Column 2
16. C	h	$400 \sin 38°$

Maintaining Skills

The following skills and concepts are reviewed:

Adding and subtracting rational expressions

Solving rational equations

Solving inequalities

MAINTAINING SKILLS

Simplify each expression.

Example $\frac{x}{r} - \frac{t}{y} = \frac{x(y)}{r(y)} - \frac{t(r)}{y(r)} = \frac{xy - tr}{ry}$

1. $\frac{2}{q} + \frac{1}{3}$ $\frac{6+q}{3q}$

2. $\frac{2}{x} + \frac{3}{2y}$ $\frac{3x+4y}{2xy}$

3. $\frac{2}{p} - \frac{p+q}{q}$ $\frac{2q - p^2 - pq}{pq}$

4. $\frac{2}{5} - 4x$ $\frac{2-20x}{5}$

5. $\frac{x}{x-3} - \frac{3}{x-3}$ 1

6. $\frac{x-3}{x^2-4} + \frac{x}{x+2}$ $\frac{x^2 - x - 3}{x^2 - 4}$

Solve for *x*.

Example

$$\frac{1}{2} + \frac{2x}{x+4} = 3$$

$$2(x+4)\left(\frac{1}{2} + \frac{2x}{x+4}\right) = 3 \cdot 2\,(x+4) \qquad \textit{Multiply both sides by LCD.}$$

$$x + 4 + 4x = 6x + 24$$

$$5x + 4 = 6x + 24$$

$$-20 = x$$

7. $\frac{2}{3}x - \frac{4}{5} = R$ $\frac{15R + 12}{10}$

8. $\frac{x}{a} + \frac{x}{b} = 2$ $\frac{2ab}{a+b}, a \neq -b$

9. $\frac{2x}{a} + \frac{c}{d} = b$ $\frac{abd - ac}{2d}, d \neq 0$

10. $\frac{ax}{b} + c = d$ $\frac{bd - bc}{a}, a \neq 0$

11. $\frac{2x}{5y} = \frac{R - S}{4}$ $\frac{5Ry - 5Sy}{8}$

12. $\frac{x}{x-3} + \frac{4}{7} = -2$ $\frac{54}{25}$

Solve for *x*.

Examples

$$8 - 2x > 16 \qquad\qquad 14 + \frac{x}{3} < 9$$

$$-2x > 8 \qquad\qquad \frac{x}{3} < -5$$

$$x < -4 \qquad\qquad x < -15$$

13. $-2x > 7$ $x < -\frac{7}{2}$

14. $\frac{2}{3} - 5x > 7$ $x < -\frac{19}{15}$

15. $5 + x > -2x + 7$ $x > \frac{2}{3}$

16. $-3x - 5 < -7x - 10$ $x < -\frac{5}{4}$

17. $2x < 3x + 7$ $x > -7$

18. $\frac{x}{2} < \frac{2}{3} + 8$ $x < \frac{52}{3}$

19. Two angles are supplementary. The measure of one angle is $\frac{4}{5}$ the measure of the other angle. Find the measure of the smaller angle. 80°.

20. The measure of an angle is 46° more than the measure of its complement. Find the measure of the larger angle. 68°

OVERVIEW • Chapter 4

SUMMARY

The law of sines is proved then used to solve triangles when the measures of two angles and one side are known. Since it is not always possible to solve a triangle using the law of sines (the *ambiguous* case), the conditions for which it is possible are discussed. The law of cosines is proved and used to solve triangles when the measures of two sides and the included angle or the measures of three sides are known. Next, the law of tangents is proved and used to solve triangles when the measures of two sides and the included angle are known. A formula for the area of a triangle is developed using the law of sines. Then Heron's formula is proved and used to find the area of a triangle when the lengths of three sides are known. The chapter concludes with an introduction to the concept of a vector.

CHAPTER OBJECTIVES

- To introduce and prove the law of sines
- To use the law of sines to solve triangles when the measures of two angles and one side are known
- To solve a triangle when the measures of two sides and an angle opposite one of them are given
- To introduce and prove the law of cosines
- To use the law of cosines to solve triangles when the measures of two sides and the included angle or the measures of three sides are given
- To introduce and prove the law of tangents
- To use the law of tangents to solve triangles when the measures of two sides and the included angle are known
- To find the area of a triangle when the measures of two sides and the included angle are known
- To use the law of sines to find the area of a triangle when the measures of one side and two angles are known
- To use Heron's formula to find the area of a triangle and the length of an altitude when the lengths of three sides are known
- To find the magnitudes of the x- and y-components of a vector, and the sum of two vectors
- To find the measure of the direction angle between two vectors

CHAPTER HIGHLIGHTS

The *theme* of Chapter 4 is robotics. Among the chapter's special features is an analysis of the types of motion involved in the mobility of industrial robots.

APPLICATIONS

The law of sines, the law of cosines, and the law of tangents are all applied to the fields of surveying and navigation. Problems in surveying are also presented in the lessons involving the area of a triangle. Several applications from physics and aviation arise in Lesson 4.7, which involve vectors in a plane.

TECHNOLOGY

Calculator

Use of calculators is strongly recommended for problems which involve the trigonometric laws. Vector problems are greatly facilitated by the use of calculators as well.

Computer

Computer programs may be used for problems that involve any of the trigonometric laws. A computer may also be used when discussing the area of a triangle and Heron's formula.

RESOURCES

Teacher's Resource Book

- Teaching Aid 4
- Transparencies 7 and 8

ASSIGNMENT GUIDE Meeting Student Needs

STUDENT TEXT				TEACHER'S RESOURCE BOOK	
Chapter Content	**Basic**	**Average**	**Enriched**	**P**	**E**
4.1 The Law of Sines	D: 160/1–25 odd, 37	D: 160/7–33 odd, 37	D: 161/11–39 odd	1	2
4.2 Law of Sines: The Ambiguous Case	D: 166/1–23 odd, 35 R: 160/4, 16, 60	D: 166/9–31 odd, 35 R: 161/10, 16, 24	D: 166/13–35 odd R: 161/14, 16, 24	3	4
4.3 The Law of Cosines	D: 172/1–23 odd, 35 R: 166/4, 12, 18	D: 172/7–31 odd, 35, 37 R: 166/10, 18, 26	D: 173/17–39 odd R: 166/12, 18, 26	5	6
4.4 The Law of Tangents	D: 179/1–17 odd, 23 R: 172/4, 12, 18 180/TY	D: 179/3–23 odd R: 172/10, 14, 20 180/TY	D: 179/5–23 odd R: 172/16, 20, 24 180/TY	7	8
4.5 The Area of Triangles	D: 184/1–19 odd, 33 R: 179/2, 8, 14	D: 184/11–29 odd, 33 R: 179/6, 12, 16	D: 184/13–35 odd R: 179/6, 12, 16	9	10
4.6 Heron's Formula	Omit	D: 189/7–25 odd R: 184/12, 18, 22	D: 189/7–25 odd R: 184/16, 18, 24	11	12
4.7 Vectors in the Plane	D: 195/1–27 odd R: 184/2, 10, 18 197/TY	D: 195/3, 9, 13–39 odd, 45 R: 189/8, 14, 18 197/TY	D: 195/3, 9, 15–45 odd R: 189/10, 16, 18 197/TY	13	14

D = Daily R = Review TY = Test Yourself P = Practice E = Enrichment

	STUDENT TEXT				TEACHER'S RESOURCE BOOK	
Review and Testing	Test Yourself	180, 197	College Ent. Exam Rev.	203	Tests	
	Chapter Sum. and Rev.	200	Cumulative Review	204	• Quizzes	33–36
	Chapter Test	202			• Chapter Test (Form A)	37–38
					• Chapter Test (Form B)	39–40
Special Features	Math Club Activity	162	Career	185	Applications—Chapter 4	15–16
	Challenge	167	Challenge	189	Critical Thinking	3
	Trigonometry in Three Dimensions	174	Application	198	Alg. and Geom. Review	13–16
					Technology	4

4 Oblique Triangles

Robotics is the use of mechanical devices to perform tasks which are normally performed by humans. Robots are programmed to use linear and rotary (angular) motion to conduct routine, heavy, dangerous, or repetitive tasks. They can even be programmed to play chess!

BACKGROUND

Once only a popular topic in science fiction, robotics now contributes greatly to industrial growth. Sophisticated programs which govern robotic motion make use of the simple trigonometric functions studied in this course.

LESSON PLAN

Vocabulary
Law of sines

Materials/Manipulatives
Scientific calculators

BACKGROUND

In the Preview, measurement activities involving a scalene triangle are used to suggest the law of sines.

4.1 The Law of Sines

Objectives: To introduce and prove the law of sines
To use the law of sines to solve triangles when the measures of two angles and one side are known

In Chapter 3, you learned to solve *right* triangles. In this chapter, you will find the unknown sides and angles of *oblique* (nonright) triangles.

Preview

A triangle is called *scalene* if no two of its sides have the same length. Draw a scalene triangle with no right angles on a full sheet of paper. Label the angles A, B, and C and the sides opposite these angles a, b, and c, respectively.

1. Use a ruler to measure a, b, and c and a protractor to measure $\angle A$, $\angle B$, and $\angle C$. Answers may vary.
2. Use a calculator to compute the following three ratios: $\frac{\sin A}{a}$, $\frac{\sin B}{b}$, $\frac{\sin C}{c}$. Answers may vary.
3. What appears to be true about the three ratios you calculated in Exercise 2? They are equal.
4. Repeat the activity with a second triangle. What can you conclude? They are equal.

The activity in the Preview suggests a relationship called the *law of sines*.

Law of Sines

For any $\triangle ABC$ in which a, b, and c are the lengths of the sides opposite the angles with measures A, B, and C, respectively,

$$\frac{\sin A}{a} = \frac{\sin B}{b} = \frac{\sin C}{c}$$

The law of sines can also be expressed in the following alternate form:

$$\frac{a}{\sin A} = \frac{b}{\sin B} = \frac{c}{\sin C}$$

To prove the law of sines, consider two cases, one in which all angles are acute, and one in which one angle is *obtuse* (has a measure between 90° and 180°). Let A, B, and C be the angles of any scalene triangle, and let a, b, and c be the sides opposite these angles. Draw the altitude (height) h_1 from C to side AB or to the extension of AB. Draw the altitude h_2 from A to side BC.

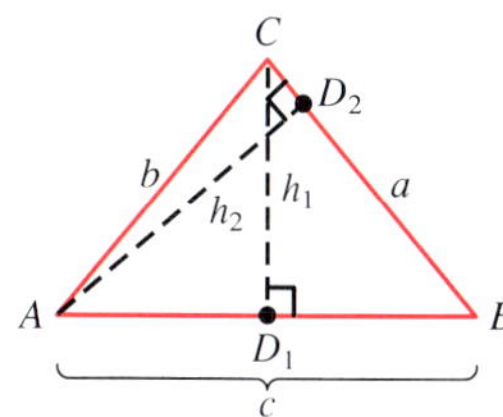

All three angles are acute (case 1).

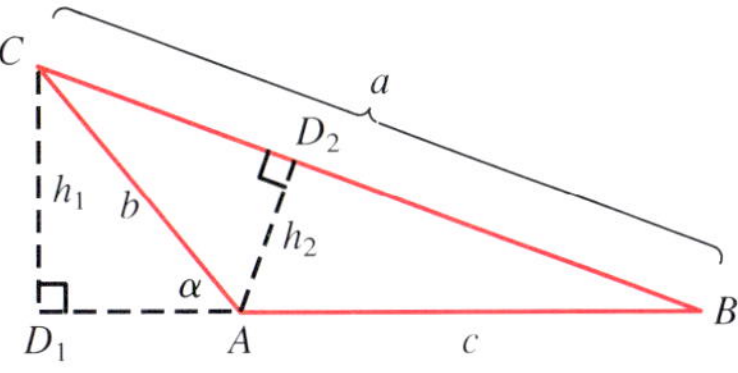

$\angle A$ is obtuse (case 2).

In the acute triangle (case 1), $\sin A = \dfrac{h_1}{b}$ and $\sin B = \dfrac{h_1}{a}$.

Thus, $h_1 = b \sin A$ and $h_1 = a \sin B$.

$b \sin A = a \sin B$ — *Substitution*

$\dfrac{\sin A}{a} = \dfrac{\sin B}{b}$ — *Divide both sides by ab.*

Similarly, $\sin B = \dfrac{h_2}{c}$ and $\sin C = \dfrac{h_2}{b}$. Thus, $h_2 = c \sin B$ and $h_2 = b \sin C$.

$c \sin B = b \sin C$ — *Substitution*

$\dfrac{\sin B}{b} = \dfrac{\sin C}{c}$ — *Divide both sides by bc.*

Therefore, $\dfrac{\sin A}{a} = \dfrac{\sin B}{b} = \dfrac{\sin C}{c}$.

In the obtuse triangle (case 2), $\sin B = \dfrac{h_1}{a}$ and $\sin \alpha = \sin(180° - A) = \dfrac{h_1}{b}$.

Thus, $h_1 = a \sin B$ and $h_1 = b \sin(180° - A)$.

$b \sin(180° - A) = a \sin B$ — *Substitution*

$b \sin A = a \sin B$ — *sin (180° − A) = sin A* *(See top of page 158.)*

$\dfrac{\sin A}{a} = \dfrac{\sin B}{b}$ — *Divide both sides by ab.*

Similarly, $\sin B = \dfrac{h_2}{c}$ and $\sin C = \dfrac{h_2}{b}$. Thus, $h_2 = c \sin B$ and $h_2 = b \sin C$.

$c \sin B = b \sin C$ — *Substitution*

$\dfrac{\sin B}{b} = \dfrac{\sin C}{c}$ — *Divide both sides by bc.*

Therefore, $\dfrac{\sin A}{a} = \dfrac{\sin B}{b} = \dfrac{\sin C}{c}$.

TEACHING SUGGESTIONS

- Emphasize that the law of sines may be used to solve a triangle when two angles and any one side of the triangle are given.
- Encourage students to use their scientific calculators to do any necessary calculations.

Critical Thinking

Analysis Ask students to consider an equilateral triangle, an isosceles triangle, and a right triangle to explain whether it is necessary to use the law of sines to solve them. Using the definitions and properties of each triangle, students should conclude that it is not necessary to use the law of sines to solve any of these triangles. They should point out that each angle of an equilateral triangle is 60°, and all sides of an equilateral triangle have the same length. Students should also note that a right triangle can be solved by using right triangle trigonometry and the Pythagorean theorem. Similarly, these methods can be used to solve an isosceles triangle if the altitude from the vertex angle is drawn.

CHALKBOARD EXAMPLES

- **For Example 1**

  Solve each triangle *ABC*. Express the lengths of sides to two significant digits and angle measures to the nearest degree.

1. $\angle A = 42°$, $\angle B = 61°$, $c = 65$
   $\angle C = 77°$, $a = 45$, $b = 58$
2. $\angle B = 72°$, $\angle C = 22°$, $a = 17$
   $\angle A = 86°$, $b = 16$, $c = 6.4$

- **For Example 2**

  Find the unknown sides for each triangle. Give the lengths of the sides to three significant digits and the angle measure to the nearest tenth of a degree.

3.

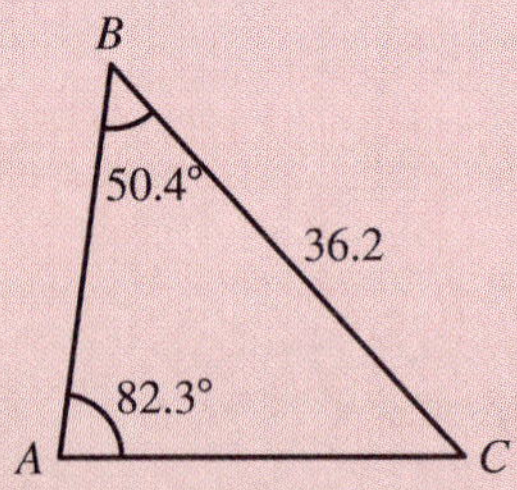

$b = 28.1$, $c = 26.8$, $\angle C = 47.3°$

4.

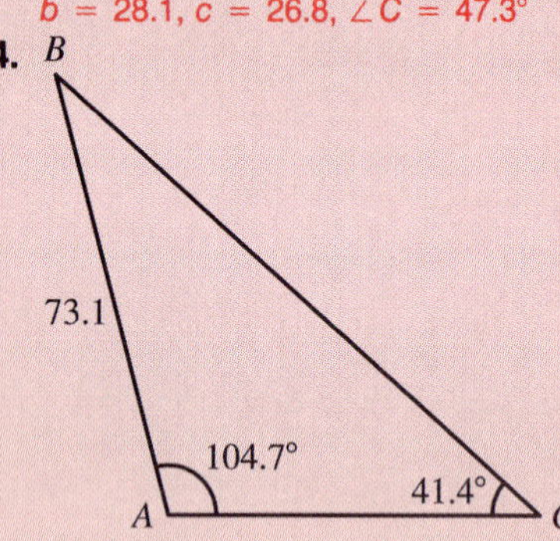

$a = 107$, $b = 61.7$, $\angle B = 33.9°$

The proof of the law of sines uses the following trigonometric identity:

$$\sin(180° - \theta) = \sin\theta$$

To see why this is an identity, refer to the figure at the right. Angle θ intersects a circle of radius r at point (x, y). Similarly, $180° - \theta$ intersects the circle at point $(-x, y)$. Since $\sin\theta$ and $\sin(180° - \theta)$ are each equal to $\frac{y}{r}$, $\sin(180° - \theta) = \sin\theta$.

The figure can also be used to prove the following identity, which will be used later in this chapter.

$$\cos(180° - \theta) = -\cos\theta$$

You will be asked to prove this identity in the exercises.

The law of sines can be used to find the unknown sides of a triangle when the measures of two angles and an included side are known.

EXAMPLE 1 **Solve $\triangle ABC$, pictured at the right, if $\angle A = 63°$, $\angle B = 49°$, and $c = 78$. Give the lengths of the sides to two significant digits and the angle measure to the nearest degree.**

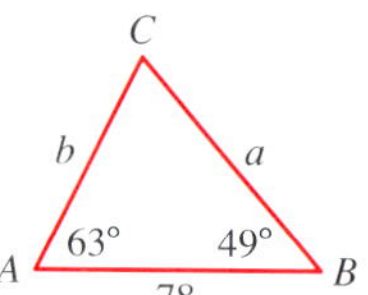

$\angle C = 180° - (63° + 49°) = 68°$ *Find $\angle C$.*

$$\frac{\sin A}{a} = \frac{\sin C}{c} \quad \text{and} \quad \frac{\sin B}{b} = \frac{\sin C}{c} \qquad \textit{Use the law of sines.}$$

$$\frac{\sin 63°}{a} = \frac{\sin 68°}{78} \qquad \frac{\sin 49°}{b} = \frac{\sin 68°}{78}$$

$$78 \sin 63° = a \sin 68° \qquad 78 \sin 49° = b \sin 68°$$

$$\frac{78 \sin 63°}{\sin 68°} = a \qquad \frac{78 \sin 49°}{\sin 68°} = b \qquad \textit{Calculation-ready form}$$

$$75 = a \qquad 63 = b$$

Therefore, $\angle A = 63°$, $\angle B = 49°$, $\angle C = 68°$, $a = 75$, $b = 63$, and $c = 78$.

The law of sines can also be used when the measures of two angles and the side opposite an angle are given.

EXAMPLE 2 **Find the unknown sides for the triangle pictured at the right. Give the lengths of the sides to three significant digits and the angle measure to the nearest tenth of a degree.**

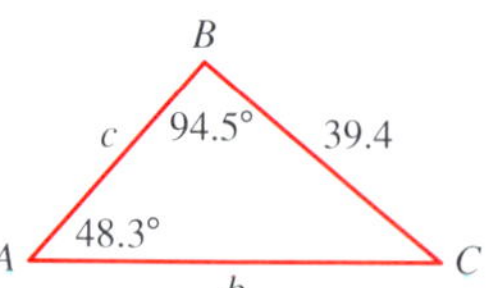

$\angle C = 180° - (94.5° + 48.3°) = 37.2°$ *Find $\angle C$.*

$$\frac{\sin A}{a} = \frac{\sin B}{b} \quad \text{and} \quad \frac{\sin A}{a} = \frac{\sin C}{c}$$

$$\frac{\sin 48.3°}{39.4} = \frac{\sin 94.5°}{b} \qquad \frac{\sin 48.3°}{39.4} = \frac{\sin 37.2°}{c}$$

$$b \sin 48.3° = 39.4 \sin 94.5° \qquad c \sin 48.3° = 39.4 \sin 37.2°$$

$$b = \frac{39.4 \sin 94.5°}{\sin 48.3°} \qquad c = \frac{39.4 \sin 37.2°}{\sin 48.3°}$$

$$b = 52.6 \qquad c = 31.9$$

Therefore, $\angle A = 48.3°$, $\angle B = 94.5°$, $\angle C = 37.2°$, $a = 39.4$, $b = 52.6$, and $c = 31.9$.

The law of sines can be combined with techniques from earlier chapters to solve real-world problems.

EXAMPLE 3 **From a point A, the angle of elevation to the top of a tree is 38°. From a point B 25 ft closer to the tree, the angle of elevation to the top is 48°. How far is it from point B to T? How tall is the tree? Express your answers to the nearest foot.**

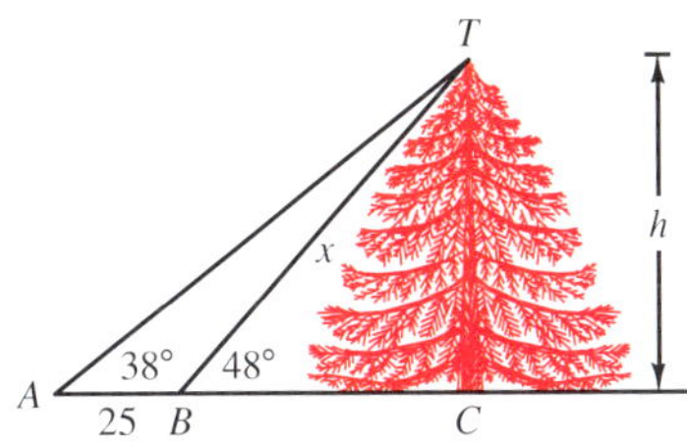

$\angle ABT = 180° - 48° = 132°$ *$\angle ABT$ and $\angle CBT$ are supplementary.*

$\angle ATB = 180° - (132° + 38°) = 10°$

$\frac{\sin 10°}{25} = \frac{\sin 38°}{x}$ *Use the law of sines.*

$x \sin 10° = 25 \sin 38°$

$x = \frac{25 \sin 38°}{\sin 10°}$ *Calculation-ready form*

$= 88.6363$ *Do not round to the nearest whole number yet, since the value of x is used in calculating h.*

$\sin 48° = \frac{h}{x}$ *Definition of sine*

$h = x \sin 48°$

$= 88.6363(\sin 48°)$ *On a calculator, simply retain the displayed value of x (88.63632829) and multiply by $\sin 48°$.*

$= 65.8696$

The tree is approximately 66 ft high, and it is approximately 89 ft from point B to the top of the tree.

- **For Example 3**
  5. From a point A, the angle of elevation to the top of a building is 50°. From a point B 11 m closer to the building, the angle of elevation of the top is 63°. How far is it from point B to point C at the top of the building? How tall is the building? Express your answers to the nearest meter. 37 m; 33 m
  6. From a point A, the angle of elevation to the top of a tree is 45°. From a point B 16 ft closer to the tree, the angle of elevation of the top is 72°. How far is it from point B to the top of the tree? How tall is the tree? Express your answers to the nearest foot. 25 ft; 24 ft

Common Error

- Students often use incorrect ratios. Emphasize $\frac{\sin A}{a} = \frac{\sin B}{b} = \frac{\sin C}{c}$.
- See *Teacher's Resource Book* for additional remediation.

LESSON FOLLOW-UP

Discussion

List various combinations of given information about a triangle. Which combination(s) do not give unique solutions? Why? *AAA, ASA, etc.; SSA, AAA; answers may vary.*

Assignment Guide

See p. 154B for assignments.

Math Club Activity

The activity is based on the fact that the bisectors of the base angles of an isosceles triangle form another isosceles triangle.

Lesson Quiz

1. Find the measure of the unknown angle in the triangle.

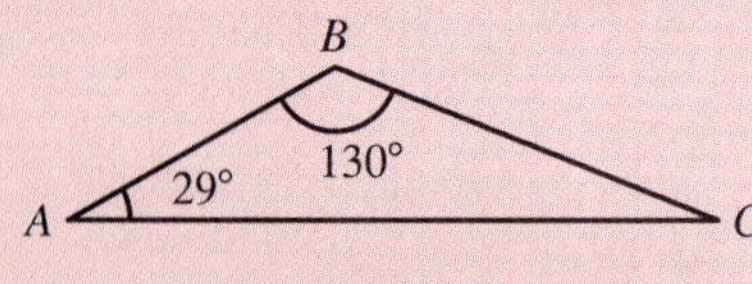

21°

2. Find a to the nearest whole number if $\dfrac{\sin 46^\circ}{a} = \dfrac{\sin 75^\circ}{11}$. 8
3. Find A to the nearest degree if $\dfrac{\sin A}{12} = \dfrac{\sin 36^\circ}{15}$. 28°
4. In triangle ABC, if $\angle A = 25^\circ$, $\angle C = 47^\circ$, and $b = 15$; find a. 7
5. In triangle ABC, if $\angle B = 47^\circ$, $\angle A = 25^\circ$, and $a = 11$; find b. 19

Enrichment

Prove: $(a - b)(\sin B) = b(\sin A - \sin B)$

$$\frac{\sin A}{a} = \frac{\sin B}{b}$$

$$\frac{a}{b} = \frac{\sin A}{\sin B}$$

$$\frac{a}{b} - \frac{b}{b} = \frac{\sin A}{\sin B} - \frac{\sin B}{\sin B}$$

$$\frac{a - b}{b} = \frac{\sin A - \sin B}{\sin B}$$

$$(a - b)(\sin B) = b(\sin A - \sin B)$$

CLASS EXERCISES

Find the measures of the unknown angle or angles in each triangle.

1. (85°, 54°, x) 41°

2.

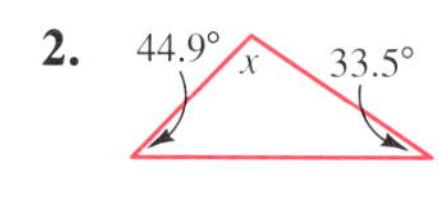

101.6°

3. (42°, 65°, x, y) $x = 23°$; $y = 25°$

4.

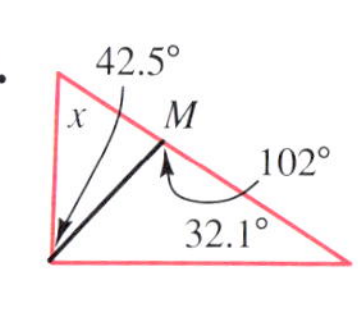

59.5°

Solve each triangle for the indicated side. Leave the answer in calculation-ready form.

5. $a = \dfrac{18 \sin 46^\circ}{\sin 29^\circ}$

6. $b = \dfrac{20 \sin 60^\circ}{\sin 48^\circ}$

7. $c = \dfrac{41 \sin 30^\circ}{\sin 58^\circ}$

8. $c = \dfrac{75 \sin 58^\circ}{\sin 92^\circ}$

For discussion

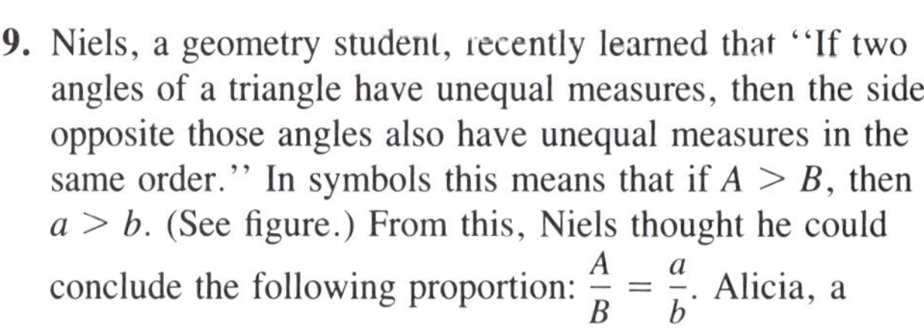

9. Niels, a geometry student, recently learned that "If two angles of a triangle have unequal measures, then the sides opposite those angles also have unequal measures in the same order." In symbols this means that if $A > B$, then $a > b$. (See figure.) From this, Niels thought he could conclude the following proportion: $\dfrac{A}{B} = \dfrac{a}{b}$. Alicia, a trigonometry student, disagreed with Niels. Who is right? Why?

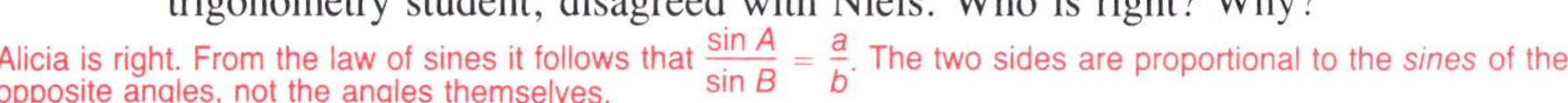

Alicia is right. From the law of sines it follows that $\dfrac{\sin A}{\sin B} = \dfrac{a}{b}$. The two sides are proportional to the *sines* of the opposite angles, not the angles themselves.

PRACTICE EXERCISES

Solve each triangle for the indicated side. Express your answer to two significant digits.

A

1. $b = 18$

2.

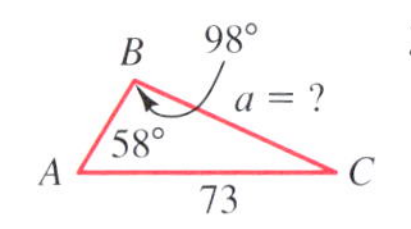

$a = 63$

3.

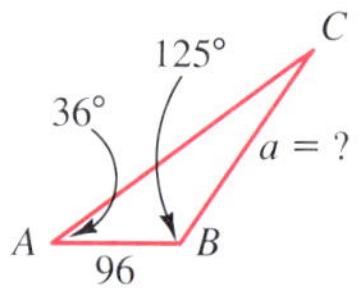

$a = 170$

4. $b = 70$

5. $\angle A = 39^\circ$, $\angle B = 42^\circ$, $c = 47$; find a. 30
6. $\angle A = 41^\circ$, $\angle B = 57^\circ$, $c = 52$; find b. 44
7. $\angle B = 72^\circ$, $\angle C = 31^\circ$, $a = 103$; find b. 100
8. $\angle B = 34^\circ$, $\angle C = 71^\circ$, $a = 115$; find b. 67
9. $\angle A = 48^\circ$, $\angle B = 38^\circ$, $b = 49$; find c. 79

10. $\angle A = 35°$, $\angle B = 56°$, $a = 51$; find c. 89

11. $\angle A = 128°$, $\angle C = 19°$, $a = 47$; find c. 19

12. $\angle B = 119°$, $\angle C = 21°$, $b = 59$; find c. 24

13. $\angle A = 49.7°$, $\angle B = 48.6°$, $a = 31$; find b. 30

14. $\angle A = 48.7°$, $\angle B = 40.2°$, $b = 29$; find a. 34

Solve each triangle PQR. Give angle measures to the nearest degree and lengths to two significant digits.

15. $p = 18$, $\angle Q = 46°$, $\angle R = 39°$ $\angle P = 95°$, $q = 13$, $r = 11$

16. $p = 24$, $\angle Q = 51°$, $\angle R = 38°$ $\angle P = 91°$, $q = 19$, $r = 15$

17. $q = 48$, $\angle P = 63°$, $\angle R = 51°$ $\angle Q = 66°$, $p = 47$, $r = 41$

18. $q = 75$, $\angle P = 42°$, $\angle R = 20°$ $\angle Q = 118°$, $p = 57$, $r = 29$

Solve each triangle PQR. Express the lengths of the sides to three significant digits and angle measures to the nearest tenth of a degree.

B 19. $\angle P = 32.6°$, $\angle R = 46.9°$, $r = 115$ $\angle Q = 100.5°$, $q = 155$, $p = 84.9$

20. $\angle P = 45.8°$, $\angle R = 32.6°$, $p = 113$ $\angle Q = 101.6°$, $q = 154$, $r = 84.9$

21. $\angle P = 54.2°$, $\angle Q = 45.9°$, $r = 76.1$ $\angle R = 79.9°$, $p = 62.7$, $q = 55.5$

22. $\angle P = 76.7°$, $\angle Q = 29.3°$, $r = 87.0$ $\angle R = 74.0°$, $p = 88.1$, $q = 44.3$

23. $\angle Q = 113.4°$, $\angle P = 27.5°$, $p = 56.3$ $\angle R = 39.1°$, $r = 76.9$, $q = 112$

24. $\angle Q = 129.7°$, $\angle P = 23.8°$, $p = 112$ $\angle R = 26.5°$, $r = 124$, $q = 214$

25. $\angle R = 54.9°$, $\angle Q = 110.3°$, $q = 73.2$ $\angle P = 14.8°$, $p = 19.9$, $r = 63.9$

26. $\angle R = 21.7°$, $\angle Q = 97.5°$, $q = 85.3$ $\angle P = 60.8°$, $p = 75.1$, $r = 31.8$

27. $\angle P = 47.3°$, $\angle Q = 65.2°$, $p = 96.4$ $\angle R = 67.5°$, $r = 121$, $q = 119$

28. $\angle P = 55.9°$, $\angle Q = 73.8°$, $p = 73.4$ $\angle R = 50.3°$, $r = 68.2$, $q = 85.1$

29. $\angle Q = 132.7°$, $\angle P = 28.1°$, $p = 67.4$ $\angle R = 19.2°$, $r = 47.1$, $q = 105$

30. $\angle Q = 31.7°$, $\angle P = 42.9°$, $p = 87.6$ $\angle R = 105.4°$, $r = 124$, $q = 67.6$

C 31. Is it true that for any $\triangle ABC$, $\frac{a}{b} = \frac{\sin A}{\sin B}$? Justify your answer. See side column.

32. If $\sin A = \sin B$ in $\triangle ABC$, when will $A = B$? Justify your answer. See side column.

33. Prove that the law of sines holds for a right triangle. See side column.

34. Prove that for any $\triangle ABC$, $\frac{a - b}{a + b} = \frac{\sin A - \sin B}{\sin A + \sin B}$. See side column.

35. Prove that $\cos(180° - \theta) = -\cos\theta$ is an identity. See side column.

Applications

Surveying In Exercises 36–40, give angle measures to the nearest degree and lengths to two significant digits, unless otherwise specified.

36. From two points P and Q that are 140 ft apart, the lines of sight to a flagpole across a river make angles of 79° and 58°, respectively, with the line joining P and Q. What are the distances from P and Q to the flagpole? 170 ft; 200 ft

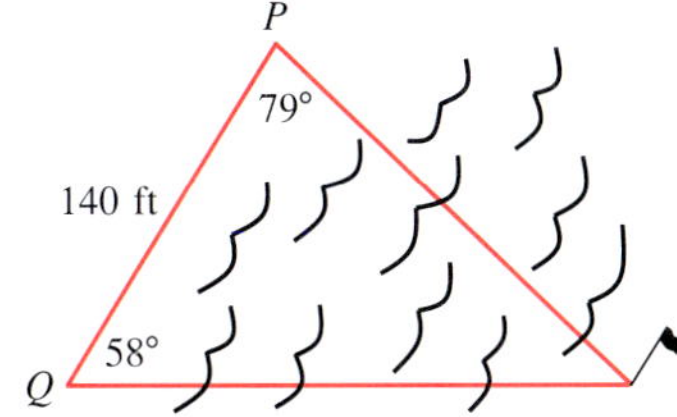

Additional Answers

31. yes; $\frac{\sin A}{a} = \frac{\sin B}{b}$; $b \sin A = a \sin B$; $\frac{b \sin A}{\sin B} = a$; $\frac{\sin A}{\sin B} = \frac{a}{b}$

32. When $\sin A = \sin B$ and there is a given value for A, there are two possible values for B, which are symmetric about the y-axis. The value of B which is in the same quadrant as A will be equal to A.

33. For right triangle ABC with right angle C, $\sin A = \frac{a}{c}$, $\sin B = \frac{b}{c}$, $\frac{\sin A}{a} = \frac{\frac{a}{c}}{a} = \frac{1}{c}$; $\frac{\sin B}{b} = \frac{\frac{b}{c}}{b} = \frac{1}{c}$; $\frac{\sin C}{c} = \frac{\sin 90}{c} = \frac{1}{c}$.

34. $\frac{\sin B}{b} = \frac{\sin A}{a}$; $a \sin B = b \sin A$; $2a \sin B = 2b \sin A$; $(a + a) \sin B = (b + b) \sin A$; $(a + a + b - b) \sin B = (b + b + a - a) \sin A$; $((a + b) + (a - b)) \sin B = ((a + b) - (a - b)) \sin A$; $(a + b) \sin B + (a - b) \sin B = (a + b) \sin A - (a - b) \sin A$; $(a - b) \sin B + (a - b) \sin A = (a + b) \sin A - (a + b) \sin B$; $(a - b)(\sin A + \sin B) = (a + b)(\sin A - \sin B)$; $\frac{a - b}{a + b} = \frac{\sin A - \sin B}{\sin A + \sin B)}$

35. $\cos(180° - \theta) = \cos(90° + 90° - \theta) = \cos(90° - (\theta - 90°)) = \sin(\theta - 90°) = -\cos\theta$

Teacher's Resource Book
Practice—Chapter 4, p. 1
Enrichment—Chapter 4, p. 2

37. Suppose that a parcel of land is triangular, with vertices A and B on the roadway and the third vertex marked at point C. A surveyor measures the distance from A to B and finds that it is 245.8 ft. The lines of sight from A and B to C make angles of 79.46° and 51.67°, respectively, with the line from A to B. Find the measure of angle C and the lengths of sides AC and BC. Give the angle measure to the nearest hundredth of a degree and the lengths to the nearest tenth of a foot. 256.0 ft; 320.8 ft; 48.87°

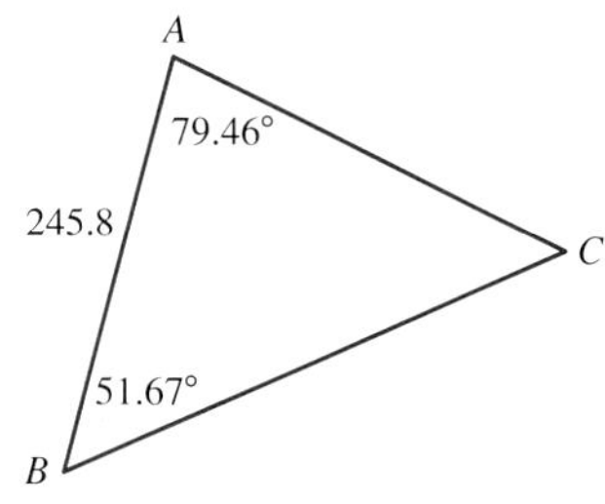

38. As you walk on a straight level path toward a mountain, the measure of the angle of elevation to the peak from one point is 33°. From a point 1000 ft closer, the angle of elevation is 35°. How high is the mountain? 9000 ft

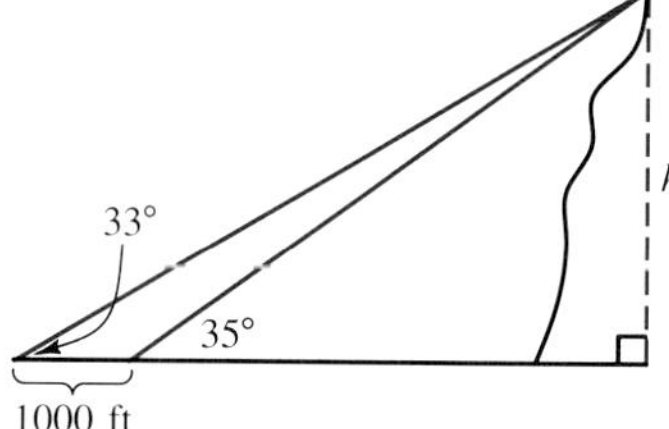

39. A flagpole stands on the edge of the bank of a river. From a point on the opposite bank directly across from the flagpole, the measure of the angle of elevation to the top of the pole is 25°. From a point 200 ft further away and in line with the pole and the first point, the measure of the angle of elevation to the top of the pole is 21°. Draw a diagram. Then find the distance across the river. 930 ft

40. You measure the angle of elevation to an airplane as 46.3°. At the same time your friend, who is 600 ft closer to a point directly beneath the plane, measures the angle of elevation as 47.5°. Draw a diagram. Then find the altitude of the plane. 15,000 ft

105°; The triangle need not be isosceles. If $\angle B = 2x$ and $\angle C = 2y$, then $2x + 2y + 30 = 180$. Therefore, $x + y = 75$. $(x + y) + M = 180°$. Thus, $M = 105°$.

MATH CLUB ACTIVITY

The measure of $\angle A$ is 30°. What is the measure of $\angle M$, the angle formed by the bisectors of angles B and C? (The triangle shown appears to be isosceles. Is this necessarily so? Explain your reasoning.) See above

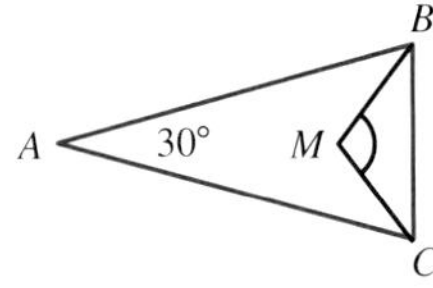

4.2 Law of Sines: The Ambiguous Case

Objective: To solve a triangle when the measures of two sides and an angle opposite one of them are given

For some triangles, the law of sines must be used with care.

Preview

The arm of a power shovel is attached to the shovel's beam at a point 60 ft from the base of the power shovel. The distance from the beam to the bottom of the shovel is 18 ft. A triangle is sometimes formed by the beam, the shovel arm, and the ground. If the angle α between the beam and the ground is too large, the shovel will not touch the ground and there is no triangle (Figure 1). If the shovel arm can touch the ground at only one point, then only one triangle is determined (Figure 2). If the shovel can touch the ground at two points, then two triangles are determined and earth can be picked up as the shovel moves from one point to the other (Figure 3).

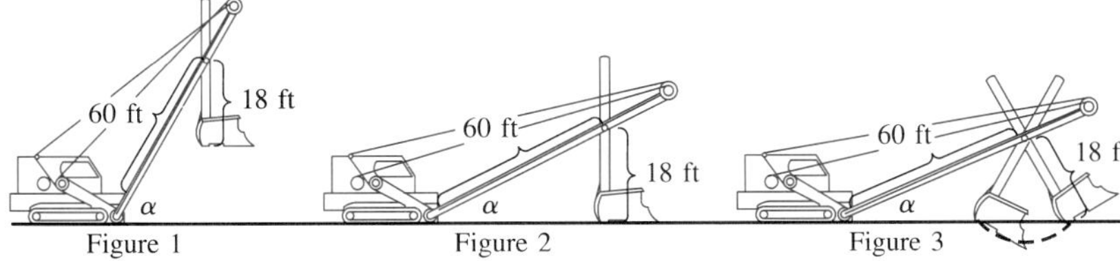

Figure 1 Figure 2 Figure 3

1. Refer to the power shovel above. Given sides of 60 ft and 18 ft and an angle α of 15°, how many triangles can be constructed? 2
2. How many triangles can be constructed if $\alpha = 50°$? *Hint*: Recall that $-1 \leq \sin \alpha \leq 1$. none
3. If the bottom of the shovel touches the ground at a right angle, what is the measure of angle α, to the nearest hundredth of a degree? 17.46°
4. If you are given two sides of a triangle and an angle opposite one side, is a unique triangle determined? sometimes

When the measures of *two sides and the angle opposite one of them* are given, you may be able to use the law of sines to find the measure of another angle of the triangle. Two angle measures, one, or none may satisfy the given value of the sine ratio. This situation is called the *ambiguous case*, since it is not always possible to determine a unique triangle under these conditions.

LESSON PLAN

Materials/Manipulatives
Scientific calculators

BACKGROUND

In the Preview, a situation analogous to the ambiguous case of the law of sines is presented. In this case, the angle formed is acute.

TEACHING SUGGESTIONS

- Emphasize that students should make a drawing of the triangle before solving. This should help determine how many solutions exist.
- Stress that when given two sides and the angle opposite one of those sides, it does not always determine a unique solution.
- Encourage students to use their scientific calculators to do any necessary calculations.

Critical Thinking

Analysis Ask students to find the number of solutions that exist given only the values of a, b, and angle A, as indicated below, without making a drawing.

$\angle A < 90°$

$a < b \sin A$ no solutions

$a = b \sin A$ one solution

$a > b \sin A$ two solutions

$b \sin A < a < b$ two solutions

$\angle A \geq 90°$

$a \leq b$ no solutions

$a > b$ one solution

CHALKBOARD EXAMPLES

- **For Example 1**

  Solve each triangle ABC. Round lengths to two significant digits and angle measures to the nearest degree.

  1. $b = 10, c = 14, \angle B = 45°$

  $\angle C = 82°, \angle A = 53°, a = 11$, or $\angle C = 98°, \angle A = 37°, a = 8.5$

  2. $\angle A = 40°, b = 14, a = 11$

  $\angle B = 55°, \angle C = 85°, c = 17$, or $\angle B = 125°, \angle C = 15°, c = 4.4$

- **For Example 2**

  Solve each triangle DEF, if possible. Round lengths to two significant digits and angle measures to the nearest degree.

  3. $d = 25, e = 40, \angle D = 50°$

  no solution

  4. $\angle D = 49°, d = 15, e = 27$

  no solution

EXAMPLE 1 **Solve $\triangle ABC$, where $a = 9.1$, $b = 12$, and $\angle A = 35°$. Round lengths to two significant digits and angle measures to the nearest degree.**

$$\frac{\sin 35°}{9.1} = \frac{\sin B}{12} \quad \textit{Use the law of sines.}$$

$$\frac{12 \sin 35°}{9.1} = \sin B \quad \textit{Calculation-ready form}$$

$$0.7564 = \sin B$$

$$\angle B = 49° \quad \text{or} \quad \angle B = 180° - 49° = 131°$$

There are two possible solutions to $\triangle ABC$, which are pictured below.

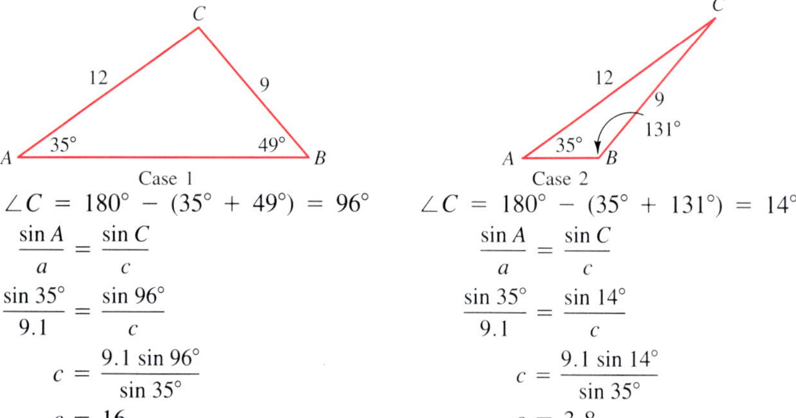

Case 1:

$$\angle C = 180° - (35° + 49°) = 96°$$

$$\frac{\sin A}{a} = \frac{\sin C}{c}$$

$$\frac{\sin 35°}{9.1} = \frac{\sin 96°}{c}$$

$$c = \frac{9.1 \sin 96°}{\sin 35°}$$

$$c = 16$$

Case 2:

$$\angle C = 180° - (35° + 131°) = 14°$$

$$\frac{\sin A}{a} = \frac{\sin C}{c}$$

$$\frac{\sin 35°}{9.1} = \frac{\sin 14°}{c}$$

$$c = \frac{9.1 \sin 14°}{\sin 35°}$$

$$c = 3.8$$

One solution is $\angle A = 35°$, $\angle B = 49°$, $\angle C = 96°$, $a = 9.1$, $b = 12$, and $c = 16$. Another solution is $\angle A = 35°$, $\angle B = 131°$, $\angle C = 14°$, $a = 9.1$, $b = 12$, and $c = 3.8$

Sometimes a triangle has no solution.

EXAMPLE 2 **Solve $\triangle ABC$, where $a = 25$, $b = 46$, and $\angle A = 37°$.**

$$\frac{\sin 37°}{25} = \frac{\sin B}{46} \quad \textit{Use the law of sines.}$$

$$\frac{46 \sin 37°}{25} = \sin B \quad \textit{Calculation-ready form}$$

$$1.1073 = \sin B$$

There is no angle whose sine is equal to 1.1073 because $-1 \leq \sin \theta \leq 1$. Therefore, there is no triangle that satisfies the given conditions.

Below is a summary of the ambiguous case (given two sides and the angle opposite one of them).

1. If $\angle A$ is an acute angle and $a < b$, there are three possibilities.

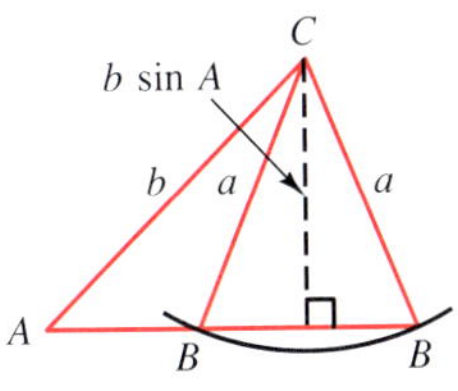

Two solutions
$a > b \sin A$

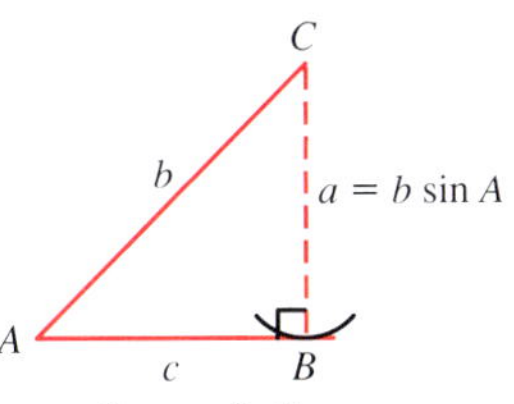

One solution
$a = b \sin A$

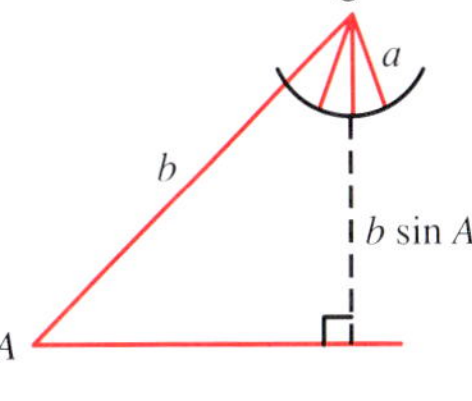

No solution
$a < b \sin A$

2. If $\angle A$ is an acute angle and $a \geq b$, then there is exactly one solution.

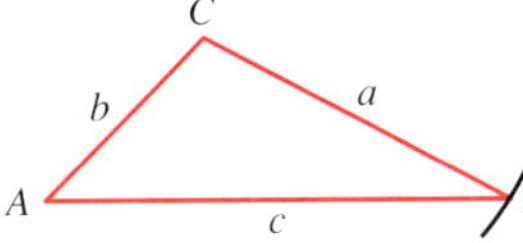

3. If $\angle A$ is an obtuse or right angle, there are two possibilities:

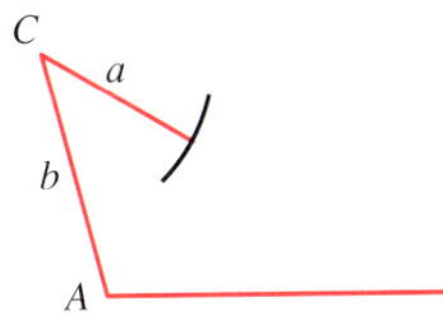

No solution
$a \leq b$

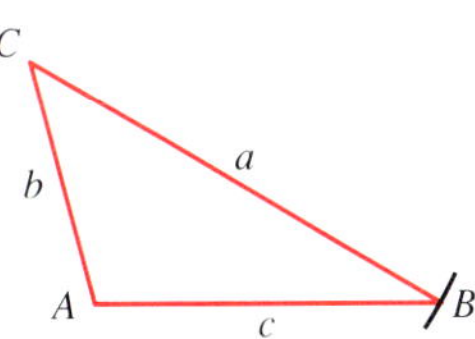

One solution
$a > b$

EXAMPLE 3 **How many solutions are there in each case?**

a. $\angle C = 47°$, $c = 20$, $a = 12$

b. $\angle C = 97°$, $c = 45$, $a = 39$

c. $\angle A = 114°$, $a = 21$, $b = 32$

a.

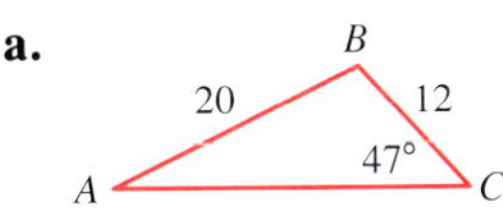

One solution, since C is acute and $c \geq a$.

b.

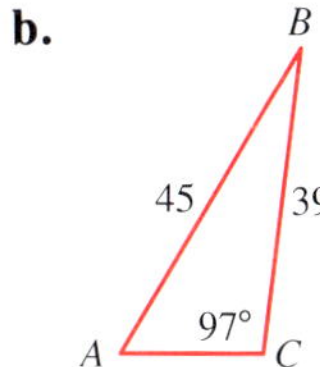

One solution; C is obtuse and $c > a$.

c.

No solution, since A is obtuse and $a \leq b$.

- **For Example 3**
  Determine the number of solutions in each case.
  5. $\angle A = 52°$, $a = 19$, $b = 30$
  no solution
  6. $\angle C = 112°$, $b = 51$, $c = 85$
  one solution

Common Error

- Students often fail to realize that more than one solution exists. These sketches may help.

$a < b$
$A < 90°$

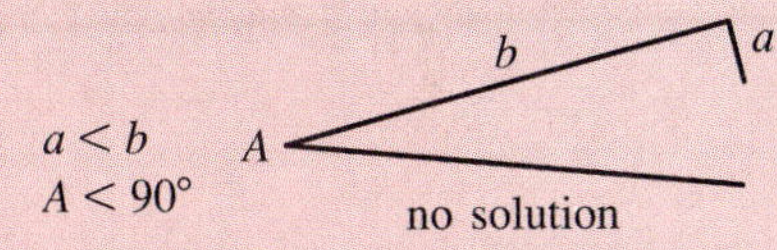

no solution

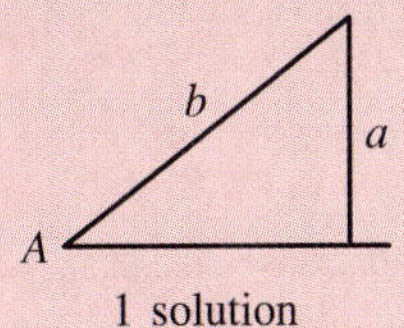

1 solution

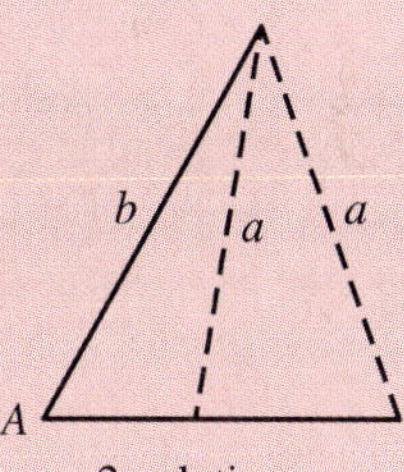

2 solutions

$a \leq b$
$A > 90°$

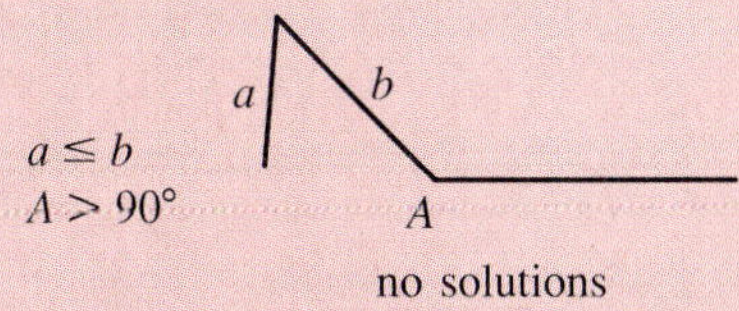

no solutions

$a > b$
$A > 90°$

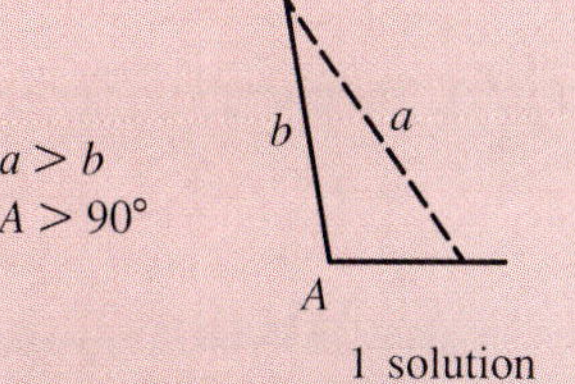

1 solution

- See *Teacher's Resource Book* for additional remediation.

LESSON FOLLOW-UP

Discussion

Show how to find the perimeter of an isosceles triangle with a base of 26 cm and a vertex angle of 46°.

$\sin 23° = \frac{13}{x}$

$x = \frac{13}{\sin 23°}$

$x = 33.27$

$P = 2(33.27) + 26$

$P = 93$ cm

Assignment Guide

See p. 154B for assignments.

Lesson Quiz

Determine how many solutions exist.

1. $\angle A = 36°, a = 17, b = 19$ two
2. $\angle A = 42°, a = 21, b = 11$ one
3. $\angle A = 112°, a = 11, b = 16$ none

Solve each triangle ABC. Round all lengths to two significant digits and all angle measures to the nearest degree.

4. $\angle A = 33°, a = 19, c = 12$
   $\angle C = 20°, \angle B = 127°, b = 28$
5. $\angle A = 27°, a = 15, b = 21$
   $\angle B = 39°, \angle C = 114°, c = 30$, or $\angle B = 141°, \angle C = 12°, c = 6.9$

Enrichment

The sides of a rhombus measure 12 cm. The measure of $\angle CBA$ is 40°. Find the lengths of the diagonals.

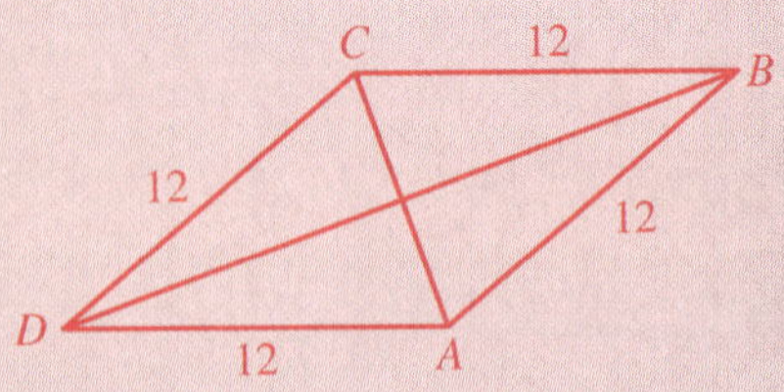

$\angle DCB = 140°; \angle ACB = 70°; \frac{\sin 40°}{AC} = \frac{\sin 70°}{12}; AC = 8.2$ cm; $\angle DBC = 20°$; $\frac{\sin 20°}{12} = \frac{\sin 140°}{BD}; BD = 22.6$ cm

CLASS EXERCISES

Find the number of solutions in each case. When in doubt, carefully draw a sketch.

1. $\angle A = 94°, a = 25, b = 15$ 1
2. $\angle A = 84°, a = 25, b = 15$ 1
3. $\angle A = 80°, a = 20, b = 20$ 1
4. $\angle A = 40°, a = 19, b = 20$ 2
5. $\angle A = 30°, a = 10, b = 20$ 1
6. $\angle A = 65°, a = 10, b = 20$ 0
7. $\angle A = 40°, a = 10, b = 20$ 0
8. $\angle A = 150°, a = 10, b = 20$ 0
9. $\angle A = 45°, a = 10, b = 10\sqrt{2}$ 1
10. $\angle A = 60°, a = 10\sqrt{3}, b = 20$ 1

PRACTICE EXERCISES

Determine how many solutions exist. When either one or two solutions exist, solve the triangle or triangles. In Exercises 1–10, round all lengths to two significant digits and all angle measures to the nearest degree. In Exercises 11–28, round all lengths to three significant digits and all angle measures to the nearest tenth of a degree. The number of solutions are shown below. For solutions, see side column on page 167.

A

1. $\angle A = 67°, a = 18, b = 20$ 0
2. $\angle A = 71°, a = 37, b = 40$ 0
3. $\angle A = 32°, a = 7, b = 10$ 2
4. $\angle A = 29°, a = 15, b = 19$ 2
5. $\angle A = 87°, a = 47, b = 50$ 0
6. $\angle A = 79°, a = 52, b = 55$ 0
7. $\angle A = 113°, a = 49, b = 54$ 0
8. $\angle A = 110°, a = 76, b = 85$ 0
9. $\angle A = 37°, a = 49, b = 54$ 2
10. $\angle A = 40°, a = 75, b = 85$ 2
11. $\angle A = 59.8°, a = 80.8, b = 73.9$ 1
12. $\angle A = 69.8°, a = 74.5, b = 21.3$ 1
13. $\angle A = 31.9°, a = 30.6, b = 37.9$ 2
14. $\angle A = 29.8°, a = 28.6, b = 35.8$ 2
15. $\angle A = 85.8°, a = 23.9, b = 26.4$ 0
16. $\angle A = 76.4°, a = 27.3, b = 29.0$ 0

B

17. $\angle C = 47.1°, b = 15.3, c = 11.9$ 2
18. $\angle C = 51.6°, b = 32.4, c = 28.0$ 2
19. $\angle B = 36.3°, b = 46.3, c = 51.2$ 2
20. $\angle B = 41.2°, b = 83.2, c = 76.2$ 1
21. $\angle B = 54.3°, a = 62.5, b = 29.6$ 0
22. $\angle B = 58.7°, a = 118.6, b = 62.4$ 0
23. $\angle C = 108.7°, a = 51.2, c = 54.3$ 1
24. $\angle C = 103.4°, a = 89.4, c = 98.4$ 1
25. $\angle C\ 123°, b = 106.9, c = 104.3$ 0
26. $\angle C = 115°, b = 54.8, c = 53.4$ 0
27. $\angle C\ 23.47°, a = 26.49, c = 20.5$ 2
28. $\angle C = 29.7°, a = 78.92, c = 58.9$ 2

In Exercises 29–33, give angle measures to the nearest degree.

C **29.** One angle of a triangle measures 51°, an adjacent side is 49 cm, and the opposite side is 46 cm. What are the measures of the other two angles?
$\angle B = 56°$, $\angle C = 73°$ or $\angle B = 124°$, $\angle C = 5°$

30. One of the angles of a triangle is to be obtuse. Another angle is to measure 73°, with an adjacent side of 54 cm and an opposite side of 100 cm. Does such a triangle exist? If so, what are the measures of the other two angles? does not exist

31. One of the angles of a triangle measures 31°. An adjacent side measures 40 cm, and the opposite side is twice as long as the other adjacent side. What are the measures of the other two angles? 15°, 134°

32. One of the angles of a triangle measures 42°. An adjacent side measures 40 cm, and the opposite side is three times as long as the other adjacent side. What are the measures of the other two angles? 13°, 125°

33. If $\angle A$ is acute and the measures of sides a and b are known, show that two triangles will satisfy these conditions if and only if $b \sin A < a < b$.
See side column.

Applications

In Exercises 34–35, give angle measures to the nearest degree.

34. Engineering If a pole has one 62-ft guy wire that makes an angle of 39° with the ground, and a second 50-ft guy wire is available for the opposite side of the pole, what angle measure will the second wire make with the ground? 51°

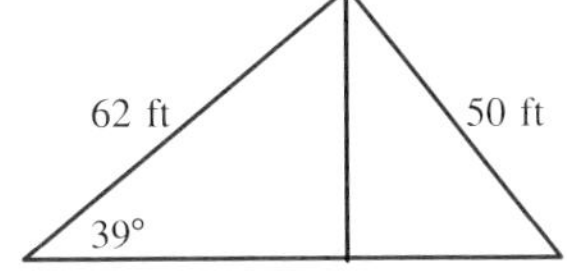

35. Surveying Two roads intersect at an obtuse angle. Two points are chosen: A, 0.15 mi from the intersection, and B, 0.23 mi from the intersection. If the line of sight from A to B makes an angle of 42° with one road at point A, what is the measure of the obtuse angle between the roads? 112°

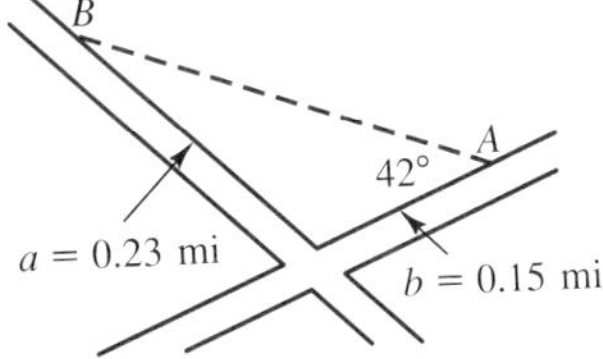

CHALLENGE

On Ms. Berger's trigonometry test, $\frac{1}{2}$ of the triangles had one solution, $\frac{1}{3}$ had two solutions, and 15 had no solution. How many triangles were on the test?
90 if triangles with no solution are counted, 75 otherwise

Teacher's Resource Book
Practice—Chapter 4, p. 3
Enrichment—Chapter 4, p. 4

Additional Answers

3. $\angle B = 131°$, $\angle C = 17°$, $c = 3.9$ or $\angle B = 49°$, $\angle C = 99°$, $c = 13$
4. $\angle B = 38°$, $\angle C = 113°$, $c = 28$ or $\angle B = 142°$, $\angle C = 9°$, $c = 4.8$
9. $\angle B = 42°$, $\angle C = 101°$, $c = 80$ or $\angle B = 138°$, $\angle C = 5°$, $c = 7.1$
10. $\angle B = 47°$, $\angle C = 93°$, $c = 120$ or $\angle B = 133°$, $\angle C = 7°$, $c = 14$
11. $\angle B = 52.2°$, $\angle C = 68.0°$, $c = 86.7$
12. $\angle B = 15.6°$, $\angle C = 94.6°$, $c = 79.1$
13. $\angle B = 40.9°$, $\angle C = 107.2°$, $c = 55.3$ or $\angle B = 139.1°$, $\angle C = 9.0°$, $c = 9.1$
14. $\angle B = 38.5°$, $\angle C = 111.7°$, $c = 53.5$ or $\angle B = 141.5°$, $\angle C = 8.7°$, $c = 8.7$
17. $\angle B = 70.4°$, $\angle A = 62.5°$, $a = 14.4$ or $\angle B = 109.6°$, $\angle A = 23.3°$, $a = 6.43$
18. $\angle B = 65.1°$, $\angle A = 63.3°$, $a = 31.9$ or $\angle B = 114.9°$, $\angle A = 13.5°$, $a = 8.34$
19. $\angle C = 40.9°$, $\angle A = 102.8°$, $a = 76.3$ or $\angle C = 139.1°$, $\angle A = 4.6°$, $a = 6.27$
20. $\angle C = 37.1°$, $\angle A = 101.7°$, $a = 124$
23. $\angle A = 63.3°$, $\angle B = 8.0°$, $b = 7.98$
24. $\angle A = 62.1°$, $\angle B = 14.5°$, $b = 25.3$
27. $\angle A = 31.0°$, $\angle B = 125.5°$, $b = 41.9$ or $\angle A = 149.0°$, $\angle B = 7.5°$, $b = 6.75$
28. $\angle A = 41.6°$, $\angle B = 108.7°$, $b = 113$ or $\angle A = 138.4°$, $\angle B = 11.9°$, $b = 24.5$
33. $\frac{a}{\sin A} = \frac{b}{\sin B}$, $b \sin A = a \sin B$; since $\sin B < 1$, $a \sin B < a \cdot 1$, so $b \sin A \leq a$. Sin $B = 1$ only when B is a right angle, in which case, only one triangle exists, due to HL. Thus, $b \sin A < a$ when B is not a right triangle. If $b = a$, then $\angle A = \angle B$ and only one triangle exists, due to ASA. If $b < a$, then $\angle B < \angle A$, so $\angle B$ must be acute, and there can be only one triangle since in a unit circle only an angle and its supplement have the same sine when restricted to quadrants I and II. Since only one solution exists when $b = a$ and when $b < a$, the only possible condition for two solutions is when $b > a$. Therefore, only $b \sin A < a < b$ can produce two solutions.

LESSON PLAN

Vocabulary
Law of cosines

Materials/Manipulatives
Scientific calculators

BACKGROUND

In the Preview, a problem involving the distances between three objects is presented. Students are asked to discuss how the distances could be found.

4.3 The Law of Cosines

Objectives: To introduce and prove the law of cosines
To use the law of cosines to solve triangles when the measures of two sides and the included angle or the measures of three sides are given

The law of sines cannot be used to solve every triangle. If the measures of *two sides and the included angle* or the measures of *all three sides* are given, another formula called the *law of cosines* is used.

Preview

Mrs. McDonald, Mr. Irizarry, and Ms. Rothstein have cottages on a lake, as shown in the diagram. Suppose you have measured the distances from the McDonald cottage to the Irizarry cottage and from the Irizarry cottage to the Rothstein cottage, as shown.

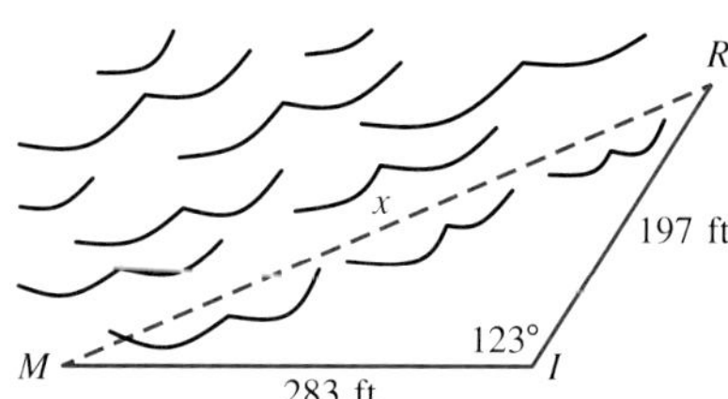

1. Describe how you could use a scale drawing to find the distance from the McDonald cottage to the Rothstein cottage. Measure, using appropriate scale ratios.
2. To use the law of sines, what additional information would you need? measure of angle M or measure of angle R
3. If the measure of $\angle I$ were 90° instead of 123°, would you need additional information? Explain your answer. No; you would use the Pythagorean theorem.

If you know the lengths of sides a and b in a triangle ABC, and if the included angle is a right angle, you can find side c using the Pythagorean theorem, $c^2 = a^2 + b^2$. If the included angle is not a right angle, you can use a generalization of this theorem called the *law of cosines*.

Law of Cosines

For any triangle ABC, where a, b, and c are the lengths of the sides opposite the angles with measures A, B, and C, respectively,

$$a^2 = b^2 + c^2 - 2bc \cos A$$
$$b^2 = a^2 + c^2 - 2ac \cos B$$
$$c^2 = a^2 + b^2 - 2ab \cos C$$

To prove the law of cosines it is necessary to consider two cases, one in which the included angle is acute and another in which the included angle is obtuse.

In the figure, $\angle A$, $\angle B$, and $\angle C$ represent the angles of any acute triangle, and a, b, and c, respectively, represent the lengths of the sides opposite these angles. Let h be the length of an altitude that separates $\triangle ABC$ into two right triangles with common side AD. Using the Pythagorean theorem and the cosine ratio, you can derive a relationship among a, b, c, and the measure of $\angle C$. Note that since x is the measure of segment DC, $a - x$ is the measure of segment BD.

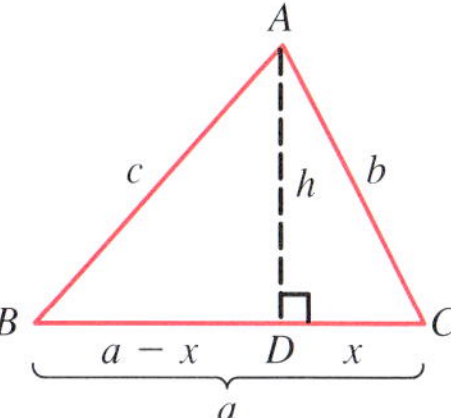

$c^2 = (a - x)^2 + h^2$	*Apply the Pythagorean theorem to $\triangle ABD$.*
$= a^2 - 2ax + x^2 + h^2$	
$= a^2 - 2ax + b^2$	*In $\triangle ADC$, $b^2 = h^2 + x^2$, so substitute.*
$= a^2 - 2a(b \cos C) + b^2$	*$\cos C = \frac{x}{b}$, so $x = b \cos C$.*
$= a^2 + b^2 - 2ab \cos C$	

Therefore, $c^2 = a^2 + b^2 - 2ab \cos C$.

You can derive similar expressions for a^2 and b^2. In the exercises, you will be asked to prove the case for an included angle that is obtuse.

EXAMPLE 1 **Find the length of side c for the given triangle.**

$\angle C = 60°$
$a = 10$
$b = 14$

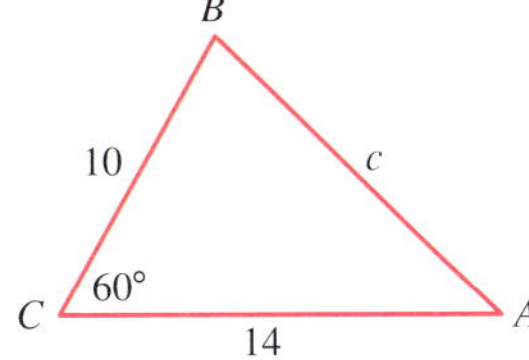

$c^2 = a^2 + b^2 - 2ab \cos C$	
$= 10^2 + 14^2 - 2(10)(14) \cos 60°$	*Calculation-ready form*
$= 156$	
$c = 12$	*To two significant digits*

Therefore, the length of side c is approximately 12 units.

In most cases, you will need a calculator or table to find the cosine of the given angle. Once you have used the law of cosines to find the missing side, you can use the law of sines to complete the solution.

TEACHING SUGGESTIONS

- Emphasize that the law of cosines is used when two sides and the included angle or three sides of a triangle are given. These are the situations where the law of sines cannot be used.
- Point out that it is usually necessary to use the law of cosines only once when solving a triangle.
- Encourage students to use their scientific calculators to do any necessary calculations.

Critical Thinking

Analysis Ask students to determine whether the law of sines or the law of cosines should be used first to solve each triangle below and to give a reason for each answer.

$C = 29°$, $a = 18$, $b = 13$
Law of cosines; two sides and the included angle are given.

$A = 32°$, $a = 8$, $b = 6$
Law of sines; two sides and the angle opposite one of the sides are given.

$a = 11$, $b = 6$, $c = 14$
Law of cosines; three sides are given.

CHALKBOARD EXAMPLES

- **For Example 1**

  Find the length of side c to two significant digits for the given triangle.

  1. $a = 3$, $b = 2$, $\angle C = 141°$ 4.7
  2. $a = 6$, $b = 9$, $\angle C = 35°$ 5.3

- **For Example 2**

  Solve triangle ABC. Compute the length of the side to three significant digits and the angle measures to the nearest tenth of a degree.

  3. $a = 17.4, c = 10.2, \angle B = 69.3°$
  $b = 16.8, \angle C = 34.6°, \angle A = 76.1°$

  4. $a = 12.1, c = 33.3, \angle B = 87.8°$
  $b = 35.0, \angle A = 20.2°, \angle C = 71.9°$

- **For Example 3**

  Solve triangle ABC. Round angle measures to the nearest degree.

  5. $a = 3, b = 5, c = 7$
  $\angle B = 38°, \angle A = 22°, \angle C = 120°$

  6. $a = 16, b = 13, c = 24$
  $\angle A = 38°, \angle B = 30°, \angle C = 112°$

EXAMPLE 2 **Solve $\triangle ABC$ if $\angle B = 98.1°$, $a = 17.2$, and $c = 21.5$. Compute the length of the side to three significant digits and the angle measures to the nearest tenth of a degree.**

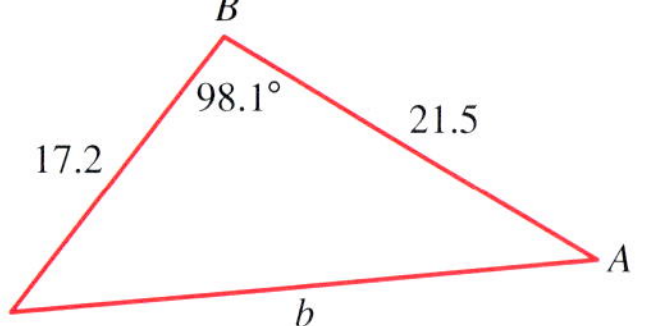

$$\begin{aligned} b^2 &= a^2 + c^2 - 2ac\cos B \\ &= (17.2)^2 + (21.5)^2 - 2(17.2)(21.5)\cos 98.1° \\ &= 862.3006 \\ b &= 29.3650 \end{aligned}$$

$$\frac{\sin C}{c} = \frac{\sin B}{b} \qquad \textit{Use the law of sines.}$$

$$\frac{\sin C}{21.5} = \frac{\sin 98.1°}{29.3650}$$

$$\sin C = \frac{21.5 \sin 98.1°}{29.3650}$$

On a calculator, retain the displayed value of b to use in the calculation of sin C.

$$= 0.7249$$

$$\angle C = 46.5°$$

$$\angle A = 180° - (98.1° + 46.5°) = 35.4°$$

Thus, $\angle A = 35.4°$, $\angle C = 46.5°$, and $b = 29.4$.

If the third form of the law of cosines, $c^2 = a^2 + b^2 - 2ab\cos C$, is solved for $\cos C$, it yields an expression for $\cos C$ in terms of a, b, and c, the three sides of the triangle.

$$\begin{aligned} c^2 &= a^2 + b^2 - 2ab\cos C \\ 2ab\cos C &= a^2 + b^2 - c^2 \\ \cos C &= \frac{a^2 + b^2 - c^2}{2ab} \end{aligned}$$

Similarly, you can show that $\cos A = \dfrac{b^2 + c^2 - a^2}{2bc}$ and $\cos B = \dfrac{a^2 + c^2 - b^2}{2ac}$.

If the lengths of all three sides of a triangle are given, you can find two of the missing angle measures by applying this form of the law of cosines twice. The third angle measure can be found using the fact that the sum of the measures of the angles of a triangle is 180°.

EXAMPLE 3 **Solve $\triangle ABC$ if $a = 43$, $b = 39$, and $c = 51$. Round angle measures to the nearest degree.**

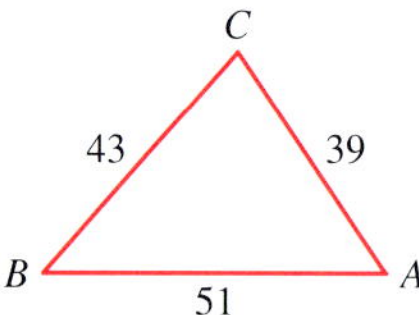

$$\cos A = \frac{b^2 + c^2 - a^2}{2bc}$$

$$= \frac{39^2 + 51^2 - 43^2}{2(39)(51)} \qquad \textit{Calculation-ready form}$$

$$= 0.5714$$

$$\angle A = 55°$$

$$\cos B = \frac{a^2 + c^2 - b^2}{2ac}$$

$$= \frac{43^2 + 51^2 - 39^2}{2(43)(51)} \qquad \textit{Calculation-ready form}$$

$$= 0.6678$$

$$\angle B = 48°$$

$$\angle C = 180° - (\angle A + \angle B) \qquad A + B + C = 180°$$

$$= 180° - (55° + 48°) \qquad \textit{Calculation-ready form}$$

$$= 77°$$

Therefore, $\angle A = 55°$, $\angle B = 48°$, and $\angle C = 77°$, to the nearest degree.

The law of cosines can be used to solve real-world problems.

EXAMPLE 4 **A field is triangular in shape with sides 473, 512 and 734. To the nearest tenth of a degree, what is the measure of the angle between the sides measuring 512 and 734?**

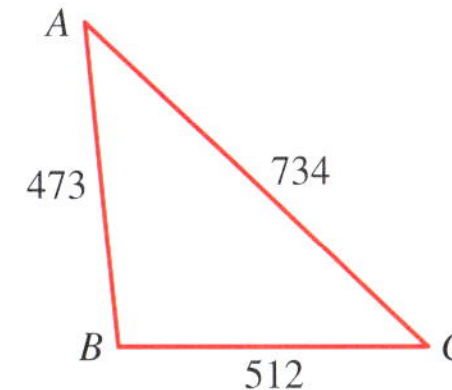

$$\cos C = \frac{734^2 + 512^2 - 473^2}{2(734)(512)}$$

$$= 0.7679$$

$$\angle C = 39.8° \qquad \textit{To the nearest tenth of a degree}$$

The following table indicates which law to use when solving triangles.

Given Information	Appropriate Law
Three sides	law of cosines
Two sides and the included angle	law of cosines
Two sides and an angle opposite one side (ambiguous case)	law of sines
One side and two angles	law of sines

- **For Example 4**
  A field is triangular in shape with sides of lengths 83 m, 120 m, and 165 m.
  7. To the nearest tenth of a degree, what is the measure of the angle formed by the sides measuring 120 m and 165 m? 28.7°
  8. To the nearest tenth of a degree, what is the measure of the angle formed by the sides measuring 83 m and 120 m? 107.3°

Common Error

- Students often make sign errors when substituting for cosine. Emphasize that cosine is negative for angle measures between 90° and 180°.
- See *Teacher's Resource Book* for additional remediation.

LESSON FOLLOW-UP

Discussion

Show how to find the lengths of the sides of a parallelogram if the diagonals are 12 cm and 17 cm long and intersect at a 35° angle.

$\angle DEC = 35°$; $\angle AED = 145°$;
$(DC)^2 = (8.5)^2 + (6)^2 - 2(8.5)(6)\cos 35°$
$DC = 4.97$ cm;
$(AD)^2 = (8.5)^2 + (6)^2 - 2(8.5)(6)\cos 145°$
$AD = 13.8$ cm

Assignment Guide

See p. 154B for assignments.

Trigonometry in 3-Space

Students are asked to prove a three-dimensional version of the law of cosines using certain given information.

Lesson Quiz

1. Write an equation that can be used to find the indicated side. Do not solve it.

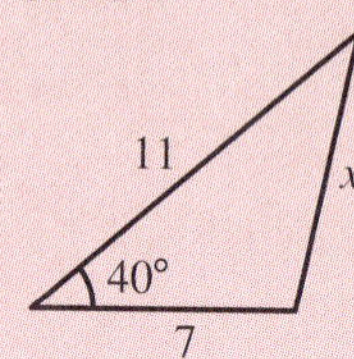

$x = \sqrt{(7)^2 + (11)^2 - 2(7)(11)(\cos 40°)}$

2. Write an expression to find the cosine of the unknown angle. Do not complete the computation.

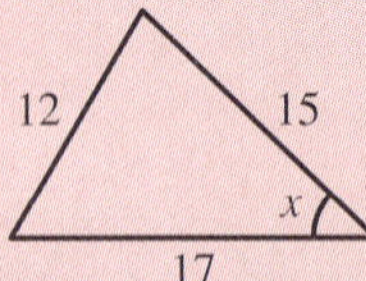

$\cos x = \dfrac{(15)^2 + (17)^2 - (12)^2}{2(15)(17)}$

3. Given triangle ABC where $\angle A = 47°$, $c = 12$, and $b = 17$, find a. 12
4. Given triangle ABC where $a = 5$, $c = 11$, and $b = 7$, find $\angle C$. 132°

CLASS EXERCISES

Write an equation to solve for the indicated side. Do not solve the equation.

1. **2.**

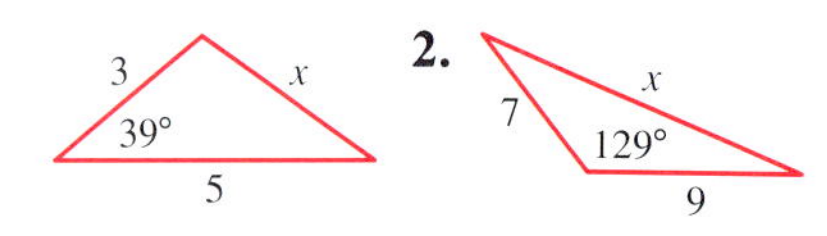

3.

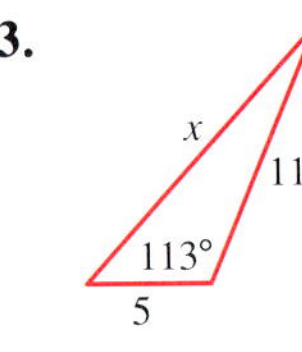

4.

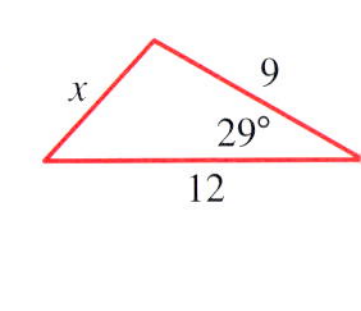

1. $x^2 = 3^2 + 5^2 - (2)(3)(5)\cos 39°$
2. $x^2 = 7^2 + 9^2 - (2)(7)(9)\cos 129°$
3. $x^2 = 5^2 + 11^2 - (2)(5)(11)\cos 113°$
4. $x^2 = 9^2 + 12^2 - (2)(9)(12)\cos 29°$

Write an expression to find the cosine of the unknown angle. Do not complete the computation.

5.

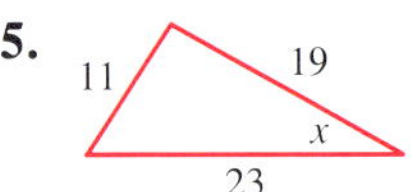

6.

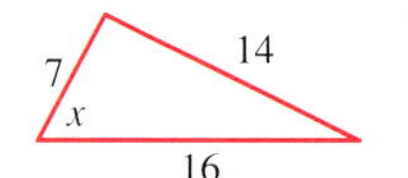

7.

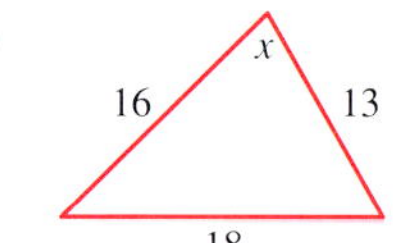

8.

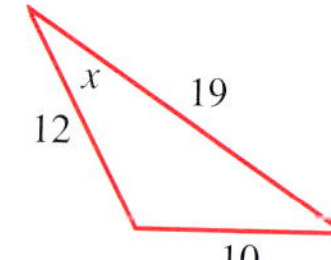

5. $\cos x = \dfrac{19^2 + 23^2 - 11^2}{2(19)(23)}$
6. $\cos x = \dfrac{7^2 + 16^2 - 14^2}{2(7)(16)}$
7. $\cos x = \dfrac{16^2 + 13^2 - 18^2}{2(16)(13)}$
8. $\cos x = \dfrac{12^2 + 19^2 - 10^2}{2(12)(19)}$

PRACTICE EXERCISES

Solve for the length of the missing side of each triangle. Round your answer to two significant digits.

A **1.**

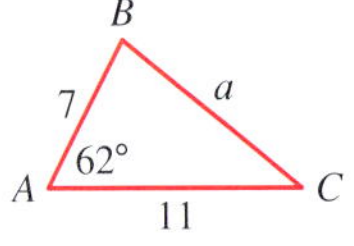

10

2.

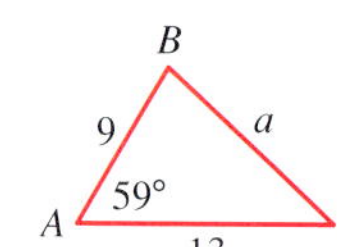

11

3.

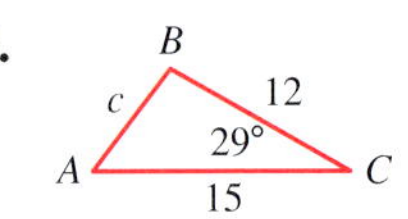

7.4

4.

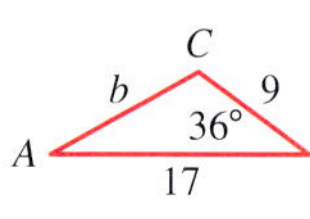

11

5. $\angle C = 115°$, $a = 11$, $b = 21$ 28
6. $\angle C = 113°$, $a = 13$, $b = 23$ 31
7. $\angle A = 32°$, $b = 23$, $c = 47$ 30
8. $\angle A = 34°$, $b = 24$, $c = 46$ 29
9. $\angle B = 31°$, $a = 17$, $c = 14$ 8.8
10. $\angle B = 32°$, $a = 15$, $c = 18$ 9.5

Solve each triangle for the specified angle measure. Round your answer to the nearest degree.

11. $a = 11$, $b = 14$, $c = 17$; $\angle A$ 40°
12. $a = 12$, $b = 16$, $c = 19$; $\angle A$ 39°
13. $a = 23$, $b = 43$, $c = 31$; $\angle B$ 105°
14. $a = 21$, $b = 42$, $c = 31$; $\angle B$ 106°
15. $a = 12$, $b = 12$, $c = 17$; $\angle C$ 90°
16. $a = 17$, $b = 17$, $c = 24$; $\angle C$ 90°

Solve each triangle PQR. Unless otherwise directed, round lengths to two significant digits and angle measures to the nearest degree.

B **17.** $\angle R = 30°$, $p = 18$, $q = 16$
$r = 9.0$, $\angle Q = 63°$, $\angle P = 87°$

18. $\angle R = 45°$, $p = 13$, $q = 19$
$r = 13$, $\angle P = 47°$, $\angle Q = 88°$

19. $\angle P = 83°$, $r = 43$, $q = 51$
$p = 63$, $\angle R = 43°$, $\angle Q = 54°$

20. $\angle P = 77°$, $r = 76$, $q = 49$
$p = 81$, $\angle R = 67°$, $\angle Q = 36°$

21. $\angle Q = 113°$, $p = 27$, $r = 43$
$q = 59$, $\angle P = 25°$, $\angle R = 42°$

22. $\angle Q = 129°$, $p = 45$, $r = 71$
$q = 105$, $\angle P = 19°$, $\angle R = 32°$

23. $p = 15$, $q = 19$, $r = 23$
$\angle R = 84°$, $\angle P = 41°$, $\angle Q = 55°$

24. $p = 27$, $q = 33$, $r = 41$
$\angle R = 86°$, $\angle Q = 53°$, $\angle P = 41°$

25. $p = 310$, $q = 250$, $r = 160$
$\angle R = 31°$, $\angle Q = 53°$, $\angle P = 96°$

26. $p = 200$, $q = 410$, $r = 280$
$\angle R = 38°$, $\angle Q = 116°$, $\angle P = 26°$

27. $p = 104.3$, $q = 135.7$, $r = 154.6$ (Give angle measures to the nearest hundredth of a degree.)
$\angle P = 41.47°$, $\angle Q = 59.51°$, $\angle R = 79.02°$

28. $p = 65.5$, $q = 92.7$, $r = 114$ (Give angle measures to the nearest tenth of a degree.)
$\angle R = 90.5°$, $\angle Q = 54.4°$, $\angle P = 35.1°$

C **29.** Prove that the Pythagorean theorem is a special case of the law of cosines. See side column.

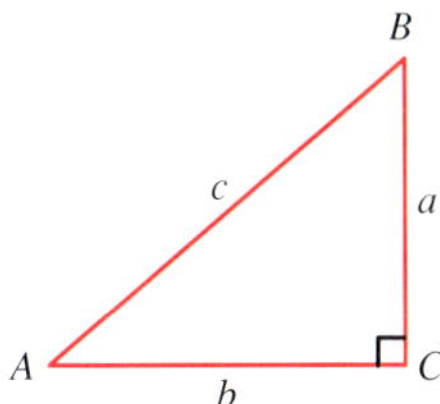

30. Given quadrilateral $ABCD$ with $AB = 47$, $BC = 31$, $CD = 61$, $\angle B = 117°$, and $\angle C = 76°$, determine AD. *Hint*: This is a three-step process. 41

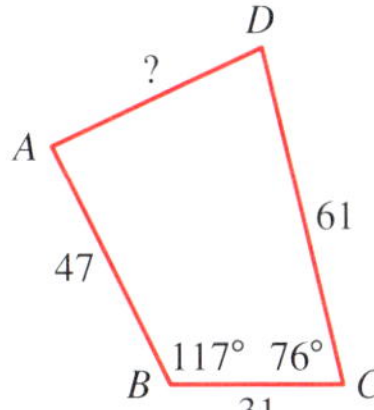

31. A parallelogram has adjacent sides of measure a and b and diagonals of measure p and q. Use the law of cosines to show that $2a^2 + 2b^2 = p^2 + q^2$.

32. Derive the law of cosines for the case in which $\angle C$ is an obtuse angle. *Hint*: Recall that $\cos(180° - C) = -\cos C$. See page 482.

33. Show that in $\triangle ABC$, if $\angle C$ is acute, $c < \sqrt{a^2 + b^2}$, and if $\angle C$ is obtuse, $c > \sqrt{a^2 + b^2}$. See page 482.

Applications

Surveying

34. A triangular field is 452 ft on one side and 572 ft on another. The sides meet in an angle of 67.1°. Find the length of the third side, to the nearest foot. 575 ft

35. If a triangular parcel of land has sides of lengths 541 ft, 429 ft, and 395 ft, what are the measures of the angles between the sides, to the nearest tenth of a degree? 82.0°, 51.7°, 46.3°

Enrichment

The basepaths in baseball are 90 ft. The center fielder stands 300 ft from second base in straight center field. How far is he from third base? $a^2 = (90)^2 + (300)^2 - (2)(90)(300)(\cos 135°) = 369$ ft

Additional Answers

29. The law of cosines states that for any triangle ABC, $c^2 = a^2 + b^2 - 2ab \cos C$. But if ABC is a right triangle, choosing C as the right angle would give $c^2 = a^2 + b^2 - 2ab \cos 90°$; $c^2 = a^2 + b^2 - 2ab\,(0)$; $c^2 = a^2 + b^2$.

31. By law of cosines: $q^2 = a^2 + b^2 - 2ab \cos \angle URS$ and $p^2 = a^2 + b^2 - 2ab \cos \angle RST$. Since consecutive angles of a parallelogram are supplementary (i.e. $180° - \angle URS = \angle RST$), $\cos \angle URS = -\cos \angle RST$. Therefore, $q^2 + p^2 = 2a^2 + 2b^2 - 2ab\,(-\cos$ $2ab \cos \angle URS$, $q^2 + p^2 = 2a^2 + 2b^2 + 2ab \cos \angle URS - 2ab \cos \angle URS$ $q^2 + p^2 = 2a^2 + 2b^2$.

Teacher's Resource Book
Practice—Chapter 4, p. 5
Enrichment—Chapter 4, p. 6

36. The east end of Silver Pond is 5.6 mi from the intersection of Hennick Road and Nutt Road, and the west end is 3.2 mi from the intersection. If the two roads are straight and intersect at an angle of 74°, what is the distance from the west end to the east end of the pond, to the nearest tenth of a mile? 6 mi

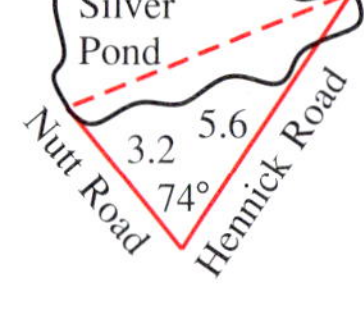

37. A parcel of land has sides measuring 175 ft, 234 ft, 295 ft, and 415 ft, and the angle between the sides of lengths 234 ft and 295 ft has measure 137.1°. What is the measure of the angle opposite this angle, to the nearest tenth of a degree? 106.0°

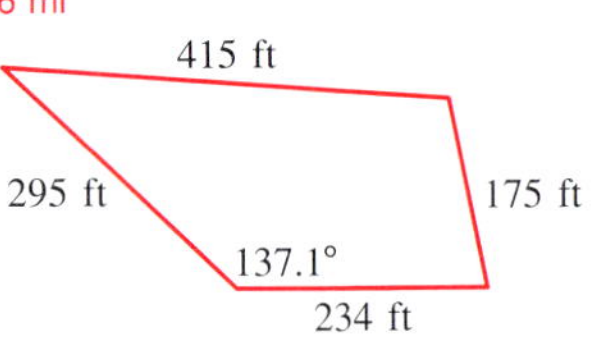

Navigation

38. Two roads intersect at an angle of 102.1°. Your friend's mailbox is 476 ft from the intersection. Your mailbox is on the other road and is 615 ft from the intersection. How far is it from your mailbox to your friend's, to the nearest foot? 853 ft

39. A plane leaves the airport and travels due east at 540 mi/h. Another plane leaves the airport $\frac{1}{4}$ h later and travels 20° east of north at the rate of 575 mi/h. To the nearest ten miles, how far apart are they $\frac{1}{2}$ h after the second plane leaves? 410 mi

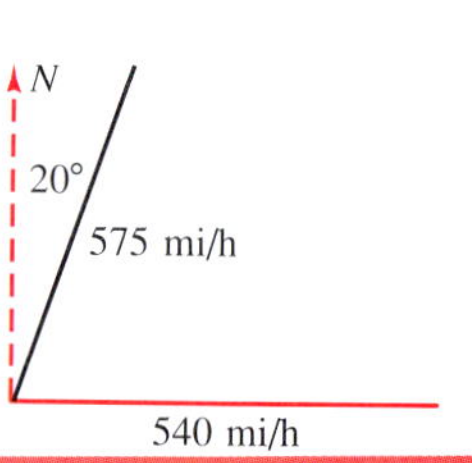

TRIGONOMETRY IN THREE DIMENSIONS

The law of cosines was proved on page 169 for two dimensions. Given the following information and figure, prove a three-dimensional version of the law of cosines.

Given: $PQRS$ is a right tetrahedron such that PS is perpendicular to plane PQR, $PS = h$, the area of $\triangle PQR = C$, and the measure of $\angle RPQ = \gamma$. Let the areas of the lateral faces $\triangle PSQ$, $\triangle PSR$, and $\triangle RSQ$ be A, B, and D, respectively.

Prove: $D^2 = A^2 + B^2 + C^2 - 2AB \cos \gamma$
See page 482.

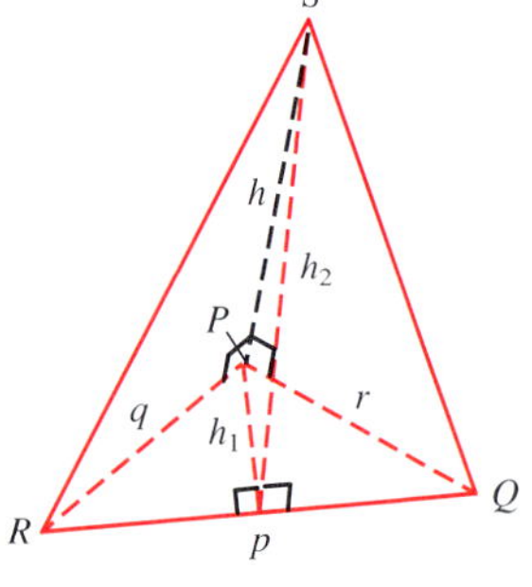

4.4 The Law of Tangents

Objectives: To introduce the law of tangents
To use the law of tangents to solve triangles when the measures of two sides and the included angle are known

If you are given two sides of a triangle and the included angle, you can use the law of cosines to find the third side and then use it again to find the other angles. In this lesson, you will learn another way to compute the angles.

Preview

To solve a system of two linear equations in two variables, add multiples of one equation to another to obtain one equation in one variable. Solve this equation, substitute the known value into one of the original equations, and then solve to find the second value.

EXAMPLE **Solve the following system for x and y:**

$$\begin{cases} 2x + y = 17 \\ x - y = 4 \end{cases}$$

$3x = 21$ *Add the two equations.*
$x = 7$

$7 - y = 4$
$-y = 4 - 7$
$y = 3$ *Substitute 7 for x in one of the equations.*

Solve each system of equations.

1. $\begin{cases} x - y = 10 \\ x + y = 4 \end{cases}$ $x = 7, y = -3$
2. $\begin{cases} 2x - y = 3 \\ x - y = 7 \end{cases}$ $x = -4, y = -11$
3. $\begin{cases} x - 2y = 3 \\ x + y = 15 \end{cases}$ $x = 11, y = 4$
4. $\begin{cases} 2x + 3y = 16 \\ 3x + 4y = 7 \end{cases}$ $x = -43, y = 34$

Another law that can be used to solve triangles is the *law of tangents*.

Law of Tangents

For any $\triangle ABC$ in which a, b, and c are the lengths of the sides opposite angles with measures A, B, and C, respectively,

$$\frac{a-b}{a+b} = \frac{\tan\frac{1}{2}(A-B)}{\tan\frac{1}{2}(A+B)} \qquad \frac{c-a}{c+a} = \frac{\tan\frac{1}{2}(C-A)}{\tan\frac{1}{2}(C+A)} \qquad \frac{b-c}{b+c} = \frac{\tan\frac{1}{2}(B-C)}{\tan\frac{1}{2}(B+C)}$$

LESSON PLAN

Vocabulary
Law of tangents

Materials/Manipulatives
Scientific calculators

BACKGROUND

In the Preview, solving systems of equations is reviewed. This skill is employed when using the law of tangents to solve a triangle.

TEACHING SUGGESTIONS

- Emphasize when the law of tangents may be used:
  a. When given two sides and the included angle.
  b. When given two angles and the sum of sides opposite them.
- Encourage students to use their scientific calculators to do any necessary calculations.

CHALKBOARD EXAMPLES

- **For Example 1**

  Find, to the nearest degree, the measures of the other two angles of each triangle. Then find the measure of the other side to two significant digits.

  1. $a = 14$, $b = 7$, $\angle C = 42°$
  $\angle A = 110°$, $\angle B = 28°$, $c = 10$

  2. $c = 21$, $a = 15$, $\angle B = 44°$
  $\angle C = 90°$, $\angle A = 46°$, $b = 15$

The law of tangents is used to determine the other angles when two sides and an included angle are known. It generates a system of two equations in two variables, which are solved to give the missing angles. The derivation of the law involves certain identities that are developed in a later chapter. Therefore, the law is presented here without proof.

EXAMPLE 1 **Find, to the nearest degree, the measures of the other two angles of a triangle in which $a = 13$, $b = 10$, and $\angle C = 58°$. Then find the measure of side c to two significant digits.**

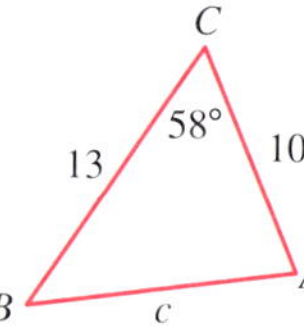

Since $\angle C = 58°$, $A + B = 180° - 58° = 122°$.

$$\frac{13 - 10}{13 + 10} = \frac{\tan \frac{1}{2}(A - B)}{\tan \frac{1}{2}(122°)}$$ *Use the law of tangents.*

$$\frac{3}{23} = \frac{\tan \frac{1}{2}(A - B)}{\tan 61°}$$

$$\tan \frac{1}{2}(A - B) = \frac{3 \tan 61°}{23}$$ *Calculation-ready form*

$$= 0.2353$$

$$\frac{1}{2}(A - B) = 13.24°$$

$$A - B = 26.48°$$

Thus, $A + B = 122°$ and $A - B = 26.48°$.

$$\begin{aligned} A - B &= 26.48° \\ A + B &= 122° \\ \hline 2A &= 148.48° \\ A &= 74.24° \\ B &= 122° - 74.24° = 47.76° \end{aligned}$$ *Add the two equations.*

$$\frac{\sin 74.24°}{13} = \frac{\sin 58°}{c}$$ *Use the law of sines.*

$$c = \frac{13 \sin 58°}{\sin 74.24°}$$

$$c = 11.46$$

Therefore, $\angle A = 74°$, $\angle B = 48°$, and $c = 11$.

If the measures of two angles and the sum of the lengths of the sides opposite them are known, the law of tangents can be used to determine the measures of the sides.

EXAMPLE 2 **In $\triangle ABC$, the sum of sides a and b is 142, $\angle A = 48°$, and $\angle B = 32°$. Find sides a and b. Then find $\angle C$ and side c. Express lengths to two significant digits and the angle measure to the nearest degree.**

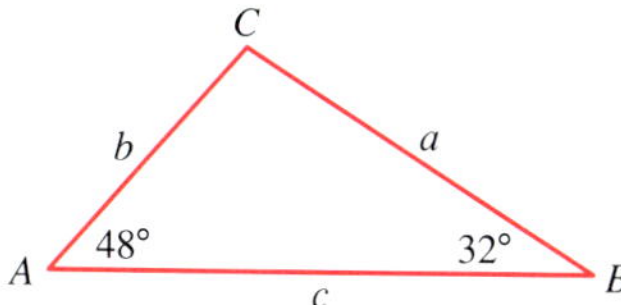

You know that $a + b = 142$.

$$\frac{a - b}{142} = \frac{\tan \frac{1}{2}(48° - 32°)}{\tan \frac{1}{2}(48° + 32°)}$$ *Use the law of tangents.*

$$\frac{a - b}{142} = \frac{\tan 8°}{\tan 40°}$$

$$a - b = \frac{142 \tan 8°}{\tan 40°}$$ *Calculation-ready form*

$$= 23.7836$$

Thus, $a + b = 142$ and $a - b = 23.7836$.

$$\begin{aligned} a - b &= 23.7836 \\ a + b &= 142 \\ \hline 2a &= 165.7836 \\ a &= 82.8918 \\ b &= 142 - 82.8918 = 59.1082 \end{aligned}$$ *Add the two equations.*

$$\angle C = 180° - (48° + 32°) = 100°$$

$$\frac{\sin 48°}{82.8918} = \frac{\sin 100°}{c}$$ *Use the law of sines.*

$$c = \frac{82.8918 \sin 100°}{\sin 48°}$$

$$c = 109.8473$$

Therefore, $\angle C = 100°$, $a = 83$, $b = 59$, and $c = 110$.

The law of tangents can be used to solve surveying problems if two sides and the included angle of a triangular parcel have been measured.

- **For Example 2**

Express lengths to two significant digits and angle measures to the nearest degree.

3. In triangle ABC, the sum of sides a and b is 112, $\angle A = 56°$, and $\angle B = 44°$. Find sides a and b. Then find $\angle C$ and side c.
   $a = 61$, $b = 51$, $\angle C = 80°$, $c = 72$
4. In triangle ABC, find b and c if the sum of sides b and c is 96, $\angle B = 85°$, and $\angle C = 67°$. Then find $\angle A$ and side a.
   $b = 50$, $c = 46$, $\angle A = 28°$, $a = 24$

- **For Example 3**
  5. In triangle ABC, $\angle B = 41°$, $c = 46$ m and $a = 32$ m. Find the measures of $\angle A$ and $\angle C$ to the nearest tenth of a degree. $\angle A = 43.9°$, $\angle C = 95.1°$
  6. In triangle ABC, $\angle A = 68.4°$, $b = 196$, and $c = 164$. Find the measures of $\angle B$ and $\angle C$ to the nearest tenth of a degree. $\angle B = 63.3°$, $\angle C = 48.3°$

Common Error

- Some students may reverse the signs in the formula. Emphasize that the proper order of signs is $A - B$ over $A + B$ and $a - b$ over $a + b$.
- See *Teacher's Resource Book* for additional remediation.

LESSON FOLLOW-UP

Discussion

Suppose triangle ABC has three sides of equal length. Explain whether or not it is necessary to use the law of tangents to solve the triangle. No, each angle of an equilateral triangle is 60°.

Critical Thinking

Analysis Ask students whether they would prefer to use the law of cosines or the law of tangents to solve a triangle for which two sides and the included angle are given and to give a reason for their answer. Answers may vary.

Assignment Guide

See p. 154B for assignments.

Additional Answers

1. 0.3939; 1.2131
2. −0.0787; 0.7954
3. 0.9657; 1.4826
4. 0.8693; 1.4826
5. 0.1944; 4.3315
6. 0.3057; 1.7321

EXAMPLE 3 **A triangular lot has two sides of length 248 ft and 379 ft, and the angle between them is 73.5°. The third side contains thick brush so that a line of sight along it cannot be established. What are the measures of the other two angles of the parcel, to the nearest tenth of a degree?**

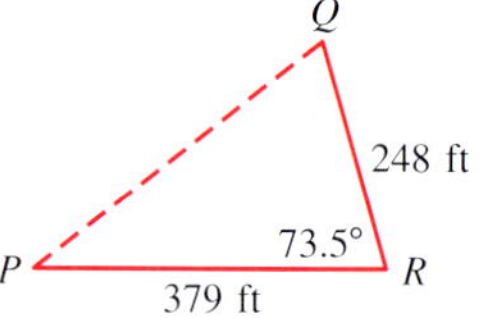

$Q + P = 180° - 73.5° = 106.5°$

$$\frac{379 - 248}{379 + 248} = \frac{\tan \frac{1}{2}(Q - P)}{\tan \frac{1}{2}(106.5°)}$$ *Use the law of tangents.*

$$\frac{131}{627} = \frac{\tan \frac{1}{2}(Q - P)}{\tan 53.25°}$$

$$\tan \frac{1}{2}(Q - P) = \frac{131 \tan 53.25°}{627}$$ *Calculation-ready form*

$$= 0.2798$$

$$\frac{1}{2}(Q - P) = 15.63°$$
$$Q - P = 31.26°$$

Thus, $Q + P = 106.5°$ and $Q - P = 31.26°$.

$$\begin{aligned} Q - P &= 31.26° \\ Q + P &= 106.5° \\ \hline 2Q &= 137.76° \\ Q &= 68.88° \\ P &= 106.5° - 68.88° = 37.62° \end{aligned}$$ *Add the two equations.*

Thus, $\angle P = 37.6°$ and $\angle Q = 68.9°$.

CLASS EXERCISES See side column.

1. If $\angle A = 72°$ and $\angle B = 29°$, find $\tan \frac{1}{2}(A - B)$ and $\tan \frac{1}{2}(A + B)$.
2. If $\angle A = 34°$ and $\angle B = 43°$, find $\tan \frac{1}{2}(A - B)$ and $\tan \frac{1}{2}(A + B)$.
3. If $\angle A = 100°$ and $\angle C = 12°$, find $\tan \frac{1}{2}(A - C)$ and $\tan \frac{1}{2}(A + C)$.
4. If $\angle A = 97°$ and $\angle C = 15°$, find $\tan \frac{1}{2}(A - C)$ and $\tan \frac{1}{2}(A + C)$.
5. If $\angle B = 88°$ and $\angle C = 66°$, find $\tan \frac{1}{2}(B - C)$ and $\tan \frac{1}{2}(B + C)$.
6. If $\angle B = 77°$ and $\angle C = 43°$, find $\tan \frac{1}{2}(B - C)$ and $\tan \frac{1}{2}(B + C)$.

PRACTICE EXERCISES

Use the law of tangents to find the measures of the unknown angles of $\triangle ABC$. Then use the law of sines to find the length of the unknown side. Express angle measures to the nearest degree and lengths to two significant digits.

A

1. $\angle C = 47°, a = 46, b = 19$ $\angle A = 110°, \angle B = 23°, c = 36$
2. $\angle C = 51°, a = 71, b = 27$ $\angle A = 108°, \angle B = 21°, c = 58$
3. $\angle C = 104°, a = 27, b = 15$ $\angle A = 51°, \angle B = 25°, c = 34$
4. $\angle C = 97°, a = 36, b = 19$ $\angle A = 57°, \angle B = 26°, c = 43$
5. $\angle A = 43°, b = 61, c = 47$ $\angle B = 87°, \angle C = 50°, a = 42$
6. $\angle A = 51°, b = 93, c = 87$ $\angle B = 68°, \angle C = 61°, a = 78$
7. $\angle A = 123°, b = 15, c = 13$ $\angle B = 31°, \angle C = 26°, a = 25$
8. $\angle A = 114°, b = 28, c = 25$ $\angle B = 35°, \angle C = 31°, a = 44$
9. $\angle B = 26°, a = 45, c = 34$ $\angle A = 108°, \angle C = 46°, b = 21$
10. $\angle B = 56°, a = 52, c = 32$ $\angle A = 86°, \angle C = 38°, b = 43$
11. $\angle B = 111°, a = 87, c = 54$ $\angle A = 44°, \angle C = 25°, b = 120$
12. $\angle B = 154°, a = 96, c = 82$ $\angle A = 14°, \angle C = 12°, b = 170$

Solve $\triangle ABC$. In Exercises 13–16, express angle measures to the nearest degree and lengths to two significant digits. In Exercises 17–20, express angle measures to the nearest tenth of a degree and lengths to three significant digits.

B

13. $\angle A = 47°, \angle B = 29°, a + b = 49$ $a = 29, b = 20, c = 39, \angle C = 104°$
14. $\angle A = 37°, \angle B = 62°, a + b = 76$ $a = 31, b = 45, c = 51, \angle C = 81°$
15. $\angle A = 94°, \angle C = 22°, a + b = 114$ $a = 60, b = 54, c = 23, \angle B = 64°$
16. $\angle A = 42°, \angle C = 101°, a + b = 213$ $a = 110, b = 100, c = 160, \angle B = 37°$
17. $\angle A = 43.6°, \angle B = 21.7°, a + b = 92.3$ $a = 60.1, b = 32.2, c = 79.2, \angle C = 114.7°$
18. $\angle A = 92.7°, \angle B = 11.6°, a + b = 27.4$ $a = 22.8, b = 4.59, c = 22.1, \angle C = 75.7°$
19. $\angle B = 36.7°, \angle C = 48.9°, b + c = 104.3$ $a = 77.0, b = 46.1, c = 58.2, \angle A = 94.4°$
20. $\angle B = 103.4°, \angle C = 21.8°, b + c = 67.5$ $a = 41.0, b = 48.9, c = 18.6, \angle A = 54.8°$

C

21. Use the law of tangents to show that in $\triangle ABC$, if $A > B$, then $a > b$. See side column.
22. Use the law of tangents to show that if two angles of a triangle are congruent, then the triangle is isosceles. See side column.

Applications

23. **Surveying** A surveyor wishes to calculate the distance between two signs. The distance from the surveyor to the first sign is 70 m, and the distance to the second sign is 100 m. The angle between the two lines of sight measures 56°. Find the distance between the two signs, to the nearest meter. 84 m

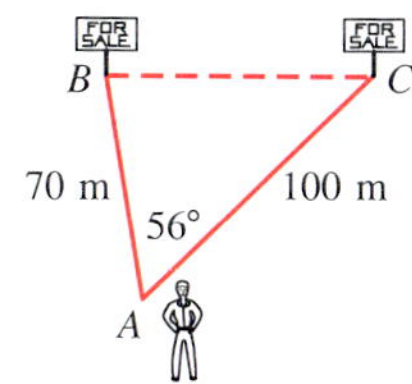

Test Yourself

See *Teacher's Resource Book, Tests*, pp. 33–34.

Lesson Quiz

1. In triangle ABC, $b = 5$, $a = 12$, and $\angle C = 78°$. Find $\angle A$ and $\angle B$. 78°; 24°
2. In triangle ABC, $a + b = 12$, $\angle A = 50°$, and $\angle B = 33°$. Find a and b. 7; 5

Additional Answers

21. $\dfrac{a-b}{a+b} = \dfrac{\tan\frac{1}{2}(A-B)}{\tan\frac{1}{2}(A+B)}$.

If $A > B$, then $A - B > 0$, $\frac{1}{2}(A - B) > 0$, so $\tan\frac{1}{2}(A - B) > 0$. Also $A + B > 0$, so $\frac{1}{2}(A + B) > 0$. Therefore, $\tan\frac{1}{2}(A + B) > 0$. Since both numerator and denominator are positive, $\dfrac{\tan\frac{1}{2}(A-B)}{\tan\frac{1}{2}(A+B)} > 0$. Thus, $\dfrac{a-b}{a+b} > 0$. $(a + b)\dfrac{(a-b)}{a+b} > 0\,(a + b)$, $a - b > 0$, $a > b$.

22. Suppose $\angle A \cong \angle B$, then $A - B = 0$ and $\frac{1}{2}(A - B) = 0$. Therefore $\dfrac{a-b}{a+b} = \dfrac{\tan 0}{\tan\frac{1}{2}(A+B)} = 0$. $\dfrac{a-b}{a+b} = 0$, $a - b = 0$, so $a = b$ and $\triangle ABC$ is isosceles.

Teacher's Resource Book
Practice—Chapter 4, p. 7
Enrichment—Chapter 4, p. 8

24. Geometry The sides of a parallelogram are 12 ft and 8 ft, and each of the larger angles measures 125°. Find the length of the shorter diagonal of the parallelogram, to the nearest foot. 10 ft

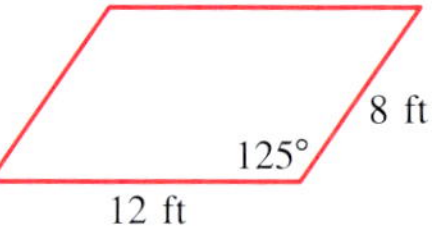

TEST YOURSELF

In all exercises, give angle measures to the nearest degree and lengths to two significant digits.

Solve each triangle.

4.1

1. $\angle A = 33°$, $\angle B = 56°$, $a = 15$
   $b = 23$, $c = 28$, $\angle C = 91°$
2. $\angle A = 62°$, $\angle B = 48°$, $b = 21$
   $a = 25$, $c = 27$, $\angle C = 70°$
3. $\angle A = 43°$, $\angle C = 50°$, $c = 12$
   $a = 11$, $b = 16$, $\angle B = 87°$
4. $\angle B = 65°$, $\angle C = 50°$, $c = 14$
   $a = 17$, $b = 17$, $\angle A = 65°$

4.2

5. How many triangles are there with angle A equal to 95°, adjacent side 23, and opposite side 31? If there are any, solve them. 1; $\angle B = 48°$, $\angle C = 37°$, $c = 19$
6. How many triangles are there with angle A equal to 47°, adjacent side 14, and opposite side 12? If there are any, solve them. 2; $\angle B = 59°$, $\angle C = 74°$, $c = 16$ or $\angle B = 121°$, $\angle C = 12°$, $c = 3.4$

Solve each triangle.

4.3

7. $\angle C = 84°$, $a = 42$, $b = 50$
   $\angle A = 42°$, $\angle B = 54°$, $c = 62$
8. $\angle A = 78°$, $b = 75$, $c = 48$
   $\angle B = 66°$, $\angle C = 36°$, $a = 80$
9. $a = 32$, $b = 25$, $c = 17$
   $\angle A = 97°$, $\angle B = 51°$, $\angle C = 32°$
10. a = 19, $b = 40$, $c = 27$
    $\angle A = 24°$, $\angle B = 120°$, $\angle C = 36°$
11. How many triangles are there with an angle of 51°, adjacent side 27, and opposite side 11? If there are any, solve them. none
12. If a triangular plot of land has sides of 476 m, 512 m, and 544 m, what are the measures of the three angles? 53°, 60°, 67°

Use the law of tangents to solve $\triangle ABC$.

4.4

13. $\angle C = 120°$, $a = 20$, $b = 35$
    $c = 48$, $\angle A = 21°$, $\angle B = 39°$
14. $\angle B = 118°$, $a = 10$, $c = 17$
    $b = 23$, $\angle A = 22°$, $\angle C = 40°$

Solve $\triangle ABC$.

15. $\angle A = 11°$, $\angle B = 118°$, $a + c = 39$
    $a = 7.7$, $b = 36$, $c = 31$, $\angle C = 51°$
16. $\angle A = 95°$, $\angle C = 21°$, $a + b = 110$
    $a = 58$, $b = 52$, $c = 21$, $\angle B = 64°$
17. The lengths of two sides of a parallelogram are 8 cm and 6 cm, and each of the smaller angles measures 78°. Find the length of the longer diagonal of the parallelogram, to the nearest centimeter.
    11 cm

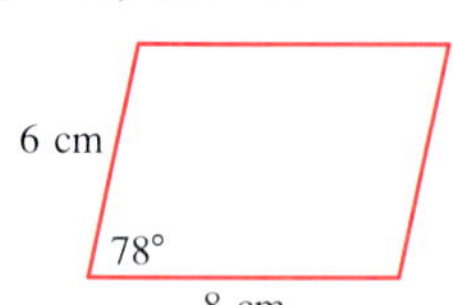

4.5 The Area of a Triangle

Objectives: To find the area of a triangle when the measures of two sides and the included angle are known
To use the law of sines to find the area of a triangle when the measures of one side and two angles are known

You have learned how to find the missing parts of a triangle when you know three sides, two sides and the included angle, or two angles and any side. Since a triangle is completely determined in these cases, its area is also determined.

Preview

The area K of a triangle is given by the formula $K = \frac{1}{2}bh$, where b is the length of one side and h is the length of the altitude to that side.

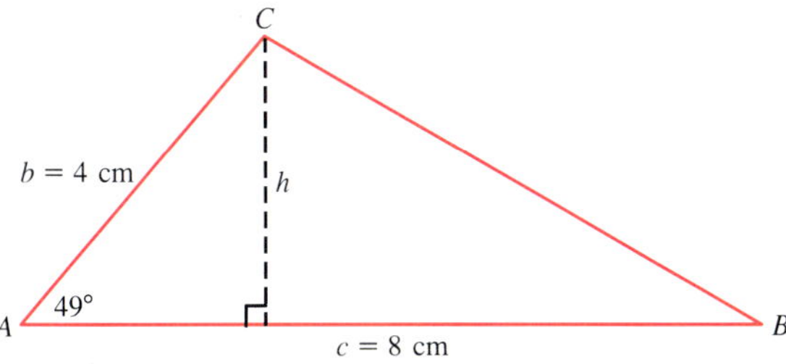

1. Refer to $\triangle ABC$ above. Use a centimeter ruler to measure the length of the altitude to side AB. Then compute the area of the triangle. $h = 3$ cm; $K = 12$ cm^2
2. Use trigonometry to find the length of the altitude from C to side AB. Then compute the area again. $h = 4 \sin 49° = 3$; $K = 12$ cm^2

For each triangle, use the methods given in Exercises 1 and 2 to find *h* and the area *K*.

3. 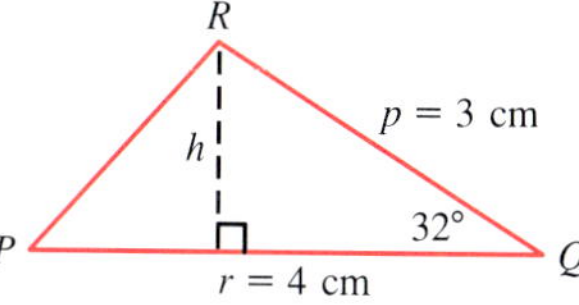

$h = 1.6$ cm; $K = 3.2$ cm^2

4. 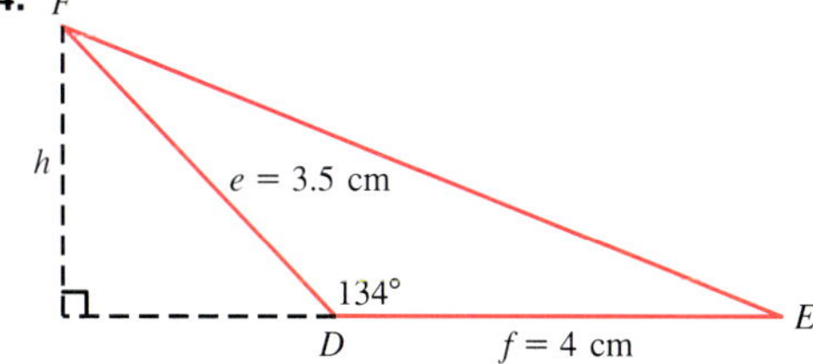

$h = 2.5$ cm; $K = 5$ cm^2

LESSON PLAN

Materials/Manipulatives
Scientific calculators

BACKGROUND

In the Preview, the formula $K = \frac{1}{2}bh$ for finding the area of a triangle is reviewed. This formula will be used in the lesson to develop trigonometric formulas for area.

TEACHING SUGGESTIONS

- Restate the formula $K = \frac{1}{2}ab \sin C$ as "the area of a triangle is one-half the product of the lengths of two sides and the sine of the included angle." Emphasize that the formula applies to any triangle: acute, right, or obtuse.
- Encourage students to use their scientific calculators to do any necessary calculations.

CHALKBOARD EXAMPLES

- **For Example 1**
  1. In $\triangle ABC$, $\angle C = 49°$, $a = 16$ cm, and $b = 25$ cm. Find the area of the triangle to the nearest ten square centimeters. 150 cm^2
  2. In $\triangle ABC$, $\angle B = 81°$, $a = 6$, and $c = 11$ in. Find the area of the triangle to the nearest ten square inches. 30 $in.^2$

The formula $K = \frac{1}{2}bh$ presented in the Preview can be transformed into other area formulas. Assume that sides a and b are given, and that h is the altitude to side AC. Consider the case in which $\angle A$ is acute.

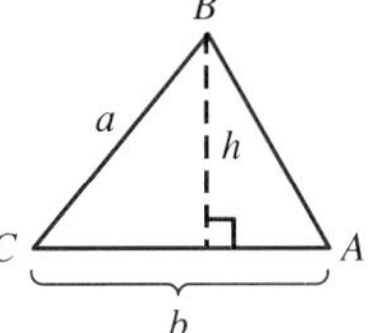

Then, $K = \frac{1}{2}bh$

But, $h = a \sin C$ $\sin C = \frac{h}{a}$

Therefore, $K = \frac{1}{2}ab \sin C$ *Substitute.*

You can write two other expressions for area involving the sine of $\angle A$ or $\angle B$. The results would be the same if an angle of the triangle is obtuse. The proof of the case in which one angle is obtuse is left as an exercise.

> The area K of any triangle ABC is given by any one of these formulas:
>
> $K = \frac{1}{2}bc \sin A \qquad K = \frac{1}{2}ac \sin B \qquad K = \frac{1}{2}ab \sin C$

EXAMPLE 1 **In $\triangle ABC$, $\angle C = 57°$, $a = 31$ cm, and $b = 42$ cm. Find the area of the triangle to the nearest ten square centimeters.**

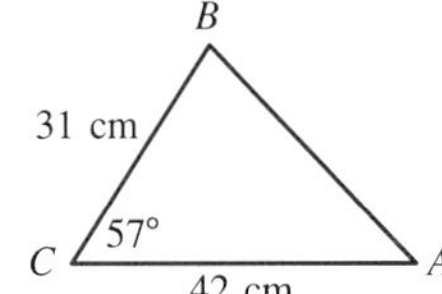

$K = \frac{1}{2}ab \sin C$

$= \frac{1}{2}(31)(42) \sin 57°$ *Calculation-ready form*

$= 546$

The area is approximately 550 cm^2.

The preceding formula can be used in conjunction with the law of sines to derive another area formula that can be used when at least two angles and one side are known.

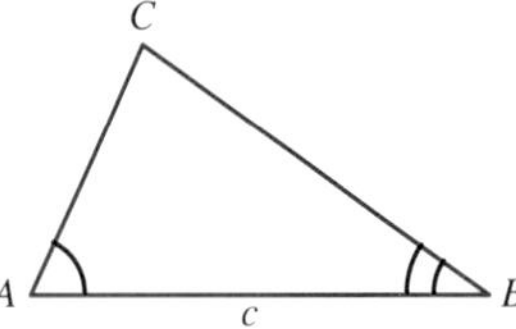

Assume that $\angle A$, $\angle B$, and c are known. Then $\angle C = 180° - (\angle A + \angle B)$.

$$\frac{b}{\sin B} = \frac{c}{\sin C} \qquad \textit{Use the law of sines to find b.}$$

$$b = \frac{c \sin B}{\sin C}$$

$$K = \left(\frac{1}{2}\right)\left(\frac{c \sin B}{\sin C}\right)(c)(\sin A) \qquad \textit{Substitute in the area formula } K = \frac{1}{2}bc \sin A.$$

$$K = \frac{c^2 \sin A \sin B}{2 \sin C}$$

It can also be shown that $K = \dfrac{a^2 \sin B \sin C}{2 \sin A}$ and $K = \dfrac{b^2 \sin A \sin C}{2 \sin B}$ are area formulas for $\triangle ABC$.

EXAMPLE 2 **Given $\angle A = 61°$, $\angle B = 78°$, and $c = 37$ in., compute the area of the triangle to the nearest ten square inches.**

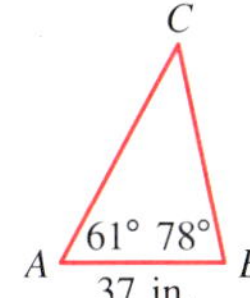

$\angle C = 180° - (61° + 78°) = 41°$

$$K = \frac{c^2 \sin A \sin B}{2 \sin C}$$

$$K = \frac{(37)^2 (\sin 61°)(\sin 78°)}{2 \sin 41°} \qquad \textit{Calculation-ready form}$$

$K = 893 \text{ in.}^2$ The area is approximately 890 in.^{2}

CLASS EXERCISES

Substitute values into the appropriate area formula, but do not solve.

1.

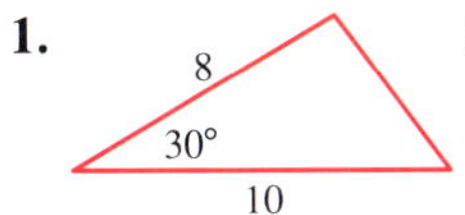

$K = \frac{1}{2}(8)(10) \sin 30°$

2.

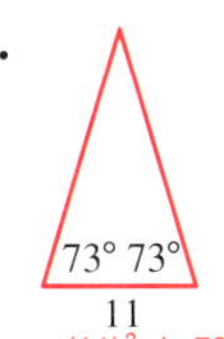

$K = \dfrac{(11)^2 \sin 73° \sin 73°}{2 \sin 34°}$

3.

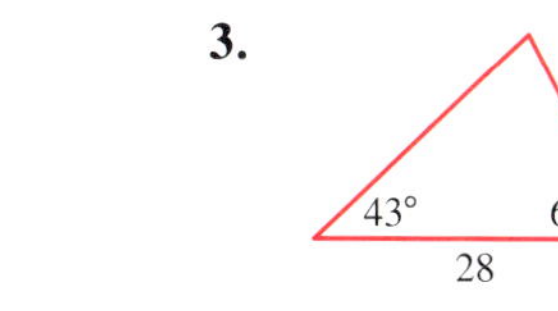

$K = \dfrac{(28)^2 \sin 43° \sin 62°}{2 \sin 75°}$

4.

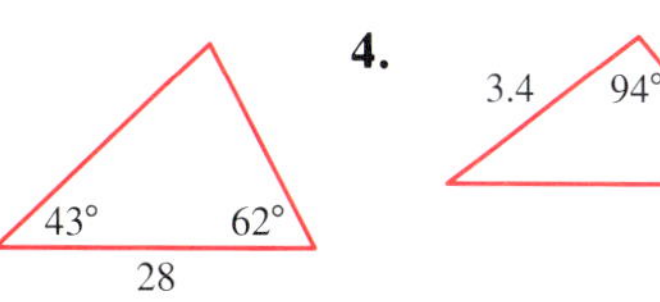

$K = \frac{1}{2}(3.4)(2.7) \sin 94°$

Find the area of the given triangle to the nearest ten square units.

5.

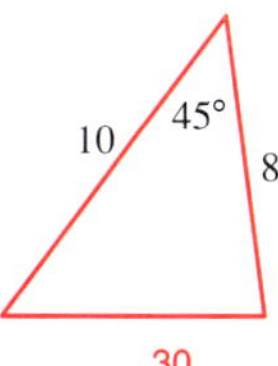

30

6.

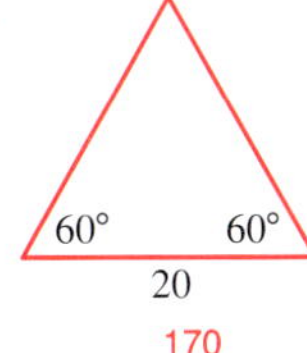

170

7.

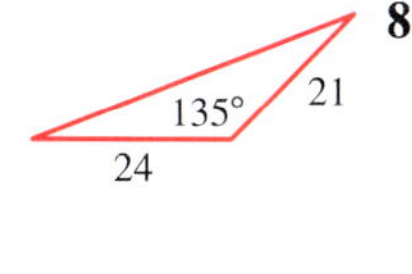

180

8. 30° 120° 28

340

- **For Example 2**
  3. Given $\angle A = 47°$, $\angle B = 69°$, and $c = 19$ cm, compute the area of the triangle to the nearest ten square centimeters. 140 cm^2
  4. Given $\angle B = 27°$, $\angle C = 64°$, and $a = 30$ cm, compute the area of the triangle to the nearest ten square centimeters. 180 cm^2

Common Error

- When finding the area of a triangle, some students may become confused when given two angles and the included side. An alternate approach would be to have those students use the law of sines to find a second side length and then use the formula $K = \frac{1}{2}ab \sin C$.
- See *Teacher's Resource Book* for additional remediation.

LESSON FOLLOW-UP

Discussion

Show how to find the exact area of a parallelogram with side lengths 18 and 22 and an angle of 45°.

$K = (18)(22) \sin 45°$

$= (18)(22)\left(\frac{\sqrt{2}}{2}\right)$

$= 198\sqrt{2} \text{ units}^2$

Critical Thinking

Synthesis Ask students to solve several problems like the one in the Enrichment section, each with a different figure and radius, and then generalize the results.

n = number of sides

r = radius of circle

$A = (n)\left(\frac{1}{2}\right)(r)(r) \sin \frac{360°}{n}$

$A = \frac{nr^2}{2} \sin\left(\frac{360°}{n}\right)$

Assignment Guide

See p. 154B for assignments.

Career: CAD (Computer-Aided Design) Technician

The occupation of a computer-aided design technician is described. Education requirements, personal qualifications, future outlook, and advancement within the field are discussed.

Lesson Quiz

Find the area of the given triangle to the nearest ten square units.

1. $h = 12$ cm, $b = 6$ cm 40 cm^2
2. $\angle A = 37°$, $b = 12$ cm, $c = 17$ 60 cm^2
3. The area of a triangular plot is 1,393,817 ft^2. Find the acreage of this plot (1 acre = 43,560 ft^2). 32 acres

Enrichment

Find the area of a regular hexagon inscribed in a circle of radius 15.

$K = \frac{1}{2}(15)(15)\sin 60°$

$K = \frac{1}{2}(15)(15)\left(\frac{\sqrt{3}}{2}\right)$

$K = \frac{225(\sqrt{3})}{4}$ (area of 1 triangle)

area of hexagon $= 6\left(\frac{225(\sqrt{3})}{4}\right)$

$= \frac{675\sqrt{3}}{2}$ units^2

PRACTICE EXERCISES

Find the area of the given triangle to the nearest ten square units.

A

1. 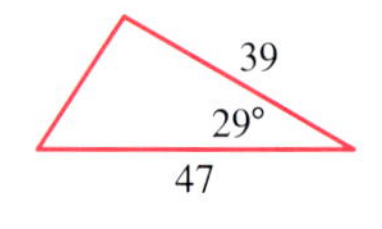

440

2. 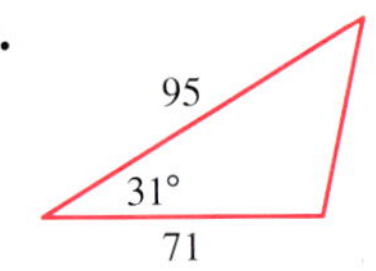

1740

3. 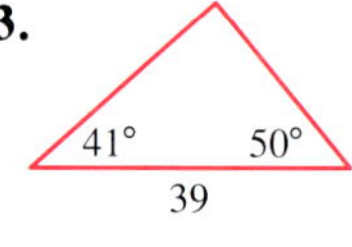

380

4. 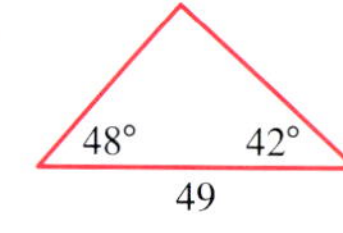

600

5. $\angle A = 92°$, $\angle B = 61°$, $b = 15$ 60

6. $\angle B = 61°$, $\angle C = 93°$, $b = 25$ 160

7. $\angle A = 110°$, $\angle B = 28°$, $c = 47$ 730

8. $\angle B = 27°$, $\angle C = 108°$, $a = 53$ 860

9. $\angle A = 34°$, $b = 11$, $c = 11$ 30

10. $\angle C = 35°$, $a = 17$, $c = 17$ 140

11. $\angle A = 71°$, $b = 21$, $c = 87$ 860

12. $\angle A = 73°$, $b = 161$, $c = 41$ 3160

13. $\angle B = 90°$, $a = 23$, $c = 47$ 540

14. $\angle A = 90°$, $b = 28$, $c = 15$ 210

15. $\angle B = 60°$, $\angle C = 60°$, $a = 10$ 40

16. $\angle B = 60°$, $\angle A = 60°$, $c = 100$ 4330

Find the area of the given triangle to the nearest square unit.

B

17. $\angle A = 87.3°$, $b = 11.3$, $c = 19.2$ 108

18. $\angle B = 86.4°$, $b = 37.7$, $c = 21.4$ 346

19. $\angle A = 41.3°$, $\angle B = 112.5°$, $c = 46.7$ 1506

20. $\angle B = 113.9°$, $\angle C = 43.9°$, $a = 57.9$ 2812

21. $\angle R = 62.4°$, $q = 23.6$, $p = 19.2$ 201

22. $\angle R = 57.3°$, $q = 17.6$, $p = 19.8$ 147

23. $\angle Q = 83.9°$, $\angle R = 46.7°$, $p = 46.7$ 1039

24. $\angle Q = 67.3°$, $\angle R = 51.2°$, $p = 31.3$ 401

25. $\angle P = 108.4°$, $\angle Q = 27.3°$, $r = 19.7$ 121

26. $\angle P = 97.8°$, $\angle Q = 34.9°$, $r = 17.3$ 115

27. $\angle P = 67.3°$, $\angle Q = 76.4°$, $p = 19.6$ 120

28. $\angle Q = 53.7°$, $\angle R = 61.9°$, $q = 23.1$ 263

C

29. Prove that $K = \frac{1}{2}ab \sin C$ is an area formula for $\triangle ABC$ when $\angle A$ is an obtuse angle (that is, when the altitude from B does not intersect segment AC). See side column on page 185.

30. Use the results of this lesson to show that the area of a parallelogram is given by $K = ab \sin \theta$, where a and b are the lengths of the adjacent sides and θ is the measure of the included angle. See side column on page 185.

31. Compute the area of trapezoid $ABCD$, if it has base angles of 70° and 45° and bases of 110 cm and 80 cm. Round your answer to the nearest hundred square centimeters. 2100 cm²

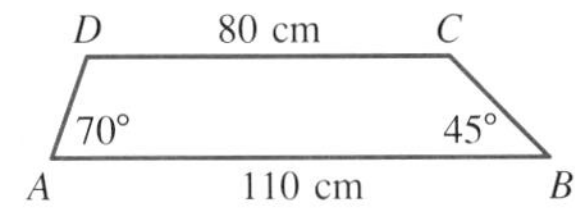

Applications

Surveying

32. To the nearest square foot, what is the area of a triangular parcel of land, if two adjacent sides measure 184 ft and 246 ft and the angle between them measures 76.0°? 21,960 ft²
33. To the nearest square foot, what is the area of a triangular pond, if two adjacent sides measure 112 ft and 96 ft and the angle between them measures 65.4°? 4888 ft²
34. What is the area, to the nearest 0.1 acre, of a triangular field that is 476 ft on one side and 723 ft on another, if the angle between these sides measures 71.3°? (1 acre = 43,560 ft²) 3.7 acres
35. What is the area, to the nearest 0.1 acre, of a triangular field that is 529 ft on one side and 849 ft on another, if the angle between these sides measures 102.7°? 5.0 acres

CAREER

CAD (Computer-Aided Design) Technician

CAD technicians create layouts, designs, and line drawings for any step in the manufacturing process, including product development, testing, and modification. They also provide assistance to architects and engineers. They use light pens to move, rotate, or zoom in on a portion of a computer display, and they edit and modify computer drawings.

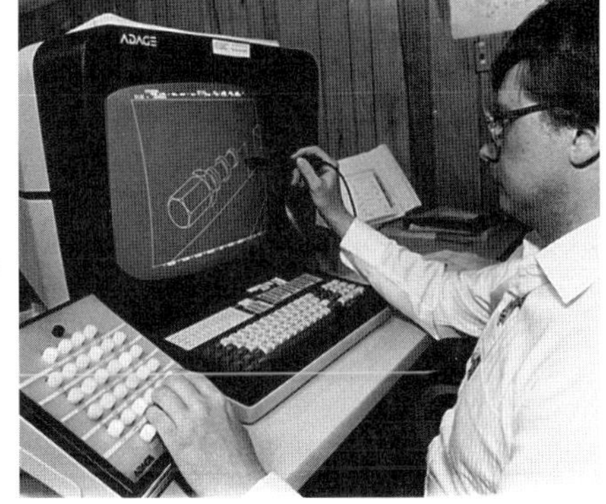

Requirements include skill in mechanical drawing and drafting, the ability to use a computer program, proficiency in mathematics, an eye for detail, and the ability to visualize spatial relationships.

CAD technicians need at least a high school education. Technical school or college study in technical graphics, architectural drafting, and product design is useful. Some companies provide on-the-job training. A bachelor's degree, experience, training, and serious effort can lead to excellent job prospects in this field and related fields like robotics.

Teacher's Resource Book
Practice—Chapter 4, p. 9
Enrichment—Chapter 4, p. 10

Additional Answers

29.

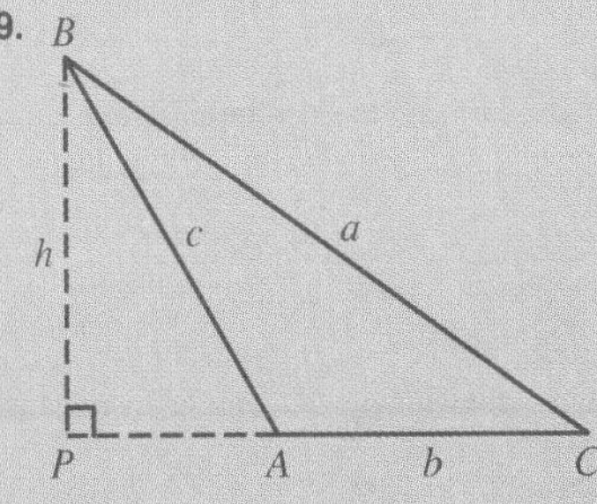

$K = \frac{1}{2}bh$

$h = a \sin C$, since $\sin C = \frac{h}{a}$.

Therefore, $K = \frac{1}{2}ab \sin C$.

30.

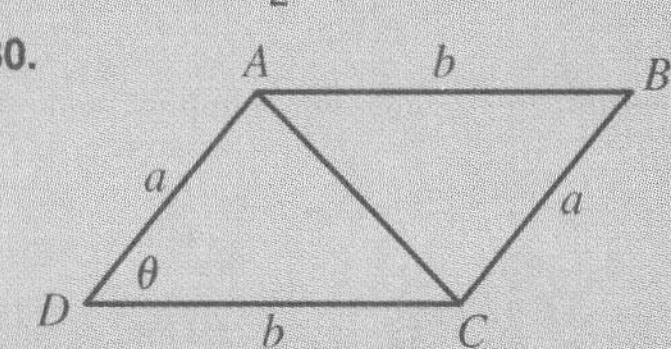

Let K = area of ▱$ABCD$

Then, $K = 2(\text{area } \triangle ADC)$

$= 2\left(\frac{1}{2}ab \sin \theta\right)$

$= ab \sin \theta$

LESSON PLAN

Vocabulary
Heron's formula

Materials/Manipulatives
Calculators

BACKGROUND

In the Preview, students are asked to apply formulas from previous lessons to a situation where three side lengths are given and the area of a triangle is to be found.

4.6 Heron's Formula

Objectives: To use Heron's formula to find the area of a triangle when the lengths of the three sides are known
To use Heron's formula to find the length of an altitude of a triangle when the lengths of the three sides are known

You know how to find the area of a triangle when the measure of at least one angle is known. It is also possible to find the area of a triangle when only the lengths of the three sides are known. Indeed, in real-life settings it is often easier to measure the sides of a triangle than it is to measure altitudes or angles.

Preview

A city block is triangular in shape. The frontage along High Street is 423 ft, that along Conover Street is 764 ft, and that along First Street is 524 ft. A developer wishes to find the area of the plot.

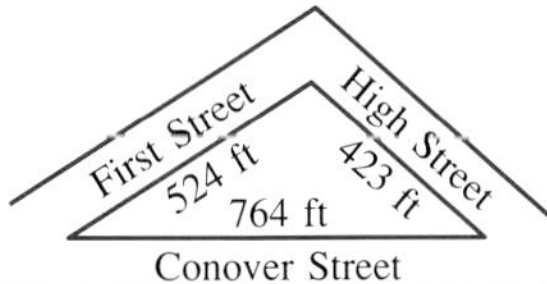

1. What additional information would he need to find the area using one of the formulas from the previous lesson? 1 angle
2. What law or laws could he use to find that information? law of cosines
3. Find, to three significant digits, the area of the triangle using a formula from the previous lesson. 106,000 ft^2

If you know the measures of three sides of a triangle, you can find the measure of an angle using the law of cosines. You can then apply a formula from the previous lesson to find the area of the triangle. However, *Heron's formula* often involves fewer calculations.

Heron's Formula

If a, b, and c are measures of the sides of a triangle, then the area K of the triangle is given by

$$K = \sqrt{s(s-a)(s-b)(s-c)}, \quad \text{where } s = \frac{a+b+c}{2}$$

The quantity s is called the *semiperimeter* of a triangle.

To prove Heron's formula, start with the area formula $K = \frac{1}{2}bc \sin A$.

$$K = \frac{1}{2}bc \sin A$$

$$K^2 = \frac{1}{4}b^2c^2 \sin^2 A \qquad \textit{Square both sides.}$$

$$K^2 = \frac{1}{4}b^2c^2(1 - \cos^2 A) \qquad \textit{Replace } \sin^2 A \textit{ with } 1 - \cos^2 A.$$

$$K^2 = \left[\frac{bc}{2}(1 + \cos A)\right]\left[\frac{bc}{2}(1 - \cos A)\right] \qquad \textit{Factor.}$$

Using the law of cosines, replace $\cos A$ with $\frac{b^2 + c^2 - a^2}{2bc}$.

$$K^2 = \left[\frac{bc}{2}\left(1 + \frac{b^2 + c^2 - a^2}{2bc}\right)\right]\left[\frac{bc}{2}\left(1 - \frac{b^2 + c^2 - a^2}{2bc}\right)\right]$$

$$K^2 = \left[\frac{2bc + b^2 + c^2 - a^2}{4}\right]\left[\frac{2bc - b^2 - c^2 + a^2}{4}\right] \qquad \textit{Multiply and simplify.}$$

$$K^2 = \left[\frac{(b^2 + 2bc + c^2) - a^2}{4}\right]\left[\frac{a^2 - (b^2 - 2bc + c^2)}{4}\right] \qquad \textit{Regroup terms.}$$

$$K^2 = \left[\frac{(b + c)^2 - a^2}{4}\right]\left[\frac{a^2 - (b - c)^2}{4}\right] \qquad b^2 + 2bc + c^2 = (b + c)^2,\ b^2 - 2bc + c^2 = (b - c)^2$$

$$K^2 = \left[\frac{b + c + a}{2}\right]\left[\frac{b + c - a}{2}\right]\left[\frac{a - b + c}{2}\right]\left[\frac{a + b - c}{2}\right] \qquad \textit{Factor.}$$

Let $s = \frac{a + b + c}{2}$. The formula then simplifies to

$$K^2 = s(s - a)(s - b)(s - c)$$

Thus, the area of $\triangle ABC$ is given by

$$K = \sqrt{s(s - a)(s - b)(s - c)}, \text{ where } s = \frac{a + b + c}{2}$$

Before the development of calculators, Heron's formula required a great deal of time-consuming computation. Calculators make the use of the formula more practical.

EXAMPLE 1 **Find the area of a triangle with sides 23 cm, 26 cm, and 31 cm, to the nearest ten square centimeters.**

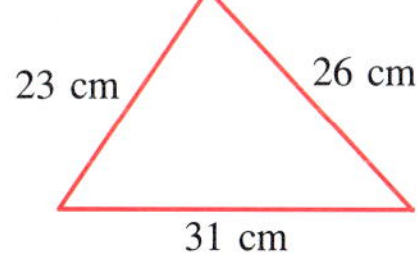

$s = \frac{23 + 26 + 31}{2} = 40$ *Find s, where* $s = \frac{a + b + c}{2}$.

$K = \sqrt{40(40 - 23)(40 - 26)(40 - 31)}$ *Use* $K = \sqrt{s(s - a)(s - b)(s - c)}$.

$K = \sqrt{40(17)(14)(9)}$

$K = 293$ The area is approximately 290 cm^2.

TEACHING SUGGESTIONS

- Emphasize that s is the semiperimeter or the sum of the sides divided by two.
- Students may use calculators to complete any necessary computations.

CHALKBOARD EXAMPLES

- **For Example 1**

  Find the area of the given triangle to the nearest ten square units.

  1. $a = 6$ cm, $b = 9$ cm, $c = 13$ cm 20 cm^2

  2. $a = 4$ ft, $b = 7$ ft, $c = 9$ ft 10 ft^2

- **For Example 2**

  Round answers to the nearest unit.

  3. Find the altitude from A to side a if $a = 16$ cm, $b = 18$ cm, and $c = 22$ cm. 18 cm

  4. Find the altitude from B to side b if $a = 6$ ft, $b = 9$ ft, $c = 11$ ft. 6 ft

Common Error

- Students may make computational errors. If students are using a calculator, consider checking answers twice.
- See *Teacher's Resource Book* for additional remediation.

LESSON FOLLOW-UP

Discussion

Use Heron's formula to develop a general formula for the area of an equilateral triangle.

Let z be the length of a side of the triangle.

$$s = \frac{3z}{2}$$

$$k = \sqrt{s(s-a)(s-b)(s-c)}$$

$$k = \sqrt{\frac{3z}{2}\left(\frac{3z}{2} - z\right)\left(\frac{3z}{2} - z\right)\left(\frac{3z}{2} - z\right)}$$

$$k = \sqrt{\frac{3z}{2}\left(\frac{z}{2}\right)\left(\frac{z}{2}\right)\left(\frac{z}{2}\right)} = \sqrt{\frac{3z^4}{16}} = \frac{z^2\sqrt{3}}{4}$$

Critical Thinking

Analysis Ask students to list formulas for finding the area of a triangle and the sets of information for which each could be used. Answers may vary.

Assignment Guide

See p. 154B for assignments.

Challenge

A challenge problem requiring students to identify right triangles in which the perimeter is n units and the area is n square units is considered. Encourage students to develop a computer program to help them find the correct solution(s).

Lesson Quiz

1. The perimeter of $\triangle ABC$ is 74 cm. If $a = 27$ cm and $c = 31$ cm, find b. 16 cm
2. For $\triangle XYZ$, calculate s, the semiperimeter.

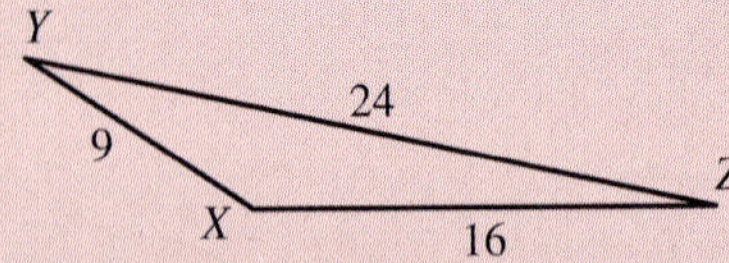

24.5 units

3. Compute the area of $\triangle PQR$ if $p = 6$, $q = 8$, and $r = 10$. 24 units^2
4. In $\triangle PQR$ of Exercise 3 above, find the length of the altitude from Q to side PR to the nearest centimeter. 6 cm

Heron's formula can also be used to find the length of the altitude of a triangle, if the measures of three sides are known.

EXAMPLE 2 **Find the length of the altitude from A to side BC, if $a = 24$ ft, $b = 14$ ft, and $c = 18$ ft. Round the answers to the nearest foot.**

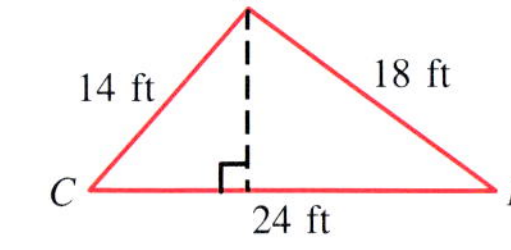

$$s = \frac{24 + 14 + 18}{2} = 28$$

$K = \sqrt{28(28-24)(28-14)(28-18)}$ *Use Heron's formula.*

$K = \sqrt{28(4)(14)(10)}$ *Calculation-ready form*

$K = 125.2$

Next, use $K = \frac{1}{2}ah$, where $K = 125.2$ and the base, a, is 24.

$125.2 = \frac{1}{2}(24)(h)$ *$K = \frac{1}{2}ah$*

$h = 10.4$

The altitude from A to side BC is approximately 10 ft long.

CLASS EXERCISES

Substitute values into the appropriate area formula, but do not solve.

1. 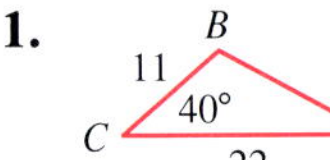

$K = \frac{1}{2}(11)(22)\sin 40°$

2. 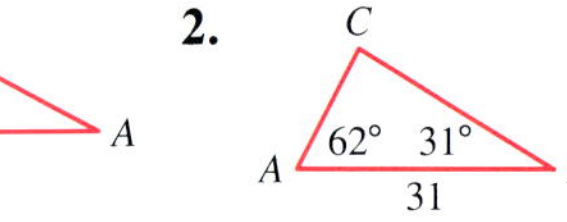

$K = \frac{(31)^2 \sin 62° \sin 31°}{2 \sin 87°}$

3. 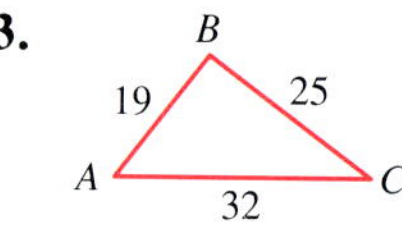

$K = \sqrt{38(38-19)(38-25)(38-32)}$

4. 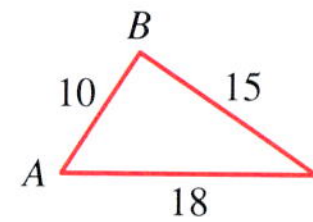

$K = \sqrt{21.5(21.5-10)(21.5-15)(21.5-18)}$

5. 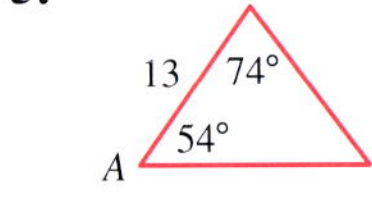

$K = \frac{(13)^2 \sin 74° \sin 54°}{2 \sin 52°}$

6. 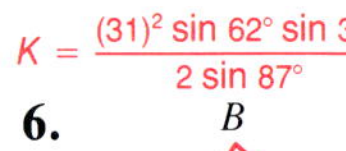

$K = \frac{1}{2}(14)(18)\sin 83°$

7. 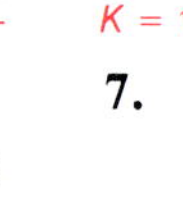

$K = \frac{(37)^2 \sin 29° \sin 98°}{2 \sin 53°}$

8. B 93°, 41, 61°, A, C

$K = \frac{(41)^2 \sin 93° \sin 61°}{2 \sin 26°}$

PRACTICE EXERCISES

Find the area of the given triangle to the nearest whole square unit.

A

1. 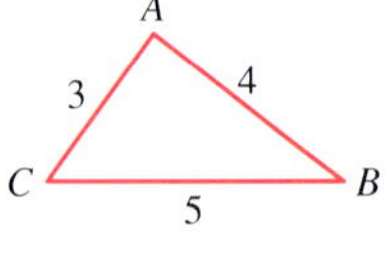

6

2. 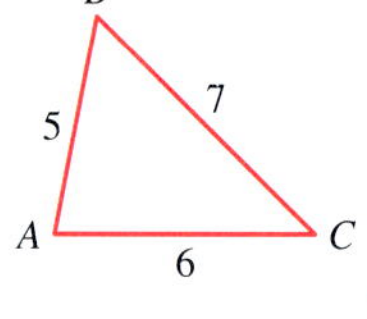

15

Find the area of the given triangle to the nearest ten square units.

3.

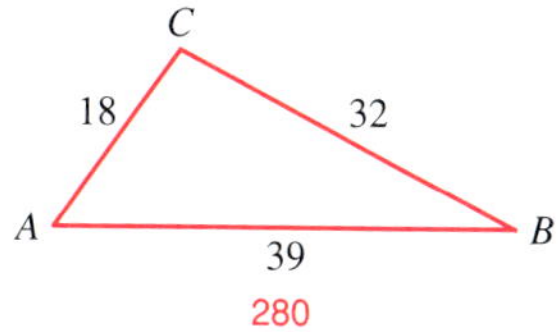

280

4.

460

5. $a = 31, b = 23, c = 14$ 150

6. $a = 61, b = 23, c = 47$ 480

7. $a = 22, b = 25, c = 30$ 270

8. $a = 40, b = 45, c = 50$ 850

9. $a = 12, b = 12, c = 12$ 60

10. $a = 20, b = 20, c = 20$ 170

11. $a = 18, b = 24, c = 30$ 220

12. $a = 10, b = 24, c = 26$ 120

Find the area of the given triangle to the nearest hundred square units.

B 13. $a = 127, b = 246, c = 291$ 15,500

14. $a = 221, b = 119, c = 246$ 13,100

15. $p = 176, q = 187, r = 176$ 13,900

16. $p = 267, q = 283, r = 296$ 34,300

Find the length of the altitude from vertex A to side BC for each triangle to the nearest unit.

17. $a = 10, b = 12, c = 16$ 12

18. $a = 15, b = 22, c = 18$ 18

19. $a = 48, b = 26, c = 50$ 26

20. $a = 38, b = 42, c = 16$ 16

C **Find the area of the given quadrilateral to the nearest hundred square units.**

21.

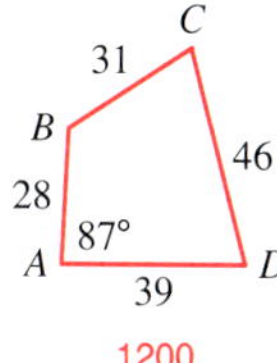

1200

22.

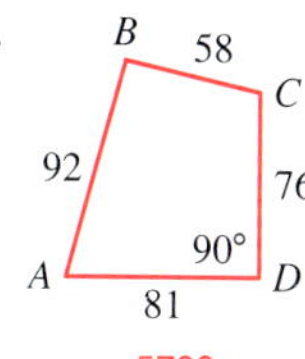

5700

23.

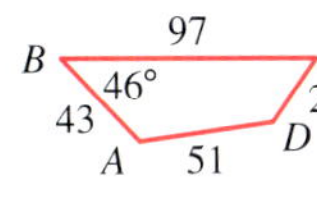

2000

24.

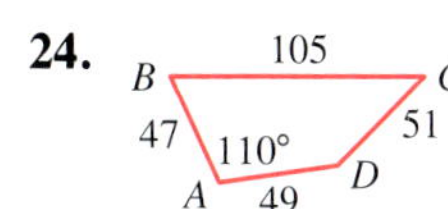

3000

Applications

25. **Construction** Find the area of a triangular patio that has sides 18 ft, 20 ft, and 22 ft, to the nearest ten square feet. 170 ft²

26. **Landscaping** Find the area of a triangular flower garden with sides 11 ft, 19 ft, and 18 ft, to the nearest ten square feet. 100 ft²

CHALLENGE

Find one or more right triangles with perimeter n units and area n square units.
(5, 12, 13); (6, 8, 10)

Teacher's Resource Book

Practice—Chapter 4, p. 11

Enrichment—Chapter 4, p. 12

LESSON PLAN

Vocabulary
Direction angle
Equivalent vectors
Initial point
Magnitude
Opposite vectors
Resultant
Scalar
Standard position
Terminal point
Vector
Vector quantity
x-component
y-component

BACKGROUND

In the Preview, students are asked to list quantities that have magnitude and direction, characteristics that also apply to vectors.

4.7 Vectors in the Plane

Objectives: To find the magnitudes of the *x*- and *y*-components of a vector
To find the sum of two vectors
To find the measure of the angle between two vectors

Many quantities can be described by their magnitude and direction.

Preview

The velocity of an airplane traveling at 620 mi/h due east has both magnitude (speed) and direction. A 20-N force exerted at the end of a plank in order to move a large rock has both magnitude and direction.

Answers may vary.

1. Name some quantities other than velocity and force that have both magnitude and direction.
2. Name some quantities that have only magnitude.
3. Name some quantities that have only direction.
4. Name some quantities that have neither magnitude nor direction.

Any quantity that has both magnitude and direction is called a **vector quantity.** A vector quantity may be represented by a directed line segment, or **vector.** The length of any vector is proportional to the magnitude of the corresponding vector quantity. The direction of the vector represents the direction of the vector quantity.

In the figure at the right, vector *AB*, written $\overrightarrow{AB}$, has the same magnitude and direction as $\overrightarrow{CD}$. The vectors are said to be **equivalent.** This fact can be denoted by $\overrightarrow{AB} = \overrightarrow{CD}$. *A* and *C* are called the **initial points** of the vectors, and *B* and *D* are called the **terminal points.**

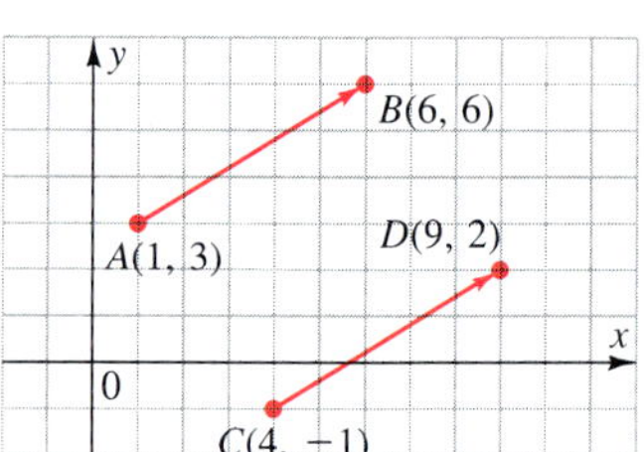

A vector can also be represented by a single letter such as $\vec{V}$. The norm, or **magnitude,** of a vector $\vec{V}$ is represented as $\|\vec{V}\|$ and is found by using the distance formula.

EXAMPLE 1 **Draw an arrow representing the vector from $A(-1, 3)$ to $B(3, 6)$, and find its magnitude.**

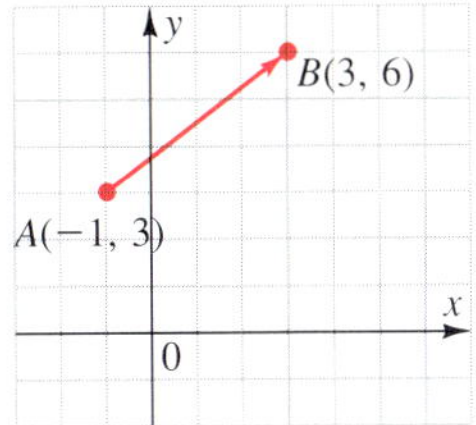

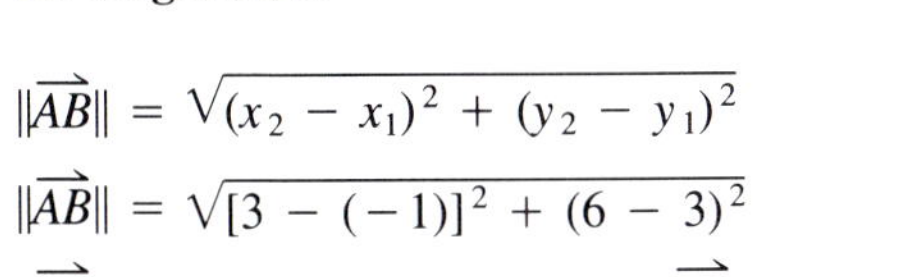

$\|\overrightarrow{AB}\| = \sqrt{(x_2 - x_1)^2 + (y_2 - y_1)^2}$

$\|\overrightarrow{AB}\| = \sqrt{[3 - (-1)]^2 + (6 - 3)^2}$

$\|\overrightarrow{AB}\| = 5$ The magnitude of $\overrightarrow{AB}$ is 5.

If a vector is placed on a standard coordinate plane with its tail, or initial point, at the origin, the vector is said to be in **standard position.** Assume that the endpoint of $\overrightarrow{OC}$ has coordinates (a, b). By the distance formula, $\|\overrightarrow{OC}\| = \|\vec{V}\| = \sqrt{a^2 + b^2}$.

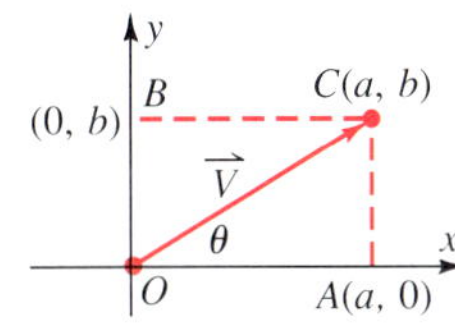

Furthermore, if θ is the measure of the angle between the vector and the positive x-axis, called the **direction angle** of $\overrightarrow{OC}$, then $\frac{a}{\|\vec{V}\|} = \cos\theta$ and $\frac{b}{\|\vec{V}\|} = \sin\theta$.

Vectors $\overrightarrow{OA}$ and $\overrightarrow{OB}$ are the ***x*-** and ***y*-components,** respectively, of $\overrightarrow{OC}$. If you know the magnitude and direction angle of a vector, you can determine the magnitude of its x-component and of its y-component.

EXAMPLE 2 **Find the magnitudes of the x- and y-components of $\overrightarrow{OC}$, given that the measure of the direction angle is 60° and $\|\overrightarrow{OC}\| = 10$.**

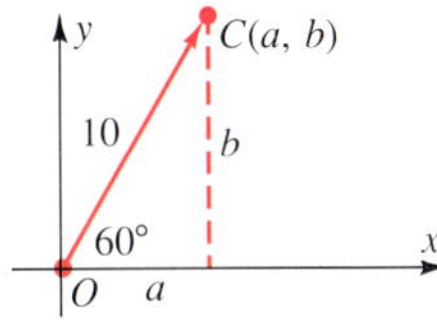

$a = 10\cos 60° = 5$ The magnitude of the x-component is 5.

$b = 10\sin 60° = 5\sqrt{3}$ The magnitude of the y-component is $5\sqrt{3}$.

The sum, or **resultant,** of two vectors is also a vector. Vectors $\vec{U}$ and $\vec{V}$ can be added using *parallelogram* addition or *tail-to-head* (triangular) addition. Each method yields the same resultant vector $\vec{W}$, as illustrated below.

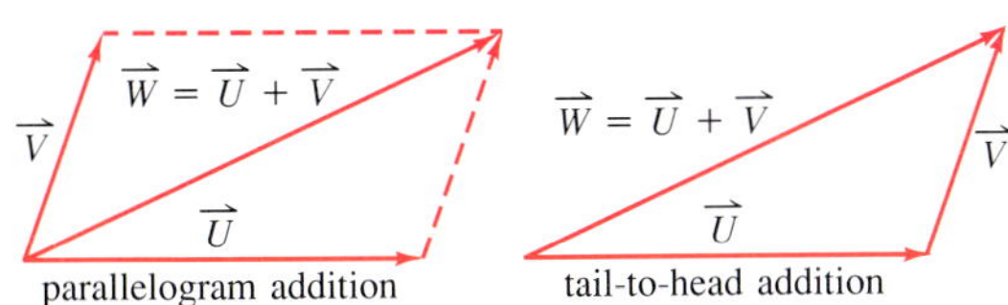

parallelogram addition tail-to-head addition

Note that in parallelogram addition, vectors $\vec{U}$ and $\vec{V}$ have the same initial point. In tail-to-head addition, the head (terminal point) of $\vec{U}$ is the same as the tail of $\vec{V}$.

TEACHING SUGGESTIONS

- Ask students to draw vectors before finding resultants.
- Stress that the resultant is the diagonal of a parallelogram and that the law of cosines can be used to find the magnitude of the resultant.
- Emphasize that the symbol $\|\ \|$ does not mean absolute value but the magnitude of a vector.
- Emphasize that expressing vectors in component form simplifies addition of vectors.

CHALKBOARD EXAMPLES

- **For Example 1**

  Draw each vector $\overrightarrow{AB}$ and find its magnitude.

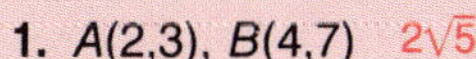

1. $A(2,3)$, $B(4,7)$ $2\sqrt{5}$

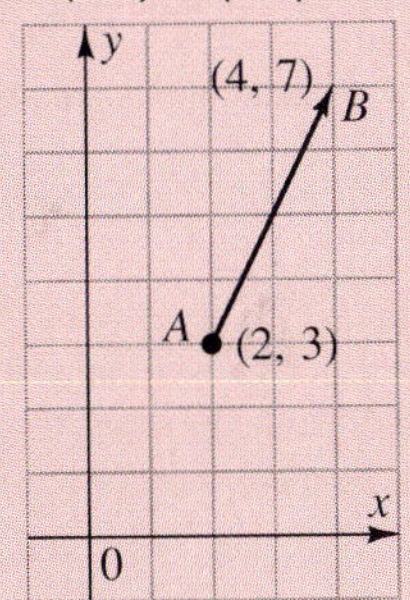

2. $A(-3,-1)$, $B(5,5)$ 10

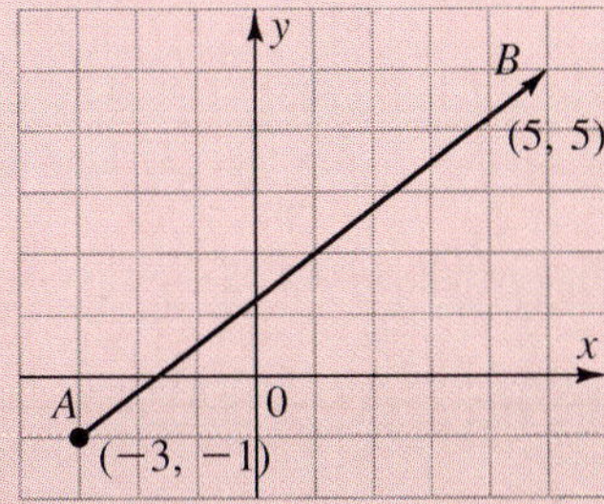

- **For Example 2**

  Find the magnitudes of the x- and y-components of $\overrightarrow{OC}$.

3. $\|\overrightarrow{OC}\| = 5, \theta = 30°$ $\frac{5\sqrt{3}}{2}; \frac{5}{2}$

4. $\|\overrightarrow{OC}\| = 8, \theta = 45°$ $4\sqrt{2}; 4\sqrt{2}$

- **For Example 3**

5. Find $\overrightarrow{AB} + \overrightarrow{BC}$ using tail-to-head addition.

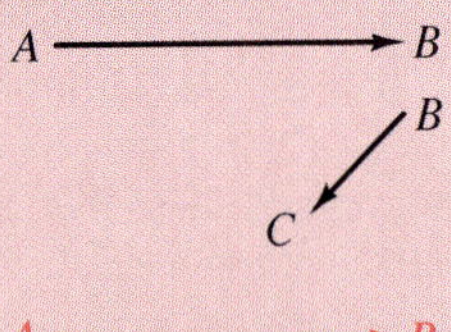

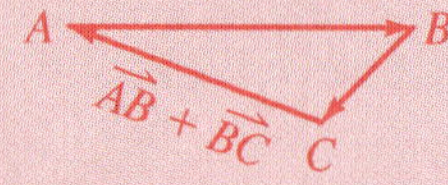

6. Find $\overrightarrow{AB} + \overrightarrow{AC}$ using parallelogram addition.

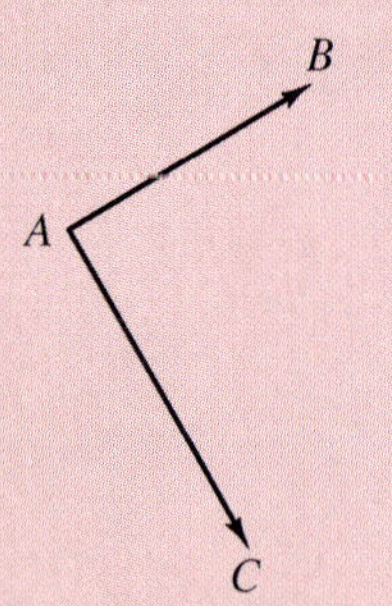

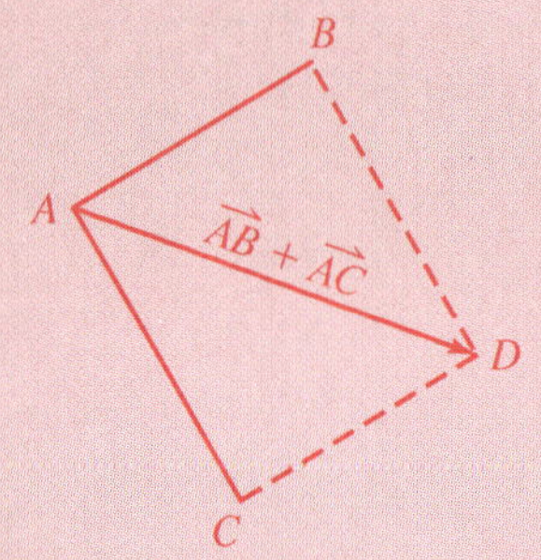

EXAMPLE 3 **Find each of the following:**

a. $\overrightarrow{AB} + \overrightarrow{AC}$, by parallelogram addition **b.** $\overrightarrow{U} + \overrightarrow{V}$, by tail-to-head addition

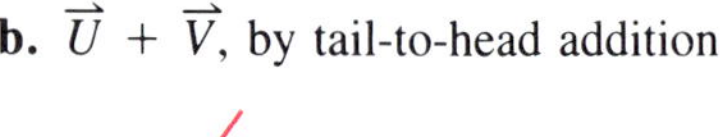

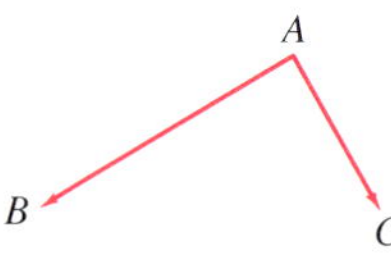

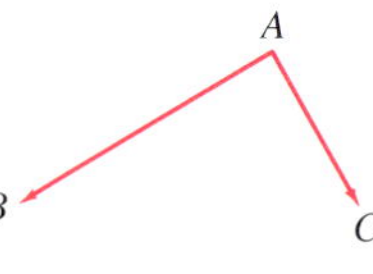

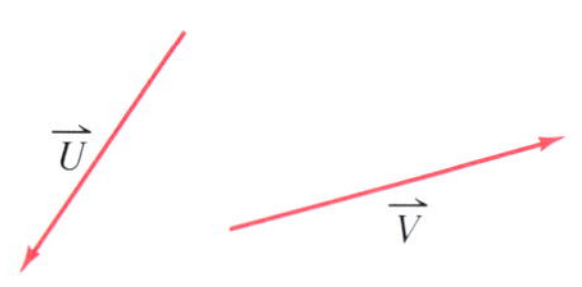

a.

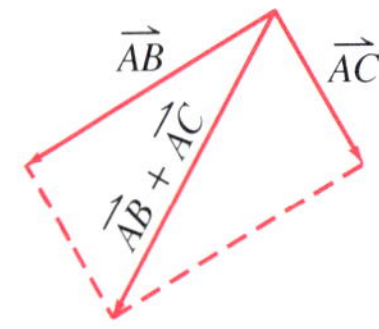

b.

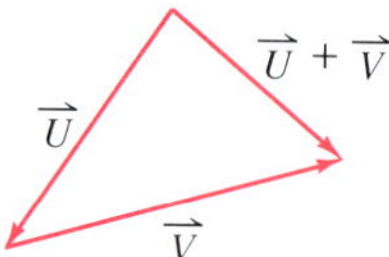

In the figure at the right, all three vectors are parallel and have the same magnitude. However, only $\overrightarrow{AB}$ and $\overrightarrow{CD}$ are equivalent, since the direction of $\overrightarrow{EF}$ is opposite to that of the other two vectors. The relationship between $\overrightarrow{EF}$ and $\overrightarrow{AB}$ is expressed as $\overrightarrow{EF} = -\overrightarrow{AB}$ or $\overrightarrow{AB} = -\overrightarrow{EF}$. $\overrightarrow{AB}$ and $\overrightarrow{EF}$ are said to be **opposite vectors.**

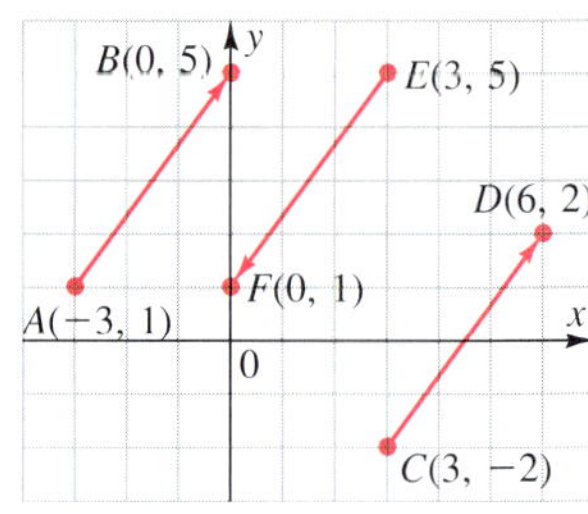

It is possible to subtract a vector from a vector. If $\overrightarrow{U}$ and $\overrightarrow{V}$ are vectors, and $-\overrightarrow{V}$ is the opposite of $\overrightarrow{V}$, then the vector difference $\overrightarrow{U} - \overrightarrow{V}$ is defined as follows: $\overrightarrow{U} - \overrightarrow{V} = \overrightarrow{U} + (-\overrightarrow{V})$.

It is also possible to multiply a real number, or **scalar,** by a vector. To do so, multiply the scalar by the magnitude of the vector.

EXAMPLE 4 **Use the vectors shown to draw each of the following.**

a. $\overrightarrow{U} - \overrightarrow{V}$ **b.** $2\overrightarrow{V}$ **c.** $-2\overrightarrow{U}$

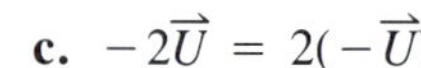

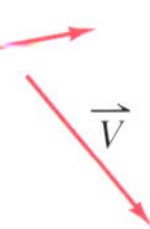

a. Draw $-\overrightarrow{V}$, then $\overrightarrow{U} + (-\overrightarrow{V})$. **b.** **c.** $-2\overrightarrow{U} = 2(-\overrightarrow{U})$

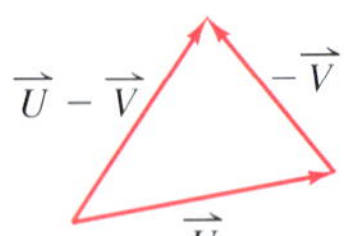

$2\overrightarrow{V}$

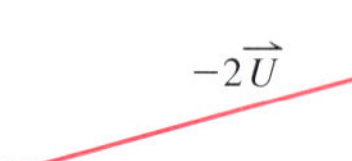

The magnitude and direction of the sum or difference of two vectors can be determined from the magnitudes and directions of the vectors. Let $\vec{U}$ and $\vec{V}$ be vectors, and let θ be the measure of the angle between them. (If you know the directions of both vectors, you can easily compute θ.) By the law of cosines,

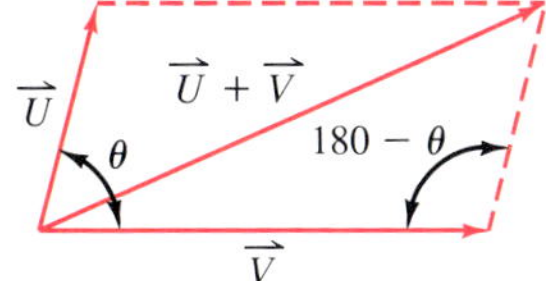

$$\|\vec{U} + \vec{V}\|^2 = \|\vec{U}\|^2 + \|\vec{V}\|^2 - 2\|\vec{U}\|\|\vec{V}\|\cos(180° - \theta)$$

$$\|\vec{U} + \vec{V}\|^2 = \|\vec{U}\|^2 + \|\vec{V}\|^2 + 2\|\vec{U}\|\|\vec{V}\|\cos\theta \qquad \textit{cos (180° − θ) = −cos θ}$$

EXAMPLE 5 **Find the magnitude and direction of the sum $\vec{W}$ of two vectors $\vec{U}$ and $\vec{V}$ of magnitudes 6.9 and 13, respectively, if the angle between them measures 48°. Round the magnitude to the nearest whole number and the direction to the nearest degree.**

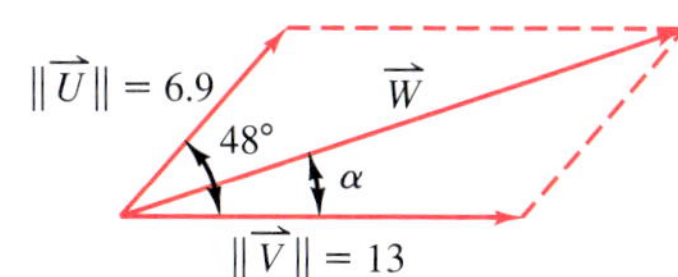

$$\|\vec{U} + \vec{V}\|^2 = \|\vec{U}\|^2 + \|\vec{V}\|^2 + 2\|\vec{U}\|\|\vec{V}\|\cos\theta \qquad \textit{Use the above equation.}$$

$$\|\vec{W}\|^2 = 6.9^2 + 13^2 + 2(6.9)(13)\cos 48°$$

$$\|\vec{W}\|^2 = 336.6520$$

$$\|\vec{W}\| = 18.3481 \qquad \text{The magnitude of } \vec{W} \text{ is 18.}$$

$$\cos\alpha = \frac{\|\vec{W}\|^2 + \|\vec{V}\|^2 - \|\vec{U}\|^2}{2\|\vec{W}\|\|\vec{V}\|} \qquad \textit{Use the formula derived from the law of cosines to find } \alpha.$$

$$\cos\alpha = \frac{18.3481^2 + 13^2 - 6.9^2}{2(18.3481)(13)} = 0.9602$$

$$\alpha = 16.22° \qquad \text{The angle between } \vec{V} \text{ and } \vec{W} \text{ is 16°.}$$

The figures below represent the flight of an airplane in still air and in the presence of wind from the south. In both cases the plane's engines provide *air velocity*, which is a vector quantity. The magnitude and direction of the air velocity are called the *air speed* and *heading*, respectively. Notice that the heading is measured clockwise from the north. The *ground velocity* is the vector sum of the air velocity and wind velocity.

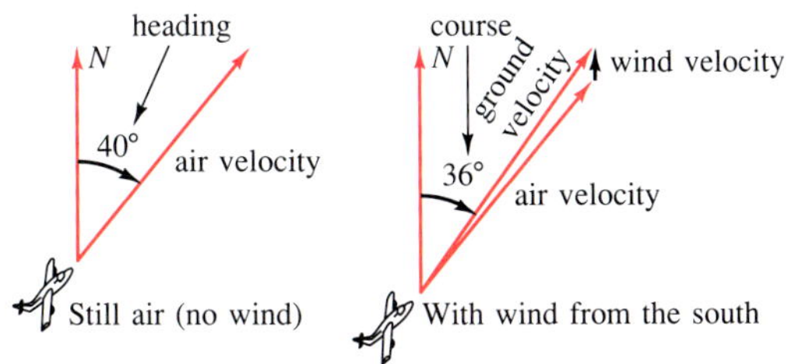

- **For Example 4**
  Use the vectors shown below to draw each of the following:

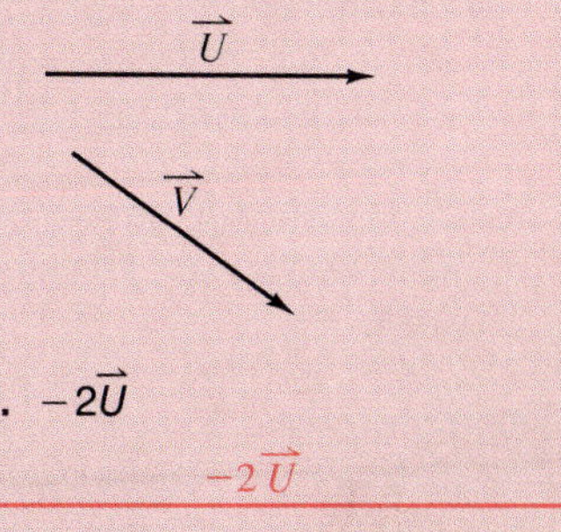

7. $-2\vec{U}$

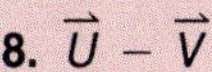

8. $\vec{U} - \vec{V}$

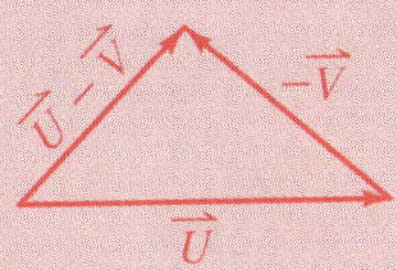

- **For Example 5**
  Find the magnitude of the sum $\vec{W}$ of two vectors $\vec{U}$ and $\vec{V}$ if the angle between them is 55°. Round answers to the nearest whole number.

9. $\|\vec{U}\| = 6$, $\|\vec{V}\| = 9$ 13

10. $\|\vec{U}\| = 8$, $\|\vec{V}\| = 15$ 21

- **For Example 6**

Find the ground speed to two significant digits and the course to the nearest degree.

11. Air speed: 270 mi/h; heading: 55°; wind speed: 25 mi/h from north 260 mi/h; 60°

12. Air speed; 400 mi/h; heading: 50°; wind speed: 20 mi/h from north 390 mi/h; 52°

Common Errors

- When drawing two vectors and their resultant, students often use the wrong diagonal. For example, they want to use $\overrightarrow{BC}$ or $\overrightarrow{CB}$ for $u + v$.

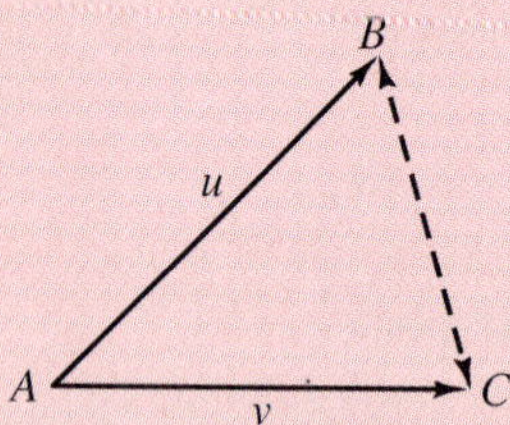

Emphasize that the correct diagonal for $u + v$ is $\overrightarrow{AD}$.

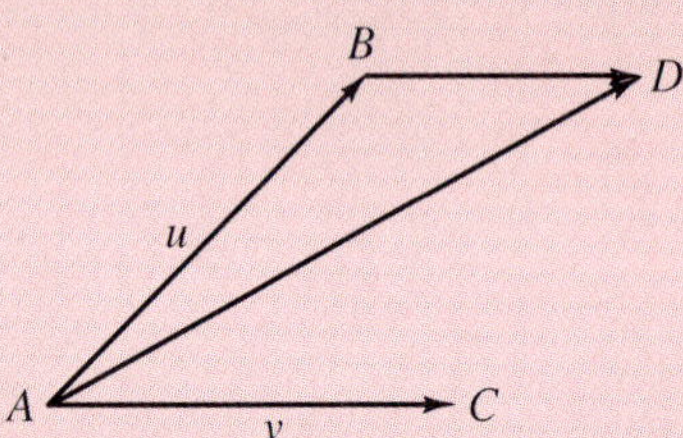

Consequently, they use the incorrect triangle ($\triangle ABC$) in problems (instead of the correct $\triangle ABD$ or $\triangle ACD$).

- See *Teacher's Resource Book* for additional remediation.

LESSON FOLLOW-UP

Assignment Guide

See p. 154B for assignments.

EXAMPLE 6 **An airplane has an air speed of 450 mi/h and a heading of 60 degrees. A wind is blowing from the north at 42 mi/h. Find the plane's ground velocity, including the *ground speed* (magnitude) to two significant digits and the *course* (direction) to the nearest degree. Draw a diagram, showing the sum of the air velocity and wind velocity as $\overrightarrow{V}$, the ground velocity.**

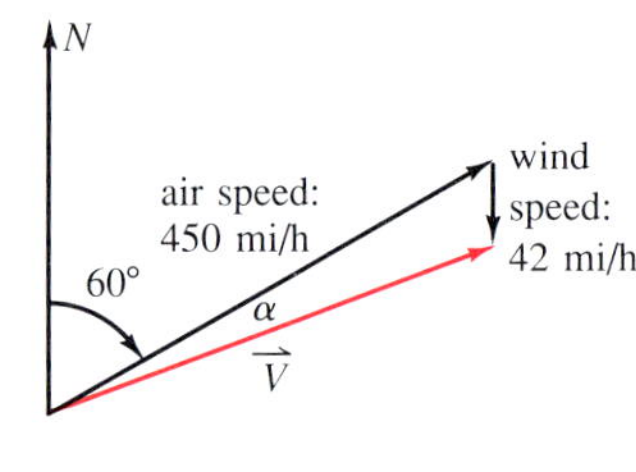

$$a^2 = b^2 + c^2 - 2bc \cos A \qquad \textit{Law of cosines}$$

$$\|\overrightarrow{V}\|^2 = 450^2 + 42^2 - 2(450)(42)\cos 60°$$

$$\|\overrightarrow{V}\|^2 = 185{,}364$$

$$\|\overrightarrow{V}\| = 430.5392 \qquad \textit{The ground speed is 430 mi/h.}$$

$$\frac{\sin A}{a} = \frac{\sin C}{c} \qquad \textit{Use the law of sines.}$$

$$\frac{\sin 60°}{430.5392} = \frac{\sin \alpha}{42}$$

$$\sin \alpha = \frac{42 \sin 60°}{430.5392}$$

$$\alpha = 4.85° \qquad \textit{To the nearest degree } \alpha = 5°.$$

The course is approximately 60° + 5° = 65°, and the ground speed is approximately 430 mi/h.

CLASS EXERCISES

Draw each vector $\overrightarrow{OC}$. Then use a ruler and protractor to find the vector's magnitude and the measure of its direction angle. See Solutions Manual for drawings.

1. $O(0, 0), C(7, 2)$ $\sqrt{53}$; 16° **2.** $O(0, 0), C(-3, 5)$ $\sqrt{34}$; 121° **3.** $O(0, 0), C(6, -2)$ $2\sqrt{10}$; 342°

Use a ruler and a protractor to draw the following vectors in standard position. Draw their sum. Measure the angle between them. See Solutions Manual for drawings.

4. $\|\overrightarrow{U}\| = 4$; $\overrightarrow{U}$ has a direction angle of 55°.
$\|\overrightarrow{V}\| = 7$; $\overrightarrow{V}$ has a direction angle of 20°. 35°

5. $\|\overrightarrow{U}\| = 9$; $\overrightarrow{U}$ has a direction angle of 45°.
$\|\overrightarrow{V}\| = 5$; $\overrightarrow{V}$ has a direction angle of −30°. 75°

6. $\|\overrightarrow{W}\| = 3.5$; $\overrightarrow{W}$ has a direction angle of 110°.
$\|\overrightarrow{Z}\| = 5.2$; $\overrightarrow{Z}$ has a direction angle of −15°. 125°

7. $\|\overrightarrow{W}\| = 9.2$; $\overrightarrow{W}$ has a direction angle of 245°.
$\|\overrightarrow{Z}\| = 7.4$; $\overrightarrow{Z}$ has a direction angle of 15°. 130°

PRACTICE EXERCISES

Draw each vector $\overrightarrow{AB}$ and find its magnitude. See Solutions Manual for drawings.

A **1.** $A(2, 5), B(7, 17)$ $\|\overrightarrow{AB}\| = 13$

2. $A(1, 3), B(5, 6)$ $\|\overrightarrow{AB}\| = 5$

3. $A(11, 5), B(-2, -6)$ $\|\overrightarrow{AB}\| = \sqrt{290}$

4. $A(-5, 7), B(2, -4)$ $\|\overrightarrow{AB}\| = \sqrt{170}$

Find the magnitudes of the *x*- and *y*-components, to the nearest whole number.

5. $\|\vec{V}\| = 15$, and $\vec{V}$ has a direction angle of 30°. 13; 8

6. $\|\vec{V}\| = 29$, and $\vec{V}$ has a direction angle of 45°. 21; 21

7. $\|\vec{V}\| = 18$, and $\vec{V}$ has a direction angle of 25°. 16; 8

8. $\|\vec{V}\| = 25$, and $\vec{V}$ has a direction angle of 43°. 18; 17

9. $\|\vec{V}\| = 12$, and $\vec{V}$ has a direction angle of 112°. −4; 11

Use the vectors shown to draw each of the following. See page 483.

10. $\vec{S} + \vec{T}$ **11.** $\overrightarrow{AB} + \overrightarrow{CD}$ **12.** $\vec{U} + \vec{V}$

13. $\overrightarrow{CD} + \overrightarrow{AB}$ **14.** $\vec{S} + \overrightarrow{CD}$ **15.** $\vec{V} - \vec{T}$

16. $2\vec{U}$ **17.** $3\vec{T}$ **18.** $-2\vec{S}$

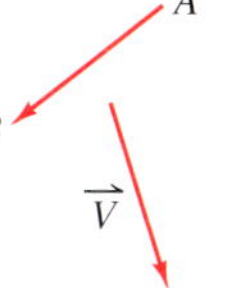

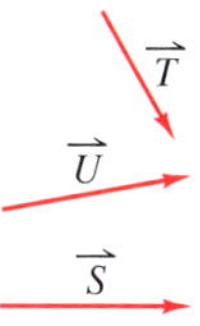

Find the magnitude and direction of the sum $\vec{W}$ of the following vectors. Express the direction as a measure of the angle between $\vec{V}$ and $\vec{W}$. In Exercises 19–24, round magnitudes to two significant digits and angle measures to the nearest degree. In Exercises 25–28, round magnitudes to three significant digits and angle measures to the nearest tenth of a degree. Assume that $\vec{V}$ lies on the positive *x*-axis.

19. $\|\vec{U}\| = 18$, $\|\vec{V}\| = 7$, and the angle between them measures 60°. 22; 44°

20. $\|\vec{U}\| = 10$, $\|\vec{V}\| = 15$, and the angle between them measures 60°. 22; 23°

21. $\|\vec{U}\| = 14$, $\|\vec{V}\| = 19$, and the angle between them measures 45°. 31; 19°

22. $\|\vec{U}\| = 21$, $\|\vec{V}\| = 26$, and the angle between them measures 45°. 43; 20°

B **23.** $\|\vec{U}\| = 6.5$, $\|\vec{V}\| = 7.2$, and the angle between them measures 150°. 3.6; 64°

24. $\|\vec{U}\| = 9.2$, $\|\vec{V}\| = 10.6$, and the angle between them measures 150°. 5.3; 60°

25. $\|\vec{U}\| = 12.4$, $\|\vec{V}\| = 19.3$, and the angle between them measures 48°. 29.1; 18.5°

26. $\|\vec{U}\| = 15.7$, $\|\vec{V}\| = 18.9$, and the angle between them measures 54°. 30.9; 24.3°

27. $\|\vec{U}\| = 32.5$, $\|\vec{V}\| = 29.4$, and the angle between them measures 114°. 33.8; 61.4°

28. $\|\vec{U}\| = 23.7$, $\|\vec{V}\| = 39.4$, and the angle between them measures 127°. 31.5; 37.0°

Test Yourself

See *Teacher's Resource Book, Tests*, pp. 35–36.

Lesson Quiz

1. Draw the vector from $A(-2,0)$ to $B(2,3)$ and find its magnitude. 5

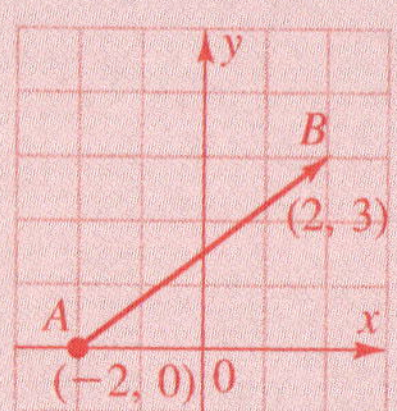

2. Find the magnitudes of the x- and y-components of OB if $\theta = 30°$ and $\|\overrightarrow{OB}\| = 12$. $6\sqrt{3}$; 6
3. Determine the magnitude of the sum of $\|\overrightarrow{U}\|$ and $\|\overrightarrow{V}\|$ if $\|\overrightarrow{U}\| = 9$ and $\|\overrightarrow{V}\| = 12$ and the angle between them is 30°. 20
4. Find the ground speed to two significant digits and the course to the nearest degree of a plane with an air speed of 425 mi/h and a heading of 45° if the wind is blowing from north at 40 mi/h. 398 mi/h; 49°

Enrichment

Use a diagram to illustrate that vector addition has the associative property.

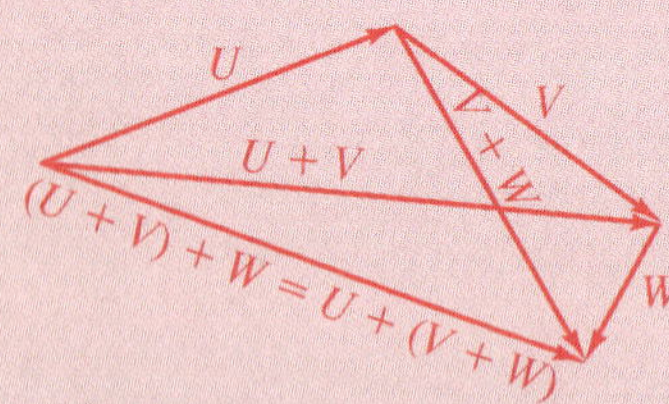

Use the vectors from Exercises 10–18 to draw each of the following.
See page 483.

29. $\overrightarrow{S} + (\overrightarrow{T} + \overrightarrow{V})$ **30.** $(\overrightarrow{S} + \overrightarrow{T}) + \overrightarrow{V}$ **31.** $2(\overrightarrow{U} + \overrightarrow{V})$ **32.** $\frac{1}{2}(\overrightarrow{S} - \overrightarrow{T})$

In Exercises 33 and 34, an airplane has the indicated air velocity and is affected by the given wind velocity. Find the ground speed to two significant digits and the course to the nearest degree.

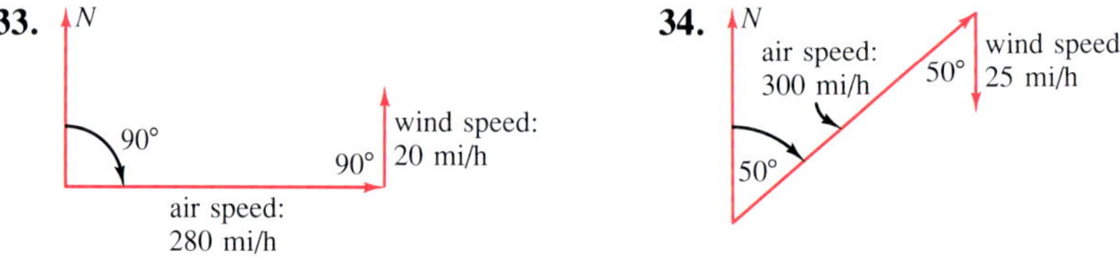

33. 280 mi/h, 86° 34. 280 mi/h, 54°

In Exercises 35–40, round magnitudes to two significant digits and angle measures to the nearest degree.

35. The air speed of an airplane is 250 mi/h, and its heading is 45 degrees. The wind is blowing from the north at 30 mi/h. Find the ground speed and course of the plane. 230 mi/h; 50°

36. The air speed of an airplane is 520 mi/h, and its heading is 110 degrees. The wind is blowing from the south at 15 mi/h. Find the ground speed and course of the plane. 520 mi/h; 108°

C **37.** A plane has a heading of 26 degrees and an air speed of 575 mi/h. The wind is blowing in a direction of 67 degrees east of north at a speed of 23 mi/h. Find the ground speed and course of the plane. 590 mi/h; 27° E of N

38. The air speed of a plane is 280 mi/h, its heading is 190 degrees, and its course is 195°. If the wind is blowing at 23 mi/h and the ground speed of the plane is 275 mi/h, in what direction is the wind blowing? 294°, or 66° W of N

39. A pilot wants to maintain a course of 40 degrees and a ground speed of 300 mi/h against a 45-mi/h wind from 20 degrees west of north. What should be his heading and air speed? 33°; 320 mi/h

40. Two planes left an airport at the same time and in still air. One flew east at 450 mi/h, and the other flew at 380 mi/h. In what direction did the second pilot fly if the two planes were 800 mi apart after 1 h? 239° or 301°

41. The magnitudes of the x- and y-components of $\overrightarrow{U}$ are a and b, respectively. The magnitudes of the x- and y-components of $\overrightarrow{V}$ are c and d, respectively. Use principles of coordinate geometry to show that the magnitudes of the x- and y-coordinates of $\overrightarrow{U} + \overrightarrow{V}$ are $a + c$ and $b + d$, respectively.
See side column on page 197.

42. Use the law of cosines to show that $\|\overrightarrow{U}\|^2 + \|\overrightarrow{V}\|^2 = \|\overrightarrow{U} + \overrightarrow{V}\|^2$ if and only if the angle between $\overrightarrow{U}$ and $\overrightarrow{V}$ is 90°. See side column on page 197.

43. The magnitudes of the x- and y-components of $\vec{U}$ are a and b, respectively. The magnitudes of the x- and y-components of $\vec{V}$ are c and d, respectively. Show that the angle between $\vec{U}$ and $\vec{V}$ is 90° if and only if $ac + bd = 0$. See side column.

Applications

In Exercises 44–45, give magnitudes to two significant digits and directions to the nearest degree.

Physics

44. Forces of 8.0 lb and 15 lb act on a body at right angles. Find the magnitude of the resultant force. 17 lb

45. Find the measure of the angle between forces of 80 lb and 70 lb, if the magnitude of the resultant force is 98 lb. 99°

46. Two forces are pushing an ice shanty along the ice. One force has a magnitude of 330 lb in a direction due east. The other force has a magnitude of 110 lb in a direction 54° east of north. What are the magnitude and direction of the resultant force? Give the magnitude to the nearest ten pounds and the direction to the nearest degree. 420 lb; 81°

TEST YOURSELF

Find the area of each triangle to the nearest ten square units.

1. $\angle C = 55°, a = 32, b = 43$ 560 **2.** $\angle B = 38°, a = 10, c = 11$ 30 **4.5**

3. $\angle A = 112°, \angle B = 23°, c = 37$ 350 **4.** $\angle B = 43°, \angle C = 29°, a = 73$ 930

Use Heron's formula to find the area of each triangle to the nearest ten square units.

5. $a = 30, b = 24, c = 13$ 150 **6.** $a = 62, b = 24, c = 45$ 440 **4.6**

7. $a = 12, b = 24, c = 27$ 140 **8.** $a = 31, b = 35, c = 40$ 520

9. Given $A(3, 4)$ and $B(-2, 7)$, what is the magnitude of $\overrightarrow{AB}$? $\sqrt{34}$ **4.7**

10. Find the magnitude of the x- and y-components of $\vec{V}$, if $\|\vec{V}\| = 10$ and $\vec{V}$ has a direction angle of 45°. $5\sqrt{2}$; $5\sqrt{2}$

11. If $\|\vec{U}\| = 9$, $\|\vec{V}\| = 11$, and the angle between $\vec{U}$ and $\vec{V}$ is 39°, find the magnitude and direction of $\vec{W}$, the sum of $\vec{U}$ and $\vec{V}$. Express the direction as a measure of the angle between $\vec{V}$ and $\vec{W}$. 18.9; 17.5°

12. An airplane has an air speed of 500 mi/h and a heading of 120°. A wind is blowing from the north at 30 mi/h. Find, to two significant digits, the plane's ground speed and course. 520 mi/h; 123°

Teacher's Resource Book
Practice—Chapter 4, p. 13
Enrichment—Chapter 4, p. 14

Additional Answers

41.

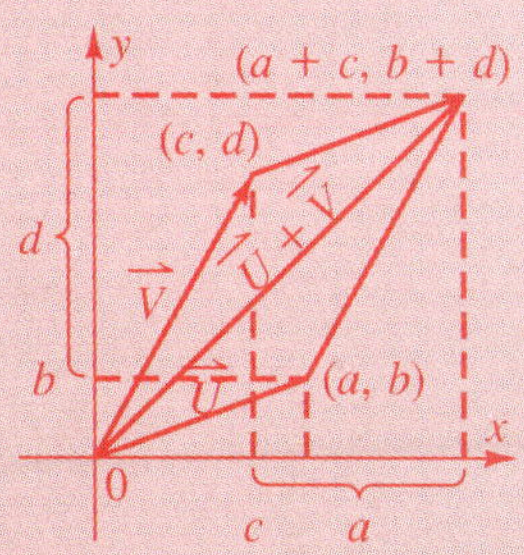

Given $\vec{U}$ with components (a, b) and $\vec{V}$ with components (c, d). Construct parallelogram with $\vec{U} + \vec{V}$ as diagonal. Project one component from each of four sides onto axes. Add x-components. Add y-components. Components of $\vec{U} + \vec{V}$ are $(a + c, b + d)$.

42. $\vec{U}$, $\vec{V}$, and $\vec{U} + \vec{V}$ always form a triangle unless $\vec{U}$ and $\vec{V}$ are collinear. So Law of cosines gives: $\vec{V}$ $|\vec{U} + \vec{V}|^2 = |\vec{U}|^2 + |\vec{V}|^2 + 2\,|\vec{U}| \cdot |\vec{V}| \cos A$, where $\angle A$ is the angle formed by $\vec{U}$ and $\vec{V}$. Thus, $|\vec{U} + \vec{V}|^2 = |\vec{U}|^2 + |\vec{V}|^2$ only if $+ 2\,|\vec{U}|\,|\vec{V}| \cos A = 0$. Since $|\vec{U}| \neq 0$ and $|\vec{V}| \neq 0$, then $\cos A = 0$, therefore $A = 90$. So $180° - A = 90°$ = measure of $\angle$ formed by $\vec{U}$ and $\vec{V}$.

43. $\|\vec{U}\| = \sqrt{a^2 + b^2}$, $\|\vec{V}\| = \sqrt{c^2 + d^2}$, $\|\vec{U} + \vec{V}\| = \sqrt{(a + c)^2 + (b + d)^2}$, θ = measure of $\angle$ formed by $\vec{U}$ and $\vec{V}$, $\|\vec{U} + \vec{V}\|^2 = \|\vec{U}\|^2 + \|\vec{V}\|^2 - 2\,\|\vec{U}\|\,\|\vec{V}\| \cos\theta$, $(a + c)^2 + (b + d)^2 = (a^2 + b^2) + (c^2 + d^2) + 2\sqrt{a^2 + b^2} \cdot \sqrt{c^2 + d^2} \cdot \cos\theta$,

$$\frac{a^2 + 2ac + c^2 + b^2 + 2bd + d^2 - a^2 - b^2 - c^2 - d^2}{2\sqrt{a^2 + b^2} \cdot \sqrt{c^2 + d^2}} = \cos\theta,$$

$$\frac{2ac + 2bd}{2\sqrt{a^2 + b^2} \cdot \sqrt{c^2 + d^2}} = \cos\theta,$$

$$\frac{ac + bd}{\sqrt{a^2 + b^2} \cdot \sqrt{c^2 + d^2}} = \cos\theta.$$

If $\theta = 90°$, then $\cos\theta = 0$; therefore, $ac + bd = 0$. If $ac + bd = 0$, then $\cos\theta = 0$; thus $\theta = 90°$

Application

Some of the complex formulas that are used in programming the motion of industrial robots are introduced. Such formulas provide an illustration of a fascinating development in mechanical engineering.

APPLICATION: Industrial Robots

Did you know that robots are used in industry to perform routine, heavy, dangerous, or repetitive tasks? As you observe them in operation, they appear to think, but this impression is an illusion; their every move has been programmed beforehand by a human being. To understand how a robot is controlled, visualize three-dimensional space.

To perform its tasks, a robot may use both linear and rotary (angular) motion. Two types of robot configurations will be considered. *Cartesian motion* acts in straight lines for forward-backward, right-left, and up-down directions. *Cylindrical motion* acts in straight lines for forward-backward and up-down direction. Horizontal motion is angular.

Illustrated at the right is a three-dimensional *rectangular coordinate system*. The coordinates of a point $P(x, y, z)$ are ordered triples. To program robotic motion, a programmer needs to know the coordinates of two different positions. The difference between the two positions is then calculated.

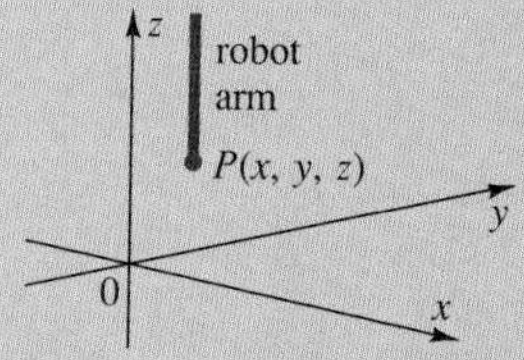

EXAMPLE 1 Suppose for a robot operating in a rectangular system, the coordinates of the first position of the end of its arm are $P_1\ (7, -2, 1)$, and the coordinates of a second position are $P_2\ (2, 4, -6)$. Calculate the distance the end of the robot's arm moves along each axis.

	P_1	P_2	Difference	
x	7	2	-5	*moved backward 5 units*
y	-2	4	6	*moved right 6 units*
z	1	-6	-7	*moved downward 7 units*

In the *cylindrical coordinate system*, a point is represented as an ordered triple (r, θ, z). In the diagram, point P is r units from the z-axis and z units directly above a point P' in the xy-plane. Angle θ is the measure of the counterclockwise acute angle between OP' and the positive x-axis.

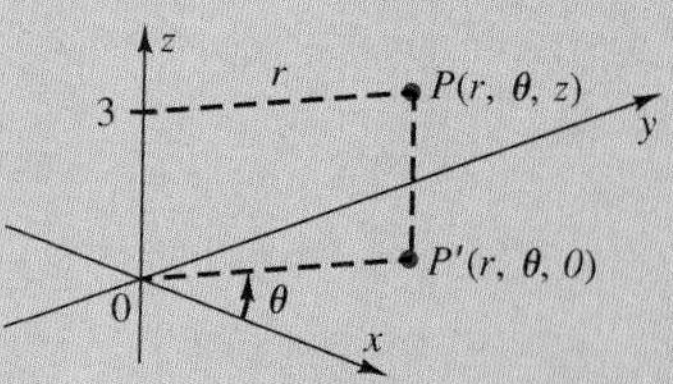

EXAMPLE 2 A robot using cylindrical coordinates has its arm rotating about the z-axis. The end of the arm is at P_1 so that θ has a value of 45° with $x_1 = 5$ and $z_1 = 8$. The arm moves to P_2 by rotating to $\theta = -30°$ with $x_2 = 4$ and $z_2 = 6$. Find the distance the robot moves along the x- and z-axes, the length of the arm at P_1 and P_2, the amount the arm length changed from P_1 to P_2, and the y-coordinate at P_2.

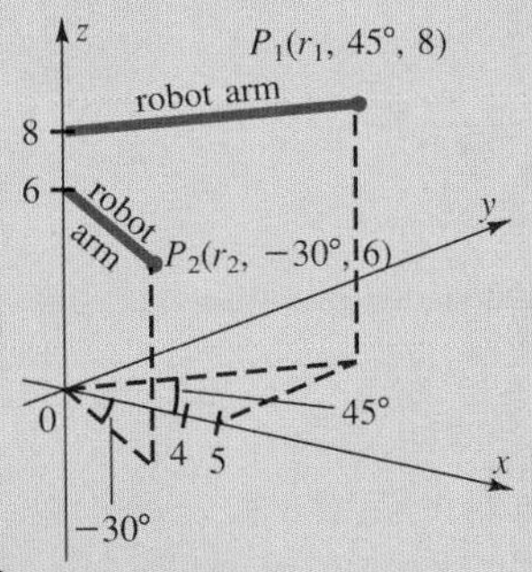

	P_1	P_2	Difference	
x	5	4	−1	*moved backward 1 unit*
θ	45°	−30°	−75°	*moved counterclockwise 75°*
z	8	6	−2	*moved downward 2 units*

arm length at P₁: $\cos 45° = \dfrac{5}{\text{arm length}}$

arm length at P₁ = 7.07 units

arm length at P₂: $\cos(-30°) = \dfrac{4}{\text{arm length}}$

arm length = 4.62 units

$\tan(-30°) = \dfrac{y_2}{4}$

$y_2 = 4\tan(-30°) = -2.31$

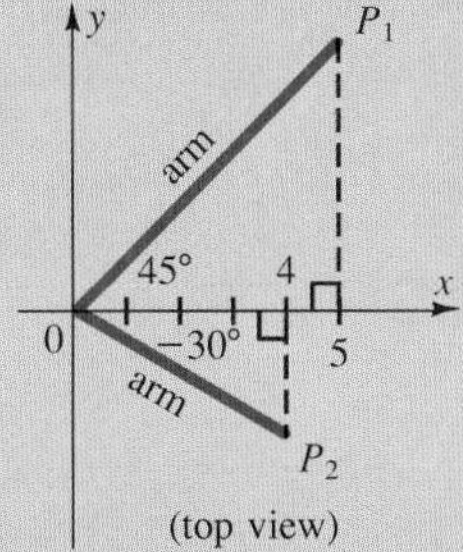

(top view)

The difference in the length of the arm between P_1 and P_2 is −2.45 units. (The arm retracted, or "shrank.") The y-coordinate of P_2 is −2.31.

Robot engineers are trying to design a robot that will direct some of its own movements. However, this goal is several years away.

EXERCISES

Suppose a robot operating in a rectangular system has the end of its arm located at $P_1(4, 6, 3)$. The arm then moves to $P_2(7, 5, 9)$.

1. Draw a diagram for the robot at P_1 and at P_2.
   See side column.
2. Calculate the distance the end of the robot's arm moves along each axis.
   x: forward 3 units; y: left 1 unit; z: upward 6 units
3. A robot using cylindrical coordinates has the end of its arm located at $P_1(r_1, 48°, -2)$ with $x_1 = 9$. The arm then rotates to $P_2(r_2, 76°, 8)$ with $x_2 = 5$. Find the distance the end of the robot's arm moves along the x- and z-axes, the length of the arm at P_1 and P_2, the amount the arm changed from P_1 to P_2, and the y-coordinate at P_2. x: backward 4 units; z: upward 10 units; arm length at P_1: 13.45 units; arm length at P_2: 20.67 units; 7.22 units increase; y_2: 20.05

Additional Answers

1.

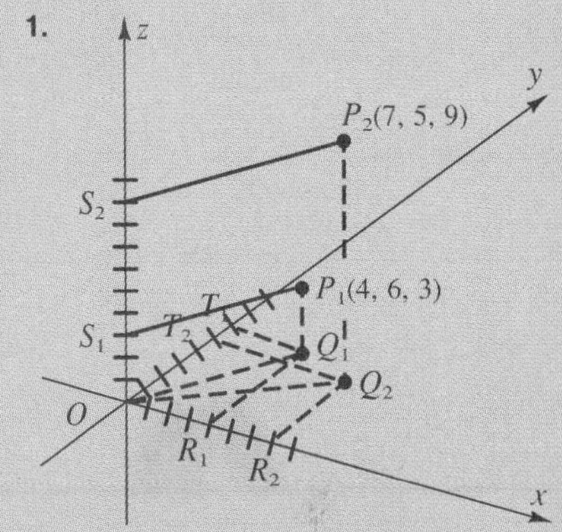

CHAPTER 4 SUMMARY AND REVIEW

Vocabulary

ambiguous case (164)
components of a vector (191)
direction of a vector (191)
direction angle of a vector (191)
initial point of a vector (190)
Heron's formula (186)
law of cosines (168)
law of sines (156)
magnitude of a vector (190)
law of tangents (175)
opposite vectors (192)
parallelogram addition (191)
resultant of a vector (191)
scalar (192)
standard position of a vector (191)
tail-to-head addition (191)
terminal point of a vector (190)
vector (190)

In all exercises, round lengths to two significant digits and angle measures to the nearest degree, unless otherwise stated.

Law of Sines The law of sines can be used to solve triangles when two angles and any side are given. **4.1**

$$\frac{\sin A}{a} = \frac{\sin B}{b} = \frac{\sin C}{c}$$

1. In $\triangle ABC$, find b if $c = 26$, $\angle B = 49°$, and $\angle C = 53°$. 24.6

2. Solve $\triangle ABC$ if $a = 41$, $\angle B = 28°$, and $\angle C = 120°$. $b = 36.3$, $c = 67.0$, $\angle A = 32°$

Law of Sines: The Ambiguous Case When the measures of two sides and an angle opposite one of them are given, the law of sines must be used with care, since *two* triangles, *one* triangle, or *no* triangles may be determined. **4.2**

Solve $\triangle ABC$.

3. $\angle A = 56°, b = 12, a = 11$ $c = 2$, $\angle B = 115°$, $\angle C = 9°$ or $c = 11$, $\angle B = 65°$, $\angle C = 59°$

4. $\angle B = 125°, b = 13, c = 20$ does not exist

5. $\angle A = 99°, a = 25, b = 15$ $c = 18$, $\angle B = 36°$, $\angle C = 45°$

6. $\angle C = 105°, a = 11, c = 14$ $b = 6$, $\angle A = 49°$, $\angle B = 26°$

Law of Cosines The law of cosines can be used to solve a triangle when two sides and the included angle or three sides are given. **4.3**

$$a^2 = b^2 + c^2 - 2bc \cos A \qquad \cos A = \frac{b^2 + c^2 - a^2}{2bc}$$

$$b^2 = a^2 + c^2 - 2ac \cos B \qquad \cos B = \frac{a^2 + c^2 - b^2}{2ac}$$

$$c^2 = a^2 + b^2 - 2ab \cos C \qquad \cos C = \frac{a^2 + b^2 - c^2}{2ab}$$

Solve $\triangle ABC$.

7. $\angle C = 58°, a = 12, b = 16$
$c = 14, \angle A = 47°, \angle B = 75°$

8. $a = 25, b = 33, c = 28$
$\angle A = 48°, \angle B = 77°, \angle C = 56°$

Law of Tangents When two sides and the included angle are given, the law of tangents may allow you to solve the triangle with fewer calculations. 4.4

$$\frac{a-b}{a+b} = \frac{\tan \frac{1}{2}(A-B)}{\tan \frac{1}{2}(A+B)} \qquad \frac{c-a}{c+a} = \frac{\tan \frac{1}{2}(C-A)}{\tan \frac{1}{2}(C+A)} \qquad \frac{b-c}{b+c} = \frac{\tan \frac{1}{2}(B-C)}{\tan \frac{1}{2}(B+C)}$$

Solve $\triangle ABC$.

9. $\angle B = 68°, b = 14, c = 18$
does not exist

10. $\angle A = 52°, \angle B = 33°, a + b = 50$
$a = 30, b = 20, c = 38, \angle C = 95°$

Area of a Triangle The following formulas can be used to find the area of a triangle when two sides and the included angle are known: 4.5

$$K = \frac{1}{2}ab \sin C \qquad K = \frac{1}{2}bc \sin A \qquad K = \frac{1}{2}ac \sin B$$

These formulas are used when two angles and the included side are known:

$$K = \frac{c^2 \sin A \sin B}{2 \sin C} \qquad K = \frac{a^2 \sin B \sin C}{2 \sin A} \qquad K = \frac{b^2 \sin A \sin C}{2 \sin B}$$

Find the area of $\triangle ABC$ to the nearest ten square units.

11. $\angle A = 48°, b = 24, c = 18$ 160

12. $\angle A = 53°, \angle B = 100°, c = 20$ 350

Heron's Formula When three sides of a triangle are given, Heron's formula can be used to find the area of the triangle and the altitude. 4.6

$$K = \sqrt{s(s-a)(s-b)(s-c)}, \text{ where } s = \frac{a+b+c}{2}$$

13. Find the area of $\triangle ABC$ to the nearest unit, if $a = 24$, $b = 28$, and $c = 35$. 334

14. Find the length of the altitude to side BC in $\triangle ABC$, to the nearest unit, if $a = 30$, $b = 48$, and $c = 60$. 48

Vectors in the Plane A vector has direction and magnitude. Vectors can be added or subtracted using either parallelogram addition or tail-to-head addition. To find the sum of two vectors, use the law of cosines. 4.7

15. Find the magnitudes of the x- and y-components of $\vec{V}$, if $\vec{V}$ has a direction angle of 135° and $\|\vec{V}\| = 20$. $10\sqrt{2}$, $10\sqrt{2}$

16. The air speed of an airplane is 400 mi/h, and its heading is 150°. A wind is blowing from the south at 20 mi/h. Find the plane's ground speed and course.
382.8 mi/h; 148.5°

See *Teacher's Resource Book, Tests*, pp. 37–40.

CHAPTER TEST

Solve $\triangle ABC$. Give lengths to two significant digits and angle measures to the nearest degree.

1. $\angle A = 38°, b = 19, c = 23$
$a = 14, \angle B = 56°, \angle C = 86°$

2. $a = 12, b = 17, c = 21$
$\angle A = 35°, \angle B = 54°, \angle C = 91°$

3. $\angle A = 37°, \angle B = 28°, c = 18$
$a = 12, b = 9.3, \angle C = 115°$

4. $\angle B = 62°, a = 14, c = 23$
$b = 21, \angle A = 37°, \angle C = 81°$

5. $\angle A = 49°, b = 23, c = 19$
$a = 18, \angle B = 77°, \angle C = 54°$

6. $a = 15, b = 19, c = 22$
$\angle A = 42°, \angle B = 58°, \angle C = 80°$

7. $\angle A = 51°, \angle B = 63°, c = 26$
$a = 22, b = 25, \angle C = 66°$

8. $\angle A = 103°, a = 49, b = 35$
$c = 27, \angle B = 44°, \angle C = 33°$

9. $\angle B = 37°, a = 44, b = 38$
$c = 62, \angle A = 44°, \angle C = 99°$ or $c = 8, \angle A = 136°, \angle C = 7°$

10. $\angle C = 10°, a = 15, c = 40$
$b = 56, \angle A = 4°, \angle B = 166°$

11. $a = 12, b = 16, c = 19$
$\angle A = 39°, \angle B = 57°, \angle C = 84°$

12. $a = 21, b = 17, c = 15$
$\angle A = 82°, \angle B = 53°, \angle C = 45°$

Determine the area of each triangle to the nearest ten square units.

13. $\angle B = 82°, a = 12, c = 15$ 90

14. $\angle A = 73°, b = 42, c = 67$ 1350

15. $\angle A = 46°, \angle B = 37°, c = 43$ 400

16. $\angle A = 51°, \angle C = 29°, b = 32$ 200

Use Heron's formula to find the area of each triangle to the nearest ten square units.

17. $a = 12, b = 14, c = 19$ 80

18. $a = 15, b = 16, c = 11$ 80

19. $a = 14, b = 16, c = 21$ 110

20. $a = 24, b = 27, c = 31$ 310

Find the altitude to side BC in $\triangle ABC$, to the nearest unit.

21. $a = 32, b = 50, c = 64$ 49

22. $a = 18, b = 21, c = 16$ 16

23. Find the magnitude of the x- and y-components of $\vec{U}$, if $\vec{U}$ has a direction angle of 45° and $\|\vec{U}\| = 40$. magnitude of x-component is $20\sqrt{2}$; magnitude of y-component is $20\sqrt{2}$

24. Find the magnitude and direction of $\vec{U}$ and $\vec{V}$, if $\|\vec{U}\| = 10$, $\|\vec{V}\| = 15$, and if the angle between $\vec{U}$ and $\vec{V}$ measures 110°. Express the direction of $\vec{W}$ as the measure of the angle between $\vec{V}$ and $\vec{W}$. 14.9; 39°

25. An airplane has an air speed of 600 mi/h and a heading of 70°. A wind is blowing from the south at 40 mi/h. Find, to two significant digits, the plane's ground speed and course. 610 mph; 66.5°

Challenge

Prove that for any $\triangle ABC$, $a = b\cos C + c\cos B$. See side column.

Additional Answer

Challenge

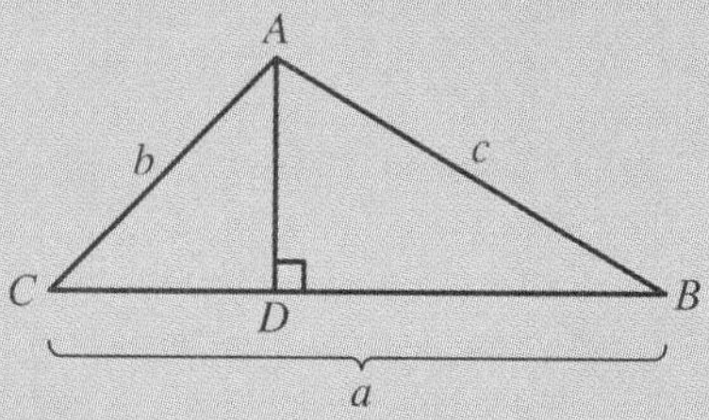

Construct $AD \perp CB$.

$\overline{CD} = b\cos C$

$\overline{DB} = c\cos B$

$a = \overline{CD} + \overline{DB}$

$a = b\cos C + c\cos B$

COLLEGE ENTRANCE EXAM REVIEW

Select the best choice for each question.

1. In the triangle shown, $\angle A = 30°$,
C $\angle B = 65°$, and $a = 5$. Find b.

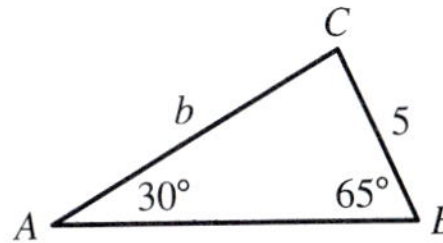

A. $\frac{\sin 65°}{c \cos 30°}$ **B.** $\frac{\cos 65°}{5 \sin 65°}$
C. $\frac{5 \sin 65°}{\sin 30°}$ **D.** $\frac{\sin 30°}{5 \sin 65°}$
E. $\frac{\sin 30°}{\cos 65°}$

2. The expression $|2x - 1| \leq 9$ is
D equivalent to
A. $4 \leq x \leq 5$ **B.** $x \leq 5$
C. $-4 \leq x \leq 10$ **D.** $-4 \leq x \leq 5$
E. $-8 \leq x \leq 10$

3. $8x^2 - 3x = 1$ has
B **A.** one real number solution
B. two real number solutions
C. one real number solution and one complex number solution
D. two complex number solutions
E. one complex number solution

4. A painter can paint a house in 3 hr.
C The painter's assistant can complete the task in 6 hr. How long would it take to paint the house if the painter and the assistant worked together?

A. $4\frac{1}{2}$ hr **B.** 5 hr **C.** 2 hr
D. 3 hr **E.** 1 hr

5. If $3p - a = 6p - d$, then p equals
A **A.** $\frac{d - a}{3}$ **B.** $\frac{a - d}{3}$ **C.** $3a + d$
D. $d - \frac{1}{3}a$ **E.** $\frac{d + a}{3}$

6. In $\triangle ABC$, if $a = 5$, $c = 9$, and
B $\cos B = \frac{1}{15}$, find b.
A. 15 **B.** 10 **C.** 9
D. 20 **E.** No solution

7. Find the area of an equilateral
B triangle if the measure of one side is 6.
A. 15 **B.** $9\sqrt{3}$ **C.** 18
D. $36\sqrt{3}$ **E.** $6\sqrt{3}$

8. The area of parallelogram
E $ABCD$ is

A. $\frac{2\sqrt{2}}{15}$
B. $\frac{15\sqrt{2}}{4}$

B C 3 30° A $5\sqrt{2}$ D

C. $\frac{4\sqrt{2}}{15}$ **D.** $\frac{15\sqrt{3}}{2}$ **E.** $\frac{15\sqrt{2}}{2}$

9. Find the area of a square whose
A side measures $3\sqrt{7}$.
A. 63 **B.** $6\sqrt{7}$ **C.** $9\sqrt{7}$
D. 42 **E.** 23

10. Which of the following is not
B a function?

I. $y = 3x$
II. $x = -6$
III. $x = 5y$

A. I only **B.** II only **C.** III only
D. I and II only **E.** I and III only

11. $\cos 45° \sin 30° - \tan 45° =$
C **A.** $\frac{\sqrt{2} - 2}{2}$ **B.** $\frac{-\sqrt{2} - 2}{2}$
C. $\frac{\sqrt{2} - 4}{4}$ **D.** $\frac{\sqrt{6} - 4}{4}$
E. $\frac{-\sqrt{6} - 4}{4}$

203

See *Teacher's Resource Book*, Cumulative Test, pp. 41–48.

Additional Answers

4. $\dfrac{\dfrac{1}{\sin\theta}+\dfrac{1}{\cos\theta}}{\dfrac{1}{\sin\theta}-\dfrac{1}{\cos\theta}}=$

$\left(\dfrac{\sin\theta\cos\theta}{\sin\theta\cos\theta}\right)\dfrac{\dfrac{1}{\sin\theta}+\dfrac{1}{\cos\theta}}{\dfrac{1}{\sin\theta}-\dfrac{1}{\cos\theta}}=$

$\dfrac{\cos\theta+\sin\theta}{\cos\theta-\sin\theta}$

10.

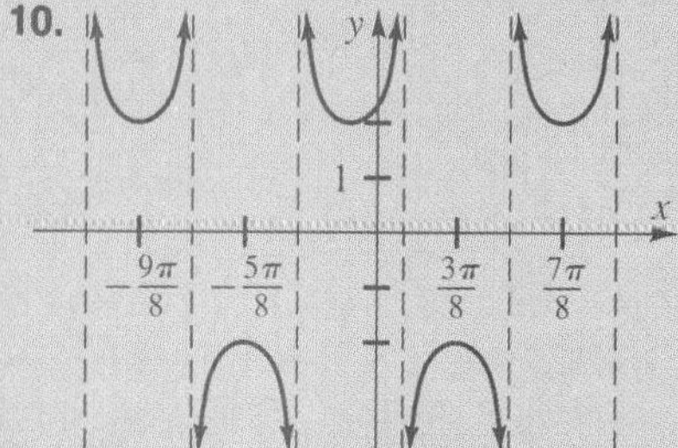

12.

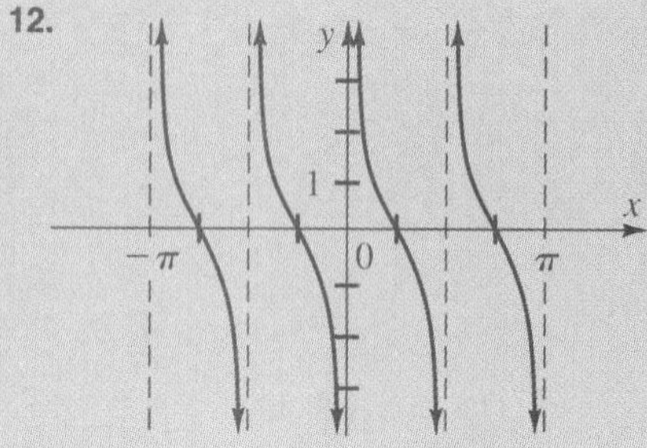

15. $(\sin\theta+\cos\theta+1)(\sin\theta+\cos\theta-1)=\sin^2\theta+\sin\theta\cos\theta-\sin\theta+\sin\theta\cos\theta+\cos^2\theta-\cos\theta+\sin\theta+\cos\theta-1=\sin^2\theta+\cos^2\theta+2\sin\theta\cos\theta-1=1+2\sin\theta\cos\theta-1=2\sin\theta\cos\theta$

CUMULATIVE REVIEW

1. Use the law of cosines to solve $\triangle ABC$ given $a = 11$, $b = 14$, and $\angle C = 34°$. Round the length of side c to two significant digits and the lengths of angle measures to the nearest degree. $\angle A = 52°$, $\angle B = 94°$, $c = 7.9$ **4.2**

2. Determine which real number(s) must be excluded from the domain of $f(x) = \dfrac{2}{x-3}$ $x \neq 3$ **1.1**

3. Express 132°18′25″ in decimal degrees to the nearest hundredth of a degree. 132.31° **1.4**

4. Prove $\dfrac{\dfrac{1}{\sin\theta}+\dfrac{1}{\cos\theta}}{\dfrac{1}{\sin\theta}-\dfrac{1}{\cos\theta}}=\dfrac{\cos\theta+\sin\theta}{\cos\theta-\sin\theta}$. See side column. **3.6**

5. Find $\cos(-48°)$ if $\cos 48° = 0.6691$. 0.6691 **2.2**

6. Find the area of the triangle with sides 5m, 7m, and 3m to the nearest square meter. 6 m^2 **4.6**

7. Given right triangle ABC with $\angle C = 90°$, $a = 11$ cm, and $c = 21$ cm, find b to the nearest centimeter. 18 cm **3.1**

A belt on a lawnmower runs a pulley of radius 3 in. at 100 rpm. Find **1.5**

8. the angular velocity of the pulley in radians per second $\frac{10\pi}{3}$ rad/s

9. the linear velocity of the belt in inches per second 10π in./s

10. Graph $y = 2\sec\left(x + \frac{\pi}{8}\right)$ over a two-period interval. See side column. **2.7**

11. Check graphically to see whether or not $2\cos^2 x = 1 + \cos 2x$ is an identity. Identity. **3.7**

12. Sketch the graph of $y = 3\cot 2x$ over the interval $-\pi \leq x \leq \pi$. See side column. **2.6**

13. Find the magnitude and direction of the sum $\vec{W}$ of two vectors $\vec{U}$ and $\vec{V}$ of magnitudes 6 and 1, respectively, if the angle between them measures 62°. Round the magnitude to the nearest whole number and the direction to the nearest degree. 7; 54° **4.7**

14. State whether or not the relation {(1, 6), (2, 3), (1, 7)} is a function. not a function **1.1**

15. Convert $(\sin\theta + \cos\theta + 1)(\sin\theta + \cos\theta - 1)$ to $2\sin\theta\cos\theta$. See side column. **3.5**

16. Using the law of cosines, find the length of a to two significant digits given $\angle A = 37°$, $b = 10$, and $c = 14$. 8.5 4.3

17. If $0° \le \theta \le 360°$, find the value(s) of θ that satisfy $\cos\theta = -1$ and $\tan\theta = 0$. 180° 1.8

18. A cat playtoy hangs at rest from a spring. It is then pulled down 12 inches and released. The toy bounces back and forth through its rest position making one full cycle every 4 s. Using the sine function as a model, write an equation for the function and then sketch its graph. $y = 12 \sin \frac{\pi}{2}(t - 1)$ 2.8

19. Verify the following identity for the given angle measure: $\sin^2 \frac{\pi}{6} \cos \frac{\pi}{6} \sec \frac{\pi}{6} = 1 - \cos^2 \frac{\pi}{6}$ See side column. 3.4

20. Compute to two significant digits the unknown sides for $\triangle ABC$ if $\angle C = 42°$, $\angle B = 87°$, and $c = 21$ in. $a = 24$ in.; $b = 31$ in. 4.1

21. Determine whether the function $y = x^2 - 5$ is symmetric with respect to the x-axis, to the y-axis, to the origin, to more than one of these, or to none of these. y-axis 2.1

22. Graph $y = \cos x - \sin 2x$ over the interval. $-2\pi \le x \le 2\pi$. See side column. 2.5

23. Given $\angle D = 58°$, $\angle F = 77°$, and $d = 35$ in., compute the area of $\triangle DEF$ to the nearest 10 square inches. 500 in.^{2} 4.5

24. A hip roof is in the form of four congruent isosceles triangles. If each triangle has a 30-ft base and a height of 12 ft, find the measure of each base angle. 39° 3.3

25. Let θ be an angle in standard position. In which quadrant(s) can θ lie when $\cos\theta > 0$ and $\sin\theta < 0$? Quadrant IV 1.6

26. Determine the amplitude and period for the function $y = -2\sin(-2x)$. Then sketch its graph over the interval $-\pi \le x \le \pi$. See side column. 2.3

27. Find the exact values of the other five trigonometric functions for an angle in standard position in quadrant II whose cosine is $-\frac{4}{5}$. $\frac{3}{5}$; $-\frac{3}{4}$; $\frac{5}{3}$; $-\frac{5}{4}$; $-\frac{4}{3}$ 1.7

28. Determine the amplitude, period, phase shift, and vertical shift for the function $y = -2\cos(2x + 2\pi) - 3$. 2; π; π left; 3 down 2.4

29. The shadow of a tower is 200 ft when the sun's elevation is 75.8°. Find the height of the tower to three significant digits. 790 ft 3.2

30. Find θ, where $0° \le \theta \le 360°$, to the nearest tenth of one degree given that $\cos\theta = -0.5253$ and $\sin\theta$ is positive. 121.7° 1.9

31. Use the law of tangents to find the unknown angles of $\triangle ABC$ to the nearest degree if $\angle B = 65°$, $a = 32$, and $c = 17$. $\angle A = 83°$, $\angle C = 32°$ 4.4

Additional Answers

19. $\sin^2 \frac{\pi}{6} \cos \frac{\pi}{6} \sec \frac{\pi}{6} = 1 - \cos^2 \frac{\pi}{6}$

$\left(\frac{1}{2}\right)^2 \left(\frac{\sqrt{3}}{2}\right)\left(\frac{2\sqrt{3}}{3}\right) = 1 - \left(\frac{\sqrt{3}}{2}\right)^2$

$\frac{1}{4} \cdot 1 = 1 - \frac{3}{4}; \frac{1}{4} = \frac{1}{4}$

22.

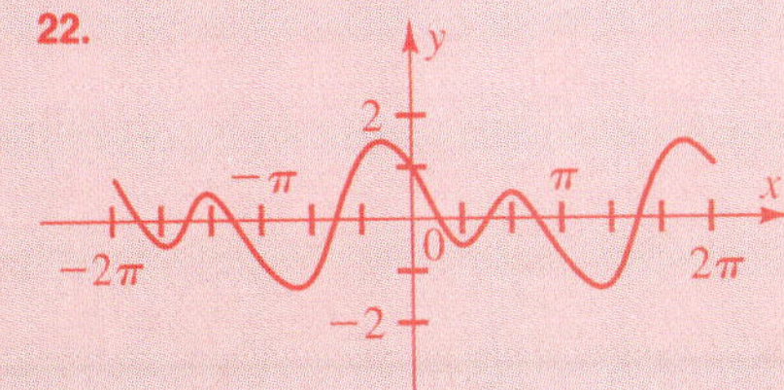

26. Amplitude: 2; period: π

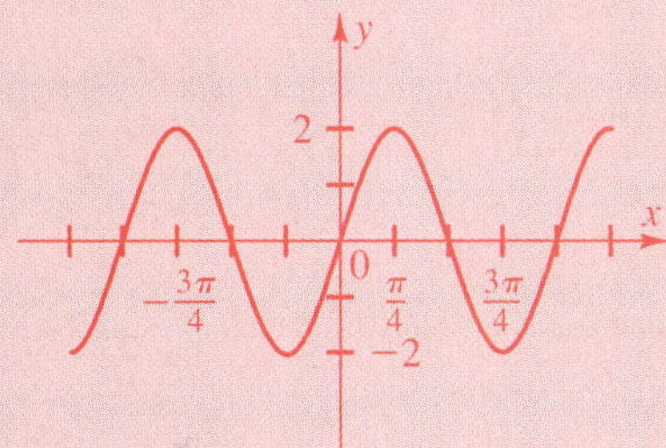

Additional Answers

32. $\dfrac{\cot^2\alpha}{\sec^2\alpha} =$

$\dfrac{\frac{\cos^2\alpha}{\sin^2\alpha}}{\frac{1}{\cos^2\alpha}} =$

$\dfrac{\sin^2\alpha}{\sin^2\alpha}\cdot\dfrac{\frac{\cos^2\alpha}{\sin^2\alpha}}{\frac{1}{\cos^2\alpha}} =$

$\dfrac{\cos^2\alpha}{\frac{\sin^2\alpha}{\cos^2\alpha}} =$

$\dfrac{\cos^2\alpha}{\tan^2\alpha}$

34. $\angle B = 41.4°$; $\angle C = 102.6°$; $c = 80$
$\angle B = 138.6°$; $\angle C = 5.4°$; $c = 7.7$

42. $\dfrac{1}{1 - \sin\alpha} + \dfrac{1}{1 + \sin\alpha} =$

$\dfrac{1 + \sin\alpha + 1 - \sin\alpha}{(1 - \sin\alpha)(1 + \sin\alpha)} =$

$\dfrac{2}{1 - \sin^2\alpha} = \dfrac{2}{\cos^2\alpha}$

48. $\sec^2\alpha - \tan^2\alpha = \dfrac{r^2}{x^2} - \dfrac{y^2}{x^2}$

$= \dfrac{r^2 - y^2}{x^2}$; but by the Pythagorean theorem, $r^2 - y^2 = x^2$, so $\sec^2\alpha - \tan^2\alpha = \dfrac{x^2}{x^2} = 1.$

50. $\dfrac{1 - \sin\beta}{\cos\beta} =$

$\left(\dfrac{1 + \sin\beta}{1 + \sin\beta}\right)\left(\dfrac{1 - \sin\beta}{\cos\beta}\right) =$

$\dfrac{1 - \sin^2\beta}{(1 + \sin\beta)(\cos\beta)} =$

$\dfrac{\cos^2\beta}{(1 + \sin\beta)(\cos\beta)} =$

$\dfrac{\cos\beta}{1 + \sin\beta}$

32. Prove $\dfrac{\cot^2\alpha}{\sec^2\alpha} = \dfrac{1 - \sin^2\alpha}{\tan^2\alpha}$. See side column. 3.6

33. Find x if the distance between (2, 2) and (x, 10) is 10. $x = 8$ or $x = -4$ 1.2

34. Determine the number of solutions to $\triangle ABC$ if $\angle A = 36°$, $a = 48$, and $b = 54$. If one or more solutions exist, solve the triangle or triangles. See side column. 4.2

Determine whether each function is even, odd, or neither even nor odd. 2.1

35. $g(x) = 4x^3 + 2x^2$ neither

36. $f(x) = 0.5x^2 - 3$ even

37. $h(x) = -2x^3 + x$ odd

38. A sailboat on Lake Michigan is 3.7 miles due north of buoy A. Buoy B is 9.3 miles due west of buoy A. How far apart are the sailboat and buoy B? 10 miles 3.3

39. Name three angles (two positive and one negative) that are coterminal with an angle of $-57°$. 303°, 663°, −417° 1.3

If $\angle A = 65°$ and $\angle B = 32°$, compute 4.4

40. $\tan\frac{1}{2}(A - B)$ 0.2962

41. $\tan\frac{1}{2}(A + B)$ 1.1303

42. Write the expression $\dfrac{1}{1 - \sin\alpha} + \dfrac{1}{1 + \sin\alpha}$ in terms of $\cos\alpha$. See side column. 3.5

Use the fact that sine and cosine are periodic functions and cofunctions to determine the following values. 2.2

43. If $\cos 67° = 0.3907$, find $\sin 23°$. 0.3907

44. If $\sin 192° = -0.2079$, find $\sin 912°$. −0.2079

45. If $\sin 33° = 0.5446$, find $\sin(-327°)$. 0.5446

46. If $\cos 160° = -0.9397$, find $\cos(-200°)$. −0.9397

47. Give the measure of the reference angle θ' for $\theta = 344°$. 16° 1.8

48. Use the definitions of the trigonometric functions to prove $\sec^2\alpha - \tan^2\alpha = 1$. See side column. 3.4

49. What is the area, in acres, of a triangular lot that is 296 ft on one side and 242 ft on an adjacent side, if the angle between these two sides measures 37.5°? (1 acre = 43,560 ft²) $\frac{1}{2}$ acre 4.5

50. Prove $\dfrac{1 - \sin\beta}{\cos\beta} = \dfrac{\cos\beta}{1 + \sin\beta}$. See side column. 3.6

51. Find the measure of the reference angle θ' for $\theta = 1227°$ 33° 1.8

52. A ladder rests against a building at a point that is 39 ft from the ground. If the ladder makes a 50° angle with the ground, what is the length of the ladder? 51 ft 3.1

OVERVIEW • Chapter 5

SUMMARY

Developing sum and difference identities for cosine, sine, and tangent of a sum or difference of two angles is taught to provide a smooth transition to double-angle, half-angle, and product/sum identities later in the chapter. Proving a number of different types of identities is introduced utilizing Pythagorean, reciprocal, and other basic identities.

After this chapter is completed, students should be able to prove all types of sum, difference, double-angle, half-angle, or product/sum identities. They should be able to apply their knowledge of these identities to simplifying complicated trigonometry expressions and proving other trigonometric identities.

CHAPTER OBJECTIVES

- To develop and use formulas for the cosine of a sum or difference of two angle measures
- To develop and use formulas for the sine of a sum or difference of two angle measures
- To develop and use formulas for the tangent of a sum or difference of two angle measures
- To develop and use double-angle identities
- To develop and use half-angle identities
- To develop and use product/sum identities

CHAPTER HIGHLIGHTS

The *theme* of Chapter 5 is aviation. Special features in the chapter include an application of trigonometry to standard-rate balanced turns used by airline pilots.

APPLICATIONS

Trigonometric identities, including sum and difference identities, double-angle identities and half-angle identities, are applied to a variety of fields. Such applications are generally geometric in nature, dealing with topics like surveying, indirect measurement, and determining slopes. An application of trigonometry in tidal motion is presented in Lesson 5.4.

TECHNOLOGY

Calculator

Calculators are helpful for verifying trigonometric identities, and in particular, for verifying double-angle and half-angle identities.

Computer

A program that computes the sine of an angle by using the sum formula and randomly generated angles is given in Lesson 5.2. Another program is presented which evaluates a product/sum formula, given two angle measures.

RESOURCES

Teacher's Resource Book

- Teaching Aid 5
- Transparencies 9 and 10

ASSIGNMENT GUIDE Meeting Student Needs

STUDENT TEXT				TEACHER'S RESOURCE BOOK	
Chapter Content	**Basic**	**Average**	**Enriched**	**P**	**E**
5.1 Cosine: Sum and Difference Identities	D: 212/1–19 odd, 31	D: 212/5, 9–27 odd, 31	D: 212/7, 11–31 odd	1	2
5.2 Sine: Sum and Difference Identities	D: 218/1–19 odd R: 212/4, 10, 18	D: 218/5, 9–27 odd R: 212/4, 14, 20	D: 218/5, 11, 15, 17–31 odd R: 212/8, 14, 20	3	4
5.3 Tangent: Sum and Difference Identities	D: 223/1–19 odd R: 218/2, 14, 18 225/TY	D: 223/5, 9–29 odd, 33 R: 218/6, 10, 18 225/TY	D: 223/7, 11, 15, 17–33 odd R: 218/8, 12, 20 225/TY	5	6
5.4 Double-Angle Identities	D: 228/1–21 odd, 33 R: 223/4, 10, 18	D: 229/9–29 odd, 33 R: 223/6, 10, 18	D: 229/11–33 odd R: 223/8, 12, 20	7	8
5.5 Half-Angle Identities	D: 235/1–23 odd R: 229/6, 14, 20	D: 235/7–29 odd, 33 R: 229/10, 18, 20	D: 235/9–33 odd R: 229/12, 18, 24	9	10
5.6 Product/Sum Identities	Omit	D: 240/3, 7, 11, 15–27 odd R: 235/8, 18, 26 241/TY	D: 240/3, 7, 11, 15–29 odd R: 235/12, 22, 28 241/TY	11	12

D = Daily R = Review TY = Test Yourself P = Practice E = Enrichment

	STUDENT TEXT				TEACHER'S RESOURCE BOOK	
Review and Testing	Test Yourself	225, 241	College Ent. Exam Rev.	247	Tests	
	Chapter Sum. and Rev.	244	Maintaining Skills	248	• Quizzes	49–52
	Chapter Test	246			• Chapter Test (Form A)	53–54
					• Chapter Test (Form B)	55–56
Special Features	Extra	213, 219	Trigonometry in Geography	236	Applications—Chapter 5	13
	Trigonometry in Tidal Motion	230	Application	242	Critical Thinking	4
					Alg. and Geom. Review	17–20
					Technology	5

5 Trigonometric Identities

Galaxies are aggregates of stars, dust, and gas that are held together by gravity. Scientists classify them into elliptical or spiral galaxies, depending upon their shape and appearance. The Milky Way (pictured above) is the galaxy in which our solar system is located.

BACKGROUND

The twentieth century has witnessed man's mastery of flight. The exploration of space provides a venue for continued growth in aviation. Students may be interested in researching some of the scientific concepts that were used in the development of flight.

LESSON PLAN

BACKGROUND

In the Preview, 45°-45°-90° and 30°-60°-90° triangle relationships are used to find the values of the trigonometric functions of special angles.

Additional Answers

Answers are in the following order: sin; cos; tan; csc; sec; cot.

1. $-\frac{\sqrt{2}}{2}, -\frac{\sqrt{2}}{2}, 1, -\sqrt{2}, -\sqrt{2}, 1$

2. $-\frac{1}{2}, \frac{\sqrt{3}}{2}, -\frac{\sqrt{3}}{3}, -2, \frac{2\sqrt{3}}{3}, -\sqrt{3}$

3. $-\frac{\sqrt{3}}{2}, -\frac{1}{2}, \sqrt{3}, -\frac{2\sqrt{3}}{3}, -2, \frac{\sqrt{3}}{3}$

4. $-\frac{\sqrt{3}}{2}, \frac{1}{2}, -\sqrt{3}, -\frac{2\sqrt{3}}{3}, 2, -\frac{\sqrt{3}}{3}$

5. $\frac{\sqrt{2}}{2}, -\frac{\sqrt{2}}{2}, -1, \sqrt{2}, -\sqrt{2}, -1$

6. $\frac{1}{2}, -\frac{\sqrt{3}}{2}, -\frac{\sqrt{3}}{3}, 2, -\frac{2\sqrt{3}}{3}, -\sqrt{3}$

5.1 Cosine: Sum and Difference Identities

Objective: To develop and use formulas for the cosine of a sum or difference of two angle measures

Before introducing the sum and difference identities, the trigonometric functions of angles whose measures are multiples of 30°, 45°, or 60° are reviewed.

Preview

Recall that the lengths of the sides of a 45°-45°-90° triangle are in the ratio $1:1:\sqrt{2}$, where $\sqrt{2}$ corresponds to the hypotenuse. The lengths of the sides of a 30°-60°-90° triangle are in the ratio $1:\sqrt{3}:2$, where 2 corresponds to the hypotenuse, and $\sqrt{3}$ to the side opposite the 60° angle. You can use these facts and reference angles to find the values of the trigonometric functions of special angles

EXAMPLE **Find the exact values of the six trigonometric functions of 120°.**

The reference angle is 60°. Let $r = 2$.
Then $x = -1$ and $y = \sqrt{3}$.

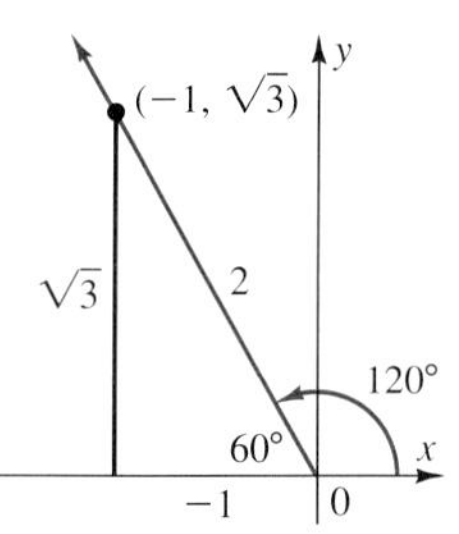

$$\sin 120° = \frac{\sqrt{3}}{2} \qquad \csc 120° = \frac{2}{\sqrt{3}} = \frac{2\sqrt{3}}{3}$$

$$\cos 120° = \frac{-1}{2} = -\frac{1}{2} \qquad \sec 120° = \frac{2}{-1} = -2$$

$$\tan 120° = \frac{\sqrt{3}}{-1} = -\sqrt{3} \qquad \cot 120° = \frac{-1}{\sqrt{3}} = -\frac{\sqrt{3}}{3}$$

Find the exact values of the six trigonometric functions of each angle.
See side column.

1. 225° **2.** 330° **3.** 240° **4.** −60° **5.** $\frac{3\pi}{4}$ **6.** $-\frac{7\pi}{6}$

In previous chapters, the identities that were presented involved only one angle (usually called θ). Now, identities that involve two angles, such as α and β, will be introduced. The first is the *difference identity for cosine*.

Difference Identity for Cosine

$$\cos(\alpha - \beta) = \cos\alpha\cos\beta + \sin\alpha\sin\beta$$

To prove this identity, place angles α and β in standard position, as shown in the figure at the right. The terminal sides of angles α and β intersect the unit circle O at points D and F, respectively, and the measure of central angle DOF is $\alpha - \beta$. Note that the coordinates of points D and F are

$$D(\cos \alpha, \sin \alpha) \qquad F(\cos \beta, \sin \beta)$$

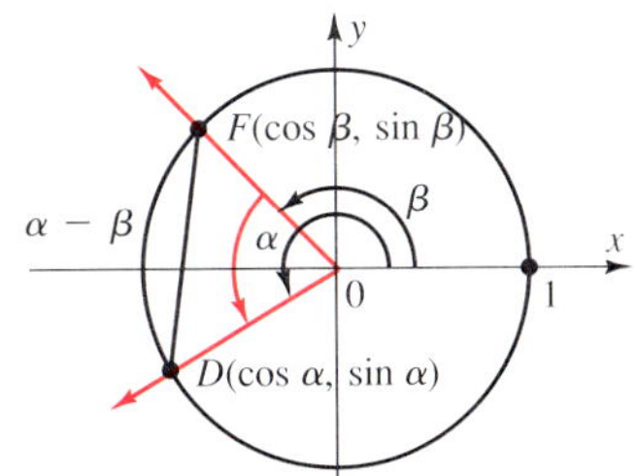

The figure at the right shows $\angle AOC$, with measure $\alpha - \beta$, in standard position. Note that the coordinates of points A and C are

$$A(1, 0) \qquad C(\cos(\alpha - \beta), \sin(\alpha - \beta))$$

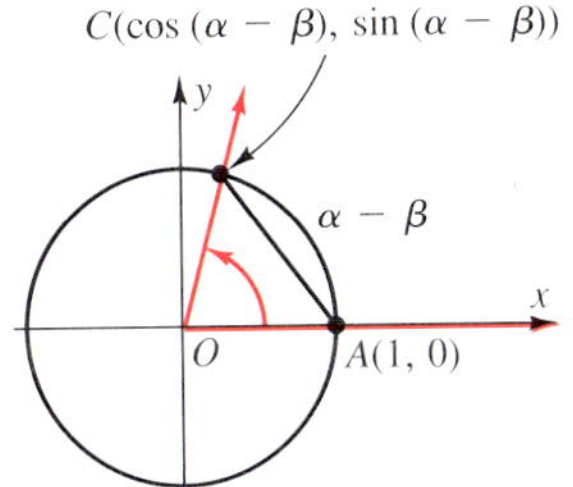

Since $\angle AOC$ and $\angle DOF$ are congruent, $\overline{AC}$ and $\overline{DF}$ are congruent and $AC = DF$. Use the distance formula to express AC and DF in terms of the coordinates of points A, C, D, and F.

First, find AC.

$$\begin{aligned} AC &= \sqrt{[\cos(\alpha - \beta) - 1]^2 + [\sin(\alpha - \beta) - 0]^2} && \textit{Distance formula} \\ &= \sqrt{\cos^2(\alpha - \beta) - 2\cos(\alpha - \beta) + 1 + \sin^2(\alpha - \beta)} \\ &= \sqrt{\cos^2(\alpha - \beta) + \sin^2(\alpha - \beta) + 1 - 2\cos(\alpha - \beta)} \\ &= \sqrt{1 + 1 - 2\cos(\alpha - \beta)} && \sin^2\theta + \cos^2\theta = 1 \\ &= \sqrt{2 - 2\cos(\alpha - \beta)} \end{aligned}$$

Then, find DF.

$$\begin{aligned} DF &= \sqrt{(\cos\alpha - \cos\beta)^2 + (\sin\alpha - \sin\beta)^2} && \textit{Distance formula} \\ &= \sqrt{\cos^2\alpha - 2\cos\alpha\cos\beta + \cos^2\beta + \sin^2\alpha - 2\sin\alpha\sin\beta + \sin^2\beta} \\ &= \sqrt{\cos^2\alpha + \sin^2\alpha + \cos^2\beta + \sin^2\beta - 2\cos\alpha\cos\beta - 2\sin\alpha\sin\beta} \\ &= \sqrt{1 + 1 - 2\cos\alpha\, 2\cos\beta - 2\sin\alpha\sin\beta} && \sin^2\theta + \cos^2\theta = 1 \\ &= \sqrt{2 - 2\cos\alpha\cos\beta - 2\sin\alpha\sin\beta} \end{aligned}$$

Since $AC = DF$, the proof is completed as follows:

$$\begin{aligned} \sqrt{2 - 2\cos(\alpha - \beta)} &= \sqrt{2 - 2\cos\alpha\cos\beta - 2\sin\alpha\sin\beta} \\ 2 - 2\cos(\alpha - \beta) &= 2 - 2\cos\alpha\cos\beta - 2\sin\alpha\sin\beta \\ -2\cos(\alpha - \beta) &= -2\cos\alpha\cos\beta - 2\sin\alpha\sin\beta \\ \cos(\alpha - \beta) &= \cos\alpha\cos\beta + \sin\alpha\sin\beta \end{aligned}$$

Note that $\cos(\alpha - \beta) = \cos[-(\alpha - \beta)] = \cos(\beta - \alpha)$, so the identity holds for any values of α and β.

TEACHING SUGGESTIONS

- It may be necessary to review the definitions of central angle of a circle, chord, congruence, the congruence symbol ≅, and the distance formula.
- Emphasize that $\cos(60 - 45)° \neq \cos 60° - \cos 45°$.
- Review the values of the trigonometric functions for the special angles: 30°, 45°, 60°, 90°, 180°, 270°, and 360°.

Critical Thinking

Discovering patterns Ask students to find values for α and β such that $\sin(\alpha + \beta) = \sin\alpha + \sin\beta$ is not an identity. Answers may vary. Example: $\sin(60° + 45°) = \sin(105°) \approx 0.9659$, which is not equal to $\sin 60° + \sin 45°$ which is equal to $0.866 + 0.7071$, or 1.5731.

CHALKBOARD EXAMPLES

- **For Example 1**

  Find the exact value of each trigonometric function.

  1. $\cos 255°$
     $\cos 255° = \cos(225° + 30°)$; $\frac{-\sqrt{6} + \sqrt{2}}{4}$
  2. $\cos \frac{13\pi}{12}$ $\cos \frac{13\pi}{12} = \cos\left(\frac{5\pi}{4} - \frac{\pi}{6}\right)$; $\frac{-\sqrt{6} - \sqrt{2}}{4}$

- **For Example 2**

  Find the exact value of each trigonometric function.

  3. $\cos 75°$ $\cos 75° = \cos(45° + 30°)$; $\frac{\sqrt{6} - \sqrt{2}}{4}$
  4. $\cos \frac{11\pi}{12}$ $\cos \frac{11\pi}{12} = \cos\left(\frac{2\pi}{3} + \frac{\pi}{4}\right)$; $\frac{-\sqrt{2} - \sqrt{6}}{4}$

EXAMPLE 1 **Find the exact value of cos 15°.**

Since $15° = 45° - 30°$, let $\alpha = 45°$ and $\beta = 30°$.

$$\begin{aligned}\cos 15° &= \cos(45° - 30°)\\ &= \cos 45° \cos 30° + \sin 45° \sin 30° \quad \textit{Difference identity for cosine}\\ &= \frac{\sqrt{2}}{2}\left(\frac{\sqrt{3}}{2}\right) + \frac{\sqrt{2}}{2}\left(\frac{1}{2}\right) = \frac{\sqrt{6}}{4} + \frac{\sqrt{2}}{4} = \frac{\sqrt{6} + \sqrt{2}}{4}\end{aligned}$$

The *sum identity for cosine* follows from the difference identity for cosine.

Sum Identity for Cosine

$$\cos(\alpha + \beta) = \cos\alpha\cos\beta - \sin\alpha\sin\beta$$

To prove the sum identity for cosine, write β as $-(-\beta)$. Then

$$\begin{aligned}\cos(\alpha + \beta) &= \cos[\alpha - (-\beta)]\\ &= \cos\alpha\cos(-\beta) + \sin\alpha\sin(-\beta)\\ &= \cos\alpha\cos\beta - \sin\alpha\sin\beta \quad \textit{cos}(-\beta) = \textit{cos}\,\beta;\ \textit{sin}(-\beta) = -\textit{sin}\,\beta\end{aligned}$$

EXAMPLE 2 **Find the exact value of $\cos \frac{7\pi}{12}$.**

Since $\frac{7\pi}{12} = \frac{4\pi}{12} + \frac{3\pi}{12} = \frac{\pi}{3} + \frac{\pi}{4}$, let $\alpha = \frac{\pi}{3}$ and $\beta = \frac{\pi}{4}$.

$$\begin{aligned}\cos\frac{7\pi}{12} &= \cos\left(\frac{\pi}{3} + \frac{\pi}{4}\right)\\ &= \cos\frac{\pi}{3}\cos\frac{\pi}{4} - \sin\frac{\pi}{3}\sin\frac{\pi}{4} \quad \textit{Sum identity for cosine}\\ &= \frac{1}{2}\left(\frac{\sqrt{2}}{2}\right) - \frac{\sqrt{3}}{2}\left(\frac{\sqrt{2}}{2}\right) = \frac{\sqrt{2}}{4} - \frac{\sqrt{6}}{4} = \frac{\sqrt{2} - \sqrt{6}}{4}\end{aligned}$$

It is important to note that $\cos(\alpha - \beta)$ is *not* equal to $\cos\alpha - \cos\beta$. To illustrate, $\cos(45° - 30°) = \frac{\sqrt{6} + \sqrt{2}}{4}$ (from Example 1), but $\cos 45° - \cos 30° = \frac{\sqrt{2}}{2} - \frac{\sqrt{3}}{2} = \frac{\sqrt{2} - \sqrt{3}}{2}$. Similarly, $\cos(\alpha + \beta)$ is *not* equal to $\cos\alpha + \cos\beta$.

EXAMPLE 3 **Find the exact values of $\cos(\alpha + \beta)$ and $\cos(\alpha - \beta)$ if $\sin \alpha = \frac{3}{5}$, $\tan \beta = -\frac{5}{12}$, $0 < \alpha < \frac{\pi}{2}$, and $\frac{\pi}{2} < \beta < \pi$.**

Find $\cos \alpha$, $\sin \beta$, and $\cos \beta$ before using the identities for cosine.

To find $\cos \alpha$, note that $\sin \alpha = \frac{3}{5}$ and α is in quadrant I. Use $y = 3$ and $r = 5$, and calculate x.

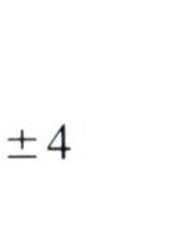

$$x = \pm\sqrt{5^2 - 3^2} = \pm\sqrt{16} = \pm 4$$

Since α is in quadrant I, $x = 4$.

Thus, $\cos \alpha = \frac{4}{5}$.

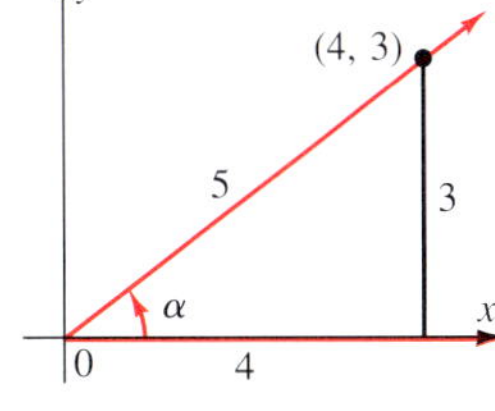

To find $\sin \beta$ and $\cos \beta$, note that $\tan \beta = -\frac{5}{12}$ and β is in quadrant II. Use $x = -12$ and $y = 5$, and calculate r.

$$r = \sqrt{(-12)^2 + 5^2} = 13$$

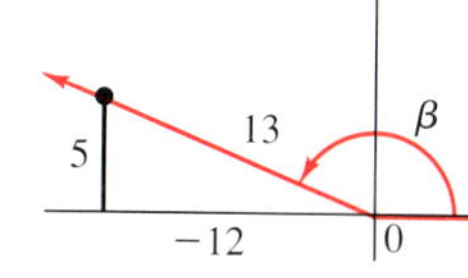

Thus, $\sin \beta = \frac{5}{13}$ and $\cos \beta = -\frac{12}{13}$.

Then, $\cos(\alpha + \beta) = \frac{4}{5}\left(-\frac{12}{13}\right) - \frac{3}{5}\left(\frac{5}{13}\right)$
$= -\frac{48}{65} - \frac{15}{65} = -\frac{63}{65}$

Also, $\cos(\alpha - \beta) = \frac{4}{5}\left(-\frac{12}{13}\right) + \frac{3}{5}\left(\frac{5}{13}\right)$
$= -\frac{48}{65} + \frac{15}{65} = -\frac{33}{65}$

The above sum and difference identities can be used to prove other identities.

EXAMPLE 4 **Prove: $\sin \theta = \cos\left(\frac{\pi}{2} - \theta\right)$**

$\sin \theta$	$\cos\left(\frac{\pi}{2} - \theta\right)$
	$\cos \frac{\pi}{2} \cos \theta + \sin \frac{\pi}{2} \sin \theta$
	$0(\cos \theta) + 1(\sin \theta)$
$=$	$\sin \theta$

Therefore, $\sin \theta = \cos\left(\frac{\pi}{2} - \theta\right)$.

- **For Example 3**
  Find the exact value of each of the following: $\sin \alpha = \frac{4}{5}$, $\tan \beta = -\frac{7}{24}$, $0 < \alpha < \frac{\pi}{2}$, and $\frac{\pi}{2} < \beta < \pi$.

  5. $\cos(\alpha + \beta)$ $-\frac{4}{5}$

  6. $\cos(\alpha - \beta)$ $-\frac{44}{125}$

- **For Example 4**
  Prove each identity.

  7. $\cos\left(\alpha - \frac{\pi}{2}\right) = \sin \alpha$

$\cos\left(\alpha - \frac{\pi}{2}\right)$	$\sin \alpha$
$\cos \alpha \cos \frac{\pi}{2} + \sin \alpha \sin \frac{\pi}{2}$	
$(\cos \alpha)(0) + (\sin \alpha)(1)$	
$\sin \alpha$	$=$

  8. $\cos(\pi + \theta) = -\cos \theta$

$\cos(\pi + \theta)$	$-\cos \theta$
$\cos \pi \cos \theta - \sin \pi \sin \theta$	
$(-1)(\cos \theta) - (0)(\sin \theta)$	
$-\cos \theta$	$=$

Common Errors

- Some students have trouble finding exact values for the cosine of an angle because they do not express the angle as the sum or difference of special angles. Stress that an angle should be expressed only in terms of 30°, 45°, 60°, and 90° $\left(\frac{\pi}{6}, \frac{\pi}{4}, \frac{\pi}{3}, \text{and } \frac{\pi}{2}\right)$ or their nonzero multiples. For example, 75° = 45° + 30° or 75° = 135° − 60°.
- Some students confuse the sum and difference identities for cosine. Stress that the sign between the terms on the right side of the equation is the opposite of the sign between the angle measures.
- See *Teacher's Resource Book* for additional remediation.

LESSON FOLLOW-UP

Discussion

If $f(x) = \cos x$, prove that

$$\frac{f(x+h) - f(x)}{h} = \cos x\left(\frac{\cos h - 1}{h}\right) - \sin x\left(\frac{\sin h}{h}\right).$$

$$\frac{f(x+h) - f(x)}{h} = \frac{\cos(x+h) - \cos x}{h} =$$

$$\frac{\cos x \cos h - \sin x \sin h - \cos x}{h} =$$

$$\frac{\cos x \cos h - \cos x - \sin x \sin h}{h} =$$

$$\frac{\cos x(\cos h - 1)}{h} - \sin x\left(\frac{\sin h}{h}\right)$$

Thus, $\frac{f(x+h) - f(x)}{h} =$

$\cos x\left(\frac{\cos h - 1}{h}\right) - \sin x\left(\frac{\sin h}{h}\right)$ is an identity if $f(x) = \cos x$.

Assignment Guide

See p. 206B for assignments.

Extra

An alternate method, using the law of cosines, for deriving the sum identity for cosine for acute angles is shown.

CLASS EXERCISES

Simplify each expression by writing it in terms of the cosine of one angle.

1. $\cos 80° \cos 20° + \sin 80° \sin 20°$ $\cos 60°$
2. $\cos 20° \cos 45° - \sin 20° \sin 45°$ $\cos 65°$
3. $\cos 30° \cos 35° - \sin 30° \sin 35°$ $\cos 65°$
4. $\cos 85° \cos 30° + \sin 85° \sin 30°$ $\cos 55°$
5. $\cos \frac{7\pi}{11} \cos \frac{2\pi}{11} - \sin \frac{7\pi}{11} \sin \frac{2\pi}{11}$ $\cos \frac{9\pi}{11}$
6. $\cos \frac{7\pi}{6} \cos \frac{5\pi}{6} + \sin \frac{7\pi}{6} \sin \frac{5\pi}{6}$ $\cos \frac{\pi}{3}$

Tell whether each of the following statements is true or false.

7. $\cos 57° = \cos(40° + 17°)$ True
8. $\cos(-20°) = \cos 30° - \cos 50°$ False
9. $\cos 65° = \cos 30° + \cos 35°$ False
10. $\cos 70° = \cos 80° - \cos 10°$ False

PRACTICE EXERCISES

Use the sum or the difference identity for cosine to find the exact value of each trigonometric function.

A

1. $\cos 75°$ $\frac{\sqrt{6} - \sqrt{2}}{4}$
2. $\cos 105°$ $\frac{\sqrt{2} - \sqrt{6}}{4}$
3. $\cos 165°$ $\frac{-\sqrt{6} - \sqrt{2}}{4}$
4. $\cos 195°$ $\frac{-\sqrt{2} - \sqrt{6}}{4}$
5. $\cos 345°$ $\frac{\sqrt{2} + \sqrt{6}}{4}$
6. $\cos 285°$ $\frac{\sqrt{6} - \sqrt{2}}{4}$
7. $\cos\left(\frac{5\pi}{3} + \frac{\pi}{4}\right)$ $\frac{\sqrt{2} + \sqrt{6}}{4}$
8. $\cos\left(\frac{7\pi}{6} + \frac{\pi}{4}\right)$ $\frac{\sqrt{2} - \sqrt{6}}{4}$

Find the exact value of $\cos(\alpha + \beta)$, given $0 < \alpha < \frac{\pi}{2}$, $0 < \beta < \frac{\pi}{2}$.

9. $\cos \alpha = \frac{4}{5}$, $\cos \beta = \frac{12}{13}$ $\frac{33}{65}$
10. $\sin \alpha = \frac{3}{5}$, $\sin \beta = \frac{5}{13}$ $\frac{33}{65}$

Find the exact value of $\cos(\alpha + \beta)$, given $\frac{\pi}{2} < \alpha < \pi$, $\frac{\pi}{2} < \beta < \pi$.

11. $\sin \alpha = \frac{7}{25}$, $\tan \beta = -\frac{3}{4}$ $\frac{3}{5}$
12. $\tan \alpha = -\frac{15}{8}$, $\sin \beta = \frac{5}{13}$ $\frac{21}{221}$

Find the exact value of $\cos(\alpha - \beta)$, given $\pi < \alpha < \frac{3\pi}{2}$, $\frac{\pi}{2} < \beta < \pi$.

13. $\cos \alpha = -\frac{7}{25}$, $\tan \beta = -\frac{3}{4}$ $-\frac{44}{125}$
14. $\cos \alpha = -\frac{5}{13}$, $\sin \beta = \frac{15}{17}$ $-\frac{140}{221}$

Find the exact value of $\cos(\alpha + \beta)$, given $\frac{3\pi}{2} < \alpha < 2\pi$, $\pi < \beta < \frac{3\pi}{2}$.

15. $\cos \alpha = \frac{15}{17}$, $\tan \beta = \frac{4}{3}$ $-\frac{77}{85}$
16. $\cos \alpha = \frac{7}{25}$, $\sin \beta = -\frac{5}{13}$ $-\frac{204}{325}$

Use the sum or the difference identity for cosine to prove each identity.

See page 483.

B

17. $\cos(360° + \alpha) = \cos \alpha$
18. $\cos(180° + \alpha) = -\cos \alpha$
19. $\cos(270° - \beta) = -\sin \beta$
20. $\cos(180° - \beta) = -\cos \beta$

21. $\cos\left(\frac{\pi}{3} - \beta\right) = \frac{\cos\beta + \sqrt{3}\sin\beta}{2}$

22. $\cos\left(\beta - \frac{\pi}{6}\right) = \frac{\sqrt{3}\cos\beta + \sin\beta}{2}$

23. $\cos\left(\frac{3\pi}{4} + \alpha\right) = -\frac{\sqrt{2}}{2}(\cos\alpha + \sin\alpha)$

24. $\cos 2\alpha = \cos^2\alpha - \sin^2\alpha$

Use the sum or the difference identity for cosine to find the exact value of each trigonometric function.

C 25. $\cos(-285°)$ $\frac{\sqrt{6}-\sqrt{2}}{4}$

26. $\cos(-105°)$ $\frac{\sqrt{2}-\sqrt{6}}{4}$

27. $\cos\left(-\frac{17\pi}{12}\right)$ $\frac{\sqrt{2}-\sqrt{6}}{4}$

28. $\cos\left(-\frac{5\pi}{12}\right)$ $\frac{\sqrt{6}-\sqrt{2}}{4}$

Prove each identity. See side column.

29. $\cos(\alpha - \beta) - \cos(\alpha + \beta) = 2\sin\alpha\sin\beta$

30. $\cos(\alpha + \beta) + \cos(\alpha - \beta) = 2\cos\alpha\cos\beta$

Applications

31. **Surveying** Part of a surveyor's diagram is shown at the right. Find the exact value of the cosine of $\angle AOD$. $\frac{\sqrt{2}-\sqrt{6}}{4}$

32. **Aviation** A jet is flying at an altitude of 12 mi, and its radar unit senses objects on the ground up to 13 mi from the plane in one direction and up to 20 mi from the plane in the opposite direction, as shown in the figure. Find the exact value of $\cos(\alpha + \beta)$. $\frac{16}{65}$

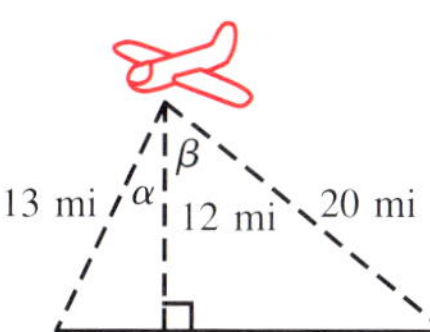

EXTRA

You can use the law of cosines, $b^2 = a^2 + c^2 - 2ac\cos(\alpha + \beta)$, to prove the sum identity for cosine for acute angles.

$$b^2 = a^2 + c^2 - 2ac\cos(\alpha + \beta)$$

$$\cos(\alpha + \beta) = \frac{a^2 + c^2 - b^2}{2ac}$$

$$\cos(\alpha + \beta) = \frac{(h^2 + y^2) + (h^2 + x^2) - (x + y)^2}{2ac}$$

$$\cos(\alpha + \beta) = \frac{2h^2 - 2xy}{2ac} = \frac{h^2 - xy}{ac} = \frac{h}{c}\left(\frac{h}{a}\right) - \frac{x}{c}\left(\frac{y}{a}\right) = \cos\alpha\cos\beta - \sin\alpha\sin\beta$$

Use the diagram at the right to derive the difference identity for cosine in a similar way. Start with this statement:

$$e^2 = f^2 + g^2 - 2fg\cos(\alpha - \beta)$$

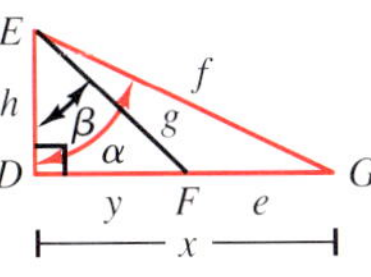

See below.

Lesson Quiz

Write each expression in terms of the cosine of one angle.

1. cos 70° cos 48° + sin 70° sin 48° cos 22°
2. cos 102° cos 16° − sin 102° sin 16° cos 118°

Use the sum or difference identity for cosine to find the exact value of each trigonometric function.

3. cos 330° $\frac{\sqrt{3}}{2}$
4. $\cos\frac{5\pi}{12}$ $\frac{\sqrt{6}-\sqrt{2}}{4}$
5. Find the exact value of $\cos(\alpha + \beta)$ if $\cos\alpha = \frac{3}{5}$, $\cos\beta = -\frac{5}{13}$, $270° < \alpha < 360°$, and $90° < \beta < 180°$. $\frac{33}{65}$
6. Find the exact value of $\cos(\alpha - \beta)$ if $\tan\alpha = \frac{3}{4}$, $\sin\beta = \frac{7}{25}$, $180° < \alpha < 270°$, and $90° < \beta < 180°$. $\frac{75}{125}$ or $\frac{3}{5}$
7. Prove: $\cos(\beta + 270°) = \sin\beta$

$\cos(\beta + 270°)$	$\sin\beta$
$\cos\beta\cos 270° - \sin\beta\sin 270°$	
$(\cos\beta)(0) - (\sin\beta)(-1)$	
$\sin\beta$	=

Teacher's Resource Book

Practice—Chapter 5, p. 1

Enrichment—Chapter 5, p. 2

Additional Answers

29. $\cos(\alpha - \beta) - \cos(\alpha + \beta) = (\cos\alpha \cdot \cos\beta + \sin\alpha \cdot \sin\beta) - (\cos\alpha \cdot \cos\beta - \sin\alpha \cdot \sin\beta) = \cos\alpha \cdot \cos\beta + \sin\alpha \cdot \sin\beta - \cos\alpha \cdot \cos\beta + \sin\alpha \cdot \sin\beta = 2\sin\alpha \cdot \sin\beta$

30. $\cos(\alpha + \beta) + \cos(\alpha - \beta) = \cos\alpha \cdot \cos\beta - \sin\alpha \cdot \sin\beta + \cos\alpha \cdot \cos\beta + \sin\alpha \cdot \sin\beta = 2\cos\alpha \cdot \cos\beta$

Extra

$e^2 = f^2 + g^2 - 2fg\cos(\alpha - \beta)$

$$\cos(\alpha - \beta) = \frac{f^2 + g^2 - e^2}{2fg} = \frac{(h^2 + x^2) + (h^2 + y^2) - (x^2 - y^2)}{2fg} = \frac{2h^2 + 2xy}{2fg} = \frac{h^2 + xy}{fg}$$

$$= \left(\frac{h}{f}\cdot\frac{h}{g}\right) + \left(\frac{x}{f}\cdot\frac{y}{g}\right) = \cos\alpha \cdot \cos\beta + \sin\alpha\sin\beta$$

LESSON PLAN

Vocabulary
Cofunction

BACKGROUND

In the Preview, cofunctions and complementary angles are discussed.

5.2 Sine: Sum and Difference Identities

Objective: To develop and use formulas for the sine of a sum or difference of two angle measures

The cosine function, the cotangent function, and the cosecant function are *cofunctions* of the sine function, the tangent function, and the secant function, respectively.

Preview

Any trigonometric function of θ, where $0° < \theta < 90°$, is equal to the cofunction of its complementary angle.

EXAMPLES

a. $\sin 50° = \cos (90° - 50°) = \cos 40°$

b. $\cot \frac{\pi}{6} = \tan \left(\frac{\pi}{2} - \frac{\pi}{6}\right) = \tan \frac{\pi}{3}$

Express each function of an angle in terms of the cofunction of its complementary angle.

1. $\sec 22.2°$ $\csc 67.8°$ **2.** $\tan 15.5°$ $\cot 74.5°$ **3.** $\csc 17°50'$ $\sec 72°10'$ **4.** $\sin 18°12'$ $\cos 71°48'$ **5.** $\cos \frac{\pi}{3}$ $\sin \frac{\pi}{6}$ **6.** $\cot \frac{\pi}{4}$ $\tan \frac{\pi}{4}$

It can be shown that the sine and the cosine are cofunctions for all angles. That is,

$$\cos (90° - \theta) = \sin \theta \quad \text{and} \quad \cos \theta = \sin (90° - \theta)$$

for *all* values of θ. The difference identity for cosine is used to prove these identities.

$$\begin{aligned} \cos (90° - \theta) &= \cos 90° \cos \theta + \sin 90° \sin \theta \\ &= 0 (\cos \theta) + 1(\sin \theta) \\ &= 0 + \sin \theta \\ &= \sin \theta \end{aligned}$$

$$\begin{aligned} \cos \theta &= \cos [90° - (90° - \theta)] \quad \textit{Substitute } 90° - (90° - \theta) \textit{ for } \theta. \\ &= \cos 90° \cos (90° - \theta) + \sin 90° \sin (90° - \theta) \\ &= 0 [\cos (90° - \theta)] + 1[\sin (90° - \theta)] \\ &= 0 + \sin (90° - \theta) \\ &= \sin (90° - \theta) \end{aligned}$$

Sum Identity for Sine

$\sin(\alpha + \beta) = \sin\alpha\cos\beta + \cos\alpha\sin\beta$

Cofunctions and the difference identity for cosine are used in the proof.

$$\begin{aligned}\sin(\alpha+\beta) &= \cos[90^\circ - (\alpha+\beta)] && \textit{\sin\theta = \cos(90^\circ - \theta)}\\ &= \cos[(90^\circ - \alpha) - \beta]\\ &= \cos(90^\circ-\alpha)\cos\beta + \sin(90^\circ-\alpha)\sin\beta\\ &= \sin\alpha\cos\beta + \cos\alpha\sin\beta\end{aligned}$$

EXAMPLE 1 **Find the exact value of sin 165°.**

Since $165^\circ = 120^\circ + 45^\circ$, let $\alpha = 120^\circ$ and $\beta = 45^\circ$.

$$\begin{aligned}\sin 165^\circ &= \sin(120^\circ + 45^\circ)\\ &= \sin 120^\circ\cos 45^\circ + \cos 120^\circ \sin 45^\circ && \textit{Sum identity for sine}\\ &= \frac{\sqrt{3}}{2}\left(\frac{\sqrt{2}}{2}\right) + \left(-\frac{1}{2}\right)\left(\frac{\sqrt{2}}{2}\right) = \frac{\sqrt{6}}{4} - \frac{\sqrt{2}}{4} = \frac{\sqrt{6}-\sqrt{2}}{4}\end{aligned}$$

Another important identity is the *difference identity for sine.*

Difference Identity for Sine

$\sin(\alpha - \beta) = \sin\alpha\cos\beta - \cos\alpha\sin\beta$

The proof follows from the sum identity for sine.

$$\begin{aligned}\sin(\alpha-\beta) &= \sin[\alpha + (-\beta)]\\ &= \sin\alpha\cos(-\beta) + \cos\alpha\sin(-\beta)\\ &= \sin\alpha\cos\beta - \cos\alpha\sin\beta && \textit{\cos(-\beta) = \cos\beta;\ \sin(-\beta) = -\sin\beta}\end{aligned}$$

EXAMPLE 2 **Find the exact value of $\sin\frac{\pi}{12}$.**

Since $\frac{\pi}{12} = \frac{4\pi}{12} - \frac{3\pi}{12} = \frac{\pi}{3} - \frac{\pi}{4}$, let $\alpha = \frac{\pi}{3}$ and $\beta = \frac{\pi}{4}$.

$$\begin{aligned}\sin\frac{\pi}{12} &= \sin\left(\frac{\pi}{3} - \frac{\pi}{4}\right)\\ &= \sin\frac{\pi}{3}\cos\frac{\pi}{4} - \cos\frac{\pi}{3}\sin\frac{\pi}{4} && \textit{Difference identity for sine}\\ &= \frac{\sqrt{3}}{2}\left(\frac{\sqrt{2}}{2}\right) - \frac{1}{2}\left(\frac{\sqrt{2}}{2}\right) = \frac{\sqrt{6}}{4} - \frac{\sqrt{2}}{4} = \frac{\sqrt{6}-\sqrt{2}}{4}\end{aligned}$$

TEACHING SUGGESTIONS

- It may be necessary to review the definition of complementary angles.
- You may wish to remind students that $\frac{\pi}{2}$ is 90°, and you may wish to review other degree and radian equivalents.

Critical Thinking

Comprehension Ask students to explain why exact values of the trigonometric functions of an angle are obtained when the sum and difference identities are used. Students should reason that when using the sum and difference identities, the given angle is expressed as the sum or difference of two angles that are multiples of 30°, 45°, 60°, or 90°, since exact values of the trigonometric functions of these angles are known.

CHALKBOARD EXAMPLES

- **For Example 1**
  Find the exact value of each trigonometric function.
  1. sin 105° $\frac{\sqrt{6}+\sqrt{2}}{4}$
  2. sin 285° $\frac{-\sqrt{6}-\sqrt{2}}{4}$
- **For Example 2**
  Find the exact value of each trigonometric function.
  3. $\sin\frac{5\pi}{12}$ $\frac{\sqrt{6}+\sqrt{2}}{4}$
  4. $\sin\frac{7\pi}{4}$ $-\frac{\sqrt{2}}{2}$

- **For Example 3**

  Find the exact value of each of the following if $\sin\alpha = \frac{4}{5}$, $\cos\beta = \frac{12}{13}$, $\frac{\pi}{2} < \alpha < \pi$, and $\frac{3\pi}{2} < \beta < 2\pi$.

  5. $\sin(\alpha + \beta)$ $\frac{63}{65}$

  6. $\sin(\alpha - \beta)$ $\frac{33}{65}$

- **For Example 4**

  Prove each identity.

  7. $\sin(\alpha - 0°) = \sin\alpha$

$\sin(\alpha - 0°)$	$\sin\alpha$
$\sin\alpha\cos 0° - \cos\alpha\sin 0°$	
$(\sin\alpha)(1) - (\cos\alpha)(0)$	
$\sin\alpha$	$=$

  8. $\sin(180° + \theta) = -\sin\theta$

$\sin(180° + \theta)$	$-\sin\theta$
$\sin 180°\cos\theta +$ $\cos 180°\sin\theta$	
$0(\cos\theta) + (-1)(\sin\theta)$	
$-\sin\theta$	$=$

Common Error

- Some students confuse the sum and difference identities for sine. Stress that the sign between the terms on the right side of the equation is the same as the sign between the angle measures on the left side.
- See *Teacher's Resource Book* for additional remediation.

Use the values of α and β from Examples 1 and 2 to verify the fact that

$$\sin(\alpha + \beta) \neq \sin\alpha + \sin\beta \qquad \text{and} \qquad \sin(\alpha - \beta) \neq \sin\alpha - \sin\beta$$

EXAMPLE 3 **Find the exact values of $\sin(\alpha + \beta)$ and $\sin(\alpha - \beta)$ if $\sin\alpha = \frac{\sqrt{3}}{2}$, $\tan\beta = -\frac{15}{8}$, $\frac{\pi}{2} < \alpha < \pi$, and $\frac{3\pi}{2} < \beta < 2\pi$.**

Find $\cos\alpha$, $\sin\beta$, and $\cos\beta$ before using the identities for sine.

To find $\cos\alpha$, note that $\sin\alpha = \frac{\sqrt{3}}{2}$ and α is in quadrant II. Use $y = \sqrt{3}$ and $r = 2$, and calculate x.

$$x = \pm\sqrt{2^2 - (\sqrt{3})^2} = \pm 1$$

Since α is in quadrant II, $x = -1$.

Thus, $\cos\alpha = -\frac{1}{2}$.

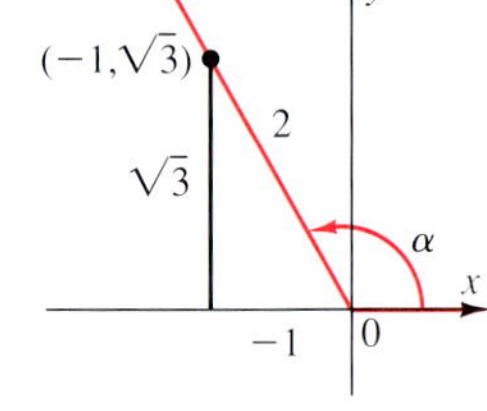

To find $\sin\beta$ and $\cos\beta$, note that $\tan\beta = -\frac{15}{8}$ and β is in quadrant IV. Use $y = -15$ and $x = 8$, and find r.

$$r = \sqrt{8^2 + (-15)^2} = 17$$

Thus, $\sin\beta = -\frac{15}{17}$ and $\cos\beta = \frac{8}{17}$.

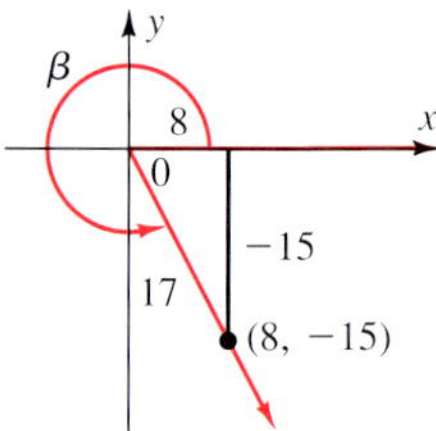

Then,
$$\sin(\alpha + \beta) = \frac{\sqrt{3}}{2}\left(\frac{8}{17}\right) + \left(-\frac{1}{2}\right)\left(-\frac{15}{17}\right) \qquad \textit{Sum identity for sine}$$
$$= \frac{8\sqrt{3}}{34} + \frac{15}{34}$$
$$= \frac{8\sqrt{3} + 15}{34}$$

$$\sin(\alpha - \beta) = \frac{\sqrt{3}}{2}\left(\frac{8}{17}\right) - \left(-\frac{1}{2}\right)\left(-\frac{15}{17}\right) \qquad \textit{Difference identity for sine}$$
$$= \frac{8\sqrt{3}}{34} - \frac{15}{34}$$
$$= \frac{8\sqrt{3} - 15}{34}$$

The sum and difference identities for sine can be used to prove other identities.

EXAMPLE 4 **Prove: $\sin(\pi + \theta) = -\sin\theta$**

$$\begin{array}{r|l} \sin(\pi + \theta) & -\sin\theta \\ \sin\pi\cos\theta + \cos\pi\sin\theta & \\ 0(\cos\theta) + (-1)(\sin\theta) & \\ -\sin\theta = & \end{array}$$

Work with the left side.

Thus, $\sin(\pi + \theta) = -\sin\theta$.

CLASS EXERCISES

Simplify each expression by writing it in terms of the sine of one angle.

1. $\sin 30° \cos 40° + \cos 30° \sin 40°$ sin 70°
2. $\sin 25° \cos 20° - \cos 25° \sin 20°$ sin 5°
3. $\sin 65° \cos 25° - \cos 65° \sin 25°$ sin 40°
4. $\sin 50° \cos 35° + \cos 50° \sin 35°$ sin 85°
5. $\sin 100° \cos 12° + \cos 100° \sin 12°$ sin 112°
6. $\sin 105° \cos 85° - \cos 105° \sin 85°$ sin 20°
7. $\sin\frac{6\pi}{7}\cos\frac{2\pi}{7} + \cos\frac{6\pi}{7}\sin\frac{2\pi}{7}$ $\sin\frac{8\pi}{7}$
8. $\sin\frac{7\pi}{9}\cos\frac{4\pi}{9} - \cos\frac{7\pi}{9}\sin\frac{4\pi}{9}$ $\sin\frac{\pi}{3}$
9. $\sin\frac{2\pi}{3}\cos\frac{\pi}{3} - \cos\frac{2\pi}{3}\sin\frac{\pi}{3}$ $\sin\frac{\pi}{3}$
10. $\sin\frac{\pi}{3}\cos\frac{4\pi}{3} + \cos\frac{\pi}{3}\sin\frac{4\pi}{3}$ $\sin\frac{5\pi}{3}$
11. $\sin\frac{\pi}{3}\cos\frac{\pi}{6} + \cos\frac{\pi}{3}\sin\frac{\pi}{6}$ $\sin\frac{\pi}{2}$
12. $\sin\frac{\pi}{3}\cos\frac{\pi}{6} - \cos\frac{\pi}{3}\sin\frac{\pi}{6}$ $\sin\frac{\pi}{6}$
13. $\sin\frac{5\pi}{4}\cos\frac{\pi}{2} - \cos\frac{5\pi}{4}\sin\frac{\pi}{2}$ $\sin\frac{3\pi}{4}$
14. $\sin\frac{5\pi}{4}\cos\frac{\pi}{2} + \cos\frac{5\pi}{4}\sin\frac{\pi}{2}$ $\sin\frac{7\pi}{4}$

Tell whether each of the following statements is true or false.

15. $\sin 75° = \sin 50° \cos 25° - \cos 50° \sin 25°$ false
16. $\sin 70° = \sin 80° \cos 10° - \cos 80° \sin 10°$ true
17. $\sin 40° = \sin 50° - \sin 10°$ false
18. $\sin 35° = \sin 20° + \sin 15°$ false
19. $\sin 50° = \sin 70° \cos 20° - \cos 70° \sin 20°$ true
20. $\sin 105° = \sin 90° \cos 15° - \cos 90° \sin 15°$ false
21. $\sin 120° = \sin 70° \cos 50° + \sin 70° \cos 50°$ false
22. $\sin 25° = \sin 10° \cos 15° + \sin 15° \cos 10°$ true

LESSON FOLLOW-UP

Discussion

What is the value of $\sin^2(\alpha + \beta) + \cos^2(\alpha + \beta)$? Justify your answer.

1; $\sin^2(\alpha + \beta) + \cos^2(\alpha + \beta) = [\sin(\alpha + \beta)]^2 + [\cos(\alpha + \beta)]^2 = (\sin\alpha\cos\beta + \cos\alpha\sin\beta)^2 + (\cos\alpha\cos\beta - \sin\alpha\sin\beta)^2 = \sin^2\alpha\cos^2\beta + 2\sin\alpha\cos\alpha\sin\beta\cos\beta + \cos^2\alpha\sin^2\beta + \cos^2\alpha\cos^2\beta - 2\sin\alpha\cos\alpha\sin\beta\cos\beta + \sin^2\alpha\sin^2\beta = \sin^2\alpha\sin^2\beta + \sin^2\alpha\cos^2\beta + \cos^2\alpha\sin^2\beta + \cos^2\alpha\cos^2\beta = \sin^2\alpha(\cos^2\beta + \sin^2\beta) + \cos^2\alpha(\cos^2\beta + \sin^2\beta) = \sin^2\alpha + \cos^2\alpha = 1$

Assignment Guide

See p. 206B for assignments.

Extra

An alternate method, using the unit circle, for deriving the sum identity for sine for acute angles is discussed.

Lesson Quiz

Write each expression in terms of the sine of one angle.

1. $\sin 49° \cos 58° - \cos 49° \sin 58°$ $\sin(-9°)$
2. $\sin 142° \cos 30° + \cos 142° \sin 30°$ $\sin 172°$

Use the sum or difference identity for sine to find the exact value of each trigonometric function.

3. $\sin 315°$ $-\frac{\sqrt{2}}{2}$
4. $\sin 285°$ $\frac{-\sqrt{6}-\sqrt{2}}{4}$
5. Find the exact value of $\sin(\alpha + \beta)$ if $\sin \alpha = -\frac{5}{13}$, $\sin \beta = \frac{4}{5}$, $270° < \alpha < 360°$, and $0° < \beta < 90°$. $\frac{33}{65}$
6. Find the exact value of $\sin(\alpha - \beta)$ if $\cos \alpha = \frac{7}{25}$, $\tan \beta = \frac{12}{5}$, $270° < \alpha < 360°$, and $0° < \beta < 90°$. $-\frac{204}{325}$
7. Prove: $\sin\left(\frac{\pi}{2} - \beta\right) = \cos \beta$

$\sin\left(\frac{\pi}{2} - \beta\right)$	$\cos \beta$
$\sin\frac{\pi}{2}\cos\beta - \cos\frac{\pi}{2}\sin\beta$	
$(1)(\cos\beta) - 0(\sin\beta)$	
$\cos\beta$	$=$

PRACTICE EXERCISES

Use the sum or the difference identity for sine to find the exact value of each trigonometric function.

A

1. $\sin 15°$ $\frac{\sqrt{6}-\sqrt{2}}{4}$
2. $\sin 75°$ $\frac{\sqrt{6}+\sqrt{2}}{4}$
3. $\sin 195°$ $\frac{\sqrt{2}-\sqrt{6}}{4}$
4. $\sin 255°$ $\frac{-\sqrt{2}-\sqrt{6}}{4}$
5. $\sin 375°$ $\frac{\sqrt{6}-\sqrt{2}}{4}$
6. $\sin 345°$ $\frac{\sqrt{2}-\sqrt{6}}{4}$
7. $\sin\left(\frac{\pi}{3} + \frac{\pi}{4}\right)$ $\frac{\sqrt{2}+\sqrt{6}}{4}$
8. $\sin\left(\frac{\pi}{4} + \frac{4\pi}{3}\right)$ $\frac{-\sqrt{2}-\sqrt{6}}{4}$

Find the exact value of $\sin(\alpha + \beta)$.

9. $\sin \alpha = \frac{12}{13}$, $\sin \beta = \frac{7}{25}$, $0 < \alpha < \frac{\pi}{2}$, $0 < \beta < \frac{\pi}{2}$ $\frac{323}{325}$
10. $\cos \alpha = -\frac{3}{5}$, $\cos \beta = \frac{8}{17}$, $\frac{\pi}{2} < \alpha < \pi$, $0 < \beta < \frac{\pi}{2}$ $-\frac{13}{85}$
11. $\cos \alpha = \frac{15}{17}$, $\tan \beta = \frac{4}{3}$, $\frac{3\pi}{2} < \alpha < 2\pi$, $\pi < \beta < \frac{3\pi}{2}$ $-\frac{36}{85}$
12. $\cos \alpha = -\frac{7}{25}$, $\tan \beta = -\frac{3}{4}$, $\pi < \alpha < \frac{3\pi}{2}$, $\frac{\pi}{2} < \beta < \pi$ $\frac{3}{5}$

Find the exact value of $\sin(\alpha - \beta)$.

13. $\cos \alpha = \frac{3}{5}$, $\tan \beta = \frac{12}{5}$, $0 < \alpha < \frac{\pi}{2}$, $0 < \beta < \frac{\pi}{2}$ $-\frac{16}{65}$
14. $\cos \alpha = \frac{4}{5}$, $\tan \beta = \frac{5}{12}$, $0 < \alpha < \frac{\pi}{2}$, $0 < \beta < \frac{\pi}{2}$ $\frac{16}{65}$
15. $\cos \alpha = \frac{7}{25}$, $\sin \beta = \frac{5}{13}$, $\frac{3\pi}{2} < \alpha < 2\pi$, $0 < \beta < \frac{\pi}{2}$ $-\frac{323}{325}$
16. $\cos \alpha = -\frac{5}{13}$, $\sin \beta = \frac{15}{17}$, $\pi < \alpha < \frac{3\pi}{2}$, $\frac{\pi}{2} < \beta < \pi$ $\frac{171}{221}$

Use the sum or the difference identity for sine to prove each identity.

See page 483.

B

17. $\sin(180° - \alpha) = \sin \alpha$
18. $\sin(\alpha - 180°) = -\sin \alpha$
19. $\sin(210° + \alpha) = \frac{-\cos \alpha - \sqrt{3}\sin \alpha}{2}$
20. $\sin(270° + \beta) = -\cos \beta$
21. $\sin(45° + \alpha) = \frac{\sqrt{2}}{2}(\cos \alpha + \sin \alpha)$
22. $\sin\left(\frac{\pi}{2} + \beta\right) = \cos \beta$
23. $\sin\left(\frac{\pi}{3} - \beta\right) = \frac{1}{2}(\sqrt{3}\cos \beta - \sin \beta)$
24. $\sin(2\beta) = 2 \sin \beta \cos \beta$
25. $\sin\left(\beta - \frac{7\pi}{6}\right) = -\frac{1}{2}(\sqrt{3}\sin \beta - \cos \beta)$
26. $\sin(-\beta) = -\sin \beta$

Find the exact value of each trigonometric function using the sine identities.

C **27.** $\sin(-105°)$ $\frac{-\sqrt{2}-\sqrt{6}}{4}$ **28.** $\sin(-285°)$ $\frac{\sqrt{2}+\sqrt{6}}{4}$ **29.** $\sin\left(-\frac{5\pi}{12}\right)$ $\frac{-\sqrt{2}-\sqrt{6}}{4}$ **30.** $\sin\left(-\frac{17\pi}{12}\right)$ $\frac{\sqrt{2}+\sqrt{6}}{4}$

Prove each identity. See side column.

31. $\sin(\alpha + \beta) + \sin(\alpha - \beta) = 2 \sin \alpha \cos \beta$

32. $\sin(\alpha + \beta) - \sin(\alpha - \beta) = 2 \cos \alpha \sin \beta$

Applications

Computer In the program at the right, the user is asked to enter an angle measure m, in degrees. Line 20 uses randomly generated numbers to generate angle A and then computes angle B such that $A + B = m$. The sine of angle m is found using the sum identity for sine.

```
10 HOME: INPUT "ENTER AN ANGLE MEASURE"; M
20 A = INT(M*RND(1) + 1): B = M - A
30 HOME: PRINT "SIN("A" + "B") =
   SIN"A" COS "B" + COS "A" SIN "B :
   C = 0.017453292
40 PRINT "SIN("A" + "B") =
   "SIN(A*C)*COS(B*C) + COS(A*C)*SIN(B*C)
50 END
```

Use the program to evaluate each of the following.

33. $\sin 20°$ 0.3420 **34.** $\sin 168°$ 0.2079 **35.** $\sin 247°$ −0.9205 **36.** $\sin 312°$ −0.7431

EXTRA

Use the portion of the unit-circle diagram at the right to derive the sum identity for sine for acute angles.

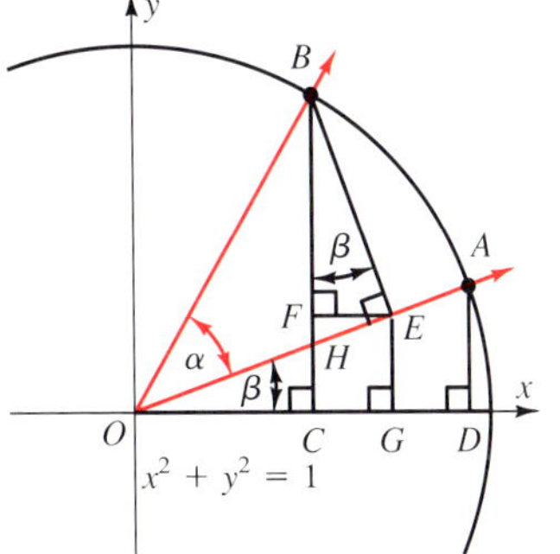

1. The measure of $\angle FBE$ is β. Why?

$\triangle OHC \sim \triangle BHE$, so corresp. $\angle$s are $\cong$

2. Complete the derivation of the sum identity for sine.

$$\begin{aligned}\sin(\alpha + \beta) &= BC \\ &= BF + FC \\ &= BE(\cos \underline{?}) + FC \quad \beta \\ &= \sin \underline{?} \cos \underline{?} + EG \quad \alpha;\ \beta \\ &= \sin \underline{?} \cos \underline{?} + OE(\sin \underline{?}) \quad \alpha;\ \beta;\ \beta \\ &= \sin \underline{?} \cos \underline{?} + \cos \underline{?} \sin \underline{?} \quad \alpha;\ \beta;\ \alpha;\ \beta\end{aligned}$$

3. Use the diagram to derive the sum identity for cosine for acute angles. See side column.

Computer

The complete form of the program for calculating the sine using a sum formula is on the disk provided with the *Teacher's Resource Book*.

Teacher's Resource Book

Practice—Chapter 5, p. 3
Enrichment—Chapter 5, p. 4

Additional Answers

31. $\sin(\alpha + \beta) + \sin(\alpha - \beta)$
$= \sin \alpha \cdot \cos \beta + \cos \alpha \cdot \sin \beta$
$+ \sin \alpha \cdot \cos \beta - \cos \alpha \cdot \sin \beta$
$= 2 \sin \alpha \cdot \cos \beta$

32. $\sin(\alpha + \beta) - \sin(\alpha - \beta)$
$= (\sin \alpha \cdot \cos \beta + \cos \alpha \cdot \sin \beta)$
$- (\sin \alpha \cdot \cos \beta - \cos \alpha \cdot \sin \beta)$
$= \sin \alpha \cdot \cos \beta + \cos \alpha \cdot \sin \beta$
$- \sin \alpha \cdot \cos \beta + \cos \alpha \cdot \sin \beta$
$= 2 \cos \alpha \cdot \sin \beta$

Extra

3. $\cos(\alpha + \beta) = OC$
$= OG - CG$
$= OE(\cos \beta) - FE$
$= \cos \alpha \cos \beta - FE$
$= \cos \alpha \cos \beta - BE(\sin \beta)$
$= \cos \alpha \cos \beta - \sin \alpha \sin \beta$

LESSON PLAN

BACKGROUND

In the Preview, conjugates are reviewed and used to rationalize denominators. This skill is used to simplify exact values obtained by using the sum or difference identity for tangent.

5.3 Tangent: Sum and Difference Identities

Objective: To develop and use formulas for the tangent of a sum or difference of two angle measures

To use tangent identities, you must rationalize denominators.

Preview

Recall the special product $(a + b)(a - b) = a^2 - b^2$, where $a + b$ and $a - b$ are conjugates. Conjugates are used to rationalize denominators.

EXAMPLE $\frac{4}{1+\sqrt{2}} = \left(\frac{4}{1+\sqrt{2}}\right)\left(\frac{1-\sqrt{2}}{1-\sqrt{2}}\right) = \frac{4-4\sqrt{2}}{1-2} = -4 + 4\sqrt{2}$

Simplify by rationalizing the denominator.

1. $\frac{2}{1-\sqrt{3}}$ $\quad -1-\sqrt{3}$
2. $\frac{\sqrt{2}}{1+\sqrt{2}}$ $\quad 2-\sqrt{2}$
3. $\frac{1-\sqrt{2}}{3+\sqrt{2}}$ $\quad \frac{5-4\sqrt{2}}{7}$
4. $\frac{\sqrt{2}-\sqrt{6}}{\sqrt{2}+\sqrt{6}}$ $\quad -2+\sqrt{3}$
5. $\frac{\sqrt{2}+3}{1-2\sqrt{2}}$ $\quad -1-\sqrt{2}$

The sum identities for sine and cosine lead to the *sum identity for tangent.*

Sum Identity for Tangent

$$\tan(\alpha + \beta) = \frac{\tan\alpha + \tan\beta}{1 - \tan\alpha\tan\beta},\quad \tan\alpha\tan\beta \neq 1$$

Therefore,

$$\tan(\alpha+\beta) = \frac{\sin(\alpha+\beta)}{\cos(\alpha+\beta)}$$

$\tan\theta = \frac{\sin\theta}{\cos\theta}$

$$= \frac{\sin\alpha\cos\beta + \cos\alpha\sin\beta}{\cos\alpha\cos\beta - \sin\alpha\sin\beta}$$

$$= \frac{\frac{\sin\alpha\cos\beta}{\cos\alpha\cos\beta} + \frac{\cos\alpha\sin\beta}{\cos\alpha\cos\beta}}{\frac{\cos\alpha\cos\beta}{\cos\alpha\cos\beta} - \frac{\sin\alpha\sin\beta}{\cos\alpha\cos\beta}}$$

Divide each term by $\cos\alpha\cos\beta$.

$$= \frac{\frac{\sin\alpha}{\cos\alpha} + \frac{\sin\beta}{\cos\beta}}{1 - \left(\frac{\sin\alpha}{\cos\alpha}\right)\left(\frac{\sin\beta}{\cos\beta}\right)} = \frac{\tan\alpha + \tan\beta}{1 - \tan\alpha\tan\beta}$$

EXAMPLE 1 **Find the exact value of tan 345°.**

Since $345° = 300° + 45°$, let $\alpha = 300°$ and $\beta = 45°$.

$$\tan 345° = \tan(300° + 45°)$$

$$= \frac{\tan 300° + \tan 45°}{1 - \tan 300° \tan 45°} \qquad \textit{Sum identity for tangent}$$

$$= \frac{-\sqrt{3} + 1}{1 - (-\sqrt{3})(1)}$$

$$= \frac{-\sqrt{3} + 1}{1 + \sqrt{3}}$$

$$= \left(\frac{1 - \sqrt{3}}{1 + \sqrt{3}}\right)\left(\frac{1 - \sqrt{3}}{1 - \sqrt{3}}\right) \qquad \textit{Simplify using the conjugate.}$$

$$= \frac{1 - 2\sqrt{3} + 3}{1 - 3} = \frac{4 - 2\sqrt{3}}{-2} = -2 + \sqrt{3}$$

The *difference identity for tangent* can be derived from the sum identity.

Difference Identity for Tangent

$$\tan(\alpha - \beta) = \frac{\tan\alpha - \tan\beta}{1 + \tan\alpha\tan\beta}, \quad \tan\alpha\tan\beta \neq -1$$

The proof makes use of the sum identity for tangent and the identity $\tan(-\beta) = -\tan\beta$. The proof of $\tan(-\beta) = -\tan\beta$ is given first.

$$\tan(-\beta) = \frac{\sin(-\beta)}{\cos(-\beta)} = \frac{-\sin\beta}{\cos\beta} = -\tan\beta$$

Then, $\tan(\alpha - \beta) = \tan[\alpha + (-\beta)]$

$$= \frac{\tan\alpha + \tan(-\beta)}{1 - \tan\alpha\tan(-\beta)} \qquad \textit{Sum identity for tangent}$$

$$= \frac{\tan\alpha - \tan\beta}{1 + \tan\alpha\tan\beta} \qquad \tan(-\beta) = -\tan\beta$$

EXAMPLE 2 **Find the exact value of $\tan\frac{5\pi}{12}$.**

Since $\frac{5\pi}{12} = \frac{8\pi}{12} - \frac{3\pi}{12} = \frac{2\pi}{3} - \frac{\pi}{4}$, let $\alpha = \frac{2\pi}{3}$ and $\beta = \frac{\pi}{4}$.

$$\tan\frac{5\pi}{12} = \tan\left(\frac{2\pi}{3} - \frac{\pi}{4}\right)$$

TEACHING SUGGESTIONS

- When discussing Example 1, be sure to review rationalizing denominators. Include single term denominators and those where you must multiply by the conjugate.
- You may wish to point out that if students memorize the identities for $\sin(\alpha + \beta)$ and $\cos(\alpha + \beta)$, they can then use $\sin(-\theta) = -\sin\theta$, $\cos(-\theta) = \cos\theta$, $\tan\theta = \frac{\sin\theta}{\cos\theta}$, and $\tan(-\theta) = -\tan\theta$ to obtain the identities for $\sin(\alpha - \beta)$, $\cos(\alpha - \beta)$, $\tan(\alpha + \beta)$, and $\tan(\alpha - \beta)$ quickly.

CHALKBOARD EXAMPLES

- **For Example 1**

  Find the exact value of each trigonometric function.

  1. tan 75° $2 + \sqrt{3}$

  2. $\tan\frac{5\pi}{3}$ $-\sqrt{3}$

- **For Example 2**

  Find the exact value of each trigonometric function.

  3. $\tan\frac{7\pi}{12}$ $-2 - \sqrt{3}$

  4. $\tan(-15°)$ $-2 + \sqrt{3}$

- **For Example 3**
  Find the exact value of each of the following if $\sin \alpha = \frac{3}{5}$, $\sin \beta = \frac{24}{25}$, $0 < \alpha < \frac{\pi}{2}$, and $\frac{\pi}{2} < \beta < \pi$.

  5. $\tan(\alpha + \beta)$ $\quad -\frac{3}{4}$

  6. $\tan(\alpha - \beta)$ $\quad -\frac{117}{44}$

- **For Example 4**
  Prove each identity.

  7. $\tan(\alpha + \pi) = \tan \alpha$

$\tan(\alpha + \pi)$	$\tan \alpha$
$\frac{\tan \alpha + \tan \pi}{1 - \tan \alpha \tan \pi}$	
$\frac{\tan \alpha + 0}{1 - \tan \alpha\,(0)}$	
$\frac{\tan \alpha}{1}$	
$\tan \alpha$	$=$

  8. $\tan\left(\frac{\pi}{4} - \alpha\right) = \frac{1 - \tan \alpha}{1 + \tan \alpha}$

$\tan\left(\frac{\pi}{4} - \alpha\right)$	$\frac{1 - \tan \alpha}{1 + \tan \alpha}$
$\frac{\tan \frac{\pi}{4} - \tan \alpha}{1 + \tan \frac{\pi}{4} \tan \alpha}$	
$\frac{1 - \tan \alpha}{1 + 1\,(\tan \alpha)}$	
$\frac{1 - \tan \alpha}{1 + \tan \alpha}$	$=$

$$\tan \frac{5\pi}{12} = \frac{\tan \frac{2\pi}{3} - \tan \frac{\pi}{4}}{1 + \tan \frac{2\pi}{3} \tan \frac{\pi}{4}} \qquad \textit{Difference identity for tangent}$$

$$= \frac{-\sqrt{3} - 1}{1 + (-\sqrt{3})(1)} = \frac{-\sqrt{3} - 1}{1 - \sqrt{3}}$$

$$= \left(\frac{-\sqrt{3} - 1}{1 - \sqrt{3}}\right)\left(\frac{1 + \sqrt{3}}{1 + \sqrt{3}}\right) = \frac{-4 - 2\sqrt{3}}{-2} = 2 + \sqrt{3}$$

Note that $\tan(\alpha + \beta) \neq \tan \alpha + \tan \beta$ and $\tan(\alpha - \beta) \neq \tan \alpha - \tan \beta$.

EXAMPLE 3 **Find the exact value of $\tan(\alpha + \beta)$ if $\sin \alpha = \frac{4}{5}$, $\cos \beta = -\frac{5}{13}$, $0 < \alpha < \frac{\pi}{2}$, and $\frac{\pi}{2} < \beta < \pi$.**

To find $\tan \alpha$, note that $\sin \alpha = \frac{4}{5}$ and α is in quadrant I. Use $y = 4$ and $r = 5$, and find x.

$$x = \pm\sqrt{5^2 - 4^2} = \pm 3$$

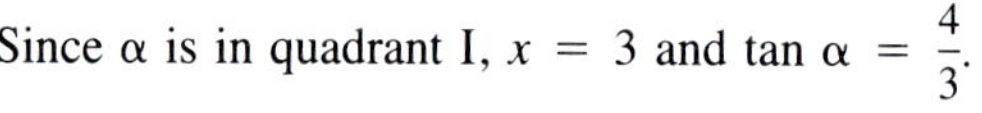
Since α is in quadrant I, $x = 3$ and $\tan \alpha = \frac{4}{3}$.

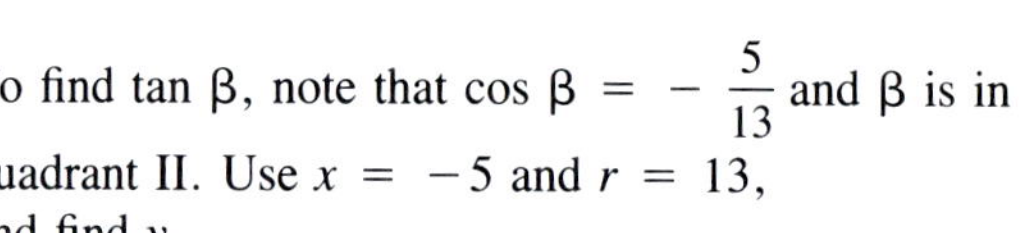
To find $\tan \beta$, note that $\cos \beta = -\frac{5}{13}$ and β is in quadrant II. Use $x = -5$ and $r = 13$, and find y.

$$y = \pm\sqrt{13^2 - (-5)^2} = \pm 12$$

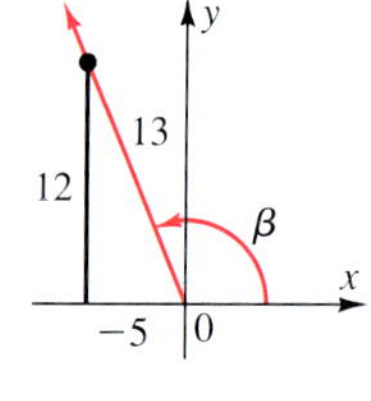

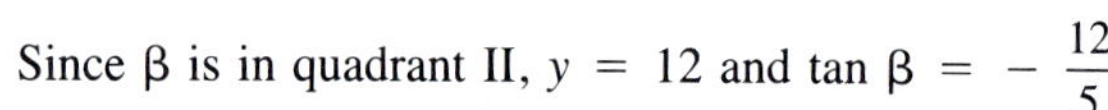
Since β is in quadrant II, $y = 12$ and $\tan \beta = -\frac{12}{5}$.

$$\tan(\alpha + \beta) = \frac{\tan \alpha + \tan \beta}{1 - \tan \alpha \tan \beta}$$

$$= \frac{\frac{4}{3} + \left(-\frac{12}{5}\right)}{1 - \left(\frac{4}{3}\right)\left(-\frac{12}{5}\right)} \qquad \textit{To simplify, multiply by } \frac{15}{15}.$$

$$= \frac{20 - 36}{15 + 48} = -\frac{16}{63}$$

The sum and difference identities can be used to prove other identities.

EXAMPLE 4 **Prove: $\tan(\pi + \theta) = \tan\theta$**

$\tan(\pi + \theta)$	$\tan\theta$
$\dfrac{\tan\pi + \tan\theta}{1 - \tan\pi\tan\theta}$	
$\dfrac{0 + \tan\theta}{1 - (0)(\tan\theta)}$	
$\dfrac{\tan\theta}{1}$	
$\tan\theta =$	

Thus, $\tan(\pi + \theta) = \tan\theta$.

CLASS EXERCISES

Write each expression in terms of the tangent of one angle.

1. $\dfrac{\tan 40° + \tan 20°}{1 - \tan 40° \tan 20°}$ tan 60°
2. $\dfrac{\tan 70° - \tan 10°}{1 + \tan 70° \tan 10°}$ tan 60°
3. $\dfrac{\tan\frac{5\pi}{9} - \tan\frac{4\pi}{9}}{1 + \tan\frac{5\pi}{9}\tan\frac{4\pi}{9}}$ $\tan\frac{\pi}{9}$
4. $\dfrac{\tan\frac{\pi}{8} + \tan\frac{\pi}{8}}{1 - \tan\frac{\pi}{8}\tan\frac{\pi}{8}}$ $\tan\frac{\pi}{4}$
5. $\dfrac{\tan\frac{\pi}{4} - \tan\frac{\pi}{4}}{1 + \tan\frac{\pi}{4}\tan\frac{\pi}{4}}$ tan 0
6. $\dfrac{\tan\frac{3\pi}{4} + \tan\frac{\pi}{4}}{1 - \tan\frac{3\pi}{4}\tan\frac{\pi}{4}}$ $\tan\pi$

Tell whether each of the following statements is true or false.

7. $\tan 100° = \dfrac{\tan 90° + \tan 10°}{1 + \tan 90° \tan 10°}$ false
8. $\tan 65° = \dfrac{\tan 40° + \tan 25°}{1 - \tan 40° \tan 25°}$ true
9. $\tan 45° = \dfrac{\tan 40° + \tan 5°}{1 - \tan 40° \tan 5°}$ true
10. $\tan 80° = \dfrac{\tan 95° - \tan 15°}{1 - \tan 95° \tan 15°}$ false
11. $\tan 75° = \dfrac{\tan 80° - \tan 5°}{1 + \tan 70° \tan 5°}$ false
12. $\tan 120° = \dfrac{\tan 60° + \tan 60°}{1 - \tan 60° \tan 60°}$ true
13. $\tan 25° = \dfrac{\tan 45° - \tan 20°}{1 + \tan 20° \tan 45°}$ true
14. $\tan 50° = \dfrac{\tan 65° - \tan 15°}{1 + \tan 15° \tan 65°}$ true

PRACTICE EXERCISES

Use the sum or the difference identity for tangent to find the exact value.

A

1. tan 255° $2 + \sqrt{3}$
2. tan 15° $2 - \sqrt{3}$
3. tan 105° $-\sqrt{3} - 2$
4. tan 285° $-\sqrt{3} - 2$
5. tan 165° $\sqrt{3} - 2$
6. tan 195° $2 - \sqrt{3}$
7. $\tan\left(\frac{5\pi}{3} + \frac{\pi}{4}\right)$ $\sqrt{3} - 2$
8. $\tan\left(\frac{4\pi}{3} + \frac{3\pi}{4}\right)$ $2 - \sqrt{3}$

LESSON FOLLOW-UP

Discussion

How could you derive the sum and difference identities for cotangent? Find an expression for $\cot(\alpha + \beta)$ and $\cot(\alpha - \beta)$. Use the sum and difference identities for sine and cosine or tangent.

$$\cot(\alpha + \beta) = \frac{\cos(\alpha + \beta)}{\sin(\alpha + \beta)} = \frac{\cos\alpha\cos\beta - \sin\alpha\sin\beta}{\sin\alpha\cos\beta + \cos\alpha\sin\beta} = \frac{\frac{\cos\alpha\cos\beta}{\sin\alpha\sin\beta} - \frac{\sin\alpha\sin\beta}{\sin\alpha\sin\beta}}{\frac{\sin\alpha\cos\beta}{\sin\alpha\sin\beta} + \frac{\cos\alpha\sin\beta}{\sin\alpha\sin\beta}} = \frac{\cot\alpha\cot\beta - 1}{\cot\beta + \cot\alpha}$$

In a similar way, $\cot(\alpha - \beta) = \frac{\cos(\alpha - \beta)}{\sin(\alpha - \beta)} = \frac{\cot\alpha\cot\beta + 1}{\cot\beta - \cot\alpha}$.

Using tangent, $\cot(\alpha + \beta) = \frac{1}{\tan(\alpha + \beta)} = \frac{1}{\frac{\tan\alpha + \tan\beta}{1 - \tan\alpha\tan\beta}} = \frac{1 - \tan\alpha\tan\beta}{\tan\alpha + \tan\beta} = \frac{1 - \frac{1}{\cot\alpha\cot\beta}}{\frac{1}{\cot\alpha} + \frac{1}{\cot\beta}} = \frac{\cot\alpha\cot\beta - 1}{\cot\beta + \cot\alpha}$; $\cot(\alpha - \beta)$ can be derived in a similar way.

Assignment Guide

See p. 206B for assignments.

Test Yourself

See *Teacher's Resource Book, Tests*, pp. 49–50.

Lesson Quiz

Write each expression in terms of the tangent of one angle.

1. $\dfrac{\tan 56^\circ + \tan 28^\circ}{1 - \tan 56^\circ \tan 28^\circ}$ $\tan 84^\circ$

2. $\dfrac{\tan 28^\circ - \tan 58^\circ}{1 - \tan 28^\circ \tan 58^\circ}$ $\tan(-30^\circ)$

Use the sum or difference identity for tangent to find the exact value of each trigonometric function.

3. $\tan \dfrac{7\pi}{12}$ $-\sqrt{3} - 2$

4. $\tan 210^\circ$ $\dfrac{\sqrt{3}}{3}$

5. Find the exact value of $\tan(\alpha + \beta)$ if $\cos\alpha = \frac{4}{5}$, $\sin\beta = -\frac{12}{13}$, $0^\circ < \alpha < 90^\circ$, and $180^\circ < \beta < 270^\circ$. $-\dfrac{63}{16}$

6. Find the exact value of $\tan(\alpha - \beta)$ if $\cos\alpha = \frac{4}{5}$, $\sin\beta = -\frac{12}{13}$, $0^\circ < \alpha < 90^\circ$, and $270^\circ < \beta < 360^\circ$. $-\dfrac{63}{16}$

7. Prove: $\tan(\beta - 180^\circ) = \tan\beta$

$\tan(\beta - 180^\circ)$	$\tan\beta$
$\dfrac{\tan\beta - \tan 180^\circ}{1 + \tan\beta \tan 180^\circ}$	
$\dfrac{\tan\beta - 0}{1 + \tan\beta(0)}$	
$\tan\beta$	$=$

Enrichment

Find an expression for $\tan 3\theta$.

$\tan 3\theta = \tan(2\theta + \theta) = \tan[(\theta + \theta) + \theta] = \dfrac{\tan(\theta + \theta) + \tan\theta}{1 - \tan(\theta + \theta)(\tan 2\theta)} = \dfrac{\dfrac{\tan\theta + \tan\theta}{1 - \tan\theta\tan\theta} + \tan\theta}{1 - \dfrac{\tan\theta + \tan\theta}{1 - \tan\theta\tan\theta}(\tan\theta)} = \dfrac{2\tan\theta + \tan\theta - \tan^3\theta}{1 - \tan^2\theta - \tan\theta(2\tan\theta)} = \dfrac{3\tan\theta - \tan^3\theta}{1 - 3\tan^2\theta}$

Find the exact value of $\tan(\alpha + \beta)$.

9. $\cos\alpha = \frac{3}{5}$, $\sin\beta = -\frac{5}{13}$, $\frac{3\pi}{2} < \alpha < 2\pi$, $\pi < \beta < \frac{3\pi}{2}$ $-\frac{33}{56}$

10. $\cos\alpha = \frac{7}{25}$, $\sin\beta = \frac{5}{13}$, $\frac{3\pi}{2} < \alpha < 2\pi$, $0 < \beta < \frac{\pi}{2}$ $-\frac{253}{204}$

11. $\tan\alpha = \frac{8}{15}$, $\cos\beta = -\frac{4}{5}$, $0 < \alpha < \frac{\pi}{2}$, $\frac{\pi}{2} < \beta < \pi$ $-\frac{13}{84}$

12. $\cos\alpha = -\frac{5}{13}$, $\tan\beta = \frac{15}{8}$, $\frac{\pi}{2} < \alpha < \pi$, $\pi < \beta < \frac{3\pi}{2}$ $-\frac{21}{220}$

Find the exact value of $\tan(\alpha - \beta)$.

13. $\cos\alpha = \frac{7}{25}$, $\cos\beta = -\frac{4}{5}$, $0 < \alpha < \frac{\pi}{2}$, $\frac{\pi}{2} < \beta < \pi$ $-\frac{117}{44}$

14. $\cos\alpha = \frac{3}{5}$, $\cos\beta = \frac{5}{13}$, $0 < \alpha < \frac{\pi}{2}$, $\frac{3\pi}{2} < \beta < 2\pi$ $-\frac{56}{33}$

15. $\tan\alpha = -\frac{4}{3}$, $\sin\beta = \frac{24}{25}$, $\frac{\pi}{2} < \alpha < \pi$, $\frac{\pi}{2} < \beta < \pi$ $\frac{44}{117}$

16. $\tan\alpha = \frac{12}{5}$, $\sin\beta = \frac{3}{5}$, $\pi < \alpha < \frac{3\pi}{2}$, $\frac{\pi}{2} < \beta < \pi$ $-\frac{63}{16}$

Use the sum or the difference identity for tangent to prove each identity. See pages 483–484.

B

17. $\tan(45^\circ - \alpha) = \dfrac{1 - \tan\alpha}{1 + \tan\alpha}$

18. $\tan(\alpha + 45^\circ) = \dfrac{1 + \tan\alpha}{1 - \tan\alpha}$

19. $\tan(180^\circ + \alpha) = \tan\alpha$

20. $\tan(180^\circ - \alpha) = -\tan\alpha$

21. $\tan(360^\circ + \alpha) = \tan\alpha$

22. $\tan(360^\circ - \alpha) = -\tan\alpha$

23. $\tan\left(\dfrac{3\pi}{4} + \alpha\right) = \dfrac{\tan\alpha - 1}{\tan\alpha + 1}$

24. $\tan\left(\dfrac{3\pi}{4} - \alpha\right) = \dfrac{\tan\alpha + 1}{\tan\alpha - 1}$

25. $\tan\left(\alpha - \dfrac{5\pi}{4}\right) = \dfrac{\tan\alpha - 1}{\tan\alpha + 1}$

26. $\tan\left(\alpha + \dfrac{7\pi}{4}\right) = \dfrac{\tan\alpha - 1}{\tan\alpha + 1}$

Find the exact value of each trigonometric function using the tangent identities.

C

27. $\tan(-15^\circ)$ $\sqrt{3} - 2$

28. $\tan(-75^\circ)$ $-2 - \sqrt{3}$

29. $\tan\left(-\dfrac{5\pi}{12}\right)$ $-2 - \sqrt{3}$

30. $\tan\left(-\dfrac{\pi}{12}\right)$ $\sqrt{3} - 2$

Prove each identity. See page 484.

31. $\tan(\beta + 45^\circ) + \tan(\beta - 45^\circ) = 2\tan 2\beta$

32. $\tan(\beta + 45^\circ) + \tan(45^\circ - \beta) = 2\left(\dfrac{1 + \tan^2\beta}{1 - \tan^2\beta}\right)$

Applications

Coordinate Geometry Recall from algebra that the slope m of a line is

$$m = \frac{\text{change in the } y\text{-coordinates}}{\text{change in the } x\text{-coordinates}}$$

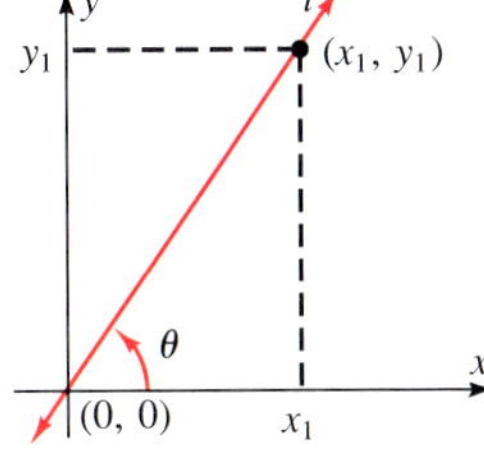

In the figure at the right, $m = \frac{y_1 - 0}{x_1 - 0} = \frac{y_1}{x_1}$.

But by definition of tangent, $\tan\theta = \frac{y_1}{x_1}$.

Therefore, the slope of line t is equal to $\tan\theta$.

33. Let θ be the acute angle formed by the intersection of lines n and p, which form angles α and β, respectively, with the x-axis. Let m_1 be the slope of n and let m_2 be the slope of p. Show that $\tan\theta = \frac{m_2 - m_1}{1 + m_1 m_2}$. See side column.

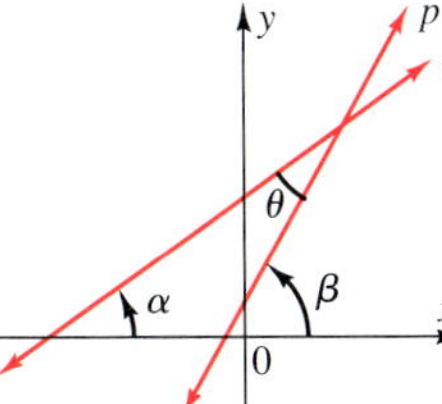

34. Use the formula in Exercise 33 to find $\tan\theta$ if $m_1 = \frac{1}{2}$ and $m_2 = \frac{2}{3}$. $\frac{1}{8}$

TEST YOURSELF

Use a sum or difference identity to find the exact value of each.

1. $\cos 75°$ $\frac{\sqrt{6}-\sqrt{2}}{4}$ **2.** $\cos 255°$ $-\frac{\sqrt{6}-\sqrt{2}}{4}$ 5.1

3. $\sin 15°$ $\frac{\sqrt{6}-\sqrt{2}}{4}$ **4.** $\sin 105°$ $\frac{\sqrt{6}+\sqrt{2}}{4}$ 5.2

5. $\tan 375°$ $2-\sqrt{3}$ **6.** $\tan 75°$ $2+\sqrt{3}$ 5.3

7. Find $\cos(\alpha + \beta)$ if $\sin\alpha = \frac{3}{5}$, $\cos\beta = \frac{3}{5}$, $0 < \alpha < \frac{\pi}{2}$, and $\frac{3\pi}{2} < \beta < 2\pi$. $\frac{24}{25}$ 5.1

8. Find $\sin(\alpha - \beta)$ if $\cos\alpha = -\frac{8}{17}$, $\sin\beta = -\frac{7}{25}$, $\frac{\pi}{2} < \alpha < \pi$, and $\pi < \beta < \frac{3\pi}{2}$. $-\frac{416}{425}$ 5.2

9. Find $\tan(\alpha - \beta)$ if $\sin\alpha = -\frac{4}{5}$, $\sin\beta = -\frac{7}{25}$, $\pi < \alpha < \frac{3\pi}{2}$, and $\frac{3\pi}{2} < \beta < 2\pi$. $\frac{117}{44}$ 5.3

Prove each identity. See side column.

10. $\cos\left(\frac{3\pi}{2} + \alpha\right) = \sin\alpha$ **11.** $\cos\left(\frac{\pi}{2} + \alpha\right) = -\sin\alpha$ 5.1

12. $\tan(\alpha - 180°) = \tan\alpha$ **13.** $\tan(360° - \beta) = -\tan\beta$ 5.3

Teacher's Resource Book
Practice—Chapter 5, p. 5
Enrichment—Chapter 5, p. 6

Additional Answers

33. $\theta = (\beta - \alpha)$
$\tan\theta = \tan(\beta - \alpha)$
$\tan\beta = m_2$
$\tan\alpha = m_1$
$\tan\theta = \frac{m_2 - m_1}{1 + m_2 m_1}$

Test Yourself

10. $\cos\left(\frac{3\pi}{2} + \alpha\right) = \cos\frac{3\pi}{2}\cdot\cos\alpha - \sin\frac{3\pi}{2}\cdot\sin\alpha = 0\cdot\cos\alpha - (-1)\sin\alpha = \sin\alpha$

11. $\cos\left(\frac{\pi}{2} + \alpha\right) = \cos\frac{\pi}{2}\cdot\cos\alpha - \sin\frac{\pi}{2}\cdot\sin\alpha = 0\cdot\cos\alpha - 1\cdot\sin\alpha = -\sin\alpha$

12. $\tan(\alpha - 180°) = \frac{\tan\alpha - \tan 180°}{1 + \tan\alpha\cdot\tan 180°} = \frac{\tan\alpha - 0}{1 + \tan\alpha\cdot 0} = \tan\alpha$

13. $\tan(360° - \beta) = \frac{\tan 360° - \tan\beta}{1 + \tan 360°\cdot\tan\beta} = \frac{0 - \tan\beta}{1 + 0\cdot\tan\beta} = -\tan\beta$

LESSON PLAN

BACKGROUND

In the Preview, a navigation method known as doubling the angle on the bow is discussed.

5.4 Double-Angle Identities

Objective: To develop and use double-angle identities

Some cases of the sum and difference identities are used so often that they are given special names. The first group consists of the *double-angle identities*.

Preview

A navigator on a ship uses a method known as *doubling the angle on the bow* to determine his position relative to a fixed point, such as a lighthouse. He begins at a point A, where the acute angle between the path of his ship and the lighthouse is α. He then measures the distance the ship moves to a point B, where the angle between the path of the ship and the lighthouse is 2α. The distance from A to B is equal to the distance from B to the lighthouse at L.

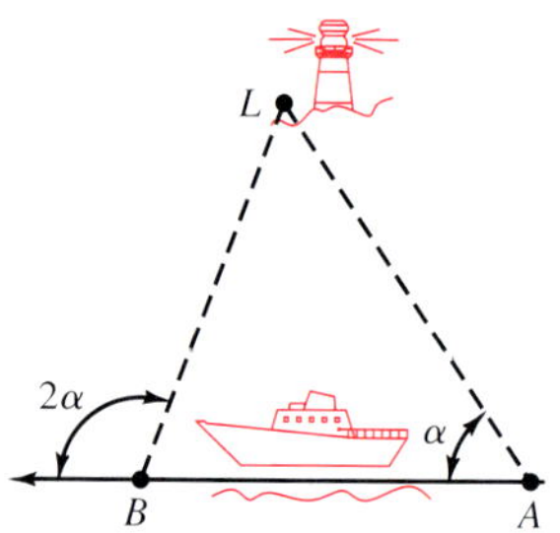

1. What is the measure of $\angle BLA$? α
2. Why is AB equal to BL? sides opposite equal angles of a triangle are equal
3. If α were 60°, and the distance from A to B were 3 mi, what would be the distance from A to L? 3 mi

The identities for $\sin(\alpha + \beta)$, $\cos(\alpha + \beta)$, and $\tan(\alpha + \beta)$ can be used to derive identities for $\sin 2\alpha$, $\cos 2\alpha$, and $\tan 2\alpha$.

$$\begin{aligned} \sin(\alpha + \beta) &= \sin\alpha\cos\beta + \cos\alpha\sin\beta \\ \sin(\alpha + \alpha) &= \sin\alpha\cos\alpha + \cos\alpha\sin\alpha && \text{Let } \beta = \alpha. \\ \sin 2\alpha &= 2\sin\alpha\cos\alpha \end{aligned}$$

$$\begin{aligned} \cos(\alpha + \beta) &= \cos\alpha\cos\beta - \sin\alpha\sin\beta \\ \cos(\alpha + \alpha) &= \cos\alpha\cos\alpha - \sin\alpha\sin\alpha && \text{Let } \beta = \alpha. \\ \cos 2\alpha &= \cos^2\alpha - \sin^2\alpha \end{aligned}$$

$$\begin{aligned} \tan(\alpha + \beta) &= \frac{\tan\alpha + \tan\beta}{1 - \tan\alpha\tan\beta} \\ \tan(\alpha + \alpha) &= \frac{\tan\alpha + \tan\alpha}{1 - \tan\alpha\tan\alpha} && \text{Let } \beta = \alpha. \\ \tan 2\alpha &= \frac{2\tan\alpha}{1 - \tan^2\alpha} \end{aligned}$$

Two alternate forms of $\cos 2\alpha$ can be derived using $\cos^2 \alpha + \sin^2 \alpha = 1$.

Alternate Form 1 (cosine form)	**Alternate Form 2 (sine form)**
$\cos 2\alpha = \cos^2 \alpha - \sin^2 \alpha$	$\cos 2\alpha = \cos^2 \alpha - \sin^2 \alpha$
$= \cos^2 \alpha - (1 - \cos^2 \alpha)$	$= (1 - \sin^2 \alpha) - \sin^2 \alpha$
$= 2 \cos^2 \alpha - 1$	$= 1 - 2 \sin^2 \alpha$

Double-Angle Identities

$\sin 2\alpha = 2 \sin \alpha \cos \alpha$

$\tan 2\alpha = \dfrac{2 \tan \alpha}{1 - \tan^2 \alpha}$, $\tan \alpha \neq \pm 1$

$\cos 2\alpha = \cos^2 \alpha - \sin^2 \alpha$

$\cos 2\alpha = 2 \cos^2 \alpha - 1$

$\cos 2\alpha = 1 - 2 \sin^2 \alpha$

EXAMPLE 1 **If $\sin \alpha = -\frac{4}{5}$ and $270° < \alpha < 360°$, find the exact values of each:**

a. $\sin 2\alpha$ **b.** $\cos 2\alpha$ **c.** $\tan 2\alpha$

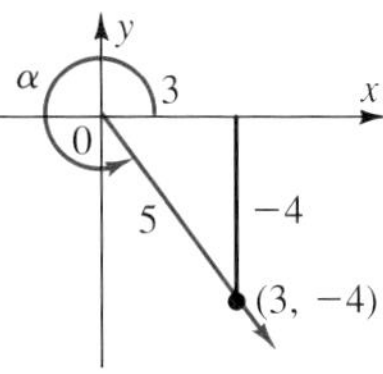

$\sin \alpha = -\frac{4}{5}$ and α is in quadrant IV.

Use $y = -4$ and $r = 5$, and find the value of x.

$$x = \pm\sqrt{(5)^2 - (-4)^2} = \pm\sqrt{9} = \pm 3$$

Since α is the quadrant IV, $x = 3$. Thus, $\cos \alpha = \frac{3}{5}$ and $\tan \alpha = -\frac{4}{3}$.

a. $\sin 2\alpha = 2\left(-\frac{4}{5}\right)\left(\frac{3}{5}\right) = -\frac{24}{25}$ *$\sin 2\alpha = 2 \sin \alpha \cos \alpha$*

b. $\cos 2\alpha = 1 - 2\left(-\frac{4}{5}\right)^2 = -\frac{7}{25}$ *$\cos 2\alpha = 1 - 2 \sin^2 \alpha$*

c. $\tan 2\alpha = \dfrac{2\left(-\frac{4}{3}\right)}{1 - \left(-\frac{4}{3}\right)^2} = \dfrac{24}{7}$ *$\tan 2\alpha = \dfrac{2 \tan \alpha}{1 - \tan^2 \alpha}$*

Formulas for multiples of angle measures follow from the double-angle identities.

EXAMPLE 2 **Derive an identity for $\cos 3\theta$ in terms of $\cos \theta$.**

$$\begin{aligned}
\cos 3\theta &= \cos (2\theta + \theta) \\
&= \cos 2\theta \cos \theta - \sin 2\theta \sin \theta \\
&= (2 \cos^2 \theta - 1) \cos \theta - (2 \sin \theta \cos \theta) \sin \theta \\
&= 2 \cos^3 \theta - \cos \theta - 2 \sin^2 \theta \cos \theta \\
&= 2 \cos^3 \theta - \cos \theta - 2 (1 - \cos^2 \theta) \cos \theta \\
&= 2 \cos^3 \theta - \cos \theta - 2 \cos \theta + 2 \cos^3 \theta \\
&= 4 \cos^3 \theta - 3 \cos \theta
\end{aligned}$$

TEACHING SUGGESTIONS

- Show the advantage of having multiple formulas for $\cos 2\alpha$. In Example 1, find $\cos 2\alpha$ using $2 \cos^2 \alpha - 1$.
- Emphasize that if students do not memorize the double-angle identities, they must be able to derive them quickly from the corresponding sum identities.

Critical Thinking

Causal explanation Ask students to explain how they can tell which of the three identities for $\cos 2\alpha$ is most advantageous to use. Students should reason that $\cos 2\alpha = \cos^2 \alpha - \sin^2 \alpha$ is most advantageous when both $\sin \alpha$ and $\cos \alpha$ are known; $\cos 2\alpha = 2 \cos^2 \alpha - 1$ is most advantageous when only $\cos \alpha$ is known; and $\cos 2\alpha = 1 - 2 \sin^2 \alpha$ is most advantageous when only $\sin \alpha$ is known.

CHALKBOARD EXAMPLES

- **For Example 1**

  If $\sin \theta = \frac{12}{13}$ and $90° < \theta < 180°$, find the exact values for each trigonometric function.

  1. $\cos 2\theta$ $-\frac{119}{169}$
  2. $\tan 2\theta$ $\frac{120}{119}$
  3. $\sin 2\theta$ $-\frac{120}{169}$

- **For Example 2**

  4. Derive an identity for $\sin 3\theta$ in terms of $\sin \theta$.

  $\sin 3\theta = \sin (2\theta + \theta) =$
  $\sin 2\theta \cos \theta + \cos 2\theta \sin \theta =$
  $(2 \sin \theta \cos \theta)\cos \theta + (\cos^2 \theta - \sin^2 \theta)\sin \theta =$
  $2 \sin \theta \cos^2 \theta + \sin \theta \cos^2 \theta - \sin^3 \theta =$
  $3 \sin \theta \cos^2 \theta - \sin^3 \theta =$
  $3 \sin \theta(1 - \sin^2 \theta) - \sin^3 \theta -$
  $3 \sin \theta - 3 \sin^3 \theta - \sin^3 \theta =$
  $3 \sin \theta - 4 \sin^3 \theta$, or
  $\sin \theta(3 - 4 \sin^2 \theta)$

- **For Example 3**
  Prove each identity.
  5. $\cot\theta + \tan\theta = 2\csc 2\theta$

$$
\begin{array}{l|r}
\cot\theta + \tan\theta & 2\csc 2\theta \\
\frac{\cos\theta}{\sin\theta} + \frac{\sin\theta}{\cos\theta} & \\
\frac{\cos^2\theta + \sin^2\theta}{\sin\theta\cos\theta} & \\
\frac{1}{\sin\theta\cos\theta} & \\
\frac{2}{2\sin\theta\cos\theta} & \\
\frac{2}{\sin 2\theta} & \\
2\csc 2\theta & =
\end{array}
$$

  6. $\cos^4\theta - \sin^4\theta = \cos 2\theta$

$$
\begin{array}{l|r}
\cos^4\theta - \sin^4\theta & \cos 2\theta \\
(\cos^2\theta - \sin^2\theta)\cdot(\cos^2\theta + \sin^2\theta) & \\
[\cos^2\theta - (1 - \cos^2\theta)](1) & \\
2\cos^2\theta - 1 & \\
\cos 2\theta & =
\end{array}
$$

Common Error

- When using the double-angle identities, some students get the incorrect sign for the trigonometric function. Emphasize that the quadrant in which θ lies determines the sign of the trigonometric function.
- See *Teacher's Resource Book* for additional remediation.

LESSON FOLLOW-UP

Critical Thinking

Application Ask students to explain how to derive an identity for $\cot 2\theta$. Then ask them to derive the identity. Students should note that since the cotangent of an angle is equal to the quotient of the cosine and sine of that angle, and the cotangent of an angle is the reciprocal of the tangent of that angle, $\cot 2\theta$ can be derived by using $\cot 2\theta = \frac{\cos 2\theta}{\sin 2\theta}$ or by using $\cot 2\theta = \frac{1}{\tan 2\theta}$.
$\cot 2\theta = \cot\theta - \csc 2\theta$

Assignment Guide

See p. 206B for assignments.

The double-angle identities can be used to prove other identities.

EXAMPLE 3 **Prove:** $\sin 2\alpha = \dfrac{2\tan\alpha}{1 + \tan^2\alpha}$

$$
\begin{array}{r|ll}
\sin 2\alpha & \frac{2\tan\alpha}{1+\tan^2\alpha} & \\
 & \frac{2\tan\alpha}{\sec^2\alpha} & \textit{1 + tan}^2\alpha = \textit{sec}^2\alpha \\
 & \frac{2\frac{\sin\alpha}{\cos\alpha}}{\frac{1}{\cos^2\alpha}} & \textit{tan}\ \alpha = \frac{\sin\alpha}{\cos\alpha};\ \textit{sec}^2\alpha = \frac{1}{\cos^2\alpha} \\
 & 2\sin\alpha\cos\alpha & \textit{Multiply numerator and denominator by } \cos^2\alpha. \\
= \sin 2\alpha & &
\end{array}
$$

Thus, $\sin 2\alpha = \dfrac{2\tan\alpha}{1+\tan^2\alpha}$.

CLASS EXERCISES

Tell whether each of the following statements is true or false.

1. $\cos 2(20°) = 2\cos^2 40° - 1$ false
2. $\cos 2(25°) = 1 - 2\sin^2 25°$ true
3. $\sin(-50°) = 2\sin(-25°)\cos(-25°)$ true
4. $\sin 2(42°) = 2\sin 84°\cos 84°$ false
5. $\cos 70° = \cos^2 35° - \sin^2 35°$ true
6. $\cos 40° = 2\cos^2 40° - 1$ false
7. $\tan 2(45°) = \dfrac{2\tan 45°}{1 - \tan^2 45°}$ true
8. $\tan(-70°) = \dfrac{2\tan(-35°)}{1 - \tan^2(-35°)}$ true

Write each expression in terms of a trigonometric function of one angle.

9. $2\sin 35°\cos 35°$ sin 70°
10. $1 - 2\sin^2 40°$ cos 80°
11. $\dfrac{2\tan 22.5°}{1 - \tan^2 22.5°}$ tan 45°
12. $2\sin 67.5°\cos 67.5°$ sin 135°

PRACTICE EXERCISES

Use the double-angle identities to find the exact value of each trigonometric function.

A

1. If $\cos\alpha = \frac{3}{5}$ and $0° < \alpha < 90°$, find $\cos 2\alpha$. $-\frac{7}{25}$
2. If $\sin\alpha = \frac{3}{5}$ and $0° < \alpha < 90°$, find $\cos 2\alpha$. $\frac{7}{25}$
3. If $\tan\alpha = \frac{4}{3}$ and $0° < \alpha < 90°$, find $\tan 2\alpha$. $-\frac{24}{7}$

4. If $\tan\alpha = \frac{3}{4}$ and $0° < \alpha < 90°$, find $\tan 2\alpha$. $\frac{24}{7}$

5. If $\sin\alpha = \frac{4}{5}$ and $90° < \alpha < 180°$, find $\sin 2\alpha$. $-\frac{24}{25}$

6. If $\cos\alpha = \frac{4}{5}$ and $270° < \alpha < 360°$, find $\sin 2\alpha$. $-\frac{24}{25}$

7. If $\sin\alpha = \frac{5}{13}$ and $90° < \alpha < 180°$, find $\sin 2\alpha$. $-\frac{120}{169}$

8. If $\cos\alpha = \frac{12}{13}$ and $270° < \alpha < 360°$, find $\sin 2\alpha$. $-\frac{120}{169}$

9. If $\sin\theta = \frac{3}{5}$ and $90° < \theta < 180°$, find $\cos 2\theta$. $\frac{7}{25}$

10. If $\sin\theta = -\frac{12}{13}$ and $180° < \theta < 270°$, find $\cos 2\theta$. $-\frac{119}{169}$

11. If $\cos\theta = \frac{5}{13}$ and $270° < \theta < 360°$, find $\tan 2\theta$. $\frac{120}{119}$

12. If $\cos\theta = \frac{4}{5}$ and $0° < \theta < 90°$, find $\tan 2\theta$. $\frac{24}{7}$

13. If $\tan\theta = -\frac{24}{7}$ and $90° < \theta < 180°$, find $\sin 2\theta$. $-\frac{336}{625}$

14. If $\tan\theta = -\frac{12}{5}$ and $90° < \theta < 180°$, find $\sin 2\theta$. $-\frac{120}{169}$

15. If $\tan\theta = -\frac{4}{3}$ and $90° < \theta < 180°$, find $\cos 2\theta$. $-\frac{7}{25}$

16. If $\tan\theta = \frac{7}{24}$ and $180° < \theta < 270°$, find $\sin 2\theta$. $\frac{336}{625}$

B 17. Derive an identity for $\sin 3\theta$ in terms of $\sin\theta$. $\sin 3\theta = 3\sin\theta - 4\sin^3\theta$

18. Derive an identity for $\tan 3\theta$ in terms of $\tan\theta$. $\tan 3\theta = \frac{3\tan\theta - \tan^3\theta}{1 - 3\tan^2\theta}$

Prove each identity. See pages 484–485.

19. $\cos 2\beta = \frac{1 - \tan^2\beta}{\sec^2\beta}$

20. $\tan\beta = \frac{2\sin\beta\cos\beta}{1 + \cos 2\beta}$

21. $\cot\beta = \frac{1 + \cos 2\beta}{2\sin\beta\cos\beta}$

22. $\sec 2\theta = \frac{1}{1 - 2\sin^2\theta}$

23. $\frac{1}{\cos\theta} = \frac{\sin 2\theta\cos\theta - \cos 2\theta\sin\theta}{\sin\theta\cos\theta}$

24. $\cot\alpha - \tan\alpha = \frac{2\cos 2\alpha}{\sin 2\alpha}$

25. $\sin^2\beta = \frac{1}{2}(1 - \cos 2\beta)$

26. $\cos^2\alpha = \frac{1}{2}(\cos 2\alpha + 1)$

27. $\sec 2\beta = \frac{\cot\beta + \tan\beta}{\cot\beta - \tan\beta}$

28. $(\sin\theta + \cos\theta)^2 = \sin 2\theta + 1$

C 29. $(\cos\theta - \sin\theta)^2 = 1 - \sin 2\theta$

30. $\cos 2\alpha = \cos^4\alpha - \sin^4\alpha$

31. $\sin 4\alpha = 4\sin\alpha\cos\alpha\cos 2\alpha$

32. $\cos 4\beta = 1 - 8\sin^2\beta\cos^2\beta$

Trigonometry in Tidal Motion

Expressing the relationship between low and high water tides in terms of the cosine function is discussed.

Lesson Quiz

Write each expression in terms of the trigonometric function of one angle.

1. $\cos^2 76° - \sin^2 76°$ $\cos 152°$
2. $2\sin 112°\cos 112°$ $\sin 224°$
3. $1 - 2\sin^2 18°$ $\cos 36°$
4. $\frac{2\tan 104°}{1 - \tan^2 104°}$ $\tan 208°$

Use a double-angle identity to find the exact value of each trigonometric function.

5. $\cos 2\left(\frac{\pi}{6}\right)$ $\frac{1}{2}$
6. $\tan 2\left(\frac{2\pi}{3}\right)$ $\sqrt{3}$
7. Use a double-angle identity to find the exact value of $\sin 2\theta$ if $\cos\theta = -\frac{4}{5}$ and θ is in quadrant II. $-\frac{24}{25}$
8. Prove: $\tan\beta = \frac{\sin 2\beta}{1 + \cos 2\beta}$

$\tan\beta$	$\frac{\sin 2\beta}{1 + \cos 2\beta}$
	$\frac{2\sin\beta\cos\beta}{1 + (2\cos^2\beta - 1)}$
	$\frac{2\sin\beta\cos\beta}{2\cos^2\beta}$
	$\frac{\sin\beta}{\cos\beta}$
=	$\tan\beta$

Enrichment

Express $\cos 2\theta$ in terms of $\tan\theta$.

$\cos 2\theta = 2\cos^2\theta - 1 = \frac{2}{\sec^2\theta} - 1 = \frac{2 - \sec^2\theta}{\sec^2\theta} = \frac{2 - (\tan^2\theta + 1)}{\tan^2\theta + 1} = \frac{1 - \tan^2\theta}{\tan^2\theta + 1}$

Teacher's Resource Book
Practice—Chapter 5, p. 7
Enrichment—Chapter 5, p. 8

Applications

33. Indirect Measurement A pole casts a shadow of 25 ft at one time and a shadow of 10 ft at a later time when the angle of elevation is twice as large. Find the height h of the pole, to the nearest foot. 11 ft

34. Indirect Measurement The World Trade Center in New York City casts a 3260-ft shadow at one time and a 1350-ft shadow at a later time when the angle of elevation is twice as large. Find the height x of the World Trade Center, to the nearest foot. 1351 ft

Trigonometry in Tidal Motion

The gravitational effect of the moon and the sun on a large body of water causes tides. The water level is lowest at low tide and highest at high tide, and tides repeat every 12.4 hours. The *height of the tide* at a given time is the difference between the height of the water at that time and the height at low tide. The relationship between the height of the tide and the time elapsed since the last high tide can sometimes be approximated using the cosine function

$$y = \frac{a}{2} + \frac{a}{2}\cos\left(\frac{2\pi x}{12.4}\right)$$

where y is the height of the tide in meters, a is the number of meters by which high tide exceeds low tide, and x is the number of hours elapsed since the most recent high tide.

1. High tide at Tampa, Florida exceeds low tide by about 1 meter. Complete the second row of the tide table below by substituting 1 for a in the equation above. Round each entry to the nearest 0.1 meter.

x = time elapsed since last high tide (in hours)	0	1	2	3	4	5	6.2	7	8	9	10	11	12.4
y = height of the tide (in meters)	1	0.9	0.8	0.5	0.3	0.1	0	0	0.2	0.4	0.7	0.9	1

2. Use the information in the table in Exercise 1 to graph the ordered pairs (x, y). Join the points in a smooth curve to complete a graph showing the approximate heights of the tide at Tampa during a 12.4-hour period. Compare this graph with the graph of $y = \cos x$. See below.

2.

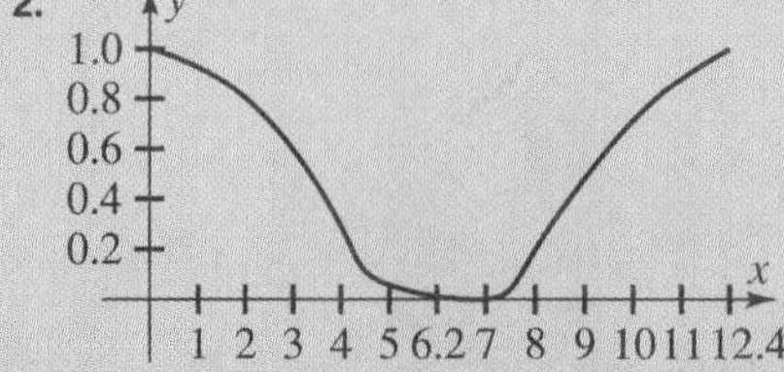

5.5 Half-Angle Identities

Objective: To develop and use half-angle identities

A sonic boom can be described in terms of a function of a half-angle.

Preview

Supersonic speed is described using a Mach number, named after Ernst Mach, an Austrian physicist who contributed to the study of sound. A plane traveling at the speed of sound (about 742.82 mph at 32°F at sea level) is traveling at Mach 1. When the speed of a plane is greater than that of sound (greater than Mach 1), a listener in the area will hear a sonic boom. The Mach number M is defined as

$$M = \frac{\text{speed of plane}}{\text{speed of sound}}$$

A sonic boom is created by sound waves that form a cone with vertex angle θ. Mathematicians have shown that if $M > 1$, then $\sin\left(\frac{\theta}{2}\right) = \frac{1}{M}$.

1. On July 28, 1976 Capt. Eldon W Joersz (USAF) flew a Lockheed SR-71A at 2193.16 mph. Calculate his Mach number to the nearest tenth. 3.0
2. If $\theta = \frac{\pi}{4}$, find the Mach number to the nearest tenth. 2.6
3. If $\theta = \frac{\pi}{8}$, find the Mach number to the nearest tenth. 5.1
4. If θ is halved, what is the effect on the Mach number? It is almost doubled.

The half-angle identities for sine and cosine can be derived by setting θ equal to $\frac{\alpha}{2}$ in the double-angle identities for cosine.

$$\cos 2\left(\frac{\alpha}{2}\right) = 1 - 2\sin^2\left(\frac{\alpha}{2}\right)$$

$$\cos \alpha = 1 - 2\sin^2\left(\frac{\alpha}{2}\right)$$

$$2\sin^2\left(\frac{\alpha}{2}\right) = 1 - \cos \alpha$$

$$\sin^2\left(\frac{\alpha}{2}\right) = \frac{1 - \cos \alpha}{2}$$

$$\sin \frac{\alpha}{2} = \pm\sqrt{\frac{1 - \cos \alpha}{2}}$$

Since $\sin \frac{\alpha}{2}$ does not have two values, the sign is either positive *or* negative, depending upon the quadrant in which $\frac{\alpha}{2}$ lies.

LESSON PLAN

Vocabulary

Mach number
Great circle
Meridians
Parallels

BACKGROUND

In the Preview, the relationship between supersonic speed and trigonometry is discussed.

TEACHING SUGGESTIONS

In the half-angle identities, the use of the $\pm$ is different from the use of the $\pm$ in algebra. For example, stress that the use of $\pm$ with the quadratic formula indicates two possible correct solutions. In trigonometry, the proper sign of each half-angle identity depends upon the quadrant in which $\frac{\alpha}{2}$ is located. The identities require either $+$ or $-$, but not both.

Critical Thinking

Application Ask students to develop an identity for $\sec\frac{\alpha}{2}$ in terms of $\sec\alpha$.

Observing that $\sec\frac{\alpha}{2} = \frac{1}{\cos\frac{\alpha}{2}}$, students should make a series of substitutions on the right side of the equation (as shown below) to arrive at the identity $\sec\frac{\alpha}{2} = \pm\sqrt{\frac{2\sec^2\alpha - 2}{\sec^2\alpha - 1}}$.

$$\sec\frac{\alpha}{2} = \frac{1}{\cos\frac{\alpha}{2}} = \frac{1}{\pm\sqrt{\frac{1+\cos\alpha}{2}}} =$$

$$\sqrt{\frac{2}{1+\cos\alpha}} = \pm\sqrt{\frac{2-2\cos\alpha}{1-\cos^2\alpha}} =$$

$$\pm\sqrt{\frac{2-\frac{2}{\sec\alpha}}{1-\frac{1}{\sec^2\alpha}}}\cdot\sqrt{\frac{\sec^2\alpha}{\sec^2\alpha}} =$$

$$\pm\sqrt{\frac{2\sec^2\alpha - 2}{\sec^2\alpha - 1}}$$

For $\cos\frac{\pi}{2}$, the sign is also determined by the quadrant in which $\frac{\alpha}{2}$ lies.

$$\cos 2\left(\frac{\alpha}{2}\right) = 2\cos^2\left(\frac{\alpha}{2}\right) - 1$$

$$\cos\alpha = 2\cos^2\left(\frac{\alpha}{2}\right) - 1$$

$$2\cos^2\left(\frac{\alpha}{2}\right) = 1 + \cos\alpha$$

$$\cos^2\left(\frac{\alpha}{2}\right) = \frac{1+\cos\alpha}{2}$$

$$\cos\frac{\alpha}{2} = \pm\sqrt{\frac{1+\cos\alpha}{2}}$$

The half-angle identity for tangent is derived using the identity $\tan\theta = \frac{\sin\theta}{\cos\theta}$.

$$\tan\frac{\alpha}{2} = \frac{\sin\frac{\alpha}{2}}{\cos\frac{\alpha}{2}}$$

$$= \frac{\pm\sqrt{\frac{1-\cos\alpha}{2}}}{\pm\sqrt{\frac{1+\cos\alpha}{2}}} = \pm\sqrt{\frac{\frac{1-\cos\alpha}{2}}{\frac{1+\cos\alpha}{2}}}$$

$$= \pm\sqrt{\frac{1-\cos\alpha}{1+\cos\alpha}}$$

There are two alternate forms for $\tan\frac{\alpha}{2}$.

$$\tan\frac{\alpha}{2} = \frac{1-\cos\alpha}{\sin\alpha},\ \sin\alpha \neq 0 \qquad \text{and} \qquad \tan\frac{\alpha}{2} = \frac{\sin\alpha}{1+\cos\alpha},\ \cos\alpha \neq -1$$

These forms are often preferred because the expressions on the right are not preceded by $\pm$ symbols; that is, each identity gives the correct sign directly. Exercises 23 and 24 ask for the derivations of these identities.

Half-Angle Identities

$$\sin\frac{\alpha}{2} = \pm\sqrt{\frac{1-\cos\alpha}{2}} \qquad \tan\frac{\alpha}{2} = \pm\sqrt{\frac{1-\cos\alpha}{1+\cos\alpha}},\ \cos\alpha \neq -1$$

$$\cos\frac{\alpha}{2} = \pm\sqrt{\frac{1+\cos\alpha}{2}} \qquad \tan\frac{\alpha}{2} = \frac{1-\cos\alpha}{\sin\alpha},\ \sin\alpha \neq 0$$

$$\tan\frac{\alpha}{2} = \frac{\sin\alpha}{1+\cos\alpha},\ \cos\alpha \neq -1$$

EXAMPLE 1 **Find the exact value of each trigonometric function:**

a. $\sin \frac{7\pi}{12}$ **b.** $\cos 165°$

a. Let $\frac{\alpha}{2} = \frac{7\pi}{12}$. Then angle $\frac{\alpha}{2}$ lies in quadrant II, where the sine is positive. Also, $\alpha = 2\left(\frac{7\pi}{12}\right) = \frac{7\pi}{6}$.

$$\sin \frac{7\pi}{12} = \sqrt{\frac{1 - \cos \frac{7\pi}{6}}{2}} \qquad \mathit{\sin \frac{\alpha}{2} = \sqrt{\frac{1 - \cos \alpha}{2}}}$$

$$= \sqrt{\frac{1 - \left(-\frac{\sqrt{3}}{2}\right)}{2}} = \sqrt{\frac{2 + \sqrt{3}}{4}} = \frac{\sqrt{2 + \sqrt{3}}}{2}$$

b. Let $\frac{\alpha}{2} = 165°$. Then angle $\frac{\alpha}{2}$ lies in quadrant II, where the cosine is negative. Also, $\alpha = 2(165°) = 330°$.

$$\cos 165° = -\sqrt{\frac{1 + \cos 330°}{2}} \qquad \mathit{\cos \frac{\alpha}{2} = -\sqrt{\frac{1 + \cos \alpha}{2}}}$$

$$= -\sqrt{\frac{1 + \frac{\sqrt{3}}{2}}{2}} = -\sqrt{\frac{2 + \sqrt{3}}{4}} = -\frac{\sqrt{2 + \sqrt{3}}}{2}$$

You can use the half-angle identities to find the value of a trigonometric function of a half-angle if you are given the value of one trigonometric function of the angle and the quadrant in which the terminal side lies.

EXAMPLE 2 **Find the exact value of $\tan \frac{\alpha}{2}$ if $\tan \alpha = -\frac{4}{3}$ and α is in quadrant IV.**

Tan $\alpha = -\frac{4}{3}$ and α is in quadrant IV. Use $y = -4$ and $x = 3$, and find r.

$$r = \sqrt{3^2 + (-4)^2} = \sqrt{25} = 5$$

Thus, $\sin \alpha = -\frac{4}{5}$ and $\cos \alpha = \frac{3}{5}$.

$$\tan \frac{\alpha}{2} = \frac{-\frac{4}{5}}{1 + \frac{3}{5}} = \frac{-\frac{4}{5}}{\frac{8}{5}} = -\frac{4}{8} = -\frac{1}{2} \qquad \mathit{\tan \frac{\alpha}{2} = \frac{\sin \alpha}{1 + \cos \alpha}}$$

Show that the other forms for $\tan \frac{\alpha}{2}$ yield the same result.

CHALKBOARD EXAMPLES

- **For Example 1**

  Find the exact value of each trigonometric function.

  1. $\cos 15°$ $\frac{\sqrt{2 + \sqrt{3}}}{2}$

  2. $\tan\left(-\frac{\pi}{6}\right)$ $-\frac{\sqrt{3}}{3}$

- **For Example 2**

  Find the exact value of each trigonometric function if $\cos \theta = -\frac{3}{5}$ and θ is in quadrant II.

  3. $\cos \frac{\theta}{2}$ $\frac{\sqrt{5}}{5}$

  4. $\sin \frac{\theta}{2}$ $\frac{2\sqrt{5}}{5}$

- **For Example 3**

  Find the exact value of each trigonometric function if $0° < \theta < 360°$.

  5. $\cos \frac{\theta}{2}$, if $\tan \theta = -2$ and θ lies in quadrant IV. $\pm\sqrt{\frac{-5 + \sqrt{5}}{10}}$

  6. $\cos \frac{\theta}{2}$, if $\cos \theta = -\frac{3}{5}$ and θ lies in quadrant III. $-\frac{\sqrt{5}}{5}$

- **For Example 4**
  Prove each identity.

7. $\cot\frac{\theta}{2} = \frac{\sin\theta}{1-\cos\theta}$

$\cot\frac{\theta}{2}$	$\frac{\sin\theta}{1-\cos\theta}$
$\frac{\cos\frac{\theta}{2}}{\sin\frac{\theta}{2}}$	
$\frac{2\sin\frac{\theta}{2}\cos\frac{\theta}{2}}{2\sin^2\frac{\theta}{2}}$	
$\frac{\sin\theta}{2\frac{1-\cos\theta}{2}}$	
$\frac{\sin\theta}{1-\cos\theta}$	$=$

8. $2\cos^2\frac{\alpha}{2} - 1 = \cos\alpha$

$2\cos^2\frac{\alpha}{2} - 1$	$\cos\alpha$
$2\left(\frac{1+\cos\alpha}{2}\right) - 1$	
$1 + \cos\alpha - 1$	
$\cos\alpha$	$=$

LESSON FOLLOW-UP

Discussion

Calculate the value of sin 75° using (a) $\sin\frac{150°}{2}$ and (b) $\sin(45° + 30°)$. Then compare the results.

a. $\sin\frac{150°}{2} = \sqrt{\frac{1-\cos 150°}{2}} = \sqrt{\frac{1+\frac{\sqrt{3}}{2}}{2}} = \frac{\sqrt{2+\sqrt{3}}}{2}$

b. $\sin(45° + 30°) =$
$\sin 45°\cos 30° + \cos 45°\sin 30°$
$\frac{\sqrt{2}}{2}\cdot\frac{\sqrt{3}}{2} + \frac{\sqrt{2}}{2}\cdot\frac{1}{2} = \frac{\sqrt{6}}{4} + \frac{\sqrt{2}}{4}$

The results are the same since $\frac{\sqrt{2+\sqrt{3}}}{2} \approx 0.9659$ and $\frac{\sqrt{6}+\sqrt{2}}{4} \approx 0.9659$.

Care must be taken to determine the correct quadrant for a half-angle.

EXAMPLE 3 **Find the exact value of $\sin\frac{\theta}{2}$ if $\cos\theta = -\frac{3}{5}$, θ lies in quadrant III, and $0° < \theta < 360°$.**

Since θ lies in quadrant III and $0° < \theta < 360°$, it must be true that $180° < \theta < 270°$. Thus, $90° < \frac{\theta}{2} < 135°$ and $\frac{\theta}{2}$ lies in quadrant II, where the sine is positive.

y, x, θ, θ/2, 0, 5, (−3, −4)

$$\sin\frac{\theta}{2} = \sqrt{\frac{1-\left(-\frac{3}{5}\right)}{2}} = \sqrt{\frac{5+3}{10}} = \sqrt{\frac{4}{5}} = \frac{2\sqrt{5}}{5}$$

$\sin\frac{\theta}{2} = \sqrt{\frac{1-\cos\theta}{2}}$

Half-angle identities can be used to prove other identities.

EXAMPLE 4 **Prove: $-\cos\theta = 2\sin^2\left(\frac{\theta}{2}\right) - 1$**

$-\cos\theta$	$2\sin^2\left(\frac{\theta}{2}\right) - 1$
	$2\left(\pm\sqrt{\frac{1-\cos\theta}{2}}\right)^2 - 1$
	$2\left(\frac{1-\cos\theta}{2}\right) - 1$
	$1 - \cos - 1$
	$= -\cos\theta$

Thus, $-\cos\theta = 2\sin^2\left(\frac{\theta}{2}\right) - 1$.

CLASS EXERCISES

Tell whether each statement is true or false.

1. $\cos 10° = \sqrt{\frac{1+\sin 5°}{2}}$ false

2. $\sin 20° = \sqrt{\frac{1-\cos 40°}{2}}$ true

3. $\tan 200° = \sqrt{\frac{\sin 100°}{1+\cos 100°}}$ false

4. $\tan 10° = \sqrt{\frac{1-\cos 20°}{\sin 20°}}$ false

5. $\tan 15° = \sqrt{\frac{1-\cos 30°}{1+\cos 30°}}$ true

6. $\tan 100° = -\sqrt{\frac{\sin 50°}{1+\cos 50°}}$ false

7. $\sin 60° = \sqrt{\frac{1+\cos 120°}{2}}$ false

8. $\cos 40° = -\sqrt{\frac{1+\cos 40°}{2}}$ false

Match each expression on the left with an equivalent one on the right.

9. $\cos 40°$ C

10. $\sin 20°$ A

11. $\tan 20°$ B

A. $\sqrt{\dfrac{1-\cos 40°}{2}}$

B. $\sqrt{\dfrac{1-\cos 40°}{1+\cos 40°}}$

C. $\sqrt{\dfrac{1+\cos 80°}{2}}$

PRACTICE EXERCISES

Use the half-angle identities to find the exact value of each function.

A

1. $\tan 105°$ $-2-\sqrt{3}$
2. $\cos 105°$ $\frac{-\sqrt{2-\sqrt{3}}}{2}$
3. $\sin 22.5°$ $\frac{\sqrt{2-\sqrt{2}}}{2}$
4. $\tan 22.5°$ $\sqrt{2}-1$
5. $\cos 67.5°$ $\frac{\sqrt{2-\sqrt{2}}}{2}$
6. $\sin 67.5°$ $\frac{\sqrt{2+\sqrt{2}}}{2}$
7. $\sin \frac{5\pi}{8}$ $\frac{\sqrt{2+\sqrt{2}}}{2}$
8. $\cos \frac{5\pi}{8}$ $\frac{-\sqrt{2-\sqrt{2}}}{2}$
9. $\cos \frac{7\pi}{8}$ $\frac{-\sqrt{2+\sqrt{2}}}{2}$
10. $\tan \frac{7\pi}{8}$ $1-\sqrt{2}$
11. $\tan \frac{\pi}{12}$ $2-\sqrt{3}$
12. $\sin \frac{\pi}{12}$ $\frac{\sqrt{2-\sqrt{3}}}{2}$

Find the exact value of each trigonometric function. Assume $0° < \theta < 360°$.

13. $\cos \frac{\theta}{2}$ if $\cos \theta = \frac{4}{5}$ and θ lies in quadrant I $\frac{3\sqrt{10}}{10}$
14. $\sin \frac{\theta}{2}$ if $\cos \theta = \frac{3}{5}$ and θ lies in quadrant I $\frac{\sqrt{5}}{5}$
15. $\cos \frac{\theta}{2}$ if $\cos \theta = \frac{1}{2}$ and θ lies in quadrant I $\frac{\sqrt{3}}{2}$
16. $\sin \frac{\theta}{2}$ if $\cos \theta = \frac{\sqrt{2}}{2}$ and θ lies in quadrant I $\frac{\sqrt{2-\sqrt{2}}}{2}$

B

17. $\tan \frac{\theta}{2}$ if $\tan \theta = 2$ and θ lies in quadrant III $\frac{-\sqrt{5}-1}{2}$
18. $\tan \frac{\theta}{2}$ if $\tan \theta = -2$ and θ lies in quadrant II $\frac{\sqrt{5}+1}{2}$
19. $\tan \frac{\theta}{2}$ if $\sin \theta = -\frac{24}{25}$ and θ lies in quadrant IV $-\frac{3}{4}$
20. $\tan \frac{\theta}{2}$ if $\sin \theta = -\frac{8}{17}$ and θ lies in quadrant IV $-\frac{1}{4}$
21. $\sin \theta$ if $\cos 2\theta = \frac{5}{13}$ and θ lies in quadrant I $\frac{2\sqrt{13}}{13}$
22. $\cos \theta$ if $\cos 2\theta = -\frac{12}{13}$ and θ lies in quadrant III $-\frac{\sqrt{26}}{26}$

Assignment Guide

See p. 206B for assignments.

Trigonometry in Geometry

Spherical trigonometry and geography are discussed in relation to a great circle.

Lesson Quiz

Write each expression in terms of the trigonometric function of one angle.

1. $\sqrt{\dfrac{1-\cos 70°}{2}}$

sin 35°

2. $\sqrt{\dfrac{1-\cos 102°}{1+\cos 102°}}$

tan 51°

Use the half-angle identities to find the exact value of each trigonometric function.

3. $\sin 105°$ $\frac{\sqrt{2+\sqrt{3}}}{2}$
4. $\cos \frac{5\pi}{6}$ $-\frac{\sqrt{3}}{2}$
5. Use a half-angle identity to find the exact value of $\tan \frac{\theta}{2}$ if $\sin \theta = \frac{3}{5}$ and θ lies in quadrant II. Assume $0° < \theta < 360°$ 3
6. Prove: $\cos \alpha = \cos^2 \frac{\alpha}{2} - \sin^2 \frac{\alpha}{2}$

$\cos \alpha$	$\cos^2 \frac{\alpha}{2} - \sin^2 \frac{\alpha}{2}$
	$\frac{1+\cos \alpha}{2} - \frac{1-\cos \alpha}{2}$
	$\frac{2\cos \alpha + 1 - 1}{2}$
=	$\cos \alpha$

Enrichment

Obtain an alternate expression for tan $\frac{\theta}{2}$ by modifying the right side of $\tan\frac{\theta}{2} = \frac{\sin\theta}{1+\cos\theta}$.

$\frac{\sin\theta}{1+\cos\theta} = \frac{\sin\theta}{1+\cos\theta}\cdot\frac{1-\cos\theta}{1-\cos\theta} = \frac{\sin\theta - \sin\theta\cos\theta}{1-\cos^2\theta} = \frac{\sin\theta - \sin\theta\cos\theta}{\sin^2\theta} = \frac{1-\cos\theta}{\sin\theta}$; thus, $\tan\frac{\theta}{2} = \frac{1-\cos\theta}{\sin\theta}$.

Teacher's Resource Book

Practice—Chapter 5, p. 9
Enrichment—Chapter 5, p. 10

Prove each identity. See page 485.

23. $\tan\frac{\alpha}{2} = \frac{\sin\alpha}{1+\cos\alpha}$

24. $\tan\frac{\alpha}{2} = \frac{1-\cos\alpha}{\sin\alpha}$

25. $\cos\alpha = 2\cos^2\left(\frac{\alpha}{2}\right) - 1$

26. $\sin\beta = 2\sin\frac{\beta}{2}\cos\frac{\beta}{2}$

27. $\csc\beta = \frac{1+\tan^2\left(\frac{\beta}{2}\right)}{2\tan\left(\frac{\beta}{2}\right)}$

28. $\tan^2\left(\frac{\alpha}{2}\right) = \frac{(1-\cos\alpha)^2}{\sin^2\alpha}$

C **29.** $\cos\alpha = \frac{\cot\frac{\alpha}{2} - \tan\frac{\alpha}{2}}{\tan\frac{\alpha}{2} + \cot\frac{\alpha}{2}}$

30. $\sec\beta = \frac{1+\tan^2\left(\frac{\beta}{2}\right)}{1-\tan^2\left(\frac{\beta}{2}\right)}$

31. $\cos 2\theta = 1 - 8\sin^2\left(\frac{\theta}{2}\right)\cos^2\left(\frac{\theta}{2}\right)$

32. $\tan^2\left(\frac{\theta}{2}\right) = 1 - \frac{2\cos\theta}{1+\cos\theta}$

Applications

33. Geometry The area A of an isosceles triangle can be expressed as $A = \left(s\sin\frac{\alpha}{2}\right)\left(s\cos\frac{\alpha}{2}\right)$, where s is the length of each of the two congruent sides, and α is the angle formed by those sides. Show that this formula is equivalent to $A = \frac{1}{2}s^2\sin\alpha$.

34. Geometry Use one of the formulas in Exercise 33 to find the area of an isosceles triangle, to the nearest square centimeter, if $s = 10$ cm and $\alpha = 48°$. 37 cm^2

TRIGONOMETRY IN GEOGRAPHY

The earth is nearly spherical, with a radius of about 3958 mi. A *great circle* is formed by a plane passing through the center O of a sphere and cutting the surface. The *equator* is a great circle on the earth, and *parallels* are smaller circles formed by planes intersecting the earth parallel to the equator.

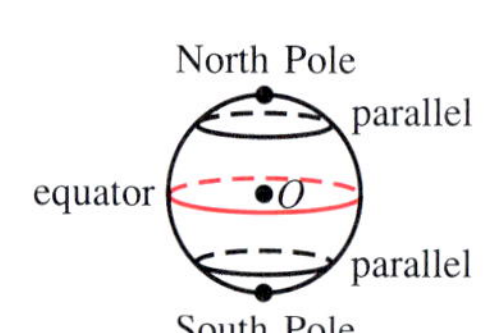

Meridians are great circles that intersect the geographic North and South Poles. In the figure at the right, $\angle XOY$ determines a unique great circle containing shorter $\widehat{XY}$.

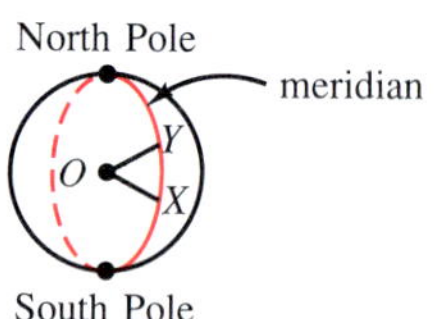

If the measure of $\angle XOY$ is 10°, what is the length of $\widehat{XY}$, to the nearest mile? 691 mi

5.6 Product/Sum Identities

Objective: To develop and use product/sum identities

By adding and subtracting sum and difference identities for sine and cosine, a new set of identities can be derived.

Preview

Planes usually fly along great-circle routes, since this is the shortest distance between two points on a sphere. A great circle is the circle formed by the intersection of a sphere and a plane through its center. Latitude and longitude can be used to find the great-circle distance between two cities, A and B, in the Northern Hemisphere using the formula

$$\cos d = \cos \theta \cos a \cos b + \sin a \sin b$$

where

θ = the positive longitudinal difference between cities A and B
a = latitude of city A
b = latitude of city B
d = degree measure of the great-circle arc between A and B ($0^\circ < d \leq 180^\circ$).

Find the degree measure, to the nearest tenth of a degree, of the great-circle arc between the given cities. Assume that the earth is spherical.

1. New York City (latitude 40.8°; longitude 74°) and Los Angeles (latitude 34.1°; longitude 118.3°) 35.4°
2. Boston (latitude 42.4°; longitude 71.1°) and Miami (latitude 25.8°; longitude 80.2°) 18.2°
3. Use the results in Exercises 1 and 2 to find the great-circle distance, to the nearest 10 mi, between each pair of cities. Assume that a 1° arc of a great circle has a length of approximately 69.1 mi. NY to LA: 2450 mi; Bos. to Miami: 1260 mi

At times it is necessary to express a trigonometric sum as a product or a product as a sum.

LESSON PLAN

BACKGROUND

In the Preview, the use of trigonometry in airline travel is discussed.

TEACHING SUGGESTIONS

- Since the product sum identities are very similar, spend time discussing their differences.
- It may be necessary to review solving systems of equations.

Critical Thinking

Analysis Ask students to approximate the maximum value of $\sqrt{3}\sin\alpha - \cos\beta$.

Students should observe that since $-1 \leq \sin\alpha \leq 1$ and $-1 \leq \cos\beta \leq 1$, the maximum occurs when $\sin\alpha = 1$ and $\cos\beta = -1$. Thus, the maximum value is approximately 3 since $\sqrt{3}(1) - (-1) \approx 2.7 \approx 3$.

CHALKBOARD EXAMPLES

- **For Example 1**
  1. Express 2 sin 60° sin 25° as a difference. cos 35° − cos 85°
  2. Express 2 cos 45° sin 30° as a difference. sin 75° − sin 15°
- **For Example 2**
  3. Express cos 16° − cos 42° as a product. 2 sin 29° sin 13°
  4. Express sin 56° − sin 28° as a product. 2 sin 42° cos 14°

Product/Sum Identities

$$2\cos\alpha\cos\beta = \cos(\alpha-\beta) + \cos(\alpha+\beta)$$
$$2\sin\alpha\sin\beta = \cos(\alpha-\beta) - \cos(\alpha+\beta)$$
$$2\sin\alpha\cos\beta = \sin(\alpha+\beta) + \sin(\alpha-\beta)$$
$$2\cos\alpha\sin\beta = \sin(\alpha+\beta) - \sin(\alpha-\beta)$$

Each of the product identities is found by adding or subtracting the sum and difference identities for sine or cosine.

$$\begin{array}{rl} \cos\alpha\cos\beta + \sin\alpha\sin\beta &= \cos(\alpha-\beta) \\ \cos\alpha\cos\beta - \sin\alpha\sin\beta &= \cos(\alpha+\beta) \\ \hline 2\cos\alpha\cos\beta &= \cos(\alpha-\beta) + \cos(\alpha+\beta) \end{array}$$

Add $\cos(\alpha+\beta)$ *to* $\cos(\alpha-\beta)$.

$$\begin{array}{rl} \cos\alpha\cos\beta + \sin\alpha\sin\beta &= \cos(\alpha-\beta) \\ \cos\alpha\cos\beta - \sin\alpha\sin\beta &= \cos(\alpha+\beta) \\ \hline 2\sin\alpha\sin\beta &= \cos(\alpha-\beta) - \cos(\alpha+\beta) \end{array}$$

Subtract $\cos(\alpha+\beta)$ *from* $\cos(\alpha-\beta)$.

The other two identities can be verified in a similar manner.

EXAMPLE 1 **Express 2 cos 40° sin 15° as a difference.**

$$\begin{aligned} 2\cos 40^\circ \sin 15^\circ &= \sin(40^\circ + 15^\circ) - \sin(40^\circ - 15^\circ) \\ &= \sin 55^\circ - \sin 25^\circ \end{aligned}$$

$2\cos\alpha\sin\beta = \sin(\alpha+\beta) - \sin(\alpha-\beta)$

The product/sum identities can also be used to express sums or differences as products.

EXAMPLE 2 **Express cos 27° + cos 61° as a product.**

Let $\alpha + \beta = 61^\circ$ and $\alpha - \beta = 27^\circ$.

$$\begin{array}{rl} \alpha + \beta &= 61^\circ \\ \alpha - \beta &= 27^\circ \\ \hline 2\alpha &= 88^\circ \\ \alpha &= 44^\circ \end{array}$$

Add the two equations and solve for α.

$44^\circ + \beta = 61^\circ$, $\beta = 17^\circ$ $\quad$ $\alpha + \beta = 61^\circ$ $\quad$ *Solve for* β.

Then use the identity $\cos(\alpha-\beta) + \cos(\alpha+\beta) = 2\cos\alpha\cos\beta$.

$$\cos 27^\circ + \cos 61^\circ = 2\cos 44^\circ \cos 17^\circ$$

Verify the fact that if $\alpha + \beta = 27^\circ$ and $\alpha - \beta = 61^\circ$ in Example 2, you obtain the same result.

There are alternate forms for the product/sum identities that can often be more easily applied. To derive these alternate identities, let $x = \alpha + \beta$ and $y = \alpha - \beta$. Add these equations to solve for α and subtract to solve for β.

$$\begin{array}{rl} x &= \alpha + \beta \\ y &= \alpha - \beta \\ \hline x + y &= 2\alpha \\ \frac{x+y}{2} &= \alpha \end{array} \qquad \begin{array}{rl} x &= \alpha + \beta \\ y &= \alpha - \beta \\ \hline x - y &= 2\beta \\ \frac{x-y}{2} &= \beta \end{array}$$

Substitute these values into the product/sum identities to obtain the alternate forms.

$$\cos x + \cos y = 2 \cos\left(\frac{x+y}{2}\right) \cos\left(\frac{x-y}{2}\right) \qquad \sin x + \sin y = 2 \sin\left(\frac{x+y}{2}\right) \cos\left(\frac{x-y}{2}\right)$$

$$\cos x - \cos y = -2 \sin\left(\frac{x+y}{2}\right) \sin\left(\frac{x-y}{2}\right) \qquad \sin x - \sin y = 2 \cos\left(\frac{x+y}{2}\right) \sin\left(\frac{x-y}{2}\right)$$

EXAMPLE 3 **Express sin 70° + sin 24° as a product.**

$$\sin x + \sin y = 2 \sin\left(\frac{x+y}{2}\right) \cos\left(\frac{x-y}{2}\right)$$

$$\sin 70^\circ + \sin 24^\circ = 2 \sin\left(\frac{70^\circ + 24^\circ}{2}\right) \cos\left(\frac{70^\circ - 24^\circ}{2}\right) \quad \textit{x = 70° and y = 24°}$$

$$= 2 \sin 47^\circ \cos 23^\circ$$

The product/sum identities can be used to prove other identities.

EXAMPLE 4 **Prove:** $\cot \theta = \dfrac{\sin 4\theta + \sin 6\theta}{\cos 4\theta - \cos 6\theta}$

$$\cot \theta \;\Big|\; \frac{\sin 4\theta + \sin 6\theta}{\cos 4\theta - \cos 6\theta}$$

$$\frac{2 \sin \frac{4\theta + 6\theta}{2} \cos \frac{4\theta - 6\theta}{2}}{-2 \sin \frac{4\theta + 6\theta}{2} \sin \frac{4\theta - 6\theta}{2}} \qquad \textit{sin x + sin y = 2 sin}\left(\frac{x+y}{2}\right) \textit{cos}\left(\frac{x-y}{2}\right); \quad \textit{cos x − cos y = −2 sin}\left(\frac{x+y}{2}\right)\textit{sin}\left(\frac{x-y}{2}\right)$$

$$\frac{2 \sin 5\theta \cos(-\theta)}{-2 \sin 5\theta \sin(-\theta)}$$

$$\frac{\cos(-\theta)}{-\sin(-\theta)}$$

$$\frac{\cos \theta}{\sin \theta} \qquad \textit{cos}(-\theta) = \textit{cos}\,\theta;\ \textit{sin}(-\theta) = -\textit{sin}\,\theta$$

$$= \cot \theta \qquad \text{Thus, } \cot \theta = \frac{\sin 4\theta + \sin 6\theta}{\cos 4\theta - \cos 6\theta}$$

- **For Example 3**

5. Express $\cos 5\pi - \cos 2\pi$ as a product. $-2 \sin \frac{7\pi}{2} \sin \frac{3\pi}{2}$
6. Express $\sin 60^\circ + \sin 44^\circ$ as a product. $2 \sin 52^\circ \cos 8^\circ$

- **For Example 4**

Prove each identity.

7. $\tan \theta = \dfrac{\sin 3\theta - \sin \theta}{\cos 3\theta + \cos \theta}$

$$\tan \theta \;\Big|\; \frac{\sin 3\theta - \sin \theta}{\cos 3\theta + \cos \theta}, \quad \frac{2 \cos 2\theta \sin \theta}{2 \cos 2\theta \cos \theta}, \quad \frac{\sin \theta}{\cos \theta}, \quad = \tan \theta$$

8. $\tan \theta = \dfrac{\cos 3\theta - \cos 5\theta}{\sin 3\theta + \sin 5\theta}$

$$\tan \theta \;\Big|\; \frac{\cos 3\theta - \cos 5\theta}{\sin 3\theta + \sin 5\theta}, \quad \frac{2 \sin 4\theta \sin \theta}{2 \sin 4\theta \cos \theta}, \quad \frac{\sin \theta}{\cos \theta}, \quad = \tan \theta$$

LESSON FOLLOW-UP

Discussion

Use the addition formulas to reduce $\sin\left(\theta + \frac{n\pi}{2}\right)$, where n is an integer, to an expression involving only $\sin\theta$ or $\cos\theta$.

Procedures may vary. $\sin\left(\theta + \frac{n\pi}{2}\right) = \pm\cos\theta$ when n is an odd integer, and $\sin\left(\theta + \frac{n\pi}{2}\right) = \sin\theta$ when n is an even integer. The sign depends on the quadrant in which $\left(\theta + \frac{n\pi}{2}\right)$ lies.

Assignment Guide

See p. 206B for assignments.

Test Yourself

See *Teacher's Resource Book*, *Tests*, pp. 51–52.

Lesson Quiz

Express each as a sum or difference.

1. $2\cos 63^\circ \sin 40^\circ$
   $\sin 103^\circ - \sin 23^\circ$
2. $2\sin 104^\circ \sin 71^\circ$
   $\cos 67^\circ - \cos 175^\circ$

Express each sum or difference as a product.

3. $\sin 14x - \sin 4x$ $2\cos 9x \sin 5x$
4. $\cos 3y + \cos 5y$ $2\cos 4y \cos y$
5. Prove: $\tan 4\alpha = \dfrac{\cos 2\alpha - \cos 6\alpha}{\sin 6\alpha - \sin 2\alpha}$

$$\tan 4\alpha \;\Bigg|\; \frac{\cos 2\alpha - \cos 6\alpha}{\sin 6\alpha - \sin 2\alpha} = \frac{2\sin 4\alpha \sin 2\alpha}{2\cos 4\alpha \sin 2\alpha} = \frac{\sin 4\alpha}{\cos 4\alpha} = \tan 4\alpha$$

Enrichment

Find the simplest alternate expression for $\tan\left(\alpha + \frac{\pi}{2}\right)$. $-\cot\alpha$

CLASS EXERCISES

Match each expression on the left with an equivalent expression on the right.

1. $2\sin 20^\circ \sin 5^\circ$ C
2. $2\sin 15^\circ \sin 10^\circ$ D
3. $2\cos 10^\circ \cos 40^\circ$ A
4. $2\sin 100^\circ \cos 50^\circ$ B

A. $\cos 50^\circ + \cos(-30^\circ)$
B. $\sin 150^\circ + \sin 50^\circ$
C. $\cos 15^\circ - \cos 25^\circ$
D. $\cos 5^\circ - \cos 25^\circ$

Express each product as a sum or difference.

5. $2\sin 50^\circ \cos 20^\circ$ $\sin 70^\circ + \sin 30^\circ$
6. $2\sin 25^\circ \sin 5^\circ$ $\cos 20^\circ - \cos 30^\circ$
7. $2\cos 40^\circ \cos 10^\circ$ $\cos 30^\circ + \cos 50^\circ$
8. $2\cos 30^\circ \sin 10^\circ$ $\sin 40^\circ - \sin 20^\circ$

Express each sum or difference as a product.

9. $\cos 70^\circ + \cos 10^\circ$ $2\cos 40^\circ \cos 30^\circ$
10. $\sin 70^\circ - \sin 10^\circ$ $2\cos 40^\circ \sin 30^\circ$
11. $\sin 4x + \sin 2x$ $2\sin 3x \cos x$
12. $\sin 10x - \sin 6x$ $2\cos 8x \sin 2x$
13. $\cos 4x - \cos 8x$ $2\sin 6x \sin 2x$
14. $\cos 12x + \cos 4x$ $2\cos 8x \cos 4x$

PRACTICE EXERCISES

Express each product as a sum or difference.

A

1. $2\sin 42^\circ \sin 18^\circ$ $\cos 24^\circ - \cos 60^\circ$
2. $2\cos 26^\circ \cos 16^\circ$ $\cos 10^\circ + \cos 42^\circ$
3. $2\cos 5x \sin 2x$ $\sin 7x - \sin 3x$
4. $2\sin 6x \sin x$ $\cos 5x - \cos 7x$
5. $2\sin\frac{\pi}{12}\cos\frac{\pi}{6}$ $\sin\frac{\pi}{4} - \sin\left(\frac{\pi}{12}\right)$
6. $2\cos\frac{\pi}{8}\sin\frac{\pi}{4}$ $\sin\frac{3\pi}{8} + \sin\left(\frac{\pi}{8}\right)$
7. $\cos\frac{3\pi}{8}\cos\frac{5\pi}{8}$ $\frac{1}{2}\left(\cos\frac{\pi}{4} + \cos\pi\right)$
8. $\sin\frac{\pi}{12}\sin\frac{\pi}{4}$ $-\frac{1}{2}\left(\cos\frac{\pi}{3} - \cos\frac{\pi}{6}\right)$

Express each sum or difference as a product.

9. $\cos 38^\circ + \cos 106^\circ$ $2\cos 72^\circ \cos 34^\circ$
10. $\sin 46^\circ + \sin 98^\circ$ $2\sin 72^\circ \cos 26^\circ$
11. $\sin 12\theta + \sin 3\theta$ $2\sin\frac{15\theta}{2}\cos\frac{9\theta}{2}$
12. $\sin 15\theta - \sin 4\theta$ $2\cos\frac{19\theta}{2}\sin\frac{11\theta}{2}$
13. $\cos\frac{6\pi}{7} - \cos\frac{2\pi}{7}$ $-2\sin\frac{4\pi}{7}\sin\frac{2\pi}{7}$
14. $\cos\frac{7\pi}{9} + \cos\frac{2\pi}{9}$ $2\cos\frac{\pi}{2}\cos\frac{5\pi}{18}$

Prove each identity. See below.

B

15. $\cot\beta = \dfrac{\cos 3\beta + \cos\beta}{\sin 3\beta - \sin\beta}$
16. $\cot\alpha = \dfrac{\cos 5\alpha + \cos 3\alpha}{\sin 5\alpha - \sin 3\alpha}$

15. $$\frac{\cos 3\beta + \cos\beta}{\sin 3\beta - \sin\beta} = \frac{2\cos\frac{3\beta+\beta}{2}\cos\frac{3\beta-\beta}{2}}{2\cos\frac{3\beta+\beta}{2}\sin\frac{3\beta-\beta}{2}} = \frac{\cos\beta}{\sin\beta} = \cot\beta$$

16. $$\frac{\cos 5\alpha + \cos 3\alpha}{\sin 5\alpha - \sin 3\alpha} = \frac{2\cos\frac{5\alpha+3\alpha}{2}\cos\frac{5\alpha-3\alpha}{2}}{2\cos\frac{5\alpha+3\alpha}{2}\sin\frac{5\alpha-3\alpha}{2}} = \frac{\cos\alpha}{\sin\alpha} = \cot\alpha$$

17. $\tan 5\beta = \dfrac{\cos 8\beta - \cos 2\beta}{\sin 2\beta - \sin 8\beta}$

18. $\tan 2\alpha = \dfrac{\cos \alpha - \cos 3\alpha}{\sin 3\alpha - \sin \alpha}$

See pages 485–486.

19. $\dfrac{1}{\tan \alpha} = \dfrac{\cos 4\alpha + \cos 2\alpha}{\sin 4\alpha - \sin 2\alpha}$

20. $\dfrac{\tan 6\beta}{\tan \beta} = \dfrac{\sin 7\beta + \sin 5\beta}{\sin 7\beta - \sin 5\beta}$

21. $-\tan \theta = \dfrac{\cos 3\theta - \cos \theta}{\sin 3\theta + \sin \theta}$

22. $\dfrac{\csc \theta}{\sec \theta} = \dfrac{\sin 5\theta + \sin 3\theta}{\cos 3\theta - \cos 5\theta}$

C 23. $-\dfrac{\csc 5\theta}{\sec 5\theta} = \dfrac{\sin 8\theta - \sin 2\theta}{\cos 8\theta - \cos 2\theta}$

24. $\dfrac{\csc 3\theta}{\sec 3\theta} = \dfrac{\cos 4\theta + \cos 2\theta}{\sin 4\theta + \sin 2\theta}$

25. $\cos^2 y - \sin^2 x = \cos(x + y)\cos(x - y)$

26. $\sin x \cos x + \sin y \cos y = \sin(x + y)\cos(x - y)$

Applications

Computer The user is asked to supply two angle values. The program then evaluates both sides of $2 \sin \alpha \cos \beta = \sin(\alpha + \beta) + \sin(\alpha - \beta)$. The results are then compared.

```
10 HOME: INPUT"ENTER TWO
   ANGLE MEASURES": A,B
20 R1 = A*0.017453292 :
   R2 = B*0.017453292
30 L = 2*SIN(R1)*COS(R2)
40 R = SIN(R1 + R2) + SIN(R1 - R2)
50 PRINT: PRINT"2 SIN A COS B",
   "=", "SIN"(A + B) + "SIN"(A - B)
60 PRINT L,R
70 END
```

27. If the angle values are the same, how could lines 30 and 40 be simplified?
L = 2 * SIN (R1) * COS (R1); R = SIN (2 * R1)

28. Name some angle values for which it is just as easy to do the calculations by hand.
Answers may vary.

TEST YOURSELF

5.4

1. Use double-angle identities to find the exact value of $\cos 2\theta$ if $\sin \theta = \frac{5}{13}$ and $90° < \theta < 180°$. $\frac{119}{169}$

2. Use double-angle identities to find the exact value of $\tan 2\theta$ if $\tan \theta = \frac{5}{4}$ and $180° < \theta < 270°$. $-\frac{40}{9}$

5.5

3. Use half-angle identities to find the exact value of $\sin 105°$. $\frac{\sqrt{2 + \sqrt{3}}}{2}$

4. Use half-angle identities to find the exact value of $\tan \frac{5\pi}{8}$. $-\sqrt{2} - 1$

5. Find the exact value of $\tan \frac{\alpha}{2}$ if $\cos \alpha = -\frac{2}{3}$ and α lies in quadrant III $0° < \alpha < 360°$. $-\sqrt{5}$

6. Find the exact value of $\cos \frac{\alpha}{2}$ if $\cos \alpha = -\frac{12}{13}$ and α lies in quadrant II $0° < \alpha < 360°$. $\frac{\sqrt{26}}{26}$

7. Prove: $\csc 2\alpha + \cot 2\alpha = \cot \alpha$
See side column.

8. Prove: $\sec 2\theta = \dfrac{\csc^2 \theta}{\csc^2 \theta - 2}$

5.6

9. Express as a product: $\sin 6x + \sin 4x$ $2 \sin 5x \cos x$

10. Express as a product: $\cos 25° - \cos 15°$ $-2 \sin 20° \sin 5°$

Computer

The complete form of the program listed in the applications is on the disk provided with the *Teacher's Resource Book*.

Teacher's Resource Book

Practice—Chapter 5, p. 11
Enrichment—Chapter 5, p. 12

Additional Answers

Test Yourself

7. $\csc 2\alpha + \cot 2\alpha = \dfrac{1}{2 \sin \alpha \cdot \cos \alpha} + \dfrac{1 - \tan^2 \alpha}{2 \tan \alpha} = \dfrac{1}{2 \sin \alpha \cdot \cos \alpha} + \dfrac{\dfrac{(\cos^2 \alpha - \sin^2 \alpha)}{\cos^2 \alpha}}{\dfrac{2 \sin \alpha}{\cos \alpha}} = \dfrac{1}{2 \sin \alpha \cdot \cos \alpha} + \dfrac{\cos^2 \alpha - \sin^2 \alpha}{2 \sin \alpha \cdot \cos \alpha} = \dfrac{1 + \cos^2 \alpha - \sin^2 \alpha}{2 \sin \alpha \cdot \cos \alpha} = \dfrac{2 \cos^2 \alpha}{2 \sin \alpha \cdot \cos \alpha} = \dfrac{\cos \alpha}{\sin \alpha} = \cot \alpha$

8. $\sec 2\theta = \dfrac{1}{\cos 2\theta} = \left(\dfrac{1}{1 - 2 \sin^2 \theta}\right) \cdot \dfrac{\left(\dfrac{1}{\sin^2 \theta}\right)}{\left(\dfrac{1}{\sin^2 \theta}\right)} = \dfrac{\dfrac{1}{\sin^2 \theta}}{\dfrac{1}{\sin^2 \theta} - 2} = \dfrac{\csc^2 \theta}{\csc^2 \theta - 2}$

Application

Students are given an introduction to the methods used for calculating the angle of bank for a *standard-rate balanced turn*. This is just one of the many applications of trigonometry in the field of aviation.

See *Teacher's Resource Book* for *Application*, Chapter 5, p. 13.
See Teacher's Resource Book for *Technology*, p. 5.

APPLICATION: Aviation

Did you know that in many professions or occupations there are certain "rules of thumb" that are approximately true and are passed down from one generation to the next generation? Some "rules of thumb" are based on legend, some on customs and folklore, and still others on mathematical principles. Airplane pilots, for example, use experience, good judgment, common sense, and some "rules of thumb." The "rule of thumb" considered here will have its basis in trigonometry.

An airline pilot, unlike a motorist in an automobile, cannot safely make sharp or sudden changes in the direction of his or her aircraft. A rate of turn must be selected that is both safe and effective. For many types of aircraft, a rate of turn called the *standard-rate balanced turn* is used. This rate is defined to be exactly 3° per second. Hence, a pilot using a standard-rate balanced turn requires 120 s to turn the aircraft through a complete revolution of 360°.

When turning the aircraft, a pilot needs to choose an *angle of bank*, that is, the tilt of the plane's wings from the horizontal. The angle is determined by both the plane's velocity and its rate of turn.

Mathematicians and pilots have developed a formula that can be used to calculate the angle of bank for a standard-rate balanced turn.

$\tan \theta = 2.3855 \times 10^{-3}\, v$ *Formula for the angle of bank, where v is the velocity of the plane in miles per hour*

EXAMPLE Suppose you are piloting an aircraft through a standard-rate balanced turn of 3° traveling 135 mi/h. What angle of bank do you need?

$$\tan \theta = 2.3855 \times 10^{-3}\, v$$
$$\tan \theta = 2.3855 \times 10^{-3} \times 135 \quad \textit{Calculation-ready form}$$
$$\tan \theta = 0.3320$$
$$\theta = 18^\circ$$

The angle of bank is approximately 18°.

The table shows the result of using the formula to find the angle of bank at different velocities for a standard-rate balanced turn.

Velocity	Angle of bank (degrees)
60	8
80	11
100	13.4
120	16
140	18.5
160	21
180	23
200	25.5

At times, a pilot may need a quick and reasonably accurate way of determining the angle of bank. A ''rule of thumb'' used to find the angle of bank for standard-rate balanced turns at various velocities is

$$\text{angle of bank} = \frac{\text{velocity in miles per hour}}{10} + 5$$

In the example, the ''rule of thumb'' would have given an angle of bank of 18.5°.

EXERCISES

Using the formula $\tan \theta = 2.3855 \times 10^{-3} v$, calculate the angle of bank for a standard-rate balanced turn at various velocities.

1. 90 mi/h 12°
2. 150 mi/h 20°
3. 250 mi/h 31°
4. 350 mi/h 40°

Using the ''rule of thumb,'' estimate the bank angle for a standard-rate balanced turn at the following velocities.

5. 110 mi/h 16°
6. 145 mi/h 19.5°
7. 300 mi/h 35°
8. 400 mi/h 45°
9. If you know that the angle of bank is 22° for a standard-rate balanced turn, use $\tan \theta = 2.3855 \times 10^{-3} v$ to find the plane's speed in miles per hour. 169 mi/h
10. The angle of bank of an airplane is 31°. Approximate the plane's speed in miles per hour using the ''rule of thumb'' formula. 260 mi/h
11. Find the amount of error that results from using the ''rule of thumb'' rather than the formula to find the measure of the angle of bank for a standard-rate balanced turn at 90 mi/h. Express this error as a percent of the angle measure calculated using the formula. 1.9°; 16%
12. Find the percent of error made by using the ''rule of thumb'' rather than the formula for the following velocities (all expressed in miles per hour): 60, 120, 150, 250, 350, 600, 700. 35%; 6.4%; 1.6%; 2.6%; 0.35%; 18%; 27%
13. Based on your results in Exercises 11 and 12, do you think that the ''rule of thumb'' is most accurate for high, intermediate, or low velocities? intermediate

CHAPTER 5 SUMMARY AND REVIEW

Vocabulary

difference identity
- for cosine (208)
- for sine (215)
- for tangent (221)

sum identity
- for cosine (210)
- for sine (215)
- for tangent (220)

other identities
- double-angle (227)
- half-angle (232)
- sum/product (238)

Cosine: Sum and Difference Identities — 5.1

$\cos(\alpha + \beta) = \cos\alpha\cos\beta - \sin\alpha\sin\beta$ $\quad$ $\cos(\alpha - \beta) = \cos\alpha\cos\beta + \sin\alpha\sin\beta$

1. Find the exact value of $\cos 255°$. $\frac{\sqrt{2} - \sqrt{6}}{4}$
2. Find the exact value of $\cos\frac{\pi}{12}$. $\frac{\sqrt{6} + \sqrt{2}}{4}$
3. Find the exact value of $\cos(\alpha - \beta)$ if $\tan\alpha = \frac{4}{3}$, $\sin\beta = -\frac{1}{2}$, $\pi < \alpha < \frac{3\pi}{2}$, and $\frac{3\pi}{2}, < \beta < 2\pi$. $\frac{3\sqrt{3} + 4}{10}$
4. Prove: $\cos(5\pi + \theta) = -\cos\theta$
   $\cos(5\pi + \theta) = \cos 5\pi \cos\theta - \sin 5\pi \sin\theta = -1 \cos\theta - 0 \cdot \sin\theta = -\cos\theta$

Sine: Sum and Difference Identities — 5.2

$\sin(\alpha + \beta) = \sin\alpha\cos\beta + \cos\alpha\sin\beta$ $\quad$ $\sin(\alpha - \beta) = \sin\alpha\cos\beta - \cos\alpha\sin\beta$

5. Find the exact value of $\sin 105°$. $\frac{\sqrt{6} + \sqrt{2}}{4}$
6. Find the exact value of $\sin\frac{5\pi}{12}$. $\frac{\sqrt{6} + \sqrt{2}}{4}$
7. Find the exact value of $\sin(\alpha + \beta)$ if $\cos\alpha = -\frac{3}{5}$, $\cos\beta = \frac{7}{25}$, $\pi < \alpha < \frac{3\pi}{2}$, and $0 < \beta < \frac{\pi}{2}$. $-\frac{4}{5}$
8. Prove: $\sin(270° + \alpha) = -\cos\alpha$
   $\sin(270° + \alpha) = \sin 270° \cos\alpha + \cos 270° \sin\alpha = -1 \cos\alpha + 0 \cdot \sin\alpha = -\cos\alpha$

Tangent: Sum and Difference Identities — 5.3

$$\tan(\alpha + \beta) = \frac{\tan\alpha + \tan\beta}{1 - \tan\alpha\tan\beta}, \quad \tan\alpha\tan\beta \neq 1$$

$$\tan(\alpha - \beta) = \frac{\tan\beta - \tan\beta}{1 + \tan\alpha\tan\beta}, \quad \tan\alpha\tan\beta \neq -1$$

9. Find the exact value of $\tan 75°$. $2 + \sqrt{3}$
10. Find the exact value of $\tan\frac{7\pi}{12}$. $-2 - \sqrt{3}$
11. Find the exact value of $\tan(\alpha - \beta)$ if $\cos\alpha = -\frac{1}{2}$, $\cos\beta = -\frac{1}{2}$, $\frac{\pi}{2} < \alpha < \pi$, and $\pi < \beta < \frac{3\pi}{2}$. $\sqrt{3}$
12. Prove: $\tan(3\pi - \alpha) = -\tan\alpha$ $\quad$ $\tan(3\pi - \alpha) = \frac{\tan 3\pi - \tan\alpha}{1 + \tan 3\pi \tan\alpha} = \frac{0 - \tan\alpha}{1 + 0 \cdot \tan\alpha} = -\tan\alpha$

244

Double-Angle Identities 5.4

$\sin 2\alpha = 2 \sin \alpha \cos \alpha$

$\tan 2\alpha = \dfrac{2 \tan \alpha}{1 - \tan^2 \alpha}$, $\tan \alpha \neq \pm 1$

$\cos 2\alpha = \cos^2 \alpha - \sin^2 \alpha$

$\cos 2\alpha = 2 \cos^2 \alpha - 1$

$\cos 2\alpha = 1 - 2 \sin^2 \alpha$

If $\sin \alpha = -\frac{3}{5}$ and α lies in quadrant IV, find the exact value of each expression.

13. $\sin 2\alpha$ $-\frac{24}{25}$ **14.** $\cos 2\alpha$ $\frac{7}{25}$ **15.** $\tan 2\alpha$ $-\frac{24}{7}$

16. Prove: $\tan \theta = \csc 2\theta - \cot 2\theta$ See side column.

Half-Angle Identities 5.5

$\sin \frac{\alpha}{2} = \pm\sqrt{\dfrac{1 - \cos \alpha}{2}}$

$\cos \frac{\alpha}{2} = \pm\sqrt{\dfrac{1 + \cos \alpha}{2}}$

$\tan \frac{\alpha}{2} = \pm\sqrt{\dfrac{1 - \cos \alpha}{1 + \cos \alpha}}$, $\cos \alpha \neq -1$

$\tan \frac{\alpha}{2} = \dfrac{\sin \alpha}{1 + \cos \alpha}$, $\cos \alpha \neq -1$

$\tan \frac{\alpha}{2} = \dfrac{1 - \cos \alpha}{\sin \alpha}$, $\sin \alpha \neq 0$

If $\sin \theta = -\frac{12}{13}$ and θ lies in quadrant III, find the exact value of each expression. Assume that $0° < \theta < 360°$.

17. $\sin \frac{\theta}{2}$ $\frac{3\sqrt{13}}{13}$ **18.** $\cos \frac{\theta}{2}$ $-\frac{2\sqrt{13}}{13}$ **19.** $\tan \frac{\theta}{2}$ $-\frac{3}{2}$

20. Prove: $\cot \frac{\alpha}{2} = \dfrac{1}{\csc \alpha - \cot \alpha}$ $\cot \frac{\alpha}{2} = \left(\dfrac{\sin \alpha}{1 - \cos \alpha}\right)\left(\dfrac{\frac{1}{\sin \alpha}}{\frac{1}{\sin \alpha}}\right) = \left(\dfrac{1}{\frac{1}{\sin \alpha} - \frac{\cos \alpha}{\sin \alpha}}\right) = \dfrac{1}{\csc \alpha - \cot \alpha}$

Product/Sum Identities 5.6

$2 \cos \alpha \cos \beta = \cos (\alpha - \beta) + \cos (\alpha + \beta)$

$2 \sin \alpha \sin \beta = \cos (\alpha - \beta) - \cos (\alpha + \beta)$

$2 \sin \alpha \cos \beta = \sin (\alpha + \beta) + \sin (\alpha - \beta)$

$2 \cos \alpha \sin \beta = \sin (\alpha + \beta) - \sin (\alpha - \beta)$

There are also alternate forms for the product/sum identities.

21. Express $2 \sin 9x \sin 4x$ as a sum. $\cos 5x - \cos 13x$

22. Express $2 \cos 50° \sin 34°$ as a sum. $\sin 84° - \sin 16°$

23. Express $\sin 73° + \sin 41°$ as a product. $2 \sin 57° \cos 16°$

24. Express $\cos (-2x) - \cos (-4x)$ as a product. $2 \sin 3x \sin x$

25. Prove: $\dfrac{\sin 2x + \sin 6x}{\sin 10x - \sin 2x} = \dfrac{\cos (-2x)}{\cos 6x}$ See side column.

26. Prove: $\tan 3\alpha = \dfrac{\cos \alpha - \cos 5\alpha}{\sin 5\alpha - \sin \alpha}$

Additional Answers

16. $\csc 2\theta - \cot 2\theta = \dfrac{1}{\sin 2\theta} - \dfrac{\cos 2\theta}{\sin 2\theta} = \dfrac{1}{2 \sin \theta \cos \theta} - \dfrac{1 - 2\sin^2 \theta}{2 \sin \theta \cos \theta} = \dfrac{2 \sin^2 \theta}{2 \sin \theta \cos \theta} = \dfrac{\sin \theta}{\cos \theta} = \tan \theta$

25. $\dfrac{\sin 2x + \sin 6x}{\sin 10x - \sin 2x} = \dfrac{2 \sin\left(\frac{2x + 6x}{2}\right) \cos\left(\frac{2x - 6x}{2}\right)}{2 \cos\left(\frac{10x + 2x}{2}\right) \sin\left(\frac{10x - 2x}{2}\right)} = \dfrac{2 \sin 4x \cdot \cos (-2x)}{2 \cos 6x \cdot \sin 4x} = \dfrac{\cos (-2x)}{\cos 6x} = \dfrac{\cos 2x}{\cos 6x}$

26. $\dfrac{\cos \alpha - \cos 5\alpha}{\sin 5\alpha - \sin \alpha} = \dfrac{-2 \sin\left(\frac{\alpha + 5\alpha}{2}\right) \sin\left(\frac{\alpha - 5\alpha}{2}\right)}{2 \cos\left(\frac{5\alpha + \alpha}{2}\right) \sin\left(\frac{5\alpha - \alpha}{2}\right)} = \dfrac{-2 \sin 3\alpha \cdot \sin (-2\alpha)}{2 \cos 3\alpha \cdot \sin 2\alpha} = \dfrac{\sin 3\alpha}{\cos 3\alpha} = \tan 3\alpha$

See *Teacher's Resource Book, Tests*, pp. 53–56.

CHAPTER TEST

Use sum and difference identities to find the value of each trigonometric function.

1. $\cos 75°$ $\frac{\sqrt{6}-\sqrt{2}}{4}$ **2.** $\sin 285°$ $\frac{-\sqrt{6}-\sqrt{2}}{4}$ **3.** $\tan 375°$ $2-\sqrt{3}$

4. Find the exact value of $\cos(\alpha+\beta)$ if $\sin\alpha = \frac{5}{13}$, $\tan\beta = \frac{7}{24}$, $0 < \alpha < \frac{\pi}{2}$, and $\pi < \beta < \frac{3\pi}{2}$. $-\frac{253}{325}$

5. Find the exact value of $\sin(\alpha-\beta)$ if $\cos\alpha = -\frac{1}{2}$, $\sin\beta = \frac{3}{5}$, $\frac{\pi}{2} < \alpha < \pi$, and $\frac{\pi}{2} < \beta < \pi$. $\frac{-4\sqrt{3}+3}{10}$

6. Find the exact value of $\tan(\alpha+\beta)$ if $\sin\alpha = -\frac{4}{5}$, $\sin\beta = \frac{\sqrt{2}}{2}$, $\pi < \alpha < \frac{3\pi}{2}$, and $\frac{\pi}{2} < \beta < \pi$. $\frac{1}{7}$

7. If $\cos\theta = \frac{7}{25}$ and $0° < \theta < 90°$, find $\sin 2\theta$. $\frac{336}{625}$

8. If $\tan\theta = -\frac{12}{5}$ and $270° < \theta < 360°$, find $\cos 2\theta$. $\frac{-119}{169}$

9. If $\cos\theta = -0.5$ and $90° < \theta < 180°$, find $\tan 2\theta$. $\sqrt{3}$

If $\cos\alpha = -\frac{3}{5}$ and α lies in quadrant III, find the exact value of each expression. Assume that $0° < \alpha < 360°$.

10. $\sin\frac{\alpha}{2}$ $\frac{2\sqrt{5}}{5}$

11. $\tan\frac{\alpha}{2}$ -2

12. Express $2\cos 7x \sin x$ as a sum. $\sin 8x - \sin 6x$

13. Express $\cos 76° - \cos 14°$ as a product. $-2\sin 45°\sin 31°$

14. Prove: $\sin 2\alpha = \frac{2}{\sec\alpha\csc\alpha}$ $\quad \frac{2}{\sec\alpha\csc\alpha} = 2\left(\frac{1}{\sec\alpha}\right)\left(\frac{1}{\csc\alpha}\right) = 2\cos\alpha\sin\alpha = \sin 2\alpha$

15. Prove: $\cos 2\alpha = \cos^4\alpha - \sin^4\alpha$ $\quad \cos^4\alpha - \sin^4\alpha = (\cos^2\alpha + \sin^2\alpha)(\cos^2\alpha - \sin^2\alpha) = 1(\cos^2\alpha - \sin^2\alpha) = \cos 2\alpha$

16. Prove: $\cot 2\beta = \frac{\cos\beta + \cos 3\beta}{\sin 3\beta + \sin\beta}$

$\frac{\cos\beta + \cos 3\beta}{\sin 3\beta + \sin\beta} = \frac{2\cos 2\beta\cos\beta}{2\sin 2\beta\cos\beta} = \frac{\cos 2\beta}{\sin 2\beta} = \cot 2\beta$

Challenge

Prove: $\left(\sin\frac{\alpha}{2} + \cos\frac{\alpha}{2}\right)^2 = 1 + \sin\alpha$ $\quad \left(\sin\frac{\alpha}{2} + \cos\frac{\alpha}{2}\right)^2 = \left(\sin^2\frac{\alpha}{2} + \cos^2\frac{\alpha}{2}\right) + \left(2\sin\frac{\alpha}{2}\cos\frac{\alpha}{2}\right) = 1 + \sin 2\left(\frac{\alpha}{2}\right) = 1 + \sin\alpha$

COLLEGE ENTRANCE EXAM REVIEW

Select the best choice for each question.

1. $\sin 15° =$
B

A. $\frac{\sqrt{3} + \sqrt{2}}{4}$ **B.** $\frac{\sqrt{6} - \sqrt{2}}{4}$

C. $\frac{\sqrt{6} - \sqrt{2}}{2}$ **D.** $\frac{\sqrt{3} - \sqrt{2}}{4}$

E. $\frac{\sqrt{2} + \sqrt{6}}{4}$

2. $\cos(\alpha + \beta)$ equals
C

A. $\cos \alpha + \cos \beta$
B. $\sin \alpha \sin \beta + \cos \alpha \cos \beta$
C. $\cos \alpha \cos \beta - \sin \alpha \sin \beta$
D. $\sin \alpha \cos \beta - \sin \beta \cos \alpha$
E. $\cos \alpha \sin \alpha + \cos \beta \sin \beta$

3. If $f(x) = x^2 + x + 1$,
D
then $f(a + b) =$

A. $a^2 + a + b^2 + b + 2$
B. $a^2 + b^2 + a + b + 1$
C. $2a^2 + 2ab + 2b^2 + 1$
D. $a^2 + 2ab + b^2 + a + b + 1$
E. $a^2 + 2ab + b^2 + 1$

4. If $\sin \theta = -\frac{3}{5}$, where $\pi < \theta < \frac{3\pi}{2}$,
C
then $\tan 2\theta =$

A. $\frac{3}{8}$ **B.** 3 **C.** $\frac{24}{7}$ **D.** $\frac{8}{3}$ **E.** $-\frac{24}{7}$

5. The equation of a circle with radius 3 and center $(2, -3)$ is
A

A. $(x - 2)^2 + (y + 3)^2 = 9$
B. $(x + 2)^2 + (y - 3)^2 = 3$
C. $(x - 3)^2 + (y + 2)^2 = 9$
D. $x^2 + y^2 = 9$
E. $(x - 2)^2 + (y - 3)^2 = 3$

6. $\cos(7\pi - x)$ is equivalent to
D

A. $\cos 7\pi$ **B.** $\sin(7\pi - x)$
C. $-\sin x$ **D.** $-\cos x$
E. $\sin 7\pi$

7. If $A = \frac{1}{2}h(b_1 + b_2)$, then $b_2 =$
E

A. $\frac{A - 2hb_1}{2h}$ **B.** $\frac{A - hb_1}{h}$
C. $A - hb_1$ **D.** $2A - 2b_1$
E. $\frac{2A - hb_1}{h}$

8. Solve for x.
A

$$\frac{6}{x^2 - 6x + 8} - \frac{x}{x - 2} = \frac{9}{x - 4}$$

A. $x = -8, x = 3$
B. $x = 6, x = 4$
C. $x = 8, x = -4$
D. $x = -6, x = 5$
E. $x = 9$

9. Evaluate: $\begin{vmatrix} 3 & -1 & 4 \\ 0 & 2 & 2 \\ 3 & 1 & -2 \end{vmatrix}$
B

A. 24 **B.** -48 **C.** -14
D. -42 **E.** -12

10. The area of a square is 25 ft^2. What is its perimeter?
B

A. 10 **B.** 20 **C.** 40 **D.** 25
E. not enough information

11. For $0 \le x \le \pi$, the following represents the graph of
C

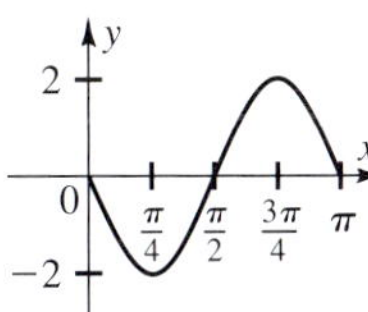

A. $y = 2 \cos x$ **B.** $y = \sin x$
C. $y = -2 \sin 2x$
D. $y = -2 \sin x$
E. $y = -2 \cos 2x$

247

Maintaining Skills

The following skills and concepts are reviewed:

- Simplifying expressions containing radicals
- Solving equations containing radicals
- Solving equations by using the quadratic formula
- Finding the inverses of functions

MAINTAINING SKILLS

Simplify.

Examples $\frac{\sqrt{3}+\sqrt{6}}{\sqrt{8}} = \frac{\sqrt{3}+\sqrt{6}}{\sqrt{8}} \cdot \frac{\sqrt{2}}{\sqrt{2}} = \frac{\sqrt{6}+\sqrt{12}}{\sqrt{16}} = \frac{\sqrt{6}+2\sqrt{3}}{4}$

$$\sqrt[3]{\frac{1}{3x}} = \sqrt[3]{\frac{1}{3x} \cdot \frac{3^2x^2}{3^2x^2}} = \frac{\sqrt[3]{9x^2}}{\sqrt[3]{27x^3}} = \frac{\sqrt[3]{9x^2}}{3x}$$

1. $\frac{2+\sqrt{3}}{\sqrt{3}}$ $\frac{2\sqrt{3}+3}{3}$
2. $\frac{5}{3\sqrt{2x}}$ $\frac{5\sqrt{2x}}{6x}$
3. $\frac{2-\sqrt{7}}{\sqrt{8}}$ $\frac{2\sqrt{2}-\sqrt{14}}{4}$
4. $\sqrt[3]{\frac{x}{y}}$ $\frac{\sqrt[3]{xy^2}}{y}$
5. $\sqrt[5]{\frac{3}{16}}$ $\frac{\sqrt[5]{6}}{2}$
6. $\sqrt[4]{\frac{2x}{y^3}}$ $\frac{\sqrt[4]{2xy}}{y}$

Solve each radical equation for x.

Example
$$\sqrt[3]{2x+3} = -3$$
$$2x+3 = -27 \quad \textit{Cube each side.}$$
$$x = -15$$

7. $\sqrt{x-3} = 5$ 28
8. $-\sqrt{x+2} = -7$ 47
9. $\sqrt[3]{x-2} = 5$ 127
10. $\sqrt{2x+4} = 8$ 30
11. $\sqrt{x^2+9} = x+1$ 4
12. $\sqrt[3]{7x+1} = 4$ 9

Solve each equation using the quadratic formula $x = \frac{-b \pm \sqrt{b^2-4ac}}{2a}$.

Example $5x^2 - 38x + 56 = 0$ *$a = 5$, $b = -38$, $c = 56$*

$$x = \frac{-(-38) \pm \sqrt{(-38)^2 - (4)(5)(56)}}{(2)(5)}$$
$$= \frac{38 \pm \sqrt{1444-1120}}{10} = \frac{38 \pm \sqrt{324}}{10} = \frac{38 \pm 18}{10} = 5.6 \text{ or } 2$$

13. $x^2 + 3x = -2$ $-1, -2$
14. $2x^2 - 7x - 15 = 0$ $\frac{-3}{2}, 5$
15. $2x^2 + 4x - 6 = 0$ $-3, 1$
16. $6x^2 + x - 1 = 0$ $\frac{-1}{2}, \frac{1}{3}$
17. $x^2 = 3x$ 0, 3
18. $x^2 - 6x + 9 = 0$ 3

Find the inverse of each function.

Example
$$y = 2x + 6$$
$$x = 2y + 6 \quad \textit{Interchange x and y.}$$
$$x - 6 = 2y \quad \textit{Solve for y.}$$
$$\tfrac{1}{2}x - 3 = y$$

19. $y = 3x - 5$ $y = \frac{1}{3}x + \frac{5}{3}$
20. $y = 4x + 3$ $y = \frac{1}{4}x - \frac{3}{4}$
21. $y = \frac{1}{2}x - \frac{1}{2}$ $y = 2x + 1$
22. $y = \frac{1}{3}x + 9$ $y = 3x - 27$

OVERVIEW • Chapter 6

SUMMARY

Initially, inverses of relations are found by interchanging the *x* and *y* components. Then the relations and their inverses are graphed and the vertical line test is used to determine if relations are functions. Inverses of the trigonometric functions are discussed and the notation for the inverses are introduced. By limiting the domains of the functions their inverses will also be functions. Notation used to indicate that the domains have been restricted or that *principal values* are required is discussed. Students solve conditional trigonometric equations involving special angle solutions as well as approximate solutions. The chapter concludes with a discussion of rotation of axes as it applies to trigonometry.

CHAPTER OBJECTIVES

- To graph functions and relations and their inverses
- To evaluate arcsin and arccos expressions
- To define the inverse sine and cosine functions
- To evaluate expressions involving Arcsine and Arccosine
- To define the inverse Tangent, Cotangent, Secant, and Cosecant functions
- To evaluate expressions involving these functions
- To define conditional equations
- To find exact solutions to trigonometric equations
- To find approximate values of solutions to trigonometric equations
- To derive and use formulas for rotations of axes
- To graph equations using rotated axes

CHAPTER HIGHLIGHTS

The *theme* of Chapter 6 is light. An application which analyzes the mathematical and scientific aspects of refraction is among the chapter's special features.

APPLICATIONS

Inverses figure prominently in the solution process for many equations and are therefore used in many real-world applications. The fields of engineering, construction, and mechanics are all sources of applications of inverses of trigonometric functions. Problems from physics are solved using the methods presented in Lessons 6.4 and 6.5.

TECHNOLOGY

Calculator
Calculators are used to find the values of inverse trigonometric functions. Students are instructed in their use for this purpose in Lessons 6.2 and 6.3.

Computer
A computer can be helpful for determining if solutions to trigonometric equations are correct. The formulas for rotation of axes are also good candidates for a computer program.

RESOURCES

Teacher's Resource Book

- Teaching Aid 6
- Transparencies 11 and 12

ASSIGNMENT GUIDE Meeting Student Needs

STUDENT TEXT				TEACHER'S RESOURCE BOOK	
Chapter Content	**Basic**	**Average**	**Enriched**	**P**	**E**
6.1 Inverse Relations and Functions	D: 253/1–27 odd, 37	D: 253/7–33 odd, 37	D: 253/9–37 odd	1	2
6.2 The Inverse Sine and Cosine Functions	D: 257/1–31 odd R: 253/2, 6, 20	D: 257/7, 13–43 odd, 49 R: 253/10, 16, 22	D: 257/7, 15–49 odd R: 253/14, 18, 30	3	4
6.3 Other Inverse Trigonometric Functions	D: 263/1–21 odd R: 257/4, 10, 26 264/TY	D: 263/3, 7–25 odd, 29 R: 257/8, 14, 34 264/TY	D: 263/3, 7, 13–29 odd R: 257/8, 16, 36 264/TY	5	6
6.4 Solving Trigonometric Equations: Using Special Angles	D: 267/1–25 odd R: 263/2, 6, 22	D: 267/3, 9, 13–35 odd, 41 R: 263/2, 14, 22	D: 267/5, 11, 17–41 odd R: 263/4, 16, 24	7	8
6.5 Trigonometric Equations: Approximate Solutions	D: 271/1–23 odd R: 267/2, 8, 14	D: 271/11–31 odd, 37 R: 267/4, 16, 22	D: 271/17–37 odd R: 267/6, 24, 34	9	10
6.6 Rotation of Axes	Omit	D: 267/5, 11–25 odd, 29 R: 271/12, 22, 28 277/TY	D: 276/7, 13–29 odd R: 271/22, 26, 30 277/TY	11	12

D = Daily R = Review TY = Test Yourself P = Practice E = Enrichment

	STUDENT TEXT				TEACHER'S RESOURCE BOOK	
Review and Testing	Test Yourself	264, 277	College Ent. Exam Rev.	283	Tests	
	Chapter Sum. and Rev.	280	Cumulative Review	284	• Quizzes	57–60
	Chapter Test	282			• Chapter Test (Form A)	61–62
					• Chapter Test (Form B)	63–64
Special Features	Extra	254	Challenge	268, 272	Applications—Chapter 6	13
	Biography	259	Application	278	Critical Thinking	5
					Alg. and Geom. Review	21–24
					Technology	6

6 Inverse Trigonometric Functions

Light is electromagnetic radiation that may be perceived by the normal human eye. Trigonometry plays an important role in understanding its path. Lightning is a visible flash of light that results from a large-scale natural electric discharge in the atmosphere.

BACKGROUND

Many forms of light radiate from a number of different sources. The distinctions between these forms of light are often described using trigonometric functions. Have students research some of the different forms that electromagnetic radiation can take and the physical laws that apply to them.

LESSON PLAN

Vocabulary
Inverse Function

BACKGROUND

In the Preview, an equation that represents donors to a public television station is used to develop the concepts of inverses.

Additional Answers

Preview

1.

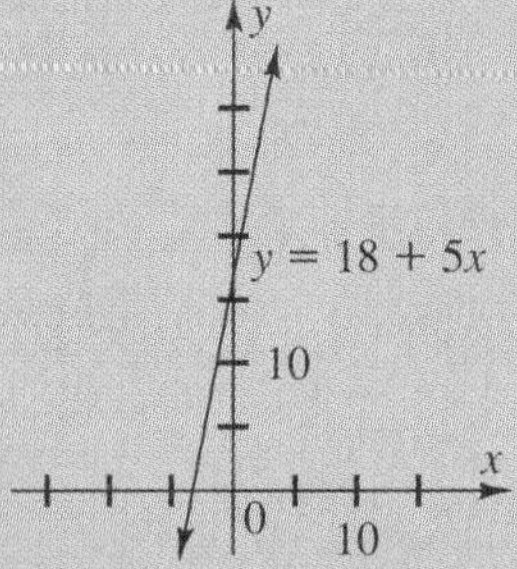

2. x: no. of people, y: no. of spots; $y = \frac{x}{5} - \frac{18}{5}$; Domain and range are interchanged.

3. The graphs are "reflected" about the line $y = x$.

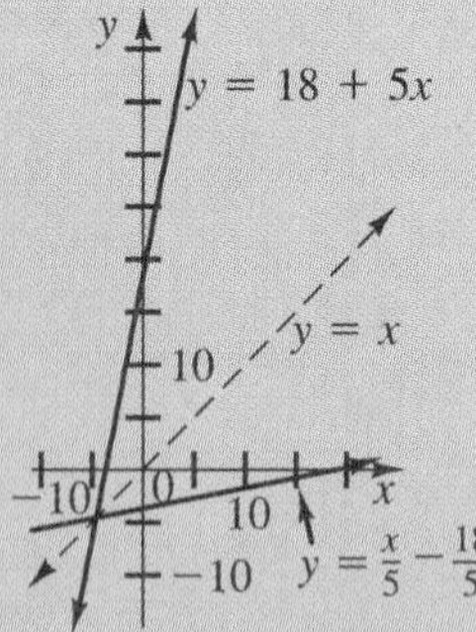

6.1 Inverse Relations and Functions

Objectives To graph functions and relations and their inverses
To evaluate expressions involving the inverse sine and inverse cosine

Recall that a relation is a set of ordered pairs, and that a function is a relation in which each element in the domain is paired with a unique element in the range.

Preview

A public television station recorded the number of people who donated to the station following fund-raising appeals. It found that the number of people, y, who gave per week could be approximated by thc cquation

$$y = 18 + 5x$$

where x is the number of advertising spots for the appeal during that week.

1. Graph the equation $y = 18 + 5x$. See side column.
2. Interchange x and y in the equation $y = 18 + 5x$ and solve the new equation for y. What do x and y represent in the new equation? How do the domain and range of the first relation compare to those of the second? See side column.
3. Graph the equation you obtained in Exercise 2 and the line $y = x$ on the same coordinate plane as you graphed $y = 18 + 5x$. Fold the paper along the line $y = x$. What assumptions can be made about the graph of $y = 18 + 5x$ and the graph of the equation you wrote for Exercise 2? See side column.

Consider the relation

$$\{(1, 2), (2, 2), (3, 4), (4, 6), (5, 8)\}$$

The domain is $\{1, 2, 3, 4, 5\}$ and the range is $\{2, 4, 6, 8\}$. The inverse of this relation is formed by interchanging the first and second elements in each pair of the relation. The result is

$$\{(2, 1), (2, 2), (4, 3), (6, 4), (8, 5)\} \qquad \textit{Inverse of relation above}$$

The domain of the inverse, $\{2, 4, 6, 8\}$, is the range of the original relation, and the range of the inverse, $\{1, 2, 3, 4, 5\}$, is the domain of the original relation. The original relation is a function, but the inverse relation is not, since the first element, 2, is paired with both 1 and 2.

When a relation is defined by an equation in two variables, its inverse can often be formed by first interchanging the variables.

EXAMPLE 1 **Find the inverse of the relation $y = \pm\sqrt{x}$, and state whether or not the inverse is a function.**

$y = \pm\sqrt{x}$	
$x = \pm\sqrt{y}$	*Interchange x and y.*
$x^2 = y$	*Square both sides and solve for y.*
$y = x^2$	*The inverse of $y = \pm\sqrt{x}$*

In this case the original relation is not a function, since there are two values of y for each nonzero value of x. However, the inverse, $y = x^2$, is a function since for each x there is only one y.

When the inverse of a function is also a function, it is called the **inverse function** of the original function. If $f(x)$ is a function, then its inverse function is denoted by $f^{-1}(x)$. Keep in mind that the symbol -1 does *not* indicate the reciprocal, $\frac{1}{f(x)}$, when it is used in this way.

EXAMPLE 2 **Find the inverse of the function $f(x) = 2x + 4$.**

$y = 2x + 4$	*Write $f(x) = 2x + 4$ as $y = 2x + 4$.*
$x = 2y + 4$	*Interchange x and y.*
$x - 4 = 2y$	*Solve for y.*
$y = \frac{x}{2} - 2$	

The inverse of $f(x) = 2x + 4$ is $f^{-1}(x) = \frac{x}{2} - 2$. It is also a function, since there is only one value of y for each value of x.

The graphs of the functions in Example 2 are shown at the right. If the coordinate plane were folded along the line $y = x$, the graphs of $f(x) = 2x + 4$ and $f^{-1}(x) = \frac{x}{2} - 2$ would coincide. Therefore, $f(x)$ and $f^{-1}(x)$ are symmetric with respect to the line $y = x$. This symmetry about $y = x$ exists for any pair of inverse functions or relations.

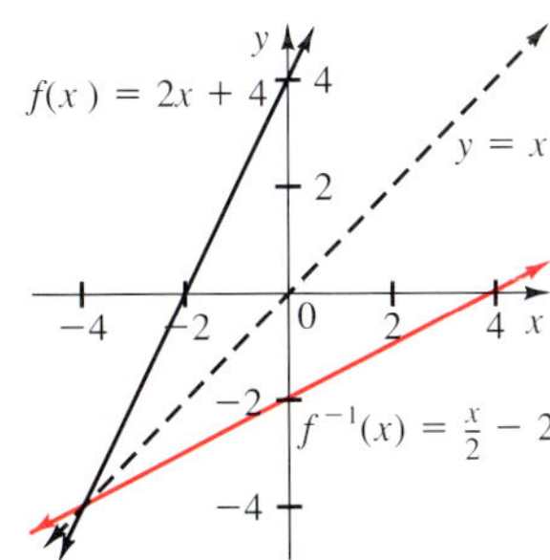

TEACHING SUGGESTIONS

- Review the definitions of domain, range, and function.
- Be sure that students understand how to find the inverse of a function. Do several examples where the domain and range are interchanged.
- Arcsin x and arccos x are sometimes written $\sin^{-1} x$ and $\cos^{-1} x$.

CHALKBOARD EXAMPLES

- **For Example 1**

Find the inverse of each relation and state whether or not the inverse is a function.

1.

x	2	4	6	8	10	12
y	3	2	1	1	2	3

(3, 2), (2, 4), (1, 6), (1, 8), (2, 10) (3, 12); the original is a function but the inverse is not.

2. $x^2 - 4 = y$

$y = \pm\sqrt{x + 4}$; the original relation is a function but the inverse is not.

- **For Example 2**

Find the inverse of each function.

3. $f(x) = 10x - 6$ $\quad f^{-1}(x) = \frac{x}{10} + \frac{3}{5}$

4. $f(x) = \frac{4}{5}x - \frac{3}{4}$ $\quad f^{-1}(x) = \frac{5}{4}x + \frac{15}{16}$

- **For Example 3**

5. Graph $y = \tan x$ and its inverse $y = \arctan x$ on the same coordinate plane.

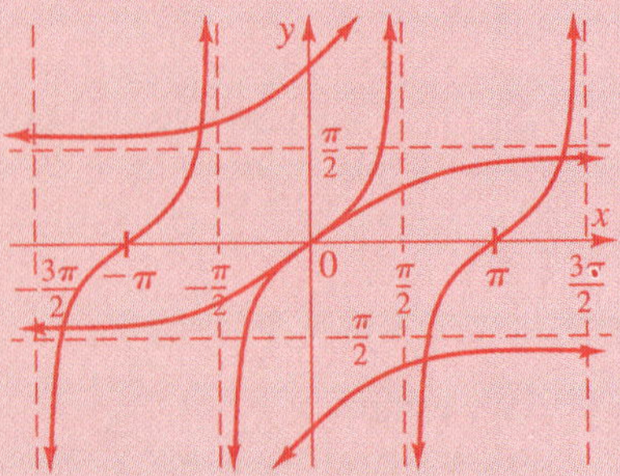

- **For Example 4**

Evaluate each expression in degrees.

6. $\arcsin \frac{1}{2}$

$y = 30° + n(360°)$ and $y = 150° + n(360°)$, where n is an integer

7. $\arctan \sqrt{3}$

$y = 60° + n(360°)$ and $y = 240° + n(360°)$, where n is an integer

Common Errors

- Students may try to add or subtract variables to interchange x and y. Stress that no algebraic manipulation is involved; the x and y values are simply interchanged.
- Students may confuse $f^{-1}(x)$ with reciprocals like 5^{-1}. Emphasize that $f^{-1}(x)$ means the inverse of $f(x)$.
- See *Teacher's Resource Book* for additional remediation.

The trigonometric functions also have inverses. To find the inverse of $y = \sin x$, you may interchange the variables to obtain $x = \sin y$, but you will not be able to solve this equation algebraically for y in terms of x. Therefore, the inverse of $y = \sin x$ is written as follows:

$$y = \sin^{-1} x \qquad \textit{Read: y is a number whose sine is x.}$$

Remember that $\sin^{-1} x$ is *not* the reciprocal of $\sin x$. The reciprocal of $\sin x$ is written as either $(\sin x)^{-1}$ or $\dfrac{1}{\sin x}$. The inverse of each of the other trigonometric functions can be formed in the same way. For example, the inverse of $y = \cos x$ is

$$y = \cos^{-1} x \qquad \textit{Read: y is a number whose cosine is x.}$$

The inverse of the sine function can be graphed by interchanging the x- and y-coordinates of points on the graph of the sine function.

EXAMPLE 3 **Graph $y = \sin x$ and $y = \sin^{-1} x$ on the same coordinate plane.**

Sketch the graph of $y = \sin x$. Then interchange the coordinates of points on the graph to find the corresponding points on the graph of $y = \sin^{-1} x$. The coordinates of some of the points are shown below.

$y = \sin x$	$y = \arcsin x$
$(-\pi, 0)$	$(0, -\pi)$
$\left(-\frac{\pi}{2}, -1\right)$	$\left(-1, -\frac{\pi}{2}\right)$
$(0, 0)$	$(0, 0)$
$\left(\frac{\pi}{2}, 1\right)$	$\left(1, \frac{\pi}{2}\right)$
$(\pi, 0)$	$(0, \pi)$
$\left(\frac{3\pi}{2}, -1\right)$	$\left(-1, \frac{3\pi}{2}\right)$
$(2\pi, 0)$	$(0, 2\pi)$

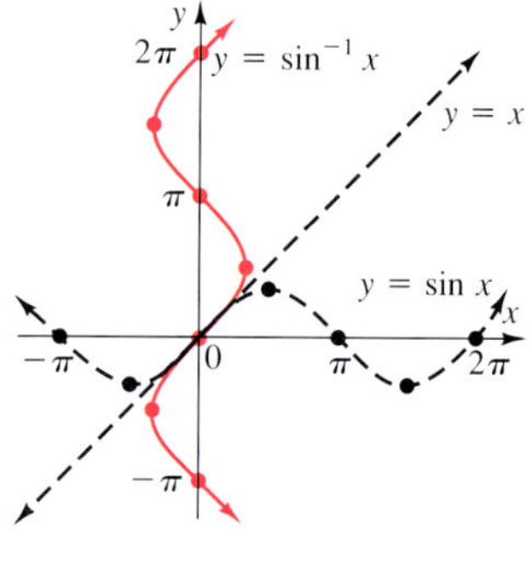

The vertical line test shows that $y = \sin^{-1} x$ is not a function. For example, there are an infinite number of values of $y = \sin^{-1} x$ for $x = 0$:

$$y = \sin^{-1} 0 = \cdots, -2\pi, -\pi, 0, \pi, 2\pi, \cdots$$

or

$$y = \sin^{-1} 0 = n\pi, \quad \text{where } n \text{ is an integer}$$

The inverse of $y = \cos x$ can be graphed in a similar manner (see Exercise 33). The relation $y = \cos^{-1} x$ is not a function, since there are an infinite number of values of y for any given value of x, where $-1 \le x \le 1$.

An alternate way of writing $\sin^{-1} x$ and $\cos^{-1} x$ is as arcsin x and arccos x. This alternate form is used in Example 4.

EXAMPLE 4 **Evaluate arccos $\frac{\sqrt{2}}{2}$ in degrees.**

Let $y = \arccos \frac{\sqrt{2}}{2}$. Then $\cos y = \frac{\sqrt{2}}{2}$.

Therefore, $y = \cdots, -315°, -45°, 45°, 315°, \cdots$

or

$y = 45° + n(360°)$ or $y = -45° + n(360°)$,

where n is an integer.

CLASS EXERCISES

Find the inverse of each relation. Is the relation a function? Is the inverse a function?

1. $\{(2, 4), (3, 6), (4, 8), (5, 10)\}$ $\{(4, 2), (6, 3), (8, 4), (10, 5)\}$; yes; yes
2. $\{(1, 3), (1, 4), (1, 5), (1, 6)\}$ $\{(3, 1), (4, 1), (5, 1), (6, 1)\}$; no; yes
3. $\{(-2, 3), (-1, 3), (0, 3), (1, 3), (2, 3)\}$ $\{(3, -2), (3, -1), (3, 0), (3, 1), (3, 2)\}$; yes; no
4. $\{(5, 3), (5, 4), (4, 4), (3, 2)\}$ $\{(3, 5), (4, 5), (4, 4), (2, 3)\}$; no; no

Find the inverse of each function.

5. $y = 4x$ $y = \frac{1}{4}x$
6. $y = x + 3$ $y = x - 3$
7. $y = -x$ $y = -x$
8. $y = x - 1$ $y = x + 1$

PRACTICE EXERCISES

Graph each relation and its inverse. See side column on page 254.

A
1. $y = x^3$
2. $y = 2x$

Find the inverse of each function. 9. $y = \frac{1}{2}x - \frac{1}{2}$ 10. $y = \frac{1}{4}x + \frac{3}{4}$

3. $y = 10x$ $y = \frac{1}{10}x$
4. $y = 6x$ $y = \frac{1}{6}x$
5. $y = x - 8$ $y = x + 8$
6. $y = x + 13$ $y = x - 13$
7. $y = -7x$ $y = -\frac{1}{7}x$
8. $y = -9x$ $y = -\frac{1}{9}x$
9. $y = 2x + 1$ See above.
10. $y = 4x - 3$ See above.
11. $y = 5x - 4$ $y = \frac{1}{5}x + \frac{4}{5}$
12. $y = 2x + 5$ $y = \frac{1}{2}x - \frac{5}{2}$
13. $y = \frac{2}{3}x + 6$ $y = \frac{3}{2}x - 9$
14. $y = \frac{2}{3}x - 5$ $y = \frac{3}{2}x + \frac{15}{2}$

Evaluate each expression in degrees.

15. $\sin^{-1} 1$ $90° + n(360°)$
16. $\cos^{-1} 0$ $90° + n(180°)$
17. $\arccos(-1)$ $180° + n(360°)$
18. $\arcsin(-1)$ $270° + n(360°)$

Evaluate each expression in radians.

B
19. $\cos^{-1} 0$ $\frac{\pi}{2} + n\pi$
20. $\sin^{-1}(-1)$ $\frac{3\pi}{2} + 2\pi n$
21. $\sin^{-1} 1$ $\frac{\pi}{2} + 2\pi n$
22. $\cos^{-1}(-1)$ $\pi + 2\pi n$
23. $\arccos(-0.5)$ $\frac{2\pi}{3} + 2\pi n$; $\frac{4\pi}{3} + 2\pi n$
24. $\arcsin\left(-\frac{\sqrt{3}}{2}\right)$ $\frac{4\pi}{3} + 2\pi n$; $\frac{5\pi}{3} + 2\pi n$
25. $\arcsin(-0.5)$ $\frac{7\pi}{6} + 2\pi n$; $\frac{11\pi}{6} + 2\pi n$
26. $\arccos \frac{\sqrt{2}}{2}$ $\frac{\pi}{4} + 2\pi n$; $\frac{7\pi}{4} + 2\pi n$

LESSON FOLLOW-UP

Discussion

Show how to find $\sin\left(\arccos \frac{12}{13}\right)$.

$\cos x = \frac{12}{13}$ $\sin x = \sqrt{1 - \cos^2 x}$

$\sin x = \sqrt{1 - \left(\frac{12}{13}\right)^2}$

$\sin x = \sqrt{1 - \frac{144}{169}}$

$\sin x = \sqrt{\frac{25}{169}}$

$\sin x = \pm \frac{5}{13}$

Critical Thinking

Analysis Ask students to show that $g(x)$ and $g^{-1}(x)$ are never perpendicular. Students should reason that since $g(x) = mx + b$, $g^{-1}(x) = \frac{1}{m}x - \frac{b}{m}$ and the product of their slopes $m \cdot \frac{1}{m} = 1$. Thus, they conclude that $g(x)$ and $g^{-1}(x)$ are never perpendicular because the product of their slopes is not -1.

Assignment Guide

See p. 248B for assignments.

Extra

The horizontal line test is a shortened method of determining whether or not the inverse of a function is also a function. Students are asked to answer questions regarding the inverses of functions in this feature.

Lesson Quiz

1. Find the inverse of the relation $y = x^2$ and tell whether or not the inverse is a function. $y = \pm\sqrt{x}$; no
2. Find the inverse of $f(x) = 2x - 5$. $y = \frac{x}{2} + \frac{5}{2}$
3. Evaluate arcsin $\frac{\sqrt{2}}{2}$ in degrees. $y = 45° + n(360°)$ or $y = 135° + n(360°)$
4. Evaluate arccos $-\frac{\sqrt{2}}{2}$ in radians. $y = \frac{3\pi}{4} + 2\pi n$ or $y = \frac{5\pi}{4} + 2\pi n$

Enrichment

A linear function is represented by $g(x) = mx + b$. For what values of m and b will $g(x) = g^{-1}(x)$ for all values of x. $m = \pm 1$ and $b = 0$

Teacher's Resource Book

Practice—Chapter 6, p. 1
Enrichment—Chapter 6, p. 2

Additional Answers

Practice Exercises

1.

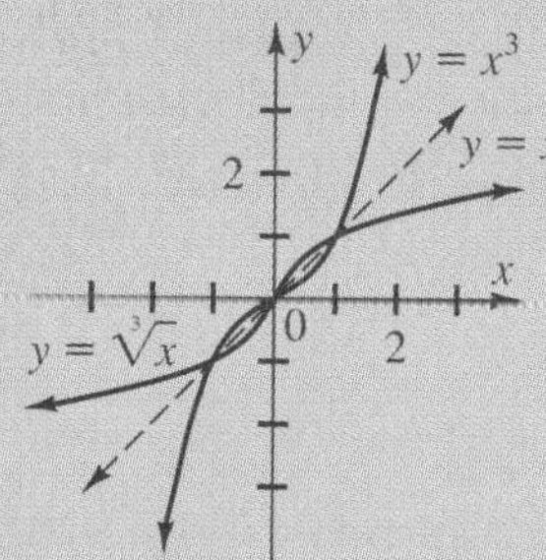

2.

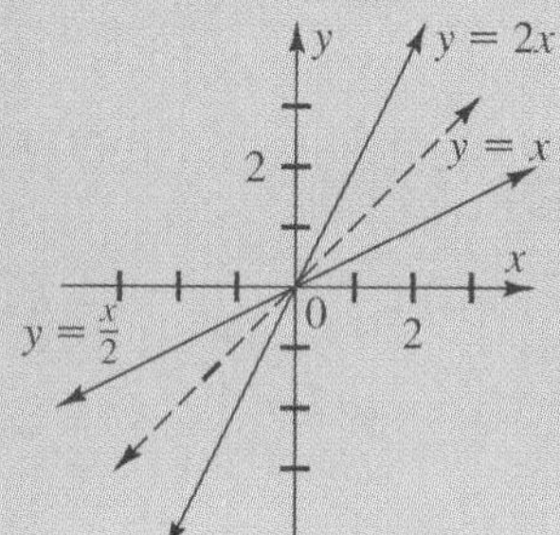

33.

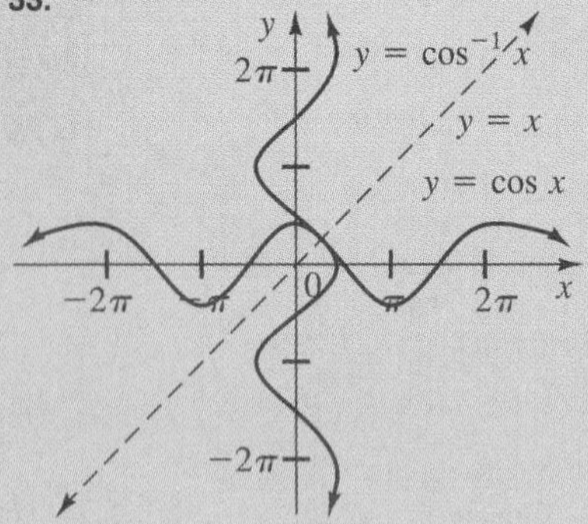

Extra

2.

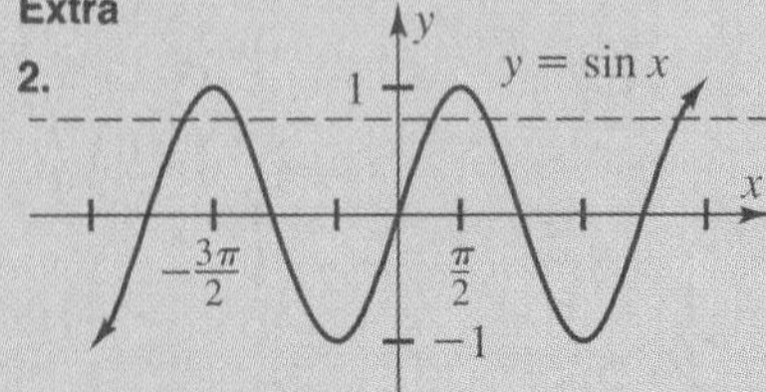

not a function

Find the inverse of each function. Graph each function and its inverse.
See page 486.

27. $y = 4x$ $y = \frac{1}{4}x$ **28.** $y = \frac{1}{2}x$ $y = 2x$ **29.** $y = x - 1$ $y = x + 1$

30. $y = x + 3$ $y = x - 3$ **31.** $y = 2x - 4$ $y = \frac{1}{2}x + 2$ **32.** $y = 3x + 5$ $y = \frac{1}{3}x - \frac{5}{3}$

C **33.** Graph $y = \cos x$ and its inverse on the same coordinate plane. See side column.

34. Given $f(x) = 3x - 24$, find $f^{-1}(x)$. Then find both $f(f^{-1}(x))$ and $f^{-1}(f(x))$. What is the result? *Hint*: To find $f(f^{-1}(x))$, use the expression you found for $f^{-1}(x)$ as a replacement for x in $3x - 24$. $f^{-1}(x) = \frac{1}{3}x + 8$; $f(f^{-1}(x)) = x$; $f^{-1}(f(x)) = x$

35. Given $f(x) = x^3 + 1$, find $f^{-1}(x)$, $f(f^{-1}(x))$ and $f^{-1}(f(x))$. What is the result?

36. Based on your results in Exercises 34 and 35, write a rule for testing to see if two functions are inverses of each other.
f and g are inverse functions if and only if $f(g(x)) = g(f(x)) = x$.

Applications

35. $f^{-1}(x) = \sqrt[3]{x - 1}$; $f(f^{-1}(x)) = x$; $f^{-1}(f(x)) = x$

37. Science The function that expresses Celsius temperature in terms of Fahrenheit temperature is given by $C(x) = \frac{5}{9}(x - 32)$. Find the inverse function that expresses Fahrenheit temperature in terms of Celsius. $F(x) = \frac{9}{5}x + 32$

38. Geometry The formula for expressing the circumference of a circle in terms of its radius is $C(r) = 2\pi r$. Find the inverse function. $R(C) = \frac{C}{2\pi}$

EXTRA

The *horizontal-line test* provides a quick way to tell whether or not the inverse of a function is also a function. Specifically, the inverse of a function is also a function if and only if no horizontal line can be found that intersects the graph of the given function in more than one point. Thus, the function defined by the equation $y = x^3 + 3x^2$ has an inverse that is not a function. However, if the domain of $y = x^3 + 3x^2$ is restricted to the interval $-2 \le x \le 0$, the restricted function's inverse is also a function.

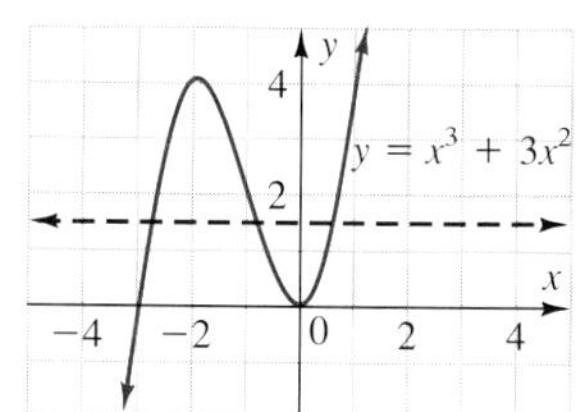

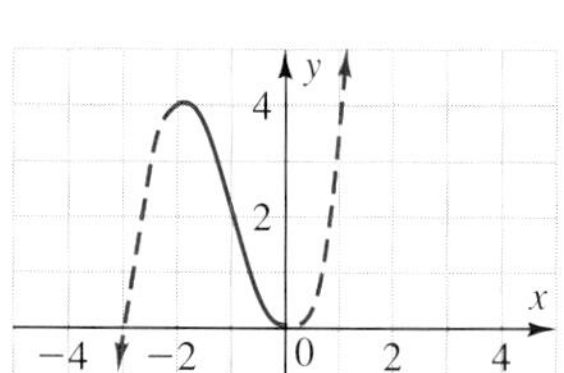

1. Describe at least two other ways to restrict the domain of $y = x^3 + 3x^2$ so that the inverse of the restricted function is also a function.
Answers may vary.
2. Sketch the graph of the function $y = \sin x$ for $-2\pi \le x \le 2\pi$. Use the horizontal-line test to determine whether or not the inverse of $y = \sin x$ is a function over that interval. See side column.

6.2 The Inverse Sine and Cosine Functions

Objectives: To define the inverse Sine and Cosine functions
To evaluate expressions involving inverse Sines and Cosines

The inverses of the sine and cosine functions are not functions, but the domains of the sine and cosine can be restricted so that their inverses *are* functions.

Preview

The exact values of the sine and cosine functions can be given for some angle measures. For example, $\cos 45° = \frac{\sqrt{2}}{2}$ and $\sin\left(-\frac{\pi}{3}\right) = -\frac{\sqrt{3}}{2}$.

Give the exact value of each sine or cosine function.

1. $\cos 0°$ 1 **2.** $\sin 30°$ $\frac{1}{2}$ **3.** $\cos 60°$ $\frac{1}{2}$ **4.** $\sin 135°$ $\frac{\sqrt{2}}{2}$ **5.** $\cos(-45°)$ $\frac{\sqrt{2}}{2}$ **6.** $\sin 2\pi$ 0

If the domain of the sine function is restricted to values from $-\frac{\pi}{2}$ through $\frac{\pi}{2}$, the inverse of this restricted function *is* a function. The domain can be restricted in other ways, but the interval $-\frac{\pi}{2} \leq x \leq \frac{\pi}{2}$ is chosen because it gives a range that is the same as that of the unrestricted function $y = \sin x$. That is, the range of the restricted sine function is $-1 \leq y \leq 1$. This function is often written $y = \text{Sin}\, x$, where the capital *S* distinguishes it from $y = \sin x$.

The inverse of $y = \text{Sin}\, x$ is the **inverse Sine** function, $y = \text{Sin}^{-1} x$. The capital *S* distinguishes this function from the relation $y = \sin^{-1} x$. The inverse Sine function is also commonly denoted by $y = \text{Arcsin}\, x$.

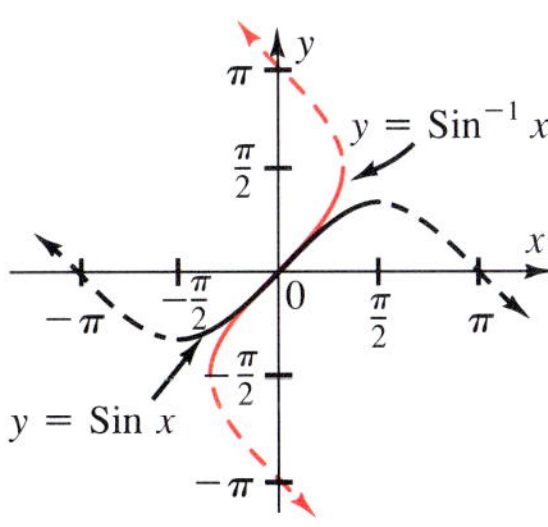

$$y = \text{Sin}^{-1} x \quad \text{if and only if} \quad \sin y = x \quad \text{and} \quad -\frac{\pi}{2} \leq y \leq \frac{\pi}{2}.$$

Thus, the domain of $y = \text{Sin}^{-1} x$ is $-1 \leq x \leq 1$ and the range is $-\frac{\pi}{2} \leq y \leq \frac{\pi}{2}$.

LESSON PLAN

Vocabulary
Inverse Cosine
Inverse Sine
Principal values

Materials/Manipulatives
Scientific calculator

BACKGROUND

In the Preview, exact values of sine and cosine for special angles are reviewed.

TEACHING SUGGESTIONS

- Emphasize the difference between $y = \text{Arcsin } x$ and $y = \text{arcsin } x$, and the difference between $y = \text{Sin } x$ and $y = \sin x$.
- Stress the definition of principal values.
- A scientific calculator is useful for approximating the values in Practice Exercises 9–16 and 29–36.

CHALKBOARD EXAMPLES

- **For Example 1**

  Find the exact value of each expression in radians.

  1. Arcsin -1 $-\frac{\pi}{2}$

  2. $\text{Cos}^{-1} \frac{\sqrt{3}}{2}$ $\frac{\pi}{6}$

- **For Example 2**

  Find the approximate value of each expression to the nearest degree.

  3. $\text{Cos}^{-1}\ 0.3090$ 72°

  4. Arcsin 0.8007 53°

The values in the range of $\text{Sin}^{-1} x$ are called the **principal values** of $\sin^{-1} x$. Similarly, if the domain of the cosine function is restricted to values of x from 0 through π, then the inverse of the restricted function is also a function. The restricted cosine function is denoted $y = \text{Cos } x$, and its inverse, the **inverse Cosine** function is denoted $y = \text{Cos}^{-1} x$ or $y = \text{Arccos } x$.

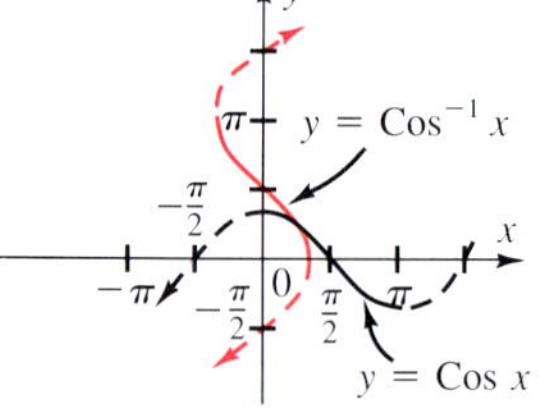

$$y = \text{Cos}^{-1} x \quad \text{if and only if} \quad \cos y = x \quad \text{and} \quad 0 \le y \le \pi.$$

Thus, the domain of $y = \text{Cos}^{-1} x$ is $-1 \le x \le 1$ and the range is $0 \le y \le \pi$. The values in the range of $\text{Cos}^{-1} x$ are called the *principal values* of $\cos^{-1} x$.

EXAMPLE 1 **Evaluate in radians: a.** Arccos $\frac{\sqrt{3}}{2}$ **b.** $\text{Sin}^{-1}\left(-\frac{\sqrt{2}}{2}\right)$

a. Let $y = \text{Arccos } \frac{\sqrt{3}}{2}$. Then $\cos y = \frac{\sqrt{3}}{2}$.

Since $\cos \frac{\pi}{6} = \frac{\sqrt{3}}{2}$ and $0 \le \frac{\pi}{6} \le \pi$, $y = \text{Arccos } \frac{\sqrt{3}}{2} = \frac{\pi}{6}$.

b. Let $y = \text{Sin}^{-1}\left(-\frac{\sqrt{2}}{2}\right)$. Then $\sin y = -\frac{\sqrt{2}}{2}$.

Since $\sin\left(-\frac{\pi}{4}\right) = -\frac{\sqrt{2}}{2}$ and $-\frac{\pi}{2} \le -\frac{\pi}{4} \le \frac{\pi}{2}$,

$y = \text{Sin}^{-1}\left(-\frac{\sqrt{2}}{2}\right) = -\frac{\pi}{4}$.

A scientific calculator can be used to find the values of inverse trigonometric functions. On some calculators, *Sin*$^{-1}$ and *Cos*$^{-1}$ keys are provided. On others, the *inv* key and the *sin* and *cos* keys are used. It is important to note that a calculator displays only the principal values, which are given in decimal form. To illustrate, in Example 1b a calculator would display the answer as -45 in degree mode or as -0.785398163 in radian mode.

EXAMPLE 2 **Find the approximate value of Sin^{-1} 0.6823 in degrees.**

Be sure the calculator is in degree mode.

$\text{Sin}^{-1}\ 0.6823 = 43°$ *Enter 0.6823 and use the* Sin^{-1} *or the* inv *and* sin *keys.*

EXAMPLE 3 **Evaluate: $\sin\left(\text{Cos}^{-1}\frac{\sqrt{3}}{2}\right)$**

Let $y = \text{Cos}^{-1}\frac{\sqrt{3}}{2}$.

Then $\cos y = \frac{\sqrt{3}}{2}$ and $y = \text{Cos}^{-1}\frac{\sqrt{3}}{2} = \frac{\pi}{6}$. $0 \le \frac{\pi}{6} \le \pi$

So, $\sin\left(\text{Cos}^{-1}\frac{\sqrt{3}}{2}\right) = \sin\frac{\pi}{6} = \frac{1}{2}$.

From the definitions of the inverse Sine and inverse Cosine functions, it is clear that sin (Arcsin x) = x and cos (Arccos x) = x. However, it is not always true that Arcsin (sin x) = x or that Arccos (cos x) = x. The next example illustrates this.

EXAMPLE 4 **Evaluate: $\text{Arcsin}\left(\sin\frac{7\pi}{4}\right)$**

$\text{Arcsin}\left(\sin\frac{7\pi}{4}\right) = \text{Arcsin}\left(-\frac{\sqrt{2}}{2}\right)$ $\sin\frac{7\pi}{4} = -\frac{\sqrt{2}}{2}$

Let $y = \text{Arcsin}\left(-\frac{\sqrt{2}}{2}\right)$. Then $\sin y = -\frac{\sqrt{2}}{2}$. The principal value is $-\frac{\pi}{4}$.

CLASS EXERCISES

For Discussion

1. Explain the difference between $y = \text{Cos}^{-1} x$ and $y = \cos^{-1} x$. See side column.
2. For what values of x is $\text{Sin}^{-1}(\sin x)$ equal to x? $-\frac{\pi}{2} \le x \le \frac{\pi}{2}$

Find each value.

3. $\text{Arcsin}\frac{1}{2}$ $\frac{\pi}{6}$
4. $\text{Arccos}\frac{1}{2}$ $\frac{\pi}{3}$
5. $\text{Arccos}\frac{\sqrt{2}}{2}$ $\frac{\pi}{4}$
6. $\text{Arcsin}\frac{\sqrt{3}}{2}$ $\frac{\pi}{3}$
7. $\text{Sin}^{-1} 0$ 0
8. $\text{Cos}^{-1} 0$ $\frac{\pi}{2}$
9. $\text{Sin}^{-1} 1$ $\frac{\pi}{2}$
10. $\text{Cos}^{-1} 1$ 0

PRACTICE EXERCISES

Evaluate each expression in radians.

A

1. $\text{Arcsin}\left(-\frac{\sqrt{3}}{2}\right)$ $-\frac{\pi}{3}$
2. $\text{Arccos}\left(-\frac{\sqrt{3}}{2}\right)$ $\frac{5\pi}{6}$
3. $\text{Arccos}\frac{1}{2}$ $\frac{\pi}{3}$
4. $\text{Arcsin}\left(-\frac{1}{2}\right)$ $-\frac{\pi}{6}$
5. $\text{Sin}^{-1}\left(-\frac{\sqrt{3}}{2}\right)$ $-\frac{\pi}{3}$
6. $\text{Cos}^{-1}\frac{\sqrt{2}}{2}$ $\frac{\pi}{4}$
7. $\text{Sin}^{-1}(-1)$ $-\frac{\pi}{2}$
8. $\text{Cos}^{-1}(-1)$ π

- **For Example 3**

Find the exact value of each expression

5. $\cos\left(\text{Sin}^{-1}\frac{\sqrt{2}}{2}\right)$ $\frac{\sqrt{2}}{2}$
6. $\sin\left(\text{Cos}^{-1}\left(\frac{\sqrt{2}}{2}\right)\right)$ $\frac{\sqrt{2}}{2}$

- **For Example 4**

Evaluate each expression in radians

7. $\text{Arcsin}\left(\sin\left(\frac{-2\pi}{3}\right)\right)$ $-\frac{\pi}{3}$
8. $\text{Arccos}\left(\cos\frac{7\pi}{6}\right)$ $\frac{5\pi}{6}$

Common Errors

- Students may confuse Arcsine with arcsine or Arccosine with arccosine. Emphasize that the capital A means that the domain has been restricted.
- Students may not read directions carefully. Stress that calculators should be set in the mode requested by the directions.
- See *Teacher's Resource Book* for additional remediation.

LESSON FOLLOW-UP

Critical Thinking

Analysis Ask students to tell how the graph of $y = \sin^{-1} x$ can be sketched using only the graphs of $y = \sin x$ and $y = x$. Students should reason that since $y = \sin^{-1} x$ is the inverse of $y = \sin x$ and since $f(x)$ and $f^{-1}(x)$ are symmetric with respect to the line $y = x$, the graph of $y = \sin^{-1} x$ can be sketched by reflecting the graph of $y = \sin x$ over the line $y = x$.

Assignment Guide

See p. 248B for assignments.

Additional Answers

Class Exercises

1. $y = \cos^{-1} x$ has an unlimited range; $y = \text{Cos}^{-1} x$ has a limited range, $0 \le y \le \pi$.

Biography: Irmgard Flugge-Lotz

A summary regarding the life and achievements of Irmgard Flugge-Lotz, a leader in the field of aerodynamics is featured.

Lesson Quiz

1. Find the approximate value of Arcsin (0.7427) in degrees. 48°

Evaluate each expression in radians.

2. Arcsin $\left(\frac{1}{2}\right)$ $\frac{\pi}{6}$
3. $\text{Cos}^{-1}\frac{\sqrt{2}}{2}$ $\frac{\pi}{4}$
4. $\text{Cos}\left(\sin^{-1}\frac{\sqrt{3}}{2}\right)$ $\frac{1}{2}$
5. $\text{Arccos}\left(\sin\frac{5\pi}{6}\right)$ $\frac{\pi}{3}$

Enrichment

Express $y = \sin(\text{Cos}^{-1} x)$ without using trigonometric or inverse trigonometric functions. $y = \sqrt{1 - x^2}$

Find the approximate value of each expression to the nearest degree.

9. Arcsin 0.8123 54°
10. Arccos 0.4927 60°
11. Arccos 0.9427 19°
12. Arcsin 0.72 46°
13. $\text{Sin}^{-1}\, 0.6142$ 38°
14. $\text{Cos}^{-1}\, 0.3129$ 72°
15. $\text{Cos}^{-1}(-0.7123)$ 135°
16. $\text{Sin}^{-1}(-0.85)$ −58°

Find the exact value of each expression.

17. $\sin(\text{Sin}^{-1}\, 0.4)$ 0.4
18. $\cos(\text{Cos}^{-1}\, 0.8667)$ 0.8667
19. $\sin\left(\text{Cos}^{-1}\frac{\sqrt{2}}{2}\right)$ $\frac{\sqrt{2}}{2}$
20. $\sin\left(\text{Cos}^{-1}\left(-\frac{1}{2}\right)\right)$ $\frac{\sqrt{3}}{2}$
21. $\cos\left(\text{Arcsin}\frac{\sqrt{3}}{2}\right)$ $\frac{1}{2}$
22. $\cos\left(\text{Arcsin}\left(-\frac{\sqrt{3}}{2}\right)\right)$ $\frac{1}{2}$
23. $\text{Arcsin}\left(\sin\frac{2\pi}{3}\right)$ $\frac{\pi}{3}$
24. $\text{Arccos}\left(\cos\left(-\frac{4\pi}{3}\right)\right)$ $\frac{2\pi}{3}$

B

25. $\text{Arcsin}\left(\sin\frac{\pi}{6}\right)$ $\frac{\pi}{6}$
26. $\text{Arccos}\left(\cos\left(-\frac{\pi}{3}\right)\right)$ $\frac{\pi}{3}$
27. $\text{Arccos}\left(\sin\frac{3\pi}{4}\right)$ $\frac{\pi}{4}$
28. $\text{Arcsin}\left(\cos\frac{\pi}{4}\right)$ $\frac{\pi}{4}$

Approximate each number to four decimal places.

29. sin (Arccos 0.4127) 0.9109
30. cos (Arcsin 0.7459) 0.6661
31. cos (Arcsin 0.6148) 0.7887
32. sin (Arccos 0.1292) 0.9916
33. $\sin[\text{Cos}^{-1}(-0.8234)]$ 0.5675
34. $\cos[\text{Sin}^{-1}(-0.6927)]$ 0.7212
35. $\cos[\text{Sin}^{-1}(-0.5123)]$ 0.8588
36. $\sin[\text{Cos}^{-1}(-0.8139)]$ 0.5810

Graph each equation. See side column on page 259.

37. $y = \text{Sin}^{-1} x + 2$
38. $y = \text{Sin}^{-1} x - 1$
39. $y = 2\,\text{Arccos}\, x$
40. $y = 3\,\text{Arccos}\, x$

Find the exact value of each expression.

C

41. $\sin\left(\text{Arccos}\frac{\sqrt{3}}{2} + \pi\right)$ $-\frac{1}{2}$
42. $\sin\left(\text{Arcsin}\frac{1}{2} - \pi\right)$ $-\frac{1}{2}$
43. $\cos\left(\text{Arccos}\frac{\sqrt{2}}{2} + \frac{\pi}{2}\right)$ $-\frac{\sqrt{2}}{2}$
44. $\cos\left(\text{Arcsin}\left(-\frac{\sqrt{3}}{2}\right) - \frac{\pi}{2}\right)$ $-\frac{\sqrt{3}}{2}$
45. If $-1 \le x \le 0$, is it true that Arcsin $x <$ Arccos x? Explain. yes
46. For what values of x is it true that Arcsin x = Arccos x? $\frac{\sqrt{2}}{2}$
47. For what values of x is it true that Arcsin $x \le$ Arccos x? $-1 \le x \le \frac{\sqrt{2}}{2}$

Applications

48. Mechanics The distance x that a slider moves when it is attached to a rod mechanism is given by $x = s - s \cos(\text{Arcsin } 0.4)$, where s is the length of the rod. Find x when s is 10 cm. 0.835

49. Engineering The ends of two vertical supports of a bridge under construction are connected by a cross-member that is 10 ft long. One end of the cross-member is attached at point A, 8 ft from the ground, and the other end is attached to point B, 2 ft from the ground. Express the measure of α, the acute angle that the cross-member makes with the right vertical support in terms of the inverse Cosine function. $\alpha = \text{Cos}^{-1}\left(\frac{3}{5}\right)$

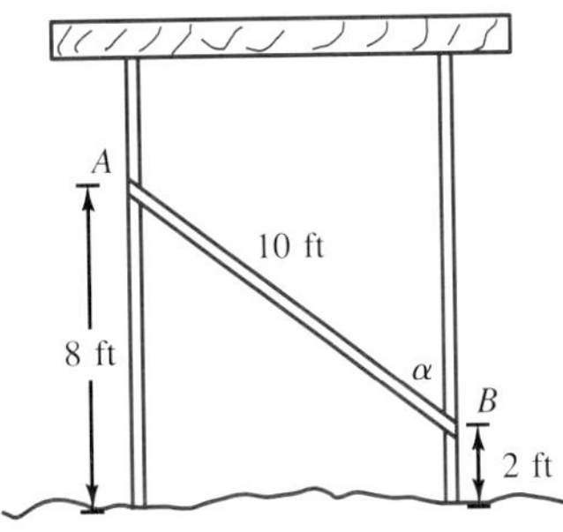

BIOGRAPHY: IRMGARD FLUGGE-LOTZ (1903–1974)

Irmgard Lotz was born in Hameln, Germany. Since her mother's family was in the construction business and her father was a mathematician, it was natural for Irmgard to develop an interest in technical subjects. In 1923, she began the study of fluid dynamics and applied mathematics at the Technische Hochschule of Hanover. After receiving her undergraduate degree in 1927, she continued her studies at Hanover and received a doctorate in engineering in 1929. Later, she was employed at the Aerodynamische Versuchsanstalt (AVA) at Göttingen as a junior research engineer, where she worked with Ludwig Prandtl, one of the founders of the science of aerodynamics.

In 1948, as Irmgard Flugge-Lotz, she and her husband, Wilhelm Flugge-Lotz, joined the faculty at Stanford University in California. She was later appointed Stanford's first female full professor of engineering mechanics, aeronautics, and astronautics. She was also appointed a fellow of the American Institute of Aeronautics and Astronautics (AIAA), becoming just the second woman so honored. The Society of Woman Engineers honored her with its Achievement Award that same year.

Teacher's Resource Book
Practice—Chapter 6, p. 3
Enrichment—Chapter 6, p. 4

Additional Answers

37.

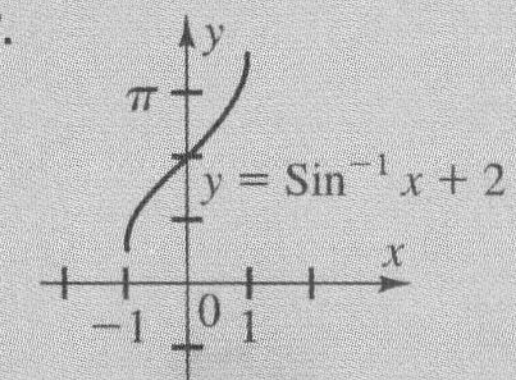

38.

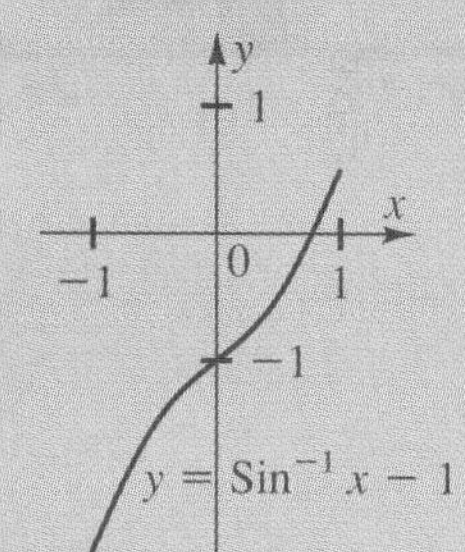

39.

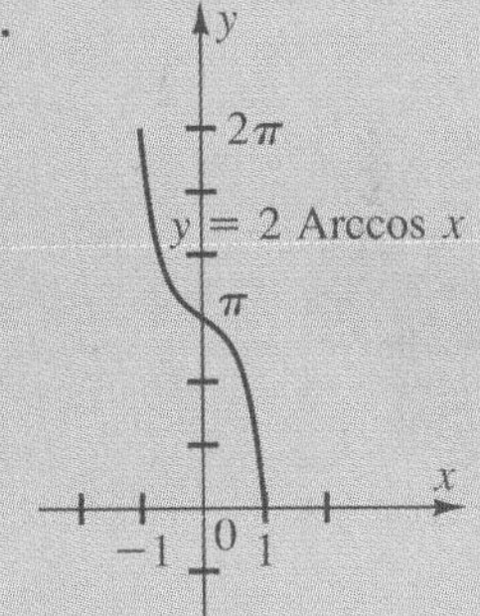

40.

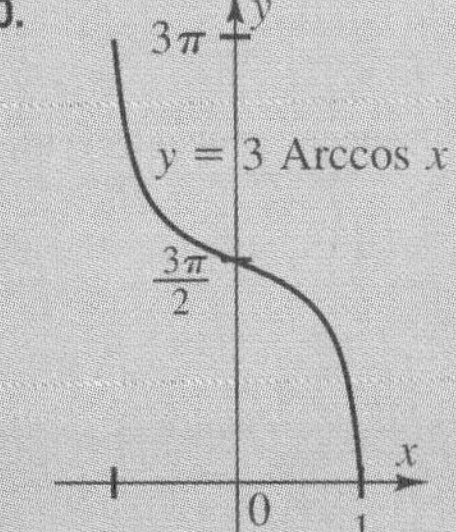

LESSON PLAN

Vocabulary
Inverse Cosecant
Inverse Cotangent
Inverse Secant
Inverse Tangent

Materials/Manipulatives
Scientific calculators

BACKGROUND

In the Preview, exact values of tangent, cotangent, secant, and cosecant for special angles are reviewed.

6.3 Other Inverse Trigonometric Functions

Objectives: To define the inverse Tangent, Cotangent, Secant, and Cosecant functions
To evaluate expressions involving these functions

You have seen that the domains of the sine and cosine functions can be restricted so that their inverses are also functions. It is also possible to restrict the domains of the other four trigonometric functions in such a way that their inverses are functions.

Preview

The exact values of the secant, cosecant, tangent, and cotangent functions can be given for some angle measures. For example, $\tan(-60°) = -\sqrt{3}$.

Give the exact value of each function.

1. $\cot 30°$ $\sqrt{3}$ **2.** $\tan(-45°)$ -1 **3.** $\sec 120°$ -2 **4.** $\csc(-30°)$ -2 **5.** $\cot 60°$ $\frac{\sqrt{3}}{3}$

6. $\tan \frac{\pi}{6}$ $\frac{\sqrt{3}}{3}$ **7.** $\tan \frac{\pi}{4}$ 1 **8.** $\cot \frac{\pi}{4}$ 1 **9.** $\sec \frac{3\pi}{4}$ $-\sqrt{2}$ **10.** $\csc\left(-\frac{\pi}{3}\right)$ $-\frac{2\sqrt{3}}{3}$

The inverses of the tangent, cotangent, secant, and cosecant functions are *not* functions.

Function	$y = \tan x$	$y = \cot x$	$y = \sec x$	$y = \csc x$
Inverse Relation	$y = \tan^{-1} x$	$y = \cot^{-1} x$	$y = \sec^{-1} x$	$y = \csc^{-1} x$

When the domains of $y = \tan x$ and $y = \cot x$ are appropriately restricted, their inverses, the **inverse Tangent** and **inverse Cotangent** are functions. The values of $y = \text{Tan}^{-1} x$ and $y = \text{Cot}^{-1} x$ are called the *principal values*. For $y = \text{Tan}^{-1} x$, the principal values are between $-\frac{\pi}{2}$ and $\frac{\pi}{2}$. For $y = \text{Cot}^{-1} x$, the principal values are between 0 and π. These domains are chosen so that $y = \text{Tan}\, x$ and $y = \text{Cot}\, x$ are continuous and have the same ranges as the unrestricted tangent and cotangent functions, respectively.

$$y = \text{Tan}^{-1} x \quad \text{if and only if} \quad \tan y = x \quad \text{and} \quad -\frac{\pi}{2} < y < \frac{\pi}{2}$$
$$y = \text{Cot}^{-1} x \quad \text{if and only if} \quad \cot y = x \quad \text{and} \quad 0 < y < \pi$$

The graphs of these functions are shown at the top of the next page.

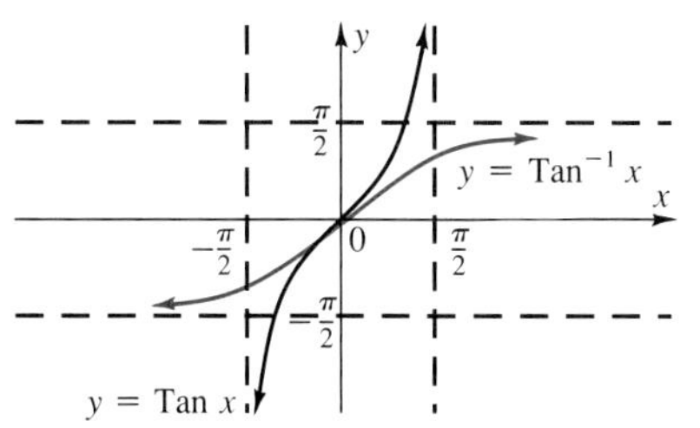

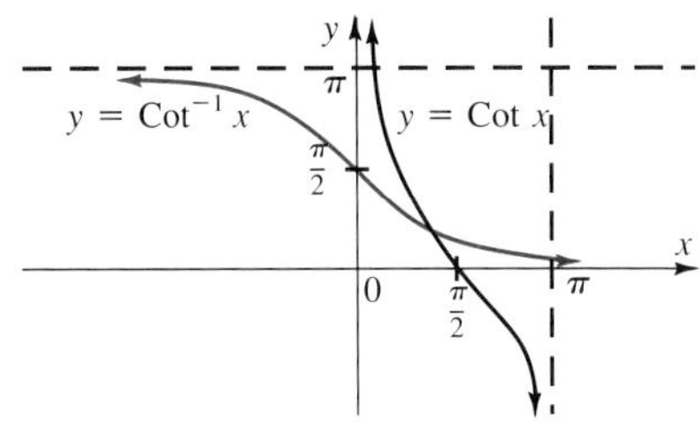

The **inverse Secant** and **inverse Cosecant** functions are the inverses of the restricted secant and cosecant functions, $y = \text{Sec } x$ and $y = \text{Csc } x$. The domains of these functions are restricted in the same way as the Sine and Cosine, respectively, except where the secant and cosecant are not defined.

$y = \text{Sec}^{-1} x$ if and only if $\sec y = x$ and $0 \leq y \leq \pi, y \neq \frac{\pi}{2}$

$y = \text{Csc}^{-1} x$ if and only if $\csc y = x$ and $-\frac{\pi}{2} \leq y \leq \frac{\pi}{2}, y \neq 0$

The graphs of these functions are shown below.

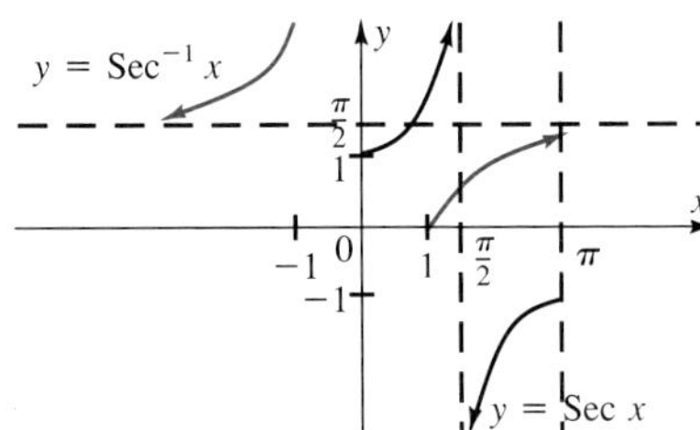

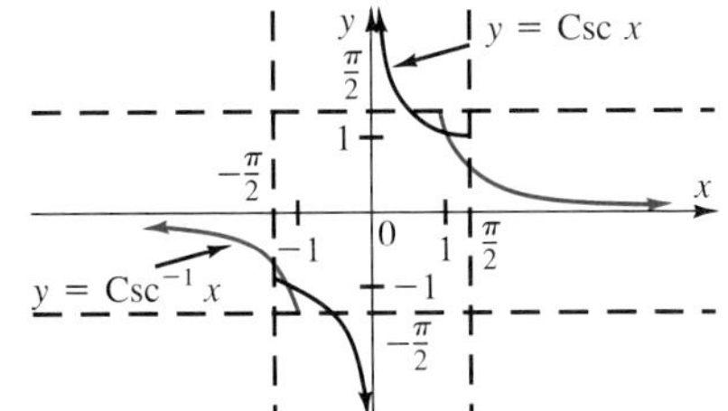

Here are the domains and ranges of the trigonometric functions and their inverses.

Function Domain Range	$y = \text{Sin } x$ $-\frac{\pi}{2} \leq x \leq \frac{\pi}{2}$ $-1 \leq y \leq 1$	$y = \text{Sin}^{-1} x$ $-1 \leq x \leq 1$ $-\frac{\pi}{2} \leq y \leq \frac{\pi}{2}$	$y = \text{Cos } x$ $0 \leq x \leq \pi$ $-1 \leq y \leq 1$	$y = \text{Cos}^{-1} x$ $-1 \leq x \leq 1$ $0 \leq y \leq \pi$
Function Domain Range	$y = \text{Tan } x$ $-\frac{\pi}{2} < x < \frac{\pi}{2}$ all real numbers	$y = \text{Tan}^{-1} x$ all real numbers $-\frac{\pi}{2} < y < \frac{\pi}{2}$	$y = \text{Cot } x$ $0 < x < \pi$ all real numbers	$y = \text{Cot}^{-1} x$ all real numbers $0 < y < \pi$
Function Domain Range	$y = \text{Sec } x$ $0 \leq x \leq \pi, x \neq \frac{\pi}{2}$ $\lvert y \rvert \geq 1$	$y = \text{Sec}^{-1} x$ $\lvert x \rvert \geq 1$ $0 \leq y \leq \pi, y \neq \frac{\pi}{2}$	$y = \text{Csc } x$ $-\frac{\pi}{2} \leq x \leq \frac{\pi}{2}, x \neq 0$ $\lvert y \rvert \geq 1$	$y = \text{Csc}^{-1} x$ $\lvert x \rvert \geq 1$ $-\frac{\pi}{2} \leq y \leq \frac{\pi}{2}, y \neq 0$

TEACHING SUGGESTIONS

- Emphasize the difference between Arctangent and arctangent, Arcsecant and arcsecant, and so forth.
- Do several examples using a calculator to find Sec^{-1} and stress the difference between the domains of $y = \cot x$ and $y = \tan x$.

Critical Thinking

Analysis Ask students to determine how the graphs of $y = \text{Tan}^{-1} x$, $y = \text{Cot}^{-1} x$, and so forth can be sketched without a table of values. Noting that $y = \text{Tan } x$ and $y = \text{Tan}^{-1} x$ are inverse functions, students should conclude that the graph of $y = \text{Tan}^{-1} x$ can be sketched by graphing $y = \tan x$ and $y = x$ on the same coordinate plane, and then reflecting $y = \tan x$ over the line $y = x$. The graphs of $y = \text{Cot}^{-1} x$, $y = \text{Sec}^{-1} x$, and $y = \text{Csc}^{-1} x$ can be sketched in a similar manner.

CHALKBOARD EXAMPLES

- **For Example 1**

  Find the exact value of each expression in radians.

  1. Arcsec $(-\sqrt{2})$ $\frac{3\pi}{4}$
  2. $\text{Csc}^{-1} 2$ $\frac{\pi}{6}$

- **For Example 2**

  Find the approximate value of each expression to the nearest degree.

  3. Arctan 3.121 72°

  4. Arctan −1.452 −55°

- **For Example 3**

  Find the approximate value of each expression to the nearest degree.

  5. Arcsec (3.3817) 73°

  6. Csc^{-1} (1.6242) 38°

- **For Example 4**

  Prove each identity.

  7. $\tan(\text{Arccot}\ \omega) = \frac{1}{\omega}$

$\tan(\text{Arccot}\ \omega)$	$\frac{1}{\omega}$
$\tan\left(\text{Arctan}\ \frac{1}{\omega}\right)$	
$\frac{1}{\omega}$	=

  8. $\text{Sin}^{-1} 1 - \text{Sin}^{-1}\left(\frac{\sqrt{2}}{2}\right) = \frac{\pi}{4}$

$\text{Sin}^{-1} 1 - \text{Sin}^{-1}\frac{\sqrt{2}}{2}$	$\frac{\pi}{4}$
$\frac{\pi}{2} - \frac{\pi}{4}$	
$\frac{\pi}{4}$	=

Common Errors

- Some students may have their calculators set in the wrong mode. Stress that the calculator mode should be the same as that stated in the directions of the problem.
- When evaluating Cot^{-1}, some students may not add 180° to get the principal value. Stress that this is necessary due to the domain differences between Cot^{-1} x and Tan^{-1} x.
- See *Teacher's Resource Book* for additional remediation.

The inverse Tangent, inverse Cotangent, inverse Secant, and inverse Cosecant may also be written as Arctangent, Arccotangent, Arcsecant, and Arccosecant. This notation is shown in Example 1.

EXAMPLE 1 **Evaluate:** Arctan (−1)

Let $y = \text{Arctan}(-1)$. Then $\tan y = -1$.

Since $\tan\left(-\frac{\pi}{4}\right) = -1$ and $-\frac{\pi}{2} < -\frac{\pi}{4} < \frac{\pi}{2}$,

$y = \text{Arctan}(-1) = -\frac{\pi}{4}$.

When there is no *Tan*$^{-1}$ key on a calculator, the *inv* key and the *tan* key are used to find approximate values of the inverse Tangent.

EXAMPLE 2 **Find the approximate value of Tan^{-1} 2.246 in degrees.**

$\text{Tan}^{-1} 2.246 = 66°$ *Use the degree mode. Enter 2.246 and use the* Tan^{-1} *key or the* inv *and* tan *keys.*

It is usually necessary to use the $\frac{1}{x}$ key to find approximate values of the inverse Cotangent, inverse Secant, and inverse Cosecant.

EXAMPLE 3 **Find the approximate value to the nearest degree:**
a. Sec^{-1} 1.843 **b.** Cot^{-1} (−1.862)

a. Since $\sec x = \frac{1}{\cos x}$ and the domain of $y = \text{Sec}\ x$ is the same as that of $y = \text{Cos}\ x$ $\left(\text{if } x \neq \frac{\pi}{2}\right)$, use the $\frac{1}{x}$ key.

$\text{Sec}^{-1} 1.843 = 57°$ *Enter 1.843 and use the* $\frac{1}{x}$ *and the* Cos^{-1} *keys, or the* $\frac{1}{x}$, inv, *and* cos *keys.*

b. Since $\cot x = \frac{1}{\tan x}$ (when both functions are defined), an angle with cotangent −1.862 has a tangent of $\frac{1}{-1.862}$. However, the domain of $y = \text{Cot}\ x$ is $0 < x < \pi$, while the domain of $y = \text{Tan}\ x$ is $-\frac{\pi}{2} < x < \frac{\pi}{2}$. When the $\frac{1}{x}$, *inv*, and *tan* keys are used to find $\text{Cot}^{-1}(-1.862)$, the calculator gives the principal value of the angle with tangent $\frac{1}{-1.862}$. To find the principal value of the angle with cotangent −1.862, you must add 180°.

$$\text{Cot}^{-1}(-1.862) = \text{Tan}^{-1}\left(\frac{1}{-1.862}\right) + 180°$$
$$= -28° + 180°$$
$$= 152°$$

The following identities follow from the definitions:

$$\tan(\text{Tan}^{-1} x) = x \qquad \cot(\text{Cot}^{-1} x) = x$$
$$\sec(\text{Sec}^{-1} x) = x \qquad \csc(\text{Csc}^{-1} x) = x$$

Identities involving inverse trigonometric functions can be proved using the definitions of the trigonometric functions and other known identities.

EXAMPLE 4 **Prove: $\csc(\text{Arcsin } v) = \frac{1}{v}$**

$\csc(\text{Arcsin } v)$	$\frac{1}{v}$
$\csc\left(\text{Arccsc } \frac{1}{v}\right)$	

$\sin x = \frac{1}{\csc x}$

$\frac{1}{v} =$ $\csc(\text{Arccsc } x) = x$

CLASS EXERCISES

Find the exact value of each expression in radians.

1. Arctan 1 $\frac{\pi}{4}$
2. $\text{Tan}^{-1}\sqrt{3}$ $\frac{\pi}{3}$
3. $\text{Cot}^{-1} 1$ $\frac{\pi}{4}$
4. Arccot $\sqrt{3}$ $\frac{\pi}{6}$
5. Arcsec 1 0
6. Arccsc 1 $\frac{\pi}{2}$
7. $\text{Csc}^{-1}\sqrt{2}$ $\frac{\pi}{4}$
8. $\text{Sec}^{-1} 2$ $\frac{\pi}{3}$

PRACTICE EXERCISES

Find the exact value of each expression in radians.

A
1. $\text{Sec}^{-1}\sqrt{2}$ $\frac{\pi}{4}$
2. $\text{Csc}^{-1}(-2)$ $-\frac{\pi}{6}$
3. Arctan $(-\sqrt{3})$ $-\frac{\pi}{3}$
4. Arccot (-1) $\frac{3\pi}{4}$

Find the approximate value of each expression to the nearest degree.

5. Arctan 2.253 66°
6. Arccot 3.341 17°
7. $\text{Sec}^{-1}(-1.242)$ 144°
8. $\text{Csc}^{-1} 1.294$ 51°

Find the exact value of each expression.

9. $\tan(\text{Tan}^{-1} 0.6)$ 0.6
10. $\cot(\text{Cot}^{-1} 0.3)$ 0.3
11. $\sec(\text{Sec}^{-1} 5)$ 5
12. $\csc(\text{Csc}^{-1} 6)$ 6
13. $\tan\left[\text{Arcsin}\left(-\frac{1}{2}\right)\right]$ $-\frac{\sqrt{3}}{3}$
14. $\sin(\text{Arctan }\sqrt{3})$ $\frac{\sqrt{3}}{2}$

B
15. $\cos[\text{Arctan}(-1)]$ $\frac{\sqrt{2}}{2}$
16. $\sin[\text{Arccot}(-1)]$ $\frac{\sqrt{2}}{2}$
17. $\sin(\text{Cot}^{-1}\sqrt{2})$ $\frac{\sqrt{3}}{3}$
18. $\cos[\text{Cot}^{-1}(-\sqrt{2})]$ $-\frac{\sqrt{6}}{3}$
19. $\sec(\text{Csc}^{-1} 4)$ $\frac{4\sqrt{15}}{15}$
20. $\sin\left(\text{Tan}^{-1}\frac{12}{5}\right)$ $\frac{12}{13}$

LESSON FOLLOW-UP

Discussion

Show how to evaluate $\tan\left[\text{Arcsin}\left(-\frac{4}{5}\right) + \text{Arctan}\frac{1}{3}\right]$

Let $a = \text{Arcsin}\left(-\frac{4}{5}\right)$ and $b = \text{Arctan}\frac{1}{3}$. Then $\tan\left[\text{Arcsin}\left(-\frac{4}{5}\right) + \text{Arctan}\frac{1}{3}\right] = \tan(a+b) = \frac{-\frac{4}{3}+\frac{1}{3}}{1-\left(-\frac{4}{3}\right)\left(\frac{1}{3}\right)} = -\frac{9}{13}$

Assignment Guide

See p. 248B for assignments.

Test Yourself

See *Teacher's Resource Book*, *Tests*, pp. 57–58.

Lesson Quiz

Find the approximate value of each expression to the nearest degree.

1. $\text{Csc}^{-1}(2.357)$ 25°
2. $\text{Tan}^{-1}(3.756)$ 75°

Find the exact value of each expression in radians.

3. Arcsec (2) $\frac{\pi}{3}$
4. Arcsec (-2) $\frac{2\pi}{3}$
5. Find the exact value of sin $(\text{Csc}^{-1}(-2))$. $-\frac{1}{2}$

Enrichment

Express $\cos(\text{Tan}^{-1} x)$ without using circular or inverse circular functions.

$\frac{1}{\sqrt{x^2+1}} = \frac{\sqrt{x^2+1}}{x^2+1}$

Teacher's Resource Book
Practice—Chapter 6, p. 5
Enrichment—Chapter 6, p. 6

Additional Answers

21. $\text{Sec}^{-1} 2 + \text{Tan}^{-1} \frac{\sqrt{3}}{3} = \frac{\pi}{3} + \frac{\pi}{6} = \frac{\pi}{2}$

22. $\text{Cot}^{-1} \frac{\sqrt{3}}{3} - \text{Csc}^{-1} 2 = \frac{\pi}{3} - \frac{\pi}{6} = \frac{\pi}{6}$

23. $\cos(\text{Arcsec } w) = \cos\left(\text{Arccos } \frac{1}{w}\right) = \frac{1}{w}$

24. $\cot(\text{Arctan } z) = \cot\left(\text{Arccot } \frac{1}{z}\right) = \frac{1}{z}$

27.

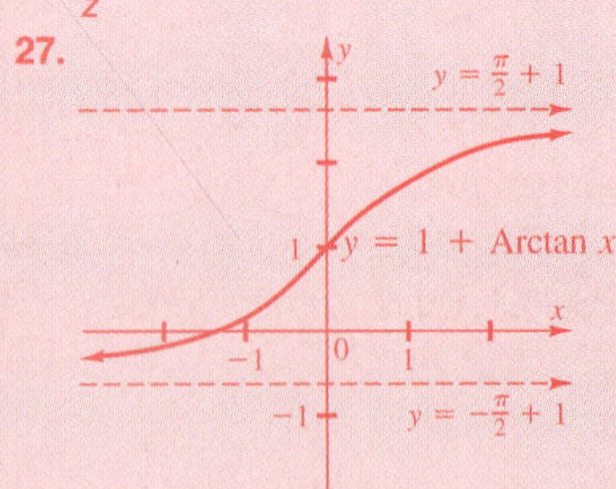

28.

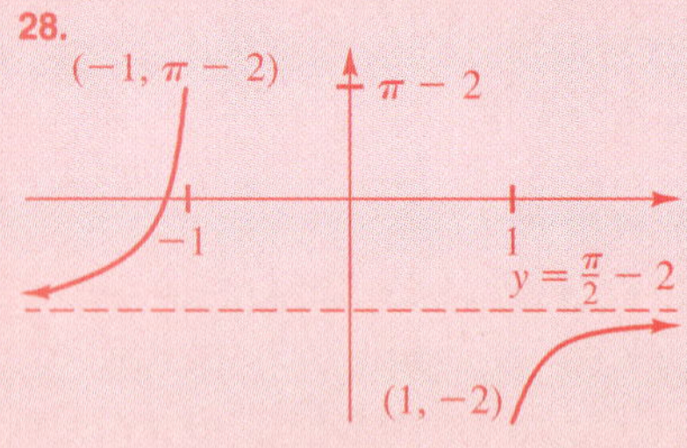

Prove. See side column.

21. $\text{Sec}^{-1} 2 + \text{Tan}^{-1} \frac{\sqrt{3}}{3} = \frac{\pi}{2}$

22. $\text{Cot}^{-1} \frac{\sqrt{3}}{3} - \text{Csc}^{-1} 2 = \frac{\pi}{6}$

23. $\cos(\text{Arcsec } w) = \frac{1}{w}$

24. $\cot(\text{Arctan } z) = \frac{1}{z}$

Find the exact value of each expression.

C 25. $\sin(\text{Tan}^{-1} 1 + \text{Csc}^{-1} 2)$ $\frac{\sqrt{6} + \sqrt{2}}{4}$

26. $\tan(\text{Tan}^{-1} 2 + \text{Cot}^{-1} 1)$ -3

Sketch each graph. See side column.

27. $y = 1 + \text{Arctan } x$

28. $y = \text{Arcsec } x - 2$

Applications

29. **Physics** The maximum angle by which a lever rotates around its pivot when it is attached to an elliptical cam is Arctan $\frac{1}{7}$. Find the measure of the angle to the nearest tenth of one degree. 8.1°

30. **Construction** A steel ball is suspended from a primary cable attached to a crane 21 ft above the ground. A second cable is used to draw the ball upward. At point A, 10 ft above the ground, the ball is released. (See the figure.) Express α in terms of the Arctangent function. Then find the value of α. $\alpha = \text{Arctan}\left(\frac{13}{11}\right)$; $\alpha = 50°$

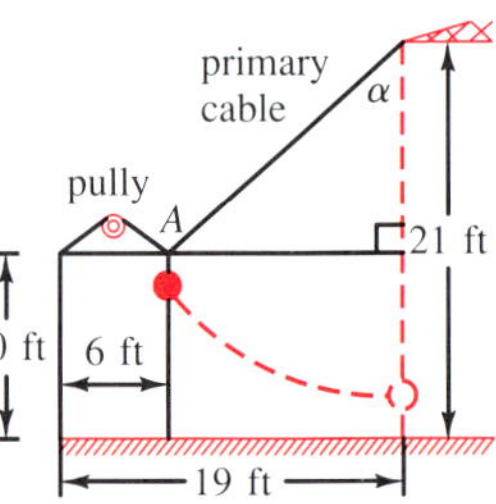

TEST YOURSELF

Evaluate each expression in radians.

1. $\arcsin \frac{1}{2}$ $\frac{\pi}{6} + 2\pi n$ or $\frac{5\pi}{6} + 2\pi n$

2. $\arccos\left(-\frac{\sqrt{2}}{2}\right)$ $\frac{3\pi}{4} + 2\pi n$ or $\frac{5\pi}{4} + 2\pi n$

3. $\arccos \frac{\sqrt{3}}{2}$ $\frac{\pi}{6} + 2\pi n$ or $-\frac{\pi}{6} + 2\pi n$

4. $\arcsin 0$ πn **6.1**

Find the approximate value of each expression to the nearest degree.

5. $\text{Sin}^{-1} 0.2463$ 14°

6. $\text{Csc}^{-1} (-1.6124)$ −38°

7. $\text{Arcsin } 0.8136$ 54° **6.2**

Approximate each number to four decimal places.

8. $\sin(\text{Cos}^{-1} 0.3262)$ 0.9453

9. $\cos(\text{Cos}^{-1} 0.7125)$ 0.7125

10. $\sin(\text{Sin}^{-1} 0.1344)$ 0.1344

Find the exact value of each expression.

11. $\sec\left(\text{Arccsc } \frac{2\sqrt{3}}{3}\right)$ 2

12. $\tan[\text{Cot}^{-1}(-1)]$ −1

13. $\csc\left(\text{Tan}^{-1} \frac{3}{4}\right)$ $\frac{5}{3}$ **6.3**

6.4

Solving Trigonometric Equations: Using Special Angles

Objectives: To define conditional equations
To find exact solutions to trigonometric equations

Trigonometric equations that are true for all values of a variable for which the equation is defined are called *trigonometric identities*. These identities can be useful in solving certain other trigonometric equations.

Preview

In order to review some of the important trigonometric identities, match each expression on the left with an equivalent expression on the right.

1. $\sin 2x$ c	**a.** $\sin x$
2. $1 + \tan^2 x$ d	**b.** $\cot^2 x$
3. $1 - \cos^2 x$ f	**c.** $2 \sin x \cos x$
4. $\tan x \cos x$ a	**d.** $\sec^2 x$
5. $\csc^2 x - 1$ b	**e.** $\sin x \cot x$
6. $\cos 2x$ g	**f.** $\sin^2 x$
7. $\cos x$ e	**g.** $\cos^2 x - \sin^2 x$

An equation that is true for at least one value of the variable, but not all values, is called a *conditional equation*. The equation $\cos x = 0$ is a conditional equation, since it is true for some values of x, such as $-\frac{3\pi}{2}, -\frac{\pi}{2}, \frac{\pi}{2}$, and $\frac{3\pi}{2}$, but not for others. The rules of algebra can often be used to solve conditional trigonometric equations.

EXAMPLE 1 **Solve $2 \sin x - 1 = 0$, where $0 \leq x < 2\pi$.**

$$2 \sin x - 1 = 0$$
$$2 \sin x = 1$$
$$\sin x = \frac{1}{2}$$

Over the interval $0 \leq x < 2\pi$, the only values of x for which $\sin x = \frac{1}{2}$ is true are $x = \frac{\pi}{6}$ and $x = \frac{5\pi}{6}$.

LESSON PLAN

BACKGROUND

In the Preview, basic trigonometric identities are reviewed. Identities are often used when solving trigonometric equations.

TEACHING SUGGESTIONS

- It may be helpful to use the graphs of the circular functions to illustrate where solutions are located.
- Emphasize that trigonometric equations are usually true for certain values of the variable but not for all values.
- Stress that solutions should always be checked in the original equation.

CHALKBOARD EXAMPLES

- **For Example 1**
  Solve each equation, where $0 \le x \le 2\pi$.
  1. $2 \cos x + 2 = 0$ π
  2. $\sin^2 x = \frac{1}{2}$ $\frac{\pi}{4}, \frac{3\pi}{4}, \frac{5\pi}{4}, \frac{7\pi}{4}$
- **For Example 2**
  Solve each equation, where $0° \le x \le 360°$.
  3. $-2 \cos^2 x + 3 \cos x = \cos^2 x + \sin^2 x$ $0°, 60°, 300°, 360°$
  4. $\cos^2 x + 3\sqrt{3} \cos x + 4 = \sin^2 x$ $150°, 210°$
- **For Example 3**
  Solve each equation, where $0 \le x < 2\pi$.
  5. $\cos 2x = \sin x + 1$ $0, \pi, \frac{7\pi}{6}, \frac{11\pi}{6}$
  6. $\tan^4 x - 4 \tan^2 x + 3 = 0$ $\frac{\pi}{4}, \frac{\pi}{3}, \frac{2\pi}{3}, \frac{3\pi}{4}, \frac{5\pi}{4}, \frac{4\pi}{3}, \frac{5\pi}{3}, \frac{7\pi}{4}$

Common Errors

- Students may not realize that an expression like $\cos \theta = 2$ will yield no solutions. Review the maximum and minimum values of the trigonometric functions.
- Students do not always check solutions. Do examples where solutions may not satisfy the original equation. For example, in the equation $\sin x = 1 - \cos x$, 270° does not check.
- See *Teacher's Resource Book* for additional remediation.

Check to show that $\frac{\pi}{6}$ and $\frac{5\pi}{6}$ are solutions.

$$2 \sin \frac{\pi}{6} - 1 \stackrel{?}{=} 0 \qquad 2 \sin \frac{5\pi}{6} - 1 \stackrel{?}{=} 0$$
$$2\left(\frac{1}{2}\right) - 1 \stackrel{?}{=} 0 \qquad 2\left(\frac{1}{2}\right) - 1 \stackrel{?}{=} 0$$
$$0 = 0 \checkmark \qquad 0 = 0 \checkmark$$

When an equation contains two or more trigonometric functions, it is sometimes helpful to express the terms using one function only.

EXAMPLE 2 **Solve $2 \cos^2 x + \sin x + 1 = 0$, where $0 \le x < 360°$.**

$$2 \cos^2 x + \sin x + 1 = 0$$
$$2(1 - \sin^2 x) + \sin x + 1 = 0 \qquad \textit{Express } \cos^2 x \textit{ in terms of } \sin x\textit{: } \cos^2 x = 1 - \sin^2 x.$$
$$2 \sin^2 x - \sin x - 3 = 0$$
$$(2 \sin x - 3)(\sin x + 1) = 0 \qquad \textit{Factor.}$$
$$2 \sin x - 3 = 0 \quad \text{or} \quad \sin x + 1 = 0$$
$$\sin x = \frac{3}{2} \quad \text{or} \quad \sin x = -1$$

Since $\sin x$ can be no greater than 1, $\sin x = \frac{3}{2}$ has no solution. If $\sin x = -1$, then $x = 270°$. Check in the original equation to show that 270° is a solution.

A double-angle identity is used to solve the equation in the next example.

EXAMPLE 3 **Solve $\sin 2x = \sqrt{3} \cos x$, where $0 \le x < 2\pi$.**

$$\sin 2x = \sqrt{3} \cos x$$
$$2 \sin x \cos x = \sqrt{3} \cos x \qquad \textit{sin } 2x = 2 \textit{ sin x cos x}$$
$$2 \sin x \cos x - \sqrt{3} \cos x = 0 \qquad \textit{Write the equation with all nonzero terms on one side.}$$
$$\cos x(2 \sin x - \sqrt{3}) = 0 \qquad \textit{Factor.}$$
$$\cos x = 0 \quad \text{or} \quad 2 \sin x - \sqrt{3} = 0$$
$$\cos x = 0 \quad \text{or} \quad \sin x = \frac{\sqrt{3}}{2}$$
$$x = \frac{\pi}{2} \text{ or } x = \frac{3\pi}{2} \quad \text{or} \quad x = \frac{\pi}{3} \text{ or } x = \frac{2\pi}{3}$$

Check in the original equation to show that $\frac{\pi}{3}, \frac{\pi}{2}, \frac{2\pi}{3}$, and $\frac{3\pi}{2}$ are solutions.

CLASS EXERCISES

Tell whether each equation is an identity or a conditional equation.

1. $\sin x = 0.5$ conditional equation
2. $\sec x \cos x = 1$ identity
3. $\tan x = 15$ conditional equation
4. $\sin x \cos x = 1$ conditional equation
5. $1 + \tan x = \cot x$ conditional equation
6. $\sin x = \tan x \cos x$ identity

Find the exact solutions to each equation for the interval $0 \le x < 2\pi$.

7. $\sin x = -\frac{1}{2}$ $\frac{7\pi}{6}; \frac{11\pi}{6}$
8. $\cos x = \frac{\sqrt{3}}{2}$ $\frac{\pi}{6}; \frac{11\pi}{6}$
9. $\sec x = 2$ $\frac{\pi}{3}; \frac{5\pi}{3}$

PRACTICE EXERCISES

Find the exact solutions to each equation for the interval $0 \le x < 2\pi$.

A

1. $54 \sin x = 27$ $\frac{\pi}{6}; \frac{5\pi}{6}$
2. $36 \cos x = 18$ $\frac{\pi}{3}; \frac{5\pi}{3}$
3. $10 \cos x = 5\sqrt{2}$ $\frac{\pi}{4}; \frac{7\pi}{4}$
4. $18 \sin x = 9\sqrt{2}$ $\frac{\pi}{4}; \frac{3\pi}{4}$
5. $4 \csc x = -8$ $\frac{7\pi}{6}; \frac{11\pi}{6}$
6. $3 \sec x = -6$ $\frac{2\pi}{3}; \frac{4\pi}{3}$

Find the exact solutions to each equation for the interval $0° \le x < 360°$.

7. $2 + 2 \cos x = 0$ 180°
8. $2 \sin x - 2 = 0$ 90°
9. $1 + \sqrt{3} \tan x = 0$ 150°; 330°
10. $\sqrt{3} + \cot x = 0$ 150°; 330°
11. $2 + \sec x = 0$ 120°; 240°
12. $2 + \sqrt{2} \csc x = 0$ 225°; 315°

Find the exact solutions to each equation for the interval $0 \le x < 2\pi$.

13. $7 \cos x + 12 = 6 \cos x + 13$ 0
14. $15 \sin x + 19 = 14 \sin x + 18$ $\frac{3\pi}{2}$
15. $4 \tan x - 5 = 5 \tan x - 4$ $\frac{3\pi}{4}; \frac{7\pi}{4}$
16. $5 \cot x + 12 = 6 \cot x + 13$ $\frac{3\pi}{4}; \frac{7\pi}{4}$
17. $4 \sin^2 x = 3$ $\frac{\pi}{3}; \frac{2\pi}{3}; \frac{4\pi}{3}; \frac{5\pi}{3}$
18. $2 \cos^2 x = 1$ $\frac{\pi}{4}; \frac{7\pi}{4}; \frac{3\pi}{4}; \frac{5\pi}{4}$
19. $\cot^2 x = 1$ $\frac{\pi}{4}; \frac{3\pi}{4}; \frac{5\pi}{4}; \frac{7\pi}{4}$
20. $\csc^2 x = 2$ $\frac{\pi}{4}; \frac{3\pi}{4}; \frac{5\pi}{4}; \frac{7\pi}{4}$

B

21. $2 \sin^2 x + 3 \sin x + 1 = 0$ $\frac{3\pi}{2}; \frac{7\pi}{6}; \frac{11\pi}{6}$
22. $2 \cos^2 x + 3 \cos x + 1 = 0$ $\frac{2\pi}{3}; \pi; \frac{4\pi}{3}$
23. $\tan^2 x = \tan x$ $0; \frac{\pi}{4}; \pi; \frac{5\pi}{4}$
24. $\cot^2 x - \cot x = 0$ $\frac{\pi}{4}; \frac{\pi}{2}; \frac{5\pi}{4}; \frac{3\pi}{2}$
25. $\tan x = \cot x$ $\frac{\pi}{4}; \frac{3\pi}{4}; \frac{5\pi}{4}; \frac{7\pi}{4}$
26. $\cos x = \sec x$ $0, \pi$
27. $2 \cos^2 x + \sin x = 2$ $0; \frac{\pi}{6}; \frac{5\pi}{6}; \pi$
28. $4 \sin^2 x - 8 \cos x + 1 = 0$ $\frac{\pi}{3}; \frac{5\pi}{3}$
29. $\sec^2 x - 2 \tan x = 0$ $\frac{\pi}{4}; \frac{5\pi}{4}$
30. $\cos 2x + 3 \cos x = -2$ $\frac{2\pi}{3}; \pi; \frac{4\pi}{3}$
31. $\cos x = \sin 2x$ $\frac{\pi}{6}; \frac{\pi}{2}; \frac{5\pi}{6}; \frac{3\pi}{2}$
32. $2 \sin x - \sin 2x = 0$ $0; \pi$
33. $\cos 2x = 2 \sin^2 x$ $\frac{\pi}{6}; \frac{5\pi}{6}; \frac{7\pi}{6}; \frac{11\pi}{6}$
34. $\sin^2 x + 3 \cos x = 3$

C

35. $2 \sin^3 x - \sin^2 x - \sin x = 0$ $0; \frac{\pi}{2}; \pi; \frac{7\pi}{6}; \frac{11\pi}{6}$
36. $2 \cos^3 x + \cos^2 x - \cos x = 0$ $\frac{\pi}{3}; \frac{\pi}{2}; \pi; \frac{3\pi}{2}; \frac{5\pi}{3}$
37. $2 \sin 2x = -\tan 2x$ See below.
38. $\cos^2 2x = \frac{1}{2} \cos 2x$ See below.
39. $\frac{\sec x}{1 + \sec x} = \frac{\sec^2 x}{2 + \sec x}$ $\frac{\pi}{4}; \frac{3\pi}{4}; \frac{5\pi}{4}; \frac{7\pi}{4}$
40. $\frac{\sin x}{\sin x - 1} = \frac{\sin^2 x}{\sin x - 3}$ $0; \pi$

37. $0; \frac{\pi}{3}; \frac{\pi}{2}; \frac{2\pi}{3}; \pi; \frac{4\pi}{3}; \frac{3\pi}{2}; \frac{5\pi}{3}$

38. $\frac{\pi}{4}; \frac{3\pi}{4}; \frac{5\pi}{4}; \frac{7\pi}{4}; \frac{\pi}{6}; \frac{5\pi}{6}; \frac{7\pi}{6}; \frac{11\pi}{6}$

LESSON FOLLOW-UP

Discussion

Explain how to solve $6 \sin x \cos x = 4 \cos x$.

Multiply both sides by $\frac{1}{2}$, set right side equal to 0, and factor. Note: multiplying both sides by $\frac{1}{\cos x}$ will eliminate solutions for $\cos x = 0$.

Critical Thinking

Analysis Ask students to develop a checklist or set of steps that could be used to solve any trigonometric equation. Answers may vary.

Assignment Guide

See p. 248B for assignments.

Challenge

A puzzle involving the use of trigonometric functions and values is featured.

Lesson Quiz

Find the exact solutions to each equation in radians, for the interval $0 \le x < 2\pi$.

1. $5\sqrt{3} \tan x - 5 = 0$ $\frac{\pi}{6}, \frac{7\pi}{6}$
2. $\cos^2 x = 2 \cos x - 1$ 0
3. $\cos x = \sin 2x$ $\frac{\pi}{6}, \frac{\pi}{2}, \frac{5\pi}{6}, \frac{3\pi}{2}$
4. $\csc^2 x + \csc x = 2$ $\frac{\pi}{2}, \frac{7\pi}{6}, \frac{11\pi}{6}$

Find the exact solutions to each equation in degrees, for the interval $0 \le x < 360°$.

5. $8 \sin x = 4$ 30°, 150°
6. $\cos 2x = 1$ 0°, 180°
7. $18 \cos^2 x = 9$ 45°, 135° 225°, 315°

Enrichment

Solve $\tan^2 2x = 3$ if $0 \le x < 2\pi$.

$\tan^2 2x = 3$

$\frac{4 \tan^2 x}{1 - 2\tan^2 x + \tan^4 x} = 3$

$3 \tan^4 x - 10 \tan^2 x + 3 = 0$

$\tan x = \pm\sqrt{3}$ or $\tan x = \pm\frac{\sqrt{3}}{3}$

$\frac{\pi}{6}, \frac{\pi}{3}, \frac{2\pi}{3}, \frac{5\pi}{6}, \frac{7\pi}{6}, \frac{4\pi}{3}, \frac{5\pi}{3}, \frac{11\pi}{6}$

Teacher's Resource Book
Practice—Chapter 6, p. 7
Enrichment—Chapter 6, p. 8

Applications

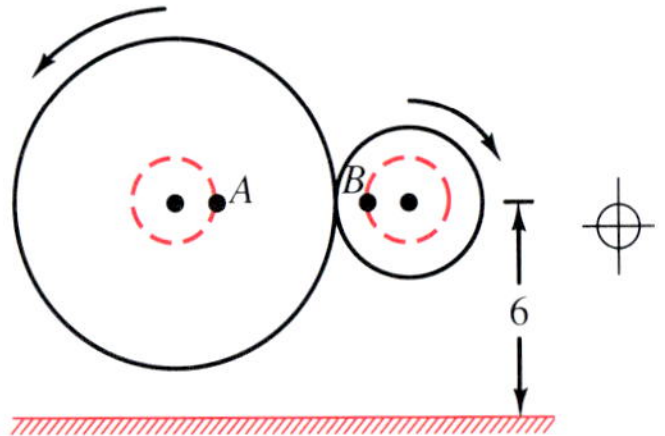

41. Physics Two friction wheels rotate in contact with each other. In the figure, points A and B are on unit circles. The elevation of A is given by the expression $6 + \sin\theta$, and the elevation of B is given by $6 + \sin 2\theta$, where θ is the angle of rotation of the larger wheel from the initial position shown. Find the values of θ, where $0 \le \theta < 2\pi$, for which the elevations of the two points are equal. $0;\ \frac{\pi}{3};\ \pi;\ \frac{5\pi}{3}$

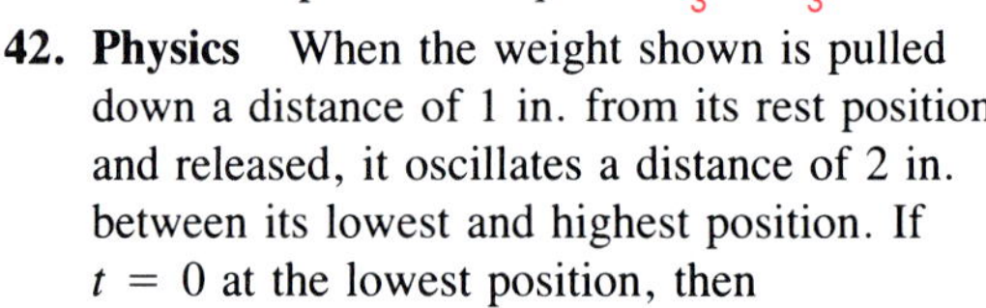

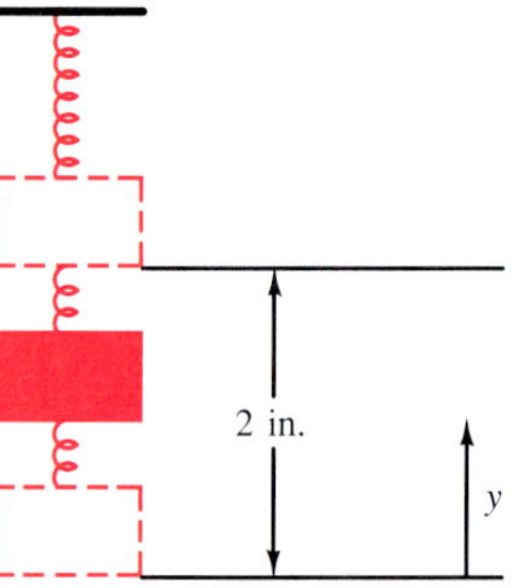

42. Physics When the weight shown is pulled down a distance of 1 in. from its rest position and released, it oscillates a distance of 2 in. between its lowest and highest position. If $t = 0$ at the lowest position, then

$$y = 1 - \cos\left(\frac{\pi t}{2}\right)$$

where y is the distance of the weight from its lowest position and t is the number of seconds elapsed. If $0 \le t < 4$, find t when $y = 1.5$. $1\frac{1}{3}$ s; $2\frac{2}{3}$ s

CHALLENGE

Find the value of each letter in **$T_1R_1I_1G\ O_1N\ O_2M\ E\ T_2R_2Y\ T_3R_3I_2V\ I_3A$**, using the information given below. Then evaluate the expression $-\frac{10}{3}$

$$\frac{(T_1T_2T_3) + (O_1O_2) + (R_1R_2R_3) + (I_1I_2I_3)}{G\,N\,M\,E\,Y\,V\,A}$$

$T_1 = \cot 30°$

$I_1 = \cos 360° + \sin 450°$

$O_1 = \tan 405° + \cot 405°$

$O_2 = \sin\left(\text{Cos}^{-1}\frac{\sqrt{3}}{2}\right)$

$E = \csc 90°$

$R_2 = \tan\left[\text{Sin}^{-1}\left(-\frac{4}{5}\right)\right]$

$T_3 = \cos\alpha$, if $\sin\alpha = \frac{1}{3}$ and $\tan\alpha > 0$

$I_2 = \cos 2\beta + 2\sin^2\beta$

$I_3 = \cos\left[\text{Cos}^{-1}\left(-\frac{\sqrt{3}}{2}\right) + \text{Sin}^{-1}\frac{1}{2}\right]$

$R_1 = \sin 30°$

$G = \sec^2\theta - \tan^2\theta$

$N = \tan 45°$

$M = \tan(-60°)$

$T_2 = \tan(\theta - \pi)$, if $\tan\theta = \sqrt{6}$

$Y = \cot\left(\text{Tan}^{-1}\frac{\sqrt{3}}{3}\right)$

$R_3 = \tan(135° + z)$, if $\tan z = \frac{1}{3}$

$V = \sin 75° \cos 75°$

$A = \tan 2\alpha$, where $\tan\alpha = \frac{1}{2}$

6.5 Trigonometric Equations: Approximate Solutions

Objective: To find approximate values of solutions to trigonometric equations

A scientific calculator can be used to approximate the values of solutions to trigonometric equations.

Preview

EXAMPLE **Find, to the nearest tenth of a degree, all values of x such that $0° \le x < 360°$ and $\sin x = 0.4826$.**

$\sin x = 0.4826$
$x = 28.9°$ — *Use the Sin^{-1} key or the inv and sin keys to obtain the principal value.*
and
$x = 180° - 28.9° = 151.1°$ — *Since sin x is positive, there is also a value of x between 90° and 180°.*

Find, to the nearest degree, all values of x such that $0° \le x < 360°$ and the given statement is true.

1. $\cos x = 0.8823$ 28°; 332°
2. $\cos x = -0.3623$ 111°; 249°
3. $\tan x = 2.544$ 69°; 249°
4. $\sin x = -0.3410$ 200°; 340°
5. $\sin x = 0.9672$ 75°; 105°
6. $\tan x = -0.2547$ 166°; 346°
7. $\sec x = 2.398$ 65°; 295°
8. $\csc x = -1.253$ 233°; 307°
9. $\cot x = -3.265$ 163°; 343°

The equations considered in the preceding lesson had exact solutions. However, solutions to most trigonometric equations must be represented as approximations.

EXAMPLE 1 **Find, to the nearest tenth of a degree, the solutions to $8 \cos x + 3 = 5 \cos x + 5$, where $0° \le x < 360°$.**

$$8 \cos x + 3 = 5 \cos x + 5$$
$$3 \cos x = 2$$
$$\cos x = \frac{2}{3}$$
$$x = 48.2°$$

Use the Cos^{-1} key or the inv and cos keys to find the principal value.

and

$$x = 360° - 48.2° = 311.8°$$

Since cos x is positive, there is also a value of x between 270° and 360°.

Check to show that 48.2° and 311.8° are approximate solutions.

LESSON PLAN

Materials/Manipulatives
Scientific calculator
Graphing calculator
Computer

BACKGROUND

In the Preview, calculators are used to solve trigonometric equations that do not have exact solutions.

TEACHING SUGGESTIONS

- Review the different methods of factoring.
- Review the quadratic formula and its use in solving equations.

Critical Thinking

Analysis List several equations on the board and ask students to describe how they would solve each one.
Answers may vary.

CHALKBOARD EXAMPLES

- **For Example 1**

  Approximate the solutions to each equation, where $0° \le x < 360°$.

  1. $7 \sin x + 13 = 10 + 3 \sin x$
  228.6°, 311.4°

  2. $5 \tan^2 x = 18$
  62.2°, 117.8°, 242.2°, 297.8°

- **For Example 2**

  Approximate the solutions to each equation, where $0° \le x < 360°$.

  3. $24 \sin^2 x + 38 \sin x + 3 = 0$
  184.7°, 355.2°

  4. $3 \tan^2 x - 22 \tan x + 7 = 0$
  18.4°, 81.9°, 198.4°, 261.9°

- **For Example 3**

  Approximate the solutions to each equation, where $0° \le x < 360°$.

  5. $5 \cos^2 x = 3 \cos x + 1$
  33°, 103.8°, 256.2°, 327°

  6. $\tan^2 x = 7 \tan x - 3$
  24.6°, 81.3°, 204.6°, 261.3°

Common Errors

- Some students may not recognize expressions that may be factored. Have those students do several exercises that require the use of factoring.
- Students may use incorrect key sequences when using a calculator to find approximate solutions. Review the correct sequences.
- See *Teacher's Resource Book* for additional remediation.

EXAMPLE 2 **Find, to the nearest tenth of a degree, the solutions to $12 \sin^2 x + 7 \sin x + 1 = 0$, where $0° \le x < 360°$.**

$$12 \sin^2 x + 7 \sin x + 1 = 0$$
$$(4 \sin x + 1)(3 \sin x + 1) = 0$$
$$4 \sin x + 1 = 0 \quad \text{or} \quad 3 \sin x + 1 = 0$$
$$\sin x = -\frac{1}{4} \quad \text{or} \quad \sin x = -\frac{1}{3}$$
$$x = -14.5° \qquad x = -19.5° \quad \textit{Principal values}$$

These values are not solutions over the interval $0° \le x < 360°$. However,

$$x = 180° - (-14.5°) = 194.5°$$
$$x = 180° - (-19.5°) = 199.5°$$
$$x = 360° + (-19.5°) = 340.5°$$
$$x = 360° + (-14.5°) = 345.5°$$

Since $\sin x = -\frac{1}{4}$ or $\sin x = -\frac{1}{3}$, x is between 180° and 360°.

Check that 194.5°, 199.5°, 340.5°, and 345.5° are approximate solutions.

If an equation cannot be solved by factoring, the quadratic formula can sometimes be used. The solutions to $ax^2 + bx + c = 0$ are given by

$$x = \frac{-b \pm \sqrt{b^2 - 4ac}}{2a}, \; a \ne 0$$

EXAMPLE 3 **Find, to the nearest tenth of a degree, the solutions to $3 \tan^2 x + 5 \tan x - 1 = 0$, where $0° \le x < 360°$.**

$$3 \tan^2 x + 5 \tan x - 1 = 0$$
$$\tan x = \frac{-5 \pm \sqrt{5^2 - 4(3)(-1)}}{2(3)} \quad \textit{Use the quadratic formula.}$$
$$\tan x = \frac{-5 + \sqrt{37}}{6} \quad \text{or} \quad \tan x = \frac{-5 - \sqrt{37}}{6}$$
$$\tan x = 0.1805 \quad \text{or} \quad \tan x = -1.847$$
$$x = 10.2° \quad \text{or} \quad x = -61.6° \quad \textit{Principal values}$$

The principal value 10.2° is between 0° and 90°. There is a value of x between 180° and 270° where $\tan x$ is also positive.

$$x = 180° + 10.2° = 190.2°$$

The principal value $-61.6°$ is not a solution, since it is not between 0° and 360°, but there are two values of x, one between 90° and 180°, the other between 270° and 360°, where $\tan x$ is also negative.

$$x = 180° + (-61.6°) = 118.4°$$
$$x = 360° + (-61.6°) = 298.4°$$

Check that 10.2°, 118.4°, 190.2°, and 298.4° are approximate solutions.

A graphing calculator or a computer can be used to determine whether solutions to a trigonometric equation are approximately correct. To illustrate, the solutions to the equation in Example 3 can be checked by graphing $y = 3\tan^2 x + 5\tan x - 1$ over the interval $0° \le x < 360°$. The points where the graph intersects the x-axis are the points where $y = 3\tan^2 x + 5\tan x - 1 = 0$. Therefore, the x-coordinates of those points are the solutions to the equation.

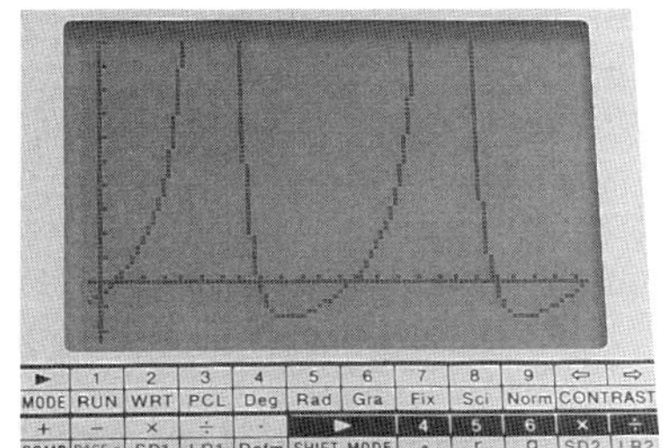

CLASS EXERCISES

Find, to the nearest degree, the solutions to each equation, where $0° \le x < 360°$.

1. $7\sin x + 8 = \sin x + 10$ 19°; 161°
2. $6\cos x - 5 = 9\cos x - 4$ 109°; 251°
3. $8\tan x + 15 = 3\tan x + 11$ 141°; 321°
4. $7\cot x + 2 = \cot x + 8$ 45°; 225°
5. $3\sec x - 7 = \sec x + 2$ 77°; 283°
6. $13\csc x + 25 = 5\csc x + 7$ 206°; 334°
7. $8\cos 2x + 3 = 2\cos 2x - 1$ 66°; 114°; 246°; 294°
8. $10\sin 2x + 5 = 5\sin 2x + 8$ 18°; 72°; 198°; 252°

PRACTICE EXERCISES

Find, to the nearest degree, the solutions to each equation, where $0° \le x < 360°$. Verify the solutions by graphing.

A

1. $8\sin x - 1 = 2\sin x + 3$ 42°; 138°
2. $18\cos x - 58 = 31\cos x - 54$ 108°; 252°
3. $16\cos x + 29 = 7\cos x + 22$ 141°; 219°
4. $73\sin x - 64 = 29\sin x - 51$ 17°; 163°
5. $25\cos^2 x - 1 = 0$ 78°; 102°; 258°; 282°
6. $36\sin^2 x - 1 = 0$ 10°; 170°; 190°; 350°
7. $6\sin^2 x - 5\sin x - 1 = 0$ 90°; 190°; 350°
8. $12\cos^2 x - 8\cos x + 1 = 0$ 60°; 80°; 280°; 300°
9. $12\cos^2 x + 7\cos x + 1 = 0$ 104°; 109°; 251°; 256°
10. $8\sin^2 x - 2\sin x - 1 = 0$ 30°; 150°; 194°; 346°
11. $8\sin^2 x + 5\sin x - 1 = 0$ 9°; 171°; 232°; 308°
12. $9\sin^2 x + 7\sin x - 2 = 0$ 13°; 167°; 270°
13. $12\cos^2 x - 11\cos x + 2 = 0$ 48°; 76°; 284°; 312°
14. $7\sin^2 x - 4\sin x - 1 = 0$ 49°; 131°; 191°; 349°
15. $\sec^2 x - 12\sec x + 2 = 0$ 85°; 275°
16. $\csc^2 x - 15\csc x + 1 = 0$ 4°; 176°
17. $\csc^2 x + 15\csc x + 8 = 0$ 184°; 356°
18. $\sec^2 x + 18\sec x + 9 = 0$ 93°; 267°

B

19. $5\sin 2x + 2 = 8\sin 2x + 1$ 10°; 80°; 190°; 260°
20. $8\cos 2x + 7 = \cos 2x + 4$ 58°; 122°; 238°; 302°
21. $10\cos 2x - 19 = 6\cos 2x - 18$ 38°; 142°; 218°; 322°
22. $22\sin 2x - 5 = 10\sin 2x + 3$ 21°; 69°; 201°; 249°
23. $8\tan 2x - 9 = 7\tan 2x + 13$ 44°; 134°; 224°; 314°
24. $12\cot 2x - 7 = 2\cot 2x - 10$ 53°; 143°; 233°; 323°
25. $35\sin^2 x - 20\sin x + 1 = 0$ 3°; 31°; 149°; 177°
26. $24\cos^2 x + 11\cos x + 1 = 0$ 97°; 109°; 251°; 263°

LESSON FOLLOW-UP

Discussion

Show how to solve $2\tan(\theta - 23°) = \sqrt{7}$ for the interval $0° < \theta < 360°$.

$$\tan(\theta - 23°) = \frac{\sqrt{7}}{2} = 1.32287$$
$$\theta - 23° = 52.9° \text{ or } 232.9°$$
$$\theta = 75.9° \text{ or } 255.9°$$

Assignment Guide

See p. 248B for assignments.

Challenge

A calculator sequence which gives the same result for any value of an acute angle is featured. Students are encouraged to make up a similar type sequence.

Lesson Quiz

Find the approximate or exact solutions for x, where $0° \le x < 360°$.

1. $12\sin^2 x = 3\sin x$
   0°, 14.5°, 165.5°, 180°
2. $13\cos x - 1 = \cos x$
   85.2°, 274.8°
3. $5\sec^2 x = 16$
   56°, 124°, 236°, 304°
4. $3\cos^2 x + 8\cos x + 1 = 0$
   97.6°, 262.4°
5. $7\tan x = 14$ 63.4°, 243.4°

Enrichment

Solve $\cos^2\theta = -\sin 2\theta$, for the interval $0° \le \theta < 360°$.

$$\cos^2\theta + \sin 2\theta = 0$$
$$\cos^2\theta + 2\sin\theta\cos\theta = 0$$
$$(\cos\theta)(\cos\theta + 2\sin\theta) = 0$$
$$\cos\theta = 0 \quad \cos\theta + 2\sin\theta = 0$$
$$\theta = 90°, 270° \quad \cos\theta = -2\sin\theta$$
$$\frac{\cos\theta}{\sin\theta} = -2$$
$$\cot\theta = -2$$
$$\theta = 153.4°, 333.4°$$

Thus, the solutions are 90°, 153.4°, 270°, 333.4°

Teacher's Resource Book
Practice—Chapter 6, p. 9
Enrichment—Chapter 6, p. 10

Additional Answers

31. 0.02; 0.10; 0.30; 0.38; 0.59; 0.67; 0.87; 0.96; 1.16; 1.24; 1.45; 1.53; 1.73; 1.81; 2.02; 2.10; 2.30; 2.38; 2.59; 2.67; 2.87; 2.96; 3.16; 3.24; 3.44; 3.53; 3.73; 3.81; 4.02; 4.10; 4.30; 4.38; 4.59; 4.67; 4.87; 4.95; 5.16; 5.24; 5.44; 5.53; 5.73; 5.81; 6.02; 6.10

32. 0.04; 0.12; 0.29; 0.36; 0.53; 0.60; 0.77; 0.84; 1.01; 1.08; 1.25; 1.32; 1.49; 1.57; 1.74; 1.81; 1.98; 2.05; 2.22; 2.29; 2.46; 2.53; 2.70; 2.77; 2.94; 3.02; 3.19; 3.26; 3.43; 3.50; 3.67; 3.74; 3.91; 3.98; 4.15; 4.22; 4.39; 4.47; 4.64; 4.71; 4.88; 4.95; 5.12; 5.19; 5.36; 5.43; 5.60; 5.67; 5.84; 5.92; 6.09; 6.16

33. 0.09; 0.44; 0.98; 1.34; 2.19; 2.54; 3.08; 3.43; 4.28; 4.63; 5.17; 5.52

34. 0.40; 1.31; 1.89; 2.97; 3.54; 4.45; 5.03; 6.11

35. no solution

36. 0.16, 0.31, 0.79, 0.94, 1.42, 1.57, 2.05, 2.17, 2.68, 2.82, 3.30, 3.45, 3.93, 4.08, 4.56, 4.71, 5.19, 5.34, 5.82, 5.96

Find, to the nearest hundredth of a radian, the solutions to each equation, where $0 \le x < 2\pi$.

27. $\sec^2 x + 7 \sec x + 10 = 0$
1.77; 2.09; 4.19; 4.51

28. $\csc^2 x + 10 \csc x + 16 = 0$
3.27; 3.67; 5.76; 6.16

29. $\csc^2 x - 9 \csc x - 10 = 0$
0.10; 3.04; 4.71

30. $\sec^2 x - 14 \sec x - 15 = 0$
1.50; 3.14; 4.78

C **31.** $8 \tan^2 11x - 17 \tan 11x + 3 = 0$
See side column.

32. $12 \cot^2 13x - 19 \cot 13x + 1 = 0$

33. $5 \cos^2 (3x + 1) + 2 \cos (3x + 1) - 1 = 0$

34. $7 \sin^2 (2x + 3) + \sin (2x + 3) - 2 = 0$

35. $\tan^2 (7x - 5) + 6 \tan (7x - 5) + 10 = 0$

36. $\cot^2 (5x + 8) - 7 \cot (5x + 8) - 12 = 0$

Applications

37. Physics A pressure wave is transmitted down a long tube, and the pressure p is given by the equation

$$p = 2 \cos [2\pi(x - 0.5t)]$$

where x is measured in meters and t in seconds. A pressure sensor is located 1 m from the opening of the tube. If the wave originates when $x = 0$ and $t = 0$, find the first value of t for which the sensor reads 0.5. 1.58 s

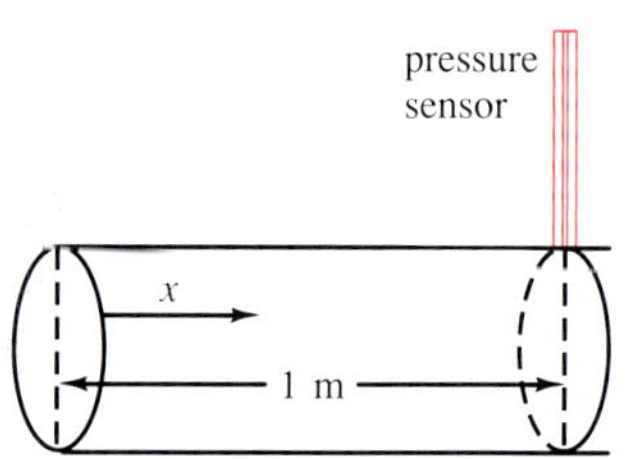

38. Physics Point A rotates on a shaft, and point B is on a slider that moves horizontally. If x is the distance moved by the horizontal slider when the wheel rotates through an angle θ, it can be shown that

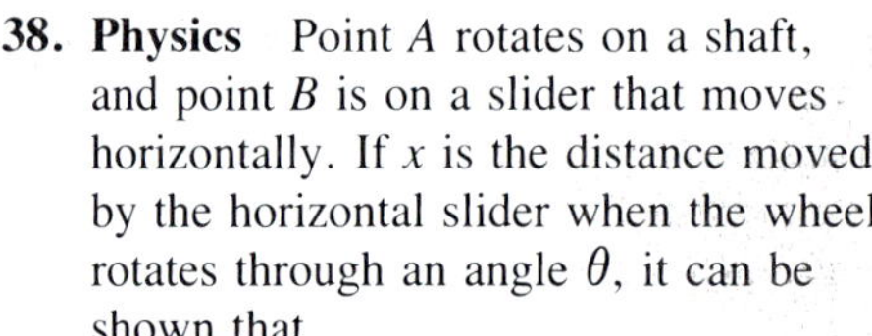

$$x = 1.54 (1 - \cos \theta) + 2 \sin^2 \theta$$

Find θ, $0° \le \theta \le 360°$, if $x = 2$.
55°; 305°

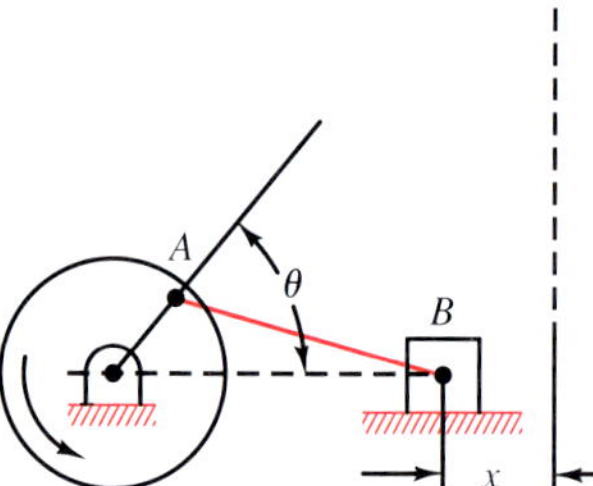

CHALLENGE

Choose any acute angle θ. Use a calculator to find $\tan \theta$, square the result, add 1, and then multiply by $\cos^2 \theta$. Explain the result. Make up another calculator sequence that always gives the same result for any value of acute angle θ.
$((\tan \theta)^2 + 1) \cos^2 \theta = \sec^2 \theta \cos^2 \theta = 1$; Answers may vary.

6.6 Rotation of Axes

Objectives: To derive and use formulas for rotations of axes
To graph equations using rotated axes

The graphs of $x^2 - y^2 = 2$ and $y = \frac{1}{x}$ can be made to coincide with one another by rotating the plane of one of the graphs about the origin.

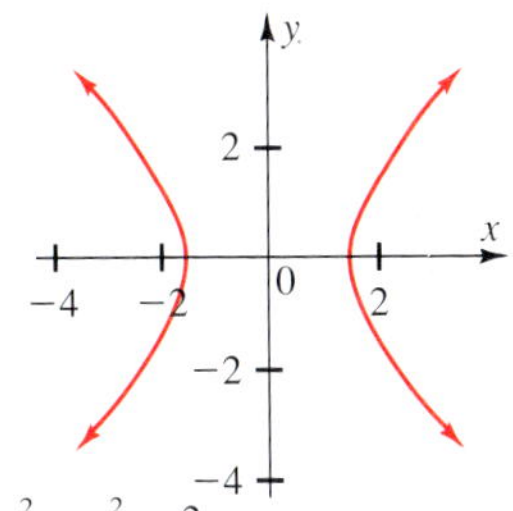

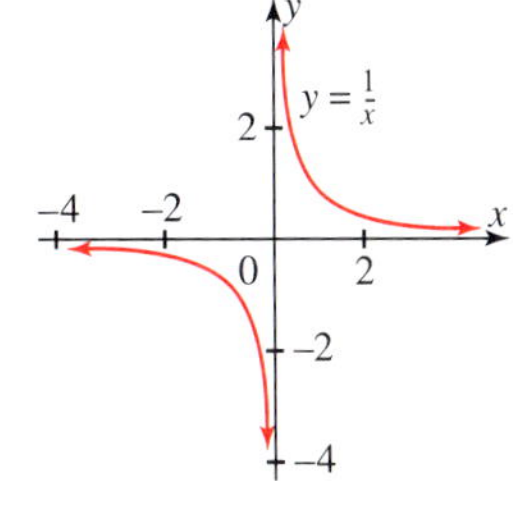

Preview

1. Copy the graph of $x^2 - y^2 = 2$ on a piece of lightweight paper, and place the copy over the graph of $y = \frac{1}{x}$ shown above so that the origins of the coordinate planes coincide. Through what angle must you rotate the graph of $x^2 - y^2 = 2$ in order to make it coincide with the graph of $y = \frac{1}{x}$? 45°

2. Draw the graphs of the equations $\frac{x^2}{9} + \frac{y^2}{4} = 1$ and $\frac{x^2}{4} + \frac{y^2}{9} = 1$. Through what angle must you rotate the first graph in order to make it coincide with the second graph? 90°

When the xy-coordinate axes are rotated about the origin through an angle θ, this motion is called a **rotation of axes** and θ is called the **angle of rotation.** The new axes are labeled the x'- and y'-axes, and each point P on the plane has two pairs of coordinates, (x, y) and (x', y'). The sum identities for sine and cosine can be used to show how the xy- and $x'y'$-coordinates of a point P are related.

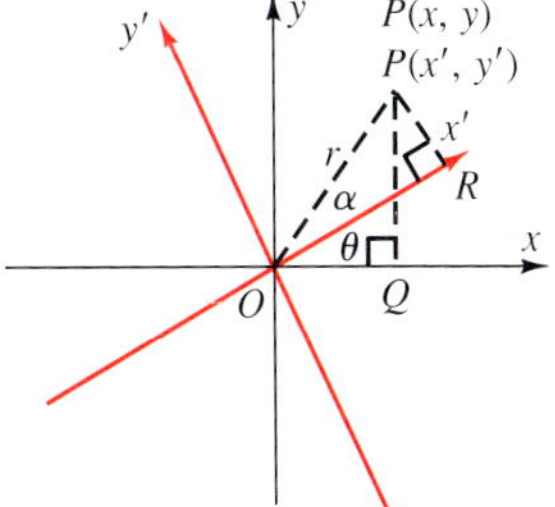

LESSON PLAN

Vocabulary
Angle of rotation
Rotation of axes

BACKGROUND

In the Preview, actual tracing of graphs is used to introduce the concept of axes rotation.

TEACHING SUGGESTIONS

- Review the sum identities for sine and cosine and solving systems of equations.
- It may be necessary to review the general equations of conic sections.

CHALKBOARD EXAMPLES

- **For Example 1**
  1. If the coordinate axes are rotated 45° about the origin, find the new coordinates of the point $(5, -2)$. $\left(\frac{3\sqrt{2}}{2}, -\frac{7\sqrt{2}}{2}\right)$
  2. If the coordinate axes are rotated 70° about the origin, find the new coordinates of the point $(3, 5)$ to the nearest tenth. $(5.7, -1.1)$

$$\cos(\theta + \alpha) = \frac{x}{r} \qquad \textit{cos } (\theta + \alpha) = \frac{OQ}{OP}$$

$$x = r\cos(\theta + \alpha)$$

$$x = r(\cos\theta\cos\alpha - \sin\theta\sin\alpha) \qquad \textit{Sum identity for cosine}$$

$$x = r\cos\theta\cos\alpha - r\sin\theta\sin\alpha$$

$$x = r\left(\frac{x'}{r}\right)\cos\theta - r\left(\frac{y'}{r}\right)\sin\theta \qquad \cos\alpha = \frac{OR}{OP} = \frac{x'}{r};\ \sin\alpha = \frac{PR}{OP} = \frac{y'}{r}$$

$$x = x'\cos\theta - y'\sin\theta$$

Similarly

$$\sin(\theta + \alpha) = \frac{y}{r} \qquad \textit{sin } (\theta + \alpha) = \frac{PQ}{OP}$$

$$y = r\sin(\theta + \alpha)$$

$$y = r(\sin\theta\cos\alpha + \cos\theta\sin\alpha) \qquad \textit{Sum identity for sine}$$

$$y = r\sin\theta\cos\alpha + r\cos\theta\sin\alpha$$

$$y = r\left(\frac{x'}{r}\right)\sin\theta + r\left(\frac{y'}{r}\right)\cos\theta \qquad \cos\alpha = \frac{OR}{OP} = \frac{x'}{r};\ \sin\alpha = \frac{PR}{OP} = \frac{y'}{r}$$

$$y = x'\sin\theta + y'\cos\theta$$

This system of equations

$$\begin{cases} x = x'\cos\theta - y'\sin\theta \\ y = x'\sin\theta + y'\cos\theta \end{cases}$$

can be solved to find x' and y' in terms of x, y, and functions of θ

$$x' = x\cos\theta + y\sin\theta$$

$$y' = -x\sin\theta + y\cos\theta$$

You will be asked to derive the above two formulas in Exercise 28.

The second pair of equations above are used to find the new coordinates of a point after the xy-plane has been rotated a given number of degrees.

EXAMPLE 1 **If the coordinate axes are rotated 30° about the origin, find the new coordinates of the point for which the old coordinates are (4, 8).**

$$x' = 4\cos 30° + 8\sin 30° \qquad x' = x\cos\theta + y\sin\theta$$

$$= 4\left(\frac{\sqrt{3}}{2}\right) + 8\left(\frac{1}{2}\right)$$

$$= 2\sqrt{3} + 4$$

$$y' = -4\sin 30° + 8\cos 30° \qquad y' = -x\sin\theta + y\cos\theta$$

$$= -4\left(\frac{1}{2}\right) + 8\left(\frac{\sqrt{3}}{2}\right)$$

$$= -2 + 4\sqrt{3}$$

The new coordinates are $(2\sqrt{3} + 4, -2 + 4\sqrt{3})$.

If an equation in x and y is given, along with an angle of rotation, an equation in terms of x' and y' can be written.

EXAMPLE 2 **A curve with equation $xy = 2$ is rotated about the origin through an angle of 45°. Find the equation of the curve in the new coordinate system.**

$$xy = 2$$
$$(x' \cos 45° - y' \sin 45°)(x' \sin 45° + y' \cos 45°) = 2$$
$$\left[x'\left(\frac{\sqrt{2}}{2}\right) - y'\left(\frac{\sqrt{2}}{2}\right)\right]\left[x'\left(\frac{\sqrt{2}}{2}\right) + y'\left(\frac{\sqrt{2}}{2}\right)\right] = 2$$
$$\frac{(x')^2}{2} - \frac{(y')^2}{2} = 2$$
$$\frac{(x')^2}{4} - \frac{(y')^2}{4} = 1$$

$x = x' \cos \theta - y' \sin \theta$; $y = x' \sin \theta + y' \cos \theta$

The choice of $\theta = 45°$ in Example 2 transformed the equation into one that can be readily recognized as that of a hyperbola. It is not usually obvious what angle of rotation should be chosen to transform an equation into one with a more familiar form. However, it is frequently helpful to eliminate an xy-term. It can be shown that an equation of the form $Ax^2 + Bxy + Cy^2 + Dx + Ey + F = 0$, where $B \neq 0$, can be transformed into an equation without an xy-term if the axes are rotated through an angle such that

$$\text{Cot } 2\theta = \frac{A - C}{B} \quad \text{or} \quad \theta = \frac{1}{2}\text{Cot}^{-1}\frac{A - C}{B}$$

EXAMPLE 3 **Determine the angle of rotation needed to transform the equation $x^2 + xy + y^2 = 3$ into an equation without an xy-term. Find the new equation and draw its graph, showing both sets of axes.**

In the equation $x^2 + xy + y^2 = 3$, $A = 1$, $B = 1$, and $C = 1$.

Therefore, $\theta = \frac{1}{2}\text{Cot}^{-1}\frac{1-1}{1}$ *Substitute in $\theta = \frac{1}{2}\text{Arccot}\frac{A - C}{B}$.*

$$= \frac{1}{2}\text{Cot}^{-1} 0$$
$$= \frac{1}{2}(90°) = 45°$$

Calculate the values of x and y in terms of x' and y'.

$$x = x' \cos 45° - y' \sin 45° \qquad x = x' \cos \theta - y' \sin \theta$$
$$= x'\left(\frac{\sqrt{2}}{2}\right) - y'\left(\frac{\sqrt{2}}{2}\right)$$
$$y = x' \sin 45° + y' \cos 45° \qquad y = x' \sin \theta + y' \cos \theta$$
$$= x'\left(\frac{\sqrt{2}}{2}\right) + y'\left(\frac{\sqrt{2}}{2}\right)$$

- **For Example 2**

3. The coordinate axes are rotated about the origin through an angle of 30°. Find the equation of a curve in the new coordinate system if the original equation is $4x^2 + 3\sqrt{3}xy + y^2 = 22$.
$\frac{(x')^2}{4} - \frac{(y')^2}{44} = 1$

4. The coordinate axes are rotated about the origin through an angle of 45°. Find the equation of a curve in the new coordinate system if the original equation is $5x^2 + 6xy + 5y^2 = 8$.
$\frac{(x')^2}{1} + \frac{(y')^2}{4} = 1$, or $(x')^2 + \frac{(y')^2}{4} = 1$

- **For Example 3**

5. Determine the angle rotation needed to transform the equation $6x^2 - \sqrt{3}xy + 5y^2 = 234$ into an equation without an xy-term. Find the new equation.
60°; $\frac{(x')^2}{52} + \frac{(y')^2}{36} = 1$

Critical Thinking

Analysis Ask students to explain why circles are not discussed in terms of axes rotation. Students should reason that circles with center at the origin are completely symmetric and therefore circles with the same radii will always coincide while those with different radii will never coincide.

Common Errors

- When determining the angle of rotation needed to eliminate an xy-term, students may use incorrect signs. Emphasize that in an expression like $6x^2 - xy - 7y^2 = 3$, $A = 6$, $B = -1$, and $C = -7$.
- See *Teacher's Resource Book* for additional remediation.

LESSON FOLLOW-UP

Discussion

Determine which conic is represented by each of the following equations.

1. $x = -2y^2 + 3y + 7$ parabola
2. $4x^2 - 8x + y^2 - 6y = -15$ ellipse
3. $\frac{x^2}{3} - \frac{y^2}{4} = 1$ hyperbola

Assignment Guide

See p. 248B for assignments.

Test Yourself

See *Teacher's Resource Book, Tests*, pp. 59–60.

Lesson Quiz

1. Find the angle of rotation needed to transform this equation into one without an xy-term. $4x^2 + 2\sqrt{3}xy + 2y^2 = 7$ 30°
2. Find the new coordinates of the point with coordinates $(5, -5)$ for a rotation of 45° about the origin. $(0, -5\sqrt{2})$
3. The xy-axes are rotated about the origin through an angle of 30°. Find the equation of a curve in the new coordinate system if the original equation is $2x^2 + \sqrt{3}xy + y^2 = 10$. $\frac{(x')^2}{4} + \frac{(y')^2}{20} = 1$
4. Determine the angle of rotation needed to transform $x^2 + \sqrt{3}xy + 2y^2 = 50$ into an equation without an xy-term. Find the new equation. 60°; $\frac{(x')^2}{20} + \frac{(y')^2}{100} = 1$

Enrichment

Given $x' = \frac{3}{5}x + \frac{4}{5}y$ and $y' = -\frac{4}{5}x + \frac{3}{5}y$, find the angle of rotation.

$\cos\theta = \frac{3}{5} = \frac{x}{r} \rightarrow x = 3$

$\sin\theta = \frac{4}{5} = \frac{y}{r} \rightarrow y = 4$

$\tan\theta = \frac{4}{3}, \theta \approx 53°$

Then substitute these expressions for x and y in the original equation. After the left side of the equation is simplified, the following equation is obtained.

$$\frac{(x')^2}{2} + \frac{(y')^2}{6} = 1$$

The new equation is clearly that of an ellipse. It is easier to recognize and to sketch the new equation than the original equation. Rotations preserve all distances and angles, so they also preserve the shapes of geometric figures. If you wish to study the geometric properties of the graph with the equation $x^2 + xy + y^2 = 3$, you may do so by studying the graph of $\frac{x^2}{2} + \frac{y^2}{6} = 1$, which is easier to sketch.

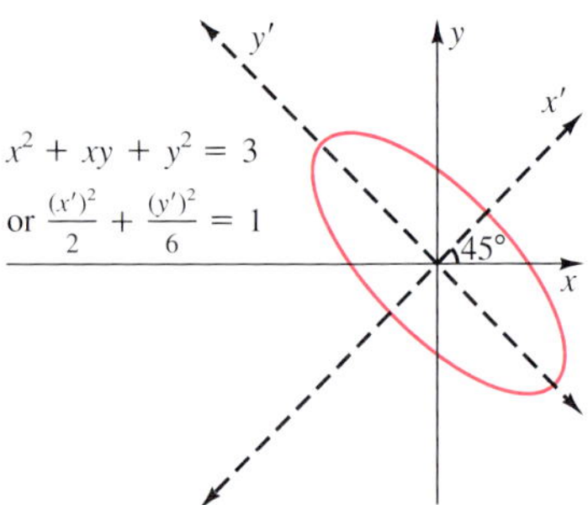

CLASS EXERCISES

Describe the shape of the graph of each equation.

1. $\frac{x^2}{9} - \frac{y^2}{25} = 4$ hyperbola
2. $\frac{x^2}{4} + \frac{y^2}{9} = 1$ ellipse
3. $\frac{y^2}{16} - \frac{x^2}{9} = 1$ hyperbola
4. $x^2 - y = 3$ parabola

Find the angle of rotation needed to transform each equation into one without an xy-term.

5. $3x^2 + 5xy + 3y^2 = 19$ 45°
6. $5x^2 - xy + 4y^2 = 8$ 67.5°
7. $10x^2 - 12xy - 2y^2 + 3y = 23$ 67.5°
8. $x^2 - xy + 3y = 9$ 13.3°

1. $(\sqrt{2}, -4\sqrt{2})$ 2. $\left(\frac{6\sqrt{3}+9}{2}, \frac{-6+9\sqrt{3}}{2}\right)$ 3. $\left(\frac{8-5\sqrt{3}}{2}, \frac{-8\sqrt{3}-5}{2}\right)$ 4. $(5\sqrt{2}, 2\sqrt{2})$

PRACTICE EXERCISES

5. $\left(\frac{-6\sqrt{3}-7}{2}, \frac{6-7\sqrt{3}}{2}\right)$ 6. $\left(\frac{-7-10\sqrt{3}}{2}, \frac{7\sqrt{3}-10}{2}\right)$ 7. $(2\sqrt{2}, \sqrt{2})$

Find the new coordinates of each point for the given angle of rotation.

A

1. $(5, -3)$, 45°
2. $(6, 9)$, 30°
3. $(8, -5)$, 60°
4. $(3, 7)$, 45°
5. $(-6, -7)$, 30°
6. $(-7, -10)$, 60°
7. $(-3, 1)$, 135°
8. $(11, -4)$, 120° See below.

The xy-axes are rotated through the given angle θ. Find an equation of the curve in the new coordinate system. See below.

9. $x^2 + 4xy + y^2 = 4; \theta = 45°$
10. $2x^2 - 3xy + 2y^2 = 8; \theta = 45°$
11. $2x^2 + 3\sqrt{3}xy - y^2 = 35; \theta = 30°$
12. $x^2 - 2\sqrt{3}xy - y^2 = 32; \theta = 60°$
13. $5x^2 + 6xy + 5y^2 = 8; \theta = 45°$
14. $3x^2 + 4xy + 3y^2 = 12; \theta = 45°$

8. $\left(\frac{-11-4\sqrt{3}}{2}, \frac{-11\sqrt{3}+4}{2}\right)$

Additional Answers

9. $3(x')^2 - (y')^2 = 4$
10. $(x')^2 + 7(y')^2 = 16$
11. $7(x')^2 - 5(y')^2 = 70$
12. $-(x')^2 + (y')^2 = 16$
13. $4(x')^2 + (y')^2 = 4$
14. $5(x')^2 + (y')^2 = 12$

Find the angle of rotation needed to transform the given equation into one without an xy-term. Find the new equation and draw its graph, showing both sets of axes. See pages 486–488.

B **15.** $5x^2 + xy + 5y^2 = 4$

16. $xy = 10$

17. $xy = -6$

18. $x^2 + 5xy + y^2 = 10$

19. $5x^2 - 6xy + 5y^2 = 30$

20. $x^2 - \sqrt{3}xy = 5$

21. $\sqrt{3}xy - y^2 = 9$

22. $19xy = -73$

23. $6x^2 - xy + 7y^2 = 4$

24. $x^2 + 2\sqrt{3}xy - y^2 = 8$

C **25.** $x^2 + 3xy - 5y^2 = 8$

26. $14x^2 - 9xy - 2y^2 = 18$

27. If only even powers of x and y appear in an equation, and the axes are rotated 180°, what can be said about the new equation?

28. Derive the formulas $x' = x\cos\theta + y\sin\theta$ and $y' = -x\sin\theta + y\cos\theta$ by solving the system of equations $\begin{cases} x = x'\cos\theta - y'\sin\theta \\ y = x'\sin\theta + y'\cos\theta \end{cases}$.

Applications

29. Engineering The equation of the shape of an elliptical cam is $13x^2 + 6\sqrt{3}xy + 7y^2 = 32$. Determine the angle of rotation needed to eliminate the xy-term from the equation. Then sketch the shape of the cam, showing both the xy- and the $x'y'$-axes. See side column.

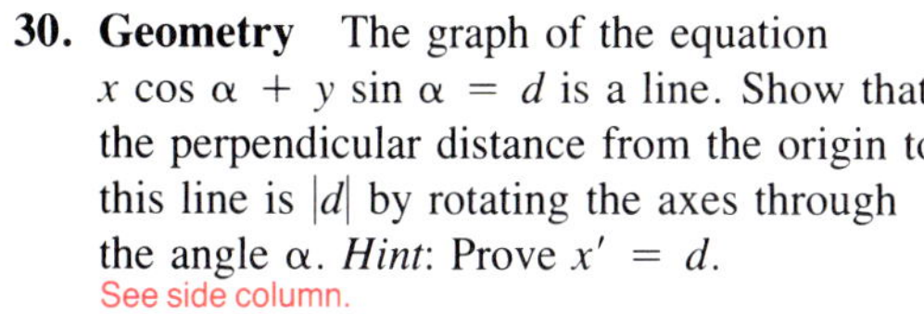

30. Geometry The graph of the equation $x\cos\alpha + y\sin\alpha = d$ is a line. Show that the perpendicular distance from the origin to this line is $|d|$ by rotating the axes through the angle α. *Hint*: Prove $x' = d$.
See side column.

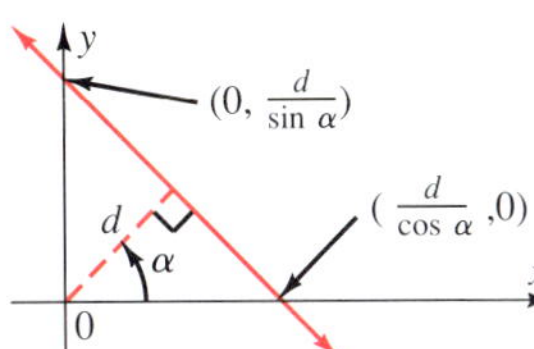

TEST YOURSELF

Find the exact solutions to each equation in radians, for $0 \le x < 2\pi$.

1. $7\tan x = 5\tan x + 2$ — $x = \frac{\pi}{4}, \frac{5\pi}{4}$

2. $\sin^2 x + \sin x = 0$ — $x = 0, \frac{3\pi}{2}, \pi$

3. $\cos^2 x + \cos x = 2$ — $x = 0$ 6.4

Approximate the solutions to each equation in degrees, for $0° \le x < 360°$.

4. $7\sin x + 2 = 3\sin x + 5$ — $x = 48.6°$, $x = 131.4°$

5. $5\tan^2 x - 4\tan x - 1 = 0$ — $x = 348.7°$, $x = 168.7°$, $x = 45°$, $x = 225°$ 6.5

6. Find the new coordinates of the point $(4, -7)$ when the axes are rotated through a 30° angle. $\left(2\sqrt{3} - \frac{7}{2}, -\frac{7\sqrt{3}}{2} - 2\right)$ 6.6

7. Determine the angle of rotation needed to transform $5x^2 + \sqrt{3}xy + 4y^2 + 3y = 9$ into an equation without an xy-term. 30°

Teacher's Resource Book
Practice—Chapter 6, p. 11
Enrichment—Chapter 6, p. 12

Additional Answers

29. $\theta = 30°$; $4(x')^2 + (y')^2 = 8$

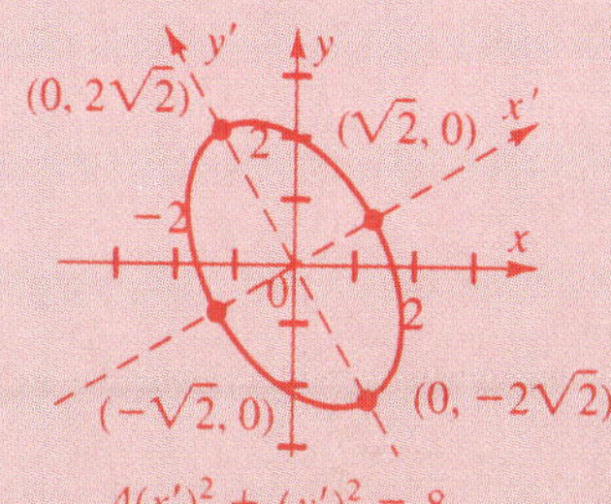

$4(x')^2 + (y')^2 = 8$

30. Substitute $x = x'\cos\alpha - y'\sin\alpha$ and $y = x'\sin\alpha + y'\cos\alpha$ into $x\cos\alpha + y\sin\alpha = d$. $\cos\alpha(x'\cos\alpha - y'\sin\alpha) + \sin\alpha(x'\sin\alpha + y'\cos\alpha) = d$; $x'(\cos^2\alpha + \sin^2\alpha) + (-y' + y')(\sin\alpha\cos\alpha) = d$; $x'(1) + 0(\sin\alpha\cos\alpha) = d$; $x' = d$

Application

A rainbow provides a vivid illustration of the concept of refraction. Snell's Law uses trigonometry to determine the effects on a ray of light as it passes from one medium to another.

See *Teacher's Resource Book* for *Application*, Chapter 6, p. 13.
See Teacher's Resource Book, *Technology*, p. 6.

APPLICATION: Rainbow

Did you know that the natural phenomenon known as a *rainbow* can only be observed when the sun is low in the sky and behind you and the rain clouds are ahead of you? Scientists and philosophers have been theorizing about rainbows since the time of Aristotle, about 2500 years ago.

Rainbows are caused by the reflection and refraction of the sun's rays striking drops of water. When an incident light ray passes at an angle from one medium (such as air) into another medium (such as water), part of the ray is *reflected* away from the boundary at an angle equal to the angle of incidence. The remainder of the ray enters the new medium, bending away from its original path. This bending, called *refraction*, arises because light rays travel more slowly in water than in a vacuum. For a ray of pure red light

$$\frac{\text{velocity in a vacuum}}{\text{velocity in water}} = 1.332$$

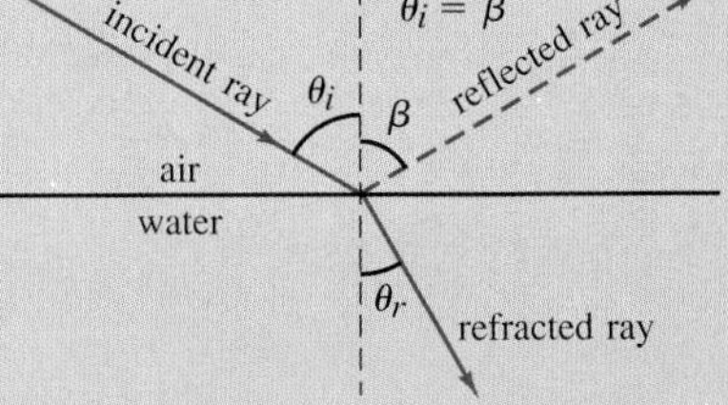

Thus, red light travels about 1.3 times as fast in a vacuum as in water. The ratio

$$\frac{\text{velocity in a vacuum}}{\text{velocity in medium}}$$

is called the *index of refraction* of the medium. The velocity of light in air and in a vacuum are virtually the same, so the index of refraction of air is one.

The path of the refracted ray depends upon both the angle of incidence and the indices of refraction of the two media involved. Willebrord Snell experimentally formulated this relationship as *Snell's Law* in about 1621.

$$n_1 \sin \theta_i = n_2 \sin \theta_r$$

where n_1 and n_2 are the indices of refraction of the first medium and the second medium, respectively, θ_i is the angle of incidence, and θ_r is the angle of refraction.

A raindrop acts like a tiny prism, reflecting back part of the sun's light and refracting the remainder. An incident ray enters the drop, rebounding (in part) inside the drop, and emerges from the drop, usually in an entirely different direction from which it began.

EXAMPLE The index of refraction of water for red light is 1.332. Use Snell's Law to find θ_{r_2}, the angle that refracted ray r_2 makes as it emerges from the raindrop shown below. The angle of incidence of ray i_1 is 30°.

$$n_1 \sin \theta_i = n_2 \sin \theta_r$$
$$1 \sin 30° = 1.332 \sin \theta_{r_1}$$
$$\frac{\sin 30°}{1.332} = \sin \theta_{r_1}$$
$$0.3754 = \sin \theta_{r_1}$$
$$\sin^{-1} 0.3754 = 22.05°$$

Thus, $\theta_{r_1} = 22.05°$.

It can be shown that θ_{r_1} equals θ_{i_2}

Therefore, $\theta_{i_2} = 22.05°$ also.

Use Snell's Law again to find θ_{r_2}

$$1.332 \sin \theta_{i_2} = 1 \sin \theta_{r_2}$$
$$1.332 \sin 22.05° = \sin \theta_{r_2}$$
$$0.5 = \sin \theta_{r_2}$$
$$\theta_{r_2} = \sin^{-1} 0.5 = 30°$$

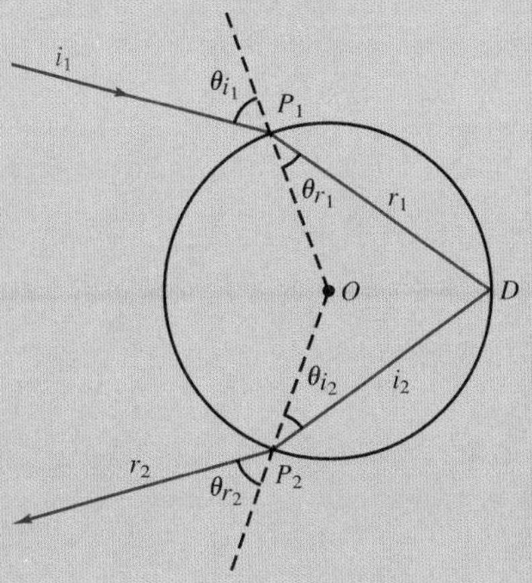

Refracted ray r_2 emerges from the raindrop at an angle of 30°.

The index of refraction of water for violet light is 1.344. So, if the refracted ray in the example had been a ray of violet light rather than red, the angle of refraction θ_{r_1} would have been a bit smaller. This would have caused r_2 to emerge from the raindrop in a slightly different direction from that of the red light ray. By sending red and violet rays into separate directions, the raindrop *disperses* the two colors. In a similar fashion, white light (a mixture of *all* colors) is dispersed by a raindrop into a full spectrum of colors. This dispersion of white light by raindrops, magnified many many times, results in the perception of a rainbow.

EXERCISES

1. An incident ray of violet light strikes a raindrop at an angle of 35°. Find the angle of refraction θ_{r_1}. 25.3°
2. In the example, θ_{i_1} and θ_{r_2} are both 30°. Is this a coincidence? Explain. See side column.
3. Suppose that $\theta_{r_2} = 21°$. Find the angle of incidence θ_{i_1} if the color of the light is red. 21°
4. Use your knowledge of geometry or trigonometry to explain why $\theta_{r_1} = \theta_{i_2}$. See side column.

Additional Answers

2. No, it is not a coincidence. If the drop is spherical, then θ_{i_1} will always equal θ_{r_2}.
4. Imagine a line segment drawn from point O to point D. By the law of reflection (at the boundary), $\angle P_1DO = \angle ODP_2 \cdot \overline{P_1O} = \overline{OP_2}$ since they are both radii of the circle. Thus, $\triangle P_1DO$ and $\triangle P_2DO$ are congruent. Since corresponding angles are congruent, $\theta_{r_1} = \theta_{i_2}$.

CHAPTER 6 SUMMARY AND REVIEW

Vocabulary

angle of rotation (273)
inverse Cosecant (261)
inverse Cosine (256)
inverse Cotangent (260)
inverse function (251)
inverse Secant (261)
inverse Sine (255)
inverse Tangent (260)
principal value (255)
rotation of axes (273)

Inverse Relations and Functions The inverse of $y = \sin x$ is $y = \sin^{-1} x$, and the inverse of $y = \cos x$ is $y = \cos^{-1} x$. 6.1

1. Find the inverse of the function $y = x - \frac{3}{4}$. $y = x + \frac{3}{4}$

Evaluate each expression in radians.

2. arccos 0.5 $\frac{\pi}{3} + 2\pi k$, where k is an integer and $\frac{5\pi}{3} + 2\pi k$, where k is an integer
3. arcsin 1 $\frac{\pi}{2} + 2\pi k$, where k is an integer

The Inverse Sine and Cosine Functions The appropriately restricted sine and cosine functions, $y = \operatorname{Sin} x$ and $y = \operatorname{Cos} x$, have inverses that are also functions. The inverse functions are denoted $y = \operatorname{Sin}^{-1} x$ (or $y = \operatorname{Arcsin} x$) and $y = \operatorname{Cos}^{-1} x$ (or $y = \operatorname{Arccos} x$). 6.2

$y = \operatorname{Sin}^{-1} x$	domain: $-1 \le x \le 1$	range: $-\frac{\pi}{2} \le y \le \frac{\pi}{2}$
$y = \operatorname{Cos}^{-1} x$	domain: $-1 \le x \le 1$	range: $0 \le y \le \pi$

4. Find the exact value of $\operatorname{Cos}^{-1}\left(-\frac{\sqrt{2}}{2}\right)$ in radians. $\frac{3\pi}{4}$
5. Find the exact value of cos (Arcsin 0.5). $\frac{\sqrt{3}}{2}$
6. Find the approximate value of $\operatorname{Sin}^{-1} 0.4765$ to the nearest degree. 28°

Other Inverse Trigonometric Functions If the domains of the tangent, cotangent, secant, and cosecant functions are appropriately restricted, their inverses are also functions. 6.3

$y = \operatorname{Tan}^{-1} x$	domain: all real numbers	range: $-\frac{\pi}{2} < y < \frac{\pi}{2}$
$y = \operatorname{Cot}^{-1} x$	domain: all real numbers	range: $0 < y < \pi$
$y = \operatorname{Sec}^{-1} x$	domain: $\lvert x\rvert \ge 1$	range: $0 \le y \le \pi$, $y \ne \frac{\pi}{2}$
$y = \operatorname{Csc}^{-1} x$	domain: $\lvert x\rvert \ge 1$	range: $-\frac{\pi}{2} \le y \le \frac{\pi}{2}$, $y \ne 0$

7. Find the exact value of $\operatorname{Tan}^{-1}(-1)$ in radians. $-\frac{\pi}{4}$

8. Find the approximate value of Arccsc (-4.321) to the nearest degree. $-13°$

9. Find the exact value of tan [Arcsec (-2)]. $-\sqrt{3}$

Solving Trigonometric Equations: Using Special Angles Many trigonometric equations can be solved using algebraic properties and known identities. 6.4

10. $x = \frac{\pi}{3}, x = \frac{5\pi}{3}$ 11. $x = \frac{\pi}{6}, x = \frac{5\pi}{6}, x = \frac{3\pi}{2}$

Find the exact solutions to each equation for the interval $0 \le x < 2\pi$.

10. $12 \sec x + 15 = 10 \sec x + 19$

11. $2 \sin^2 x + \sin x - 1 = 0$

12. $2 \csc^2 x + 3 \csc x - 2 = 0$ $x = \frac{7\pi}{6}, x = \frac{11\pi}{6}$

13. $\sin 2x + \cos x = 2 \cos x$ $x = \frac{\pi}{6}, x = \frac{\pi}{2}, x = \frac{5\pi}{6}, x = \frac{3\pi}{2}$

Trigonometric Equations: Approximate Solutions When the solutions to a trigonometric equation are not exact, a calculator can be used. 6.5

Find, to the nearest degree, the solutions to each equation for the interval $0° \le x < 360°$.

14. $8 \cos x + 9 = 2 \cos x + 11$ $x = 71°, x = 289°$

15. $5 \cos^2 x - \cos x - 1 = 0$ $x = 56°, x = 111°, x = 249°, x = 304°$

Rotations of Axes If the x- and y-axes are rotated through an angle θ, a new $x'y'$-coordinate system is obtained. The coordinates (x, y) and (x', y') of a point are related by the following two systems of equations: 6.6

$$x = x' \cos \theta - y' \sin \theta \qquad x' = x \cos \theta + y \sin \theta$$
$$y = x' \sin \theta + y' \cos \theta \qquad y' = -x \sin \theta + y \cos \theta$$

An equation of the form $Ax^2 + Bxy + Cy^2 + Dx + Ey + F = 0$, where $B \neq 0$, can be transformed into an equation without an xy-term if the axes are rotated through an angle $\theta = \frac{1}{2} \text{Cot}^{-1} \frac{A - C}{B}$.

Find the new coordinates of each point for the given angle of rotation.

16. $(22, -17)$, $45°$ $\left(\frac{5\sqrt{2}}{2}, -\frac{39\sqrt{2}}{2}\right)$

17. $(15, 19)$, $120°$ $\left(\frac{19\sqrt{3} - 15}{2}, \frac{-15\sqrt{3} - 19}{2}\right)$

The xy-axes are rotated through the given angle θ. Find an equation of the curve in the new coordinate system.

18. $5x^2 + 3xy - 6y^2 = 11$; $\theta = 90°$ $-6(x')^2 - 3x'y' + 5(y')^2 = 11$

19. $4x^2 + 5xy - 2y^2 = 9$; $\theta = 90°$ $4(y')^2 - 5x'y' - 2(x')^2 = 9$

Find the angle of rotation needed to transform the given equation into one without an xy-term. Find the new equation and draw its graph, showing both the xy- and $x'y'$-axes. See side column for graphs.

20. $x^2 + xy + y^2 = 6$ $45°$; $3(x')^2 + (y')^2 = 12$

21. $xy = 8$ $45°$; $(x')^2 - (y')^2 = 16$

Additional Answers

20.

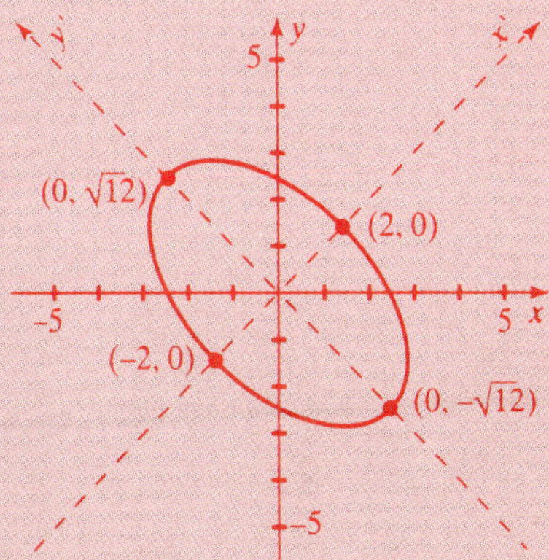

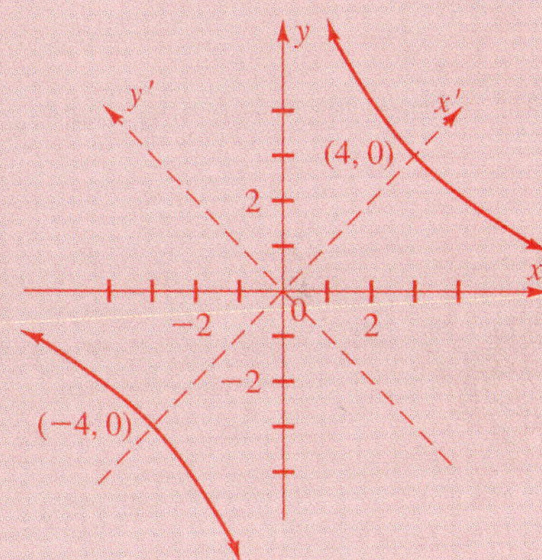

See *Teacher's Resource Book, Tests*, pp. 61–64.

Additional Answers

15.

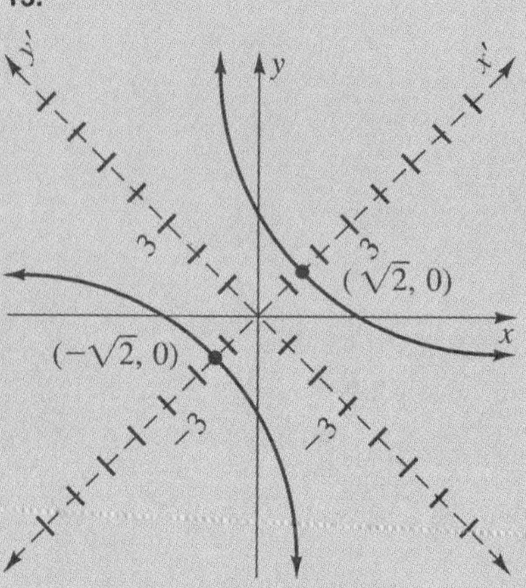

CHAPTER TEST

1. Find the inverse of the function $y = 10x - 7$. $y = \frac{x+7}{10}$
2. Evaluate $\arcsin\left(-\frac{\sqrt{2}}{2}\right)$ in degrees. $225° + (360 \cdot k)°$ or $315° + (360 \cdot k)°$, where k is an integer
3. Evaluate $\arccos \frac{\sqrt{3}}{2}$ in radians. $\pm\frac{\pi}{6} + 2\pi k$, where k is an integer
4. Find the exact value of $\text{Sin}^{-1}\, 1$ in radians. $\frac{\pi}{2}$
5. Find the exact value of $\text{Tan}^{-1}(-1)$ in degrees. $-45°$
6. Find the exact value of csc [Arccsc(-2)]. -2
7. Find the exact value of $\sin\left[\text{Arccos}\left(-\frac{1}{2}\right)\right]$. $\frac{\sqrt{3}}{2}$

Find the approximate value of each expression to the nearest degree.

8. Arcsin 0.6193 38°
9. $\text{Tan}^{-1}(-4.562)$ $-78°$

Find the exact solutions to each equation in radians, for the interval $0 \le x < 2\pi$.

10. $13 \tan x - 9 = 11 \tan x - 11$
$x = \frac{3\pi}{4}$, $x = \frac{7\pi}{4}$
11. $2 \cos^2 x - \cos x - 1 = 0$
$x = 0$, $x = \frac{2\pi}{3}$, $x = \frac{4\pi}{3}$

Find, to the nearest degree, the solutions to each equation for the interval $0° \le x < 360°$.

12. $7 \sin x - 9 = 2 \sin x - 7$
$x = 24°$, $x = 156°$
13. $3 \tan^2 x - 4 \tan x - 7 = 0$
$x = 67°$, $x = 135°$, $x = 247°$, $x = 315°$
14. If the x- and y-axes are rotated about the origin through an angle of 120°, what are the coordinates of the point $(-5, 7)$ in the new coordinate system?
15. Determine the angle of rotation that will transform $x^2 + 5xy + y^2 = 7$ into an equation without an xy-term. Find the new equation and draw its graph, showing both the xy- and $x'y'$-axes. See side column.

$7(x')^2 - 3(y')^2 = 14$

14. $\left(\frac{7\sqrt{3}+5}{2}, \frac{5\sqrt{3}-7}{2}\right)$

CHALLENGE

If $-1 \le x \le 1$, which of the following expressions is equal to sin (2 Arcsin x)? c

a. $2\sqrt{1-x^2}$ **b.** $2x(1-x^2)$ **c.** $2x\sqrt{1-x^2}$ **d.** $2x$

COLLEGE ENTRANCE EXAM REVIEW

In each item you are to compare a quantity in Column 1 with a quantity in Column 2. Write the letter of the correct answer from these choices:

A. The quantity in Column 1 is greater than the quantity in Column 2.
B. The quantity in Column 2 is greater than the quantity in Column 1.
C. The quantity in Column 1 is equal to the quantity in Column 2.
D. The relationship cannot be determined from the information given.

Notes: Information centered over both columns refers to one or both of the quantities being compared. A symbol that appears in both columns has the same meaning in each column. All variables represent real numbers. Most figures are not drawn to scale.

	Column 1	Column 2
1. D	$x^2 + 2x - 5$	6
	$\sin^2 x - 1 = 0,\ 0° \le x < 360°$	
2. A	x	$45°$
	$y = \text{Arcsin}\left(-\frac{1}{2}\right)$	
3. C	y	$-\frac{\pi}{6}$
	$y = \sin\left(\text{Arcsin}\,\frac{2}{5}\right)$	
4. A	y	$\frac{1}{5}$
	$5x - 2y = 8$	
5. B	The slope of the line	3
6. C	$\dfrac{9x^2 - 15x}{3x}$	$3x - 5$
7. D	$\dfrac{x}{y}$	xy
	$x - y = 7$ $2x + y = 5$	
8. B	y	x

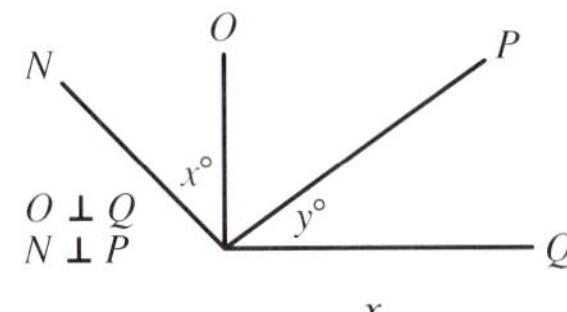

	Column 1	Column 2
9. A	y	$\frac{x}{2}$

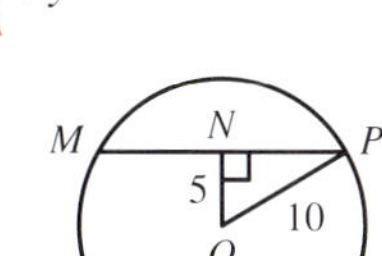

	Column 1	Column 2
10. B	$MN + NP$	$20\sqrt{3}$
11. D	The area of a circle with radius 2	The area of a circle with radius $\frac{x}{2}$
	$y = 4 \sin 2x$	
12. C	The maximum value of the function	The amplitude of the function
	$y = \text{Sin}^{-1} x$	
13. C	$\cos y$	$\sqrt{1 - x^2}$

283

See *Teacher's Resource Book, Tests*, pp. 65–68.

Additional Answers

6. $\sin 2\alpha \cot \alpha =$

$2 \sin \alpha \cos \alpha \cdot \dfrac{\cos \alpha}{\sin \alpha} = 2 \cos^2 \alpha;$

$1 + \cos 2\alpha = 1 + \cos^2 \alpha - \sin^2 \alpha = \cos^2 \alpha + \cos^2 \alpha = 2 \cos^2 \alpha$

$\therefore \sin 2\alpha \cot \alpha = 1 + \cos^2 \alpha$

7. 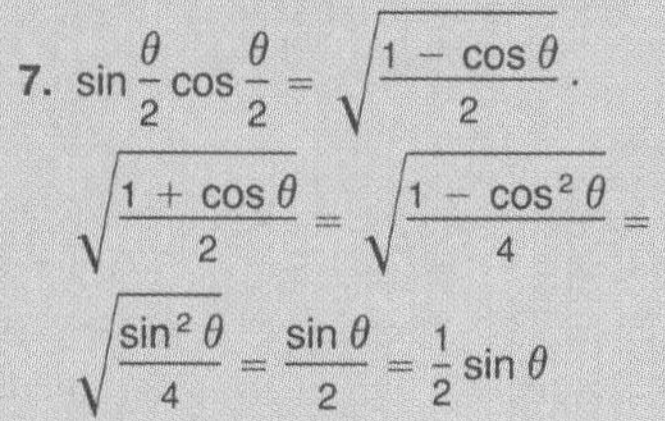

$\sin \frac{\theta}{2} \cos \frac{\theta}{2} = \sqrt{\frac{1 - \cos \theta}{2}} \cdot \sqrt{\frac{1 + \cos \theta}{2}} = \sqrt{\frac{1 - \cos^2 \theta}{4}} = \sqrt{\frac{\sin^2 \theta}{4}} = \frac{\sin \theta}{2} = \frac{1}{2} \sin \theta$

8. 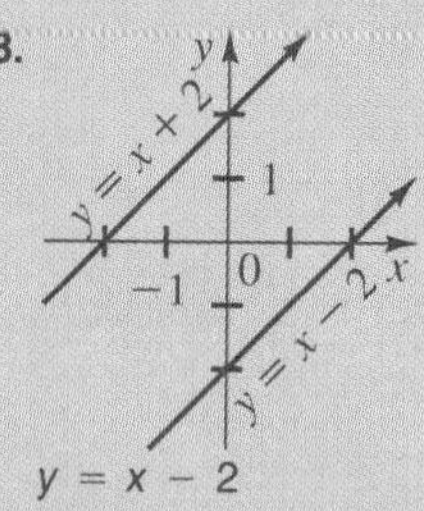

9. 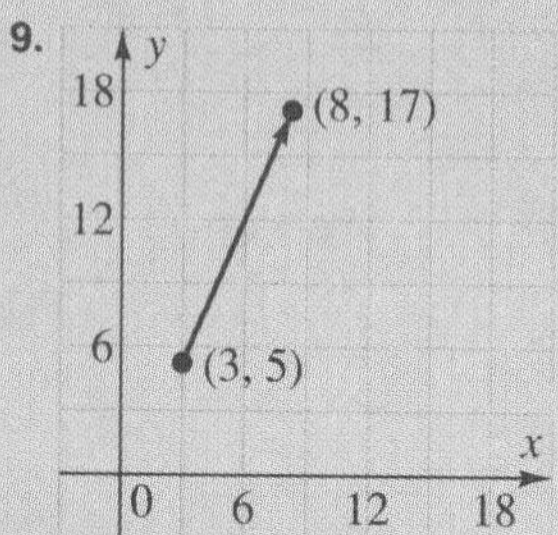

12. 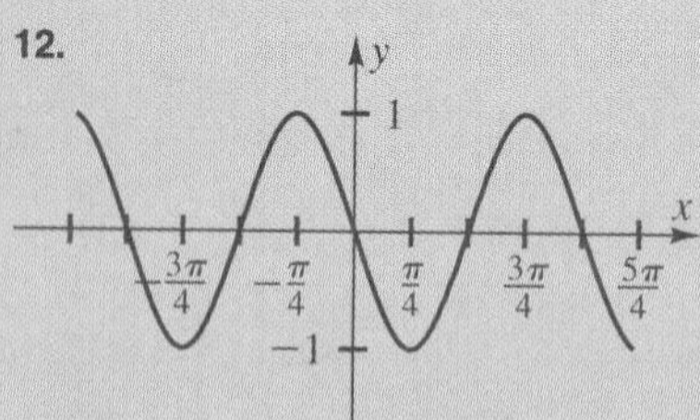

CUMULATIVE REVIEW

1. Express $\cos 52° + \cos 88°$ as a product. $2 \cos 70° \cos 18°$ — 5.6

Check graphically to see whether or not each equation is an identity. — 3.7

2. $\cos x \sin 2x - \sin x \cos 2x = \sin x$ identity

3. $\tan^2 \alpha + \cot^2 \alpha = \dfrac{\tan \alpha}{\cot \alpha} - \dfrac{\cot \alpha}{\tan \alpha}$ not an identity

Find the exact solutions to each equation in radians, for the interval $0 \le x \le 2\pi$. — 6.4

4. $\sqrt{2} \cos x - 1 = 0$ $\frac{\pi}{4}, \frac{7\pi}{4}$

5. $\sin^2 x - 5 \sin x = -4$ $\frac{\pi}{2}$

Prove each identity. See side column. — 5.4, 5.5

6. $\sin 2\alpha \cot \alpha = 1 + \cos 2\alpha$

7. $\sin \frac{\theta}{2} \cos \frac{\theta}{2} = \frac{1}{2} (\sin \theta)$

8. Find the inverse of the function $y = x + 2$. Graph the function and its inverse. See side column. — 6.1

9. Draw an arrow representing the vector from $A(3, 5)$ to $B(8, 17)$ and find its magnitude. $\|\overrightarrow{AB}\| = 13$; See side column. — 4.7

10. Evaluate $\sin (\alpha + \beta)$ if $\cos \alpha = \frac{4}{5}$ and $\tan \beta = \frac{5}{12}$, $0 < \alpha < \frac{\pi}{2}$, $0 < \beta < \frac{\pi}{2}$. $\frac{56}{65}$ — 5.2

11. Find the exact value of $\sin \left(\text{Arccos} \left(-\frac{\sqrt{3}}{2}\right)\right)$. $\frac{1}{2}$ — 6.2

12. Sketch the graph of the function $y = \cos \left(2x + \frac{\pi}{2}\right)$. See side column.

Approximate the solutions to each equation, where $0° \le x < 360°$. — 6.5

13. $2 \sin 3x - 1 = 0$ 10°, 50°, 130°, 170°, 250°, 290°

14. $2 \csc^2 x + 10 \csc x + 11 = 0$ 197°, 218°, 322°, 343°

15. Find the area, in acres, of a triangular plot of land with sides 475 ft, 628 ft, and 710 ft. (*Note*: 1 acre = 43,560 ft^2) 3.36 — 4.6

16. Express $\dfrac{\tan 85° + \tan 52°}{1 - \tan 85° \tan 52°}$ in terms of a single tangent. $\tan 137°$ — 5.3

17. $-\frac{7\pi}{12}$ is an angle in standard position. Find the measure of its reference angle θ'. $\frac{5\pi}{12}$ — 1.8

18. Express $2 \cos \frac{\pi}{6} \sin \frac{\pi}{12}$ as a sum or difference. $\sin \frac{\pi}{4} - \sin \frac{\pi}{12}$ — 5.6

OVERVIEW • Chapter 7

SUMMARY

The chapter begins with a development of the relationship between the Cartesian coordinate system and the polar coordinate system. Students convert coordinates from one system to the other before graphing polar equations of a circle, limaçon, cardioid, and rose. Complex numbers are introduced and the operations of addition, subtraction, multiplication, and division for complex numbers are defined. Expressing complex numbers in polar form, as well as changing the polar form to rectangular form, are introduced and used to multiply and divide complex numbers expressed in polar form. Finally, DeMoivre's theorem and the Complex Roots Theorem are developed and used to evaluate powers and roots of complex numbers and to solve and graph equations of the form $x^4 + 81 = 0$.

CHAPTER OBJECTIVES

- To define polar coordinates
- To convert rectangular coordinates to polar coordinates and polar coordinates to rectangular coordinates
- To graph polar equations
- To introduce the complex numbers
- To define addition and subtraction of complex numbers
- To find the product and quotient of two complex numbers
- To express complex numbers in polar form
- To multiply and divide complex numbers expressed in polar form
- To use DeMoivre's theorem to evaluate powers of complex numbers
- To find the *n* distinct roots of a complex number

CHAPTER HIGHLIGHTS

The *theme* of Chapter 7 is fractals. Special features in the chapter include a description of some of the computer technology that was used in the development of the Mandelbrot set.

APPLICATIONS

The applications for complex numbers are of interest in that they illustrate the practical uses for even the most abstract mathematical concepts. Many of the real-world problems in Chapter 7 are from physics. In particular, imaginary numbers are used in problems involving electric circuitry to relate measures of resistance, reactance, and impedance.

TECHNOLOGY

Calculator

Calculators are helpful for converting from polar coordinates to rectangular coordinates and vice versa. Many calculators possess a feature that performs these operations.

Computer

Programs which compute the product of complex numbers in standard and polar form are presented. The technology feature on fractals exposes students to a program that uses the complex plane to generate the Mandelbrot set.

RESOURCES

Teacher's Resource Book

- Teaching Aid 7
- Transparencies 13 and 14

ASSIGNMENT GUIDE Meeting Student Needs

STUDENT TEXT				TEACHER'S RESOURCE BOOK	
Chapter Content	**Basic**	**Average**	**Enriched**	**P**	**E**
7.1 Polar Coordinates	D: 290/1–35 odd	D: 290/7–41 odd, 45	D: 290/7–45 odd	1	2
7.2 Graphs of Polar Equations	D: 296/1–23 odd R: 290/2, 14, 24, 30	D: 296/1, 3, 9–29 odd, 33 R: 290/8, 16, 26, 30	D: 296/1, 5, 11–35 odd R: 290/10, 16, 28, 32	3	4
7.3 Sums and Differences of Complex Numbers	D: 301/1–37 odd, 57 R: 296/2, 4, 14	D: 301/11–49 odd, 57 R: 296/2, 4, 16	D: 301/13, 19–59 odd R: 296/2, 6, 16	5	6
7.4 Products and Quotients of Complex Numbers	D: 305/1–39 odd, 57 R: 301/4, 22, 36 307/TY	D: 305/13–51 odd, 57 R: 301/14, 24, 42 307/TY	D: 305/15–59 odd R: 301/20, 34, 44 307/TY	7	8
7.5 Complex Numbers in Polar Form	D: 311/1–25 odd, 41 R: 305/2, 20, 44	D: 311/5, 9, 15–35 odd, 41 R: 305/14, 24, 32	D: 311/7, 11, 19–43 odd R: 305/18, 28, 36	9	10
7.6 Multiplying and Dividing Complex Numbers in Polar Form	D: 315/1–19 odd R: 311/2, 14, 22	D: 316/5, 9, 15–29 odd R: 311/8, 16, 24	D: 316/9, 15–33 odd R: 311/12, 20, 30	11	12
7.7 DeMoivre's Theorem	D: 320/1–19 odd, 29 R: 315/2, 12, 18	D: 321/7–25 odd, 29 R: 316/6, 16, 20	D: 321/11–29 odd R: 316/10, 14, 20	13	14
7.8 Roots of Complex Numbers	Omit	D: 325/9–35 odd R: 321/6, 20, 24 326/TY	D: 325/9, 13, 17–39 odd R: 321/12, 22, 24 326/TY	15	16

D = Daily R = Review TY = Test Yourself P = Practice E = Enrichment

	STUDENT TEXT				TEACHER'S RESOURCE BOOK	
Review and Testing	Test Yourself	307, 326	College Ent. Exam Rev.	331	Tests	
	Chapter Sum. and Rev.	328	Maintaining Skills	332	• Quizzes	69–72
	Chapter Test	330			• Chapter Test (Form A)	73–74
					• Chapter Test (Form B)	75–76
Special Features	Trigonometry and Pascal's Triangle	291	Historical Note	312	Applications—Chapter 7	17
			Challenge	317	Critical Thinking	6
	Extra	297	Extra	321	Alg. and Geom. Review	25–28
	Biography	302	Technology	327	Technology	7

7 Complex Numbers

Benoit B. Mandelbrot developed the concept of fractals, which are curves that have sections that are small-scale replicas of the whole. A computer can use complex numbers to generate pictures of fractals.

BACKGROUND

"Nature exhibits not simply a higher degree but an altogether different level of complexity." This quote is attributed to Benoit B. Mandelbrot, whose work with fractals deals with complex natural shapes such as coastlines and wood grains without reducing them to the simple forms of Euclidean geometry.

LESSON PLAN

Vocabulary
Polar axis
Polar coordinates
Polar coordinate system
Pole

Materials/Manipulatives
Calculators

BACKGROUND

In the Preview, the distance formula for finding the distance between the origin and a point on the rectangular coordinate plane is reviewed.

7.1 Polar Coordinates

Objectives: To define polar coordinates
To convert rectangular coordinates to polar coordinates and polar coordinates to rectangular coordinates

Points on the Cartesian, or rectangular, coordinate plane are represented by ordered pairs of real numbers (x, y).

Preview

The distance d between the origin (0, 0) and any point with *rectangular coordinates* (x, y) is given by the distance formula

$$d = \sqrt{(x - 0)^2 + (y - 0)^2} = \sqrt{x^2 + y^2}$$

Find the distance from the origin to each point.

1. $(5, -9)$ $\sqrt{106}$ **2.** $(0, 4)$ 4 **3.** $(-1, -1)$ $\sqrt{2}$ **4.** $(-4, 0)$ 4

5. $(-\sqrt{7}, 1)$ $2\sqrt{2}$ **6.** $(-1, \sqrt{11})$ $2\sqrt{3}$ **7.** $(-\sqrt{2}, \sqrt{3})$ $\sqrt{5}$ **8.** $(3\sqrt{5}, 2\sqrt{7})$ $\sqrt{73}$

There is another way to locate points in the plane. In the **polar coordinate system,** the reference system consists of a point O, called the **pole** (or origin), and a ray called the **polar axis,** with O its initial point.

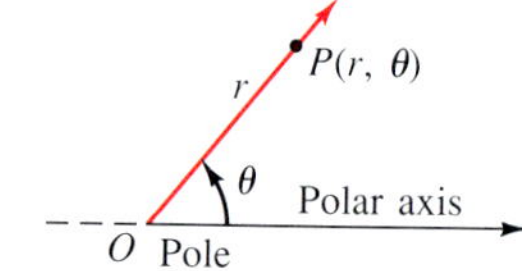

The **polar coordinates** of a point P in the polar coordinate system are an ordered pair (r, θ), where $|r|$ is the distance from the pole to P and θ is the measure of the angle from the polar axis to ray OP. The pole, O, has polar coordinates $(0, \theta)$, where θ is arbitrary. Angle θ can be measured in degrees or in radians.

EXAMPLE 1 **Graph point Q with polar coordinates (2, 30°).**

Sketch a 30° angle in standard position. Use the polar axis as its initial side.

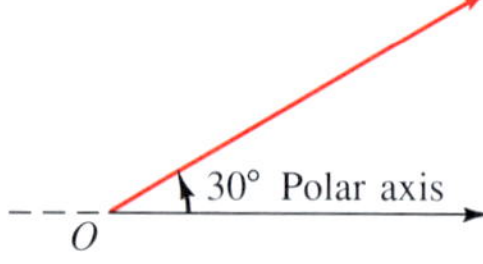

Then locate on the ray the point that is 2 units from the pole.

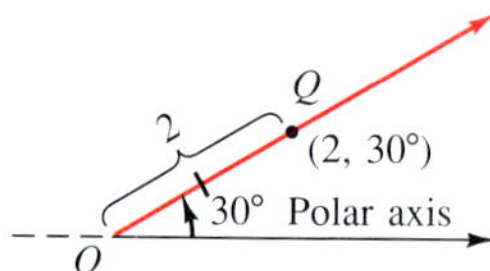

Both r and θ can be either positive or negative. If r is *negative*, the point (r, θ) is $|r|$ units from the pole on the ray opposite the terminal side of the angle with measure θ. If θ is *negative*, the angle is measured in a clockwise direction from the polar axis. If θ is in standard position, then a point P with coordinates (r, θ) is located as follows, where ray OT makes an angle θ with the polar axis.

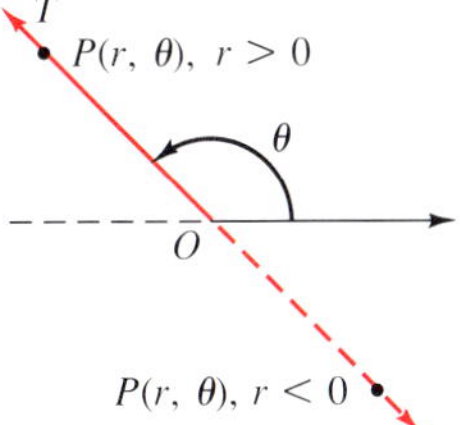

If $r > 0$, P lies on ray OT at a distance r from O.
If $r < 0$, P lies on the ray opposite ray OT at a distance $|r|$ from O.
If $r = 0$, P is the pole O.

In general, $(-r, \theta) = (r, \theta \pm 180°)$, if θ is in degrees, or $(-r, \theta) = (r, \theta \pm \pi)$, if θ is in radians.

EXAMPLE 2 **Graph the point D with polar coordinates $(-3, -60°)$.**

Draw $\theta = -60°$ in standard position. Since r is negative, locate the point $|-3|$ units from the pole on the ray *opposite* the terminal side of the angle. Note that D can also be represented by $(3, -60° + 180°) = (3, 120°)$ or by $(3, -60° - 180°) = (3, -240°)$.

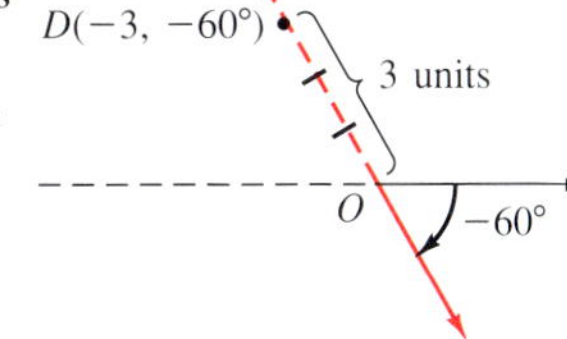

In the rectangular coordinate system, each point is represented by a unique ordered pair. In the polar system, points may be named with an infinite number of pairs of polar coordinates. For example, the polar coordinates $(3, 105°)$, $(-3, -75°)$, $(-3, 285°)$, and $(3, -255°)$ all represent point P.

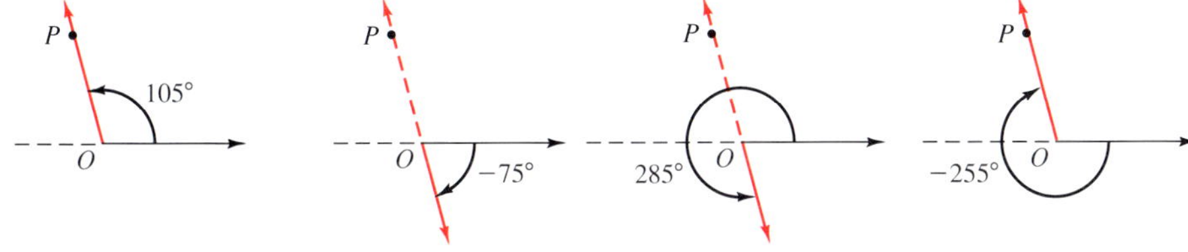

Generally, any point (r, θ) can be represented by $(r, \theta + n \cdot 360°)$ or $(r, \theta + 2\pi n)$, where n is an integer.

EXAMPLE 3 **If $-2\pi \le \theta \le 2\pi$, name four pairs of polar coordinates for point R.**

Point R is represented by

$\left(5, \frac{3\pi}{4}\right)$, $\left(-5, -\frac{\pi}{4}\right)$, $\left(-5, \frac{7\pi}{4}\right)$, or $\left(5, -\frac{5\pi}{4}\right)$.

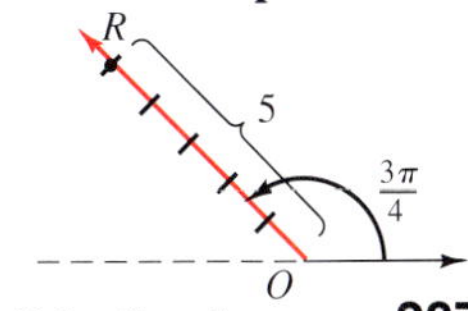

TEACHING SUGGESTIONS

- Point out that θ in (r, θ) can be measured in degrees or radians.
- Emphasize that a point in the rectangular coordinate system is represented by a unique ordered pair of coordinates while an unlimited number of polar coordinates can represent the same point in the polar coordinate system.
- Encourage students to use calculators for appropriate problems.

Critical Thinking

Analysis Ask students to compare and contrast the rectangular and polar coordinate systems. When comparing the rectangular and polar coordinate systems, students should note that: both systems provide means of locating points in a plane; both systems contain an origin; the point in each system is represented by ordered pairs of real numbers. When contrasting the systems, students should note that the rectangular coordinate system consists of two axes while the polar coordinate system consists of an origin, called the pole, and a ray called the polar axis; both coordinates of a point in the rectangular system represent distances while in the polar system one coordinate represents distance and the other represents an angle measure.

CHALKBOARD EXAMPLES

- **For Example 1**
  1. Graph the point P that has polar coordinates $\left(3, \frac{\pi}{4}\right)$.

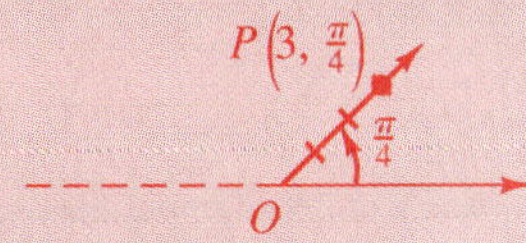

- **For Example 2**

  Graph the point P with the given coordinates.

  2. $P(-2, -30°)$

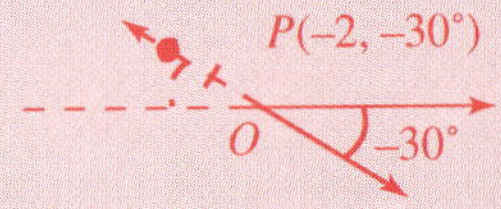

  3. $P(-3, -45°)$

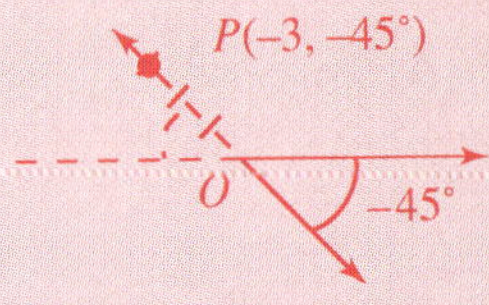

- **For Example 3**

  If $-2\pi \le \theta \le 2\pi$, name four different pairs of polar coordinates that represent point T in each figure.

  4.

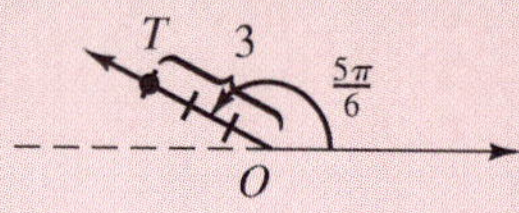

  $\left(3, \frac{5\pi}{6}\right), \left(3, -\frac{7\pi}{6}\right), \left(-3, -\frac{\pi}{6}\right), \left(-3, \frac{11\pi}{6}\right)$

  5.

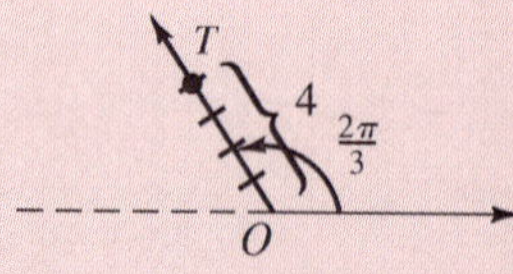

  $\left(4, \frac{2\pi}{3}\right), \left(4, -\frac{4\pi}{3}\right), \left(-4, -\frac{\pi}{3}\right), \left(-4, \frac{5\pi}{3}\right)$

If a rectangular coordinate system is superimposed upon a polar coordinate system and the units are assumed to be the same, then for any point P, a correspondence can be established between its rectangular coordinates (x, y) and its polar coordinates (r, θ).

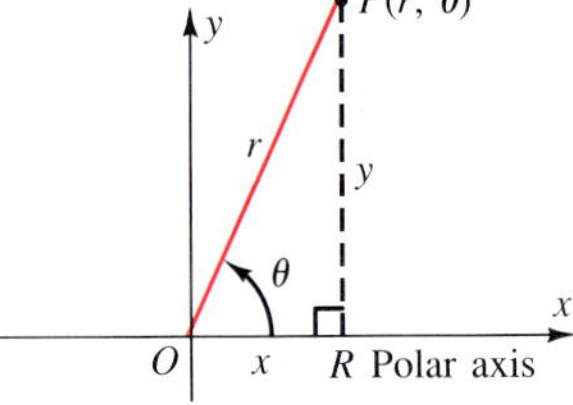

Notice that the two origins coincide. Also, the positive x-axis of the rectangular system coincides with the polar axis of the polar system. In triangle OPR, $\cos\theta = \frac{x}{r}$ and $\sin\theta = \frac{y}{r}$. Therefore, $x = r\cos\theta$ and $y = r\sin\theta$.

> The rectangular coordinates (x, y) of a point P that has polar coordinates (r, θ) can be found using the following conversion formulas:
>
> $$x = r\cos\theta \qquad y = r\sin\theta$$

EXAMPLE 4 **Find the rectangular coordinates of the following points:**

a. $A(4, -50°)$ **b.** $B\left(-5, \frac{2\pi}{3}\right)$

a. $x = r\cos\theta$ $\quad$ $y = r\sin\theta$

$= 4\cos(-50°)$ $\quad$ $= 4\sin(-50°)$

$= 2.57$ *To the nearest hundredth* $= -3.06$

The rectangular coordinates of A are $(2.57, -3.06)$.

b. $x = r\cos\theta$ $\quad$ $y = r\sin\theta$

$= -5\cos\frac{2\pi}{3} = \frac{5}{2}$ $\quad$ $= -5\sin\frac{2\pi}{3} = -\frac{5\sqrt{3}}{2}$

The rectangular coordinates of B are $\left(\frac{5}{2}, -\frac{5\sqrt{3}}{2}\right)$.

When a point is identified with rectangular coordinates (x, y), a pair of polar coordinates (r, θ) for the point can be found using the Pythagorean theorem and the Arctangent function. The principal values of the inverse tangent relation are between $-\frac{\pi}{2}$ and $\frac{\pi}{2}$, so if the point is in the first or fourth quadrant, $\theta = \text{Arctan}\,\frac{y}{x}$. If the point is in the second or third quadrant, add π radians (or 180°) to $\text{Arctan}\,\frac{y}{x}$.

For $x < 0$, $\theta = \text{Arctan}\,\frac{y}{x} + \pi$

A pair of polar coordinates (r, θ) of a point named by the rectangular coordinates (x, y) can be found using the following formulas:

$$r = \sqrt{x^2 + y^2}$$
$$\theta = \text{Arctan}\,\frac{y}{x}, \text{ if } x > 0$$
$$\theta = \text{Arctan}\,\frac{y}{x} + \pi \left(\text{or Arctan}\,\frac{y}{x} + 180°\right), \text{ if } x < 0$$
$$\theta = \frac{\pi}{2} \text{ (or } 90°\text{), if } x = 0 \text{ and } y > 0$$
$$\theta = \frac{3\pi}{2} \text{ (or } 270°\text{), if } x = 0 \text{ and } y < 0$$
$$\theta = \text{any real number, if } x = 0 \text{ and } y = 0$$

EXAMPLE 5 **Find polar coordinates of point C with rectangular coordinates $(-2, 2)$.**

$$r = \sqrt{(-2)^2 + 2^2} = \sqrt{8} = 2\sqrt{2} \qquad r = \sqrt{x^2 + y^2}$$

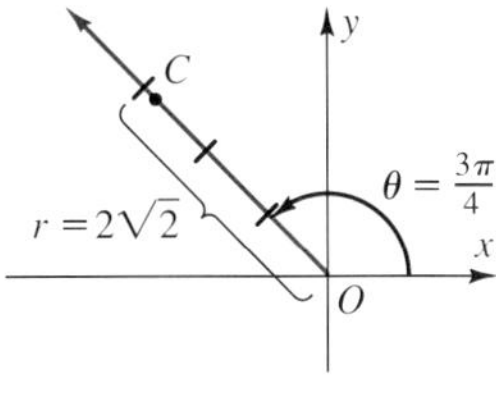

Since $x < 0$, $\theta = \text{Arctan}\left(\frac{2}{-2}\right) + \pi$
$= \text{Arctan}\,(-1) + \pi$
$= -\frac{\pi}{4} + \pi = \frac{3\pi}{4}$

Polar coordinates of C are $\left(2\sqrt{2}, \frac{3\pi}{4}\right)$, or $(2\sqrt{2}, 135°)$.

As you know, a point represented by a unique pair of rectangular coordinates (x, y) can be represented by an infinite number of pairs of polar coordinates. When the equation $r = \sqrt{x^2 + y^2}$ is used, a *nonnegative* value of r is always obtained. It is possible to represent the same point using $r = -\sqrt{x^2 + y^2}$ and an appropriate value of θ. To illustrate, show that point C in Example 5 could also be represented by the polar coordinates $(-2\sqrt{2}, -45°)$.

You can change equations from rectangular to polar form and the reverse.

EXAMPLE 6 **a. Express $y = 5$ in polar-coordinate form.**
b. Express $r = 3 \cos \theta$ in rectangular form.

a.
$$y = 5$$
$$r \sin \theta = 5 \quad \textit{Replace } y \textit{ with } r \sin \theta.$$
$$r = \frac{5}{\sin \theta}$$

- **For Example 4**
  Find the rectangular coordinates of the following points.
  6. $Q\left(-3, \frac{\pi}{4}\right)$ $\left(-\frac{3\sqrt{2}}{2}, -\frac{3\sqrt{2}}{2}\right)$
  7. $S\left(2, \frac{5\pi}{6}\right)$ $(-\sqrt{3}, 1)$
- **For Example 5**
  Find the polar coordinates for each point with the given rectangular coordinates.
  8. $(-3, 3)$ $(3\sqrt{2}, 135°)$, or $\left(3\sqrt{2}, \frac{3\pi}{4}\right)$
  9. $(4, 5)$ $(\sqrt{41}, 51°)$, or $\left(\sqrt{41}, \frac{17\pi}{60}\right)$
- **For Example 6**
  10. Express $x = 3$ in polar coordinate form. $r = 3 \sec \theta$
  11. Express $r = 2 \sin \theta$ in rectangular form. $x^2 + y^2 = 2y$

Common Errors

- Some students have difficulty plotting (r, θ) if one of the coordinates, or both, is negative. Emphasize that when $r < 0$, r is measured on the ray opposite the terminal side of θ and that when $\theta < 0°$, or $\theta < 0$ radians, θ is measured in a clockwise direction from the pole.
- When changing from rectangular coordinates to polar coordinates, some students have difficulty finding θ. Refer those students to the formulas on text p. 289.
- See *Teacher's Resource Book* for additional remediation.

LESSON FOLLOW-UP

Discussion

Give an example of something that functions on the principle of polar coordinates. Answers may vary; for example, radar (origin is your position).

Assignment Guide

See p. 284B for assignments.

Lesson Quiz

1. Plot $(-2, -120°)$.

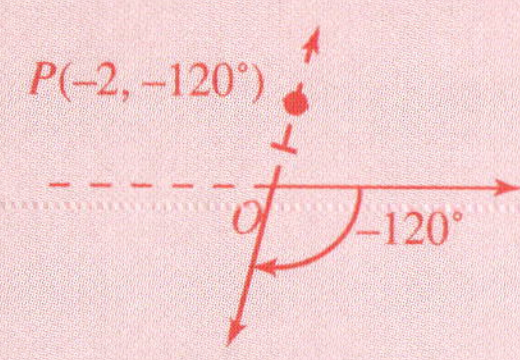

Find the rectangular coordinates of each point with the given polar coordinates.

2. $M(4, 150°)$ $(-3.46, 2)$
3. $J(-1, -50°)$ $(-0.64, 0.77)$

Find the polar coordinates of each point with the given rectangular coordinates.

4. $B(\sqrt{3}, 1)$ $(2, 30°)$, or $\left(2, \frac{\pi}{6}\right)$
5. $A(-2, 2)$ $(2\sqrt{2}, 135°)$, or $\left(2\sqrt{2}, \frac{3\pi}{4}\right)$
6. Express $x^2 = 6y$ in polar coordinate form. $r = 6 \sec \theta \tan \theta$
7. Express $r = 7 \sec \theta$ in rectangular form. $x = 7$

Enrichment

Find the equation of the line through (0,0) and $\left(3, \frac{2\pi}{3}\right)$ in polar coordinate form. $\left(3, \frac{2\pi}{3}\right) = (-1.5, 2.6)$ in rectangular form; $y - 2.6 = -\frac{2.6}{1.5}(x + 1.5)$; $y = -1.73x$; $r \sin \theta = -1.73 r \cos \theta$; $\sin \theta = -1.73 \cos \theta$; $\tan \theta = -1.73$; $\theta = \frac{2\pi}{3}$

b.

$$r = 3 \cos \theta$$

$$r^2 = 3 r \cos \theta \qquad \textit{Multiply both sides by } r.$$

$$x^2 + y^2 = 3x \qquad \textit{Replace } r^2 \textit{ with } x^2 + y^2 \textit{ and } r \cos \theta \textit{ with } x.$$

CLASS EXERCISES

Estimate polar coordinates for each point T.

1.

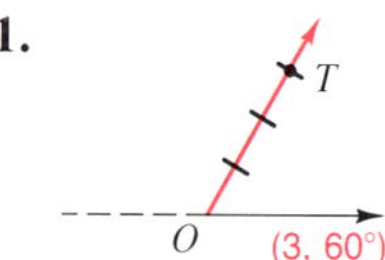

(3, 60°)

2.

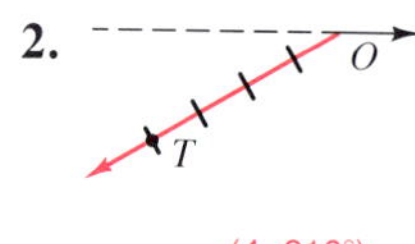

(4, 210°)

3.

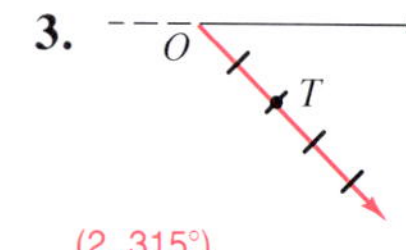

(2, 315°)

4.

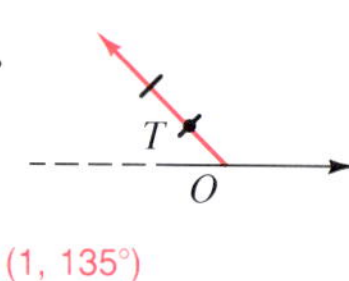

(1, 135°)

Find the rectangular coordinates of each point.

5. $(3, 90°)$ (0, 3) **6.** $(5, 270°)$ (0, −5) **7.** $(-3, 180°)$ (3, 0) **8.** $(-1, 0°)$ (−1, 0)

Find polar coordinates for each point. Give θ to the nearest degree.

9. $(5, -5)$ $(5\sqrt{2}, 315°)$ **10.** $(-6, 6)$ $(6\sqrt{2}, 135°)$ **11.** $(3, 4)$ (5, 53°) **12.** $(-4, 3)$ (5, 143°)

PRACTICE EXERCISES

Plot each point with the given polar coordinates. See page 488.

A

1. $(3, 45°)$ **2.** $(2, 60°)$ **3.** $(-2, 270°)$ **4.** $(-1, 180°)$

5. $(4, -420°)$ **6.** $(1, -435°)$ **7.** $(0, 300°)$ **8.** $\left(-4, -\frac{5\pi}{6}\right)$

9. $\left(-3, -\frac{2\pi}{3}\right)$ **10.** $\left(0, \frac{7\pi}{4}\right)$ **11.** $\left(3, \frac{5\pi}{2}\right)$ **12.** $\left(3, \frac{\pi}{2}\right)$

If $-360° \leq \theta \leq 360°$, give four pairs of polar coordinates for each point P.

13. **14.** **15.** **16.**

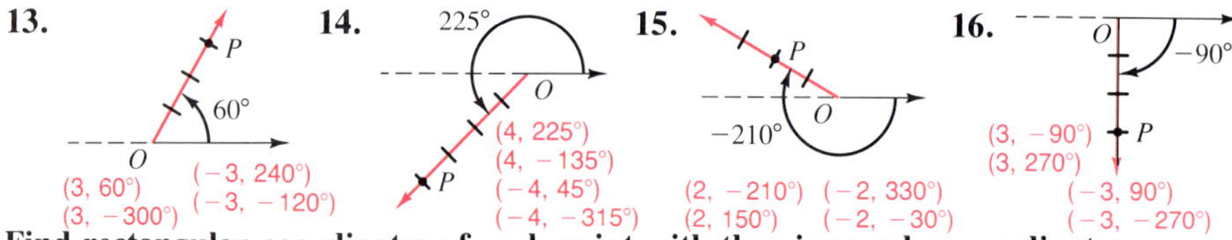

Find rectangular coordinates of each point with the given polar coordinates.

17. $(0, 90°)$ (0, 0) **18.** $(0, 180°)$ (0, 0) **19.** $(2, 130°)$ (−1.29, 1.53)

20. $(3, 80°)$ (0.52, 2.95) **21.** $(3, -70°)$ (1.03, −2.82) **22.** $(2, -115°)$ (−0.85, −1.8

Find polar coordinates for each point with the given rectangular coordinates.

B

23. $(0, -1)$ (1, 270°) **24.** $(0, 2)$ (2, 90°) **25.** $(2, -2)$ $(2\sqrt{2}, 315°)$

26. $(-2, -2)$ $(2\sqrt{2}, 225°)$ **27.** $(-3, \sqrt{3})$ $(2\sqrt{3}, 150°)$ **28.** $(3, -\sqrt{3})$ $(2\sqrt{3}, 330°)$

Write each equation in polar-coordinate form.

29. $x = 4$ $r = 4 \sec \theta$ **30.** $x^2 + y^2 = 49$ $r = 7$ **31.** $x^2 + y^2 = 36$ $r = 6$ **32.** $5x - 2y = 1$ $r = \frac{1}{5 \cos \theta - 2 \sin \theta}$

Write each equation in rectangular-coordinate form.

33. $r = 10 \sin \theta$ $x^2 + y^2 = 10y$ **34.** $r = -6 \cos \theta$ $x^2 + y^2 = -6x$ **35.** $r = 6 \csc \theta$ $y = 6$ **36.** $r = 3 \sec \theta$ $x = 3$

Write each equation in polar-coordinate form.

C **37.** $4x^2 + 4y^2 = -3y$ $r = -\frac{3}{4} \sin \theta$ **38.** $3x^2 + 3y^2 = -7x$ $r = -\frac{7}{3} \cos \theta$

39. $y^2 + (x - 6)^2 = 36$ $r = 12 \cos \theta$ **40.** $x^2 + (y - 5)^2 = 25$ $r = 10 \sin \theta$

Write each equation in rectangular-coordinate form.

41. $r = \frac{1}{2 + \sin \theta}$ $4x^2 + 3y^2 + 2y = 1$ **42.** $r = \frac{1}{3 - \cos \theta}$ $8x^2 + 9y^2 - 2x = 1$

43. $r = 6 \tan \theta \sec \theta$ $x^2 = 6y$ **44.** $r = 8 \cot \theta \csc \theta$ $y^2 = 8x$

Applications

45. Physics The radiation pattern of an antenna is given by the equation $r = 50(1 + \cos \theta)$. Write the equation in rectangular-coordinate form. $x^2 + y^2 = 50\sqrt{x^2 + y^2} + 50x$

46. Physics A satellite has an orbit with the equation $r = \frac{200{,}000}{43 - 2 \cos \theta}$. Write the equation in rectangular-coordinate form. $1845x^2 + 1849y^2 - 800{,}000x = 40{,}000{,}000{,}000$

TRIGONOMETRY AND PASCAL'S TRIANGLE

Pascal's triangle, a special triangular array of numbers, appears in algebra, in set theory, and in probability. When $\tan (n\theta)$ is expressed in terms of powers of $\tan \theta$, Pascals's triangle appears again.

	Coefficients
$\tan \theta = \frac{\mathbf{1} \tan^1 \theta}{\mathbf{1}}$	1 1
$\tan 2\theta = \frac{\mathbf{2} \tan^1 \theta}{\mathbf{1} - \mathbf{1} \tan^2 \theta}$	1 2 1
$\tan 3\theta = \frac{\mathbf{3} \tan^1 \theta - \mathbf{1} \tan^3 \theta}{\mathbf{1} - \mathbf{3} \tan^2 \theta}$	1 3 3 1
$\tan 4\theta = \frac{\mathbf{4} \tan^1 \theta - \mathbf{4} \tan^3 \theta}{\mathbf{1} - \mathbf{6} \tan^2 \theta + \mathbf{1} \tan^4 \theta}$	1 4 6 4 1

Note the following patterns: Odd powers of $\tan \theta$ appear in the numerator. Even powers of $\tan \theta$ appear in the denominator. The signs alternate in numerator and denominator. Pascal's triangle is formed by alternately selecting coefficients from the denominator and numerator. For $n = 5$, express $\tan (n\theta)$ and the coefficients of Pascal's triangle. 1, 5, 10, 10, 5, 1; $\tan 5\theta = \frac{5 \tan^1 \theta - 10 \tan^3 \theta + 1 \tan^5 \theta}{1 - 10 \tan^2 \theta + 5 \tan^4 \theta}$

Teacher's Resource Book

Practice—Chapter 7, p. 1

Enrichment—Chapter 7, p. 2

LESSON PLAN

Vocabulary
Cardioid
Limaçon
Rose

Materials/Manipulatives
Polar coordinate graph paper

BACKGROUND

In the Preview, types of microphones are discussed. The graphs of their pick-up patterns are similar to graphs of cardioids and lemniscates.

Additional Answers

Preview

1. 90° 60° 180° 0° 270°

2. 90° 45° 180° 0° 270°

3. 90° 180° 0° 270°

7.2 Graphs of Polar Equations

Objective: To graph polar equations

Graphs of polar equations depict many everyday occurrences, such as the manner in which microphones pick up sound.

Preview

Microphones are designed to pick up sound in a variety of patterns. *Uni*directional microphones pick up sound from one direction, whereas *bi*directional microphones pick up sound from two directions.

The polar-coordinate diagrams below show the pickup patterns of a unidirectional microphone and a bidirectional microphone. In each case, the microphone is placed at the pole. The two parts, or *leaves*, of the pickup pattern in the second diagram are arranged as far away from one another as possible in order to avoid interference.

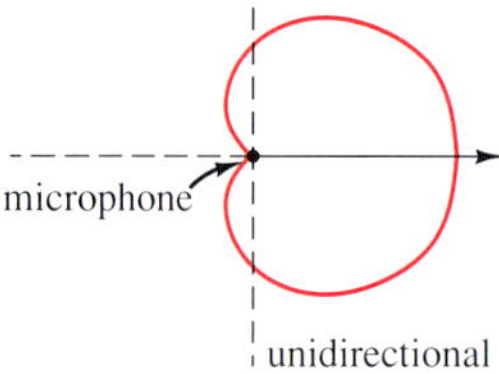

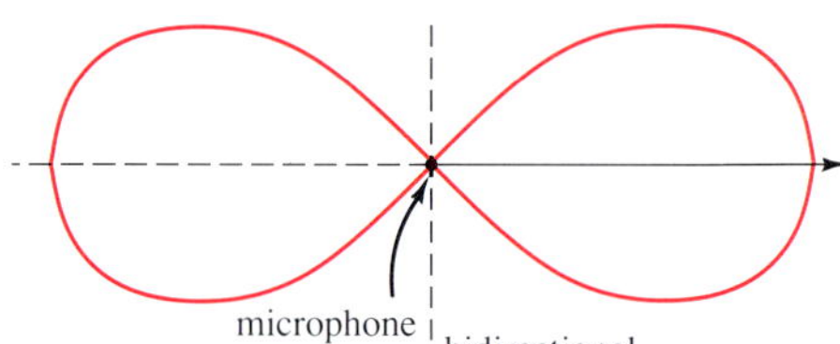

Draw a pickup pattern for each type of microphone. Assume the element of the microphone is placed at the pole to avoid interference.
See side column.

1. A *tri*directional microphone
2. A *quadri*directional microphone
3. An *omni*directional microphone (from all directions)

A *polar equation* involves polar coordinates, and a *polar graph* is a graph of the set of all points (r, θ) that satisfy a given polar equation.

EXAMPLE 1 **Graph $\theta = 60°$.**

In the equation $\theta = 60°$, the variable r can take on any real value. Thus, the graph is a straight line through the origin that forms an angle of 60° with the polar axis. Note that on this line the points for which $r > 0$ lie in quadrant I, while the points for which $r < 0$ are in quadrant III.

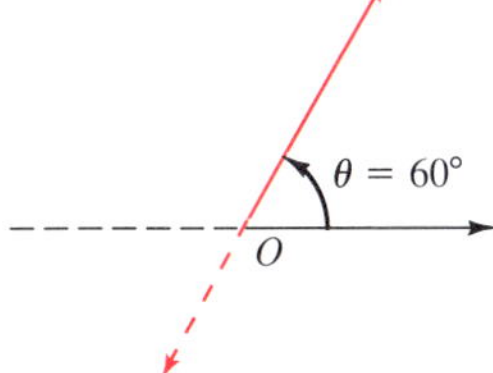

Another familiar graph is shown in Example 2.

EXAMPLE 2 **Graph $r = 4$. Then express the equation $r = 4$ using rectangular coordinates.**

In the equation $r = 4$, the variable θ can have any real value. Thus, the graph is a circle with center (0,0) and radius 4.

$$r = \sqrt{x^2 + y^2}$$

$$4 = \sqrt{x^2 + y^2} \quad \textit{Substitute 4 for r.}$$

$$16 = x^2 + y^2 \quad \textit{Square each side.}$$

$x^2 + y^2 = 16$ is the equation of a circle with center (0, 0) and radius 4.

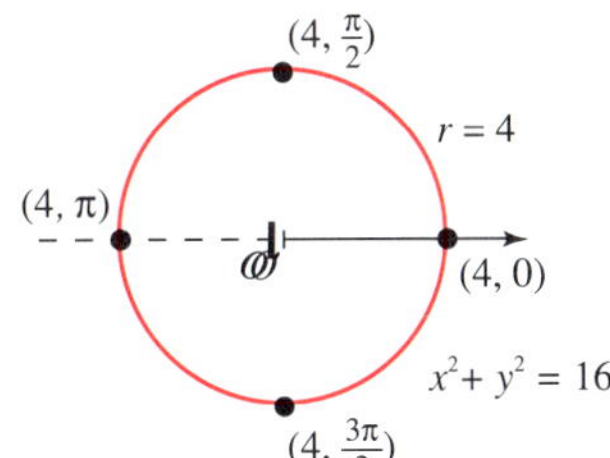

It is helpful to set up a table of values before graphing a polar equation. Let θ range from 0° to 360° and find corresponding values for r.

EXAMPLE 3 **Graph $r = 4 \cos \theta$. Then express the equation in rectangular form.**

First construct a table of values, $0° \leq \theta \leq 360°$. Round $4 \cos \theta$ to the nearest tenth. Then graph the points and connect them with a smooth curve.

θ	$r = 4 \cos \theta$
0°	4
30°	3.5
45°	2.8
60°	2
90°	0
120°	−2
135°	−2.8
150°	−3.5
180°	−4

θ	$r = 4 \cos \theta$
210°	−3.5
225°	−2.8
240°	−2
270°	0
300°	2
315°	2.8
330°	3.5
360°	4

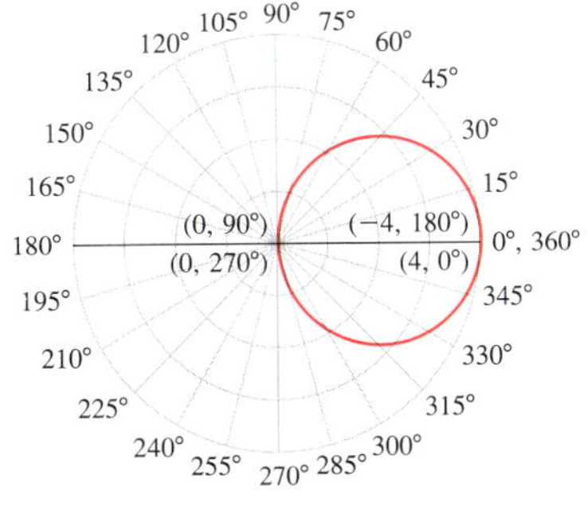

TEACHING SUGGESTIONS

- You may wish to have students make a table comparing the equations of limaçons, cardioids, and roses.

Limaçon	Cardioid
$r = a \pm b \cos \theta$	$r = a \pm a \cos \theta$
$r = a \pm b \sin \theta$	$r = a \pm a \sin \theta$

Rose
$r = a \cos n\theta$
$r = a \sin n\theta$

- When graphing polar equations, point out that it is necessary to plot many points in order to determine the exact shape of a graph.

CHALKBOARD EXAMPLES

- **For Example 1**

1. Graph $\theta = -\dfrac{5\pi}{4}$.

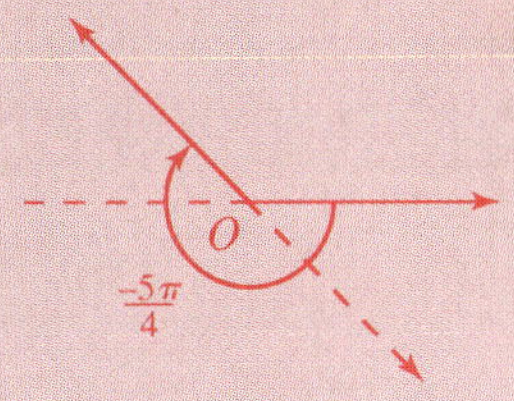

- **For Example 2**

2. Graph $r = 1$. Then express $r = 1$ using rectangular coordinates.

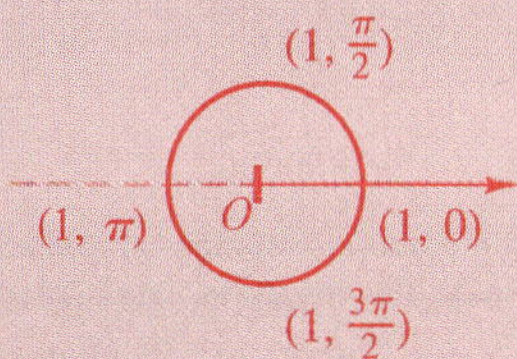

$x^2 + y^2 = 1$

- **For Example 3**

  3. Graph $r = -4 \cos \theta$. Then express the equation in rectangular form.

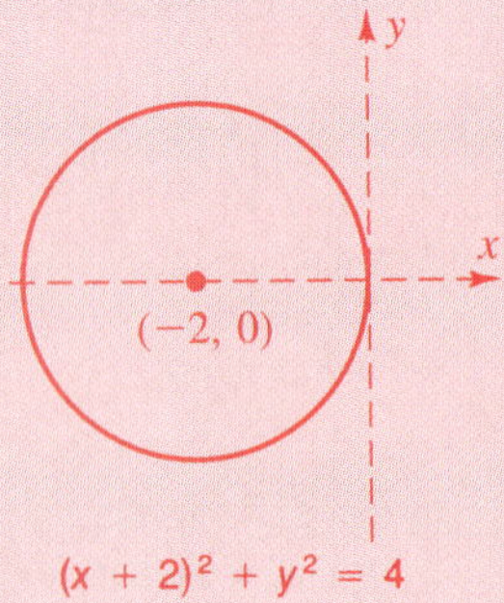

- **For Example 4**

  4. Graph: $r = 1 + 4 \sin \theta$

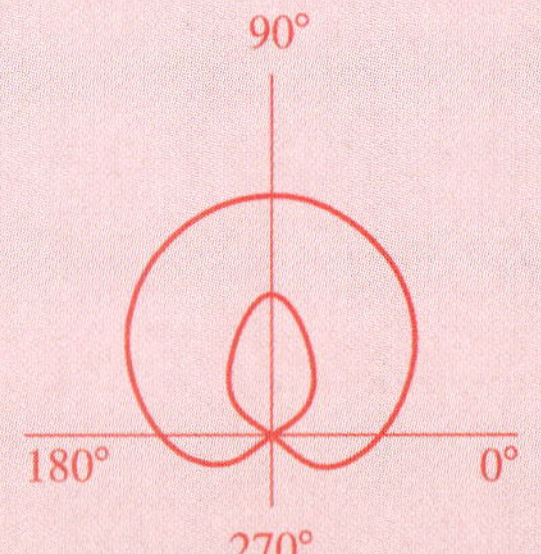

5. Graph: $r = 1 - 3 \cos \theta$

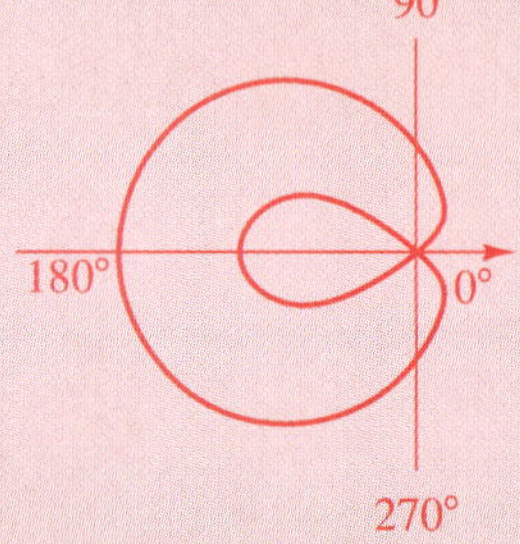

- **For Example 5**

  6. Graph: $r = 3 - 3 \sin \theta$

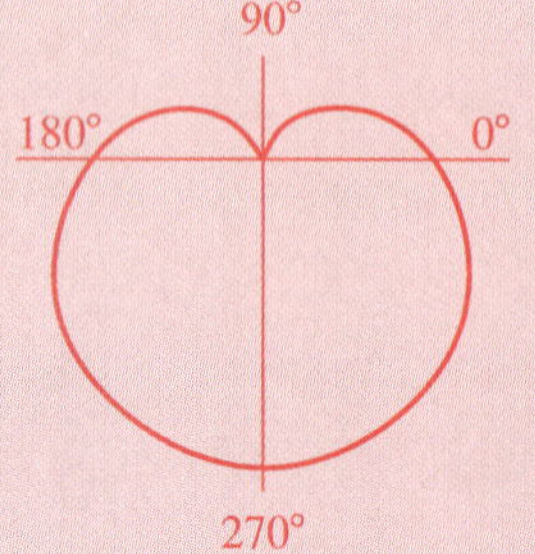

Notice that a pattern develops in the table of values for $r = 4 \cos \theta$. The point $(4, 0°)$ is the same as the point $(-4, 180°)$, the point $(3.5, 30°)$ is the same as the point $(-3.5, 210°)$, and so on. The graph appears to be a circle with center at $(2, 0)$. To check, express the equation in rectangular form.

$$r = 4 \cos \theta$$
$$r^2 = 4r \cos \theta \qquad \textit{Multiply each side by r.}$$
$$x^2 + y^2 = 4x \qquad \textit{Substitute } x^2 + y^2 \textit{ for } r^2 \textit{ and } x \textit{ for } r \cos \theta.$$
$$x^2 - 4x + y^2 = 0$$
$$x^2 - 4x + 4 + y^2 = 4 \qquad \textit{Complete the square on x.}$$
$$(x - 2)^2 + y^2 = 4$$

This is the equation of a circle with center $(2, 0)$ and radius 2.

In general, the graph of any polar equation of the form $r = 2a \cos \theta$ is a circle with radius a and center $(a, 0)$. Two other types of curves will be studied in this lesson. The first is the **limaçon.** Limaçons are curves represented by polar equations of the form $r = a \pm b \cos \theta$ or $r = a \pm b \sin \theta$.

If $a < b$, the graph has an extra loop.

If $a = b$, the graph is heart-shaped and is called a **cardioid.**

If $a > b$, the graph has no extra loop.

EXAMPLE 4 **Graph $r = 1 + 4 \cos \theta$.**

$1 + 4 \cos \theta = 0$ *Let $r = 0$ in $1 + 4 \cos \theta = r$.*

$\theta = 104.48°$ or $\theta = 255.52°$

Then graph the points. Since $1 < 4$ (that is, $a < b$), the graph has an extra loop.

θ	r	θ	r
0°	5	210°	−2.5
30°	4.5	225°	−1.8
45°	3.8	240°	−1
60°	3	256°	0
90°	1	270°	1
104°	0	300°	3
120°	−1	315°	3.8
135°	−1.8	330°	4.5
150°	−2.5	360°	5
180°	−3		

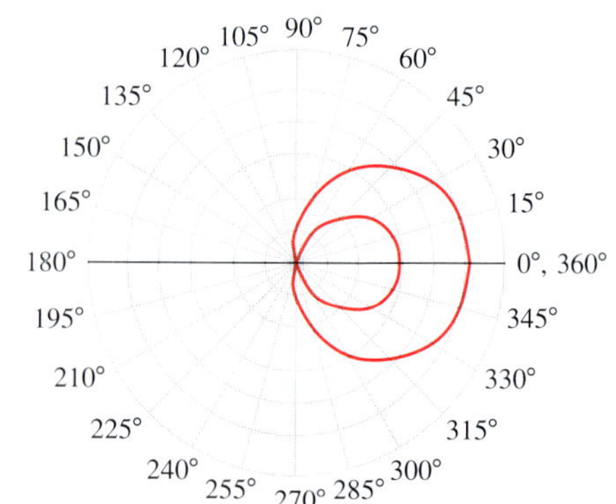

The graph of a cardioid (with $a = b$) is shown in the next example.

EXAMPLE 5 **Graph $r = 4 + 4\cos\theta$.**

θ	r	θ	r
0	8	$\frac{7\pi}{6}$	0.5
$\frac{\pi}{6}$	7.5	$\frac{5\pi}{4}$	1.2
$\frac{\pi}{4}$	6.8	$\frac{4\pi}{3}$	2
$\frac{\pi}{3}$	6	$\frac{3\pi}{2}$	4
$\frac{\pi}{2}$	4	$\frac{5\pi}{3}$	6
$\frac{2\pi}{3}$	2	$\frac{7\pi}{4}$	6.8
$\frac{3\pi}{4}$	1.2	$\frac{11\pi}{6}$	7.5
π	0		

Another special type of curve is the **rose.** Rose curves are represented by polar equations of the form $r = a\cos n\theta$ or $r = a\sin n\theta$, where n is a positive integer. If n is odd, the rose has n leaves. If n is even, the rose has $2n$ leaves.

EXAMPLE 6 **Graph $r = 4\cos 2\theta$.**

θ	$r = 4\cos 2\theta$	θ	$r = 4\cos 2\theta$
0°	4	210°	2
30°	2	225°	0
45°	0	240°	−2
60°	−2	270°	−4
90°	−4	300°	−2
120°	−2	315°	0
135°	0	330°	2
150°	2	360°	4
180°	4		

Notice that the rose has $2(2) = 4$ leaves.

CLASS EXERCISES

Match each equation with a description on the right.

1. $r = 3 + 3\sin\theta$ c
2. $r = 2$ e
3. $r = 3\cos 2\theta$ d
4. $r = 3 + 2\sin\theta$ b
5. $r = 2 - 4\cos\theta$ a
6. $r = -2\sin 2\theta$ d
7. $r = 3(1 + 2\sin\theta)$ a
8. $r = 6(1 + \cos\theta)$ c

a. Limaçon with an extra loop
b. Limaçon with no extra loop
c. Cardioid
d. Rose
e. Circle

- **For Example 6**

  7. Graph: $r = 4\sin 2\theta$

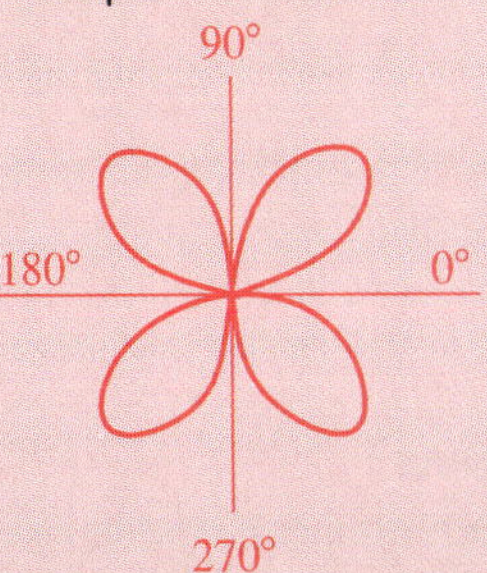

Common Errors

- Some students have difficulty plotting a polar equation correctly. Having those students classify the curve before constructing a table of values may be helpful.
- Some students obtain incorrect values when constructing a table of values. Have those students review the trigonometric functions for the special angles.
- See *Teacher's Resource Book* for additional remediation.

LESSON FOLLOW-UP

Discussion

Explain how to solve this system of polar equations:

$$\begin{cases} r = 4 \\ r = 4\cos\theta \end{cases}$$

Graph both equations on the same coordinate system. Then check all points of intersection to determine which, if any, are solutions of the system. In this case the solutions are (4, 0°), (4, 360°), (4, 720°), . . .

Critical Thinking

Synthesis Ask students to discuss problems which may arise when solving systems of polar equations by graphing. Recalling that points in the polar coordinate system are not uniquely determined, students note that points of intersection are not always solutions of the system of polar equations.

Assignment Guide

See p. 284B for assignments.

Extra

The graph of a lemniscate is shown in several rotations along with an activity showing that the sine function is simply a rotation of the cosine function.

Lesson Quiz

Match each equation on the left with a description on the right.

1. $r = 2 - 2 \sin \theta$ B
2. $r = 3 \cos 4\theta$ C
3. $r = 9 \cos \theta$ A
4. $r = 4 - 2 \cos \theta$ E
5. $r = 3 + 5 \sin \theta$ D

A. Circle
B. Cardioid
C. Rose
D. Limaçon with a loop
E. Limaçon without a loop

Graph each polar equation.

6. $r = 2 + 4 \sin \theta$

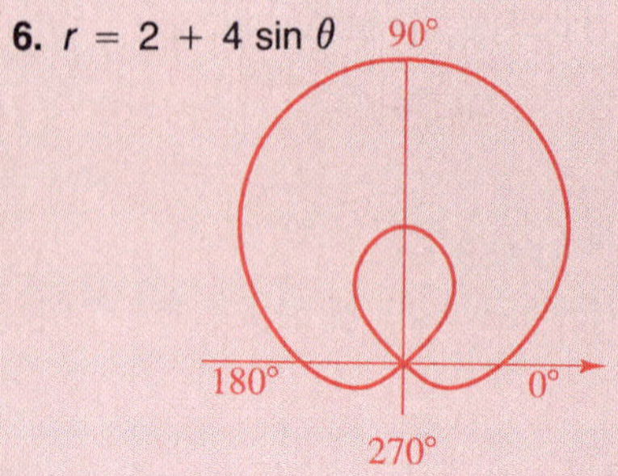

7. $r = 3 - 3 \cos \theta$

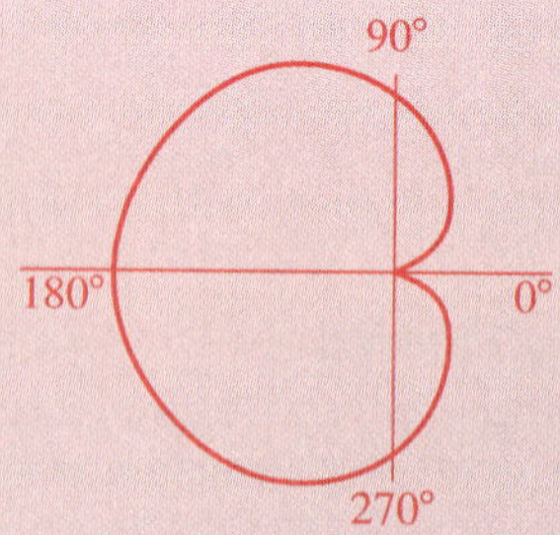

PRACTICE EXERCISES

Graph each polar equation. See side column on page 297.

A **1.** $\theta = 135°$ **2.** $\theta = -\frac{2\pi}{3}$

Graph each polar equation. Then express the equation in rectangular form.

3. $r = 5$ $x^2 + y^2 = 25$ **4.** $r = 3$ $x^2 + y^2 = 9$ **5.** $r = -2$ $x^2 + y^2 = 4$ **6.** $r = -4$ $x^2 + y^2 = 16$

See pages 488–489 for graphs.

Graph each polar equation. Then express the equation in rectangular form. See page 489 for graphs.

7. $r = 6 \cos \theta$ $(x - 3)^2 + y^2 = 9$
8. $r = 8 \cos \theta$ $(x - 4)^2 + y^2 = 16$
9. $r = -8 \cos \theta$ $(x + 4)^2 + y^2 = 16$
10. $r = -6 \cos \theta$ $(x + 3)^2 + y^2 = 9$
11. $r = 8 \sin \theta$ $x^2 + (y - 4)^2 = 16$
12. $r = 4 \sin \theta$ $x^2 + (y - 2)^2 = 4$

Graph each equation. See pages 489–491.

13. $r = 4 + 3 \sin \theta$ (limaçon)
14. $r = 5 + 2 \sin \theta$ (limaçon)
15. $r = 6 - 4 \cos \theta$ (limaçon)
16. $r = 7 - 3 \cos \theta$ (limaçon)
B **17.** $r = 4 + 4 \sin \theta$ (cardioid)
18. $r = 3 + 3 \cos \theta$ (cardioid)
19. $r = 3 - 3 \cos \theta$ (cardioid)
20. $r = 2 - 2 \sin \theta$ (cardioid)
21. $r = 1 - \cos \theta$ (cardioid)
22. $r = 1 + \sin \theta$ (cardioid)
23. $r = 6 \sin 2\theta$ (four-leaved rose)
24. $r = \cos 2\theta$ (four-leaved rose)
25. $r = 4 \cos 3\theta$ (three-leaved rose)
26. $r = \sin 5\theta$ (five-leaved rose)
27. $r = 5 \cos 4\theta$ (eight-leaved rose)
28. $r = 2 \sin 4\theta$ (eight-leaved rose)

Another special curve is called the *lemniscate*. Graph each polar equation to find the shape of a lemniscate.

C **29.** $r^2 = \cos 2\theta$ **30.** $r^2 = \sin 2\theta$

The *Spiral of Archimedes* is also a special curve. Graph each polar equation to find the shape of an Archimedean spiral.

31. $r = \theta$ **32.** $r = -3\theta$

Applications

33. Physics The orbit of a satellite has the equation $r = \frac{2}{2 + \cos \theta}$. Sketch the graph of its orbit. Is the path of the orbit a circle, a parabola, an ellipse, or a hyperbola? ellipse

34. Physics The path of a comet has the equation $r = \frac{2}{1 - \sin \theta}$. Sketch the graph of its path. Is the path a circle, a parabola, an ellipse, or a hyperbola? parabola

35. Coordinate Geometry Find the points of intersection of the graphs of $r = \cos \theta$ and $r = \sin \theta$. the pole, $\left(\frac{\sqrt{2}}{2}, 45^\circ\right)$

EXTRA

Rotations of Polar Graphs The equation for a *lemniscate* is of the form $r^2 = a^2 \cos 2\theta$ or $r^2 = a^2 \sin 2\theta$. A lemniscate always has two leaves, each of length a.

The graph below is a lemniscate with the equation $r^2 = 25 \cos 2\theta$. A table of values is shown. Note that the values of r in the table are found by taking the square root of each side of the equation.

$$r^2 = 25 \cos 2\theta$$
$$r = \pm 5\sqrt{\cos 2\theta}$$

θ	$r = \pm 5\sqrt{\cos 2\theta}$
0°	±5
15°	±4.7
30°	±3.5
45°	0
45° < θ < 135°	undefined
135°	0
150°	±3.5
165°	±4.7
180°	±5

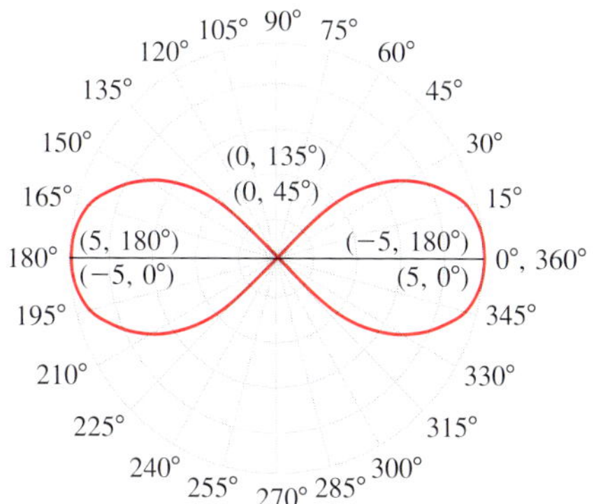

1. Construct a table of values of the equation $r^2 = 25 \cos 2(\theta - 45^\circ)$. Then graph the equation. See page 491.

2. Through how many degrees would the graph of $r^2 = 25 \cos 2\theta$ (shown above) have to be rotated in a counterclockwise direction in order to obtain the graph you drew for Exercise 1? 45°

3. Construct a table of values for $r^2 = 25 \sin 2\theta$. Then graph the equation. How does this graph compare with that of $r^2 = 25 \cos 2(\theta - 45^\circ)$? equivalent; See page 491.

4. To explain your results in Exercise 3, show that $r^2 = 25 \sin 2\theta$ is equivalent to $r^2 = 25 \cos 2(\theta - 45^\circ)$. *Hint*: Start with $r^2 = 25 \sin 2\theta$ and use the fact that $\sin 2\theta = \cos(90^\circ - 2\theta)$. See page 491.

Enrichment

Graph: $r = \sin \theta \cos^2 \theta$

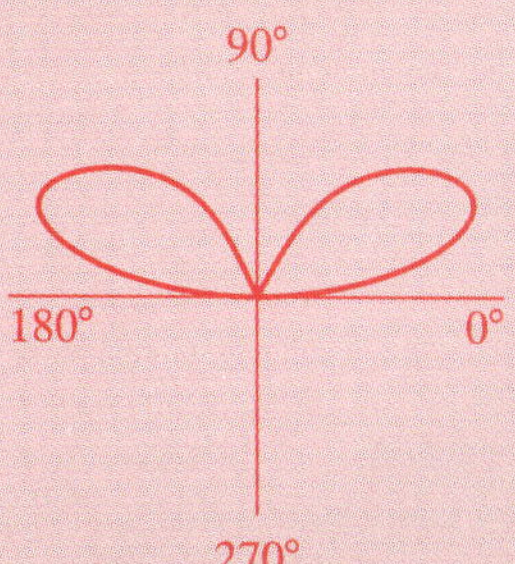

Teacher's Resource Book

Practice—Chapter 7, p. 3

Enrichment—Chapter 7, p. 4

Additional Answers

Practice Exercises

1.

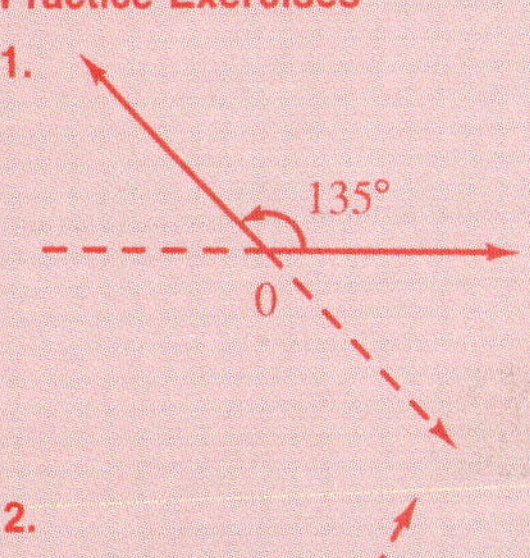

2.

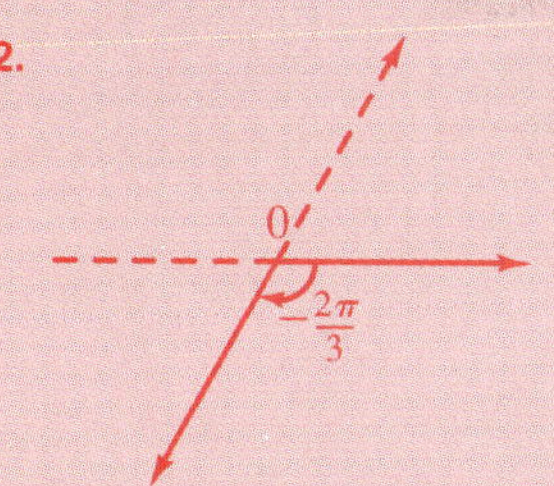

33.

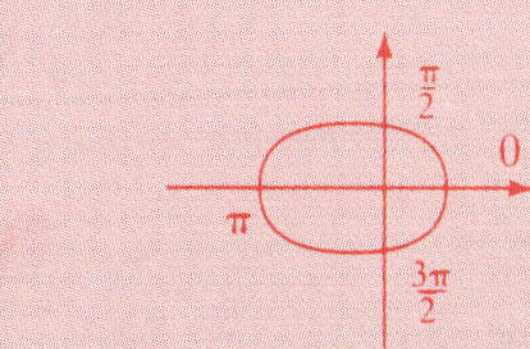

34.

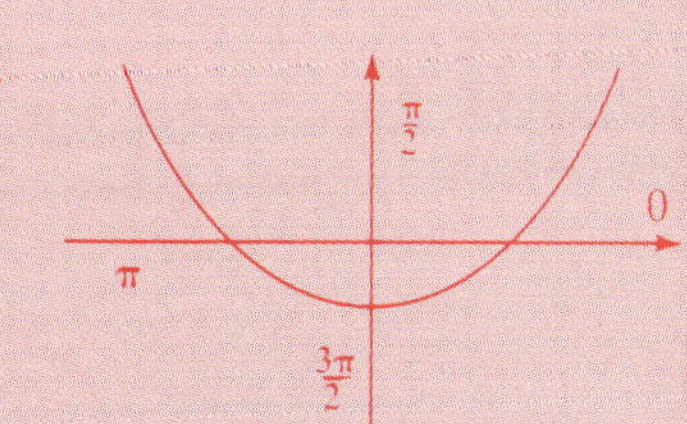

LESSON PLAN

Vocabulary

Complex number
Imaginary number

BACKGROUND

In the Preview, determining which of the given equations do not have a solution in a specified set of numbers is reviewed to motivate a discussion of the complex number system.

7.3 Sums and Differences of Complex Numbers

Objectives: To introduce the complex numbers
To define addition and subtraction of complex numbers

The solutions to many equations are members of the set of real numbers, which includes the integers, the rational numbers, and the irrational numbers. However, a new set of numbers, called the *complex numbers*, must be introduced before certain types of equations can be solved.

Preview

Determine which of the equations at the right do *not* have a solution that is a member of the given set of numbers.

a. $2x - 4 = 3$ **b.** $x^2 = 4$
c. $3x^2 = 4$ **d.** $x^2 + 4 = 2$
e. $3x + 4 = 2$ **f.** $x^2 + 4 = 3$
g. $\frac{x^3}{2} + 4 = 0$ **h.** $x^2 + \frac{3}{4} = 1$

1. The set of integers a, c, d, e, f, h
2. The set of rational numbers c, d, f
3. The set of real numbers d, f

The equations $x^2 + 1 = 0$ and $x^2 = -1$ are equivalent. Since the square root of a negative number is not a real number, the equation $x^2 = -1$ has no solution in the real-number system. In order to solve an equation of this type, the **imaginary number *i*** is defined by the equation

$$i = \sqrt{-1}$$

It follows that $i^2 = -1$. Therefore, the solutions of $x^2 = -1$ are $x = i$ and $x = -i$. Using imaginary numbers, the square root of a negative number can be expressed as the product of a real number and the number i. For example, $\sqrt{-4} = \sqrt{4}\sqrt{-1} = 2\sqrt{-1} = 2i$. For any *positive* real number b,

$$\sqrt{-b} = \sqrt{b}\sqrt{-1} = \sqrt{b}\, i = i\sqrt{b}$$

It is customary to write $i\sqrt{b}$ rather than $\sqrt{b}i$ to avoid confusion with $\sqrt{bi}$.

EXAMPLE 1 **Simplify:** **a.** $\sqrt{-9}$ **b.** $\sqrt{-24}$

a. $\sqrt{-9} = i\sqrt{9} = i(3) = 3i$

b. $\sqrt{-24} = i\sqrt{24} = i\sqrt{4}\sqrt{6} = i(2)\sqrt{6} = 2i\sqrt{6}$

A **complex number** is a number that can be written in the form $a + bi$, where a and b are real numbers and $i = \sqrt{-1}$. The *real* part of a complex number $a + bi$ is a, and the *imaginary* part is bi. A complex number $a + bi$ is *imaginary* if $b \neq 0$, *pure imaginary* if $a = 0$ and $b \neq 0$, and *real* if $b = 0$.

Complex Numbers

Imaginary Numbers	Real Numbers
$2 + 3i$, $6 - i$, $1-\sqrt{-8}$; Pure Imaginary Numbers: $-2i\sqrt{5}$	19, -5, $9.\overline{3}$, 0, $\frac{13}{3}$, 10.2; Irrational Numbers: $0.212212221\ldots$, $\pi\sqrt{7}$

The powers of i follow a cyclic pattern. A few of these powers are listed below.

$i^1 = i$	$i^5 = i^4 \cdot i = 1i = i$
$i^2 = -1$	$i^6 = i^4 \cdot i^2 = 1(-1) = -1$
$i^3 = i^2 \cdot i = -i$	$i^7 = i^4 \cdot i^3 = 1(-i) = -i$
$i^4 = i^2 \cdot i^2 = (-1)(-1) = 1$	$i^8 = i^4 \cdot i^4 = 1(1) = 1$

Notice that the numbers above repeat in cycles of four. Any higher integral powers of i can be factored into powers of i^4 and a single factor of i, i^2, or i^3. Since $i^4 = 1$, every integral power of i is equal to i, -1, $-i$, or 1.

EXAMPLE 2 **Simplify:** **a.** i^{18} **b.** i^{223}

a. $i^{18} = i^{16} \cdot i^2 = (i^4)^4 i^2 = (1)^4(-1) = 1(-1) = -1$

b. $i^{223} = i^{220} \cdot i^3 = (i^4)^{55} \cdot i^3 = (1)^{55}(-i) = (1)(-i) = -i$

Two complex numbers are equal if and only if their real parts are equal and their imaginary parts are equal. That is,

$$a + bi = c + di \quad \text{if and only if} \quad a = c \text{ and } b = d$$

Complex-number arithmetic has many properties that are similar to real-number arithmetic. For example, the definition of addition for complex numbers uses the commutative, associative, and distributive properties.

Addition of Complex Numbers For all real numbers a, b, c, and d

$$(a + bi) + (c + di) = (a + c) + (b + d)i$$

Notice that when two complex numbers are added, the real parts are grouped and then added, as are the imaginary parts.

TEACHING SUGGESTIONS

- Stress that since $\sqrt{ab} = \sqrt{a}\sqrt{b}$ does not hold true if a and b are negative, the definition $i = \sqrt{-1}$ was introduced.
- Emphasize that adding and subtracting complex numbers is similar to adding and subtracting polynomials by grouping similar, or like, terms.
- Show that the distributive property can be used when subtracting complex numbers. For example, $(1 + 7i) - (2 + i) = 1 + 7i - 2 - i = -1 + 6i$.

Critical Thinking

Analysis Ask students to list the first 16 powers of i with their simplified forms, then ask them to determine the value of i^n working only with the exponent. Observing that every positive integral power of i is equal to one of the values $1, -1, -i$, or i, students should reason that i^n is equal to one of the values i or $-i$ when n is a positive odd integer, and i^n is equal to one of the values -1 or 1 when n is a positive even integer.

CHALKBOARD EXAMPLES

- **For Example 1**
  Simplify:
  1. $\sqrt{-48}$ $4i\sqrt{3}$
  2. $\sqrt{-18}$ $3i\sqrt{2}$
- **For Example 2**
  Simplify:
  3. i^{20} 1
  4. i^{37} i

- **For Example 3**
  Find the sum and simplify:
  5. $(-4 - 5i) + (2 + i)$ $-2 - 4i$
  6. $(5 + 2i) + (-7 - 7i)$
  $-2 - 5i$
- **For Example 4**
  Find the difference and simplify:
  7. $(-8 - 2i) - (-1 - 3i)$
  $-7 + i$
  8. $(9 - 2i) - (-3 + 4i)$ $12 - 6i$

Common Errors

- Students often simplify $-5i^2$ to -5, forgetting that $i^2 = -1$. Emphasize the values of the powers of i.
- Some students make sign mistakes when subtracting complex numbers. Remind those students that each term of the complex number being subtracted must be multiplied by -1 before it is added to the other number.
- See *Teacher's Resource Book* for additional remediation.

LESSON FOLLOW-UP

Discussion

How could you graph complex numbers on a rectangular coordinate system? Use the x-axis as the real axis and the y-axis as the imaginary axis. Then $a + bi$ would have coordinates (a, b).

Assignment Guide

See p. 284B for assignments.

EXAMPLE 3 **Find the sum and simplify:**
a. $(3 + 2i) + (5 + 7i)$ **b.** $(-4 + 7i) + (4 - 7i)$

a. $(3 + 2i) + (5 + 7i) = (3 + 5) + (2 + 7)i = 8 + 9i$

b. $(-4 + 7i) + (4 - 7i) = (-4 + 4) + (7 - 7)i = 0 + 0i = 0$

As suggested in part b of Example 3, every complex number has a unique *additive inverse*. To subtract complex number $c + di$ from $a + bi$, add the additive inverse of $c + di$, which is $-(c + di)$ or $(-c - di)$.

> **Subtraction of Complex Numbers** For all real numbers a, b, c, and d
>
> $$(a + bi) - (c + di) = (a + bi) + (-c - di) = (a - c) + (b - d)i$$

EXAMPLE 4 **Find the difference and simplify: $(11 + 5i) - (-7 - 6i)$**

$$\begin{aligned}(11 + 5i) - (-7 - 6i) &= (11 + 5i) + (7 + 6i)\\ &= (11 + 7) + (5 + 6)i\\ &= 18 + 11i\end{aligned}$$

CLASS EXERCISES

For Discussion

1. Is the set of complex numbers closed under addition? That is, is the sum of two complex numbers always a complex number? yes
2. Is the set of imaginary numbers closed under addition? no

Simplify.

3. $\sqrt{-81}$ $9i$ **4.** $\sqrt{-100}$ $10i$ **5.** $\sqrt{-144}$ $12i$ **6.** $\sqrt{-121}$ $11i$

7. $(2 + 3i) + (4 + 6i)$ $6 + 9i$ **8.** $(1 + 7i) + (2 + i)$ $3 + 8i$

9. $(7 + 4i) - (5 + 3i)$ $2 + i$ **10.** $(9 + 2i) - (6 + i)$ $3 + i$

In which set(s) of numbers listed at the right does each of the following numbers belong?

Complex
Imaginary
Pure Imaginary
Real
Rational
Irrational

11. $-3i$ complex, imaginary, pure imaginary **12.** 19.131131113 . . . complex, real, irrational

13. 17 complex, real, rational **14.** $7 + 3i$ complex, imaginary

15. $6 - 2i$ complex, imaginary **16.** $6i$ complex, imaginary, pure imaginary

17. i^2 complex, real, rational **18.** i^4 complex, real, rational

19. $\sqrt{7}$ complex, real, irrational **20.** 7.125 complex, real, rational

PRACTICE EXERCISES

Simplify.

A

1. $\sqrt{-25}$ $5i$ **2.** $\sqrt{-9}$ $3i$ **3.** $\sqrt{-\frac{1}{49}}$ $\frac{1}{7}i$ **4.** $\sqrt{-\frac{36}{25}}$ $\frac{6}{5}i$

5. $\sqrt{-28}$ $2i\sqrt{7}$ **6.** $\sqrt{-20}$ $2i\sqrt{5}$ **7.** $\sqrt{-150}$ $5i\sqrt{6}$ **8.** $\sqrt{-180}$ $6i\sqrt{5}$

9. $\sqrt{-9} + \sqrt{-9}$ $6i$ **10.** $\sqrt{-7} + \sqrt{-7}$ $2i\sqrt{7}$ **11.** $\sqrt{-4} - \sqrt{-8}$ $2i - 2i\sqrt{2}$ **12.** $\sqrt{-9} - \sqrt{-18}$ $3i - 3i\sqrt{2}$

13. $\sqrt{-16} + \sqrt{-49}$ $11i$ **14.** $\sqrt{-4} - \sqrt{-125}$ $2i - 5i\sqrt{5}$ **15.** i^{25} i **16.** i^{37} i

17. i^{104} 1 **18.** i^{84} 1 **19.** i^{246} -1 **20.** i^{263} $-i$

Find each sum or difference and simplify.

21. $(6 + 9i) + (8 + 5i)$ $14 + 14i$ **22.** $(7 + 10i) + (4 + 6i)$ $11 + 16i$

23. $(-5 - 3i) + (2 - 7i)$ $-3 - 10i$ **24.** $(9 - 11i) + (-14 - 2i)$ $-5 - 13i$

25. $(7 + 15i) - (14 - i)$ $-7 + 16i$ **26.** $(9 + 16i) - (12 - 4i)$ $-3 + 20i$

27. $(-4 - 7i) - (10 - 5i)$ $-14 - 2i$ **28.** $(-14 - 6i) + (14 - 5i)$ $-11i$

B

29. $(-12 - \sqrt{-15}) + (-5 + \sqrt{-15})$ -17 **30.** $(-8 - \sqrt{-9}) - (6 - \sqrt{-9})$ -14

31. $(12 + \sqrt{-5}) - (12 - \sqrt{-6})$ $i\sqrt{5} + i\sqrt{6}$ **32.** $(15 + \sqrt{-49}) - (9 - \sqrt{-1})$ $6 + 8i$

33. $(2 - 3\sqrt{-72}) + (-4 + 2\sqrt{-18})$ $-2 - 12i\sqrt{2}$ **34.** $(-5 + \sqrt{-12}) - (7 + 2\sqrt{-27})$ $-12 - 4i\sqrt{3}$

Let n be a positive integer and compute the following.

35. i^{4n} 1 **36.** i^{4n+1} i **37.** i^{4n+3} $-i$ **38.** i^{4n+4} 1

Solve each equation for x.

39. $(3 - 4i) + x = (-7 + i)$ $-10 + 5i$ **40.** $(-5 + 3i) + x = (-9 - 4i)$ $-4 - 7i$

41. $(-4 + 6i) + x = (-1 - 3i)$ $3 - 9i$ **42.** $(7 - 9i) + x = (-3 + 5i)$ $-10 + 14i$

43. $(-5 - 4i) + x = 2i$ $5 + 6i$ **44.** $(7 - 2i) + x = 4i$ $-7 + 6i$

Find the real values of x and y that make each statement true.

45. $5x + yi = -10 + 7i$ $x = -2;\ y = 7$ **46.** $6x + yi = -24 + 4i$ $x = -4;\ y = 4$

47. $4x + 3yi = 7i$ $x = 0;\ y = \frac{7}{3}$ **48.** $6x + 4yi = 15i$ $x = 0;\ y = \frac{15}{4}$

C

49. Determine whether $x = 3 - 2i$ is a solution of the equation $7 - 5i - x = 4 + 3i$. no

50. Determine whether $(-2 + 5i, 9 + i)$ is a solution of the equation $3x + 2y = 17i$. no

Biography: Abraham DeMoivre

A brief biography of DeMoivre is presented and students are requested to research DeMoivre's contributions in mathematical areas other than trigonometry.

Lesson Quiz

Simplify:

1. $\sqrt{-16}$ $4i$

2. $\sqrt{-45}$ $3i\sqrt{5}$

3. $\sqrt{-144} + \sqrt{-64}$ $20i$

4. i^{19} $-i$

5. i^{12} 1

6. $(4 + 7i) + (6 + 3i)$ $10 + 10i$

7. $(-8 + 2i) + (1 - 5i)$ $-7 - 3i$

8. $(9 + 2i) - (6 + i)$ $3 + i$

9. $(-4 - i) - (3 - 5i)$ $-7 + 4i$

Solve for x.

10. $(8 - 3i) + x = 5i$ $x = -8 + 8i$

Enrichment

Prove that addition of complex numbers is associative.

$[(a + bi) + (c + di)] + (e + fi) =$
$[(a + c) + (b + d)i] + (e + fi) =$
$(a + c + e) + (b + d + f)i;$
$(a + bi) + [(c + di) + (e + fi)] =$
$(a + bi) + [(c + e) + (d + f)i] =$
$(a + c + e) + (b + d + f)i$

Teacher's Resource Book

Practice—Chapter 7, p. 5
Enrichment—Chapter 7, p. 6

Additional Answers

58. $x + yi + (w + zi + u + vi) =$
$x + yi + w + zi + u + vi =$
$x + w + u + yi + zi + vi =$
$x + w + u + (y + z + v)i$
$(x + yi + w + zi) + u + vi =$
$x + yi + w + zi + u + vi =$
$x + w + u + yi + zi + vi =$
$x + w + u + (y + z + v)i$
so, $x + yi + (w + zi + u + vi) =$
$(x + yi + w + zi) + u + vi$

Simplify.

51. $(9 + i^9) + (6 - 3i^3)$ $15 + 4i$

52. $(-2 + i^7) + (3 + 2i^2)$ $-1 - i$

53. $(7 - i^8) - (5 - 3i^7)$ $1 - 3i$

54. $(6 - i^6) - (-2 + 2i^{22})$ 11

55. $(3 + i^{118}) - (-2 - i^{68})$ 5

56. $(-1 - 5i^{14}) - (-3 + 17i^{17})$ $7 - 17i$

Applications

57. Algebra The solutions of a quadratic equation are $(5 + 6i)$ and $(5 - 6i)$. Find the sum of the two solutions. 10

58. Algebra Using $x + yi$, $w + zi$, and $u + vi$, show that the associative property holds for the addition of complex numbers. See side column.

Electricity In an electrical circuit, current can be obstructed by resistors, capacitors, and inductances. Resistance R (in ohms) from resistors is represented by a real number. Reactance X (in ohms), from capacitors X_C and inductances X_L, is represented by a pure imaginary number. Impedance Z, which is the total effective resistance, is the sum of the resistances and reactances.

$$Z = R + Xi \qquad \text{where } X = X_L - X_C$$

The magnitude of the impedance is $|Z|$ (in ohms) where $|Z| = \sqrt{R^2 + X^2}$.

59. A circuit has a resistance of 8 ohms and a reactance of 6 ohms. Find the magnitude of the impedance. 10 ohms

60. In a circuit, the resistance is 8.055 ohms and the reactance from capacitors is 3.701 ohms and from inductances 9.099 ohms. Find the magnitude of the impedance. 9.696 ohms

BIOGRAPHY: Abraham De Moivre

Abraham De Moivre was born in France in 1667, studied in Paris, and moved to London in 1688. He lived there until his death in 1754. As his writings attest, he was interested in a wide variety of mathematical topics, including trigonometry. His best friends included Isaac Newton and other notable mathematicians of that time. De Moivre was elected a fellow of the prestigious Royal Society, a member of the Berlin Academy of Sciences, and a foreign associate of the Paris Academy of Sciences.

Research the life of Abraham De Moivre and investigate some areas of mathematics, other than trigonometry, in which he made contributions.

7.4 Products and Quotients of Complex Numbers

Objective: To find the product and the quotient of two complex numbers

Multiplying complex numbers is similar to multiplying binomials.

Preview

The *FOIL* method can be used to multiply two binomials. FOIL means to add the products of the *F*irst, the *O*uter, the *I*nner, and the *L*ast terms.

EXAMPLE **Multiply: $(x + 2)(3x - 7)$**

$$\begin{aligned}(x + 2)(3x - 7) &= \overset{F}{x(3x)} + \overset{O}{x(-7)} + \overset{I}{2(3x)} + \overset{L}{2(-7)} \\ &= 3x^2 - 7x + 6x - 14 \\ &= 3x^2 - x - 14\end{aligned}$$

Multiply.

1. $(x - 9)(x + 4)$ $x^2 - 5x - 36$ **2.** $(2x - 3)(3x - 5)$ $6x^2 - 19x + 15$ **3.** $(x + 2y)(3x + 5y)$ $3x^2 + 11xy + 10y^2$

4. $(0.5a - 1)(0.2a + 4)$ $0.1a^2 + 1.8a - 4$ **5.** $(n + \sqrt{2})(n - \sqrt{2})$ $n^2 - 2$ **6.** $(2x - \sqrt{3})(3x - \sqrt{3})$ $6x^2 - 5x\sqrt{3} + 3$

To multiply two complex numbers in the form $a + bi$, use the FOIL method and the fact that $i^2 = -1$.

EXAMPLE 1 **Find the product and simplify:**

a. $(a + bi)(c + di)$ **b.** $(3 + 2i)(5 - 4i)$

c. $\sqrt{-4}\,(\sqrt{-16})$ **d.** $(5 - 3i)(5 + 3i)$

a. $$\begin{aligned}(a + bi)(c + di) &= a(c) + a(di) + bi(c) + bi(di) \\ &= ac + adi + bci + bdi^2 \\ &= ac + bd(-1) + (ad + bc)i = (ac - bd) + (ad + bc)i\end{aligned}$$

b. $$\begin{aligned}(3 + 2i)(5 - 4i) &= 3(5) + 3(-4i) + 2i(5) + 2i(-4i) \\ &= 15 + (-12i) + 10i + (-8i^2) \\ &= 15 - 2i - 8(-1) = 23 - 2i\end{aligned}$$

c. $\sqrt{-4}\,(\sqrt{-16}) = i\sqrt{4}\,(i\sqrt{16}) = 2i(4i) = 8i^2 = 8(-1) = -8$

d. $$\begin{aligned}(5 - 3i)(5 + 3i) &= 5(5) + 5(3i) + [-3i(5)] + [-3i(3i)] \\ &= 25 + 15i + (-15i) + (-9i^2) \\ &= 25 - 9(-1) = 25 + 9 = 34\end{aligned}$$

LESSON PLAN

Vocabulary
Conjugates

BACKGROUND

In the Preview, the FOIL method for multiplying binomials is reviewed. This method will be used when multiplying complex numbers.

TEACHING SUGGESTIONS

- Compare multiplying complex numbers to multiplying binomials, emphasizing that complex numbers be in i form before this operation is performed.
- Emphasize that multiplying conjugates is similar to multiplying $(a + b)$ by $(a - b)$.
- It may be necessary to do several examples on rationalizing the denominator of a complex number.

Critical Thinking

Analysis Ask students to find the additive inverse of α and the multiplicative inverse of α if $\alpha = a + bi$. Using the definitions for additive and multiplicative inverses and substitution, students should conclude that the additive inverse of $\alpha = -\alpha = -(a + bi) = -a - bi$ and the multiplicative inverse of $\alpha = \frac{1}{\alpha} = \frac{1}{a + bi} = \left(\frac{1}{a + bi}\right)\left(\frac{a - bi}{a - bi}\right) = \frac{a - bi}{a^2 + b^2}$.

CHALKBOARD EXAMPLES

- **For Example 1**

  Find the product and simplify:

  1. $\sqrt{-1}\,(\sqrt{-1})$ -1
  2. $(\sqrt{-8})\,(-3 - 2i)$ $4\sqrt{2} - 6i\sqrt{2}$
  3. $(5 - 4i)\,(2 + 3i)$ $22 + 7i$
  4. $(7 + 2i)\,(7 - 2i)$ 53
  5. $(3 - 2i)^2$ $5 - 12i$

- **For Example 2**

  Find the reciprocal and simplify:

  6. i $-i$
  7. $3 + i$ $\frac{3 - i}{10}$ or $\frac{3}{10} - \frac{1}{10}i$
  8. $-7i$ $\frac{i}{7}$ or $0 + \frac{1}{7}i$
  9. $8 - 3i$ $\frac{8 + 3i}{73}$ or $\frac{8}{73} + \frac{3i}{73}$

In part (c) of Example 1, note that $\sqrt{-4} \cdot \sqrt{-16} \neq \sqrt{(-4)(-16)}$. If $x \geq 0$ and $y \geq 0$, then $\sqrt{x} \cdot \sqrt{y} = \sqrt{xy}$. However, $\sqrt{-x} \cdot \sqrt{-y} \neq \sqrt{(-x)(-y)}$. Therefore, square roots of negative numbers should be expressed in terms of i before they are multiplied.

In part d of Example 1, the complex numbers $5 - 3i$ and $5 + 3i$ are called *conjugates*. In general, complex numbers of the form $a + bi$ and $a - bi$ are called **conjugates** of each other. They differ only in the sign of the imaginary part. The product of a complex number and its conjugate is a real number, since

$$\begin{aligned}(a + bi)(a - bi) &= a^2 - abi + abi - b^2i^2 \\ &= a^2 + 0 - b^2(-1) \\ &= a^2 + b^2\end{aligned}$$

Conjugates can be used to simplify the reciprocal of a complex number.

EXAMPLE 2 **Find the reciprocal of $3 - i$ and simplify.**

$$\begin{aligned}\frac{1}{3 - i} &= \left(\frac{1}{3 - i}\right)\left(\frac{3 + i}{3 + i}\right) \\ &= \frac{3 + i}{3^2 + 1^2} \\ &= \frac{3 + i}{9 + 1} \\ &= \frac{3 + i}{10}, \text{ or } \frac{3}{10} + \frac{1}{10}i\end{aligned}$$

Multiply the numerator and denominator by the conjugate of the denominator.
$(a + bi)(a - bi) = a^2 + b^2$

The quotient of two complex numbers can be simplified using the conjugate of the denominator, as illustrated in Example 3.

EXAMPLE 3 **Find the quotient and simplify: a. $\frac{a + bi}{c + di}$ b. $\frac{4 + 5i}{3 + 2i}$**

a.
$$\begin{aligned}\frac{a + bi}{c + di} &= \left(\frac{a + bi}{c + di}\right)\left(\frac{c - di}{c - di}\right) \\ &= \frac{ac - adi + bci - bdi^2}{c^2 + d^2} \\ &= \frac{ac - adi + bci - (bd)(-1)}{c^2 + d^2} \\ &= \frac{ac - adi + bci + bd}{c^2 + d^2} \\ &= \frac{(ac + bd) + (bc - ad)i}{c^2 + d^2}, \text{ or } \frac{ac + bd}{c^2 + d^2} + \frac{bc - ad}{c^2 + d^2}i\end{aligned}$$

b. $\frac{4+5i}{3+2i} = \left(\frac{4+5i}{3+2i}\right)\left(\frac{3-2i}{3-2i}\right)$

$$= \frac{12 - 8i + 15i - 10i^2}{3^2 + 2^2}$$

$$= \frac{12 + 7i - 10(-1)}{9 + 4}$$

$$= \frac{22 + 7i}{13}, \text{ or } \frac{22}{13} + \frac{7}{13}i$$

Many of the properties of real-number arithmetic hold for complex numbers. For example, the addition and multiplication of complex numbers are commutative, associative, and distributive.

CLASS EXERCISES

For Discussion

1. Is the set of complex numbers closed under multiplication? yes
2. Is the set of imaginary numbers closed under multiplication? no

Multiply.

3. $i(2 - i)$ $1 + 2i$
4. $i(3 + i)$ $-1 + 3i$
5. $(2 + i)(3 + 2i)$ $4 + 7i$
6. $(4 + i)(1 - 2i)$ $6 - 7i$
7. $\sqrt{-100} \cdot \sqrt{-64}$ -80
8. $\sqrt{-36} \cdot \sqrt{-100}$ -60

State the conjugate of each complex number and find the product of each number and its conjugate.

9. $6 + i$ $6 - i;\ 37$
10. $2 - 4i$ $2 + 4i;\ 20$
11. $15i$ $-15i;\ 225$
12. 12 $12;\ 144$
13. $-i$ $i;\ 1$

PRACTICE EXERCISES

Multiply.

A

1. $i(4 - 2i)$ $2 + 4i$
2. $i(-3 + 6i)$ $-6 - 3i$
3. $(2 + i)(3 - 2i)$ $8 - i$
4. $(-4 - 2i)(-2 + 3i)$ $14 - 8i$
5. $(-7 - 4i)(-3 - 3i)$ $9 + 33i$
6. $(-6 - 2i)(-2 - 5i)$ $2 + 34i$
7. $\sqrt{-6} \cdot \sqrt{-600}$ -60
8. $\sqrt{-10} \cdot \sqrt{-640}$ -80
9. $\sqrt{-3} \cdot \sqrt{-8}$ $-2\sqrt{6}$
10. $\sqrt{-2} \cdot \sqrt{-16}$ $-4\sqrt{2}$
11. $(4 + 5i)^2$ $-9 + 40i$
12. $(6 - 5i)^2$ $11 - 60i$
13. $(4 - 3i)^2$ $7 - 24i$
14. $(3 - 5i)^2$ $-16 - 30i$
15. $(2 + 3i)(2 - 3i)$ 13
16. $(4 - 5i)(4 + 5i)$ 41
17. $(\sqrt{5} + i)(\sqrt{5} - 4i)$ $9 - 3i\sqrt{5}$
18. $(\sqrt{3} - 5i)(\sqrt{3} + i)$ $8 - 4i\sqrt{3}$

- **For Example 3**

  Find the quotient and simplify:

  10. $\frac{1}{1+i}$ $\frac{1-i}{2}$ or $\frac{1}{2} - \frac{1}{2}i$
  11. $\frac{2+i}{2-i}$ $\frac{3+4i}{5}$ or $\frac{3}{5} + \frac{4}{5}i$
  12. $\frac{6+12i}{1+i}$ $9 + 3i$
  13. $\frac{3-i}{3+i}$ $\frac{4-3i}{5}$ or $\frac{4}{5} - \frac{3}{5}i$
  14. $\frac{3+5i}{-2-i}$ $\frac{-11-7i}{5}$ or $-\frac{11}{5} - \frac{7}{5}i$

Common Errors

- When multiplying complex numbers, some students forget that $i^2 = -1$ and therefore simplify the product incorrectly. Emphasize that complex numbers must be in i form before multiplying them. Have those students review the powers of i.
- When dividing complex numbers, students may fail to rationalize correctly. Emphasize that numerator and denominator must be multiplied by the conjugate of the denominator.
- See *Teacher's Resource Book* for additional remediation.

LESSON FOLLOW-UP

Assignment Guide

See p. 284B for assignments.

Test Yourself

See *Teacher's Resource Book, Tests*, pp. 69–70.

Lesson Quiz

Multiply:

1. $i(7 - 3i)$ $3 + 7i$
2. $(3 + i)(-4 - 2i)$ $-10 - 10i$
3. $(4 + 3i)^2$ $7 + 24i$
4. Find the reciprocal of $6 - 2i$ and simplify: $\frac{3 + i}{20}$ or $\frac{3}{20} + \frac{1}{20}i$
5. Find the quotient and simplify: $\frac{1 - 3i}{1 + 2i}$ $-1 - i$

Enrichment

Simplify: $\frac{1}{(2 - i)(3 + 8i)}$

$\frac{1}{(2 - i)(3 + 8i)} = \frac{1}{14 + 13i} \cdot \frac{14 - 13i}{14 - 13i} = \frac{14 - 13i}{365}$ or $\frac{14}{365} - \frac{13}{365}i$

Additional Answers

22. $\frac{-5 - 2i}{29}$
24. $\frac{-5 + 3i}{34}$
39. $\frac{-6 + 13i}{5}$
45. $\frac{1 \pm i\sqrt{6}}{2}$

Find the reciprocal and simplify.

19. $5 + i$ $\frac{5 - i}{26}$
20. $6 - i$ $\frac{6 + i}{37}$
21. $3 - 4i$ $\frac{3 + 4i}{25}$
22. $-5 + 2i$ See side column.
23. $-6 - 3i$ $\frac{-2 + i}{15}$
24. $-5 - 3i$ See side column.
25. $-4i$ $\frac{i}{4}$
26. $-5i$ $\frac{i}{5}$
27. $3i$ $\frac{-i}{3}$
28. $4i$ $\frac{-i}{4}$

Find each quotient and simplify.

B

29. $\frac{1 + i}{1 - i}$ i
30. $\frac{1 - i}{1 + i}$ $-i$
31. $\frac{3 + 4i}{3 - 4i}$ $\frac{-7 + 24i}{25}$
32. $\frac{5 + 7i}{5 - 7i}$ $\frac{-12 + 35i}{37}$
33. $\frac{2 + 6i}{-3 - i}$ $\frac{-6 - 8i}{5}$
34. $\frac{3 + 7i}{-4 - i}$ $\frac{-19 - 25i}{17}$
35. $\frac{-4 - 5i}{-2 + 9i}$ $\frac{-37 + 46i}{85}$
36. $\frac{-6 - 5i}{-3 + 7i}$ $\frac{-17 + 57}{58}$

Solve for x.

37. $(1 - 5i)x = i$ $\frac{-5 + i}{26}$
38. $(2 + 3i)x = i$ $\frac{3 + 2i}{13}$
39. $(2 + i)x + 5 - i = 3i$ See side column.
40. $ix + 3 - 5i = 4$ $5 - i$
41. $ix + 2 = 3i$ $3 + 2i$
42. $(1 - i)x - 1 = i$ i

Solve each equation. Recall that the quadratic formula is $x = \frac{-b \pm \sqrt{b^2 - 4ac}}{2a}$.

43. $x^2 - 2x + 5 = 0$ $1 \pm 2i$
44. $x^2 - 2x + 19 = 0$ $1 \pm 3i\sqrt{2}$
45. $4x^2 - 4x + 7 = 0$ See side column.
46. $2x^2 - x + 3 = 0$ $\frac{1 \pm i\sqrt{23}}{4}$
47. $x^2 + 26 = 0$ $\pm i\sqrt{26}$
48. $2x^2 + 5 = 0$ $\frac{\pm i\sqrt{10}}{2}$

Simplify.

C

49. $(3 + 2i)^3$ $-9 + 46i$
50. $(3 - 2i)^4$ $-119 - 120i$
51. $(4 + \sqrt{-49}) \div (-\sqrt{-1})^{13}$ $-7 + 4i$
52. $(7 - \sqrt{-7}) \div (\sqrt{-1})^{11}$ $\sqrt{7} + 7i$
53. Find x and y such that $(2 + xi)(3 - 6i) = 30 + yi$. $x = 4;\ y = 0$
54. Find x and y such that $\frac{x + 2i}{1 - 3i} = -\frac{1}{5} + yi$. $x = 4;\ y = \frac{7}{5}$

Prove each statement. *Hint*: Let $a + bi$ and $c + di$ represent the two numbers.

55. The sum of the conjugates of two complex numbers equals the conjugate of their sum. See side column on page 307.
56. The product of the conjugates of two complex numbers equals the conjugate of their product. See side column on page 307.

Applications

Physics The voltage in an electric circuit can be represented by the formula $V = IZ$, where I is the current (in amps) and Z is the impedance (in ohms).

57. If the current in a circuit is $(3 + 5i)$ amps and the impedance is $(6 - 2i)$ ohms, find the voltage. $28 + 24i$ volts

58. If the voltage in a circuit is $(20 + 2i)$ volts, and the current is $(2 + i)$ amps, find the impedance. $\frac{42 - 16i}{5}$ ohms

59. The total voltage of one circuit is $(10 + 5i)$ volts. The total voltage of a second circuit is $(6 - 2i)$ volts. Find the ratio of the total voltage of the first circuit to that of the second circuit. Simplify your answer. $\frac{5 + 5i}{4}$

60. Computer This program prints the product of two complex numbers of the form $A + Bi$ and $C + Di$. The user supplies the values of A, B, C, and D. Use the program to check your answers for Exercises 1–18.

```
10 HOME
20 INPUT "ENTER THE VALUES OF
   A, B, C, AND D"; A, B, C, D
30 F = A*C
40 M = A*D + C*B
50 L = B*D*-1
60 PRINT "THE PRODUCT IS" F + L" + "M"I"
70 END
```

TEST YOURSELF

Find the rectangular coordinates of each point with the given polar coordinates.

1. $(-3, 150°)$ (2.6, −1.5) **2.** $(2, -40°)$ (1.53, −1.29) 7.1

Find polar coordinates for each point with the given rectangular coordinates.

3. $(\sqrt{3}, -1)$ (2, 330°) **4.** $(2, 0)$ (2, 0°)

Write each equation in polar-coordinate form.

5. $y = 5$ $r = 5 \csc \theta$ **6.** $x^2 = 5y$ $r = 5 \tan \theta \sec \theta$

Write each equation in rectangular-coordinate form.

7. $r = 5 \sec \theta$ $x = 5$ **8.** $r = 3$ $x^2 + y^2 = 9$

Graph each polar equation. See side column.

9. $r = 3 + 4 \cos \theta$ **10.** $r = -2 \sin 2\theta$ 7.2

Simplify.

11. $\sqrt{-40}$ $2i\sqrt{10}$ **12.** $\sqrt{-75}$ $5i\sqrt{3}$ **13.** i^{19} $-i$ **14.** i^{36} 1 7.3

Find each sum or difference and simplify.

15. $(10 + 4i) + (12 - 3i)$ $22 + i$ **16.** $(-7 + 2i) - (-8 + 5i)$ $1 - 3i$

Find each product or quotient and simplify.

17. $(-9 + i)(2 - 3i)$ $-15 + 29i$ **18.** $\frac{2 + 4i}{1 - 3i}$ $-1 + i$ 7.4

Teacher's Resource Book
Practice—Chapter 7, p. 7
Enrichment—Chapter 7, p. 8

Additional Answers

55. sum of the conjugates: $(a - bi) + (c - di) = a - bi + c - di = a + c - bi - di = (a + c) - (b + d)i$; conjugate of the sum: $(a + bi) + (c + di) = a + bi + c + di = a + c + bi + di = (a + c) + (b + d)i$; conjugate: $(a + c) - (b + d)i$

56. product of the conjugates: $(a - bi)(c - di) = ac - adi - bci + bdi^2 = (ac - bd) - adi - bci = (ac - bd) - (ad + bc)i$; conjugate of the product: $(a + bi)(c + di) = ac + adi + bci + bdi^2 = ac - bd + adi + bci = (ac - bd) + (ad + bc)i$; conjugate: $(ac - bd) - (ad + bc)i$

Additional Answers

9.

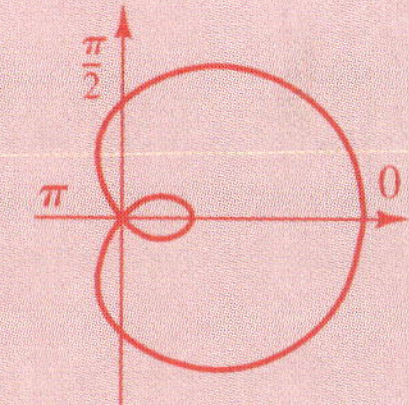

10.

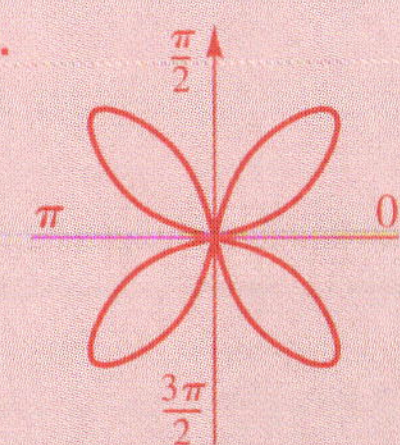

LESSON PLAN

Vocabulary
Argument
Complex plane
Modulus
Polar form of a complex number
Rectangular form of a complex number

BACKGROUND

In the Preview, the mathematical contribution of Jean-Robert Argand is discussed.

7.5 Complex Numbers in Polar Form

Objective: To express complex numbers in polar form

It is useful to be able to plot complex numbers on a coordinate plane.

Preview

The *Argand diagram* on which complex numbers are graphed (pictured at the right) was named after Jean-Robert Argand (1768–1822), a little-known Swiss bookkeeper. He wrote a short book on the geometric representation of complex numbers and received credit for the idea, even though De Moivre, Euler, and Gauss had previously thought of complex numbers as points on the plane.

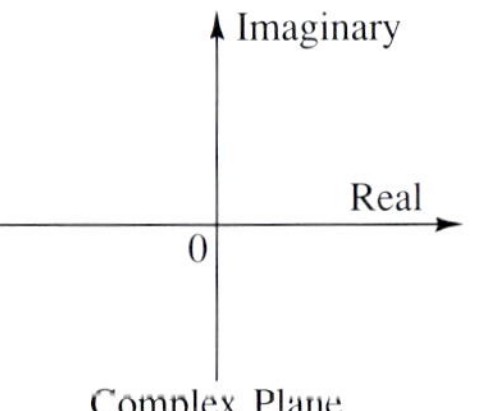

Complex Plane

1. Why do you think Argand chose the horizontal axis as the real axis, and the vertical axis as the imaginary axis? Answers may vary.
2. Can you think of any other men or women who made significant contributions to science, mathematics, or society outside their field of study? Answers may vary.

When a complex number $a + bi$ is represented by a point in the plane, the real part is associated with the horizontal axis, and the imaginary part is associated with the vertical axis. Thus, the complex number $a + bi$ is represented by the ordered pair (a, b) and plotted in the usual manner.

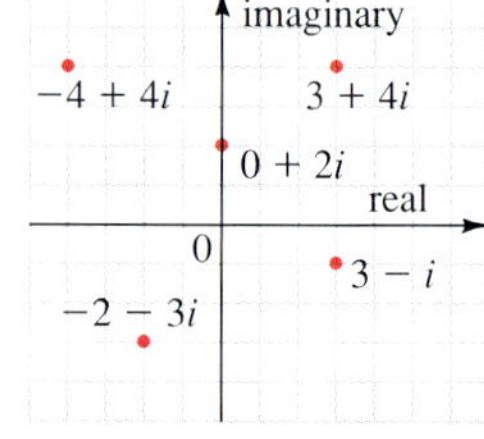

Each point in the **complex plane** determines a unique directed line segment, or vector, that begins at the origin and ends at that point. The vector OP represents $a + bi$, or (a, b), and has both length (magnitude) and direction. The length of a vector z is denoted by $\|z\|$. If $z = a + bi$, then

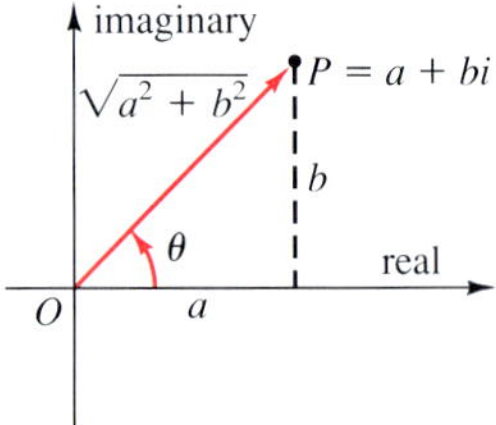

$$\|z\| = \|a + bi\| = \sqrt{a^2 + b^2}$$

The vector OP forms an angle θ with the real axis, and the measure of θ is called the **argument** of the complex number. Since the principal value of Arctan θ is between $-\frac{\pi}{2}$ and $\frac{\pi}{2}$, the value of θ is found as follows:

$$\theta = \text{Arctan}\,\frac{b}{a} \quad \text{if } a > 0$$

$$\theta = \text{Arctan}\,\frac{b}{a} + \pi \quad \left(\text{or Arctan}\,\frac{b}{a} + 180°\right) \quad \text{if } a < 0$$

$$\theta = \frac{\pi}{2} \quad (\text{or } 90°) \quad \text{if } a = 0 \text{ and } b > 0$$

$$\theta = \frac{3\pi}{2} \quad (\text{or } 270°) \quad \text{if } a = 0 \text{ and } b < 0$$

$$\theta = \text{any real number} \quad \text{if } a = 0 \text{ and } b = 0$$

EXAMPLE 1 **Find $\|z\|$ and θ for each vector z. Then graph the vector.**

a. $z = 3 + 4i$ **b.** $z = -4$

a. $\|z\| = \|3 + 4i\|$
$= \sqrt{3^2 + 4^2}$ $\quad |a + bi| = \sqrt{a^2 + b^2}$
$= 5$

$\theta = \text{Arctan}\,\frac{4}{3}$ $\quad a > 0$
$= 53°$

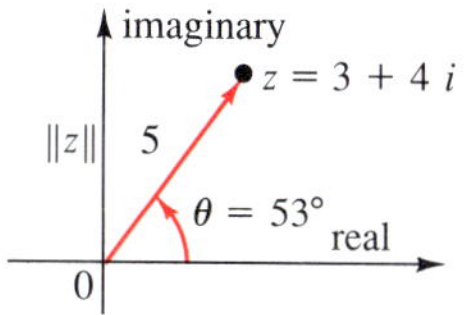

b. $\|z\| = \|-4 + 0i\|$
$= \sqrt{(-4)^2 + 0^2}$
$= 4$

$\theta = \text{Arctan}\,\frac{0}{-4} + 180°$ $\quad a < 0$
$= 180°$

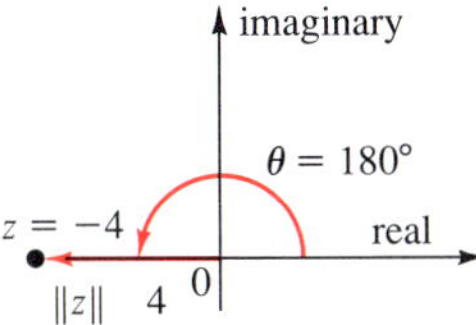

The definition of the magnitude of a complex number is consistent with the definition of the absolute value of a real number. The magnitude of a complex number is simply its distance from the origin.

The complex number $a + bi$ is said to be in **rectangular form.** A useful form, called the **polar form,** or *trigonometric form*, of a complex number, can be derived using vector OP. Recall that vector OP has length $r = \sqrt{a^2 + b^2}$ and it forms an angle θ with the real axis. In the triangle shown at the right, $\cos\theta = \frac{a}{r}$, $\sin\theta = \frac{b}{r}$, and $\tan\theta = \frac{b}{a}$. Therefore, $a = r\cos\theta$ and $b = r\sin\theta$. Substituting these values for a and b yields

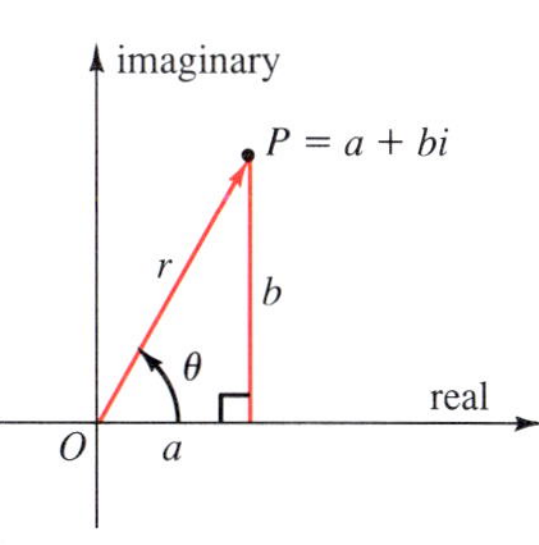

$$a + bi = r\cos\theta + (r\sin\theta)i = r(\cos\theta + i\sin\theta)$$

TEACHING SUGGESTIONS

- Emphasize that the argument of a complex number (θ) is chosen to be the smallest nonnegative angle. The value of θ is determined by the values of a and b.
- Emphasize the differences between rectangular form and polar form of a complex number.

Critical Thinking

Analysis Ask students to evaluate $\|\alpha\| \cdot \|r\|$ and $\|\alpha r\|$ for $\alpha = 1 - i$ and $r = 3i$ and for several other values of α and r. Then ask them to draw a conclusion about $\|\alpha\| \cdot \|r\|$ and $\|\alpha r\|$ for any α and r. Evaluating $\|\alpha\| \cdot \|r\|$, students obtain $\|\alpha\| \cdot \|r\| = \|1 - i\| \cdot \|3i\| = \sqrt{2} \cdot 3$, or $3\sqrt{2}$. Evaluating $\|\alpha r\|$, students obtain $\|(1 - i)(3i)\| = \|3i + 3\| = \sqrt{18}$, or $3\sqrt{2}$. After evaluating $\|\alpha\| \cdot \|r\|$ and $\|\alpha r\|$ for other values of α and r, students should conclude that for all α and r, $\|\alpha\| \cdot \|r\| = \|\alpha r\|$.

CHALKBOARD EXAMPLES

- **For Example 1**

Find $\|z\|$ and θ for each vector z. Then graph z.

1. $z = -5 + 12i$ $\quad$ 13; 112.6°

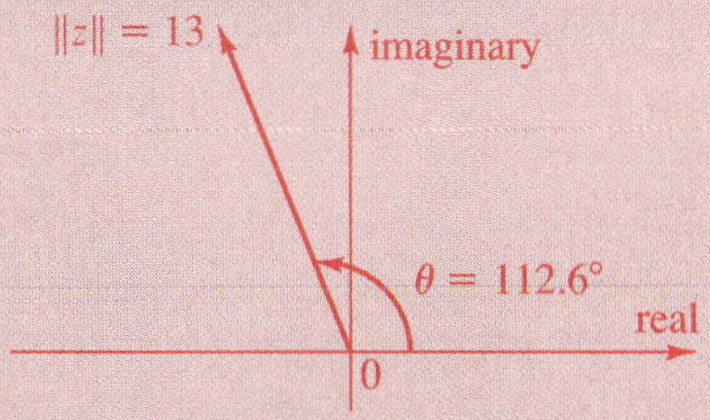

2. $z = -7i$ $\quad$ 7; 270°

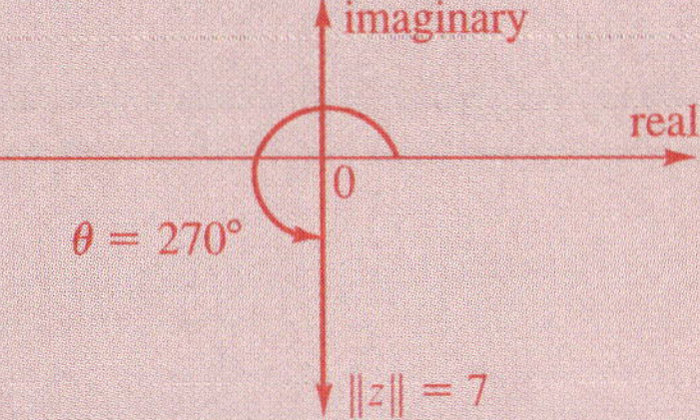

- **For Example 2**

  Express each complex number in polar form.

  3. $\sqrt{3} + i$ $2(\cos 30° + i \sin 30°)$

  4. $\sqrt{6} - i\sqrt{2}$

  $2\sqrt{2}(\cos 330° + i \sin 330°)$

  5. $1.035 - 3.864i$

  $4(\cos 285° + i \sin 285°)$

  6. $3i$ $3(\cos 90° + i \sin 90°)$

- **For Example 3**

  Express each complex number in rectangular form.

  7. $5\left(\cos\frac{3\pi}{4} + i\sin\frac{3\pi}{4}\right)$

  $\frac{-5\sqrt{2}}{2} + \frac{5i\sqrt{2}}{2}$

  8. $3(\cos 240° + i \sin 240°)$

  $-\frac{3}{2} - \frac{3\sqrt{3}}{2}i$

Common Error

- When expressing complex numbers in polar form and vice versa, some students do not use the correct values for sine and cosine. Have those students review the values of the trigonometric functions for the special angles.
- See *Teacher's Resource Book* for additional remediation.

The **polar form** of the complex number $a + bi$ is

$$r(\cos\theta + i\sin\theta) \quad \text{where } r = \sqrt{a^2 + b^2} \text{ and } \tan\theta = \frac{b}{a},\ a \neq 0$$

The quantity r is called the **modulus** and θ is called the *argument*, as mentioned earlier. $r(\cos\theta + i\sin\theta)$ is sometimes written as $r \text{ cis } \theta$.

$r(\cos\theta + i \sin\theta)$
$r(\text{c} \qquad i\,\text{s} \quad \theta)$

EXAMPLE 2 **Express each complex number in polar form:**

a. $-1 + i\sqrt{3}$ **b.** $-5 - 3i$ **c.** $-5i$

a. $r = \sqrt{(-1)^2 + (\sqrt{3})^2} = 2$ $\quad a = -1, b = \sqrt{3}$ $\quad \theta = \text{Arctan}\left(\frac{\sqrt{3}}{-1}\right) + \pi = \frac{2\pi}{3}$, or $120°$

Thus, $-1 + i\sqrt{3} = 2(\cos 120° + i \sin 120°)$.

b. $r = \sqrt{(-5)^2 + (-3)^2} = \sqrt{34}$ $\quad a = -5, b = -3$ $\quad \theta = \text{Arctan}\left(\frac{-3}{-5}\right) + \pi = 211°$

Thus, $-5 - 3i = \sqrt{34}(\cos 211° + i \sin 211°)$.

c. Since $a = 0$ and $b = -5$, the number is on the imaginary axis.

$r = \sqrt{0^2 + (-5)^2} = 5$

$\theta = \frac{3\pi}{2}$, or $270°$ $\quad a = 0, b < 0$

Thus, $-5i = 5\left(\cos\frac{3\pi}{2} + i\sin\frac{3\pi}{2}\right)$.

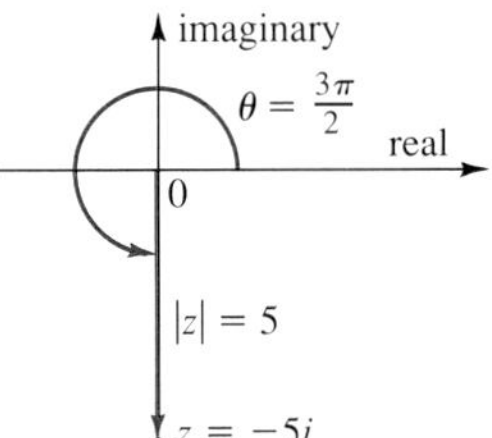

To change a complex number from polar to rectangular form, find a and b.

EXAMPLE 3 **Express each complex number in rectangular form:**

a. $4(\cos 225° + i \sin 225°)$ **b.** $6\left(\cos\frac{2\pi}{3} + i\sin\frac{2\pi}{3}\right)$

a. $a = 4\cos 225° = 4\left(-\frac{\sqrt{2}}{2}\right) = -2\sqrt{2}$ $\quad r = 4, \theta = 225°$ $\quad b = 4\sin 225° = 4\left(-\frac{\sqrt{2}}{2}\right) = -2\sqrt{2}$

Thus, $4(\cos 225° + i \sin 225°) = -2\sqrt{2} - 2i\sqrt{2}$.

b. $a = 6 \cos \frac{2\pi}{3}$ $\quad r = 6, \theta = \frac{2\pi}{3}$ $\quad b = 6 \sin \frac{2\pi}{3}$

$= 6\left(-\frac{1}{2}\right) = -3$ $\qquad = 6\left(\frac{\sqrt{3}}{2}\right) = 3\sqrt{3}$

Thus, $6\left(\cos \frac{2\pi}{3} + i \sin \frac{2\pi}{3}\right) = -3 + 3i\sqrt{3}$.

Some calculators have a rectangular-to-polar and polar-to-rectangular conversion feature. Consult your owner's manual for instructions.

CLASS EXERCISES

Express each complex number as an ordered pair and then graph it.

1. $-3 - 4i$ $(-3, -4)$ **2.** $-4 + i$ $(-4, 1)$ **3.** $3 - 3i$ $(3, -3)$ **4.** $4i$ $(0, 4)$ **5.** -2 $(-2, 0)$

Express each complex number in polar form. See side column.

6. $-2 + 2i$ **7.** $-3 + 3i\sqrt{3}$ **8.** $-5 - 5i$ **9.** 5 **10.** $5i$

PRACTICE EXERCISES

Calculate $\|z\|$ and θ. Then graph each complex number.

A **1.** $z = -1 - i$ $\|z\| = \sqrt{2}; \theta = 225°$ **2.** $z = -1 + i$ $\|z\| = \sqrt{2}; \theta = 135°$ **3.** $z = -2 + 2i$ $\|z\| = 2\sqrt{2}; \theta = 135°$ **4.** $z = 2 - 2i$ $\|z\| = 2\sqrt{2}; \theta = 315°$

5. $z = \sqrt{3} - i\sqrt{3}$ $\|z\| = \sqrt{6}; \theta = 315°$ **6.** $z = -\sqrt{5} + i\sqrt{5}$ $\|z\| = \sqrt{10}; \theta = 135°$ **7.** $z = 4i$ $\|z\| = 4; \theta = 90°$ **8.** $z = -2i$ $\|z\| = 2; \theta = 270°$

9. $z = -5$ $\|z\| = 5; \theta = 180°$ **10.** $z = 4$ $\|z\| = 4, \theta = 0°$ **11.** $z = -5 + 5i\sqrt{5}$ $\|z\| = 5\sqrt{6}; \theta = 114°$ **12.** $z = \sqrt{2} - 2i\sqrt{2}$ $\|z\| = \sqrt{10}; \theta = 297°$

Express each complex number in polar form.

13. $1 - i$ $\sqrt{2}(\cos 315° + i \sin 315°)$ **14.** $1 + i$ $\sqrt{2}(\cos 45° + i \sin 45°)$ **15.** $\sqrt{3} + i$ $2(\cos 30° + i \sin 30°)$ **16.** $1 - i\sqrt{3}$ $2(\cos 300° + i \sin 300°)$

17. 4 $4(\cos 0° + i \sin 0°)$ **18.** $-6i$ $6(\cos 270° + i \sin 270°)$ **19.** $2\sqrt{3} - 2i$ $4(\cos 330° + i \sin 330°)$ **20.** $2 - 2i\sqrt{3}$ $4(\cos 300° + i \sin 300°)$

Express each complex number in rectangular form.

B **21.** $3(\cos 45° + i \sin 45°)$ $\frac{3\sqrt{2}}{2} + \frac{3i\sqrt{2}}{2}$ **22.** $2(\cos 30° + i \sin 30°)$ $\sqrt{3} + i$

23. $\cos(-60°) + i \sin(-60°)$ $\frac{1}{2} - \frac{i\sqrt{3}}{2}$ **24.** $\cos(-240°) + i \sin(-240°)$ $-\frac{1}{2} + \frac{i\sqrt{3}}{2}$

25. $5(\cos 30° + i \sin 30°)$ $\frac{5\sqrt{3}}{2} + \frac{5i}{2}$ **26.** $4(\cos 60° + i \sin 60°)$ $2 + 2i\sqrt{3}$

27. $2[\cos(-45°) + i \sin(-45°)]$ $\sqrt{2} - i\sqrt{2}$ **28.** $3[\cos(-30°) + i \sin(-30°)]$ $\frac{3\sqrt{3}}{2} - \frac{3i}{2}$

29. $\cos 0 + i \sin 0$ $1 + 0i$ **30.** $3(\cos \pi + i \sin \pi)$ $-3 + 0i$

31. $3 \text{ cis}\left(\frac{2\pi}{3}\right)$ $\frac{-3}{2} + \frac{3i\sqrt{3}}{2}$ **32.** $5 \text{ cis}\left(\frac{3\pi}{4}\right)$ $\frac{-5\sqrt{2}}{2} + \frac{5i\sqrt{2}}{2}$ **33.** $4 \text{ cis}\left(-\frac{3\pi}{4}\right)$ $-2\sqrt{2} - 2i\sqrt{2}$ **34.** $2 \text{ cis}\left(-\frac{5\pi}{6}\right)$ $-\sqrt{3} - i$

LESSON FOLLOW-UP

Discussion

Express the equation $x^2 + y^2 = 36$ in polar form.

Substitute $r \cos \theta$ for x and $r \sin \theta$ for y:

$x^2 + y^2 = 36$

$(r \cos \theta)^2 + (r \sin \theta)^2 = 36$

$r^2 \cos^2 \theta + r^2 \sin^2 \theta = 36$

$r^2(\cos^2 \theta + \sin^2 \theta) = 36$

$r^2 = 36$

$r = 6$

Assignment Guide

See p 284B for assignments.

Historical Note

A short history of complex numbers in chart form is featured. The chart highlights 1st through 19th century contributors.

Lesson Quiz

1. Calculate $|z|$ for $3 - 4i$. 5

Express each complex number in polar form.

2. $1 + i\sqrt{3}$ $2(\cos 60° + i \sin 60°)$

3. $4i$ $4(\cos 90° + i \sin 90°)$

Express each complex number in rectangular form.

4. $6(\cos 135° + i \sin 135°)$ $-3\sqrt{2} + 3i\sqrt{2}$

5. $5(\cos 180° + i \sin 180°)$ -5

6. $\cos \frac{5\pi}{6} + i \sin \frac{5\pi}{6}$ $-\frac{\sqrt{3}}{2} + \frac{1}{2}i$

Additional Answers

Class Exercises

6. $2\sqrt{2}(\cos 135° + i \sin 135°)$ **7.** $6(\cos 120° + i \sin 120°)$ **8.** $5\sqrt{2}(\cos 225° + i \sin 225°)$ **9.** $5(\cos 0° + i \sin 0°)$ **10.** $5(\cos 90° + i \sin 90°)$

Enrichment

Express $\frac{4 - 4i}{2\sqrt{3} - 2i}$ in polar form.

$\frac{4 - 4i}{2\sqrt{3} - 2i} = \frac{2 - 2i}{\sqrt{3} - i} \cdot \frac{\sqrt{3} + i}{\sqrt{3} + i} =$

$\frac{(2\sqrt{3} + 2) + (2i - 2i\sqrt{3})}{4} =$

$\frac{(\sqrt{3} + 1) + (1 - \sqrt{3})i}{2} = \frac{2.732 - 0.732i}{2} =$

$1.366 - 0.366i$; $r =$

$r = \sqrt{(1.366)^2 + (0.366)^2} = \sqrt{2}$; $\theta =$

Arctan $\frac{-0.366}{1.366} = 345°$; thus, $\frac{4 - 4i}{2\sqrt{3} - 2i} =$

$\sqrt{2}(\cos 345° + i \sin 345°)$ in polar form; alternate solution: $\frac{4 - 4i}{2\sqrt{3} - 2i} =$

$\frac{4\sqrt{2}(\cos 315° + i \sin 315°)}{4(\cos 330° + i \sin 330°)} =$

$\sqrt{2}[\cos(-15°) + i \sin(-15°)] =$

$\sqrt{2}(\cos 345° + i \sin 345°)$

Teacher's Resource Book

Practice—Chapter 7, p. 9

Enrichment—Chapter 7, p. 10

Additional Answers

35. $\sqrt{z \cdot \bar{z}} = \sqrt{(a + bi)(a - bi)} =$
$\sqrt{a^2 - abi + abi - b^2i^2} =$
$\sqrt{a^2 + b^2} = \|z\|$

36. $\|\bar{z}\| = |a - bi| = \sqrt{a^2 + (-b)^2} =$
$\sqrt{a^2 + b^2} = \|z\|$

37. $\|2z\| = |2a + 2bi| =$
$\sqrt{(2a)^2 + (2b)^2} = \sqrt{4a^2 + 4b^2} =$
$\sqrt{4(a^2 + b^2)} = 2\sqrt{a^2 + b^2} = 2\|z\|$

Given $z = a + bi$ and $\bar{z} = a - bi$, prove each of the following statements.
See side column.

C **35.** $\|z\| = \sqrt{z \cdot \bar{z}}$ **36.** $\|z\| = \|\bar{z}\|$ **37.** $2\|z\| = \|2z\|$

38. Let $z_1 = a + bi$ and $z_2 = c + di$. Prove that $\|z_1 + z_2\| \leq \|z_1\| + \|z_2\|$.
See page 491.

39. The inequality in Exercise 38 is called the *triangle inequality*. Relate z_1 and z_2 to two sides of a triangle and explain why this title is appropriate. Show that $\|z_2 - z_1\|$ is the distance from z_2 to z_1 in the complex number plane.
See page 491.

Applications

Geometry Complex numbers can be added on the complex number plane using the parallelogram law for vector addition. If $a + bi$ and $c + di$ are represented as vectors drawn from the origin to points (a, b) and (c, d), their sum is represented by the diagonal of the parallelogram.

Find each sum graphically. Check algebraically.

40. $4 + i$, $-4 + 5i$ $0 + 6i$

41. $3 + 2i$, $-2 + 4i$ $1 + 6i$

42. $6 - i$, $3 - 2i$ $9 - 3i$

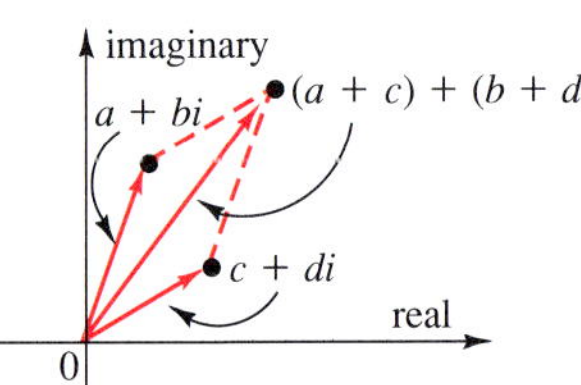

HISTORICAL NOTE

This table gives some of the high points in the history of complex numbers.

Century	Mathematicians	Contributions
1st	Heron	First wrote the square root of a negative number
16th	Jerome Cardan Rafael Bombelli	Used square roots of negative numbers to solve certain equations
17th	René Descartes	Coined the terms *real* and *imaginary*
	Abraham De Moivre	Introduced the important relation $(\cos \theta + i \sin \theta)^n = \cos n\theta + i \sin n\theta$
18th	Leonhard Euler	Denoted $\sqrt{-1}$ by i
19th	Karl Friedrich Gauss	Introduced the term *complex number*
	Augustin Louis Cauchy	Introduced the terms *conjugate* and *modulus*
	William Rowan Hamilton	Defined complex numbers as pairs of real numbers

Find out about other high points in the history of complex numbers.

7.6 Multiplying and Dividing Complex Numbers in Polar Form

Objective: To multiply and divide complex numbers in polar form

It is sometimes necessary to find an angle coterminal with a given angle when the product or quotient of complex numbers is computed.

Preview

Coterminal angles are angles in standard position whose terminal sides coincide. Their measures differ by an integral multiple of 360°. Thus, angles coterminal with angle θ have measures equal to $n \cdot 360° + \theta$, where n is an integer.

Find an angle between 0° and 360° that is coterminal with each angle.

1. $-75°$ 285° **2.** 395° 35° **3.** 700° 340° **4.** $-550°$ 170° **5.** 1025° 305° **6.** $-900°$ 180°

The addition or subtraction of two complex numbers expressed in rectangular form is a straightforward procedure. However, the multiplication or division of two complex numbers can require many steps. If the complex numbers are expressed in polar form, multiplication and division may be less complicated.

Let z_1 and z_2 be complex numbers, where $z_1 = r_1(\cos\theta_1 + i\sin\theta_1)$ and $z_2 = r_2(\cos\theta_2 + i\sin\theta_2)$. Then

$$\begin{aligned}
z_1z_2 &= r_1(\cos\theta_1 + i\sin\theta_1)\cdot r_2(\cos\theta_2 + i\sin\theta_2)\\
&= r_1r_2(\cos\theta_1 + i\sin\theta_1)(\cos\theta_2 + i\sin\theta_2)\\
&= r_1r_2(\cos\theta_1\cos\theta_2 + i\cos\theta_1\sin\theta_2 + i\sin\theta_1\cos\theta_2 + i^2\sin\theta_1\sin\theta_2)\\
&= r_1r_2(\cos\theta_1\cos\theta_2 + i\cos\theta_1\sin\theta_2 + i\sin\theta_1\cos\theta_2 - \sin\theta_1\sin\theta_2)\\
&= r_1r_2[(\cos\theta_1\cos\theta_2 - \sin\theta_1\sin\theta_2) + i(\cos\theta_1\sin\theta_2 + \sin\theta_1\cos\theta_2)]\\
&= r_1r_2[\cos(\theta_1 + \theta_2) + i\sin(\theta_1 + \theta_2)]
\end{aligned}$$

Use the sum identities for sine and cosine.

The product of two complex numbers $z_1 = r_1(\cos\theta_1 + i\sin\theta_1)$ and $z_2 = r_2(\cos\theta_2 + i\sin\theta_2)$ is given by the formula

$$z_1z_2 = r_1r_2[\cos(\theta_1 + \theta_2) + i\sin(\theta_1 + \theta_2)]$$

Notice that the modulus of the product of two complex numbers is the product of the two moduli. The argument of the product is the sum of the two arguments.

LESSON PLAN

BACKGROUND

In the Preview, the method for finding coterminal angles is reviewed. This skill may be necessary when multiplying or dividing complex numbers.

TEACHING SUGGESTIONS

- It may be necessary to review the terms modulus and argument.
- Show, or have students show, that multiplying and dividing complex numbers expressed in polar form is easier than multiplying or dividing complex numbers expressed in rectangular form and then changing to polar form.
- Emphasize that θ in $r(\cos\theta + i\sin\theta)$ should be the smallest positive angle measure between 0° and 360°.

Critical Thinking

Analysis Ask students to decide which form of complex numbers (rectangular or polar) is more advantageous and ask them to give reasons to support their choice. Answers may vary.

CHALKBOARD EXAMPLES

- **For Example 1**

  Find each product and simplify.

  1. $2(\cos 32° + i\sin 32°) \cdot 3(\cos 105° + i\sin 105°)$ $6(\cos 137° + i\sin 137°)$

  2. $\sqrt{2}(\cos 187° + i\sin 187°) \cdot 5(\cos 205° + i\sin 205°)$ $5\sqrt{2}(\cos 32° + i\sin 32°)$

- **For Example 2**

  Find each quotient and simplify.

  3. $\dfrac{12(\cos 240° + i\sin 240°)}{3(\cos 32° + i\sin 32°)}$ $4(\cos 208° + i\sin 208°)$

  4. $14\left(\cos\dfrac{3\pi}{4} + i\sin\dfrac{3\pi}{4}\right) \div 7\left(\cos\dfrac{3\pi}{4} + i\sin\dfrac{3\pi}{4}\right)$ 2

EXAMPLE 1 **Find the product and simplify: $3(\cos 27° + i\sin 27°) \cdot 5(\cos 82° + i\sin 82°)$**

$$
\begin{aligned}
3(\cos 27° + i\sin 27°) \cdot 5(\cos 82° + i\sin 82°) &= 3 \cdot 5[\cos(27° + 82°) + i\sin(27° + 82°)] \\
&= 15(\cos 109° + i\sin 109°)
\end{aligned}
$$

The procedure for the division of two complex numbers in polar form is very similar to the procedure for multiplication. Let z_1 and z_2 be complex numbers, where $z_1 = r_1(\cos\theta_1 + i\sin\theta_1)$ and $z_2 = r_2(\cos\theta_2 + i\sin\theta_2)$. Then

$$
\begin{aligned}
\frac{z_1}{z_2} &= \frac{r_1(\cos\theta_1 + i\sin\theta_1)}{r_2(\cos\theta_2 + i\sin\theta_2)} \\
&= \frac{r_1(\cos\theta_1 + i\sin\theta_1)}{r_2(\cos\theta_2 + i\sin\theta_2)} \cdot \left(\frac{\cos\theta_2 - i\sin\theta_2}{\cos\theta_2 - i\sin\theta_2}\right) \\
&= \frac{r_1}{r_2}\left[\frac{\cos\theta_1\cos\theta_2 - i\cos\theta_1\sin\theta_2 + i\sin\theta_1\cos\theta_2 - i^2\sin\theta_1\sin\theta_2}{\cos^2\theta_2 - i^2\sin^2\theta_2}\right] \\
&= \frac{r_1}{r_2}\left[\frac{(\cos\theta_1\cos\theta_2 + \sin\theta_1\sin\theta_2) + i(\sin\theta_1\cos\theta_2 - \cos\theta_1\sin\theta_2)}{\cos^2\theta_2 + \sin^2\theta_2}\right] \\
&= \frac{r_1}{r_2}[\cos(\theta_1 - \theta_2) + i\sin(\theta_1 - \theta_2)]
\end{aligned}
$$

$\cos^2\theta_2 + \sin^2\theta_2 = 1$; $\cos\theta_1\cos\theta_2 + \sin\theta_1\sin\theta_2 = \cos(\theta_1 - \theta_2)$; $\sin\theta_1\cos\theta_2 - \cos\theta_1\sin\theta_2 = \sin(\theta_1 - \theta_2)$

> The quotient of two complex numbers $z_1 = r_1(\cos\theta_1 + i\sin\theta_1)$ and $z_2 = r_2(\cos\theta_2 + i\sin\theta_2)$ is given by the formula
>
> $$\frac{z_1}{z_2} = \frac{r_1}{r_2}[\cos(\theta_1 - \theta_2) + i\sin(\theta_1 - \theta_2)],\ z_2 \neq 0$$

Notice that the modulus of the quotient of two complex numbers is the quotient of their moduli, and the argument of the quotient is the difference of their arguments.

EXAMPLE 2 **Find the quotient and simplify: $\dfrac{9(\cos 54° + i\sin 54°)}{18(\cos 77° + i\sin 77°)}$**

$$
\begin{aligned}
\frac{9(\cos 54° + i\sin 54°)}{18(\cos 77° + i\sin 77°)} &= \frac{9}{18}[\cos(54° - 77°) + i\sin(54° - 77°)] \\
&= \frac{1}{2}[\cos(-23°) + i\sin(-23°)] \\
&= \frac{1}{2}(\cos 337° + i\sin 337°)
\end{aligned}
$$

A 337° angle is coterminal with a −23° angle.

You may be required to express a complex number in polar form before using the formulas discussed in this lesson.

EXAMPLE 3 **Calculate the quotient $\dfrac{3 + 5i}{2 + 4i}$ using the formulas in this lesson.**

First express $3 + 5i$ and $2 + 4i$ in polar form.

For $3 + 5i$: $r = \sqrt{3^2 + 5^2} = \sqrt{34}$, $\theta = \text{Arctan}\,\frac{5}{3} = 59°$

For $2 + 4i$: $r = \sqrt{2^2 + 4^2} = 2\sqrt{5}$, $\theta = \text{Arctan}\,\frac{4}{2} = 63°$

$3 + 5i = \sqrt{34}\,(\cos 59° + i \sin 59°)$ $\qquad$ $2 + 4i = 2\sqrt{5}\,(\cos 63° + i \sin 63°)$

Therefore,
$$\frac{3 + 5i}{2 + 4i} = \frac{\sqrt{34}\,(\cos 59° + i \sin 59°)}{2\sqrt{5}\,(\cos 63° + i \sin 63°)}$$
$$= \frac{\sqrt{170}}{10}[\cos (59° - 63°) + i \sin (59° - 63°)]$$
$$= \frac{\sqrt{170}}{10}[\cos (-4°) + i \sin (-4°)] = \frac{\sqrt{170}}{10}[\cos 356° + i \sin 356°]$$

CLASS EXERCISES

Find r and θ for each product or quotient.

1. $2(\cos 4° + i \sin 4°) \cdot 3(\cos 10° + i \sin 10°)$ $r = 6;\ \theta = 14°$
2. $6(\cos 30° + i \sin 30°) \cdot 5(\cos 4° + i \sin 4°)$ $r = 30;\ \theta = 34°$
3. $3(\cos 15° + i \sin 15°) \cdot 6(\cos 2° + i \sin 2°)$ $r = 18;\ \theta = 17°$
4. $2(\cos 150° + i \sin 150°) \cdot 6(\cos 2° + i \sin 2°)$ $r = 12;\ \theta = 152°$
5. $16(\cos 80° + i \sin 80°) \div 2(\cos 100° + i \sin 100°)$ $r = 8;\ \theta = 340°$
6. $30(\cos 70° + i \sin 70°) \div 5(\cos 10° + i \sin 10°)$ $r = 6;\ \theta = 60°$
7. $\dfrac{14(\cos 60° + i \sin 60°)}{2(\cos 50° + i \sin 50°)}$ $r = 7;\ \theta = 10°$
8. $\dfrac{21(\cos 90° + i \sin 90°)}{3(\cos 110° + i \sin 110°)}$ $r = 7;\ \theta = 340°$

PRACTICE EXERCISES

Find each product or quotient.

A

1. $2(\cos 70° + i \sin 70°) \cdot 3(\cos 60° + i \sin 60°)$ $6(\cos 130° + i \sin 130°)$
2. $4(\cos 80° + i \sin 80°) \cdot 3(\cos 50° + i \sin 50°)$ $12(\cos 130° + i \sin 130°)$
3. $4\left(\cos \frac{2\pi}{3} + i \sin \frac{2\pi}{3}\right) \cdot 6\left(\cos \frac{\pi}{4} + i \sin \frac{\pi}{4}\right)$ $24\left(\cos \frac{11\pi}{12} + i \sin \frac{11\pi}{12}\right)$
4. $2\left(\cos \frac{5\pi}{6} + i \sin \frac{5\pi}{6}\right) \cdot 4\left(\cos \frac{\pi}{3} + i \sin \frac{\pi}{3}\right)$ $8\left(\cos \frac{7\pi}{6} + i \sin \frac{7\pi}{6}\right)$

- **For Example 3**

Calculate each quotient using the formulas in this lesson.

5. $\dfrac{1 - i}{\sqrt{3} + i}$ $\frac{\sqrt{2}}{2}(\cos 285° + i \sin 285°)$
6. $(2 + 2i) \div (1 - i\sqrt{3})$ $\sqrt{2}(\cos 105° + i \sin 105°)$

LESSON FOLLOW-UP

Discussion

If $z = 4\left(\cos \frac{\pi}{4} + i \sin \frac{\pi}{4}\right)$, how would you find z^3? Find z^3. Multiply $4\left(\cos \frac{\pi}{4} + i \sin \frac{\pi}{4}\right)$ by itself then multiply the result by $4\left(\cos \frac{\pi}{4} + i \sin \frac{\pi}{4}\right)$; $z^3 = \left[4\left(\cos \frac{\pi}{4} + i \sin \frac{\pi}{4}\right) \cdot 4\left(\cos \frac{\pi}{4} + i \sin \frac{\pi}{4}\right)\right] \cdot \left[4\left(\cos \frac{\pi}{4} + i \sin \frac{\pi}{4}\right)\right] = \left[16\left(\cos \frac{\pi}{2} + i \sin \frac{\pi}{2}\right)\right] \cdot \left[4\left(\cos \frac{\pi}{2} + i \sin \frac{\pi}{4}\right)\right] = 64\left(\cos \frac{3\pi}{4} + i \sin \frac{3\pi}{4}\right)$

Assignment Guide

See p. 284B for assignments.

5. $7(\cos 63° + i \sin 63°) \cdot 12(\cos 89° + i \sin 89°)$ $84(\cos 152° + i \sin 152°)$

6. $6(\cos 48° + i \sin 48°) \cdot 15(\cos 94° + i \sin 94°)$ $90(\cos 142° + i \sin 142°)$

7. $10(\cos 176° + i \sin 176°) \cdot 5(\cos 216° + i \sin 216°)$ $50(\cos 32° + i \sin 32°)$

8. $9(\cos 214° + i \sin 214°) \cdot 11(\cos 215° + i \sin 215°)$ $99(\cos 69° + i \sin 69°)$

9. $7\left(\cos \frac{5\pi}{2} + i \sin \frac{5\pi}{2}\right) \cdot 8\left(\cos \frac{3\pi}{4} + i \sin \frac{3\pi}{4}\right)$ $56\left(\cos \frac{5\pi}{4} + i \sin \frac{5\pi}{4}\right)$

10. $6\left(\cos \frac{7\pi}{3} + i \sin \frac{7\pi}{3}\right) \cdot 9\left(\cos \frac{5\pi}{6} + i \sin \frac{5\pi}{6}\right)$ $54\left(\cos \frac{7\pi}{6} + i \sin \frac{7\pi}{6}\right)$

11. $8(\cos 30° + i \sin 30°) \div 2(\cos 10° + i \sin 10°)$ $4(\cos 20° + i \sin 20°)$

12. $12(\cos 40° + i \sin 40°) \div 4(\cos 15° + i \sin 15°)$ $3(\cos 25° + i \sin 25°)$

13. $20(\cos 86° + i \sin 86°) \div 5(\cos 61° + i \sin 61°)$ $4(\cos 25° + i \sin 25°)$

14. $36(\cos 104° + i \sin 104°) \div 9(\cos 79° + i \sin 79°)$ $4(\cos 25° + i \sin 25°)$

15. $7\left(\cos \frac{2\pi}{3} + i \sin \frac{2\pi}{3}\right) \div 2\left(\cos \frac{\pi}{6} + i \sin \frac{\pi}{6}\right)$ $\frac{7}{2}\left(\cos \frac{\pi}{2} + i \sin \frac{\pi}{2}\right)$

16. $8\left(\cos \frac{3\pi}{4} + i \sin \frac{3\pi}{4}\right) \div 5\left(\cos \frac{\pi}{2} + i \sin \frac{\pi}{2}\right)$ $\frac{8}{5}\left(\cos \frac{\pi}{4} + i \sin \frac{\pi}{4}\right)$

Calculate each product or quotient in polar form. Check by repeating the calculations using the rectangular form of the numbers. 19. $\frac{10}{7}(\cos 150° + i \sin 150°)$

B 17. $8(\cos 20° + i \sin 20°) \cdot 2(\cos 25° + i \sin 25°)$ $16(\cos 45° + i \sin 45°)$

18. $7(\cos 60° + i \sin 60°) \cdot 4(\cos 135° + i \sin 135°)$ $28(\cos 195° + i \sin 195°)$

19. $\dfrac{10(\cos 330° + i \sin 330°)}{7(\cos 180° + i \sin 180°)}$

20. $\dfrac{9(\cos 240° + i \sin 240°)}{5(\cos 120° + i \sin 120°)}$ $\frac{9}{5}$(cos 120° + i sin 12

21. $(3 + 4i)(\sqrt{3} - i)$ $10(\cos 23° + i \sin 23°)$

22. $(4\sqrt{3} - 4i)(3 + 3i)$ $24\sqrt{2}$(cos 15° + i sin 15

23. $(3 + i)(2i)$ $2\sqrt{10}(\cos 108° + i \sin 108°)$

24. $(6 + 2i)(-i)$ $2\sqrt{10}(\cos 288° + i \sin 288°)$

26. $2\sqrt{10}(\cos 108° + i \sin 108°)$

25. $\dfrac{3 + i}{2i}$

26. $\dfrac{6 + 2i}{-i}$

27. $\dfrac{3 + 4i}{\sqrt{3} - i}$

28. $\dfrac{4\sqrt{3} - 4i}{3 + 3i}$

25. $\frac{\sqrt{10}}{2}(\cos 288° + i \sin 288°)$ $\frac{5}{2}(\cos 83° + i \sin 83°)$ $\frac{4\sqrt{2}}{3}(\cos 285° + i \sin 285°)$

C 29. Use repeated multiplication to show that $\text{cis}\ \frac{\pi}{3}$ is a solution of $z^6 = 1$. See side column.

30. Use repeated multiplication to show that cis 45° is a solution of $z^8 = 1$.

The polar form of $z = a + bi$ is $z = r \text{ cis } \theta$. The conjugate of z, denoted $\bar{z}$, is $a - bi$. Prove each of the following statements. See side column on page 317.

31. $\bar{z} = r \text{ cis }(-\theta)$

32. $\frac{1}{z} = \frac{1}{r} \text{ cis }(-\theta)$

33. $r^2 = z \cdot \bar{z}$

Challenge

Given complex numbers A, B, and C in polar form and similar triangles AOC and DOB, students are challenged to show that $D = A \cdot B$.

Lesson Quiz

Find each product.

1. $4(\cos 43° + i \sin 43°) \cdot 3(\cos 32° + i \sin 32°)$ $12(\cos 75° + i \sin 75°)$
2. $5(\cos 340° + i \sin 340°) \cdot 6(\cos 160° + i \sin 160°)$ $30(\cos 140° + i \sin 140°)$

Find each quotient.

3. $24(\cos 114° + i \sin 114°) \div 8(\cos 58° + i \sin 58°)$ $3(\cos 56° + i \sin 56°)$
4. $7(\cos 127° + \sin 127°) \div 3(\cos 241° + i \sin 241°)$ $\frac{7}{3}(\cos 246° + i \sin 246°)$
5. Calculate the product $6(\cos 70° + i \sin 70°) \cdot 3(\cos 50° + i \sin 50°)$ in polar form, then express the answer in rectangular form. $18(\cos 120° + i \sin 120°)$; $-9 + 9i\sqrt{3}$

Enrichment

Given: $A = -\frac{1}{2} + \frac{i\sqrt{3}}{2}$

Prove: $A^4 = A$

$A = -\frac{1}{2} + \frac{i\sqrt{3}}{2} = \cos 120° + i \sin 120°$
$A^4 = [(\cos 120° + i \sin 120°) \cdot (\cos 120° + i \sin 120°)] \cdot [(\cos 120° + i \sin 120°) \cdot (\cos 120° + i \sin 120°)]$; $A^4 = (\cos 240° + i \sin 240°) \cdot (\cos 240° + i \sin 240°)$; $A^4 = \cos 480° + i \sin 480°$; $A^4 = \cos 120° + i \sin 120°$; $A^4 = A$

Additional Answers

29. $\left(\text{cis}\frac{\pi}{3}\right)\left(\text{cis}\frac{\pi}{3}\right)\left(\text{cis}\frac{\pi}{3}\right)\left(\text{cis}\frac{\pi}{3}\right)\left(\text{cis}\frac{\pi}{3}\right)\left(\text{cis}\frac{\pi}{3}\right) = 1 \text{ cis}\left(\frac{\pi}{3} + \frac{\pi}{3} + \frac{\pi}{3} + \frac{\pi}{3} + \frac{\pi}{3} + \frac{\pi}{3}\right) = \text{cis } 2\pi = \cos 2\pi + i \sin 2\pi = 1 + 0 = 1$

30. (cis 45°) (cis 45°) (cis 45°) (cis 45°) (cis 45°) (cis 45°) (cis 45°) (cis 45°) = cis (45 + 45 + 45 + 45 + 45 + 45 + 45 + 45) = cis 360° = cos 360° + i sin 360° = 1 + 0 = 1

Applications

Computer The program below can be used to find the product of two complex numbers in polar form. The user supplies the values of A, B, W, and S in the expressions A cis W and B cis S.

34. If 4 cis 225° and 6 cis 245° are multiplied, the computer output will be 24 cis 470°. Write an equivalent expression, using a coterminal angle. 24 cis 110°

35. If 2 cis 40° and 6 cis 40° are multiplied using this program, predict the output. Find the rectangular form of the product. 12 cis 80°; $2.08 + 11.82i$

```
10 HOME
20 INPUT "ENTER THE VALUE OF A
   IN A CIS W";A
30 INPUT "ENTER W";W
40 INPUT "ENTER THE VALUE OF B
   IN B CIS S";S
50 INPUT "ENTER S";S
60 HOME: PRINT"THE PRODUCT OF
   ("A" CIS "W") AND ("B" CIS "S") IS"
70 PRINT A*B "CIS" W + S
80 END
```

The program calculates the quotient of A cis W and B cis S. The values of A, B, W, and S are supplied by the user.

36. If 2 cis 30° is divided by 4 cis 150°, find the program output. Write an equivalent expression using a coterminal angle with a positive measure. $\frac{1}{2}$ cis (−120°); $\frac{1}{2}$ cis 240°

37. Suppose that $5 + 4i$ is to be divided by $3 + 2i$. Express the numbers in polar form and predict the program output. 1.78 cis (4.97°)

```
10 HOME
20 INPUT "ENTER THE VALUE OF A
   IN A CIS W";A
30 INPUT "ENTER W";W
40 INPUT "ENTER THE VALUE OF B
   IN B CIS S";S
50 INPUT "ENTER S";S
60 HOME: PRINT "THE QUOTIENT OF
   ("A" CIS "W") AND ("B" CIS "S") IS"
70 PRINT A/B "CIS" W - S
80 END
```

CHALLENGE

Given: A, B, C, and D are complex numbers.

$$A = r_1(\cos \alpha + i \sin \alpha)$$
$$B = r_2(\cos \beta + i \sin \beta)$$
$$C = 1(\cos 0° + i \sin 0°) = 1$$

Also, triangle AOC is similar to triangle DOB. See page 491.

Show that $D = A \cdot B$.

Hint: Let the modulus of D be r_3. Find a relationship between r_1, r_2, and r_3.

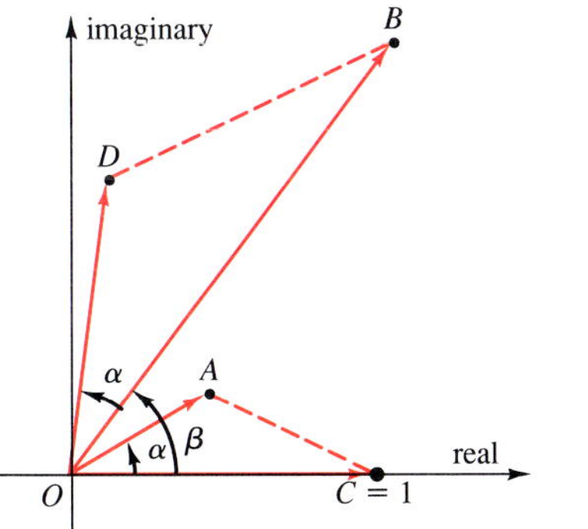

Teacher's Resource Book
Practice—Chapter 7, p. 11
Enrichment—Chapter 7, p. 12

Additional Answers

31. $z = a + bi = r \text{ cis } \theta$; $\sqrt{a^2 + b^2} = r$ and $\frac{a}{r} = \cos \theta$, $\frac{b}{r} = \sin \theta$. If $\bar{z} = a - bi = r_1 \text{ cis } \theta_1$, then $r_1 = \sqrt{a^2 + (-b)^2} = r$ and $\frac{a}{r} = \cos \theta_1$; $-\frac{b}{r} = \sin \theta_1$. From the unit circle notice that $\theta_1 = -\theta$. Therefore, $r_1 \text{ cis } \theta_1 = r \text{ cis } (-\theta)$

32. $\frac{1}{z} = \frac{1}{a + bi} \cdot \frac{a - bi}{a - bi} = \frac{a - bi}{a^2 + b^2} = \frac{r \text{ cis } (-\theta)}{r^2} = \frac{1}{r} \text{ cis } (-\theta)$

33. $z = r(\cos \theta + i \sin \theta)$; $\bar{z} = r[\cos(-\theta) + i \sin(-\theta)]$; $z \cdot \bar{z} = r(\cos \theta + i \sin \theta) \cdot r[\cos(-\theta) + i \sin(-\theta)] = r \cdot r[\cos(\theta + (-\theta)) + i \sin(\theta + (-\theta))] = r^2(\cos 0 + i \sin 0) = r^2(1 + 0) = r^2$

LESSON PLAN

BACKGROUND

In the Preview, the angle-sum and double-angle identities are used to derive identities for sine and cosine of 3θ and 4θ.

Additional Answers

1. $\cos^3\theta - 3\sin^2\theta\cos\theta$
2. $4\sin\theta\cos\theta(\cos^2\theta - \sin^2\theta)$
3. $\cos^4\theta - 6\cos^2\theta\sin^2\theta + \sin^4\theta$

7.7 De Moivre's Theorem

Objective: To use De Moivre's theorem to evaluate powers of complex numbers

De Moivre's theorem will enable you to find powers of a complex number expressed in polar form and to derive certain identities using complex numbers in polar form.

Preview

The angle-sum and double-angle identities can be used to derive identities for multiples of an angle measure.

EXAMPLE **Derive an identity for $\sin 3\theta$ in terms of $\sin\theta$ and $\cos\theta$.**

$$\begin{aligned}\sin 3\theta &= \sin(2\theta + \theta)\\ &= \sin 2\theta\cos\theta + \cos 2\theta\sin\theta\\ &= (2\sin\theta\cos\theta)\cos\theta + (\cos^2\theta - \sin^2\theta)\sin\theta\\ &= 2\sin\theta\cos^2\theta + \cos^2\theta\sin\theta - \sin^3\theta\\ &= 3\cos^2\theta\sin\theta - \sin^3\theta\end{aligned}$$

Derive an identity for each expression in terms of $\sin\theta$ and $\cos\theta$.

See side column.

1. $\cos 3\theta$
2. $\sin 4\theta$
3. $\cos 4\theta$

To square a complex number expressed in polar form, multiply the number by itself using the product rule from the last lesson.

$$\begin{aligned}[r(\cos\theta + i\sin\theta)]^2 &= [r(\cos\theta + i\sin\theta)][r(\cos\theta + i\sin\theta)]\\ &= r\cdot r[\cos(\theta+\theta) + i\sin(\theta+\theta)]\\ &= r^2(\cos 2\theta + i\sin 2\theta)\end{aligned}$$

Notice that 2 is both the exponent of r and the coefficient of θ.

To cube a complex number expressed in polar form, use the number as a factor three times.

$$\begin{aligned}[r(\cos\theta + i\sin\theta)]^3 &= [r(\cos\theta + i\sin\theta)][r(\cos\theta + i\sin\theta)][r(\cos\theta + i\sin\theta)]\\ &= [r(\cos\theta + i\sin\theta)][r^2(\cos 2\theta + i\sin 2\theta)]\\ &= r^3[\cos(\theta + 2\theta) + i\sin(\theta + 2\theta)]\\ &= r^3(\cos 3\theta + i\sin 3\theta)\end{aligned}$$

Similarly, $[r(\cos \theta + i \sin \theta)]^4 = r^4(\cos 4\theta + i \sin 4\theta)$, and so forth. *De Moivre's theorem* provides a general formula for finding any power of a complex number in polar form.

De Moivre's Theorem If z is a complex number in polar form, and n is a positive integer, then

$$z^n = [r(\cos \theta + i \sin \theta)]^n = r^n(\cos n\theta + i \sin n\theta)$$

EXAMPLE 1 **Evaluate $[2(\cos 72° + i \sin 72°)]^5$. Then express the answer in rectangular form.**

$$\begin{aligned}[2(\cos 72° + i \sin 72°)]^5 &= 2^5[\cos 5(72°) + i \sin 5(72°)]^5 && \textit{Apply De Moivre's theorem.}\\ &= 32(\cos 360° + i \sin 360°)\\ &= 32[1 + i(0)] = 32\end{aligned}$$

In order to apply De Moivre's theorem, complex numbers must be in polar form.

EXAMPLE 2 **Find $(1 + i\sqrt{3})^4$. Express the answer in rectangular form.**

First write $1 + i\sqrt{3}$ in polar form.

$$r = \sqrt{1^2 + (\sqrt{3})^2} = 2 \qquad \theta = \text{Arctan}\,\frac{\sqrt{3}}{1} = 60°$$

Therefore, $1 + i\sqrt{3} = 2(\cos 60° + i \sin 60°)$

$$\begin{aligned}\text{Then,}\quad (1 + i\sqrt{3})^4 &= [2(\cos 60° + i \sin 60°)]^4\\ &= 2^4[\cos 4(60°) + i \sin 4(60°)] && \textit{Apply De Moivre's theorem.}\\ &= 16(\cos 240° + i \sin 240°)\\ &= 16\left[-\frac{1}{2} + \left(-\frac{\sqrt{3}}{2}\right)i\right] = -8 - 8i\sqrt{3}\end{aligned}$$

De Moivre's theorem also applies when n is a negative integer, if $a + bi \neq 0$.

EXAMPLE 3 **Find $(-1 + i)^{-4}$. Express the answer in rectangular form.**

First write $-1 + i$ in polar form. $r = \sqrt{(-1)^2 + 1^2} = \sqrt{2}$ and $\theta = \text{Arctan}\,\frac{1}{-1} + \pi = \frac{3\pi}{4}$, so $-1 + i = \sqrt{2}\left(\cos \frac{3\pi}{4} + i \sin \frac{3\pi}{4}\right)$.

$$\begin{aligned}\text{Then,}\quad (-1 + i)^{-4} &= (\sqrt{2})^{-4}\left[\cos(-4)\left(\frac{3\pi}{4}\right) + i \sin(-4)\left(\frac{3\pi}{4}\right)\right]\\ &= \frac{1}{4}[\cos(-3\pi) + i \sin(-3\pi)] = \frac{1}{4}[-1 + i(0)] = -\frac{1}{4}\end{aligned}$$

TEACHING SUGGESTIONS

- To emphasize the value of De Moivre's theorem, ask students to simplify $(3 + 2i)^6$, change the result to polar form, and then simplify using De Moivre's theorem.
- Emphasize that directions should be read carefully, since the form of an answer depends on the directions.
- Stress that De Moivre's theorem is only applicable to complex numbers expressed in polar form.
- Note that there are three forms of $\cos 2\theta$. Therefore, students may obtain different answers when deriving identities for $\sin n\theta$ or $\cos n\theta$, where n is an integer.

Critical Thinking

Analysis Ask students to discuss De Moivre's theorem and its implications on mathematical operations with complex numbers. Answers may vary.

CHALKBOARD EXAMPLES

- **For Example 1**

  Evaluate each power. Express the answer in rectangular form.

  1. $[2(\cos 15° + i \sin 15°)]^4$ $8 + 8i\sqrt{3}$

  2. $[3(\cos 36° + i \sin 36°)]^5$ $-243 + 0i$

- **For Example 2**

  Find each power. Express the answer in rectangular form.

  3. $(1 + i)^5$ $-4 + -4i$

  4. $(3 - 4i)^3$ $-117 - 44i$

- **For Example 3**

  Find each power. Express the answer in rectangular form.

  5. $(\sqrt{3} + i)^{-4}$ $-\frac{1}{32} - \frac{i\sqrt{3}}{32}$

  6. $(-1)^{-5}$ $-1 + 0i$

- **For Example 4**

  7. Use De Moivre's theorem to derive an identity for $\sin 3\theta$ in terms of $\cos \theta$ and $\sin \theta$.
  $\sin 3\theta = 3 \cos^2 \theta \sin \theta - \sin^3 \theta$

Common Error

- Some students evaluate exponential expressions incorrectly. Having those students use calculators should eliminate this mistake.
- See *Teacher's Resource Book* for additional remediation.

LESSON FOLLOW-UP

Discussion

Find the multiplicative inverse of $3\sqrt{3} + 3i$ using DeMoivre's theorem and compare the answer to the one found by simplifying $\frac{1}{3\sqrt{3} + 3i}$.

$\frac{1}{3\sqrt{3} + 3i} = (3\sqrt{3} + 3i)^{-1} =$
$(6 \text{ cis } 30°)^{-1} = \frac{1}{6} \text{cis}(-30°) =$
$\frac{1}{6}(\cos 330° + i \sin 330°)$; $\frac{1}{3\sqrt{3} + 3i} \cdot$
$\frac{3\sqrt{3} - 3i}{3\sqrt{3} - 3i} = \frac{\sqrt{3}}{12} - \frac{i}{12} =$
$\frac{1}{6}(\cos 330° + i \sin 330°)$

Assignment Guide

See p. 284B for assignments.

Extra

Solving equations involving cosines is discussed.

The result in Example 3 can be checked as follows:

$$\begin{aligned}(-1 + i)^{-4} &= \frac{1}{(-1 + i)^4} \\ &= \frac{1}{(-1 + i)^2(-1 + i)^2} \\ &= \frac{1}{(1 - 2i + i^2)(1 - 2i + i^2)} \\ &= \frac{1}{(-2i)(-2i)} = \frac{1}{4i^2} = \frac{1}{-4} = -\frac{1}{4}\end{aligned}$$

Trigonometric identities that involve the representation of $\cos n\theta$ and $\sin n\theta$ in terms of $\cos \theta$ and $\sin \theta$, where n is a positive integer, can be derived using De Moivre's theorem.

EXAMPLE 4 **Derive the identities for $\cos 2\theta$ and $\sin 2\theta$ in terms of $\cos \theta$ and $\sin \theta$.**

$$(\cos \theta + i \sin \theta)^2 = \cos 2\theta + i \sin 2\theta \quad \textit{De Moivre's theorem}$$
$$\cos^2\theta + 2i \cos \theta \sin \theta + i^2\sin^2\theta = \cos 2\theta + i \sin 2\theta$$
$$(\cos^2\theta - \sin^2\theta) + 2i \cos \theta \sin \theta = \cos 2\theta + i \sin 2\theta$$

Now use the last equation above and the definition of equality for two complex numbers.

$$\cos^2\theta - \sin^2\theta = \cos 2\theta \quad \textit{The real parts are equal.}$$
$$2 \cos \theta \sin \theta = \sin 2\theta \quad \textit{The imaginary parts are equal.}$$

CLASS EXERCISES

Match each expression on the left with an equivalent expression on the right.

1. $[2(\cos 10° + i \sin 10°)]^6$ c
2. $[4(\cos 10° + i \sin 10°)]^3$ e
3. $[3(\cos 15° + i \sin 15°)]^4$ a
4. $[9(\cos 15° + i \sin 15°)]^2$ f
5. $(\cos 12° + i \sin 12°)^9$ d
6. $[3(\cos 54° + i \sin 54°)]^2$ b

a. $81(\cos 60° + i \sin 60°)$
b. $9(\cos 108° + i \sin 108°)$
c. $64(\cos 60° + i \sin 60°)$
d. $\cos 108° + i \sin 108°$
e. $64(\cos 30° + i \sin 30°)$
f. $81(\cos 30° + i \sin 30°)$

PRACTICE EXERCISES

Apply De Moivre's theorem to evaluate each power. Express your answers in rectangular form.

A

1. $[3(\cos 40° + i \sin 40°)]^3$ $\frac{-27}{2} + \frac{27\sqrt{3}}{2}i$
2. $[4(\cos 30° + i \sin 30°)]^5$ $-512\sqrt{3} + 512i$
3. $[5(\cos 15° + i \sin 15°)]^3$ $\frac{125\sqrt{2}}{2} + \frac{125\sqrt{2}}{2}i$
4. $[5(\cos 45° + i \sin 45°)]^3$ $-\frac{125\sqrt{2}}{2} + \frac{125\sqrt{2}}{2}i$

5. $\{2[\cos(-20°) + i\sin(-20°)]\}^3$ $4 - 4i\sqrt{3}$

6. $\{2[\cos(-10°) + i\sin(-10°)]\}^3$ $4\sqrt{3} - 4i$

7. $(\cos 30° + i\sin 30°)^3$ i

8. $(\cos 45° + i\sin 45°)^6$ $-i$

9. $[6(\cos 200° + i\sin 200°)]^3$ $-108 - 108i\sqrt{3}$

10. $[3(\cos 240° + i\sin 240°)]^3$ 27

11. $(2 - 2i)^4$ -64

12. $(1 - i\sqrt{3})^5$ $16 + 16i\sqrt{3}$

13. $(\sqrt{3} + i)^3$ $8i$

14. $(-1 + i)^4$ -4

B 15. $(-1 + i)^6$ $8i$

16. $(2\sqrt{2} - 2i\sqrt{2})^4$ -256

17. $(3 - 3i)^3$ $-54 - 54i$

18. $(-\sqrt{3} - i)^3$ $-8i$

19. i^{-5} $-i$

20. $(-i)^{-4}$ 1

21. $(2 \text{ cis } 60°)^{-3}$ $-\frac{1}{8}$

22. $(4 \text{ cis } 30°)^{-4}$ $-\frac{1}{512} - \frac{\sqrt{3}}{512}i$

Use De Moivre's theorem to derive an identity for each expression in terms of sin θ and cos θ.

23. $\cos 3\theta$ $4\cos^3\theta - 3\cos\theta$

24. $\sin 3\theta$ $3\sin\theta - 4\sin^3\theta$

C 25. $\sin 4\theta$ $4\cos^3\theta\sin\theta - 4\cos\theta\sin^3\theta$

26. $\cos 4\theta$ $\cos^4\theta + \sin^4\theta - 6\cos^2\theta\sin^2\theta$

Apply De Moivre's theorem and then simplify. Express answers in rectangular form.

27. $\dfrac{(1 + i)^3}{1 + i\sqrt{3}}$ $\dfrac{-1 + \sqrt{3}}{2} + \left(\dfrac{1 + \sqrt{3}}{2}\right)i$

28. $\dfrac{(2 - 2i)^4}{-1 - i\sqrt{3}}$ $+16 - 16i\sqrt{3}$

Applications

29. **Algebra** Show that $\cos 45° + i\sin 45°$ is a solution of $x^4 + x^2 = -1 + i$. See side column.

30. **Algebra** Show that $\cos\frac{\pi}{6} + i\sin\frac{\pi}{6}$ is a solution of $x^{12} + x^6 + x^3 - i = 0$.

EXTRA

To solve some equations involving cosines, use the fact that $\cos a = \cos b$ if and only if $a = b + k(360°)$ or $a = -b + k(360°)$, where k is an integer.

EXAMPLE Solve: $\cos(660° - 3x) = \cos 2x$

$660° - 3x = 2x + k(360°)$ or $660° - 3x = -2x + k(360°)$

$\dfrac{660° - k(360°)}{5} = x$ or $660° - k(360°) = x$

Substitute 1 for k to obtain two solutions.

$$x = \frac{660° - 1(360°)}{5} = \frac{300°}{5} = 60° \qquad x = 660° - 1(360°) = 300°$$

1. Use $k = 2$ and $k = 3$ to find additional solutions of $\cos(660° - 3x) = \cos 2x$. $-12°$, $-60°$, $-84°$, $-420°$
2. Find at least four solutions of the equation $\cos(150° - 2x) = \cos 4x$, using the method outlined above. $-35°$, $105°$, $-95°$, $285°$

Lesson Quiz

1. Evaluate $(\sqrt{2} + i\sqrt{2})^4$. Express answer in both polar and rectangular form.
   $16(\cos 180° + i\sin 180°)$; $-16 + 0i$
2. Evaluate $\left(\dfrac{1}{4} + \dfrac{\sqrt{3}i}{4}\right)^{-3}$. Express the answer in rectangular form.
   $-8 + 0i$

Apply DeMoivre's theorem to evaluate each power. Then express the answer in rectangular form.

3. $[2(\cos 60° + i\sin 60°)]^3$
   $8(\cos 180° + i\sin 180°)$; $-8 + 0i$
4. $\left[3\left(\cos\dfrac{\pi}{8} + i\sin\dfrac{\pi}{8}\right)\right]^4$
   $81\left(\cos\dfrac{\pi}{2} + i\sin\dfrac{\pi}{2}\right)$; $0 + 81i$
5. Evaluate $\dfrac{(1 - i)^4}{(2\sqrt{3} + 2i)^3}$. Express the answer in rectangular form.
   $\dfrac{1}{16}i$; or $0 + \dfrac{1}{16}i$

Enrichment

Evaluate $(4 + 2i)^{\frac{2}{3}}$.

$(4 + 2i)^{\frac{2}{3}} =$
$[(4.47)^{\frac{1}{3}}(\cos 26.6° + i\sin 26.6°)^{\frac{1}{3}}]^2 =$
$[1.65(\cos 8.87° + i\sin 8.87°)]^2 =$
$2.72(\cos 17.74° + i\sin 17.74°)$
$2.72(\cos 257.74° + i\sin 257.74°)$
$2.72(\cos 137.74° + i\sin 137.74°)$

Teacher's Resource Book

Practice—Chapter 7, p. 13
Enrichment—Chapter 7, p. 14

Additional Answers

29. $(\text{cis } 45°)^4 + (\text{cis } 45°)^2 = \text{cis } 180° + \text{cis } 90° = (\cos 180° + i\sin 180°) + (\cos 90° + i\sin 90°) = (-1 + 0i) + (0 + i) = -1 + i$

30. $\left(\text{cis}\dfrac{\pi}{6}\right)^{12} + \left(\text{cis}\dfrac{\pi}{6}\right)^6 + \left(\text{cis}\dfrac{\pi}{6}\right)^3 - i = \text{cis } 2\pi + \text{cis } \pi + \text{cis}\dfrac{\pi}{2} - i = (1 + 0i) + (-1 + 0i) + (0 + i) - i = i - i = 0$

LESSON PLAN

BACKGROUND

In the Preview, finding the real solutions of equations of degree two and higher is reviewed.

7.8 Roots of Complex Numbers

Objective: To find the n distinct roots of a complex number

To find any possible real nth roots of a number a, you can solve an equation of the form $x^n - a = 0$, or $x^n = a$.

Preview

The real solutions of a quadratic equation such as $x^2 - 4 = 0$ can be found by factoring or by taking square roots.

$$\begin{aligned} x^2 - 4 &= 0 \\ (x + 2)(x - 2) &= 0 \quad \textit{Factor.} \\ x + 2 = 0 \quad &\text{or} \quad x - 2 = 0 \\ x = -2 \quad &\text{or} \quad x = 2 \end{aligned}$$

$$\begin{aligned} x^2 - 4 &= 0 \\ x^2 &= 4 \\ x &= \pm 2 \quad \textit{Take the square roots.} \\ x &= -2 \text{ or } x = 2 \end{aligned}$$

Find the real solution(s) of each equation.

1. $x^2 - 169 = 0$ 13, −13 **2.** $x^2 - 12 = 0$ $2\sqrt{3}, -2\sqrt{3}$ **3.** $x^3 - 8 = 0$ 2

4. $x^3 + 27 = 0$ −3 **5.** $x^4 - 256 = 0$ 4, −4 **6.** $x^5 + 1 = 0$ −1

Consider the cubic (third degree) equation $x^3 = 64$. If the cube root of each side is taken, the real root $x = 4$ is obtained. There are, however, two other roots of 64 that are complex numbers. The *complex roots theorem*, which can be shown to follow from De Moivre's theorem, is used to find all of the nth roots of a complex number.

Complex Roots Theorem If n is a positive integer, then $a + bi = r(\cos\theta + i\sin\theta)$ has n distinct roots, which are given by

$$\sqrt[n]{r}\left[\cos\frac{1}{n}(\theta + k\cdot 360^\circ) + i\sin\frac{1}{n}(\theta + k\cdot 360^\circ)\right] \quad \text{or}$$

$$\sqrt[n]{r}\left[\cos\frac{1}{n}(\theta + 2k\pi) + i\sin\frac{1}{n}(\theta + 2k\pi)\right] \quad \text{where}$$

$k = 0, 1, 2, \ldots, n - 1$

EXAMPLE 1 **Find the three complex roots of 64. That is, find all of the complex roots of $x^3 = 64$. Express the roots in rectangular form.**

$64 = 64 + 0i = 64(\cos 0° + i \sin 0°)$ *Express 64 in polar form.*

$$x^3 = 64(\cos 0° + i \sin 0°)$$

$$x = \sqrt[3]{64}\left[\cos \tfrac{1}{3}(0° + k \cdot 360°) + i \sin \tfrac{1}{3}(0° + k \cdot 360°)\right]$$ *$\sqrt[3]{64}$ is the real cube root of 64.*

$$= 4[\cos (0° + k \cdot 120°) + i \sin (0° + k \cdot 120°)]$$

For $k = 0$: $x = 4[\cos (0° + 0 \cdot 120°) + i \sin (0° + 0 \cdot 120°)]$
$= 4(\cos 0° + i \sin 0°) = 4(1 + 0) = 4$

For $k = 1$: $x = 4[\cos (0° + 1 \cdot 120°) + i \sin (0° + 1 \cdot 120°)]$
$= 4(\cos 120° + i \sin 120°) = 4\left(-\frac{1}{2} + i\frac{\sqrt{3}}{2}\right) = -2 + 2i\sqrt{3}$

For $k = 2$: $x = 4[\cos (0° + 2 \cdot 120°) + i \sin (0° + 2 \cdot 120°)]$
$= 4(\cos 240° + i \sin 240°) = 4\left[\left(-\frac{1}{2} + i\left(-\frac{\sqrt{3}}{2}\right)\right)\right] = -2 - 2i\sqrt{3}$

Thus, the three cube roots of 64 are 4, $-2 + 2i\sqrt{3}$, and $-2 - 2i\sqrt{3}$. Check to show that each number, raised to the third power, equals 64.

The complex roots theorem can be used to solve equations of higher degree.

EXAMPLE 2 **Solve: $x^4 + 81 = 0$**

$x^4 + 81 = 0$ is equivalent to $x^4 = -81$. Express -81, or $-81 + 0i$, in polar form.

$$r = \sqrt{(-81)^2 + 0^2} = 81 \qquad \theta = \text{Arctan}\,\frac{0}{-81} + \pi = \pi$$

Thus, $-81 = 81(\cos \pi + i \sin \pi)$

and $x = \sqrt[4]{81}\left[\cos \frac{1}{4}(\pi + 2k\pi) + i \sin \frac{1}{4}(\pi + 2k\pi)\right]$

$$= 3\left[\cos\left(\frac{\pi}{4} + \frac{2k\pi}{4}\right) + i \sin\left(\frac{\pi}{4} + \frac{2k\pi}{4}\right)\right]$$

For $k = 0$: $x = 3\left(\cos \frac{\pi}{4} + i \sin \frac{\pi}{4}\right) = 3\left(\frac{\sqrt{2}}{2} + i\frac{\sqrt{2}}{2}\right) = \frac{3\sqrt{2}}{2} + \frac{3i\sqrt{2}}{2}$

For $k = 1$: $x = 3\left(\cos \frac{3\pi}{4} + i \sin \frac{3\pi}{4}\right) = 3\left(-\frac{\sqrt{2}}{2} + i\frac{\sqrt{2}}{2}\right) = -\frac{3\sqrt{2}}{2} + \frac{3i\sqrt{2}}{2}$

For $k = 2$: $x = 3\left(\cos \frac{5\pi}{4} + i \sin \frac{5\pi}{4}\right) = 3\left(-\frac{\sqrt{2}}{2} - i\frac{\sqrt{2}}{2}\right) = -\frac{3\sqrt{2}}{2} - \frac{3i\sqrt{2}}{2}$

For $k = 3$: $x = 3\left(\cos \frac{7\pi}{4} + i \sin \frac{7\pi}{4}\right) = 3\left(\frac{\sqrt{2}}{2} - i\frac{\sqrt{2}}{2}\right) = \frac{3\sqrt{2}}{2} - \frac{3i\sqrt{2}}{2}$

To check, raise each of the four roots to the fourth power.

TEACHING SUGGESTIONS

- Emphasize that the coefficient in the complex roots theorem is the *n*th root of *r* not $\frac{r}{n}$.
- While discussing the examples, point out that the greatest *k*-value is always one less than the number of roots that are to be found.

Critical Thinking

Analysis Ask students to discuss the impact of the complex roots theorem on solving equations. Answers may vary.

CHALKBOARD EXAMPLES

- **For Example 1**
  1. Find the two complex roots of i or solve $x^2 = i$. Express the roots in rectangular form.
     $\frac{\sqrt{2}}{2} + i\frac{\sqrt{2}}{2}, -\frac{\sqrt{2}}{2} - i\frac{\sqrt{2}}{2}$
  2. Find the four complex roots of 256 or solve $x^4 = 256$. Express the roots in rectangular form.
     $4 + 0i, 0 + 4i, -4 + 0i, 0 - 4i$
- **For Example 2**
  Solve each equation.
  3. $x^3 - 8 = 0$ $\quad 2 + 0i, -1 + i\sqrt{3}, -1 - \sqrt{3}$
  4. $x^4 + 1 = 0$ $\quad \frac{\sqrt{2}}{2} + i\frac{\sqrt{2}}{2}, -\frac{\sqrt{2}}{2} + i\frac{\sqrt{2}}{2}, -\frac{\sqrt{2}}{2} - i\frac{\sqrt{2}}{2}, \frac{\sqrt{2}}{2} - i\frac{\sqrt{2}}{2}$

- **For Example 3**
  Graph the roots of each equation.
  5. $x^4 = 16$

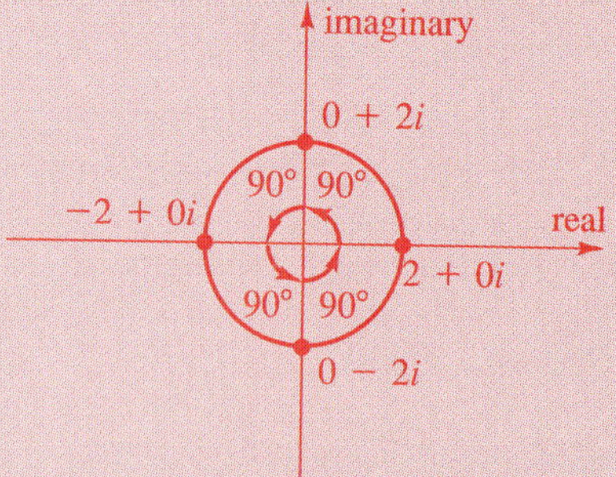

6. $x^3 = -1$

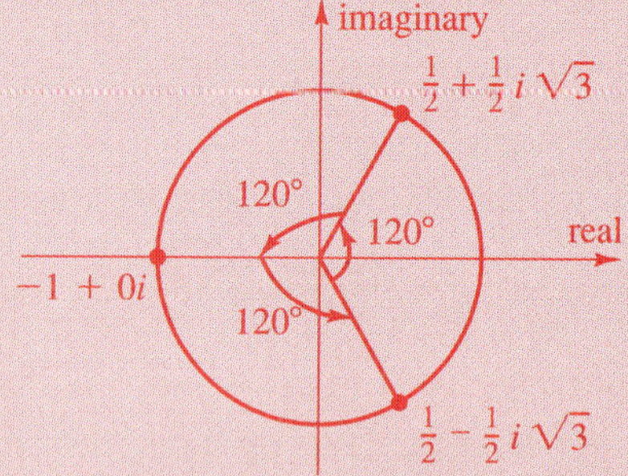

- **For Example 4**
  Find the indicated roots. Express the roots in polar form.
  7. the cube roots of $1 + i$
  $\sqrt[6]{2}(\cos 15° + i \sin 15°)$,
  $\sqrt[6]{2}(\cos 135° + i \sin 135°)$,
  $\sqrt[6]{2}(\cos 255° + i \sin 255°)$
  8. the fifth roots of i
  $\cos 18° + i \sin 18°$,
  $\cos 90° + i \sin 90°$,
  $\cos 162° + i \sin 162°$,
  $\cos 234° + i \sin 234°$,
  $\cos 306° + i \sin 306°$

Common Error

- When using the complex roots theorem, some students begin with a k-value of 1. Emphasize that the values of k range from 0 through $n - 1$.
- See *Teacher's Resource Book* for additional remediation.

Notice that the roots of $x^4 + 81 = 0$ are cyclic in nature. In fact, when graphed on the complex plane, the nth roots of a complex number are equally spaced around a circle. The figure at the right shows the four fourth roots of -81.

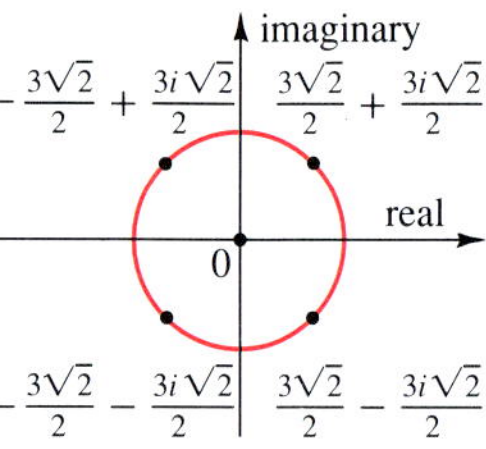

If one nth root of a complex number is known, then all the nth roots can be graphed on the complex plane as follows:

Step 1 With its center at the origin, draw a circle with radius $\sqrt[n]{r}$.
Step 2 Graph the known root.
Step 3 To locate the other roots, divide the circle into n arcs of equal length. The measure of the angle between the consecutive roots is $\frac{360°}{n}$, or $\frac{2\pi}{n}$.

EXAMPLE 3 **Graph the five fifth roots of 32.**

Step 1 Since $\sqrt[5]{32} = 2$, a circle with radius 2 should be drawn.
Step 2 Graph the first root, $2 + 0i$.
Step 3 The five roots are equally spaced $\frac{360°}{5} = 72°$ apart.

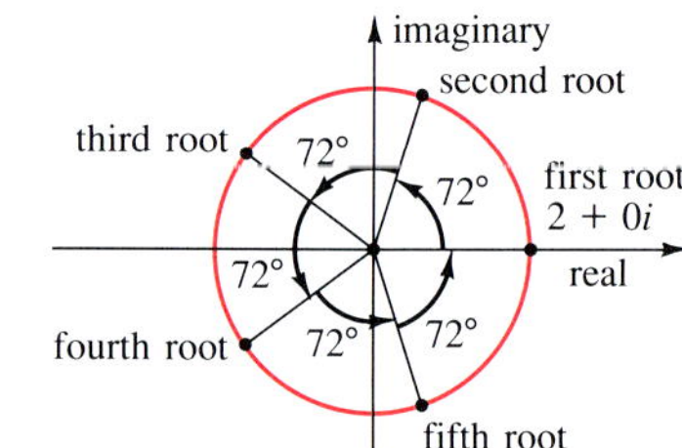

The complex roots theorem can also be used to find the roots of a complex number, $a + bi$, where $b \neq 0$.

EXAMPLE 4 **Find the five fifth roots of $-1 - i$. Express your answers in polar form.**

First express $-1 - i$ in polar form.

$$r = \sqrt{(-1)^2 + (-1)^2} = \sqrt{2} \qquad \theta = \text{Arctan}\,\frac{-1}{-1} + 180° = 225°$$

Thus, $\quad -1 - i = \sqrt{2}(\cos 225° + i \sin 225°)$

The five fifth roots are given by

$$\sqrt[10]{2}\left[\cos\left(\frac{225°}{5} + \frac{k \cdot 360°}{5}\right) + i \sin\left(\frac{225°}{5} + \frac{k \cdot 360°}{5}\right)\right] \quad \text{or}$$

$$\sqrt[10]{2}\,[\cos(45° + k \cdot 72°) + i \sin(45° + k \cdot 72°)]$$

For $k = 0$: $\quad x = \sqrt[10]{2}(\cos 45° + i \sin 45°)$
For $k = 1$: $\quad x = \sqrt[10]{2}(\cos 117° + i \sin 117°)$
For $k = 2$: $\quad x = \sqrt[10]{2}(\cos 189° + i \sin 189°)$
For $k = 3$: $\quad x = \sqrt[10]{2}(\cos 261° + i \sin 261°)$
For $k = 4$: $\quad x = \sqrt[10]{2}(\cos 333° + i \sin 333°)$

The five fifth roots of $-1 - i$

CLASS EXERCISES

Given the number n of nth roots of a complex number, give the measure of the angle between any two consecutive roots on the complex plane.

1. $n = 3$ 120° **2.** $n = 4$ 90° **3.** $n = 5$ 72° **4.** $n = 6$ 60° **5.** $n = 8$ 45° **6.** $n = 9$ 40°

7. Show that the nth *roots of unity* (nth roots of 1) are of the form $\cos \dfrac{k \cdot 360°}{n} + i \sin \dfrac{k \cdot 360°}{n}$. See side column.

8. Find the cube roots of unity. $1 + 0i, -\frac{1}{2} + \frac{\sqrt{3}}{2}i, -\frac{1}{2} - \frac{\sqrt{3}}{2}i$

PRACTICE EXERCISES

Find the indicated roots. Express the roots in rectangular form. See side column on page 326.

A

1. the square roots of $9i$

2. the square roots of $16i$

3. the cube roots of 27

4. the cube roots of -27

5. the cube roots of $-64i$

6. the cube roots of $-8i$

7. the fourth roots of $-1 - i\sqrt{3}$

8. the cube roots of $1 - i$

9. the fourth roots of 16

10. the cube roots of $-27i$

11. the square roots of 1

12. the fourth roots of 1

13. the sixth roots of 1

14. the eighth roots of 1

Graph the roots of each equation on the complex plane. See solutions manual.

15. $x^4 = 1$

16. $x^{10} = 1$

17. $x^3 = 27$

18. $x^6 = 64$

19. $x^5 = -243$

20. $x^3 = -125$

Solve each equation. See pages 491–492.

B

21. $x^3 - 27 = 0$

22. $x^3 + 64 = 0$

23. $x^5 + 1 = 0$

24. $x^4 - 1 = 0$

25. $x^3 + 8 = 0$

26. $x^3 + 27 = 0$

27. $16x^4 = i$

28. $x^4 = -i$

29. $x^3 - (32 + 32i\sqrt{3}) = 0$

30. $x^4 = -1 - i\sqrt{3}$

31. $x^5 - 32 = 0$

32. $x^5 + 32 = 0$

33. $x^6 = -1$

34. $x^6 = 1$

C

35. $x^3 + \left(\frac{1}{2} - \frac{i\sqrt{3}}{2}\right) = 0$

36. $x^3 - (1 + i\sqrt{3}) = 0$

LESSON FOLLOW-UP

Discussion

What would you have to do before you could solve the equation $Z^3 = 2i$? Solve the equation. Write both sides in the form $r(\cos\theta + i \sin\theta)$
$Z^3 = 2i$; $(a + bi)^3 = 2i$
$[r(\cos\theta + i\sin\theta)]^3 = 2(\cos 90° + i \sin 90°)$;
$r^3(\cos 3\theta + i \sin 3\theta) = 1(\cos 90° + i\sin 90°)$; $r^3 = 2(\cos 30° + i \sin 30°)$; $r = \sqrt[3]{2}(\cos 30° + i \sin 30°)^{\frac{1}{3}}$;
$\sqrt[3]{2}(\cos 30° + i \sin 30°)$,
$\sqrt[3]{2}(\cos 150° + i \sin 150°)$,
$\sqrt[3]{2}(\cos 270° + i \sin 270°)$

Assignment Guide

See p. 284B for assignments.

Test Yourself

See *Teacher's Resource Book*, *Tests*, pp. 71–72.

Lesson Quiz

1. Find the square roots of $8 + 8i\sqrt{3}$. Express the roots in rectangular form. $2\sqrt{3} + 2i, -2\sqrt{3} - 2i$

2. One cube root of z is $3(\cos 20° + i \sin 20°)$. Give the polar form of the other cube roots of z. $3(\cos 140° + i \sin 140°)$, $3(\cos 260° + i \sin 260°)$

3. Find all the fourth roots of -16. $\sqrt{2} + i\sqrt{2}, -2 + i\sqrt{2}, -2 - i\sqrt{2}, \sqrt{2} - i\sqrt{2}$

4. Solve for z: $z^2 = 2 + 2i\sqrt{3}$ $z = \sqrt{3} + i, z = -\sqrt{3} - i$

Additional Answers

7. $x^n = 1$; $\sqrt[n]{1}\left[\cos\frac{1}{n}(0° + k \cdot 360°) + i \sin\frac{1}{n}(0° + k \cdot 360°)\right] = \cos\left(\frac{k \cdot 360°}{n}\right) + i \sin\left(\frac{k \cdot 360°}{n}\right)$

Enrichment

If $A = 2 - i$ show that A is a root of the equation $z^4 = -7 - 24i$.

Express $2 - i$ in polar form, then use DeMoivre's theorem to find $(2 - i)^4$: $(2 - i)^4 =$ $[\sqrt{5}(\cos 333.4° + i \sin 333.4°)]^4 =$ $25(\cos 253.6° + i \sin 253.6°)$ Express $7 - 24i$ in polar form, then compare answers: $7 - 24i =$ $25(\cos 253.7° + i \sin 253.7°)$. Since the answers are the same, A, $2 - i$, is a root of the equation $z^4 = -7 - 24i$.

Teacher's Resource Book

Practice—Chapter 7, p. 15

Enrichment—Chapter 7, p. 16

Additional Answers

Practice Exercises

1. $\frac{3\sqrt{2}}{2} + \frac{3\sqrt{2}i}{2}; -\frac{3\sqrt{2}}{2} - \frac{3\sqrt{2}i}{2}$
2. $2\sqrt{2} + 2i\sqrt{2}; -2\sqrt{2} - 2i\sqrt{2}$
3. $3, -\frac{3}{2} + \frac{3\sqrt{3}i}{2}, -\frac{3}{2} - \frac{3\sqrt{3}i}{2}$
4. $-3; \frac{3}{2} + \frac{3\sqrt{3}}{2}i; \frac{3}{2} - \frac{3\sqrt{3}}{2}i$
5. $4i, 2\sqrt{3} - 2i, -2\sqrt{3} - 2i$
6. $2i, -\sqrt{3} - i, \sqrt{3} - i$
7. $\frac{\sqrt[4]{2}}{2} + \frac{\sqrt[4]{18}}{2}i; -\frac{\sqrt[4]{18}}{2} + \frac{\sqrt[4]{2}}{2}i; -\frac{\sqrt[4]{2}}{2} - \frac{\sqrt[4]{18}}{2}i; \frac{\sqrt[4]{18}}{2} - \frac{\sqrt[4]{2}}{2}i$
8. $-\frac{\sqrt[3]{4}}{2} - \frac{\sqrt[3]{4}}{2}i;$ $\sqrt[6]{2}\left(\frac{\sqrt{2} - \sqrt{6}}{4} + \frac{\sqrt{2} + \sqrt{6}}{4}i\right);$ $\sqrt[6]{2}\left(\frac{\sqrt{2} + \sqrt{6}}{4} + \frac{\sqrt{2} - \sqrt{6}}{4}i\right)$
9. $2, -2, 2i, -2i$
10. $0 + 3i, \frac{-3\sqrt{3}}{2} - \frac{3}{2}i, \frac{3\sqrt{3}}{2} - \frac{3}{2}i$
11. $1; -1$
12. $1; -1; i; -i$
13. $\frac{1}{2} + \frac{\sqrt{3}i}{2}; -\frac{1}{2} + \frac{\sqrt{3}i}{2}; -\frac{1}{2} - \frac{\sqrt{3}i}{2}; \frac{1}{2} - \frac{\sqrt{3}i}{2}; 1; -1$
14. $\frac{\sqrt{2}}{2} + \frac{\sqrt{2}i}{2}; \frac{-\sqrt{2}}{2} + \frac{\sqrt{2}i}{2}; \frac{-\sqrt{2}}{2} + \frac{\sqrt{2}i}{2}; \frac{\sqrt{2}}{2} - \frac{\sqrt{2}i}{2}; 1, -1, i, -i$

Find the indicated roots.

37. $(1 + i)^{\frac{4}{3}}$

38. $\left(\frac{27\sqrt{2}}{2} + \frac{27i\sqrt{2}}{2}\right)^{\frac{2}{3}}$

39. $(4\sqrt{2} - 4i\sqrt{2})^{\frac{2}{3}}$

40. $(-8 + 8i\sqrt{3})^{\frac{3}{4}}$

Applications

Computer This program calculates N nth roots of a complex number R cis W. The values of N, R, and W are supplied by the user. 3 cis 40°; 3 cis 160°; 3 cis 280°

```
10  HOME : INPUT "ENTER THE NUMBER OF
    ROOTS ";N
20  INPUT "ENTER THE VALUE OF R ";R:
    INPUT "ENTER THE VALUE OF W ";W
30  PRINT : PRINT "THE "N" ROOTS OF
    "R" (CIS "W") ARE: "
40  FOR K = 0 TO N - 1
50  PRINT : PRINT  R ^ (1 / N);"(COS ";
    (1/N) * (W + K * 360);" + i SIN ";
    (1/N) * (W + K * 360),")"
60  NEXT K
70  END
```

41. Use the program to find the 3rd roots of 27 cis 120°.

42. Use the program to find the 6th roots of 64 cis 120°. See below

43. Use the program to find the fifth roots of unity. 1 cis 0°; 1 cis 72°; 1 cis 144°; 1 cis 216°; 1 cis 288°

TEST YOURSELF

Express each complex number in polar form. 7.5

1. $3 - 3i$ $3\sqrt{2}(\cos 315° + i \sin 315°)$

2. $5 + 5i$ $5\sqrt{2}(\cos 45° + i \sin 45°)$

Express each complex number in rectangular form.

3. $4(\cos 135° + i \sin 135°)$ $-2\sqrt{2} + 2\sqrt{2}i$

4. $2(\cos 60° + i \sin 60°)$ $1 + \sqrt{3}i$

Find each product or quotient. 7.6

5. $4(\cos 194° + i \sin 194°) \cdot 10(\cos 160° + i \sin 160°)$ $40(\cos 354° + i \sin 354°)$

6. $9(\cos 107° + i \sin 107°) \div 3(\cos 62° + i \sin 62°)$ $3(\cos 45° + i \sin 45°)$

Evaluate each power. Express your answers in rectangular form. 7.7

7. $[4(\cos 10° + i \sin 10°)]^{12}$ $-8{,}388{,}608 + 14{,}529{,}495i$

8. $(6 + 6i)^4$ -5184

Find the indicated roots. Express the roots in rectangular form. 7.8

9. the cube roots of -27 $\frac{3}{2} + \frac{3\sqrt{3}}{2}i, -3 + 0i, \frac{3}{2} - \frac{3\sqrt{3}}{2}i$

10. the fourth roots of 16 $2 + 0i, -2 + 0i, 0 - 2i, 0 + 2i$

42. 2 cis 20°; 2 cis 80°; 2 cis 140°; 2 cis 200°; 2 cis 260°; 2 cis 320°

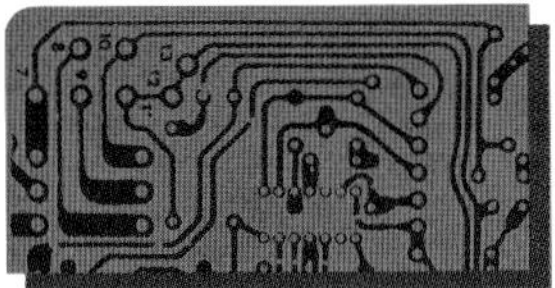

TECHNOLOGY: Fractals

Did you know that fractal shapes occur in many natural phenomena such as a coastline, a mountain range, a fern, and a galaxy? In the 1970s, Benoit B. Mandelbrot developed the concept of a fractal curve, which he defined as a curve that is not straight, yet has the property that its parts are small-scale replicas of the whole.

The computer program shown draws the Mandelbrot set. To write complex numbers in a form the computer can use, $0 + 0i$ is represented by A = 0: B = 0 (line 70). To find the next complex number, recall that

$$\begin{aligned}(a + bi)^2 &= (a + bi)(a + bi)\\ &= a^2 + 2abi - b^2 = (a^2 - b^2) + 2abi\end{aligned}$$

Thus the new value of a is NA = A*A − B*B − K (line 90), where K is a constant.

If the modulus of the resulting complex number is greater than $\sqrt{10}$, then the point is shaded in a color depending on the number of times the process was repeated. If the modulus of the number does not exceed $\sqrt{10}$ after 20 repetitions, then the point stays black. For example, let $K = -2$.

```
10 N = 20:M = 150:R = 10
20 X1 = - 1.6:Y1 = - 2:X2 = 2.4:Y2 = 2
30 HGR
40 FOR P = 1 TO M
50 FOR Q = 1 TO M
60 K = X1 + (X2 - X1) * P / M:
L = Y1 + (Y2 - Y1) * Q / M
70 A = 0:B = 0
80 FOR I = 1 TO N
90 NA = A * A - B * B - K
100 NB = 2 * A * B - L
110 A = NA:B = NB
120 IF (A * A + B * B < = R) THEN
GOTO 170
130 C = INT (((I / 7) - INT (I / 7)) * 7)
140 IF C = 4 THEN C = 6
150 HCOLOR = C: HPLOT P,Q
160 GOTO 180
170 NEXT I
180 NEXT Q: NEXT P
```

First iteration, $a + bi = 0 + 0i$.
new value of A: $0^2 - 0^2 + 2 = 2$
new value of B: $2(0)(0) + 2 = 2$
modulus: $\sqrt{2^2 + 2^2} = \sqrt{8} < \sqrt{10}$

Second iteration, $a + bi = 2 + 2i$.
new value of a: $2^2 - 2^2 + 2 = 2$
new value of b: $2(2)(2) + 2 = 10$
modulus: $\sqrt{2^2 + 10^2} = \sqrt{104} > \sqrt{10}$

Therefore, the process stops after the second iteration. In this program, the point is shaded green when the process stops after 2 iterations. The program takes some time to run since the calculations are complicated and the computer must go through this process for every dot on the screen.

EXERCISES

1. Determine how many iterations will be performed for $K = 1$ and $K = -1$. 3; 2
2. Can you think of any values of K for which the process would not terminate before at least 20 iterations have been performed? K = 0 or fractional values of K

Technology

A program which generates the Mandelbrot set, an important fractal image, is presented. An interesting aspect of the Mandelbrot set is that continued magnifications of sections of the image reveal small-scale replicas of the set.

See *Teacher's Resource Book* for *Application*, Chapter 7, p. 17.
See *Teacher's Resource Book*, *Technology*, p. 7.

Additional Answers

1.

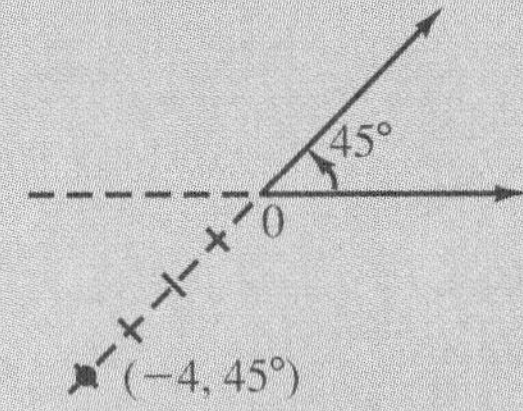

4.

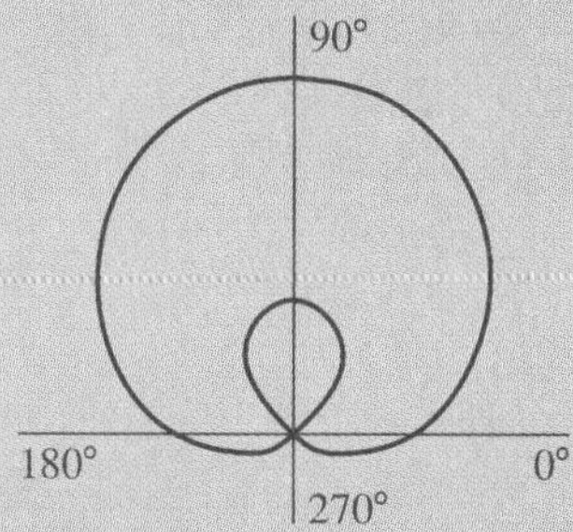

5.

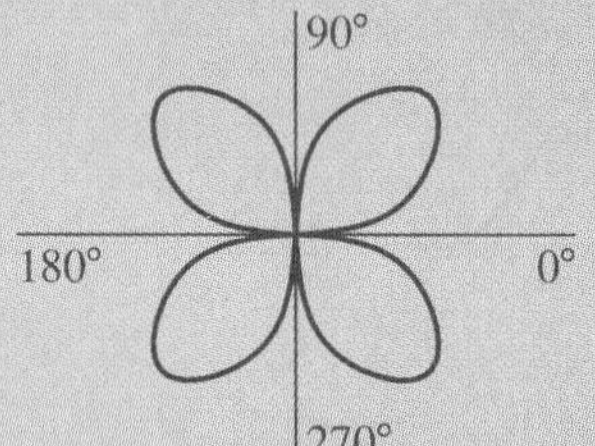

6.

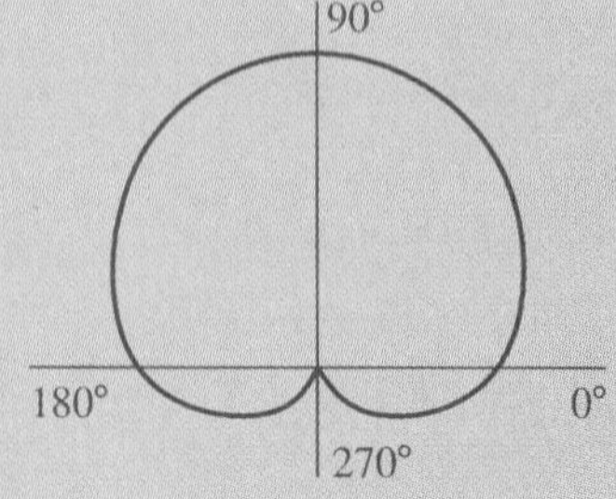

CHAPTER 7 SUMMARY AND REVIEW

Vocabulary

argument (309)	modulus (310)
cardioid (294)	polar axis (286)
cis θ (310)	polar coordinates (286)
complex number (299)	polar coordinate system (286)
complex plane (308)	polar form of a complex number (309)
complex roots theorem (322)	pole (286)
conjugate (304)	pure imaginary number (299)
De Moivre's theorem (319)	rectangular coordinates (286)
imaginary number (298)	rectangular form of a complex number (309)
limaçon (294)	rose (295)

Polar Coordinates A point on the plane has rectangular coordinates of the form (x, y) and polar coordinates of the form (r, θ). To change from polar to rectangular coordinates, use the formulas $x = r\cos\theta$ and $y = r\sin\theta$. To change from rectangular to polar coordinates, use $r = \sqrt{x^2 + y^2}$, $\theta = \text{Arctan}\,\frac{y}{x}$ (if $x > 0$), and $\theta = \text{Arctan}\,\frac{y}{x} + 180°$ (if $x < 0$). **7.1**

1. Graph the point $(-4, 45°)$ on the polar coordinate system. See side column.
2. Find the rectangular coordinates of the point $(2, 45°)$. $(\sqrt{2}, \sqrt{2})$
3. Find the polar coordinates of the point $(1, \sqrt{3})$. $(2, 60°)$

Graphs of Polar Equations Some curves have special names. **7.2**

Circle	$r = 2a\cos\theta$
Limaçon (with extra loop)	$r = a \pm b\cos\theta$ or $r = a \pm b\sin\theta$, $a < b$
Limaçon (without extra loop)	$r = a \pm b\cos\theta$ or $r = a \pm b\sin\theta$, $a > b$
Cardioid	$r = a \pm a\cos\theta$ or $r = a \pm a\sin\theta$
Rose	$r = a\cos n\theta$ or $r = a\sin n\theta$, n is a positive integer

Graph each equation. See side column.

4. $r = 1 + 2\sin\theta$
5. $r = -6\sin 2\theta$
6. $r = 2 + 2\sin\theta$

Sums and Differences of Complex Numbers The definitions of addition and subtraction of two complex numbers $a + bi$ and $c + di$, are: **7.3**

$$(a + bi) + (c + di) = (a + c) + (b + d)i$$
$$(a + bi) - (c + di) = (a + bi) + (-c - di) = (a - c) + (b - d)i$$

Simplify.

7. $\sqrt{-20}$ $2i\sqrt{5}$

8. $(-4 + 3i) + (9 - 11i)$ $5 - 8i$

9. $(-2 - i) - (6 - 8i)$ $-8 + 7i$

Products and Quotients of Complex Numbers To multiply two complex numbers, use the FOIL method and the fact that $i^2 = -1$. The two complex numbers $a + bi$ and $a - bi$ are conjugates, and the quotient of two complex numbers can be simplified by multiplying the dividend and the divisor by the conjugate of the divisor. 7.4

10. Multiply and simplify: $(-1 - 3i)(4 + 5i)$ $11 - 17i$

11. Divide and simplify: $(5 + 6i) \div (-3 - 2i)$ $\frac{-27 - 8i}{13}$

Complex Numbers in Polar Form The polar form of a complex number $a + bi$ is $r(\cos \theta + i \sin \theta)$, where $r = \sqrt{a^2 + b^2}$ and $\tan \theta = \frac{b}{a}$, $a \neq 0$. The rectangular form of a complex number can be found using the formulas $a = r \cos \theta$ and $b = r \sin \theta$. 7.5

12. Express in polar form: $3 + 3i$ $3\sqrt{2}(\cos 45° + i \sin 45°)$

13. Express in rectangular form: $6(\cos 150° + i \sin 150°)$ $-3\sqrt{3} + 3i$

Multiplying and Dividing Complex Numbers in Polar Form The product and the quotient of two complex numbers $z_1 = r_1(\cos \theta_1 + i \sin \theta_1)$ and $z_2 = r_2(\cos \theta_2 + i \sin \theta_2)$ are given by the following formulas: 7.6

$$z_1 z_2 = r_1 r_2[\cos(\theta_1 + \theta_2) + i \sin(\theta_1 + \theta_2)]$$

$$\frac{z_1}{z_2} = \frac{r_1}{r_2}[\cos(\theta_1 - \theta_2) + i \sin(\theta_1 - \theta_2)]$$

14. Multiply: $12(\cos 30° + i \sin 30°) \cdot 6(\cos 150° + i \sin 150°)$ $72(\cos 180° + i \sin 180°)$

15. Divide: $12(\cos 100° + i \sin 100°) \div 6(\cos 40° + i \sin 40°)$ $2(\cos 60° + i \sin 60°)$

De Moivre's Theorem If z is a complex number in polar form, and n is a positive integer, then $z^n = r^n(\cos n\theta + i \sin n\theta)$. 7.7

Evaluate each power. Express your answers in rectangular form. 17. $\frac{15625}{2} + \frac{15625\sqrt{3}}{2}i$

16. $[3(\cos 15° + i \sin 15°)]^2$ $\frac{9\sqrt{3}}{2} + \frac{9}{2}i$

17. $[5(\cos 10° + i \sin 10°)]^6$

Roots of Complex Numbers The complex roots theorem states that if n is a positive integer, then $a + bi = r(\cos \theta + i \sin \theta)$ has n distinct roots, given by $\sqrt[n]{r}\left[\cos \frac{1}{n}(\theta + k \cdot 360°) + i \sin \frac{1}{n}(\theta + k \cdot 360°)\right]$, where $k = 0, 1, 2, \ldots, n - 1$.

18. Find the cube roots of -1. $-1 + 0i, \frac{1}{2} - \frac{\sqrt{3}}{2}i, \frac{1}{2} + \frac{\sqrt{3}}{2}i$

19. Find the fourth roots of 625. $5 + 0i, -5 + 0i, 0 - 5i, 0 + 5i$

See *Teacher's Resource Book, Tests*, pp. 73–76.

Additional Answers

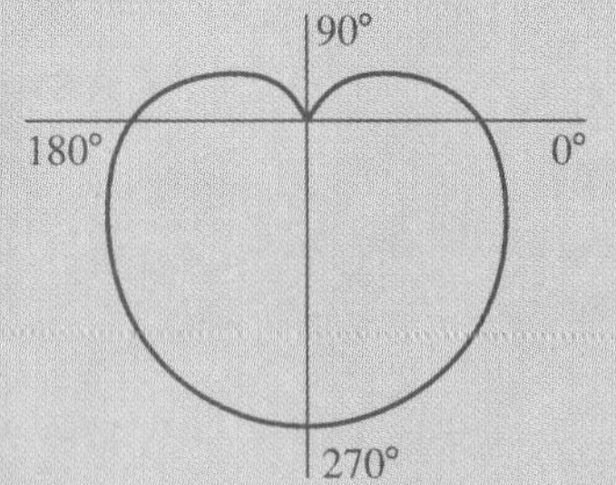

CHAPTER TEST

Find the rectangular coordinates of each point.

1. $(2, -135°)$ $(-\sqrt{2}, -\sqrt{2})$

2. $(-3, 120°)$ $\left(\frac{3}{2}, -\frac{3\sqrt{3}}{2}\right)$

Find the polar coordinates of each point.

3. $(-3, -3)$ $(3\sqrt{2}, 225°)$

4. $(-1, \sqrt{3})$ $(2, 120°)$

5. Graph the polar equation $r = 2 - 2 \sin \theta$. See side column.

6. Simplify: i^{60} 1

7. Find the sum: $(6 + 3i) + (11 - 4i)$ $17 - i$

8. Find the difference: $(2 - 5i) - (-4 - i)$ $6 - 4i$

9. Multiply: $(9 - 6i)(-1 + i)$ $-3 + 15i$

10. Divide: $(2 + 3i) \div (5 - i)$ $\frac{7 + 17i}{26}$

Express each complex number in polar form.

11. $-\sqrt{3} + i$ $2(\cos 150° + i \sin 150°)$

12. $4 + 4i$ $4\sqrt{2}(\cos 45° + i \sin 45°)$

13. Express $2(\cos 225° + i \sin 225°)$ in rectangular form. $-\sqrt{2} - i\sqrt{2}$

Multiply.

14. $8(\cos 96° + i \sin 96°) \cdot 4(\cos 129° + i \sin 129°)$ $32(\cos 225° + i \sin 225°)$

15. $3(\cos 264° + i \sin 264°) \cdot 12(\cos 126° + i \sin 126°)$ $36(\cos 30° + i \sin 30°)$

Divide.

16. $16(\cos 199° + i \sin 199°) \div 2(\cos 49° + i \sin 49°)$ $8(\cos 150° + i \sin 150°)$

17. $20(\cos 214° + i \sin 214°) \div 4(\cos 274° + i \sin 274°)$ $5(\cos 300° + i \sin 300°)$

Calculate each power. Express your answers in rectangular form.

18. $[4(\cos 20° + i \sin 20°)]^3$ $32 + 32\sqrt{3}\,i$

19. $[2(\cos 60° + i \sin 60°)]^4$ $-8 - 8\sqrt{3}\,i$

20. Find the two square roots of $25i$. $\frac{5\sqrt{2}}{2} + \frac{5i\sqrt{2}}{2}, -\frac{5\sqrt{2}}{2} - \frac{5i\sqrt{2}}{2}$

Challenge

Find the points of intersection of the graphs of $r = 6 \cos 2\theta$ and $r = 6 \sin \theta$.
(3, 30°); (3, 150°); (−6, 270)

COLLEGE ENTRANCE EXAM REVIEW

Select the best choice for each question.

1. Solve for x: $2x^2 - 4x + 3 = 0$
C
A. $4 \pm 2i\sqrt{2}$ **B.** $-4 \pm i\sqrt{2}$
C. $\dfrac{2 \pm i\sqrt{2}}{2}$ **D.** $\dfrac{4 \pm i\sqrt{7}}{4}$
E. $\dfrac{2 \pm i\sqrt{7}}{4}$

2. $(-3 + 2i)(5 - 5i) =$
B
A. $5 + 5i$ **B.** $-5(1 - 5i)$
C. $5 - 5i$ **D.** $-5(1 + 5i)$
E. $-5(1 + i)$

3. Which of the following is (are) true for the function f defined by
B

$$f(x) = \begin{cases} x^2 + 2 & \text{if } x > 0 \\ -5 & \text{if } x = 0 \\ x + 1 & \text{if } x < 0 \end{cases}$$

I. $f(-1) < f(0)$
II. x can be any real number.
III. $f(4) = 16$

A. I only **B.** II only **C.** III only
D. I and II only **E.** II and III only

4. $4 - 4i$ is equivalent to
D
A. $2\sqrt{2}\left[\cos\left(-\frac{\pi}{4}\right) + i \sin\left(-\frac{\pi}{4}\right)\right]$
B. $\sqrt{2}\left[\cos\left(-\frac{\pi}{4}\right) + i \sin\left(-\frac{\pi}{4}\right)\right]$
C. $\sqrt{6}\left[\cos\left(-\frac{\pi}{3}\right) + i \sin\left(-\frac{\pi}{3}\right)\right]$
D. $4\sqrt{2}\left[\cos\left(-\frac{\pi}{4}\right) + i \sin\left(-\frac{\pi}{4}\right)\right]$
E. $4\sqrt{2}\left[\cos\left(-\frac{\pi}{6}\right) + i \sin\left(-\frac{\pi}{6}\right)\right]$

5. $i^{36} \cdot i^{25} =$
C
A. 1 **B.** $-i$ **C.** i **D.** -1 **E.** 0

6. Evaluate $(-\sqrt{3} + i)^6$.
B
A. $3 - 2i$ **B.** -64 **C.** $12 - 12i$
D. 36 **E.** 54

7. $\dfrac{2 + i}{5 - 2i} =$
A
A. $\frac{8}{29} + \frac{9}{29}i$ **B.** $8 + 9i$
C. $-\frac{8}{29} + 9i$ **D.** $-\frac{8}{29} - \frac{1}{29}i$
E. $8 + \frac{9}{29}i$

8. Find x.
B
A. 4 **B.** 2
C. 5 **D.** 3
E. 6

36
6
10
x

9. Which of the following represents the equation of the line which contains the point $(5, -2)$ and is parallel to the line $-3x + y = 2$?
E
A. $y = 3x + 2$ **B.** $y = -3x - 2$
C. $y = -3x - 2$ **D.** $y = 3x - 1$
E. $y = 3x - 17$

10. Write the complex number $5\left(\cos\frac{\pi}{3} + i \sin\frac{\pi}{3}\right)$ in the form $a + bi$.
D
A. $5 + 5i$ **B.** $\frac{5\sqrt{3}}{2} + \frac{5}{2}i$
C. $-5 - 5i$ **D.** $\frac{5}{2} + \frac{5\sqrt{3}}{2}i$
E. $-\frac{5\sqrt{3}}{2} - \frac{5}{2}i$

11. Three-fourths of a number is 10 less than twice the number. Find the number.
D
A. 5 **B.** 12 **C.** 4 **D.** 8 **E.** 24

12. The total surface area S of a cylinder is $2\pi r^2 + 2\pi rh$. Find the radius r if $S = 30\pi$ and $h = 2$.
C
A. 5 **B.** -5 **C.** 3 **D.** -3 **E.** 15

331

Maintaining Skills

The following skills and concepts are reviewed:

Raising a constant to a power

Simplifying expressions involving exponents

Order of operations

Evaluating polynomial expressions for given values

Using scientific notation

MAINTAINING SKILLS

Simplify.

Examples $-2^4 = -1(2^4) = -1(2)(2)(2)(2) = -16$
$(-2)^4 = (-2)(-2)(-2)(-2) = 16$

1. -3^5 -243 **2.** $(-4)^3$ -64 **3.** $-\left(\frac{1}{2}\right)^4$ $-\frac{1}{16}$

4. $-\left(\frac{1}{4}\right)^2$ $-\frac{1}{16}$ **5.** $\left(-\frac{1}{4}\right)^2$ $\frac{1}{16}$ **6.** 0.2^3 0.008

Simplify.

Example $(5x^2y^4)^3(x^2y)^2 = (5)^3(x^2)^3(y^4)^3(x^2)^2(y)^2 = 125x^6y^{12}x^4y^2 = 125x^{10}y^{14}$

7. $\frac{6x^5}{2x^2}$ $3x^3$ **8.** $\frac{9x^9y^6}{3x^2y^{-1}}$ $3x^7y^7$ **9.** $\left(\frac{x}{y^{-2}}\right)^3$ x^3y^6

10. $(-2xy)(3x^2y)$ $-6x^3y^2$ **11.** $(3a^2b^3)^{-3}(27a^8b^{10})$ a^2b **12.** $(x^{-4})\left(\frac{3x^{10}y}{y^{-2}}\right)$ $3x^6y^3$

Evaluate each expression.

Example $5(2)^4 - 10 \div 2 = 5(16) - 10 \div 2 = 80 - 5 = 75$

13. $(10 \div 2)(3 + 4)$ 35 **14.** $10 \div [2(3 + 4)]$ $\frac{5}{7}$ **15.** $3 - 2(5)$ -7

16. $(3 - 2)(5)$ 5 **17.** $(2 + 3)^2 \div 5 + 1$ 6 **18.** $(2 + 3)^2 \div (5 + 1)$ $\frac{25}{6}$

Evaluate each expression for $x = -2$ and $y = 3$.

Example $x^2 - 4y$
$(-2)^2 - 4(3) = 4 - 12 = -8$

19. $x^2 - y$ 1 **20.** $xy - 2y^2$ -24 **21.** $\frac{x}{y} + \frac{y}{2}$ $\frac{5}{6}$

22. $-3x - 2y$ 0 **23.** $5x^2 + 2y$ 26 **24.** $-3x^2 - 2y^2$ -30

Express each number in scientific notation.

Example $0.000000247 = 2.47 \times 10^{-7}$

25. 625000 6.25×10^5 **26.** 0.0000387 3.87×10^{-5} **27.** 63800000 6.38×10^7

28. 0.000375 3.75×10^{-4} **29.** 37.2×10^6 3.72×10^7 **30.** 0.625×10^4 6.25×10^3

OVERVIEW • Chapter 8

SUMMARY

The chapter reviews concepts of exponential functions and logarithmic functions from second-year algebra with emphasis on real numbers as exponents and bases. Both common and natural logarithms are discussed. Working with the concepts of exponential and logarithmic expressions necessitates the use of a scientific calculator. Exponential and logarithmic equations are introduced and used to solve real-world problems involving exponential growth and decay.

CHAPTER OBJECTIVES

- To evaluate expressions containing real-number exponents
- To solve exponential equations
- To define exponential functions
- To graph exponential functions
- To define and graph logarithmic functions
- To solve logarithmic functions
- To state the properties of logarithms
- To use the properties to express logarithms in expanded form and to solve logarithmic equations
- To evaluate common and natural logarithms
- To solve exponential equations
- To change the base of a logarithmic expression
- To use exponential and logarithmic equations to solve real world problems

CHAPTER HIGHLIGHTS

The *theme* of Chapter 8 is seismology. This theme is elaborated upon in a discussion of the Richter scale, included among the chapter's special features.

APPLICATIONS

Exponential functions and logarithms are used to solve problems that deal with growth and decay. Archaeology, biology, optics, and business are only some of the disciplines to which the concepts in this chapter are applied. Lesson 8.6 provides problems involving population growth, carbon-14 dating, and compounding interest on investments.

TECHNOLOGY

Calculator

A calculator is used to evaluate expressions involving rational exponents and compute natural exponential functions and common and natural logarithms. Students are instructed in the use of a calculator for these purposes in Lessons 8.1, 8.2, 8.3, and 8.5.

Computer

Chapter 8 provides a program which uses the Richter scale to compare the respective magnitudes of different earthquakes.

RESOURCES

Teacher's Resource Book.

- Teaching Aid 8
- Transparencies 15 and 16

ASSIGNMENT GUIDE Meeting Student Needs

STUDENT TEXT					TEACHER'S RESOURCE BOOK	
Chapter Content		Basic	Average	Enriched	P	E
8.1	Real Exponents	Omit	Omit	D: 336/3, 9, 17–41 odd	1	2
8.2	Exponential Functions	Omit	Omit	D: 341/3, 7, 17–33 odd R: 336/4, 10, 30	3	4
8.3	Logarithmic Functions	Omit	Omit	D: 345/7, 15, 21–59 odd R: 341/4, 16, 24 346/TY	5	6
8.4	Properties of Logarithms	Omit	Omit	D: 351/17–67 odd R: 345/8, 16, 28	7	8
8.5	Evaluating Logarithms and Solving Exponential Equations	Omit	Omit	D: 357/5, 11, 17, 23–65 odd R: 351/16, 30, 46	9	10
8.6	Applications: Exponential and Logarithmic Equations	Omit	Omit	D: 363/5–19 odd R: 357/24, 44, 50 364/TY	11	12

D = Daily R = Review TY = Test Yourself P = Practice E = Enrichment

	STUDENT TEXT				TEACHER'S RESOURCE BOOK	
Review and Testing	Test Yourself	346, 364	College Ent. Exam Rev.	369	Tests	
	Chapter Sum. and Rev.	366	Cumulative Review	370	• Quizzes	77–80
	Chapter Test	368			• Chapter Test (Form A)	81–82
					• Chapter Test (Form B)	83–84
Special Features	Challenge	337	Math Club Activity	358	Applications—Chapter 8	13
	Extra	341	Technology	365	Critical Thinking	7
	Trigonometry in Chemistry	352			Alg. and Geom. Review	29–32
					Technology	8

8 Exponential and Logarithmic Functions

Seismology is the study of earthquakes and the mechanical properties of the earth. A seismogram is a record made by a seismograph. Its period and amplitude reflect the intensity, direction, and duration of any movement of the ground.

BACKGROUND

The earth is in constant motion. This motion can take the form of a devastating earthquake or a slight tremor, perceptible only to precision instruments. To incorporate the wide range of intensity of seismological forces scientists use the Richter scale, a measuring system that is logarithmic in nature.

LESSON PLAN

Vocabulary
Exponential equation

Materials/Manipulatives
Scientific Calculators

BACKGROUND

In the Preview, the properties of rational exponents are reviewed.

Critical Thinking

Analysis Ask students to describe how they would proceed to evaluate the expression $4^{(3 + \sqrt{2})}$.
Answers may vary.

8.1 Real Exponents

Objectives: To evaluate expressions containing real number exponents
To solve exponential equations

Before introducing real number exponents, it will be helpful to review the properties of rational number exponents.

Preview

If a and b are positive real numbers, and m and n are rational numbers, then the following properties hold true:

$$a^m a^n = a^{m+n} \qquad \frac{a^m}{a^n} = a^{m-n} \qquad (a^m)^n = a^{mn} \qquad (ab)^m = a^m b^m$$

$$\left(\frac{a}{b}\right)^m = \frac{a^m}{b^m} \qquad a^{-m} = \frac{1}{a^m} \qquad a^{\frac{1}{n}} = \sqrt[n]{a} \qquad a^{\frac{m}{n}} = \sqrt[n]{a^m} = (\sqrt[n]{a})^m$$

EXAMPLE **Simplify:** **a.** $(2x)^{-3}$ **b.** $(27x^6)^{\frac{2}{3}}$ **c.** $(81x)^{\frac{1}{2}}$

a. $(2x)^{-3} = (2)^{-3}(x)^{-3} = \left(\frac{1}{8}\right)\left(\frac{1}{x^3}\right) = \frac{1}{8x^3}$

b. $(27x^6)^{\frac{2}{3}} = (27)^{\frac{2}{3}}(x^6)^{\frac{2}{3}} = (3)^2x^4 = 9x^4$

c. $(81x)^{\frac{1}{2}} = (81)^{\frac{1}{2}}(x)^{\frac{1}{2}} = 9\sqrt{x}$

Simplify.

1. $(2x^3)^2$ $4x^6$ **2.** $(3^2)(3^4)$ 729 **3.** $(-32)^{\frac{2}{5}}$ 4 **4.** $81^{-\frac{1}{4}}$ $\frac{1}{3}$ **5.** $(3x)^{-3}$ $\frac{1}{27x^3}$ **6.** $4^{-2.5}$ $\frac{1}{32}$

The properties of rational exponents can also be applied to real number exponents.

EXAMPLE 1 **Simplify:** **a.** $(2^{\sqrt{30}})^{\sqrt{120}}$ **b.** $2^{-\sqrt{3}} \cdot 2^{\sqrt{48}}$

a. $(2^{\sqrt{30}})^{\sqrt{120}} = 2^{\sqrt{3600}} = 2^{60}$

b. $2^{-\sqrt{3}} \cdot 2^{\sqrt{48}} = 2^{-\sqrt{3}} \cdot 2^{4\sqrt{3}} = 2^{3\sqrt{3}}$

It is customary to say that an expression is in simplest form when every exponent is positive.

EXAMPLE 2 **Simplify and express with positive exponents:**

a. $\dfrac{-32x^{\frac{4}{5}}y^{\frac{1}{6}}}{4x^{\frac{1}{5}}y^{\frac{5}{6}}}$ **b.** $\dfrac{x^{3\sqrt{2}}y^{-2\sqrt{3}}}{x^{4\sqrt{2}}y^{-4\sqrt{3}}}$ **c.** $(x^{-\sqrt{8}}y^{\sqrt{50}})^{\sqrt{2}}$

a. $\dfrac{-32x^{\frac{4}{5}}y^{\frac{1}{6}}}{4x^{\frac{1}{5}}y^{\frac{5}{6}}} = -8(x^{\frac{4}{5}-\frac{1}{5}})(y^{\frac{1}{6}-\frac{5}{6}})$

$= -8x^{\frac{3}{5}}y^{-\frac{4}{6}}$

$= -8x^{\frac{3}{5}}y^{-\frac{2}{3}}$

$= \dfrac{-8x^{\frac{3}{5}}}{y^{\frac{2}{3}}}$

b. $\dfrac{x^{3\sqrt{2}}y^{-2\sqrt{3}}}{x^{4\sqrt{2}}y^{-4\sqrt{3}}} = (x^{3\sqrt{2}-4\sqrt{2}})(y^{-2\sqrt{3}-(-4\sqrt{3})})$

$= x^{-\sqrt{2}}y^{2\sqrt{3}}$

$= \dfrac{y^{2\sqrt{3}}}{x^{\sqrt{2}}}$

c. $(x^{-\sqrt{8}}y^{\sqrt{50}})^{\sqrt{2}} = x^{-\sqrt{16}}y^{\sqrt{100}}$

$= x^{-4}y^{10}$

$= \dfrac{y^{10}}{x^4}$

You can also use a scientific calculator to evaluate expressions containing real number exponents. The y^x key makes this possible. To enter the exponent, use the parentheses keys or the memory.

EXAMPLE 3 **Evaluate the following to the nearest thousandth:**

a. $248^{\frac{1}{5}}$ **b.** $3^{\frac{2}{7}}$ **c.** $4^{-\sqrt{6}}$ **d.** $2^{\sqrt{2}+\sqrt{3}}$

a. $248^{\frac{1}{5}} = 3.012$ *Use the y^x key. Enter the exponent in parentheses: (1 ÷ 5) or use the memory.*

b. $3^{\frac{2}{7}} = 1.369$ *Use the y^x key. Enter the exponent in parentheses or use the memory.*

c. $4^{-\sqrt{6}} = 0.034$ *Use the y^x key, the $\sqrt{\ }$ key, and the +/− key.*

d. $2^{\sqrt{2}+\sqrt{3}} = 8.854$ *Use the y^x key. Enter the exponent in parentheses or use the memory.*

TEACHING SUGGESTIONS

- Review the properties of exponents starting with whole-number exponents. Then discuss the effect of enlarging the domain to integral exponents, rational-number exponents, and finally real-number exponents.
- Students will find their scientific calculators helpful when evaluating numbers raised to real-number powers involving radicals.

CHALKBOARD EXAMPLES

- **For Example 1**

Simplify.

1. $4^{\sqrt{12}} \cdot 4^{\sqrt{3}}$ $4^{3\sqrt{3}}$
2. $(3^{\sqrt{10}})^{\sqrt{20}}$ $3^{10\sqrt{2}}$
3. $2^{-\sqrt{8}}$ $\dfrac{1}{2^{2\sqrt{2}}}$

- **For Example 2**

Simplify and express with positive exponents.

4. $\dfrac{16x^{\frac{3}{4}}y^{\frac{1}{5}}}{4x^{\frac{1}{4}}y^{\frac{4}{5}}}$ $\dfrac{4x^{\frac{1}{2}}}{y^{\frac{3}{5}}}$
5. $\dfrac{n^{\sqrt{2}}m^{3\sqrt{2}}}{n^{2\sqrt{2}}m^{5\sqrt{2}}}$ $\dfrac{1}{n^{\sqrt{2}}m^{2\sqrt{2}}}$
6. $(w^{\sqrt{3}}y^{-\sqrt{27}})^{2\sqrt{3}}$ $\dfrac{w^6}{y^{18}}$

- **For Example 3**

Evaluate the following to the nearest thousandths.

7. $243^{\frac{1}{5}}$ 3.000
8. $2^{\frac{3}{8}}$ 1.297
9. $5^{\frac{2}{3}\sqrt{3}}$ 6.414
10. $3^{(\sqrt{5}-\sqrt{2})}$ 2.467

- **For Example 4**

Solve for x.

11. $32 = 4^x$ $\frac{5}{2}$
12. $9^{4x+1} = 3^{2x+3}$ $\frac{1}{6}$

Common Error

- Some students do not write expressions with the same base in an exponential equation. Remind those students that in the equation $a^x = b^y$, $x = y$ if and only if $a = b$.
- See *Teacher's Resource Book* for additional remediation.

LESSON FOLLOW-UP

Discussion

Evaluate the use of the definition "An exponent shows how many times the base is used as a factor." Answers may vary.

Assignment Guide

See p. 332B for assignments.

Challenge

To solve the problem, students may use calculators to evaluate each expression and compare results or they may write each expression in terms of the same base then compare results.

Lesson Quiz

Simplify.

1. $4^{\sqrt{45}} \cdot 4^{\sqrt{5}}$ $4^{4\sqrt{5}}$
2. $\dfrac{n^{3\sqrt{2}} m^{-2\sqrt{2}}}{n^{\sqrt{2}} m^{2\sqrt{2}}}$ $\dfrac{n^{2\sqrt{2}}}{m^{4\sqrt{2}}}$
3. $(w^{\sqrt{6}} y^{\sqrt{8}})^{\sqrt{2}}$ $w^{2\sqrt{3}} y^4$

Evaluate the following to the nearest thousandth.

4. $243^{\frac{2}{3}}$ 38.941
5. $2^{\sqrt{3}}$ 3.322
6. $5^{0.268}$ 1.539
7. $3^{2\sqrt{3}\sqrt{2}}$ 217.474

Solve for x.

8. $81 = 3^x$ 4
9. $4^{4x+1} = 8^{2x+5}$ $\frac{13}{2}$
10. $(2^x)^x = 16$ ± 2

Enrichment

Solve for x.

$2^{4x} = (2\sqrt{2})^{\frac{3}{4}}$ $\quad 2^{4x} = (2^{\frac{3}{2}})^{\frac{3}{4}}, x = \frac{9}{32}$

An equation in which the variable is in the exponent is called an **exponential equation.** Examples of exponential equations are $3^x = 81$ and $2^x = \frac{1}{16}$.

To solve exponential equations, recall that for $b > 0$, $b \neq 1$, $b^m = b^n$ if and only if $m = n$. You can use this fact to solve an exponential equation by first expressing both sides of the equation in terms of the same base.

EXAMPLE 4 **Solve for x:**

a. $81 = 27^x$ **b.** $25^{n-1} = \left(\frac{1}{5}\right)^{4n\ \ 1}$

a. $81 = 27^x$

$$\begin{aligned} 3^4 &= (3^3)^x \\ 3^4 &= 3^{3x} \\ 4 &= 3x \\ \frac{4}{3} &= x \end{aligned}$$

b. $25^{n-1} = \left(\frac{1}{5}\right)^{4n-1}$

$$\begin{aligned} (5^2)^{n-1} &= (5^{-1})^{4n-1} \\ 5^{2n-2} &= 5^{-4n+1} \\ 2n - 2 &= -4n + 1 \\ 6n &= 3 \\ n &= \frac{1}{2} \end{aligned}$$

CLASS EXERCISES

Simplify.

1. $2^{\sqrt{7}} \cdot 2^{\sqrt{28}}$ $2^{3\sqrt{7}}$
2. $(3^{\sqrt{2}})^{\sqrt{18}}$ 729
3. $5^{-\sqrt{5}} \cdot 5^{\sqrt{45}}$ $5^{2\sqrt{5}}$

Simplify and express with positive exponents.

4. $(3x^{0.7})(81^{-0.5}x^{0.3})$ $\frac{x}{3}$
5. $\dfrac{x^{5\sqrt{2}}y^{-3\sqrt{5}}}{x^{2\sqrt{2}}y^{\sqrt{5}}}$ $\dfrac{x^{3\sqrt{2}}}{y^{4\sqrt{5}}}$
6. $(x^{-\sqrt{3}}y)^{\sqrt{12}}$ $\dfrac{y^{2\sqrt{3}}}{x^6}$

Evaluate the following to the nearest thousandth.

7. $5^{0.75}$ 3.344
8. $6^{\sqrt{7}}$ 114.497
9. $12^{\sqrt{14}}$ 10912.556

Solve for x.

10. $7^4 = 7^{2x+1}$ $\frac{3}{2}$
11. $8^{x-1} = 16^{3x}$ $-\frac{1}{3}$
12. $3^{4x-3} = 27^{2x-4}$ $\frac{9}{2}$

PRACTICE EXERCISES

Simplify.

A

1. $5^{\sqrt{6}} \cdot 5^{\sqrt{216}}$ $5^{7\sqrt{6}}$
2. $6^{\sqrt{12}} \cdot 6^{\sqrt{243}}$ $6^{11\sqrt{3}}$
3. $(4^{\sqrt{5}})^{\sqrt{180}}$ 4^{30}
4. $(3^{-\sqrt{8}})^{-\sqrt{800}}$ 3^{80}

Simplify and express with positive exponents.

5. $2^{\sqrt{200}} \cdot 2^{-\sqrt{8}}$ $2^{8\sqrt{2}}$
6. $3^{\sqrt{5}} \cdot 3^{-\sqrt{500}}$ $\frac{1}{3^{9\sqrt{5}}}$
7. $(3^{\sqrt{8}})^{-\sqrt{800}}$ $\frac{1}{3^{80}}$
8. $(4^{-\sqrt{128}})^{\sqrt{5}}$ $\frac{1}{4^{8\sqrt{10}}}$
9. $\dfrac{15x^{5\sqrt{3}}y^{-\sqrt{5}}}{45x^{\sqrt{3}}y^{\sqrt{5}}}$ $\frac{x^{4\sqrt{3}}}{3y^{2\sqrt{5}}}$
10. $\dfrac{25x^{2\sqrt{2}}y^{7\sqrt{7}}}{15x^{5\sqrt{2}}y^{2\sqrt{7}}}$ $\frac{5y^{5\sqrt{7}}}{3x^{3\sqrt{2}}}$
11. $(x^{-\sqrt{162}}y^{\sqrt{242}})^{\sqrt{2}}$ $\frac{y^{22}}{x^{18}}$
12. $(x^{\sqrt{32}}y^{-\sqrt{200}})^{\sqrt{8}}$ $\frac{x^{16}}{y^{40}}$

Evaluate each of the following to the nearest thousandth.

13. $6^{\frac{3}{7}}$ 2.155
14. $5^{\frac{5}{9}}$ 2.445
15. $7^{\frac{6}{11}}$ 2.890
16. $3^{\frac{6}{7}}$ 2.564
17. $6^{-\frac{6}{7}}$ 0.215
18. $5^{-\frac{5}{9}}$ 0.409
19. $7^{-\frac{6}{11}}$ 0.346
20. $3^{-\frac{6}{7}}$ 0.390

B

21. $3^{\sqrt{2}+\sqrt{3}}$ 31.707
22. $4^{\sqrt{3}+\sqrt{2}}$ 78.386
23. $2^{\sqrt{5}+\sqrt{6}}$ 25.733
24. $6^{\sqrt{7}+\sqrt{3}}$ 2550.305

Solve for *x*.

25. $3^x = 3^{2x-3}$ 3
26. $5^4 = 25^{2x}$ 1
27. $6^{x+4} = 6^{2x-1}$ 5
28. $9^x = 9^{2x-1}$ 1
29. $5^{x-2} = 5^{3x+1}$ $-\frac{3}{2}$
30. $3^{3x} = 3^{5x+2}$ -1
31. $2^{x+3} = 4^{x-5}$ 13
32. $25^{x-1} = 125^x$ -2
33. $4^{2x+1} = 8^{x+5}$ 13
34. $8^{2x-1} = 16^x$ $\frac{3}{2}$

C

35. $\left(\frac{1}{2}\right)^{x+2} = 16^{-x}$ $\frac{2}{3}$
36. $\left(\frac{1}{5}\right)^{x+3} = 25^{-x}$ 3
37. $(3a)^x = (9a^2)^{x+2}$ -4
38. $(4a)^x = (16a^2)^{x+2}$ -4
39. $3^{\sqrt{x}+4} = 27^{\sqrt{x}}$ 4
40. $5^{\sqrt{x}+8} = 125^{\sqrt{x}}$ 16

Applications

41. **Biology** The number N of bacteria in a particular solution can be determined by the formula $N = N_0 2^t$, where t is the time measured in hours, and N_0 is the number of bacteria present at time $t = 0$. Find N if $N_0 = 20{,}000$ and $t = \frac{3}{7}$ h. 26,918 to the nearest whole number

42. **Archaeology** The number A of milligrams (mg) of carbon-14 in a fossil can be determined by the formula $A = A_0 2^{-0.75t}$, where t is the time in years, and A_0 is the number of milligrams of carbon-14 present at time $t = 0$. Find the amount of carbon-14 left after 5 years if $A_0 = 150$. 11.15 mg

CHALLENGE

Which is greater, $16^{\sqrt{75}}$ or $32^{\sqrt{48}}$? They are equal.

Teacher's Resource Book

Practice—Chapter 8, p. 1
Enrichment—Chapter 8, p. 2

LESSON PLAN

Vocabulary
Exponential function
Compound interest formula

Materials/Manipulatives
Scientific calculators
Graphing calculators or computer

BACKGROUND

In the Preview, the compound interest formula is used to review the use of exponents in formulas prior to introducing exponential functions.

8.2 Exponential Functions

Objectives: To define exponential functions
To graph exponential functions

Exponential functions are very useful for expressing growth and decay in mathematical terms.

Preview

If you invest an amount of money in an account in which interest is compounded, the total amount A in the account at the end of one year is

$$A = P\left(1 + \frac{r}{n}\right)^n$$

where n is the number of times per year that interest is compounded, r is the yearly rate of interest, and P is the principal invested.

EXAMPLE **Mary Ann invests $1000 at 8% in an account in which interest is compounded twice per year. How much money is in her account at the end of one year?**

$$A = P\left(1 + \frac{r}{n}\right)^n$$
$$= 1000\left(1 + \frac{0.08}{2}\right)^2$$
$$= 1000(1.04)^2 \quad \textit{Calculation-ready form}$$
$$= 1081.60 \quad \textit{To the nearest cent}$$

There is $1081.60 in her account at the end of one year.

Find A to the nearest cent under each of the following conditions.

1. $P = \$1000$, $r = 8\%$, $n = 3$ $1082.15
2. $P = \$1000$, $r = 5\%$, $n = 4$ $1050.95
3. $P = \$12{,}000$, $r = 6\%$, $n = 4$ $12,736.36
4. $P = \$100{,}000$, $r = 7.5\%$, $n = 4$ $107,713.59
5. How much more interest will Mary Ann earn on $1000 invested at 8% if the interest is compounded monthly rather than twice per year? $1.40
6. How much more interest will she earn if the interest in Exercise 5 is compounded daily (assume 365 days per year) rather than twice per year? $1.68

An **exponential function** is a function in which the variable is in the exponent. If x and b are real numbers such that $b > 0$ and $b \neq 1$, then $f(x) = b^x$ is an exponential function with base b. Examples of exponential functions include $y = 3^x$, $f(x) = 6^x$, and $f(x) = 2^x$.

EXAMPLE 1 **Graph the function $y = 2^x$.**

Begin by constructing a table of values. Write nonintegral powers of 2 in decimal form. Then plot the points and join them with a smooth curve.

x	−3	−2	−1	0	1	2	3
y	0.125	0.25	0.5	1	2	4	8

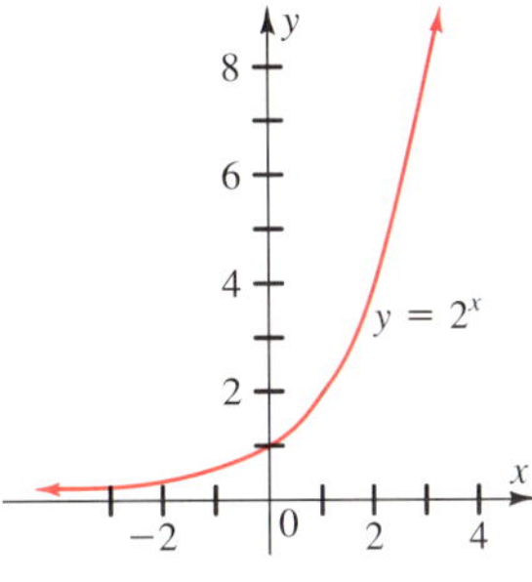

Since the base of the function $y = 2^x$ is positive, the function is never negative. The x-axis is a horizontal asymptote. Any *horizontal line* intersects the graph in at most one point, so $y = 2^x$ is a *one-to-one function*.

Note that irrational numbers are included in the domain and range of $y = 2^x$. Its graph is a smooth curve including a point such as $(\sqrt{3}, 2^{\sqrt{3}})$. The approximate location of $(\sqrt{3}, 2^{\sqrt{3}})$ is shown on the graph at (1.7, 3.3).

The value of y in Example 1 *increases* rapidly as x increases. Functions of this form are often said to increase *exponentially*. Some functions exhibit great *decreases* as x increases in value. The next example shows the graph of an exponential function in which the base is a positive number less than 1.

EXAMPLE 2 **Graph the function $y = \left(\frac{1}{3}\right)^x$.**

Write nonintegral powers of $\frac{1}{3}$ in decimal form.
Then plot the points and join them with a smooth curve.

x	−3	−2	−1	0	1	2	3
y	27	9	3	1	0.333	0.111	0.037

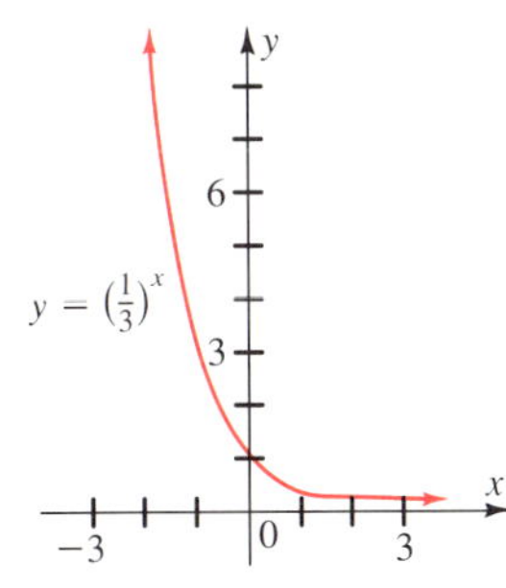

The x-axis is again a horizontal asymptote. Since any horizontal line intersects the graph in at most one point, $y = \left(\frac{1}{3}\right)^x$ is a one-to-one function.

In Example 1, the graph approaches the negative x-axis. In Example 2, the graph approaches the positive x-axis. In general, the graph of $y = b^x$ rises to the right for $b > 1$, and to the left for $0 < b < 1$.

TEACHING SUGGESTIONS

- Stress the idea that the graph of $f(x) = a^x$ is a function and has all the properties of a function.
- Students should use their scientific calculators to find the decimal values for non-integral powers.
- Students may use a graphing calculator or a computer to display graphs.

CHALKBOARD EXAMPLES

- **For Example 1**

  1. Graph the function $y = 3^x$.

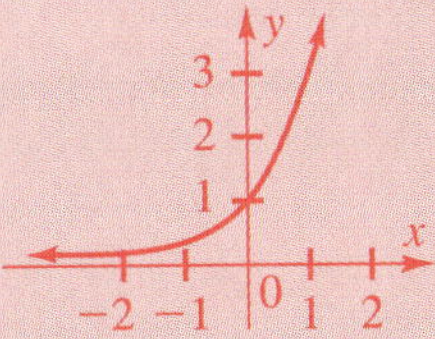

- **For Example 2**

  2. Graph the function $y = \left(\frac{1}{2}\right)^x$.

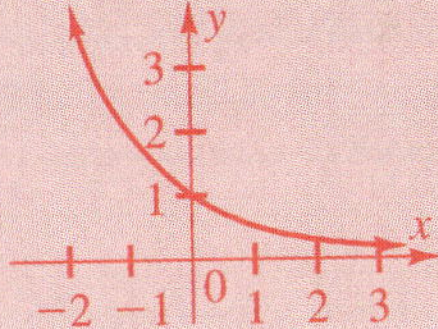

- **For Example 3**

  Find each value to three decimal places.

  3. $e^{0.2}$ 1.221 4. $e^{2.1}$ 8.166

- **For Example 4**

  5. Graph the function $f(x) = e^{-x}$.

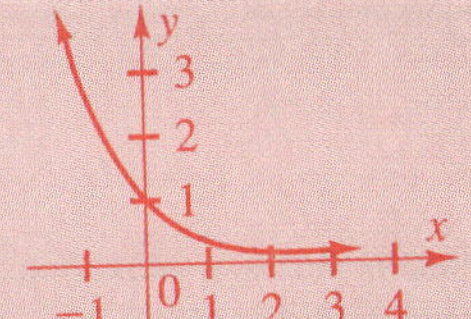

LESSON FOLLOW-UP

Critical Thinking

Analysis Point out that the function $y = a^x$ is defined for $a > 0$ and $a \neq 1$. Then ask students to describe the graph of $y = a^x$ for $a = 0$, for $a = 1$, and for $a < 0$. Then ask them to tell in which case, if any, $y = a^x$ is a function. Using the definitions for a^x for $a = 0$, $a = 1$, or $a < 0$, students should conclude that the graph of $y = a^x$ is the point (0, 0) when $a = 0$ and x is any real number and the graph of $y = a^x$ is the point (1, 1) when $a = 1$ and x is any real number. When $a < 0$ and x is any real number, the points on the graph of $y = a^x$ are alternately located above and below the x-axis. Students should also conclude that $y = a^x$ is a function in each case but is not a one-to-one function when $a < 0$.

Assignment Guide

See p. 332B for assignments.

Extra

To solve the problem students must first determine that each stack of paper in the problem contains 2^{n-1} sheets of computer paper.

Properties of Exponential Functions

- The domain is the set of all real numbers.
- The range is the set of positive real numbers.
- The y-intercept of the graph is 1.
- The x-axis is an asymptote of the graph.
- The function is one-to-one.

In advanced mathematics, science, and finance, the exponential function with base e is used frequently. The number e is irrational and can be approximated by 2.7182818. Many scientific calculators have an e^x key. To use most such calculators to evaluate e^x, enter the exponent and then press the e^x key.

EXAMPLE 3 **Find the value of: a.** e^3 **b.** $e^{0.4}$ **c.** $e^{-1.2}$
Round your answers to three decimal places.

a. $e^3 = 20.086$ **b.** $e^{0.4} = 1.492$ **c.** $e^{-1.2} = 0.301$

The function $f(x) = e^x$ is called the **natural exponential function.** You can use the e^x key to generate a table of values.

EXAMPLE 4 **Graph the function $f(x) = e^x$.**

Begin by constructing a table of values.

x	-3	-2	-1	0	1	2	3
$f(x)$	0.05	0.14	0.37	1	2.72	7.39	20.09

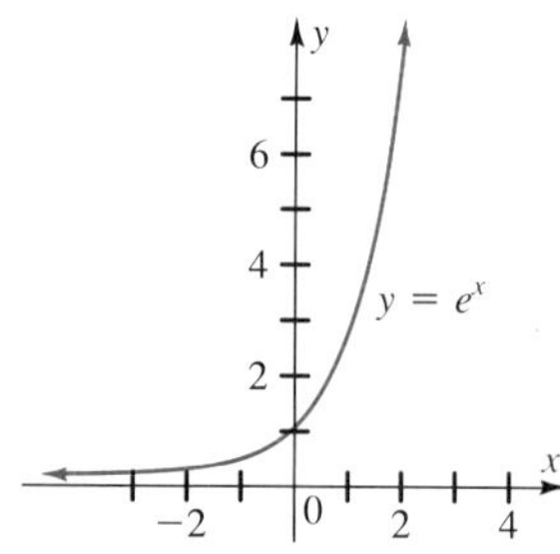

CLASS EXERCISES

Evaluate each of the following for the function $f(x) = 3^x$.

1. $f(5)$ 243 **2.** $f(-5)$ 0.004 **3.** $f(0.5)$ 1.732 **4.** $f(-0.5)$ 0.577

5. Graph on the same set of axes: $f(x) = 2^x, f(x) = \left(\frac{1}{2}\right)^x, f(x) = 2^{-x}$. See side column on page 341.

6. How do the graphs of $f(x) = 2^x$ and $f(x) = \left(\frac{1}{2}\right)^x$ differ? $f(x) = 2^x$ is increasing; $f(x) = \left(\frac{1}{2}\right)^x$ is decreasing.

For Discussion

7. The exponential function is defined for $b > 0$ and $b \neq 1$. Describe the graph if $b = 0$. *x*-axis if $b = 1$. the line $y = 1$

PRACTICE EXERCISES

Evaluate each of the following for the function $f(x) = 4^x$.

A 1. $f(2)$ 16 2. $f(-3)$ 0.016 3. $f(0.5)$ 2 4. $f(-0.2)$ 0.758

Evaluate each of the following for the function $f(x) = e^x$.

5. $f(4)$ 54.598 6. $f(-4)$ 0.018 7. $f(1.4)$ 4.055 8. $f(-4.2)$ 0.015

Graph each function. See pages 492–493.

9. $f(x) = 4^x$ 10. $f(x) = 7^x$ 11. $f(x) = 5^x$ 12. $f(x) = 6^x$

13. $f(x) = 5^{-x}$ 14. $f(x) = 10^{-x}$ 15. $f(x) = 2^{-x}$ 16. $f(x) = 4^{-x}$

B 17. $f(x) = \left(\frac{1}{2}\right)^x$ 18. $f(x) = \left(\frac{1}{5}\right)^x$ 19. $f(x) = \left(\frac{1}{4}\right)^x$ 20. $f(x) = \left(\frac{1}{10}\right)^x$

21. $f(x) = 4^{2x}$ 22. $f(x) = 3^{2x}$ 23. $f(x) = 2^{2x+1}$ 24. $f(x) = 3^{2x-1}$

25. $f(x) = \left(\frac{1}{2}\right)^{2x}$ 26. $f(x) = \left(\frac{1}{3}\right)^{2x}$ 27. $f(x) = e^{-x}$ 28. $f(x) = e^{2x}$

C 29. $f(x) = 2^{x^2}$ 30. $f(x) = 2^{-x^2}$

Find the base of an exponential function, $y = b^x$, containing the points.

31. $\left(-2, \frac{1}{16}\right)$ 4 32. $\left(\frac{1}{2}, 2\right)$ 4

Applications

33. **Algebra** If a number is squared and raised to a certain power, the result is the same number raised to a power that is one more than three times the power of the number being squared. What is the power? −1

34. **Optics** The optical intensity of objects can be found using the formula $I = 10^{-d}$. Graph the function and determine the intensity when $d = 0.8$. 0.158 See side column.

EXTRA

One sheet of paper is 0.005 in. thick. Suppose you make stacks of computer paper such that each stack has twice as many sheets as the previous stack. How tall would the 30th stack be if the first stack has only 1 sheet? 42.4 miles

Lesson Quiz

Evaluate each of the following for the function $f(x) = 5^x$. Round answers to three decimal places.

1. $f(0.5)$ 2.236 2. $f(-2)$ 0.04 3. $f(3.6)$ 328.316

Evaluate each of the following for the function $f(x) = e^x$. Round to three decimal places.

4. $f(0.5)$ 1.649 5. $f(-3.2)$ 0.041

Graph each function.

6. $f(x) = e^x$

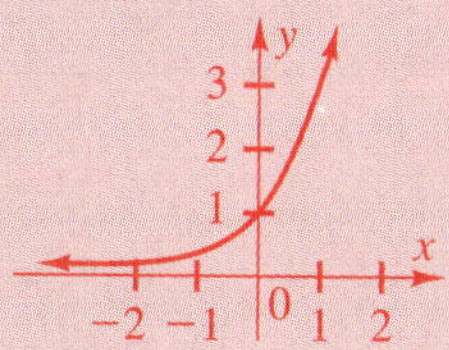

7. $f(x) = 3^{-x}$

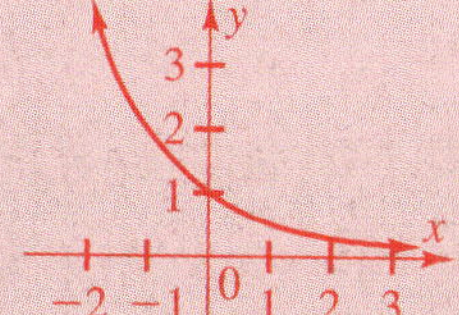

Enrichment

Describe the graph of the function $f(x) = 2^x - 3$. It is the graph of $y = 2^x$ shifted down 3 units.

Teacher's Resource Book

Practice—Chapter 8, p. 3
Enrichment—Chapter 8, p. 4

Additional Answers

Class Exercises

5.

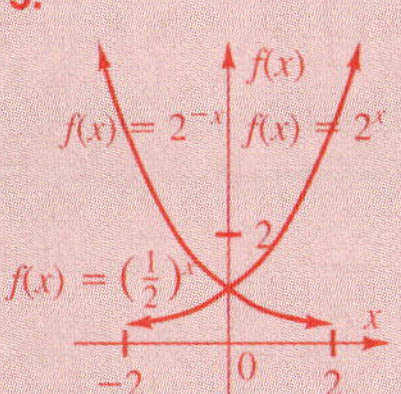

Applications

34.

LESSON PLAN

Vocabulary
Logarithmic function

Materials/Manipulatives
Scientific calculators
Graphs of exponential functions on actetate sheet

BACKGROUND

In the Preview, graphing a function and its inverse on the same set of axes and determining whether or not the inverse is a function is reviewed.

Additional Answers

Preview

1.

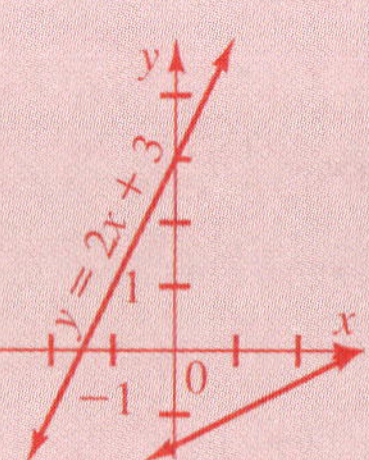

2.

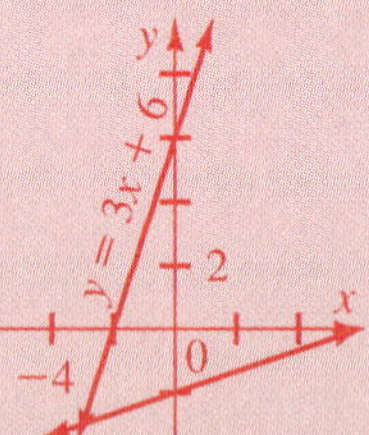

3.

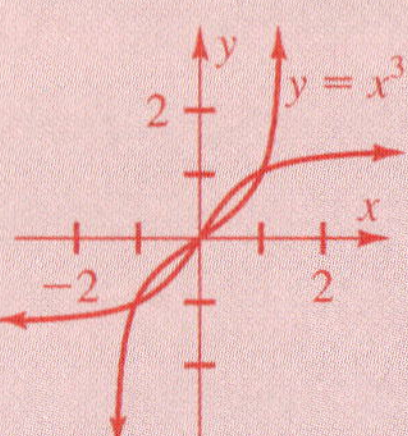

4.

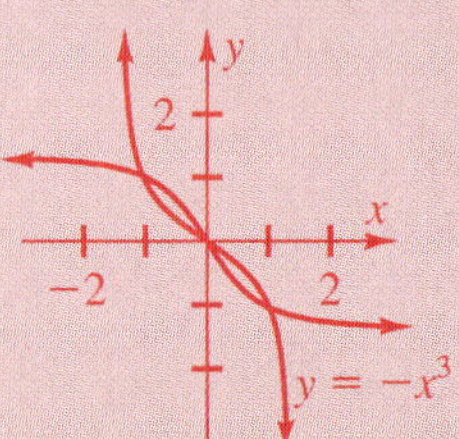

5.

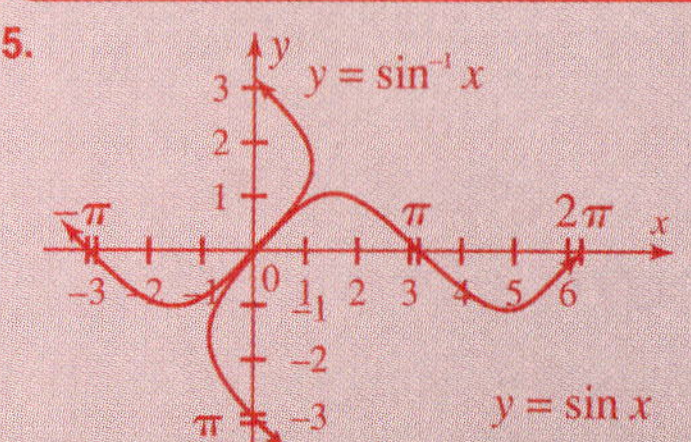

6.

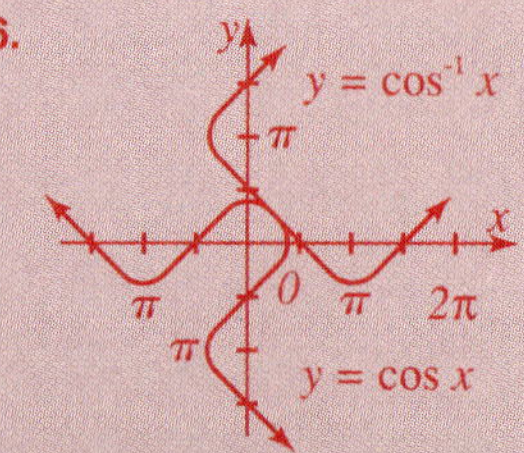

8.3 Logarithmic Functions

Objectives: To define and graph logarithmic functions
To solve logarithmic equations

The concentration of hydrogen ions in a solution is called the pH of the solution. The pH is defined using a special function called a logarithm. The logarithmic function is the inverse of the exponential function.

Preview

Recall that the inverse of a function $y = f(x)$ can be found by interchanging the x- and y-coordinates and then solving for y. As you learned when studying the trigonometric functions, not all functions have inverses that are also functions.

EXAMPLE **Graph the function $y = x^2$ and its inverse on the same set of axes. Is the inverse also a function?**

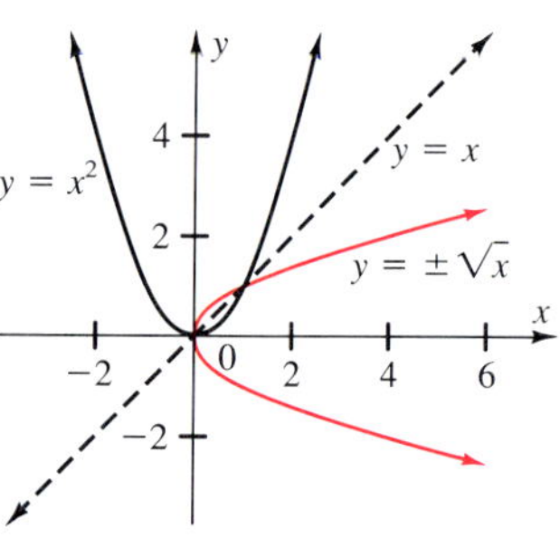

$y = x^2$

$x = y^2$ *Interchange x and y.*

$\pm\sqrt{x} = y$ *Solve for y.*

Since for each positive value of x there is more than one value of y, $y = \pm\sqrt{x}$ is *not* a function.

Note that the graph of $y = x^2$ is symmetric to the graph of its inverse, $y = \pm\sqrt{x}$, about the line $y = x$. Recall that the graphs of *all* functions are symmetric to their inverses about the line $y = x$.

Graph each function and its inverse on the same set of axes. Tell whether or not each inverse is also a function. See side column.

1. $y = 2x + 3$ yes
2. $y = 3x + 6$ yes
3. $y = x^3$ yes
4. $y = -x^3$ yes
5. $y = \sin x$ no
6. $y = \cos x$ no

The exponential function $y = b^x$ is a one-to-one function. Therefore, its inverse, $x = b^y$, is also a function. In order to write the inverse function $x = b^y$ in the form $y = f(x)$, the *logarithmic function* is defined.

For all positive real numbers x and b, $b \neq 1$, the inverse of the exponential function $y = b^x$ is the **logarithmic function**:

$y = \log_b x$ *Read: "y equals the logarithm to the base b of x" or "y equals log base b of x."*

$y = \log_b x$ if and only if $x = b^y$.

Recall that the graph of a function is symmetric to the graph of its inverse about $y = x$. Thus, the graph of $y = \log_b x$ is symmetric to that of $y = b^x$ about $y = x$.

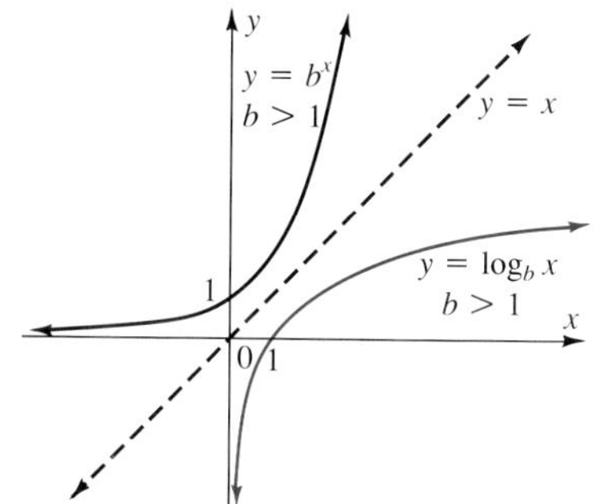

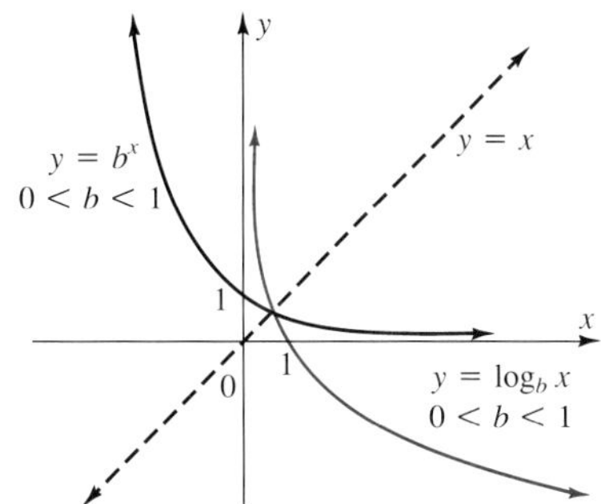

For the logarithmic function with $b > 1$, notice that as x increases, y increases. As x decreases, y decreases and the curve approaches the y-axis, its asymptote. For the logarithmic function with $0 < b < 1$, as x increases, y decreases. As x decreases, y increases and the curve approaches the y-axis, its asymptote.

Properties of Logarithmic Functions

- The domain is the set of positive real numbers.
- The range is the set of all real numbers.
- The x-intercept of the graph is 1.
- The y-axis is an asymptote of the graph.
- The function is one-to-one.

EXAMPLE 1 **Graph $y = \log_3 x$.**

Recall that $y = \log_3 x$ is equivalent to $x = 3^y$. Choose values for y and calculate values of x.

x	$\frac{1}{27}$	$\frac{1}{9}$	$\frac{1}{3}$	1	3	9	27
y	-3	-2	-1	0	1	2	3

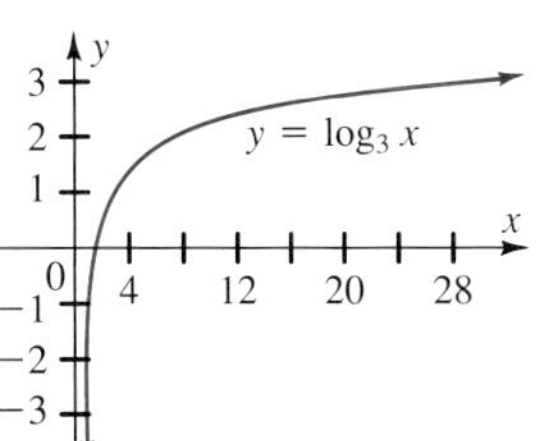

You can express an equation given in logarithmic form in exponential form.

TEACHING SUGGESTIONS

To be sure that students understand why the exponential and logarithmic functions are smooth curves, have them use some nonintegral values for the exponents.

Critical Thinking

Application Ask students to draw the graph of a relation that is not a function whose inverse is a function and then to list the properties that relations of this type would have that would be consistent across the group. Answers may vary.

CHALKBOARD EXAMPLES

- **For Example 1**

1. Graph $y = \log_5 x$.

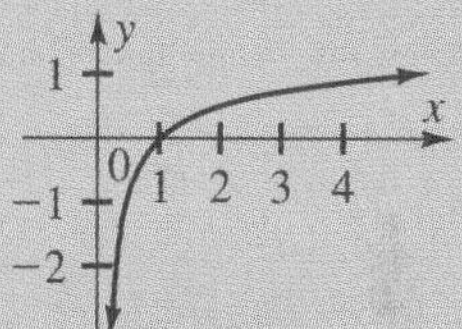

- **For Example 2**

Express in exponential form.

2. $\log_4 16 = 2$ $4^2 = 16$
3. $\log_3 64 = 4$ $4^3 = 64$
4. $\log_5 0.04 = -2$ $5^{-2} = 0.04$
5. $\log_{25} 5 = 0.5$ $25^{0.5} = 5$

- **For Example 3**

Express in logarithmic form.

6. $5^3 = 125$ $\log_5 125 = 3$
7. $5^{-2} = 0.04$ $\log_5 0.04 = -2$
8. $36^{\frac{3}{2}} = 216$ $\log_{36} 216 = \frac{3}{2}$
9. $8^{-\frac{2}{3}} = \frac{1}{4}$ $\log_8 \frac{1}{4} = -\frac{2}{3}$

- **For Example 4**

Evaluate.

10. $\log_2 32$ 5

11. $\log_{16} 8$ $\frac{3}{4}$

12. $\log_5 0.008$ -3

13. $\log_{25} 125$ $\frac{3}{2}$

- **For Example 5**

Solve.

14. $\log_x 49 = 2$ 7

15. $\log_3 x = 4$ 81

16. $\log_9 x = -2$ $\frac{1}{81}$

17. $\log_x 5 = 0.2$ 3125

Common Error

- Some students forget that a logarithm is an exponent. Having those students change from logarithmic form to exponential form and from exponential to logarithmic form will help establish the relationship.
- See *Teacher's Resource Book* for additional remediation.

Additional Answers

Class Exercises

1. $\log_3 x$ increases more rapidly than $\log_{10} x$.

EXAMPLE 2 **Express in exponential form:**

a. $\log_4 64 = 3$ **b.** $\log_2 32 = 5$ **c.** $\log_5 0.04 = -2$ **d.** $\log_{36} 6 = 0.5$

a. $\log_4 64 = 3$, $4^3 = 64$ **b.** $\log_2 32 = 5$, $2^5 = 32$ **c.** $\log_5 0.04 = -2$, $5^{-2} = 0.04$ **d.** $\log_{36} 6 = 0.5$, $36^{0.5} = 6$

You can also convert an expression given in exponential form to logarithmic form.

EXAMPLE 3 **Express in logarithmic form:**

a. $7^2 = 49$ **b.** $5^4 = 625$ **c.** $16^{\frac{5}{4}} = 32$ **d.** $27^{-\frac{4}{3}} = \frac{1}{81}$

a. $7^2 = 49$, $\log_7 49 = 2$ **b.** $5^4 = 625$, $\log_5 625 = 4$ **c.** $16^{\frac{5}{4}} = 32$, $\log_{16} 32 = \frac{5}{4}$ **d.** $27^{-\frac{4}{3}} = \frac{1}{81}$, $\log_{27} \frac{1}{81} = -\frac{4}{3}$

Some logarithms can be evaluated by setting them equal to a variable and then rewriting the resulting equation in exponential form.

EXAMPLE 4 **Evaluate:** **a.** $\log_3 81$ **b.** $\log_5 \frac{1}{25}$ **c.** $\log_2 2\sqrt{2}$

a. $\log_3 81$

Let $x = \log_3 81$

$3^x = 81$

$3^x = 3^4$

$x = 4$

b. $\log_5 \frac{1}{25}$

Let $x = \log_5 \frac{1}{25}$

$5^x = \frac{1}{25}$

$5^x = 5^{-2}$

$x = -2$

c. $\log_2 2\sqrt{2}$

Let $x = \log_2 2\sqrt{2}$

$2^x = 2\sqrt{2}$

$2^x = 2^{\frac{3}{2}}$

$x = \frac{3}{2}$

Some logarithmic equations can be solved if they are converted to exponential form. If the base is not an integer, use a scientific calculator.

EXAMPLE 5 **Solve:** **a.** $\log_{100} x = \frac{5}{2}$ **b.** $\log_5 125 = x$ **c.** $\log_x 32 = 3$

a. $\log_{100} x = \frac{5}{2}$

$100^{\frac{5}{2}} = x$

$100{,}000 = x$

b. $\log_5 125 = x$

$5^x = 125$

$5^x = 5^3$

$x = 3$

c. $\log_x 32 = 3$

$x^3 = 32$

$x = 32^{\frac{1}{3}}$

$x = 3.17$

CLASS EXERCISES

For Discussion

1. How do the graphs of $y = \log_3 x$ and $y = \log_{10} x$ differ? See side column, page 344.

Express in logarithmic form.

2. $2^6 = 64$ $\log_2 64 = 6$
3. $16^{0.5} = 4$ $\log_{16} 4 = 0.5$
4. $10^{-3} = 0.001$ $\log_{10} 0.001 = -3$

Express in exponential form.

5. $\log_6 36 = 2$ $6^2 = 36$
6. $\log_{25} 5 = 0.5$ $25^{0.5} = 5$
7. $\log_{10} 0.01 = -2$ $10^{-2} = 0.01$

Evaluate.

8. $\log_{25} 5$ $\frac{1}{2}$
9. $\log_{10} 100$ 2
10. $\log_{10} 0.001$ -3

Solve.

11. $\log_x 625 = 4$ 5
12. $\log_4 x = 3$ 64
13. $\log_x 16 = 4$ 2

PRACTICE EXERCISES

Express in exponential form.

A

1. $\log_8 64 = 2$ $8^2 = 64$
2. $\log_6 36 = 2$ $6^2 = 36$
3. $\log_3 27 = 3$ $3^3 = 27$
4. $\log_4 256 = 4$ $4^4 = 256$
5. $\log_{36} 6 = 0.5$ $36^{0.5} = 6$
6. $\log_{81} 9 = 0.5$ $81^{0.5} = 9$
7. $\log_2 \frac{1}{16} = -4$ $2^{-4} = \frac{1}{16}$
8. $\log_3 \frac{1}{81} = -4$ $3^{-4} = \frac{1}{81}$

Express in logarithmic form.

9. $4^3 = 64$ $\log_4 64 = 3$
10. $6^3 = 216$ $\log_6 216 = 3$
11. $49^{\frac{1}{2}} = 7$ $\log_{49} 7 = \frac{1}{2}$
12. $343^{\frac{1}{3}} = 7$ $\log_{343} 7 = \frac{1}{3}$
13. $27^{-\frac{2}{3}} = \frac{1}{9}$ $\log_{27} \frac{1}{9} = -\frac{2}{3}$
14. $16^{-\frac{1}{4}} = \frac{1}{2}$ $\log_{16} \frac{1}{2} = -\frac{1}{4}$
15. $7^{\frac{3}{2}} = 7\sqrt{7}$ $\log_7 7\sqrt{7} = \frac{3}{2}$
16. $13^{\frac{3}{2}} = 13\sqrt{13}$ $\log_{13} 13\sqrt{13} = \frac{3}{2}$

Evaluate.

17. $\log_{10} 1000$ 3
18. $\log_{10} 1$ 0
19. $\log_{10} 0.1$ -1
20. $\log_{10} 0.0001$ -4
21. $\log_4 0.5$ $-\frac{1}{2}$
22. $\log_8 16$ $\frac{4}{3}$
23. $\log_9 27$ $\frac{3}{2}$
24. $\log_{16} 2$ $\frac{1}{4}$

Graph. See below.

25. $y = \log_4 x$
26. $y = \log_3 x$
27. $y = \log_{\frac{1}{4}} x$
28. $y = \log_{\frac{1}{3}} x$

Solve.

B

29. $\log_4 64 = y$ 3
30. $\log_x 16 = 4$ 2
31. $\log_6 z = 3$ 216
32. $\log_9 27 = y$ $\frac{3}{2}$
33. $\log_x 18 = 1$ 18
34. $\log_3 z = -2$ $\frac{1}{9}$
35. $\log_3 243 = y$ 5
36. $\log_x 10 = 0.5$ 100

LESSON FOLLOW-UP

Discussion

What is the value of x in the expression $\log_3 3^2 = x$? 2

Assignment Guide

See p. 332B for assignments.

Test Yourself

See *Teacher's Resource Book, Tests*, pp. 77–78.

Lesson Quiz

1. Graph $y = \log_2 x$.

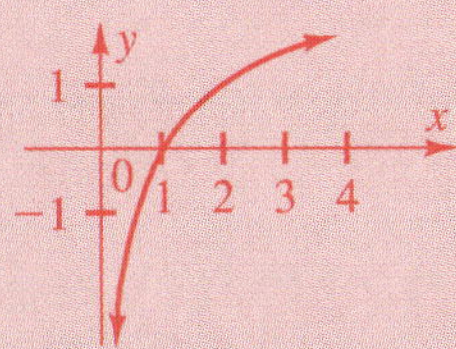

Express in exponential form.

2. $\log_4 64 = 3$ $4^3 = 64$
3. $\log_5 0.04 = -2$ $5^{-2} = 0.04$

Express in logarithmic form.

4. $8^3 = 512$ $\log_8 512 = 3$
5. $4^{-2} = 0.0625$ $\log_4 0.0625 = -2$

Evaluate.

6. $\log_3 81$ 4
7. $\log_{25} 5$ $\frac{1}{2}$ or 0.5
8. $\log_{10} 0.001$ -3
9. $\log_4 0.5$ $-\frac{1}{2}$

Solve.

10. $\log_x 125 = 3$ 5
11. $\log_2 x = 4$ 16
12. $\log_8 x = -2$ $\frac{1}{64}$
13. $\log_x 17 = 1$ 17
14. $\log_x 21 = 2$ 4.5826
15. $\log_x 15 = 2.4$ 3.0906

25.
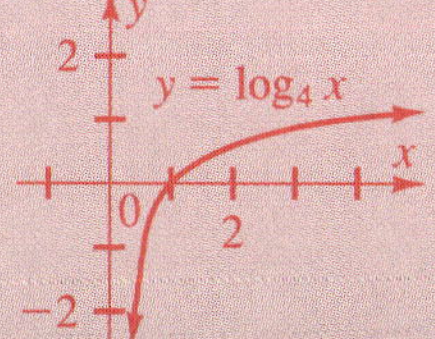

26.
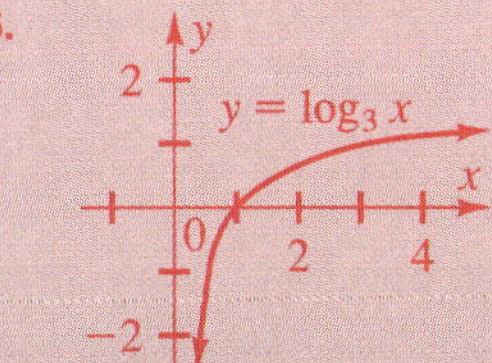

27.
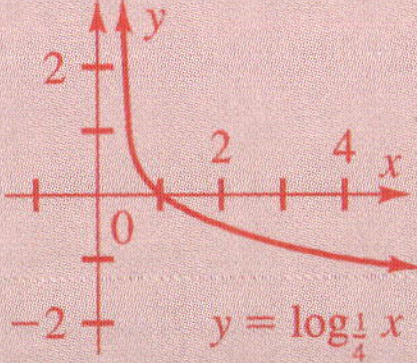

28.
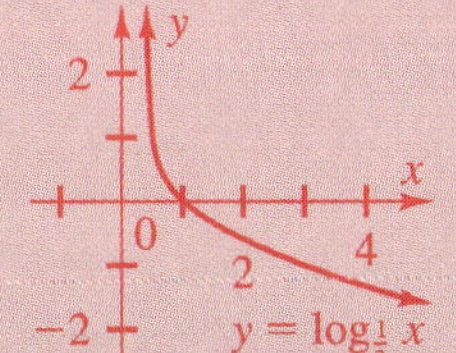

Teacher's Resource Book
Practice—Chapter 8, p. 5
Enrichment—Chapter 8, p. 6

Additional Answers

Practice Exercises

49.

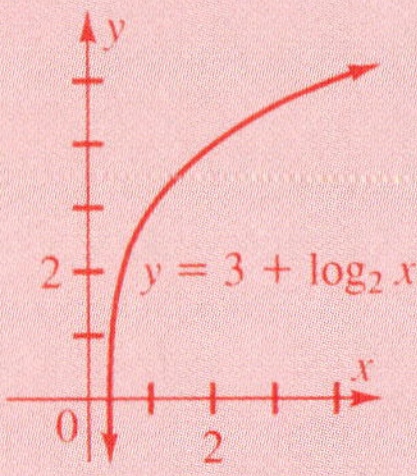

$D = \{x: x > 0\}$
$R = \{y: y \text{ is a real number}\}$

50.

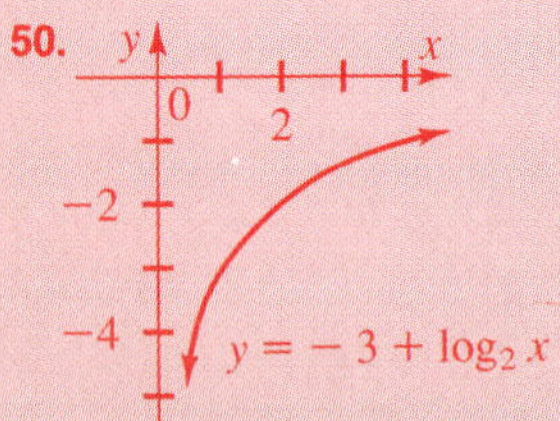

$D = \{x: x > 0\}$
$R = \{y: y \text{ is a real number}\}$

51.

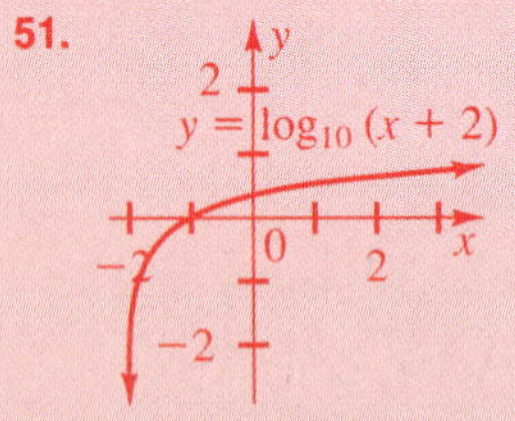

$D = \{x: x > -2\}$
$R = \{y: y \text{ is a real number}\}$

52.

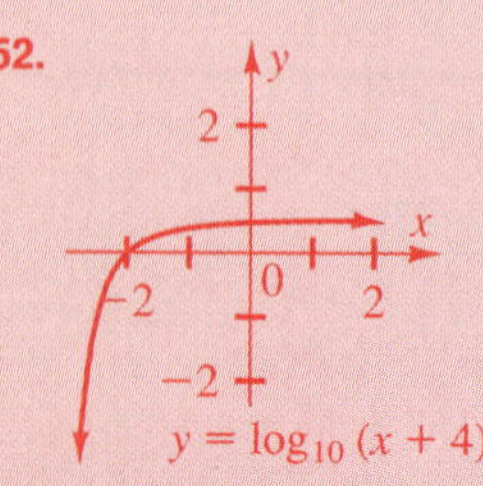

$D = \{x: x > -4\}$
$R = \{y: y \text{ is a real number}\}$

37. $\log_7 z = 0.5$ $\sqrt{7}$ **38.** $\log_{10} 10 = y$ 1 **39.** $\log_x 4 = \frac{1}{3}$ 64 **40.** $\log_{\frac{1}{2}} z = 8$ $\frac{1}{256}$

41. $\log_x 10 = 3.8$ 1.83 **42.** $\log_x 18 = 2.1$ 3.96 **43.** $\log_x 24 = 0.3$ 39,875.32 **44.** $\log_x 36 = 0.28$ 361,595.67

45. $\log_x 9 = 14$ 1.17 **46.** $\log_x 17 = 5.4$ 1.69 **47.** $\log_x 4.2 = 0.7$ 7.77 **48.** $\log_x 12 = 0.19$ 478,521.32

Graph and state the domain and range. See side column.

C **49.** $y = 3 + \log_2 x$ **50.** $y = -3 + \log_2 x$

51. $y = \log_{10}(x + 2)$ **52.** $y = \log_{10}(x + 4)$

Solve.

53. $\log_{\frac{2}{5}} \frac{4}{25} = \frac{x}{4}$ 8 **54.** $\log_{\frac{3}{5}} \frac{27}{125} = 2x$ $\frac{3}{2}$ **55.** $\log_{\frac{3}{8}} \frac{9}{64} = 2x + 2$ 0

56. $\log_{\frac{5}{8}} \frac{125}{512} = 2x - 7$ 5 **57.** $\log_{\frac{2}{7}} \frac{8}{343} = 3x - 6$ 3 **58.** $\log_{\frac{3}{7}} \frac{81}{2401} = 6x - 4$ $\frac{4}{3}$

Applications

59. Physics The electric current I in a particular circuit is given by $\log_2 I = -t$. Write this formula in exponential form. $2^{-t} = I$

60. Computer Use a graphing calculator or a computer to graph the function $y = \log_{10} x$. Let the domain be the interval $0 \le x \le 20$ in increments of 1. Use the graph to evaluate $\log_{10} 1.9$ and $\log_{10} 5.3$. $\log_{10} 1.9 = 0.2788$; $\log_{10} 5.3 = 0.7243$

TEST YOURSELF

Evaluate each of the following to the nearest thousandth.

1. $7^{\sqrt{11}}$ 635.145 **2.** $6^{-\frac{2}{9}}$ 0.672 **8.1**

Solve each exponential equation.

3. $5^{2x+1} = 125^{x-1}$ 4 **4.** $\left(\frac{1}{4}\right)^{x+3} = 2^{-x}$ −6

5. Graph $f(x) = 3^{-x}$ and $f(x) = 3^x$ on the same set of axes. See side column. **8.2**

6. Evaluate $6^{1.2}$ and $e^{-0.3}$. $6^{1.2} = 8.586$; $e^{-0.3} = 0.741$

7. Express $4^{-3} = \frac{1}{64}$ in logarithmic form and $\log_{49} 7 = \frac{1}{2}$ in exponential form. **8.3**
$\log_4 \frac{1}{64} = -3$; $49^{\frac{1}{2}} = 7$

Solve.

8. $\log_4 1024 = x$ 5 **9.** $\log_5 \frac{1}{25} = x$ −2 **10.** $\log_x 11 = 0.5$ 121 **11.** $\log_{10} x = 6$ 1,000,000

12. Graph $y = \log_5 x$ and $y = \log_{\frac{1}{5}} x$ on the same set of axes. See side column.

Test Yourself

5.

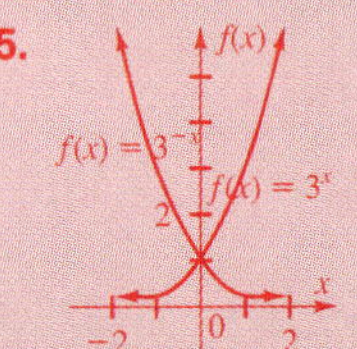

12.

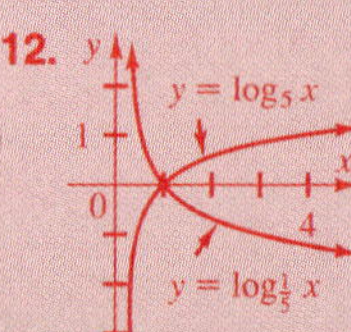

8.4 Properties of Logarithms

Objectives: To state the properties of logarithms
To use these properties to express logarithms in expanded form and to solve logarithmic equations

Loudness is a sensation that is perceived differently by different people. It is related to intensity, which is a physically measurable quantity. The human ear can detect sounds over a very wide range of intensities. However, what a human ear perceives as loudness is not directly proportional to the intensity of the sound. Therefore, a logarithmic scale is used to specify sound intensity.

Preview

The **intensity level** (β), measured in decibels (dB), of any sound is related to its **intensity** (I), measured in watts/meter^2 (W/m^2), using the following formula:

$$\beta = 10 \log_{10} \frac{I}{I_0}$$

where $I_0 = 1 \times 10^{-12}$ W/m^2, which is the minimum intensity audible to an average person and is called the *threshold of hearing.*

EXAMPLE **Find the intensity level in decibels of the rustle of leaves, which has intensity 1×10^{-11} W/m^2.**

$$\begin{aligned} \beta &= 10 \log_{10} \frac{10^{-11}}{10^{-12}} \\ &= 10 \log_{10} 10 \\ &= 10(1) \\ &= 10 \text{ dB} \end{aligned}$$

1. Find the intensity level in dB of a whisper, which has intensity 1×10^{-10} W/m^2. 20 dB
2. Find the intensity level of a quiet radio, which has intensity 1×10^{-8} W/m^2. 40 dB
3. As the intensity is increased from 1×10^{-10} W/m^2 to 1×10^{-8} W/m^2 (by a factor of 100), the intensity level increases by how many decibels? 20 dB
4. How much more intense would a 70-dB sound be than a 50-dB sound? 100 times as intense

LESSON PLAN

Vocabulary
Expanded form
Logarithmic equation
Properties of logarithms

BACKGROUND

In the Preview, a formula for finding the intensity level of any sound in decibels is introduced and used to solve problems.

TEACHING SUGGESTIONS

- Emphasize that the properties of logarithms hold only for logarithms with the same base.
- When solving logarithmic equations, point out that the expressions in each member should be expressed in the most concise form before the definition of logarithms is used to rewrite the equation.

Critical Thinking

Casual Explanation Ask students what they think is the reason why the base of a logarithm cannot be one. Students should reason that since a logarithm is an exponent and that one to any power is one, the base of a logarithm cannot be one because it does not produce a unique answer.

CHALKBOARD EXAMPLES

- **For Example 1**

  Express in expanded form.

  1. $\log_2 (x + 1)(x - 3)$ $\log_2 (x + 1) + \log_2 (x - 3)$
  2. $\log_{10} \dfrac{2s - 1}{s + 3}$ $\log_{10} (2s - 1) - \log_{10} (s + 3)$
  3. $\log_{10} k^4$ $4 \log_{10} k$
  4. $\log_7 \sqrt[5]{b}$ $\frac{1}{5} \log_7 b$

The properties of logarithms follow directly from the properties of exponents, because logarithmic functions and exponential functions are inverses. Recall that logarithms are defined only for positive real numbers and that the base of a logarithm can be any positive real number except 1.

Properties of Logarithms

If M, N, and b ($b \neq 1$) are positive real numbers, and r is any real number, then

$$\log_b MN = \log_b M + \log_b N$$
$$\log_b \frac{M}{N} = \log_b M - \log_b N$$
$$\log_b N^r = r \log_b N$$

The expression on the right side of each of the above equations is called the *expanded form* of the logarithm on the left side.

Each of these properties can be derived from the properties of exponents.

To derive the first property, $\log_b MN = \log_b M + \log_b N$, let

$\log_b M = u$ and $\log_b N = v$

Then $b^u = M$ and $b^v = N$

$MN = b^u b^v = b^{u+v}$

$\log_b MN = u + v$ *Definition of logarithms*

Therefore $\log_b MN = \log_b M + \log_b N$ *Substitute $\log_b M$ for u and $\log_b N$ for v.*

You will be asked to derive the other two properties in the exercises.

EXAMPLE 1 **Express in expanded form:**

a. $\log_4 xy$ **b.** $\log_{10} \dfrac{I}{I_0}$ **c.** $\log_3 x^4$ **d.** $\log_5 \sqrt[4]{y}$

a. $\log_4 xy = \log_4 x + \log_4 y$

b. $\log_{10} \dfrac{I}{I_0} = \log_{10} I - \log_{10} I_0$

c. $\log_3 x^4 = 4 \log_3 x$

d. $\log_5 \sqrt[4]{y} = \log_5 y^{\frac{1}{4}} = \dfrac{1}{4} \log_5 y$

It is often necessary to use more than one property of logarithms when expanding an expression.

EXAMPLE 2 **Express in expanded form:** **a.** $\log_4 x^3 y^2$ **b.** $\log_5 \sqrt{\frac{x}{y^5}}$

a. $\log_4 x^3 y^2 = \log_4 x^3 + \log_4 y^2$ *$\log_b MN = \log_b M + \log_b N$*

$= 3 \log_4 x + 2 \log_4 y$ *$\log_b N^r = r \log_b N$*

b. $\log_5 \sqrt{\frac{x}{y^5}} = \log_5 \left(\frac{x}{y^5}\right)^{\frac{1}{2}}$

$= \frac{1}{2} \log_5 \frac{x}{y^5}$ *$\log_b N^r = r \log_b N$*

$= \frac{1}{2} (\log_5 x - \log_5 y^5)$ *$\log_b \frac{M}{N} = \log_b M - \log_b N$*

$= \frac{1}{2} (\log_5 x - 5 \log_5 y)$ *$\log_b N^r = r \log_b N$*

If the terms of a logarithmic expression in expanded form have the same base, you can rewrite the expression as a single logarithm.

EXAMPLE 3 **Express as a single logarithm:**
a. $3 \log_3 x + \log_3 y$ **b.** $2 \log_4 x + \log_4 3 - \log_4 y$

a. $3 \log_3 x + \log_3 y = \log_3 x^3 + \log_3 y$ *$r \log_b N = \log_b N^r$*

$= \log_3 x^3 y$ *$\log_b M + \log_b N = \log_b MN$*

b. $2 \log_4 x + \log_4 3 - \log_4 y = \log_4 x^2 + \log_4 3 - \log_4 y$ *$r \log_b N = \log_b N^r$*

$= \log_4 3x^2 - \log_4 y$ *$\log_b M + \log_b N = \log_b MN$*

$= \log_4 \frac{3x^2}{y}$ *$\log_b M - \log_b N = \log_b \frac{M}{N}$*

Since logarithmic functions and exponential functions are inverses, the following properties are true:

$$\log_b b^x = x \quad \text{and} \quad b^{\log_b x} = x$$

Two other important properties of logarithms are

$\log_b b = 1$ *Recall that $b^1 = b$.*

$\log_b 1 = 0$ *Recall that $b^0 = 1$.*

You can use these properties to evaluate logarithmic expressions.

EXAMPLE 4 **Evaluate the following expressions:**
a. $3 \log_5 5 - \log_5 25$ **b.** $\log_4 16^2 + \log_4 8^3$

a. $3 \log_5 5 - \log_5 25 = 3(1) - \log_5 25$ *$\log_b b = 1$*

$= 3 - \log_5 5^2$

$= 3 - 2 = 1$ *$\log_b b^x = x$*

- **For Example 2**

Express in expanded form.

5. $\log_7 (a^5 y^2)^4$ $4 (5 \log_7 a + 2 \log_7 y)$
6. $\log_{10} \frac{2s}{t}$ $\log_{10} 2s - \log_{10} t = \log_{10} 2 + \log_{10} s - \log_{10} t$
7. $\log_8 \frac{(w - 1)^3}{z + 5}$ $3 \log_8 (w - 1) - \log_8 (z + 5)$

- **For Example 3**

Express as a single logarithm.

8. $\log_2 (c - 4) - \log_2 (c + 7)$ $\log_2 \frac{c - 4}{c + 7}$
9. $\log_{10} (3b + 1) + 3 \log_{10} (5b - 2)$ $\log_{10} (3b + 1)(5b - 2)^3$

- **For Example 4**

Evaluate the following expressions.

10. $\log_2 32 - \log_2 0.5$ 6
11. $\log_3 9 + 3 \log_3 \frac{1}{3}$ -1
12. $\log_5 25^2 + \log_5 5^3 - \log_5 5$ 6

- **For Example 5**
  Solve and check.
  13. $\log_4 2x + \log_4 3x = \log_4 294$ 7
  14. $\log_6 (x - 3) + \log_6 (x + 2) = 1$ 4

Common Error

- Some students multiply logarithms when given a product and divide logarithms when given a quotient. Remind those students that the properties of logarithms are as those for exponents.
- See *Teacher's Resource Book* for additional remediation.

Additional Answers

Class Exercises

1. The log of a quotient is the difference of the logs, *not* the quotient of the logs.
2. It should be $\frac{1}{2}(\log_3 2 + \log_3 x^2)$
6. The difference of the logs is equal to the log of the quotient, *not* the quotient of the logs.

b.
$$\begin{aligned} \log_4 16^2 + \log_4 8^3 &= 2\log_4 16 + 3\log_4 8 && \log_b N^r = r\log_b N \\ &= 2\log_4 4^2 + 3\log_4 4^{\frac{3}{2}} \\ &= 2(2) + 3\left(\frac{3}{2}\right) && \log_b b^x = x \\ &= 4 + \frac{9}{2} \\ &= \frac{17}{2} \end{aligned}$$

Since the logarithmic function is continuous and one-to-one, every positive real number has a unique logarithm to the base b. Therefore,

$$\log_b N = \log_b M \qquad \text{if and only if} \qquad N = M$$

This property can be used to solve *logarithmic equations*. It is important to check all possible solutions, since the logarithm of a variable expression is defined only when the expression is positive.

EXAMPLE 5 **Solve and check:**
a. $\log_3 4x + \log_3 x = \log_3 144$ **b.** $\log_5 (x^2 - 25) - \log_5 (x - 5) = 2$

a.
$$\begin{aligned} \log_3 4x + \log_3 x &= \log_3 144 \\ \log_3 4x^2 &= \log_3 144 \\ 4x^2 &= 144 \\ x^2 &= 36 \\ x &= \pm 6 \end{aligned}$$

The solution is 6.

Check: Reject -6 because logarithms are only defined for positive numbers.
$$\begin{aligned} \log_3 4x + \log_3 x &= \log_3 144 \\ \log_3 4(6) + \log_3 6 &\stackrel{?}{=} \log_3 144 \\ \log_3 24 + \log_3 6 &\stackrel{?}{=} \log_3 144 \\ \log_3 144 &\stackrel{?}{=} \log_3 144 \checkmark \end{aligned}$$

b.
$$\begin{aligned} \log_5 (x^2 - 25) - \log_5 (x - 5) &= 2 \\ \log_5 \frac{x^2 - 25}{x - 5} &= 2 \\ \frac{(x + 5)(x - 5)}{x - 5} &= 25 \\ x + 5 &= 25 \\ x &= 20 \end{aligned}$$

The solution is 20.

Check:
$$\begin{aligned} \log_5 (x^2 - 25) - \log_5 (x - 5) &= 2 \\ \log_5 [20^2 - 25] - \log_5 (20 - 5) &\stackrel{?}{=} 2 \\ \log_5 (400 - 25) - \log_5 15 &\stackrel{?}{=} 2 \\ \log_5 375 - \log_5 15 &\stackrel{?}{=} 2 \\ \log_5 25 &\stackrel{?}{=} 2 \\ 2 &\stackrel{?}{=} 2 \checkmark \end{aligned}$$

CLASS EXERCISES

For Discussion

Tell whether each equation is true or false. If it is false, explain why.
See side column.

1. $\log_5 \frac{x}{y} = \frac{\log_5 x}{\log_5 y}$ F
2. $\log_3 \sqrt{2x^2} = \frac{1}{2}\log_3 2 + \log_3 x^2$ F

See side column.

3. $\log_4 \sqrt[n]{x^m} = \frac{m}{n} \log_4 x$ T

4. $\log_4 \sqrt[5]{5x} = \frac{1}{5} \log_4 5 + \frac{1}{5} \log_4 x$ T

5. $\log_2 x^3y^4 = 3 \log_2 x + 4 \log_2 y$ T

6. $\frac{\log_3 x}{\log_3 y} = \log_3 x - \log_3 y$ F

7. $\log_6 \sqrt[5]{2y} = \frac{1}{5} \log_6 2y$ T

8. $\log_3 9 + \log_3 27 = 2 + 3$ T

PRACTICE EXERCISES

Express in expanded form.

A

1. $\log_3 xy$ $\log_3 x + \log_3 y$
2. $\log_4 xz$ $\log_4 x + \log_4 z$
3. $\log_5 2x$ $\log_5 2 + \log_5 x$
4. $\log_2 3y$ $\log_2 3 + \log_2 y$
5. $\log_2 \frac{x}{z}$ $\log_2 x - \log_2 z$
6. $\log_3 \frac{y}{x}$ $\log_3 y - \log_3 x$
7. $\log_3 \frac{x}{5}$ $\log_3 x - \log_3 5$
8. $\log_4 \frac{10}{y}$ $\log_4 10 - \log_4 y$
9. $\log_3 x^2$ $2 \log_3 x$
10. $\log_2 x^3$ $3 \log_2 x$
11. $\log_6 \sqrt[5]{x}$ $\frac{1}{5} \log_6 x$
12. $\log_7 \sqrt[8]{y}$ $\frac{1}{8} \log_7 y$
13. $\log_2 x^4y^3$ $4 \log_2 x + 3 \log_2 y$
14. $\log_3 x^5y^2$ $5 \log_3 x + 2 \log_3 y$
15. $\log_2 (3x)^3$ $3 \log_2 3 + 3 \log_2 x$
16. $\log_3 (4x)^4$ $4 \log_3 4 + 4 \log_3 x$
17. $\log_3 \sqrt[3]{4}$ $\frac{1}{3} \log_3 4$
18. $\log_2 \sqrt[5]{6}$ $\frac{1}{5} \log_2 6$
19. $\log_5 xyz$ $\log_5 x + \log_5 y + \log_5 z$
20. $\log_6 2xq$ $\log_6 2 + \log_6 x + \log_6 q$

Express as a single logarithm.

21. $\log_2 x + \log_2 y$ $\log_2 xy$
22. $\log_4 w + \log_4 z$ $\log_4 wz$
23. $\log_5 z + 2 \log_5 w - 3 \log_5 x$ $\log_5 \frac{zw^2}{x^3}$
24. $3 \log_n 4x - (\log_n 5y + \log_n 6z)$ $\log_n \frac{32x^3}{15yz}$

Evaluate.

25. $\log_2 8 + \log_2 16$ 7
26. $3 \log_3 81 - \log_3 27$ 9
27. $\log_5 25 + \log_5 125$ 5
28. $\log_8 16 + \log_8 256$ 4
29. $\log_3 27 - 6 \log_3 9$ −9
30. $\log_2 64 - 7 \log_2 4$ −8

Solve and check.

31. $\log_{10} (x^2 + 36) = \log_{10} 100$ ±8
32. $\log_{10} (x^2 - 25) = \log_{10} 144$ ±13

Express in expanded form. See side column, page 352.

B

33. $\log_3 \sqrt{\frac{x^2}{y}}$
34. $\log_4 \sqrt{\frac{x}{y^2}}$
35. $\log_2 \sqrt{\frac{x^2}{y^3}}$
36. $\log_3 \sqrt{\frac{x^3}{y^2}}$
37. $\log_4 3x^2y^4$
38. $\log_5 6x^3y^3$
39. $\log_3 (2xy)^2$
40. $\log_4 (4yz)^3$
41. $\log_b \sqrt[4]{x^5}$
42. $\log_b \sqrt[5]{x^4}$
43. $\log_b \sqrt[3]{x^2y^4}$
44. $\log_b \sqrt{7x^5}$
45. $\log_b \frac{3\sqrt{x}}{y^4}$
46. $\log_b \frac{2x^2}{\sqrt{y}}$
47. $\log_b \frac{6x^4y}{\sqrt[3]{z}}$
48. $\log_b \frac{2x^2y^2}{\sqrt[5]{z}}$

LESSON FOLLOW-UP

Discussion

What conditions must exist in order to write a logarithmic expression as a single logarithm? The terms in the logarithmic expression must have the same base.

Trigonometry in Chemistry

A formula for measuring the acidity (pH) of a solution is presented.

Assignment Guide

See p. 332B for assignments.

Lesson Quiz

Express in expanded form.

1. $\log_5 \frac{(x - 2)(x + 3)}{2x - 4}$ $[\log_5 (x - 2) + \log_5 (x + 3)] - \log_5 (2x - 4)$
2. $\log_5 (a^4b^3)^2$ $2(4 \log_5 a + 3 \log_5 b)$
3. $\log_{10} \frac{3w}{4z}$ $\log_{10} 3w - \log_{10} 4z = \log_{10} 3 + \log_3 w - (\log_{10} 4 + \log_{10} z)$

Express as a single logarithm.

4. $\log_2 (a - 4) - \log_2 (2a + 7)$ $\log_2 \frac{a-4}{2a+7}$
5. $\log_{10} (3b - 1) + 4 \log_{10} (5b + 2)$ $\log_{10} (3b - 1)(5b + 2)$

Evaluate.

6. $\log_2 16 + 3 \log_2 4$ 10
7. $\log_4 256 + \log_4 64 - \log_4 4$ 6

Solve and check.

8. $\log_4 n + \log_4 8 = 2$ 2
9. $\log_6 (z + 2) + \log_6 (z - 3) = 1$ 4
10. $3 \log_7 k - \log_7 k = 2$ $\sqrt{7}$

Enrichment

Solve.

$\log_2 x + \log_2 (2x + 1) = 0$ $\frac{1}{2}, -1$

Teacher's Resource Book

Practice—Chapter 8, p. 7
Enrichment—Chapter 8, p. 8

Additional Answers

33. $\log_3 x - \frac{1}{2}\log_3 y$

34. $\frac{1}{2}\log_4 x - \log_4 y$

35. $\log_2 x - \frac{3}{2}\log_2 y$

36. $\frac{3}{2}\log_3 x - \log_3 y$

37. $\log_4 3 + 2\log_4 x + 4\log_4 y$

38. $\log_5 6 + 3\log_5 x + 3\log_5 y$

39. $\log_3 4 + 2\log_3 x + 2\log_3 y$

40. $\log_4 64 + 3\log_4 y + 3\log_4 z$

41. $\frac{5}{4}\log_b x$

42. $\frac{4}{5}\log_b x$

43. $\frac{2}{3}\log_b x + \frac{4}{3}\log_b y$

44. $\frac{1}{2}\log_b 7 + \frac{5}{2}\log_b x$

45. $\log_b 3 + \frac{1}{2}\log_b x - 4\log_b y$

46. $\log_b 2 + 2\log_b x - \frac{1}{2}\log_b y$

47. $\log_b 6 + 4\log_b x + \log_b y - \frac{1}{3}\log_b z$

48. $\log_b 2 + 2\log_b x + 2\log_b y - \frac{1}{5}\log_b z$

57. $\frac{1}{3}\log_b x + \frac{3}{2}\log_b y - \frac{3}{4}\log_b z$

58. $\frac{3}{2}\log_b x + 4\log_b y - \frac{2}{3}\log_b z$

59. $\frac{5}{2}\log_2 y - \log_2 z - 2\log_2 x$

60. $\frac{1}{3}\log_b x - \log_b 6 - \frac{2}{3}\log_b y - \frac{2}{3}\log_b z$

65. $\log_b M = x$; $\log_b N = y$; $M = b^x$; $N = b^y$ $\frac{M}{N} = \frac{b^x}{b^y} = b^{x-y}$; $\log_b \frac{M}{N} = x - y$; $\log_b \frac{M}{N} = \log_b M - \log_b N$

66. $\log_b N = x$; $N = b^x$; $N^r = (b^x)^r = b^{xr}$; $\log_b N^r = \log_b b^{xr} = xr$; $\log_b N^r = r\log_b N$

352

Express as a single logarithm.

49. $\frac{1}{2}\log_4 x^2 + \frac{1}{3}\log_4 y$ $\log_4 x\sqrt[3]{y}$

50. $\frac{1}{5}\log_4 x^2 + \frac{1}{2}\log_4 y^3$ $\log_4 \sqrt[5]{x^2}\sqrt{y^3}$

51. $\frac{1}{3}\log_4 x - \frac{1}{2}\log_4 y$ $\log_4 \frac{\sqrt[3]{x}}{\sqrt{y}}$

52. $\log_3 x - \frac{1}{4}\log_3 y$ $\log_3 \frac{x}{\sqrt[4]{y}}$

Solve and check.

53. $\log_6 n - \log_6 3 = 2$ 108

54. $\log_5 n - \log_5 4 = 2$ 100

55. $\log_5 4x + \log_5 x = \log_5 100$ 5

56. $\log_2 9x + \log_2 x = \log_2 900$ 10

Express in expanded form. See side column.

C 57. $\log_b \frac{(\sqrt[3]{x})(\sqrt{y^3})}{\sqrt[4]{z^3}}$

58. $\log_b \frac{(\sqrt{x^3})(y^4)}{\sqrt[3]{z^2}}$

59. $\log_2 \frac{\sqrt{x^2y^5}}{zx^3}$

60. $\log_b \frac{\sqrt[3]{xyz}}{6yz}$

Solve and check.

61. $\log_3 (x + 1) - \log_3 x = \log_3 4$ $\frac{1}{3}$

62. $\log_4 x - \log_4 (x - 1) = \log_4 5$ $\frac{5}{4}$

63. $\log_7 (x - 5) + \log_7 (x + 1) = 1$ 6

64. $\log_2 (9x + 5) - \log_2 (x^2 - 1) = 2$ 3

65. Derive $\log_b \frac{M}{N} = \log_b M - \log_b N$ using properties of exponents. See side column.

66. Derive $\log_b N^r = r\log_b N$ using properties of exponents. See side column.

Applications

67. **Physics** Expand the formula for intensity level $\beta = 10\log_{10}\frac{I}{I_0}$. $\beta = 10(\log_{10} I - \log_{10} I_0)$

68. **Physics** Find the intensity level of an indoor rock concert with intensity 1. 120 dB

TRIGONOMETRY IN CHEMISTRY

Chemists use the pH scale (values range from 0 to 14) to measure the acidity or alkalinity of a solution. The hydrogen ion concentration of a solution determines whether it is an acid (with a pH below 7), a base (with a pH above 7), or neutral (with a pH equal to 7). The pH of a solution is determined by the formula $\text{pH} = \log_{10}\frac{1}{H^+}$, where H^+ is the concentration of hydrogen ions in gram atoms per liter.

Find the pH of a solution if $H^+ = 10^{-5}$. 5

8.5 Evaluating Logarithms and Solving Exponential Equations

Objectives: To evaluate common and natural logarithms
To solve exponential equations
To change the base of a logarithmic expression

Logarithms were developed to simplify complicated calculations. Since our number system is based on powers of 10, base-10 logarithms, or **common logarithms,** were tabulated. In order to find the value of the common logarithm of a number using a table, the number must be expressed in scientific notation. Calculators also use scientific notation to display very large and very small numbers.

Preview

A positive number of the form $a \times 10^b$, where $1 \leq a < 10$ and b is an integer, is in scientific notation.

Decimal form	Scientific notation
6,100,000	6.1×10^6
0.000123	1.23×10^{-4}

Express each number in scientific notation.

1. 6120 6.12×10^3
2. 160,000,000 1.6×10^8
3. 0.0525 5.25×10^{-2}
4. 0.0000028 2.8×10^{-6}

Express each number in decimal form.

5. 4.05×10^4 40,500
6. 2.4×10^{-2} 0.024
7. 1.02×10^{11} 102,000,000,000
8. 7.13×10^{-5} 0.0000713

It is customary to omit the base when writing a common logarithm.

$$\log N = \log_{10} N$$

You can use the log key on a scientific calculator to find the common logarithm of a number. Although most logarithms displayed by a calculator are approximations, the = symbol, rather than the ≈ symbol, is customarily used.

EXAMPLE 1 **Find each logarithm to four decimal places:**
a. log 340 **b.** log 34 **c.** log 3.4 **d.** log 0.34 **e.** log 0.034

a. log 340 = 2.5315 **b.** log 34 = 1.5315 **c.** log 3.4 = 0.5315
d. log 0.34 = −0.4685 **e.** log 0.034 = −1.4685

LESSON PLAN

Vocabulary
Antilogarithm
Change-of-base formula
Common logarithm
Natural logarithm

Materials/Manipulatives
Scientific calculators

BACKGROUND

In the Preview, expressing a number in scientific notation and expressing a number in decimal form are reviewed.

Critical Thinking

Analysis The base of the logarithmic function is not defined for $b = 1$. What do you think the reason for this could be? one to any power is one

TEACHING SUGGESTIONS

- Students should use their scientific calculators to find the logarithms of numbers.
- Be sure that students understand the relationship between the values of the logarithmic expressions in Example 1 on text p. 353.

CHALKBOARD EXAMPLES

- **For Example 1**

  Find each logarithm to four decimal places.

  1. log 562 2.7497
  2. log 21.8 1.3385
  3. log 4.61 0.6637
  4. log 0.00041 −3.3872

- **For Example 2**

  Find x to four significant digits.

  5. $\log x = 1.6749$ 47.30
  6. $\log x = 15.23$ 1.698×10^{15}
  7. $\log x = -1.6073$ 0.0247

- **For Example 3**

  Find each logarithm to four decimal places.

  8. ln 453 6.1159
  9. ln 62.82 4.1403
  10. ln 0.234 −1.4524
  11. ln 2.7182 1.0000

Note that common logarithms are of base 10. Therefore, the common logarithms of numbers that differ by a factor of 10, such as those in Example 1, differ by 1. Thus, log 340 is between 2 and 3, log 34 is between 1 and 2, log 3.4 is between 0 and 1, log 0.34 is between −1 and 0, and log 0.034 is between −2 and −1. The common logarithm of a number can be expressed as the sum of an integer, called the *characteristic*, and a nonnegative number less than 1, called the *mantissa* (0.5315 in this example).

If you know $\log x$, you can find x. The number x is called the *antilogarithm*, or *antilog*, of $\log x$.

Recall that 10^x is the inverse of $\log_{10} x$. If your calculator has a 10^x key, you can use it to find the antilog. If your calculator does not have a 10^x key, you can use the inverse key and the log key.

EXAMPLE 2 **Find x to four significant digits:**

a. $\log x = 3.233$ **b.** $\log x = 13.06$

a. $\log x = 3.233$ *Enter 3.233. Use the inverse and log keys, or the 10^x key.*

$x = 1710$ *To four significant digits*

b. $\log x = 13.06$ *Enter 13.06. Use the inverse and log keys, or the 10^x key.*

$x = 1.148 \times 10^{13}$ *To four significant digits. Your calculator uses scientific notation to express this number.*

Natural logarithms are to the base e. They are of great importance in the study of calculus and in other areas of mathematics. Just as the symbol *log x* is used for common logarithms, the symbol *ln x* is used for natural logarithms.

$$\ln x = \log_e x$$ *ln is read "el en."*

You can use the ln key on a scientific calculator to find the natural logarithm of a number.

EXAMPLE 3 **Find each logarithm to four decimal places:**

a. ln 367 **b.** ln 0.023

a. $\ln 367 = 5.9054$ *Enter 367 and use the ln key.*

b. $\ln 0.023 = -3.7723$ *Enter 0.023 and use the ln key.*

The inverse of $\log_e x$, or $\ln x$, is e^x. If your calculator has an e^x key, you can use it to find the antilog of a natural logarithm. If your calculator does not have an e^x key, you can use the inverse key and the ln key.

EXAMPLE 4 **Find x to four significant digits:**

a. $\ln x = 3.2335$ **b.** $\ln x = -28.62$

a. $\ln x = 3.2335$ *Enter 3.2335. Use the inverse and ln keys, or the e^x key.*

$x = 25.37$ *To four significant digits*

b. $\ln x = -28.62$ *Enter 28.62. Use the +/− key. Then use the inverse and ln keys, or the e^x key.*

$x = 3.720 \times 10^{-13}$ *To four significant digits. Your calculator uses scientific notation to express this number.*

A scientific calculator makes it easier to evaluate complicated logarithmic expressions. There is often more than one correct way to make entries. You may have to use the parentheses keys or the memory feature. Use your owner's manual and experiment with the keys to determine the sequences of keys that yield the correct answers in Example 5 below.

EXAMPLE 5 **Evaluate. Round each answer to four decimal places.**

a. $5(\ln 2.3 + \ln 18)$ **b.** $\frac{\log 34}{\log 8 + \log 3}$

a. $5(\ln 2.3 + \ln 18) = 18.6164$ *Use the parentheses keys: $5 \times (\ln 2.3 + \ln 18)$.*

b. $\frac{\log 34}{\log 8 + \log 3} = 1.1096$ *Use the parentheses keys: $\log 34 \div (\log 8 + \log 3)$.*

Logarithms can be used to solve exponential equations in which both sides cannot be written in terms of the same base, as the following example illustrates.

EXAMPLE 6 **Find x to four significant digits:**

a. $3^x = 8$ **b.** $4^{x+2} = 15$

a.

$3^x = 8$

$\log 3^x = \log 8$ *Take the log of both sides.*

$x \log 3 = \log 8$ *$\log_b N^r = r \log_b N$*

$x = \frac{\log 8}{\log 3}$ *Calculation-ready form*

$x = 1.893$

Check: Use the y^x key on your calculator to show that $3^{1.893}$ is approximately equal to 8.

- **For Example 4**

Find x to four significant digits.

12. $\ln x = 0.2345$ 1.264
13. $\ln x = 4.0758$ 58.90
14. $\ln x = -3.7679$ 0.0231

- **For Example 5**

Evaluate. Give the answer to four decimal places.

15. $\log 34.5 + \log 0.237 - \log 789$ −1.9845
16. $3(\ln 45.1 - \ln 3.2)$ 7.9372
17. $(\log 723)(\log 25.78)$ 4.0351
18. $\frac{\log 8.93 + \log 93.4}{\log 3.21}$ 5.7674

- **For Example 6**

Find x to four significant digits.

19. $5^x = 12$ 1.544
20. $3^{x+2} = 15$ 0.4650

- **For Example 7**
  Use common logarithms to find each logarithm to four decimal places.
  21. $\log_4 7$ 1.4037
  22. $\log_3 8$ 1.8928

LESSON FOLLOW-UP

Discussion

If you do not remember the change-of-base formula, how could you calculate $\log_7 9$? Write $\log_7 9$ as an exponential equation, $7^x = 9$, then solve the equation.

Assignment Guide

See p. 332B for assignments.

b.

$$\begin{aligned} 4^{x+2} &= 15 \\ \log 4^{x+2} &= \log 15 && \textit{Take the log of both sides.} \\ (x + 2)\log 4 &= \log 15 && \log_b N^r = r \log_b N \\ x \log 4 + 2 \log 4 &= \log 15 && \textit{Distributive property} \\ x \log 4 &= \log 15 - 2 \log 4 \\ x &= \frac{\log 15 - 2 \log 4}{\log 4} && \textit{Calculation-ready form} \\ x &= -0.0466 \end{aligned}$$

Check: Use the y^x key on your calculator to show that $4^{-0.0466+2}$ is approximately equal to 15.

The method demonstrated in Example 7 *changes* the base of an exponential expression. If you know the logarithm of a number to one base (the *reference base*), you can use the *change-of-base formula* to find the logarithm of that number to some other base.

Change-of-Base Formula

$$\log_b x = \frac{\log_a x}{\log_a b} \qquad \textit{The reference base is a.}$$

Let $n = \log_b x$

Then

$$\begin{aligned} b^n &= x && \textit{Definition of logarithm} \\ \log_a b^n &= \log_a x && \textit{Take } \log_a \textit{ of both sides.} \\ n \log_a b &= \log_a x && \log_b N^r = r \log_b N \\ n &= \frac{\log_a x}{\log_a b} \\ \log_b x &= \frac{\log_a x}{\log_a b} && \textit{Replace } n \textit{ with } \log_b x. \end{aligned}$$

Since a scientific calculator can be used to approximate values of common logarithms, you can replace a with 10 in the change-of-base formula.

EXAMPLE 7 **Use common logarithms to find $\log_6 5$ to four decimal places.**

$$\begin{aligned} \log_6 5 &= \frac{\log_{10} 5}{\log_{10} 6} && \log_b x = \frac{\log_{10} x}{\log_{10} b} \\ &= \frac{\log 5}{\log 6} \\ &= 0.8982 \end{aligned}$$

CLASS EXERCISES

For Discussion

1. Would the results of Example 6 be different if you took the natural log of both sides of the equation instead of the common log? Explain.
See side column, page 358.

Find each logarithm to four decimal places.

2. log 456 2.6590
3. log 46.27 1.6653
4. log 0.45 −0.3468
5. ln 32.1 3.4689
6. ln 0.345 −1.0642
7. ln 68.234 4.2229
8. $\log_3 7$ 1.7712
9. $\log_2 2.4$ 1.2630
10. $\log_5 11.1$ 1.4955

Find x to four significant digits.

11. $\log x = 3.2341$ 1714
12. $\log x = -1.7823$ 0.0165
13. $\log x = 0.2891$ 1.946
14. $\ln x = 0.3679$ 1.445
15. $\ln x = -3.2738$ 0.0379
16. $\ln x = 2.9845$ 19.78

PRACTICE EXERCISES

Find each logarithm to four decimal places.

A

1. log 792 2.8987
2. log 26.36 1.4209
3. log 0.37 −0.4318
4. log 312.4 2.4947
5. log 0.0034 −2.4685
6. log 0.0000468 −4.3298
7. ln 56.2 4.0289
8. ln 0.123 −2.0956
9. ln 32.567 3.4833
10. ln 0.345 −1.0642
11. ln 0.000425 −7.7634
12. ln 1.239 0.2143

Find x to four significant digits.

13. $\log x = 2.3176$ 207.8
14. $\log x = -3.1842$ 0.0006543
15. $\log x = 0.3345$ 2.160
16. $\log x = 1.3924$ 24.68
17. $\log x = -1.3657$ 0.0431
18. $\log x = -3.5743$ 0.0002665
19. $\ln x = 0.3679$ 1.445
20. $\ln x = 3.2738$ 26.41
21. $\ln x = 2.9845$ 19.78
22. $\ln x = -2.5691$ 0.0766
23. $\ln x = 0.8734$ 2.395
24. $\ln x = -1.2363$ 0.2905

Evaluate. Round each answer to four decimal places.

25. log 3 + log 7 1.3222
26. log 153 − log 34 0.6532
27. 6(log 3.2 + log 4.5) 6.9502
28. $\dfrac{\log 5}{\log 7}$ 0.8271
29. $\dfrac{5(\log 13)}{\log 4}$ 9.2511
30. $\dfrac{\log 45.1}{3(\log 8)}$ 0.6106

Find x to four significant digits.

B

31. $\log x = 12.2481$ 1.771×10^{12}
32. $\log x = 21.2934$ 1.965×10^{21}
33. $\log x = -13.4218$ 3.786×10^{-14}
34. $\ln x = -23.5347$ 6.012×10^{-11}
35. $\ln x = 10.4321$ 3.393×10^{4}
36. $\ln x = 23.2634$ 1.268×10^{10}

Lesson Quiz

Find each logarithm to four decimal places.

1. log 487 2.6875
2. log 34.2 1.5340
3. log 0.0612 −1.2132
4. log 56,000 4.7482
5. ln 51.3 3.9377
6. ln 3456 8.1479

Find x to four significant digits.

7. $\log x = 3.2351$ 1718
8. $\log x = -1.6073$ 0.0247
9. $\ln x = 2.4156$ 11.20

Evaluate. Give the answer to four decimal places.

10. log 54.5 + log 0.252 − log 289 −1.3231
11. 2(ln 43.1 − ln 5.2) 4.2297
12. (log 231)(log 32.13) 3.5618

Find x to four significant digits.

13. $4^x = 12$ 1.792
14. $5^{x+2} = 15$ −0.3174
15. Use common logarithms to find $\log_4 2$. 0.5

Enrichment

Find the value of N without using your calculator: $\log N = \left(\log 7 + \frac{1}{2}\log 3\right) - \frac{1}{2}\left(\log 4 + \frac{1}{3}\log 27\right)$

$$\log N = \frac{\log(7 \cdot 3^{\frac{1}{2}})}{\log[4(27)^{\frac{1}{3}}]^{\frac{1}{2}}}$$

$$\log N = \frac{\log(7 \cdot 3^{\frac{1}{2}})}{\log[4(3)]^{\frac{1}{2}}}$$

$$\log N = \frac{\log 7 \cdot 3^{\frac{1}{2}}}{\log 2 \cdot 3^{\frac{1}{2}}}$$

$$N = \frac{7}{2}$$

Teacher's Resource Book
Practice—Chapter 8, p. 9
Enrichment—Chapter 8, p. 10

Additional Answers

Class Exercises

1. No. The results would be the same because the change of base formula does not require that the reference base be ten, and so base *e* may be used.

Solve and check. Round answers to four significant digits.

37. $2^x = 9$ 3.170 **38.** $3^x = 10$ 2.096 **39.** $4^x = 23$ 2.262

40. $5^x = 27$ 2.048 **41.** $2^{2x} = 19$ 2.124 **42.** $3^{2x} = 20$ 1.363

43. $3^{x+2} = 11$ 0.1827 **44.** $7^{x+3} = 17$ −1.544 **45.** $7^{x-1} = 19$ 2.513

Use common logarithms to find each logarithm to four decimal places.

46. $\log_3 11$ 2.1827 **47.** $\log_4 9$ 1.5850 **48.** $\log_6 86$ 2.4860

49. $\log_5 65$ 2.5937 **50.** $\log_{0.8} 70$ −19.0393 **51.** $\log_{2.8} 75$ 4.1933

C **52.** $\log_{1.2} 7$ 10.6730 **53.** $\log_{\sqrt{3}} 17$ 5.1578 **54.** $\log_{\sqrt{2}} 38$ 10.4959

Solve and check. Round answers to four significant digits.

55. $6.2^{2x-1} = 74.1$ 1.680 **56.** $2.9^{\sqrt{2x}+1} = 87.6$ 5.123 **57.** $2.5^{\sqrt{2x}+2} = 90.5$ 4.254

Evaluate. Give each answer to four decimal places.

58. $\dfrac{(\log 22)^2}{2 + \log 17}$ 0.5578 **59.** $\sqrt{\dfrac{2 + \log 6}{1 + \log 12}}$ 1.1559 **60.** $\dfrac{1 + 3 \ln 3}{3 + \sqrt{\ln 3}}$ 1.0612

Applications

61. Astronomy The limiting magnitude L of an optical telescope with a lens of diameter d, in inches, is $L = 8.8 + 5.1 \log d$. Find the limiting magnitude of a 6-in. telescope. 12.77

62. Business Solve $1200 = 1000\left(1 + \dfrac{0.05}{4}\right)^{4t}$ for t to determine how many years it will take \$1000, deposited at 5% interest compounded quarterly, to grow to \$1200. (Assume no withdrawals or additional deposits are made.) 3.669 yr

63. Business Solve $1200 = 1000\left(1 + \dfrac{0.05}{12}\right)^{12t}$ for t to determine how many years it will take \$1000, deposited at 5% interest compounded monthly, to grow to \$1200. (Assume no withdrawals or additional deposits are made.) 3.654 yr

64. Business Solve $1200 = 1000e^{0.05t}$ for t to determine how many years it will take \$1000, deposited at 5% interest compounded continuously, to grow to \$1200. 3.646 yr

MATH CLUB ACTIVITY

How can common logarithms be used to determine the number of digits in a number of the form n^{n^n}, where n is an integer? Use your answer to find the number of digits in 4^{4^4}. Add 1 to the characteristic of $n^n \log n$; 155 digits in the number 4^{4^4}.

8.6 Applications: Exponential and Logarithmic Equations

Objectives: To use exponential and logarithmic equations to solve real-world problems

Exponential and logarithmic equations can be used to describe exponential growth and decay, learning curves, and logistic growth. These equations are used by scientists, economists, and sociologists. Exponential equations also can be used to model the interest accumulated on bank deposits or loans.

Preview

Interest may be compounded periodically. For example, it may be compounded semiannually, quarterly, or daily. The following formula gives the total amount in a savings account after a given period of time.

Compound Interest Formula

$$A = P\left(1 + \frac{r}{n}\right)^{nt}$$

A = total amount (interest + principal), in dollars
P = principal invested, in dollars
r = rate of annual interest, expressed as a decimal
n = number of times per year the interest is compounded
t = number of years the principal is invested

If interest is compounded *continuously*, it can be shown that the formula becomes $A = Pe^{rt}$.

EXAMPLE **How much money will Wai Ming have at the end of 5 yr if he deposits $1000 at 9% interest compounded semiannually?**

$$A = P\left(1 + \frac{r}{n}\right)^{nt}$$

$$= 1000\left(1 + \frac{0.09}{2}\right)^{2(5)} = \$1552.97 \qquad \textit{Calculation-ready form}$$

1. How much money will Wai Ming have if interest is compounded quarterly? $1560.51
2. How much money will he have if interest is compounded monthly? $1565.68
3. How much money will he have if interest is compounded daily? $1568.22
4. How much money will he have if interest is compounded continuously? $1568.31
5. Do your answers to questions 3 and 4 differ greatly? Why? No; Answers may vary.
6. If interest is compounded monthly, approximately how many years will it take the investment to increase to $2500? 10 yr

LESSON PLAN

Vocabulary
Exponential decay formula

Materials/Manipulatives
Scientific calculators

BACKGROUND

In the Preview, the compound interest formula is used to find the total amount of money in an account for different methods of compounding interest: quarterly, semiannually, monthly, daily, and continuously.

Critical Thinking

Analysis The function $y = a^x$ is defined for $a > 0$. Describe the graph if $a = 0$. Describe the graph if $a < 0$. Would it still be a function? If $a = 0$, it would be the line $y = 0$. If $a < 0$, points would alternate between the first and second quadrants, depending on whether x is odd or even. It would be a function.

TEACHING SUGGESTIONS

- Most of this section consists of evaluating formulas using logarithms. If students follow the examples carefully they should not have any trouble.
- Encourage students to use their scientific calculators to find the logarithms of numbers and to do any other necessary calculations.

CHALKBOARD EXAMPLES

- **For Example 1**

  The growth rate for a certain culture of bacteria is calculated using the formula $B = 1000(10)^{\frac{t}{48}}$.

  1. How many bacteria will there be after 6 hours? 1334
  2. How long will it take for there to be 15,000 bacteria in the culture? 56 h

The number of bacteria in a culture increases exponentially.

EXAMPLE 1 **The growth rate for a certain bacterial culture can be calculated using the formula $B = 1000\,(2)^{\frac{t}{48}}$, where B is the number of bacteria and t is the elapsed time in hours.**

a. How many bacteria will be present after 4 hours?

b. How long will it take, to the nearest hour, for there to be 12,000 bacteria in the culture?

a.
$$\begin{aligned} B &= 1000(2)^{\frac{t}{48}} \\ &= 1000(2)^{\frac{4}{48}} && \textit{Calculation-ready form} \\ &= 1059 \end{aligned}$$

There will be approximately 1059 bacteria present after 4 h.

b.
$$\begin{aligned} B &= 1000(2)^{\frac{t}{48}} \\ 12{,}000 &= 1000(2)^{\frac{t}{48}} \\ 12 &= 2^{\frac{t}{48}} \\ \log 12 &= \log 2^{\frac{t}{48}} && \textit{Take the log of both sides.} \\ \log 12 &= \frac{t}{48}\log 2 && \log_b N^r = r\log_b N \\ \frac{48\log 12}{\log 2} &= t && \textit{Calculation-ready form} \\ 172 &= t \end{aligned}$$

It will take approximately 172 hours for there to be 12,000 bacteria in the culture.

It is possible to approximate the age of fossils by measuring the amount of carbon-14, a radioactive isotope, that remains inside them, and then applying the *exponential decay formula*.

Exponential Decay Formula

$$A = A_0 2^{-\frac{t}{k}}$$

A = present amount of the radioactive isotope

A_0 = original amount of the radioactive isotope, measured in the same units as A

t = time it takes to reduce original amount of the isotope to present amount

k = half-life of the isotope, measured in the same units as t

The half-life is the average time required for one-half of the atoms of a sample of a radioactive substance to decay.

EXAMPLE 2 **A fossil that originally contained 100 mg of carbon-14 now contains 75 mg of the isotope. Determine the approximate age of the fossil, to the nearest 100 years, if the half-life of carbon-14 is 5570 yr.**

$$A = A_0 2^{-\frac{t}{k}}$$

$$75 = 100(2)^{-\frac{t}{5570}}$$

$$0.75 = 2^{-\frac{t}{5570}}$$

$$\log 0.75 = \log 2^{-\frac{t}{5570}} \qquad \textit{Take the log of both sides.}$$

$$\log 0.75 = -\frac{t}{5570}\log 2 \qquad \log_b N^r = r \log_b N$$

$$\frac{5570 \log 0.75}{\log 2} = -t \qquad \textit{Calculation-ready form}$$

$$2311.76 = t$$

The fossil is approximately 2300 years old.

Sociologists use exponential equations to describe the spread of a rumor.

EXAMPLE 3 **In a town of 15,000 people, the spread of a rumor that the local transit company would go on strike was such that t hours after the rumor started, $f(t)$ persons heard the rumor, where experience over time has shown that**

$$f(t) = \frac{15{,}000}{1 + 7499e^{-0.8t}}$$

a. How many people started the rumor?
b. How many people heard the rumor after 5 hours?
c. How many hours does it take for 14,000 people to hear the rumor?

a. $f(t) = \dfrac{15{,}000}{1 + 7499e^{-0.8t}}$

$f(0) = \dfrac{15{,}000}{1 + 7499e^{0}}$

$= 2$

Two people started the rumor.

b. $f(t) = \dfrac{15{,}000}{1 + 7499e^{-0.8t}}$

$f(5) = \dfrac{15{,}000}{1 + 7499e^{-0.8(5)}}$

$= 108$

After 5 hours, 108 people heard the rumor.

- **For Example 2**

  3. A fossil that originally contained 100 mg of carbon-14 now contains 65 mg of the isotope. Determine the age of the fossil, to the nearest 100 yr, if the half-life of carbon-14 is 5570 yr. 3500 yr

- **For Example 3**

  4. Using the data in Example 3, determine the number of people that heard the rumor after 10 h. 4267

Common Error

- Some students may substitute incorrectly into the formula, omit a negative sign, and so forth. Emphasize the importance of checking for careless errors before attempting any calculations.
- See *Teacher's Resource Book* for additional remediation.

LESSON FOLLOW-UP

Discussion

Why were logarithms used to solve the problems in this lesson? Answers may vary.

Assignment Guide

See p. 332B for assignments.

Test Yourself

See *Teacher's Resource Book*, *Tests*, pp. 79–80.

Lesson Quiz

1. The growth rate for a particular bacterial culture can be calculated using the formula $B = 1200(9)^{\frac{t}{48}}$ (t in hours). How many bacteria will be present after 6 h? 1579
2. How many hours will it take for 12,000 bacteria to be present in Exercise 1? 50 h
3. A bone found by an archaeologist contains 75 g of carbon-14. If the bone originally contained 100 g of the isotope, what is the approximate age of the fossil, to the nearest 100 yr, if the half-life of carbon-14 is 5570 yr? Use the formula $A = A_0 2^{\frac{t}{k}}$. 2300 yr

c.

$$f(t) = \frac{15{,}000}{1 + 7499e^{-0.8t}}$$

$$14{,}000 = \frac{15{,}000}{1 + 7499e^{-0.8t}}$$

$$14{,}000 + 14{,}000(7499e^{-0.8t}) = 15{,}000$$

$$104{,}986{,}000e^{-0.8t} = 1000$$

$$e^{-0.8t} = \frac{1000}{104{,}986{,}000}$$

$$\ln e^{-0.8t} = \ln\left(\frac{1000}{104{,}986{,}000}\right) \quad \textit{Take ln of both sides.}$$

$$-0.8t \ln e = \ln\left(\frac{1000}{104{,}986{,}000}\right) \quad \log_b N^r = r \log_b N$$

$$-0.8t = \ln\left(\frac{1000}{104{,}986{,}000}\right) \quad \ln e = 1$$

$$t = 14.45$$

It takes approximately 14 hours for 14,000 people to hear the rumor.

CLASS EXERCISES

For Discussion

1. Why was the natural logarithm (ln) taken of both sides of the equation in Example 3(c) instead of the common logarithm (log)? The base is e.

Solve.

2. If the bacteria culture in Example 1 is allowed to grow for 12 h, how many bacteria will be present? 1189
3. A radioactive isotope has a half-life of 5 days. How many days will it take for a 10 g sample to decay to 4 g? 6.6 d
4. Jones Savings and Loan pays 7% interest compounded semiannually. If \$2500 is deposited for 10 yr, how much money will be in the account at the end of that time period? \$4974.47
5. Approximately how many years will it take for \$1600 to increase to \$2100 if it is deposited in a bank that pays 7.5% interest compounded quarterly? 4 yr

PRACTICE EXERCISES

Solve. Round answers to the nearest whole number unless otherwise told.

A 1. The growth rate for a particular bacterial culture can be calculated using the formula $B = 900(2)^{\frac{t}{50}}$, where B is the number of bacteria and t is the elapsed time in hours. How many bacteria will be present after 5 h? 965

2. Approximately how many hours will it take for there to be 9000 bacteria present in the culture in Exercise 1? 166 h

3. How many hours will it take for there to be 18,000 bacteria present in the culture in Exercise 1? 216 h

4. A dinosaur bone was found to have 80 g of carbon-14. If the bone originally contained 200 g of the isotope, determine the approximate age of the bone to the nearest 100 years. (Carbon-14 has a half-life of 5570 yr.) 7400 yr

5. A fossil that originally contained 150 mg of carbon-14 now contains 85 mg of the isotope. Determine the approximate age of the fossil to the nearest 100 years. 4600 yr

6. A shell fossil now contains 60 mg of carbon-14. If there were originally 150 mg of the isotope in the fossil, determine the approximate age of the fossil, to the nearest 100 years. 7400 yr

7. A radioactive isotope has a half-life of 4 days. How many days will it take for a 15-g sample of this isotope to decay to 3 g? 9 d

8. A particular isotope has a half-life of 25 days. How many days will it take for a 8-mg sample to decay to 2 mg? 50 d

In Exercises 9–12, round answers to the nearest tenth.

9. Dan invests $5000 at 8% annual interest compounded monthly. How many years will it take him to earn $2000 interest? 4.2 yr

10. Debbie invests $4000 at 7% annual interest compounded semiannually. How many years will it take her to earn $3000 interest? 8.1 yr

11. Allison invests $3400 at 8.5% annual interest compounded quarterly. How many years will it take her money to double in value? 8.2 yr

12. Eric invests $2500 at 10.5% annual interest compounded monthly. How many years will it take his money to double in value? 6.6 yr

B 13. If Allison invests the same amount of money at the same interest rate as in Exercise 11 compounded *continuously*, how many years will it take for her investment to double in value? 8.2 yr

14. Mr. Kakutani found 12 mg of a radioactive isotope in a soil sample. After 5 h, only 8.2 mg of the isotope remained. Determine the approximate half-life of the isotope. 9 h

15. Ms. Carmenza discovered 15 mg of an isotope in a fossil. After 10 h, only 13.2 mg of the isotope remained. Determine the approximate half-life of the isotope. 54 h

16. In the town of Middlebury, the spread of a virus was such that t weeks after its outbreak, $f(t)$ persons had contracted it, where $f(t) = \frac{12{,}000}{1 + 599e^{-0.7t}}$. How many people had the virus at its outbreak? 20

Teacher's Resource Book
Practice—Chapter 8, p. 11
Enrichment—Chapter 8, p. 12

17. In Exercise 16, how many weeks will it take for 3000 people to contract the virus? 8 wk

18. In Exercise 16, how many weeks will it take for 6000 people to contract the virus? 9 wk

C 19. As a person's experience increases, competence at completing a task increases rapidly at first, and then slows down as additional experience is gained. This phenomena can be represented by a *learning curve*. After t hours of typing, José can type $f(t) = 80(1 - e^{-0.025t})$ words per minute. How many words per minute can he be expected to type after 10 h of practice? 18

20. In Exercise 19, how many words per minute can José type after 25 h of practice? 37

TEST YOURSELF

Express as a single logarithm.

1. $\log_4 d + 3 \log_4 2e - 3 \log_4 f$ $\log_4 \frac{8de^3}{f^3}$ **8.4**
2. $(\log_5 2x + 3 \log_5 y) - \log_5 3z$ $\log_5 \frac{2xy^3}{3z}$

Express in expanded form.

3. $\log_b 2xy^2 z^3$ $\log_b 2 + \log_b x + 2 \log_b y + 3 \log_b z$
4. $\log_3 \frac{8x^3}{y^2}$ $\log_3 8 + 3 \log_3 x - 2 \log_3 y$

Find each logarithm to four decimal places.

5. log 329 2.5172
6. log 0.042 −1.3768 **8.5**
7. ln 0.321 −1.1363
8. ln 24.3 3.1905

Find x to four significant digits.

9. $\log x = 2.5642$ 366.6
10. $\log x = 2.1226$ 132.6
11. $\ln x = 0.2639$ 1.302
12. $\ln x = -1.1728$ 0.3095

13. Use common logarithms to find $\log_5 8$ to four decimal places. 1.2920
14. Evaluate 3(log 60.2 + log 7.5) − log 21.3 to four decimal places. 6.6356 **8.6**
15. Sally invests $9000 at 7.5% annual interest compounded quarterly. How many years will it take for her investment to increase to $11,500? 3.3 yr
16. Kareem found 25 mg of an isotope in a fossil. After 12 h, only 20 mg of the isotope remained. Find the approximate half-life of the isotope. 37 h

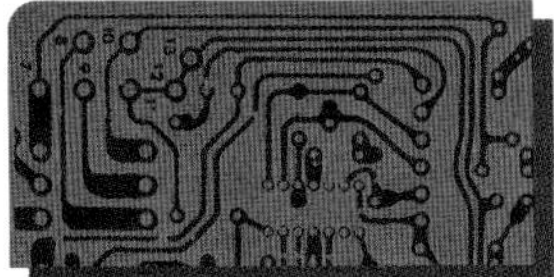

TECHNOLOGY: The Richter Scale

Did you know that the Richter scale is used to measure the relative intensity of an earthquake? Created in 1935 by American seismologist Charles F. Richter, the scale describes the amount of energy released at the center of the quake.

```
10 INPUT "WEAKER QUAKE? ";R1
20 INPUT "STRONGER QUAKE? ";R2
30 D = 10 ^ R2 / 10 ^ R1
40 PRINT "THE QUAKE MEASURING
   "R2" IS "D" TIMES AS POWERFUL
   AS THE QUAKE WHICH
   MEASURES "R1"."
```

Because the Richter scale is logarithmic, the energy released increases by powers of 10 in relation to the Richter numbers. For example, an earthquake of magnitude 7 is 10 times more powerful than one of magnitude 6 and 10^2 or 100 times more powerful than one of magnitude 5. Thus, an earthquake of magnitude 8 is not twice as powerful as one of magnitude 4, but 10^4 or 10,000 times more powerful.

The table below lists the probable effects of earthquakes of various magnitudes:

Magnitude	Intensity Level	Probable Effect
1	10^1	Detectable only by
2	10^2	seismographic instruments
3	10^3	Slight damage
4	10^4	within a
5	10^5	limited area
6	10^6	Moderate damage
7	10^7	Major damage
8	10^8	Catastrophic damage

EXERCISES

Use the computer program above to answer the following questions.

1. Use common logarithms to derive the formula on line 30. See side column.
2. Is it true that an earthquake of magnitude 2.5 is half as intense as an earthquake of magnitude 5? No, it is 0.003 times as intense.
3. Compare the difference between a quake which measures 3.4 and one which measures 3.9, and the difference between a quake which measures 7.4 and one which measures 7.9. Is this result surprising? See side column.
4. The highest Richter magnitude ever recorded was 8.9. How many times more powerful was this quake than the 1988 Armenian quake of magnitude 6.9? 100

Technology.

Students are introduced to the Richter scale, which is used to gauge the magnitude of earthquakes. This device, crucial in the science of seismology, demonstrates the use of logarithms in a real-world situation.

The complete form of the computer program that uses the Richter scale is on the disk provided with the *Teacher's Resource Book.*

See *Teacher's Resource Book* for *Application*, Chapter 8, p. 13.

See *Teacher's Resource Book, Technology*, p. 8.

Additional Answers

1. $\log D = R_2 - R_1$;
   $D = 10^{R_2 - R_1}$;
   $D = \frac{10^{R_2}}{10^{R_1}}$

3. The quake measuring 3.9 is greater than the quake measuring 3.4 by about 5400, while the quake measuring 7.9 is greater than the one measuring 7.4 by about 54 million. In both cases, however, the quake with the higher magnitude is about 3.16 times greater than the one with the lesser magnitude.

Additional Answers

8.

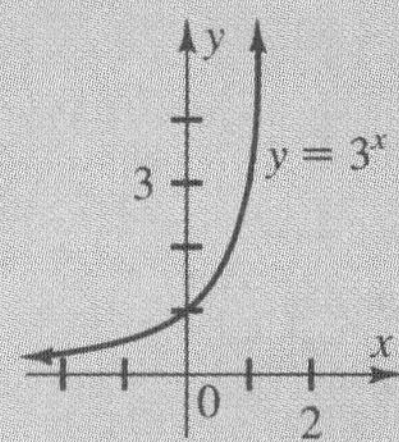

CHAPTER 8 SUMMARY AND REVIEW

Vocabulary

change-of-base formula (356)
common logarithm (353)
compound interest formula (359)
exponential decay formula (360)
exponential function (339)
logarithmic function (343)
natural exponential function (340)
natural logarithm (354)

Real Exponents Expressions containing real number exponents can be simplified using the properties of exponents and evaluated using the y^x key on a scientific calculator. 8.1

An equation in which the variable is in the exponent is called an exponential equation. Some exponential equations can be solved by first expressing both sides of the equation in terms of the same base.

Evaluate the following to the nearest thousandth.

1. $81^{\frac{1}{11}}$ 1.491

2. $25^{\sqrt{3}}$ 263.817

Solve for x.

3. $5^4 = 5^{2x+1}$ $\frac{3}{2}$

4. $4^{x-1} = 8^{3x}$ $-\frac{2}{7}$

Exponential Functions An exponential function is a function in which the variable is in the exponent. If x and b are real numbers such that $b > 0$ and $b \neq 1$, then $f(x) = b^x$ is an exponential function. 8.2

Evaluate each of the following for the function $f(x) = 3^x$.

5. $f(5)$ 243

6. $f(-2)$ $\frac{1}{9}$

7. $f(0.5)$ 1.732

8. Graph $f(x) = 3^x$. See side column.

Logarithmic Functions The expression $y = \log_b x$ means y is the exponent to which b is raised to produce x. The logarithmic function is the inverse of the exponential function. 8.3

Express in logarithmic form.

9. $3^4 = 81$ $\log_3 81 = 4$

10. $49^{0.5} = 7$ $\log_{49} 7 = 0.5$

Express in exponential form.

11. $\log_4 64 = 3$ $4^3 = 64$

12. $\log_9 3 = 0.5$ $9^{0.5} = 3$

Evaluate.

13. $\log_{81} 9$ $\frac{1}{2}$

14. $\log_{10} 0.01$ -2

Solve for x.

15. $\log_x 125 = 3$ 5

16. $\log_6 x = 3$ 216

17. $\log_3 27 = x$ 3

18. $\log_x 50 = 4$ 2.66

Properties of Logarithms 8.4

$$\log_b MN = \log_b M + \log_b N$$
$$\log_b \frac{M}{N} = \log_b M - \log_b N$$
$$\log_b N^r = r \log_b N$$

19. Express $\log_4 \frac{x^5 y^2}{z}$ in expanded form. $5 \log_4 x + 2 \log_4 y - \log_4 z$

20. Express $2 \log_n 3s - (\log_n 4t + 3 \log_n v)$ as a single logarithm. $\log_n \frac{9s^2}{4tv^3}$

Solve and check.

21. $\log_2 6 + \log_2 3 - \log_2 n = \log_2 9$ 2

22. $\log_{10} (x - 2) + \log_{10} (x + 1) = 1$ 4

Evaluating Logarithms and Solving Exponential Equations 8.5

Common (base 10) and natural (base e) logarithms can be computed using a scientific calculator. If you know the common or natural logarithm of a number, you can find the number. Logarithms can be used to solve exponential equations. The change-of-base formula is $\log_b x = \frac{\log_a x}{\log_a b}$.

Evaluate to four decimal places.

23. $\log 47.65$ 1.6781

24. $\ln 0.056$ -2.8824

Find x to four significant digits.

25. $\log x = -3.1252$ 0.0007495

26. $\ln x = 1.5378$ 4.654

27. Evaluate $2(\log 51.3 + \log 7.8) - \log 32$ to four decimal places. 3.6993

28. Use common logarithms to evaluate $\log_3 8$ to four decimal places. 1.8928

29. Use natural logarithms to evaluate $\log_4 7$ to four decimal places. 1.4037

30. Joanie invests \$8000 at 6.75% annual interest compounded semiannually. How many years will it take her investment to double in value? 10.4 yr 8.6

31. A radioactive isotope has a half-life of 4 days. How many days will it take for a 12-g sample to decay to 3 g? 8 d

See *Teacher's Resource Book, Tests*, pp. 81–84.

CHAPTER TEST

Evaluate to the nearest thousandth.

1. $(64)^{\frac{7}{9}}$ 25.398

2. $6^{\sqrt{5}}$ 54.954

3. $e^{1.2}$ 3.320

Solve for x.

4. $7^{x+4} = 49^{2x-1}$ 2

5. $8^{x+2} = 16^{2x}$ $\frac{6}{5}$

Evaluate.

6. $\log_5 125$ 3

7. $\log_{27} 3$ $\frac{1}{3}$

8. $\log_8 16$ $\frac{4}{3}$

Solve for x.

9. $\log_3 x = 6$ 729

10. $\log_2 x = 4$ 16

11. $\log_x 21 = 3$ 2.759

12. $\log_x 8 = 1.6$ 3.668

13. Express $\log_2 4m + 3 \log_2 n - 3 \log_2 p^3$ as a single logarithm. $\log_2 \frac{4mn^3}{p^9}$

14. Express $\log_7 \frac{8x^3 y}{z}$ in expanded form. $\log_7 8 + 3 \log_7 x + \log_7 y - \log_7 z$

Find each logarithm to four decimal places.

15. $\log 26.37$ 1.4211

16. $\ln 0.1572$ −1.8502

Find x to four significant digits.

17. $\log x = 2.6134$ 410.6

18. $\ln x = -2.5124$ 0.0811

19. Use common logarithms to find $\log_5 9$ to four decimal places. 1.3652

20. The growth rate for a particular bacterial culture can be calculated using the formula $B = 1000(2)^{\frac{t}{40}}$, where B is the number of bacteria and t is the elapsed time in hours. How many hours will it take for there to be 8000 bacteria present? 120 h

Challenge

A rare diamond was purchased in 1930 for \$1500, and its value has doubled every 10 yr since that time. Let $f(t)$ represent its value t yr after its purchase. Find an equation for $f(t)$ and use it to find the value of the diamond in 1990.

$f(t) = 1500 \cdot 2^{\frac{t}{10}}$; \$96,000.00

COLLEGE ENTRANCE EXAM REVIEW

In each item you are to compare a quantity in Column 1 with a quantity in Column 2. Write the letter of the correct answer from these choices:

A. The quantity in Column 1 is greater than the quantity in Column 2.
B. The quantity in Column 2 is greater than the quantity in Column 1.
C. The quantity in Column 1 is equal to the quantity in Column 2.
D. The relationship cannot be determined from the information given.

Notes: Information centered over both columns refers to one or both of the quantities being compared. A symbol that appears in both columns has the same meaning in each column. All variables represent real numbers. Most figures are not drawn to scale.

	Column 1	Column 2
	$f(x) = 2^x$	
1. B	$f(0)$	$f(0.5)$
2. B	$\log_5 1$	$\log_{64} 8$
3. C	$\log (MN)$	$\log M + \log N$
4. A	$\log (8^2)$	$\log \frac{8}{2}$
	$\log_5 \sqrt[8]{5^7} = x$	
5. A	x	0.25
	$-6x < 2x - 16$	
6. A	x	2
	$5^x = 7$	
7. A	x	$\log 7$
	$f(x) = \left(\frac{1}{3}\right)^{-x}$	
8. A	$f(1)$	$f(-1)$
	$\log_x 0.04 = -2$	
9. C	5	x

	Column 1	Column 2
	a: $3x - 2y = 4$ b: $4x + y = -2$	
10. A	slope of line a	slope of line b
	$x < y$	
11. D	$\frac{x^2}{y}$	$\frac{y}{x^2}$
	$3^{x+2} = 27^x$	
12. B	x	2
	$\log (x - 3) + \log x = 1$	
13. B	x	6
14. C	$\frac{1}{3} \log 27$	$\log \sqrt{9}$
15. C	$\ln e$	1
16. D	10^x	e^x
17. D	$\frac{\log M}{\log N}$	$\log M - \log N$
	$x > y$	
18. D	$\log x$	$\log \frac{1}{y}$

See *Teacher's Resource Book, Tests*, pp. 85–88.

Additional Answers

4. $\frac{\cot^2\alpha + 1}{\sec^2\alpha} = \frac{\csc^2\alpha}{\sec^2\alpha} = \frac{\frac{1}{\sin^2\alpha}}{\frac{1}{\cos^2\alpha}} = \frac{1}{\sin^2\alpha} \cdot \frac{\cos^2\alpha}{1} = \frac{\cos^2\alpha}{\sin^2\alpha} = \cot^2\alpha$

5. 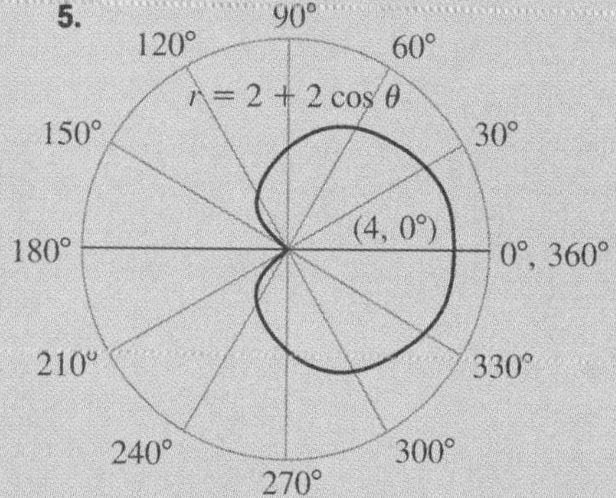

Table of values:

$(4, 0)\ \left(2, \frac{\pi}{2}\right),\ (0, \pi)\ \left(2, \frac{3\pi}{2}\right)\ \left(3, \frac{\pi}{3}\right)$
$\left(3, \frac{5\pi}{3}\right)\ \left(1, \frac{2\pi}{3}\right)\ \left(1, \frac{4\pi}{3}\right)$

20. 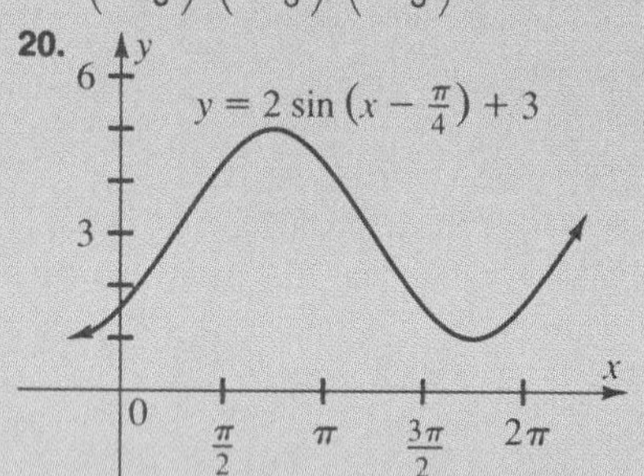

CUMULATIVE REVIEW

Evaluate each of the following for the function $f(x) = e^{x+2}$.

1. $f(3)$ 148.413
2. $f(-3.1)$ 0.333 **8.2**
3. Find the three cube roots of -8. Express your answers in rectangular form. $-2 + 0i,\ 1 + i\sqrt{3},\ 1 - i\sqrt{3}$ **7.8**
4. Prove $\frac{\cot^2\alpha + 1}{\sec^2\alpha} = \cot^2\alpha$. See side column. **3.6**
5. Construct a table of values and graph the equation $r = 2 + 2\cos\theta$. See side column. **7.2**
6. Evaluate $\arccos\left(-\frac{\sqrt{2}}{2}\right)$ in radians. $\frac{3\pi}{4} \pm 2\pi k$ or $\frac{5\pi}{4} \pm 2\pi k$ **6.1**

Find each sum or difference.

7. $(7 + 2i) - (5 + i)$ $2 + i$
8. $(4 - i\sqrt{2}) + (-3 - 2i\sqrt{2})$ $1 - 3i\sqrt{2}$ **7.3**

Find each logarithm to four decimal places.

9. ln 159.7 5.0733
10. log 0.0798 −1.0980 **8.5**

Evaluate each of the following using a double-angle identity.

11. $\sin 2\left(\frac{\pi}{4}\right)$ 1
12. $\tan 2(60°)$ $-\sqrt{3}$ **5.4**

Express in logarithmic form.

13. $6^4 = 1296$ $\log_6 1296 = 4$
14. $32^{-\frac{1}{5}} = \frac{1}{2}$ $\log_{32}\frac{1}{2} = -\frac{1}{5}$ **8.3**

Find each product or quotient and simplify.

15. $5(\cos 42° + i\sin 42°) \cdot 2(\cos 78° + i\sin 78°)$ $-5 + 5i\sqrt{3}$ **7.6**
16. $10(\cos 90° + i\sin 90°) \div 5(\cos 45° + i\sin 45°)$ $\sqrt{2} + i\sqrt{2}$
17. Find the length of the arc intercepted by a central angle of $\theta = \frac{\pi}{2}$ and a radius of 8 in. Express the answer in terms of π. 4π in. **1.5**
18. Solve and check $2^x = 7$. 2.81 **8.5**
19. Use the law of sines to find $\angle B$ given $\angle A = 112°$, $a = 23$, and $b = 13$. 31.6° **4.2**
20. Sketch the graph of the function $y = 2\sin\left(x - \frac{\pi}{4}\right) + 3$ over the interval $0 \le x < 2\pi$. See side column. **2.4**

OVERVIEW • Chapter 9

SUMMARY

The *n*th term formulas for arithmetic and geometric sequences are developed and then used to solve a variety of problems. These formulas are used as building blocks in the discussion of arithmetic and geometric series. Partial sums are used to develop the sum formula for an infinite geometric series, and limits aid in the discussion of convergent and divergent series. Sine, cosine, and exponential functions are defined in terms of power series, and the power series e^x and e^{-x} are then used to define the hyperbolic functions. The similarities between the trigonometric identities and the hyperbolic identities are discussed and used to prove other identities.

CHAPTER OBJECTIVES

- To develop and use a formula to find specified terms of an arithmetic sequence
- To find arithmetic means
- To develop and use a formula to find specified terms of a geometric sequence
- To find geometric means
- To use sigma notation to represent the sum of a series
- To develop and use formulas to find a partial sum of an arithmetic series
- To develop and use formulas to find a partial sum of a geometric series
- To determine whether or not a series is arithmetic or geometric
- To develop and use a formula to find the sum of a convergent geometric series
- To express repeating decimals as rational numbers in fraction form
- To define power series
- To estimate values of circular functions using power series
- To define the hyperbolic functions in terms of e^x and e^{-x} and to express them as power series

CHAPTER HIGHLIGHTS

The *theme* of Chapter 9 is spirals. One of the chapter's special features examines natural occurrences of Archimedean spirals and the mathematical formulas which govern them.

APPLICATIONS

The mathematical concepts involved in sequences and series are applied to a great number of real-world situations, many of which are financial in nature. An important use of geometric series in statistics, the standard normal probability curve, is introduced in Lesson 9.6. The use of hyperbolic functions in engineering is also addressed, and a catenary curve is defined.

TECHNOLOGY

Calculator

Calculators are used extensively when dealing with sequences and series. Finding partial sums, arithmetic and geometric means, specific terms, common ratios, and infinite sums are all made easier with the use of a calculator.

Computer

A computer may be used for many of the operations that arise in dealing with sequences and series. Chapter 9 provides a computer program for generating Archimedean spirals in an extension of the chapter's theme.

RESOURCES

Teacher's Resource Book

- Teaching Aid 9
- Transparencies 17 and 18

ASSIGNMENT GUIDE Meeting Student Needs

STUDENT TEXT				TEACHER'S RESOURCE BOOK	
Chapter Content	**Basic**	**Average**	**Enriched**	**P**	**E**
9.1 Arithmetic Sequences	D: 378/1–29 odd, 49	D: 378/3, 9, 17–43 odd, 49	D: 378/5, 11, 17, 23–49 odd	1	2
9.2 Geometric Sequences	D: 383/1–27 odd, 41 R: 378/2, 8, 20	D: 383/3, 9, 13–37 odd, 41 R: 378/4, 10, 28	D: 383/5, 11, 15–43 odd R: 378/6, 12, 30	3	4
9.3 Arithmetic Series	D: 389/1–31 odd, 49 R: 383/2, 8, 20	D: 389/5, 15–43 odd, 49 R: 383/10, 16, 26	D: 389/7,19–51 odd R: 383/12, 18, 28	5	6
9.4 Geometric Series	D: 395/1–29 odd, 45 R: 389/4, 14, 26 397/TY	D: 396/9–39 odd, 45 R: 389/16, 26, 32 397/TY	D: 396/15–47 odd R: 390/18, 28, 34 397/TY	7	8
9.5 Infinite Geometric Series	D: 402/1–25 odd, 37 R: 395/4, 14, 24	D: 402/9–33 odd, 37, 39 R: 396/14, 24, 28	D: 402/9–41 odd R: 396/14, 26, 30	9	10
9.6 Power Series and Trigonometric Functions	D: 407/1–27 odd R: 402/6, 14, 20	D: 407/3, 7, 11–37 odd, 41, 43 R: 402/10, 16, 20	D: 407/3, 7, 11–47 odd R: 402/12, 18, 22	11	12
9.7 Hyperbolic Functions	Omit	D: 412/3, 9–21 odd, 25 R: 407/18, 26, 30 413/TY	D: 412/3, 9–25 odd R: 407/20, 28, 32 413/TY	13	14

D = Daily R = Review TY = Test Yourself P = Practice E = Enrichment

	STUDENT TEXT				TEACHER'S RESOURCE BOOK	
Review and Testing	Test Yourself	394, 411	College Ent. Exam Rev.	417	Tests	
	Chapter Sum. and Rev.	414	Cumulative Review	418	• Quizzes	85–88
	Chapter Test	416			• Chapter Test (Form A)	89–90
					• Chapter Test (Form B)	91–92
Special Features	Extra	377	Historical Note	401	Applications—Chapter 9	15
	Trigonometry in Finance	383	Challenge	406	Critical Thinking	8
	Extra	389	Technology	412	Alg. and Geom. Review	33–36
					Technology	9

9

Sequences and Series

Spirals often occur in nature. They are used as a symbol or ornament in the world of art. The spiral staircase shown above is indeed a piece of artwork. Archimedean spirals are characterized by arithmetic sequences, and logarithmic spirals by geometric sequences.

BACKGROUND

Spirals manifest themselves in many real-world phenomena. Examination of these spirals reveals that their behavior can be described by mathematical sequences and series, such as the Fibonacci sequence. You may wish to have students research various examples of spirals and the sequences and series which govern them.

LESSON PLAN

Vocabulary
Arithmetic mean
Arithmetic sequence
Common difference
Explicitly
Finite sequence
Infinite sequence
Recursively
Term

BACKGROUND

In the Preview, relations and functions as well as domain and range are reviewed. Sequences will be defined in terms of functions.

9.1 Arithmetic Sequences

Objectives: To develop and use a formula to find specified terms of an arithmetic sequence
To find arithmetic means

In order to define a *sequence*, it is helpful to recall the definitions of a relation and a function.

Preview

A *relation* is a set of ordered pairs. Set T, shown below, is a relation.

$$T = \{(1, 2), (2, 4), (3, 6), (4, 8)\}$$

The domain of this relation is $\{1, 2, 3, 4\}$ and the range is $\{2, 4, 6, 8\}$. A relation is a *function* if for every element of the domain, there is one and only one corresponding element of the range. Thus, the relation T is also a function. T can be described by the rule $f(x) = 2x$, where x is a member of the set $\{1, 2, 3, 4\}$.

Find the domain and range of each relation and state whether or not the relation is a function. 3. $D = \{x: x \neq 0\}$
$R = \{y \neq 0\}$; function

1. $\{(2, 4), (3, 4), (5, 4), (6, 4)\}$
$D = \{2, 3, 5, 6\}$, $R = \{4\}$; function

2. $\{(-2, 1), (1, 3), (4, 1), (5, -2), (-2, 0)\}$
$D = \{-2, 1, 4, 5\}$, $R = \{1, 3, -2, 0\}$; not a function

3. $f(x) = \dfrac{1}{x}$

4. $f(x) = x^2$
$D = \{\text{real numbers}\}$
$R = \{y: y \geq 0\}$; function

5. $f(x) = \pm x$
$D = \{\text{real numbers}\}$
$R = \{\text{real numbers}\}$; not

6. $f(x) = -2x^2 + 1$
$D = \{\text{real numbers}\}$
$R = \{y: y \leq 1\}$; function

The following table shows the value, in pennies, of from 1 to 8 nickels.

number of nickels	1	2	3	4	5	6	7	8
value in pennies	5	10	15	20	25	30	35	40

The set of ordered pairs $\{(1, 5), (2, 10), \ldots, (8, 40)\}$ follows a pattern and is an example of a *sequence*. A **sequence** is a function whose domain is a set of consecutive positive integers. Each number in the range of a sequence is called a *term*. Using symbols, the first term is denoted by a_1, the second term by a_2, and the nth term, or **general term,** by a_n. The nickels-and-pennies sequence can also be written as $5, 10, 15, \ldots, 40$, where $a_1 = 5$, $a_2 = 10$, and so on. This is an example of a **finite sequence,** since it has a last term.

A sequence can be defined **explicitly** by expressing a_n as a function of n, where n is a positive integer. An explicit formula for the nickels-and-pennies sequence is $a_n = 5n$, where n is a member of the set $\{1, 2, 3, 4, 5, 6, 7, 8\}$. That is, the general term of the sequence is $5n$.

A sequence can also be defined *recursively* by stating the first term, a_1, and a rule for obtaining the $(n + 1)$th term from the preceding term, a_n. A **recursive formula** for the nickels-and-pennies sequence is $a_1 = 5$, $a_{n+1} = a_n + 5$, where n is a member of $\{1, 2, 3, 4, 5, 6, 7, 8\}$.

An **infinite sequence** does not have a last term. However, any desired number of consecutive terms can be found if the sequence is defined recursively.

EXAMPLE 1 **The following infinite sequence is defined recursively: $a_1 = 7$ and $a_{n+1} = a_n + 2$. Write the first five terms.**

$a_1 = 7$
$a_2 = 7 + 2 = 9$ $\quad a_2 = a_1 + 2$
$a_3 = 9 + 2 = 11$ $\quad a_3 = a_2 + 2$
$a_4 = 11 + 2 = 13$ $\quad a_4 = a_3 + 2$
$a_5 = 13 + 2 = 15$ $\quad a_5 = a_4 + 2$

The first five terms of the sequence are 7, 9, 11, 13, and 15.

An **arithmetic sequence,** or an *arithmetic progression*, is a sequence of numbers in which the difference between any two successive terms is a constant. This constant is called the **common difference.** You can find the common difference, d, by subtracting any two consecutive terms. That is,

$$d = a_{n+1} - a_n$$

This formula can be written in a different way to obtain a recursive formula for an arithmetic sequence. If the first term, a_1, is given, then

$$a_{n+1} = a_n + d$$

EXAMPLE 2 **Find the next three terms of the infinite arithmetic sequence 18, 24, 30, ...**

Since $24 - 18 = 6$, the common difference is 6.
Then, $30 + 6 = 36$ $\quad a_{n+1} = a_n + 6$
$36 + 6 = 42$
$42 + 6 = 48$ $\quad$ The next three terms are 36, 42, and 48.

Consider the terms of the sequence in Example 2.

a_1	a_2	a_3	a_4	a_5	a_6
18	24	30	36	42	48
18 + 0(6)	18 + 1(6)	18 + 2(6)	18 + 3(6)	18 + 4(6)	18 + 5(6)
$a_1 + 0(d)$	$a_1 + 1(d)$	$a_1 + 2(d)$	$a_1 + 3(d)$	$a_1 + 4(d)$	$a_1 + 5(d)$

TEACHING SUGGESTIONS

- To emphasize the characteristics of arithmetic sequences, write several on the chalkboard and ask students to identify the first term, common difference, and so on.
- Emphasize that the domain of a sequence refers to the set of positive integers and the range of the sequence refers to the numbers which are the terms of the sequence.

Critical Thinking

Analysis Ask students to describe the graph of an arithmetic sequence using the term numbers as the domain and the actual terms as the range. Then ask them to predict the slope of the graph. After graphing several arithmetic sequences, students should note that each graph will be a straight line and that the slope of each graph will be the common difference of the terms of the sequence.

CHALKBOARD EXAMPLES

- **For Example 1**

  The following sequences are defined recursively. Write the first 5 terms of each sequence.

  1. $a_1 = 5$, $a_{n+1} = a_n + 3$
  5, 8, 11, 14, 17

  2. $a_1 = -2$, $a_{n+1} = a_n - 4$
  −2, −6, −10, −14, −18

- **For Example 2**

  Find the next three terms of each arithmetic sequence.

  3. 33, 29, 25 21, 17, 13

  4. −7, −10, −13
  −16, −19, −22

- **For Example 3**

  Write the general term a_n of each arithmetic sequence.

  5. $-7, -4, -1, \ldots$ $a_n = 3n - 10$

  6. $8, 4, 0, \ldots$ $a_n = 12 - 4n$

- **For Example 4**

  Find the given term of each arithmetic sequence.

  7. $a_1 = 5, d = 4; a_6$ 25

  8. $a_1 = -2, d = 5; a_{21}$ 98

- **For Example 5**

  9. The 8th term of an arithmetic sequence is 24 and the 15th term is 45. Find a_1 and d. $a_1 = 3, d = 3$

  10. The 5th term of an arithmetic sequence is -1 and the 12th term is 34. Find a_1 and d. $a_1 = -21, d = 5$

In general, an explicit formula for the general term, a_n, of an arithmetic sequence is

$$a_n = a_1 + (n - 1)d$$

where a_1 is the first term and d is the common difference.

EXAMPLE 3 **Find the general term a_n of the arithmetic sequence $-8, -3, 2, \ldots$.**

$a_1 = -8$ and $d = -3 - (-8) = 5$

$a_n = -8 + (n - 1)(5)$ — *$a_n = a_1 + (n - 1)d$*

$a_n = -8 + 5n - 5$

$a_n = 5n - 13$

The formula for the general term a_n can be used to find any specified term of a sequence if the first term and the common difference are known.

EXAMPLE 4 **Find the sixth term of the arithmetic sequence if $a_1 = 11$ and $d = 4$.**

$a_6 = 11 + (6 - 1)(4)$ — *Substitute $a_1 = 11$, $n = 6$, and $d = 4$ in the formula $a_n = a_1 + (n - 1)d$.*

$= 11 + 20 = 31$ — The sixth term is 31.

If two terms of an arithmetic sequence are known, a_1 and d can be found by using the formula for the general term twice and then solving the resulting system of equations.

EXAMPLE 5 **The tenth term of an arithmetic sequence is 77 and the sixteenth term is 119. Find a_1 and d.**

$a_n = a_1 + (n - 1)d$ — *Use the formula for the nth term twice.*

$77 = a_1 + (10 - 1)d$ — *Let $a_{10} = 77$ and $n = 10$.*

$119 = a_1 + (16 - 1)d$ — *Let $a_{16} = 119$ and $n = 16$.*

$77 = a_1 + 9d$

$119 = a_1 + 15d$ — *Simplify each equation and solve the system of equations.*

$-77 = -a_1 - 9d$ — *Multiply both sides of the first equation by -1.*

$119 = a_1 + 15d$

$42 = 6d$ — *Add.*

$7 = d$ — *Solve for d.*

$77 = a_1 + 9(7)$ — *Substitute $d = 7$ in one of the original equations and solve for a_1.*

$77 = a_1 + 63$

$14 = a_1$ — Thus, $a_1 = 14$ and $d = 7$.

The terms between any two nonconsecutive terms of an arithmetic sequence are called **arithmetic means.** In the sequence 5, 12, 19, 26, 33, . . . , the terms 12, 19, and 26 are the arithmetic means between 5 and 33. A single arithmetic mean between any two given numbers is called *the arithmetic mean* or the *average* of the two numbers. For all real numbers a and b, the arithmetic mean is $\frac{a+b}{2}$.

Any finite number of arithmetic means can be inserted between two numbers.

EXAMPLE 6 **Insert five arithmetic means between 13 and −11.**

Let $a_1 = 13$ and $a_7 = -11$. The sequence is 13, a_2, a_3, a_4, a_5, a_6, −11.

$a_7 = a_1 + (7 - 1)d$ *Use the formula $a_n = a_1 + (n - 1)d$, where $n = 7$.*

$-11 = 13 + 6d$ *Substitute: $a_7 = -11$, $a_1 = 13$.*

$-24 = 6d$

$-4 = d$ *Solve for d.*

Therefore, $a_2 = 13 + (-4) = 9$ $a_{n+1} = a_n + d$

$a_3 = 9 + (-4) = 5$

$a_4 = 5 + (-4) = 1$

$a_5 = 1 + (-4) = -3$

$a_6 = -3 + (-4) = -7$

The five arithmetic means between 13 and −11 are 9, 5, 1, −3, and −7.

For any arithmetic sequence in which a_n is the general term, a_1 is the first term, d is the common difference, and n is a positive integer

$$a_n = a_1 + (n - 1)d$$
$$a_{n+1} = a_n + d \quad \text{or} \quad d = a_{n+1} - a_n$$

CLASS EXERCISES

Consider the arithmetic sequence 3, 8, 13, ... Find each value.

1. d 5 **2.** a_1 3 **3.** a_3 13 **4.** a_4 18 **5.** a_5 23 **6.** a_6 28

Determine whether or not each sequence is arithmetic. If it is, find the common difference and write the general term, a_n.

7. 15, 18, 21 yes; $d = 3$; $a_n = 3n + 12$ **8.** 14, 16, 19 no **9.** 10, 5, 10 no **10.** −5, −3, −1 yes; $d = 2$; $a_n = 2n - 7$

11. 7, 11, 16 no **12.** 1, 2, 3 yes; $d = 1$; $a_n = n$ **13.** −1, −2, −3 yes; $d = -1$; $a_n = -n$ **14.** 4, 0, 4 no

- **For Example 6**

11. Insert four arithmetic means between −6 and −26. −10, −14, −18, −22

12. Insert three arithmetic means between −12 and 8. −7, −2, 3

LESSON FOLLOW-UP

Discussion

What is the next term of the sequence 8, 5, 2, . . .? Give a reason for your answer. If the sequence is assumed to be arithmetic, then the fourth term is −1. If the sequence is not assumed to be arithmetic, then the fourth term is 1 since an expression for the nth term of the sequence is $|3n - 11|$.

Assignment Guide

See p. 370B for assignments.

Extra

The study of relationships among numbers is referred to as number theory. Arithmetic sequences are often helpful in this study.

Lesson Quiz

Consider the arithmetic sequence 5, 9, 13, ... Give the value of each.

1. d 4
2. a_1 5
3. a_5 21

Determine whether each sequence is arithmetic. If it is, find the common difference and write the general term a_n.

4. 4, 2, 4 no
5. 7, 8, 9 yes; $a_n = 6 + n$
6. 8, 4, 0 yes; $a_n = 12 - 4n$
7. Find a_9 if $a_1 = 23$ and $d = -3.4$. -4.2
8. Insert three arithmetic means between 19 and -5. 13, 7, 1
9. Using the formula $a_n = a_1 + (n - 1)d$, find n if 88 is the nth term of 4, 7, 10, 13 29
10. Kathy received \$2 per week for an allowance at age 8. If her allowance increased \$1.50 each year, how much did she get when she was 17 years old? \$15.50

Enrichment

The harmonic mean of two numbers s and t is x if $\frac{1}{s}, \frac{1}{x}$, and $\frac{1}{t}$ are terms of an arithmetic sequence. Find x in terms of s and t.

$$\frac{1}{x} - \frac{1}{s} = \frac{1}{t} - \frac{1}{x}$$

$$\frac{s - x}{sx} = \frac{x - t}{tx}$$

$$\frac{tx(s - x)}{sx} = x - t$$

$$ts - tx = sx - st$$

$$-sx - tx = -2st$$

$$-x(s + t) = -2st$$

$$x = \frac{2st}{s + t}$$

PRACTICE EXERCISES

The following sequences are defined recursively. Give the first five terms of each sequence.

A

1. $a_{n+1} = a_n + 4$; $a_1 = 5$ 5, 9, 13, 17, 21
2. $a_{n+1} = a_n + 6$; $a_1 = 3$ 3, 9, 15, 21, 27
3. $a_{n+1} = a_n - 3$; $a_1 = 4$ 4, 1, -2, -5, -8
4. $a_{n+1} = a_n - 4$; $a_1 = 6$ 6, 2, -2, -6, -10
5. $a_{n+1} = a_n + 6$; $a_1 = -17$ -17, -11, -5, 1, 7
6. $a_{n+1} = a_n + 3$; $a_1 = -1$ -1, 2, 5, 8, 11

List the next four terms of each arithmetic sequence. Then, find the general term, a_n.

7. 7, 11, 15, . . . 19, 23, 27, 31; $a_n = 4n + 3$
8. 3, 6, 9, . . . 12, 15, 18, 21; $a_n = 3n$
9. $-32, -28, -24, \ldots$ $-20, -16, -12, -8$; $a_n = 4n -$
10. $\frac{1}{2}, \frac{3}{2}, \frac{5}{2}, \ldots$ 10. $\frac{7}{2}, \frac{9}{2}, \frac{11}{2}, \frac{13}{2}$; $a_n = n - \frac{1}{2}$
11. $1, -4, -9, \ldots$ $-14, -19, -24, -29$; $a_n = -5n + 6$
12. $-2.5, 1.5, 5.5, \ldots$ 9.5, 13.5, 17.5, 21.5; $a_n = 4n -$

Find the given term of each arithmetic sequence.

13. $a_1 = 5, d = 5; a_7$ 35
14. $a_1 = 4, d = 3; a_3$ 10
15. $a_1 = \frac{1}{3}, d = -\frac{2}{3}; a_5$ -3
16. $a_1 = 15, d = -2; a_8$ 1
17. $a_1 = 3, d = -2.5; a_{11}$ -22
18. $a_1 = 0.8, d = 1; a_{15}$ 14.8

Find the first term and the common difference for each arithmetic sequence, using the given terms.

19. $a_3 = 11, a_8 = 26$ $a_1 = 5$; $d = 3$
20. $a_5 = 24, a_{12} = 45$ $a_1 = 12$; $d = 3$
21. $a_6 = -19, a_8 = -29$ $a_1 = 6$; $d = -5$
22. $a_7 = -23, a_{14} = -37$ $a_1 = -11$; $d = -2$
23. $a_{21} = 13, a_{27} = 16$ $a_1 = 3$; $d = \frac{1}{2}$
24. $a_{50} = 5.4, a_{75} = 7.9$ $a_1 = 0.5$; $d = 0.1$

B

25. Insert three arithmetic means between 6 and 46. 16, 26, 36
26. Insert six arithmetic means between 18 and 53. 23, 28, 33, 38, 43, 48
27. Insert four arithmetic means between 6 and -6. 3.6, 1.2, -1.2, -3.6
28. Insert three arithmetic means between -8 and 1. -5.75, -3.50, -1.25
29. Insert two arithmetic means between 6.2 and 7.8 $\frac{101}{15}, \frac{109}{15}$
30. Insert two arithmetic means between -5.4 and -2.6. $-\frac{67}{15}, -\frac{53}{15}$

Find the indicated term in each arithmetic sequence.

31. a_{12} of $-16, -13, -10, \ldots$ 17
32. a_{21} of 10, 7, 4, . . . -50
33. a_{32} of 4, 7, 10, . . . 97
34. a_6 of $8, 3, -2, \ldots$ -17
35. a_{12} of $\frac{3}{4}, \frac{3}{2}, \frac{9}{4}, \ldots$ 9
36. a_{10} of $\frac{5}{3}, \frac{6}{3}, \frac{7}{3}, \ldots$ $\frac{14}{3}$
37. In the arithmetic sequence $-9, -2, 5, \ldots$, which term is 131? $n = 21$
38. In the arithmetic sequence $-3, 2, 7, \ldots$, which term is 142? $n = 30$

39. In the arithmetic sequence 7, 2, -3, . . . , which term is -28? $n = 8$

40. In the arithmetic sequence $\frac{9}{4}$, 2, $\frac{7}{4}$, . . . , which term is $-\frac{17}{4}$? $n = 27$

C **41.** Find three arithmetic means between a and b. $\frac{3a + b}{4}, \frac{a + b}{2}, \frac{a + 3b}{4}$

42. The arithmetic mean of -22 and another number is -8. Find the other number. 6

43. The last term of an arithmetic sequence is 207, the common difference is 3, and the number of terms is 14. What is the first term of the sequence? 168

44. The fifth term of an arithmetic sequence is 19 and the eleventh term is 43. Find the first term and the sixty-seventh term of the sequence. $a_1 = 3$; $a_{67} = 267$

45. The fifth term of an arithmetic sequence is 3 and the fifteenth term is 8. What is the general term of the sequence? $a_n = \frac{1}{2}n + \frac{1}{2}$

46. The third term of an arithmetic sequence is 12 and the thirteenth term is 7. What is the general term of the sequence? $a_n = -\frac{n}{2} + \frac{27}{2}$

Find a_1 and d in terms of r and s for each arithmetic sequence.

47. $a_4 = r$; $a_9 = s$ $a_1 = \frac{8r - 3s}{5}$; $d = \frac{s - r}{5}$

48. $a_6 = r + s$; $a_{11} = 25$ $a_1 = 2r + 2s - 25$; $d = \frac{25 - r - s}{5}$

Applications

49. Consumerism In 1970, a family membership in a country club cost \$650. This cost increased \$35 each year since 1970. How much will a 1990 membership cost? \$1350

50. Business A department store chain had 112 franchises in 1978 and opened 8 new franchises each year thereafter. How many franchises were there at the end of 1985? 168

EXTRA

Number theory is the study of relationships among numbers, particularly positive integers. Some number theory problems deal with sequences.

1. In an arithmetic sequence of three positive integers, can the sum of the terms be a prime number? *Hint*: Let a_1, $a_1 + d$, and $a_1 + 2d$ represent the three terms of the sequence, where d is the common difference. If the sum is divisible by an integer other than 1, it is not prime. no
2. In an arithmetic sequence of five positive integers, can the sum of the terms be a prime number? no
3. Is it possible to have an arithmetic sequence of six positive integers whose sum is not divisible by 6? If so, give an example. Yes; answers may vary.

Teacher's Resource Book

Practice—Chapter 9, p. 1
Enrichment—Chapter 9, p. 2

LESSON PLAN

Vocabulary
Common ratio
Geometric mean
Geometric sequence
Mean proportional

Materials/Manipulatives
Calculators

BACKGROUND

In the Preview, a sequence in which the terms decrease by a constant factor is discussed. This discussion will aid in the development of the formulas used in a geometric sequence.

9.2 Geometric Sequences

Objectives: To develop and use a formula to find specified terms of a geometric sequence
To find geometric means

In many sequences, the terms increase or decrease by a constant factor rather than by a constant addend.

Preview

The Singletons have just purchased a car for \$15,000. The car is expected to depreciate by 20 percent per year. Therefore, the value of the car each year will be 0.8 times its value from the previous year.

At the end of the first year, the car will be worth

\$15,000(0.8) = \$12,000

At the end of the second year, the car will be worth

\$12,000(0.8) = \$15,000$(0.8)^2$ = \$9600

Find the value of the car, to the nearest dollar, at the end of each year.

1. year 3 \$7680 **2.** year 4 \$6144 **3.** year 5 \$4915 **4.** year 8 \$2517

A **geometric sequence,** or *geometric progression*, is a sequence in which the ratio of any two consecutive terms is a constant. This constant is called the **common ratio.** To illustrate, in the geometric sequence 1, 3, 9, 27, . . . , each term is 3 times the preceding term. That is, the common ratio is $\frac{3}{1} = \frac{9}{3} = \frac{27}{9} = 3$.

In general, the common ratio r can be found by dividing any term in the sequence by the preceding term. Thus,

$$r = \frac{a_{n+1}}{a_n}$$

Note that this formula can be written in a different way to give a recursive formula for a geometric sequence for which the first term, a_1, is known.

$$a_{n+1} = a_n r$$

EXAMPLE 1 **Write the first six terms of the geometric sequence with first term 2 and common ratio 5.**

Begin with 2 and multiply by 5 to get each succeeding term.

$a_1 = 2$
$a_2 = 2(5) = 10$ $\quad a_2 = a_1r$
$a_3 = 10(5) = 50$ $\quad a_3 = a_2r$
$a_4 = 50(5) = 250$ $\quad a_4 = a_3r$
$a_5 = 250(5) = 1250$ $\quad a_5 = a_4r$
$a_6 = 1250(5) = 6250$ $\quad a_6 = a_5r$

Each term of a geometric sequence can be written as the product of the first term, a_1, and a power of r.

Term		
1	a_1	$= a_1$
2	a_2	$= a_1r$
3	$a_3 = a_2r = a_1r \cdot r$	$= a_1r^2$
4	$a_4 = a_3r = a_1r^2 \cdot r$	$= a_1r^3$
5	$a_5 = a_4r = a_1r^3 \cdot r$	$= a_1r^4$

An explicit formula for the *n*th term, or general term, of a geometric sequence, where a_1 is the first term and r is the common ratio, is

$$a_n = a_1r^{n-1}$$

EXAMPLE 2 **Write the general term, a_n, of the geometric sequence 3, 12, 48, ...**

$a_1 = 3$ and $r = \frac{12}{3} = 4$
$a_n = 3(4)^{n-1}$ $\quad a_n = a_1r^{n-1}$

You may wish to use a calculator when you work with geometric sequences.

EXAMPLE 3 **Find the sixth term of a geometric sequence with first term 3 and common ratio 4**

$a_6 = 3(4)^{6-1}$ *Use the formula $a_n = a_1r^{n-1}$, where $a_1 = 3$, $n = 6$, $r = 4$.*
$= 3(4)^5$ *Calculation-ready form*
$= 3072$

It is possible to find the first term and the common ratio of a geometric sequence if any two terms are given.

TEACHING SUGGESTIONS

- Give students examples of several geometric sequences and ask them to find the common ratio.
- Give students the first term (a_1) and the common ratio (r) and ask them to write a specified number of terms of the sequence.
- Encourage students to use calculators when solving the problems in this lesson.

Critical Thinking

Contrasting Ask students to find x in the arithmetic sequence 3, x, 9 and in the geometric sequence 15, x, 25. Then ask them to explain the difference between arithmetic and geometric means. 6; $\pm \frac{1}{3}\sqrt{15}$; arithmetic means are unique; geometric means are not unique.

CHALKBOARD EXAMPLES

- **For Example 1**
  Write the first five terms of each geometric sequence.
  1. 1st term 3, common ratio 6 3, 18, 108, 648, 3888
  2. 1st term -2, common ratio -3 $-2, 6, -18, 54, -162$
- **For Example 2**
  Write the general term a_n of each geometric sequence.
  3. 3, 6, 12, . . . $a_n = 3(2)^{n-1}$
  4. 5, -15, 45, . . . $a_n = 5(-3)^{n-1}$
- **For Example 3**
  Find the specified term of each geometric sequence.
  5. $_1 = 3, r = -2; a_5$ 48
  6. $a_1 = -7, r = 2; a_9$ -1792

- **For Example 4**

7. The second term of a geometric sequence is 10 and the sixth term is 160. Find the first term and the common ratio.
$a_1 = \pm 5, r = \pm 2$

8. The third term of a geometric sequence is 45 and the fifth term is 405. Find the first term and the common ratio.
$a_1 = 5, r = \pm 3$

- **For Example 5**

9. Insert three geometric means between 2 and 162.
6, 18, 54 or −6, 18, −54

10. Insert three geometric means between 3125 and 5.
625, 125, 25 or −625, 125, −25

EXAMPLE 4 **The third term of a geometric sequence is 32 and the fifth term is 128. Find the first term and the common ratio.**

$32 = a_1r^2$ *Use the formula $a_n = a_1r^{n-1}$, where $a_3 = 32$ and $n = 3$.*

$128 = a_1r^4$ *Now let $a_5 = 128$ and $n = 5$.*

$\frac{128}{32} = \frac{a_1r^4}{a_1r^2}$ *Solve the system of equations by dividing.*

$4 = r^2$

$\pm 2 = r$

$32 = a_1(2)^2$ $\quad 32 = a_1(-2)^2$ *Use one of the original equations to solve for a_1.*

$32 = 4a_1$ $\quad 32 = 4a_1$

$8 = a_1$ $\quad 8 = a_1$ Thus $a_1 = 8$ and $r = 2$ or $r = -2$.

The terms between any two nonconsecutive terms of a geometric sequence are called **geometric means.** Any finite number of geometric means can be inserted between two given numbers.

EXAMPLE 5 **Insert four geometric means between 972 and 4.**

Let $a_1 = 972$ and $a_6 = 4$. Then, the sequence is: 972, a_2, a_3, a_4, a_5, 4.

$a_6 = a_1r^{6-1}$ *Use the formula $a_n = a_1r^{n-1}$, where $n = 6$.*

$4 = 972r^5$ *Substitute: $a_6 = 4$, $a_1 = 972$.*

$\frac{4}{972} = r^5$

$\left(\frac{4}{972}\right)^{\frac{1}{5}} = r$ *Solve for r.*

$\frac{1}{3} = r$

Thus, $a_2 = 972\left(\frac{1}{3}\right) = 324$ $\quad a_3 = 324\left(\frac{1}{3}\right) = 108$ $\quad a_{n+1} = a_nr$

$a_4 = 108\left(\frac{1}{3}\right) = 36$ $\quad a_5 = 36\left(\frac{1}{3}\right) = 12$

The geometric means are 324, 108, 36, and 12.

A single geometric mean between two numbers is called *the geometric mean*, or the **mean proportional,** between the two numbers. If $\frac{a}{m} = \frac{m}{b}$, where a and b are nonzero real numbers, then m is the mean proportional between a and b. Thus, $m^2 = ab$ and $m = \sqrt{ab}$ or $m = -\sqrt{ab}$. It is customary to let m equal $\sqrt{ab}$ when a and b are positive, and $-\sqrt{ab}$ when a and b are negative.

EXAMPLE 6 **Find the mean proportional between -12 and -300.**

a and b are negative, so $m = -\sqrt{ab}$.

$m = -\sqrt{(-12)(-300)}$ **Check:** $\frac{-12}{-60} \stackrel{?}{=} \frac{-60}{-300}$

$= -60$ $\frac{1}{5} = \frac{1}{5}$

> For any geometric sequence in which a_n is the general term, a_1 is the first term, r is the common ratio, and n is a positive integer,
>
> $$a_n = a_1 r^{n-1}$$
>
> $$a_{n+1} = a_n r \quad \text{or} \quad r = \frac{a_{n+1}}{a_n}$$

CLASS EXERCISES

Determine whether or not each sequence is geometric. If it is, find the common ratio and write the general term a_n.

1. $-3, -12, -48$ yes; $r = 4$; $(-3)(4)^{n-1}$
2. $-2, 4, -6$ no
3. $20, 10, 5$ yes; $r = \frac{1}{2}$; $20\left(\frac{1}{2}\right)^{n-1}$
4. $18, 6, 2$ yes; $r = \frac{1}{3}$; $18\left(\frac{1}{3}\right)^{n-1}$

Find the specified term of each geometric sequence.

5. $a_1 = -1, r = 3; a_4$ -27
6. $a_1 = 2, r = \frac{1}{2}; a_6$ $\frac{1}{16}$
7. $a_1 = 16, r = \frac{1}{4}; a_5$ $\frac{1}{16}$
8. $a_1 = 81, r = -\frac{1}{3}; a_8$ $-\frac{1}{27}$
9. Insert three geometric means between 2 and 162. 6, 18, 54 or -6, 18, -54
10. Find the mean proportional of 16 and 36. Check your answer. 24

PRACTICE EXERCISES

Find the first four terms of each geometric sequence.

A

1. $a_1 = 3.5, r = 3$ 3.5, 10.5, 31.5, 94.5
2. $a_1 = 3, r = -4$ 3, -12, 48, -192
3. $a_1 = 1, r = -2$ 1, -2, 4, -8
4. $a_1 = 1.5, r = 2$ 1.5, 3, 6, 12
5. $a_1 = 8, r = -\frac{1}{2}$ 8, -4, 2, -1
6. $a_1 = 10, r = -\frac{1}{5}$ 10, -2, $\frac{2}{5}$, $-\frac{2}{25}$

List the next three terms of each geometric sequence. Then, write the general term a_n.

7. $3, -9, 27, \ldots$ -81, 243, -729; $a_n = 3(-3)^{n-1}$
8. $64, -32, 16, \ldots$ -8, 4, -2; $a_n = 64\left(-\frac{1}{2}\right)^{n-1}$
9. $9, 3, 1, \ldots$ $\frac{1}{3}, \frac{1}{9}, \frac{1}{27}$; $a_n = 9\left(\frac{1}{3}\right)^{n-1}$
10. $-\frac{1}{2}, 1, -2, \ldots$ 4, -8, 16; $a_n = -\frac{1}{2}(-2)^{n-1}$
11. $-\frac{1}{9}, -\frac{1}{3}, -1, \ldots$ -3, -9, -27; $a_n = -\frac{1}{9}(3)^{n-1}$
12. $\sqrt{3}, 3, 3\sqrt{3}, \ldots$ 9, $9\sqrt{3}$, 27; $a_n = \sqrt{3}(\sqrt{3})^{n-1}$

- **For Example 6**

  Find the mean proportional between each pair of numbers.

  11. -6 and -150 -30

  12. 5 and 245 35

Common Errors

- Some students confuse arithmetic and geometric sequences. Emphasize the difference and provide practice in which students classify sequences as arithmetic or geometric.
- Some students confuse geometric means and mean proportional. Emphasize that the mean proportional is a single geometric mean between two numbers and is equal to the square root of the product of those numbers.
- See *Teacher's Resource Book* for additional remediation.

LESSON FOLLOW-UP

Discussion

How would you construct the geometric mean of two given line segments with lengths x and y?

1. On a line m, construct $\overline{ST}$ and $\overline{TW}$ with lengths x and y respectively.
2. Construct a semicircle with diameter SW.
3. Through T, construct a segment perpendicular to SW.
4. Label the point where the segment intersects the semicircle Y.
5. YT is the geometric mean of ST and TW.

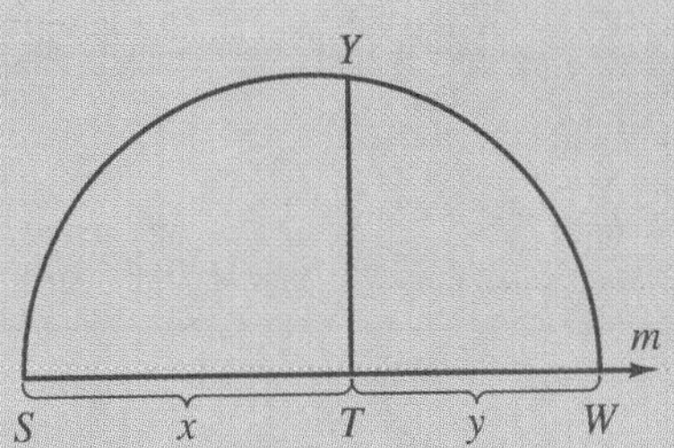

Assignment Guide

See p. 370B for assignments.

Trigonometry in Finance

The compound interest formula as an application of a geometric sequence is discussed.

Lesson Quiz

1. Find the common ratio for the sequence 6, −3, 1.5, . . . $-\frac{1}{2}$
2. Find the first four terms of the geometric sequence with $a_1 = 3$ and $r = 4$. 3, 12, 48, 192
3. Find a_1 for a geometric sequence if $a_5 = 12$ and $r = \frac{1}{3}$. 972
4. Find three geometric means between 54 and $\frac{2}{3}$. 18, 6, 2 or −18, 6, −2
5. Find the mean proportional between −49 and −9. −21
6. Suppose a new car costs $15,000 and the dealer uses a depreciation factor of 25% each year. How much would the car be worth to the nearest dollar at the end of 5 years? $3560

Enrichment

Find the mean proportional between s^2t and s^6t^9, with $s > 0$ and $t > 0$. s^4t^5

Find the specified terms of each geometric sequence.

13. $a_1 = 4, r = 6; a_3$ 144 **14.** $a_1 = 1, r = 5; a_4$ 125 **15.** $a_3 = 32, r = -\frac{1}{2}; a_6$ −4

16. $a_4 = 16, r = \frac{1}{2}; a_8$ 1 **17.** $2x, 4x, 8x, \ldots; a_7$ $128x$ **18.** $y^3, y^4, y^5, \ldots; a_{10}$ y^{12}

19. Insert two geometric means between 1 and 8. 2, 4

20. Insert five geometric means between 32 and $\frac{1}{2}$. 16, 8, 4, 2, 1 or −16, 8, −4, 2, −1

B **21.** Insert four geometric means between $\frac{1}{3}$ and 81. 1, 3, 9, 27

22. Insert three geometric means between 81 and 1. 27, 9, 3 or −27, 9, −3

23. Insert four geometric means between −3 and 96. 6, −12, 24, −48

24. Insert three geometric means between 5 and $\frac{16}{125}$. $2, \frac{4}{5}, \frac{8}{25}$ or $-2, \frac{4}{5}, -\frac{8}{25}$

Find the mean proportional between each pair of numbers.

25. −12 and −108 −36 **26.** −16 and −49 −28 **27.** 15 and 60 30 **28.** 16 and 100 40

Find the common ratio and the first term for each geometric sequence.

29. $a_5 = 16, a_8 = 2$ $r = \frac{1}{2}; a_1 = 256$ **30.** $a_8 = -21, a_{10} = -189$ $r = 3; a_1 = \frac{-7}{729}$ or $r = -3; a_1 = \frac{7}{729}$

31. $a_6 = 54, a_8 = 216$ $r = 2; a_1 = \frac{27}{16}$ or $r = -2, a_1 = -\frac{27}{16}$ **32.** $a_4 = 100, a_6 = 10{,}000$ $r = 10; a_1 = 0.1$ or $r = -10; a_1 = -\frac{1}{10}$

33. $a_7 = 6, a_{11} = \frac{3}{8}$ $r = \pm\frac{1}{2}; a_1 = 384$ **34.** $a_9 = 1.1, a_{13} = 0.00011$ $r = \pm 0.1; a_1 = 110{,}000{,}000$

C **35.** The fourth term of a geometric sequence is −10 and the sixth term is −40. Find the ninth term of the sequence. −320

36. The second term of a geometric sequence is −2 and the fifth term is 54. Find the eighth term of the sequence. −1458

37. Find the value of x so that $x - 1$, $x + 2$, and $x + 8$ are the first three terms of a geometric sequence. 4

38. Find the value of x so that $x - 4$, $x - 2$, and $x + 4$ are the first three terms of a geometric sequence. 5

39. If $a_1 = 100$ and $r = 0.01$, which term in the geometric sequence is 10^{-10}? seventh

40. If $a_1 = x^2$ and $r = x^4$, which term in the geometric sequence is x^{18}? fifth

Applications

41. Business Zoom Airlines' passenger load has been increasing by 12 percent annually. In 1980 they carried 20,500 passengers. How many passengers should they expect to carry in 1990? 63,670

42. **Business** Monica invested $50,000 in a piece of equipment. The equipment depreciates at the rate of 20 percent per year. What will be the value of the equipment, to the nearest dollar, at the end of the fifth year? $16,384

43. **Investment** Suppose an antique car you own appreciates (increases) in value at a rate of 8 percent annually. If the car was purchased for $12,000, what will it be worth during the eighth year? $22,211

44. **Geometry** A 16-in. piece of wire is cut in half. Each piece is then cut in half, and the process is repeated 8 more times. What is the size of each final piece? $\frac{1}{64}$ in.

TRIGONOMETRY IN FINANCE

The compound interest formula is $A_n = P(1 + r)^n$, where P is the original principal, n is the number of time periods for which the principal is invested, r is the interest rate per period (expressed as a decimal), and A_n is the amount in the account after n time periods. To understand how the formula is derived, study the pattern of the first few terms of the sequence. It is much like the pattern of the terms of a geometric sequence.

$A_1 = P + rP = P(1 + r)^1$

$A_2 = P(1 + r) + r[P(1 + r)] = [P(1 + r)](1 + r) = P(1 + r)^2$

$A_3 = P(1 + r)^2 + r[P(1 + r)^2] = [P(1 + r)^2](1 + r) = P(1 + r)^3$

$A_n = P(1 + r)^n$ *General term of the sequence*

To illustrate the use of this formula, consider the following problem:

If $1000 is invested at 8 percent per year, compounded quarterly, what will the value of the account be at the end of 2 years?

Since there are 4 quarters per year, $n = 2(4) = 8$.

Since the rate is 8% per year, $r = \frac{0.08}{4} = 0.02$.

$A_8 = 1000(1 + 0.02)^8$ *Use the formula $A_n = P(1 + r)^n$, where $n = 8$, $P = 1000$, and $r = 0.02$.*

$= 1000(1.02)^8$

$= 1171.659381$ The account will be worth $1171.66 to the nearest cent.

1. If $500 is invested at 7 percent per year, compounded annually, how much will the account be worth at the end of 3 years? $612.52
2. If $2000 is invested at 6 percent per year, compounded quarterly, how much will the account be worth at the end of 4 years? $2537.97
3. If $45,000 is invested at 7.5 percent per year, compounded semiannually, how much will the account be worth at the end of 5 years? $65,026.98
4. How long would it take $100 to double if it is invested at 10 percent per year, compounded annually? 7.27 yr

Teacher's Resource Book

Practice—Chapter 9, p. 3

Enrichment—Chapter 9, p. 4

LESSON PLAN

Vocabulary
Arithmetic series
Index of summation
Limits of summation
Series
Sigma notation
Summation notation

Materials/Manipulatives
Calculators

BACKGROUND

In the Preview, repetitive addition is used to solve problems involving arithmetic sequences.

9.3 Arithmetic Series

Objectives: To use sigma notation to represent the sum of a series
To develop and use formulas to find a partial sum of an arithmetic series

It is sometimes necessary to find the sum of the terms of a finite arithmetic sequence.

Preview

The bells in a tower chime as many times as the hour. From 1 AM to 12 noon, inclusive, how many times do they chime?

First, write the finite arithmetic sequence.

1, 2, 3, 4, 5, 6, 7, 8, 9, 10, 11, 12

Then add the terms of the sequence.

$1 + 2 + 3 + 4 + 5 + 6 + 7 + 8 + 9 + 10 + 11 + 12 = 78$

The bells chime 78 times from 1 AM to 12 noon, inclusive.

The front row of a triangular flower bed has 24 plants. Each successive row contains 3 more plants than the row in front of it. How many plants are there in each row?

1. 4th row 33 **2.** 6th row 39 **3.** 8th row 45 **4.** 10th row 51

5. What is the total number of plants in the first 8 rows? in the first 10 rows? 276; 375

A **series** is the indicated sum of the terms of a sequence. The indicated sum $2 + 4 + 6 + 8$ is a finite series. The indicated sum $2 + 4 + 6 + 8 + \cdots$ is an infinite series. An **arithmetic series** is the indicated sum of an arithmetic sequence. The sum of the first n terms of an arithmetic series, denoted S_n, is a **partial sum** of the series. The sum of the first five terms, S_5, of the series $2 + 4 + 6 + 8 + \cdots$ is 30, since $2 + 4 + 6 + 8 + 10 = 30$.

Sigma (Σ) notation, or **summation notation,** is a shorthand method for representing the sum of a series. The general term of a series is the same as the general term of the corresponding sequence. For the series $2 + 4 + 6 + 8 + \cdots$, the general term, a_n, is $2n$. Therefore,

$$S_5 = \sum_{n=1}^{5} 2n$$ *Read $\sum_{n=1}^{5} 2n$ as: the sum from n = 1 to 5 of 2n.*

The variable n is called the **index of summation,** the numbers 1 and 5 are called the **limits of summation,** S_5 represents the sum of the first five terms, and $2n$ is the general term. Any letter can represent the index of summation.

EXAMPLE 1 **Write $\sum_{j=3}^{6}(2j + 1)$ in expanded form and then find the sum.**

Replace j successively with 3, 4, 5, and 6. Then add.

$$\sum_{j=3}^{6}(2j + 1) = [2(3) + 1] + [2(4) + 1] + [2(5) + 1] + [2(6) + 1]$$
$$= 7 + 9 + 11 + 13 = 40$$

To express in sigma notation a partial sum given in expanded form, first find the common difference and the general term.

EXAMPLE 2 **Express in sigma notation the sum of the first six terms of the arithmetic series $-2 + (-6) + (-10) + (-14) + \cdots$.**

First, find the common difference.

$d = -6 - (-2) = -4$ *Use the formula $d = a_{n+1} - a_n$, where $a_{n+1} = -6$ and $a_n = -2$.*

Then, find the general term.

$$\begin{aligned} a_n &= -2 + (n - 1)(-4) \\ &= -2 - 4n + 4 \\ &= -2(2n - 1) \end{aligned}$$

Use the formula $a_n = a_1 + (n - 1)d$, where $a_1 = -2$ and $d = -4$.

Then, $S_6 = \sum_{n=1}^{6} -2(2n - 1)$.

If a series has a large number of terms, it is not convenient to find the sum by addition. To find a general formula for the partial sum of an arithmetic series of n terms, write S_n in two ways, as shown below, and add.

$$\begin{aligned} S_n &= a_1 + (a_1 + d) + (a_1 + 2d) + \cdots + (a_n - d) + a_n \\ S_n &= a_n + (a_n - d) + (a_n - 2d) + \cdots + (a_1 + d) + a_1 \\ \hline 2S_n &= (a_1 + a_n) + (a_1 + a_n) + (a_1 + a_n) + \cdots + (a_1 + a_n) + (a_1 + a_n) \\ 2S_n &= n(a_1 + a_n) \\ S_n &= \frac{n(a_1 + a_n)}{2} \end{aligned}$$

Since S_n has n terms, there are n terms of the form $(a_1 + a_n)$ in $2S_n$.

EXAMPLE 3 **Find S_{500}, the sum of the first 500 terms of an arithmetic series, if $a_1 = 2$ and $a_{500} = 1000$.**

$$\begin{aligned} S_{500} &= \frac{500(2 + 1000)}{2} \\ &= 250{,}500 \end{aligned}$$

Use the formula $S_n = \frac{n(a_1 + a_n)}{2}$, where $n = 500$, $a_1 = 2$, and $a_n = 1000$.

TEACHING SUGGESTIONS

- Remind students that there are two formulas for the sum of an arithmetic series and that they should use the formula which corresponds to the given information.
- Emphasize that when a series is written in sigma notation, the sum formula $\left(S_n = \frac{n(a_1 + a_n)}{2}\right)$ can be used since the first and last terms can be found easily.
- A calculator is helpful for finding partial sums.

Critical Thinking

Comparing-Contrasting Ask students to compare and contrast an arithmetic sequence and an arithmetic series. When comparing an arithmetic sequence with an arithmetic series, students should note that the sequence and the series contain the same terms and that they have the same general term. When contrasting an arithmetic sequence with an arithmetic series, students should note that a sequence is a listing of terms separated by commas, whereas a series is a sum of terms separated by addition signs.

CHALKBOARD EXAMPLES

- **For Example 1**

  Write each series in expanded form and then find the sum.

1. $\sum_{n=0}^{4} 2n$

   $\sum_{n=0}^{4} 2n = 2(0) + 2(1) + 2(2) + 2(3) + 2(4) = 20$

2. $\sum_{r=3}^{8} -r$

   $\sum_{r=3}^{8} -r = (-3) + (-4) + (-5) + (-6) + (-7) + (-8) = -33$

- **For Example 2**

  Express in sigma notation the sum of the first seven terms of each arithmetic series.

  3. $-3 + (-9) + (-15) + (-21) + \cdots$ $\sum_{n=1}^{7} -3(2n-1)$

  4. $8 + 12 + 16 + 20 + \cdots$ $\sum_{n=1}^{7} 8\left(\frac{1}{2}n + \frac{1}{2}\right)$

- **For Example 3**

  Find the indicated partial sum for each arithmetic series.

  5. $a_1 = 2$, $a_{100} = 1200$; S_{100} 60,100

  6. $a_1 = 5$, $a_{50} = 400$; S_{50} 10,125

- **For Example 4**

  Find the indicated partial sum for each arithmetic series.

  7. $\sum_{n=1}^{8} (2n-1)$ 64

  8. $\sum_{n=1}^{9} (2n+3)$ 117

- **For Example 5**

  Find the indicated partial sum for each arithmetic series.

  9. $5 + 20 + 35 + 50 + \cdots$; S_{20} 2950

  10. $(-4) + (-8) + (-12) + (-16) + \cdots$; S_{15} -480

- **For Example 6**

  Find the first five terms of each arithmetic series.

  11. $a_1 = 5$, $a_n = 73$, $S_n = 702$, $n = 18$, $d = 4$ 5, 9, 13, 17, 21

  12. $a_1 = 9$, $a_n = 163$, $S_n = 1978$, $n = 23$, $d = 7$ 9, 16, 23, 30, 37

You can also use the formula $S_n = \frac{n(a_1 + a_n)}{2}$ to find the partial sum of a series expressed in sigma notation, without writing it in expanded form.

EXAMPLE 4 **Find the sum of the arithmetic series $\sum_{n=1}^{7} (5n - 2)$.**

$a_1 = 5(1) - 2 = 3$ and $a_7 = 5(7) - 2 = 33$ *Find a_1 and a_7.*

$S_7 = \frac{7(3 + 33)}{2} = 126$ $\quad S_n = \frac{n(a_1 + a_n)}{2}$

An alternate formula for S_n can be obtained by substituting $a_1 + (n-1)d$ for a_n. This alternate formula is useful when the nth term is not given.

$$S_n = \frac{n(a_1 + a_n)}{2} = \frac{n[a_1 + a_1 + (n-1)d]}{2} = \frac{n[2a_1 + (n-1)d]}{2}$$

EXAMPLE 5 **Find the sum of the first 50 terms of the arithmetic series $10 + 85 + 160 + 235 + \cdots$.**

$d = a_{n+1} - a_n = 85 - 10 = 75$ *Find d.*

$S_{50} = \frac{50[2(10) + (50-1)(75)]}{2}$ $\quad S_n = \frac{n[2a_1 + (n-1)d]}{2}$

$= 92{,}375$

The sum of the first 50 terms is 92,375.

Specific terms of an arithmetic series can be found if the values of a_1, a_n, and S_n are known.

EXAMPLE 6 **Find the first three terms of the arithmetic series in which $a_1 = 5$, $a_n = 100$, and $S_n = 1050$.**

$1050 = \frac{n(5 + 100)}{2}$ *Solve $S_n = \frac{n(a_1 + a_n)}{2}$ for n.*

$n = \frac{2(1050)}{5 + 100} = 20$

Substitute $n = 20$ in $S_n = \frac{n[2a_1 + (n-1)d]}{2}$ and solve for d.

$$1050 = \frac{20[2(5) + (20-1)d]}{2}$$
$$1050 = 10(10 + 19d)$$
$$105 = 10 + 19d \quad \textit{Divide both sides by 10.}$$
$$5 = d$$

The first three terms are 5, 10, and 15.

The formulas for the partial sum of an arithmetic series are summarized below.

> The partial sum, S_n, of an arithmetic series is given by either of the following formulas:
>
> $$S_n = \frac{n(a_1 + a_n)}{2} \quad \text{or} \quad S_n = \frac{n[2a_1 + (n - 1)d]}{2}$$
>
> where n is the number of terms, a_1 is the first term, d is the common difference, and a_n is the nth term.

CLASS EXERCISES

Write in expanded form and then find the sum.

1. $\sum_{i=10}^{15} i$ 75
2. $\sum_{i=6}^{12} i$ 63
3. $\sum_{n=3}^{7} 2n$ 50
4. $\sum_{n=1}^{5} 3n$ 45
5. $\sum_{m=1}^{4} 2^m$ 30
6. $\sum_{m=0}^{3} 3^m$ 40
7. $\sum_{k=4}^{6} (k + 3)$ 24
8. $\sum_{k=4}^{7} (k - 2)$ 14
9. $\sum_{y=5}^{6} (y - 4)$ 3
10. $\sum_{y=3}^{4} (y + 6)$ 19
11. $\sum_{x=1}^{3} (2x - 3)$ 3
12. $\sum_{x=2}^{5} (-3x + 9)$ -6
13. Express in sigma notation the partial sum from $n = 1$ to $n = 5$ of the arithmetic series $7 + 10 + 13 + 16 + \cdots$. $\sum_{n=1}^{5} (3n + 4)$

Consider the series $\sum_{n=1}^{6} (2n - 1)$. Find the value of each partial sum.

14. S_1 1
15. S_2 4
16. S_3 9
17. S_4 16
18. S_5 25
19. S_6 36

PRACTICE EXERCISES

Write each arithmetic series in expanded form, and then find its sum.

A

1. $\sum_{k=1}^{6} 2k$ 42
2. $\sum_{k=1}^{4} 3k$ 30
3. $\sum_{j=2}^{4} (j + 3)$ 18
4. $\sum_{j=3}^{7} (2j - 1)$ 45
5. $\sum_{k=7}^{9} (k + 4)$ 36
6. $\sum_{k=9}^{13} (2k - 3)$ 95
7. $\sum_{n=4}^{6} 2n + \frac{1}{2}$ 30.5
8. $\sum_{n=5}^{8} 4n + \frac{1}{2}$ 104.5

Common Error

- Some students will try to use a sum formula that does not correspond to given information for the arithmetic series. Provide practice in which students are to identify the sum formula that is most appropriate for the stated conditions.
- See *Teacher's Resource Book* for additional remediation.

LESSON FOLLOW-UP

Discussion

Express $\sum_{x=1}^{4} x^2 + 6x + 9$ using a change of index. Since $x^2 + 6x + 9 = (x + 3)^2$, let $p = x + 3$. Therefore, $\sum_{x=1}^{4} x^2 + 6x + 9$ can be expressed as $\sum_{p=4}^{7} p^2$.

Assignment Guide

See p. 370B for assignments.

Extra

An example showing that an expression written in summation notation may often be written in different forms simply by changing the index is considered.

Lesson Quiz

1. Write $\sum_{k=3}^{7}(k+4)$ in expanded form then find its sum. $\sum_{k=3}^{7}(k+4) = 7+8+9+10+11 = 45$
2. Find S_{11} for the arithmetic series $a_1 = 4$ and $d = -3.2$ -132
3. Find the sum of the first 200 terms of an arithmetic series if $a_1 = 7$ and $a_{200} = 987$. 99,400
4. Find a_1 if $S_{40} = 8020$ and $d = 5$. 103
5. Find S_n for the series $\sum_{n=1}^{8}(2n-3)$. 48
6. A baseball 13-run pool starts at \$25. If no team scores exactly 13 runs, the kitty increases by \$15 each week. How many weeks would it take to have a kitty of \$280? 18 wk

Additional Answers

45. The series is arithmetic with $a_1 = 1$ and $d = 1$, so $S_n = \frac{n[2a_1 + (n-1)d]}{2} = \frac{n[2\cdot 1 + (n-1)1]}{2} = \frac{n(2+n-1)}{2} = \frac{n(n+1)}{2}$

46. The series is arithmetic with $a_1 = 2$ and $d = 2$, so $S_n = \frac{n[2a_1 + (n-1)d]}{2} = \frac{n[2\cdot 2 + (n-1)2]}{2} = \frac{n(4+2n-2)}{2} = \frac{n(2n+2)}{2} = \frac{2n(n+1)}{2} = n(n+1)$

47. The series is arithmetic with $a_1 = 1$ and $d = 2$, so $S_n = \frac{n[2a_1 + (n-1)d]}{2} = \frac{n[2\cdot 1 + (n-1)2]}{2} = \frac{n(2+2n-2)}{2} = \frac{n(2n)}{2} = n^2$

Find the indicated partial sum for each arithmetic series.

9. $a_1 = 2, a_{200} = 200; S_{200}$ 20,200
10. $a_1 = 5, a_{30} = 100; S_{30}$ 1575
11. $a_1 = 1, a_{10} = -20; S_{10}$ -95
12. $a_1 = 2, a_{20} = -40; S_{20}$ -380
13. $a_1 = 73, d = -3; S_{15}$ 780
14. $a_1 = 91, d = -4; S_{12}$ 828
15. $a_1 = -6, d = 0.2; S_{10}$ -51
16. $a_1 = 12, d = 0.4; S_6$ 78
17. $9 + 11 + 13 + \cdots; S_{10}$ 180
18. $-9 + (-6) + (-3) + \cdots; S_{16}$ 216
19. $-11 + (-15) + (-19) + \cdots; S_{12}$ -396
20. $3 + 9 + 15 + \cdots; S_{30}$ 2700
21. $\sum_{n=1}^{10} 4n$ 220
22. $\sum_{n=1}^{10}(2n+4)$ 150
23. $\sum_{j=1}^{6}(j-7)$ -21
24. $\sum_{j=1}^{10}(30-j)$ 245

Express in sigma notation the sum of the first six terms of each arithmetic sequence.

B

25. $3 + 7 + 11 + 15 + \cdots$ $\sum_{n=1}^{6}(4n-1)$
26. $-5 + (-3) + (-1) + 1 + \cdots$ $\sum_{n=1}^{6}(2n-7)$
27. $8 + 3 + (-2) + (-7) + \cdots$ $\sum_{n=1}^{6}(-5n+13)$
28. $-1 + (-8) + (-15) + \cdots$ $\sum_{n=1}^{6}(-7n+6)$

Find the indicated term for each arithmetic sequence.

29. $a_1 = -2, S_{10} = 205; a_{10}$ 43
30. $a_1 = -4, S_{15} = 885; a_{15}$ 122
31. $a_1 = 10, S_{50} = -12{,}975; a_{50}$ -529
32. $a_1 = -12, S_{60} = -7800; a_{60}$ -248
33. $S_{20} = 40, d = 4; a_{20}$ 40
34. $S_{30} = 120, d = 6; a_{30}$ 91
35. $S_{100} = -200, d = -2; a_{100}$ -101
36. $S_{200} = -400, d = -4; a_{200}$ -400

Find the first three terms of each arithmetic series.

37. $a_1 = 6, a_n = 306, S_n = 1716$ 6, 36, 66
38. $a_1 = 7, a_n = 139, S_n = 876$ 7, 19, 31
39. $n = 25, a_{25} = 156, S_{25} = 2175$ 18, 23.75, 29.5
40. $n = 14, a_{14} = 199, S_{14} = 1239$ $-22, -5, 12$

C

41. For an arithmetic series with $a_n = -23$, $d = -4$, and $S_n = -72$, find a_1 and n. $a_1 = 5$; $n = 8$
42. For an arithmetic series with $a_n = 45$, $d = 5$, and $S_n = 210$, find a_1 and n. $a_1 = 15$; $n = 7$, or $a_1 = -10$; $n = 12$
43. For an arithmetic series with $a_1 = -11$, $S_n = -32$, and the fourth term is -5, find d and n. $d = 2$; $n = 4$ or $n = 8$
44. For an arithmetic series with $a_1 = 6$, $S_n = -120$, and the third term is -6, find d and n. $d = -6$; $n = 8$
45. Prove: $1 + 2 + 3 + \cdots + n = \frac{n(n+1)}{2}$ See side column.

48. The series is arithmetic with $a_1 = 4$ and $d = 4$, so $S_n = \frac{n[2a_1 + (n-1)d]}{2} = \frac{n[2\cdot 4 + (n-1)4]}{2} = \frac{4[2n + n(n-1)]}{2} = 4\cdot\frac{n^2+n}{2}$. From problem 45, $\frac{n^2+n}{2} = \sum_{j=1}^{n} j$ so $4\cdot\frac{n^2+n}{2} = 4\sum_{j=1}^{n} j$

46. Prove: $2 + 4 + 6 + \cdots + 2n = n(n + 1)$ See side column.

47. Prove: $1 + 3 + 5 + \cdots + (2n - 1) = n^2$

48. Prove: $\sum_{j=1}^{n} 4j = 4 \sum_{j=1}^{n} j$

Applications

49. **Building Design** The balcony of a theater has 12 rows of seats. The last row contains 8 seats, and each of the other rows contains one more seat than the row behind it. How many seats are there in the balcony? 162

50. **Sports** During the first hour of her climb, a mountain climber ascends 675 ft. Each hour after that, she ascends 50 ft less than in the preceding hour. When will she be 4500 ft above her starting point? 10 h, 18 h

51. **Business** A contractor agreed that if a job was not done by a certain date, he would pay a \$1000 penalty for the first day of delay, and for each day after that, he would pay \$50 more than for the preceding day. How many penalty days did he use if his penalty was \$10,800? 9

52. **Number Theory** Find the sum of the integers from 1 through 100. 5050

EXTRA

An expression given in sigma notation may often be written in a different form by changing the index. Consider the expression $\sum_{k=1}^{5}(3k + 7)$.

$$3k + 7 = 3k + 6 + 1 = 3(k + 2) + 1$$

Let $i = k + 2$. Then, the general term $3k + 7$ can be written $3i + 1$.
If the index is changed to i, the limits of summation must also be changed.

For $k = 1$, $3(1) + 7 = 3i + 1$; $9 = 3i$; $3 = i$

For $k = 5$, $3(5) + 7 = 3i + 1$; $22 = 3i + 1$; $7 = i$

Thus, the series $\sum_{k=1}^{5}(3k + 7)$ may be expressed as $\sum_{i=3}^{7}(3i + 1)$.

Write each expression in a different form, using the index j, where $j = k + 4$.

1. $\sum_{k=1}^{4}(2k + 8)$ $\sum_{j=5}^{8} 2j$
2. $\sum_{k=1}^{6}(k + 5)$ $\sum_{j=5}^{10} (j + 1)$
3. $\sum_{k=3}^{8}(k^2 + 8k + 16)$ $\sum_{j=7}^{12} j^2$

Teacher's Resource Book
Practice—Chapter 9, p. 5
Enrichment—Chapter 9, p. 6

LESSON PLAN

Vocabulary
Geometric series

Materials/Manipulatives
Calculators

BACKGROUND

In the Preview, repetitive addition is used to solve problems involving geometric sequences.

9.4 Geometric Series

Objectives: To develop and use formulas to find a partial sum of a geometric series
To determine whether or not a series is arithmetic or geometric

A geometric series can be formed from a geometric sequence, just as an arithmetic series can be formed from an arithmetic sequence.

Preview

In a large corporation, there are five levels of management: President, Vice President, Regional Vice President, Regional Manager, and Local Manager. There is 1 president, and there are 4 vice presidents, 16 regional vice presidents, and so on. The sequence is geometric. How many people are on the management team?

First, write the geometric sequence.

$$1, 4, 16, 64, 256$$

Then, add the terms of the sequence.

$$1 + 4 + 16 + 64 + 256 = 341$$

There are 341 people on the management team.

Steven will receive a 5 percent salary increase at the end of each year. His beginning salary is $10,000. How much will he earn during each year?

1. 2nd year $10,500 **2.** 3rd year $11,025 **3.** 4th year $11,576.25 **4.** 5th year $12,155.06 **5.** 6th year $12,762.82

6. How much money will he earn in all during 6 years on the job? $68,019.13

A **geometric series** is the indicated sum of the terms of a geometric sequence. The series $1 + 3 + 9 + 27 + 81 + 243$ is finite. The series $1 + 4 + 16 + \cdots$ is infinite. A partial sum, S_n, of a geometric series is the sum of the first n terms of the series.

$$S_n = a_1 + a_1r + a_1r^2 + \cdots + a_1r^{n-1}$$

To find a formula for the partial sum of a geometric series, multiply S_n by r and subtract the resulting product from S_n.

$$\begin{array}{rcl} S_n &=& a_1 + a_1r + a_1r^2 + \cdots + a_1r^{n-1} \\ rS_n &=& \quad\; a_1r + a_1r^2 + \cdots + a_1r^{n-1} + a_1r^n \\ \hline S_n - rS_n &=& a_1 + 0 + 0 + \cdots + 0 - a_1r^n \end{array}$$

$$S_n(1 - r) = a_1 - a_1r^n$$

$$S_n = \frac{a_1 - a_1r^n}{1 - r}, r \neq 1 \qquad \textit{Divide each side by } 1 - r.$$

$$S_n = \frac{a_1(1 - r^n)}{1 - r}, r \neq 1 \qquad \textit{Factor.}$$

EXAMPLE 1 **Find the sum of the first six terms of the geometric series $2 + 8 + 32 + 128 + \cdots$.**

$$S_6 = \frac{2(1 - 4^6)}{1 - 4} = 2730 \qquad \textit{Use } S_n = \frac{a_1(1 - r^n)}{1 - r}, n = 6, a_1 = 2, \textit{and } r = 4.$$

You can find the first term if you know one partial sum and the common ratio.

EXAMPLE 2 **For a geometric series, find a_1 if $S_{10} = -2046$ and $r = 2$.**

$$-2046 = \frac{a_1(1 - 2^{10})}{1 - 2} \qquad S_n = \frac{a_1(1 - r^n)}{1 - r}$$

$$2046 = a_1(1 - 2^{10}) \qquad \textit{Multiply both sides by } -1.$$

$$\frac{2046}{1 - 2^{10}} = a_1 \qquad \textit{Calculation-ready form}$$

$$-2 = a_1$$

Recall that the general term of a geometric sequence is $a_n = a_1r^{n-1}$. Multiplying both sides of this equation by r yields $ra_n = a_1r^n$. This new equation can be used, along with algebraic manipulation, to find an alternate formula for S_n. If $r \neq 1$, then

$$S_n = \frac{a_1(1 - r^n)}{1 - r} = \frac{a_1 - a_1r^n}{1 - r} = \frac{a_1 - ra_n}{1 - r} \qquad \textit{Substitute } ra_n \textit{ for } a_1r^n.$$

This formula is useful when the first and nth terms are known.

EXAMPLE 3 **Find the sum of the first four terms of a geometric series, where $a_1 = 343$, $a_4 = -1$, and $r = -\frac{1}{7}$.**

$$S_4 = \frac{343 - \left(-\frac{1}{7}\right)(-1)}{1 - \left(-\frac{1}{7}\right)} = \frac{343 - \frac{1}{7}}{\frac{8}{7}} = \frac{2401 - 1}{8} = 300 \qquad \textit{Use } S_n = \frac{a_1 - ra_n}{1 - r}.$$

TEACHING SUGGESTIONS

- Before developing the general formulas, it might be wise to do several numerical examples.
- Emphasize that $r \neq 1$ in the formulas since the formulas are undefined for that particular value.
- Stress that the given information dictates which sum formula to use for a geometric series.
- A calculator is helpful for finding the partial sum of a geometric series.

Critical Thinking

Analysis Ask students to explain whether the sum of the geometric series $1 + 2 + 4 + \cdots + (2)^{n-1}$ can be determined. Students should note that the sum of the series cannot be determined because there is no largest value of n.

CHALKBOARD EXAMPLES

- **For Example 1**

  Find the sum of the first seven terms of each geometric series.

  1. 3, 12, 48, . . . 16,383

  2. 0.2, 1, 5, 25, . . . 3906.2

- **For Example 2**

  For each geometric series, find the value of a_1.

  3. $S_6 = 2730$, $r = 4$ 2

  4. $S_8 = 22{,}960$, $r = 3$ 7

- **For Example 3**

  Find the sum of the first five terms of each geometric series.

  5. $a_1 = 256$, $a_5 = 16$, $r = -\frac{1}{2}$ 176

  6. $a_1 = 729$, $a_5 = 9$, $r = -\frac{1}{3}$ 549

- **For Example 4**

7. Find $\sum_{n=1}^{6} (3n + 1)$. 69

8. Find $\sum_{n=1}^{8} (2n + 5)$. 112

- **For Example 5**

Express in sigma notation the sum of the first eight terms of each geometric series.

9. $5 + (-15) + 45 + (-135) + \cdots$ $\sum_{n=1}^{8} 5(-3)^{n-1}$

10. $48 + 24 + 12 + \cdots$ $\sum_{n=1}^{8} 48\left(\frac{1}{2}\right)^{n-1}$

Geometric series can also be written using sigma notation. For example, the finite series $6 + 12 + 24 + 48 + 96 + 192 + 384 + 768$ can be written $\sum_{n=1}^{8} 6(2)^{n-1}$. In this series, $a_1 = 6$, $r = 2$, and $n = 8$. Before you can find the sum of a series given in sigma notation, you must first determine whether it is an arithmetic or a geometric series.

EXAMPLE 4 **Find $\sum_{n=1}^{5} (4n - 1)$.**

$$\sum_{n=1}^{5} (4n - 1) = [4(1) - 1] + [4(2) - 1] + [4(3) - 1] + [4(4) - 1] + [4(5) - 1]$$
$$= 3 + 7 + 11 + 15 + 19$$

This series is arithmetic, since the sequence 3, 7, 11, 15, 19 is arithmetic. Each succeeding term is found by adding 4 to the previous term. Because this is an arithmetic series, use $S_n = \frac{n(a_1 + a_n)}{2}$ to find the sum.

$$S_5 = \frac{5(3 + 19)}{2} \qquad n = 5, a_1 = 3, a_n = 19$$
$$= 55$$

If a geometric series is given in expanded form, it can be expressed in sigma notation. First, find the common ratio and the general term.

EXAMPLE 5 **Express in sigma notation the sum of the first eight terms of the geometric series $7 + 14 + 28 + 56 + \cdots$.**

First, find the common ratio.

$$r = \frac{14}{7} = 2 \qquad \text{Use } a_{n+1} = 14 \text{ and } a_n = 7.$$

Then, find the general term.

$$a_n = 7(2)^{n-1} \qquad a_1 = 7, r = 2$$

Then, $S_8 = \sum_{n=1}^{8} 7(2)^{n-1}$

It is possible to find one partial sum of a geometric series, given another partial sum and the common ratio. To do this, apply the formula for the partial sum twice.

EXAMPLE 6 **Find S_{10} for the geometric sequence in which $S_6 = 728$ and $r = 3$.**

$728 = \frac{a_1(1 - 3^6)}{1 - 3}$ *Use $S_n = \frac{a_1(1 - r^n)}{1 - r}$ to find a_1; $n = 6$, $S_n = 728$, and $r = 3$.*

$\frac{-1456}{1 - 3^6} = a_1$

$2 = a_1$

$S_{10} = \frac{2(1 - 3^{10})}{1 - 3} = 59{,}048$ *Use the same formula to find S_{10}; $n = 10$, $a_1 = 2$, and $r = 3$.*

The partial sum, S_n, of a geometric series is given by either of the following formulas:

$$S_n = \frac{a_1(1 - r^n)}{1 - r} \quad r \neq 1 \qquad \text{or} \qquad S_n = \frac{a_1 - ra_n}{1 - r} \quad r \neq 1$$

where n is the number of terms, a_1 is the first term, r is the common ratio, and a_n is the nth term.

CLASS EXERCISES

Consider the geometric series $1 + 2 + 4 + 8 + \cdots$. Find each value.

1. r 2
2. S_1 1
3. S_2 3
4. S_4 15
5. S_8 255
6. S_{12} 4095

Find r for each geometric series. Then, find the indicated partial sum.

7. $3 + 6 + 12 + \cdots$; S_4 2; 45
8. $10 + 100 + 1000 + \cdots$; S_5 10; 111,110
9. $-32 + 16 + (-8) + \cdots$; S_6 $-\frac{1}{2}$; -21
10. $-81 + 27 + (-9) + \cdots$; S_4 $-\frac{1}{3}$; -60

PRACTICE EXERCISES

Find the indicated partial sum for each geometric series.

A

1. $a_1 = 6, r = 4$; S_4 510
2. $a_1 = 625, r = \frac{2}{5}$; S_5 1031
3. $a_1 = 16, r = -\frac{1}{2}$; S_5 11
4. $a_1 = 243, r = -\frac{1}{3}$; S_6 182
5. $a_1 = 12, a_5 = 972, r = 3$; S_5 1452
6. $a_1 = 4, a_6 = -972, r = -3$; S_6 -728
7. $a_1 = -4, a_6 = \frac{1}{8}, r = -\frac{1}{2}$; S_6 -2.625
8. $a_1 = 125, a_5 = \frac{1}{25}, r = -\frac{1}{5}$; S_5 104.2

- **For Example 6**
  Find the indicated partial sum for each geometric series.
  11. $S_4 = 765$, $a_1 = 9$, $r = 4$; S_6 12,285
  12. $S_3 = 186$, $a_1 = 6$, $r = 5$; S_8 585,936

Common Error

- Some students confuse the formulas used with an arithmetic series with those used for a geometric series. Be sure to emphasize their similarities and differences.
- See *Teacher's Resource Book* for additional remediation.

LESSON FOLLOW-UP

Discussion

Express the series $1 + 3 + 9 + 27 + 81 + 243$ in base-3 notation and then find the sum using base-3 arithmetic. 11111; 364

Assignment Guide

See p. 370B for assignments.

Test Yourself

See *Teacher's Resource Book, Tests*, pp. 85–86.

Lesson Quiz

Consider the geometric series 48 + 24 + 12 + 6 + 3.

Find:

1. r $\frac{1}{2}$
2. S_2 72
3. S_4 90
4. For a geometric series, find S_7 if $a_1 = 128$ and $r = -\frac{1}{2}$. 86
5. Find S_{10} for the geometric series $2 + 6 + 18 + \cdots$. 59,048
6. Find the value of a_1 for the geometric series in which $a_n = 450$, $r = 5$, and $S_n = 562.464$. 0.144
7. John is promised an 8% annual return on an investment. If his initial deposit is \$8500, how much will he have, to the nearest dollar, at the end of 5 yr? \$12,489

Enrichment Problem

If $-1 < x < 1$ and y is a very large positive integer, estimate the value of x^y. 0

9. $a_1 = 2, a_5 = \frac{1}{8}, r = -\frac{1}{2}$; S_5 1.375

10. $a_1 = -343, a_4 = 1, r = -\frac{1}{7}$; S_4 -300

11. $1 + 2 + 4 + \cdots$; S_{10} 1023

12. $1875 + 375 + 75 + \cdots$; S_5 2343

13. $2 + (-6) + 18 + \cdots$; S_8 -3280

14. $3 + (-6) + 12 + \cdots$; S_7 129

15. $4 + 4 + 4 + \cdots$; S_{15} 60

16. $8 + 8 + 8 + \cdots$; S_{13} 104

17. $1024 + 256 + 64 + \cdots$; S_8 1365.3125

18. $2 + 8 + 32 + \cdots$; S_9 174, 762

19. $\sum_{n=1}^{4} 2^n$ 30

20. $\sum_{n=1}^{4} 2(3)^n$ 240

21. $\sum_{n=1}^{5} 4^{n-1}$ 341

22. $\sum_{n=1}^{5} 2^{n+2}$ 248

For each geometric series, find the value of a_1.

B

23. $a_n = 420, r = \frac{3}{2}, S_n = 1270$ -5

24. $a_n = 243, r = 3, S_n = 302$ 125

25. $S_n = -172, r = -2, n = 7$ -4

26. $S_n = -1785, r = -2, n = 8$ 21

Determine whether each series is arithmetic or geometric and find its sum.

27. $\sum_{n=1}^{6} (3n + 5)$ arithmetic; 93

28. $\sum_{n=1}^{4} (6n - 2)$ arithmetic; 52

29. $\sum_{k=1}^{8} 4^k$ geometric; 87, 380

30. $\sum_{n=1}^{6} 2(2)^{n+2}$ geometric; 1008

Express in sigma notation the sum of the first 8 terms of each geometric series.

31. $3 + 1 + \frac{1}{3} + \frac{1}{9} + \cdots$ $\sum_{k=1}^{8} 3\left(\frac{1}{3}\right)^{k-1}$

32. $\frac{1}{16} + \frac{1}{4} + 1 + 4 + \cdots$ $\sum_{n=1}^{8} \frac{1}{16}(4)^{n-1}$

33. $2 + (-10) + 50 + (-250) + \cdots$ $\sum_{k=1}^{8} 2(-5)^{k-1}$

34. $120 + (-60) + 30 + (-15) + \cdots$ $\sum_{n=1}^{8} 120\left(-\frac{1}{2}\right)^{n-1}$

Find the indicated partial sum for each geometric series.

35. $S_8 = 2040, r = 2$; S_6 504

36. $S_6 = 63, r = -\frac{1}{2}$; S_3 72

37. $S_4 = 240, r = 3$; S_7 6558

38. $S_5 = -110, r = -2$; S_8 850

Find the missing values for each geometric series.

C

39. If $a_1 = 3$, $a_n = 192$, and $S_n = 129$, find r and n. $r = -2$; $n = 7$

40. If $a_1 = 6$, $r = 2$, and $a_n = 3072$, find S_n and n. $S_n = 6138$; $n = 10$

Find the indicated partial sum.

41. $x + 3x^3 + 9x^5 + \cdots$; S_6 $\frac{x - 729x^{13}}{1 - 3x^2}$

42. $xy + x^2y^4 + x^3y^7 + \cdots$; S_6 $\frac{xy - x^7y^{19}}{1 - xy^3}$

43. $\sum_{n=1}^{6} y(x^2)^{n+1}$ $\frac{y(x^4 - x^{16})}{1 - x^2}$

44. $\sum_{n=1}^{5} y^2(x^3)^n$ $\frac{y^2(x^3 - x^{18})}{1 - x^3}$

Applications

45. Construction A pile driver is used to drive a pipe into the ground. The first blow drives the pipe down 20 cm. If each succeeding blow drives the pipe down 0.4 times as far as the preceding blow, how far has the pipe been driven after 4 blows? 32.48 cm

46. Entertainment A club sponsor needs to contact all of the club members by telephone to inform them of a change in plans. The sponsor calls 2 members, each of whom calls 2 other members not previously called, and so on. If all members have been contacted after this process has been repeated 6 times, how many members (excluding the sponsor) are in the club? 126

47. Finance An investment in a gas well earned a total of \$54,450 during the first 5 yr. Each year after the first, the investment earned 3 times as much as during the preceding year. How much did the investment earn during the first year? during the fifth year? first year: \$450; fifth year: \$36,450

TEST YOURSELF

1. Find a_{16} for an arithmetic sequence in which $a_1 = 120$ and $d = -\frac{1}{3}$. 115 **9.1**

2. List the next five terms of the arithmetic sequence 100, 93, 86, ... 79, 72, 65, 58, 51

3. Insert four arithmetic means between -3 and 22. 2, 7, 12, 17

4. Find a_7 for a geometric sequence in which $a_1 = 4$ and $r = -2$. 256 **9.2**

5. List the next five terms of the geometric sequence 1000, 200, 40, ... $8, \frac{8}{5}, \frac{8}{25}, \frac{8}{125}, \frac{8}{625}$

6. Insert two geometric means between 81 and 3. 27, 9

7. Find S_{15} for an arithmetic series in which $a_1 = 8$ and $d = 6$. 750 **9.3**

8. Find a_{12} for an arithmetic series in which $a_1 = -6$ and $S_{12} = 60$. 16

9. Express in sigma notation the sum of the first five terms of the arithmetic series $24 + 21 + 18 + \cdots$. $\sum_{n=1}^{5}(-3n + 27)$

Find the indicated partial sum for each geometric series.

10. $a_1 = 10, r = 2; S_{20}$ 10, 485, 750

11. $a_1 = 80, a_5 = 5; S_5$ 155 or 55 **9.4**

Determine whether each series is arithmetic or geometric, and find its sum.

12. $\sum_{n=1}^{7}(4n - 2)$ arithmetic; 98

13. $\sum_{n=1}^{6} 4^{n-2}$ geometric; 341.25

14. Express in sigma notation the sum of the first nine terms of the geometric series $\frac{1}{2} + \left(-\frac{3}{2}\right) + \frac{9}{2} + \cdots$. $\sum_{k=1}^{9}\frac{1}{2}(-3)^{k-1}$

Teacher's Resource Book

Practice—Chapter 9, p. 7
Enrichment—Chapter 9, p. 8

LESSON PLAN

Vocabulary
Converge
Diverge
Infinite geometric series
Limit

Materials/Manipulatives
Calculators

BACKGROUND

In the Preview, repetitive addition is used to solve a problem involving an infinite geometric series.

9.5 Infinite Geometric Series

Objectives: To develop and use a formula to find the sum of a convergent geometric series
To express repeating decimals as rational numbers in fraction form

You have learned how to compute the sum of a *finite* number of terms of a geometric series. However, there are many applications for which you will need to evaluate the sum of an *infinite* number of terms.

Preview

A tennis ball is tossed upward from the ground to a height of 25 ft. Each bounce is 0.8 as high as the previous bounce. The first four heights reached by the ball form the geometric sequence 25, 20, 16, 12.8. If the ball were to bounce indefinitely, the sequence would be infinite. If the sequence is infinite, it is not possible to find the total distance traveled by the ball simply by adding terms. However, you can look at some partial sums.

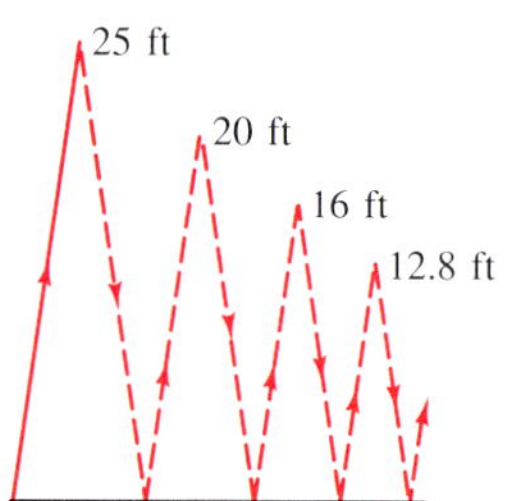

Before it bounces once, the ball travels up 25 ft and down 25 ft. Before it bounces a second time, it travels up 25(0.8) ft and down 25(0.8) ft. Let S_1 be the distance traveled before it bounces once, S_2 be the distance traveled before it bounces twice, and so on. Then,

$$S_1 = 2[25(0.8)^0] = 50$$
$$S_2 = 2[25(0.8)^0 + 25(0.8)^1] = 90$$
$$S_3 = 2[25(0.8)^0 + 25(0.8)^1 + 25(0.8)^2] = 122$$

How far has the ball traveled before it bounces the given number of times? *Hint*: Use $S_n = \frac{a_1(1 - r^n)}{1 - r}$ and then multiply by 2.

1. 5 times 168 ft
2. 10 times 223 ft
3. 15 times 241 ft
4. 25 times 249 ft
5. 50 times 250 ft
6. Do you think the distance traveled will approach a certain number as the number of bounces increases indefinitely? yes; sum approaches 250

In this lesson, a formula will be developed that will enable you to find the sum of certain types of infinite geometric series.

EXAMPLE 1 **Find the partial sums S_1, S_2, S_3, S_8, S_9, S_{10}, and S_n of the infinite geometric series $1 + \frac{1}{3} + \frac{1}{9} + \cdots$. Round to five decimal places.**

$$r = \frac{a_{n+1}}{a_n} = \frac{\frac{1}{3}}{1} = \frac{1}{3} \qquad \textit{Find } r.$$

$$S_1 = 1$$

$$S_2 = 1 + \frac{1}{3} = 1.33333$$

$$S_3 = 1 + \frac{1}{3} + \frac{1}{9} = 1.44444$$

$$S_8 = 1 + \frac{1}{3} + \frac{1}{9} + \cdots + \frac{1}{2187} = 1.49977 \qquad a_n = a_1 r^{n-1}$$

$$S_9 = 1 + \frac{1}{3} + \frac{1}{9} + \cdots + \frac{1}{6561} = 1.49992 \qquad S_n = \frac{a_1(1 - r^n)}{1 - r}$$

$$S_{10} = 1 + \frac{1}{3} + \frac{1}{9} + \cdots + \frac{1}{19{,}683} = 1.49997$$

$$S_n = 1 + \frac{1}{3} + \frac{1}{9} + \cdots + \left(\frac{1}{3}\right)^{n-1} = \frac{1\left[1 - \left(\frac{1}{3}\right)^n\right]}{1 - \frac{1}{3}} = \frac{3}{2} - \frac{3}{2}\left(\frac{1}{3}\right)^n$$

This graph represents the partial sums in Example 1. The horizontal line at 1.5 is an *asymptote*. As n gets very large, S_n approaches 1.5 as a **limit.** Since the series has a limit, it is said to *converge*, or to be *convergent*. To see why S_n approaches the limit 1.5, consider the nth partial sum

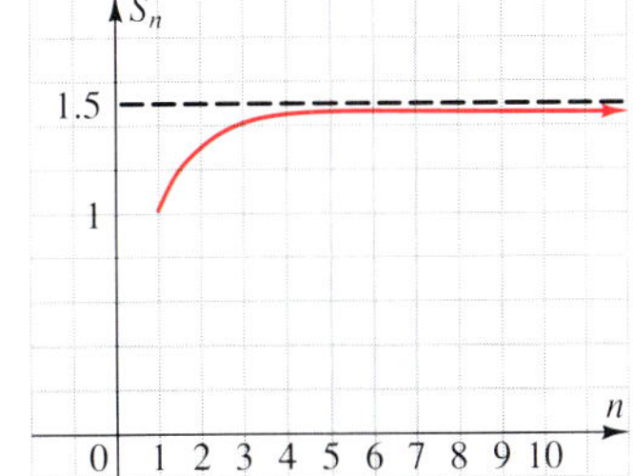

$$S_n = \frac{3}{2} - \frac{3}{2}\left(\frac{1}{3}\right)^n$$

As n gets very large, $\frac{3}{2}\left(\frac{1}{3}\right)^n$ gets very close to zero, so S_n approaches $\frac{3}{2}$, or 1.5.

In general, an infinite geometric series **converges** if the sequence of partial sums approaches a limit as n increases without bound. It can be shown that an infinite geometric series converges if and only if $|r| < 1$, where r is the common ratio. This is true because if $|r| < 1$, then r^n approaches zero as n approaches infinity.

An infinite geometric series can be expressed in sigma notation, with the upper limit infinity (∞). The series in Example 1 can be written

$$\sum_{n=1}^{\infty} \left(\frac{1}{3}\right)^{n-1}$$

TEACHING SUGGESTIONS

- As an introduction, a ruler with $\frac{1}{16}$ divisions could be used to represent this series: $4 + 2 + 1 + \frac{1}{2} + \frac{1}{4} + \frac{1}{8} + \frac{1}{16} + \cdots$.
- Stress that a convergent infinite geometric series has a common ratio between -1 and 1 and that a divergent infinite series has a common ratio less than or equal to -1 or greater than or equal to 1.
- You may wish to compare the traditional method of changing repeating decimals to rational numbers in fraction form with the infinite geometric series method.
- Calculators are especially helpful for finding the sums of infinite series.

Critical Thinking

Analysis Ask students if the following reasoning is correct:

$$A = 2^0 + 2^1 + 2^2 + 2^3 + \cdots$$
$$= 2^0 + 2(1 + 2^1 + 2^2 + \cdots)$$
$$= 2^0 + 2\left(\frac{1}{1 - 2}\right)$$
$$= 1 - 2$$
$$A = -1$$

Students should note that $r = 2$ in the series $1 + 2^2 + 2^3 + \cdots$ and that the series is divergent and therefore has no sum.

CHALKBOARD EXAMPLES

- **For Example 1**

1. Find the partial sums S_1, S_2, S_8, S_9, and S_n of the infinite geometric series $2 + 1 + \frac{1}{2} + \cdots$. Round answers to five decimal places.
$S_1 = 2$, $S_2 = 3$, $S_8 = 3.98438$
$S_9 = 3.99219$, $S_n = 4 - 4(0.5)^n$
2. Find the partial sums S_1, S_3, S_5, and S_n of the infinite geometric series $-\frac{1}{2} + \frac{1}{4} + \left(-\frac{1}{8}\right) + \cdots$. Round answers to five decimal places. $S_1 = -\frac{1}{2}$, $S_3 = -0.375$, $S_5 = -0.34375$, $S_n = -\frac{1}{3} + \frac{1}{3}(-0.5)^n$

- **For Example 2**

3. Find the partial sums S_1, S_3, S_4, and S_n for the infinite geometric series $\sum_{n=1}^{\infty} 3^n$. $S_1 = 3$, $S_3 = 39$
$S_4 = 120$, $S_n = -\frac{3}{2} + \frac{3}{2}(3)^n$
4. Find the partial sums S_1, S_2, S_5, and S_n for the infinite geometric series $\sum_{n=1}^{\infty} 3^{1-n}$. $S_1 = 1$, $S_2 = \frac{4}{3}$, $S_5 = \frac{121}{81}$, $S_n = \frac{3}{2} - \frac{3}{2}\left(\frac{1}{3}\right)^n$

EXAMPLE 2 **Find the partial sums S_1, S_2, S_3, S_4, S_5, and S_n for the infinite geometric series $\sum_{n=1}^{\infty} 2^n$.**

$$\sum_{n=1}^{\infty} 2^n = 2 + 4 + 8 + \cdots$$

$$r = \frac{a_{n+1}}{a_n} = \frac{4}{2} = 2$$

$$S_1 = 2$$
$$S_2 = 2 + 4 = 6$$
$$S_3 = 2 + 4 + 8 = 14$$
$$S_4 = 2 + 4 + 8 + 16 = 30$$
$$S_5 = 2 + 4 + 8 + 16 + 32 = 62$$

$$S_n = 2 + 4 + 8 + \cdots + 2(2)^{n-1} = \frac{2(1 - 2^n)}{1 - 2} = 2(2^n - 1)$$

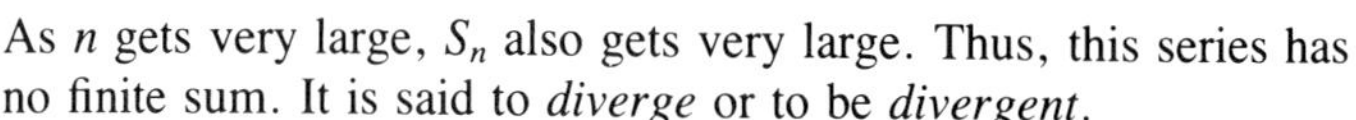

As n gets very large, S_n also gets very large. Thus, this series has no finite sum. It is said to *diverge* or to be *divergent*.

A series **diverges** if the sequence of partial sums increases or decreases without bound as n increases without bound. It can be shown that an infinite geometric series diverges if and only if $|r| \geq 1$, where r is the common ratio.

The **sum of an infinite geometric series** in which the sequence of partial sums converges to a limit S is equal to S. To develop a formula for the sum of a convergent infinite geometric series, look at the formula for the partial sum S_n.

$$S_n = \frac{a_1(1 - r^n)}{1 - r}$$

Since the series converges, $|r| < 1$. Thus, as n gets very large, r^n approaches zero as a limit. Thus,

$$S_n = \frac{a_1(1 - r^n)}{1 - r} \quad \text{approaches} \quad \frac{a_1(1 - 0)}{1 - r} = \frac{a_1}{1 - r}$$

The sum S of a convergent infinite geometric series is given by

$$S = \frac{a_1}{1 - r} \qquad |r| < 1$$

where a_1 is the first term and r is the common ratio.

EXAMPLE 3 **Find the sum of each infinite geometric series, if it exists:**

a. $\sum_{n=1}^{\infty} 4^{3-n}$ **b.** $\sum_{n=1}^{\infty} 4^{n-2}$

a. $\sum_{n=1}^{\infty} 4^{3-n} = 16 + 4 + 1 + \frac{1}{4} + \ldots$ $a_1 = 16, r = \frac{1}{4}$

Since $|r| < 1$, the series converges and the sum exists.

$$S = \frac{16}{1 - \frac{1}{4}} = \frac{64}{3}$$ *Use the formula* $S = \frac{a_1}{1 - r}$.

b. $\sum_{n=1}^{\infty} 4^{n-2} = \frac{1}{4} + 1 + 4 + 16 + \cdots$ $a_1 = \frac{1}{4}, r = 4$

Since $|r| > 1$, the series diverges and the sum does not exist.

A repeating decimal can be thought of as an infinite geometric series. Therefore, you can use the formula for the sum of an infinite geometric series to write repeating decimals as rational numbers in fraction form.

EXAMPLE 4 **Write $0.\overline{6}$ (or 0.666 . . .) as a rational number in fraction form.**

$0.\overline{6} = 0.6 + 0.06 + 0.006 + \cdots$ *Write $0.\overline{6}$ as an infinite geometric series.*

$a_1 = 0.6$ $r = 0.1$ Since $|r| < 1$, the series converges.

$$S = \frac{0.6}{1 - 0.1} = \frac{0.6}{0.9} = \frac{2}{3}$$ Therefore, $0.\overline{6} = \frac{2}{3}$.

CLASS EXERCISES

Tell whether each geometric series converges or diverges.

1. $1 + 5 + 25 + \cdots$ diverges

2. $1 + 6 + 36 + \cdots$ diverges

3. $1 + \frac{1}{5} + \frac{1}{25} + \cdots$ converges

4. $1 + \frac{1}{6} + \frac{1}{36} + \cdots$ converges

5. $48 - 36 + 27 - \cdots$ converges

6. $80 - 64 + 51.2 - \cdots$ converges

7. $-1 - 1 - 1 - \cdots$ diverges

8. $1 - 1 + 1 - \cdots$ diverges

Express each repeating decimal as a rational number in fraction form.

9. $0.\overline{3}$ $\frac{1}{3}$

10. $0.\overline{24}$ $\frac{8}{33}$

11. $0.2\overline{4}$ $\frac{11}{45}$

12. $0.\overline{032}$ $\frac{32}{999}$

13. Use the formula for the sum of an infinite geometric series to find the total distance traveled by the ball in the Preview problem on page 396. 250 ft

- **For Example 3**
  Find the sum of the each infinite series, if it exists.

  5. $\sum_{n=1}^{\infty} (-2)^{2-n}$ $-\frac{4}{3}$

  6. $\sum_{n=1}^{\infty} (-2)^{n-2}$ does not exist

- **For Example 4**
  Express each repeating decimal as a rational number in fraction form.

  7. $0.\overline{7}$ $\frac{7}{9}$

  8. $0.\overline{157}$ $\frac{157}{999}$

Common Error

- Some students confuse convergent geometric series and divergent geometric series. Emphasize that in a convergent series, $|r| < 1$ and in a divergent series $|r| \geq 1$.
- See *Teacher's Resource Book* for additional remediation.

LESSON FOLLOW-UP

Critical Thinking

Discovering patterns Ask students to determine if the series $\frac{1}{2} + \frac{1}{6} + \frac{1}{12} + \frac{1}{20} + \frac{1}{30} + \cdots$ is geometric, and if so to find a rule to represent the nth term. Writing the terms as $\frac{1}{1 \cdot 2} + \frac{1}{2 \cdot 3} + \frac{1}{3 \cdot 4} + \cdots + \frac{1}{n(n+1)}$, students should note that the series does not have a common ratio and therefore is not geometric.

Assignment Guide

See p. 370B for assignments.

Historical Note

This feature provides an historic look at famous 17th–19th century mathematicians and their contributions to the idea of limit, the foundation for calculus.

Lesson Quiz

State whether each geometric series converges or diverges.

1. $\frac{1}{16} + \frac{1}{8} + \frac{1}{4} + \cdots$ diverges
2. $400 + 100 + 25 + \cdots$ converges

Determine the sum S_n, if it exists, for each geometric series. Indicate if the sum does not exist.

3. $1 + \frac{1}{8} + \frac{1}{64} + \cdots$ $\frac{8}{7}$
4. $0.1 + 0.3 + 0.9 + \cdots$ does not exist
5. Write $0.\overline{18}$ as a rational number. $\frac{2}{11}$
6. If $a_1 = 20$ and $S_n = 80$, find r. $\frac{3}{4}$

Enrichment

Express $0.\overline{bba}$ as a common fraction. $N = \frac{bba}{999}$

PRACTICE EXERCISES

Determine the sum S, if it exists, for each geometric series. If the sum does not exist, so state.

A

1. $1 + \frac{1}{7} + \frac{1}{49} + \cdots$ $\frac{7}{6}$
2. $1 + \frac{2}{7} + \frac{4}{49} + \cdots$ $\frac{7}{5}$
3. $16 + 8 + 4 + \cdots$ 32
4. $20 + 10 + 5 + \cdots$ 40
5. $2 + 2 + 2 + \cdots$ does not exist
6. $2 - 2 + 2 - \cdots$ does not exist
7. $4 + 5 + \frac{25}{4} + \cdots$ does not exist
8. $3 + 4 + \frac{16}{3} + \cdots$ does not exist
9. $-7 + 1 - \frac{1}{7} + \cdots$ $-\frac{49}{8}$
10. $-10 + 1 - 0.1 + \cdots$ $-\frac{100}{11}$
11. $0.8 + 0.4 + 0.2 + \cdots$ $\frac{8}{5}$
12. $0.2 + 0.4 + 0.8 + \cdots$ does not exist

Write each repeating decimal as a rational number in fraction form.

13. $0.\overline{4}$ $\frac{4}{9}$
14. $0.\overline{8}$ $\frac{8}{9}$
15. $0.\overline{21}$ $\frac{7}{33}$
16. $0.\overline{12}$ $\frac{4}{33}$
17. $0.\overline{123}$ $\frac{41}{333}$
18. $0.\overline{456}$ $\frac{152}{333}$

Determine the sum S, if it exists, for each geometric series. If the sum does not exist, so state.

B

19. $\sum_{n=1}^{\infty} \left(\frac{1}{2}\right)^{n-1}$ 2
20. $\sum_{n=1}^{\infty} \left(\frac{1}{3}\right)^{n}$ $\frac{1}{2}$
21. $\sum_{n=1}^{\infty} \frac{3}{4}(3)^{n-1}$ does not exist
22. $\sum_{n=1}^{\infty} \frac{1}{2}(2)^{n+1}$ does not exist

Write each repeating decimal as a rational number in fraction form.

23. $0.4\overline{1}$ $\frac{37}{90}$
24. $0.4\overline{23}$ $\frac{419}{990}$
25. $5.\overline{27}$ $\frac{58}{11}$
26. $6.\overline{37}$ $\frac{631}{99}$

Find the specified value for each infinite geometric series.

27. $a_1 = 12, S = 30$; r $\frac{3}{5}$
28. $a_1 = 40, S = 200$; r $\frac{4}{5}$
29. $r = -0.2, S = -50$; a_1 -60
30. $r = -0.5, S = -28$; a_1 -42

Determine whether each statement is true or false. If it is false, give a counterexample (a specific case in which the statement does not hold).

C

31. If the terms of an infinite geometric series alternate between positive and negative, then the series must be divergent. false; Answers may vary.
32. If the terms of an infinite geometric series are all positive, then the series must be convergent. false; Answers may vary.
33. If the sum of an infinite geometric series is twice the first term, then the common ratio is $\frac{1}{2}$. true
34. If the common ratio of an infinite geometric series is 0.9999, then the series converges. true
35. If the sum of an infinite geometric series is positive, all the terms must be positive. false; Answers may vary.

36. If the sum of an infinite geometric series is negative, all the terms must be negative. false; Answers may vary.

Applications

37. Physics The end of the pendulum of a clock travels 40 cm on its first swing. On each succeeding swing, it travels 0.8 as far as it did on the preceding one. How far will the end of the pendulum travel before coming to rest? 200 cm

38. Physics A broken metronome travels 20 in. on its first swing. On each succeeding swing it travels $\frac{7}{8}$ as far as it did on the preceding one. How far will the metronome travel before coming to rest? 160 in.

39. Physics A ball tossed up 22 ft rebounds to $\frac{9}{10}$ of the height from which it falls on each bounce. What is the total distance (up and down) it will travel before coming to rest? 440 ft

40. Physics After its release, a weather balloon rises 100 ft in the first minute. Each minute thereafter, the balloon rises only 80 percent as far as in the previous minute. What is the maximum height the balloon will reach? 500 ft

41. Geometry Triangle *ABC* is equilateral with each side of length 40 cm. The midpoints of the sides are joined to form a second equilateral triangle *DEF*, and so on. If this midpoint-joining process continues without end, find the sum of the perimeters of all the triangles formed. 240 cm

HISTORICAL NOTE

Limits It took several centuries to formulate precisely the concept of a limit, upon which calculus is based. A number of the best mathematical minds of each century worked with the idea before an acceptable definition was devised.

Century	Mathematicians	Contributions
17th	Isaac Newton	Developed and used infinite series
	Gottfried Wilhelm Leibniz	Invented convenient symbolism and developed rules of procedure
18th	Leonhard Euler Joseph Louis Lagrange	Tried to clarify basic limit ideas
19th	Augustin-Louis Cauchy	Developed somewhat satisfactory definitions of limits
	Karl Weierstrass	Gave more precise definitions of the limit of a function and the limit of a series

Read about one of the men listed in the table and report on his additional accomplishments in mathematics and other fields. Answers may vary.

Teacher's Resource Book
Practice—Chapter 9, p. 9
Enrichment—Chapter 9, p. 10

LESSON PLAN

Vocabulary
Eüler's formula
Exponential function
Power series

Materials/Manipulatives
Calculators

BACKGROUND

In the Preview, simplifying factorials is discussed. This skill is necessary to evaluate power series.

9.6 Power Series and Trigonometric Functions

Objectives: To define power series
To estimate values of trigonometric functions using power series

Work with power series requires the simplification of expressions written in factorial notation.

Preview

Recall that $n!$ is read *n factorial* and means $n(n - 1) \ldots 3 \cdot 2 \cdot 1$. Therefore, $6! = 6 \cdot 5 \cdot 4 \cdot 3 \cdot 2 \cdot 1 = 720$. Note that $0! = 1$ by definition.

EXAMPLE **Simplify:** $\frac{10!}{3!\,5!}$

$$\frac{10!}{3!\,5!} = \frac{10 \cdot 9 \cdot 8 \cdot 7 \cdot 6 \cdot 5\,!}{3!\,5!} = \frac{10 \cdot 9 \cdot 8 \cdot 7 \cdot 6}{3 \cdot 2} = 10 \cdot 9 \cdot 8 \cdot 7 = 5040$$

Simplify each expression.

1. $\frac{8!}{4!}$ 1680 **2.** $\frac{12!}{6!}$ 665,280 **3.** $\frac{5!\,6!}{4!}$ 3600 **4.** $\frac{n!}{(n - 1)!}$ n **5.** $\frac{(n + 1)!}{n!}$ $n + 1$

The series previously considered had constant terms. Now consider series with variable terms, such as $1 + x + x^2 + \cdots + x^n + \cdots$. By substituting different values for x, a whole family of series can be generated. This series is an example of a *power series*. Expressions of the following form are called **power series** in x, where x is a variable and $a_0, a_1, a_2, \ldots, a_n, \ldots$ are constants.

$$\sum_{n=0}^{\infty} a_n x^n = a_0x^0 + a_1x^1 + a_2x^2 + \cdots + a_{n-1}x^{n-1} + \cdots$$

Power series can be used to represent certain functions, such as $\sin x$ and $\cos x$, where x is in *radians*. In fact, calculators and computers use power series to evaluate such functions.

$$\sin x = x - \frac{x^3}{3!} + \frac{x^5}{5!} - \frac{x^7}{7!} + \cdots \qquad \cos x = 1 - \frac{x^2}{2!} + \frac{x^4}{4!} - \frac{x^6}{6!} + \cdots$$

The sum of the first six terms of either expansion will provide a reasonable approximation, and it is helpful to use a calculator to find that sum. However, there are times when six terms are not enough to give a sufficiently precise approximation, and then more terms must be used.

EXAMPLE 1 **Approximate each value using the first six terms of the appropriate power series expansion. Round each answer to four decimal places.** **a.** $\cos\frac{\pi}{6}$ **b.** $\sin 23°$

a. $\cos\frac{\pi}{6} = 1 - \frac{\left(\frac{\pi}{6}\right)^2}{2!} + \frac{\left(\frac{\pi}{6}\right)^4}{4!} - \frac{\left(\frac{\pi}{6}\right)^6}{6!} + \frac{\left(\frac{\pi}{6}\right)^8}{8!} - \frac{\left(\frac{\pi}{6}\right)^{10}}{10!}$ *$\cos x = 1 - \frac{x^2}{2!} + \frac{x^4}{4!} - \cdots$*

$= 1 - \frac{\left(\frac{\pi}{6}\right)^2}{2} + \frac{\left(\frac{\pi}{6}\right)^4}{24} - \frac{\left(\frac{\pi}{6}\right)^6}{720} + \frac{\left(\frac{\pi}{6}\right)^8}{40320} - \frac{\left(\frac{\pi}{6}\right)^{10}}{3628800}$

$= 0.8660$ Use the cosine key on your calculator to verify that $\cos\frac{\pi}{6} = 0.8660$, correct to four decimal places.

b. $23° = 0.4014$ radians, to four decimal places   *Express 23° in radians.*

$\sin 0.4014 = 0.4014 - \frac{(0.4014)^3}{3!} + \frac{(0.4014)^5}{5!} - \frac{(0.4014)^7}{7!} + \frac{(0.4014)^9}{9!} - \frac{(0.4014)^{11}}{11!}$

$\sin x = x - \frac{x^3}{3!} + \frac{x^5}{5!} - \frac{x^7}{7!} + \cdots$

$= 0.4014 - \frac{(0.4014)^3}{6} + \frac{(0.4014)^5}{120} - \frac{(0.4014)^7}{5040} + \frac{(0.4014)^9}{362880} - \frac{(0.4014)^{11}}{39916800}$

$= 0.3907$ Use the sine key on your calculator to verify that $\sin 23° = 0.3907$, correct to four decimal places.

An important function in mathematics is the *exponential function*, e^x. Recall that e is the irrational number that is used as the base for natural logarithms. It can be shown that for all x

$$e^x = 1 + \frac{x}{1!} + \frac{x^2}{2!} + \frac{x^3}{3!} + \cdots = \sum_{n=0}^{\infty} \frac{x^n}{n!}$$

EXAMPLE 2 **Using $x = 1$, approximate the value of e using the first six terms of the power series for e^x. Round your answer to three decimal places.**

$e^1 = 1 + \frac{1}{1!} + \frac{1^2}{2!} + \frac{1^3}{3!} + \frac{1^4}{4!} + \frac{1^5}{5!}$ *$e^x = 1 + \frac{x}{1!} + \frac{x^2}{2!} + \frac{x^3}{3!} + \frac{x^4}{4!} + \frac{x^5}{5!}$*

$e = 1 + 1 + \frac{1}{2} + \frac{1}{6} + \frac{1}{24} + \frac{1}{120} = 2.717$

Use a calculator to show that the value of e to three decimal places is 2.718. The difference is due to the fact that only six terms are used.

TEACHING SUGGESTIONS

- Emphasize that each value of the variable in a power series generates a unique series.
- Point out that the arithmetic when evaluating a power series is very tedious, and therefore a calculator is almost a necessity.
- Emphasize that the number of terms used in a power series determines the degree of accuracy and that using a finite number of terms may produce a very poor approximation.
- Stress that angle values used in a power series must be in radians.

Critical Thinking

Analysis Ask students why the power series for sine and cosine are expressed in radians and to use the fact that $1° = \frac{\pi}{180}$ to express one of the series in degrees.
Answers may vary.

CHALKBOARD EXAMPLES

- **For Example 1**
  Approximate each value using the first four terms of the appropriate power series. Round each answer to four decimal places.

  1. $\sin\frac{\pi}{2}$ 0.9998
  2. $\cos 1°$ 0.9998
  3. $\cos 23°$ 0.9205

- **For Example 2**
  Approximate each value using the first six terms of the power series for e^x. Round each answer to three decimal places.

  4. $e^{\frac{3}{2}}$ 4.398
  5. $e^{\frac{4}{5}}$ 2.220

- **For Example 3**

  Find the exact value of each expression using Eüler's formula

  6. $e^{-\frac{3\pi i}{2}}$ i

  7. $e^{2\pi i}$ 1

- **For Example 4**

  8. Prove: $\cos x = \dfrac{e^{ix} + e^{-ix}}{2}$.

$\cos x = \dfrac{e^{ix} + e^{-ix}}{2}$

$\left|\begin{array}{l} \dfrac{\cos x + i\sin x + \cos x - i\sin x}{2} \\ \dfrac{2\cos x}{2} \end{array}\right.$

$\cos x = \cos x$

Common Error

- Students often confuse the power series for sine with that for cosine. Emphasize the differences between them.
- See *Teacher's Resource Book* for additional remediation.

LESSON FOLLOW-UP

Discussion

Evaluate $\sqrt[3]{e}$ using the first four terms of a power series.

$\sqrt[3]{e} = e^{\frac{1}{3}}$

$= 1 + \frac{1}{3} + \frac{1}{18} + \frac{1}{162}$

$= 1 + 0.\overline{3} + 0.0\overline{5} + 0.0061$

$= 1.3949$

Assignment Guide

See p. 370B for assignments.

The Swiss mathematician Leonhard Euler (1707–1783) discovered a formula relating the sine, cosine, and exponential functions. **Euler's formula** is derived by replacing x with ix in the series for e^x, where $i = \sqrt{-1}$.

$$e^{ix} = 1 + ix + \frac{(ix)^2}{2!} + \frac{(ix)^3}{3!} + \frac{(ix)^4}{4!} + \frac{(ix)^5}{5!} + \cdots + \frac{(ix)^{n-1}}{(n-1)!} + \cdots$$

$$= 1 + ix + \frac{i^2x^2}{2!} + \frac{i^3x^3}{3!} + \frac{i^4x^4}{4!} + \frac{i^5x^5}{5!} + \cdots + \frac{i^{n-1}x^{n-1}}{(n-1)!} + \cdots$$

$$= 1 + ix - \frac{x^2}{2!} - i\frac{x^3}{3!} + \frac{x^4}{4!} + i\frac{x^5}{5!} - \frac{x^6}{6!} - i\frac{x^7}{7!} + \cdots \qquad i^2 = -1, i^3 = -i, i^4 = 1,$$

$$= \left(1 - \frac{x^2}{2!} + \frac{x^4}{4!} - \frac{x^6}{6!} + \cdots\right) + i\left(x - \frac{x^3}{3!} + \frac{x^5}{5!} - \frac{x^7}{7!} + \cdots\right) \qquad \textit{Group real and imaginary terms.}$$

$$= \cos x + i\sin x \qquad 1 - \frac{x^2}{2!} + \frac{x^4}{4!} - \cdots = \cos x;\ x - \frac{x^3}{3!} + \frac{x^5}{5!} - \cdots = \sin x$$

Thus, Euler's formula is $e^{ix} = \cos x + i\sin x$.

EXAMPLE 3 **Find the exact value of $e^{\frac{3\pi i}{2}}$ using Euler's formula.**

$$e^{\frac{3\pi i}{2}} = \cos\frac{3\pi}{2} + i\sin\frac{3\pi}{2} = 0 + i(-1) = -i \qquad e^{ix} = \cos x + i\sin x$$

Another identity can be derived by substituting $-x$ for x in Euler's formula.

$$e^{i(-x)} = \cos(-x) + i\sin(-x) = \cos x + (-i\sin x) \qquad \cos(-x) = \cos x;$$
$$e^{-ix} = \cos x - i\sin x \qquad \sin(-x) = -\sin x$$

Euler's formula and this identity can be used to prove other identities.

EXAMPLE 4 **Prove: $\sin x = \dfrac{e^{ix} - e^{-ix}}{2i}$**

$$\sin x \;\left|\; \begin{array}{l} \dfrac{e^{ix} - e^{-ix}}{2i} \\ \dfrac{\cos x + i\sin x - (\cos x - i\sin x)}{2i} \\ \dfrac{2i\sin x}{2i} \end{array}\right.$$

$$= \sin x \qquad \text{Thus, } \sin x = \frac{e^{ix} - e^{-ix}}{2i}$$

CLASS EXERCISES

Use the appropriate power series to write the first six terms needed to estimate each value. Do not evaluate the sum. See side column.

1. $\cos\frac{\pi}{7}$ **2.** $\cos 3.6$ **3.** $\sin 2.4$ **4.** $\sin\frac{2\pi}{5}$ **5.** $e^{1.3}$ **6.** e^{π}

Additional Answers

1. $1 - \dfrac{\left(\frac{\pi}{7}\right)^2}{2!} + \dfrac{\left(\frac{\pi}{7}\right)^4}{4!} - \dfrac{\left(\frac{\pi}{7}\right)^6}{6!} + \dfrac{\left(\frac{\pi}{7}\right)^8}{8!} - \dfrac{\left(\frac{\pi}{7}\right)^{10}}{10!}$

2. $1 - \dfrac{(3.6)^2}{2!} + \dfrac{(3.6)^4}{4!} - \dfrac{(3.6)^6}{6!} + \dfrac{(3.6)^8}{8!} - \dfrac{(3.6)^{10}}{10!}$

3. $2.4 - \dfrac{(2.4)^3}{3!} + \dfrac{(2.4)^5}{5!} - \dfrac{(2.4)^7}{7!} + \dfrac{(2.4)^9}{9!} - \dfrac{(2.4)^{11}}{11!}$

4. $\dfrac{2\pi}{5} - \dfrac{\left(\frac{2\pi}{5}\right)^3}{3!} + \dfrac{\left(\frac{2\pi}{5}\right)^5}{5!} - \dfrac{\left(\frac{2\pi}{5}\right)^7}{7!} + \dfrac{\left(\frac{2\pi}{5}\right)^9}{9!} - \dfrac{\left(\frac{2\pi}{5}\right)^{11}}{11!}$

5. $1 + \dfrac{1.3}{1!} + \dfrac{(1.3)^2}{2!} + \dfrac{(1.3)^3}{3!} + \dfrac{(1.3)^4}{4!} + \dfrac{(1.3)^5}{5!}$

6. $1 + \dfrac{\pi}{1!} + \dfrac{\pi^2}{2!} + \dfrac{\pi^3}{3!} + \dfrac{\pi^4}{4!} + \dfrac{\pi^5}{5!}$

Find the exact value of each expression using Euler's formula.

7. $e^{2\pi i}$ 1 8. $e^{-\pi i}$ −1 9. $e^{\frac{\pi i}{4}}$ $\frac{\sqrt{2}}{2} + \frac{i\sqrt{2}}{2}$

PRACTICE EXERCISES

Approximate each value using the first six terms of the appropriate power series. Round each answer to four decimal places.

A 1. $\sin \frac{\pi}{4}$ 0.7071 2. $\cos \frac{\pi}{4}$ 0.7071 3. $\cos \frac{\pi}{6}$ 0.8660 4. $\sin \pi$ −0.0004

5. $\cos \frac{2\pi}{3}$ −0.5000 6. $\sin \frac{3\pi}{4}$ 0.7071 7. $\sin 1.5$ 0.9975 8. $\cos 0.5$ 0.8776

Approximate each value using the first six terms of the power series for e^x. Round each answer to three decimal places.

9. $e^{0.1}$ 1.105 10. $e^{0.2}$ 1.221 11. $e^{\frac{1}{3}}$ 1.396 12. $e^{\frac{1}{2}}$ 1.649 13. $e^{-\frac{1}{2}}$ 0.607 14. e^{-1} 0.367

Find the exact value of each expression using Euler's formula.

15. $e^{\pi i}$ −1 16. $e^{\frac{\pi i}{2}}$ i 17. $e^{\frac{3\pi i}{4}}$ $-\frac{\sqrt{2}}{2} + \frac{i\sqrt{2}}{2}$ 18. $e^{3\pi i}$ −1 19. $e^{-4\pi i}$ 1 20. $e^{-\frac{7\pi i}{4}}$ $\frac{\sqrt{2}}{2} + \frac{i\sqrt{2}}{2}$

Approximate each value using the first six terms of the appropriate power series. Remember to change degree measure to radian measure. Round each answer to four decimal places.

B 21. $\cos 180°$ −1.0018 22. $\sin 180°$ −0.0004 23. $\sin 112°$ 0.9272 24. $\cos 124°$ −0.5592

25. $\cos 246°$ −0.4810 26. $\sin 240°$ −0.8841 27. $\sin 162°$ 0.3089 28. $\cos 176°$ −0.9990

Prove each identity. See page 493.

29. $\cos x = \frac{e^{ix} + e^{-ix}}{2}$ 30. $\tan x = \frac{e^{ix} - e^{-ix}}{i(e^{ix} + e^{-ix})}$ 31. $\csc x = \frac{2i}{e^{ix} - e^{-ix}}$

32. $\sec x = \frac{2}{e^{ix} + e^{-ix}}$ 33. $\cot x = \frac{i(e^{ix} + e^{-ix})}{e^{ix} - e^{-ix}}$ 34. $e^{(\pi + x)i} = -e^{ix}$

C 35. Prove: $\frac{e^{ix}}{e^{iy}} = e^{(x-y)i}$

36. Use Euler's formula and double-angle formulas to show that $(e^{ix})^2 = e^{2ix}$.

37. Justify the following statement using your calculator to substitute values for n: $\left(1 + \frac{1}{n}\right)^n$ approaches e as n gets larger and larger.

Challenge

Leonard Eüler discovered the unique number e, which can be used as a log base.

Lesson Quiz

1. Use six terms of the power series for sine to evaluate $\sin \frac{\pi}{3}$ to three decimal places. 0.866
2. Use six terms of the power series for e^x to evaluate $e^{2.3}$ to three decimal places. 9.675
3. Evaluate $e^{3\pi i}$ using Eüler's formula. −1
4. Use six terms of the power series for cosine to evaluate cos 84° to four decimal places. 1.1045

Enrichment

If $z < 1$ find a power series expansion for $\frac{1}{1-z}$. $1 + z + z^2 + z^3 + \cdots$

Teacher's Resource Book
Practice—Chapter 9, p. 11
Enrichment—Chapter 9, p. 12

41. $e^{\frac{-x^2}{2}} = 1 + \left(\frac{-x^2}{2}\right) + \frac{\left(-\frac{x^2}{2}\right)^2}{2!} + \frac{\left(-\frac{x^2}{2}\right)^3}{3!} + \frac{\left(-\frac{x^2}{2}\right)^4}{4!} + \frac{\left(-\frac{x^2}{2}\right)^5}{5!} = 1 - \frac{x^2}{2} + \frac{\left(\frac{x^4}{4}\right)}{2} + \frac{\left(-\frac{x^6}{8}\right)}{6} + \frac{\left(\frac{x^8}{16}\right)}{24} + \frac{\left(-\frac{x^{10}}{32}\right)}{120} = 1 - \frac{x^2}{2} + \frac{x^4}{8} - \frac{x^6}{48} + \frac{x^8}{384} - \frac{x^{10}}{3840}$; Therefore, $y = \frac{1}{\sqrt{2\pi}} e^{\frac{-x^2}{2}} = \frac{1}{\sqrt{2\pi}}\left(1 - \frac{x^2}{2} + \frac{x^4}{8} - \frac{x^6}{48} + \frac{x^8}{384} - \frac{x^{10}}{3840}\right)$

See page 493.

38. Show that $\sin^2 x + \cos^2 x = 1$, using $\sin x = \frac{e^{ix} - e^{-ix}}{2i}$ and $\cos x = \frac{e^{ix} + e^{-ix}}{2}$.

39. Show that $1 + \cot^2 x = \csc^2 x$, using the expressions for $\cot x$ and $\csc x$ given in Exercises 31 and 33.

40. Show that $1 + \tan^2 x = \sec^2 x$, using the expressions for $\tan x$ and $\sec x$ given in Exercises 30 and 32.

Applications

Statistics The *standard normal probability curve* $y = \frac{1}{\sqrt{2\pi}} e^{-\frac{x^2}{2}}$ is of great importance in the field of statistics.

41. Show that $y = \frac{1}{\sqrt{2\pi}} e^{-\frac{x^2}{2}}$ can be approximated by this equation

$$y = \frac{1}{\sqrt{2\pi}}\left(1 - \frac{x^2}{2} + \frac{x^4}{8} - \frac{x^6}{48} + \frac{x^8}{384} - \frac{x^{10}}{3840}\right)$$ See side column.

Use the equation in Exercise 41 to find the y-coordinate of each point. Round to the nearest hundredth.

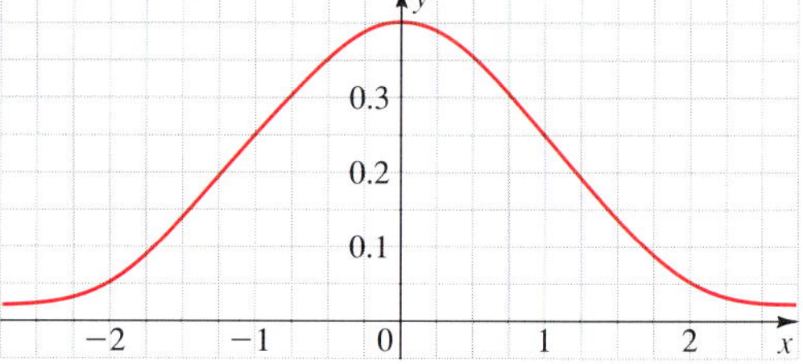

42. $A(0, \underline{?})$ 0.40
43. $B(0.25, \underline{?})$ 0.39
44. $C(-0.5, \underline{?})$ 0.35
45. $D(0.75, \underline{?})$ 0.30
46. $E(-1, \underline{?})$ 0.24
47. $F(1.25, \underline{?})$ 0.18

CHALLENGE

The letter e is used to represent the irrational number 2.7182818 (to seven decimal places). This unique number was discovered by the mathematician Leonhard Euler, who found that when n is replaced with successively greater values in the expression $\left(1 + \frac{1}{n}\right)^n$, the value of the expression approaches e as its limit. Euler also showed that e can be written as a continued fraction.

$$e = 2 + \cfrac{1}{1 + \cfrac{1}{2 + \cfrac{2}{3 + \cfrac{3}{4 + \cfrac{4}{\ddots}}}}}$$

1. Show that if the continued fraction above is evaluated through the "$3 + \frac{3}{4}$" row (that is, all rows below $3 + \frac{3}{4}$ are eliminated), the value obtained gives e correct to one decimal place. *Hint*: Start from the bottom and work up. $e = 2.7$

2. Through which row must the continued fraction be evaluated in order to obtain e correct to three decimal places? "$4 + \frac{4}{5}$" row

9.7

Hyperbolic Functions

Objective: To define the hyperbolic functions in terms of e^x and e^{-x} and to express them as power series

Hyperbolic functions are used in physics and engineering. In order to understand them, it is important to be familiar with the fundamental identities, the double-angle identities, and the sum and difference identities.

Preview

Complete each of the following identities.

1. $\frac{1}{\sin\theta} = \underline{?}$ $\csc\theta$
2. $\frac{1}{\tan\theta} = \underline{?}$ $\cot\theta$
3. $\frac{\sin\theta}{\cos\theta} = \underline{?}$ $\tan\theta$
4. $\cos^2\theta + \sin^2\theta = \underline{?}$ 1
5. $\sin(\alpha + \beta) = \underline{?}$ $\sin\alpha\cos\beta + \cos\alpha\sin\beta$
6. $\cos(\alpha + \beta) = \underline{?}$ $\cos\alpha\cos\beta - \sin\alpha\sin\beta$
7. $\cos 2\alpha = \underline{?}$ $\cos^2\alpha - \sin^2\alpha$
8. $1 + \tan^2\theta = \underline{?}$ $\sec^2\theta$
9. $\tan(\alpha - \beta) = \underline{?}$ $\frac{\tan\alpha - \tan\beta}{1 + \tan\alpha\tan\beta}$
10. $1 - 2\sin^2\theta = \underline{?}$ $\cos 2\theta$
11. $2\cos^2\beta - 1 = \underline{?}$ $\cos 2\beta$
12. $2\sin\alpha\cos\alpha = \underline{?}$ $\sin 2\alpha$

In the last lesson, you learned that $e^x = 1 + x + \frac{x^2}{2!} + \frac{x^3}{3!} + \frac{x^4}{4!} + \cdots$. If you substitute $-x$ for x in this series, the following new series is generated

$$e^{-x} = 1 + (-x) + \frac{(-x)^2}{2!} + \frac{(-x)^3}{3!} + \frac{(-x)^4}{4!} + \cdots$$

$$e^{-x} = 1 - x + \frac{x^2}{2!} - \frac{x^3}{3!} + \frac{x^4}{4!} - \cdots$$

Two new functions can be generated by adding and subtracting the power series for e^x and e^{-x}. First add.

$$e^x = 1 + x + \frac{x^2}{2!} + \frac{x^3}{3!} + \frac{x^4}{4!} + \frac{x^5}{5!} + \frac{x^6}{6!} + \cdots$$

$$e^{-x} = 1 - x + \frac{x^2}{2!} - \frac{x^3}{3!} + \frac{x^4}{4!} - \frac{x^5}{5!} + \frac{x^6}{6!} - \cdots$$

$$e^x + e^{-x} = 2 + 0 + 2\left(\frac{x^2}{2!}\right) + 0 + 2\left(\frac{x^4}{4!}\right) + 0 + 2\left(\frac{x^6}{6!}\right) + \cdots$$

$$e^x + e^{-x} = 2\left(1 + \frac{x^2}{2!} + \frac{x^4}{4!} + \frac{x^6}{6!} + \cdots\right)$$

Thus, $$\frac{e^x + e^{-x}}{2} = 1 + \frac{x^2}{2!} + \frac{x^4}{4!} + \frac{x^6}{6!} + \cdots.$$

LESSON PLAN

Vocabulary
Hyperbolic function

Materials/Manipulatives
Calculators
Computer

BACKGROUND

In the Preview, the fundamental identities are reviewed. Similar identities exist for the hyperbolic functions.

TEACHING SUGGESTIONS

- Emphasize that the power series for hyperbolic sine and hyperbolic cosine are the same as those for sine and cosine except the signs do not alternate.
- When the domain of cosh and sinh is the set of real numbers, the range of cosh is the set of all real numbers greater than or equal to 1, and the range of sinh is the set of real numbers.
- It may be necessary to develop the idea of hyperbola.
- Calculators or computers make calculations easier when evaluating hyperbolic functions.

Critical Thinking

Research Ask students to find a real-world example for which the graph of cosh x is a model. Answers may vary; for example, the Gateway Arch in St. Louis, Mo.

CHALKBOARD EXAMPLES

- **For Example 1**
  1. Evaluate cosh x and sinh x for $x = 0$. Round each answer to four decimal places. cosh (0) = 1; sinh (0) = 0
  2. Evaluate tanh x for $x = 0$. Round the answer to four decimal places. tanh (0) = 0

Recall that $\cos x = 1 - \frac{x^2}{2!} + \frac{x^4}{4!} - \frac{x^6}{6!} + \cdots$. Note that the power series representation of $\frac{e^x + e^{-x}}{2}$ is the same as that of $\cos x$, except for the fact that it does not have alternating signs. A new function, cosh x, is defined as

$$\cosh x = \frac{e^x + e^{-x}}{2}$$ *Read: hyperbolic cosine of x*

Next, subtract e^{-x} from e^x.

$$e^x = 1 + x + \frac{x^2}{2!} + \frac{x^3}{3!} + \frac{x^4}{4!} + \frac{x^5}{5!} + \cdots$$

$$e^{-x} = 1 - x + \frac{x^2}{2!} - \frac{x^3}{3!} + \frac{x^4}{4!} - \frac{x^5}{5!} + \cdots$$

$$e^x - e^{-x} = 0 + 2x + 0 + 2\left(\frac{x^3}{3!}\right) + 0 + 2\left(\frac{x^5}{5!}\right) + \cdots$$

$$e^x - e^{-x} = 2\left(x + \frac{x^3}{3!} + \frac{x^5}{5!} + \frac{x^7}{7!} + \cdots\right)$$

Thus, $$\frac{e^x - e^{-x}}{2} = x + \frac{x^3}{3!} + \frac{x^5}{5!} + \frac{x^7}{7!} + \cdots.$$

Recall that $\sin x = x - \frac{x^3}{3!} + \frac{x^5}{5!} - \frac{x^7}{7!} + \cdots$. Note that the power series representation of $\frac{e^x - e^{-x}}{2}$ is the same as that of $\sin x$, except for the fact that it does not have alternating signs. A new function, sinh x, is defined as follows:

$$\sinh x = \frac{e^x - e^{-x}}{2}$$ *Read: hyperbolic sine of x*

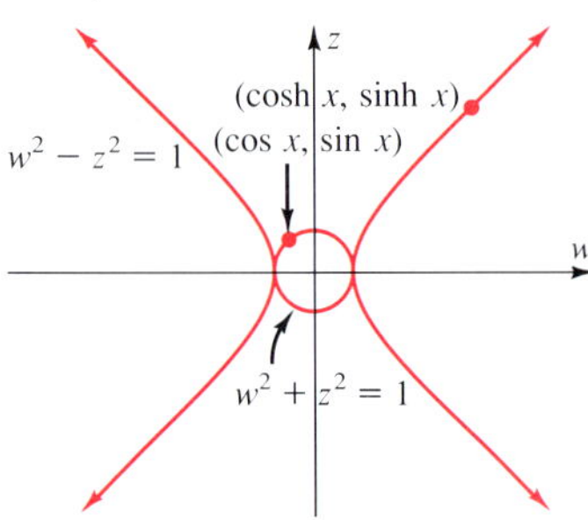

These new functions are associated with the unit hyperbola, $w^2 - z^2 = 1$, graphed on the wz-plane at the right. Note that every point on the wz-plane has coordinates of the form (w, z). For each value of x, the point (cosh x, sinh x) lies on the unit hyperbola, just as the point ($\cos x$, $\sin x$) lies on the unit circle $w^2 + z^2 = 1$.

EXAMPLE 1 **Evaluate cosh x and sinh x for $x = 1$. Round to four decimal places.**

$$\cosh x = \frac{e^x + e^{-x}}{2} \qquad \sinh x = \frac{e^x - e^{-x}}{2}$$

$$\cosh 1 = \frac{e^1 + e^{-1}}{2} \qquad \sinh 1 = \frac{e^1 - e^{-1}}{2}$$

$\cosh 1 = 1.5431$ *To four decimal places* $\sinh 1 = 1.1752$

Other similarities exist between the trigonometric functions and the **hyperbolic functions.** For example, the hyperbolic tangent, hyperbolic cotangent, hyperbolic secant, and hyperbolic cosecant functions can be defined in terms of the hyperbolic sine and cosine, just as the other trigonometric functions can be defined in terms of $\sin x$ and $\cos x$.

Trigonometric functions	**Hyperbolic functions**
$\tan x = \dfrac{\sin x}{\cos x}$	$\tanh x = \dfrac{\sinh x}{\cosh x} = \dfrac{e^x - e^{-x}}{e^x + e^{-x}}$
$\cot x = \dfrac{\cos x}{\sin x}$	$\coth x = \dfrac{\cosh x}{\sinh x} = \dfrac{e^x + e^{-x}}{e^x - e^{-x}}$
$\sec x = \dfrac{1}{\cos x}$	$\operatorname{sech} x = \dfrac{1}{\cosh x} = \dfrac{2}{e^x + e^{-x}}$
$\csc x = \dfrac{1}{\sin x}$	$\operatorname{csch} x = \dfrac{1}{\sinh x} = \dfrac{2}{e^x - e^{-x}}$

In addition, there are a number of identities for hyperbolic functions that are similar to those for the trigonometric functions.

Trigonometric identities	**Hyperbolic identities**
$\cos^2 x + \sin^2 x = 1$	$\cosh^2 x - \sinh^2 x = 1$ (since $w^2 - z^2 = 1$)
$\sin(x + y) = \sin x \cos y + \cos x \sin y$	$\sinh(x + y) = \sinh x \cosh y + \cosh x \sinh y$
$\cos(x + y) = \cos x \cos y - \sin x \sin y$	$\cosh(x + y) = \cosh x \cosh y - \sinh x \sinh y$
$\sin 2x = 2 \sin x \cos x$	$\sinh 2x = 2 \sinh x \cosh x$
$\cos 2x = 2\cos^2 x - 1$	$\cosh 2x = 2\cosh^2 x - 1$

These identities can be used to prove other identities.

EXAMPLE 2 **Prove: $-\sinh x = \sinh(-x)$**

$$-\sinh x \;\Bigg|\; \sinh(-x)$$
$$\frac{e^{-x} - e^{-(-x)}}{2}$$
$$\frac{e^{-x} - e^{x}}{2}$$
$$-\left(\frac{e^{x} - e^{-x}}{2}\right)$$
$$= -\sinh x$$

Thus, $-\sinh x = \sinh(-x)$.

- **For Example 2**
  Prove each identity.

3. $\cosh x = \cosh(-x)$

$$\cosh x = \cosh(-x)$$
$$\frac{e^x + e^{-x}}{2} \;\Bigg|\; \frac{e^{-x} + e^{-(-x)}}{2}$$
$$\frac{e^{-x} + e^{x}}{2}$$
$$= \frac{e^x + e^{-x}}{2}$$

4. $-\tanh x = \dfrac{\sinh(-x)}{\cosh(-x)}$

$$-\tanh x = \frac{\sinh(-x)}{\cosh(-x)}$$
$$-\left(\frac{e^x - e^{-x}}{e^x + e^{-x}}\right) \;\Bigg|\; \frac{\frac{e^{-x} - e^{x}}{2}}{\frac{e^{-x} + e^{x}}{2}}$$
$$\frac{e^{-x} - e^{x}}{e^{-x} + e^{x}}$$
$$-\frac{e^{x} + e^{-x}}{e^{x} + e^{-x}}$$
$$= -\left(\frac{e^x - e^{-x}}{e^x + e^{-x}}\right)$$

Common Error

- Students often confuse hyperbolic functions with the sine and cosine functions. Emphasize that hyperbolic functions are related to the unit hyperbola, while sine and cosine are related to the unit circle.
- See *Teacher's Resource Book* for additional remediation.

LESSON FOLLOW-UP

Assignment Guide

See p. 370B for assignments.

Test Yourself

See *Teacher's Resource Book*, *Tests*, pp. 87–88.

Lesson Quiz

1. Evaluate $\cosh x$ for $x = 0$. 1
2. Evaluate $\coth x$ for $x = 1$. Round your answer to 4 decimal places. 1.3130

Prove each identity.

3. $\coth x = \dfrac{\cosh x}{\sinh x}$

$$\coth x \;\Big|\; \frac{\cosh x}{\sinh x}$$

$$\frac{e^x + e^{-x}}{e^x - e^{-x}} \;\Big|\; \frac{\frac{e^x + e^{-x}}{2}}{\frac{e^x - e^{-x}}{2}} = \frac{e^x + e^{-x}}{e^x - e^{-x}}$$

4. $\cosh^2 x - \sinh^2 x = 1$

$$\cosh^2 x - \sinh^2 x \;\Big|\; 1$$

$$\left(\frac{e^x + e^{-x}}{2}\right)^2 - \left(\frac{e^x - e^{-x}}{2}\right)^2$$

$$\frac{e^{2x} + 2e^0 + e^{-2x}}{4} - \frac{e^{2x} - 2e^0 + e^{-2x}}{4}$$

$$\frac{4}{4}$$

$$1 =$$

Additional Answers

Class Exercises

7. $2\cosh^2 x - 1$

$$= 2\left(\frac{e^x + e^{-x}}{2}\right)^2 - 1 =$$

$$\frac{2(e^{2x} + e^0 + e^0 + e^{-2x})}{4} - 1 =$$

$$\frac{e^{2x} + 2 + e^{-2x} - 2}{2} =$$

$$\frac{e^{2x} + e^{-2x}}{2} = \cosh 2x$$

CLASS EXERCISES

Match each expression in Exercises 1–6 with its value listed in the second row.

1. $\sinh 0$	**2.** $\cosh 0$	**3.** $\operatorname{sech} -1$ d	**4.** $\operatorname{csch} -1$ a	**5.** $\coth 1$ e	**6.** $\tanh 1$ b
a -0.8509	**b.** 0.7616	**c.** 0	**d.** 0.6481	**e.** 1.3130	**f.** 1

7. Prove $\cosh 2x = 2\cosh^2 x - 1$ See side column.

PRACTICE EXERCISES

Evaluate each function for $x = -1$. Round each answer to four decimal places.

A **1.** $\sinh x$ -1.1752 **2.** $\cosh x$ 1.5431 **3.** $\tanh x$ -0.7616 **4.** $\coth x$ -1.3130

Prove each identity. See page 494.

5. $\cosh x = \dfrac{1}{\operatorname{sech} x}$

6. $\sinh x = \dfrac{1}{\operatorname{csch} x}$

7. $\sinh x = \dfrac{\cosh x}{\coth x}$

8. $\tanh x = \dfrac{1}{\coth x}$

9. $\cosh x = \dfrac{\coth x}{\operatorname{csch} x}$

10. $\sinh x = \dfrac{\tanh x}{\operatorname{sech} x}$

11. $\operatorname{sech} x = \dfrac{\operatorname{csch} x}{\coth x}$

12. $\cosh(-x) = \cosh x$

B **13.** $\operatorname{sech}^2 x = 1 - \tanh^2 x$

14. $\operatorname{csch}^2 x = \coth^2 x - 1$

15. $\sinh 2x = 2\sinh x \cosh x$

16. $\cosh 2x = 1 + 2\sinh^2 x$

17. $\sinh^2 x = \cosh^2 x - 1$

18. $\cosh 2x = \cosh^2 x + \sinh^2 x$

19. $e^x = \cosh x + \sinh x$

20. $e^{-2x} = (\sinh x - \cosh x)^2$

C **21.** $\sinh(x - y) = \sinh x \cosh y - \cosh x \sinh y$

22. $\cosh(x - y) = \cosh x \cosh y - \sinh x \sinh y$

23. $\cosh \dfrac{x}{2} = \sqrt{\dfrac{1 + \cosh x}{2}}$

24. $\sinh^2\left(\dfrac{x}{2}\right) = \dfrac{\cosh x - 1}{2}$

Applications

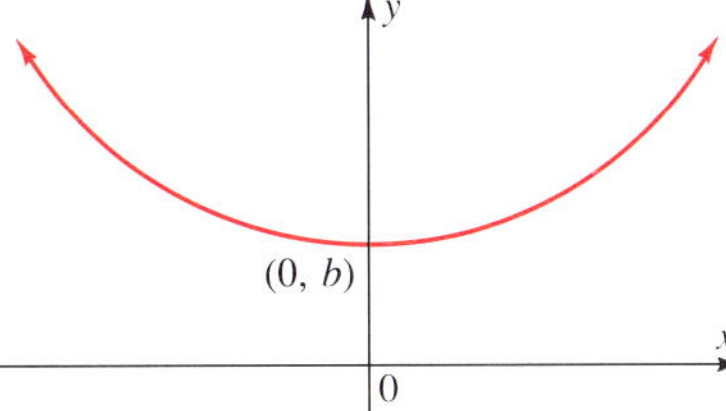

Civil Engineering Some bridge cables have the shape of a rope supported at both ends and hanging of its own weight. This curve is called a *catenary*. The catenary at the right is described by the equation $y = a \cosh x + b - a$, where b is the y-intercept and a is a quantity that depends upon forces acting on the cable.

25. For simplicity, the x-axis in the figure can be chosen so that $b = a$. In that case, what is the simplified form of the equation of the catenary? $y = a \cosh x$

26. In Exercise 25, suppose $x = 3$. What is the value of y as a function of a? $\frac{a}{2}(e^3 + e^{-3})$, or about $10a$

TEST YOURSELF

Determine the sum S, if it exists, for each infinite geometric series.

1. $1 + \frac{1}{8} + \frac{1}{64} + \cdots$ $\frac{8}{7}$

2. $\frac{1}{64} + \frac{1}{8} + 1 + \cdots$ does not exist 9.5

3. $\sum_{n=1}^{\infty} (-2)^n$ does not exist

4. $\sum_{n=1}^{\infty} \left(-\frac{1}{2}\right)^n$ $-\frac{1}{3}$

Write each repeating decimal as a fraction.

5. $0.\overline{52}$ $\frac{52}{99}$

6. $0.7\overline{4}$ $\frac{67}{90}$

Approximate each value using the first six terms of the appropriate power series. Round each answer to four decimal places.

7. $\sin \frac{5\pi}{4}$ -0.7150

8. $\cos 311°$ -0.5132

9. e^4 42.8667

10. $e^{\frac{1}{4}}$ 1.2840 9.6

Find the exact value of each expression using Euler's formula.

11. $e^{\frac{-\pi i}{4}}$ $\frac{\sqrt{2}}{2} - \frac{i\sqrt{2}}{2}$

12. $e^{\frac{-\pi i}{2}}$ $-i$

Evaluate each function for $x = 2$. Round to four decimal places.

13. $\sinh x$ 3.6269

14. $\cosh x$ 3.7622 9.7

Prove each identity. See side column.

15. $\coth x = \dfrac{\operatorname{csch} x}{\operatorname{sech} x}$

16. $\cosh x - \sinh x = e^{-x}$

Teacher's Resource Book
Practice—Chapter 9, p. 13
Enrichment—Chapter 9, p. 14

Additional Answers

15. $\coth x \;\Big|\; \dfrac{\operatorname{csch} x}{\operatorname{sech} x}$

$$\left(\frac{\frac{2}{e^x - e^{-x}}}{\frac{2}{e^x + e^{-x}}}\right)\frac{(e^x - e^{-x})(e^x + e^{-x})}{(e^x - e^{-x})(e^x + e^{-x})}$$

$$\frac{2(e^x + e^{-x})}{2(e^x - e^{-x})}$$

$$\frac{e^x + e^{-x}}{e^x - e^{-x}}$$

$= \coth x$

16. $\cosh x - \sinh x \;\Big|\; e^{-x}$

$$\frac{e^x + e^{-x}}{2} - \frac{e^x - e^{-x}}{2}$$

$$\frac{2e^{-x}}{2}$$

$e^{-x} =$

Technology

Archimedean spirals are introduced, and real-world examples of such spirals are cited. A program for generating spirals of various dimensions is presented. The complete form of this program is on the disk provided with the *Teacher's Resource Book.*

See *Teacher's Resource Book—Technology*, p. 9.
See *Teacher's Resource Book—Application*, Chapter 9, p. 15.

TECHNOLOGY: Spirals

Did you know that some spiders spin their webs in the shape of an Archimedean spiral? The grooves in a phonograph record, the coil of a spool of video or audio tape, and a coiled rope also take the form of an Archimedean spiral. In addition, these spirals are found in the plant kingdom in tomato plants, fiddlehead ferns, and sago palms.

What is a spiral, and in particular, what is an Archimedean spiral? A spiral is a curve traced by a point constantly moving around and away from a fixed point called a pole. The particular type of spiral is determined by the manner in which the curve moves away from the pole. An Archimedean spiral is a curve in which the distance from the origin of a point on the curve is directly proportional to the measure of the angle turned to reach that point. For example, if the distance from the pole is 1 when the angle is 60°, then the distance must be 2 when the angle is 120°, 3 when the angle is 180°, and so on. Thus, an Archimedean spiral increases in length at a constant rate per constant turn.

If a radius vector is superimposed on an Archimedean spiral, the spiral will intersect it at equally spaced intervals. The distances from the pole of the successive turns of the spiral along this radius vector form an arithmetic sequence: d, $2d$, $3d$, ...

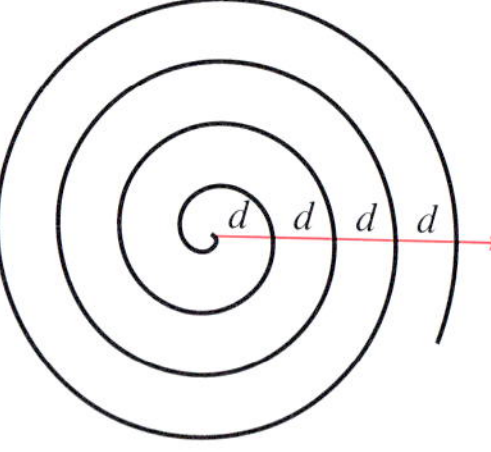

When spinning their webs, spiders are efficient at conserving energy and material while expanding over a maximum area. Spiders that spin their webs in the shape of an Archimedean spiral first create a framework of spokes that correspond to radius vectors. They then progress outward from the center, forming a spiral by guiding themselves by the previous turn and ensuring that the distance between each turn is constant.

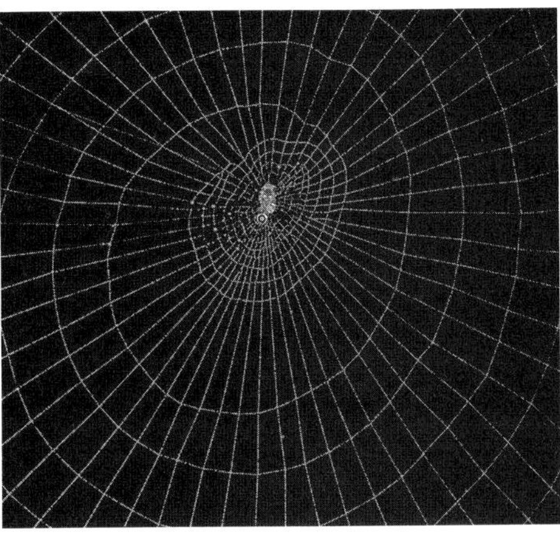

```
10 INPUT "DISTANCE "; D
20 INPUT "ANGLE (DEGREES) "; A
30 C = D / (3.1415927 * A/180)
40 HGR : HCOLOR = 2
50 HPLOT 90,0 TO 90,180
60 HPLOT 0,90 TO 180,90
70 FOR I = 0 TO 10
80 HPLOT 18 * I,87 TO 18 * I,93
90 HPLOT 87,18 * I TO 93, 18 * I
100 NEXT I: HCOLOR = 5
110 FOR I = 1 TO 20000
120 TH = I * 3.1415927 / 180
130 R = C * TH
140 X = R * COS (TH):Y = R * SIN (TH)
150 X = 18 * X + 90:Y = 18 * Y + 90
160 IF X * Y <= 0 THEN GOTO 190
170 HPLOT X,Y
180 NEXT I
190 END
```

The above computer program graphs the Archimedean spiral for a specified distance from the origin and a specified degree turn. Angle measures are entered in degrees, but the program converts them to radians. It also performs its calculations with polar coordinates, but converts the results to rectangular coordinates before graphing.

EXERCISES

1. Run the program for a distance of 1 and a turn of 360°. At what points does this spiral cross the positive x-axis? 1, 2, 3, 4, . . .

2. Run the program for a distance of 0.25 and a turn of 180°. At what points does this spiral cross the positive x-axis? 0.5, 1, 1.5, 2, . . .

3. Notice that the distances in Exercises 1 and 2 form an arithmetic sequence. What are the values of d and a_1 for Exercise 1? for Exercise 2? *Hint*: Remember that one full revolution is 360°. $d = 1$, $a_1 = 0$; $d = \frac{1}{2}$, $a_1 = 0$

4. A spider is building a web. After laying support webs, the spider makes a spiral in which the distance from the center increases by 0.1875 whenever the angle increases by 30°. How far will the spider be from the center of the web after making 4 complete revolutions? 9

5. Another spider builds a web in which the distance increases by 0.1875 whenever the angle increases by 60°. How far will this spider be from the center after making 4 complete revolutions? Explain the difference between your answers to this Exercise and Exercise 4. 4.5

6. Now increase the distance to 0.5 while keeping the angle equal to 60° as in Exercise 5. What is the effect of changing the distance without changing the angle of turn? The distance traveled in one revolution is increased.

413

CHAPTER 9 SUMMARY AND REVIEW

Vocabulary

arithmetic means (375)
arithmetic sequence (373)
arithmetic series (384)
common difference (373)
common ratio (378)
convergent series (397)
divergent series (398)
Euler's formula (404)
explicit formula (372)
finite sequence (372)
general term (372)
geometric means (380)
geometric sequence (378)
geometric series (390)
hyperbolic functions (409)
index of summation (385)
infinite sequence (373)
limit (397)
limits of summation (385)
mean proportional (380)
partial sum (384)
power series (402)
recursive formula (373)
sequence (372)
series (384)
sigma notation (384)
sum of an infinite geometric series (398)

Arithmetic Sequences If a_1 is the first term, n is the number of terms, d is the common difference, and a_n is the nth term of an arithmetic sequence, then $a_n = a_1 + (n - 1)d$. 9.1

Find the next three terms and the general term of each arithmetic sequence.

1. 10, 7, 4, . . . 1, −2, −5; $a_n = -3n + 13$
2. −0.5, 0, 0.5, . . . 1, 1.5, 2; $a_n = 0.5n - 1$

Find the indicated term of each arithmetic sequence.

3. a_{10} of −20, −11, −2, . . . 61
4. a_8 of 18, 6, −6, . . . −66
5. Find two arithmetic means between 6 and 36. 16, 26

Geometric Sequences If a_1 is the first term, n is the number of terms, r is the common ratio, and a_n is the nth term, then $a_n = a_1r^{n-1}$. 9.2

Find the next three terms and the general term of each geometric sequence.

6. 36, 12, 4, . . . $\frac{4}{3}, \frac{4}{9}, \frac{4}{27}$; $a_n = 36\left(\frac{1}{3}\right)^{n-1}$
7. 1, 3, 9, . . . 27, 81, 243; $a_n = 3^{n-1}$

Find the indicated term of each geometric sequence.

8. a_6 of 4, −8, 16, . . . −128
9. a_7 of 128, 64, 32, . . . 2
10. Find the mean proportional between −9 and −81. −27

Arithmetic Series If a_1 is the first term, n is the number of terms, d is the common difference, and a_n is the nth term, then the partial sum S_n of an arithmetic series is given by 9.3

$$S_n = \frac{n(a_1 + a_n)}{2} \quad \text{or} \quad S_n = \frac{n[2a_1 + (n - 1)d]}{2}$$

Find the indicated partial sum of each arithmetic series.

11. $-15, -10, -5, \ldots; S_{13}$ 195

12. $\sum_{m=1}^{15} (2m + 1)$ 255

13. Express in sigma notation the sum of the first six terms of the arithmetic series $-7 + (-3) + 1 + \cdots$. $\sum_{k=1}^{6} (4k - 11)$

14. If $a_8 = 100$ and $S_8 = 440$, and the series is arithmetic, find a_1. 10

Geometric Series If a_1 is the first term, n is the number of terms, r is the common ratio, and a_n is the nth term of a geometric series, then 9.4

$$S_n = \frac{a_1(1 - r^n)}{1 - r}, r \neq 1 \quad \text{or} \quad S_n = \frac{a_1 - ra_n}{1 - r}, r \neq 1$$

Find the indicated partial sum of each geometric series.

15. $135 + 45 + 15 + \cdots; S_7$ 202.4

16. $\sum_{j=1}^{6} (-3)^j$ 546

17. $a_1 = 2, a_7 = 1458; S_7$ 2186 if $r = 3$ or 1094 if $r = -3$

Infinite Geometric Series An infinite geometric series converges if and only if $|r| < 1$. If S is the sum of a convergent geometric series, then $S = \frac{a_1}{1 - r}$. 9.5

Determine the sum S, if it exists, for each infinite geometric series.

18. $100 - 10 + 1 - \cdots$ $\frac{1000}{11}$

19. $\frac{1}{2} + \frac{3}{2} + \frac{9}{2} + \cdots$ does not exist

20. $\sum_{n=1}^{\infty} 2\left(\frac{1}{2}\right)^{n+5}$ 0.0625

Power Series and Trigonometric Functions 9.6

$$\sin x = x - \frac{x^3}{3!} + \frac{x^5}{5!} - \frac{x^7}{7!} + \cdots \qquad \cos x = 1 - \frac{x^2}{2!} + \frac{x^4}{4!} - \frac{x^6}{6!} + \cdots$$

$$e^x = 1 + \frac{x}{1!} + \frac{x^2}{2!} + \frac{x^3}{3!} + \cdots \qquad e^{ix} = \cos x + i \sin x$$

Approximate each value using the first six terms of the appropriate power series. Round each answer to four decimal places.

21. $\cos \frac{3\pi}{2}$ -0.2224

22. $\sin 1.7$ 0.9917

23. $e^{0.7}$ 2.0136

24. Use Euler's formula to find the exact value of $e^{\frac{\pi i}{6}}$. $\frac{\sqrt{3}}{2} + \frac{i}{2}$

Hyperbolic Functions $\cosh x = \frac{e^x + e^{-x}}{2}$ and $\sinh x = \frac{e^x - e^{-x}}{2}$ 9.7

See side column.

25. Prove: $\text{sech } x = \frac{\tanh x}{\sinh x}$

26. Prove: $\coth^2 x = 1 + \text{csch}^2 x$

Additional Answers

25. $\frac{\tanh x}{\sinh x} = \frac{\frac{\sinh x}{\cosh x}}{\sinh x} = \frac{1}{\cosh x} = \text{sech } x$

26. $1 + \text{csch}^2 x = 1 + \left(\frac{2}{e^x - e^{-x}}\right)^2 =$

$\frac{e^{2x} + e^0 + e^0 + e^{-2x}}{e^{2x} - e^0 - e^0 + e^{-2x}} = 1 +$

$\frac{4}{e^{2x} - e^0 - e^0 + e^{-2x}} =$

$\frac{e^{2x} - 2 + e^{-2x}}{e^{2x} - 2 + e^{-2x}} + \frac{4}{e^{2x} - 2 + e^{-2x}} =$

$\frac{e^{2x} + 2 + e^{-2x}}{e^{2x} - 2 + e^{-2x}} = \left(\frac{e^x + e^{-x}}{e^x - e^{-x}}\right)^2 =$

$\coth^2 x$

See *Teacher's Resource Book, Tests*, pp. 89–92.

Additional Answers

17. $\cosh x = \frac{\sinh x}{\tanh x} = \frac{\frac{e^x - e^{-x}}{2}}{\frac{e^x - e^{-x}}{e^x + e^{-x}}} = \frac{(e^x - e^{-x})(e^x + e^{-x})}{2(e^x - e^{-x})} = \frac{e^x + e^{-x}}{2} = \cosh x$

Challenge

$\frac{1 + \tanh x \tanh y}{\tanh x - \tanh y} = \frac{1 + \frac{\sinh x \sinh y}{\cosh x \cosh y}}{\left(\frac{\sinh x}{\cosh x} - \frac{\sinh y}{\cosh y}\right)}\left(\frac{\cosh x \cosh y}{\cosh x \cosh y}\right) = \frac{\cosh x \cosh y + \sinh x \sinh y}{\sinh x \cosh y - \sinh y \cosh x} = \frac{\cosh (x - y)}{\sinh (x - y)} = \coth (x - y)$

CHAPTER TEST

1. In an arithmetic sequence, $a_1 = 24$ and $d = -0.5$. Find a_6. $\frac{43}{2}$
2. Insert two arithmetic means between -10 and 8. $-4, 2$
3. In a geometric sequence, $a_1 = 7$ and $r = 4$. Find a_9. 458, 752
4. Find the mean proportional between 2 and 18. 6
5. The sixth term of an arithmetic sequence is 14 and the 12th term is 38. Find the first term and the common difference. $a_1 = -6$; $d = 4$
6. The fourth term of a geometric sequence is -27 and the seventh term is 729. Find the first term and the common ratio. $a_1 = 1$; $r = -3$

Find the indicated partial sum for each arithmetic series.

7. $a_1 = 10, d = 7;\ S_{12}$ 582
8. $a_1 = 14, d = 12;\ S_9$ 558

Find the indicated partial sum for each geometric series.

9. $4 + (-12) + 36 + \cdots; S_{12}$ $-531,440$
10. $-7 + 14 + (-28) + \cdots; S_9$ -1197

Find the sum of each infinite geometric series, if it exists. If the sum does not exist, say so.

11. $\sum_{k=1}^{\infty} \left(\frac{4}{3}\right)^{-k}$ 3
12. $\sum_{k=1}^{\infty} \left(\frac{3}{4}\right)^{-k}$ does not exist

Approximate each value using the first six terms of the appropriate power series. Round each answer to four decimal places.

13. $\sin 2.8$ 0.3349
14. $\cos 60°$ 0.5000
15. $e^{0.3}$ 1.3499
16. Use Euler's formula to find the exact value of $e^{-6\pi i}$. 1
17. Prove: $\cosh x = \frac{\sinh x}{\tanh x}$. See side column.

Challenge

Prove: $\coth (x - y) = \frac{1 + \tanh x \tanh y}{\tanh x - \tanh y}$ See side column.

COLLEGE ENTRANCE EXAM REVIEW

Select the best choice for each question.

1. If $a_n = n^2 + 2$, what are the first four terms of the sequence?
E
A. 1, 4, 9, 16 **B.** 1, 2, 3, 4
C. 2, 5, 10, 17 **D.** 2, 4, 6, 8
E. 3, 6, 11, 18

2. $\sum_{j=2}^{5} 3j^2 - 1$ equals
B
A. 63 **B.** 158 **C.** 162
D. 109 **E.** 240

3. If $f(t) = \dfrac{2t^2 + 1}{t}$, find $f(-2)$.
C
A. $-\frac{1}{2}$ **B.** 8 **C.** -4.5
D. 7.8 **E.** 12

4. Find the 22nd term of the arithmetic sequence 8, 11, 14, 17, ...
E
A. 76 **B.** 84 **C.** 96
D. 68 **E.** 71

5. Evaluate: $\begin{vmatrix} 8 & 6 \\ 4 & -9 \end{vmatrix}$
A
A. -96 **B.** -7 **C.** 27
D. 48 **E.** -31

6. $\dfrac{x^2 - y^2}{x + 5} \div \dfrac{x - y}{x^2 + 6x + 5} =$
D
A. $x + y$ **B.** $(x + 5)(x - y)$
C. $x + 1$ **D.** $(x + y)(x + 1)$
E. $(x + y)(x - y)$

7. Find the sum of the first 12 terms of the arithmetic sequence 1, 3, 5, ...
B
A. 168 **B.** 144 **C.** 23
D. 138 **E.** 208

8. Solve $\sqrt{2x - 1} - x = -2$.
A
A. $x = 5$ **B.** $x = 5, x = 1$
C. $x = 1$ **D.** $x = -5$
E. $x = -1, x = -5$

9. Find the 8th term of the geometric sequence 4, 8, 16, 32, ...
D
A. 384 **B.** 212 **C.** 158
D. 512 **E.** 640

10. Find the sum of the geometric sequence 2, 6, 18, 54, 162.
E
A. 232 **B.** 356 **C.** 408
D. 394 **E.** 242

11. Find the sum of the infinite geometric sequence $\frac{2}{3}, \frac{8}{3}, \frac{32}{3}, \ldots$.
E
A. 1045 **B.** -66 **C.** -108
D. 440 **E.** does not exist

12. If $T = \dfrac{a + bx}{1 - x}$, $x \neq 1$, solve for x.
D
A. $\dfrac{a + b}{T}$ **B.** $T - b$ **C.** $a + T$
D. $\dfrac{T - a}{b + T}$ **E.** $\dfrac{T + b}{b}$

13. The perimeter of a rectangle is 48 ft. If the width of the rectangle is one-half its length, find the width.
D
A. 16 ft **B.** 24 ft **C.** 32 ft
D. 8 ft **E.** 40 ft

14. $\sum_{n=1}^{\infty} 2^{3-n} =$
D
A. $\frac{1}{2}$ **B.** $-\frac{1}{2}$ **C.** 4 **D.** 8 **E.** -8

15. $\cosh 2 =$
A
A. $\dfrac{e^2 + e^{-2}}{2}$ **B.** $\dfrac{e^2 - e^{-2}}{2}$
C. $\dfrac{1}{\sinh 2}$ **D.** $1 - \sinh^2 2$
E. $\dfrac{\tanh 2}{\sinh 2}$

See *Teacher's Resource Book, Cumulative Test*, pp. 93–100.

Additional Answers

5.

θ	r
0	5
$\frac{\pi}{6}$	2.5
$\frac{\pi}{4}$	0
$\frac{\pi}{3}$	-2.5
$\frac{\pi}{2}$	-5
π	5

9. $\cot 135° \cos 135° = \csc 135° - \sin 135°$,
$-1 \cdot \left(-\frac{\sqrt{2}}{2}\right) = \sqrt{2} - \frac{\sqrt{2}}{2}; \frac{\sqrt{2}}{2} = \frac{\sqrt{2}}{2}$

10. $\dfrac{2 \sin \frac{\pi}{6} \cos \frac{\pi}{6}}{1 + \cos 2\left(\frac{\pi}{6}\right)} = \dfrac{2\left(\frac{1}{2}\right)\left(\frac{\sqrt{3}}{2}\right)}{1 + \frac{1}{2}} =$
$\dfrac{\frac{\sqrt{3}}{2}}{\frac{3}{2}} = \dfrac{\sqrt{3}}{3} = \tan \dfrac{\pi}{6}$

15.

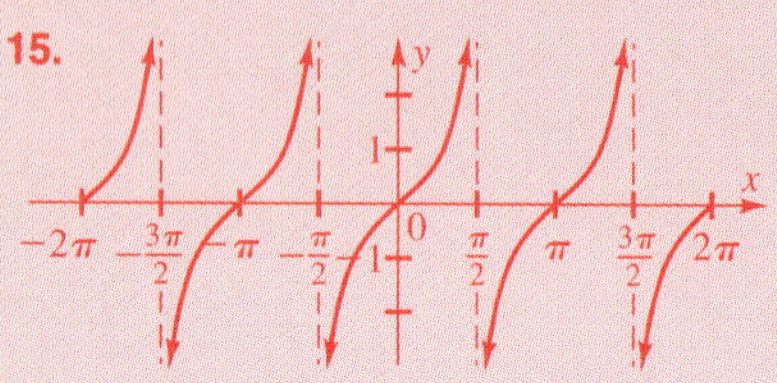

16. $\dfrac{\cos^2 \alpha - 1}{\sec \alpha} \;\Big|\; \dfrac{-\sin \alpha \sin 2\alpha}{2}$
$\dfrac{-\sin^2 \alpha}{\sec \alpha} \;\Big|\; \dfrac{-2 \sin^2 \alpha \cos \alpha}{2}$
$-\sin^2 \alpha \cos \alpha = -\sin^2 \alpha \cos \alpha$

CUMULATIVE REVIEW

Express in exponential form.

1. $\log_{125} 5 = \frac{1}{3}$ $125^{\frac{1}{3}} = 5$
2. $\log_3 \frac{1}{243} = -5$ $3^{-5} = \frac{1}{243}$ 8.3

Find the approximate value in degrees.

3. $\text{Cos}^{-1}(0.3040)$ 72°
4. $\text{Arctan}(-0.4286)$ −23° 6.3
5. Construct a table of values and then graph the equation $r = 5 \cos 2\theta$. 7.2
See side column and page 461.

Use half-angle identities to find the exact value of each trigonometric function.

6. $\tan \frac{5\pi}{8}$ $-\sqrt{2} - 1$
7. $\sin 105°$ $\frac{\sqrt{2 + \sqrt{3}}}{2}$ 5.5
8. Apply De Moivre's theorem to evaluate $(2 + 2i)^4$. Express your answer in rectangular form. $-64 + 0i$ 7.7

Verify each identity for the given angle measure. See side column.

9. $\cot 135° \cos 135° = \csc 135° - \sin 135°$
10. $\dfrac{2 \sin \frac{\pi}{6} \cos \frac{\pi}{6}}{1 + \cos 2\left(\frac{\pi}{6}\right)} = \tan \dfrac{\pi}{6}$ 3.4

Simplify.

11. $\dfrac{-27 r^{\frac{3}{4}} s^{\frac{2}{5}}}{-9 r^{\frac{1}{4}} s^{\frac{4}{5}}}$ $\dfrac{3r^{\frac{1}{2}}}{s^{\frac{2}{5}}}$
12. $(P^{\sqrt{16}} Q^{-\sqrt{12}})^{\sqrt{3}}$ $\dfrac{P^{4\sqrt{3}}}{Q^6}$ 8.1
13. Determine how many solutions exist to $\triangle ABC$, where $\angle C = 47°$, $a = 45$, and $c = 41$. Solve the triangle or triangles. 4.2
2; $\angle B = 80°$, $\angle A = 53°$, $b = 55$ or $\angle A = 127°$, $\angle B = 6°$, $b = 6$
14. Find the sum of the odd integers from 1 to 49, inclusive. 625 9.3
15. Graph $y = \tan(x + \pi)$ over the interval $-2\pi < x < 2\pi$. See side column. 2.6
16. Prove $\dfrac{\cos^2 \alpha - 1}{\sec \alpha} = \dfrac{-\sin \alpha \sin 2\alpha}{2}$. See side column. 3.6
17. Find the length of the altitude from $\angle A$ to side a for a triangle whose sides are $a = 16$, $b = 14$, and $c = 20$. 14 4.6

Find the degree measure of the angle for each rotation.

18. $\frac{25}{3}$ clockwise rotation −3000°
19. $\frac{3}{4}$ counterclockwise rotation 270° 1.3

Express each complex number in rectangular form. $-\frac{5\sqrt{2}}{2}+\frac{5i\sqrt{2}}{2}$

20. $6(\cos 330° + i \sin 330°)$ $3\sqrt{3} - 3i$

21. $5(\cos 135° + i \sin 135°)$ 7.5

22. Use the fact that the cosine function is an even function to find $\cos(-709°)$ if $\cos 709° = 0.9816$. 0.9816 2.2

Determine whether each series is arithmetic or geometric, and then find its sum.

23. $\sum_{n=1}^{6} 5^{n-1}$ geometric; 3906

24. $\sum_{k=1}^{10} (2k + 7)$ arithmetic; 180 9.4

Evaluate.

25. $\cos\left(\text{Arccos}\left(-\frac{\sqrt{2}}{2}\right)\right)$ $-\frac{\sqrt{2}}{2}$

26. $\text{Arccos}\left(\cos\left(\frac{5\pi}{4}\right)\right)$ $\frac{3\pi}{4}$ 6.2

27. Use the law of cosines to find the length of side c for $\triangle ABC$ if $\angle C = 58°$, $a = 12$, and $b = 21$. 18 4.3

Determine whether each function is even, odd, or neither even nor odd.

28. $h(x) = 3x - 4$ neither

29. $g(x) = 2x^3 - x^2$ neither 2.1

30. The entrance to a hardware store is 3.7 ft above ground level. A ramp for the handicapped from the ground level to the entrance is to be built at a 12° angle of elevation. Find the length of the ramp. 18 ft 3.2

31. Let θ be an angle in standard position. Evaluate $\cos\theta$, $\tan\theta$, $\cot\theta$, $\sec\theta$, and $\csc\theta$ if θ lies in quadrant IV and $\sin\theta = -\frac{12}{13}$. See side column. 1.7

32. Given $\angle A = 59°$, $\angle B = 76°$, and $c = 33$ cm, compute the area of $\triangle ABC$ to the nearest square centimeter. 640 cm^2 4.5

33. A ladder rests against a house at a point that is 18 ft from the ground. If the ladder makes a 45° angle with the ground, what is the length of the ladder to the nearest foot? 25 ft 3.1

34. Express in sigma notation the sum of the first 8 terms of the geometric series $3 + (-6) + 12 + (-24) + \cdots$. $\sum_{k=1}^{8} 3(-2)^{n-1}$ 9.4

35. Use the sum identity for cosine to prove $\cos(2\pi + \alpha) = \cos\alpha$. See side column. 5.1

Graph each function. See side column.

36. $f(x) = e^{-x}$

37. $g(x) = 3^x$ 8.2

38. Express $5 \cos 3x \sin x$ as a sum. $\frac{5}{2}(\sin 4x - \sin 2x)$ 5.6

Additional Answers

31. $\frac{5}{13}, -\frac{12}{5}, -\frac{5}{12}, \frac{13}{5}, -\frac{13}{12}$

35. $\cos(2\pi + \alpha) = \cos 2\pi \cos\alpha - \sin 2\pi \sin\alpha = 1 \cdot \cos\alpha - 0 \cdot \sin\alpha = \cos\alpha$

36.

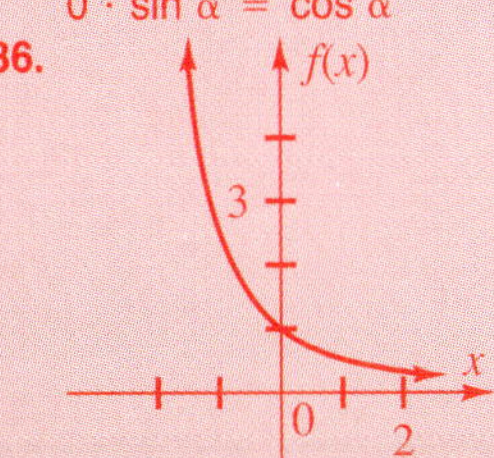

37.

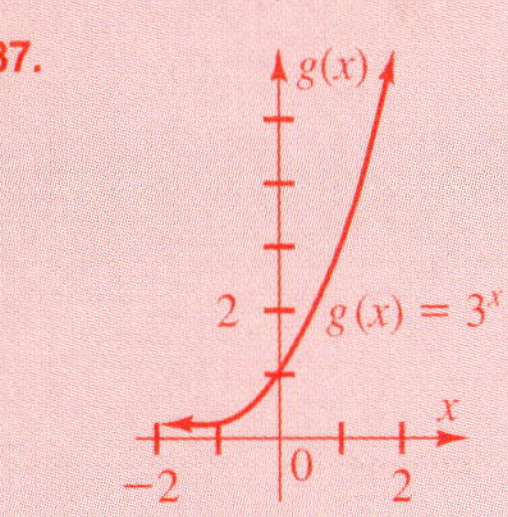

Calculate each product or quotient in polar form.

39. $6(\cos 15° + i \sin 15°) \cdot 3(\cos 30° + i \sin 30°)$ $18(\cos 45° + i \sin 45°)$ 7.6

40. $\dfrac{12(\cos 250° + i \sin 250°)}{2(\cos 70° + i \sin 70°)}$ $6(\cos 180° + i \sin 180°)$

41. Insert four arithmetic means between -9 and 11. $-5, -1, 3, 7$ 9.1

42. Use the law of sines to solve $\triangle ABC$ with $\angle B = 105°$, $\angle A = 32°$, and $a = 47.2$. Express the lengths of the sides to three significant digits and angle measures to the nearest tenth of a degree. $\angle C = 43°$, $b = 86.0$, $c = 60.7$ 4.1

Check graphically to see whether or not each equation is an identity.

43. $\dfrac{1 - \tan \alpha}{1 + \tan \alpha} = \dfrac{\cos 2\alpha}{1 + \sin 2\alpha}$ not an identity

44. $\dfrac{\sin \beta}{1 - \cos \beta} = \dfrac{1 + \cos \beta}{\sin \beta}$ identity 3.7

45. Approximate the solutions to $12 \sin^2 x - 8 \sin x + 1 = 0$ in degrees, for the interval $0° \leq x \leq 360°$. $10°, 30°, 150°, 170°$ 6.5

46. Find the length of the arc intercepted by a central angle of $\frac{3\pi}{4}$ if the radius of the circle is 120 ft. 90π ft 1.5

47. For the functions $f(x) = \sqrt{x} - 1$ and $g(x) = 2\sqrt{x}$, find $f(4) + g(9)$. 7 1.1

48. Express in sigma notation the sum of the first 8 terms of the arithmetic sequence $-6 + (-4) + (-2) + \cdots$. $\sum_{k=1}^{8} 2k - 8$ 9.3

49. The xy-axes are rotated about the origin through an angle of 60°. Find the equation of the curve $x^2 - 2\sqrt{3}\,xy - y^2 = 8$ in the new coordinate system. $-(x')^2 + (y')^2 = 4$ 6.6

50. Write the expression $\dfrac{1}{\tan \beta + \sec \beta} + \dfrac{1}{\sec \beta - \tan \beta}$ in terms of $\cos \beta$. $\dfrac{2}{\cos \beta}$ 3.5

51. How many years will it take for \$1800 to increase to \$2500, if it is deposited in an account that pays 7.25% interest, compounded monthly? 4.5 yr 8.6

52. Find the exact values of the sine and cosine of an angle θ in standard position, if the point with coordinates $(4, -4)$ lies on its terminal side. 1.6

If $\sin \alpha = \frac{\sqrt{2}}{2}$, $\tan \beta = -1$, $\frac{\pi}{2} < \alpha < \pi$, and $\frac{3\pi}{2} < \beta < 2\pi$, find the exact values for each of the following.

53. $\sin(\alpha + \beta)$ 1

54. $\sin(\alpha - \beta)$ 0 5.2

55. Solve $5^x = 37$. 2.2 8.5

56. Find the distance between the points $(2, 3)$ and $(-7, -10)$. $5\sqrt{10}$ 1.2

52. $\sin \theta = -\dfrac{\sqrt{2}}{2}$; $\cos \theta = \dfrac{\sqrt{2}}{2}$

Table of Squares, Cubes, Square and Cube Roots

N	N^2	N^3	$\sqrt{N}$	$\sqrt[3]{N}$	N	N^2	N^3	$\sqrt{N}$	$\sqrt[3]{N}$
1	1	1	1.000	1.000	**51**	2,601	132,651	7.141	3.708
2	4	8	1.414	1.260	**52**	2,704	140,608	7.211	3.733
3	9	27	1.732	1.442	**53**	2,809	148,877	7.280	3.756
4	16	64	2.000	1.587	**54**	2,916	157,464	7.348	3.780
5	25	125	2.236	1.710	**55**	3,025	166,375	7.416	3.803
6	36	216	2.449	1.817	**56**	3,136	175,616	7.483	3.826
7	49	343	2.646	1.913	**57**	3,249	185,193	7.550	3.849
8	64	512	2.828	2.000	**58**	3,364	195,112	7.616	3.871
9	81	729	3.000	2.080	**59**	3,481	205,379	7.681	3.893
10	100	1,000	3.162	2.154	**60**	3,600	216,000	7.746	3.915
11	121	1,331	3.317	2.224	**61**	3,721	226,981	7.810	3.936
12	144	1,728	3.464	2.289	**62**	3,844	238,328	7.874	3.958
13	169	2,197	3.606	2.351	**63**	3,969	250,047	7.937	3.979
14	196	2,744	3.742	2.410	**64**	4,096	262,144	8.000	4.000
15	225	3,375	3.873	2.466	**65**	4,225	274,625	8.062	4.021
16	256	4,096	4.000	2.520	**66**	4,356	287,496	8.124	4.041
17	289	4,913	4.123	2.571	**67**	4,489	300,763	8.185	4.062
18	324	5,832	4.243	2.621	**68**	4,624	314,432	8.246	4.082
19	361	6,859	4.359	2.668	**69**	4,761	328,509	8.307	4.102
20	400	8,000	4.472	2.714	**70**	4,900	343,000	8.367	4.121
21	441	9,261	4.583	2.759	**71**	5,041	357,911	8.426	4.141
22	484	10,648	4.690	2.802	**72**	5,184	373,248	8.485	4.160
23	529	12,167	4.796	2.844	**73**	5,329	389,017	8.544	4.179
24	576	13,824	4.899	2.884	**74**	5,476	405,224	8.602	4.198
25	625	15,625	5.000	2.924	**75**	5,625	421,875	8.660	4.217
26	676	17,576	5.099	2.962	**76**	5,776	438,976	8.718	4.236
27	729	19,683	5.196	3.000	**77**	5,929	456,533	8.775	4.254
28	784	21,952	5.292	3.037	**78**	6,084	474,552	8.832	4.273
29	841	24,389	5.385	3.072	**79**	6,241	493,039	8.888	4.291
30	900	27,000	5.477	3.107	**80**	6,400	512,000	8.944	4.309
31	961	29,791	5.568	3.141	**81**	6,561	531,441	9.000	4.327
32	1,024	32,768	5.657	3.175	**82**	6,724	551,368	9.055	4.344
33	1,089	35,937	5.745	3.208	**83**	6,889	571,787	9.110	4.362
34	1,156	39,304	5.831	3.240	**84**	7,056	592,704	9.165	4.380
35	1,225	42,875	5.916	3.271	**85**	7,225	614,125	9.220	4.397
36	1,296	46,656	6.000	3.302	**86**	7,396	636,056	9.274	4.414
37	1,369	50,653	6.083	3.332	**87**	7,569	658,503	9.327	4.431
38	1,444	54,872	6.164	3.362	**88**	7,744	681,472	9.381	4.448
39	1,521	59,319	6.245	3.391	**89**	7,921	704,969	9.434	4.465
40	1,600	64,000	6.325	3.420	**90**	8,100	729,000	9.487	4.481
41	1,681	68,921	6.403	3.448	**91**	8,281	753,571	9.539	4.498
42	1,764	74,088	6.481	3.476	**92**	8,464	778,688	9.592	4.514
43	1,849	79,507	6.557	3.503	**93**	8,649	804,357	9.644	4.531
44	1,936	85,184	6.633	3.530	**94**	8,836	830,584	9.695	4.547
45	2,025	91,125	6.708	3.557	**95**	9,025	857,375	9.747	4.563
46	2,116	97,336	6.782	3.583	**96**	9,216	884,736	9.798	4.579
47	2,209	103,823	6.856	3.609	**97**	9,409	912,673	9.849	4.595
48	2,304	110,592	6.928	3.634	**98**	9,604	941,192	9.899	4.610
49	2,401	117,649	7.000	3.659	**99**	9,801	970,299	9.950	4.626
50	2,500	125,000	7.071	3.684	**100**	10,000	1,000,000	10.000	4.642

Using a Table of Trigonometric Values

The table on pages 434 to 438 contains the values of the six trigonometric functions from 0°00′ to 90°00′ in 10-minute increments. The radian measure of each angle value is also listed. A portion of the table is shown below.

θ Deg.	**θ Rad.**	**Sin θ**	**Cos θ**	**Tan θ**	**Cot θ**	**Sec θ**	**Csc θ**		
19°00′	0.3316	0.3256	0.9455	0.3443	2.9042	1.0576	3.0716	1.2392	**71°00′**
10′	0.3345	0.3283	0.9446	0.3476	2.8770	1.0587	3.0458	1.2363	**50′**
20′	0.3374	0.3311	0.9436	0.3508	2.8502	1.0598	3.0206	1.2334	**40′**
30′	0.3403	0.3338	0.9426	0.3541	2.8239	1.0608	2.9957	1.2305	**30′**
40′	0.3432	0.3365	0.9417	0.3574	2.7980	1.0619	2.9713	1.2275	**20′**
50′	0.3462	0.3393	0.9407	0.3607	2.7725	1.0631	2.9474	1.2246	**10′**
20°00′	0.3491	0.3420	0.9397	0.3640	2.7475	1.0642	2.9238	1.2217	**70°00′**
10′	0.3520	0.3448	0.9387	0.3673	2.7228	1.0653	2.9006	1.2188	**50′**
		Cos θ	**Sin θ**	**Cot θ**	**Tan θ**	**Csc θ**	**Sec θ**	**θ Rad.**	**θ Deg.**

To find the value of a function for an angle between 0°00′ and 45°00′, read *down* the first column to locate the angle and across the *top* row to locate the function. The value of the function is at the intersection of that row and column. To illustrate, sin 19°30′ = 0.3338. To find the value of a function for an angle between 45°00′ and 90°00′, read *up* the last column to locate the angle and across the *bottom* row to locate the function. For instance, tan 70°50′ = 2.8770.

To find the value for a trigonometric function of an angle expressed in tenths of degrees, convert the angle measure to degrees and minutes first. To find the value for a trigonometric function of an angle expressed in radians, convert to decimal radian measure and then use the second and next to last columns of the table.

EXAMPLE 1 **Use a table to find the following:**

a. cos 42°20′ **b.** cot 54.5° **c.** $\csc \frac{\pi}{8}$

a. cos 42°20′ = 0.7392 *Use the table.*

b. cot 54.5° = cot 54°30′ *Convert to degrees and minutes.*
= 0.7133 *Use the table.*

c. $\csc \frac{\pi}{8} = \csc 0.3927$ *Convert to decimal radian measure.*
= 2.613 *Use second column of table for radians.*

To find the value for a trigonometric function of an angle greater than 90°, use a reference triangle in the appropriate quadrant. Find the value of the *reference angle,* which is the acute angle of the reference triangle that the terminal side makes with the x-axis.

EXAMPLE 2 **Use a table to find the following.** **a.** $\sin 245°40'$ **b.** $\sec \frac{23\pi}{12}$

a. $\sin 245°40' = -\sin 65°40'$ *Reference triangle is in quadrant III.*
$= -0.9112$ *Use the table.*

b. $\sec \frac{23\pi}{12} = \sec \frac{\pi}{12}$ *Reference triangle is in quadrant IV.*
$= \sec 0.2618$ *Convert to decimal radian measure.*
$= 1.0353$ *Use second column of table for radians.*

The table of trigonometric function values can also be used to solve trigonometric equations. The acute reference angle θ of a reference triangle can be found directly if the value of a function is known.

EXAMPLE 3 **Solve for θ, $0° \leq \theta < 360°$, to the nearest minute:**
a. $\tan \theta = 0.8642$ **b.** $\sin \theta = -0.5628$

a. $\tan \theta = 0.8642$
reference angle $= 40°50'$ *Use the table.*
$\theta = 40°50', 220°50'$ *Tangent is positive in quadrants I and III.*

b. $\sin \theta = -0.5616$ *Ignore negative to find reference angle.*
reference angle $= 34°10'$ *Use the table.*
$\theta = 214°10', 325°50'$ *Sine is negative in quadrants III and IV.*

EXERCISES

Use a table to find the following.

1. $\cos 9°20'$ 0.9868
2. $\sin 74°10'$ 0.9621
3. $\tan 43°50'$ 0.9601
4. $\csc 88°40'$ 1.0003
5. $\sec \frac{\pi}{18}$ 1.0154
6. $\cot \frac{5\pi}{24}$ 1.3032
7. $\sin \frac{17\pi}{9}$ −0.3420
8. $\csc \frac{23\pi}{30}$ 1.4945
9. $\tan 223°10'$ 0.9380
10. $\cos 147°20'$ −0.8418
11. $\csc 289°50'$ −1.0631
12. $\sin 315°30'$ −0.7009
13. $\sec 12°50'$ 1.0256
14. $\cot 53°40'$ 0.7355
15. $\csc 237°40'$ −1.1835
16. $\sec 302°10'$ 1.8783

Solve for θ, $0° \leq \theta < 360°$, to the nearest ten minutes.

17. $\sin \theta = 0.9013$ 64°20′; 115°40′
18. $\cot \theta = 14.301$ 4°0′; 184°0′
19. $\sec \theta = -5.4026$ 100°40′; 259°20′
20. $\tan \theta = -2.5605$ 111°20′; 291°20′
21. $\csc \theta = -1.4572$ 223°20′; 316°40′
22. $\cos \theta = 0.8843$ 27°50′; 332°10′

Linear Interpolation

The table on pages 434–438 includes angles expressed in 10-minute increments. If the measure of an angle falls between two entries, you can use a method called **linear interpolation** to find the values of trigonometric functions of that angle.

Over very small intervals, the graph of a trigonometric function is nearly a straight line. Consider the graph of $y = \sin x$ between the points $A(32°10', 0.5324)$ and $C(32°20', 0.5348)$. In order to estimate the value of $\sin x$ at a point between A and C where $x = 32°17'$, assume that the graph between points A and C is a straight line. Then, since triangles ABD and ACE are similar, corresponding sides are proportional. That is,

$$\frac{AD}{AE} = \frac{BD}{CE}$$

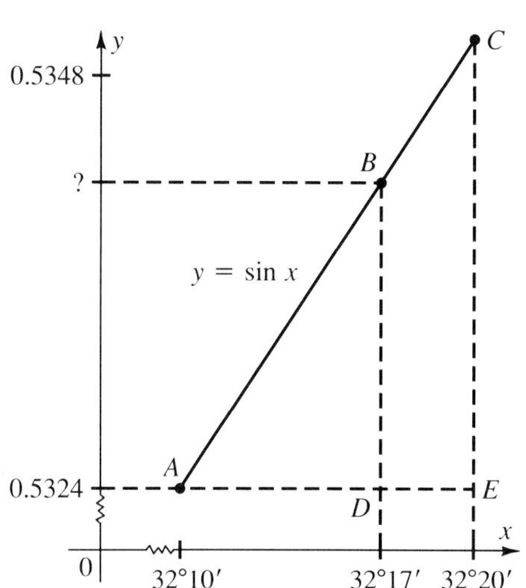

EXAMPLE 1 **Use linear interpolation to approximate sin 32°17′.**

First set up a table.

		x	$\sin x$		
$10'$	$7'$	$32°10'$	0.5324	d	0.0024
		$32°17'$	?		
		$32°20'$	0.5348		

Then write and solve a proportion.

$$\frac{7}{10} = \frac{d}{0.0024} \qquad \frac{AD}{AE} = \frac{BD}{CE}$$

$$d = \frac{7}{10}(0.0024) \approx 0.0017$$

Round to four decimal places, as in the sin table.

So, $\sin 32°17' = 0.5324 + d = 0.5324 + 0.0017 = 0.5341$

Linear interpolation can also be used to approximate the measure of an angle when the value of one of its trigonometric functions falls between entries in the table.

EXAMPLE 2 **Use linear interpolation to approximate θ if $\tan \theta = 0.4000$, $0° \le \theta \le 360°$.**

Use a trigonometric table to find the values of tangent just less than and just greater than 0.4000.

		θ	$\tan \theta$		
$10'$	d	$21°40'$	0.3973	0.0027	0.0033
		?	0.4000		
		$21°50'$	0.4006		

Write and solve a proportion.

$$\frac{d}{10'} = \frac{0.0027}{0.0033}$$

$$d = \frac{27}{33}(10') \approx 8 \qquad \textit{Round.}$$

So, $\theta = 21°40' + d = 21°40' + 8' = 21°48'$

Note that $\tan \theta$ is also positive in Quadrant III, so θ can also equal $180° + 21°48' = 201°48'$ if $\tan \theta = 0.4000$.

EXERCISES

Use linear interpolation to approximate each function.

1. $\sin 27°23'$ 0.4599
2. $\cos 31°42'$ 0.8508
3. $\tan 16°17'$ 0.2921
4. $\sec 87°21'$ 21.629
5. $\csc 12°52'$ 4.491
6. $\cot 33°47'$ 1.4947
7. $\cos -15°53'$ 0.9618
8. $\tan -25°34'$ −0.4784
9. $\sin -52°08'$ −0.7894
10. $\sec -12°05'$ 1.023
11. $\cot -82°14'$ −0.1364
12. $\csc -55°22'$ −1.215
13. $\sin 0.4111$ 0.3996
14. $\cos 0.4660$ 0.8934
15. $\tan 0.3210$ 0.3325
16. $\cos -0.6000$ 0.8253
17. $\sin -0.5302$ −0.5057
18. $\tan -0.7777$ −0.9847

Use linear interpolation to approximate θ, $0° \le \theta \le 360°$, to the nearest minute.

19. $\sin \theta = 0.3205$ 18°42′; 161°18′
20. $\cos \theta = 0.8111$ 35°48′; 324°12′
21. $\tan \theta = 1.7032$ 59°35′; 239°35′
22. $\sin \theta = -0.5107$ 210°43′; 329°17′
23. $\cos \theta = 0.2215$ 77°12′; 282°48′
24. $\tan \theta = 6.0721$ 80°39′; 260°39′
25. $\cos \theta = 0.1155$ 83°22′; 276°38′
26. $\tan \theta = 1.1617$ 49°17′; 229°17′
27. $\sin \theta = 0.5152$ 31°01′; 148°59′
28. $\tan \theta = -2.5487$ 111°25′; 291°25′
29. $\sec \theta = -5.381$ 100°43′; 259°17′
30. $\csc \theta = -1.458$ 223°18′; 316°42′
31. $\cos \theta = -0.1205$ 96°55′; 263°05′
32. $\cot \theta = -7.0123$ 171°53′; 351°53′
33. $\sin \theta = -0.2222$ 192°50′; 347°10′
34. $\csc \theta = -5.562$ 190°21′; 349°39′
35. $\sec \theta = -1.018$ 169°13′; 190°47′
36. $\cot \theta = -4.4163$ 167°14′; 347°14′

Using a Table of Common Logarithms

The table on pages 430 and 431 contains the common logarithms of numbers from 1 to 9.99. A portion of the table is shown below. The logarithm values are rounded to four decimal places and the decimal points are omitted.

N	0	1	2	3	4	5	6	7	8	9
4.1	6128	6138	6149	6160	6170	6180	6191	6201	6212	6222

To find the logarithm of a number N between 1 and 10, find the first two digits of N in the column on the left and the third digit in the top row. The logarithm of N is shown at the intersection of that row and column. To illustrate, the logarithm of 4.17 is found at the intersection of the row for 41 and the column for 7. To four decimal places, $\log 4.17 = 0.6201$.

To find the logarithm of a number greater than 10 or between 0 and 1, write the number in scientific notation and then use the properties of logarithms.

EXAMPLE 1 **Use a table to find each logarithm:**
a. $\log 4170$ **b.** $\log 0.0417$

a. $\log 4170 = \log (4.17 \times 10^3)$
$= \log 4.17 + \log 10^3$ *$\log_b MN = \log_b M + \log_b N$*
$= 0.6201 + 3$ *$\log_b b^x = x$*
$= 3.6201$ *To four decimal places*

b. $\log 0.0417 = \log (4.17 \times 10^{-2})$
$= \log 4.17 + \log 10^{-2}$
$= 0.6201 + (-2)$
$= 0.6201 - 2$, or -1.3799 *To four decimal places*

A common logarithm can be written as the sum of a *nonnegative* number less than 1, called the **mantissa,** and an integer, called the **characteristic.**

	mantissa		*characteristic*
$\log 4170 =$	0.6201	$+$	3
$\log 0.0417 =$	0.6201	$+$	(-2)

The logarithm of a number such as 0.0417 can be written in many ways. Sometimes it is more convenient to use one form than another.

$\log 0.0417 = -1.3799$	*Not* in mantissa/characteristic form	
$\log 0.0417 = 0.6201 - 2$	Mantissa: 0.6201	Characteristic: -2
$\log 0.0417 = 2.6201 - 4$	Mantissa: 0.6201	Characteristic: $2 - 4 = -2$
$\log 0.0417 = 8.6201 - 10$	Mantissa: 0.6201	Characteristic: $8 - 10 = -2$

The table of common logarithms can also be used to find a number N if $\log N$ is known. If $\log N$ is between 0 and 1, then N is between 1 and 10 and its value can be read directly from the table. For instance, if $\log N = 0.7932$, the table on page 431 shows that the logarithm closest to this value is found at the intersection of the row for 62 and the column for 1. That is, N is approximately 6.21, to three significant digits.

EXAMPLE 2 **Use a table to find N:**
a. $\log N = 5.1904$ **b.** $\log N = -3.2165$.

First write each number in mantissa/characteristic form.

a. $\log N = 5.1905$
$= 0.1905 + 5$ *Mantissa/characteristic form*
$= \log 1.55 + 5$ *Use the table.*
$= \log 1.55 + \log 10^5$
$= \log (1.55 \times 10^5)$
$= \log 155{,}000$
So, $N = 155{,}000$ *Three significant digits*

b. $\log N = -3.2165$
$= 4 + -3.2165 - 4$
$= 0.7835 - 4$
$= \log 6.07 - 4$
$= \log 6.07 + \log 10^{-4}$
$= \log (6.07 \times 10^{-4})$
$= \log 0.000607$
So, $N = 0.000607$

EXERCISES

Use a table to find each common logarithm to four decimal places.

1. log 2.58
0.4116
2. log 9.04
0.9562
3. log 16.8
1.2253
4. log 340
2.5315
5. log 72,500
4.8603
6. log 0.143
−0.8447
7. log 0.00329
−2.4828
8. log 0.0605
−1.2182

Use a table to find N to three significant digits.

9. $\log N = 0.8982$
7.91
10. $\log N = 0.5528$
3.57
11. $\log N = 1.9805$
95.6
12. $\log N = 3.6944$
4950
13. $\log N = 0.9842 - 2$
0.0964
14. $\log N = 0.1732 - 1$
0.149

Use a table to find each common logarithm to four decimal places.

15. log 300,000
5.4771
16. log 75,000,000
7.8751
17. log 0.000086
−4.0655
18. 0.0000006
−5.2218

Use a table to find N to three significant digits.

19. $\log N = 8.8131$
650,000,000
20. $\log N = 9.5265$
3,360,000,000
21. $\log N = 4.7378 - 10$
0.00000547
22. $\log N = -1.7714$
0.0169
23. $\log N = -2.3100$
0.00490
24. $\log N = 6.9272 - 10$
0.000846
25. $\log N = 11.9559$
903,000,000,000
26. $\log N = 13.0042$
10,100,000,000,000
27. $\log N = -12.5406$
0.00000000000288
28. $\log N = -10.7393$
0.0000000000182

Computation Using Common Logarithms

Common logarithms can be used to simplify computations. In this process, addition and subtraction replace the operations of multiplication and division, while multiplication replaces the operation of exponentiation.

EXAMPLE 1 **Use common logarithms from a table to evaluate:**

a. $0.00496 \times 23{,}600$ **b.** $\frac{0.000408}{7.73}$ **c.** $\sqrt[3]{0.425}$

a. Let $N = 0.00496 \times 23{,}600$

Then $\log N = \log (0.00496 \times 23{,}600)$

$= \log 0.00496 + \log 23{,}600$ *$\log MN = \log M + \log N$*

$= \log (4.96 \times 10^{-3}) + \log (2.36 \times 10^{4})$

$= (0.6955 - 3) + (4.3729)$ *Use the table.*

$= 2.0684$

$= 0.0684 + 2$ *Mantissa/characteristic form*

So, $N = 1.17 \times 10^{2}$ *Use the table again. Find the closest entry.*

$N = 117$ *Three significant digits*

b. Let $N = \frac{0.000408}{7.73}$

Then $\log N = \log \left(\frac{0.000408}{7.73}\right)$

$= \log 0.000408 - \log 7.73$ *$\log \frac{M}{N} = \log M - \log N$*

$= \log (4.08 \times 10^{-4}) - \log 7.73$

$= (0.6107 - 4) - 0.8882$ *Use the table.*

$= (1.6107 - 5) - 0.8882$ *To obtain a positive mantissa, write $0.6107 - 4$ as $1.6107 - 5$.*

$= 0.7225 - 5$ *Mantissa/characteristic form*

So, $N = 5.28 \times 10^{-5}$ *Use the table again. Find the closest entry.*

$N = 0.0000528$ *Three significant digits*

c. Let $N = \sqrt[3]{0.425} = 0.425^{\frac{1}{3}}$

Then $\log N = \log 0.425^{\frac{1}{3}}$

$= \frac{1}{3} \log 0.425$ *$\log N^{r} = r \log N$*

$= \frac{1}{3}(0.6284 - 1)$ *Use the table.*

$= \frac{1}{3}(2.6284 - 3)$ *To get an integer for the negative part of the characteristic, write $0.6284 - 1$ as $2.6284 - 3$.*

$= 0.8761 - 1$ *Mantissa/characteristic form*

So, $N = 7.52 \times 10^{-1}$ *Use the table again. Find the closest entry.*

$N = 0.752$ *Three significant digits*

In the next example, several operations are performed using common logarithms.

EXAMPLE 2 **Use common logarithms from a table to evaluate:** $\dfrac{0.0493(3.71)^4}{52.6}$

$$
\begin{aligned}
\text{Let} \qquad N &= \frac{0.0493(3.71)^4}{52.6} \\
\text{Then} \quad \log N &= \log\left(\frac{0.0493(3.71)^4}{52.6}\right) \\
&= \log 0.0493 + \log 3.71^4 - \log 52.6 \\
&= \log 0.0493 + 4 \log 3.71 - \log 52.6 \\
&= (0.6928 - 2) + 4(0.5694) - 1.7210 \\
&= 0.6928 - 2 + 2.2776 - 1.7210 \\
&= -0.7506, \quad \text{or } 0.2494 - 1 \\
\text{So,} \qquad N &= 0.178 \qquad \textit{Three significant digits}
\end{aligned}
$$

EXERCISES

Use common logarithms from a table to evaluate to three significant digits.

1. 2.96×9.74 28.8
2. 7.45×8.93 66.5
3. 0.0309×768 23.7
4. 0.00265×37.6 0.0996
5. 3.49^3 42.5
6. 6.27^4 1550
7. $\sqrt[5]{925{,}000}$ 15.6
8. $\sqrt[4]{34{,}800}$ 13.7
9. $256(4.95)^2$ 6270
10. $0.747(8.71)^2$ 56.7
11. $3760\sqrt[4]{46{,}800}$ 55,300
12. $0.0129\sqrt[3]{5080}$ 0.222
13. $\dfrac{5620}{3.85}$ 1460
14. $\dfrac{8.38}{0.125}$ 67.0
15. $\dfrac{0.714(7.36)}{247}$ 0.0213
16. $\dfrac{0.00368(2.54)}{1.13}$ 0.00827
17. $\dfrac{56.5(5.66)^4}{363{,}000}$ 0.160
18. $\dfrac{0.296(4.87)^3}{1370}$ 0.0250
19. $0.0165\sqrt[6]{7280}$ 0.0726
20. $45{,}200\sqrt[8]{0.968}$ 45,000
21. $\sqrt{26.8(127)^3}$ 7410
22. $\sqrt[3]{0.0561}(4.99)^2$ 1.12
23. $56.3(4.26)^{-6}$ 0.00942
24. $7250(36.2)^{-5}$ 0.000117
25. $\dfrac{4.72}{0.928(42.6)^2}$ 0.00280
26. $\dfrac{38.3}{2.09(0.0256)^3}$ 1,090,000
27. $\dfrac{7.66(0.0478)}{0.395(26.7)}$ 0.0347
28. $\dfrac{686(0.00261)}{0.985(49.7)}$ 0.0366
29. $\dfrac{\sqrt{21.3}}{\sqrt[7]{26{,}700}}$ 1.08
30. $\dfrac{\sqrt[5]{0.796}}{\sqrt[3]{545}}$ 0.117
31. $\left(\dfrac{\sqrt[3]{751}}{275}\right)^4$ 0.00000119
32. $\left(\dfrac{368}{\sqrt[4]{911}}\right)^3$ 300,000
33. $\sqrt{\dfrac{92.6(1.49)^3}{0.368}}$ 28.9
34. $\sqrt[3]{\dfrac{476(0.155)^2}{76.7}}$ 0.530
35. $\sqrt{\dfrac{529(1.48)^5}{1.37\sqrt[3]{0.0245}}}$ 97.2
36. $\sqrt[4]{\dfrac{29.4\sqrt[5]{0.766}}{301(0.135)^3}}$ 2.48

Table of Common Logarithms

N	0	1	2	3	4	5	6	7	8	9
1.0	0000	0043	0086	0128	0170	0212	0253	0294	0334	0374
1.1	0414	0453	0492	0531	0569	0607	0645	0682	0719	0755
1.2	0792	0828	0864	0899	0934	0969	1004	1038	1072	1106
1.3	1139	1173	1206	1239	1271	1303	1335	1367	1399	1430
1.4	1461	1492	1523	1553	1584	1614	1644	1673	1703	1732
1.5	1761	1790	1818	1847	1875	1903	1931	1959	1987	2014
1.6	2041	2068	2095	2122	2148	2175	2201	2227	2253	2279
1.7	2304	2330	2355	2380	2405	2430	2455	2480	2504	2529
1.8	2553	2577	2601	2625	2648	2672	2695	2718	2742	2765
1.9	2788	2810	2833	2856	2878	2900	2923	2945	2967	2989
2.0	3010	3032	3054	3075	3096	3118	3139	3160	3181	3201
2.1	3222	3243	3263	3284	3304	3324	3345	3365	3385	3404
2.2	3424	3444	3464	3483	3502	3522	3541	3560	3579	3598
2.3	3617	3636	3655	3674	3692	3711	3729	3747	3766	3784
2.4	3802	3820	3838	3856	3874	3892	3909	3927	3945	3962
2.5	3979	3997	4014	4031	4048	4065	4082	4099	4116	4133
2.6	4150	4166	4183	4200	4216	4232	4249	4265	4281	4298
2.7	4314	4330	4346	4362	4378	4393	4409	4425	4440	4456
2.8	4472	4487	4502	4518	4533	4548	4564	4579	4594	4609
2.9	4624	4639	4654	4669	4683	4698	4713	4728	4742	4757
3.0	4771	4786	4800	4814	4829	4843	4857	4871	4886	4900
3.1	4914	4928	4942	4955	4969	4983	4997	5011	5024	5038
3.2	5051	5065	5079	5092	5105	5119	5132	5145	5159	5172
3.3	5185	5198	5211	5224	5237	5250	5263	5276	5289	5302
3.4	5315	5328	5340	5353	5366	5378	5391	5403	5416	5428
3.5	5441	5453	5465	5478	5490	5502	5514	5527	5539	5551
3.6	5563	5575	5587	5599	5611	5623	5635	5647	5658	5670
3.7	5682	5694	5705	5717	5729	5740	5752	5763	5775	5786
3.8	5798	5809	5821	5832	5843	5855	5866	5877	5888	5899
3.9	5911	5922	5933	5944	5955	5966	5977	5988	5999	6010
4.0	6021	6031	6042	6053	6064	6075	6085	6096	6107	6117
4.1	6128	6138	6149	6160	6170	6180	6191	6201	6212	6222
4.2	6232	6243	6253	6263	6274	6284	6294	6304	6314	6325
4.3	6335	6345	6355	6365	6375	6385	6395	6405	6415	6425
4.4	6435	6444	6454	6464	6474	6484	6493	6503	6513	6522
4.5	6532	6542	6551	6561	6571	6580	6590	6599	6609	6618
4.6	6628	6637	6646	6656	6665	6675	6684	6693	6702	6712
4.7	6721	6730	6739	6749	6758	6767	6776	6785	6794	6803
4.8	6812	6821	6830	6839	6848	6857	6866	6875	6884	6893
4.9	6902	6911	6920	6928	6937	6946	6955	6964	6972	6981
5.0	6990	6998	7007	7016	7024	7033	7042	7050	7059	7067
5.1	7076	7084	7093	7101	7110	7118	7126	7135	7143	7152
5.2	7160	7168	7177	7185	7193	7202	7210	7218	7226	7235
5.3	7243	7251	7259	7267	7275	7284	7292	7300	7308	7316
5.4	7324	7332	7340	7348	7356	7364	7372	7380	7388	7396

Table of Common Logarithms

N	0	1	2	3	4	5	6	7	8	9
5.5	7404	7412	7419	7427	7435	7443	7451	7459	7466	7474
5.6	7482	7490	7497	7505	7513	7520	7528	7536	7543	7551
5.7	7559	7566	7574	7582	7589	7597	7604	7612	7619	7627
5.8	7634	7642	7649	7657	7664	7672	7679	7686	7694	7701
5.9	7709	7716	7723	7731	7738	7745	7752	7760	7767	7774
6.0	7782	7789	7796	7803	7810	7818	7825	7832	7839	7846
6.1	7853	7860	7868	7875	7882	7889	7896	7903	7910	7917
6.2	7924	7931	7938	7945	7952	7959	7966	7973	7980	7987
6.3	7993	8000	8007	8014	8021	8028	8035	8041	8048	8055
6.4	8062	8069	8075	8082	8089	8096	8102	8109	8116	8122
6.5	8129	8136	8142	8149	8156	8162	8169	8176	8182	8189
6.6	8195	8202	8209	8215	8222	8228	8235	8241	8248	8254
6.7	8261	8267	8274	8280	8287	8293	8299	8306	8312	8319
6.8	8325	8331	8338	8344	8351	8357	8363	8370	8376	8382
6.9	8388	8395	8401	8407	8414	8420	8426	8432	8439	8445
7.0	8451	8457	8463	8470	8476	8482	8488	8494	8500	8506
7.1	8513	8519	8525	8531	8537	8543	8549	8555	8561	8567
7.2	8573	8579	8585	8591	8597	8603	8609	8615	8621	8627
7.3	8633	8639	8645	8651	8657	8663	8669	8675	8681	8686
7.4	8692	8698	8704	8710	8716	8722	8727	8733	8739	8745
7.5	8751	8756	8762	8768	8774	8779	8785	8791	8797	8802
7.6	8808	8814	8820	8825	8831	8837	8842	8848	8854	8859
7.7	8865	8871	8876	8882	8887	8893	8899	8904	8910	8915
7.8	8921	8927	8932	8938	8943	8949	8954	8960	8965	8971
7.9	8976	8982	8987	8993	8998	9004	9009	9015	9020	9025
8.0	9031	9036	9042	9047	9053	9058	9063	9069	9074	9079
8.1	9085	9090	9096	9101	9106	9112	9117	9122	9128	9133
8.2	9138	9143	9149	9154	9159	9165	9170	9175	9180	9186
8.3	9191	9196	9201	9206	9212	9217	9222	9227	9232	9238
8.4	9243	9248	9253	9258	9263	9269	9274	9279	9284	9289
8.5	9294	9299	9304	9309	9315	9320	9325	9330	9335	9340
8.6	9345	9350	9355	9360	9365	9370	9375	9380	9385	9390
8.7	9395	9400	9405	9410	9415	9420	9425	9430	9435	9440
8.8	9445	9450	9455	9460	9465	9469	9474	9479	9484	9489
8.9	9494	9499	9504	9509	9513	9518	9523	9528	9533	9538
9.0	9542	9547	9552	9557	9562	9566	9571	9576	9581	9586
9.1	9590	9595	9600	9605	9609	9614	9619	9624	9628	9633
9.2	9638	9643	9647	9652	9657	9661	9666	9671	9675	9680
9.3	9685	9689	9694	9699	9703	9708	9713	9717	9722	9727
9.4	9731	9736	9741	9745	9750	9754	9759	9763	9768	9773
9.5	9777	9782	9786	9791	9795	9800	9805	9809	9814	9818
9.6	9823	9827	9832	9836	9841	9845	9850	9854	9859	9863
9.7	9868	9872	9877	9881	9886	9890	9894	9899	9903	9908
9.8	9912	9917	9921	9926	9930	9934	9939	9943	9948	9952
9.9	9956	9961	9965	9969	9974	9978	9983	9987	9991	9996

Table of Natural Logarithms (ln x)

x	0.00	0.01	0.02	0.03	0.04	0.05	0.06	0.07	0.08	0.09
1.0	0.0000	0.0100	0.0198	0.0296	0.0392	0.0488	0.0583	0.0677	0.0770	0.0862
1.1	0.0953	0.1044	0.1133	0.1222	0.1310	0.1398	0.1484	0.1570	0.1655	0.1740
1.2	0.1823	0.1906	0.1989	0.2070	0.2151	0.2231	0.2311	0.2390	0.2469	0.2546
1.3	0.2624	0.2700	0.2776	0.2852	0.2927	0.3001	0.3075	0.3148	0.3221	0.3293
1.4	0.3365	0.3436	0.3507	0.3577	0.3646	0.3716	0.3784	0.3853	0.3920	0.3988
1.5	0.4055	0.4121	0.4187	0.4253	0.4318	0.4383	0.4447	0.4511	0.4574	0.4637
1.6	0.4700	0.4762	0.4824	0.4886	0.4947	0.5008	0.5068	0.5128	0.5188	0.5247
1.7	0.5306	0.5365	0.5423	0.5481	0.5539	0.5596	0.5653	0.5710	0.5766	0.5822
1.8	0.5878	0.5933	0.5988	0.6043	0.6098	0.6152	0.6206	0.6259	0.6313	0.6166
1.9	0.6419	0.6471	0.6523	0.6575	0.6627	0.6678	0.6729	0.6780	0.6831	0.6881
2.0	0.6931	0.6981	0.7031	0.7080	0.7130	0.7178	0.7227	0.7275	0.7324	0.7372
2.1	0.7419	0.7467	0.7514	0.7561	0.7608	0.7655	0.7701	0.7747	0.7793	0.7839
2.2	0.7885	0.7930	0.7975	0.8020	0.8065	0.8109	0.8154	0.8198	0.8242	0.8286
2.3	0.8329	0.8372	0.8416	0.8459	0.8502	0.8544	0.8587	0.8629	0.8671	0.8713
2.4	0.8755	0.8796	0.8838	0.8879	0.8920	0.8961	0.9002	0.9042	0.9083	0.9123
2.5	0.9163	0.9203	0.9243	0.9282	0.9322	0.9361	0.9400	0.9439	0.9478	0.9517
2.6	0.9555	0.9594	0.9632	0.9670	0.9708	0.9746	0.9783	0.9821	0.9858	0.9895
2.7	0.9933	0.9969	1.0006	1.0043	1.0080	1.0116	1.0152	0.0188	1.0225	1.0260
2.8	1.0296	1.0332	1.0367	1.0403	1.0438	1.0473	1.0508	1.0543	1.0578	1.0613
2.9	1.0647	1.0682	1.0716	1.0750	1.0784	1.0818	1.0852	1.0886	1.0919	1.0953
3.0	1.0986	1.1019	1.1053	1.1086	1.1119	1.1151	1.1184	1.1217	1.1249	1.1282
3.1	1.1314	1.1346	1.1378	1.1410	1.1442	1.1474	1.1506	1.1537	1.1569	1.1600
3.2	1.1632	1.1663	1.1694	1.1725	1.1756	1.1787	1.1817	1.1848	1.1878	1.1909
3.3	1.1939	1.1970	1.2000	1.2030	1.2060	1.2090	1.2119	1.2149	1.2179	1.2208
3.4	1.2238	1.2267	1.2296	1.2326	1.2355	1.2384	1.2413	1.2442	1.2470	1.2499
3.5	1.2528	1.2556	1.2585	1.2613	1.2641	1.2669	1.2698	1.2726	1.2754	1.2782
3.6	1.2809	1.2837	1.2865	1.2892	1.2920	1.2947	1.2975	1.3002	1.3029	1.3056
3.7	1.3083	1.3110	1.3137	1.3164	1.3191	1.3218	1.3244	1.3271	1.3297	1.3324
3.8	1.3350	1.3376	1.3403	1.3429	1.3455	1.3481	1.3507	1.3533	1.3558	1.3584
3.9	1.3610	1.3635	1.3661	1.3686	1.3712	1.3737	1.3762	1.3788	1.3813	1.3838
4.0	1.3863	1.3888	1.3913	1.3938	1.3962	1.3987	1.4012	1.4036	1.4061	1.4085
4.1	1.4110	1.4134	1.4159	1.4183	1.4207	1.4231	1.4255	1.4279	1.4303	1.4327
4.2	1.4351	1.4375	1.4398	1.4422	1.4446	1.4469	1.4493	1.4516	1.4540	1.4563
4.3	1.4586	1.4609	1.4633	1.4656	1.4679	1.4702	1.4725	1.4748	1.4770	1.4793
4.4	1.4816	1.4839	1.4861	1.4884	1.4907	1.4929	1.4952	1.4974	1.4996	1.5019
4.5	1.5041	1.5063	1.5085	1.5107	1.5129	1.5151	1.5173	1.5195	1.5217	1.5239
4.6	1.5261	1.5282	1.5304	1.5326	1.5347	1.5369	1.5390	1.5412	1.5433	1.5454
4.7	1.5476	1.5497	1.5518	1.5539	1.5560	1.5581	1.5602	1.5623	1.5644	1.5665
4.8	1.5686	1.5707	1.5728	1.5748	1.5769	1.5790	1.5810	1.5831	1.5851	1.5872
4.9	1.5892	1.5913	1.5933	1.5953	1.5974	1.5994	1.6014	1.6034	1.6054	1.6074
5.0	1.6094	1.6114	1.6134	1.6154	1.6174	1.6194	1.6214	1.6233	1.6253	1.6273
5.1	1.6292	1.6312	1.6332	1.6351	1.6371	1.6390	1.6409	1.6429	1.6448	1.6467
5.2	1.6487	1.6506	1.6525	1.6544	1.6563	1.6582	1.6601	1.6620	1.6639	1.6658
5.3	1.6677	1.6696	1.6715	1.6734	1.6752	1.6771	1.6790	1.6808	1.6827	1.6845
5.4	1.6864	1.6882	1.6901	1.6919	1.6938	1.6956	1.6974	1.6993	1.7001	1.7029

Table of Natural Logarithms (ln x)

x	0.00	0.01	0.02	0.03	0.04	0.05	0.06	0.07	0.08	0.09
5.5	1.7047	1.7066	1.7084	1.7102	1.7120	1.7138	1.7156	1.7174	1.7192	1.7210
5.6	1.7228	1.7246	1.7263	1.7281	1.7299	1.7317	1.7334	1.7352	1.7370	1.7387
5.7	1.7405	1.7422	1.7440	1.7457	1.7475	1.7492	1.7509	1.7527	1.7544	1.7561
5.8	1.7579	1.7596	1.7613	1.7630	1.7647	1.7664	1.7682	1.7699	1.7716	1.7733
5.9	1.7750	1.7766	1.7783	1.7800	1.7817	1.7834	1.7851	1.7867	1.7884	1.7901
6.0	1.7918	1.7934	1.7951	1.7967	1.7984	1.8001	1.8017	1.8034	1.8050	1.8066
6.1	1.8083	1.8099	1.8116	1.8132	1.8148	1.8165	1.8181	1.8197	1.8213	1.8229
6.2	1.8245	1.8262	1.8278	1.8294	1.8310	1.8326	1.8342	1.8358	1.8374	1.8390
6.3	1.8406	1.8421	1.8437	1.8453	1.8469	1.8485	1.8500	1.8516	1.8532	1.8547
6.4	1.8563	1.8579	1.8594	1.8610	1.8625	1.8641	1.8656	1.8672	1.8687	1.8703
6.5	1.8718	1.8733	1.8749	1.8764	1.8779	1.8795	1.8810	1.8825	1.8840	1.8856
6.6	1.8871	1.8886	1.8901	1.8916	1.8931	1.8946	1.8961	1.8976	1.8991	1.9006
6.7	1.9021	1.9036	1.9051	1.9066	1.9081	1.9095	1.9110	1.9125	1.9140	1.9155
6.8	1.9169	1.9184	1.9199	1.9213	1.9228	1.9242	1.9257	1.9272	1.9286	1.9301
6.9	1.9315	1.9330	1.9344	1.9359	1.9373	1.9387	1.9402	1.9416	1.9430	1.9445
7.0	1.9459	1.9473	1.9488	1.9502	1.9516	1.9530	1.9544	1.9559	1.9573	1.9587
7.1	1.9601	1.9615	1.9629	1.9643	1.9657	1.9671	1.9685	1.9699	1.9713	1.9727
7.2	1.9741	1.9755	1.9769	1.9782	1.9796	1.9810	1.9824	1.9838	1.9851	1.9865
7.3	1.9879	1.9892	1.9906	1.9920	1.9933	1.9947	1.9961	1.9974	1.9988	2.0001
7.4	2.0015	2.0028	2.0042	2.0055	2.0069	2.0082	2.0096	2.0109	2.0122	2.0136
7.5	2.0149	2.0162	2.0176	2.0189	2.0202	2.0215	2.0229	2.0242	2.0255	2.0268
7.6	2.0282	2.0295	2.0308	2.0321	2.0334	2.0347	2.0360	2.0373	2.0386	2.0399
7.7	2.0412	2.0425	2.0438	2.0451	2.0464	2.0477	2.0490	2.0503	2.0516	2.0528
7.8	2.0541	2.0554	2.0567	2.0580	2.0592	2.0605	2.0618	2.0631	2.0643	2.0665
7.9	2.0669	2.0681	2.0694	2.0707	2.0719	2.0732	2.0744	2.0757	2.0769	2.0782
8.0	2.0794	2.0807	2.0819	2.0832	2.0844	2.0857	2.0869	2.0882	2.0894	2.0906
8.1	2.0919	2.0931	2.0943	2.0956	2.0968	2.0980	2.0992	2.1005	2.1017	2.1029
8.2	2.1041	2.1054	2.1066	2.1078	2.1090	2.1102	2.1114	2.1126	2.1138	2.1150
8.3	2.1163	2.1175	2.1187	2.1199	2.1211	2.1223	2.1235	2.1247	2.1258	2.1270
8.4	2.1282	2.1294	2.1306	2.1318	2.1330	2.1342	2.1353	2.1365	2.1377	2.1389
8.5	2.1401	2.1412	2.1424	2.1436	2.1448	2.1459	2.1471	2.1483	2.1494	2.1506
8.6	2.1518	2.1529	2.1541	2.1552	2.1564	2.1576	2.1587	2.1599	2.1610	2.1622
8.7	2.1633	2.1645	2.1656	2.1668	2.1679	2.1691	2.1702	2.1713	2.1725	2.1736
8.8	2.1748	2.1759	2.1770	2.1782	2.1793	2.1804	2.1815	2.1827	2.1838	2.1849
8.9	2.1861	2.1872	2.1883	2.1894	2 1905	2.1917	2.1928	2.1939	2.1950	2.1961
9.0	2.1972	2.1983	2.1994	2.2006	2.2017	2.2028	2.2039	2.2050	2.2061	2.2072
9.1	2.2083	2.2094	2.2105	2.2116	2.2127	2.2138	2.2148	2.2159	2.2170	2.2181
9.2	2.2192	2.2203	2.2214	2.2225	2.2235	2.2246	2.2257	2.2268	2.2279	2.2289
9.3	2.2300	2.2311	2.2322	2.2332	2.2343	2.2354	2.2364	2.2375	2.2386	2.2396
9.4	2.2407	2.2418	2.2428	2.2439	2.2450	2.2460	2.2471	2.2481	2.2492	2.2502
9.5	2.2513	2.2523	2.2534	2.2544	2.2555	2.2565	2.2576	2.2586	2.2597	2.2607
9.6	2.2618	2.2628	2.2638	2.2649	2.2659	2.2670	2.2680	2.2690	2.2701	2.2711
9.7	2.2721	2.2732	2.2742	2.2752	2.2762	2.2773	2.2783	2.2793	2.2803	2.2814
9.8	2.2824	2.2834	2.2844	2.2854	2.2865	2.2875	2.2885	2.2895	2.2905	2.2915
9.9	2.2925	2.2935	2.2946	2.2956	2.2966	2.2976	2.2986	2.2996	2.3006	2.3016

Table of Values of the Trigonometric Functions

θ Deg.	θ Rad.	Sin θ	Cos θ	Tan θ	Cot θ	Sec θ	Csc θ		
0°00′	0.0000	0.0000	1.0000	0.0000		1.000		1.5708	90°00′
10′	0.0029	0.0029	1.0000	0.0029	343.77	1.000	343.8	1.5679	50′
20′	0.0058	0.0058	1.0000	0.0058	171.89	1.000	171.9	1.5650	40′
30′	0.0087	0.0087	1.0000	0.0087	114.59	1.000	114.6	1.5621	30′
40′	0.0116	0.0116	0.9999	0.0116	85.940	1.000	85.95	1.5592	20′
50′	0.0145	0.0145	0.9999	0.0145	68.750	1.000	68.76	1.5563	10′
1°00′	0.0175	0.0175	0.9998	0.0175	57.290	1.000	57.30	1.5533	89°00′
10′	0.0204	0.0204	0.9998	0.0204	49.104	1.000	49.11	1.5504	50′
20′	0.0233	0.0233	0.9997	0.0233	42.964	1.000	42.98	1.5475	40′
30′	0.0262	0.0262	0.9997	0.0262	38.188	1.000	38.20	1.5446	30′
40′	0.0291	0.0291	0.9996	0.0291	34.368	1.000	34.38	1.5417	20′
50′	0.0320	0.0320	0.9995	0.0320	31.242	1.001	31.26	1.5388	10′
2°00′	0.0349	0.0349	0.9994	0.0349	28.636	1.001	28.65	1.5359	88°00′
10′	0.0378	0.0378	0.9993	0.0378	26.432	1.001	26.45	1.5330	50′
20′	0.0407	0.0407	0.9992	0.0407	24.542	1.001	24.56	1.5301	40′
30′	0.0436	0.0436	0.9990	0.0437	22.904	1.001	22.93	1.5272	30′
40′	0.0465	0.0465	0.9989	0.0466	21.470	1.001	21.49	1.5243	20′
50′	0.0495	0.0494	0.9988	0.0495	20.206	1.001	20.23	1.5213	10′
3°00′	0.0524	0.0523	0.9986	0.0524	19.081	1.001	19.11	1.5184	87°00′
10′	0.0553	0.0552	0.9985	0.0553	18.075	1.002	18.10	1.5155	50′
20′	0.0582	0.0581	0.9983	0.0582	17.169	1.002	17.20	1.5126	40′
30′	0.0611	0.0610	0.9981	0.0612	16.350	1.002	16.38	1.5097	30′
40′	0.0640	0.0640	0.9980	0.0641	15.605	1.002	15.64	1.5068	20′
50′	0.0669	0.0669	0.9978	0.0670	14.924	1.002	14.96	1.5039	10′
4°00′	0.0698	0.0698	0.9976	0.0699	14.301	1.002	14.34	1.5010	86°00′
10′	0.0727	0.0727	0.9974	0.0729	13.727	1.003	13.76	1.4981	50′
20′	0.0756	0.0756	0.9971	0.0758	13.197	1.003	13.23	1.4952	40′
30′	0.0785	0.0785	0.9969	0.0787	12.706	1.003	12.75	1.4923	30′
40′	0.0814	0.0814	0.9967	0.0816	12.251	1.003	12.29	1.4893	20′
50′	0.0844	0.0843	0.9964	0.0846	11.826	1.004	11.87	1.4864	10′
5°00′	0.0873	0.0872	0.9962	0.0875	11.430	1.004	11.47	1.4835	85°00′
10′	0.0902	0.0901	0.9959	0.0904	11.059	1.004	11.10	1.4806	50′
20′	0.0931	0.0929	0.9957	0.0934	10.712	1.004	10.76	1.4777	40′
30′	0.0960	0.0958	0.9954	0.0963	10.385	1.005	10.43	1.4748	30′
40′	0.0989	0.0987	0.9951	0.0992	10.078	1.005	10.13	1.4719	20′
50′	0.1018	0.1016	0.9948	0.1022	9.7882	1.005	9.839	1.4690	10′
6°00′	0.1047	0.1045	0.9945	0.1051	9.5144	1.006	9.567	1.4661	84°00′
10′	0.1076	0.1074	0.9942	0.1080	9.2553	1.006	9.309	1.4632	50′
20′	0.1105	0.1103	0.9939	0.1110	9.0098	1.006	9.065	1.4603	40′
30′	0.1134	0.1132	0.9936	0.1139	8.7769	1.006	8.834	1.4573	30′
40′	0.1164	0.1161	0.9932	0.1169	8.5555	1.007	8.614	1.4544	20′
50′	0.1193	0.1190	0.9929	0.1198	8.3450	1.007	8.405	1.4515	10′
7°00′	0.1222	0.1219	0.9925	0.1228	8.1443	1.008	8.206	1.4486	83°00′
10′	0.1251	0.1248	0.9922	0.1257	7.9530	1.008	8.016	1.4457	50′
20′	0.1280	0.1276	0.9918	0.1287	7.7704	1.008	7.834	1.4428	40′
30′	0.1309	0.1305	0.9914	0.1317	7.5958	1.009	7.661	1.4399	30′
40′	0.1338	0.1334	0.9911	0.1346	7.4287	1.009	7.496	1.4370	20′
50′	0.1367	0.1363	0.9907	0.1376	7.2687	1.009	7.337	1.4341	10′
8°00′	0.1396	0.1392	0.9903	0.1405	7.1154	1.010	7.185	1.4312	82°00′
10′	0.1425	0.1421	0.9899	0.1435	6.9682	1.010	7.040	1.4283	50′
20′	0.1454	0.1449	0.9894	0.1465	6.8269	1.011	6.900	1.4254	40′
30′	0.1484	0.1478	0.9890	0.1495	6.6912	1.011	6.765	1.4224	30′
40′	0.1513	0.1507	0.9886	0.1524	6.6506	1.012	6.636	1.4195	20′
50′	0.1542	0.1536	0.9881	0.1554	6.4348	1.012	6.512	1.4166	10′
9°00′	0.1571	0.1564	0.9877	0.1584	6.3138	1.012	6.392	1.4137	81°00′
		Cos θ	Sin θ	Cot θ	Tan θ	Csc θ	Sec θ	θ Rad.	θ Deg.

Table of Values of the Trigonometric Functions

θ Deg.	θ Rad.	Sin θ	Cos θ	Tan θ	Cot θ	Sec θ	Csc θ		
9°00′	0.1571	0.1564	0.9877	0.1584	6.3138	1.012	6.392	1.4137	81°00′
10′	0.1600	0.1593	0.9872	0.1614	6.1970	1.013	6.277	1.4108	50′
20′	0.1629	0.1622	0.9868	0.1644	6.0844	1.013	6.166	1.4079	40′
30′	0.1658	0.1650	0.9863	0.1673	5.9758	1.014	6.059	1.4050	30′
40′	0.1687	0.1679	0.9858	0.1703	5.8708	1.014	5.955	1.4021	20′
50′	0.1716	0.1708	0.9853	0.1733	5.7694	1.015	5.855	1.3992	10′
10°00′	0.1745	0.1736	0.9848	0.1763	5.6713	1.015	5.759	1.3963	80°00′
10′	0.1774	0.1765	0.9843	0.1793	5.5764	1.016	5.665	1.3934	50′
20′	0.1804	0.1794	0.9838	0.1823	5.4845	1.016	5.575	1.3904	40′
30′	0.1833	0.1822	0.9833	0.1853	5.3955	1.017	5.487	1.3875	30′
40′	0.1862	0.1851	0.9827	0.1883	5.3093	1.018	5.403	1.3846	20′
50′	0.1891	0.1880	0.9822	0.1914	5.2257	1.018	5.320	1.3817	10′
11°00′	0.1920	0.1908	0.9816	0.1944	5.1446	1.019	5.241	1.3788	79°00′
10′	0.1949	0.1937	0.9811	0.1974	5.0658	1.019	5.164	1.3759	50′
20′	0.1978	0.1965	0.9805	0.2004	4.9894	1.020	5.089	1.3730	40′
30′	0.2007	0.1994	0.9799	0.2035	4.9152	1.020	5.016	1.3701	30′
40′	0.2036	0.2022	0.9793	0.2065	4.8430	1.021	4.945	1.3672	20′
50′	0.2065	0.2051	0.9787	0.2095	4.7729	1.022	4.876	1.3643	10′
12°00′	0.2094	0.2079	0.9781	0.2126	4.7046	1.022	4.810	1.3614	78°00′
10′	0.2123	0.2108	0.9775	0.2156	4.6382	1.023	4.745	1.3584	50′
20′	0.2153	0.2136	0.9769	0.2186	4.5736	1.024	4.682	1.3555	40′
30′	0.2182	0.2164	0.9763	0.2217	4.5107	1.024	4.620	1.3526	30′
40′	0.2211	0.2193	0.9757	0.2247	4.4494	1.025	4.560	1.3497	20′
50′	0.2240	0.2221	0.9750	0.2278	4.3897	1.026	4.502	1.3468	10′
13°00′	0.2269	0.2250	0.9744	0.2309	4.3315	1.026	4.445	1.3439	77°00′
10′	0.2298	0.2278	0.9737	0.2339	4.2747	1.027	4.390	1.3410	50′
20′	0.2327	0.2306	0.9730	0.2370	4.2193	1.028	4.336	1.3381	40′
30′	0.2356	0.2334	0.9724	0.2401	4.1653	1.028	4.284	1.3352	30′
40′	0.2385	0.2363	0.9717	0.2432	4.1126	1.029	4.232	1.3323	20′
50′	0.2414	0.2391	0.9710	0.2462	4.0611	1.030	4.182	1.3294	10′
14°00′	0.2443	0.2419	0.9703	0.2493	4.0108	1.031	4.134	1.3265	76°00′
10′	0.2473	0.2447	0.9696	0.2524	3.9617	1.031	4.086	1.3235	50′
20′	0.2502	0.2476	0.9689	0.2555	3.9136	1.032	4.039	1.3206	40′
30′	0.2531	0.2504	0.9681	0.2586	3.8667	1.033	3.994	1.3177	30′
40′	0.2560	0.2532	0.9674	0.2617	3.8208	1.034	3.950	1.3148	20′
50′	0.2589	0.2560	0.9667	0.2648	3.7760	1.034	3.906	1.3119	10′
15°00′	0.2618	0.2588	0.9659	0.2679	3.7321	1.035	3.864	1.3090	75°00′
10′	0.2647	0.2616	0.9652	0.2711	3.6891	1.036	3.822	1.3061	50′
20′	0.2676	0.2644	0.9644	0.2742	3.6470	1.037	3.782	1.3032	40′
30′	0.2705	0.2672	0.9636	0.2773	3.6059	1.038	3,742	1.3003	30′
40′	0.2734	0.2700	0.9628	0.2805	3.5656	1.039	3.703	1.2974	20′
50′	0.2763	0.2728	0.9621	0.2836	3.5261	1.039	3.665	1.2945	10′
16°00′	0.2793	0.2756	0.9613	0.2867	3.4874	1.040	3.628	1.2915	74°00′
10′	0.2822	0.2784	0.9605	0.2899	3.4495	1.041	3.592	1.2886	50′
20′	0.2851	0.2812	0.9596	0.2931	3.4124	1.042	3.556	1.2857	40′
30′	0.2880	0.2840	0.9588	0.2962	3.3759	1.043	3.521	1.2828	30′
40′	0.2909	0.2868	0.9580	0.2994	3.3402	1.044	3.487	1.2799	20′
50′	0.2938	0.2896	0.9572	0.3026	3.3052	1.045	3.453	1.2770	10′
17°00′	0.2967	0.2924	0.9563	0.3057	3.2709	1.046	3.420	1.2741	73°00′
10′	0.2996	0.2952	0.9555	0.3089	3.2371	1.047	3.388	1.2712	50′
20′	0.3025	0.2979	0.9546	0.3121	3.2041	1.048	3.356	1.2683	40′
30′	0.3054	0.3007	0.9537	0.3153	3.1716	1.049	3.326	1.2654	30′
40′	0.3083	0.3035	0.9528	0.3185	3.1397	1.049	3.295	1.2625	20′
50′	0.3113	0.3062	0.9520	0.3217	3.1084	1.050	3.265	1.2595	10′
18°00′	0.3142	0.3090	0.9511	0.3249	3.0777	1.051	3.236	1.2566	72°00′
		Cos θ	Sin θ	Cot θ	Tan θ	Csc θ	Sec θ	θ Rad.	θ Deg.

Table of Values of the Trigonometric Functions

θ Deg.	θ Rad.	Sin θ	Cos θ	Tan θ	Cot θ	Sec θ	Csc θ		
18°00′	0.3142	0.3090	0.9511	0.3249	3.0777	1.051	3.236	1.2566	72°00′
10′	0.3171	0.3118	0.9502	0.3281	3.0475	1.052	3.207	1.2537	50′
20′	0.3200	0.3145	0.9492	0.3314	3.0178	1.053	3.179	1.2508	40′
30′	0.3229	0.3173	0.9483	0.3346	2.9887	1.054	3.152	1.2479	30′
40′	0.3258	0.3201	0.9474	0.3378	2.9600	1.056	3.124	1.2450	20′
50′	0.3287	0.3228	0.9465	0.3411	2.9319	1.057	3.098	1.2421	10′
19°00′	0.3316	0.3256	0.9455	0.3443	2.9042	1.058	3.072	1.2392	71°00′
10′	0.3345	0.3283	0.9446	0.3476	2.8770	1.059	3.046	1.2363	50′
20′	0.3374	0.3311	0.9436	0.3508	2.8502	1.060	3.021	1.2334	40′
30′	0.3403	0.3338	0.9426	0.3541	2.8239	1.061	2.996	1.2305	30′
40′	0.3432	0.3365	0.9417	0.3574	2.7980	1.062	2.971	1.2275	20′
50′	0.3462	0.3393	0.9407	0.3607	2.7725	1.063	2.947	1.2246	10′
20°00′	0.3491	0.3420	0.9397	0.3640	2.7475	1.064	2.924	1.2217	70°00′
10′	0.3520	0.3448	0.9387	0.3673	2.7228	1.065	2.901	1.2188	50′
20′	0.3549	0.3475	0.9377	0.3706	2.6985	1.066	2.878	1.2159	40′
30′	0.3578	0.3502	0.9367	0.3739	2.6746	1.068	2.855	1.2130	30′
40′	0.3607	0.3529	0.9356	0.3772	2.6511	1.069	2.833	1.2101	20′
50′	0.3636	0.3557	0.9346	0.3805	2.6279	1.070	2.812	1.2072	10′
21°00′	0.3665	0.3584	0.9336	0.3839	2.6051	1.071	2.790	1.2043	69°00′
10′	0.3694	0.3611	0.9325	0.3872	2.5826	1.072	2.769	1.2014	50′
20′	0.3723	0.3638	0.9315	0.3906	2.5605	1.074	2.749	1.1985	40′
30′	0.3752	0.3665	0.9304	0.3939	2.5386	1.075	2.729	1.1956	30′
40′	0.3782	0.3692	0.9293	0.3973	2.5172	1.076	2.709	1.1926	20′
50′	0.3811	0.3719	0.9283	0.4006	2.4960	1.077	2.689	1.1897	10′
22°00′	0.3840	0.3746	0.9272	0.4040	2.4751	1.079	2.669	1.1868	68°00′
10′	0.3869	0.3773	0.9261	0.4074	2.4545	1.080	2.650	1.1839	50′
20′	0.3898	0.3800	0.9250	0.4108	2.4342	1.081	2.632	1.1810	40′
30′	0.3927	0.3827	0.9239	0.4142	2.4142	1.082	2.613	1.1781	30′
40′	0.3956	0.3854	0.9228	0.4176	2.3945	1.084	2.595	1.1752	20′
50′	0.3985	0.3881	0.9216	0.4210	2.3750	1.085	2.577	1.1723	10′
23°00′	0.4014	0.3907	0.9215	0.4245	2.3559	1.086	2.559	1.1694	67°00′
10′	0.4043	0.3934	0.9194	0.4279	2.3369	1.088	2.542	1.1665	50′
20′	0.4072	0.3961	0.9182	0.4314	2.3183	1.089	2.525	1.1636	40′
30′	0.4102	0.3987	0.9171	0.4348	2.2998	1.090	2.508	1.1606	30′
40′	0.4131	0.4014	0.9159	0.4383	2.2817	1.092	2.491	1.1577	20′
50′	0.4160	0.4041	0.9147	0.4417	2.2637	1.093	2.475	1.1548	10′
24°00′	0.4189	0.4067	0.9135	0.4452	2.2460	1.095	2.459	1.1519	66°00′
10′	0.4218	0.4094	0.9124	0.4487	2.2286	1.096	2.443	1.1490	50′
20′	0.4247	0.4120	0.9112	0.4522	2.2113	1.097	2.427	1.1461	40′
30′	0.4276	0.4147	0.9100	0.4557	2.1943	1.099	2.411	1.1432	30′
40′	0.4305	0.4173	0.9088	0.4592	2.1775	1.100	2.396	1.1403	20′
50′	0.4334	0.4200	0.9075	0.4628	2.1609	1.102	2.381	1.1374	10′
25°00′	0.4363	0.4226	0.9063	0.4663	2.1445	1.103	2.366	1.1345	65°00′
10′	0.4392	0.4253	0.9051	0.4699	2.1283	1.105	2.352	1.1316	50′
20′	0.4422	0.4279	0.9038	0.4734	2.1123	1.106	2.337	1.1286	40′
30′	0.4451	0.4305	0.9026	0.4770	2.0965	1.108	2.323	1.1257	30′
40′	0.4480	0.4331	0.9013	0.4806	2.0809	1.109	2.309	1.1228	20′
50′	0.4509	0.4358	0.9001	0.4841	2.0655	1.111	2.295	1.1199	10′
26°00′	0.4538	0.4384	0.8988	0.4877	2.0503	1.113	2.281	1.1170	64°00′
10′	0.4567	0.4410	0.8975	0.4913	2.0353	1.114	2.268	1.1141	50′
20′	0.4596	0.4436	0.8962	0.4950	2.0204	1.116	2.254	1.1112	40′
30′	0.4625	0.4462	0.8949	0.4986	2.0057	1.117	2.241	1.1083	30′
40′	0.4654	0.4488	0.8936	0.5022	1.9912	1.119	2.228	1.1054	20′
50′	0.4683	0.4514	0.8923	0.5059	1.9768	1.121	2.215	1.1025	10′
27°00′	0.4712	0.4540	0.8910	0.5095	1.9626	1.122	2.203	1.0996	63°00′
		Cos θ	Sin θ	Cot θ	Tan θ	Csc θ	Sec θ	θ Rad.	θ Deg.

Table of Values of the Trigonometric Functions

θ Deg.	θ Rad.	Sin θ	Cos θ	Tan θ	Cot θ	Sec θ	Csc θ		
27°00′	0.4712	0.4540	0.8910	0.5095	1.9626	1.122	2.203	1.0996	63°00′
10′	0.4741	0.4566	0.8897	0.5132	1.9486	1.124	2.190	1.0966	50′
20′	0.4771	0.4592	0.8884	0.5169	1.9347	1.126	2.178	1.0937	40′
30′	0.4800	0.4617	0.8870	0.5206	1.9210	1.127	2.166	1.0908	30′
40′	0.4829	0.4643	0.8857	0.5243	1.9074	1.129	2.154	1.0879	20′
50′	0.4858	0.4669	0.8843	0.5280	1.8940	1.131	2.142	1.0850	10′
28°00′	0.4887	0.4695	0.8829	0.5317	1.8807	1.133	2.130	1.0821	62°00′
10′	0.4916	0.4720	0.8816	0.5354	1.8676	1.134	2.118	1.0792	50′
20′	0.4945	0.4746	0.8802	0.5392	1.8546	1.136	2.107	1.0763	40′
30′	0.4974	0.4772	0.8788	0.5430	1.8418	1.138	2.096	1.0734	30′
40′	0.5003	0.4797	0.8774	0.5467	1.8291	1.140	2.085	1.0705	20′
50′	0.5032	0.4823	0.8760	0.5505	1.8165	1.142	2.074	1.0676	10′
29°00′	0.5061	0.4848	0.8746	0.5543	1.8040	1.143	2.063	1.0647	61°00′
10′	0.5091	0.4874	0.8732	0.5581	1.7917	1.145	2.052	1.0617	50′
20′	0.5120	0.4899	0.8718	0.5619	1.7796	1.147	2.041	1.0588	40′
30′	0.5149	0.4924	0.8704	0.5658	1.7675	1.149	2.031	1.0559	30′
40′	0.5178	0.4950	0.8689	0.5696	1.7556	1.151	2.020	1.0530	20′
50′	0.5207	0.4975	0.8675	0.5735	1.7437	1.153	2.010	1.0501	10′
30°00′	0.5236	0.5000	0.8660	0.5774	1.7321	1.155	2.000	1.0472	60°00′
10′	0.5265	0.5025	0.8646	0.5812	1.7205	1.157	1.990	1.0443	50′
20′	0.5294	0.5050	0.8631	0.5851	1.7090	1.159	1.980	1.0414	40′
30′	0.5323	0.5075	0.8616	0.5890	1.6977	1.161	1.970	1.0385	30′
40′	0.5352	0.5100	0.8601	0.5930	1.6864	1.163	1.961	1.0356	20′
50′	0.5381	0.5125	0.8587	0.5969	1.6753	1.165	1.951	1.0327	10′
31°00′	0.5411	0.5150	0.8572	0.6009	1.6643	1.167	1.942	1.0297	59°00′
10′	0.5440	0.5175	0.8557	0.6048	1.6534	1.169	1.932	1.0268	50′
20′	0.5469	0.5200	0.8542	0.6088	1.6426	1.171	1.923	1.0239	40′
30′	0.5498	0.5225	0.8526	0.6128	1.6319	1.173	1.914	1.0210	30′
40′	0.5527	0.5250	0.8511	0.6168	1.6212	1.175	1.905	1.0181	20′
50′	0.5556	0.5275	0.8496	0.6208	1.6107	1.177	1.896	1.0152	10′
32°00′	0.5585	0.5299	0.8480	0.6249	1.6003	1.179	1.887	1.0123	58°00′
10′	0.5614	0.5324	0.8465	0.6289	1.5900	1.181	1.878	1.0094	50′
20′	0.5643	0.5348	0.8450	0.6330	1.5798	1.184	1.870	1.0065	40′
30′	0.5672	0.5373	0.8434	0.6371	1.5697	1.186	1.861	1.0036	30′
40′	0.5701	0.5398	0.8418	0.6412	1.5597	1.188	1.853	1.0007	20′
50′	0.5730	0.5422	0.8403	0.6453	1.5497	1.190	1.844	0.9977	10′
33°00′	0.5760	0.5446	0.8387	0.6494	1.5399	1.192	1.836	0.9948	57°00′
10′	0.5789	0.5471	0.8371	0.6536	1.5301	1.195	1.828	0.9919	50′
20′	0.5818	0.5495	0.8355	0.6577	1.5204	1.197	1.820	0.9890	40′
30′	0.5847	0.5519	0.8339	0.6619	1.5108	1.199	1.812	0.9861	30′
40′	0.5876	0.5544	0.8323	0.6661	1.5013	1.202	1.804	0.9832	20′
50′	0.5905	0.5568	0.8307	0.6703	1.4919	1.204	1.796	0.9803	10′
34°00′	0.5934	0.5592	0.8290	0.6745	1.4826	1.206	1.788	0.9774	56°00′
10′	0.5963	0.5616	0.8274	0.6787	1.4733	1.209	1.781	0.9745	50′
20′	0.5992	0.5640	0.8258	0.6830	1.4641	1.211	1.773	0.9716	40′
30′	0.6021	0.5664	0.8241	0.6873	1.4550	1.213	1.766	0.9687	30′
40′	0.6050	0.5688	0.8225	0.6916	1.4460	1.216	1.758	0.9657	20′
50′	0.6080	0.5712	0.8208	0.6959	1.4370	1.218	1.751	0.9628	10′
35°00′	0.6109	0.5736	0.8192	0.7002	1.4281	1.221	1.743	0.9599	55°00′
10′	0.6138	0.5760	0.8175	0.7046	1.4193	1.223	1.736	0.9570	50′
20′	0.6167	0.5783	0.8158	0.7089	1.4106	1.226	1.729	0.9541	40′
30′	0.6196	0.5807	0.8141	0.7133	1.4019	1.228	1.722	0.9512	30′
40′	0.6225	0.5831	0.8124	0.7177	1.3934	1.231	1.715	0.9483	20′
50′	0.6254	0.5854	0.8107	0.7221	1.3848	1.233	1.708	0.9454	10′
36°00′	0.6283	0.5878	0.8090	0.7265	1.3764	1.236	1.701	0.9425	54°00′
		Cos θ	Sin θ	Cot θ	Tan θ	Csc θ	Sec θ	θ Rad.	θ Deg.

Table of Values of the Trigonometric Functions

θ Deg.	θ Rad.	Sin θ	Cos θ	Tan θ	Cot θ	Sec θ	Csc θ		
36°00′	0.6283	0.5878	0.8090	0.7265	1.3764	1.236	1.701	0.9425	54°00′
10′	0.6312	0.5901	0.8073	0.7310	1.3680	1.239	1.695	0.9396	50′
20′	0.6341	0.5925	0.8056	0.7355	1.3597	1.241	1.688	0.9367	40′
30′	0.6370	0.5948	0.8039	0.7400	1.3514	1.244	1.681	0.9338	30′
40′	0.6400	0.5972	0.8021	0.7445	1.3432	1.247	1.675	0.9308	20′
50′	0.6429	0.5995	0.8004	0.7490	1.3351	1.249	1.668	0.9279	10′
37°00′	0.6458	0.6018	0.7986	0.7536	1.3270	1.252	1.662	0.9250	53°00′
10′	0.6487	0.6041	0.7969	0.7581	1.3190	1.255	1.655	0.9221	50′
20′	0.6516	0.6065	0.7951	0.7627	1.3111	1.258	1.649	0.9192	40′
30′	0.6545	0.6088	0.7934	0.7673	1.3032	1.260	1.643	0.9163	30′
40′	0.6574	0.6111	0.7916	0.7720	1.2954	1.263	1.636	0.9134	20′
50′	0.6603	0.6134	0.7898	0.7766	1.2876	1.266	1.630	0.9105	10′
38°00′	0.6632	0.6157	0.7880	0.7813	1.2799	1.269	1.624	0.9076	52°00′
10′	0.6661	0.6180	0.7862	0.7860	1.2723	1.272	1.618	0.9047	50′
20′	0.6690	0.6202	0.7844	0.7907	1.2647	1.275	1.612	0.9018	40′
30′	0.6720	0.6225	0.7826	0.7954	1.2572	1.278	1.606	0.8988	30′
40′	0.6749	0.6248	0.7808	0.8002	1.2497	1.281	1.601	0.8959	20′
50′	0.6778	0.6271	0.7790	0.8050	1.2423	1.284	1.595	0.8930	10′
39°00′	0.6807	0.6293	0.7771	0.8098	1.2349	1.287	1.589	0.8901	51°00′
10′	0.6836	0.6316	0.7753	0.8146	1.2276	1.290	1.583	0.8872	50′
20′	0.6865	0.6338	0.7735	0.8195	1.2203	1.293	1.578	0.8843	40′
30′	0.6894	0.6361	0.7716	0.8243	1.2131	1.296	1.572	0.8814	30′
40′	0.6923	0.6383	0.7698	0.8292	1.2059	1.299	1.567	0.8785	20′
50′	0.6952	0.6406	0.7679	0.8342	1.1988	1.302	1.561	0.8756	10′
40°00′	0.6981	0.6428	0.7660	0.8391	1.1918	1.305	1.556	0.8727	50°00′
10′	0.7010	0.6450	0.7642	0.8441	1.1847	1.309	1.550	0.8698	50′
20′	0.7039	0.6472	0.7623	0.8491	1.1778	1.312	1.545	0.8668	40′
30′	0.7069	0.6494	0.7604	0.8541	1.1708	1.315	1.540	0.8639	30′
40′	0.7098	0.6517	0.7585	0.8591	1.1640	1.318	1.535	0.8610	20′
50′	0.7127	0.6539	0.7566	0.8642	1.1571	1.322	1.529	0.8581	10′
41°00′	0.7156	0.6561	0.7547	0.8693	1.1504	1.325	1.524	0.8552	49°00′
10′	0.7185	0.6583	0.7528	0.8744	1.1436	1.328	1.519	0.8523	50′
20′	0.7214	0.6604	0.7509	0.8796	1.1369	1.332	1.514	0.8494	40′
30′	0.7243	0.6626	0.7490	0.8847	1.1303	1.335	1.509	0.8465	30′
40′	0.7272	0.6648	0.7470	0.8899	1.1237	1.339	1.504	0.8436	20′
50′	0.7301	0.6670	0.7451	0.8952	1.1171	1.342	1.499	0.8407	10′
42°00′	0.7330	0.6691	0.7431	0.9004	1.1106	1.346	1.494	0.8378	48°00′
10′	0.7359	0.6713	0.7412	0.9057	1.1041	1.349	1.490	0.8348	50′
20′	0.7389	0.6734	0.7392	0.9110	1.0977	1.353	1.485	0.8319	40′
30′	0.7418	0.6756	0.7373	0.9163	1.0913	1.356	1.480	0.8290	30′
40′	0.7447	0.6777	0.7353	0.9217	1.0850	1.360	1.476	0.8261	20′
50′	0.7476	0.6799	0.7333	0.9271	1.0786	1.364	1.471	0.8232	10′
43°00′	0.7505	0.6820	0.7314	0.9325	1.0724	1.367	1.466	0.8203	47°00′
10′	0.7534	0.6841	0.7294	0.9380	1.0661	1.371	1.462	0.8174	50′
20′	0.7563	0.6862	0.7274	0.9435	1.0599	1.375	1.457	0.8145	40′
30′	0.7592	0.6884	0.7254	0.9490	1.0538	1.379	1.453	0.8116	30′
40′	0.7621	0.6905	0.7234	0.9545	1.0477	1.382	1.448	0.8087	20′
50′	0.7650	0.6926	0.7214	0.9601	1.0416	1.386	1.444	0.8058	10′
44°00′	0.7679	0.6947	0.7193	0.9657	1.0355	1.390	1.440	0.8029	46°00′
10′	0.7709	0.6967	0.7173	0.9713	1.0295	1.394	1.435	0.7999	50′
20′	0.7738	0.6988	0.7153	0.9770	1.0235	1.398	1.431	0.7970	40′
30′	0.7767	0.7009	0.7133	0.9827	1.0176	1.402	1.427	0.7941	30′
40′	0.7796	0.7030	0.7112	0.9884	1.0117	1.406	1.423	0.7912	20′
50′	0.7825	0.7050	0.7092	0.9942	1.0058	1.410	1.418	0.7883	10′
45°00′	0.7854	0.7071	0.7071	1.0000	1.0000	1.414	1.414	0.7854	45°00′
		Cos θ	Sin θ	Cot θ	Tan θ	Csc θ	Sec θ	θ Rad.	θ Deg.

ANSWERS TO SELECTED EXERCISES

Chapter 1 Trigonometric Functions

Practice Exercises, pages 5–7 **1.** $D = \{1, 2, 5, 7, 16\}$; $R = \{0, 1, 3, 5, 9\}$ **3.** $D = \{-2, -1, 0, 1\}$; $R = \{-10, -1, 8, 11\}$ **5.** $D = \{-3, 2, 4\}$; $R = \{1, 2, 3\}$ **7.** $D = \{4, 6, 8, 10\}$; $R = \{25\}$ **9.** not a function **11.** function **13.** function **15.** not a function **17.** function **19.** not a function **21.** 25 **23.** -2 **25.** -4 **27.** -1.6 **29.** $D = \{-1, 5, 8\}$; $R = \{-10, -1, 3, 8\}$; not a function **31.** $D = \{x: x \text{ is a real number}\}$; $R = \{f(x): f(x) \geq 0\}$; function **33.** $D = \{x: x \text{ is a real number}\}$; $R = \{f(x): f(x) \text{ is real number}\}$; function **35.** $D = \{x: x \leq 0\}$; $R = \{y: y \text{ is a real number}\}$; not a function **37.** $D = \{x: x \leq 0, x \neq -1\}$; $R = \{y: y \geq 0, y \neq 1\}$; function **39.** 3 **41.** 24 **43.** 4 **45.** $x \neq 0$ **47.** $x \neq \pm 7$ **49.** $-\frac{1}{9}$ **51.** -81 **53.** 3 **55.** -26 **57.** 95°F **59.** 54 in.^{2}

Practice Exercises, pages 10–11 **1.** 5 **3.** $7\sqrt{2}$ **5.** $6\sqrt{2}$ **7.** $\sqrt{37}$ **9.** 5 **11.** $\sqrt{10}$ **13.** 7 **15.** 11 **17.** $2\sqrt{15}$ **19.** $\sqrt{2.5}$ **21.** $3\sqrt{2}$ **23.** $3\sqrt{3}$ **25.** $2\sqrt{a^2 + b^2}$ **27.** $x^2 + y^2 = 25$ **29.** $x^2 + y^2 = 45$ **31.** $x^2 + y^2 = 27$ **33.** $x = 9$ or $x = -7$ **35.** 24 units^2 **37.** $(-1, -6)$ **39.** $3\sqrt{5}$ mi **41.** 13 ft **43.** 74 ft

Practice Exercises, pages 15–16

1.

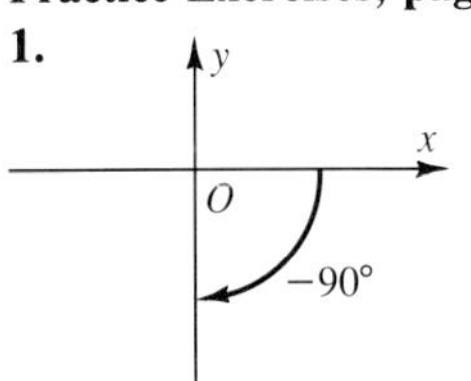

3.

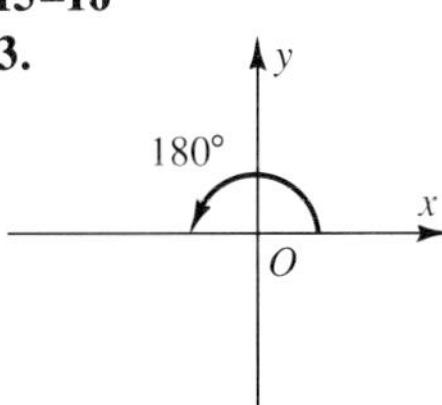

5.

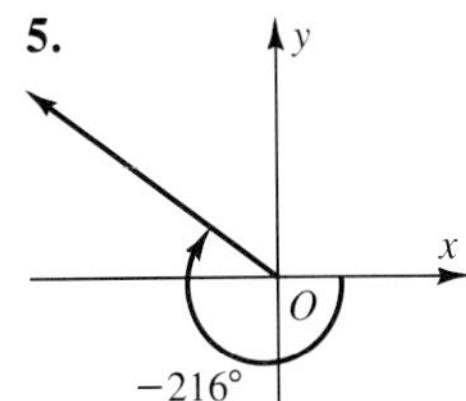

7.

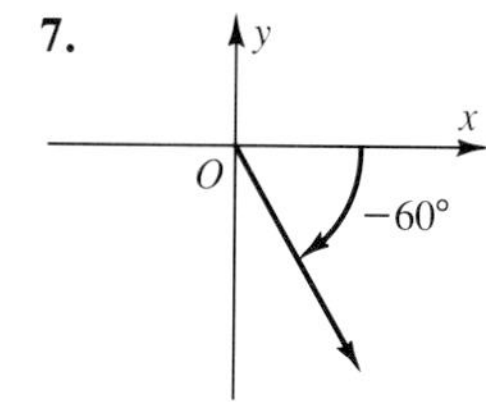

9.

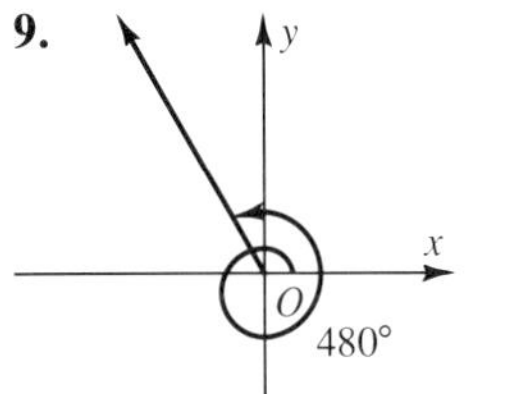

11.

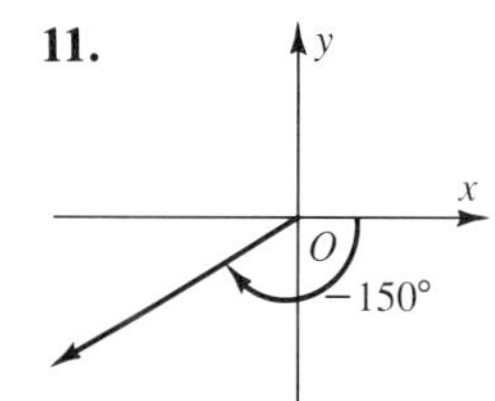

21. 835° **23.** 115° **25.** $-2909°$ **27.** 763° **29.** $-150°$ **31.** 378° **33.** 702° **35.** 570° **37.** $-410°$ **39.** 260° **41.** 174° **43.** 180° **45.** 350° **47.** 3 **49.** 18 **51.** 12,000°

Practice Exercises, pages 21–22 **1.** II **3.** III **5.** IV **7.** 27°30′ **9.** 20°12′ **11.** 134°18′ **13.** 135.5° **15.** 220.3° **17.** 25.2° **19.** $\frac{\pi}{3}$ **21.** $\frac{2\pi}{3}$ **23.** $\frac{5\pi}{6}$ **25.** $\frac{3\pi}{10}$ **27.** $\frac{67\pi}{36}$ **29.** $-\frac{23\pi}{36}$ **31.** 240° **33.** 75° **35.** $-130°$ **37.** 210° **39.** $-135°$ **41.** 300° **43.** 21°33′36″ **45.** 160°6′36″ **47.** $-313°31'48''$ **49.** $-381°52'12''$ **51.** 18.25° **53.** 315.80° **55.** $-102.03°$ **57.** $-515.25°$ **59.** $\frac{5\pi}{2}$ **61.** $-\frac{9\pi}{4}$ **63.** $-\frac{23\pi}{4}$ **65.** 945° **67.** $-1080°$ **69.** $-1200°$ **71.** 576° **73.** $-\frac{5\pi}{3}$ **75.** $\frac{5\pi}{3}$ **77.** $-\frac{19\pi}{6}$ **79.** $\frac{11\pi}{2}$ **81.** $-\frac{17\pi}{4}$ **83.** $\frac{11\pi}{6}$ **85.** $-30°$; $-540°$ **87.** $-90°$; $-\frac{\pi}{2}$

Practice Exercises, pages 26–28 **1.** 5π in. **3.** $\frac{25\pi}{2}$ in. **5.** 45π in. **7.** π rad **9.** $\frac{2\pi}{3}$ rad **11.** $\frac{6\pi}{5}$ rad **13.** $\frac{\pi}{2}$ rad/min **15.** $\frac{2\pi}{15}$ rad/min **17.** $\frac{\pi}{10}$ rad/min **19.** 200π cm/s **21.** 75π cm/s

23. $\frac{\pi}{30}$ rad/s **25.** $\frac{8\pi}{3}$ rad/min **27.** $\frac{24\pi}{5}$ rad/min

29. $1033\frac{1}{3}$ mph **31.** 890 rpm **33.** 8π ft/s; 8π ft/s

Test Yourself, page 28 **1.** $D = \{1, 2, 4, 6\}$; $R = \{3, 5, 7\}$; function **3.** $\frac{1}{7}$ **5.** undefined **7.** II

9. I **11.** $\frac{5\pi}{12}$ **13.** 12π m **15.** 84π in./s

Practice Exercises, pages 32–33

1. $\frac{4}{5}; \frac{3}{5}$

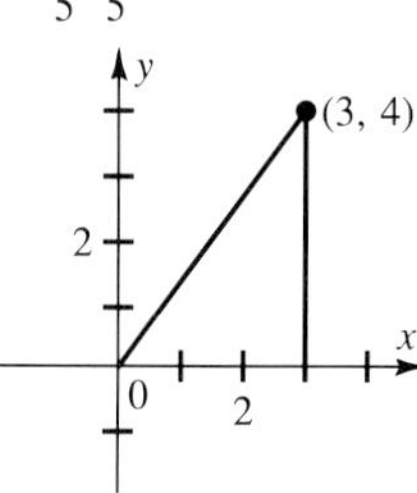

3. $-\frac{12}{13}; -\frac{5}{13}$

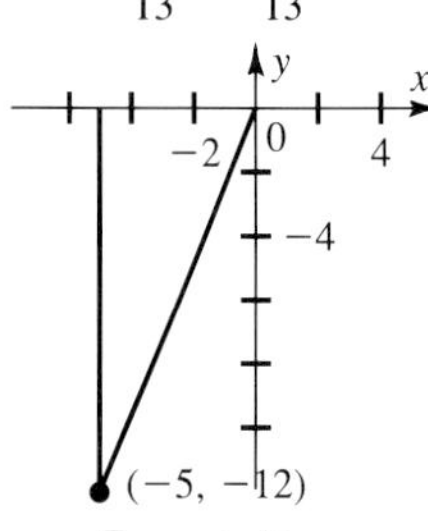

5. 0; −1

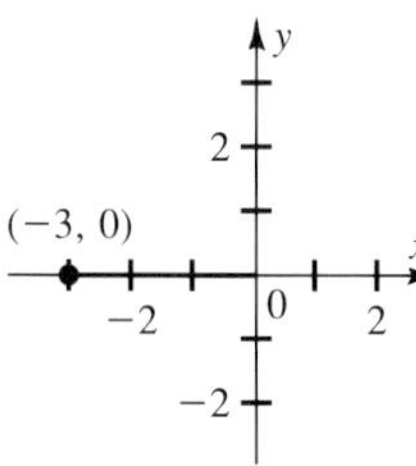

7. $\frac{\sqrt{5}}{5}; -\frac{2\sqrt{5}}{5}$

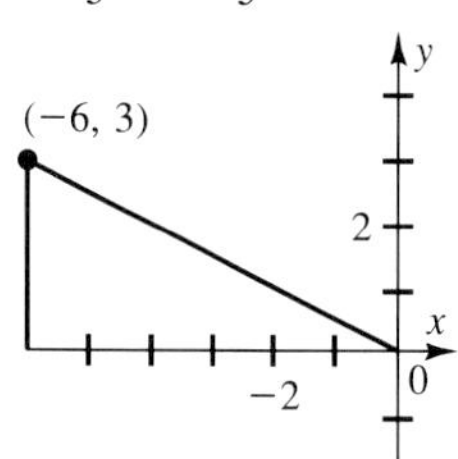

9. $\frac{3}{5}$ **11.** $\frac{5}{13}$ **13.** $-\frac{1}{2}$ **15.** $-\frac{4}{5}$

17. $\frac{7\sqrt{74}}{74}; \frac{5\sqrt{74}}{74}$

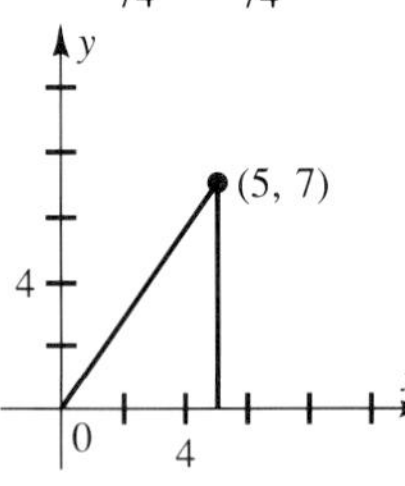

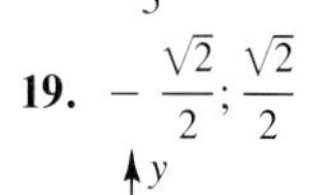

19. $-\frac{\sqrt{2}}{2}; \frac{\sqrt{2}}{2}$

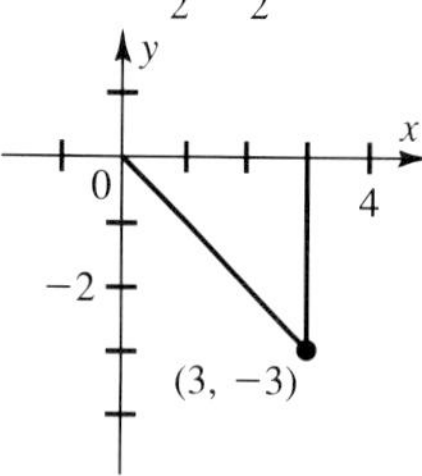

21. $\frac{5\sqrt{34}}{34}; -\frac{3\sqrt{34}}{34}$

(−6, 10)

23. $\frac{\sqrt{3}}{2}$ **25.** $-\frac{1}{2}$ **27.** $-\frac{4\sqrt{6}}{11}$

29. I; II **31.** I; III

Practice Exercises, pages 37–38 Answers are in the order: sin; cos; tan; csc; sec; cot.

1. $\frac{4}{5}; \frac{3}{5}; \frac{4}{3}; \frac{5}{4}; \frac{5}{3}; \frac{3}{4}$

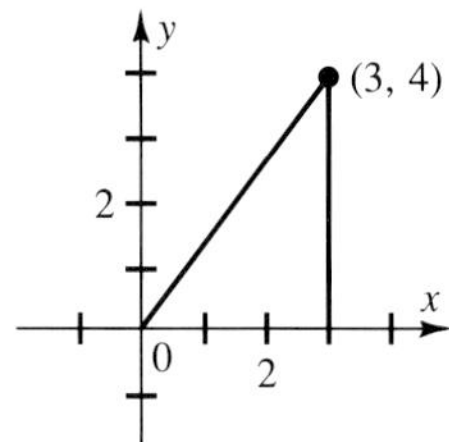

3. $-\frac{5}{13}; \frac{12}{13}; -\frac{5}{12}; -\frac{13}{5}; \frac{13}{12}; -\frac{12}{5}$

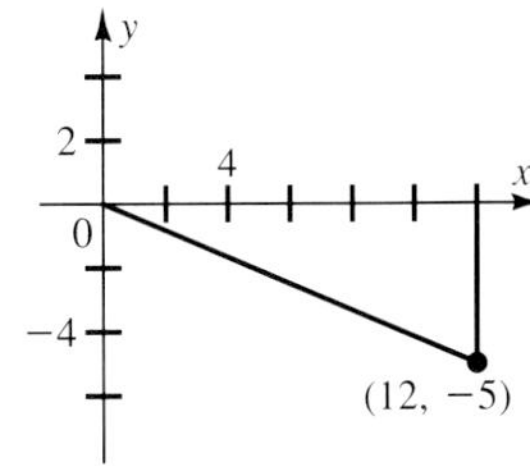

5. $-\frac{\sqrt{2}}{2}; -\frac{\sqrt{2}}{2}$; 1; $-\sqrt{2}$; $-\sqrt{2}$; 1

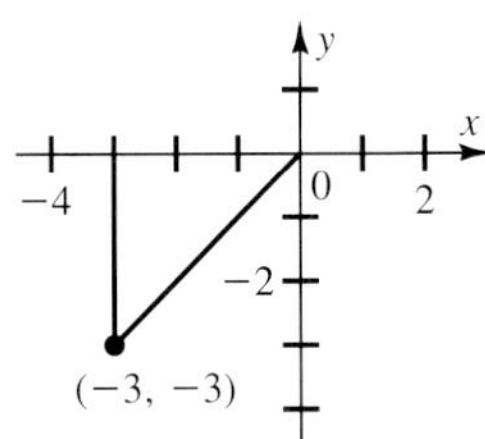

7. $-\frac{\sqrt{21}}{7}; \frac{2\sqrt{7}}{7}; -\frac{\sqrt{3}}{2}; -\frac{\sqrt{21}}{3}; \frac{\sqrt{7}}{2}; -\frac{2\sqrt{3}}{3}$

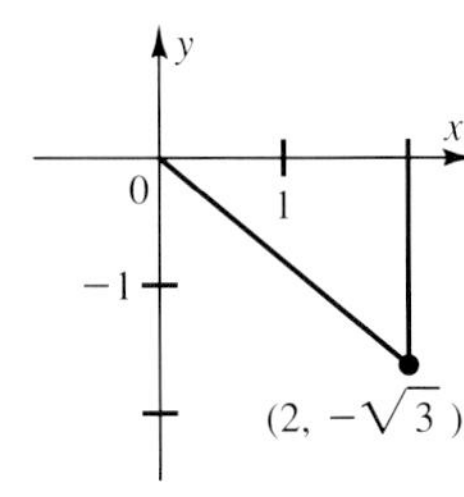

9. $-\frac{24}{25}; \frac{7}{25}; -\frac{24}{7}; -\frac{25}{24}; \frac{25}{7}; -\frac{7}{24}$

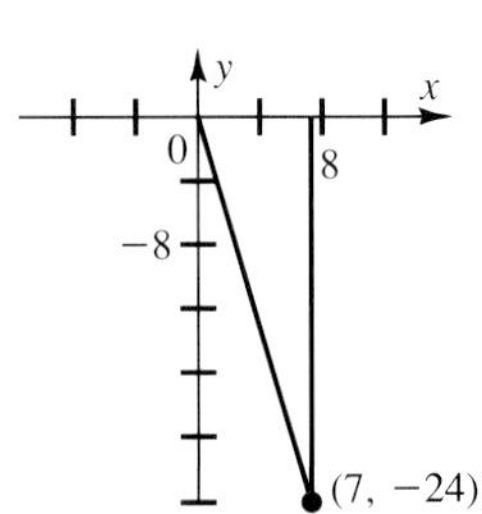

11. $\frac{\sqrt{3}}{2}$; $-\frac{1}{2}$; $-\sqrt{3}$; $\frac{2\sqrt{3}}{3}$; -2; $-\frac{\sqrt{3}}{3}$

13. I, IV **15.** IV **17.** II, IV **19.** III, IV
21. $-\frac{3}{5}$; $\frac{4}{3}$; $-\frac{5}{4}$; $-\frac{5}{3}$; $\frac{3}{4}$ **23.** $\frac{4}{5}$; $\frac{3}{5}$; $\frac{5}{4}$; $\frac{5}{3}$; $\frac{3}{4}$
25. $-\frac{\sqrt{3}}{2}$; $-\frac{1}{2}$; $\sqrt{3}$; $-\frac{2\sqrt{3}}{3}$; $\frac{\sqrt{3}}{3}$ **27.** $-\frac{7}{25}$; $-\frac{24}{25}$; $\frac{7}{24}$; $-\frac{25}{7}$; $-\frac{25}{24}$ **29.** $\frac{\sqrt{7}}{4}$; $-\frac{3}{4}$; $-\frac{\sqrt{7}}{3}$
31. $-\frac{\sqrt{3}}{2}$; $\frac{1}{2}$; $-\sqrt{3}$ **33.** $-\frac{12}{13}$; $-\frac{5}{13}$; $\frac{12}{5}$
35. $(-8, 6)$ **37.** $(1, -1)$ **39.** $(4\sqrt{2}, -4\sqrt{2})$
41. tangent **43.** $\frac{2}{5}$

Practice Exercises, pages 44–45 **1.** $\frac{1}{2}$ **3.** $\frac{\sqrt{3}}{3}$
5. $\frac{2\sqrt{3}}{3}$ **7.** $\frac{\sqrt{2}}{2}$ **9.** 1 **11.** $\sqrt{2}$ **13.** $\frac{\sqrt{3}}{2}$ **15.** $\sqrt{3}$
17. 2 **19.** 9° **21.** 55° **23.** 52° **25.** 70°
Answers to 27–45 are in the order: sin; cos; tan; csc; sec; cot.
27. $\frac{1}{2}$; $-\frac{\sqrt{3}}{2}$; $-\frac{\sqrt{3}}{3}$; 2; $-\frac{2\sqrt{3}}{2}$; $-\sqrt{3}$
29. $\frac{\sqrt{2}}{2}$; $-\frac{\sqrt{2}}{2}$; -1; $\sqrt{2}$; $-\sqrt{2}$; -1 **31.** $-\frac{\sqrt{2}}{2}$; $-\frac{\sqrt{2}}{2}$; 1; $-\sqrt{2}$; $-\sqrt{2}$; 1 **33.** $-\frac{1}{2}$; $-\frac{\sqrt{3}}{2}$; $\frac{\sqrt{3}}{3}$; -2; $-\frac{2\sqrt{3}}{3}$; $\sqrt{3}$ **35.** $\frac{1}{2}$; $\frac{\sqrt{3}}{2}$; $\frac{\sqrt{3}}{3}$; 2; $\frac{2\sqrt{3}}{3}$; $\sqrt{3}$
37. $\frac{\sqrt{2}}{2}$; $-\frac{\sqrt{2}}{2}$; -1; $\sqrt{2}$; $-\sqrt{2}$; -1 **39.** $-\frac{1}{2}$; $\frac{\sqrt{3}}{2}$; $-\frac{\sqrt{3}}{3}$; -2; $\frac{2\sqrt{3}}{3}$; $-\sqrt{3}$ **41.** $\frac{\sqrt{3}}{2}$; $-\frac{1}{2}$; $-\sqrt{3}$; $\frac{2\sqrt{3}}{3}$, -2; $-\frac{\sqrt{3}}{3}$ **43.** $-\frac{\sqrt{3}}{2}$; $\frac{1}{2}$; $-\sqrt{3}$; $-\frac{2\sqrt{3}}{3}$; 2; $-\frac{\sqrt{3}}{3}$ **45.** $-\frac{\sqrt{3}}{2}$; $-\frac{1}{2}$; $\sqrt{3}$; $-\frac{2\sqrt{3}}{3}$; -2; $\frac{\sqrt{3}}{3}$
47. $\frac{\pi}{3}$; $\frac{5\pi}{3}$ **49.** $\frac{3\pi}{4}$; $\frac{7\pi}{4}$ **51.** $\frac{\pi}{4}$; $\frac{7\pi}{4}$ **53.** $\frac{5\pi}{6}$; $\frac{11\pi}{6}$
55. 78° **57.** 7° **59.** 45° **61.** 150° **63.** 240°
65. 270° **67.** 315° **69.** 26 ft

Practice Exercises, pages 50–51 **1.** 0.3190 **3.** 9.0579 **5.** 0.9755 **7.** 0.3179 **9.** 21.2049 **11.** 0.8090 **13.** 0.8660 **15.** -1 **17.** 2.8291 **19.** 1.1343 **21.** -1.2134 **23.** 1.0122 **25.** 51.1°, 128.9° **27.** 34.1°, 214.1° **29.** 52.0°, 308.0° **31.** -0.7630 **33.** 2.8556 **35.** -0.3714 **37.** -1.5959 **39.** 0.9749 **41.** 3.4057 **43.** -0.2260 **45.** 1.4142 **47.** 23.6°, 156.4° **49.** 12.7°, 192.7° **51.** 351.4°, 171.4° **53.** 48.3° **55.** 235.5° **57.** 328.2° **59.** 102.0° **61.** $\pi \div 180$

Test Yourself, page 51 **1.** $\cos\theta = -\frac{1}{2}$
Answers to 3–5 are in the order: sin; cos; tan; csc; sec; cot.
3. $-\frac{12}{13}$; $-\frac{5}{13}$; $\frac{12}{5}$; $-\frac{13}{12}$; $-\frac{13}{5}$; $\frac{5}{12}$ **5.** $\frac{\sqrt{95}}{12}$; $-\frac{7}{12}$; $-\frac{\sqrt{95}}{7}$; $\frac{12\sqrt{95}}{95}$; $-\frac{12}{7}$; $-\frac{7\sqrt{95}}{95}$ **7.** 48°
9. $\frac{\pi}{4}$ **11.** -0.5329 **13.** 2.3048

Summary and Review, pages 54–55 **1.** $D = \{1, 5\}$; $R = \{3, 7, 9\}$; no **3.** 10 **7.** 495° **9.** 4π in. **11.** $-\frac{8}{17}$ **13.** 0; -1; 0; undefined; -1; undefined **15.** 0.7071; -1.2314

Maintaining Skills, page 58 **1.** -10 **3.** $-\frac{7}{4}$
5. $-\frac{5}{2}$ **7.** $(2x + 3)(2x - 3)$ **9.** $(2 - x)(4 + 2x + x^2)$ **11.** $(x - 3y)(x^2 + 3xy + 9y^2)$ **13.** 0, $\sqrt{2}$, $-\sqrt{2}$ **15.** 0, 2, -1 **17.** 0, 5, -1

19.

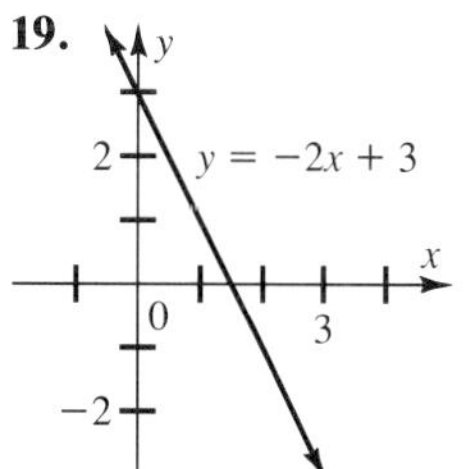

21.

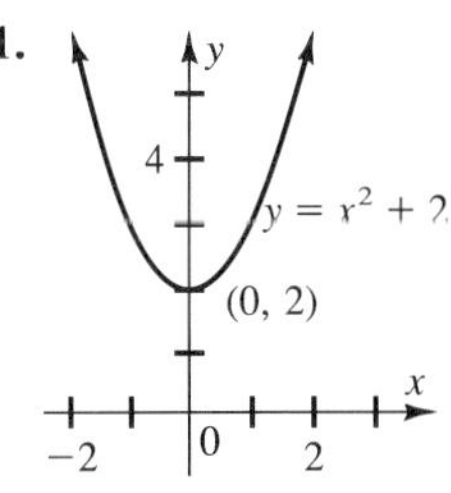

23.

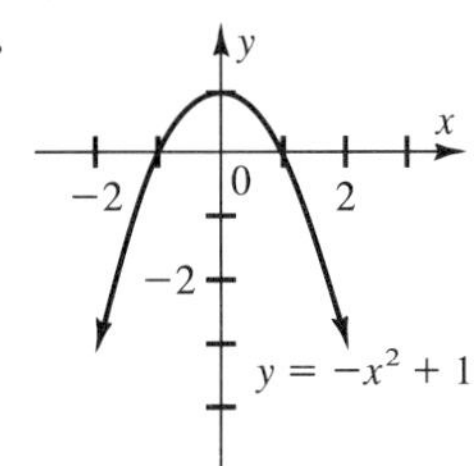

Chapter 2 Graphing Trigonometric Functions

Practice Exercises, pages 64–66 **1.** yes; yes; 2 **3.** yes; no **5.** yes; yes; 4 **7.** period 2; $0 \le x \le 8$

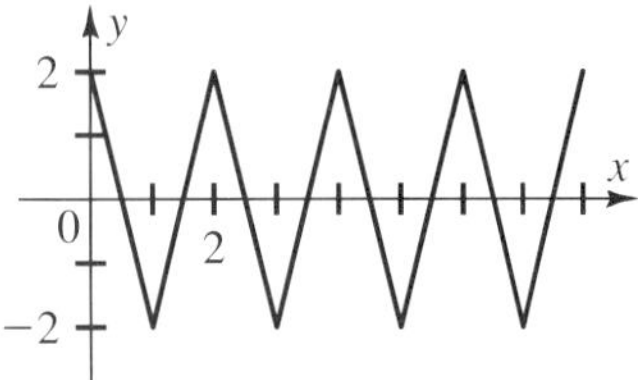

9. period 1; $-2 \le x \le 3$

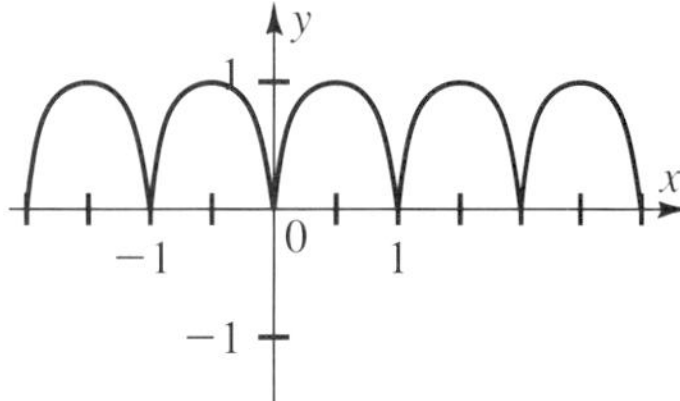

11. period 2; $-4 \le x \le 6$

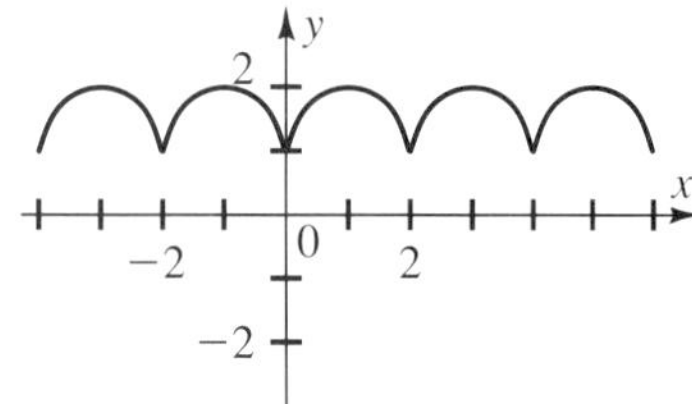

13. even **15.** odd **17.** even **19.** odd **21.** even **23.** odd **25.** y-axis **27.** none **29.** origin

31.

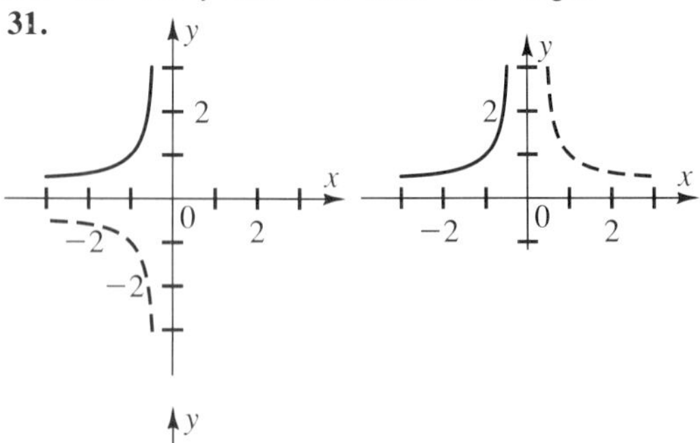

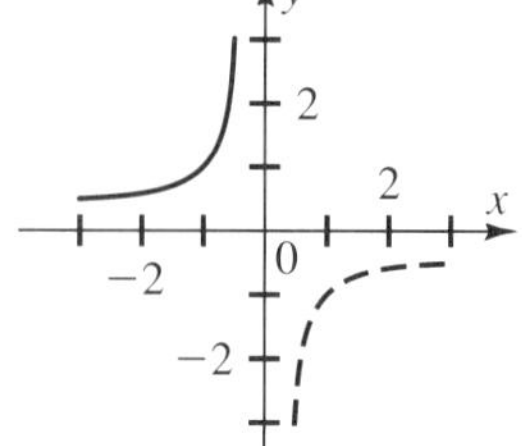

33.

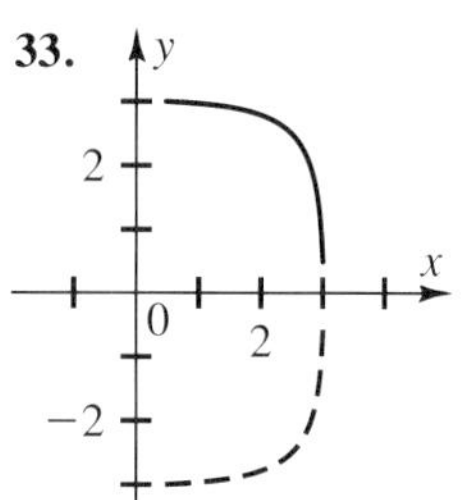

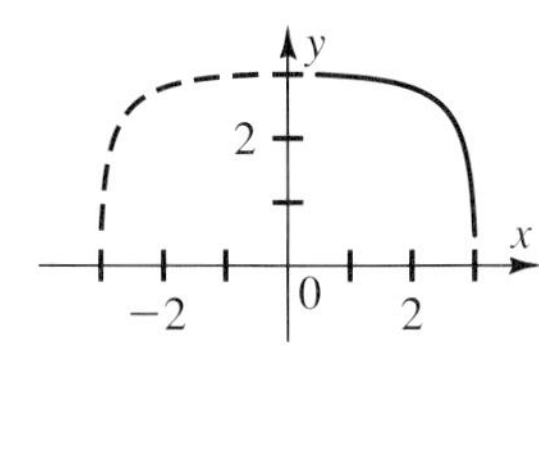

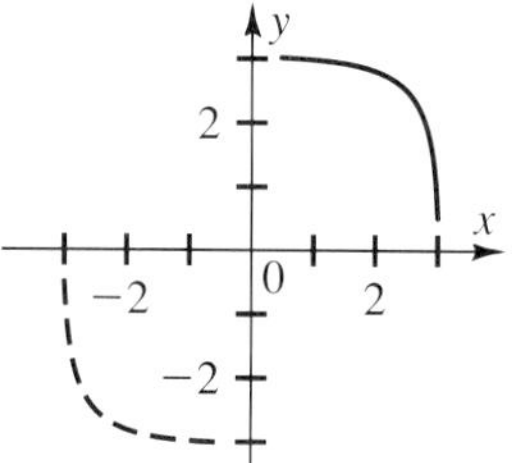

35.

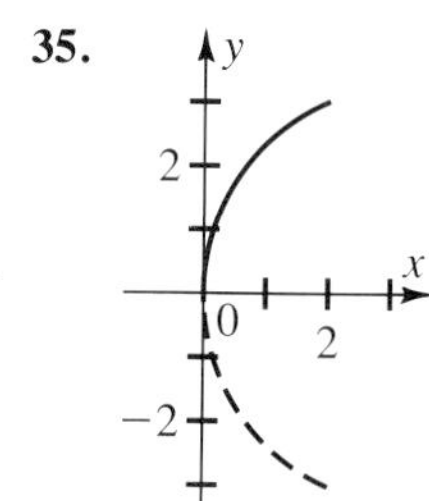

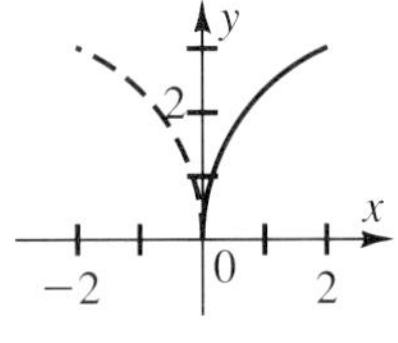

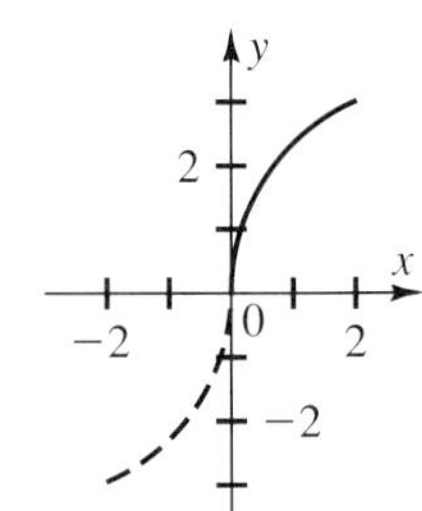

37. 31–2nd, 3rd; 33–2nd, 3rd; 35–2nd, 3rd
39. $x^2 + y^2 = 36$; y-axis, x-axis; origin

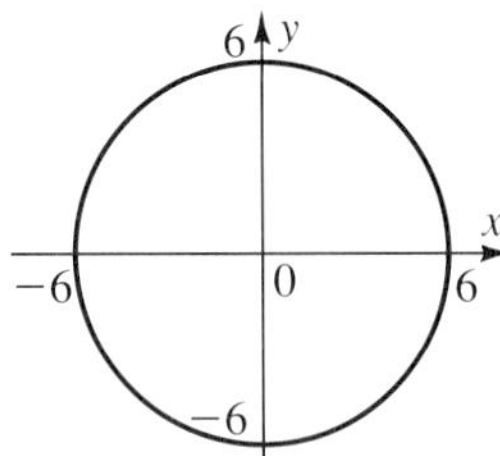

41.

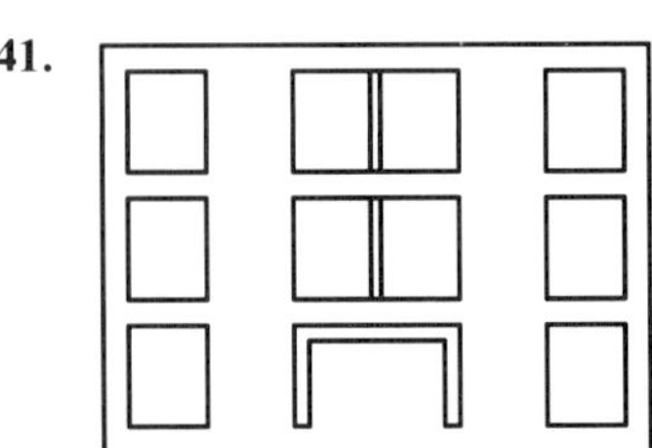

Practice Exercises, pages 71–72 **1.** cos 62° **3.** sec 5° **5.** csc 63° **7.** tan 37° **9.** −0.5299 **11.** 0.6691 **13.** 0.8480 **15.** 0.8090 **17.** 0.4695 **19.** −0.6691

21. 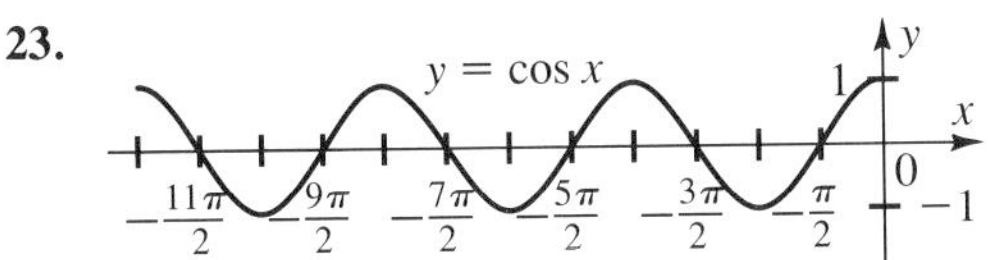

23. 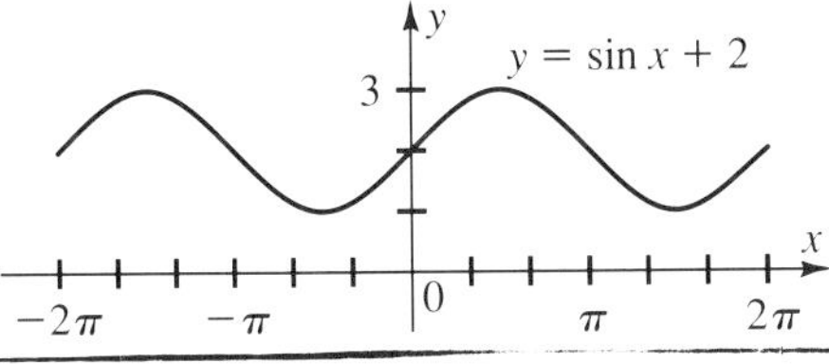

25. 0.3746 **27.** 0.4540 **29.** 0.6293
31. −0.7314 **33.** 0.6561

35. 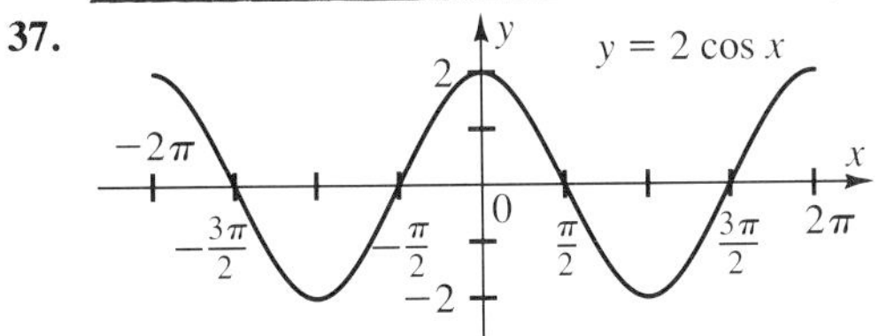

37. 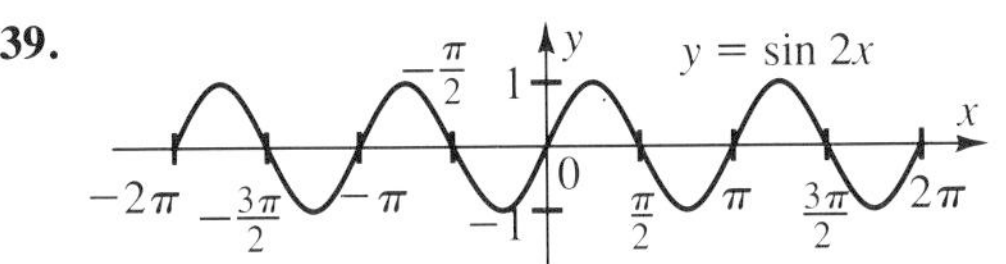

39. 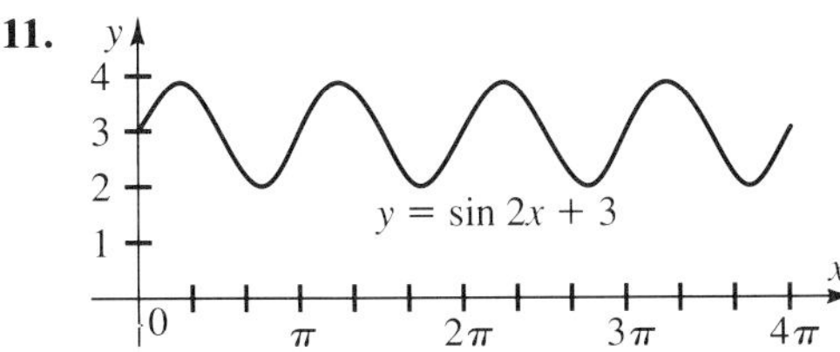

41. For $0 \leq x \leq 0.6$, $x \approx \sin x$, to the nearest tenth

Practice Exercises, pages 82–83 **1.** 2; $\frac{2\pi}{3}$; none; none **3.** 3; $\frac{\pi}{2}$; none; none **5.** 1; π; none; down 5 **7.** $\frac{1}{4}$; π; none; none **9.** 4; $\frac{2\pi}{3}$; right $\frac{\pi}{9}$; up 2

11. 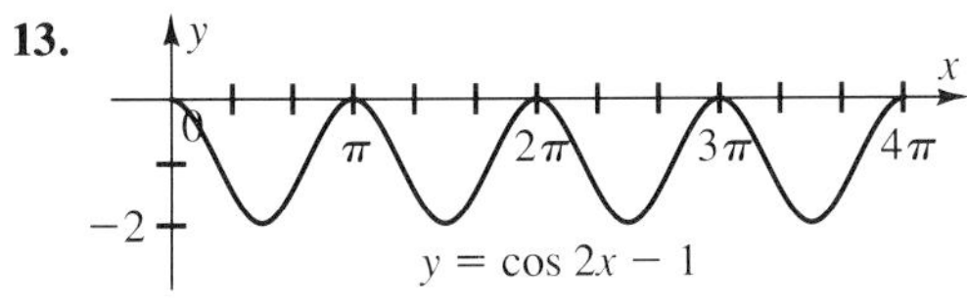

13. 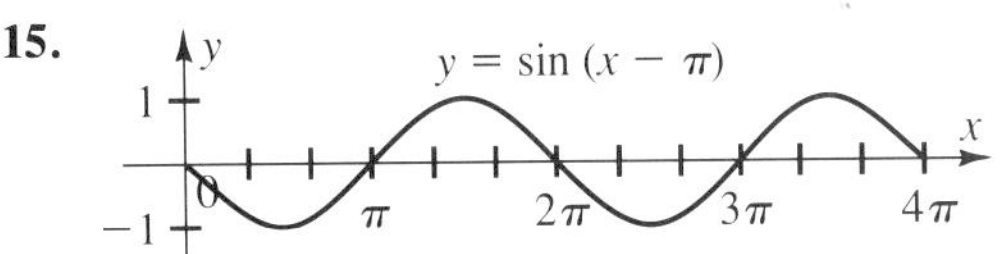

15. 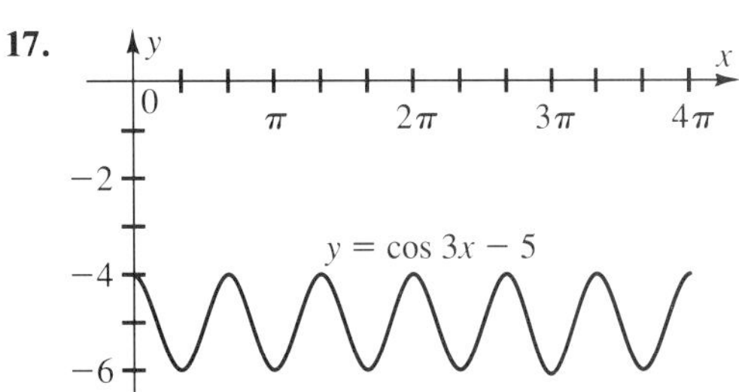

17. 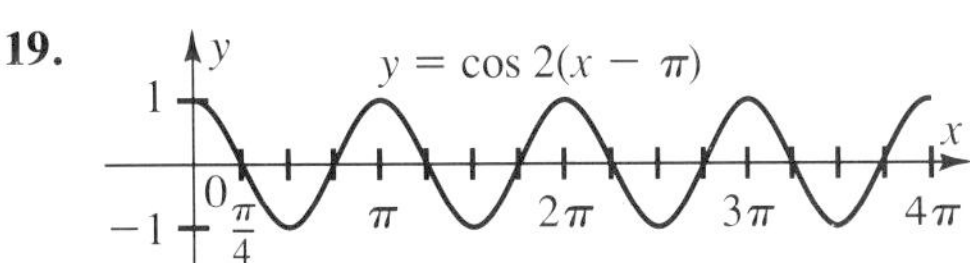

19. 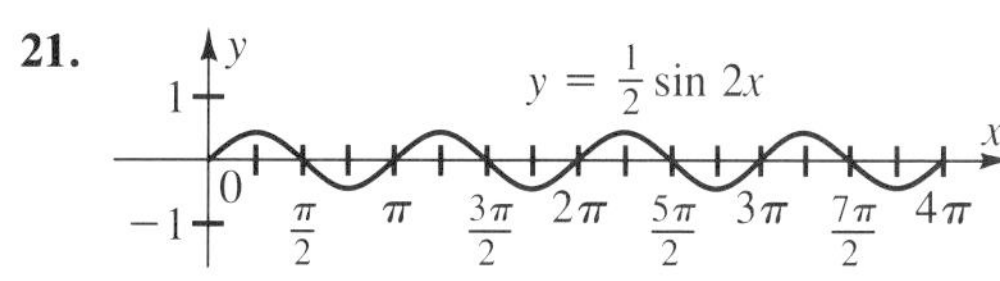

21. 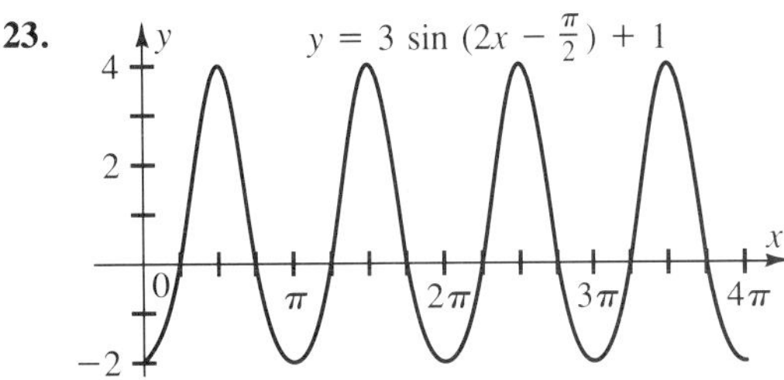

23. 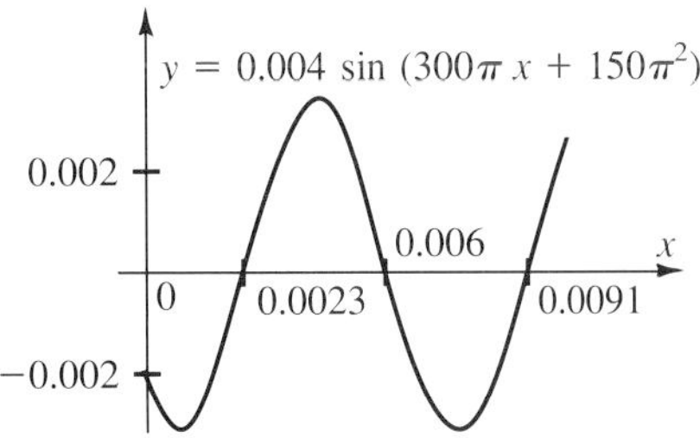

25. $y = 4 \sin 2x - 3$ **27.** $y = \frac{3}{5} \cos \frac{4}{3}x + \frac{22}{5}$ **29.** $y = \frac{1}{25} \sin 2(196\pi x)$ **31.** 0.004; $\frac{1}{150}$; $\frac{\pi}{2}$

Test Yourself, page 83 **1.** even **3.** odd **5.** x-axis **7.** −0.8572 **9.** 5; 2π **11.** 4; 6π **13.** 2; 2π; $\frac{\pi}{2}$ right; 5 up **15.** 1; π; $\frac{3\pi}{2}$ left; 1 down

17.

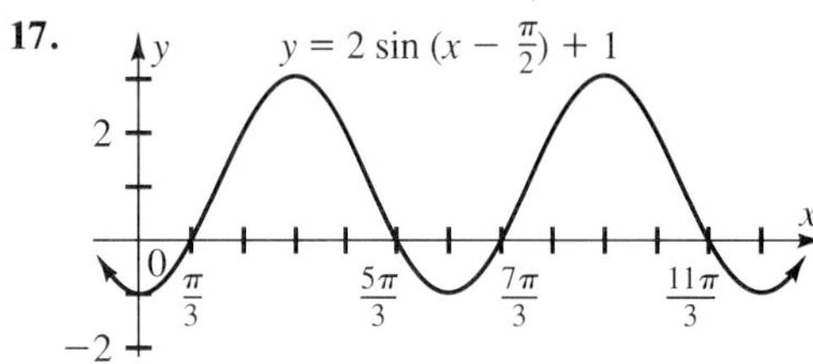

Practice Exercises, pages 86–87

1.

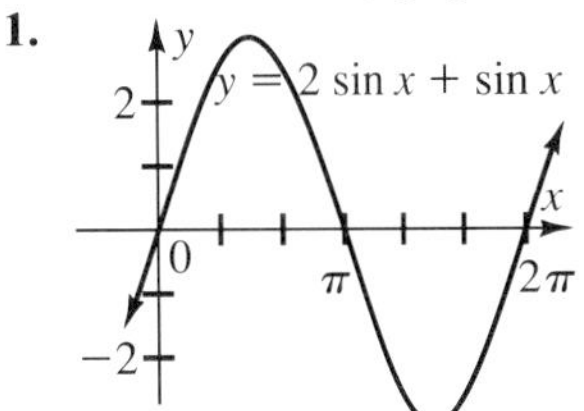

3.

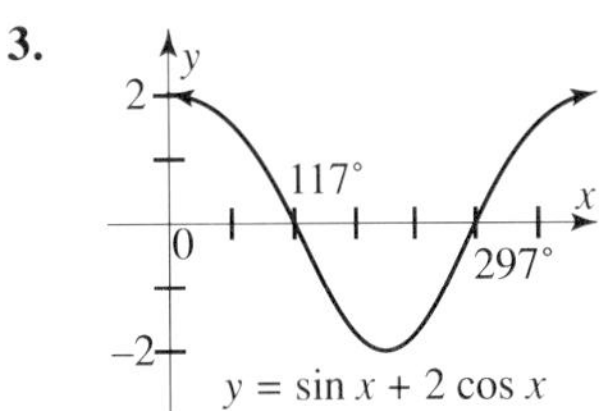

5.

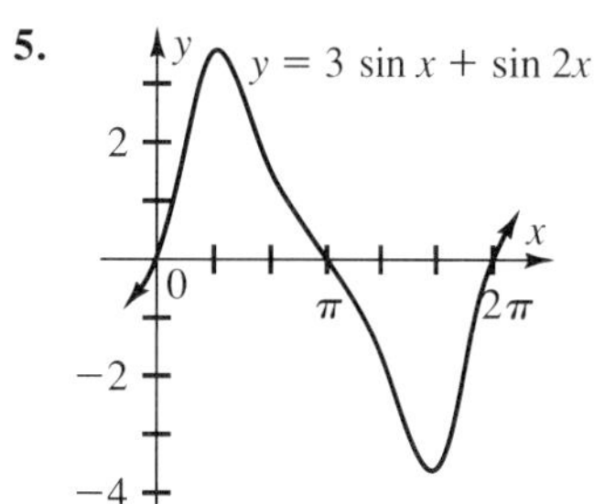

7.

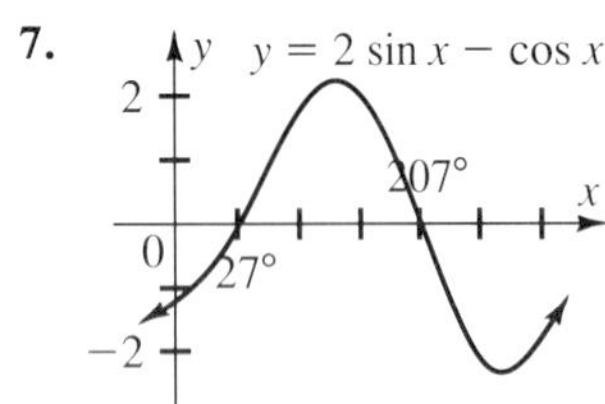

9.

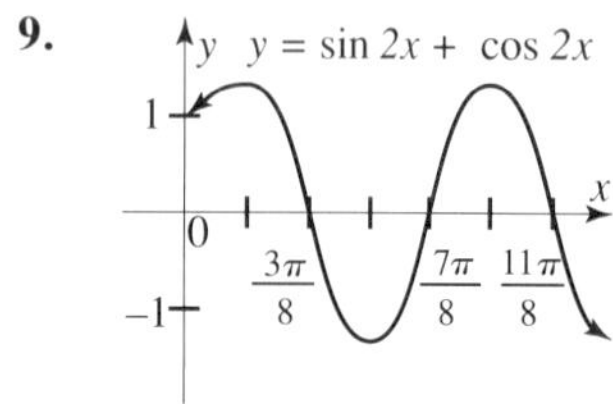

11.

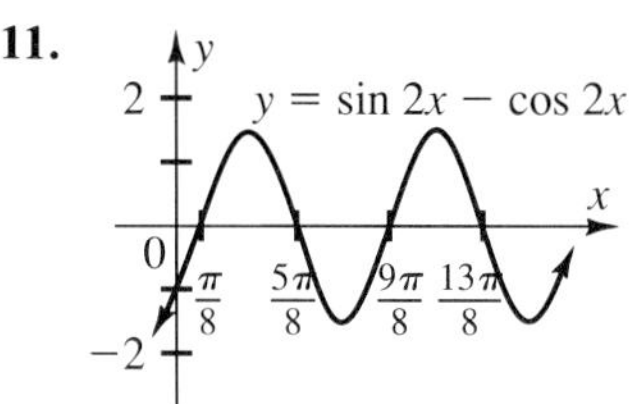

13.

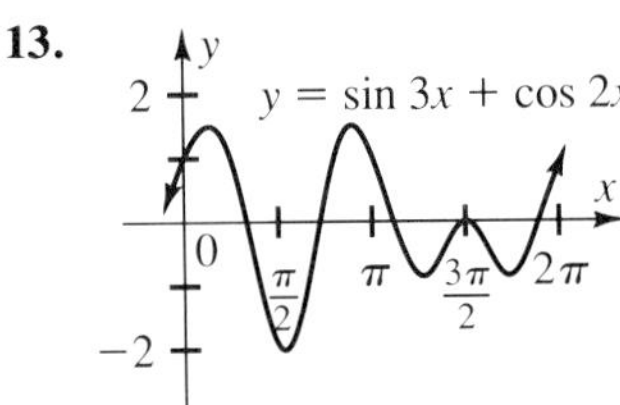

15.

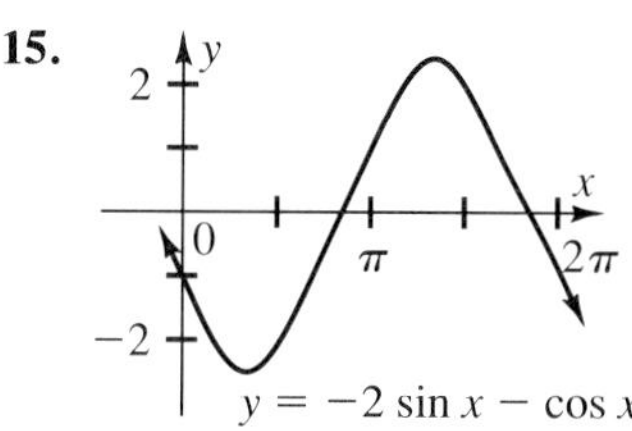

17.

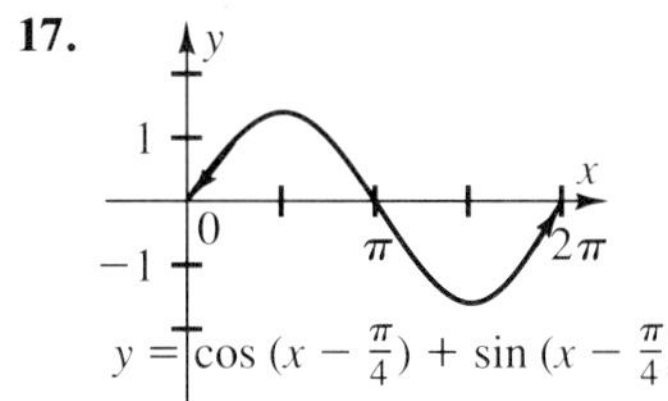

19.

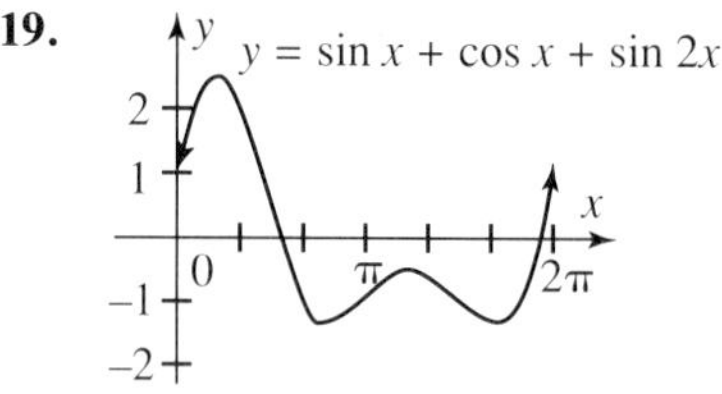

21.

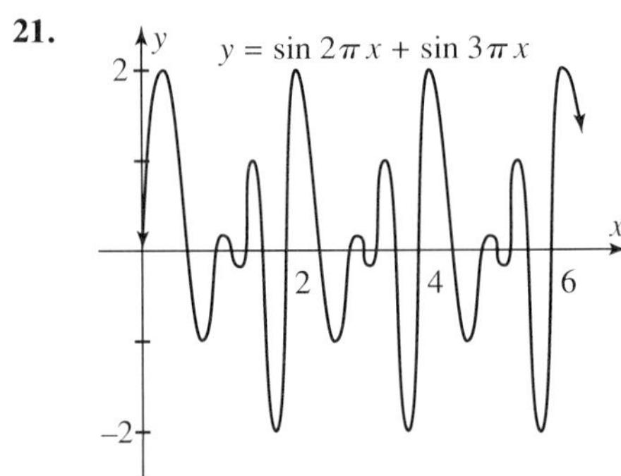

Practice Exercises, pages 92–93

1. $\frac{\pi}{3}$; π right **3.** π; $\frac{\pi}{4}$ right **5.** π; $\frac{2\pi}{3}$ left

7. $\frac{\pi}{2}$; π left; 3 up **9.** π; $\frac{4\pi}{3}$ right; 1 down

11. π; $\frac{2\pi}{3}$ left; 2 down **13.** π; $\frac{\pi}{6}$ right; 2 up

15.

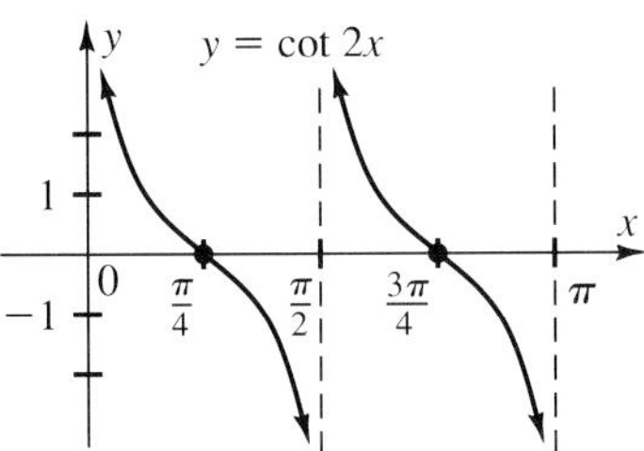

17.

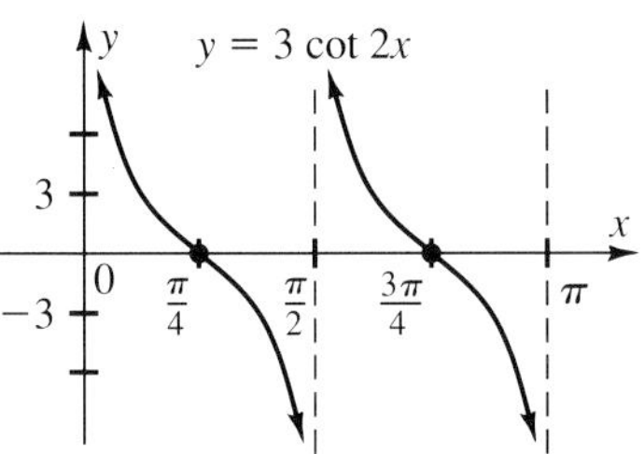

19.

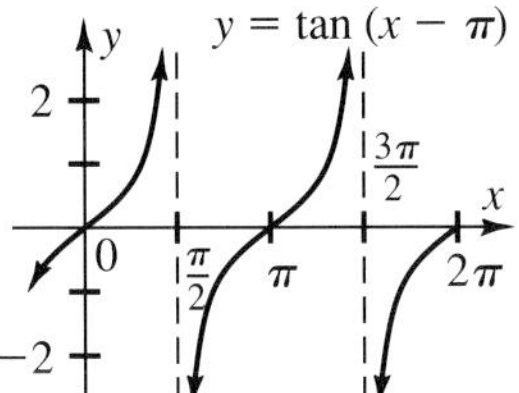

21.

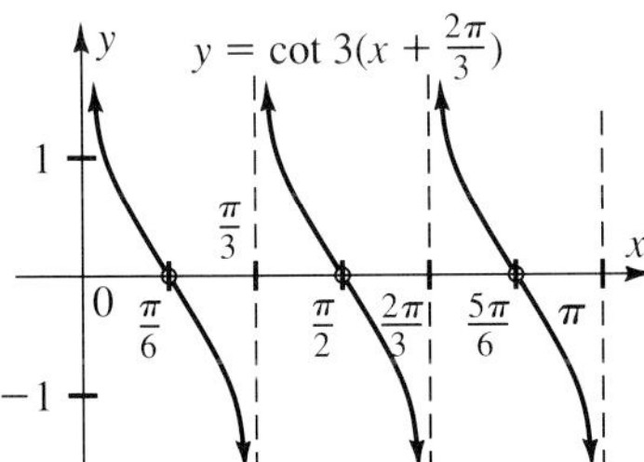

23.

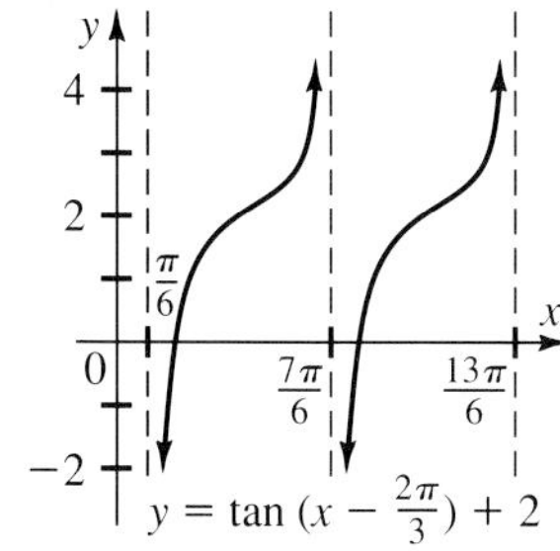

25.

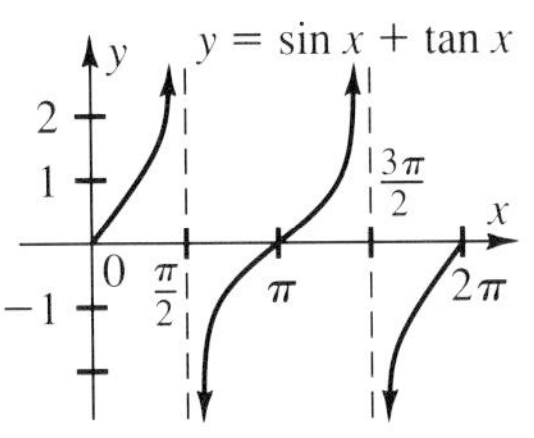

27.

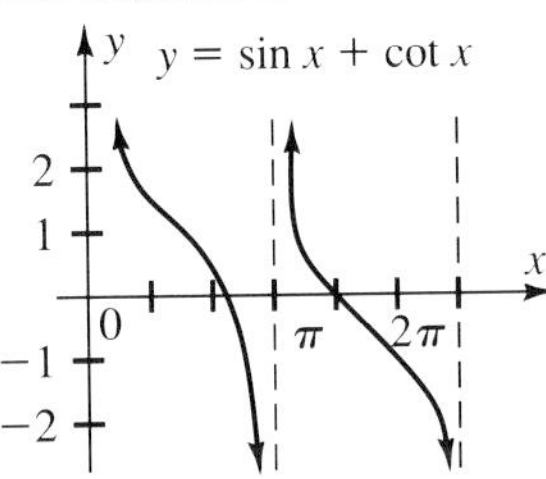

Practice Exercises, pages 98–99 **1.** $\frac{2\pi}{3}$; π right; none **3.** 2π; $\frac{\pi}{4}$ right; none **5.** 2π; $\frac{2\pi}{3}$ left; none

7. π; π left; 3 up **9.** 2π; $\frac{4\pi}{3}$ right; 1 down

11. $\frac{2\pi}{3}$; $\frac{2\pi}{9}$ left; 2 down **13.** π; $\frac{\pi}{4}$ left; 1 down

15.

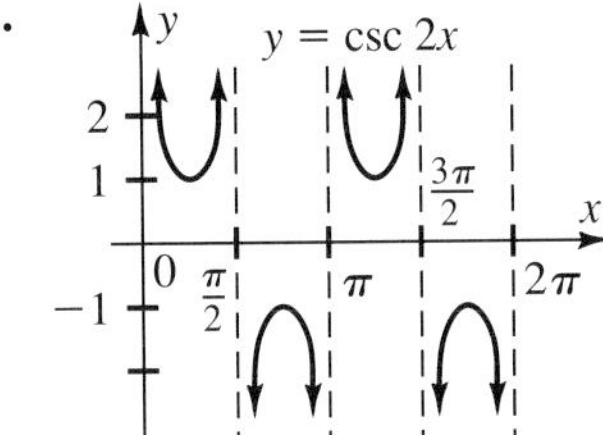

17.

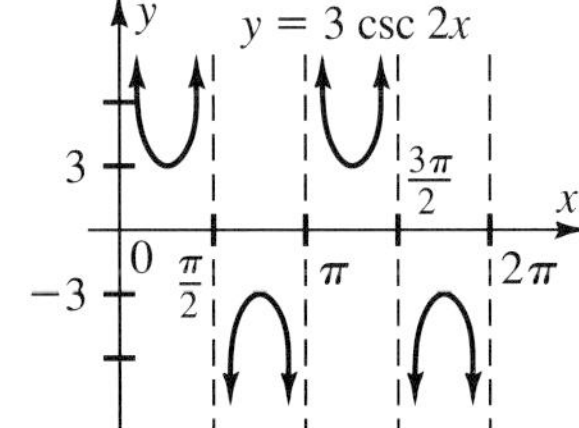

19.

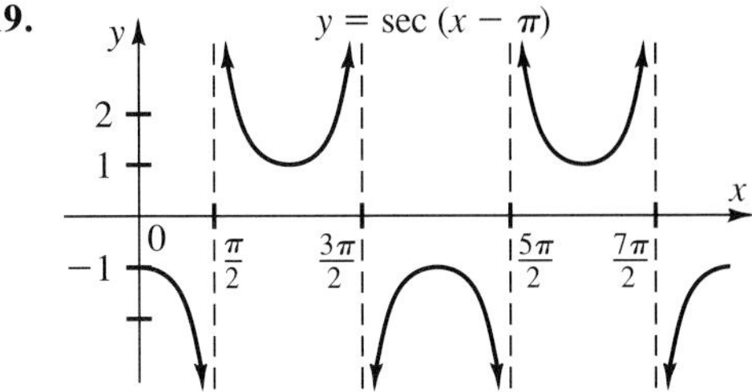

21. $y = \csc 2(x + \frac{2\pi}{3})$

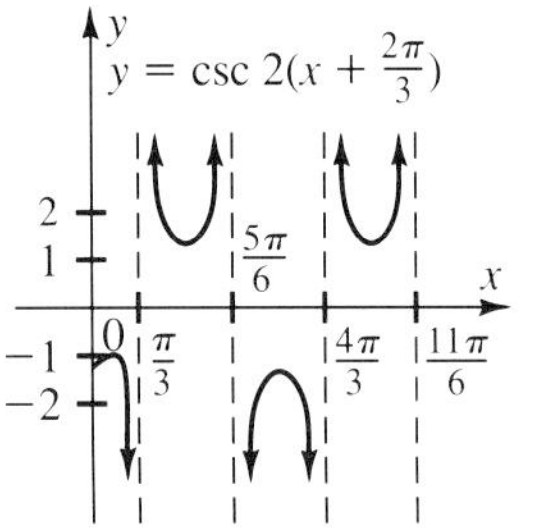

23. $y = \sec (2x - \pi)$

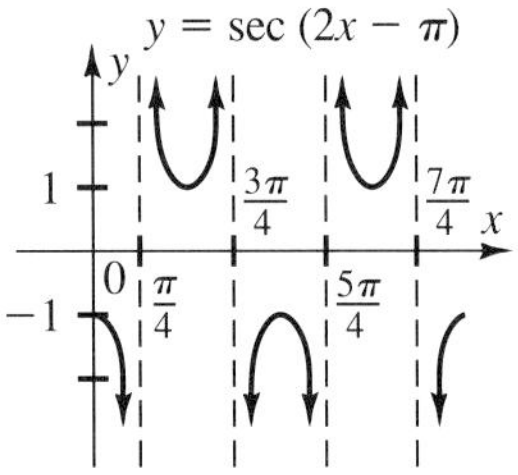

25. $y = \sin x + \sec x$

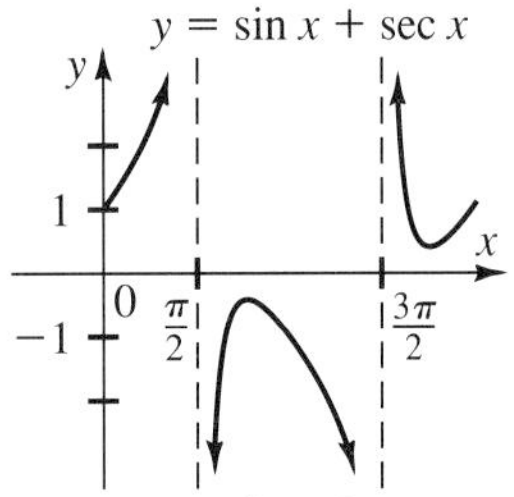

27. $y = \sin x + \csc x$

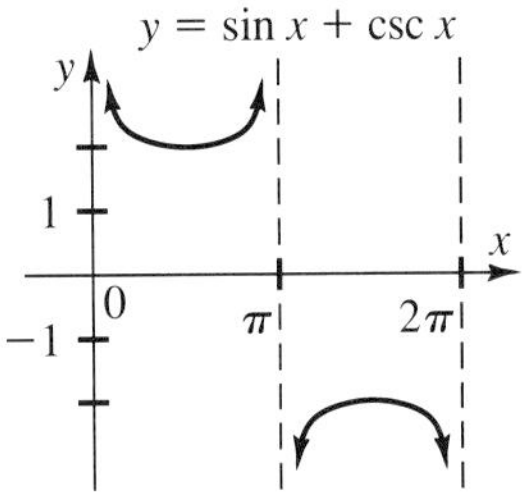

Practice Exercises, pages 103–104

1. $y = 3 \sin 2\pi\left(t - \frac{1}{4}\right)$ **3.** $y = 5 \sin \pi\left(t - \frac{1}{2}\right)$

5. $y = 5 \sin 4\pi\left(t - \frac{1}{8}\right)$ **7.** $y = 20 \sin \frac{\pi}{2}(t - 1)$

9. $y = 20 \sin 8\pi\left(t - \frac{1}{16}\right)$ **11.** $y = \sin (528\ \pi t)$

13. $y = 3.2 \sin \frac{\pi}{6.2}(t + 3.1)$ **15.** $x = 6 \sin \frac{\pi}{2} t$

Test Yourself, page 104

1.

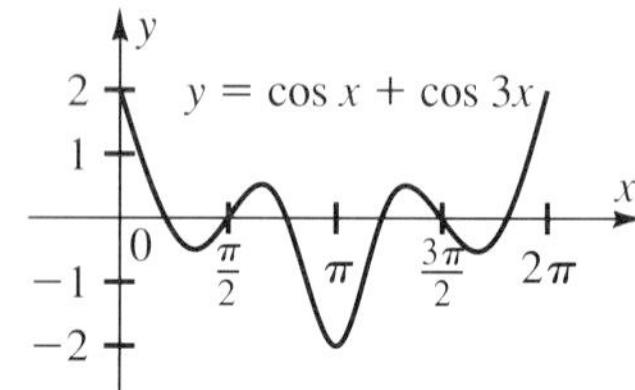

3. $\frac{\pi}{4}$; π left; none **5.** π; $\frac{\pi}{4}$ left; 1 down

7. 2π; $\frac{\pi}{2}$ left; 1 up **9.** π; $\frac{\pi}{4}$ left; 3 down

11.

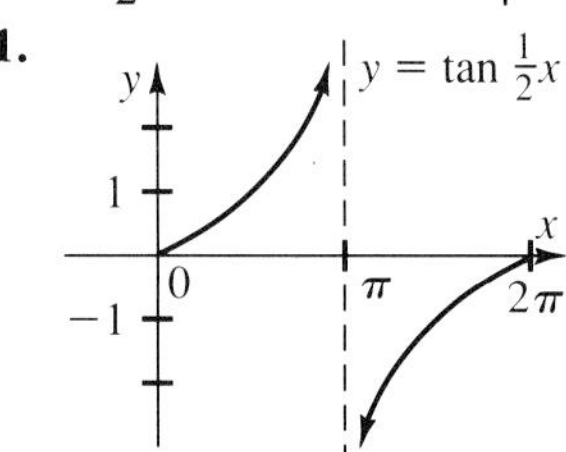

13.

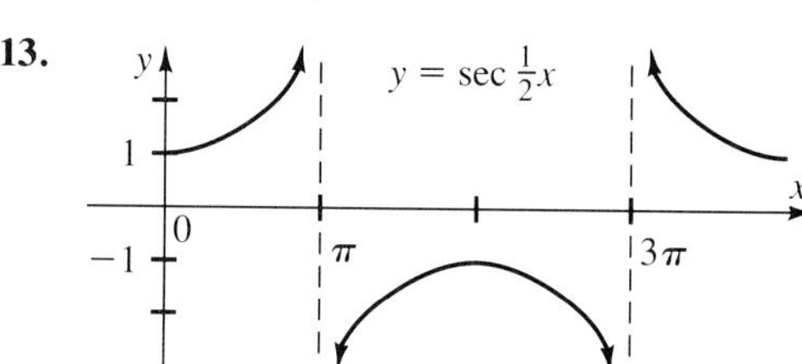

15. $y = 8 \sin \frac{\pi}{2}(t - 1)$

Summary and Review, pages 106–107 **1.** yes; period = 6, y-axis symmetry **3.** -0.8192

5. 4; $\frac{2\pi}{3}$ **7.** 3; π; π left; 4 down

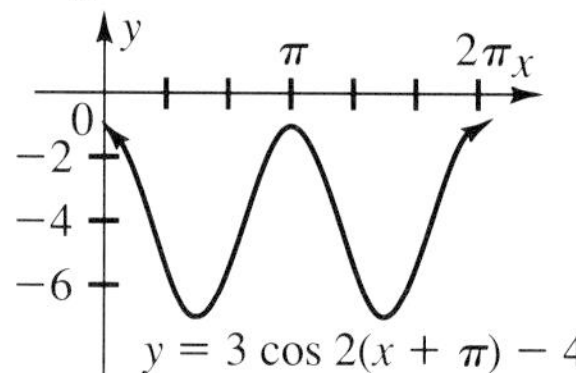

9.

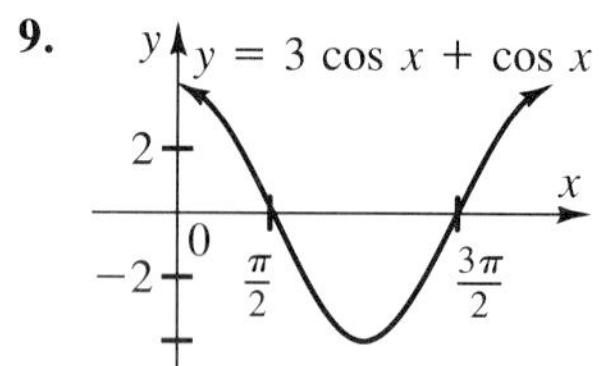

11. $\frac{\pi}{3}$; none; none

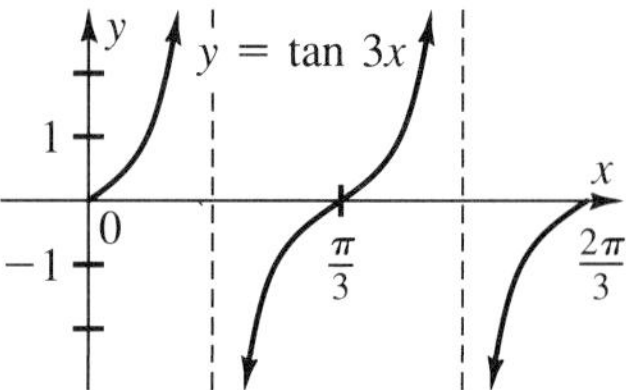

13. $\frac{2\pi}{3}$; none; none

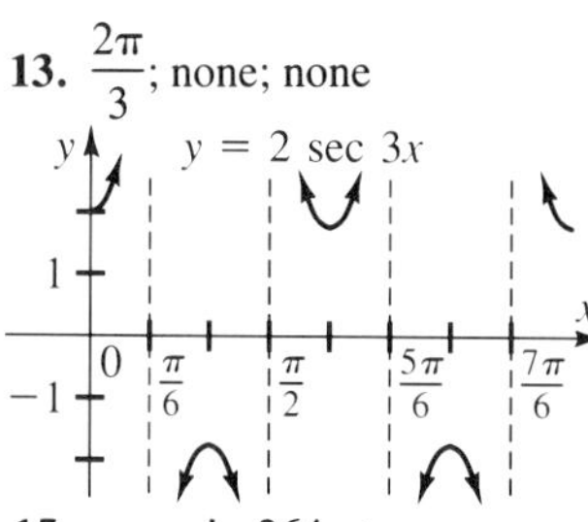

15. $y = \sin 264\pi t$

Cumulative Review, page 110

1.

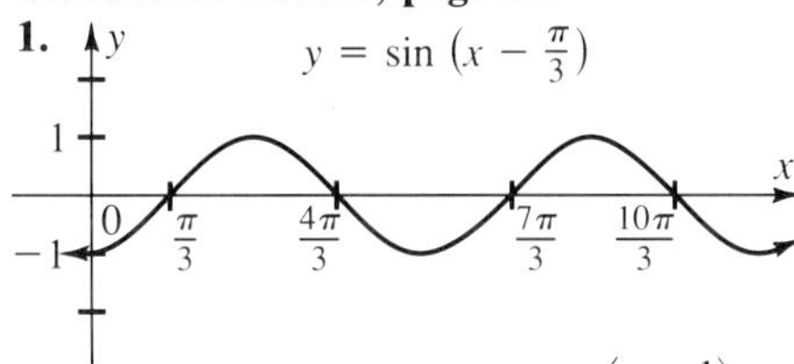

3. even 5. no 7. $y = 6 \sin \pi\left(t - \frac{1}{2}\right)$

9. $\sin 41°$

11. $y = \sin x - \cos x$

13. 24π ft/s 15. $\frac{7\pi}{4}$ 17. 330° 19. 48.8° and 311.2°

Chapter 3 Solving Right Triangles

Practice Exercises, pages 116–117 **1.** $\angle A = 65°$; $a = 30$; $c = 33$ **3.** $\angle A = 73°$; $a = 21$; $b = 6.4$ **5.** $\angle A = 57°$; $\angle B = 33°$; $c = 6.6$ **7.** $\angle B = 32°$; $a = 23$; $b = 14$ **9.** $\angle A = 74.9°$; $a = 10.3$; $b = 2.79$ **11.** $\angle B = 27°$; $b = 5.6$; $c = 12$ **13.** $\angle A = 47.5°$; $b = 172$; $c = 255$ **15.** $\angle B = 53°19'$; $b = 25.94$; $c = 32.34$ **17.** $\angle A = 17°32'$; $b = 268.5$; $c = 281.6$ **19.** $\angle A = 42.53°$; $\angle B = 47.47°$; $a = 16.16$ **21.** $\angle A = 52.68°$; $\angle B = 37.32°$; $c = 23.45$ **23.** $\angle A = 54.09°$; $\angle B = 35.91°$; $b = 1182$ Answers to 25–27 are in the order: sin; cos; tan; csc; sec; cot.

25. $\frac{3}{t}; \frac{\sqrt{t^2-9}}{t}; \frac{3\sqrt{t^2-9}}{t^2-9}; \frac{t}{3}; \frac{t\sqrt{t^2-9}}{t^2-9}; \frac{\sqrt{t^2-9}}{3}$

27. $\frac{\sqrt{t^2-1}}{t}; \frac{1}{t}; \sqrt{t^2-1}; \frac{t\sqrt{t^2-1}}{t^2-1}; t; \frac{\sqrt{t^2-1}}{t^2-1}$

29. 86 ft **31.** 16 ft

Practice Exercises, pages 121–123 **1.** 22 ft **3.** 89 ft **5.** 39° **7.** 4590 ft **9.** 1089 ft **11.** 147 ft **13.** 150 ft **15.** 55 m **17.** 13.91° **19.** 53 m

Practice Exercises, pages 126–129 **1.** 124°; 64 km **3.** 221 mi **5.** 916 ft **7.** 95 yd **9.** 61° **11.** 19.8 mi **13.** 145 mi **15.** 216 ft **17.** 198.0 mi

Test Yourself, page 129 **1.** $\angle B = 47°$; $a = 13$; $b = 14$ **3.** $\angle A = 38°$; $\angle B = 52°$; $c = 24$ **5.** 20° **7.** 149°; 37 mi **9.** 92 m

Practice Exercises, pages 133–135

9. $1 + (\sqrt{3})^2 = (2)^2$ **11.** $\left(\frac{\sqrt{2}}{2}\right)^2 + \left(\frac{\sqrt{2}}{2}\right)^2 = 1$

13. $-\sqrt{3} = \frac{-\frac{\sqrt{3}}{2}}{\frac{1}{2}}$ **15.** $-\frac{\sqrt{3}}{2} = -\left(\frac{\sqrt{3}}{2}\right)$

17. $1 - \left(\frac{\sqrt{2}}{2}\right)^2 = \left(-\frac{\sqrt{2}}{2}\right)^2$ **19.** $\frac{1}{\cos\theta} = \frac{1}{\frac{x}{r}} = \frac{r}{x} = \sec\theta$ **21.** $\frac{\cos\theta}{\sin\theta} = \frac{\frac{x}{r}}{\frac{y}{r}} = \frac{x}{y} = \cot\theta$ **23.** $1 + \tan^2\theta = 1 + \left(\frac{y}{x}\right)^2 = \frac{x^2+y^2}{x^2} = \frac{r^2}{x^2} = \sec^2\theta$

25. $\sin\theta\csc\theta = \left(\frac{y}{r}\right)\left(\frac{1}{\frac{y}{r}}\right) = \left(\frac{y}{r}\right)\left(\frac{r}{y}\right) = 1$

27. $1 - \sin^2\theta = 1 - \left(\frac{y^2}{r^2}\right) = \frac{r^2-y^2}{r^2} = \frac{x^2}{r^2} = \cos^2\theta$ **29.** $\left(\frac{\sqrt{2}}{2}\right)(1) = \frac{\sqrt{2}}{2}$

31. $\left(\frac{\sqrt{2}}{2}\right)\left(-\frac{\sqrt{2}}{2}\right)(-1) = 1 - \left(-\frac{\sqrt{2}}{2}\right)^2$

33. $41^2 = 40^2 + 9^2$; $1681 = 1681$

35. $\left(\frac{41}{40}\right)\left(\frac{9}{41}\right)\left(\frac{40}{9}\right) = 1$; $1 = 1$

37. 0.9903; 20.0 m **41.** It prevents division by 0.

Practice Exercises, pages 137–139 **1.** $1 - \sin^2\theta$ **3.** $\sec^2\beta - 1$ **5.** $\frac{(\sin\alpha - 5)(\sin\alpha + 5)}{(\sin\alpha + 5)(\sin\alpha + 5)}$

7. $\frac{(\cos\alpha - 7)(\cos\alpha - 7)}{(\cos\alpha - 7)(\cos\alpha + 7)}$

9. $\dfrac{(\tan\theta - 3)(\tan^2\theta + 3\tan\theta + 9)}{(\tan\theta - 3)(\tan\theta + 3)}$
11. $\dfrac{(\cot\theta + 2)(\cot^2\theta - 2\cot\theta + 4)}{(\cot\theta + 2)(\cot\theta - 2)}$
13. $\csc^2\theta - \cot^2\theta$ **15.** $\dfrac{\sin^2\theta}{\cos^2\theta}$ **17.** $\dfrac{\sin\theta}{1 - \sin^2\theta}$
19. $\dfrac{2}{\sin\theta}$ **21.** $\cos^2\theta$ **23.** $\dfrac{2}{\cos\theta}$ **25.** $\tan\theta$
27. $\dfrac{1}{\tan\theta}$ **29.** $\sin^2\theta + \sin\theta\cos\theta + \cos^2\theta =$
$\sin\theta\cos\theta + 1$ **31.** $a^2 - b^2$

Practice Exercises, pages 143–144
1. $\dfrac{1 - \cos^2\theta}{\cos^2\theta} = \dfrac{\sin^2\theta}{\cos^2\theta} = \left(\dfrac{\sin\theta}{\cos\theta}\right)^2$
3. $\dfrac{\cos\alpha}{1 - \sin^2\alpha} = \dfrac{\cos\alpha}{\cos^2\alpha} = \dfrac{1}{\cos\alpha}$
5. $\dfrac{1}{\tan\alpha} = \dfrac{\cot\alpha}{1} = \dfrac{\cot\alpha}{\sin^2\alpha + \cos^2\alpha}$
7. $\dfrac{1 + \cos\mu}{\cos\mu} = \dfrac{1}{\cos\mu} + \dfrac{\cos\mu}{\cos\mu} = \sec\mu + 1$
9. $\dfrac{\sin^2\alpha}{1 - \sin^2\alpha} = \dfrac{\sin^2\alpha}{\cos^2\alpha} = \tan^2\alpha = \sec^2\alpha - 1$
11. $\sin^2\theta\cos\theta\sec\theta = \sin^2\theta(\cos\theta)\left(\dfrac{1}{\cos\theta}\right) =$
$\sin^2\theta = 1 - \cos^2\theta$ **13.** $1 - \dfrac{1}{\sec^2\beta} =$
$1 - \cos^2\beta = \sin^2\beta$

Practice Exercises, pages 146–147 **1.** identity **3.** identity **5.** identity **7.** identity **9.** identity **11.** not an identity **13.** identity **15.** not an identity **17.** identity **19.** not an identity **21.** identity **23.** not an identity **25.** identity **27.** not an identity **29.** $4\cos\theta + 3\sin\theta = 5\cos(\theta - 36°52')$

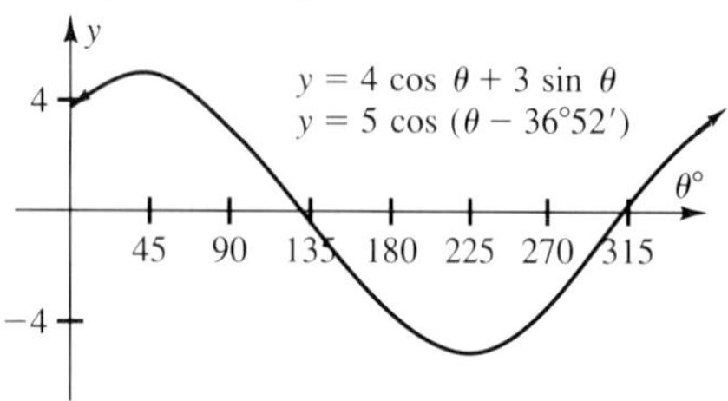

Test Yourself, page 147
1. $\left(-\dfrac{\sqrt{3}}{3}\right)^2 = \left(\dfrac{2\sqrt{3}}{3}\right)^2 - 1$
3. $\dfrac{\cot\theta}{\sin\theta}(\sec\theta - \cos\theta) =$
$\dfrac{\dfrac{\cos\theta}{\sin\theta}}{\sin\theta}\left(\dfrac{1}{\cos\theta} - \cos\theta\right) = \dfrac{\cos\theta}{\sin^2\theta}\left(\dfrac{1 - \cos^2\theta}{\cos\theta}\right) =$
$\left(\dfrac{\cos\theta}{\sin^2\theta}\right)\left(\dfrac{\sin^2\theta}{\cos\theta}\right) = 1$ **7.** identity

Summary and Review, pages 150–151 **1.** $\angle A = 69°$; $b = 65.3$; $c = 182$ **3.** 21° **5.** 213°; 44.6 km **7.** 680 ft **9.** $\dfrac{\sqrt{3}}{3} = \dfrac{-\frac{1}{2}}{-\frac{\sqrt{3}}{2}}$ **11.** $\dfrac{\cos^2\theta}{\sin^2\theta}$

Maintaining Skills, page 154
1. $\dfrac{6 + q}{3q}$ **3.** $\dfrac{2q - p^2 - pq}{pq}$ **5.** 1 **7.** $\dfrac{15R + 12}{10}$
9. $\dfrac{abd - ac}{2d}$ **11.** $\dfrac{5RY - 5SY}{8}$ **13.** $x < -\dfrac{7}{2}$
15. $x > \dfrac{2}{3}$ **17.** $x > -7$ **19.** 80°

Chapter 4 Oblique Triangles

Practice Exercises, pages 160–162 **1.** $b = 18$ **3.** $a = 170$ **5.** 30 **7.** 100 **9.** 79 **11.** 19 **13.** 30 **15.** $\angle P = 95°$, $q = 13$, $r = 11$ **17.** $\angle Q = 66°$, $p = 47$, $r = 41$ **19.** $\angle Q = 100.5°$, $q = 155$, $p = 84.9$ **21.** $\angle R = 79.9°$, $p = 62.7$, $q = 55.5$ **23.** $\angle R = 39.1°$, $r = 76.9$, $q = 112$ **25.** $\angle P = 14.8°$, $p = 19.9$, $r = 63.9$ **27.** $\angle R = 67.5°$, $r = 121$, $q = 119$ **29.** $\angle R = 19.2°$, $r = 47.1$, $q = 105$ **31.** yes
33. $\sin A = \dfrac{a}{c}$, $\sin B = \dfrac{b}{c}$, $\sin C = 1$; $\dfrac{a}{c} \div a = \dfrac{b}{c} \div b = \dfrac{1}{c}$ **37.** 48.87°; 256.0 ft; 320.8 ft
39. 930 ft

Practice Exercises, pages 166–167 **1.** 0 **3.** 2; $\angle B = 131°$, $\angle C = 17°$, $c = 3.9$ or $\angle B = 49°$, $\angle C = 99°$, $c = 13$ **5.** 0 **7.** 0 **9.** 2; $\angle B = 42°$, $\angle C = 101°$, $c = 80$ or $\angle B = 138°$, $\angle C = 5°$, $c = 7.1$ **11.** 1; $\angle B = 52.2°$, $\angle C = 68.0°$, $c = 86.7$ **13.** 2; $\angle B = 40.9°$, $\angle C = 107.2°$, $c = 55.3$ or $\angle B = 139.1°$, $\angle C = 9.0°$, $c = 9.1$ **15.** 0 **17.** 2; $\angle B = 70.4°$, $\angle A = 62.5°$, $a = 14.4$ or $\angle B = 109.6°$, $\angle A = 23.3°$, $a = 6.43$ **19.** 2; $\angle C = 40.9°$, $\angle A = 102.8°$, $a = 76.3$ or $\angle C = 139.1°$, $\angle A = 4.6°$, $a = 6.27$ **21.** 0 **23.** 1; $\angle A = 63.3°$, $\angle B = 8.0°$, $b = 7.98$ **25.** 0

27. 2; $\angle A = 31.0°$, $\angle B = 125.5°$, $b = 41.9$ or $\angle A = 149.0°$, $\angle B = 7.5°$, $b = 6.75$ **29.** $\angle B = 56°$, $\angle C = 73°$ or $\angle B = 124°$, $\angle C = 5°$ **31.** 15°, 134° **35.** 112°

Practice Exercises, pages 172–174 **1.** 10 **3.** 7.4 **5.** 28 **7.** 30 **9.** 8.8 **11.** 40° **13.** 105° **15.** 90° **17.** $r = 9.0$, $\angle Q = 63°$, $\angle P = 87°$ **19.** $p = 63$, $\angle R = 43°$, $\angle Q = 54°$ **21.** $q = 59$, $\angle P = 25°$, $\angle R = 42°$ **23.** $\angle R = 84°$, $\angle P = 41°$, $\angle Q = 55°$ **25.** $\angle R = 31°$, $\angle Q = 53°$, $\angle P = 96°$ **27.** $\angle P = 41.47°$, $\angle Q = 59.51°$, $\angle R = 79.02°$ **35.** 82.0°, 51.7°, 46.3° **37.** 106.0° **39.** 410 mi

Practice Exercises, pages 179–180 **1.** $\angle A = 110°$, $\angle B = 23°$, $c = 36$ **3.** $\angle A = 51°$, $\angle B = 25°$, $c = 34$ **5.** $\angle B = 87°$, $\angle C = 50°$, $a = 42$ **7.** $\angle B = 31°$, $\angle C = 26°$, $a = 25$ **9.** $\angle A = 108°$, $\angle C = 46°$, $b = 21$ **11.** $\angle A = 44°$, $\angle C = 25°$, $b = 120$ **13.** $\angle C = 104°$, $a = 29$, $b = 20$, $c = 39$ **15.** $\angle B = 64°$, $a = 60$, $b = 54$, $c = 23$ **17.** $\angle C = 114.7°$, $a = 60.1$, $b = 32.2$, $c = 79.2$ **19.** $\angle A = 94.4$, $a = 77.0$, $b = 46.1$, $c = 58.2$ **23.** 84 m

Test Yourself, page 180 **1.** $\angle C = 91°$, $b = 23$, $c = 28$ **3.** $\angle B = 87°$, $a = 11$, $b = 16$ **5.** 1; $\angle B = 48°$, $\angle C = 37°$, $c = 19$ **7.** $\angle A = 42°$, $\angle B = 54°$, $c = 62$ **9.** $\angle A = 97°$, $\angle B = 51°$, $\angle C = 32°$ **11.** 0 **13.** $\angle A = 21°$, $\angle B = 39°$, $c = 48$, **15.** $\angle C = 51°$, $a = 7.7$, $b = 36$, $c = 31$ **17.** 11 cm

Practice Exercises, pages 184–185 **1.** 440 **3.** 380 **5.** 60 **7.** 730 **9.** 30 **11.** 860 **13.** 540 **15.** 40 **17.** 108 **19.** 1506 **21.** 201 **23.** 1039 **25.** 121 **27.** 120 **31.** 2100 cm^2 **33.** 4888 ft^2 **35.** 5.0 acres

Practice Exercises, pages 188–189 **1.** 6 **3.** 280 **5.** 150 **7.** 270 **9.** 60 **11.** 220 **13.** 15,500 **15.** 13,900 **17.** 12 **19.** 26 **21.** 1200 **23.** 2000 **25.** 170 ft^2

Practice Exercises, pages 195–197

1. $\|\overrightarrow{AB}\| = 13$

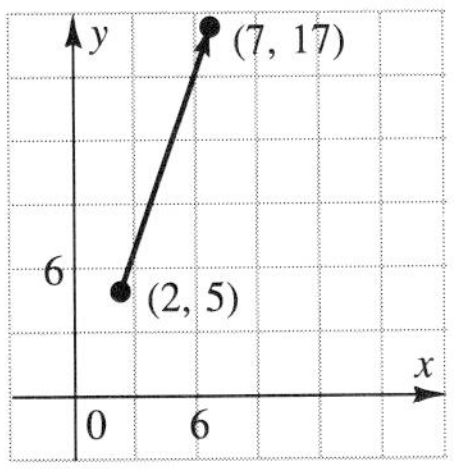

3. $\|\overrightarrow{AB}\| = \sqrt{290}$

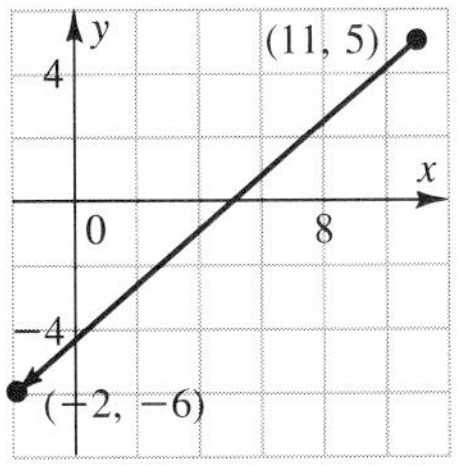

5. 13; 8 **7.** 16; 8 **9.** −4; 11

11.

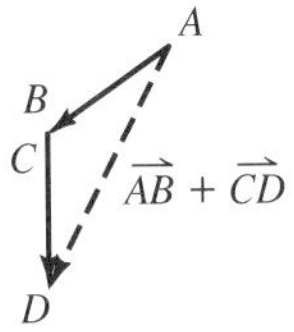

13.

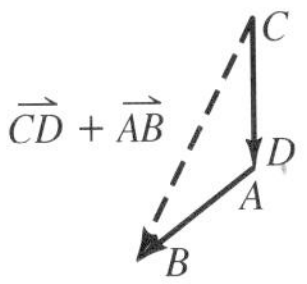

15. $\overrightarrow{V} - \overrightarrow{T}$

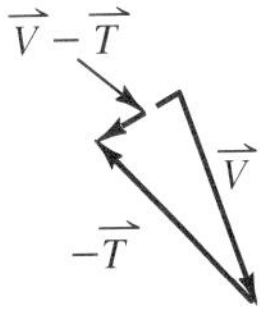

17.

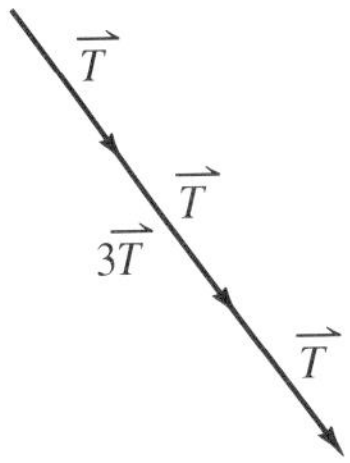

19. 22; 44° **21.** 31; 19° **23.** 3.6; 64° **25.** 29.1; 18.5° **27.** 33.8; 61.4°

29.

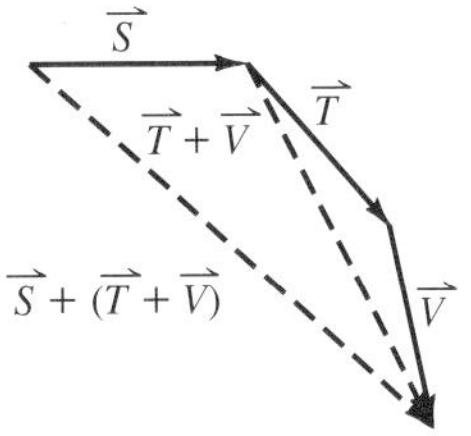

31.

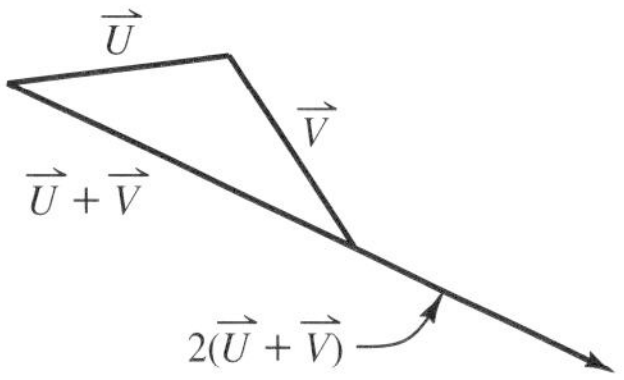

33. 280 mi/h; 86° **35.** 230 mi/h; 50° **37.** 590 mi/h; 27° E of N **39.** 320 mi/h; 33°

Test Yourself, page 197 **1.** 560 $in.^2$ **3.** 350 cm^2 **5.** 150 **7.** 140 **9.** $\sqrt{34}$ **11.** 18.9; 17.5°

Summary and Review, pages 200–201 **1.** 24.6 **3.** $c = 2$, $\angle B = 115°$, $\angle C = 9°$ or $c = 11$, $\angle B = 65°$, $\angle C = 59°$ **5.** $\angle B = 36°$, $\angle C = 45°$, $c = 18$ **7.** $\angle A = 47°$, $\angle B = 75°$, $c = 14$ **9.** triangle is not determined **11.** 160 **13.** 334 **15.** $10\sqrt{2}$; $10\sqrt{2}$

Cumulative Review, pages 204–206 **1.** $\angle A = 52°$, $\angle B = 94°$, $c = 7.9$ **3.** 132.31° **5.** 0.6691 **7.** 18 cm **9.** 10π in./s **11.** yes **13.** 7; 54° **17.** 180° **19.** $\left(\frac{1}{2}\right)^2\left(\frac{\sqrt{3}}{2}\right)\left(\frac{2\sqrt{3}}{3}\right) = 1 - \left(\frac{\sqrt{3}}{2}\right)^2$; $\left(\frac{1}{4}\right)1 = 1 - \left(\frac{3}{4}\right)$; $\frac{1}{4} = \frac{1}{4}$ **21.** y-axis

23. 500 in.^{2} **25.** quadrant IV **27.** $\sin\theta = \frac{3}{5}$; $\tan\theta = -\frac{3}{4}$; $\csc\theta = \frac{5}{3}$; $\sec\theta = -\frac{5}{4}$; $\cot\theta = -\frac{4}{3}$ **29.** 790 ft **31.** $\angle A = 83°$; $\angle C = 32°$ **33.** $x = 8$ or $x = -4$ **35.** neither **37.** odd **39.** 303°; 663°; −417° **41.** 1.1303 **43.** 0.3907 **45.** 0.5446 **47.** 16° **49.** $\frac{1}{2}$ acre

Chapter 5 Trigonometric Identities

Practice Exercises, pages 212–213

1. $\frac{\sqrt{6} - \sqrt{2}}{4}$ **3.** $\frac{-\sqrt{6} - \sqrt{2}}{4}$ **5.** $\frac{\sqrt{2} + \sqrt{6}}{4}$ **7.** $\frac{\sqrt{2} + \sqrt{6}}{4}$ **9.** $\frac{33}{65}$ **11.** $\frac{3}{5}$ **13.** $-\frac{44}{125}$ **15.** $-\frac{77}{85}$ **17.** $\cos(360° + \alpha) = \cos 360° \cos\alpha - \sin 360° \sin\alpha = 1(\cos\alpha) - 0(\sin\alpha) = \cos\alpha$ **19.** $\cos(270° - \beta) = \cos 270° \cos\beta + \sin 270° \sin\beta = 0(\cos\beta) + (-1)(\sin\beta) = -\sin\beta$ **21.** $\cos\left(\frac{\pi}{3} - \beta\right) = \cos\frac{\pi}{3}\cos\beta + \sin\frac{\pi}{3}\sin\beta = \frac{1}{2}\cos\beta + \frac{\sqrt{3}}{2}\sin\beta = \frac{\cos\beta + \sqrt{3}\sin\beta}{2}$ **23.** $\cos\left(\frac{3\pi}{4} + \alpha\right) = \cos\frac{3\pi}{4}\cos\alpha - \sin\frac{3\pi}{4}\sin\alpha = -\frac{\sqrt{2}}{2}\cos\alpha - \frac{\sqrt{2}}{2}\sin\alpha = -\frac{\sqrt{2}}{2}(\cos\alpha + \sin\alpha)$ **25.** $\frac{\sqrt{6} - \sqrt{2}}{4}$ **27.** $\frac{\sqrt{2} - \sqrt{6}}{4}$ **29.** $\cos(\alpha - \beta) - \cos(\alpha + \beta) = \cos\alpha\cos\beta + \sin\alpha\sin\beta - [\cos\alpha\cos\beta - \sin\alpha\sin\beta] = \cos\alpha\cos\beta + \sin\alpha\sin\beta - \cos\alpha\cos\beta + \sin\alpha\sin\beta = 2\sin\alpha\sin\beta$ **31.** $\frac{\sqrt{2} - \sqrt{6}}{4}$

Practice Exercises, pages 218–219

1. $\frac{\sqrt{6} - \sqrt{2}}{4}$ **3.** $\frac{\sqrt{2} - \sqrt{6}}{4}$ **5.** $\frac{\sqrt{6} - \sqrt{2}}{4}$ **7.** $\frac{\sqrt{2} + \sqrt{6}}{4}$ **9.** $\frac{323}{325}$ **11.** $-\frac{36}{85}$ **13.** $-\frac{16}{65}$ **15.** $-\frac{323}{325}$ **17.** $\sin(180° - \alpha) = \sin 180° \cos\alpha - \cos 180° \sin\alpha = 0(\cos\alpha) - (-1)(\sin\alpha) = \sin\alpha$ **19.** $\sin(210° + \alpha) = \sin 210° \cos\alpha + \cos 210° \sin\alpha = \left(-\frac{1}{2}\right)\cos\alpha + \left(-\frac{\sqrt{3}}{2}\right)\sin\alpha = \frac{-\cos\alpha - \sqrt{3}\sin\alpha}{2}$ **21.** $\sin(45° + \alpha) = \sin 45° \cos\alpha + \cos 45° \sin\alpha = \frac{\sqrt{2}}{2}\cos\alpha + \frac{\sqrt{2}}{2}\sin\alpha = \frac{\sqrt{2}}{2}(\cos\alpha + \sin\alpha)$ **23.** $\sin\left(\frac{\pi}{3} - \beta\right) = \sin\frac{\pi}{3}\cos\beta - \cos\frac{\pi}{3}\sin\beta = \frac{\sqrt{3}}{2}\cos\beta - \frac{1}{2}\sin\beta = \frac{1}{2}(\sqrt{3}\cos\beta - \sin\beta)$ **25.** $\sin\left(\beta - \frac{7\pi}{6}\right) = \sin\beta\cos\frac{7\pi}{6} - \cos\beta\sin\frac{7\pi}{6} = \sin\beta\left(-\frac{\sqrt{3}}{2}\right) - \cos\beta\left(-\frac{1}{2}\right) = -\frac{1}{2}(\sqrt{3}\sin\beta - \cos\beta)$ **27.** $\frac{-\sqrt{2} - \sqrt{6}}{4}$ **29.** $\frac{-\sqrt{2} - \sqrt{6}}{4}$ **31.** $\sin(\alpha + \beta) + \sin(\alpha - \beta) = \sin\alpha\cos\beta + \cos\alpha\sin\beta + \sin\alpha\cos\beta - \cos\alpha\sin\beta = 2\sin\alpha\sin\beta$ **33.** 0.3420 **35.** −0.9205

Practice Exercises, pages 223–224

1. $2 + \sqrt{3}$ **3.** $-2 - \sqrt{3}$ **5.** $\sqrt{3} - 2$ **7.** $\sqrt{3} - 2$ **9.** $-\frac{33}{56}$ **11.** $-\frac{13}{84}$ **13.** $-\frac{117}{44}$ **15.** $\frac{44}{117}$ **17.** $\tan(45° - \alpha) = \frac{\tan 45° - \tan\alpha}{1 + \tan 45° \tan\alpha} = \frac{1 - \tan\alpha}{1 + \tan\alpha}$ **19.** $\tan(180° + \alpha) = \frac{\tan 180° + \tan\alpha}{1 - \tan 180° \tan\alpha} = \frac{0 + \tan\alpha}{1 - 0(\tan\alpha)} = \tan\alpha$ **21.** $\tan(360° + \alpha) = \frac{\tan 360° + \tan\alpha}{1 - \tan 360° \tan\alpha} = \frac{0 + \tan\alpha}{1 - 0(\tan\alpha)} = \tan\alpha$ **23.** $\tan\left(\frac{3\pi}{4} + \alpha\right) = \frac{\tan\frac{3\pi}{4} + \tan\alpha}{1 - \tan\frac{3\pi}{4}\tan\alpha} = \frac{-1 + \tan\alpha}{1 - (-1)\tan\alpha} = \frac{\tan\alpha - 1}{\tan\alpha + 1}$

25. $\tan\left(\alpha - \frac{5\pi}{4}\right) = \frac{\tan\alpha - \tan\frac{5\pi}{4}}{1 + \tan\alpha\tan\frac{5\pi}{4}} =$

$\dfrac{\tan \alpha - 1}{1 + (\tan \alpha)(1)} = \dfrac{\tan \alpha - 1}{\tan \alpha + 1}$ **27.** $\sqrt{3} - 2$
29. $-2 - \sqrt{3}$

Test Yourself, page 225

1. $\dfrac{\sqrt{6} - \sqrt{2}}{4}$ **3.** $\dfrac{\sqrt{6} - \sqrt{2}}{4}$ **5.** $2 - \sqrt{3}$ **7.** $\dfrac{24}{25}$
9. $\dfrac{117}{44}$ **11.** $\cos\left(\dfrac{\pi}{2} + \alpha\right) = \cos\dfrac{\pi}{2}\cos \alpha - \sin\dfrac{\pi}{2}\sin \alpha = 0(\cos \alpha) - 1(\sin \alpha) = -\sin \alpha$

13. $\tan(360° - \beta) = \dfrac{\tan 360° - \tan \beta}{1 + \tan 360° \tan \beta} = \dfrac{0 - \tan \beta}{1 + 0(\tan \beta)} = -\tan \beta$

Practice Exercises, pages 228–230

1. $-\dfrac{7}{25}$ **3.** $-\dfrac{24}{7}$ **5.** $-\dfrac{24}{25}$ **7.** $-\dfrac{120}{169}$
9. $\dfrac{7}{25}$ **11.** $\dfrac{120}{119}$ **13.** $-\dfrac{336}{625}$ **15.** $-\dfrac{7}{25}$
17. $\sin 3\theta = 3 \sin \theta - 4 \sin^3 \theta$
19. $\dfrac{1 - \tan^2 \beta}{\sec^2 \beta} = \dfrac{1 - \dfrac{\sin^2 \beta}{\cos^2 \beta}}{\dfrac{1}{\cos^2 \beta}} \cdot \dfrac{\cos^2 \beta}{\cos^2 \beta}$
$= \cos^2 \beta - \sin^2 \beta = \cos 2\beta$
21. $\dfrac{1 + \cos 2\beta}{2 \sin \beta \cos \beta} = \dfrac{2 \cos^2 \beta}{2 \sin \beta \cos \beta} = \dfrac{\cos \beta}{\sin \beta} = \cot \beta$
33. 11 ft

Practice Exercises, pages 235–236 **1.** $-2 - \sqrt{3}$
3. $\dfrac{\sqrt{2 - \sqrt{2}}}{2}$ **5.** $\dfrac{\sqrt{2 - \sqrt{2}}}{2}$ **7.** $\dfrac{\sqrt{2 + \sqrt{2}}}{2}$
9. $\dfrac{-\sqrt{2 + \sqrt{2}}}{2}$ **11.** $2 - \sqrt{3}$ **13.** $\dfrac{3\sqrt{10}}{10}$
15. $\dfrac{\sqrt{3}}{2}$ **17.** $\dfrac{-\sqrt{5} - 1}{2}$ **19.** $\dfrac{3}{4}$ **21.** $\dfrac{2\sqrt{13}}{13}$
33. $\left(s \sin\dfrac{\alpha}{2}\right)\left(s \cos\dfrac{\alpha}{2}\right) = s^2 \sqrt{\dfrac{1 - \cos \alpha}{2}} \cdot \sqrt{\dfrac{1 + \cos \alpha}{2}} = s^2 \sqrt{\dfrac{1 - \cos^2 \alpha}{4}} = s^2\left(\dfrac{\sqrt{\sin^2 \alpha}}{2}\right) = \dfrac{1}{2} s^2 \sin \alpha$

Practice Exercises, pages 240–241 **1.** $\cos 24° - \cos 60°$ **3.** $\sin 7x - \sin 3x$ **5.** $\sin\dfrac{\pi}{4} - \sin\dfrac{\pi}{12}$

7. $\dfrac{1}{2}\left(\cos\dfrac{\pi}{4} + \cos \pi\right)$
9. $2 \cos 72° \cos 34°$ **11.** $2 \sin\dfrac{15\theta}{2} \cos\dfrac{9\theta}{2}$
13. $2 \sin\dfrac{4\pi}{7} \sin\dfrac{2\pi}{7}$

Test Yourself, pages 241–242 **1.** $\dfrac{119}{169}$
3. $\dfrac{\sqrt{2 + \sqrt{3}}}{2}$ **5.** $-\sqrt{5}$ **9.** $2 \sin 5x \cos x$

Summary and Review, pages 244–245

1. $\dfrac{\sqrt{2} - \sqrt{6}}{4}$ **3.** $\dfrac{-3\sqrt{3} + 4}{10}$ **5.** $\dfrac{\sqrt{6} + \sqrt{2}}{4}$
7. $-\dfrac{4}{5}$ **9.** $2 + \sqrt{3}$ **11.** $\sqrt{3}$ **13.** $-\dfrac{24}{25}$ **15.** $-\dfrac{24}{7}$
17. $\dfrac{3\sqrt{13}}{13}$ **19.** $-\dfrac{3}{2}$ **21.** $\cos 5x - \cos 13x$
23. $2 \sin 57° \sin 16°$

Maintaining Skills, page 248

1. $\dfrac{2\sqrt{3} + 3}{3}$ **3.** $\dfrac{2\sqrt{2} - \sqrt{14}}{4}$ **5.** $\dfrac{\sqrt[5]{6}}{2}$ **7.** 28
9. 127 **11.** 4 **13.** $-1, -2$ **15.** $-3, 1$
17. 0, 3 **19.** $y = \dfrac{1}{3}x + \dfrac{5}{3}$ **21.** $y = 2x + 1$

Chapter 6 Inverse Trigonometric Functions

Practice Exercises, pages 253–254

1.

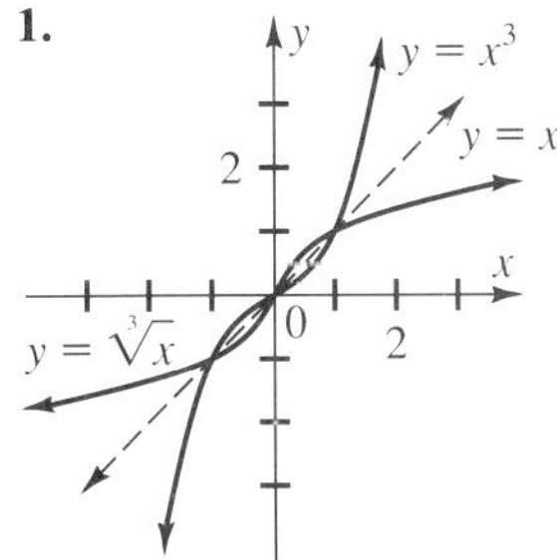

3. $y = \dfrac{1}{10}x$ **5.** $y = x + 8$ **7.** $y = -\dfrac{1}{7}x$
9. $y = \dfrac{1}{2}x - \dfrac{1}{2}$ **11.** $y = \dfrac{1}{5}x + \dfrac{4}{5}$
13. $y = \dfrac{3}{2}x - 9$ **15.** $90° + n(360°)$

17. $180° + n(360°)$ **19.** $\frac{\pi}{2} + \pi n$

21. $\frac{\pi}{2} + 2\pi n$ **23.** $\frac{2\pi}{3} + 2\pi n$ or $\frac{4\pi}{3} + 2\pi n$ **25.** $\frac{7\pi}{6} + 2\pi n$ or $\frac{11\pi}{6} + 2\pi n$

27.

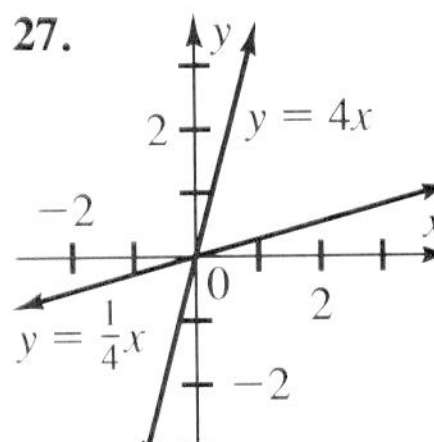

29.

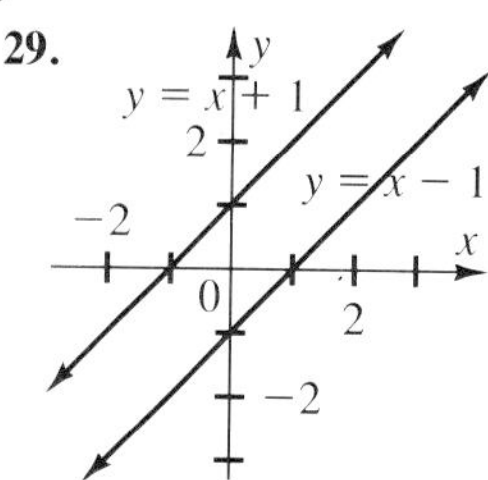

33.

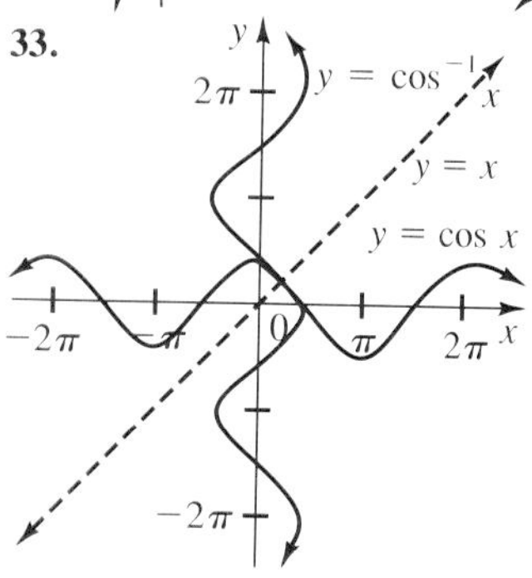

35. $f^{-1}(x) = \sqrt[3]{x-1}; f(f^{-1}(x)) = x; f^{-1}(f(x)) = x$

37. $F(x) = \frac{9}{5}x + 32$

Practice Exercises, pages 257–259

1. $-\frac{\pi}{3}$ **3.** $\frac{\pi}{3}$ **5.** $-\frac{\pi}{3}$ **7.** $-\frac{\pi}{2}$ **9.** 54° **11.** 19°

13. 38° **15.** 135° **17.** 0.4 **19.** $\frac{\sqrt{2}}{2}$ **21.** $\frac{1}{2}$

23. $\frac{\pi}{3}$ **25.** $\frac{\pi}{6}$ **27.** $\frac{\pi}{4}$ **29.** 0.9109 **31.** 0.7887

33. 0.5675 **35.** 0.8588

37.

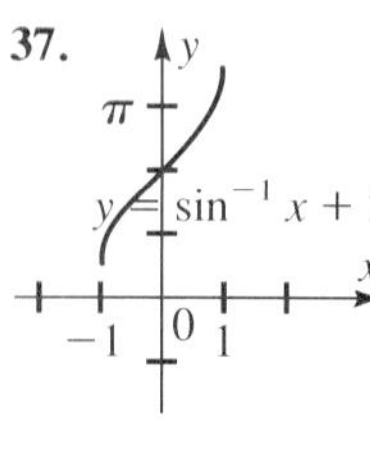

39.

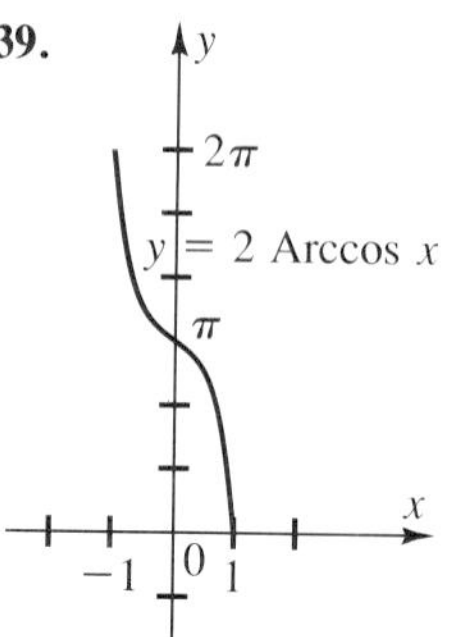

41. $-\frac{1}{2}$ **43.** $-\frac{\sqrt{2}}{2}$ **45.** yes **47.** $-1 \le x \le \frac{\sqrt{2}}{2}$

Practice Exercises, pages 263–264

1. $\frac{\pi}{4}$ **3.** $-\frac{\pi}{3}$ **5.** 66° **7.** 144° **9.** 0.6 **11.** 5

13. $-\frac{\sqrt{3}}{3}$ **15.** $\frac{\sqrt{2}}{2}$ **17.** $\frac{\sqrt{3}}{3}$ **19.** $\frac{4\sqrt{15}}{15}$

21. $\sec^{-1} 2 + \tan^{-1}\frac{\sqrt{3}}{3} = \frac{\pi}{3} + \frac{\pi}{6} = \frac{\pi}{2}$

23. $\cos(\text{Arcsec } w) = \cos\left(\text{Arccos }\frac{1}{w}\right) = \frac{1}{w}$

25. $\frac{\sqrt{6} + \sqrt{2}}{4}$ **27.**

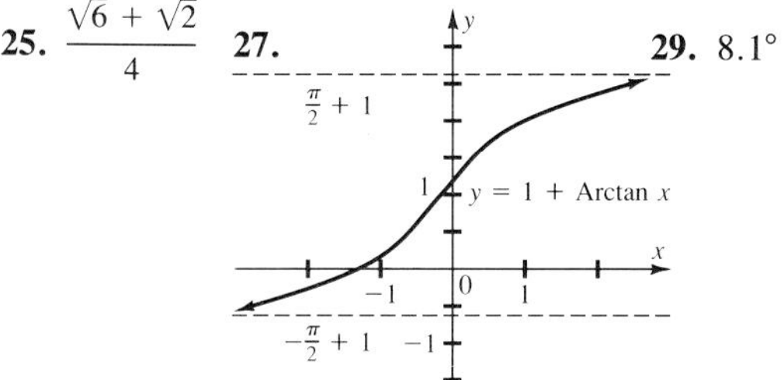

29. 8.1°

Test Yourself, page 264

1. $\frac{\pi}{6} + 2\pi n$ or $\frac{5\pi}{6} + 2\pi n$ **3.** $\frac{\pi}{6} + 2\pi n$ or $-\frac{\pi}{6} + 2\pi n$ **5.** 14° **7.** 54° **9.** 0.7125 **11.** 2 **13.** $\frac{5}{3}$

Practice Exercises, pages 267–268

1. $\frac{\pi}{6}; \frac{5\pi}{6}$ **3.** $\frac{\pi}{4}; \frac{7\pi}{4}$ **5.** $\frac{7\pi}{6}; \frac{11\pi}{6}$ **7.** 180°

9. 150°; 330° **11.** 120°; 240° **13.** 0 **15.** $\frac{3\pi}{4}; \frac{7\pi}{4}$

17. $\frac{\pi}{3}; \frac{2\pi}{3}; \frac{4\pi}{3}; \frac{5\pi}{3}$ **19.** $\frac{\pi}{4}; \frac{3\pi}{4}; \frac{5\pi}{4}; \frac{7\pi}{4}$ **21.** $\frac{3\pi}{2}; \frac{7\pi}{6}; \frac{11\pi}{6}$ **23.** $0; \frac{\pi}{4}; \pi; \frac{5\pi}{4}$ **25.** $\frac{\pi}{4}; \frac{3\pi}{4}; \frac{5\pi}{4}; \frac{7\pi}{4}$

27. $0; \frac{\pi}{6}; \frac{5\pi}{6}; \pi$ **29.** $\frac{\pi}{4}; \frac{5\pi}{4}$ **31.** $\frac{\pi}{6}; \frac{\pi}{2}; \frac{5\pi}{6}; \frac{3\pi}{2}$

33. $\frac{\pi}{6}; \frac{5\pi}{6}; \frac{7\pi}{6}; \frac{11\pi}{6}$ **35.** $0; \frac{\pi}{2}; \pi; \frac{7\pi}{6}; \frac{11\pi}{6}$

37. $0; \frac{\pi}{3}; \frac{\pi}{2}; \frac{2\pi}{3}; \pi; \frac{4\pi}{3}; \frac{3\pi}{2}; \frac{5\pi}{3}$ **39.** $\frac{\pi}{4}; \frac{3\pi}{4}; \frac{5\pi}{4}; \frac{7\pi}{4}$

41. $0; \frac{\pi}{3}; \pi; \frac{5\pi}{3}$

Practice Exercises, pages 271–272 **1.** 42°; 138° **3.** 141°; 219° **5.** 78°; 102°; 258°; 282° **7.** 90°; 190°; 350° **9.** 104°; 109°; 251°; 256° **11.** 9°; 171°; 232°; 308° **13.** 48°; 76°; 284°; 312° **15.** 85°; 275° **17.** 184°; 356° **19.** 10°; 80°; 190°; 260° **21.** 38°; 142°; 218°; 322° **23.** 44°; 134°; 224°; 314° **25.** 3°; 31°; 149°; 177° **27.** 1.77; 2.09; 4.19; 4.51 **29.** 0.10; 3.04; 4.71 **33.** 0.09; 0.44; 0.98; 1.34; 2.19; 2.54; 3.08; 3.43; 4.28; 4.63; 5.17; 5.52 **35.** no solution **37.** 1.58 *s*

Practice Exercises, pages 276–277

1. $(\sqrt{2}, -4\sqrt{2})$ **3.** $\left(\frac{8 - 5\sqrt{3}}{2}, \frac{-8\sqrt{3} - 5}{2}\right)$

5. $\left(\frac{-6\sqrt{3} - 7}{2}, \frac{6 - 7\sqrt{3}}{2}\right)$ **7.** $(2\sqrt{2}, \sqrt{2})$

9. $3(x')^2 - (y')^2 = 4$ **11.** $7(x')^2 - 5(y')^2 = 70$
13. $4(x')^2 + (y')^2 = 4$
15. $11(x')^2 + 9(y')^2 = 8$

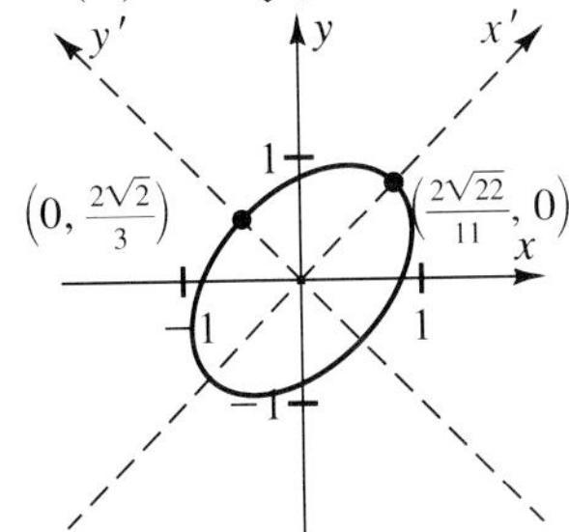

17.

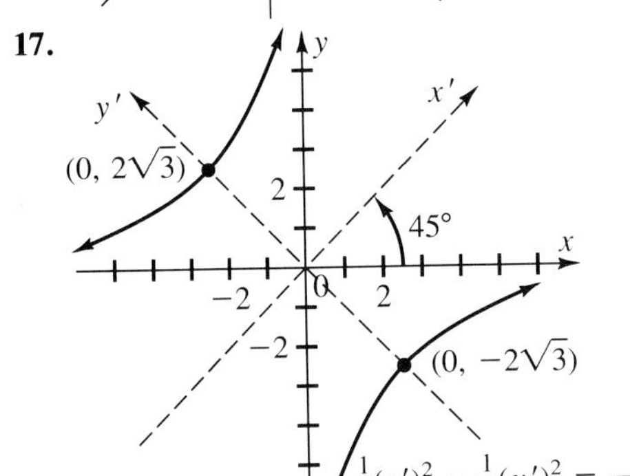

19.

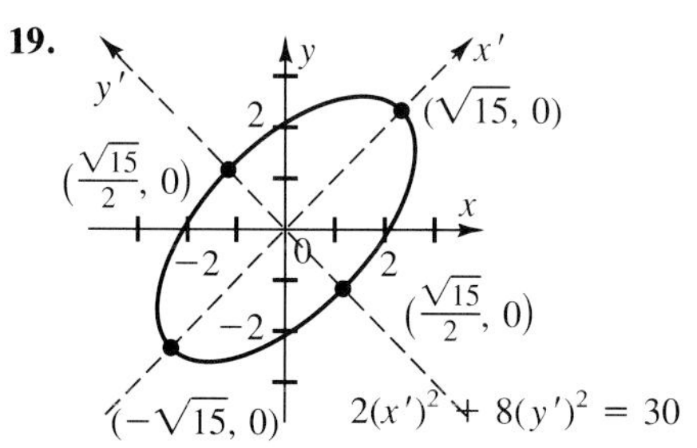

25. $1.35(x')^2 - 5.35(y')^2 = 8$

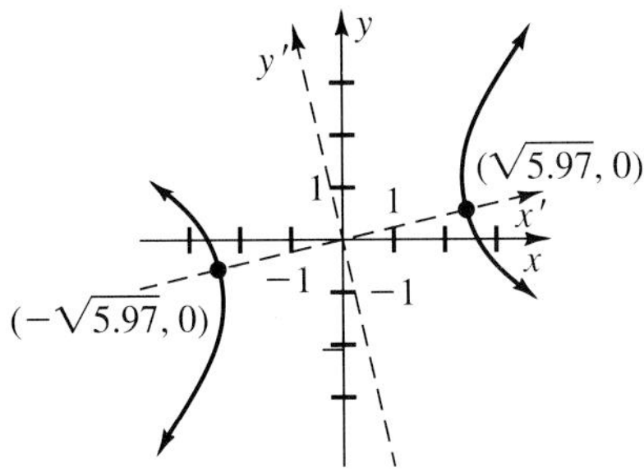

27. The new equation is equivalent to the original.
29. $\theta = 30°$; $4(x')^2 + (y')^2 = 8$

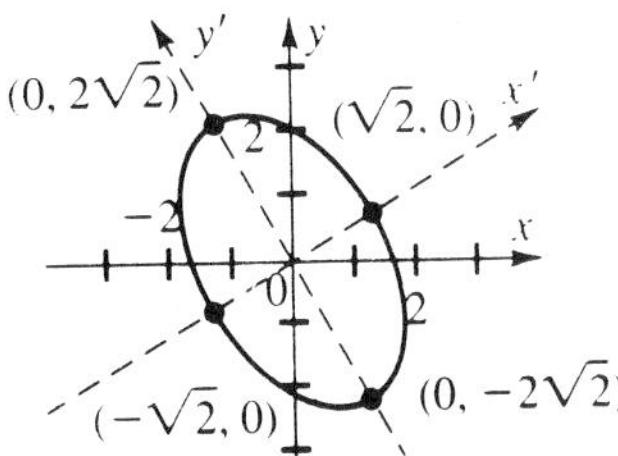

Test Yourself, page 277 **1.** $\frac{\pi}{4}; \frac{5\pi}{4}$ **3.** $x = 0$;
5. 45°; 168.7°; 225°; 348.7° **7.** 30°

Summary and Review, pages 280–281

1. $y = x + \frac{3}{4}$ **3.** $\frac{\pi}{2} + 2\pi k$, k is an integer

5. $\frac{\sqrt{3}}{2}$ **7.** $-\frac{\pi}{4}$ **9.** $-\sqrt{3}$ **11.** $\frac{\pi}{6}; \frac{5\pi}{6}; \frac{3\pi}{2}$

13. $\frac{\pi}{6}; \frac{\pi}{2}; \frac{5\pi}{6}; \frac{3\pi}{2}$ **15.** 56°; 111°; 249°; 304°

17. $\left(\frac{19\sqrt{3} - 15}{2}, \frac{-15\sqrt{3} - 19}{2}\right)$

19. $4(y')^2 - 5x'y' - 2(x')^2 = 9$
21. $\theta = 45°$;

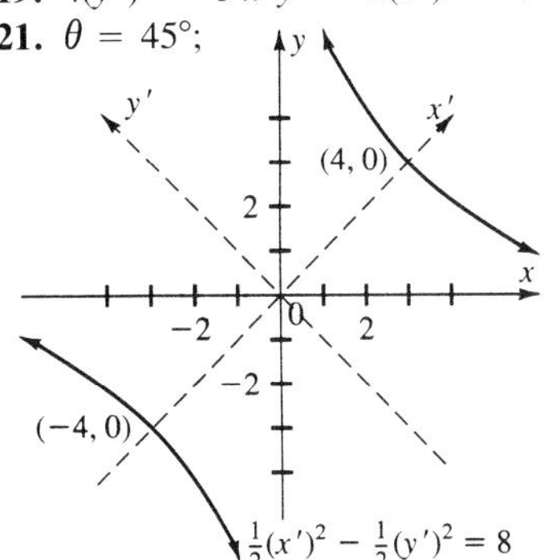

Cumulative Review, page 284

1. 2 cos 70° cos 18° **5.** $\frac{\pi}{2}$

9. $\|\overrightarrow{AB}\| = 13$

11. $\frac{1}{2}$ **13.** 10°; 50°; 130°; 170°; 250°; 290°
15. 3.36 **17.** 75°

Answers

Chapter 7 Complex Numbers

Practice Exercises, pages 290–291

1.

3

(3, 45°)

45°

Polar axis

0

3.

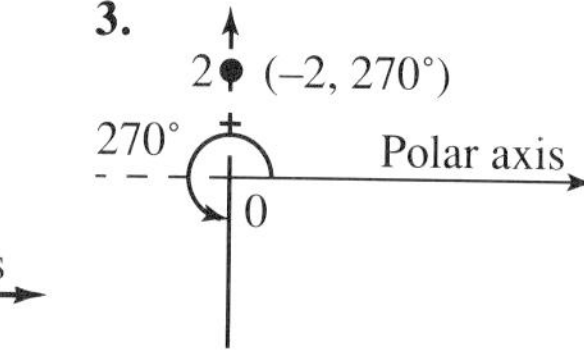

5. −420°

0

Polar axis

4

(4, −420°)

7.

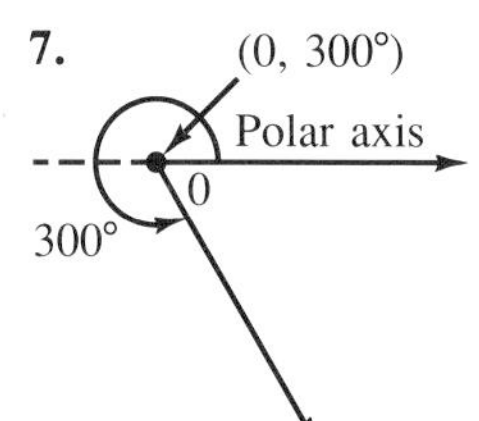

9.

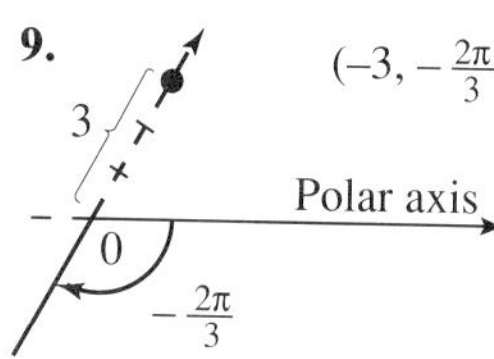

11.

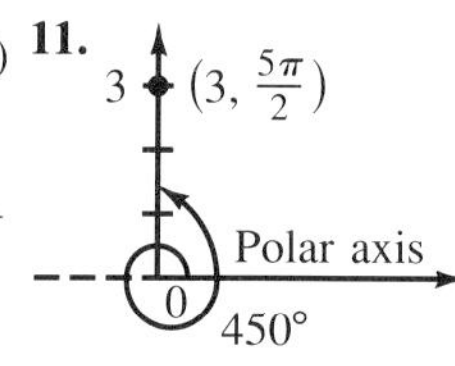

13. $(3, 60°)$; $(3, -300°)$; $(-3, 240°)$; $(-3, -120°)$
15. $(2, -210°)$; $(2, 150°)$; $(-2, 330°)$; $(-2, -30°)$
17. $(0, 0)$ **19.** $(-1.29, 1.53)$ **21.** $(1.03, -2.82)$
23. $(1, 270°)$ **25.** $(2\sqrt{2}, 315°)$ **27.** $(2\sqrt{3}, 150°)$
29. $r = 4 \sec \theta$ **31.** $r = 6$ **33.** $x^2 + y^2 = 10y$
35. $y = 6$ **37.** $r = -\frac{3}{4} \sin \theta$ **39.** $r = 12 \cos \theta$
41. $4x^2 + 3y^2 + 2y = 1$ **43.** $x^2 = 6y$
45. $x^2 + y^2 = 50\sqrt{x^2 + y^2} + 50x$

Practice Exercises, pages 296–297

1. **3.**

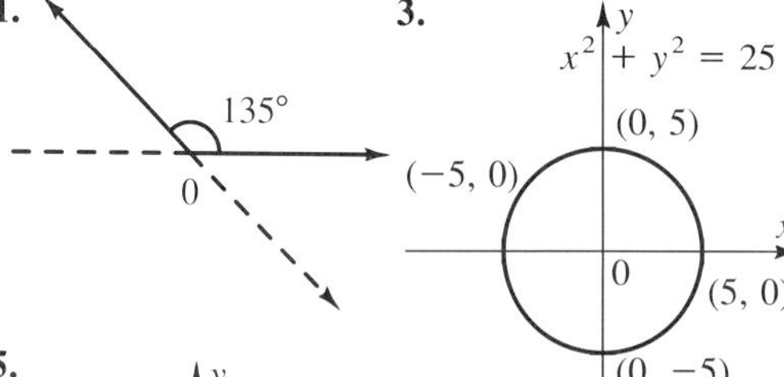

5.

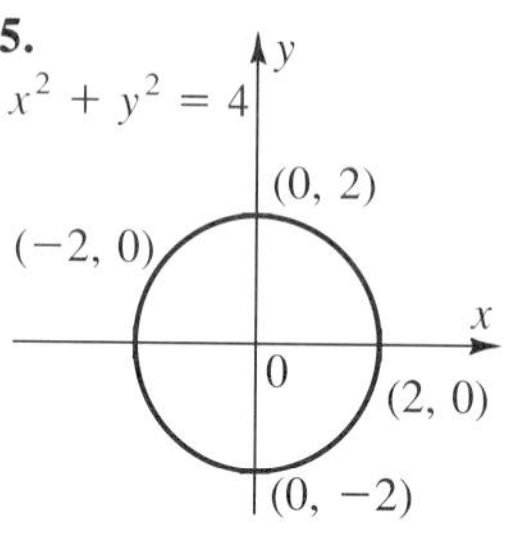

7.

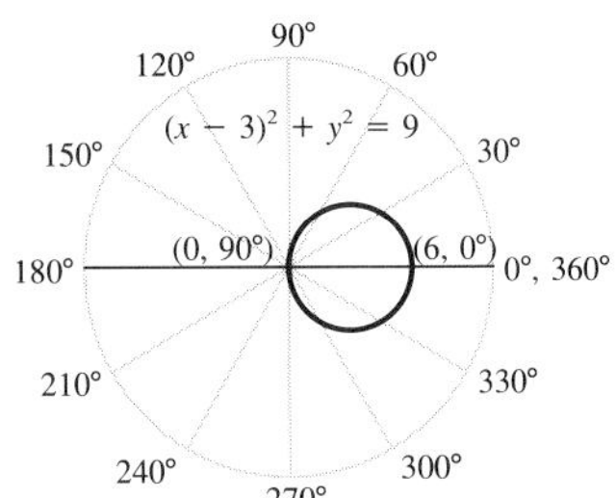

9.

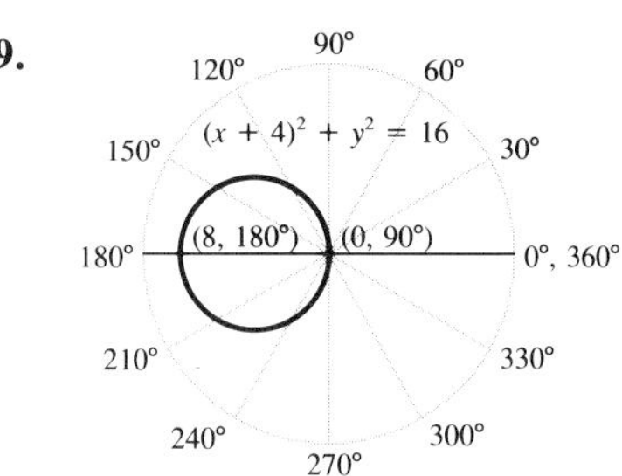

11.

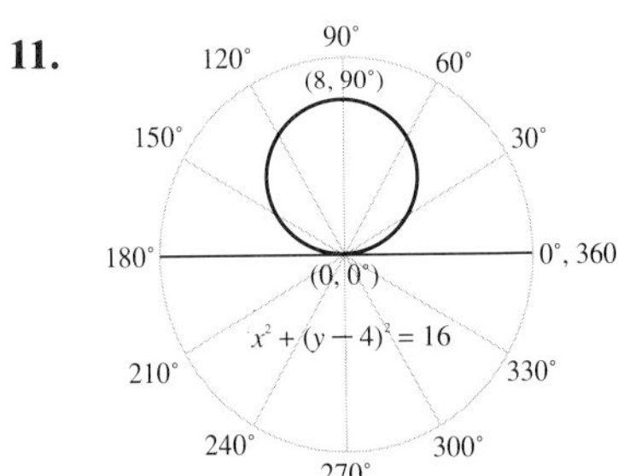

13.

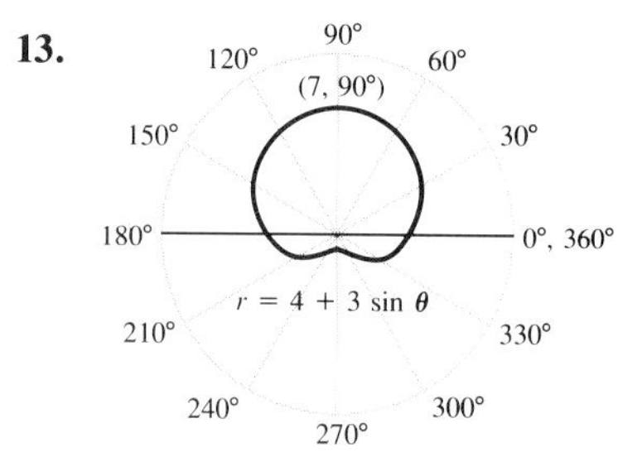

15.

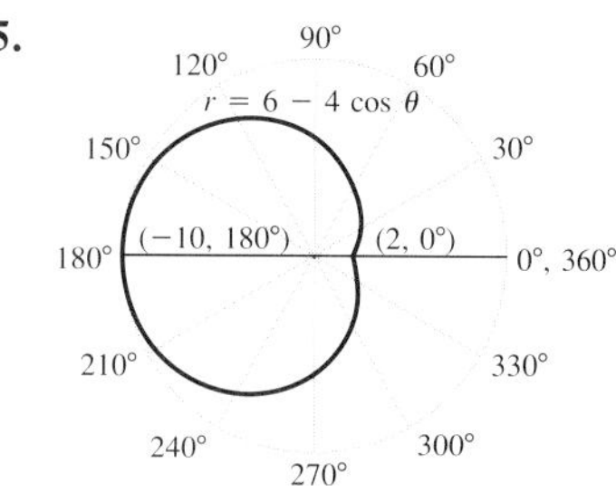

17.

$r = 4 + 4 \sin \theta$ (8, 90°)

19.

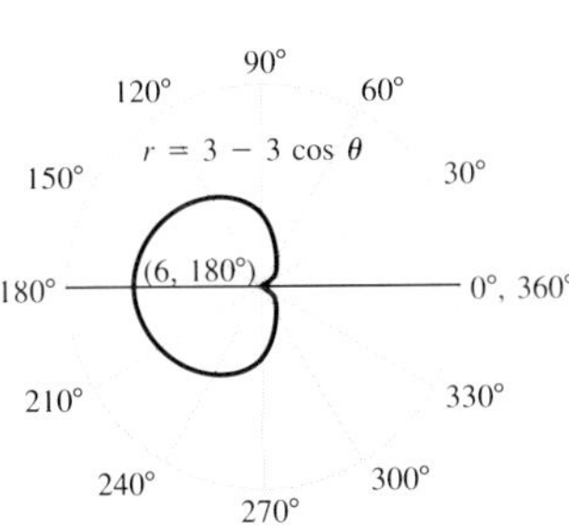

21.

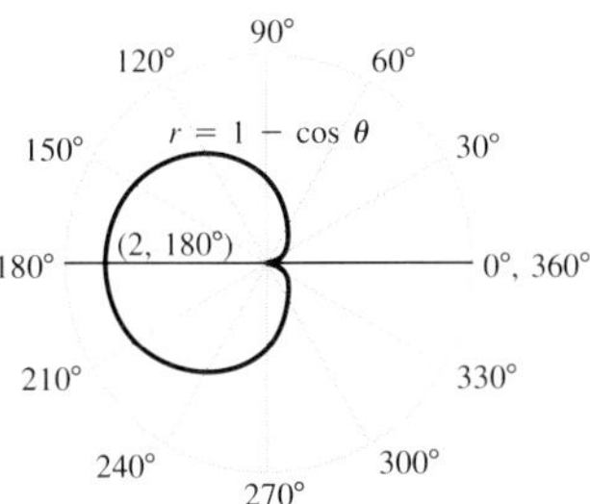

23.

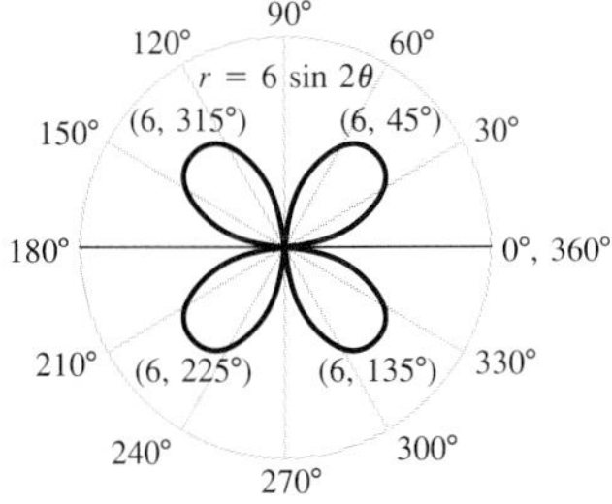

25.

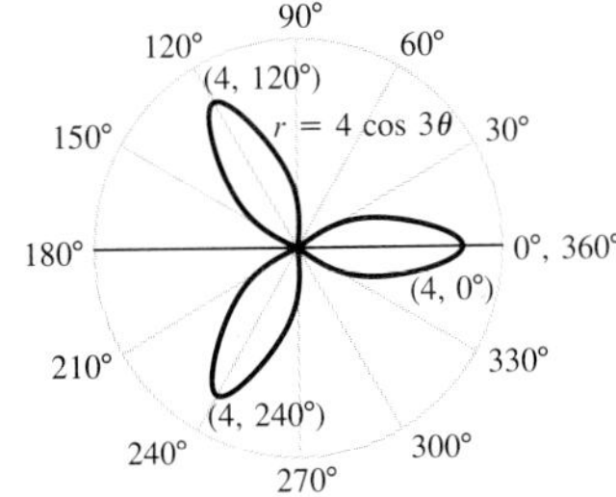

27.

$r = 5 \cos 4\theta$ (5, 90°) (5, 180°) (5, 0°) (5, 270°)

29.

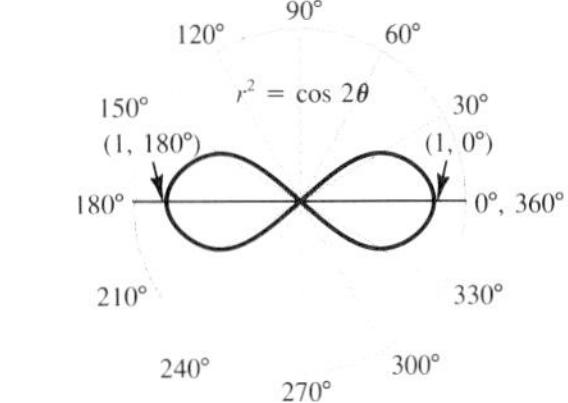

33. ellipse **35.** the pole; $\left(\frac{\sqrt{2}}{2}, 45°\right)$

Practice Exercises, pages 301–302

1. $5i$ **3.** $\frac{1}{7}i$ **5.** $2i\sqrt{7}$ **7.** $5i\sqrt{6}$ **9.** $6i$
11. $2i - 2i\sqrt{2}$ **13.** $11i$ **15.** i **17.** 1 **19.** -1
21. $14 + 14i$ **23.** $-3 - 10i$ **25.** $-7 + 16i$
27. $-14 - 2i$ **29.** -17 **31.** $i\sqrt{5} + i\sqrt{16}$
33. $-2 - 12i\sqrt{2}$ **35.** 1 **37.** $-i$ **39.** $-10 + 5i$
41. $3 - 9i$ **43.** $5 + 6i$ **45.** $x = -2; y = 7$
47. $x = 0; y = \frac{7}{3}$ **49.** no **51.** $15 + 4i$
53. $1 - 3i$ **55.** 5 **57.** 10 **59.** 10 ohms

Practice Exercises, pages 305–307 **1.** $2 + 4i$
3. $8 - i$ **5.** $9 + 33i$ **7.** -60 **9.** $-2\sqrt{6}$
11. $-9 + 40i$ **13.** $7 - 24i$ **15.** 13
17. $9 - 3i\sqrt{5}$ **19.** $\frac{5 - i}{26}$ **21.** $\frac{3 + 4i}{25}$ **23.** $\frac{-2 + i}{15}$
25. $\frac{i}{4}$ **27.** $-\frac{i}{3}$ **29.** i **31.** $\frac{-7 + 24i}{25}$
33. $\frac{-6 - 8i}{5}$ **35.** $\frac{-37 + 46i}{85}$ **37.** $\frac{-5 + i}{26}$
39. $\frac{-6 + 13i}{5}$ **41.** $3 + 2i$ **43.** $1 \pm 2i$
45. $\frac{1 \pm i\sqrt{6}}{2}$ **47.** $\pm i\sqrt{26}$ **49.** $-9 + 46i$
51. $-7 + 4i$ **53.** $x = 4; y = 0$ **55.** $(a - bi) + (c - di) = (a + c) + (-bi - di) = (a + c) - (b + d)i$ **57.** $28 + 24i$ volts **59.** $\frac{5 + 5i}{4}$

Test Yourself, page 307 **1.** (2.6, −1.5)
3. (2, 330°) **5.** $r = 5 \csc \theta$ **7.** $x = 5$

Test Yourself, page 307 **1.** (2.6, −1.5) **3.** (2, 330°) **5.** $r = 5 \csc \theta$ **7.** $x = 5$
9.

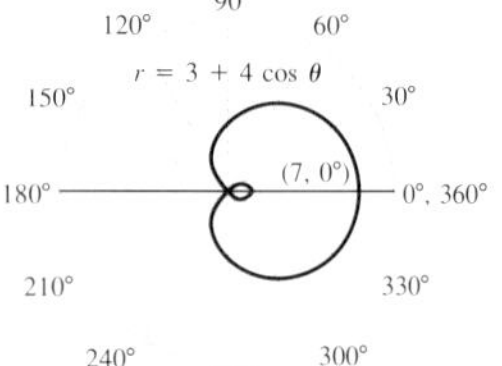

11. $2i\sqrt{10}$ **13.** $-i$ **15.** $22 + i$ **17.** $-15 + 29i$

Practice Exercises, pages 311–312

1.

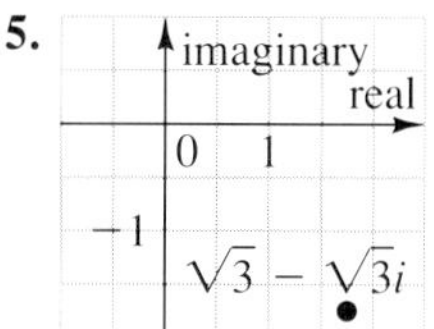

3.

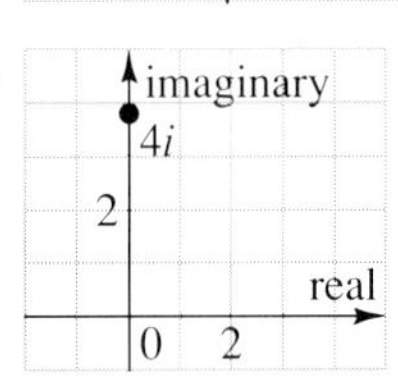

5.

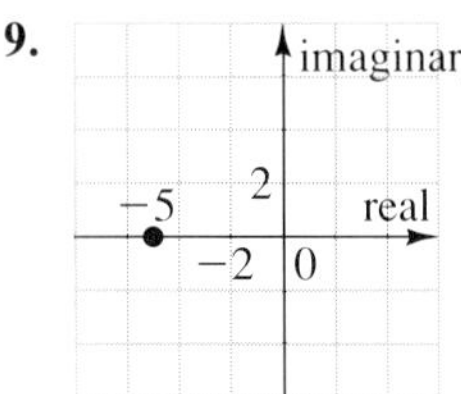

7.

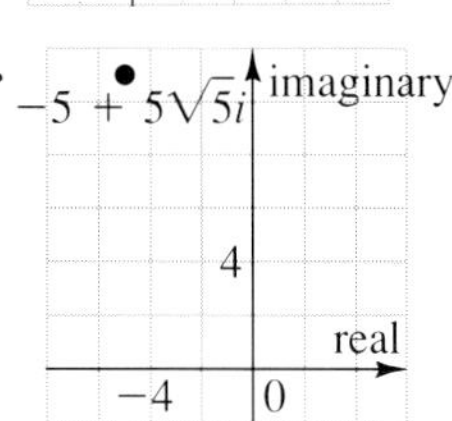

9.

11.

13. $\sqrt{2}(\cos 315° + i \sin 315°)$
15. $2(\cos 30° + i \sin 30°)$ **17.** $4(\cos 0° + i \sin 0°)$ **19.** $4(\cos 330° + i \sin 330°)$

21. $\frac{3\sqrt{2}}{2} + \frac{3\sqrt{2}}{2}i$ **23.** $\frac{1}{2} - \frac{\sqrt{3}}{2}i$ **25.** $\frac{5\sqrt{3}}{2} + \frac{5}{2}i$

27. $\sqrt{2} - \sqrt{2}i$ **29.** $1 + 0i$ **31.** $-\frac{3}{2} + \frac{3\sqrt{3}}{2}i$

33. $-2\sqrt{2} - 2\sqrt{2}i$ **41.**

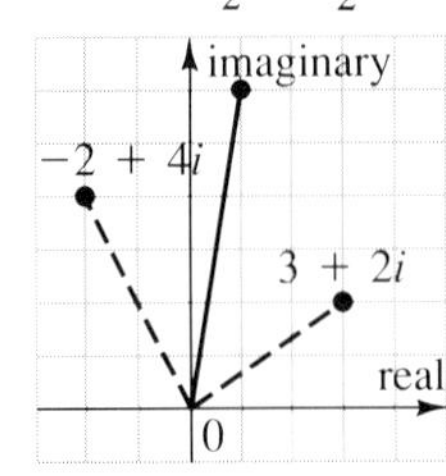

Practice Exercises, pages 315–317

1. $6(\cos 130° + i \sin 130°)$ **3.** $24\left(\cos \frac{11\pi}{12} + i \sin \frac{11\pi}{12}\right)$ **5.** $84(\cos 152° + i \sin 152°)$

7. $50(\cos 32° + i \sin 32°)$ **9.** $56\left(\cos \frac{5\pi}{4} + i \sin \frac{5\pi}{4}\right)$ **11.** $4(\cos 20° + i \sin 20°)$

13. $4(\cos 25° + i \sin 25°)$ **15.** $\frac{7}{2}\left(\cos \frac{\pi}{2} + i \sin \frac{\pi}{2}\right)$

17. $16(\cos 45° + i \sin 45°)$ **19.** $\frac{10}{7}(\cos 150° + i \sin 150°)$ **21.** $10(\cos 23° + i \sin 23°)$
23. $2\sqrt{10}(\cos 108° + i \sin 108°)$

25. $\frac{\sqrt{10}}{2}(\cos 288° + i \sin 288°)$ **27.** $\frac{5}{2}(\cos 83° + i \sin 83°)$ **35.** 12 cis 80°; $2.08 + 11.82i$
37. 1.78 cis(4.97°)

Practice Exercises, pages 320–321

1. $-\frac{27}{2} + \frac{27\sqrt{3}}{2}i$ **3.** $\frac{125\sqrt{2}}{2} + \frac{125\sqrt{2}}{2}i$

5. $4 - 4\sqrt{3}i$ **7.** i **9.** $-108 - 108i\sqrt{3}$
11. -64 **13.** $8i$ **15.** $8i$ **17.** $-54 - 54i$

19. $-i$ **21.** $-\frac{1}{8}$ **23.** $4 \cos^3 \theta - 3 \cos \theta$

25. $4 \cos^3 \theta \sin \theta - 4 \cos \theta \sin^3 \theta$

27. $\frac{-1 + \sqrt{3}}{2} + \frac{1 + \sqrt{3}}{2}i$

Practice Exercises, pages 325–326

1. $\frac{3\sqrt{2}}{2} + \frac{3\sqrt{2}}{2}i; -\frac{3\sqrt{2}}{2} - \frac{3\sqrt{2}}{2}i$ **3.** $3; -\frac{3}{2} + \frac{3\sqrt{3}}{2}i; -\frac{3}{2} - \frac{3\sqrt{3}}{2}i$ **5.** $2\sqrt{3} - 2i; 4i; -2\sqrt{3} - 2i$

7. $\frac{\sqrt[4]{18}}{2} - \frac{\sqrt[4]{2}}{2}i; \frac{\sqrt[4]{2}}{2} + \frac{\sqrt[4]{18}}{2}i; -\frac{\sqrt[4]{18}}{2} + \frac{\sqrt[4]{2}}{2}i; -\frac{\sqrt[4]{2}}{2} - \frac{\sqrt[4]{18}}{2}i$ **9.** $2 + 0i; 0 + 2i; -2 + 0i; 0 - 2i$

11. $1; -1$ **13.** $1 + 0i; \frac{1}{2} + \frac{\sqrt{3}}{2}i; -\frac{1}{2} + \frac{\sqrt{3}}{2}i; -1 + 0i; -\frac{1}{2} - \frac{\sqrt{3}}{2}i; \frac{1}{2} - \frac{\sqrt{3}}{2}i$

15. $1 + 0i; 0 + i; -1 + 0i; 0 - i$

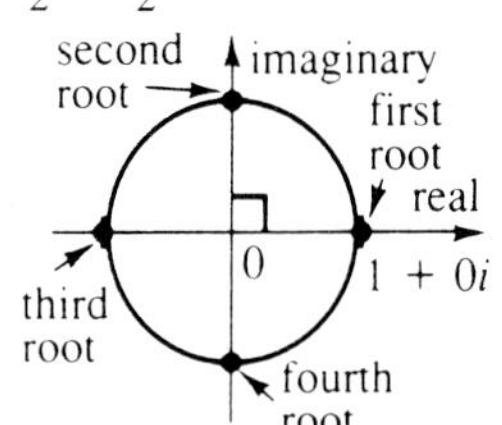

17. $-\frac{3}{2}+\frac{3\sqrt{3}}{2}i$;
$3+0i$;
$-\frac{3}{2}-\frac{3\sqrt{3}}{2}i$

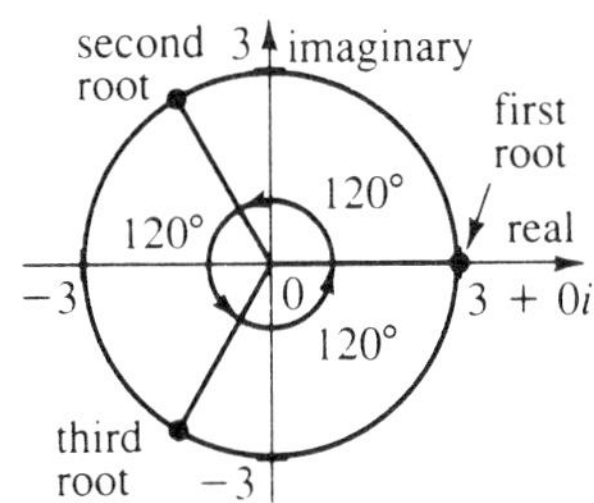

19. $3\cos 36° + 3i\sin 36°$; $3\cos 108° + 3i\sin 108°$; $-3+0i$; $3\cos 252° + 3i\sin 252°$; $3\cos 324° + 3i\sin 324°$

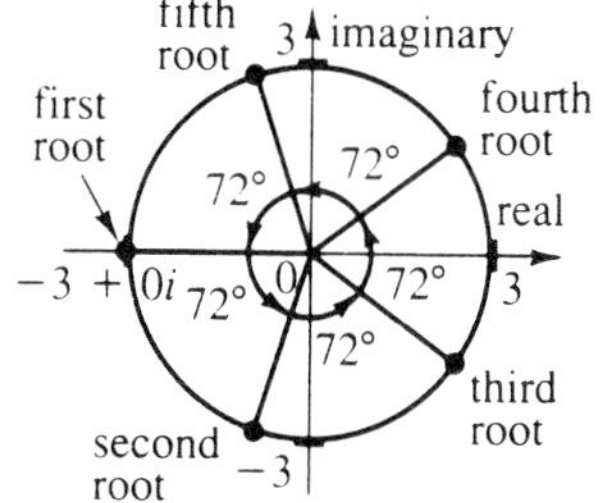

21. $-\frac{3}{2}+\frac{3\sqrt{3}}{2}i$;
$3+0i$;
$-\frac{3}{2}-\frac{3\sqrt{3}}{2}i$

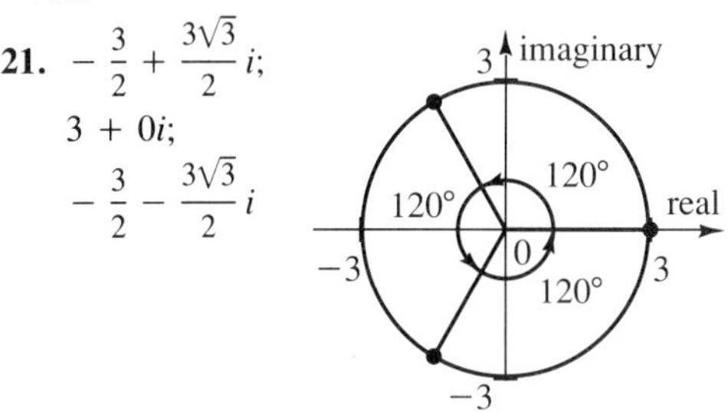

23. $\cos 36° + i\sin 36°$, $\cos 108° + i\sin 108°$; $\cos 180° + i\sin 180°$; $\cos 252° + i\sin 252°$; $\cos 324° + i\sin 324°$

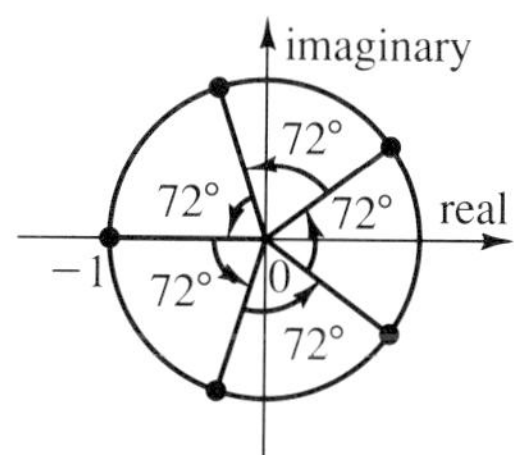

25. $1+i\sqrt{3}$; $-2+0i$; $1-i\sqrt{3}$

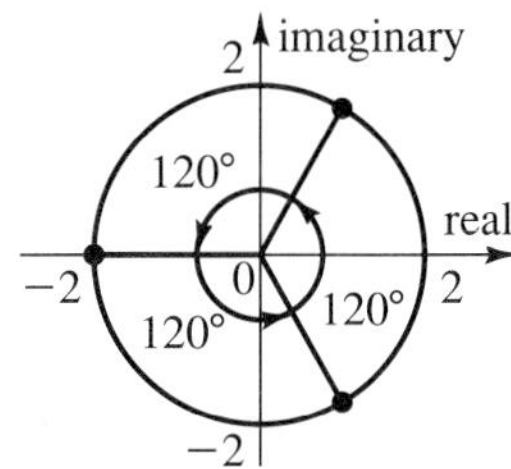

27. $\frac{\sqrt{2+\sqrt{2}}}{4}+\frac{\sqrt{2-\sqrt{2}}}{4}i$; $-\frac{\sqrt{2-\sqrt{2}}}{4}+\frac{\sqrt{2+\sqrt{2}}}{4}i$; $-\frac{\sqrt{2+\sqrt{2}}}{4}-\frac{\sqrt{2-\sqrt{2}}}{4}i$; $\frac{\sqrt{2-\sqrt{2}}}{4}-\frac{\sqrt{2+\sqrt{2}}}{4}i$

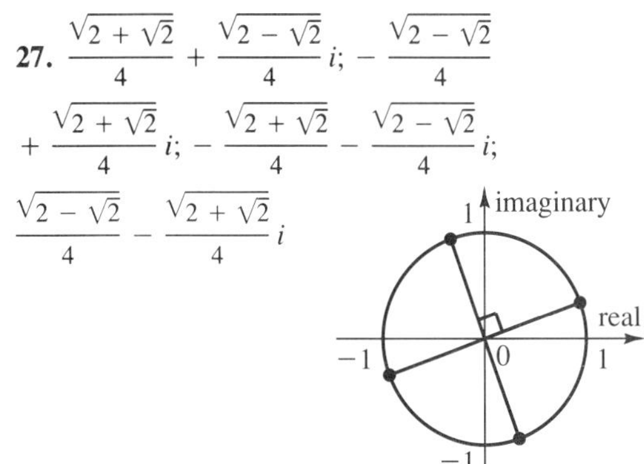

29. $4\cos 20° + 4i\sin 20°$;
$4\cos 140° + 4i\sin 140°$;
$4\cos 260° + 4i\sin 260°$

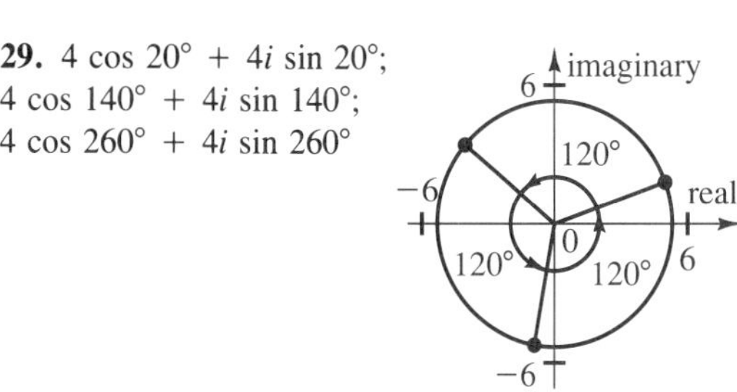

31. $2+0i$; $2\cos 72° + 2i\sin 72°$; $2\cos 144° + 2i\sin 144°$; $2\cos 216° + 2i\sin 216°$; $2\cos 288° + 2i\sin 288°$

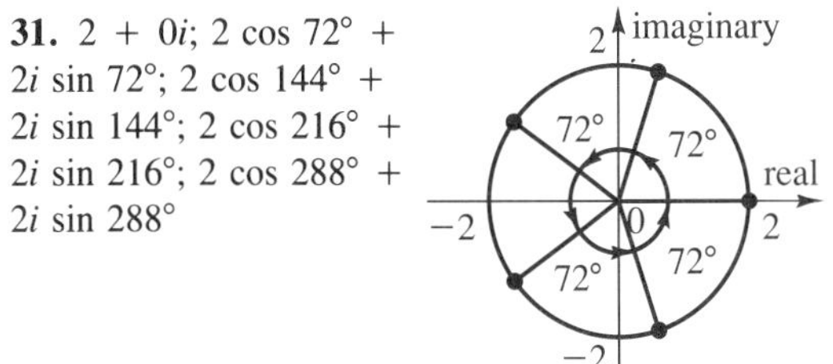

33. $\frac{\sqrt{3}}{2}+\frac{1}{2}i$; $0+i$;
$-\frac{\sqrt{3}}{2}+\frac{1}{2}i$; $-\frac{\sqrt{3}}{2}-\frac{1}{2}i$;
$0-i$; $\frac{\sqrt{3}}{2}-\frac{1}{2}i$

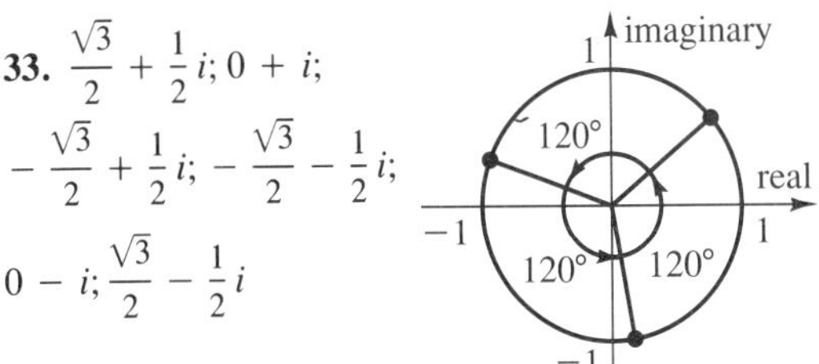

35. $\cos 40° + i\sin 40°$;
$\cos 160° + i\sin 160°$;
$\cos 280° + i\sin 280°$

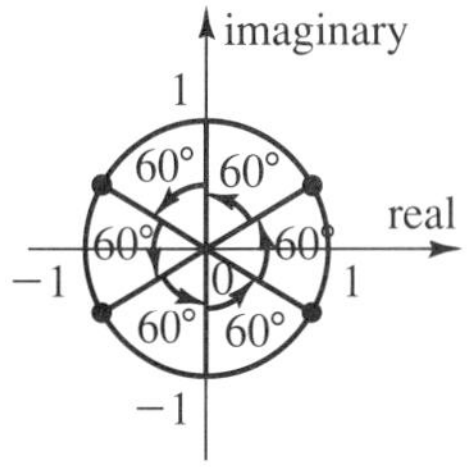

37. $\frac{\sqrt[3]{4}}{2}+\frac{\sqrt[6]{432}}{2}i$; $-\sqrt[3]{4}$; $\frac{\sqrt[3]{4}}{2}-\frac{\sqrt[6]{432}}{2}i$

39. $2\sqrt{3} - 2i$; $4i$; $-2\sqrt{3} - 2i$

41. $27^{\frac{1}{3}} \text{ cis } \left(\frac{120}{3} + k\frac{360}{3}\right)$ **43.** 2 cis 20; 2 cis 80; 2 cis 140; 2 cis 200; 2 cis 260; 2 cis 320

Test Yourself, page 326 **1.** $3\sqrt{2}$ (cos 315° + i sin 315°) **3.** $-2\sqrt{2} + 2\sqrt{2}\,i$ **5.** 40 (cos 354° + i sin 354°) **7.** $-8{,}388{,}608 + 14{,}529{,}495i$

9. $\frac{3}{2} + \frac{3\sqrt{3}}{2}i$; $-3 + 0i$; $\frac{3}{2} - \frac{3\sqrt{3}}{2}i$

Summary and Review, pages 328–329

1.

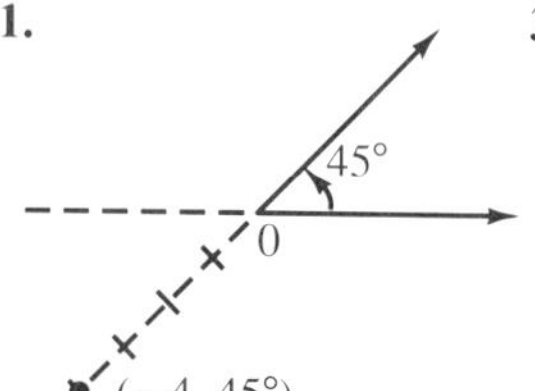

3. (2, 60°)

5.

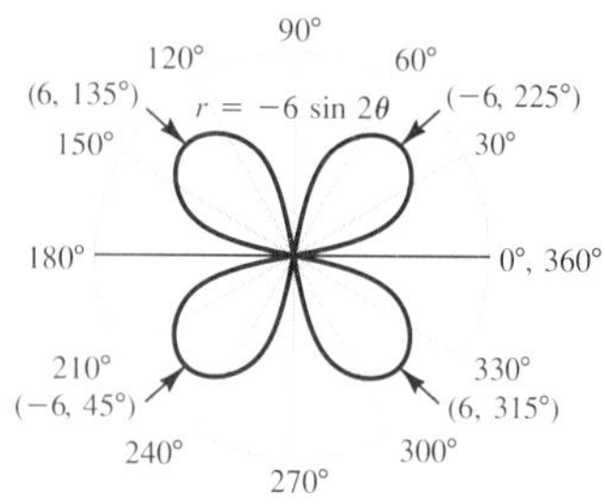

7. $2i\sqrt{5}$ **9.** $-8 + 7i$ **11.** $\frac{-27 - 8i}{13}$

13. $-3\sqrt{3} + 3i$ **15.** 2 (cos 60° + i sin 60°)

17. $\frac{15625}{2} + \frac{15625\sqrt{3}}{2}i$ **19.** $5 + 0i$; $-5 + 0i$; $0 - 5i$; $0 + 5i$

Maintaining Skills, page 332 **1.** -243 **3.** $-\frac{1}{16}$

5. $\frac{1}{16}$ **7.** $3x^3$ **9.** x^3y^6 **11.** a^2b **13.** 35

15. -7 **17.** 6 **19.** 1 **21.** $\frac{5}{6}$ **23.** 26

25. 6.25×10^5 **27.** 6.38×10^7 **29.** 3.72×10^7

Chapter 8 Exponential and Logarithmic Functions

Practice Exercises, pages 336–337

1. $5^{7\sqrt{6}}$ **3.** 4^{30} **5.** $2^{8\sqrt{2}}$ **7.** $\frac{1}{3^{80}}$ **9.** $\frac{x^{4\sqrt{3}}}{3y^{2\sqrt{5}}}$

11. $\frac{y^{22}}{x^{18}}$ **13.** 2.155 **15.** 2.890 **17.** 0.215

19. 0.346 **21.** 31.707 **23.** 25.733 **25.** 3

27. 5 **29.** $-\frac{3}{2}$ **31.** 13 **33.** 13 **35.** $\frac{2}{3}$

37. -4 **39.** 4 **41.** 26,918

Practice Exercises, page 341

1. 16 **3.** 2 **5.** 54.598 **7.** 4.055

9.

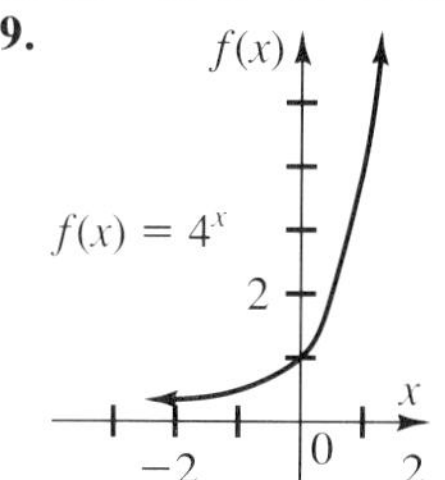

11.

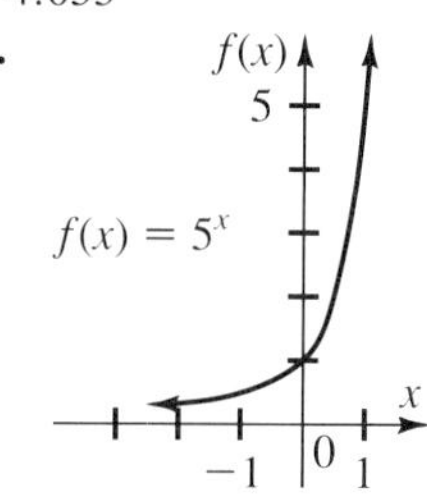

13.

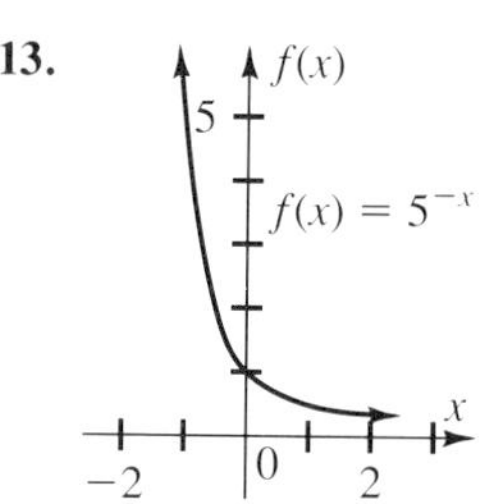

15.

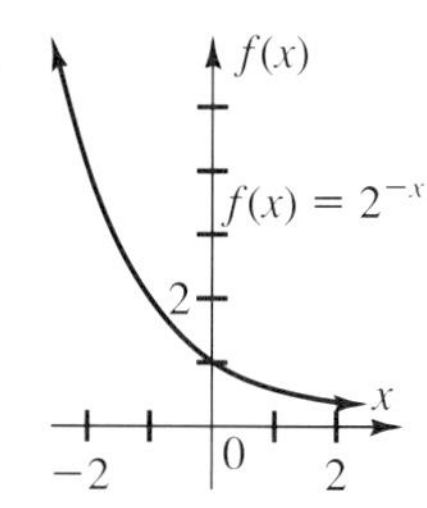

17.

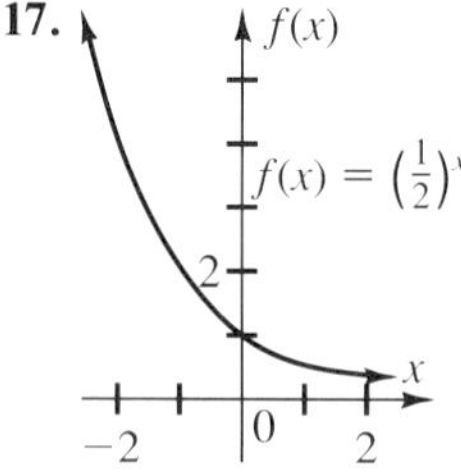

19.

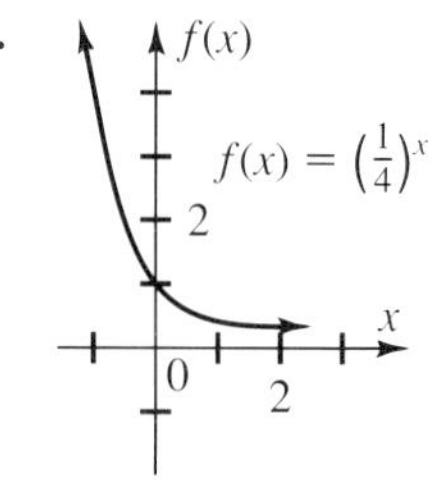

21.

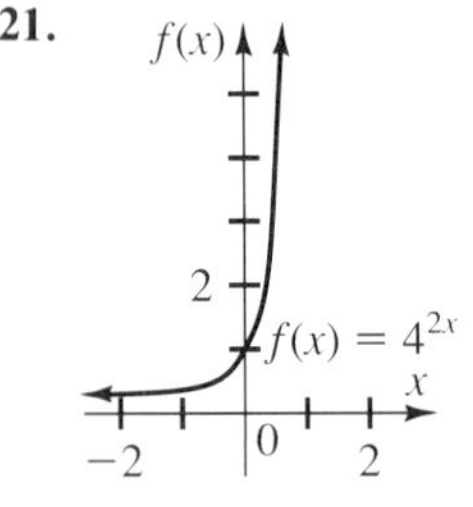

23.

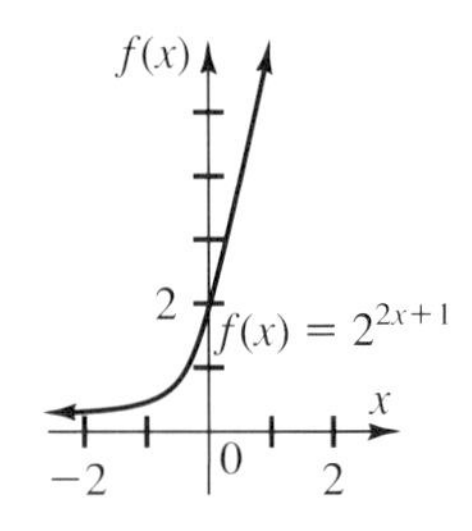

25.

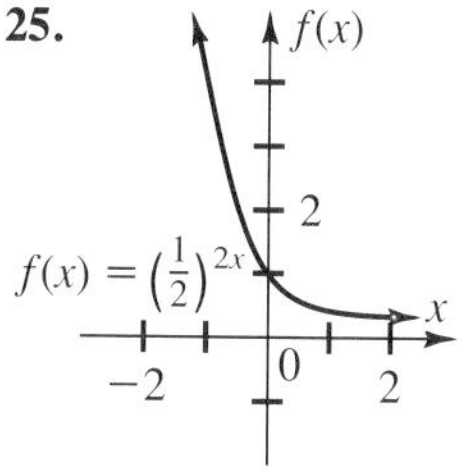

27.

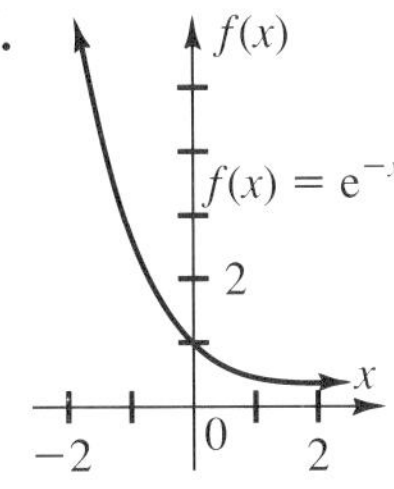

29.

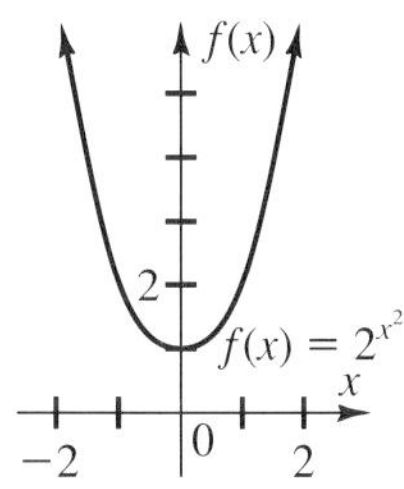

31. 4 **33.** −1

Practice Exercises, pages 345–346 **1.** $8^2 = 64$ **3.** $3^3 = 27$ **5.** $36^{0.5} = 6$ **7.** $2^{-4} = \frac{1}{16}$ **9.** $\log_4 64 = 3$ **11.** $\log_{49} 7 = \frac{1}{2}$ **13.** $\log_{27} \frac{1}{9} = -\frac{2}{3}$ **15.** $\log_7 7\sqrt{7} = \frac{3}{2}$ **17.** 3 **19.** −1 **21.** $-\frac{1}{2}$ **23.** $\frac{3}{2}$

25.

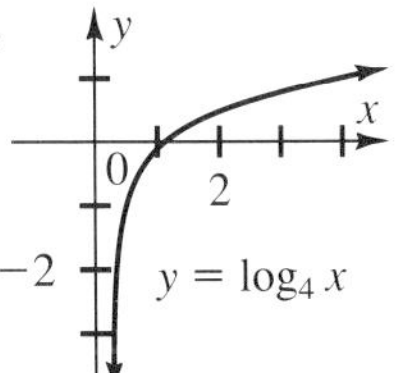

27.

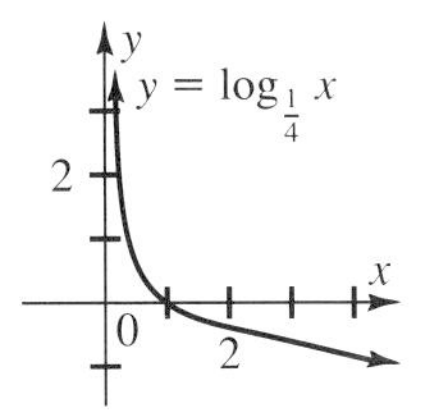

29. 3 **31.** 216 **33.** 18 **35.** 5 **37.** $\sqrt{7}$ **39.** 64 **41.** 1.83 **43.** 39,875.32 **45.** 1.17 **47.** 7.77 **49.** $D = \{x: x > 0\}$; $R = \{y: y$ is a real number$\}$ **51.** $D = \{x: x > -2\}$; $R = \{y: y$ is a real number$\}$

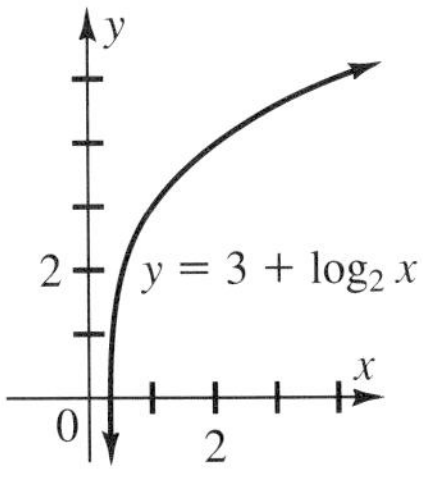

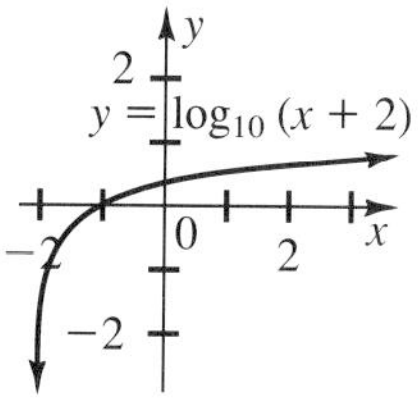

53. 8 **55.** 0 **57.** 3 **59.** $2^{-t} = I$

Test Yourself, page 346 **1.** 635.145 **3.** 4 **7.** $\log_4 \frac{1}{64} = -3$; $49^{\frac{1}{2}} = 7$ **9.** −2 **11.** 1,000,000

Practice Exercises, pages 351–352 **1.** $\log_3 x + \log_3 y$ **3.** $\log_5 2 + \log_5 x$ **5.** $\log_2 x - \log_2 z$ **7.** $\log_3 x - \log_3 5$ **9.** $2 \log_3 x$ **11.** $\frac{1}{5} \log_6 x$ **13.** $4 \log_2 x + 3 \log_2 y$ **15.** $3 \log_2 3 + 3 \log_2 x$ **17.** $\frac{1}{3} \log_3 4$ **19.** $\log_5 x + \log_5 y + \log_5 z$ **21.** $\log_2 xy$ **23.** $\log_5 \frac{zw^2}{x^3}$ **25.** 7 **27.** 5 **29.** −9 **31.** ±8 **33.** $\log_3 x - \frac{1}{2} \log_3 y$ **35.** $\log_2 x - \frac{3}{2} \log_2 y$ **37.** $\log_4 3 + 2 \log_4 x + 4 \log_4 y$ **39.** $2 \log_3 2 + 2 \log_3 x + 2 \log_3 y$ **41.** $\frac{5}{4} \log_b x$ **43.** $\frac{2}{3} \log_b x + \frac{4}{3} \log_b y$ **45.** $\log_b 3 + \frac{1}{2} \log_b x - 4 \log_b y$ **47.** $\log_b 6 + 4 \log_b x + \log_b y - \frac{1}{3} \log_b z$ **49.** $\log_4 x\sqrt[3]{y}$ **51.** $\log_4 \frac{\sqrt[3]{x}}{\sqrt{y}}$ **53.** 108 **55.** 5 **57.** $\frac{1}{3} \log_b x + \frac{3}{2} \log_b y - \frac{3}{4} \log_b z$ **59.** $-2 \log_2 x + \frac{5}{2} \log_2 y - \log_2 z$ **61.** $\frac{1}{3}$ **63.** 6 **67.** $\beta = 10(\log_{10} I - \log_{10} I_0)$

Practice Exercises, pages 357–358 **1.** 2.8987 **3.** −0.4318 **5.** −2.4685 **7.** 4.0289 **9.** 3.4833 **11.** −7.7634 **13.** 207.8 **15.** 2.160 **17.** 0.0431 **19.** 1.445 **21.** 19.78 **23.** 2.395 **25.** 1.3222 **27.** 6.9502 **29.** 9.2511 **31.** 1.771×10^{12} **33.** 3.786×10^{-14} **35.** 3.393×10^4 **37.** 3.170 **39.** 2.262 **41.** 2.124 **43.** 0.1827 **45.** 2.513 **47.** 1.5850 **49.** 2.5937 **51.** 4.1933 **53.** 5.1578 **55.** 1.680 **57.** 4.254 **59.** 1.1559 **61.** 12.77 **63.** 3.654 yr

Practice Exercises, pages 362–364 **1.** 965 **3.** 216 h **5.** 4600 yr **7.** 9 d **9.** 4.2 yr **11.** 8.2 yr **13.** 8.2 yr **15.** 54 h **17.** 8 wk **19.** 18

Test Yourself, page 364
1. $\log_4 \frac{8de^3}{f^3}$ **3.** $\log_b 2x + 2 \log_b y + 3 \log_b z$ **5.** 2.5172 **7.** −1.1363 **9.** 366.6 **11.** 1.302 **13.** 1.2920 **15.** 3.3 yr

Summary and Review, pages 366–367 **1.** 1.491 **3.** $\frac{3}{2}$ **5.** 243 **7.** 1.732 **9.** $\log_3 81 = 4$ **11.** $4^3 = 64$ **13.** $\frac{1}{2}$ **15.** 5 **17.** 3 **19.** $5 \log_4 x + 2 \log_4 y - \log_4 z$ **21.** 2 **23.** 1.6781 **25.** 0.0007495 **27.** 3.6993 **29.** 1.4037

Cumulative Review, page 370 **1.** 148.413 **3.** $-2 + 0i$, $1 + i\sqrt{3}$, $1 - i\sqrt{3}$ **5.**

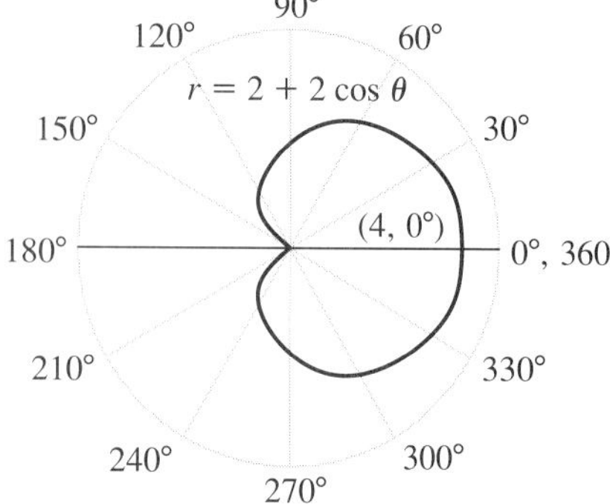

7. $2 + i$ **9.** 5.0733 **11.** 1 **13.** $\log_6 1296 = 4$ **15.** $-5 + 5i\sqrt{3}$ **17.** 4π in. **19.** 31.6°

Chapter 9 Sequences and Series

Practice Exercises, pages 376–377 **1.** 5, 9, 13, 17, 21 **3.** 4, 1, −2, −5, −8 **5.** −17, −11, −5, 1, 7 **7.** 19, 23, 27, 31; $a_n = 4n + 3$ **9.** −20, −16, −12, −8; $a_n = 4n - 36$ **11.** −14, −19, −24, −29; $a_n = -5n + 6$ **13.** 35 **15.** −3 **17.** −22 **19.** $a_1 = 5; d = 3$ **21.** $a_1 = 6; d = -5$ **23.** $a_1 = 3; d = \frac{1}{2}$ **25.** 16, 26, 36 **27.** 3.6, 1.2, −1.2, −3.6 **29.** $\frac{101}{15}, \frac{109}{15}$ **31.** 17 **33.** 97 **35.** 9 **37.** $n = 21$ **39.** $n = 8$ **41.** $\frac{3a + b}{4}, \frac{a + b}{2}, \frac{a + 3b}{4}$ **43.** 168 **45.** $a_n = \frac{1}{2}n + \frac{1}{2}$ **47.** $a_1 = \frac{8r - 3s}{5}$; $d = \frac{s - r}{5}$ **49.** $1350

Practice Exercises, pages 381–383 **1.** 3.5, 10.5, 31.5, 94.5 **3.** 1, −2, 4, −8 **5.** 8, −4, 2, −1 **7.** −81, 243, −729; $a_n = 3(-3)^{n-1}$ **9.** $\frac{1}{3}, \frac{1}{9}, \frac{1}{27}$; $a_n = 9\left(\frac{1}{3}\right)^{n-1}$ **11.** −3, −9, −27; $a_n = -\frac{1}{9}(3)^{n-1}$ **13.** 144 **15.** −4 **17.** $128x$ **19.** 2, 4 **21.** 1, 3, 9, 27 **23.** 6, −12, 24, −48 **25.** −36 **27.** 30 **29.** $r = \frac{1}{2}; a_1 = 256$ **31.** $r = 2; a_1 = \frac{27}{16}$ or $r = -2; a_1 = -\frac{27}{16}$ **33.** $r = \pm\frac{1}{2}; a_1 = 384$ **35.** −320 **37.** 4 **39.** seventh **41.** 63,670 **43.** $22,211

Practice Exercises, pages 387–389 **1.** 42 **3.** 18 **5.** 36 **7.** 31.5 **9.** 20,200 **11.** −95 **13.** 780 **15.** −51 **17.** 180 **19.** −396 **21.** 220 **23.** −21 **25.** $\sum_{n=1}^{6} (4n - 1)$ **27.** $\sum_{n=1}^{6} (-5n + 13)$ **29.** 43 **31.** −529 **33.** 40 **35.** −101 **37.** 6, 36, 66 **39.** 18, 23.75, 29.5 **41.** $a_1 = 5; n = 8$ **43.** $d = 2; n = 4$ **49.** 162 **51.** $9d$

Practice Exercises, pages 393–395 **1.** 510 **3.** 11 **5.** 1452 **7.** −2.625 **9.** 1.375 **11.** 1023 **13.** −3280 **15.** 60 **17.** 1365.3125 **19.** 30 **21.** 341 **23.** −5 **25.** −4 **27.** arithmetic; 93 **29.** geometric; 87,380 **31.** $\sum_{k=1}^{8} 3\left(\frac{1}{3}\right)^{k-1}$ **33.** $\sum_{k=1}^{8} 2(-5)^{k-1}$ **35.** 504 **37.** 6558 **39.** $r = -2; n = 7$ **41.** $\frac{x - 729x^{13}}{1 - 3x^2}$ **43.** $\frac{y(x^4 - x^{16})}{1 - x^2}$ **45.** 32.48 cm **47.** $450; $36,450

Test Yourself, page 395 **1.** 115 **3.** 2, 7, 12, 17 **5.** $8, \frac{8}{5}, \frac{8}{25}, \frac{8}{125}, \frac{8}{625}$ **7.** 750 **9.** $\sum_{n=1}^{5} (-3n + 27)$ **11.** 155 or 55 **13.** geometric; 341.25

Practice Exercises, pages 400–401 **1.** $\frac{7}{6}$ **3.** 32 **5.** does not exist **7.** does not exist **9.** $-\frac{49}{8}$ **11.** $\frac{5}{4}$ **13.** $\frac{4}{9}$ **15.** $\frac{7}{33}$ **17.** $\frac{41}{333}$ **19.** 2 **21.** does not exist **23.** $\frac{37}{90}$ **25.** $\frac{58}{11}$ **27.** $\frac{3}{5}$ **29.** −60 **31.** false **33.** true **35.** false **37.** 200 cm **39.** 440 ft **41.** 240 cm

Practice Exercises, pages 405–406 **1.** 0.7072 **3.** 0.8660 **5.** −0.5184 **7.** 0.9975 **9.** 1.105 **11.** 1.396 **13.** 0.607 **15.** −1 **17.** $-\frac{\sqrt{2}}{2} + \frac{\sqrt{2}}{2}i$

19. 1 **21.** -1.0018 **23.** 0.9272 **25.** -0.4810 **27.** 0.3089 **43.** 0.39 **45.** 0.30 **47.** 0.18

Practice Exercises, pages 410–411 **1.** -1.1752 **3.** -0.7616 **5.** $\frac{1}{\text{sech } x} = \frac{1}{\frac{1}{\cosh x}} = \cosh x$

7. $\frac{\cosh x}{\coth x} = \frac{\cosh x}{\frac{\cosh x}{\sinh x}} = \sinh x$ **9.** $\frac{\coth x}{\text{csch } x} = \frac{\frac{\cosh x}{\sinh x}}{\frac{1}{\sinh x}} = \cosh x$ **11.** $\frac{\text{csch } x}{\coth x} = \frac{\frac{1}{\sinh x}}{\frac{\cosh x}{\sinh x}} = \frac{1}{\cosh x} = \text{sech } x$ **13.** $1 - \tanh^2 x = 1 - \frac{e^{2x} - 2 + e^{-2x}}{e^{2x} + 2 + e^{-2x}} = \frac{e^{2x} + 2 + e^{-2x} - e^{2x} + 2 - e^{-2x}}{e^{2x} + 2 + e^{-2x}} = \frac{4}{e^{2x} + 2 + e^{-2x}} = \left(\frac{2}{e^x + e^{-x}}\right)^2 = \text{sech}^2 x$

15. $2 \sinh x \cosh x = 2\left(\frac{e^x - e^{-x}}{2}\right)\left(\frac{e^x + e^{-x}}{2}\right) = 2\left(\frac{e^{2x} - e^{-2x}}{4}\right) = \frac{e^{2x} - e^{-2x}}{2} = \sinh 2x$

17. $\cosh^2 x - 1 = \left(\frac{e^x + e^{-x}}{2}\right)^2 - 1 = \frac{e^{2x} - 2 + e^{-2x}}{4} = \left(\frac{e^x - e^{-x}}{2}\right)^2 = \sinh^2 x$

19. $\cosh x + \sinh x = \frac{e^x + e^{-x}}{2} + \frac{e^x - e^{-x}}{2} = \frac{2e^x}{2} = e^x$ **25.** $y = a \cosh x$

Test Yourself, page 411 **1.** $\frac{8}{7}$ **3.** does not exist **5.** $\frac{52}{99}$ **7.** -0.715 **9.** 42.8667 **11.** $\frac{\sqrt{2}}{2} - \frac{\sqrt{2}}{2}i$ **13.** 3.6269

Summary and Review, pages 414–415
1. 1, -2, -5; $a_n = -3n + 13$ **3.** 61 **5.** 16, 26 **7.** 27, 81, 243; $a_n = 3^{n-1}$ **9.** 2 **11.** 195 **13.** $\sum_{k=1}^{6} (4k - 11)$ **15.** 202.4 **17.** 2186 or 1094 **19.** does not exist **21.** -0.2224 **23.** 2.0136 **25.** $\frac{\tanh x}{\sinh x} = \frac{\frac{\sinh x}{\cosh x}}{\sinh x} = \frac{1}{\cosh x} = \text{sech } x$

Cumulative Review, pages 418–420
1. $125^{\frac{1}{3}} = 5$ **3.** 72°
5.

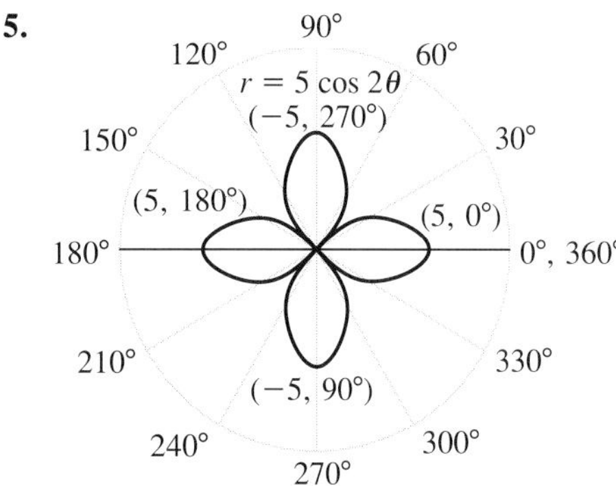

7. $\frac{\sqrt{6} + \sqrt{2}}{4}$ **9.** $(-1)\left(-\frac{\sqrt{2}}{2}\right) = \sqrt{2} - \frac{\sqrt{2}}{2}$; $\frac{\sqrt{2}}{2} = \frac{\sqrt{2}}{2}$ **11.** $\frac{3r^{\frac{1}{2}}}{s^{\frac{2}{5}}}$ **13.** 2; $\angle A = 53°$, $\angle B = 80°$, $b = 55$ or $\angle A = 6°$, $\angle B = 127°$, $b = 6$
15. $y = \tan(x + \pi)$

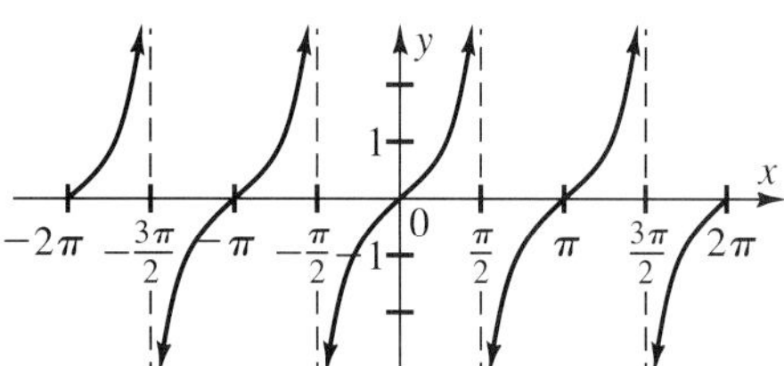

17. 14 **19.** 270° **21.** $-\frac{5\sqrt{2}}{2} + \frac{5\sqrt{2}}{2}i$ **23.** geometric; 3906 **25.** $-\frac{\sqrt{2}}{2}$ **27.** 18 **29.** neither **31.** $\frac{5}{13}$; $-\frac{12}{5}$; $-\frac{5}{12}$; $\frac{13}{5}$; $-\frac{13}{12}$
33. 25 ft **35.** $\cos(2\pi + \alpha) = \cos 2\pi \cos \alpha - \sin 2\pi \sin \alpha = (1) \cos \alpha - (0) \sin \alpha = \cos \alpha$

37.

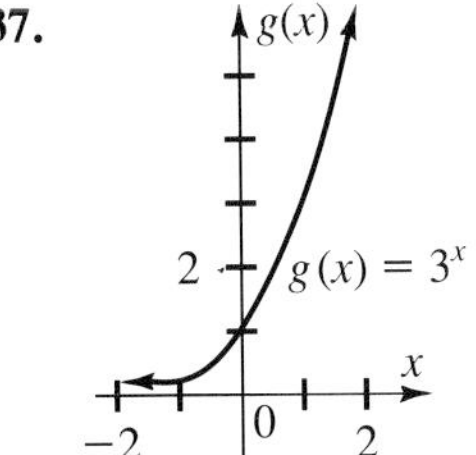

39. $9\sqrt{2} + 9\sqrt{2}\,i$ **41.** $-5, -1, 3, 7$
43. identity **45.** 10°, 30°, 150°, 170° **47.** 7
49. $\frac{(y')^2}{4} - \frac{(x')^2}{4} = 1$ **51.** 4.5 yr **53.** 1 **55.** 2.2

GLOSSARY

The explanations given in this glossary include definitions and brief descriptions of the key terms used in this book.

addition of ordinates (p. 85) A method for sketching the graph of a function that is the sum or difference of two trigonometric functions which involve adding or subtracting the two y-values for each x-value.

ambiguous case (p. 164) When the measures of two sides and the angle opposite one of them are given, it is not always possible to determine a unique triangle using the law of sines and a given value of the sine ratio.

amplitude (p. 73) Half the difference of the maximum and minimum values of a function.

angle of depression (p. 118) The angle formed between a horizontal line (the horizon) and an observer's line of sight to an object below the horizon.

angle of elevation (p. 118) The angle formed between a horizontal line (the horizon) and an observer's line of sight to an object above the horizon.

angular displacement (p. 24) The angle θ through which $\overrightarrow{OA}$ rotates where O is the axis of an object and A is a point on the edge of the object.

angular velocity (p. 24) The angular displacement per unit of time: $w = \frac{\theta}{t}$, where w represents angular velocity.

angle (p. 273) Formed by rotating a ray about its vertex from one position, called the initial side, to another, called its terminal side.

argument (p. 309) The measure of the angle formed by $\overrightarrow{OP}$ with the real axis in a complex plane.

arithmetic means (p. 377) The terms between any two nonconsecutive terms of an arithmetic sequence.

arithmetic sequence (p. 375) An arithmetic progression; a sequence of numbers in which the difference between any two successive terms is a constant.

arithmetic series (p. 386) The indicated sum of an arithmetic sequence.

asymptote (p. 88) A line that a graph approaches but never crosses.

cardioid (p. 294) A heart-shaped curve represented by polar equations of the form $r = a \pm b \cos \theta$ or $r = a \pm b \sin \theta$ in which $a = b$.

central angle (p. 17) An angle whose vertex is the center of a circle.

change-of-base formula (p. 356) $\log_b x = \frac{\log_a x}{\log_a b}$, where a is the reference base.

circular function (p. 31) The trigonometric functions are sometimes called circular functions.

cofunctions (p. 67) The sine and cosine, tangent and cotangent, and secant and cosecant, are pairs of cofunctions.

common difference (p. 375) The constant, d, between any two successive terms in an arithmetic sequence; $d = a_{n-1} - a_n$.

common logarithm (p. 353) A base-10 logarithm denoted $\log x$.

common ratio (p. 380) The constant, r, which is the ratio of any two consecutive terms in a geometric sequence; $r = \frac{a_n + 1}{a_n}$.

complementary angles (p. 67) Two acute angles the sum of whose measures is 90°.

complex number (p. 299) A number that can be written in the form $a + bi$, where a and b are real numbers and $i = \sqrt{-1}$.

complex plane (p. 308) A coordinate plane for plotting complex numbers of the form $a + bi$, where the horizontal axis is associated with the real part and the vertical axis is associated with the imaginary part.

components of a vector (p. 191) The x- and y-component of a vector in standard position such that the magnitude of x equals the product of the magnitude of the vector and the cosine of its directional angle and y equals the product of the magnitude of the vector and the sine of its directional angle.

conjugates (p. 304) Two complex numbers, such as $3 + 5i$ and $3 - 5i$, that differ only in the sign of the imaginary part.

convergent series (p. 399) An infinite geometric series whose sequence of partial sums approaches a limit as n increases without bound.

coordinate plane (p. 3) A rectangular coordinate system in which each point is associated with exactly one ordered pair in a relation.

cosecant (p. 34) If $P(x, y)$ is any point other than the origin on the terminal side of an angle θ in standard position, $\csc \theta = \frac{r}{y}$, $y \neq 0$.

cosine (p. 29, 34) The function defined by the set of ordered pairs (θ, x); if $P(x, y)$ is any point other than the origin on the terminal side of an angle θ in standard position, $\cos \theta = \frac{x}{r}$, where $r = \sqrt{x^2 + y^2}$.

cotangent (p. 34) If $P(x, y)$ is any point other than the origin on the terminal side of an angle θ in standard position, $\cot \theta = \frac{x}{y}$, $y \neq 0$.

coterminal angles (p. 13) Angles in standard position, whose terminal sides coincide.

cycle (p. 60) The part of the graph of a periodic function that is completed in each period.

degree (p. 13) A unit of angle measure equal to $\frac{1}{360}$ revolution.

direction angle of a vector (p. 191) The angle between a vector in standard position and the positive x-axis.

divergent series (p. 400) An infinite geometric series whose sequence of partial sums increase or decrease without bound as n increases without bound.

domain (p. 2) The set of replacements for the first variable in a relation or function.

equivalent vectors (p. 190) Vectors that have the same magnitude and the same direction.

even function (p. 62) A function in which $f(-x) = f(x)$ for every value in the domain of the function.

explicit formula (p. 374) A formula that defines a sequence by expressing the nth term, a_n, as a function of n.

exponent function (p. 339) A function in which the variable is in the exponent as in $f(x) = b^x$, $y = 3^x$, and $f(x) = 4^x$.

finite sequence (p. 374) A sequence that has a last term.

frequency (p. 100) For a sinusoidal graph the reciprocal of the period, which is $\frac{|b|}{2\pi}$.

function (p. 3) A relation in which each member of the domain is paired with exactly one member of the range.

fundamental identities (p. 133) The reciprocal, Pythagorean, ratio, and odd-even identities.

geometric means (p. 382) The terms between any two nonconsecutive terms of a geometric sequence.

geometric sequence (p. 380) A geometric progression; a sequence in which the ratio of any two consecutive terms is a constant.

geometric series (p. 392) The indicated sum of the terms of a geometric sequence.

general term (p. 374) The nth term, denoted a^n, in a sequence.

hyperbolic functions (p. 410–411) The trigonometric functions defined in terms of the exponential function and associated with the unit hyperbola, $\sinh x = \frac{e^x - e^{-x}}{2}$, $\cosh x = \frac{e^x + e^{-x}}{2}$, and $\tanh x = \frac{e^x - e^{-x}}{e^x + e^{-x}}$.

imaginary number i (p. 298) The square root of negative one, that is, $i = \sqrt{-1}$.

infinite sequence (p. 375) A sequence that does not have a last term.

initial point of a vector (p. 190) The point A in $\overrightarrow{AB}$; the tail of a vector.

initial side (p. 12) The position from which a ray is rotated about the vertex of an angle.

inverse function (p. 251) If the inverse of a function is also a function, it is called the inverse function.

law of cosines (p. 168) For any triangle ABC, where a, b, and c are the lengths of the sides opposite the angles with measures A, B, and C, respectively, $a^2 = b^2 + c^2 - 2\,bc \cos A$, $b^2 = a^2 + c^2 - 2\,ac \cos B$, and $c^2 = a^2 + b^2 - 2\,ab \cos C$.

law of sines (p. 156) For any triangle ABC in which a, b, and c are the lengths of the sides opposite the angles with measures A, B, and C, respectively, $\frac{\sin A}{a} = \frac{\sin B}{b} = \frac{\sin C}{c}$.

law of tangents (p. 175) For any triangle ABC in which a, b, and c are the lengths of the sides opposite angles with measures A, B, and C, respectively,

$$\frac{a - b}{a + b} = \frac{\tan \frac{1}{2}(A - B)}{\tan \frac{1}{2}(A + B)},$$

$$\frac{c - a}{c + a} = \frac{\tan \frac{1}{2}(C - A)}{\tan \frac{1}{2}(C + A)}, \text{ and}$$

$$\frac{b - c}{b + c} = \frac{\tan \frac{1}{2}(B - C)}{\tan \frac{1}{2}(B + C)}.$$

limaçon (p. 294) A curve represented by polar equations of the form $r = a + b \cos \theta$ or $r = a + b \sin \theta$.

limit (p. 399) A number that the partial sums of an infinite geometric series approach but never reach as the number of terms summed increases without bound.

linear velocity (p. 24) The linear distance traveled by a point on a revolving ray per unit of time: $v = w \cdot r$.

logarithmic function (p. 343) The function $y = \log_b x$ which is the inverse of the exponential function $y = b^x$.

magnitude of a vector (p. 190) The magnitude, or norm, of V is represented as $\|V\|$ and is found by using the distance formula.

mean proportional (p. 382) A single geometric mean between two numbers in a geometric sequence; $m = \sqrt{ab}$ or $m = -\sqrt{ab}$, where a and b are real numbers and m is the mean proportional.

minute (p. 17) A subdivision of a degree, where $1° = 60'$ (60 min).

modulus (p. 310) The quantity r in $r(\cos \theta + i \sin \theta)$.

natural exponential function (p. 340) The function $f(x) = e^x$, where $e \approx 2.7182818$.

natural logarithm (p. 354) A base e logarithm denoted $\ln x$.

odd-even identities (p. 132) Identities that can be derived from the definitions of odd and even functions: $\sin(-\theta) = -\sin \theta$, $\cos(-\theta) = \cos \theta$, and $\tan(-\theta) = -\tan \theta$.

odd function (p. 62) A function in which $f(-x) = -f(x)$ for every value in the domain of the function.

origin (p. 3) The point of intersection of the axes in a coordinate plane.

period (p. 60) For a periodic function $f(x)$, the smallest positive number p, such that $f(x + p) = f(x)$ whenever $f(x + p)$ and $f(x)$ are defined, is the period of the function.

phase shift (p. 78) The distance that a graph is shifted right or left.

polar coordinates (p. 286) The coordinates (r, θ) of a point P in the polar coordinate system, where $|r|$ is the distance from the pole and θ is the measure of the angle from the polar axis to $\overrightarrow{OP}$.

polar form of a complex number (p. 309) $a + bi = r(\cos \theta + i \sin \theta)$, where $r = \sqrt{a^2 + b^2}$ and $\tan \theta = \frac{b}{a}$, $a \neq 0$.

power series (p. 404) An expression of the form $\sum_{n=0}^{\infty} a_n x^n$, where x is a variable and a_0, a_1, a_2, . . . , a_n, . . . are constants.

principal values (p. 255) The values in the range of an inverse function.

Pythagorean theorem (p. 8) a and b are the lengths of the legs of a right triangle and c is the length of its hypotenuse if and only if $c^2 = a^2 + b^2$.

quadrantal angle (p. 39) An angle whose terminal side is on the x- or y-axis.

radian (p. 18) A unit of angle measure; a central angle of 1 radian intercepts an arc that has the same length as the radius of the circle.

range (p. 2) The set of replacements for the second variable in a relation or function.

reciprocal functions (p. 35) Sine and cosecant, cosine and secant, and tangent and cotangent are pairs of reciprocal functions.

recursive formula (p. 375) A formula that defines a sequence by stating the first term, a_1, and a rule for obtaining the $(n + 1)$th term from the preceding term, a_n.

reference angle (p. 42) The smallest positive acute angle determined by the x-axis and the terminal side of θ.

reference triangle (p. 30) A right triangle formed when a perpendicular is drawn from any point on the terminal side of an angle to the x-axis.

resultant vector (p. 191) A vector that is the sum of two vectors.

rose (p. 295) A curve $r = a \cos n\theta$ or $r = a \sin n\theta$, where n is a positive integer, consisting of n leaves if n is odd and $2n$ leaves if n is even.

rotation of axes (p. 273) A new $x'y'$-coordinate system formed by rotating the xy-coordinate axes about the origin.

scalar (p. 192) A real number used to multiply a vector.

secant (p. 34) If $P(x, y)$ is any point other than the origin on the terminal side of an angle θ in standard position, $\sec \theta = \frac{r}{x}$, $x \neq 0$.

second (p. 17) A subdivision of a minute, where $1' = 60''$ (60 s).

sequence (p. 372) A function whose domain is a set of consecutive positive integers.

series (p. 386) The indicated sum of the terms of a sequence.

sine (p. 29, 34) The function defined by the set of ordered pairs (θ, y); if $P(x, y)$ is any point other than the origin on the terminal side of an angle θ in standard position, $\sin \theta = \frac{y}{r}$, where $r = \sqrt{x^2 + y^2}$.

standard position (p. 12) When the vertex of an angle is at the origin of the coordinate plane and its initial side coincides with the positive x-axis, the angle is in standard position.

sum of an infinite geometric series (p. 400) The limit approached by the sequence of partial sums of an infinite geometric series; $S = \frac{a_1}{1 - r}$, $|r| < 1$ where a_1 is the first term and r is the common ratio.

tangent (p. 34) If $P(x, y)$ is any point other than the origin on the terminal side of an angle θ in standard position, $\tan \theta = \frac{y}{x}$, $x \neq 0$.

terminal side (p. 12) The side of an angle that determines the amount of rotation of the initial side of the angle.

trigonometric functions (p. 34) Sine, cosine, tangent, cosecant, secant, and cotangent are trigonometric functions.

unit circle (p. 29) A circle with a radius of 1.

vector (p. 190) A directed line segment that represents a vector quantity.

vertical shift (p. 80) The distance that a graph is shifted up or down.

INDEX

Accuracy (of a measurement), 114
Addition
  of complex numbers, 299-300
  of ordinates, 85
  parallelogram, 191-192
  tail-to-head, 191-192
  of vectors, 191-192
Air speed, 193
Air velocity, 193-194
Algebra
  applications, 65, 66, 72, 302, 321, 341, 358
  logical reasoning, 135
Ambiguous case, 163-165
Amplitude, 73-75
Angle(s), 12-14
  acute, 42
  of bank, 242
  central, 17
  complementary, 67
  coterminal, 13-14
  degree measure of, 13-14, 17-20
  of depression, 118-120
  direction, of a vector, 191
  of elevation, 118-119
  of incidence, 278-279
  initial side of, 12
  negative measure, 12
  positive measure, 12
  quadrantal, 39
  radian measure of, 18-20
  reference, 42-43
  refraction, 278-279
  rotation, 12-14
  of rotation, 273
  standard position of, 12
  terminal side of, 12
  vertex, 12
Angular displacement, 24
Angular velocity, 24-26
  symbol for, 24
Antilogarithm, 354
Applications
  angles of elevation and depression, 118-120
  angular and linear velocity, 23-26
  aviation, 242-243
  compact discs, 148-149
  construction, 125
  exponential and logarithmic equations, 359-362
  harmonic motion, 100-102
  industrial robots, 198-199
  magnetic resonance imaging, 105
  navigation, 124-125
  rainbow, 278-279
  space exploration, 52-53
  surveying, 125
Arc 17-18
  great-circle, 237
  length of intercepted, 23
Archimedean spiral, 296, 412-413
Architecture, applications, 66, 117
Area
  of a parallelogram, 123
  of a rectangle, 123
  of a triangle, 181-183, 186-187
Argand diagram, 308
Argument, 309
Arithmetic means, 375
Arithmetic sequence, 373-375
  application, 414-415
  common difference of, 373
  general term of, 374
Arithmetic series, 384-387
Asymptotes, 88-91
Aviation, applications, 135, 213, 242-243
Axes, rotation of, 273-276
Axis
  imaginary, 308
  polar, 286
  real, 308
  x-, 3
  y-, 3

Bearing, 124
Binary notation, 148-149
Biography
  Flugge-Lotz, Irmgard, 259
  De Moivre, Abraham, 302
  Young, Grace Chisholm, 99
Bombelli, Rafael, 312

Calculator
  approximate solutions to trigonometric equations, 269-271
  area of triangle, 182-183
  calculation-ready form, 113, 125, 158-159, 164, 169, 171, 176-178, 182-183, 188, 242, 338, 355-356, 359-361, 379
  challenge, 272
  check identities, 145
  compound interest problems, 338, 359
  decimal degrees and, 18, 114
  degree mode, 46-49
  evaluate exponential functions, 340
  evaluate inverse trigonometric functions, 256, 262-263
  evaluate logarithms, 353-355
  evaluate trigonometric functions, 46-49, 112
  geometric series, 379
  graphing, 70, 84, 93, 145, 271, 346
  Heron's formula, 187-188
  polar form and, 311
  power series, 402-403
  radian mode, 46-48
  real number exponents and, 335
  solve exponential equations, 355-356, 359-361
  solve right triangles, 113-115, 125
  solve triangles, 158-159, 164, 169-171, 176-178
Calendar, trigonometry and, 87
Cardan, Jerome, 312
Cardioid, 294-295
Career
  air traffic control specialist, 72
  CAD technician, 185
Cartesian coordinate system, 7
Cartesian motion, 198
Catenary, 410
Cauchy, Augustin Louis, 312, 401
Central angle, 17
Challenge, 38, 45, 56, 77, 108, 123, 152, 167, 189, 202, 246, 268, 272, 282, 317, 330, 337, 368, 406, 416
Change-of-base formula, 356
Chapter Review (*see* Reviews)
Chapter Test (*see* Tests)
Characteristic, 354
Chemistry, trigonometry in, 352

Circle, 17-20
arcs of, 17-18, 23
central angle of, 17
circumference of, 17
equation of, 9
great, 236, 237
unit, 29
Circular functions, 31
Circumference, 17
Cis θ, 316
Cofunction(s), 67-68
College Entrance Exam Review (*see* Reviews)
Common difference, 373
Common logarithm(s), 353-354
characteristic of, 354
computation using, 428-429
evaluating, 353-356
mantissa of, 354
table of, 430-431
using a table of, 426-427
Common ratio, 378
Compact discs (CD), 148-149
Complementary angles, 67
Complex number(s), 299
absolute value of, 309
addition of, 299-300
additive inverse, 300
Argand diagram, 308
argument of, 309
associative property for, 299, 305
commutative property for, 299, 305
conjugates, 304
cubed, 318
distributive property for, 299, 305
division of, 304-305, 314-315
equality of, 299
modulus of, 310
multiplication of, 303-304, 313-314
polar form of, 309-310
powers of, 318-320
reciprocal of, 304
rectangular form of, 309
roots of, 322-324
squared, 318
subtraction of, 300
trigonometric form of, 309
as vectors, 308-309
using computer programs containing, 307, 316-317, 326, 327
Complex plane, 308
Complex roots theorem, 322
Compound interest, 338, 359, 383
Computer (*see also* Technology)
applications, 33, 51, 135, 139, 144, 219, 241, 307, 317, 326, 346
check identities, 145-146
check solutions to equations, 271
complex numbers, 307, 316-317, 326, 327
draw Mandelbrot set, 327
evaluate product/sum identity, 241
evaluate trigonometric expressions, 139, 144
evaluate trigonometric functions, 33, 51, 135, 219
graph Archimedean spiral, 412-413
graph logarithmic functions, 346, 365
graphing, 145-146, 271, 327, 346, 365, 412-413
Conditional equation, 265-266
Conjugates, 304
Construction, applications, 11, 38, 45, 117, 125, 189, 264, 395
Continued fraction, 406
Convergent series, 397
Coordinate geometry, applications, 225, 297
Coordinate plane, 3
angles in, 12-14
distance between two points in, 8-9
graphing functions in, 3-4
graphing relations in, 3
Coordinate system(s)
application, 198-199
Cartesian, 7
cylindrical, 198-199
polar, 286-290
rectangular, 3
three-dimensional rectangular, 198-199
in space, 11, 198-199
Coriolis force, 12
Cos θ, 30
Cos x, 67
$\text{Cos}^{-1}\, x$, 252
Cosecant function(s), 34-36
graphs of, 94-97
hyperbolic, 409
inverse, 261-263
properties of, 94-97
range of, 96
Cosh x, 408
Cosine function(s), 29-31
amplitude of, 73-74
definition, 30
difference identity for, 208-211
domain of, 31
graphs of, 67-81
hyperbolic, 408-409
inverse, 256-257
maximum value of, 73
minimum value of, 73
period of, 69, 73-75
phase shift, 78-81
properties of, 67-70
range of, 73
sum identity for, 210-211
vertical shift, 80-81
Cotangent function(s), 34-36
graphs of, 88-91
hyperbolic, 409
inverse, 260-263
properties of, 88-91
range of, 90
Cot θ, 34
Cot x, 67
$\text{Cot}^{-1}\, x$, 260
Coterminal angles, 13-14
Coth x, 409
Course (direction), 194
Course (in navigation), 124
Csc θ, 34
Csc x, 69
$\text{Csc}^{-1}\, x$, 261
Csch x, 409
Cumulative Review (*see* Reviews)
Cycle, 60
Cylindrical coordinate system, 198-199
Cylindrical motion, 198

Decibel, 347
Degree, 13
decimal, 18
minutes, 17
seconds, 17
Degree measure(s), 13-14, 17-20
converting to radian measure, 19
equivalent radian measures, 20
De Moivre, Abraham, 302, 312
De Moivre's theorem, 318-320
Descartes, René, 7, 312
Difference identity
for cosine, 208-211
for sine, 215-217
for tangent, 221-223
Direction angle of a vector, 191
Discontinuous function, 89

Displacement, 101
angular, 24
Distance formula, 8-9
and equation of a circle, 9
for two points in space, 11
Divergent series, 398
Division
of complex numbers, 304-305
in polar form, 314-315
Domain, 2
Double-angle identities, 226-228

***e*, 340, 403, 406**
Eight-leaved rose, 295-296
Electricity, application, 302
Ellipse, 276
Engineering, applications, 11, 99, 134, 167, 259, 277, 410
Equation(s)
of a circle, 9
conditional, 265-266
cubic, 322
exponential, 336, 359-362
logarithmic, 350, 359-362
polar, 292-295
quadratic, 78
trigonometric, 140-146, 265-271
Equator, 236
Equivalent trigonometric expressions, 136-137
Equivalent vectors, 190
Euler, Leonhard, 312, 401, 406
Euler's formula, 404
Even function, 61-62
Exponential decay formula, 360
Exponential equation(s), 336
applications, 359-362
solving, 336, 355-356
Exponential form, 343-344
Exponential function(s), 338-340, 403-404
graphs of, 339-340
natural, 340
properties of, 340
using calculator to evaluate, 340
Extra, 11, 16, 93, 139, 213, 219, 254, 297, 321, 341, 377, 389

Factorial notation, 139, 402
Finance
application, 395
trigonometry in, 383
Finite sequence, 372
Five-leaved rose, 295-296
Flugge-Lotz, Irmgard, 259
FOIL method, 303
Formula(s)
angle of bank, 242-243
angular velocity, 24
arc length, 23
area of a triangle, 181-183
arithmetic sequence, 375
change-of-base, 356
circumference of a circle, 17
compound interest, 338
difference identity for cosine, 208
difference identity for sine, 215
difference identity for tangent, 221
distance, 8
double-angle identities, 227
Euler's, 404
exponential decay, 360
geometric sequence, 381
half-angle identities, 232
Heron's, 186
intensity level, 347
linear velocity, 24
nautical mile, 46
partial sum of arithmetic series, 387
partial sum of geometric series, 393
pH of a solution, 353
product/sum identites, 238
quadratic, 270
radian measure of central angle, 18
sum identity for cosine, 210
sum identity for sine, 215
sum identity for tangent, 220
sum of infinite geometric series, 398
Fourier, Joseph, 66
Four-leaved rose, 295-296
Fractals, 327
Frequency, 73, 100
Function(s), 3
circular, 31
domain of, 2
exponential, 338-340, 403-404
even, 61-62
graphs of, 3-4 (*see also* Graph)
hyperbolic, 407-409
inverse, 251-253
logarithmic, 342-344
notation, 4
odd, 61-62
one-to-one, 342
period of, 60
periodic, 60-61
range of, 2
reciprocal, 35-36
trigonometric, 34-36 (*see also* Trigonometric function)
value of, 4-5
vertical line test, 4
Fundamental identities, 130-133

Gauss, Karl Friedrich, 312
Gauss units, 105
General term of a sequence, 372
Geography, trigonometry in, 236
Geometric means, 380
Geometric sequence, 378-381
common ratio of, 378
general term of, 379
Geometric series, 390-393
infinite, 396-399
Geometry
applications, 7, 117, 180, 236, 254, 277, 312, 383, 401
Graph(s)
by addition of ordinates, 85
computer generated, 145-146, 271, 327, 346, 365, 412-413
of cosecant function, 94-97
of cosine function, 67-81
of cotangent function, 88-91
designs, 99
of equations using rotated axes, 273-276
of even function, 62
of exponential function, 339-340
graphing calculator, 70, 84, 93, 145, 271, 346
of inverse functions, 251-252
of logarithmic function, 343-344, 346, 365
of natural exponential function, 340
of odd function, 62
of periodic function, 61
polar, 292-295, 297
of a relation, 3
of secant function, 94-97
of sine function, 67-81
sinusoidal, 100-102
of sound wave, 149
symmetry of, 62-63
of tangent function, 88-91, 93
of trigonometric equations, 145-146, 271
Great circle, 236, 237
Greek alphabet, 16
Ground speed, 194
Ground velocity, 193-194

Half-angle identities, 231-234
Hamilton, William Rowan, 312
Harmonic motion, simple, 100-102
Heading, 193
Heron, 312
Heron's formula, 186-188
Hipparchus, 66
Historical note
complex numbers, 312
computing devices, 144
contributions to trigonometry, 66
Descartes and the coordinate plane, 7
limits, 401
Pythagorean theorem, 117
Horizontal line test, 254
Hyperbola, 88, 275
unit, 408
Hyperbolic functions, 407-409
Hyperbolic identities, 409

***i*, 298**
Identity(ies)
difference, for cosine, 208-211
difference, for sine, 215-217
difference, for tangent, 221-223
double-angle, 226-228
fundamental, 130-133
graphing to check, 145-146
half-angle, 231-234
odd-even, 132
product/sum, 237-239, 241
proving, 130-132, 140-142
Pythagorean, 131
ratio, 131
reciprocal, 130
sum, for cosine, 210-211
sum, for sine, 215-217
sum, for tangent, 220-223
trigonometic, 130-133, 208-239
Imaginary
axis, 308
number, 298
Index of refraction, 278
Index of summation, 385, 389
Industrial robots, 198-199
Infinite geometric series, 396-399
convergent, 397
divergent, 398
and repeating decimals, 399
sum of convergent, 398-399
Infinite sequence, 373
Initial side of an angle, 12
Intensity level, 347
Interpolation, linear, 424-425
Inverse function(s), 251-253
cosecant, 261-263
cosine, 256-257
cotangent, 260-263
domains, 261
horizontal line test, 254
graphs of, 251-252
ranges, 261
secant, 261-263
sine, 255-257
tangent, 260-263
trigonometric, 252-263
using calculator to evaluate, 256, 262-263
Inverse of a relation, 250-251
Isosceles right triangle, 39

Lagrange, Joseph Louis, 401
Latitude, 237
Law of cosines, 168-171, 174
Law of sines, 156-159
ambiguous case, 163-165
Law of tangents, 175-178
Learning curve, 364
Lemniscate, 296, 297
Leibniz, Gottfried Wilhelm, 401
Limaçon, 294
Limits, 397, 401
Limits of summation, 385, 389
Linear motion, 198
Linear interpolation, 424-425
Linear velocity, 24-26
Line of sight, 118
Logarithm(s),
change-of-base formula, 356
common, 353-354
computation using common, 428-429
evaluating, 353-356
expanded form, 348-349
natural, 354-355
properties of, 348-350
solving exponential equations with, 355-356
Logarithmic equations, 350, 352
applications, 359-362
Logarithmic form, 343-344
Logarithmic function(s), 342-344
application, 365
graphs of, 343-344, 346, 365
properties of, 343
Logical reasoning, 135
Longitude, 237

Mach number, 231
Machine shop, trigonometry in, 33
Magnetic Resonance Imaging (MRI), 105
Magnitude
of the impedance, 302
of a vector, 190-191
Maintaining Skills (*see* Reviews)
Mandelbrot set, 327
Mantissa, 354
Math Club Activity, 162, 358
Mean(s)
arithmetic, 375
geometric, 380
proportional, 380
Measure of an angle
degree, 13-14, 17-20
negative, 12
positive, 12
radian, 18-20
Measure of an arc, 17
Meridians, 236
Meteorology, applications, 7, 22
Minutes, 17
Modulus, 310
Motion
Cartesian, 198
cylindrical, 198
linear, 198
rotary, 24, 198
simple harmonic, 100-102
tidal, 102, 230
Multiplication
of complex numbers, 303-304
in polar form, 313-314
of vector by scalar, 192
Music, applications, 82-83, 86

Natural exponential function, 340
Natural logarithm(s), 354-355
symbol for, 354
table of, 432-433
Nautical mile, 46
Navigation, applications, 11, 124-125, 174
Newton, Isaac, 401
Normal probability curve, 406
Notation
binary, 148-149
factorial, 139, 402
function, 4
scientific, 353
sigma, 384-387
summation, 384

Oblique triangle(s)
area of, 181-183
solving, 156-178

Odd-even identities, 132
Odd function, 61-62
Omega (ω), 24
Opposite vectors, 192
Ordered pair, 2-3
Ordered triplet, 11, 198
Origin, 3
Origin symmetry, 62-63
Oscilloscope, 67, 100

Parabola, 52
Parallelogram addition, 191-192
Parallels, 236
Partial sum(s)
  of arithmetic series, 384-387
  of geometric series, 390-393
Pascal's triangle, trigonometry and, 291
Period, 60
Periodic function(s), 60-61
  amplitude of, 73-75
pH, 342, 352
Phase shift, 78-81
Physics, applications, 77, 147, 197, 264, 268, 272, 291, 296-297, 306-307, 346, 352, 401
Pi (π), 19
Pitch, 73
Plane
  complex, 308
  coordinate, 3
  slice, 105
  vectors in, 190-194
  wz-, 408
Polar axis, 286
Polar coordinate system, 286-290
Polar coordinates, 286-290
  converting to rectangular coordinates, 289
Polar equations, 292-295
Polar form
  of complex numbers, 309-310
  division in, 314-315
  multiplication in, 313-314
Polar graphs, 292-295
  rotations of, 297
Pole, 286
Power series, 402-404
  and hyperbolic functions, 407-409
Powers
  of complex numbers, 318-320
  logarithms of, 348-349
Principal values, 255-256, 260
Product(s)
  of complex numbers, 303-304, 313-314
  of conjugates, 304
  logarithms of, 348-349
  of reciprocal trigonometric functions, 35
  of vector by scalar, 192
Product/sum identities, 237-239, 241
Ptolemy, Claudius, 66
Pulleys, 25-26
Pure imaginary number, 299
Pythagoras, 17
Pythagorean identities, 131
Pythagorean theorem, 8, 117

Quadrantal angle, 39
Quadrants, 3
Quotient(s)
  of complex numbers, 304-305
  logarithms of, 348-349

Radian, 18
Radian measure(s), 18-20
  converting to degree measure, 19
  equivalent degree measures, 20
Radius vector, 30
Rainbow, 278-279
Range, 2
Ratio
  common, 378
  identities, 131
  right triangle, 113
  trigonometric, 112
Real axis, 308
Real number exponents, 334-336
  properties of, 334
Reciprocal(s)
  of complex numbers, 304
  identities, 130
  properties, 35
  trigonometric functions, 35-36
Rectangular coordinate system, 3
  three-dimensional, 198-199
Rectangular coordinates, 286
  converting to polar coordinates, 288
Rectangular form of complex numbers, 309
Regiomontanus (Johann Müller), 66
Reference angle, 42-43
Reference triangle, 30-31
Refraction, 278-279
Relation, 2-3
  domain of, 2
  graphs of, 3
  inverse of, 250-251
  range of, 2
Repeating decimals, 399
Resultant of a vector, 191
Reviews
  Chapter Summary and Review, 54-55, 106-107, 150-151, 200-201, 244-245, 280-281, 328-329, 366-367, 414-415
  College Entrance Exam, 57, 109, 153, 203, 247, 283, 331, 369, 417
  Cumulative, 110, 204-206, 284, 370, 418-420
  Maintaining Skills, 58, 154, 248, 332
Richter scale, 365
Right triangle(s)
  applications, 118-120, 124-125
  isosceles, 39
  ratios, 113
  solving, 112-115
Roots of complex numbers, 322-324
Rose, 295
Rotary motion, 24, 198
Rotation
  angles of, 273
  of axes, 273-276
  of polar graphs, 297

Scalar, 192
Scalene triangle, 156
Scientific notation, 353
Sec θ, 34
Sec x, 67
$\text{Sec}^{-1} x$, 261
Secant function(s), 34-36
  graphs of, 94-97
  hyperbolic, 409
  inverse, 261-263
  properties of, 94-97
  range of, 95
Sech x, 409
Seconds, 17
Semiperimeter, 186
Sequence(s), 372
  arithmetic, 373-375, 412-413
  defined explicitly, 372
  defined recursively, 373
  finite, 372
  general term of, 372
  geometric, 378-381
  infinite, 373
  number theory and, 377
  term of, 372
Self-test (*see* Tests, Test Yourself)

Series, 384
arithmetic, 384-387
convergent, 397
divergent, 398
finite, 384
geometric, 390-393
infinite, 384
infinite geometric, 396-399
limit, 397
partial sum of, 384-393
power, 402-404
sigma notation, 384-385
sum of, 398-399
Shift
phase, 78-81
vertical, 80-81
Sigma notation, 384-387
Significant digits, 113-114
Simple harmonic motion, 100-102
Sin θ, 30
Sin x, 67
$\text{Sin}^{-1} x$, 252
Sine function(s), 29-31
amplitude of, 73-74
definition, 30
difference identity for, 215-217
domain of, 31
graphs of, 67-81
hyperbolic, 408-409
inverse, 255-257
maximum value of, 73
minimum value of, 73
period of, 69, 73-75
phase shift, 78-81
properties of, 67-70
range of, 73
sum identity for, 215-217
vertical shift, 80-81
Sinh x, 408
Sinusoidal graphs, 100-102
Slope of a line, 225
Snell's Law, 278-279
Solving triangle(s)
ambiguous case, 163-165
law of cosines, 168-171
law of sines, 156-159
law of tangents, 175-178
oblique, 156-178
right, 112-115
using calculator, 113-115, 125, 158-159, 164, 169-171, 176-178
Sound waves, 73, 82-83, 148-149
Space
coordinate systems in, 11, 198-199
distance between two points in, 11
four-dimensional, 11
trigonometry in three dimensions, 174
Space exploration, 52-53
Speed of sound, 231
Spiral, 412
Archimedean, 296, 412-413
Standard position
of an angle, 12
of a vector, 191
Standard-rate balanced turn, 242
Standarized Test (*see* Reviews, College Entrance Exam)
Subtraction
of complex numbers, 300
of vectors, 192
Sum
of an infinite geometric series, 398-399
of two trigonometric functions, 84-85
of two vectors, 191-192
Sum identity
for cosine, 210-211
for sine, 215-217
for tangent, 220-223
Summary and Review (*see* Reviews)
Summation notation, 384
Surveying
applications, 38, 125, 161-162, 167, 173-174, 179, 185, 213
trigonometry in, 22
Symbols, 439
Symmetric, 60
Symmetry of a graph, 62-63

Table(s)
of common logarithms, 426-427, 430-431
of cube roots, 421
of cubes, 421
of natural logarithms, 432-433
of square roots, 421
of squares, 421
of values of trigonometric functions, 422-423, 434-438
using, 422-423, 426-427
Tail-to-head addition, 191-192
Tan θ, 34
Tan x, 67
$\text{Tan}^{-1} x$, 260
Tangent function(s), 34-36
difference identify for, 221-223
discontinuous, 89
graphs of, 88-91, 93
hyperbolic, 409
inverse, 260-263
Pascal's triangle and, 291
properties of, 88-91
range of, 89
sum identity for, 220-223
Tahn x, 409
Technology (*see also* Computer and Calculator)
fractals, 327
Richter scale, 365
spirals, 412-413
Term of a sequence, 372
general, 372
Terminal side of an angle, 12
Tests
Chapter Test, 56, 108, 152, 202, 246, 282, 330, 368, 416
Test Yourself, 28, 51, 83, 104, 129, 147, 180, 197, 225, 241, 264, 277, 307, 326, 346, 364, 395, 411
Theorem(s)
complex roots, 322
De Moivre's, 319
Pythagorean, 8, 117
Theta (θ), 14
Three-leaved rose, 295-296
Tidal motion, 102, 230
Transit, 22, 125
Triangle(s)
area of, 181-183, 186-187
Heron's formula, 186-188
oblique, 156
perimeter of, 189
reference, 30-31
right (*see* Right triangle)
scalene, 156
semiperimeter of, 186
similar, 30
solving, 156-178
Trigonometric equation(s)
applications, 105, 242-243
approximate solutions, 269-271
conditional, 265-266
graphs of, 145-146, 271
proving as identity, 140-142
solving, 265-266
Trigonometric expression(s), 130-131
complex, 140
writing equivalent, 136-137
using computer programs to evaluate, 139, 144

Trigonometric function(s), 34-36
amplitude of, 73-75
application, 52-53
cofunctions, 67-68
cosecant, 34-36
cosine, 29-31
cotangent, 34-36
domains, 261
evaluating, 35-49
frequency of, 73, 100
geometric interpretations of, 38
graphs of, 67-98
inverse, 252-263
period of, 73-75
and power series, 402-404
properties of, 67-70, 88-97
ranges, 261
reciprocal, 35-36
and right triangle ratios, 113
secant, 34-36
sine, 29-31
of special angles, 39-43
sum of two, 84-85
table of values of, 434-438
tangent, 34-36
using calculator to evaluate, 46-49, 112
using computer programs to evaluate, 33, 51, 135, 219
Trigonometric identity(ies), 130-133, 208-239
Trigonometric ratios, 112
Trigonometry
and the calendar, 87
in chemistry, 352
in finance, 383
in geography, 236
in the machine shop, 33
and Pascal's triangle, 291
in surveying, 22
in three dimensions, 174
in tidal motion, 230

Unit(s)
decibel, 347
degree, 17-18
gauss, 105
nautical mile, 46
radian, 18-19
rpm, 24
Unit circle, 29
diagram, 219
Unit hyperbola, 408

Values of trigonometric functions
table of, 434-438
using a table of, 422-423
Vector(s), 190-194
air velocity, 193-194
application, 105
difference of, 192
direction of, 190
direction angle of, 191
equivalent, 190
ground velocity, 193-194
initial point of, 190
magnitude of, 190-191
measure of angle between two, 193
multiply scalar by, 192
opposite, 192
quantity, 190
radius, 30
resultant, 191
standard position of, 191
sum of two, 191-192
terminal point of, 190
x-component of, 191
y-component of, 191
Velocity
air, 193-194
angular, 24-26
ground, 193-194
linear, 24-26
sum of, 193-194
Vertex of an angle, 12
Vertical line test, 4
Vertical shift, 80-81
Viète, François, 66
Vocabulary, 54, 106, 150, 200, 244, 280, 328, 366, 414

Weierstrass, Karl, 401

***x*-axis, 3**
x-axis symmetry, 62-63
x-component of a vector, 191

***y*-axis, 3**
y-axis symmetry, 62-63
y-component of a vector, 191
Young, Grace Chisholm, 99

Photo Credits

Chapter 1 **1:** NASA. **2:** Richard Megna/Fundamental Photographs. **7:** The Bettmann Archive. **12:** Fundamental Photos. **22:** Larry Smith/DPI. **23:** DPI. **24:** Ken Karp. **52:** NASA/OPC.
Chapter 2 **59:** Courtesy of Nancy Grigsby, MR Systems Engineer, GE Medical Systems. **67:** Richard Megna/Fundamental Photographs. **72:** George Goodwin/Monkmeyer Press. **93:** Jon Lei/OPC; Jon Lei/OPC. **99:** Roger and Silvia Wiegand. **105:** Courtesy of Nancy Grigsby, MR Systems Engineer, GE Medical Systems.
Chapter 3 **111:** Dan McCoy/Rainbow. **117:** The Bettmann Archive. **144:** Kip Peticolas/Fundamental Photos. **145:** John Lei/OPC. **146:** John Lei/OPC. **148:** John Lei/OPC.
Chapter 4 **155:** Stock Boston/Aronson Photographics. **185:** Ray Ellis/Photo Researchers. **190:** Douglas Aircraft Company, Inc. **198:** Hank Morgan/Rainbow.
Chapter 5 **207:** American Museum of Natural History. **230:** Elihu Blotnick/OPC. **231:** Photri. **242:** Stock Market/Ted Mahieu.
Chapter 6 **249:** Gary Ladd Photography. **250:** Frank Siteman/Taurus Photos. **259:** Society of Women Engineers. **271:** John Lei/OPC. **278:** Stock Boston/Peter Menzel.
Chapter 7 **285:** Courtesy of Professor Benoit Mandelbrot. **292:** Michael Dalton/Fundamental Photographs. **302:** Granger Collection. **327:** Courtesy of Professor Benoit Mandelbrot.
Chapter 8 **333:** Stock Boston/Donald Dietz. **338:** Audrey Gottlieb/Monkmeyer Press. **347:** Bernard Wolf/OPC.
Chapter 9 **373:** Mark Antmarr/Stock Market. **380:** Robert J. Capece/Monkmeyer Press. **386:** G. Zimbel/Monkmeyer Press. **392:** Randy Matusow/Monkmeyer Press. **415:** Stock Boston/Frank Siteman.

Additional Answers

Chapter 2

Practice Exercises, pages 64–66

7.

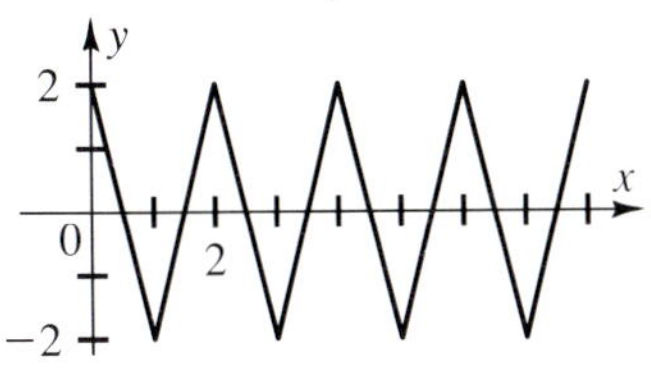

8.

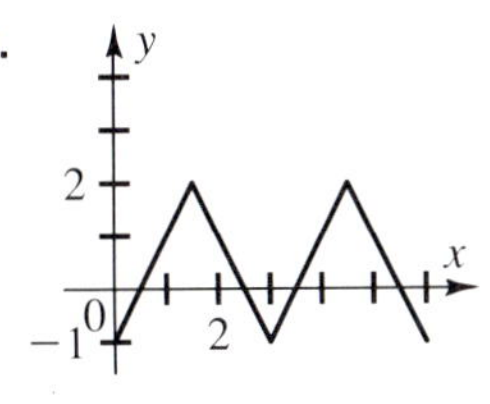

9.

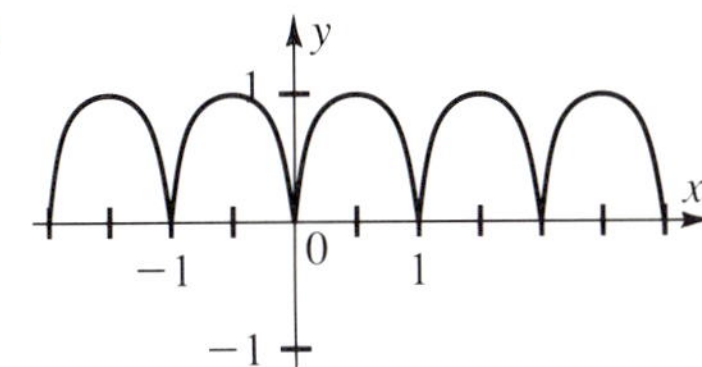

10.

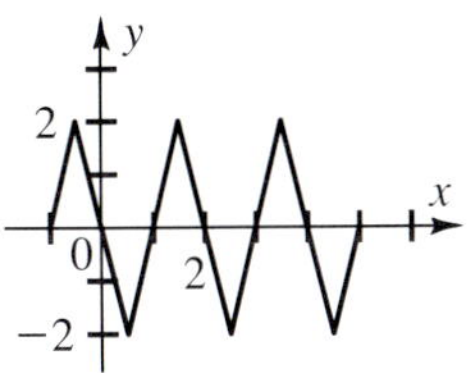

11.

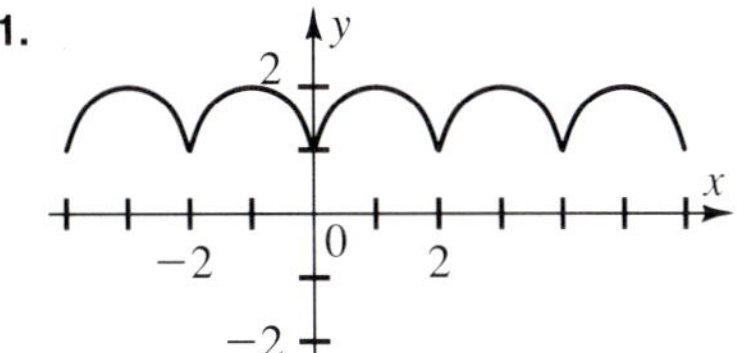

12.

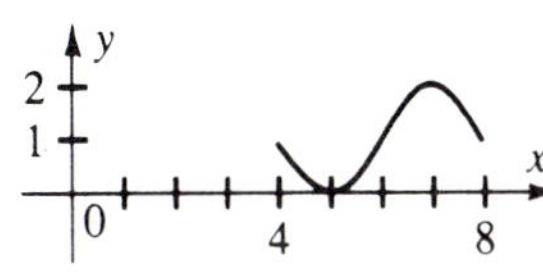

31.

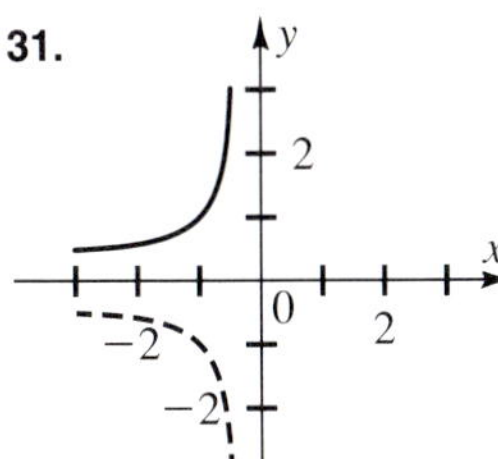

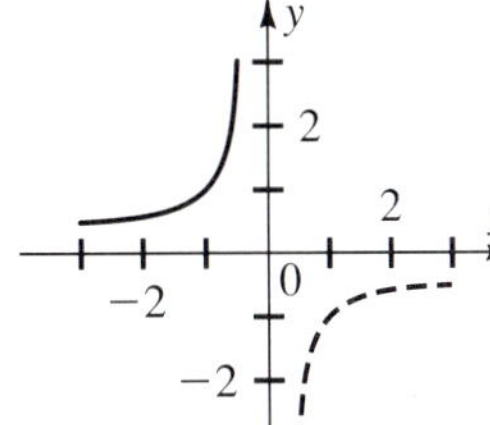

32.

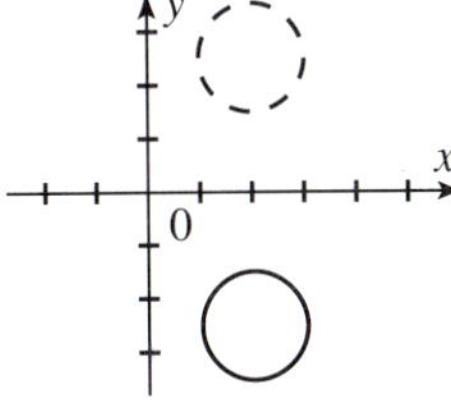

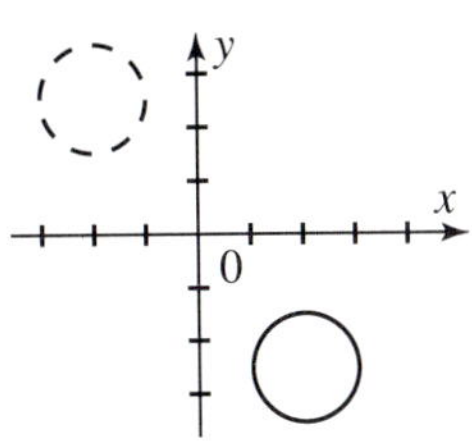

33.

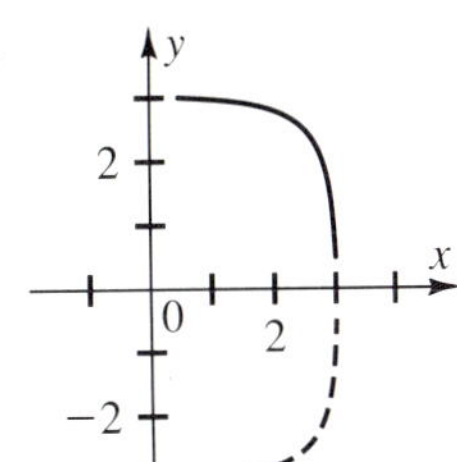

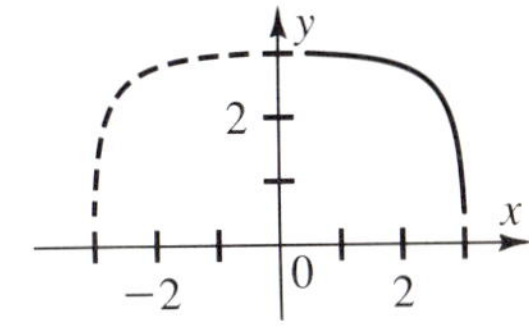

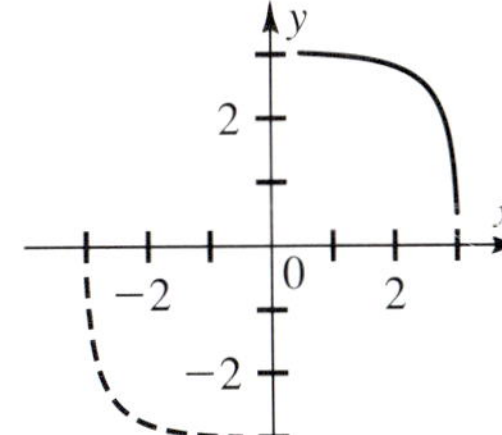

34.

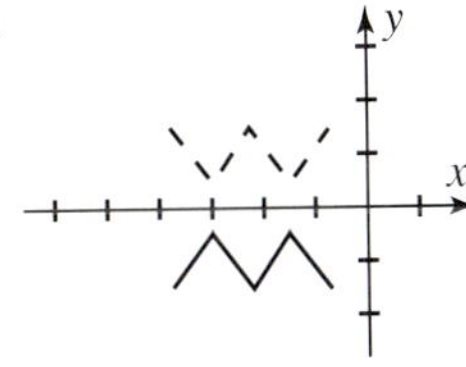

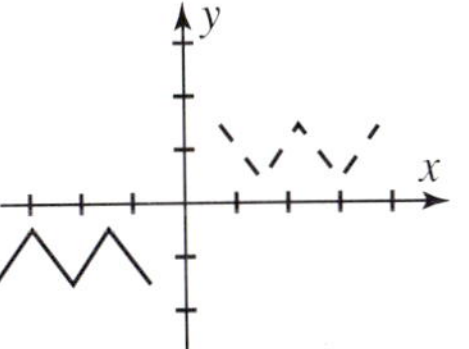

35.

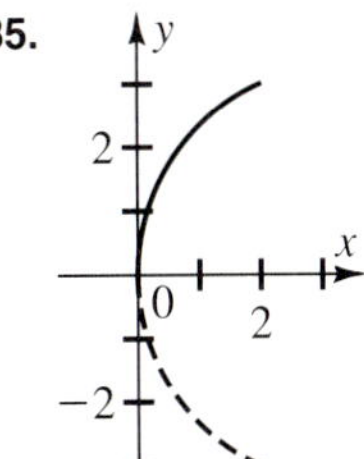

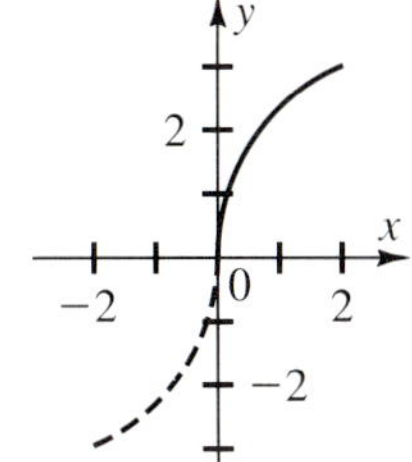

36.

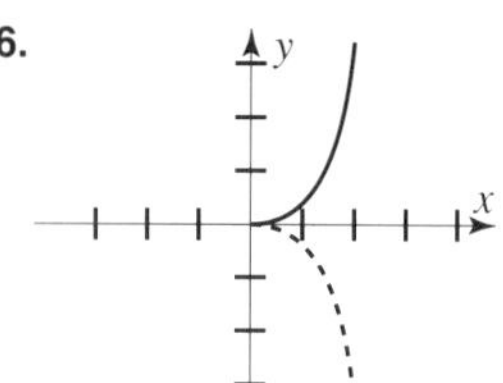

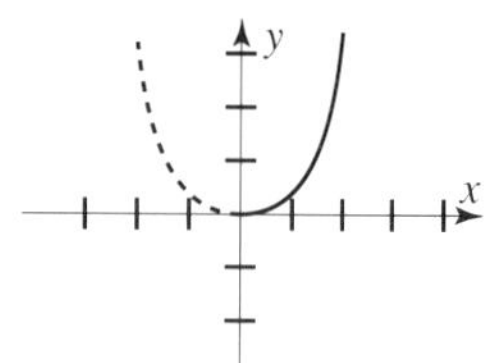

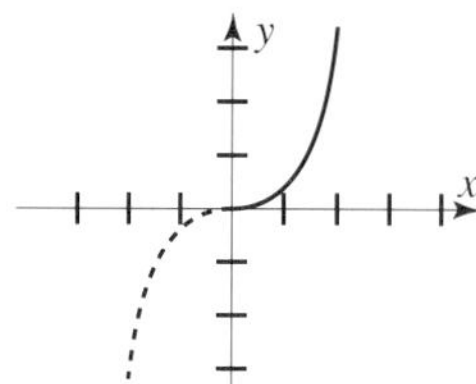

Practice Exercises, pages 71–72

21.

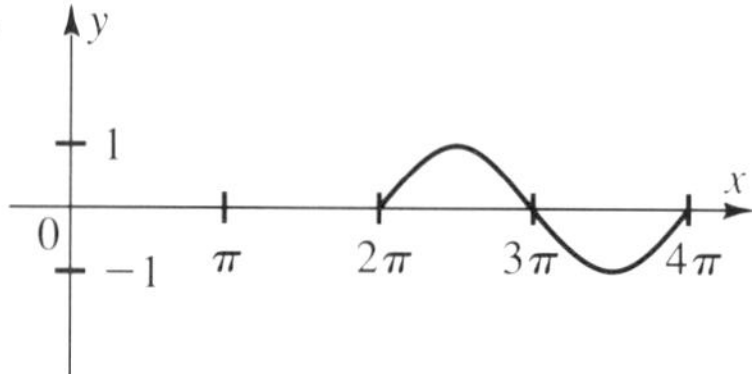

22.

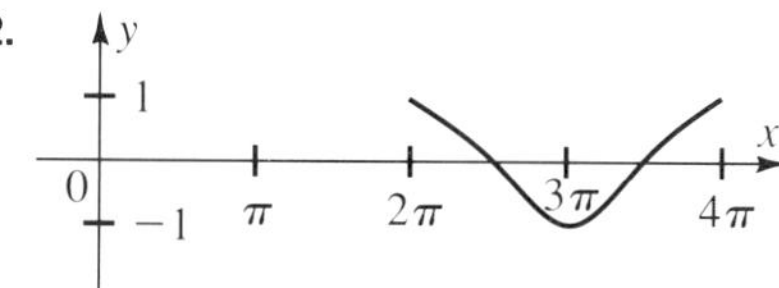

23.

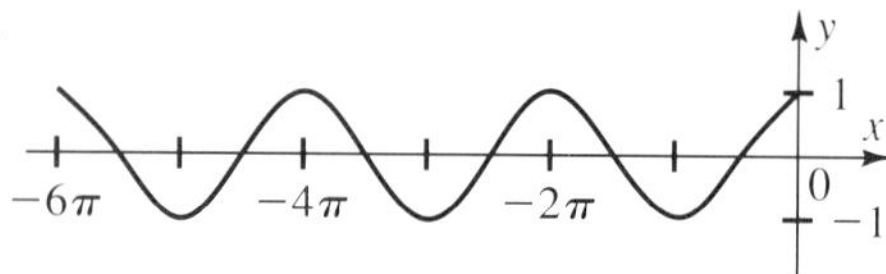

24.

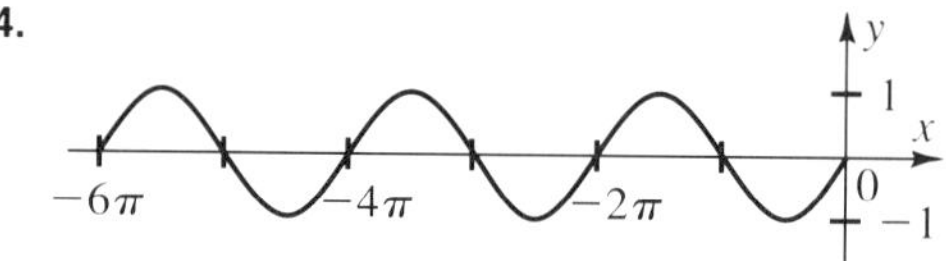

35.

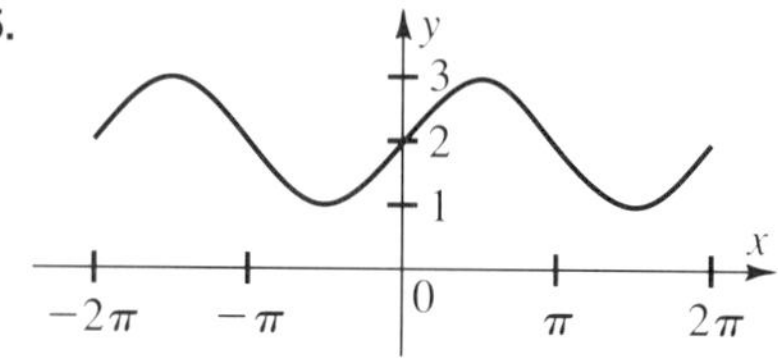

36.

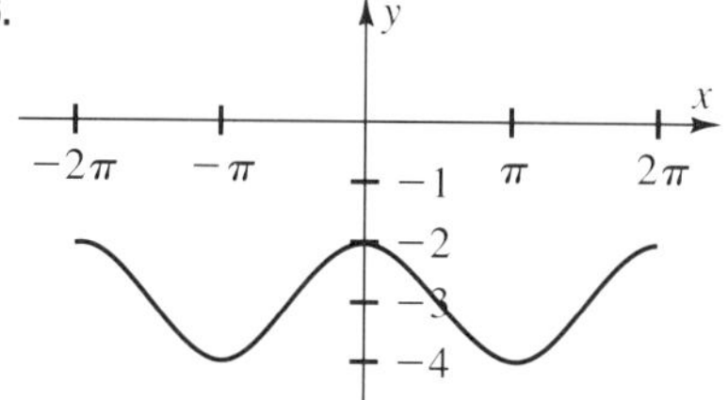

37.

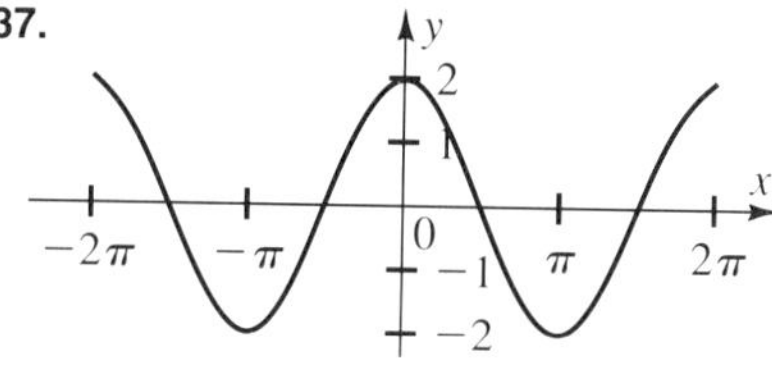

38.

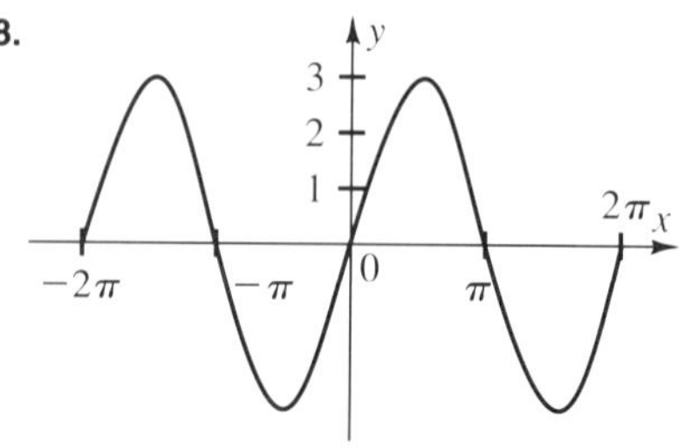

39.

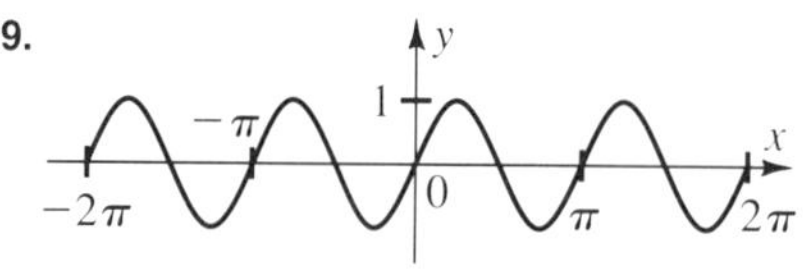

40.

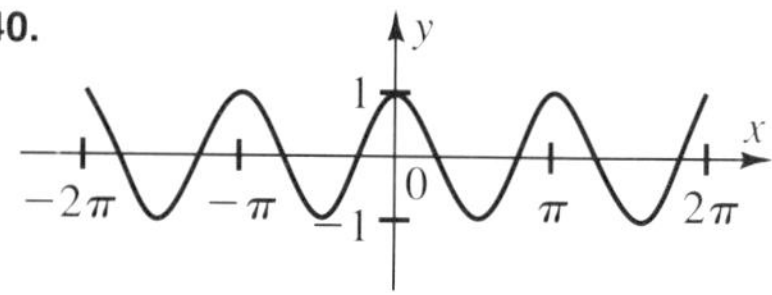

41.

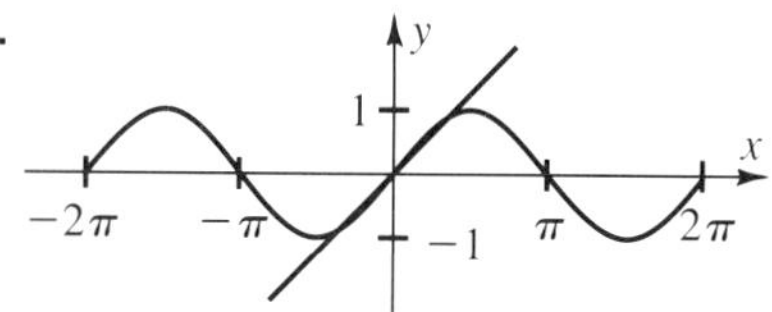

42.

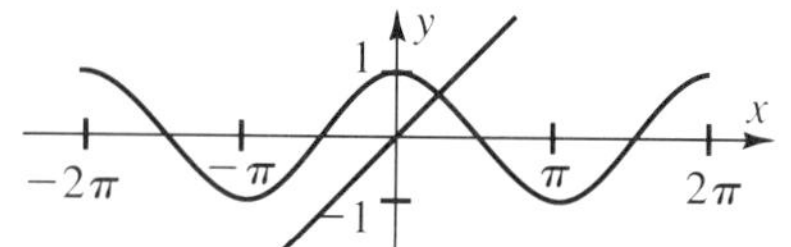

Practice Exercises, pages 82–83

11. 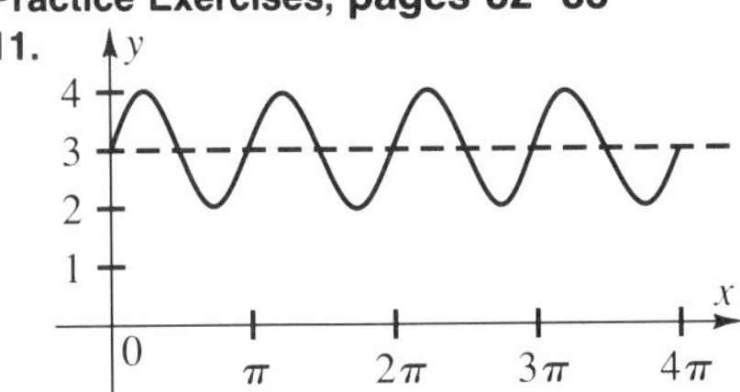

12. 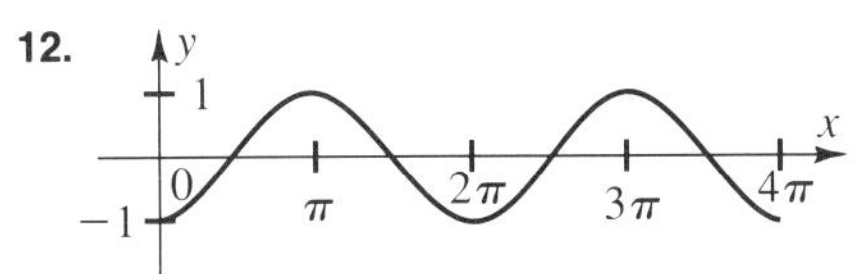

13. 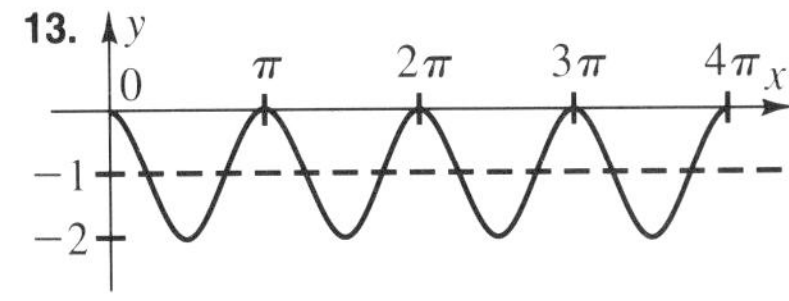

14 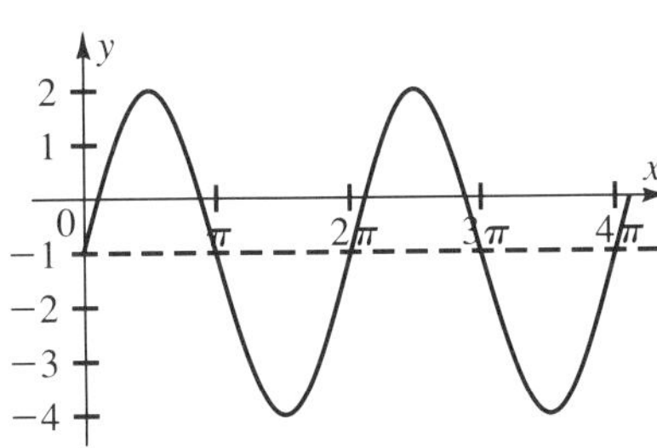

15. 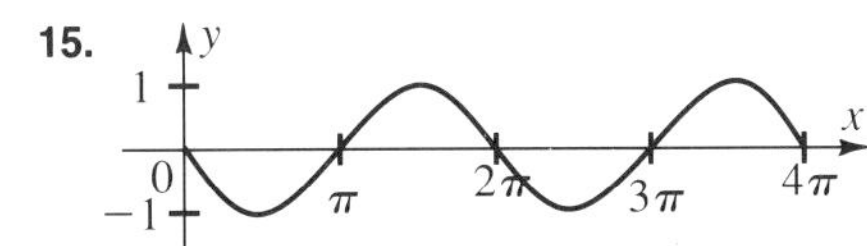

16. 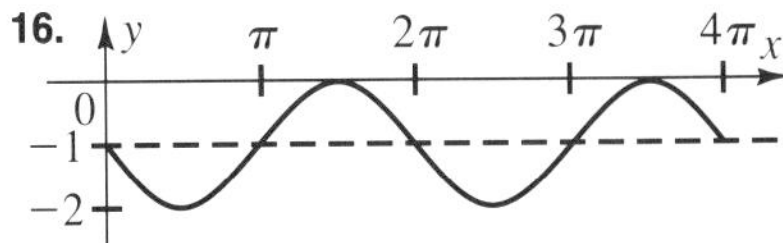

17. 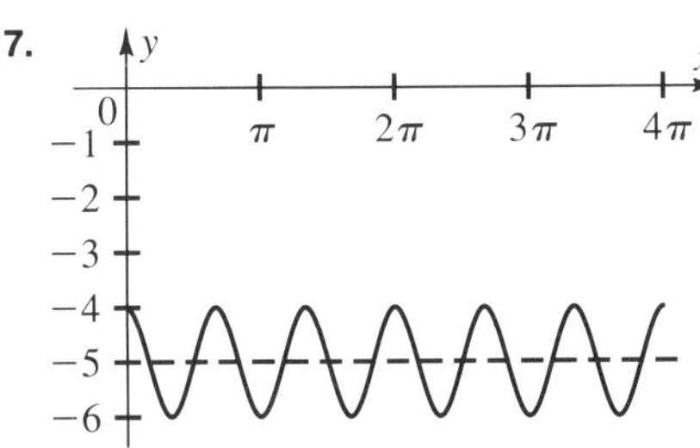

18. 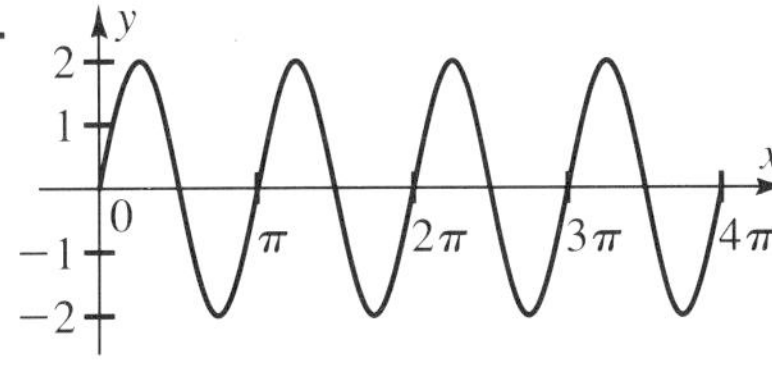

19. 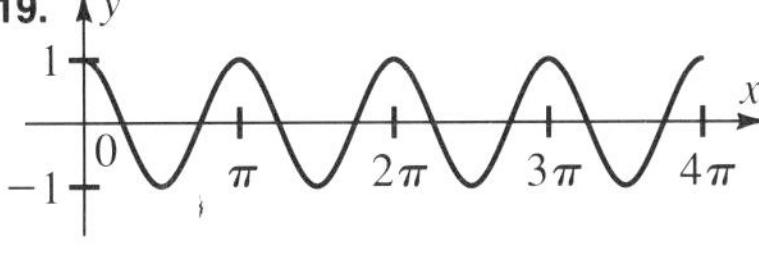

20. 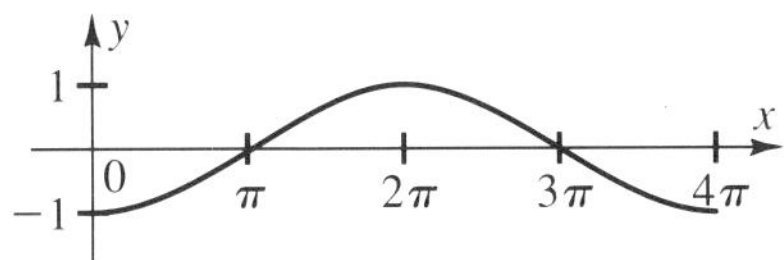

21. 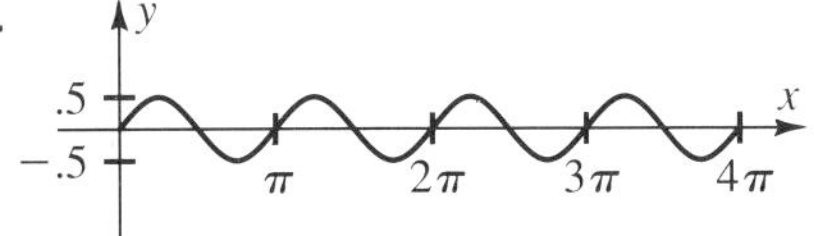

22. 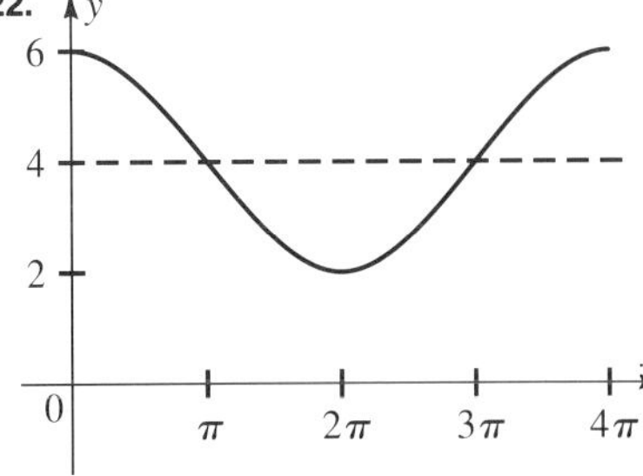

23. 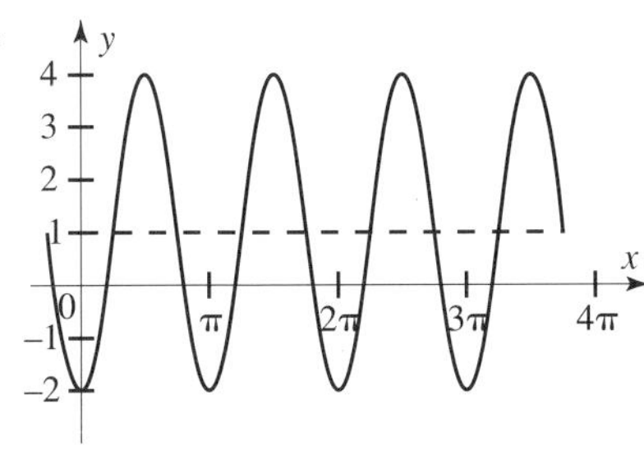

24. 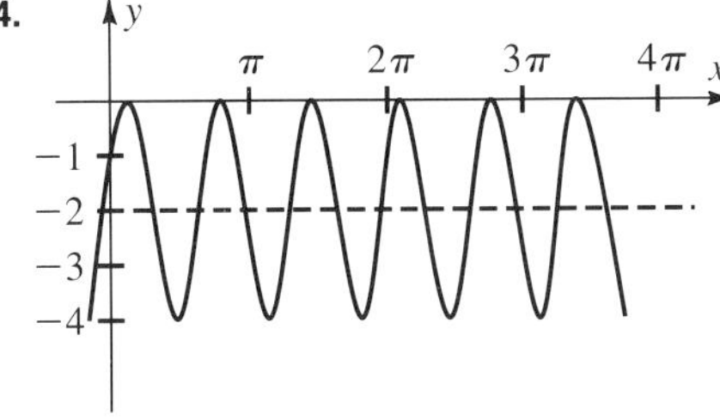

Test Yourself, page 83

17.

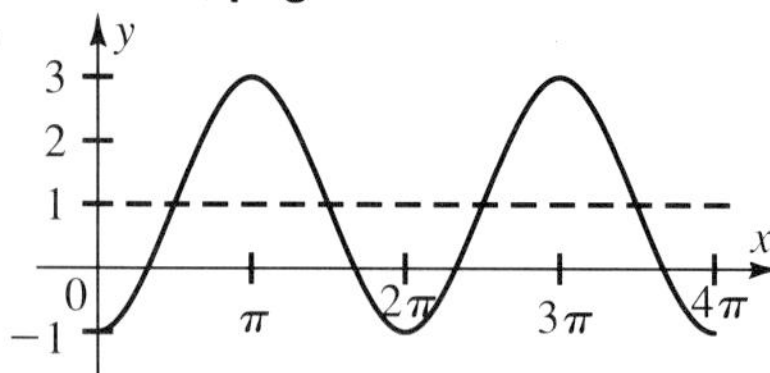

18.

Practice Exercises, page 86

1.

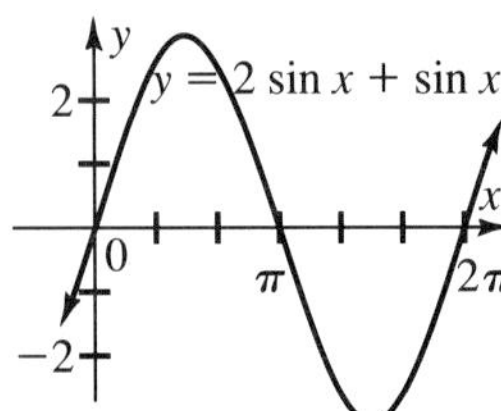

2.

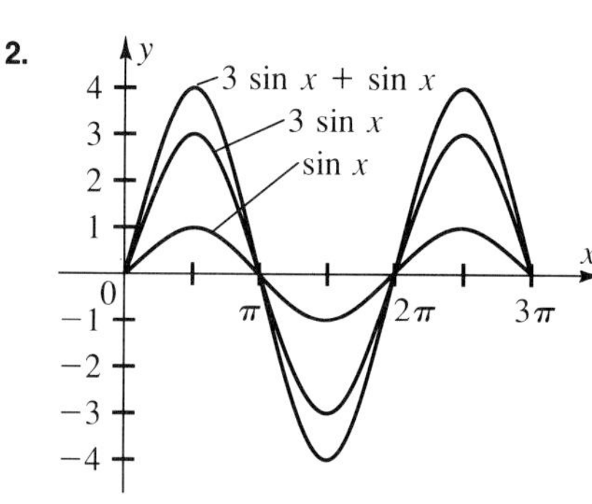

3.

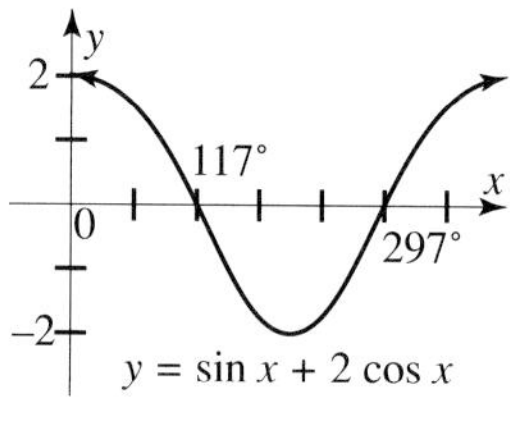

4.

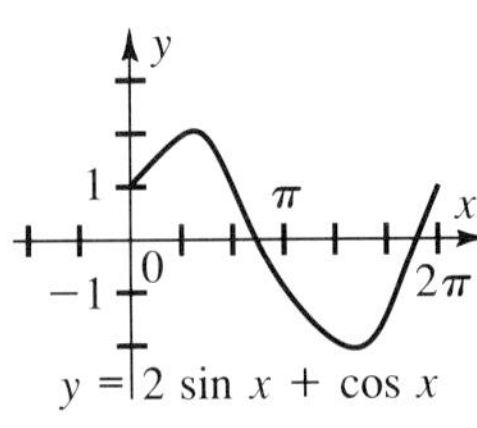

5.

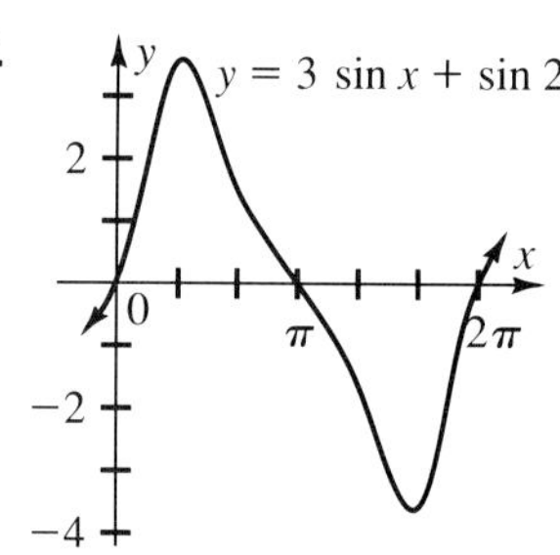

6.

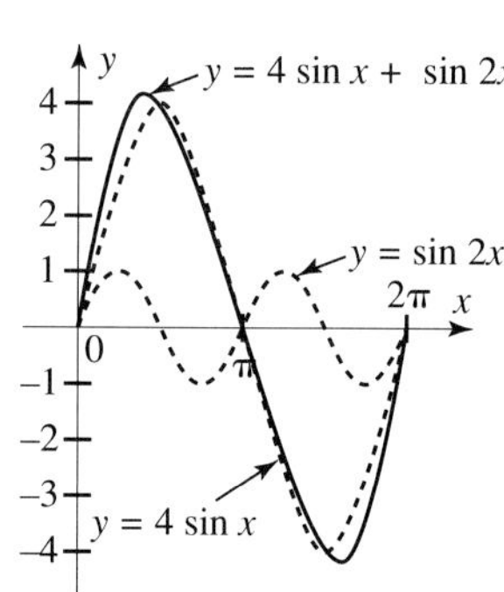

7.

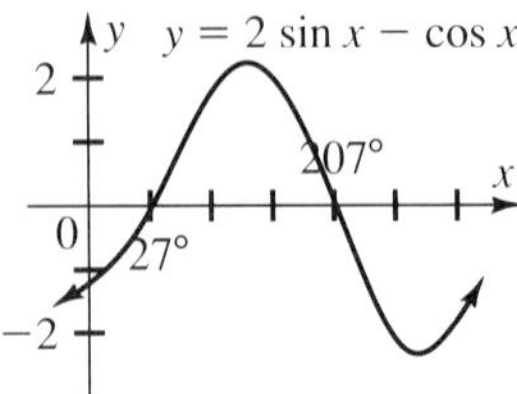

8.

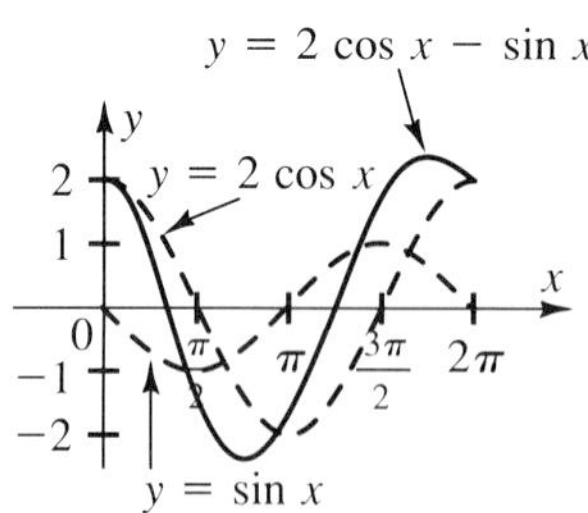

9.

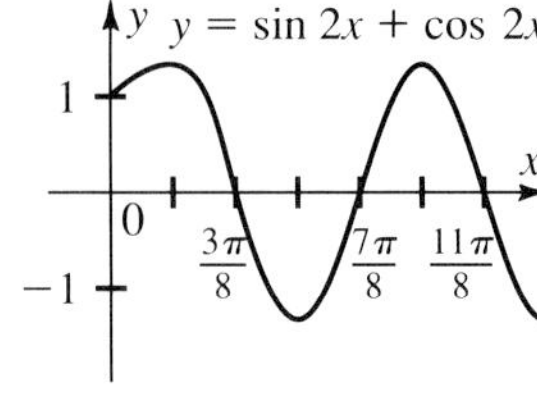

10.

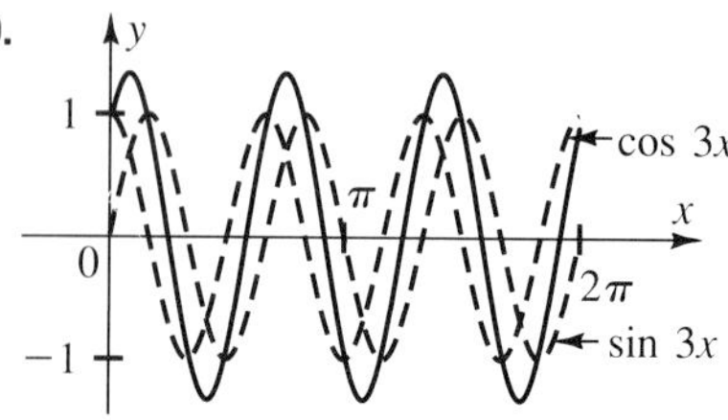

11.

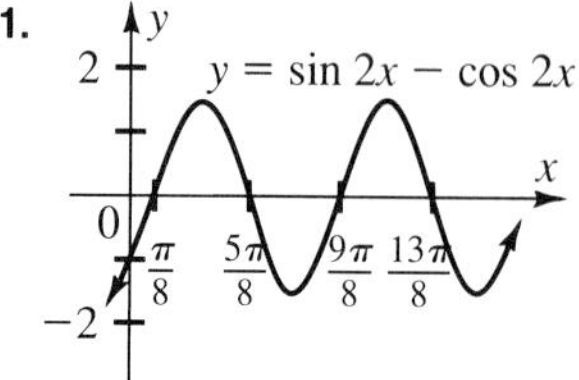

12.

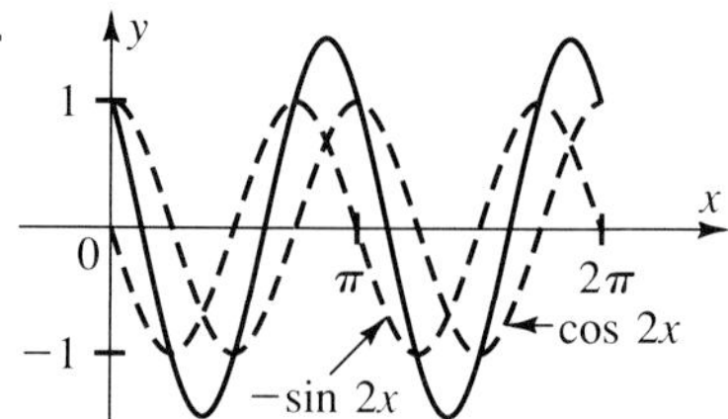

13.
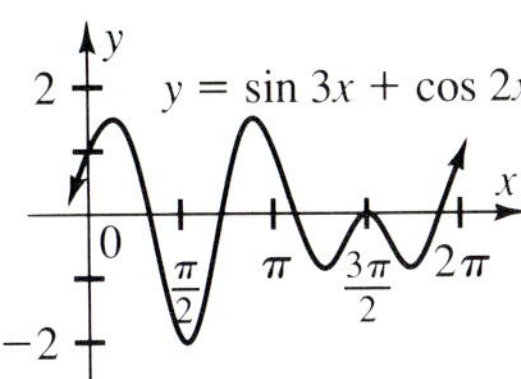

14.
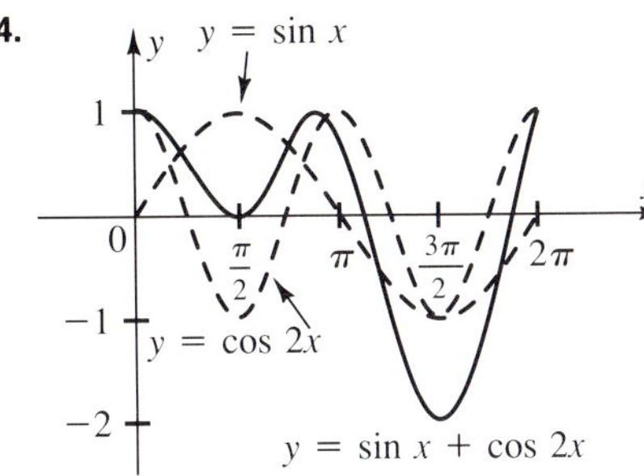

15.
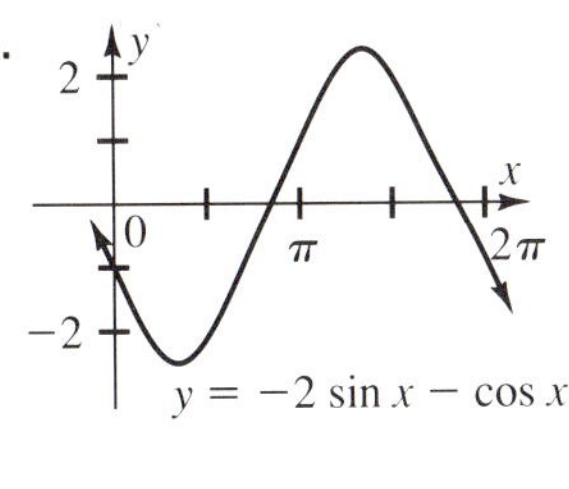

16.
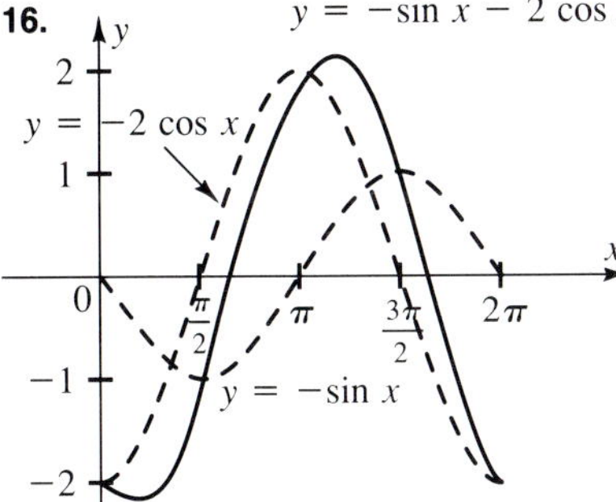

17.
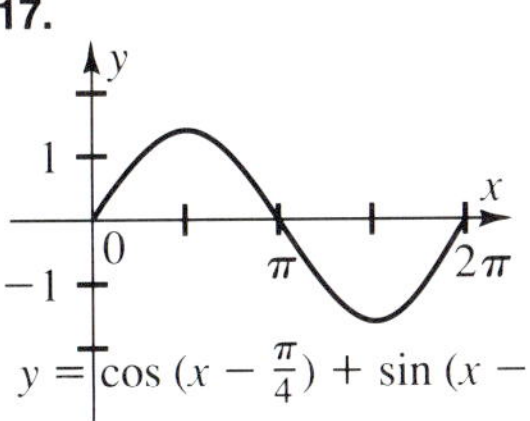

18.
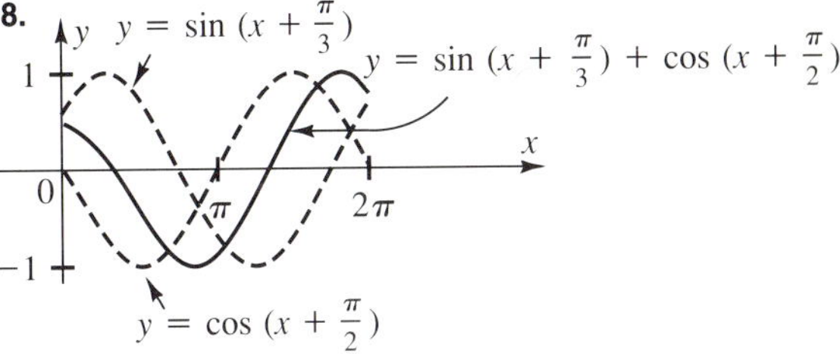

19.
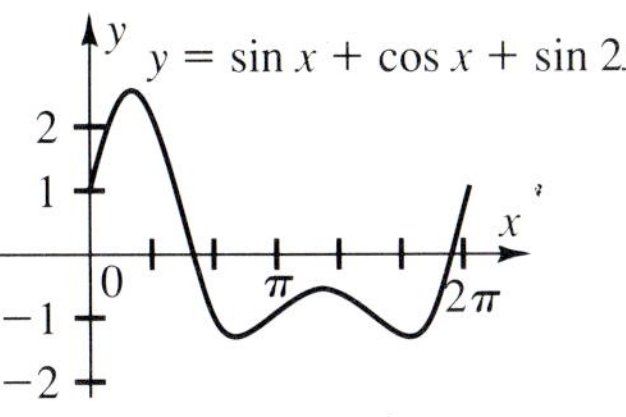

20.
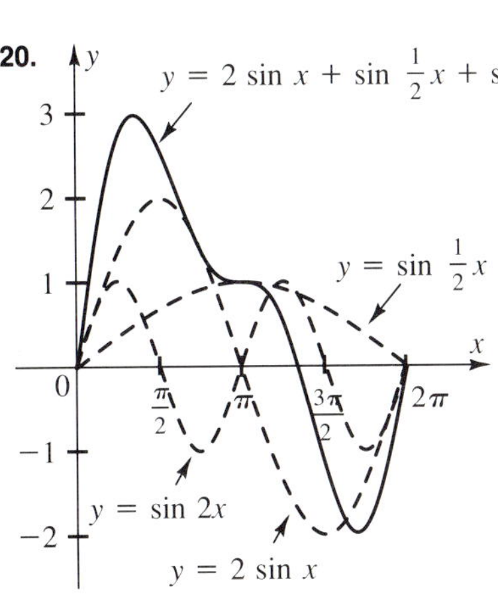

21.
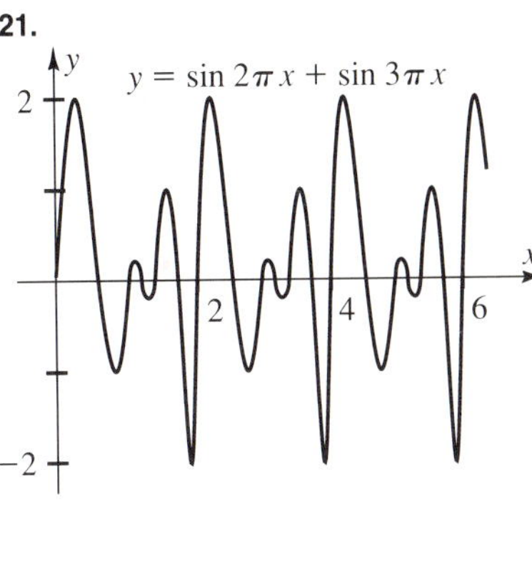

22.
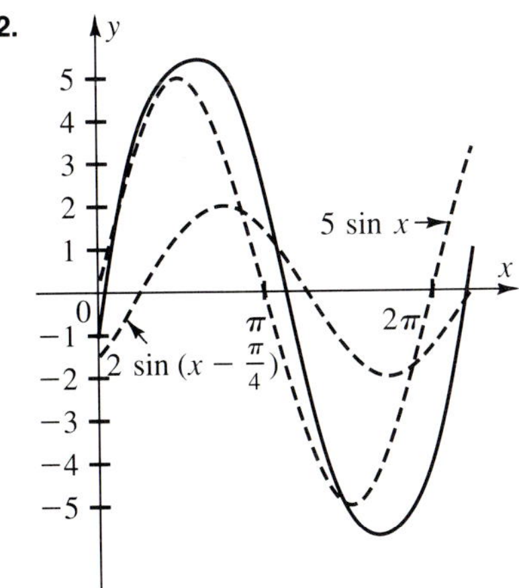

Preview, page 88

1.
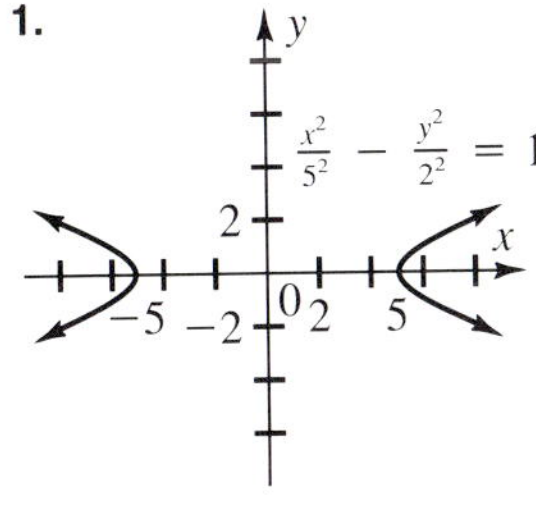

2.
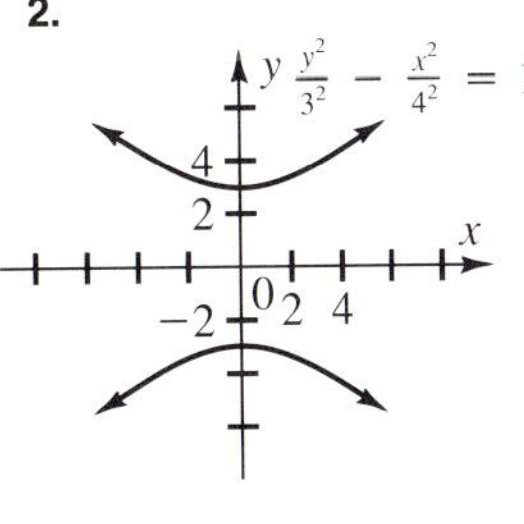

3.
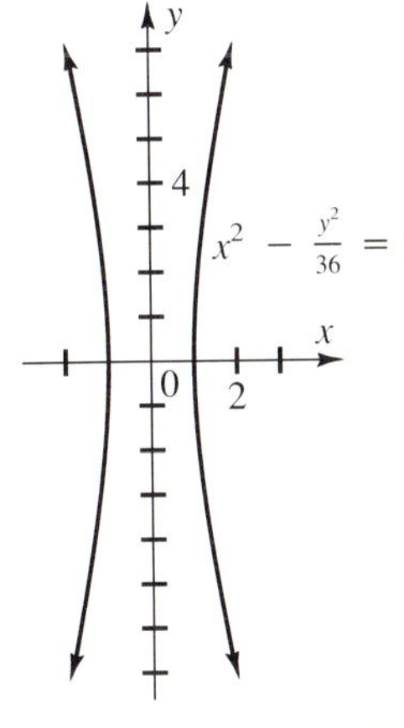

Answers

Practice Exercises, page 92

15.

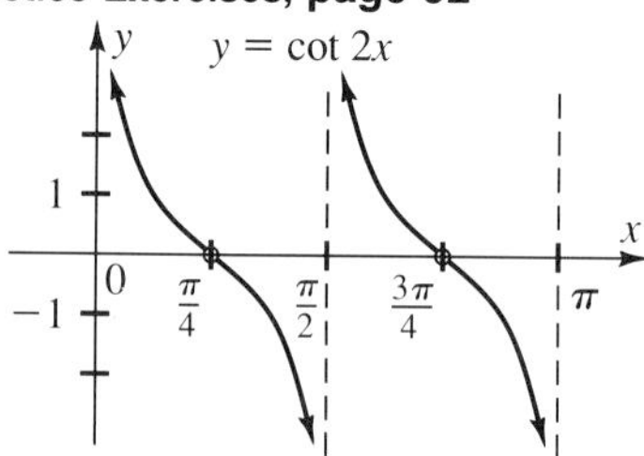

16.

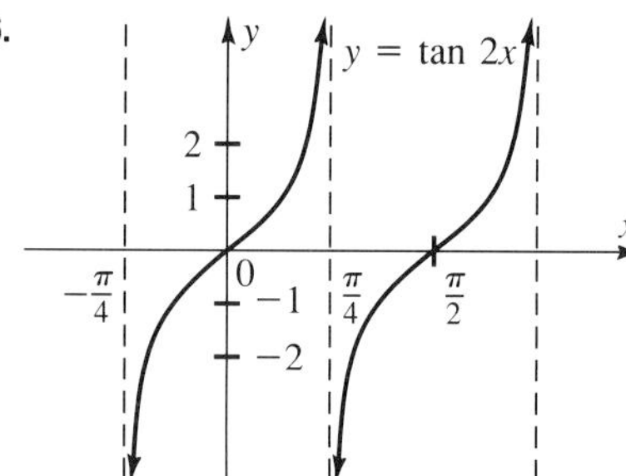

17.

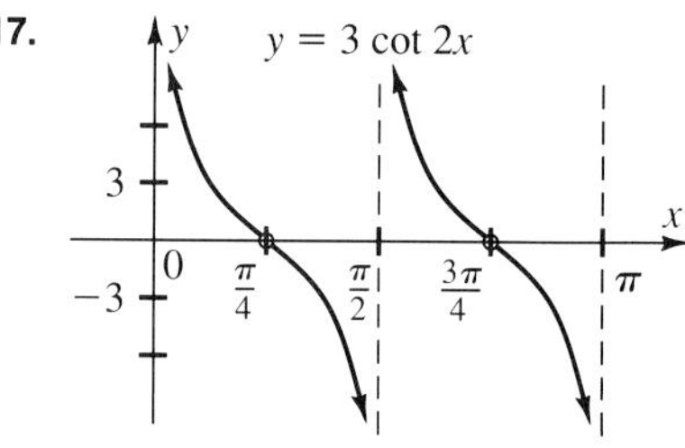

18.

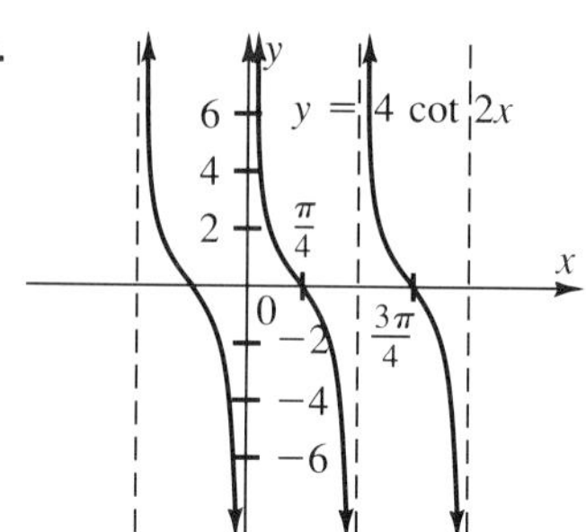

19.

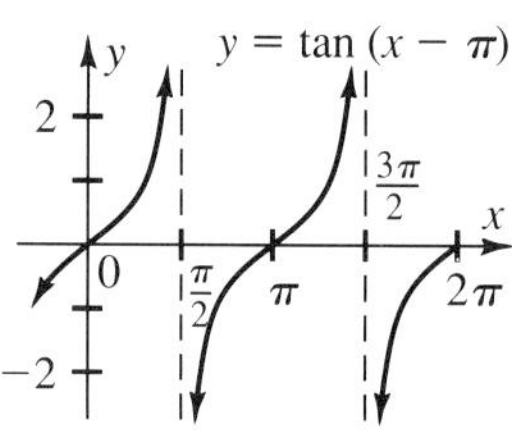

20.

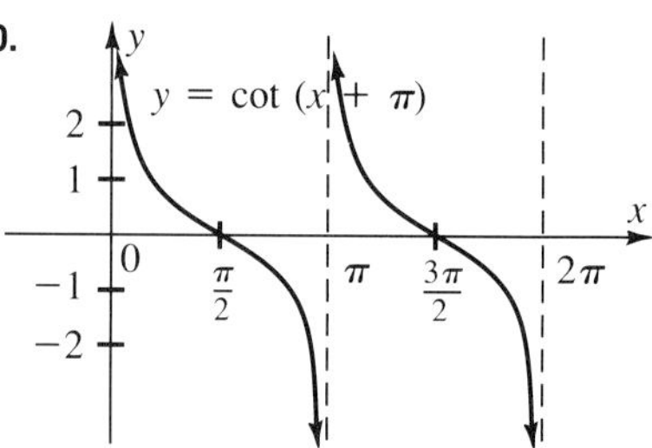

21.

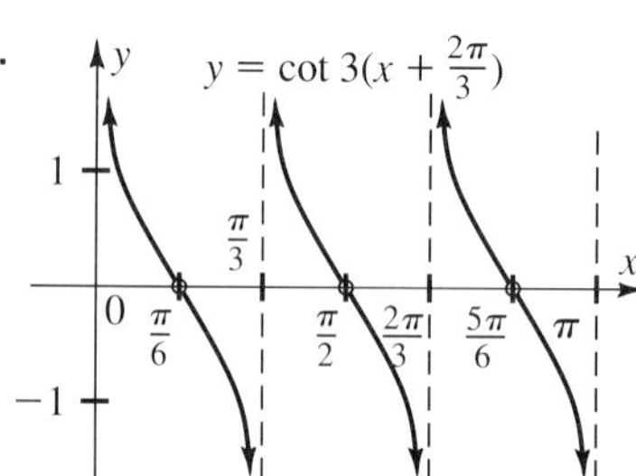

22.

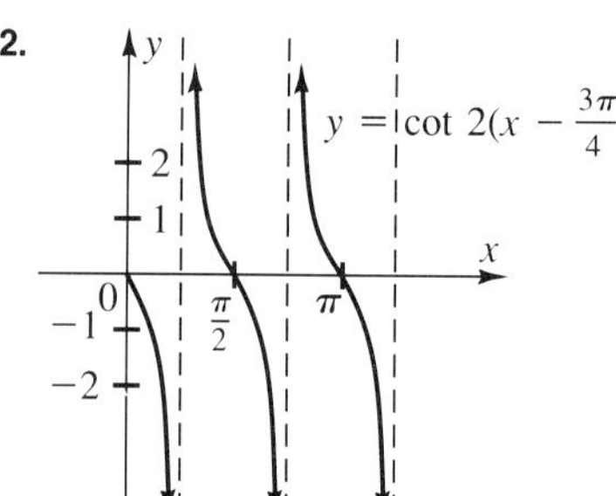

23.

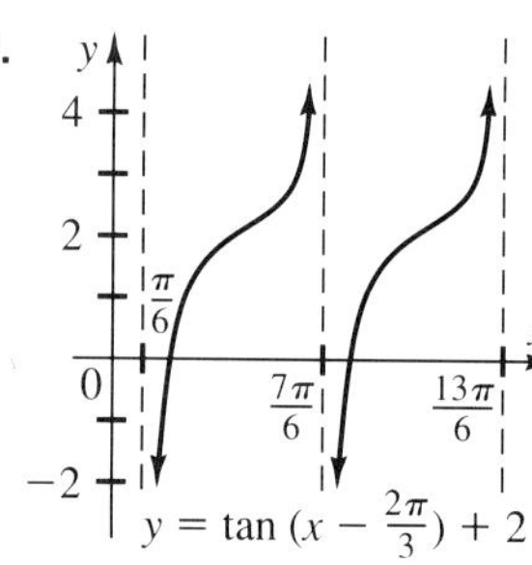

24.

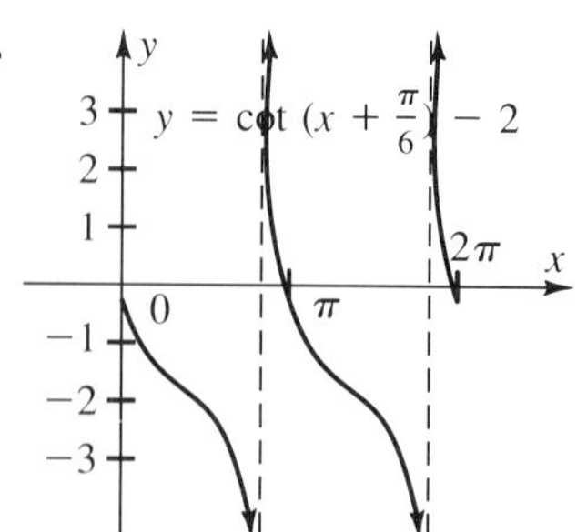

25.

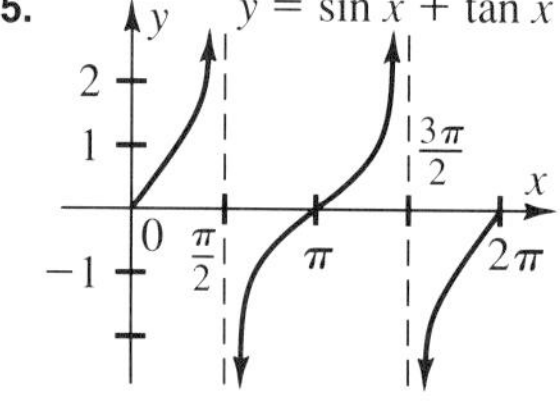

26.

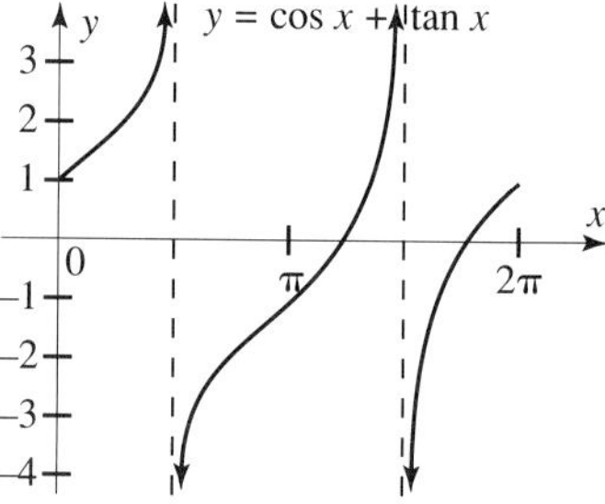

27.

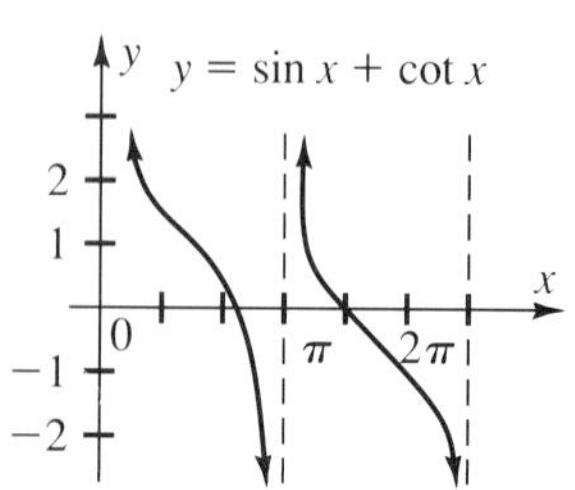

28.

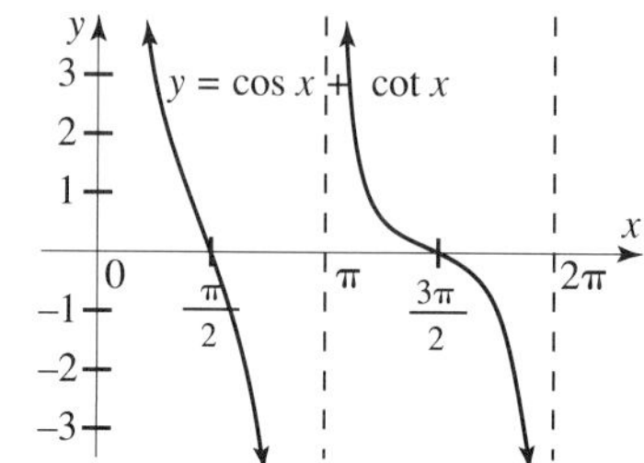

Practice Exercises, pages 98–99

15.

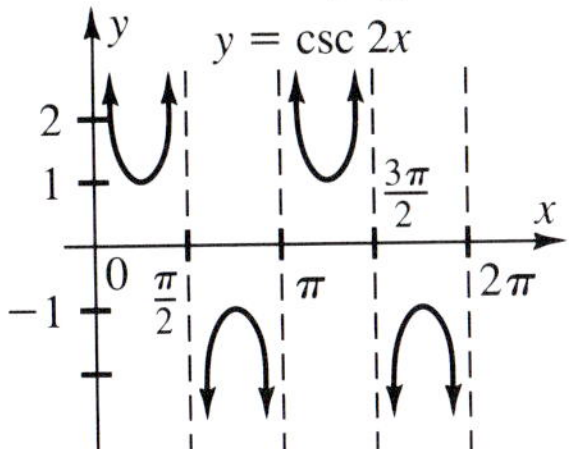

16.

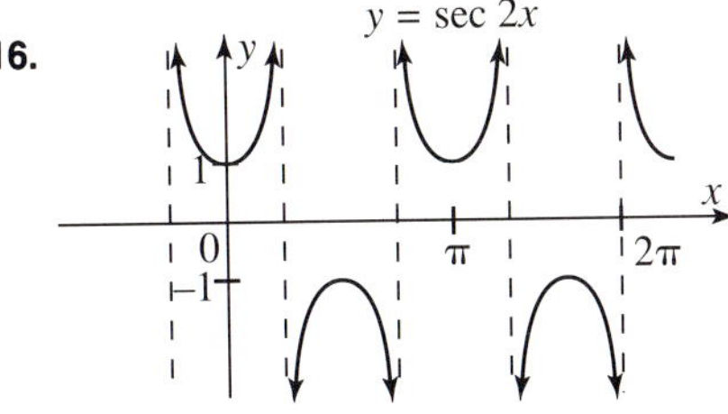

17.

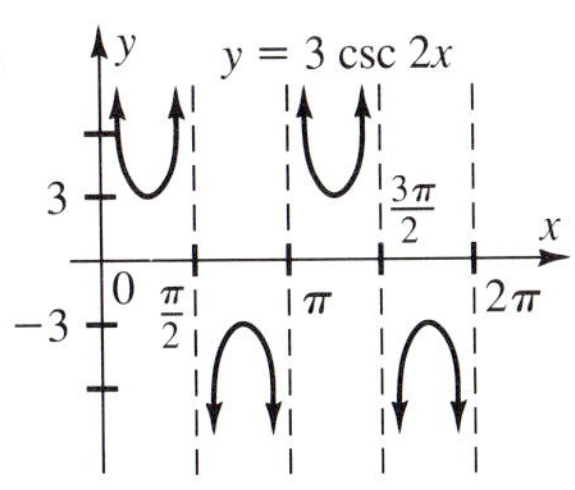

18.

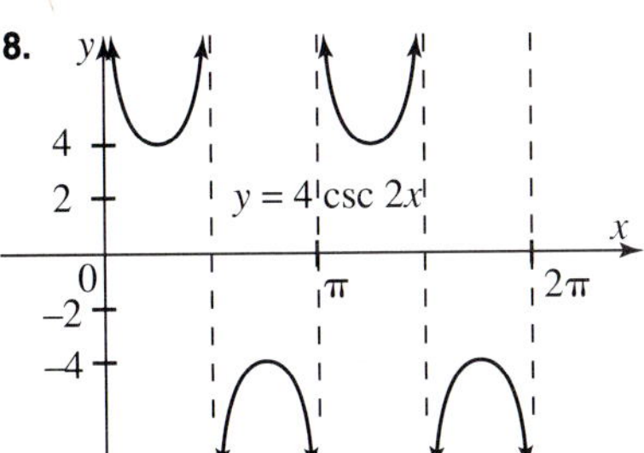

19.

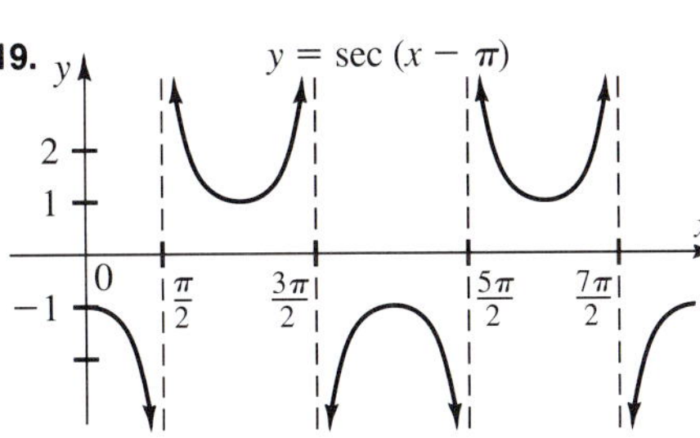

20.

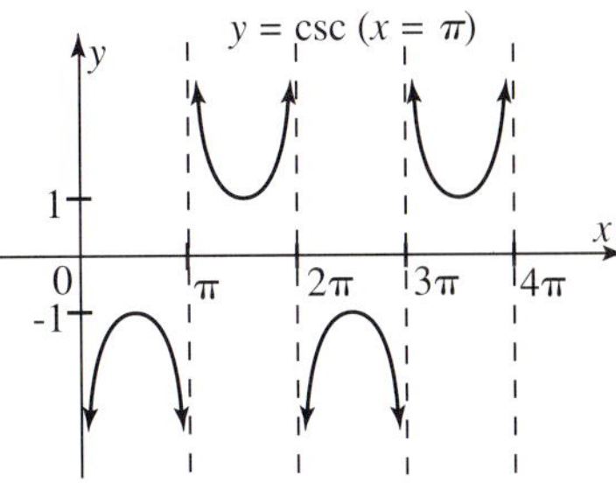

21.

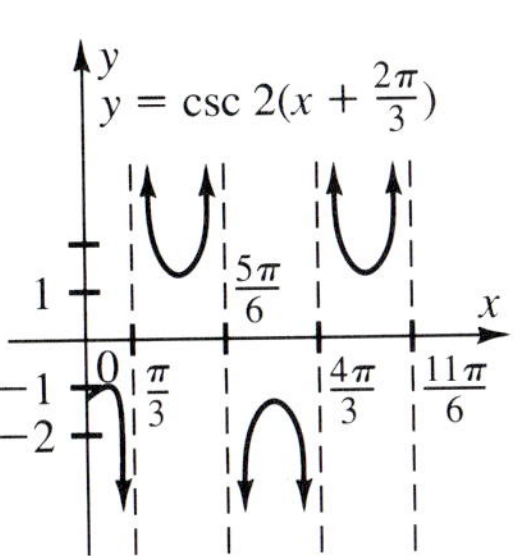

22.

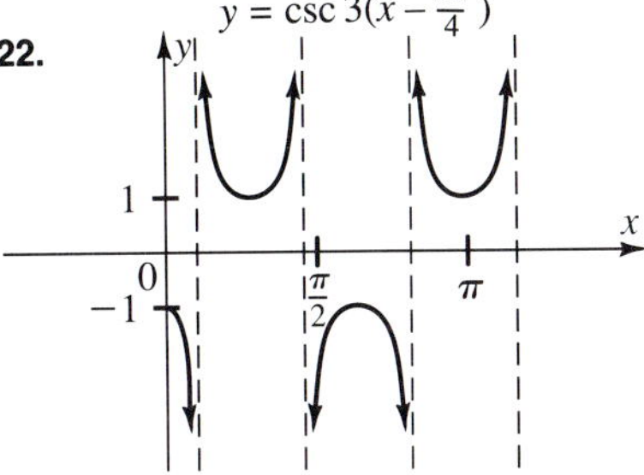

23.

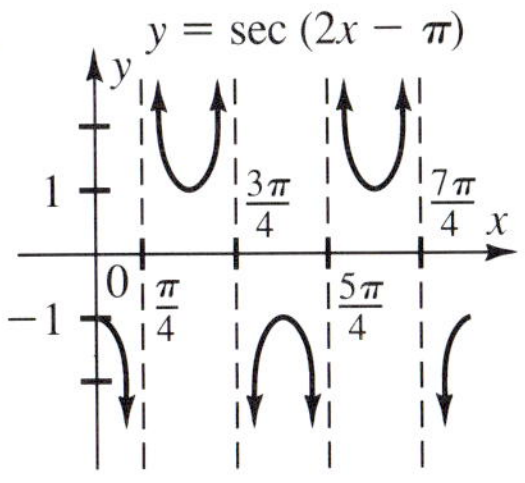

24.

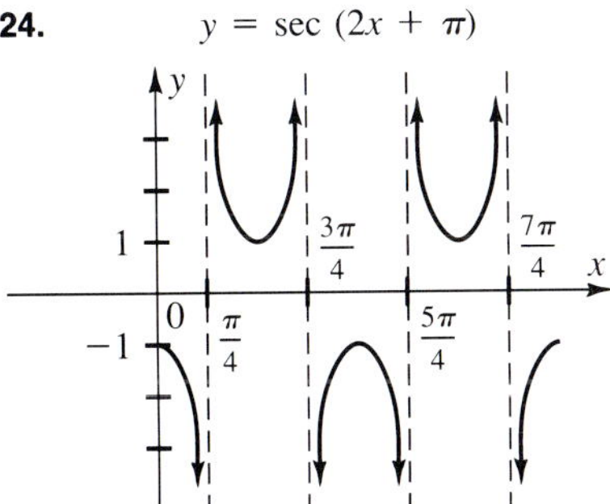

25.

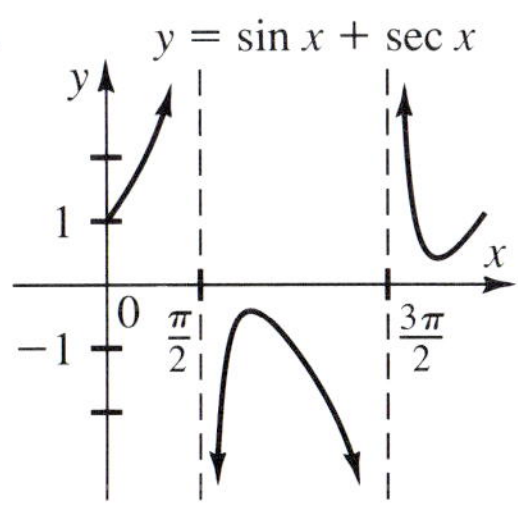

26.

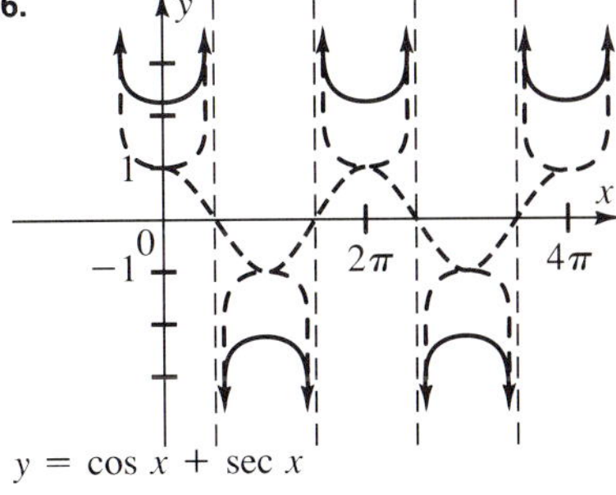

27.

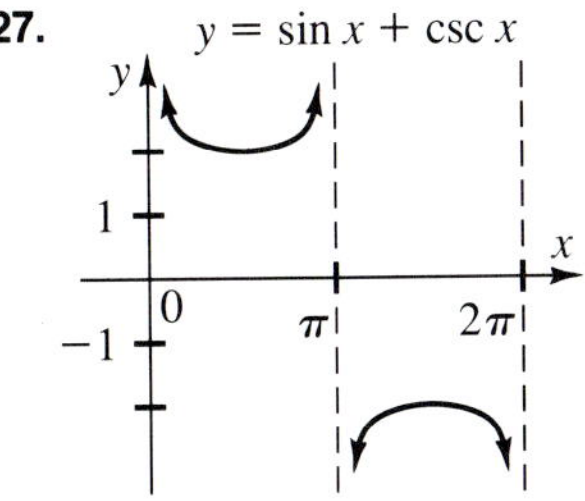

28.

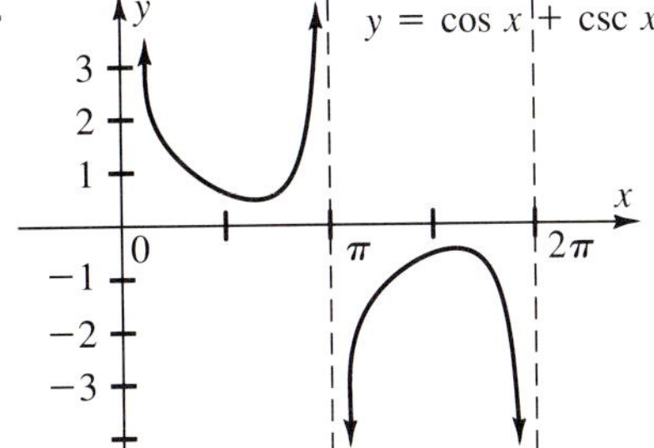

Test Yourself, page 104

1.

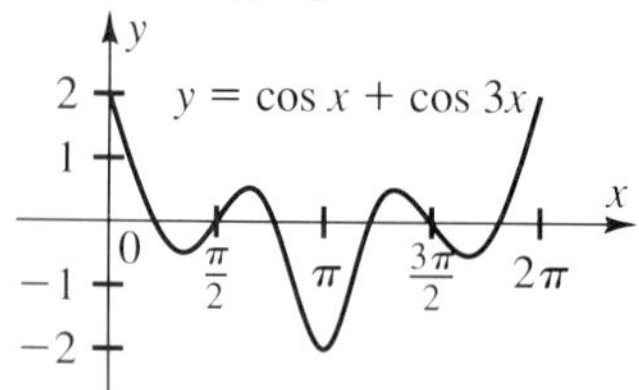

2.

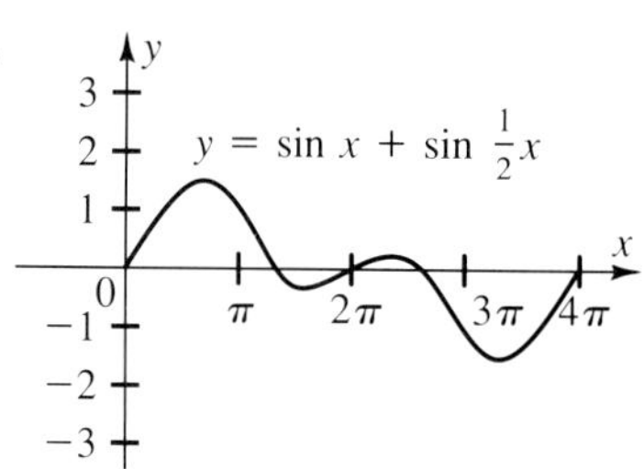

11.

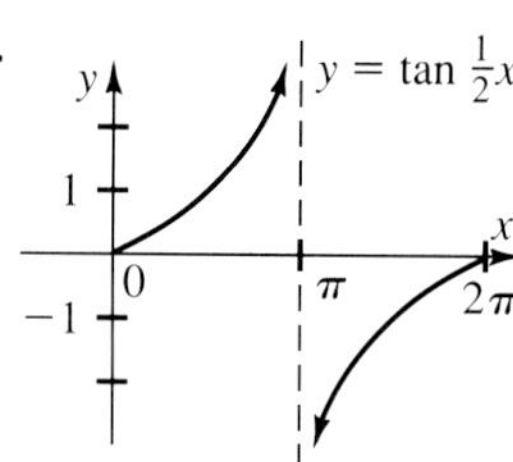

12.

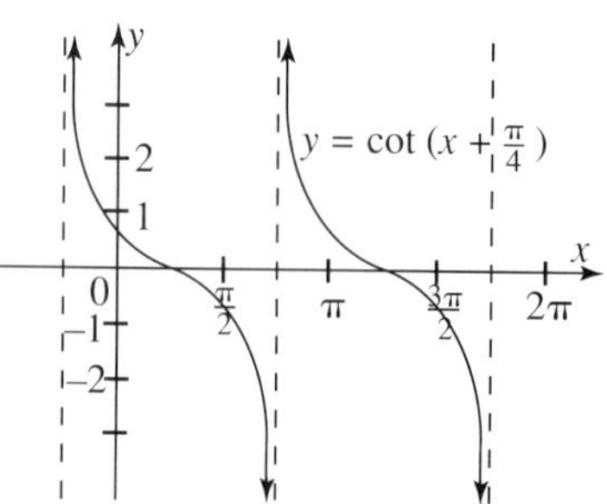

13.

$y = \sec \frac{1}{2}x$

14.

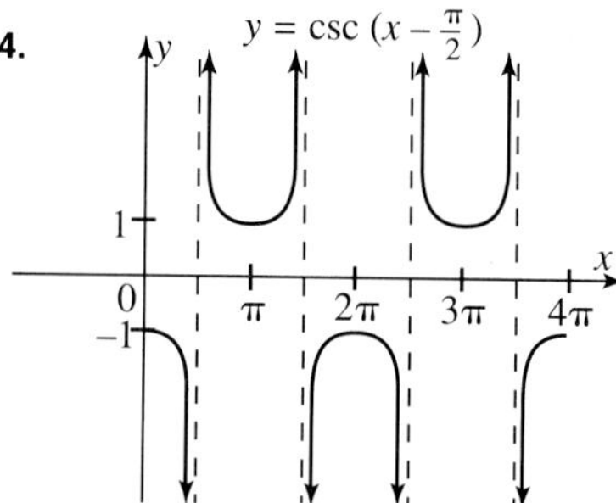

Summary and Review, page 107

7.

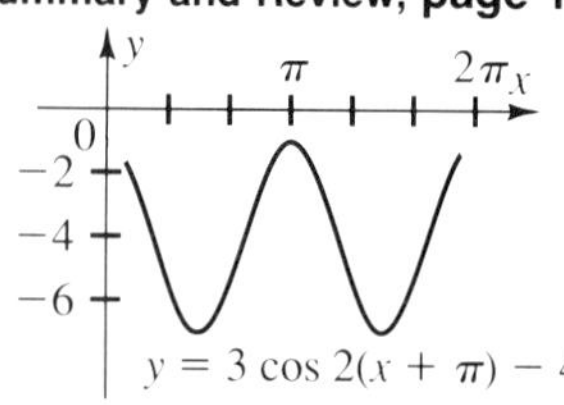

8.

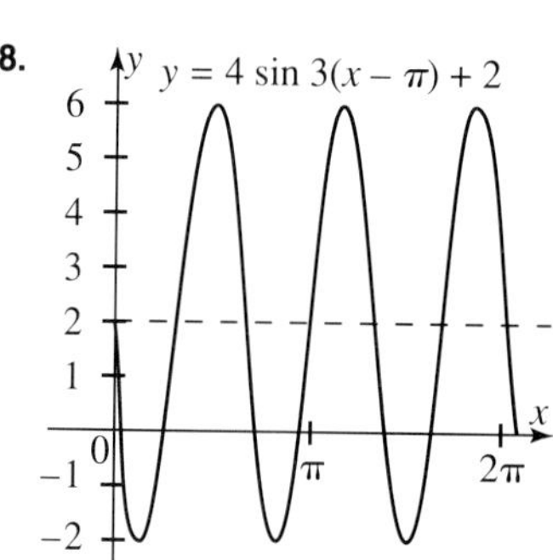

9.

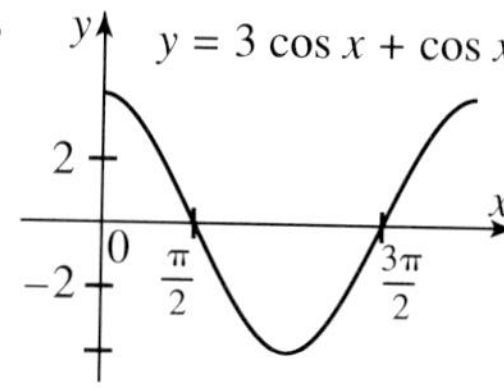

10.

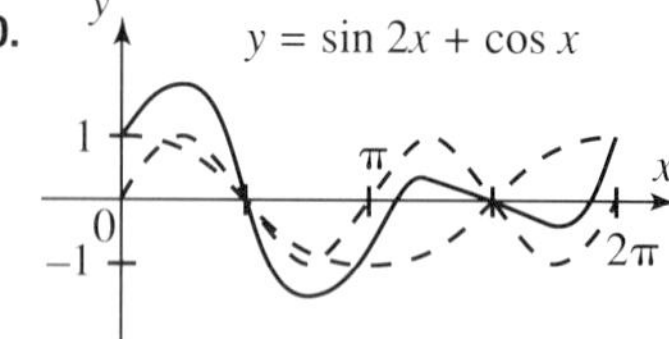

11.

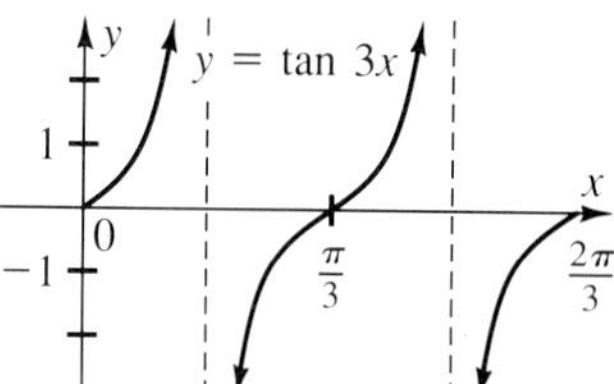

12.

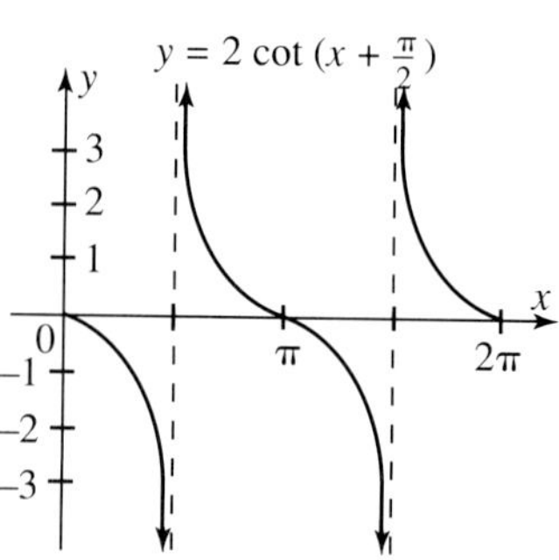

13.

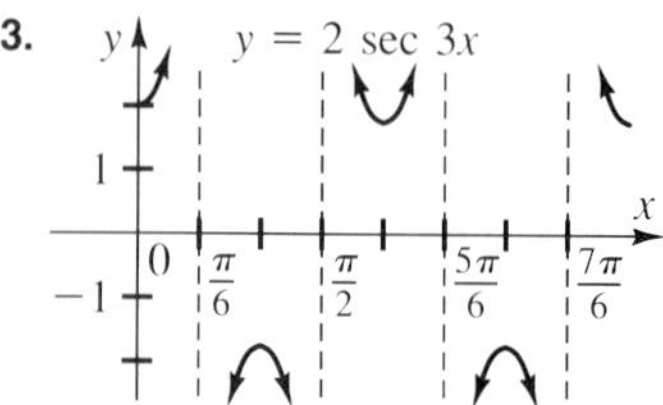

14.

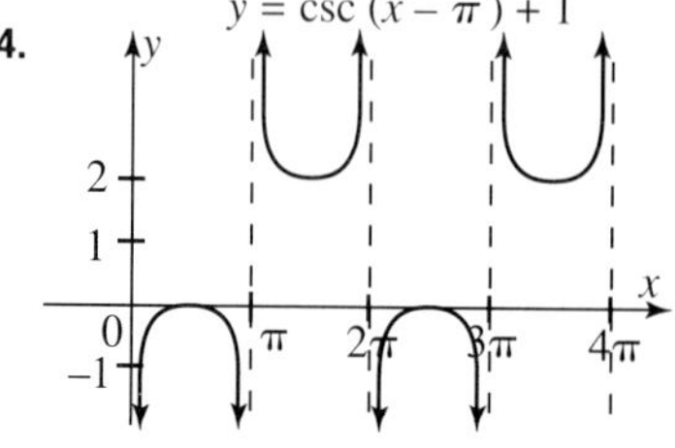

Chapter Test, page 108

10.

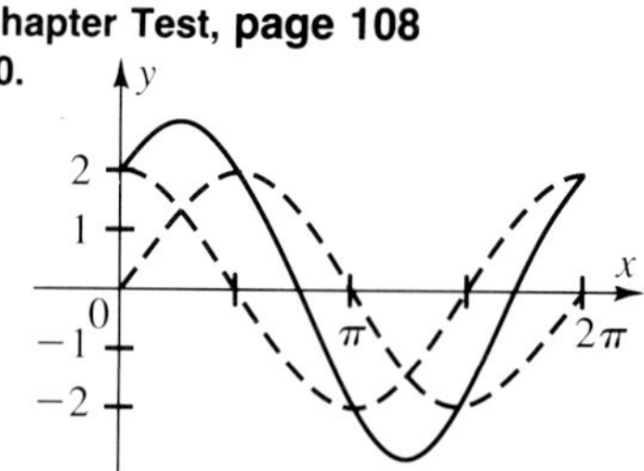

Chapter 3

Practice Exercises, page 143

1. $\frac{1-\cos^2\theta}{\cos^2\theta} = \frac{\sin^2\theta}{\cos^2\theta} = \left(\frac{\sin\theta}{\cos\theta}\right)^2$ **2.** $\frac{\sin^2\theta+\cos^2\theta}{\cos^2\theta} = \frac{1}{\cos^2\theta} = \left(\frac{1}{\cos\theta}\right)^2$ **3.** $\frac{\cos\alpha}{1-\sin^2\alpha} = \frac{\cos\alpha}{\cos^2\alpha} = \frac{1}{\cos\alpha}$

4. $\frac{1-\cos^2\lambda}{\sin\lambda} = \frac{\sin^2\lambda}{\sin\lambda} = \sin\lambda$ **5.** $\frac{\cot\alpha}{\sin^2\alpha+\cos^2\alpha} = \frac{\cot\alpha}{1} = \cot\alpha = \frac{1}{\tan\alpha}$ **6.** $\frac{1+\cot^2\alpha}{\csc^2\alpha} = \frac{\csc^2\alpha}{\csc^2\alpha} = 1$

7. $\frac{1+\cos\mu}{\cos\mu} = \frac{1}{\cos\mu}+\frac{\cos\mu}{\cos\mu} = \sec\mu+1$ **8.** $\frac{\csc^2\mu-1}{\csc^2\mu} = \frac{\csc^2\mu}{\csc^2\mu}-\frac{1}{\csc^2\mu} = 1-\sin^2\mu = \cos^2\mu$ **9.** $\frac{\sin^2\alpha}{1-\sin^2\alpha} = \frac{\sin^2\alpha}{\cos^2\alpha} = \tan^2\alpha = \sec^2\alpha-1$ **10.** $\frac{\csc\alpha-\sec\alpha}{\sec\alpha} = \frac{\csc\alpha}{\sec\alpha}-1 = \cot\alpha-1$ **11.** $\sin^2\theta\cdot\cos\theta\sec\theta = \sin^2\theta\cdot 1 = 1-\cos^2\theta$ **12.** $\sin\phi+\cos\phi\cdot\cot\phi = \frac{\sin^2\phi}{\sin\phi}+\frac{\cos^2\phi}{\sin\phi} = \frac{\sin^2\phi+\cos^2\phi}{\sin\phi} = \frac{1}{\sin\phi}$ **13.** $1-\frac{1}{\sec^2\beta} = 1-\cos^2\beta = \sin^2\beta$ **14.** $\frac{\cos\beta}{\sec\beta}-\frac{\cot\beta}{\tan\beta} = \cos^2\beta-\cot^2\beta = \cos^2\beta\left(1-\frac{1}{\sin^2\beta}\right) = \cos^2\beta(1-\csc^2\beta) = \cos^2\beta(-\cot^2\beta) = -\cos^2\beta\cot^2\beta$ **15.** $\frac{\cos\lambda}{\sec\lambda+\tan\lambda} = \frac{\cos\lambda}{\frac{1}{\cos\lambda}+\frac{\sin\lambda}{\cos\lambda}} = \frac{\cos^2\lambda}{1+\sin\lambda} = \frac{1-\sin^2\lambda}{1+\sin\lambda} = \frac{(1+\sin\lambda)(1-\sin\lambda)}{1+\sin\lambda} = 1-\sin\lambda$

16. $\frac{1+\cot^2\alpha}{\csc\alpha} = \frac{\csc^2\alpha}{\csc\alpha} = \csc\alpha$ **17.** $\cos^2\beta+\cot^2\beta+\sin^2\beta = 1+\cot^2\beta = \csc^2\beta$ **18.** $\frac{1}{\tan\beta+\cot\beta} = \frac{1}{\frac{\sin\beta}{\cos\beta}+\frac{\cos\beta}{\sin\beta}} = \frac{\sin\beta\cos\beta}{\sin^2\beta+\cos^2\beta} = \sin\beta\cos\beta$

19. $\cos^4\theta-\sin^4\theta = (\cos^2\theta-\sin^2\theta)(\cos^2\theta+\sin^2\theta) = \cos^2\theta-\sin^2\theta = \cos^2\theta-(1-\cos^2\theta) = 2\cos^2\theta-1$ **20.** $(\sin\lambda-\tan\lambda)^2 = \left(\sin\lambda-\frac{\sin\lambda}{\cos\lambda}\right)^2 = \left(\frac{\sin\lambda\cos\lambda-\sin\lambda}{\cos\lambda}\right)^2 = \left(\frac{\sin\lambda(\cos\lambda-1)}{\cos\lambda}\right)^2 = \frac{\sin^2\lambda(\cos\lambda-1)^2}{\cos^2\lambda} = \tan^2\lambda(\cos\lambda-1)^2$ **21.** $\frac{\cos\lambda}{1+\sin\lambda}+\frac{\cos\lambda}{1-\sin\lambda} = \frac{\cos\lambda(1-\sin\lambda)+\cos\lambda(1+\sin\lambda)}{1-\sin^2\lambda} = \frac{\cos\lambda-\cos\lambda\sin\lambda+\cos\lambda+\cos\lambda\sin\lambda}{\cos^2\lambda} = \frac{2\cos\lambda}{\cos^2\lambda} = \frac{2}{\cos\lambda}$

22. $\frac{\sec\lambda}{\cos\lambda}-1 = \frac{1}{\cos^2\lambda}-\frac{\cos^2\lambda}{\cos^2\lambda} = \frac{1-\cos^2\lambda}{\cos^2\lambda} = \frac{\sin^2\lambda}{\cos^2\lambda}$

23. $\frac{\sin^4\theta-1}{\cos^2\theta} = \frac{(\sin^2\theta-1)(\sin^2\theta+1)}{\cos^2\theta} = \frac{(\sin^2\theta-1)(\sin^2\theta+1)}{-1(\sin^2\theta-1)} = -(\sin^2\theta+1) = -\sin^2\theta-1 = -(1-\cos^2\theta)-1 = -1+\cos^2\theta-1 = \cos^2\theta-2$

24. $\frac{\sin\theta-1}{1-\sin^2\theta} = \frac{\sin\theta-1}{(1+\sin\theta)(1-\sin\theta)} = \frac{-1}{1+\sin\theta} = \frac{-1}{1+\frac{1}{\csc\theta}} = \frac{-1}{\frac{\csc\theta+1}{\csc\theta}} = \frac{\csc\theta}{-\csc-1}$ **25.** $\tan^2\alpha+\cos^2\alpha-1 = \tan^2\alpha+(\cos^2\alpha-1) = \tan^2\alpha+(-\sin^2\alpha) = \tan^2\alpha-\sin^2\alpha$ **26.** $\frac{1}{\csc\lambda-\cot\lambda}+\frac{1}{\csc\lambda+\cot\lambda} = \frac{\csc\lambda+\cot\lambda+\csc\lambda-\cot\lambda}{\csc^2\lambda-\cot^2\lambda} = \frac{2\csc\lambda}{1} = 2\csc\lambda$ **27.** $(\tan\alpha-\sec\alpha)^2 = \tan^2\alpha-2\tan\alpha\cdot\sec\alpha+\sec^2\alpha = \frac{\sin^2\alpha}{\cos^2\alpha}-\frac{2\sin\alpha}{\cos\alpha}\cdot\frac{1}{\cos\alpha}+\frac{1}{\cos^2\alpha} = \frac{\sin^2\alpha-2\sin\alpha+1}{\cos^2\alpha} = \frac{1-2\sin\alpha+\sin^2\alpha}{1-\sin^2\alpha} = \frac{(1-\sin\alpha)^2}{(1-\sin\alpha)(1+\sin\alpha)} = \frac{1-\sin\alpha}{1+\sin\alpha}$ **28.** $(\sin\theta+\cos\theta)^2\cdot\tan\theta = (\sin^2\theta+2\sin\theta\cos\theta+\cos^2\theta)(\tan\theta) = (1+2\sin\theta\cdot\cos\theta)\tan\theta = \tan\theta+2\sin\theta\cos\theta\tan\theta = \tan\theta+2\sin^2\theta$

29. $(1+\sin\theta+\cos\theta)(1-\sin\theta-\cos\theta) = 1+\sin\theta+\cos\theta-\sin\theta-\sin^2\theta-\sin\theta\cos\theta-\cos\theta-\sin\theta\cos\theta-\cos^2\theta = 1-(\sin^2\theta+\cos^2\theta+2\sin\theta\cos\theta) = 1-(1+2\cdot\sin\theta\cos\theta) = -2\sin\theta\cos\theta$ **30.** $\frac{\sec^2\theta-\tan^2\theta+\tan\theta}{\cos\theta} = \frac{1+\tan^2\theta-\tan^2\theta+\tan\theta}{\cos\theta} = \frac{1+\tan\theta}{\cos\theta} = \sec\theta\cdot(\tan\theta+1) = \sec\theta\left(\frac{\sin\theta}{\cos\theta}+1\right) = \sec\theta\left(\sin\theta\cdot\frac{1}{\cos\theta}+1\right) = \sec\theta(\sin\theta\sec\theta+1)$ **31.** $(\cos\alpha)(\sec^3\alpha+\cot\alpha) = \sec^2\alpha+\cos\alpha\cot\alpha = \sec^2\alpha+\frac{\cos^2\alpha}{\sin\alpha} = \frac{\tan^2\alpha}{\sin^2\alpha}+\frac{\sin\alpha}{\tan^2\alpha} = \frac{\tan^2\alpha}{1-\cos^2\alpha}+\frac{\sin\alpha}{\sec^2\alpha-1}$ **32.** $\frac{\cot\alpha}{1-\sin^2\alpha}+\frac{\cos\alpha}{\csc^2\alpha-1} = \frac{\cot\alpha}{\cos^2\alpha}+\frac{\cos\alpha}{\cot^2\alpha} = \frac{\cot^3\alpha+\cos^3\alpha}{\cos^2\alpha\cot^2\alpha} = \frac{\cot^3\alpha+\cos^3\alpha}{\frac{\cos^4\alpha}{\sin^2\alpha}} = (\cot^3\alpha+\cos^3\alpha)\cdot\frac{\sin^2\alpha}{\cos^4\alpha} = \frac{1}{\sin\alpha\cos\alpha}+\frac{\sin^2\alpha}{\cos\alpha} = \frac{1}{\cos\alpha}\left(\frac{1}{\sin\alpha}+\sin^2\alpha\right) = (\sec\alpha)(\sin^2\alpha+\csc\alpha)$

33. $\sqrt{(5\cos\theta+12\sin\theta)^2+(12\cos\theta-5\sin\theta)^2} = \sqrt{25(\cos^2\theta+\sin^2\theta)+144(\cos^2\theta+\sin^2\theta)} = \sqrt{25+144} = \sqrt{169} = 13$ **34.** $\frac{-2\sin\alpha\cos\alpha}{1-\sin\alpha-\cos\alpha} = \frac{(-1)2\sin\alpha\cos\alpha}{1-\sin\alpha-\cos\alpha} = \frac{(-1)[1+2\sin\alpha\cos\alpha-1]}{1-\sin\alpha-\cos\alpha} = \frac{(-1)[(\sin^2\alpha+2\sin\alpha\cos\alpha+\cos^2\alpha)-1]}{1-\sin\alpha-\cos\alpha} = \frac{(-1)[(\sin\alpha+\cos\alpha)^2-1]}{1-\sin\alpha-\cos\alpha} =$

$$\frac{(-1)(\sin\alpha+\cos\alpha+1)(\sin\alpha+\cos\alpha-1)}{1-\sin\alpha-\cos\alpha}=$$
$$\frac{(\sin\alpha+\cos\alpha+1)(1-\sin\alpha-\cos\alpha)}{1-\sin\alpha-\cos\alpha}=1+\sin\alpha+\cos\alpha$$

35. $\frac{2\sin\theta\cos\theta}{\sin\theta+\cos\theta-1}=\frac{[2\sin\theta\cos\theta]}{[\sin\theta+\cos\theta-1]}\cdot$
$\frac{[(\sin\theta+\cos\theta)+1]}{[(\sin\theta+\cos\theta)+1]}=\frac{2\sin\theta\cos\theta(\sin\theta+\cos\theta+1)}{(\sin\theta+\cos\theta)^2-1}=$
$\frac{2\sin\theta\cos\theta(\sin\theta+\cos\theta+1)}{\sin^2\theta+2\sin\theta\cos\theta+\cos^2\theta-1}=$
$\frac{2\sin\theta\cos\theta(\sin\theta+\cos\theta+1)}{1+2\sin\theta\cos\theta-1}=$
$\frac{2\sin\theta\cos\theta(\sin\theta+\cos\theta+1)}{2\sin\theta\cos\theta}=\sin\theta+\cos\theta+1$

36. $\frac{\sin\alpha-\cos\alpha-1}{\sin\alpha+\cos\alpha-1}=$
$\frac{[(\sin\alpha-\cos\theta)-1][(\sin\alpha+\cos\alpha)+1]}{[(\sin\alpha+\cos\alpha)-1][(\sin\alpha+\cos\alpha)+1]}=$
$\frac{(\sin\alpha-\cos\alpha)(\sin\alpha+\cos\alpha)+(\sin\alpha-\cos\alpha)-(\sin\alpha+\cos\alpha)-1}{(\sin\alpha+\cos\alpha)^2-1}=$
$\frac{\sin^2\alpha-\cos^2\alpha+\sin\alpha-\cos\alpha-\sin\alpha-\cos\alpha-1}{\sin^2\alpha+2\sin\alpha\cos\alpha+\cos^2\alpha-1}=$
$\frac{\sin^2\alpha-1-\cos^2\alpha-2\cos\alpha}{1+2\sin\alpha\cos\alpha-1}=$
$\frac{(1-\cos^2\alpha)-1-\cos^2\alpha-2\cos\alpha}{2\sin\alpha\cos\alpha}=\frac{-2\cos^2\alpha-2\cos\alpha}{2\sin\alpha\cos\alpha}=$
$\frac{-2\cos\alpha(\cos\alpha+1)}{2\sin\alpha\cos\alpha}=-\frac{\cos\alpha+1}{\sin\alpha}$ **37.** Answers may vary. **38.** 90°, 270°; $\tan x$ is undefined for $x=90°$ and $x=270°$. **39.** $\cos x+\sin x\tan x=\cos x+(\sin x)\left(\frac{\sin x}{\cos x}\right)=$
$\frac{\cos^2x+\sin^2x}{\cos x}=\frac{1}{\cos x}=\sec x$

Chapter 3

Preview, page 145

1.

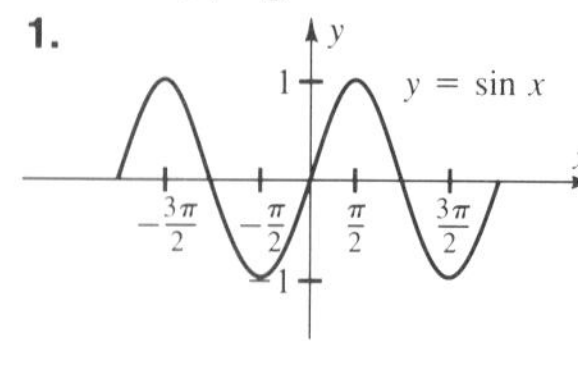

2.

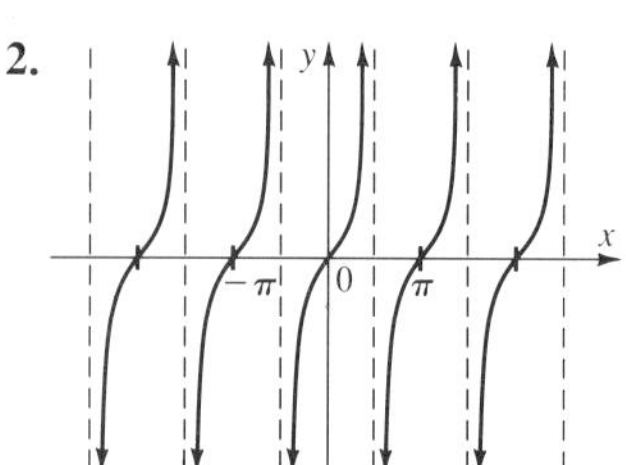

3.

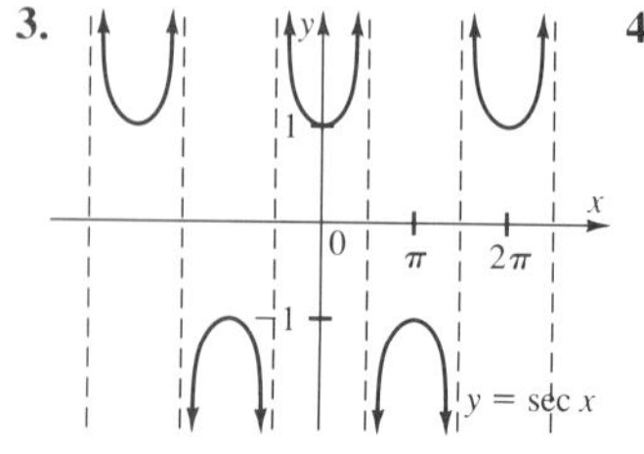

4.

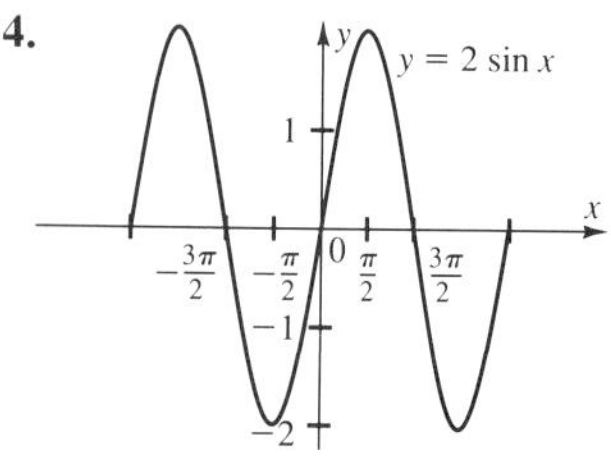

5.

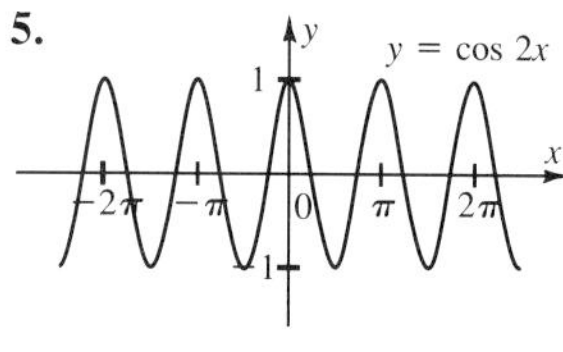

6.

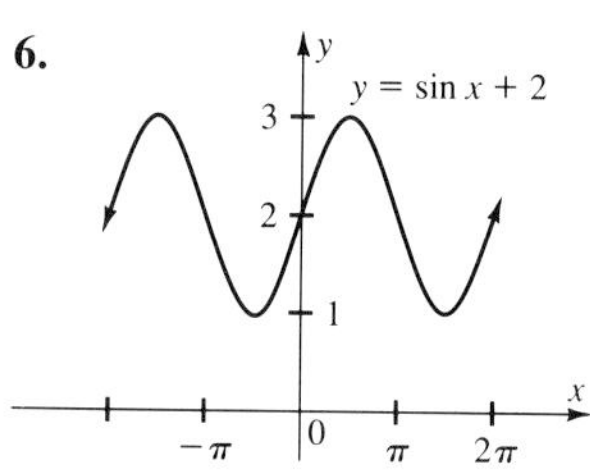

Chapter 4

Practice Exercises, pages 172–174

32. $h^2+(x+a)^2=c^2$, $h^2+x^2+2ax+a^2=c^2$. Since $h^2+x^2=b^2$ and $\frac{x}{b}=\cos(180°-C)$, $x=b\cos(180°-C)$, $x=b(-\cos C)$, $x=-b\cos C$. So, $c^2=h^2+x^2+2ax+a^2$, $c^2=b^2+2a(-b\cos C)+a^2$. Therefore, $c^2=a^2+b^2-2ab\cos C$ **33.** If $\angle C$ is acute, $\cos\angle C>0$. Since $a>0$ and $b>0$, $-2ab<0$. Therefore: $-2ab\cdot\cos\angle C<0\cdot-2ab$, $-2ab\cos\angle C<0$. Adding a^2+b^2 to each side: $a^2+b^2-2ab\cos\angle C<0+a^2+b^2$, $c^2<a^2+b^2$, $c<\sqrt{a^2+b^2}$. Similarly if $\angle C$ is obtuse, $\cos\angle C<0$, so $-2ab\cos\angle C>0$. Therefore: $a^2+b^2-2ab\cos\angle C>a^2+b^2$, $c^2>a^2+b^2$, $c>\sqrt{a^2+b^2}$.

Feature, page 174

The areas of the faces of the tetrahedron are: $A=\frac{1}{2}hr$, $B=\frac{1}{2}hr$, $C=\frac{1}{2}qr\sin\gamma$, and $D=\frac{1}{2}h_2p$. By the law of cosines $p=\sqrt{q^2+r^2-2qr\cos\gamma}$. Another equation for the area of triangle C is $C=\frac{1}{2}h_1p$. Therefore $\frac{1}{2}h_1p=\frac{1}{2}qr\sin\gamma$. Solving for h_1 gives $h_1=\frac{qr\sin\gamma}{p}$. Substituting for p gives $h_1=\frac{qr\sin\gamma}{\sqrt{q^2+r^2-2qr\cos\gamma}}$. Therefore $h_1{}^2=\frac{q^2r^2\sin^2\gamma}{q^2+r^2-2qr\cos\gamma}$. By the Pythagorean theorem, $h_2{}^2=h^2+h_1{}^2$. $A^2+B^2+C^2-2AB\cos\gamma=\frac{1}{4}h^2r^2+\frac{1}{4}h^2q^2+\frac{1}{4}qr\sin\gamma-2\left(\frac{1}{4}h^2qr\cos\gamma\right)$; $D^2=\frac{1}{4}h_2{}^2p^2=\frac{1}{4}(h^2+h_1{}^2)(q^2+r^2-2qr\cos\gamma)=\frac{1}{4}\left(h^2+\frac{q^2r^2\sin^2\gamma}{q^2+r^2-2qr\cos\gamma}\right)(q^2+r^2-2qr\cos\gamma)=\frac{1}{4}(h^2q^2+h^2r^2-2h^2qr\cos\gamma+q^2r^2\sin^2\gamma)=\frac{1}{4}h^2r^2+\frac{1}{4}h^2q^2+\frac{1}{4}q^2r^2\sin^2\gamma-2\left(\frac{1}{4}h^2qr\cos\gamma\right)$. Therefore $D^2=A^2+B^2+C^2-2AB\cos\gamma$.

Practice Exercises, pages 195–197

10.

11.

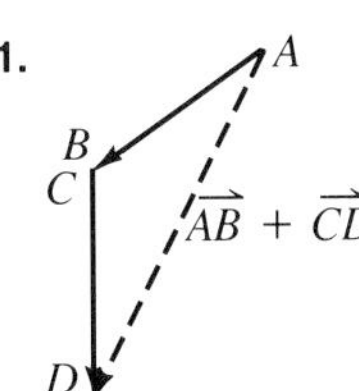

12.

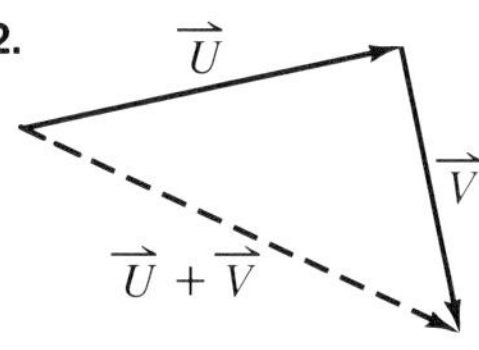

13.

14.

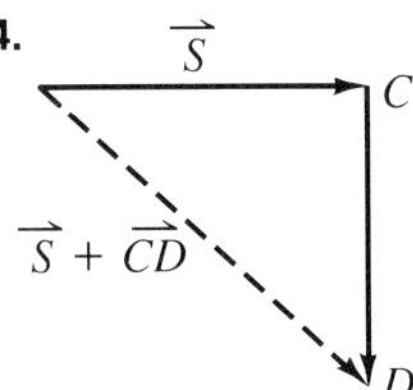

15.

16.

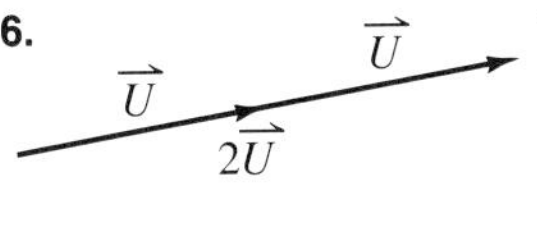

17.

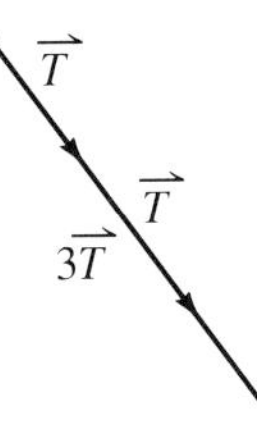

18.

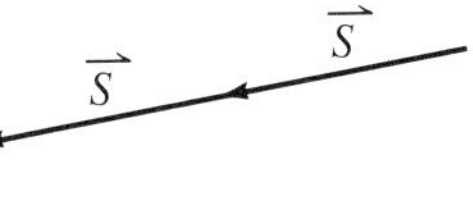

29.

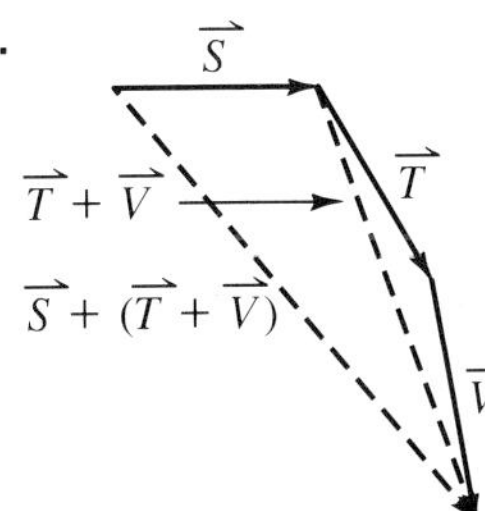

30.

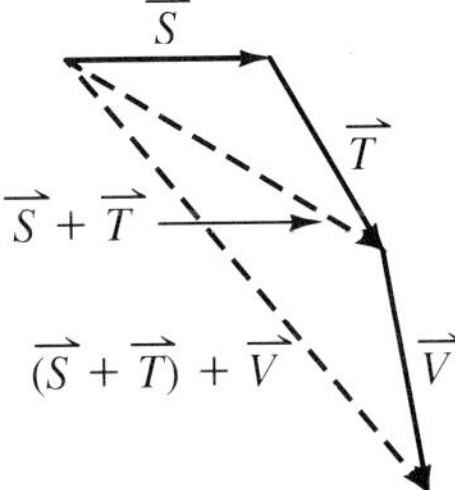

31.

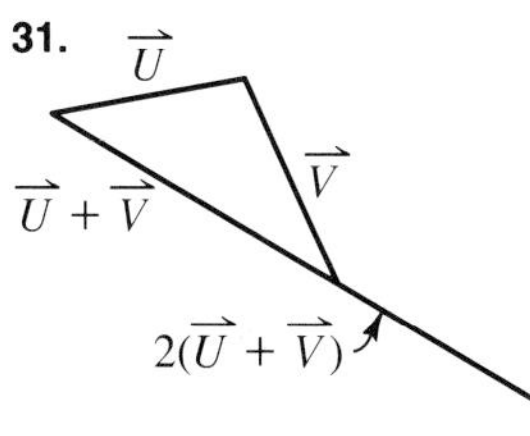

32.

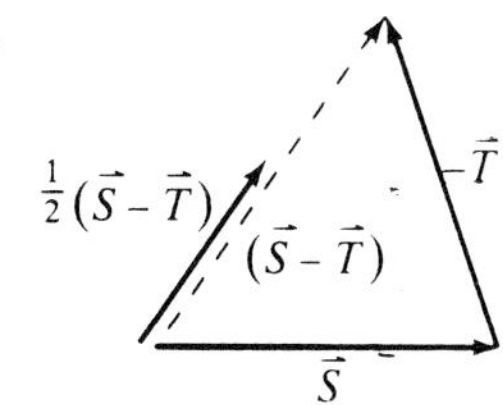

Chapter 5

Practice Exercises, pages 212–213

17. $\cos(360° + \alpha) = \cos 360° \cos\alpha - \sin 360° \sin\alpha = 1 \cdot \cos\alpha - 0(\sin\alpha) = \cos\alpha$ **18.** $\cos(180° + \alpha) = \cos 180° \cdot \cos\alpha - \sin 180° \sin\alpha = (-1)\cos\alpha - 0(\sin\alpha) = -\cos\alpha$ **19.** $\cos(270° - \beta) = \cos 270° \cos\beta + \sin 270° \sin\beta = 0 \cdot (\cos\beta) + (-1)\sin\beta = -\sin\beta$ **20.** $\cos(180° - \beta) = \cos 180° \cos\beta + \sin 180° \sin\beta = (-1)\cos\beta + 0(\sin\beta) = -\cos\beta$ **21.** $\cos\left(\frac{\pi}{3} - \beta\right) = \cos\frac{\pi}{3}\cos\beta + \sin\frac{\pi}{3}\sin\beta = \frac{1}{2}\cos\beta + \frac{\sqrt{3}}{2}\sin\beta = \frac{\cos\beta + \sqrt{3}\sin\beta}{2}$ **22.** $\cos\left(\beta - \frac{\pi}{6}\right) = \cos\beta\cos\frac{\pi}{6} + \sin\beta\sin\frac{\pi}{6} = \cos\beta \cdot \frac{\sqrt{3}}{2} + \frac{1}{2}\sin\beta = \frac{\sqrt{3}\cos\beta + \sin\beta}{2}$ **23.** $\cos\left(\frac{3\pi}{4} + \alpha\right) = \cos\frac{3\pi}{4}\cos\alpha - \sin\frac{3\pi}{4}\sin\alpha = -\frac{\sqrt{2}}{2}\cos\alpha - \frac{\sqrt{2}}{2}\sin\alpha = -\frac{\sqrt{2}}{2}(\cos\alpha + \sin\alpha)$ **24.** $\cos 2\alpha = \cos(\alpha + \alpha) = \cos\alpha\cos\alpha - \sin\alpha\sin\alpha = \cos^2\alpha - \sin^2\alpha$

Practice Exercises, pages 218–219

17. $\sin(180° - \alpha) = \sin 180° \cdot \cos\alpha - \cos 180° \cdot \sin\alpha = 0 \cdot \cos\alpha - (-1)\sin\alpha = \sin\alpha$ **18.** $\sin(\alpha - 180°) = \sin\alpha \cdot \cos 180° - \cos\alpha \cdot \sin 180° = \sin\alpha(-1) - \cos\alpha \cdot 0 = -\sin\alpha$ **19.** $\sin(210° + \alpha) = \sin 210° \cdot \cos\alpha + \cos 210° \cdot \sin\alpha = -\frac{1}{2}\cos\alpha - \frac{\sqrt{3}}{2}\sin\alpha = \frac{-\cos\alpha - \sqrt{3}\sin\alpha}{2}$
20. $\sin(270° + \beta) = \sin 270° \cdot \cos\beta + \cos 270° \cdot \sin\beta = -1 \cdot \cos\beta + 0 \cdot \sin\beta = -\cos\beta$ **21.** $\sin(45° + \alpha) = \sin 45° \cdot \cos\alpha + \cos 45° \cdot \sin\alpha = \frac{\sqrt{2}}{2}\cos\alpha + \frac{\sqrt{2}}{2}\sin\alpha = \frac{\sqrt{2}}{2}(\cos\alpha + \sin\alpha)$ **22.** $\sin\left(\frac{\pi}{2} + \beta\right) = \sin\frac{\pi}{2} \cdot \cos\beta + \cos\frac{\pi}{2} \cdot \sin\beta = 1 \cdot \cos\beta + 0 \cdot \sin\beta = \cos\beta$

23. $\sin\left(\frac{\pi}{3} - \beta\right) = \sin\frac{\pi}{3} \cdot \cos\beta - \cos\frac{\pi}{3} \cdot \sin\beta = \frac{\sqrt{3}}{2}\cos\beta - \frac{1}{2}\sin\beta = \frac{1}{2}(\sqrt{3}\cos\beta - \sin\beta)$ **24.** $\sin(2\beta) = \sin(\beta + \beta) - \sin\beta \cdot \cos\beta + \cos\beta \cdot \sin\beta = 2\sin\beta \cdot \cos\beta$ **25.** $\sin\left(\beta - \frac{7\pi}{6}\right) = \sin\beta \cdot \cos\frac{7\pi}{6} - \cos\beta \cdot \sin\frac{7\pi}{6} = \sin\beta\left(-\frac{\sqrt{3}}{2}\right) - \cos\beta\left(-\frac{1}{2}\right) = -\frac{\sqrt{3}}{2}\sin\beta + \frac{1}{2}\cos\beta = -\frac{1}{2}(\sqrt{3}\sin\beta - \cos\beta)$ **26.** $\sin(-\beta) = \sin(0° - \beta) = \sin 0° \cdot \cos\beta - \cos 0° \cdot \sin\beta = 0\cos\beta - 1 \cdot \sin\beta = -\sin\beta$

Practice Exercises, page 224

17. $\tan(45° - \alpha) = \frac{\tan 45° - \tan\alpha}{1 + \tan 45° \cdot \tan\alpha} = \frac{1 - \tan\alpha}{1 + \tan\alpha}$

18. $\tan(\alpha + 45°) = \dfrac{\tan\alpha + \tan 45°}{1 - \tan\alpha \cdot \tan 45°} = \dfrac{\tan\alpha + 1}{1 - \tan\alpha}$ **19.** $\tan(180° + \alpha) = \dfrac{\tan 180° + \tan\alpha}{1 - \tan 180° \cdot \tan\alpha} = \dfrac{0 + \tan\alpha}{1 - 0\tan\alpha} = \tan\alpha$ **20.** $\tan(180° - \alpha) = \dfrac{\tan 180° - \tan\alpha}{1 + \tan 180° \cdot \tan\alpha} = \dfrac{0 - \tan\alpha}{1 + 0\tan\alpha} = -\tan\alpha$ **21.** $\tan(360° + \alpha) = \dfrac{\tan 360° + \tan\alpha}{1 - \tan 360° \tan\alpha} = \dfrac{0 + \tan\alpha}{1 - 0 \cdot \tan\alpha} = \tan\alpha.$

22. $\tan(360° - \alpha) = \dfrac{\tan 360° - \tan\alpha}{1 + \tan 360° \tan\alpha} = \dfrac{0 - \tan\alpha}{1 + 0 \cdot \tan\alpha} = -\tan\alpha$ **23.** $\tan\left(\dfrac{3\pi}{4} + \alpha\right) = \dfrac{\tan\dfrac{3\pi}{4} + \tan\alpha}{1 - \tan\dfrac{3\pi}{4} \cdot \tan\alpha} = \dfrac{-1 + \tan\alpha}{1 - (-1)\tan\alpha} = \dfrac{\tan\alpha - 1}{\tan\alpha + 1}$ **24.** $\tan\left(\dfrac{3\pi}{4} - \alpha\right) = \dfrac{\tan\dfrac{3\pi}{4} - \tan\alpha}{1 + \tan\dfrac{3\pi}{4} \cdot \tan\alpha} = \dfrac{-1 - \tan\alpha}{1 + (-1)\tan\alpha} = \left(\dfrac{-1 - \tan\alpha}{1 - \tan\alpha}\right)\left(\dfrac{-1}{-1}\right) = \dfrac{1 + \tan\alpha}{-1 + \tan\alpha} = \dfrac{\tan\alpha + 1}{\tan\alpha - 1}$

25. $\tan\left(\alpha - \dfrac{5\pi}{4}\right) = \dfrac{\tan\alpha - \tan\dfrac{5\pi}{4}}{1 + \tan\alpha \cdot \tan\dfrac{5\pi}{4}} = \dfrac{\tan\alpha - 1}{1 + \tan\alpha\,(1)} = \dfrac{\tan\alpha - 1}{\tan\alpha + 1}$ **26.** $\tan\left(\alpha + \dfrac{7\pi}{4}\right) = \dfrac{\tan\alpha + \tan\dfrac{7\pi}{4}}{1 - \tan\alpha \cdot \tan\dfrac{7\pi}{4}} = \dfrac{\tan\alpha + (-1)}{1 - \tan\alpha\,(-1)} = \dfrac{\tan\alpha - 1}{\tan\alpha + 1}$

31. $\tan(\beta + 45°) + \tan(\beta - 45°) = \dfrac{\tan\beta + \tan 45°}{1 - \tan\beta \cdot \tan 45°} + \dfrac{\tan\beta - \tan 45°}{1 + \tan\beta \cdot \tan 45°} = \dfrac{\tan\beta + 1}{1 - \tan\beta} + \dfrac{\tan\beta - 1}{1 + \tan\beta} = \left(\dfrac{\tan\beta + 1}{1 - \tan\beta}\right)\left(\dfrac{1 + \tan\beta}{1 + \tan\beta}\right) + \left(\dfrac{\tan\beta - 1}{1 + \tan\beta}\right)\left(\dfrac{1 - \tan\beta}{1 - \tan\beta}\right) = \dfrac{\tan\beta + \tan^2\beta + 1 + \tan\beta + \tan\beta - \tan^2\beta - 1 + \tan\beta}{1 - \tan^2\beta} = \dfrac{4\tan\beta}{1 - \tan^2\beta} = \dfrac{2\tan\beta + 2\tan\beta}{1 - \tan\beta \cdot \tan\beta} = 2\tan(\beta + \beta) = 2\tan(2\beta)$

32. $\tan(\beta + 45°) + \tan(45° - \beta) = \dfrac{\tan\beta + \tan 45°}{1 - \tan\beta \cdot \tan 45°} + \dfrac{\tan 45° - \tan\beta}{1 + \tan 45° \cdot \tan\beta} = \dfrac{\tan\beta + 1}{1 - \tan\beta} + \dfrac{1 - \tan\beta}{1 + \tan\beta} = \left(\dfrac{\tan\beta + 1}{1 - \tan\beta}\right)\left(\dfrac{1 + \tan\beta}{1 + \tan\beta}\right) + \left(\dfrac{1 - \tan\beta}{1 + \tan\beta}\right)\left(\dfrac{1 - \tan\beta}{1 - \tan\beta}\right) = \dfrac{\tan\beta + \tan^2\beta + 1 + \tan\beta + 1 - \tan\beta - \tan\beta + \tan^2\beta}{1 - \tan^2\beta} = \dfrac{2 + 2\tan^2\beta}{1 - \tan^2\beta} = 2\left(\dfrac{1 + \tan^2\beta}{1 - \tan^2\beta}\right)$ **33.** Since the sum of the angles of any triangle is 180°, $\theta + \alpha + (180° - \beta) = 180°$; $\theta = 180° - \alpha - (180° - \beta) = -\alpha + \beta = \beta - \alpha$. So $\theta = \beta - \alpha$. Thus, $\tan\theta = \tan(\beta - \alpha) = \dfrac{\tan\beta - \tan\alpha}{1 + \tan\beta \cdot \tan\alpha}$. But $\tan\beta = m_2$ and $\tan\alpha = m_1$, so we have $\dfrac{m_2 - m_1}{1 + m_2 \cdot m_1}$

Practice Exercises, page 229

19. $\dfrac{1 - \tan^2\beta}{\sec^2\beta} = \dfrac{1 - \dfrac{\sin^2\beta}{\cos^2\beta}}{\dfrac{1}{\cos^2\beta}} \cdot \dfrac{\cos^2\beta}{\cos^2\beta} = \cos^2\beta - \sin^2\beta = \cos 2\beta$

20. $\tan\beta = \dfrac{2\sin\beta\cos\beta}{1 + \cos 2\beta} = \dfrac{2\sin\beta\cos\beta}{2\cos^2\beta + 1 - 1} = \dfrac{\sin\beta}{\cos\beta} = \tan\beta$

21. $\cot\beta = \dfrac{1 + \cos 2\beta}{2\sin\beta\cos\beta} = \dfrac{2\cos^2\beta + 1 - 1}{2\sin\beta\cos\beta} = \dfrac{\cos\beta}{\sin\beta} = \cot\beta$

22. $\sec 2\theta = \dfrac{1}{1 - 2\sin^2\theta} = \dfrac{1}{\cos 2\theta} = \sec 2\theta$

23. $\dfrac{1}{\cos\theta} = \dfrac{\sin 2\theta\cos\theta - \cos 2\theta\sin\theta}{\sin\theta\cos\theta} = \dfrac{2\sin\theta\cos^2\theta - (2\cos^2\theta - 1)\sin\theta}{\sin\theta\cos\theta} = \dfrac{\sin\theta}{\sin\theta\cos\theta} = \dfrac{1}{\cos\theta}$

24. $\cot\alpha - \tan\alpha = \dfrac{2\cos 2\alpha}{\sin 2\alpha} = \dfrac{2(\cos^2\alpha - \sin^2\alpha)}{2\sin\alpha\cos\alpha} = \dfrac{\cos^2\alpha}{\sin\alpha\cos\alpha} - \dfrac{\sin^2\alpha}{\sin\alpha\cos\alpha} = \dfrac{\cos\alpha}{\sin\alpha} - \dfrac{\sin\alpha}{\cos\alpha} = \cot\alpha - \tan\alpha$

25. $\sin^2\beta = \dfrac{1}{2}(1 - \cos 2\beta) = \dfrac{1}{2}[1 - (2\cos^2\beta - 1)] = \dfrac{1}{2}(-2\cos^2\beta + 2) = -\cos^2\beta + 1 = \sin^2\beta$ **26.** $\cos^2\alpha = \dfrac{1}{2}(\cos 2\alpha + 1) = \dfrac{1}{2}(1 - 2\sin^2\alpha + 1) = \dfrac{1}{2}(2 - 2\sin^2\alpha) = 1 - \sin^2\alpha = \cos^2\alpha$ **27.** $\sec 2\beta = \dfrac{\cot\beta + \tan\beta}{\cot\beta - \tan\beta} = \dfrac{\dfrac{\cos\beta}{\sin\beta} + \dfrac{\sin\beta}{\cos\beta}}{\dfrac{\cos\beta}{\sin\beta} - \dfrac{\sin\beta}{\cos\beta}} \cdot \dfrac{\sin\beta\cos\beta}{\sin\beta\cos\beta} = \dfrac{\cos^2\beta + \sin^2\beta}{\cos^2\beta - \sin^2\beta} = \dfrac{1}{\cos^2\beta - \sin^2\beta} = \dfrac{1}{\cos 2\beta} = \sec 2\beta$ **28.** $(\sin\theta + \cos\theta)^2 = \sin 2\theta + 1 = \sin^2\theta + 2\sin\theta\cos\theta + \cos^2\theta = 1 + 2\sin\theta\cos\theta = 1 + \sin 2\theta$ **29.** $(\cos\theta - \sin\theta)^2 = 1 - \sin 2\theta = \cos^2\theta - 2\sin\theta\cos\theta + \sin^2\theta = 1 - 2\sin\theta\cos\theta = 1 - \sin 2\theta$ **30.** $\cos 2\alpha = \cos^4\alpha - \sin^4\alpha = (\cos^2\alpha + \sin^2\alpha)(\cos^2\alpha - \sin^2\alpha) = (\cos^2\alpha - \sin^2\alpha) = \cos 2\alpha$ **31.** $\sin 4\alpha = 4\sin\alpha\cos\alpha\cos 2\alpha = \sin(2\alpha + 2\alpha) = \sin 2\alpha\cos 2\alpha + \cos 2\alpha\sin 2\alpha = 2(2\sin\alpha\cos\alpha)(\cos 2\alpha) = 4\sin\alpha\cos\alpha\cos 2\alpha$ **32.** $\cos 4\beta = 1 - 8\sin^2\beta\cos^2\beta = \cos(2\beta + 2\beta) = \cos 2\beta\cos 2\beta - \sin 2\beta\sin 2\beta = (1 - 2\sin^2\beta)(2\cos^2\beta - 1) - (2\sin\beta\cos\beta)(2\sin\beta\cos\beta) = 2\cos^2\beta - 1 - 4\sin^2\beta\cos^2\beta + 2\sin^2\beta - 4\sin^2\beta\cos^2\beta = 2 - 1 - 4\sin^2\beta\cos^2\beta - 4\sin^2\beta\cos^2\beta = 1 - 8\sin^2\beta\cos^2\beta$

Practice Exercises, page 236

23. $\tan\frac{\alpha}{2} = \dfrac{\sin\frac{\alpha}{2}}{\cos\frac{\alpha}{2}} = \dfrac{\sqrt{\dfrac{1-\cos\alpha}{2}}}{\sqrt{\dfrac{1+\cos\alpha}{2}}} = \sqrt{\dfrac{1-\cos\alpha}{1+\cos\alpha}} \cdot$

$\dfrac{\sqrt{1+\cos\alpha}}{\sqrt{1+\cos\alpha}} = \dfrac{\sqrt{(1-\cos\alpha)(1+\cos\alpha)}}{\sqrt{(1+\cos\alpha)^2}} = \dfrac{\sqrt{1-\cos^2\alpha}}{1+\cos\alpha} =$

$\dfrac{\sqrt{\sin^2\alpha}}{1+\cos\alpha} = \dfrac{\sin\alpha}{1+\cos\alpha}$ **24.** $\tan\frac{\alpha}{2} =$

$\sqrt{\dfrac{1-\cos\alpha}{1+\cos\alpha}} \cdot \dfrac{\sqrt{1-\cos\alpha}}{\sqrt{1-\cos\alpha}} = \dfrac{\sqrt{(1-\cos\alpha)^2}}{\sqrt{1-\cos^2\alpha}} =$

$\dfrac{1-\cos\alpha}{\sqrt{\sin^2\alpha}} = \dfrac{1-\cos\alpha}{\sin\alpha}$ **25.** $2\cos^2\left(\frac{\alpha}{2}\right) - 1 =$

$2\left(\sqrt{\dfrac{1+\cos\alpha}{2}}\right)^2 - 1 = 1 + \cos\alpha - 1 = \cos\alpha$

26. $2\sin\frac{\beta}{2}\cos\frac{\beta}{2} = 2\sqrt{\dfrac{1-\cos\beta}{2}} \cdot \sqrt{\dfrac{1+\cos\beta}{2}} =$

$2 \cdot \dfrac{\sqrt{1-\cos^2\beta}}{2} = 2\dfrac{\sqrt{\sin^2\beta}}{2} = 2 \cdot \dfrac{\sin\beta}{2} = \sin\beta$

27. $\dfrac{1+\tan^2\left(\frac{\beta}{2}\right)}{2\tan\left(\frac{\beta}{2}\right)} = \dfrac{1+\left(\dfrac{1-\cos\beta}{\sin\beta}\right)^2}{2\left(\dfrac{1-\cos\beta}{\sin\beta}\right)} =$

$\dfrac{1+\dfrac{1-2\cos\beta+\cos^2\beta}{\sin^2\beta}}{\dfrac{2-2\cos\beta}{\sin\beta}} = \dfrac{\sin^2\beta + 1 - 2\cos\beta + \cos^2\beta}{\sin^2\beta} \cdot$

$\dfrac{\sin\beta}{2-2\cos\beta} = \dfrac{1}{\sin\beta} = \csc\beta$ **28.** $\tan^2\left(\frac{\alpha}{2}\right) = \left(\dfrac{1-\cos\alpha}{\sin\alpha}\right)^2 =$

$\dfrac{(1-\cos\alpha)^2}{\sin^2\alpha}$ **29.** $\dfrac{\cot\left(\frac{\alpha}{2}\right) - \tan\left(\frac{\alpha}{2}\right)}{\tan\left(\frac{\alpha}{2}\right) + \cot\left(\frac{\alpha}{2}\right)} =$

$\dfrac{\dfrac{1+\cos\alpha}{\sin\alpha} - \dfrac{1-\cos\alpha}{\sin\alpha}}{\dfrac{1-\cos\alpha}{\sin\alpha} + \dfrac{1+\cos\alpha}{\sin\alpha}} = \dfrac{\dfrac{2\cos\alpha}{\sin\alpha}}{\dfrac{2}{\sin\alpha}} = \dfrac{2\cos\alpha}{2} =$

$\cos\alpha$ **30.** $\dfrac{1+\tan^2\left(\frac{\beta}{2}\right)}{1-\tan^2\left(\frac{\beta}{2}\right)} = \dfrac{1+\left(\dfrac{1-\cos\beta}{\sin\beta}\right)^2}{1-\left(\dfrac{1-\cos\beta}{\sin\beta}\right)^2} =$

$\dfrac{1+\dfrac{1-2\cos\beta+\cos^2\beta}{\sin^2\beta}}{1-\dfrac{1-2\cos\beta+\cos^2\beta}{\sin^2\beta}} = \dfrac{\sin^2\beta + 1 - 2\cos\beta + \cos^2\beta}{\sin^2\beta - 1 + 2\cos\beta - \cos^2\beta} =$

$\dfrac{2-2\cos\beta}{\sin^2\beta} \cdot \dfrac{\sin^2\beta}{-2\cos^2\beta + 2\cos\beta} = \dfrac{2-2\cos\beta}{(2-2\cos\beta)(\cos\beta)} =$

$\dfrac{1}{\cos\beta} = \sec\beta$ **31.** $1 - 8\sin^2\left(\frac{\theta}{2}\right)\cos^2\left(\frac{\theta}{2}\right) =$

$1 - 8\left(\dfrac{1-\cos\theta}{2}\right)\left(\dfrac{1+\cos\theta}{2}\right) = 1 - 8\left(\dfrac{1-\cos^2\theta}{4}\right) =$

$1 - 8\left(\dfrac{\sin^2\theta}{4}\right) = 1 - 2\sin^2\theta = \cos 2\theta$

32. $1 - \dfrac{2\cos\theta}{1+\cos\theta} = \dfrac{1+\cos\theta-2\cos\theta}{1+\cos\theta} =$

$\dfrac{1-\cos\theta}{1+\cos\theta} = \tan^2\frac{\theta}{2}$ **33.** $A = \left(s\sin\frac{\alpha}{2}\right)\left(s\cos\frac{\alpha}{2}\right) =$

$s^2\left(\sqrt{\dfrac{1-\cos\alpha}{2}}\right)\left(\sqrt{\dfrac{1+\cos\alpha}{2}}\right) = s^2\left(\sqrt{\dfrac{1-\cos^2\alpha}{4}}\right) =$

$s^2\left(\dfrac{\sqrt{\sin^2\alpha}}{2}\right) = s^2\left(\dfrac{\sin\alpha}{2}\right) = \dfrac{1}{2}s^2\sin\alpha$

Practice Exercises, pages 240–241

17. $\dfrac{\cos 8\beta - \cos 2\beta}{\sin 2\beta - \sin 8\beta} = \dfrac{-2\sin\left(\dfrac{8\beta+2\beta}{2}\right)\sin\left(\dfrac{8\beta-2\beta}{2}\right)}{2\cos\left(\dfrac{2\beta+8\beta}{2}\right)\sin\left(\dfrac{2\beta-8\beta}{2}\right)} =$

$\dfrac{-2\sin 5\beta\sin 3\beta}{2\cos 5\beta\sin(-3\beta)} = \dfrac{-2\sin 5\beta\sin 3\beta}{-2\cos 5\beta\sin 3\beta} = \dfrac{\sin 5\beta}{\cos 5\beta} = \tan 5\beta$

18. $\dfrac{\cos\alpha - \cos 3\alpha}{\sin 3\alpha - \sin\alpha} = \dfrac{-2\sin\left(\dfrac{\alpha+3\alpha}{2}\right)\sin\left(\dfrac{\alpha-3\alpha}{2}\right)}{2\cos\left(\dfrac{3\alpha+\alpha}{2}\right)\sin\left(\dfrac{3\alpha-\alpha}{2}\right)} =$

$\dfrac{-2\sin 2\alpha\sin(-\alpha)}{2\cos 2\alpha\sin\alpha} = \dfrac{2\sin 2\alpha\sin\alpha}{2\cos 2\alpha\sin\alpha} = \dfrac{\sin 2\alpha}{\cos 2\alpha} = \tan 2\alpha$

19. $\dfrac{\cos 4\alpha + \cos 2\alpha}{\sin 4\alpha - \sin 2\alpha} = \dfrac{2\cos\left(\dfrac{4\alpha+2\alpha}{2}\right)\cos\left(\dfrac{4\alpha-2\alpha}{2}\right)}{2\cos\left(\dfrac{4\alpha+2\alpha}{2}\right)\sin\left(\dfrac{4\alpha-2\alpha}{2}\right)} =$

$\dfrac{2\cos 3\alpha\cos\alpha}{2\cos 3\alpha\sin\alpha} = \dfrac{\cos\alpha}{\sin\alpha} = \cot\alpha = \dfrac{1}{\tan\alpha}$

20. $\dfrac{\sin 7\beta + \sin 5\beta}{\sin 7\beta - \sin 5\beta} = \dfrac{2\sin\left(\dfrac{7\beta+5\beta}{2}\right)\cos\left(\dfrac{7\beta-5\beta}{2}\right)}{2\cos\left(\dfrac{7\beta+5\beta}{2}\right)\sin\left(\dfrac{7\beta-5\beta}{2}\right)} =$

$\dfrac{2\sin 6\beta\cos\beta}{2\cos 6\beta\sin\beta} = \tan 6\beta\cot\beta = \dfrac{\tan 6\beta}{\tan\beta}$

21. $\dfrac{\cos 3\theta - \cos\theta}{\sin 3\theta + \sin\theta} = \dfrac{-2\sin\left(\dfrac{3\theta+\theta}{2}\right)\sin\left(\dfrac{3\theta-\theta}{2}\right)}{2\sin\left(\dfrac{3\theta+\theta}{2}\right)\cos\left(\dfrac{3\theta-\theta}{2}\right)} =$

$\dfrac{-2\sin 2\theta\sin\theta}{2\sin 2\theta\cos\theta} = \dfrac{-\sin\theta}{\cos\theta} = -\tan\theta$

22. $\dfrac{\sin 5\theta + \sin 3\theta}{\cos 3\theta - \cos 5\theta} = \dfrac{2\sin\left(\dfrac{5\theta+3\theta}{2}\right)\cos\left(\dfrac{5\theta-3\theta}{2}\right)}{-2\sin\left(\dfrac{3\theta+5\theta}{2}\right)\sin\left(\dfrac{3\theta-5\theta}{2}\right)} =$

$\dfrac{2\sin 4\theta\cos\theta}{-2\sin 4\theta\sin(-\theta)} = \dfrac{-\cos\theta}{\sin(-\theta)} = \dfrac{\cos\theta}{\sin\theta} = \dfrac{\dfrac{1}{\sec\theta}}{\dfrac{1}{\csc\theta}} = \dfrac{\csc\theta}{\sec\theta}$

23. $\frac{\sin 8\theta - \sin 2\theta}{\cos 8\theta - \cos 2\theta} = \frac{2\cos\left(\frac{8\theta + 2\theta}{2}\right)\sin\left(\frac{8\theta - 2\theta}{2}\right)}{-2\sin\left(\frac{8\theta + 2\theta}{2}\right)\sin\left(\frac{8\theta - 2\theta}{2}\right)} =$

$\frac{2\cos 5\theta \sin 3\theta}{-2\sin 5\theta \sin 3\theta} = -\frac{\cos 5\theta}{\sin 5\theta} = \frac{\frac{1}{\sec 5\theta}}{\frac{1}{\csc 5\theta}} = -\frac{\csc 5\theta}{\sec 5\theta}$

24. $\frac{\cos 4\theta + \cos 2\theta}{\sin 4\theta + \sin 2\theta} = \frac{2\cos\left(\frac{4\theta + 2\theta}{2}\right)\cos\left(\frac{4\theta - 2\theta}{2}\right)}{2\sin\left(\frac{4\theta + 2\theta}{2}\right)\cos\left(\frac{4\theta - 2\theta}{2}\right)} =$

$\frac{2\cos 3\theta \cos\theta}{2\sin 3\theta \cos\theta} = \frac{\cos 3\theta}{\sin 3\theta} = \frac{\frac{1}{\sec 3\theta}}{\frac{1}{\csc 3\theta}} = \frac{\csc 3\theta}{\sec 3\theta}$

25. $\cos(x + y)\cos(x - y) = \frac{1}{2}[\cos(x + y - x + y) +$ $\cos(x + y + x - y)] = \frac{1}{2}[\cos 2y + \cos 2x] = \frac{1}{2}$ $[(2\cos^2 y - 1) + (1 - 2\sin^2 x)] = \frac{1}{2}[2\cos^2 y - 2\sin^2 x] =$ $\cos^2 y - \sin^2 x$ 26. $\sin(x + y)\cos(x - y) = \frac{1}{2}[\sin(x +$ $y + x - y) + \sin(x + y - x + y)] = \frac{1}{2}[\sin 2x + \sin 2y] =$ $\frac{1}{2}[2\sin x\cos x + 2\sin y\cos y] = \sin x\cos x + \sin y\cos y$

Chapter 6

Practice Exercises, pages 253–254

27.

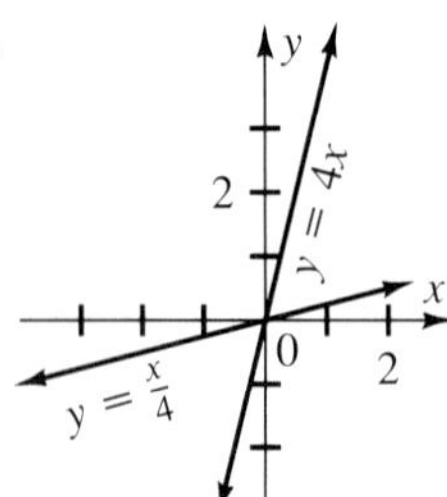

28.

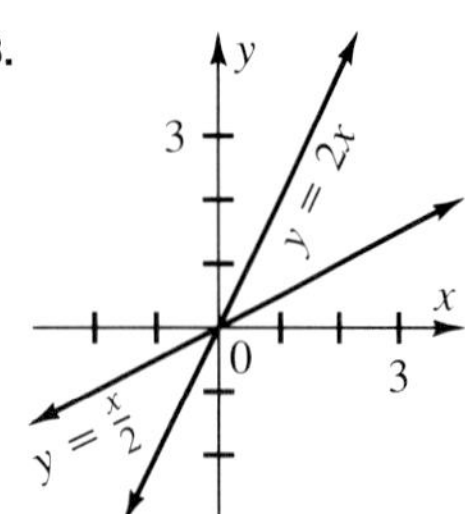

29.

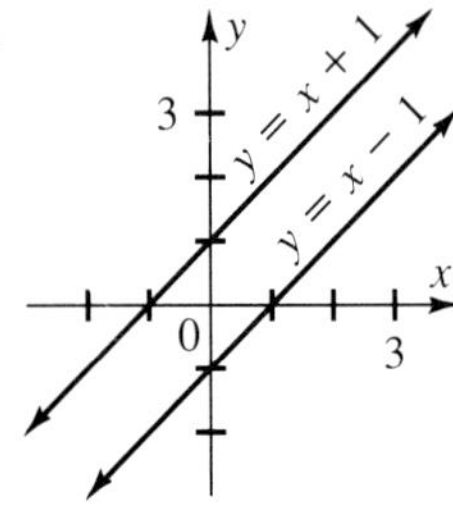

30.

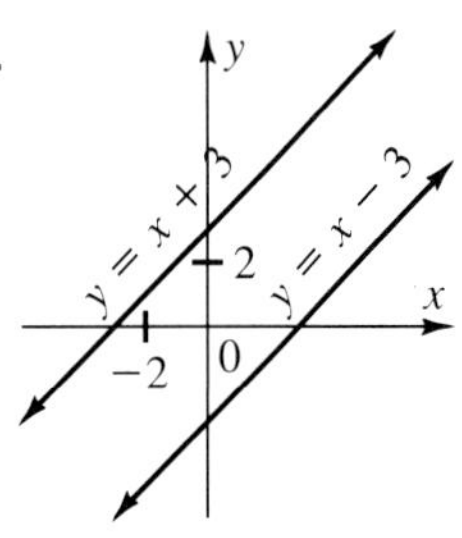

31.

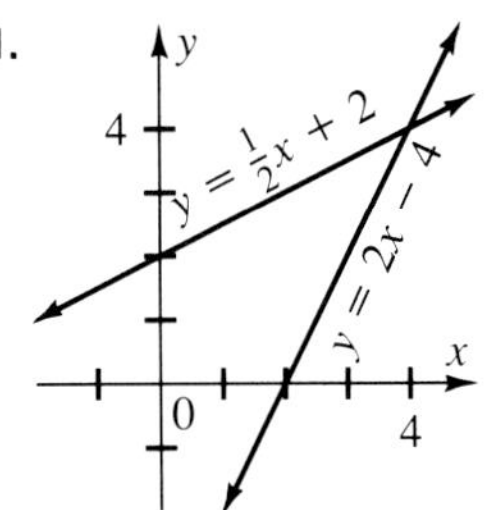

32.

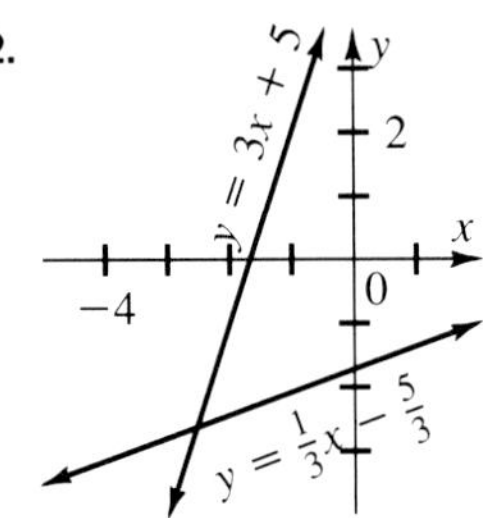

Practice Exercises, pages 276–277

15. 45°; $11(x')^2 + 9(y')^2 = 8$

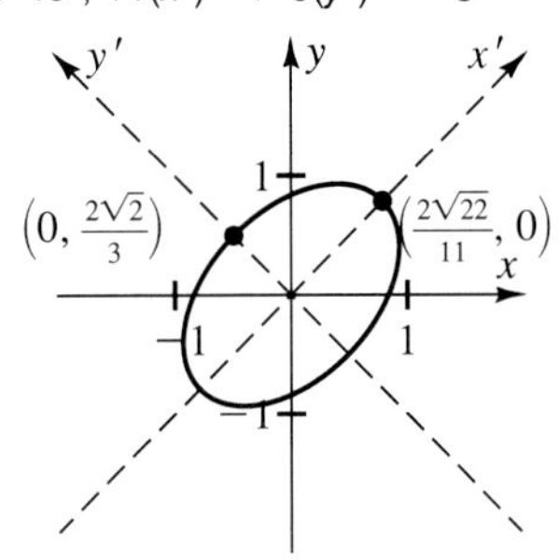

16. 45°; $\frac{(x')^2}{20} - \frac{(y')^2}{20} = 1$

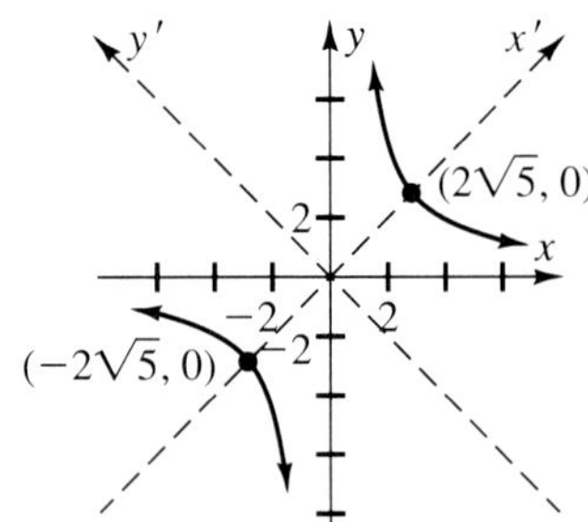

17. 45°; $\frac{(y')^2}{12} - \frac{(x')^2}{12} = 1$

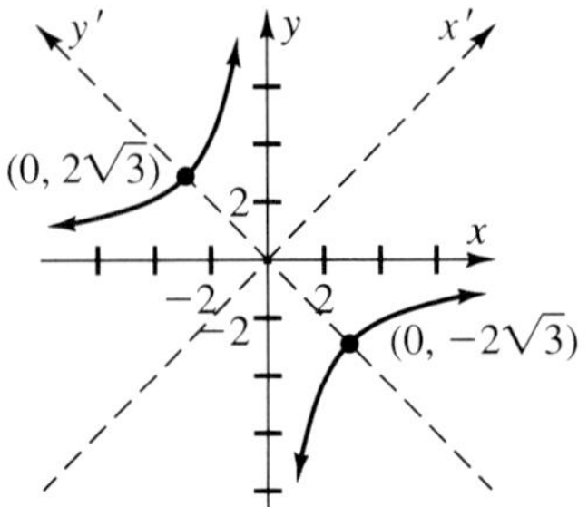

18. $45°$; $7(x')^2 - 3(y')^2 = 20$

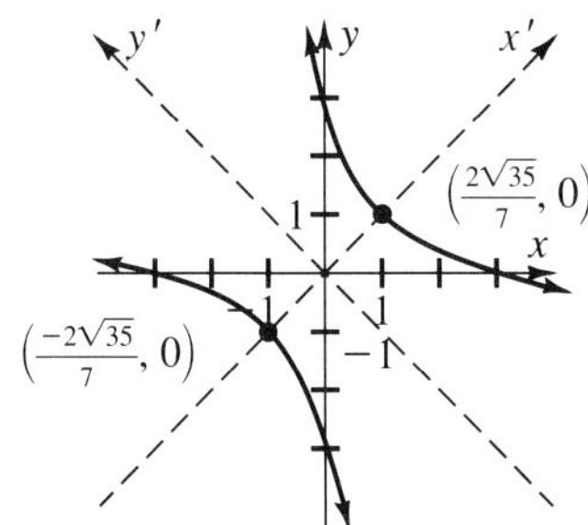

19. $45°$; $(x')^2 + 4(y')^2 = 15$

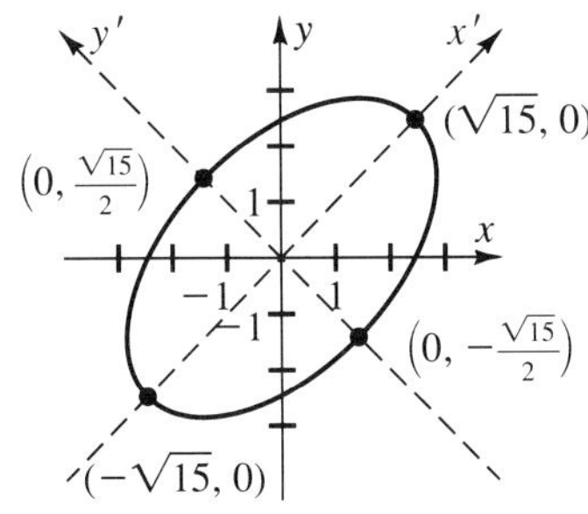

20. $60°$; $3(y')^2 - (x')^2 = 10$

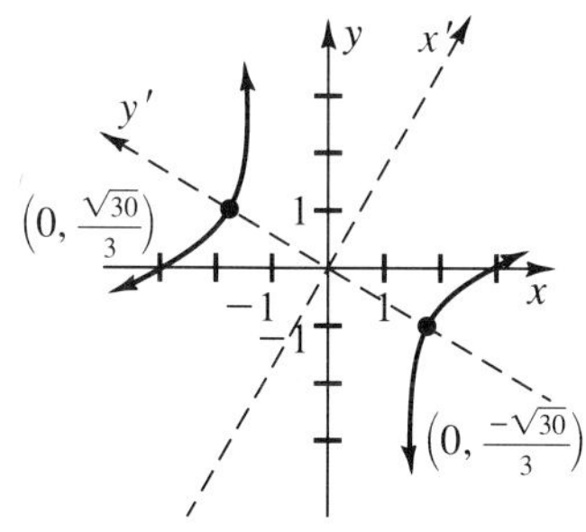

21. $30°$; $\frac{(x')^2}{18} - \frac{(y')^2}{6} = 1$

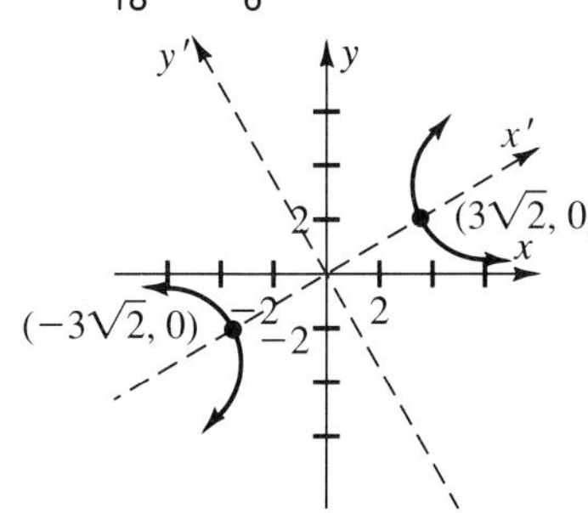

22. $45°$; $19(y')^2 - 19(x')^2 = 146$

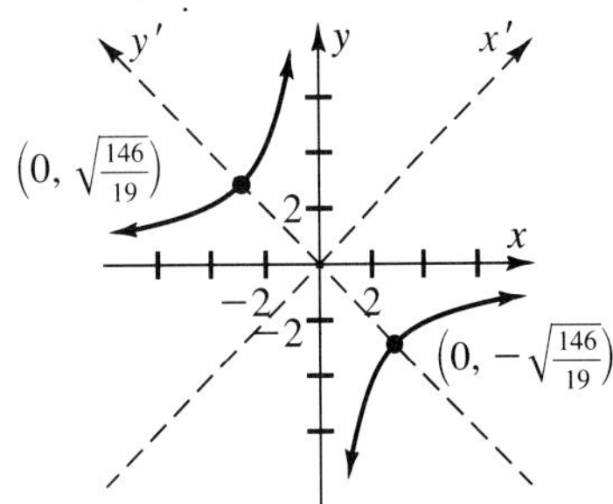

23. $22.5°$; $1.45(x')^2 + 1.82(y')^2 = 1$

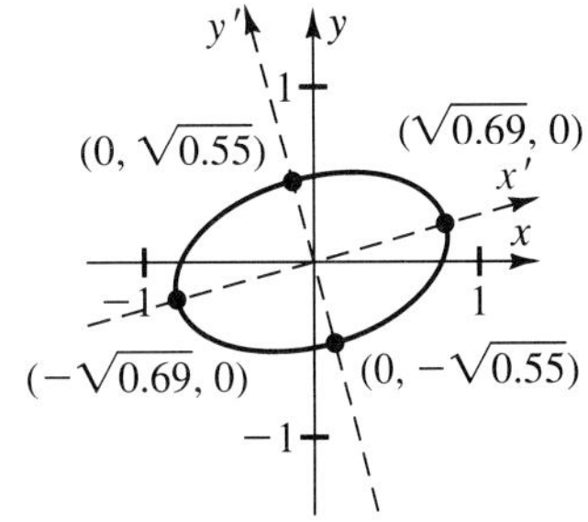

24. $30°$; $\frac{(x')^2}{4} - \frac{(y')^2}{4} = 1$

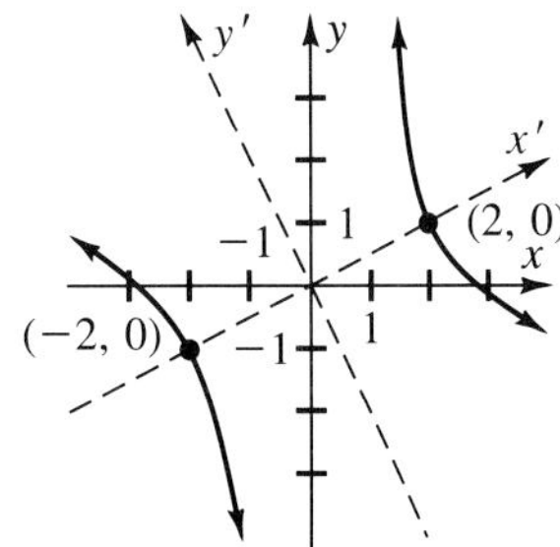

25. $13°$; $0.17(x')^2 - 0.67(y')^2 = 1$

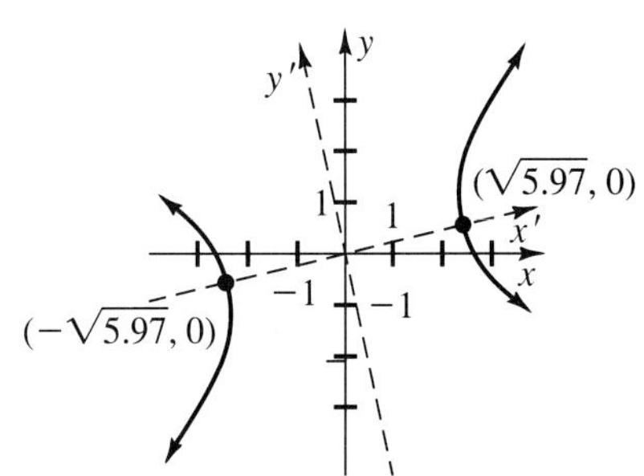

26. 75°; $0.84(y')^2 - 0.18(x')^2 = 1$

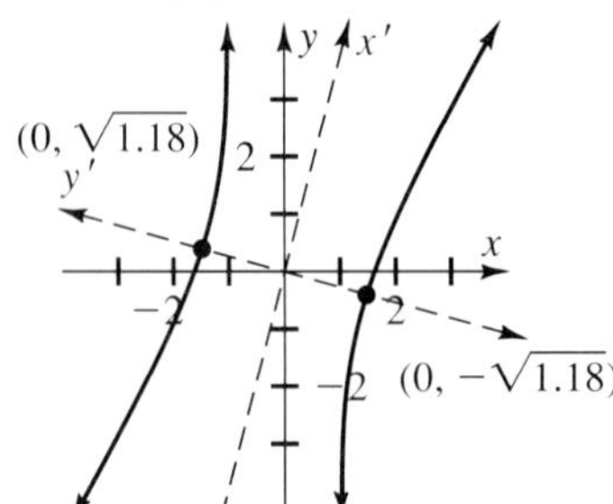

27. $\theta = 180°$; $x = x' \cos 180° - y' \sin 180°$, $x = -x'$; $y = x' \sin 180° + y' \cos 180°$, $y = y'$; If only even powers of x and y appear in the equation: $(x)^{2n} = (-x')^{2n} = (x')^{2n}$; $(y)^{2n} = (-y')^{2n} = (y')^{2n}$; Therefore, the new equation will be the same as the old one. **28.** $x = x' \cos \theta - y' \sin \theta$, $x \cos \theta = x' \cos^2 \theta - y' \sin \theta \cos \theta$; $y = x' \sin \theta + y' \cos \theta$, $y \sin \theta = x' \sin^2 \theta + y' \sin \theta \cos \theta$. Thus $x \cos \theta + y \sin \theta = x' (\cos^2 \theta + \sin^2 \theta) = x'$. $x = x' \cos \theta - y' \sin \theta$, $-x \sin \theta = -x' \sin \theta \cos \theta + y' \sin^2 \theta$; $y = x' \sin \theta + y' \cos \theta$, $y \cos \theta = x' \sin \theta \cos \theta + y' \cos^2 \theta$. Thus $y \cos \theta - x \sin \theta = y' (\sin^2 \theta + \cos^2 \theta) = y'$

Chapter 7

Practice Exercises, page 290

1.

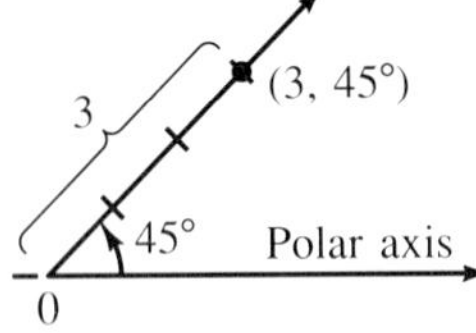

2.

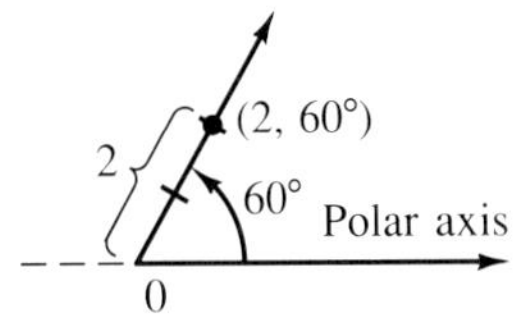

3.

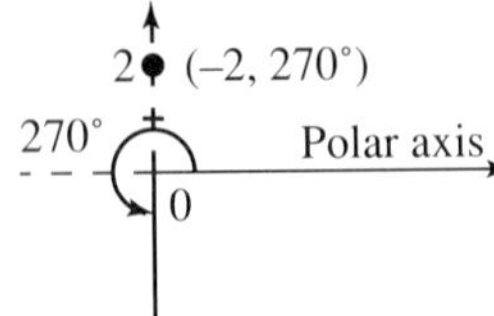

4.

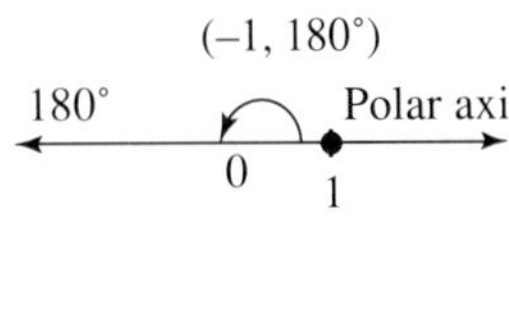

5.

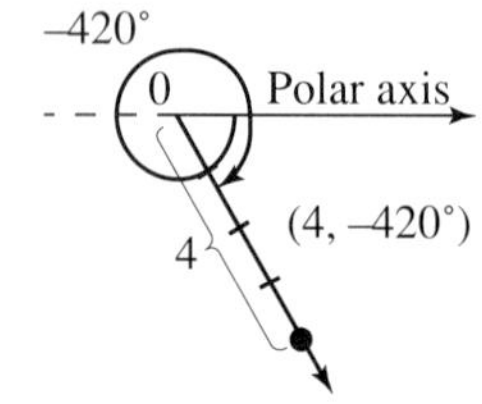

6.

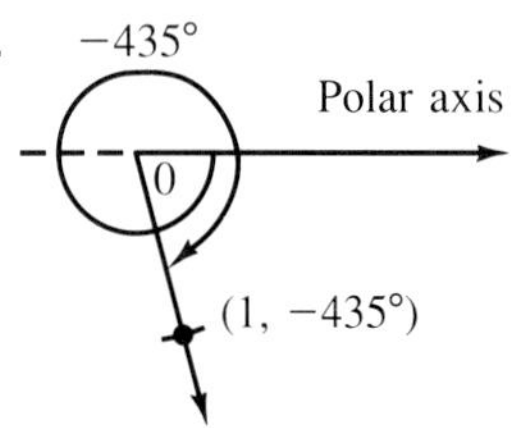

7.

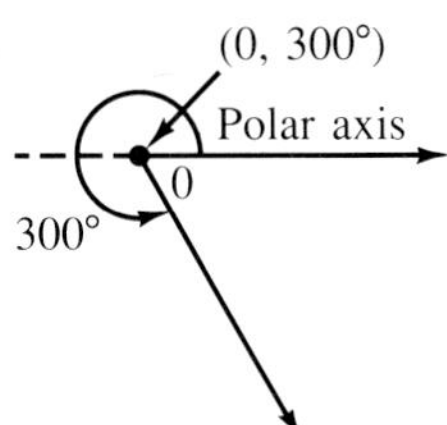

8.

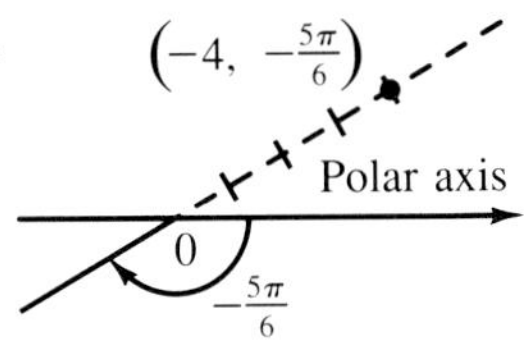

9.

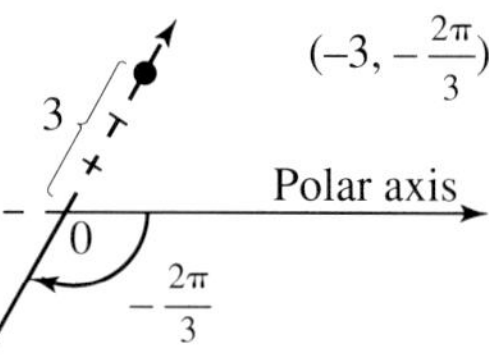

10.

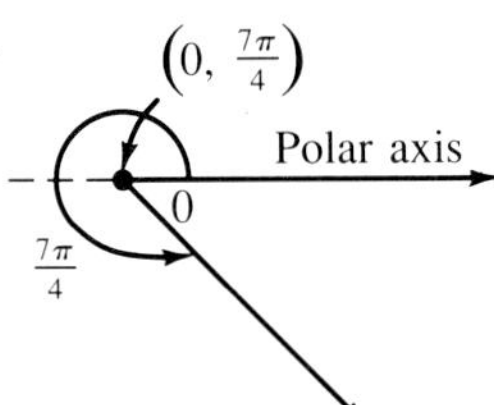

11.

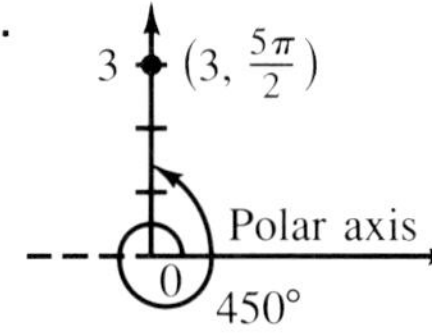

12.

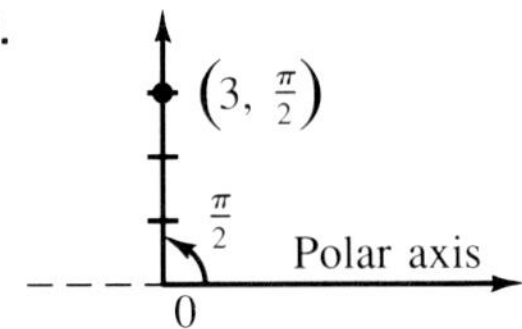

Practice Exercises, page 296

3.

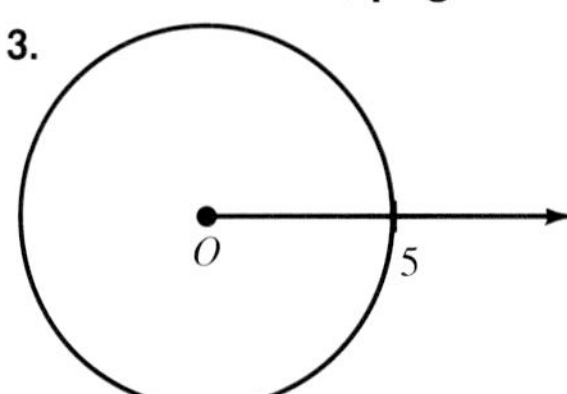

4.

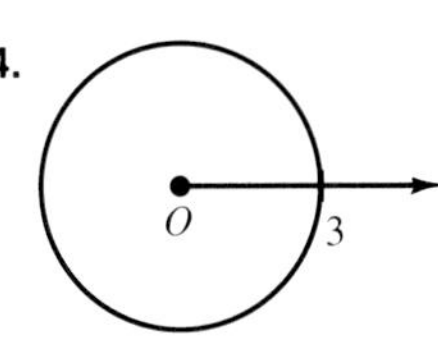

5.

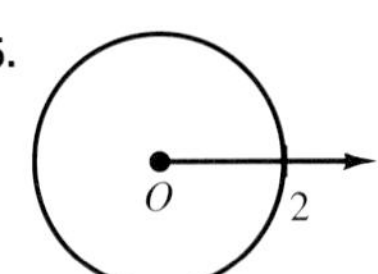

6.

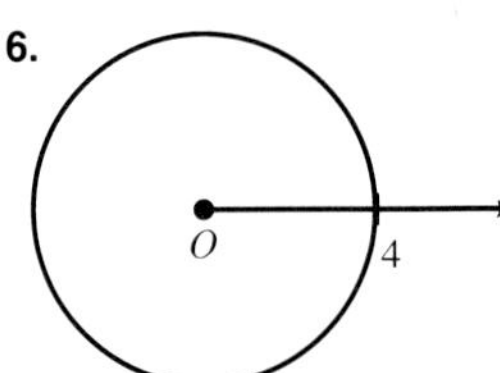

7.

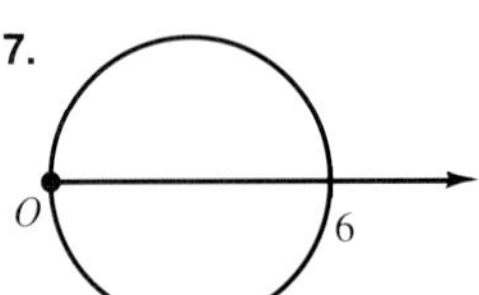

8.

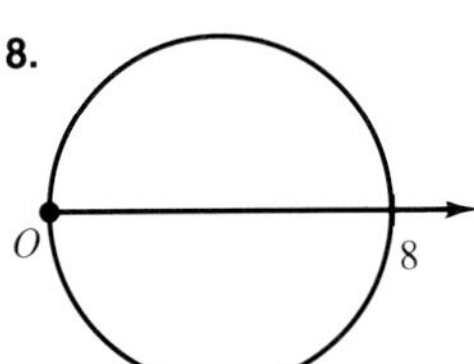

9.

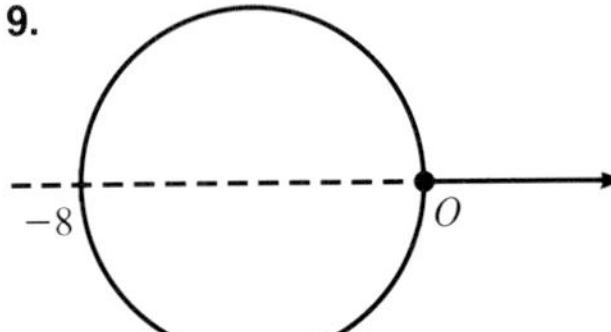

10.

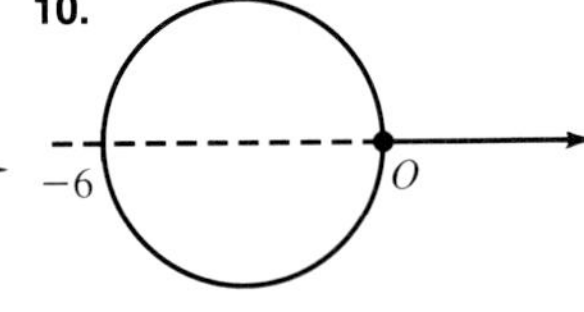

11.

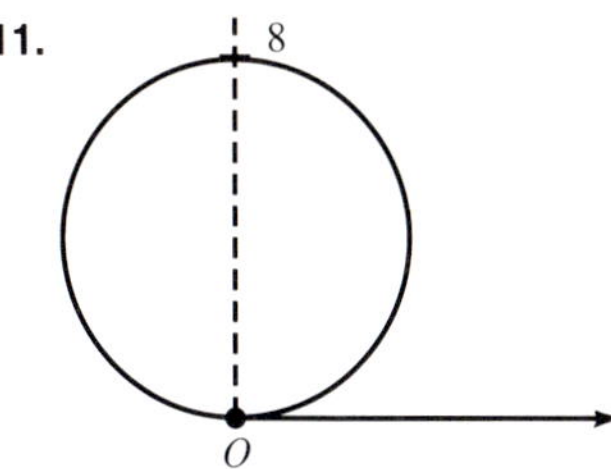

12.

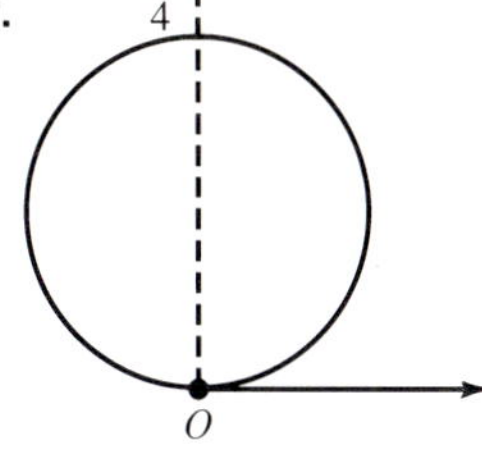

16.

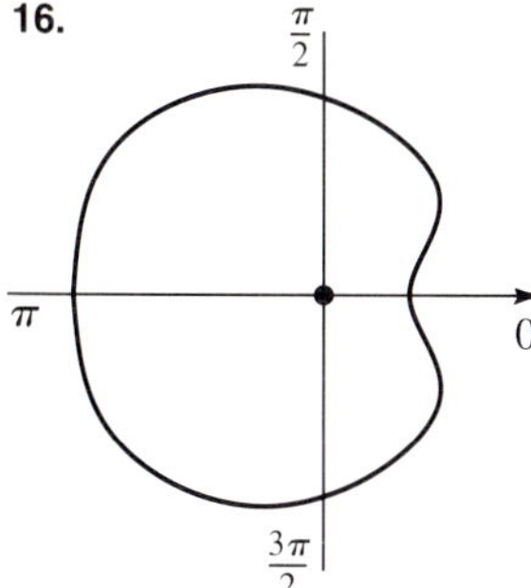

13.

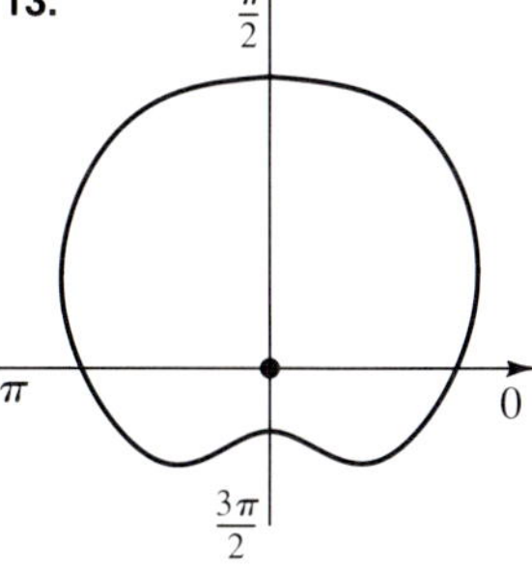

17.

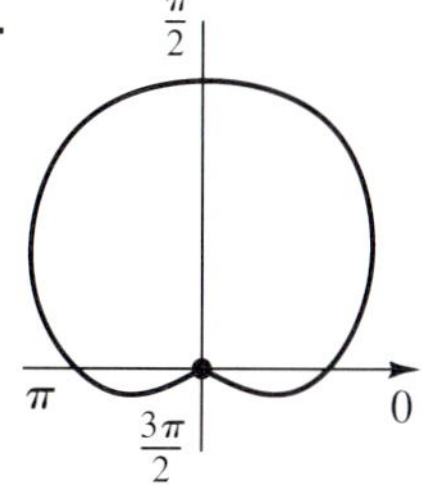

18.

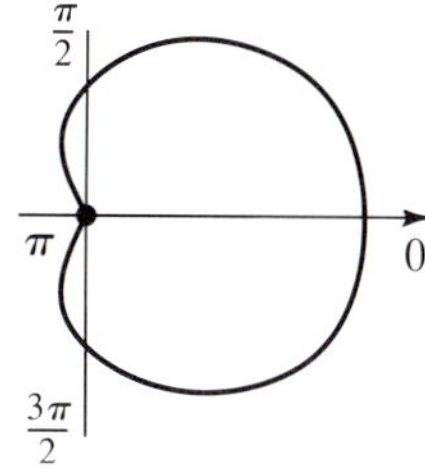

14.

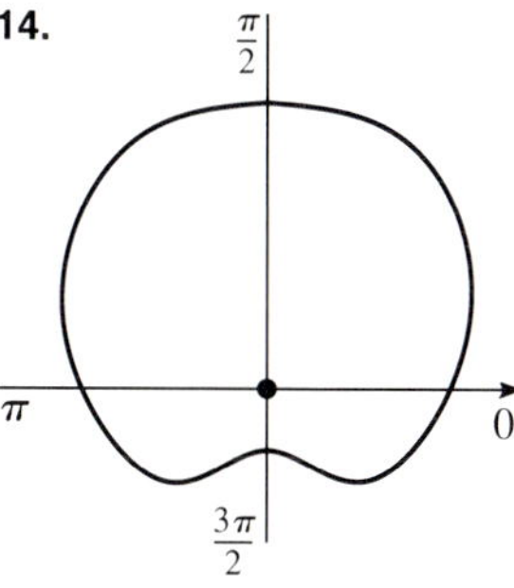

19.

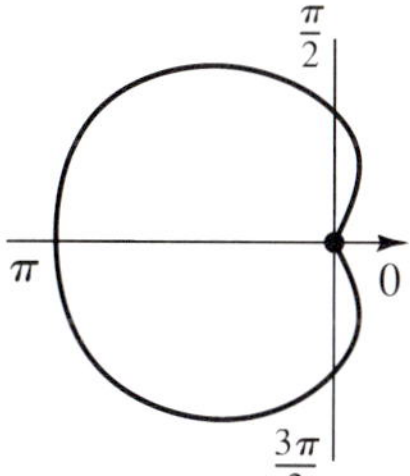

15.

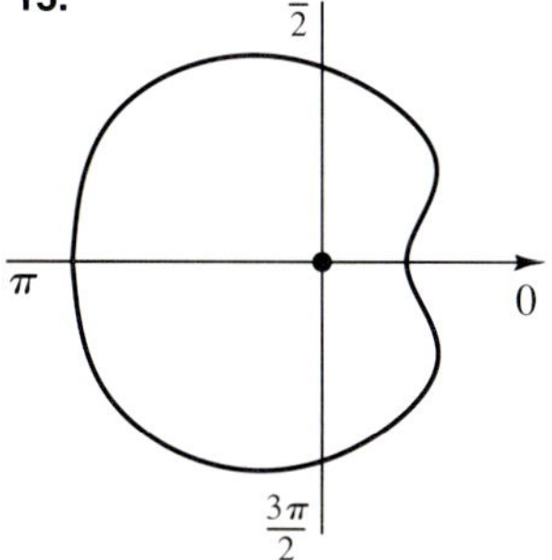

20.

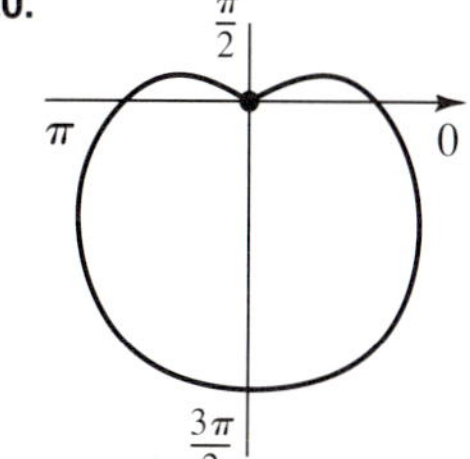

21.

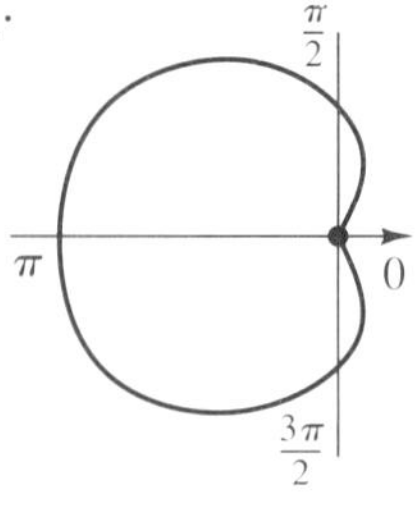

22.

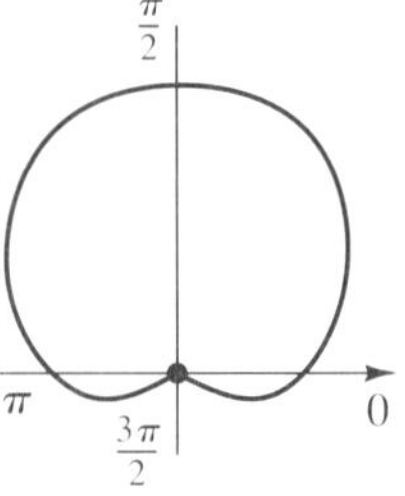

23.

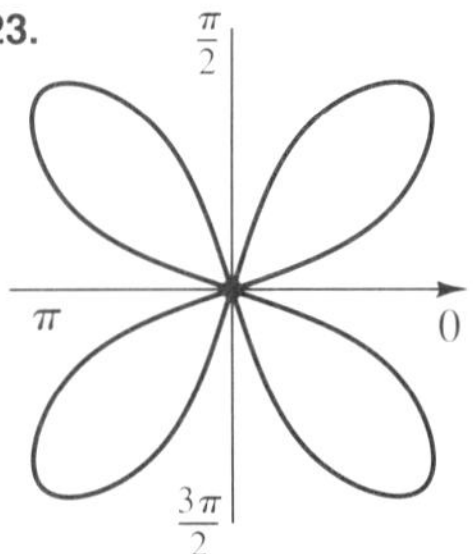

24.

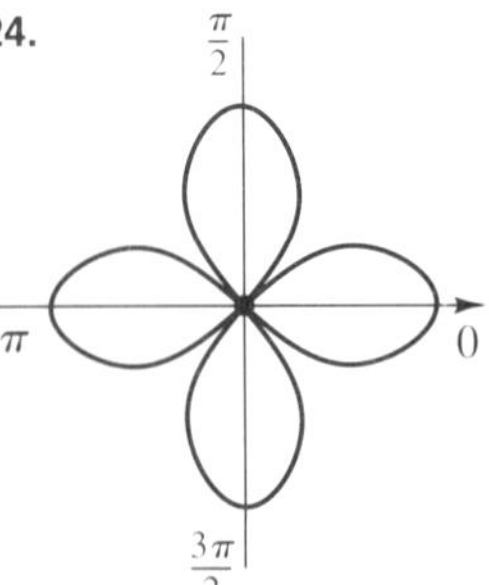

25.

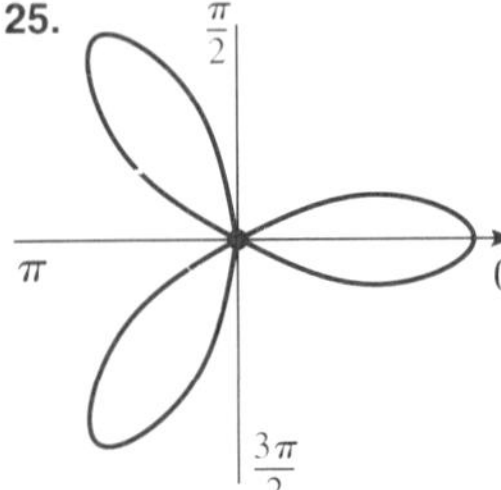

26.

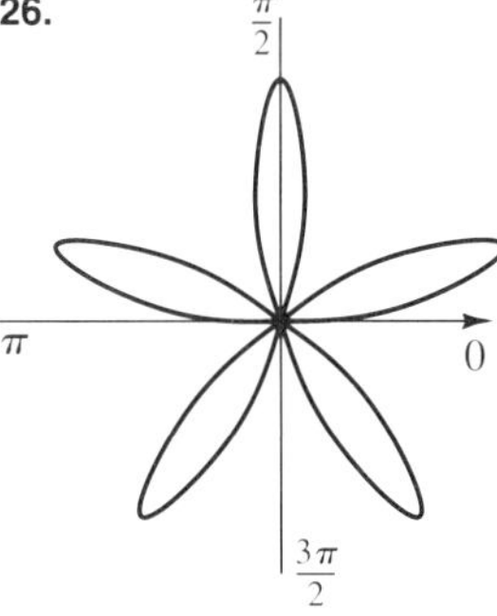

27.

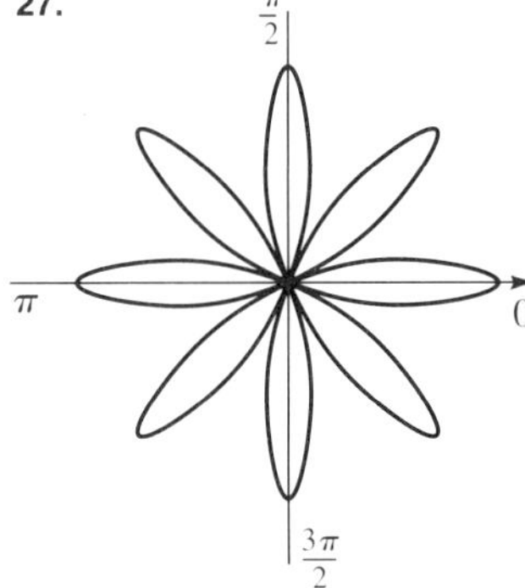

28.

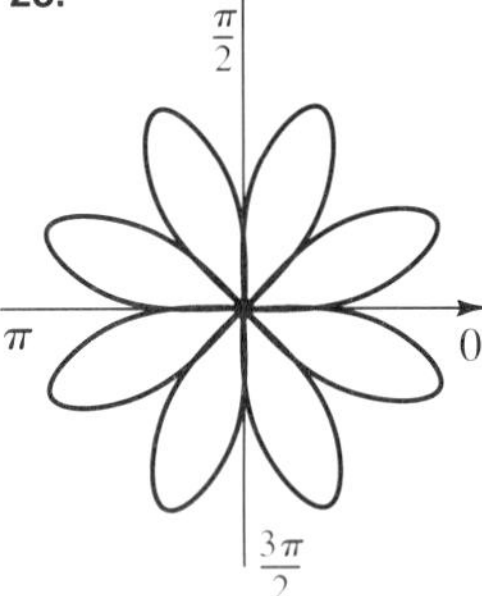

29. 30.

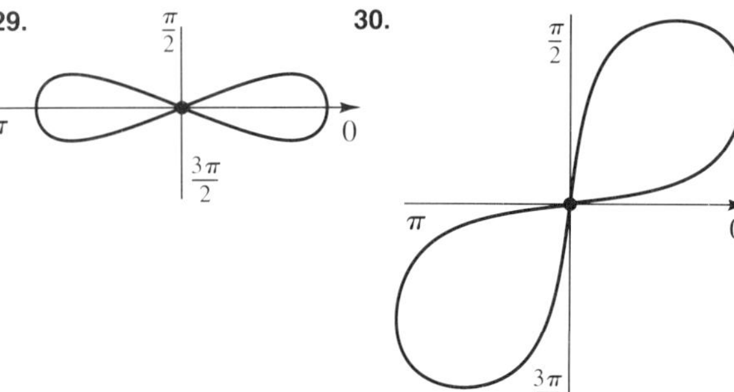

31.

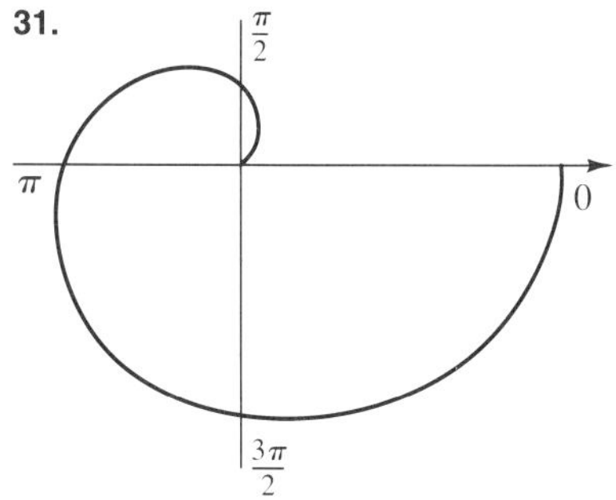

32.

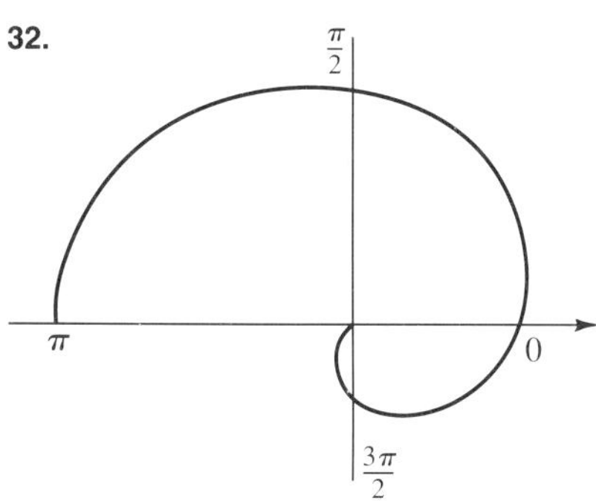

Extra, page 297

1.

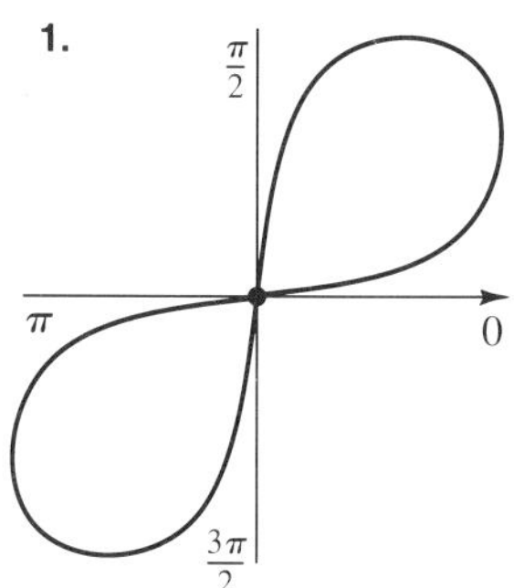

Table of values:
(0, 0); (5, 45°); (0, 90°); (0, 180°); (5, 225°); (0, 270°); (−5, 45°); (−5, 225°)

3.

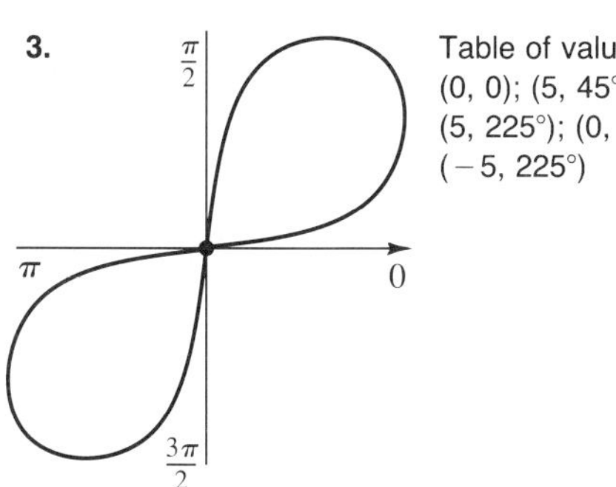

Table of values:
(0, 0); (5, 45°); (0, 90°); (0, 180°); (5, 225°); (0, 270°); (−5, 45°); (−5, 225°)

4. $r^2 = 25 \sin 2\theta$; $\sin 2\theta = \cos (90° - 2\theta) = \cos (-2)(\theta - 45°)$; But $\cos (-x) = \cos x$, so $\sin 2\theta = \cos 2(\theta - 45°)$. Therefore $r^2 = 25 \cos 2(\theta - 45°)$.

Practice Exercises, page 312

38. $z_1 = a + bi = r_1 (\cos \theta_1 + \mathrm{i} \sin \theta_1)$; $z_2 = c + di = r_2 (\cos \theta_2 + \mathrm{i} \sin \theta_2)$; $|z_1| = \sqrt{a^2 + b^2} = \sqrt{r_1^2 \cos^2 \theta + r_1^2 \sin^2 \theta} = \sqrt{r_1^2 (\cos^2 \theta_1 + \sin^2 \theta_1)} = \sqrt{r_1^2} = r_1$. Similarly, $|z_2| = r_2$. So $|z_1| + |z_2| = r_1 + r_2$; $z_1 + z_2 = (a + c) + (b + d)\mathrm{i} = (r_1 \cos \theta_1 + r_2 \cos \theta_2) + (r_1 \sin \theta_1 + r_2 \sin \theta_2)\mathrm{i}$; $|z_1 + z_2| = \sqrt{(r_1 \cos \theta_1 + r_2 \cos \theta_2)^2 + (r_1 \sin \theta_1 + r_2 \sin \theta_2)^2} = \sqrt{r_1^2 + r_2^2 + 2r_1r_2 \cos (\theta_2 - \theta_1)}$, $2r_1r_2 \geq 2r_1r_2 \cos (\theta_2 - \theta_1)$ because $\cos (\theta_2 - \theta_1) \leq 1$, $r_1^2 + r_2^2 + 2r_1r_2 \geq r_1^2 + r_2^2 + 2r_1r_2 \cos (\theta_2 - \theta_1)$, $r_1 + r_2 \geq \sqrt{r_1^2 + r_2^2 + 2r_1r_2 \cos (\theta_2 - \theta_1)}$. Thus, $|z_1| + |z_2| \geq |z_1 + z_2|$, $|z_1 + z_2| \leq |z_1 + z_2|$. **39.** Let $|z_1|$, $|z_2|$, and $|z_1 + z_2|$ represent the lengths of the sides. The sum of the lengths of any two sides of a triangle is greater than the length of the third side. $z_2 - z_1 = (c - a) + (d - b)\mathrm{i}$; $|z_2 - z_1| = \sqrt{(c - a)^2 + (d - b)^2} = \sqrt{(x_2 - x_1)^2 + (y_2 - y_1)^2}$

Practice Exercises, page 317

Challenge $A = r_1 (\cos \alpha + i \sin \alpha)$, $B = r_2 (\cos \beta + i \sin \beta)$, $C = 1 (\cos 0° + i \sin 0°)$, $D = r_3 [\cos (\alpha + \beta) + i \sin (\alpha + \beta)]$

Since $\triangle AOC \sim \triangle DOB$, $\frac{OA}{OC} = \frac{OD}{OB}$. Using substitution,

$\frac{r_1 (\cos \alpha + i \sin \alpha)}{1} = \frac{r_3 [\cos (\alpha + \beta) + i \sin (\alpha + \beta)]}{r_2 (\cos \beta + i \sin \beta)}$;

$r_1r_2 [\cos (\alpha + \beta) + i \sin (\alpha + \beta)] = r_3 [\cos (\alpha + \beta) + i \sin (\alpha + \beta)]$, $r_1r_2 = r_3$. Therefore, $D = A \cdot B$.

Practice Exercises, pages 325–326

15. $1 + 0i$, $0 + i$, $-1 + 0i$, $0 - i$ **16.** $1 + 0i$, $\cos 36° + i \sin 36°$, $\cos 72° + i \sin 72°$, $\cos 108° + i \sin 108°$, $\cos 144° + i \sin 144°$, $-1 + 0i$, $\cos 216° + i \sin 216°$, $\cos 252° + i \sin 252°$, $\cos 288° + i \sin 288°$, $\cos 324° + i \sin 324°$

17. $\frac{-3}{2} + \frac{3\sqrt{3}}{2}i$, $3 + 0i$, $\frac{-3}{2} - \frac{3\sqrt{3}}{2}i$ **18.** $2 + 0i$; $1 + \sqrt{3}i$; $-1 + \sqrt{3}i$; $-2 + 0i$; $-1 - \sqrt{3}i$; $1 - \sqrt{3}i$ **19.** $3(\cos 36° + i \sin 36°)$, $3 (\cos 108° + i \sin 108°)$, $-3 + 0i$, $3(\cos 252° + i \sin 252°)$, $3(\cos 324° + i \sin 324°)$ **20.** $\frac{5}{2} + \frac{5\sqrt{3}i}{2}$, $-5 + 0i$, $\frac{5}{2} - \frac{5\sqrt{3}i}{2}$ **21.** 3, $\frac{-3}{2} + \frac{3\sqrt{3}i}{2}$, $\frac{-3}{2} - \frac{3\sqrt{3}i}{2}$ **22.** $2 + 2\sqrt{3}i$, $-4 + 0i$, $2 - 2\sqrt{3}i$ **23.** $\cos 36° + i \sin 36°$, $\cos 108° + i \sin 108°$, $-1 + 0i$, $\cos 252° + i \sin 252°$, $\cos 324° + i \sin 324°$ **24.** $0 + i$, $-1 + 0i$, $0 - i$, $1 + 0i$ **25.** $1 + i\sqrt{3}$, $-2 + 0i$, $1 - i\sqrt{3}$ **26.** $\frac{3}{2} + \frac{3\sqrt{3}}{2}i$, $-3 + 0i$, $\frac{3}{2} - \frac{3\sqrt{3}}{2}i$

27. $\frac{\sqrt{2+\sqrt{2}}}{4} + \frac{\sqrt{2-\sqrt{2}}}{4}i, \frac{-\sqrt{2-\sqrt{2}}}{4} + \frac{\sqrt{2+\sqrt{2}}}{4}i,$
$\frac{-\sqrt{2+\sqrt{2}}}{4} - \frac{\sqrt{2-\sqrt{2}}}{4}i, \frac{\sqrt{2-\sqrt{2}}}{4} - \frac{\sqrt{2+\sqrt{2}}}{4}i$
28. $\frac{\sqrt{2-\sqrt{2}}}{2} + \frac{\sqrt{2+\sqrt{2}}}{2}i, \frac{-\sqrt{2+\sqrt{2}}}{2} + \frac{\sqrt{2-\sqrt{2}}}{2}i,$
$\frac{-\sqrt{2-\sqrt{2}}}{2} - \frac{\sqrt{2+\sqrt{2}}}{2}i, \frac{\sqrt{2+\sqrt{2}}}{2} - \frac{\sqrt{2-\sqrt{2}}}{2}i$
29. 4(cos 20° + *i* sin 20°), 4(cos 140° + *i* sin 140°), 4(cos 260° + *i* sin 260°) **30.** $\frac{\sqrt[4]{2}}{2} + \frac{\sqrt[4]{18}}{2}i, \frac{-\sqrt[4]{18}}{2} + \frac{\sqrt[4]{2}}{2}i, \frac{-\sqrt[4]{2}}{2} - \frac{\sqrt[4]{18}}{2}i, \frac{\sqrt[4]{18}}{2} - \frac{\sqrt[4]{2}}{2}i$ **31.** 2 + 0*i*, 2(cos 72° + *i* sin 72°), 2(cos 144° + *i* sin 144°), 2(cos 216° + *i* sin 216°), 2(cos 288° + *i* sin 288°) **32.** 2(cos 36° + *i* sin 36°), 2(cos 108° + *i* sin 108°), −2 + 0*i*, 2(cos 252° + *i* sin 252°), 2(cos 324° + *i* sin 324°)
33. $\frac{\sqrt{3}}{2} + \frac{i}{2}, 0 + i, \frac{-\sqrt{3}}{2} + \frac{i}{2}, \frac{-\sqrt{3}}{2} - \frac{i}{2}, 0 - i, \frac{\sqrt{3}}{2} - \frac{i}{2}$ **34.** $1 + 0i, \frac{1}{2} + \frac{\sqrt{3}}{2}i, -\frac{1}{2} + \frac{\sqrt{3}}{2}i, -1 + 0i, -\frac{1}{2} - \frac{\sqrt{3}}{2}i, \frac{1}{2} - \frac{\sqrt{3}}{2}i$ **35.** cos 40° + *i* sin 40°, cos 160° + *i* sin 160°, cos 280° + *i* sin 280° **36.** $\sqrt[3]{2}$ (cos 20° + *i* sin 20°), $\sqrt[3]{2}$ (cos 140° + *i* sin 140°), $\sqrt[3]{2}$ (cos 260° + *i* sin 260°)
37. $\frac{\sqrt[3]{4}}{2} + \frac{\sqrt[6]{432}}{2}i, -\sqrt[3]{4}, \frac{\sqrt[3]{4}}{2} - \frac{\sqrt[6]{432}}{2}i$ **38.** $\frac{9\sqrt{3}}{2} + \frac{9}{2}i, \frac{-9\sqrt{3}}{2} + \frac{9}{2}i, -9i$ **39.** $2\sqrt{3} - 2i, 4i, -2\sqrt{3} - 2i$
40. $8i, -8, -8i, 8$

Chapter 8

Practice Exercises, pages 340–341

9.

$f(x) = 4^x$

10.

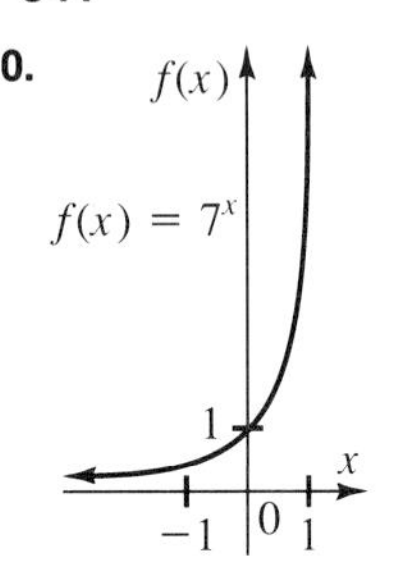

11.

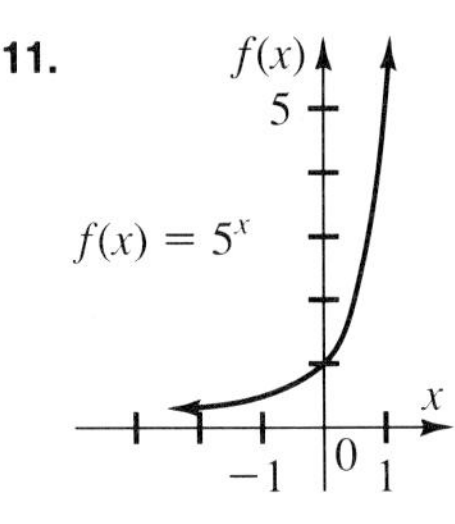

12.

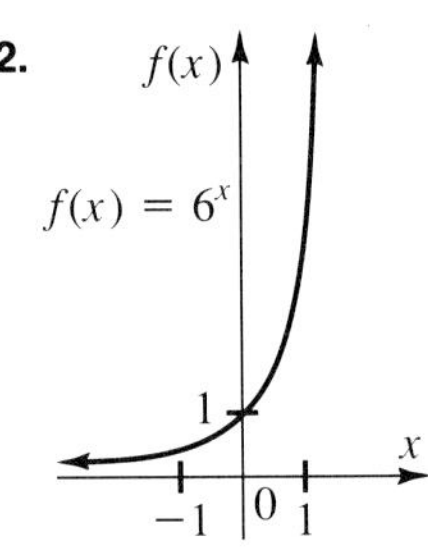

13.

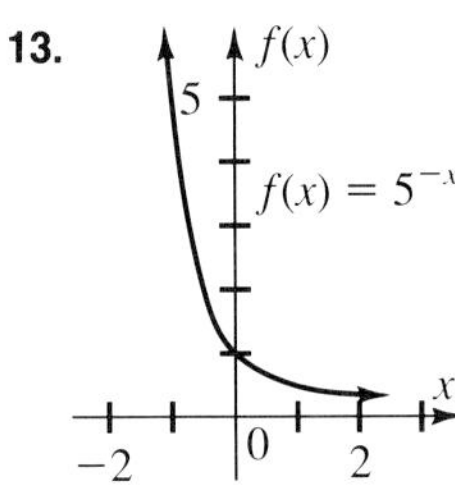

14.

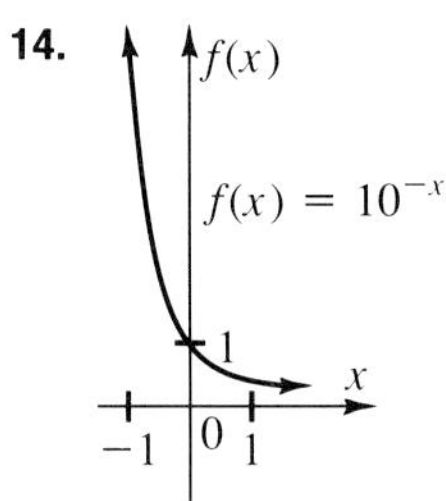

15.

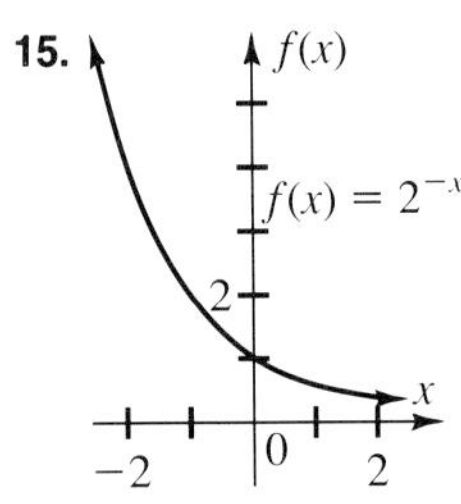

16.

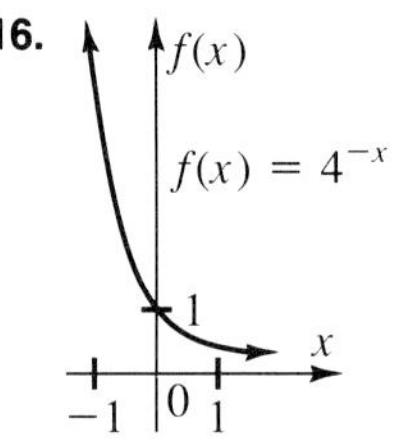

17.

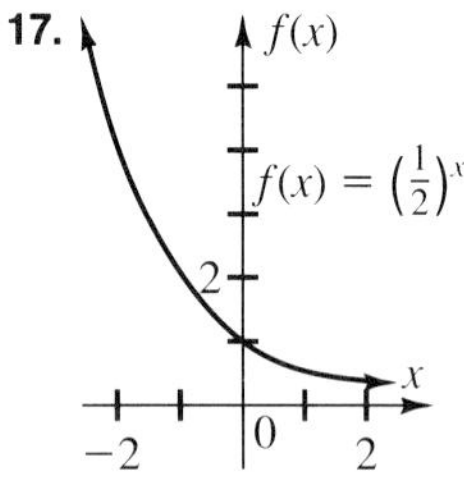

18.

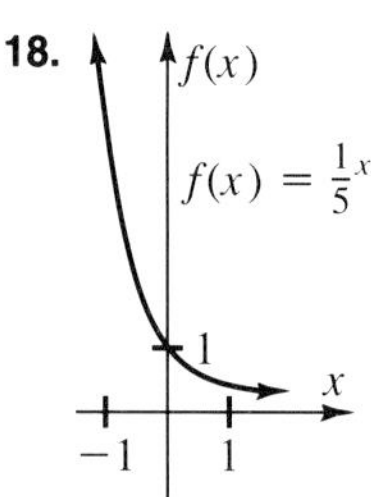

19.

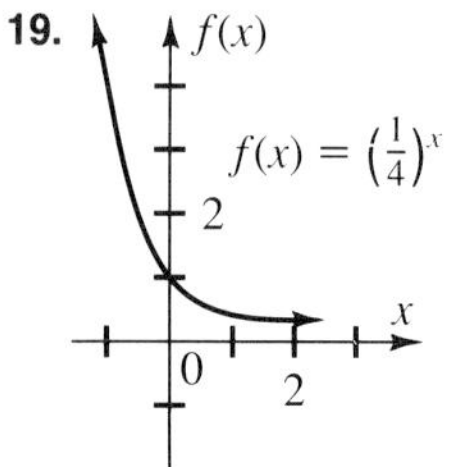

20.

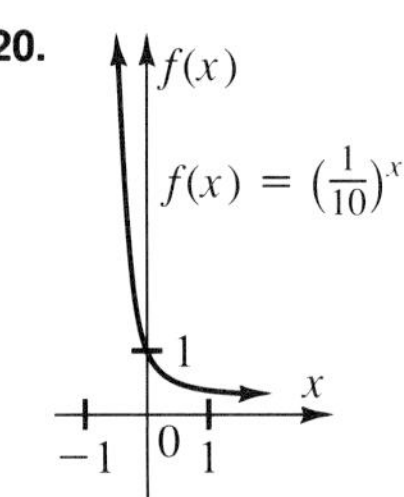

21.

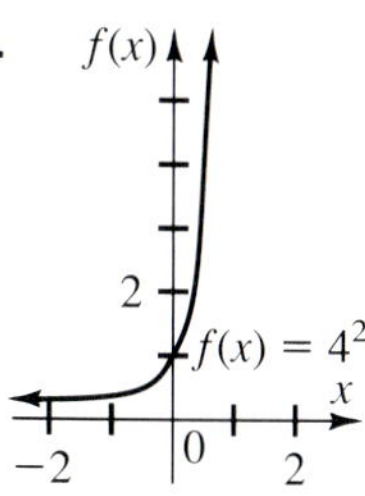

22.

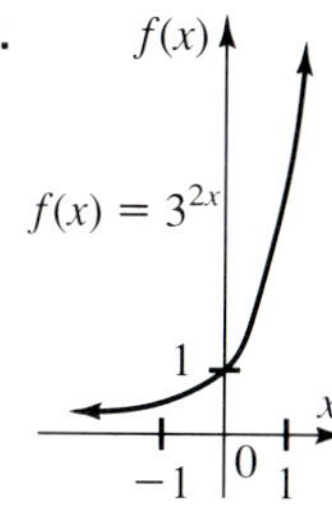

23.

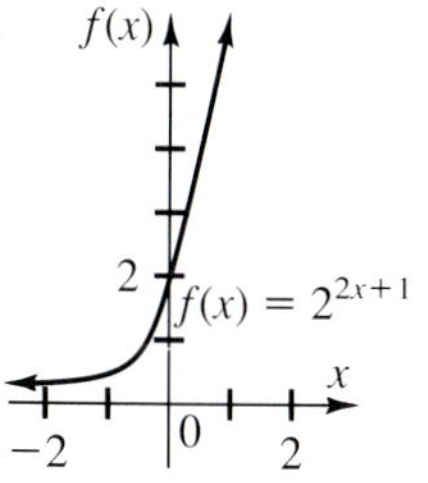

24.

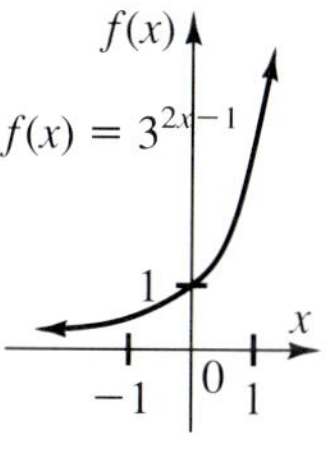

25.

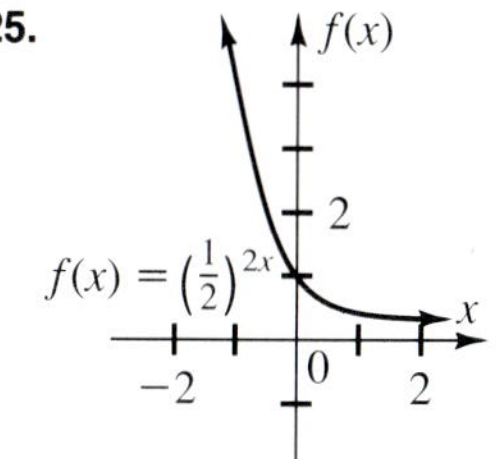

26.

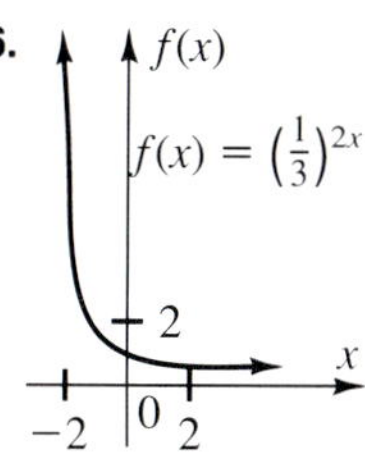

27.

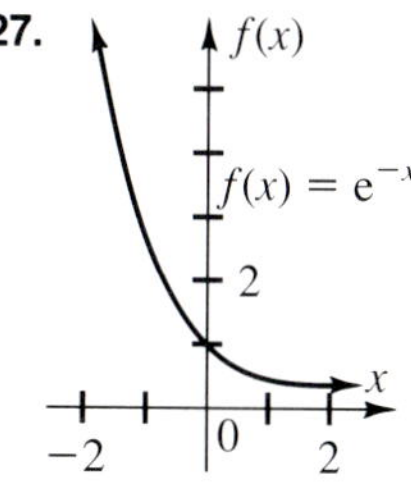

28.

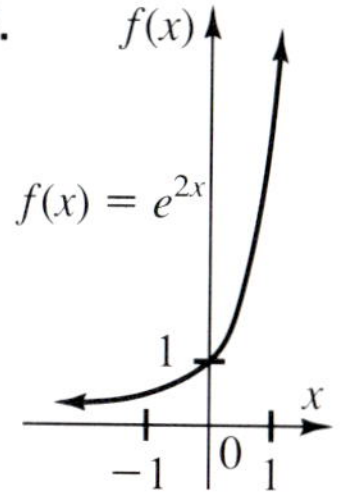

29.

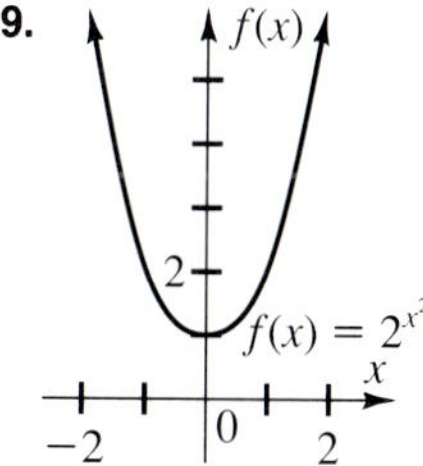

30.

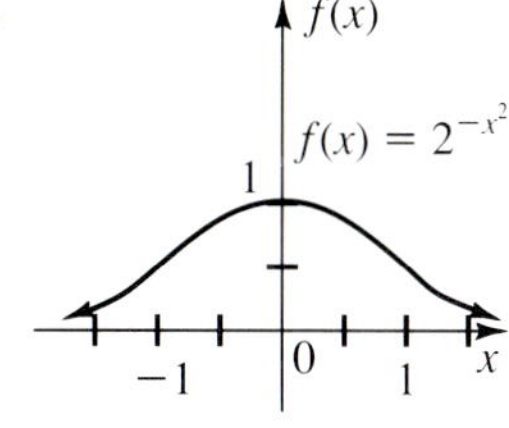

Chapter 9

Practice Exercises, pages 405–406

29. $\frac{e^{ix}+e^{-ix}}{2} = \frac{\cos x + i\sin x + \cos x - i\sin x}{2} = \frac{2\cos x}{2} = \cos x$

30. $\frac{e^{ix}-e^{-ix}}{i(e^{ix}+e^{-ix})} = \frac{\cos x + i\sin x - (\cos x - i\sin x)}{i(\cos x + i\sin x + \cos x - i\sin x)} = \frac{\cos x + i\sin x - \cos x + i\sin x}{i(2\cos x)} = \frac{2i\sin x}{2i\cos x} = \frac{\sin x}{\cos x} = \tan x$

31. $\frac{2i}{e^{ix}-e^{-ix}} = \frac{2i}{\cos x + i\sin x - (\cos x - i\sin x)} = \frac{2i}{\cos x + i\sin x - \cos x + i\sin x} = \frac{2i}{2i\sin x} = \frac{1}{\sin x} = \csc x$

32. $\frac{2}{e^{ix}+e^{-ix}} = \frac{2}{\cos x + i\sin x + \cos x - i\sin x} = \frac{2}{2\cos x} = \frac{1}{\cos x} = \sec x$

33. $\frac{i(e^{ix}+e^{-ix})}{e^{ix}-e^{-ix}} = \frac{i(\cos x + i\sin x + \cos x - i\sin x)}{\cos x + i\sin x - (\cos x - i\sin x)} = \frac{i(2\cos x)}{2i\sin x} = \frac{\cos x}{\sin x} = \cot x$

34. $e^{(\pi+x)i} = \cos(\pi + x) + i\sin(\pi + x) = -\cos x - i\sin x = -(\cos x + i\sin x) = -e^{ix}$

35. $\frac{e^{ix}}{e^{iy}} = \frac{\cos x + i\sin x}{\cos y + i\sin y} = \frac{(\cos x + i\sin x)}{(\cos y + i\sin y)} \cdot \frac{(\cos y - i\sin y)}{(\cos y - i\sin y)} = \frac{\cos x\cos y - i^2\sin x\sin y + i(\sin x\cos y - \cos y\sin y)}{\cos^2 y - i^2\sin^2 y} = \frac{\cos x\cos y + \sin x\sin y + i(\sin(x-y))}{\cos^2 y + \sin^2 y} = \cos(x-y) + i\sin(x-y) = e^{(x-y)i}$

36. $(e^{ix})^2 = (\cos x + i\sin x)^2 = \cos^2 x + 2i\cos x\sin x + i^2\sin^2 x = \cos^2 x - \sin^2 x + i(2\sin x\cos x) = \cos 2x + i\sin 2x = e^{2xi}$

37.

n	$\left(1+\frac{1}{n}\right)^n$
2	2.25
5	2.48832
10	2.593742
50	2.691588
100	2.704814
1,000	2.716924
10,000	2.718146
1,000,000	2.718280

38. $\sin^2 x + \cos^2 x = \left(\frac{e^{ix}-e^{-ix}}{2i}\right)^2 + \left(\frac{e^{ix}+e^{-ix}}{2}\right)^2 = \frac{e^{2ix}-e^0-e^0+e^{-2ix}}{-4} + \frac{e^{2ix}+e^0+e^0+e^{-2ix}}{4} = \frac{-e^{2ix}+1+1-e^{-2ix}+e^{2ix}+1+1+e^{-2ix}}{4} = \frac{4}{4} = 1$

39. $1 + \cot^2 x = \csc^2 x$; $1 + \left[\frac{i(e^{ix}+e^{-ix})}{e^{ix}-e^{-ix}}\right]^2 = \left(\frac{2i}{e^{ix}-e^{-ix}}\right)^2$;
$1 + \frac{i^2(e^{ix}+e^{-ix})^2}{(e^{ix}-e^{-ix})^2} = \frac{-4}{e^{2ix}-e^0-e^0+e^{-2ix}}$;
$1 + \frac{-1(e^{2ix}+e^0+e^0+e^{-2ix})}{(e^{2ix}-e^0-e^0+e^{-2ix})} = \frac{-4}{e^{2ix}-2+e^{-2ix}}$;
$1 + \frac{-e^{2ix}-2-e^{-2ix}}{e^{2ix}-2+e^{-2ix}} = \frac{-4}{e^{2ix}-2+e^{-2ix}}$;
$\frac{e^{2ix}-2+e^{-2ix}}{e^{2ix}-2+e^{-2ix}} + \frac{-e^{2ix}-2-e^{-2ix}}{e^{2ix}-2+e^{-2ix}} = \frac{-4}{e^{2ix}-2+e^{-2ix}}$;
$\frac{-4}{e^{2ix}-2+e^{-2ix}} = \frac{-4}{e^{2ix}-2+e^{-2ix}}$

40. $1 + \tan^2 x = \sec^2 x$; $1 + \left[\frac{e^{ix}-e^{-ix}}{i(e^{ix}+e^{-ix})}\right]^2 = \frac{2}{(e^{ix}+e^{-ix})^2}$;

$$1 + \frac{e^{2ix} - e^0 - e^0 + e^{-2ix}}{-1(e^{2ix} + e^0 + e^0 + e^{-2ix})} = \frac{4}{e^{2ix} + e^0 + e^0 + e^{-2ix}};$$

$$1 + \frac{e^{2ix} - 2 + e^{-2ix}}{-e^{2ix} - 2 - e^{-2ix}} = \frac{4}{e^{2ix} + 2 + e^{-2ix}};$$

$$\frac{-e^{2ix} - 2 - e^{-2ix}}{-e^{2ix} - 2 - e^{-2ix}} + \frac{e^{2ix} - 2 + e^{-2ix}}{-e^{2ix} - 2 - e^{-2ix}} = \frac{4}{e^{2ix} + 2 + e^{-2ix}};$$

$$\frac{-4}{-e^{2ix} - 2 - e^{-2ix}} = \frac{4}{e^{2ix} + 2 + e^{-2ix}};$$

$$\frac{4}{e^{2ix} - 2 - e^{-2ix}} = \frac{4}{e^{2ix} + 2 + e^{-2ix}}$$

Practice Exercises, pages 410–411

5. $\frac{1}{\operatorname{sech} x} = \frac{1}{\frac{1}{\cosh x}} = \cosh x$ **6.** $\frac{1}{\operatorname{csch} x} = \frac{1}{\frac{1}{\sinh x}} = \sinh x$

7. $\frac{\cosh x}{\coth x} = \frac{\cosh x}{\frac{\cosh x}{\sinh x}} = \sinh x$ **8.** $\frac{1}{\coth x} = \frac{1}{\frac{\cosh x}{\sinh x}} = \frac{\sinh x}{\cosh x} = \tanh x$

9. $\frac{\coth x}{\operatorname{csch} x} = \frac{\frac{\cosh x}{\sinh x}}{\frac{1}{\sinh x}} = \cosh x$ **10.** $\frac{\tanh x}{\operatorname{sech} x} = \frac{\frac{\sinh x}{\cosh x}}{\frac{1}{\cosh x}} = \sinh x$

11. $\frac{\operatorname{csch} x}{\coth x} = \frac{\frac{1}{\sinh x}}{\frac{\cosh x}{\sinh x}} = \frac{1}{\cosh x} = \operatorname{sech} x$

12. $\cosh(-x) = \frac{e^{-x} + e^{-(-x)}}{2} = \frac{e^{-x} + e^{x}}{2} = \cosh x$

13. $1 - \tanh^2 x = 1 - \frac{e^{2x} - 2 + e^{-2x}}{e^{2x} + 2 + e^{-2x}} = \frac{e^{2x} + 2 + e^{-2x} - e^{2x} + 2 - e^{-2x}}{e^{2x} + 2 + e^{-2x}} = \frac{4}{e^{2x} + 2 + e^{-2x}} = \left(\frac{2}{e^x + e^{-x}}\right)^2 = \operatorname{sech}^2 x$

14. $\coth^2 x - 1 = \left(\frac{e^x + e^{-x}}{e^x - e^{-x}}\right)^2 - 1 = \frac{e^{2x} + e^0 + e^0 + e^{-2x}}{e^{2x} - e^0 - e^0 + e^{-2x}} - 1 = \frac{e^{2x} + 2 + e^{-2x}}{e^{2x} - 2 + e^{-2x}} - \frac{e^{2x} - 2 + e^{-2x}}{e^{2x} - 2 + e^{-2x}} = \frac{4}{e^{2x} - 2 + e^{-2x}} = \left(\frac{2}{e^x - e^{-x}}\right)^2 = \operatorname{csch}^2 x$

15. $2 \sinh x \cosh x = 2\left(\frac{e^x - e^{-x}}{2}\right)\left(\frac{e^x + e^{-x}}{2}\right) = 2\left(\frac{e^{2x} - e^{-2x}}{4}\right) = \frac{e^{2x} - e^{-2x}}{2} = \sinh 2x$

16. $1 + 2\sinh^2 x = 1 + 2\left(\frac{e^x - e^{-x}}{2}\right)^2 = 1 + 2\left(\frac{e^{2x} - 2e^0 + e^{-2x}}{4}\right) = 1 + 2\left(\frac{e^{2x} - 2 + e^{-2x}}{4}\right) = 1 + \left(\frac{e^{2x} - 2 + e^{-2x}}{2}\right) = 1 - 1 + \frac{e^{2x} + e^{-2x}}{2} = \cosh 2x$

17. $\cosh^2 x - 1 = \left(\frac{e^x + e^{-x}}{2}\right)^2 - 1 = \frac{e^{2x} - 2 + e^{-2x}}{4} = \left(\frac{e^x - e^{-x}}{2}\right)^2 = \sinh^2 x$

18. $\cosh^2 x + \sinh^2 x = \left(\frac{e^x + e^{-x}}{2}\right)^2 + \left(\frac{e^x - e^{-x}}{2}\right)^2 = \frac{e^{2x} + e^0 + e^0 + e^{-2x}}{4} + \frac{e^{2x} - e^0 - e^0 + e^{-2x}}{4} = \frac{2e^{2x} + 2e^{-2x}}{4} = \frac{e^{2x} + e^{-2x}}{2} = \cosh 2x$

19. $\cosh x + \sinh x = \frac{e^x + e^{-x}}{2} + \frac{e^x - e^{-x}}{2} = \frac{e^x + e^{-x} + e^x - e^{-x}}{2} = \frac{2e^x}{2} = e^x$

20. $(\sinh x - \cosh x)^2 = \left(\frac{e^x - e^{-x}}{2} - \frac{e^x + e^{-x}}{2}\right)^2 = \left(\frac{-2e^{-x}}{2}\right)^2 = (-e^{-x})^2 = e^{-2x}$

21. $\sinh x \cdot \cosh y - \cosh x \cdot \sinh y = \frac{e^x - e^{-x}}{2} \cdot \frac{e^y + e^{-y}}{2} - \frac{e^x + e^{-x}}{2} \cdot \frac{e^y - e^{-y}}{2} = \frac{e^{x+y} + e^{x-y} - e^{-x+y} - e^{-x-y}}{4} - \frac{e^{x+y} - e^{x-y} + e^{-x+y} - e^{-x-y}}{4} = \frac{2e^{x-y} - 2e^{-x+y}}{4} = \frac{e^{x-y} - e^{-(x-y)}}{2} = \sinh(x - y)$

22. $\cosh x \cdot \cosh y - \sinh x \cdot \sinh y = \frac{e^x + e^{-x}}{2} \cdot \frac{e^y + e^{-y}}{2} - \frac{e^x - e^{-x}}{2} \cdot \frac{e^y - e^{-y}}{2} = \frac{e^{x+y} + e^{x-y} + e^{-x+y} + e^{-x-y}}{4} - \frac{e^{x+y} - e^{x-y} - e^{-x+y} + e^{-x-y}}{4} = \frac{2e^{x-y} + 2e^{-x+y}}{4} = \frac{e^{x-y} + e^{-(x-y)}}{2} = \cosh(x - y)$

23. $\sqrt{\frac{1 + \cosh x}{2}} = \sqrt{\frac{1 + \frac{e^x + e^{-x}}{2}}{2}} = \sqrt{\frac{\frac{2}{2} + \frac{e^x + e^{-x}}{2}}{2}} = \sqrt{\frac{e^x + 2 + e^{-x}}{4}} = \sqrt{\frac{\left(e^{\frac{x}{2}} + e^{-\frac{x}{2}}\right)^2}{4}} = \frac{e^{\frac{x}{2}} + e^{-\frac{x}{2}}}{2} = \cosh\frac{x}{2}$

24. $\sin^2\left(\frac{x}{2}\right) = \left(\frac{e^{\frac{x}{2}} - e^{-\frac{x}{2}}}{2}\right)^2 = \frac{e^x + e^{-x} - 2}{4} = \frac{\frac{e^x + e^{-x}}{2} - 1}{2} = \frac{\cosh x - 1}{2}$